NUMERICAL MODELS IN GEOMECHANICS
NUMOG VI

PROCEEDINGS OF THE SIXTH INTERNATIONAL SYMPOSIUM ON NUMERICAL MODELS IN GEOMECHANICS – NUMOG VI / MONTREAL / QUEBEC / CANADA / 2-4 JULY 1997

Numerical Models in Geomechanics

NUMOG VI

Edited by

S. PIETRUSZCZAK
McMaster University, Hamilton, Ontario, Canada

G. N. PANDE
University of Wales Swansea, UK

A. A. BALKEMA / ROTTERDAM / BROOKFIELD / 1997

The texts of the various papers in this volume were set individually by typists under the supervision of each of the authors concerned.

Published by
A.A. Balkema, P.O. Box 1675, 3000 BR Rotterdam, Netherlands (Fax: +31.10.4135947)
A.A. Balkema Publishers, Old Post Road, Brookfield, VT 05036-9704, USA (Fax: 802.2763837)

ISBN 90 5410 886 X

Printed in the Netherlands

Numerical Models in Geomechanics, Pietruszczak & Pande (eds) © 1997 Balkema, Rotterdam, ISBN 90 5410 886 X

Table of contents

3 *Modelling of reinforced soil*

4 *Modelling of transient/coupled problems*

6.2 *Piles and foundations*

6.3 *Slopes, embankments and dams*

6.4 *Other applications*

7 *Supplement*

Numerical Models in Geomechanics, Pietruszczak & Pande (eds) © 1997 Balkema, Rotterdam, ISBN 90 5410 886 X

Preface

Numerical models are being increasingly used for the solution of complex geotechnical engineering problems and also for verification of the results obtained from conventional analyses, specially in the cases where the validity of the latter is questionable. In the last two decades rapid progress has been made in many areas of Computational Geomechanics. A series of International Symposia 'Numerical Models in Geomechanics' was launched in 1982 with the main aim of encouraging an exchange of views between researchers and practising engineers on various aspects of numerical analysis. Encouraged by the success of earlier Symposia in the series held at Zürich, Switzerland (1982); Ghent, Belgium (1986); Niagara Falls, Canada (1989); Swansea, UK (1992) and Davos, Switzerland (1995), the sixth Symposium was organised at Montreal, Canada in 1997. The NUMOG series has now been well established and is perceived as an important event in the calendar of geotechnical researchers and engineers.

This book contains more than hundred papers which were selected for presentation at the Montreal symposium. The choice of papers was made on the basis of about 130 abstracts offered and subsequent review of over 120 completed manuscripts. The papers are organised in six sections:

1. Constitutive relations for geological materials: Formulation and verification;
2. Instability and strain localization in geomaterials;
3. Modelling of reinforced soil;
4. Modelling of transient/coupled problems;
5. Numerical algorithms: Formulation and performance;
6. Application of numerical techniques to practical problems;

They are generally of a high standard and contain a wealth of information which would be useful for researchers as well as practising engineers.

We are grateful to the members of the Technical Advisory Committee, consisting of: F.Darve (France), R.de Borst (Netherlands), W.D.L.Finn (Canada), J.Ghaboussi (USA), P.V.Lade (USA), R.Nova (Italy), F.Oka (Japan), H.B.Poorooshasb (Canada), B.A.Schrefler (Italy), A.P.S.Selvadurai (Canada), I.M.Smith (UK) and H.R.Thomas (UK) for their helpful suggestions and comments.

S.Pietruszczak
G.N.Pande
Swansea, April 1997

1 Constitutive relations for geomaterials: Formulation and verification

Numerical Models in Geomechanics, Pietruszczak & Pande (eds)© 1997 Balkema, Rotterdam, ISBN 90 5410 886 X

Coupled dilatancy-compaction model for homogeneous and localized deformation of granular materials

R.G.Wan & P.J.Guo
Department of Civil Engineering, University of Calgary, Alb., Canada

ABSTRACT: This paper describes a stress-dilatancy based constitutive model which addresses both stress level and void ratio sensitivities including microstructure dependencies for describing bulk material response of granular materials. Rowe's stress dilatancy concepts are extended so that the plastic dissipation energy equation reflects the aforementioned issues. The resulting dilatancy rule is then used in conjunction with a yield function and a flow rule in the framework of classical plasticity theory. Unlike many other models, the evolution law controlling dilatancy with plastic deformations couples both dilatancy and compaction mechanisms. This coupling allows for a more consistent prediction of the granular material behaviour with respect to dependencies on stress level, void ratio and strain localization as illustrated by numerical predictions for triaxial compression and biaxial tests on sands.

1 INTRODUCTION

Contrary to common belief that a granular material entirely derives its strength from pure friction sustained at the grain boundaries, it has been shown by Oda and Konishi (1974) that microstructure seems to play a dominant role in determining strength, while particle friction would merely provide a stabilizing effect.

The formation and evolution of microstructure lead to known macroscopic phenomena such as dilatancy and compaction which are normally both pressure, void ratio and loading direction sensitive. For instance, the deformation pattern can shift from strain hardening to strain softening, thus involving non-homogeneous deformations and strain localization, see Vardoulakis (1980).

To set the scene for this study, it is important to recall Rowe's stress dilatancy theory (1962) upon which most interesting aspects of granular material behaviour are predicated. The theory describes how the geometrical interlocking of the particles influences the strength of the material, and thus relates stress ratio to a dilatancy factor which quantifies a geometrical effect. Furthermore, Oda (1978), and more recently Ueng and Lee(1990), incorporated microstructure in their stress-dilatancy formulations by distinguishing between slipping and non-slipping grains. The former grains would contribute to dilatancy, while the latter would account for the microstructure build up and thus strength of the granular material.

Stress-dilatancy equations under different forms have been successfully used as a flow rule in plasticity calculations, Vermeer (1982). However, features such as density and pressure sensitivities in granular material response have not been given due consideration. To the authors' knowledge, the only relatively satisfactory work which addresses such issues are due to Bauer and Wu (1993), and Gudehus (1996) who use hypoplasticity concepts which preclude the notion of yield surfaces. It is basically assumed that the rate of Cauchy granular stress in the granular material is related to its deformation rate and void ratio by means of a functional. In order to investigate material instability in the form of shear band localization, classical bifurcation analysis can be performed on the above constitutive models. It is found that the accuracy of the results largely depends on the basic ingredients of the constitutive models, and in some cases, the use of vertex plasticity is even advocated, Bardet (1991) and Han and Drescher (1993).

While the present study is mainly focussed on the consistent description of stress and void ratio dependencies in Rowe's stress dilatancy equation for the case of monotonic loading in particular, it is also proposed to include microstructural aspects in the formulation. A special dissipation function embedded into the dilatancy rule is used in conjunction with a yield function and a flow rule in the proposed elasto-plastic constitutive model. For describing plastic deformation, a plastic evolution law is introduced which couples both dilatancy and compaction mechanisms. In fact, this coupling allows for a more consistent prediction of the granular material behaviour with respect to dependencies on stress level and void ratio.

For illustration purposes, drained compression triaxial tests results computed at different stress and initial void ratio levels are in very good agreement with published experimental data. Some drained biaxial tests in which strains localize into a shear band are also simulated for different initial void ratio and stress levels. Predictions of shear band inclination and the strain at localization are in excellent concordance with published experimental results. It is noted that the latter were not successfully modelled by other investigators.

2 PROPOSED MODEL

2.1 *Modified stress-dilatancy equation*

Consider a granular material being subjected to principal stresses σ_1 and σ_3 leading to volumetric strains ϵ_v and shear strains γ as a result of principal strains ε_1 and ε_3. By minimizing the energy dissipation developed at particle boundaries due to frictional sliding, Rowe proposed a stress dilatancy equation which couples the dilatancy factor, $D = 1 - d\epsilon_v^p/d\varepsilon_1^p$, to effective stress ratio, $R = \sigma_1/\sigma_3$, so as $R = K_{cv}\,D$. The factor $K_{cv}=\tan^2(\pi/4+\varphi_{cv}/2)$ is a material constant while φ_{cv} represents the friction angle at constant volume deformation. When a mobilized dilation angle ψ_m, defined as $\sin\psi_m = -d\epsilon_v^p/d\gamma^p$, is introduced together with some mobilized friction angle φ_m, Rowe's stress dilatancy equation takes the classical form

$$\sin\psi_m = \frac{\sin\varphi_m - \sin\varphi_{cv}}{1 - \sin\varphi_m \sin\varphi_{cv}} \tag{1}$$

In the original derivation of the above equation, no information regarding stress level, density of the grain assembly and its evolution with deformation history were considered. In view of addressing the issue of density on an average sense, a first order modification of the stress-dilatancy equation involves introducing a factor based on the current void ratio e relative to the critical void ratio e_{cr}, i.e.

$$\sin\psi_m = \frac{\sin\varphi_m - (e/e_{cr})^\alpha \sin\varphi_{cv}}{1 - (e/e_{cr})^\alpha \sin\varphi_m \sin\varphi_{cv}} \tag{2}$$

in which α is a material parameter to be determined. The validation of Eq. (2) hinges upon a modified energy dissipation equation which embeds a so-called state parameter related to the current void ratio and the critical one. In fact, the energy based factor K_{cv} in the original Rowe's equation is made void ratio sensitive so that the modified stress dilatancy relationship becomes

$$\left(\frac{\sigma_1}{\sigma_3}\right)_m = K_{cv}^*\,D\,; \qquad K_{cv}^* = \frac{1 + (e/e_{cr})^\alpha \sin\varphi_{cv}}{1 - (e/e_{cr})^\alpha \sin\varphi_{cv}} \tag{3}$$

It is obvious that void ratio dependent K_{cv}^* reverts to the original K_{cv} when the void ratio is fixed to the critical void ratio, i.e. $(e/e_{cr})^\alpha = 1$ or $\alpha = 0$. The modification of the original Rowe's equation and its relevance to classical sands can be found in Wan and Guo (1996). It is worth noting that the critical void ratio e_{cr} also varies with stress level, i.e.

$$e_{cr} = e_{cr0} \exp\left[-\left(\frac{p}{h_{cr}}\right)^{n_{cr}}\right] \tag{4}$$

where e_{cr0} is the critical void ratio at very small confining stress, n_{cr} is an exponent number and h_{cr} some parameter.

Ideally, microstructural aspects should also be reflected in Eq. (3) since a certain amount of energy must be accounted for towards creating microstructure and modifying it relative to the direction and level of loading. To illustrate this point, an extreme example refers to sustained compaction (as opposed to dilation) of a dense granular material largely caused by combined effects of internal structure development and external loading conditions. Different approaches can be used to consider microstructural effects and their impact on dilatancy. First, the structure of granular assemblies can be described by abandoning the notion of a scalar void ratio e and replacing it by a directional void ratio $e(\mathbf{n})$ along a characteristic direction $\mathbf{n}$ in space. A first order decomposition of $e(\mathbf{n})$ involves the introduction of Ω_{ij} a so-called fabric tensor of order two such that $e(\mathbf{n}) = e_m(1 + n_i\Omega_{ij}n_j)$, see Pietruszczak and Krucinski (1989) for instance. It is obvious that if $\Omega_{ij} = \delta_{ij}$ (Kronecker delta), $e(\mathbf{n})$ corresponds to an isotropic spatial distribution of void ratio valued at $2e_m$.

The use of directional void ratio is not enough to define accurately the microstructure of granular materials. The concept must be supplemented with the introduction of two more parameters, namely coordination number and contact force distribution. The former which is of geometrical nature can be incorporated as a kinematical constraint into the dilatancy factor D, while the latter related to statical stress quantities can be used in conjunction with parameter K in Eq. (3). Moreover, the nature of the distributions relative to the external loading conditions seems to be a crucial issue which also needs to be addressed.

2.2 *Plastic deformations*

The particular relevance of the modified stress-dilatancy equation (3) becomes clear when it is used as a flow rule to describe the shearing of granular material within the classical theory of plasticity. Thus, plastic strain increments $d\boldsymbol{\varepsilon}_s^p$ are calculated from a non-associated flow rule, i.e. $d\boldsymbol{\varepsilon}_s^p = d\lambda \partial Q_s / \partial \boldsymbol{\sigma}$, in which $d\lambda$ is the plastic multiplier and Q_s is a plastic potential function consistent with Eq. (3). The plastic potential function takes the form of

$$Q_s = (\sigma_1 - \sigma_3) - (\sigma_1 + \sigma_3) \sin \psi_m \tag{5}$$

in which the parameter ψ_m refers to the mobilized dilation angle at a certain stress level. A positive dilation angle corresponds to a shear induced increase in volume while a negative angle means shear induced contraction.

The onset of plastic flow is determined by using two yield surfaces: one for the shearing mechanism and another one for compaction in hydrostatic state. For the shearing mechanism, it is assumed that plastic shear yielding corresponds to a continuous mobilization of the frictional angle which results into an isotropic expansion of the yield surface in the stress space until it reaches an ultimate failure surface. At a given yielding state, the expression of the shear yield surface is

$$F_s = (\sigma_1 - \sigma_3) - (\sigma_1 + \sigma_3) \sin \varphi_m = 0 \tag{6}$$

in which φ_m is the mobilized frictional angle. The ultimate failure surface takes the same form as in the above Eq. (6) with φ_m replaced by the peak friction angle φ_p or φ_{cv}, the friction angle at constant volume, depending on the case.

With regards to describing the compaction mechanism, a vertical cut-off cap surface which also moves in the stress space is used for simplicity, i.e., $F_c = p - p_0 = 0$, with p_0 referring to the consolidation pressure as a function of compaction.

2.3 *Hardening-softening laws*

2.3.1 *Coupled hardening-softening for shearing mechanism:* During deviatoric loading history, it is postulated that both volumetric strains (void ratio) and plastic deviatoric strains control hardening and softening according to a simple function given as

$$\sin \varphi_m = \frac{\gamma^p f_d(e)}{a + \gamma^p} \sin \varphi_{cv}\,; \qquad f_d(e) = (e/e_{cr})^{-\beta} \tag{7}$$

with a, β as constants. Equation (7) basically represents a hyperbolic variation of mobilized friction angle with plastic shear as well as total volumetric strains (deviatoric and hydrostatic pressure induced) by virtue of the void ratio function f_d. It is also evident that, depending upon whether the current void ratio e is denser or looser of critical, the factor f_d will adjust the functional representation of $\sin \varphi_m$ so as to make it evolve into either a softening or hardening trend.

2.3.2 *Cap hardening during hydrostatic compaction process:* For describing the hydrostatic consolidation process, the cap surface grows (hardens) isotropically in the stress space as irrecoverable volumetric plastic strains increase. The evolution of void ratio (elastic and plastic components) with mean stress p is governed by an exponential law, i.e., $e = e_0 \exp\left[-(p/h_l)^m\right]$, in which h_l is a modulus and m an exponent. In the unloading path, the rebound curve is considered to be also an exponential such that $de/e = -(n/h_u)\,(p/h_u)^{n-1}\,dp$, where h_u and n are unloading parameters.

2.4 *Constitutive equations*

The incremental form of the stress-strain equations ($\Delta\sigma_{ij} = C_{ijkl}\,\Delta\varepsilon_{ij}$) for the proposed constitutive model can be typically obtained by applying plastic consistency conditions together with the flow rule described earlier. For the case of two dimensions such as biaxial stress conditions, the fourth order constitutive tensor emerges as $C_{ijkl} = G(L^e_{ijkl} + L^p_{ijkl})$ where

$$L^e_{ijkl} = (\Theta - 1)\delta_{ij}\delta_{kl} + 2\delta_{ik}\delta_{jl} \tag{8}$$

$$L^p_{ijkl} = -\frac{<1>}{H}\left(\frac{s_{ij}}{\tau} + \Theta\,\beta_D\,\delta_{ij}\right)\left(\frac{s_{kl}}{\tau} + \Theta\,\mu\,\delta_{kl}\right) \tag{9}$$

The operator $< . >$ represents a switch function, and H is a plastic hardening modulus defined as $H = 1 + h + \Theta\beta_D\mu$; $h = ph_t/G; h_t = d\mu/d\gamma^p; \Theta = (1+v)/(1-v)$; $\beta_D = \sin\psi_m$; $\mu = \sin\varphi_m$; $\tau = \sqrt{s_{ij}s_{ij}/2}$; s_{ij} is the deviatoric stress tensor. It is worthwhile to note that the shear modulus G which appears in the expression of C_{ijkl} varies non-linearly with mean stress p according to $G = G_0\sqrt{p}(2.17-e)^2/(1+e)$ while the Poisson ratio v is kept constant.

2.5 *Strain localization*

The condition for strain localization into a shear band is well established as it can be obtained from equilibrium of stress rates across a possible shear band. For example, in the case of continuous bifurcation and small material rotations, the condition is given as $\det(\mathbf{n}\ \mathbf{C}\ \mathbf{n}) = 0$. The orientation $\mathbf{n}$ of the shear band is calculated as real roots of $\det(\mathbf{n}\ \mathbf{C}\ \mathbf{n}) = 0$. In the case of biaxial stress, the localization condition for flow plasticity theory can be written as

$$\left(C - a_1 C_{1122} - a_2 C_{2211}\right)^2 - 4a_1a_2C_{1111}C_{2222} = 0 \tag{10}$$

$$\begin{aligned} C &= C_{1111}C_{2222} - C_{1122}C_{2211}; \\ a_{1,2} &= \frac{1}{2}C_{1212} \pm \sqrt{\frac{s_{ij}s_{ij}}{2}} \end{aligned}$$

The shear band inclination θ is finally obtained as the roots of

$$\tan\theta = \pm\sqrt[4]{(a_2 C_{2222})/(a_1 C_{1111})} \quad (11)$$

As for the deformation theory which expresses stresses to total strains, different equations must be derived in order to express strain localization conditions. The adaptation of the proposed constitutive model to deformation theory simply involves expressing frictional parameter $\mu(e,\gamma)$ and dilatancy parameter $\beta_D(e,\gamma)$ to total strain and void ratio. The condition for strain localization and the corresponding shear band inclination have been derived by Vardoulakis and Graf (1985). These are

$$[2h_t - h_s(1+\mu\beta_D) - \mu(\mu+\beta_D)]^2$$

$$-(h_s^2-\mu^2)(1-\mu^2)(1-\beta_D^2) = 0; \quad h_s = \mu/\gamma \quad (12)$$

$$\tan\theta = \pm\sqrt[4]{\frac{(1+\mu)(1+\beta_D)(h_s-\mu)}{(1-\mu)(1-\beta_D)(h_s+\mu)}} \quad (13)$$

3 NUMERICAL RESULTS

3.1 *Drained triaxial compression*

It is attempted to reproduce numerically the experimental results published by Lee and Seed (1967) for drained triaxial compression of Sacramento River Sand. Material model constants as derived from tests are listed in Table 1.

Table 1. Constants for Sacramento River sand

$G_0 = 3900$ KPa	$v = 0.25$	
$a = 0.011$	$\alpha = 1.2828$	$\beta = 1.25$
$m = 0.8324$	$h_u = 111.78$ MPa	$n = 0.7898$
$h_{cr} = 22.139$ MPa	$n_{cr} = 0.7075$	$e_{cr0} = 1.03$
$\varphi_{cv} = 34^0$	$h_l = 63.90$ MPa	

3.1.1 *Influence of void ratio and stress level:* Figure 1 shows that both strain softening and volume dilation are correctly captured at the low range of confining pressures for dense Sacramento River sand. Also, at a high confining pressure of 2 MPa, strain softening is almost suppressed and the volumetric response is virtually compressive even though the sand was initially at a dense state, i.e. $e_0 = 0.61$. Figure 2 refers the case of loose Sacramento River sand where strain softening is less prominent and the volumetric strain remains mainly contractant. However, at low confining pressure of 100 kPa, some dilation with strain softening is captured even though the sand is initially loose at $e_0 = 0.87$.

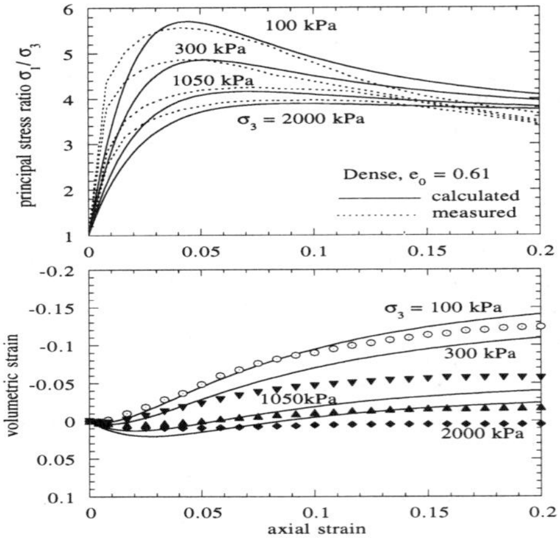

Figure 1. Comparison between calculated and measured behaviour of dense Sacramento River sand in drained triaxial compression tests

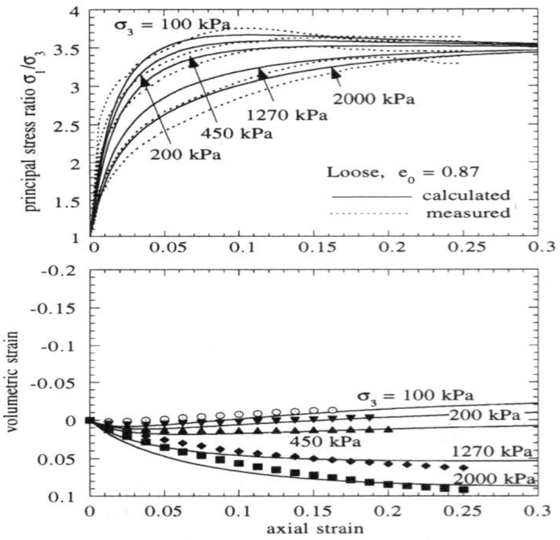

Figure 2. Comparison between calculated and measured behaviour of loose Sacramento River sand in drained triaxial compression tests

A closer look at Figs. 1 and 2 reveals that the model predicts consistently more compaction at small strains and more dilation at large strains than in the experimental results. These discrepancies are mainly due to the fact that in the experiments, deformations are not uniform, especially in the large strain range where strain localization takes place. Hence, due to non-uniformities of deformation fields, the measured volumetric strains are average ones. On the other hand, the numerical simulations assumed uniformity of deformations. It is also noted that the model, due to the nature of its formulation, naturally captures the maximum mobilized friction angle and its variation with stress level.

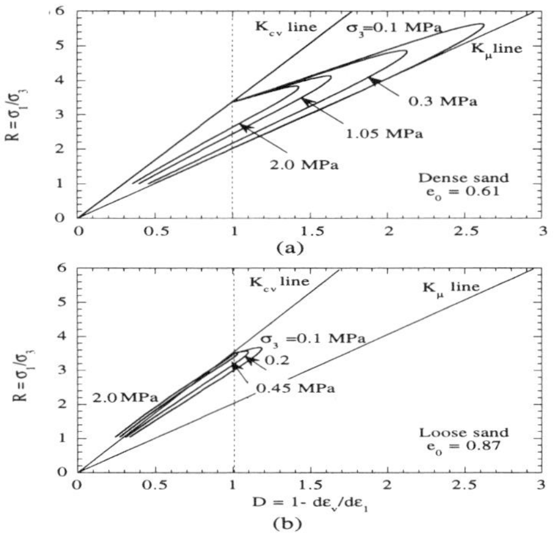

Figure 3. Computed stress-dilatancy plots for Sacramento River sand: (a) dense (b) loose

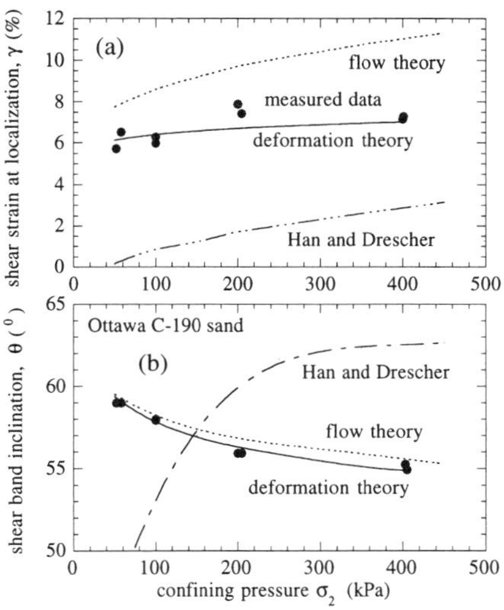

Figure 4. (a) Shear strains at localization (b) Shear band inclination

3.1.2 *Dilatancy plots:* The evolution of stress-dilatancy with deformation history is plotted in Fig. 3. It is clearly shown that for the same confining pressure, the sand may have different strengths with respect to dilatancy factor D, hence illustrating the geometrical constraint imposed by dilatancy. The dilatancy plots obtained from the model indeed describe different energy dissipation curves depending on void ratio and stress level, which are in agreement with experimental trends. This aspect of describing dilatancy is in contrast with the original Rowe's relationship which assumes a unique energy line.

3.2 *Strain localization in biaxial tests*

Spontaneous shear band formation under plane strain conditions has been experimentally investigated by Vardoulakis (1980) for Karlsruhe sand, and more recently by Han and Drescher (1993) for Ottawa C-190 sand in a special biaxial strain localization apparatus.

3.2.1 *Influence of stress level:* Han and Drescher (1993) performed biaxial tests on dry Ottawa C-190 sand at confining pressures of 50, 100, 200 and 400 kPa at fixed initial void ratio e_0 of about 0.5. From available test results, material parameters for the proposed constitutive model were deduced, and are listed in Table 2.

Strain localization predictions based on plastic flow and deformation theories were made for different confining pressures. Figures 4a, b show predictions of shear strain at localization conditions and shear band inclinations for different confining pressures. Overall, it is found that the model predictions for shear band (both flow and deformation theories) are in very good agreement with experimental data, see Fig. 4 b in particular. Also, are plotted in Figs. 4a, b plastic flow theory predictions made by Han and Drescher (1993). The latter predict incorrectly an increase in shear band inclination with increasing confining pressure, which is in contradiction with experimental results.

Table 2. Parameters for Ottawa C-190 sand

$G_0 = 2.9$ MPa	$v = 0.2$	$\alpha = 0$
$a = 0.0085$	$\beta = 1.65$	$\varphi_{cv} = 28^0$
$e_{cr0} = 0.78$	$p_0 = 10$ kPa	$\lambda_{cr} = 0.0425$
$m = 0.3885$	$h_l = 284$ MPa	

3.2.2 *Influence of void ratio:* The effect of initial void ratio on shear band formation was experimentally investigated by Vardoulakis (1980) for dry Karlsruhe sand specimens. Biaxial tests were performed at a variety of initial void ratios, i.e. $e_0 = 0.56 - 0.75$, and a range of confining pressures ($\sigma_{22} = 42$ kPa-230 kPa). Table 3 gives the material parameters which have been deduced from experimental data published in Vardoulakis and Graf (1985) for the same sand.

Table 3. Parameters for Karlsruhe sand

$G_0 = 6.9$ MPa	$v = 0.25$	$\alpha = 0$
$a = 0.0085$	$\beta = 2.5$	$\varphi_{cv} = 34^0$
$e_{cr0} = 0.85$	$p_0 = 10$ kPa	$\lambda_{cr} = 0.041$
$m = 0.7474$	$h_l = 92$ MPa	

The calculated shear strains at localization are shown in Fig. 5a for various initial void ratios at relatively constant confining pressures. No experimental data were available for comparison, but the calculations indicate that the looser the sand is initially, the higher the shear strain at localization. In simple terms, for the above low confining pressures, the occurrence of strain localization becomes less probable for initially loose states. Also, for the same loose states subjected at higher confining pressures, calculations seem to indicate that the sand would be more conducive to strain localization. Fig. 5a shows some discrepancy between the flow and deformation theory predictions, especially for the looser range of initial void ratios. This is probably due to the difference in the description of the stress strain curves using the two theories.

Shear band inclination measurements were available for comparison with model predictions. The latter compare favourably with the experimental data, see Fig. 5b.

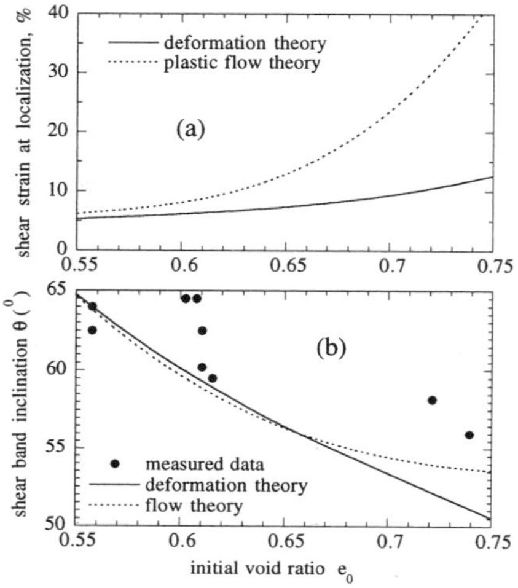

Figure 5. (a) Shear strain versus initial void ratio (b) Shear band inclination versus void ratio

4 CONCLUSIONS

This paper demonstrates that a judicious modification of the original Rowe's stress-dilatancy equation proves to be very effective in capturing void ratio and stress level dependencies of granular material behaviour. Also, rather than using complicated models, the modified dilatancy equation fits nicely in the formulation of classical rate-independent plasticity constitutive models. Thus, the stress and volumetric responses exhibited by sand over a large range of density and stress levels can be rationally modelled using a consistent set of material parameters. Furthermore, the model describes accurately shear band inclinations and shear strain at localization for sand samples subjected to different confining pressures and initial void ratios without the need to use vertex-like plasticity.

Further developments are currently being made on the model with regard to effects of microstructure and external loading variation on stress-dilatancy. To this end, ideas discussed earlier in the paper with regard to directional void ratio and contact force distribution will have to be implemented. As a result, the yield surfaces will also have to be modified in order to reflect anisotropy.

5 ACKNOWLEDGEMENTS

The financial support of the Natural Science and Engineering Research Council of Canada is gratefully acknowledged.

REFERENCES

Bardet, J. P. 1991. Orientation of shear bands in frictional soils. *J. Eng. Mech,* 117 (7): 1466-1484.

Bauer, E. and Wu, W. 1993. A hypoplastic model for granular soils under cyclic loading. *Proc. Int. Workshop on Modelling Approaches to Plasticity:* 247-258, Elsevier

Gudehus, G. A. 1996. Comprehensive constitutive equation for granular materials. *Soils and Foundations,* 36 (1): 1-12.

Han, C. and Drescher, A. 1993. Shear bands in biaxial tests on dry coarse sand. *Soils and Foundations,* 33 (1): 118-132.

Lee, K. L. and Seed, H. B. 1967. Drained characteristics of sands. *J. Soil Mech. Found. Div., ASCE,* 93, (SM6): 117-141.

Oda, M. & Konishi, J. 1974. Microscopic deformation mechanism of granular material in simple shear. *Soils and Foundations,* 14 (4): 25-38.

Pietruszczak, S. and Krucinski, S. 1989. Description of anisotropic response of clays using a tensorial measure of structural disorder, *Mech. of Materials,* 8: 237-249.

Rowe, P. W. 1962. The Stress-dilatancy relation for static equilibrium of an assembly of particles in contact. *Proc. of Royal. Soc. A,* 269: 500-527.

Ueng, T. S. and Lee. C. J. 1990. Deformation of sand under shear-particulate approach. *J. Geotech. Eng.,* 116: 1625-1640.

Vardoulakis, I. 1980. Shear band inclination and modulus of sand in biaxial tests. *Int. J. Num. Anal. Meth. Geomech,* 4: 103-119.

Vardoulakis, I. and Graf, B. 1985. Calibration of constitutive models for granular materials using data from biaxial experiments. *Geotechnique,* 35(3): 299-317.

Vermeer, P. A. 1982. A five constant model unifying well-established concepts. *Constitutive Relations for Soils*: 175-197. Rotterdam: Balkema

Wan, R. G. and Guo, P. J. 1996. A pressure and density dependent dilatancy model for granular materials. Submitted to *Soils and Foundations.*

Numerical Models in Geomechanics, Pietruszczak & Pande (eds) © 1997 Balkema, Rotterdam, ISBN 90 5410 886 X

Rotational kinematic hardening model for sand

Poul V. Lade
The Johns Hopkins University, Baltimore, Md., USA
Sinan Inel
Dames & Moore, Los Angeles, Calif., USA

ABSTRACT: To capture the behavior of soil during large stress reversals, a new kinematic hardening mechanism is proposed which incorporates rotation and intersection of yield surfaces to achieve a consistent and physically rational fit with experimentally observed soil behavior during large stress reversals. An existing elasto-plastic model with isotropic hardening is used as the basic framework to which the rotational kinematic hardening mechanism has been added. The new combined model preserves the behavior of the isotropic hardening model under monotonic loading conditions, and the extension from isotropic to rotational kinematic hardening is accomplished without introducing new material parameters. Comparisons show that the proposed model can capture the behavior of sand during large stress reversals in the triaxial plane with reasonable accuracy.

1 INTRODUCTION

Modeling the behavior of soils during large stress reversals and large changes in stress involving unloading and reloading is required to capture the behavior of earth structures and soil-structure interaction problems under static as well as dynamic loading conditions. Large stress reversals and stress rotations occur in the ground during most construction projects due to removal and addition of soil and structures. General three-dimensional stress reversals occur during earthquakes and due to fluctuating loads caused by pile driving, wind, waves, and other transient events.

To capture the soil behavior during large stress reversals, a new kinematic hardening model is developed. An existing elasto-plastic model with isotropic hardening is used as the basic framework to which the rotational kinematic hardening mechanism is added. Analyses of data from triaxial tests (Lade and Boonyachut 1982) indicate that plastic yielding occurs during unloading and reloading, and the directions of the plastic strain increment vectors tend towards a pattern of behavior, which may be described by a combination of isotropic and kinematic hardening with rotational evolution of the plastic potential and the yield surface. Thus, the model involves rotation and intersection of yield surfaces to achieve a consistent and physically rational fit with experimentally observed behavior in both triaxial and octahedral planes. The existing, isotropic work hardening law is shown to apply to isotropic as well as kinematic behavior of sand. The behavior of the proposed plastic potential is verified by studying the behavior of sand during large stress reversals on the basis of results from an experimental program incorporating conventional triaxial tests. Experiments were performed with various simple and complex stress paths involving large changes in stress during unloading and reloading, and the results are evaluated in terms of directions of plastic strain increment vectors superimposed on the stress space. Predicted plastic strain increment directions are compared with experimental data and it is shown that the proposed model can capture soil behavior with reasonable accuracy within the scatter of test results. The capability of the proposed model is examined by comparing predictions with experimental data for simple and complex stress paths involving large stress reversals in the triaxial plane. Comparisons show that the proposed model can capture the behavior of sand during large stress reversals in the triaxial plane with reasonable accuracy.

2 ISOTROPIC SINGLE HARDENING MODEL

The basis for the rotational kinematic hardening model presented here is an existing isotropic single hardening model (Kim and Lade 1988, Lade and Kim 1988a and b). This elasto-plastic model employs a nonassociated flow rule derived from a plastic potential whose shape in principal stress space resembles an asymmetric cigar with smoothly rounded triangular cross-sections in octahedral planes. The model also incorporates a single isotropic yield surface that is expressed as a contour of constant plastic work as measured from the stress origin. The resulting yield surface is shaped as an asymmetric teardrop with cross-sections similar to those of the plastic potential surface. As the plastic work increases, the isotropic yield surface inflates until the current stress point reaches the failure surface. The hardening relation between the yield criterion and the plastic work is described by a monotonically increasing function whose slope decreases with increasing plastic work. Beyond failure the yield surface contracts isotropically with increasing amount of plastic work.

3 ROTATIONAL KINEMATIC HARDENING

The essence of the rotational kinematic hardening model is presented in Fig. 1. In addition to rotation around the origin, the kinematic yield surface intersects the isotropic yield surface. During primary loading the isotropic teardrop-shaped yield surface is active. When stress reversal occurs, the pseudo-hydrostatic axis defining the position of the rotated kinematic surface passes through the stress reversal point. In this manner, both the orientation and the size of the kinematic surface is determined by the stress reversal point. The kinematic surface rotates in such a manner that it will merge into the isotropic yield surface when primary loading is encountered again. This pattern is accomplished if the pseudo-hydrostatic axis always rotates towards the true hydrostatic axis, and the kinematic surface expands such that its tip moves towards the tip of the isotropic surface. The region between the kinematic and isotropic surfaces defines an elastic domain which evolves as the kinematic surface rotates and expands in accordance with the stress reversal path. Thus, the model involves a combination of isotropic and kinematic hardening theories such that it preserves the behavior of the isotropic hardening model under monotonic loading conditions. The extension from isotropic to rotational kinematic hardening is accomplished entirely within the mathematical framework and without introducing new material parameters.

4 ROTATING PLASTIC POTENTIAL

To study whether this rotational kinematic hardening hypothesis is reasonable, the results of triaxial tests involving large stress reversals were analyzed. The relative magnitudes of the plastic strain increments may be characterized by vector directions superimposed on the triaxial plane. The directions of these vectors may be used to study the location of the plastic potential surface to which they should be perpendicular, as indicated in Fig. 2. The isotropic yield surface is attached to the plastic potential surface, and it expands monotonically as the stress point moves in the stress space.

Examples of the plastic strain increment vectors obtained according to the proposed rotational hypothesis are presented in Figs. 3 and 4. Predicted and measured plastic strain increment directions for a series of simple triaxial compression tests with several unloading-reloading cycles are shown in Fig. 3. Immediately following a stress reversal, the strains are mainly elastic, whereas the plastic strain components increase as the current stress point moves away from the point of stress reversal. The experimental plastic strain increment vectors are therefore determined towards the end of each straight section of the stress paths, where the larger

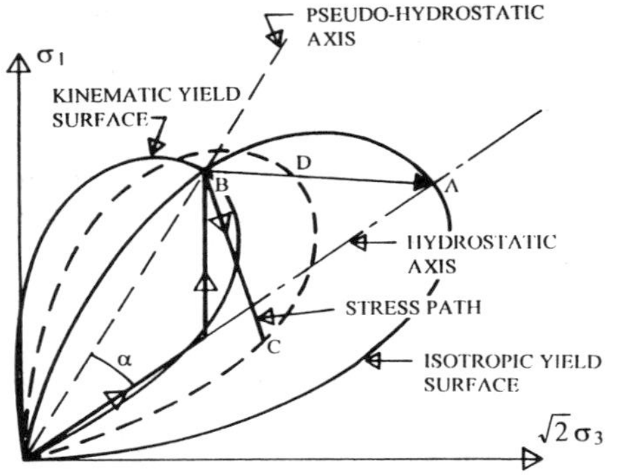

Fig. 1. Proposed modeling hypothesis consisting of rotation and growth of kinematic yield surface towards the isotropic yield surface upon stress reversal at point B.

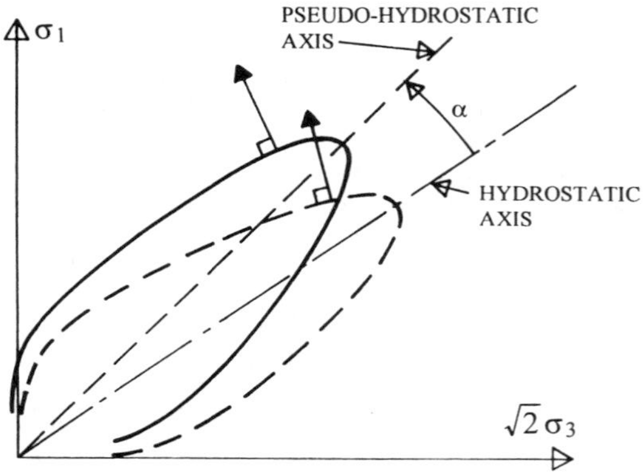

Fig. 2. Proposed modeling hypothesis consisting of rotation of plastic potential and plastic strain increment vector in triaxial plane.

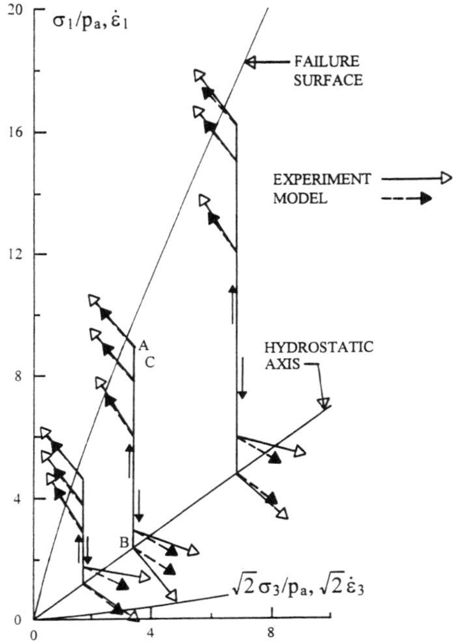

Fig. 3. Stress paths with superimposed plastic strain increment vectors in triaxial plane for unloading and reloading in conventional triaxial compression tests on loose Santa Monica Beach sand.

plastic strains provide more reliable evaluations. Note that although some of the stress states are in the region of compression, the strain increment directions for unloading point downwards similar to those observed for extension tests. For reloading, the vectors point upwards as they do for primary loading in compression.

The directions of the experimental and predicted vectors compare favorably. Although careful measurements were made in the tests performed for this investigation, these measurements do contain errors due to slight variations in void ratios between specimens, corrections for membrane penetration and sand penetration into the lubricated ends, etc. In addition, the quality of the calculated plastic strains employed in the study depends on the model and soil parameters used to compute the elastic strains, which were subtracted from the total strains measured during large stress reversals. Therefore, the experimental vector directions can be expected to show considerable scatter, especially when the strains are small. This is demonstrated in Fig. 3, where the experimental vector directions are on either side of the predicted vector directions. The vector directions predicted by the model are consistent and they do not show scatter, but they may show systematic deviations to one side of the experimental vector directions. However, it appears

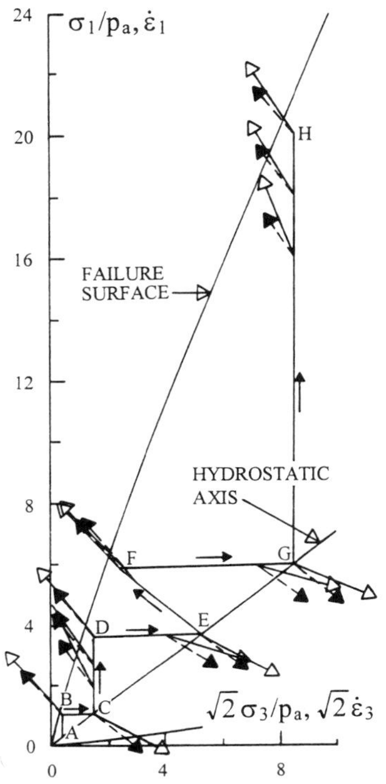

Fig. 4. Complex stress path with generally increasing mean normal stress and with superimposed plastic strain increment vectors in triaxial plane for test on loose Santa Monica Beach sand.

that the two sets of vector directions are reasonably close together, and the differences in vector directions shown in Fig. 3 are considered to reflect scatter rather than systematic deviations.

Fig. 4 shows the results of a test with a complex stress path. This test involves generally increasing mean normal stress, and the stress path is designed to explore plastic soil behavior in various directions and to see if the proposed model is able to capture the behavior of soil for such a complex stress path. Comparisons between experimental and predicted vector directions are favorable and generally within the scatter of the test results for such stress paths.

Considering the small strains measured for some stress paths, the inevitable variations from one test to another, as well as the corrections to the test data, it is clear that some scatter in the plastic strain increment directions occur. The largest amounts of scatter were associated with the stress paths producing the smallest strains. Within the scatter of the test results, the proposed model involving rotation of the plastic potential and the attached yield surface is shown to capture the behavior of sand during large stress reversals with reasonable accuracy.

5 ISOTROPIC AND KINEMATIC WORK-HARDENING

To calculate the magnitude of the plastic strain increments during large stress reversals, a work-hardening relation is required. For this purpose, the isotropic work-hardening law is extended to cover kinematic work-hardening by proposing that the already developed empirical relation between the yield criterion and its corresponding plastic work state is universally valid for both isotropic and kinematic hardening. Consequently, as the kinematic yield surface rotates and expands during a large stress reversal, a unique plastic work value may be computed at each step of its evolution.

The procedure is schematically depicted in Figs. 2 and 5. As shown in Fig. 2, the plastic strain increment direction is calculated using the plastic potential in the rotated frame of reference. Fig. 5 depicts the calculation of plastic strain increment magnitude using the increment of plastic work generated by the expansion of the yield surface. As indicated in the figure, the plastic strain increment magnitude is proportional to the difference in size of the initial and the current kinematic yield surfaces rotated back to the hydrostatic axis. Plastic strains can therefore be calculated both during primary loading and during stress reversals without introducing additional soil parameters.

For any load step, the increment of plastic work is determined by the slope of the work-hardening curve. Immediately after a stress reversal, the plastic modulus is quite high, and as loading continues, the modulus decreases, thus generating increasing amounts of plastic work and strain. In this manner, the value of the overall modulus is smoothly transformed from initially elastic to fully plastic behavior.

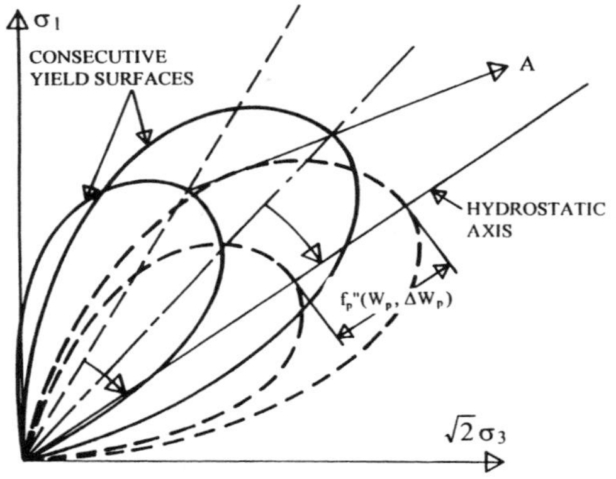

Fig. 5. Determination of magnitude of plastic work increment generated by expansion of kinematic yield surface rotated to position of isotropic yield surface.

6 SUBSEQUENT STRESS REVERSALS

Following the first stress reversal indicated at point B in Fig. 1, the stress state may continue to move towards the isotropic yield surface beyond point C (in Fig. 1), and the kinematic yield surface will merge with the isotropic surface as the stress state reaches this surface. Alternatively, another stress reversal may occur at point C. In fact, the stress reversal path within the isotropic yield surface may consist of any combination of minor reversals forming corners or cycles of unloading and reloading. For each such reversal, a progression of new kinematic surfaces must develop. This may easily be accomplished by forming consecutive kinematic surfaces according to the same logic used for the initial stress reversal. Eventually, such an approach will create a system of kinematic surfaces which are nested in the sense that each must grow towards its previously created counterpart. Since it is not desirable to keep in memory a large number of surfaces, and experimental observations show that soil tends to forget previous yielding mechanisms after sufficient yielding is activated by the latest surface, it was found appropriate to remember only one surface prior to the current kinematic surface. This surface will be referred to as the memory surface. Therefore, upon a new stress reversal at point C within the isotropic yield surface, a new kinematic yield surface is created as shown in Fig. 6. The previous kinematic yield surface is now referred to as the memory surface, with its tip at point D. The new kinematic surface will evolve towards the isotropic yield surface along the bilinear path, C-D-A, shown in Fig. 6. If a new kinematic surface is created, due to another stress reversal within the yield surface, the previous kinematic surface becomes the memory surface and the previous memory surface is erased.

The kinematic surface is related to the memory surface in the same manner as the memory surface is

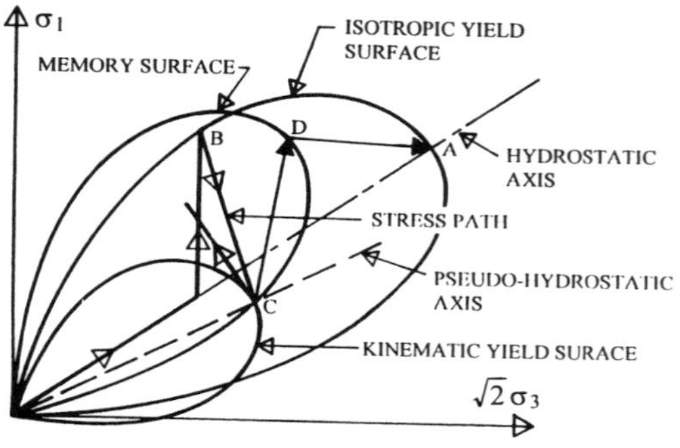

Fig. 6. Creation of new kinematic yield surface due to second stress reversal at point C. Old kinematic yield surface becomes memory surface.

related to the isotropic yield surface. The kinematic yield surface evolves in accordance with the stress reversal path and it eventually merges with the memory surface. In turn, the memory surface is activated and it subsequently evolves and merges with the isotropic yield surface. This is supported by experimental results.

7 COMPARISON OF EXPERIMENTAL DATA AND PREDICTIONS

The stress-strain behavior of sand observed in conventional triaxial laboratory tests were predicted to evaluate the proposed model. By comparing measured and predicted stress-strain curves, the model is tested in all aspects and conclusions may be made on its validity. The testing program consisting of triaxial tests with large stress reversals on loose, cylindrical specimens of Santa Monica Beach Sand was used for comparison. The model parameters required for the isotropic model were determined from conventional triaxial compression tests and employed in the predictions presented below.

Examples of predictions are compared with measured responses in Figs. 7, 8 and 9. Fig. 7 shows the prediction of a triaxial compression test at a constant confining pressure of 117.7 kPa (1.2 kg/cm^2) involving four consecutive cycles of unloading to and reloading from the hydrostatic stress state. This test corresponds to the stress path with the lowest confining pressure in Fig. 3. The model prediction captures the observed stress-strain behavior during each cycle. Note also that the volumetric strain behavior during stress reversals is captured quite accurately.

Fig. 8 shows the prediction of a triaxial compression-extension test at a constant confining pressures of 235.4 kPa (2.4 kg/cm^2). The cycle of unloading and reloading is captured quite accurately. The volumetric strain behavior is slightly off in magnitude, yet the qualitative behavior of the stress-strain relation is followed closely. Closer observation shows that the deviation in the magnitude of volumetric strains is accumulated in the initial primary loading portion of the test. If shifted, the predicted volumetric strain curve would fit the measured volumetric strain behavior with reasonable accuracy. Therefore, this deviation in magnitude may indicate some inconsistency in the measured stress-strain response or in the employed soil parameters rather than an inadequacy in the model.

Fig. 9 shows predictions for a triaxial test involving a more complicated stress path with generally decreasing mean normal stress. This test corresponds to the stress path shown in Fig. 4. The model prediction closely matches actual soil behavior along this complicated stress path. The prediction of volumetric strain behavior is almost

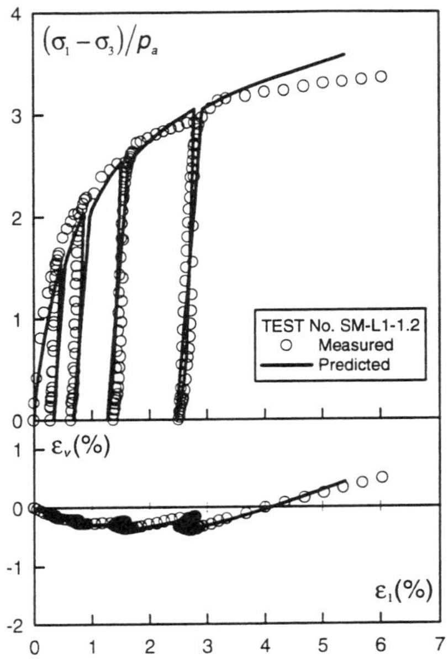

Fig. 7. Comparison of measured and predicted stress-strain and volume change relations for triaxial compression test on loose Santa Monica Beach sand.

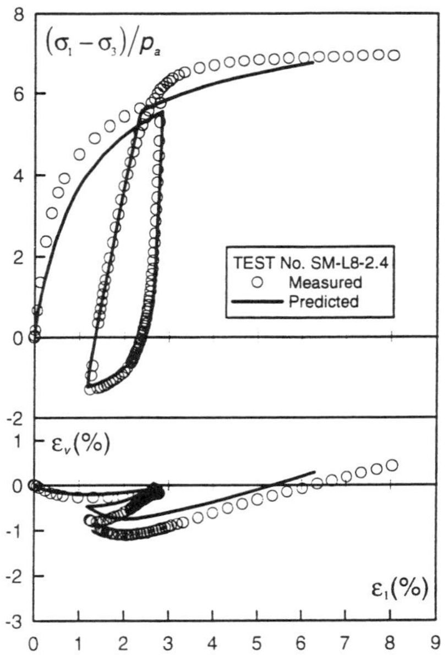

Fig. 8. Comparison of measured and predicted stress-strain and volume change relations for triaxial compression-extension test on loose Santa Monica Beach sand.

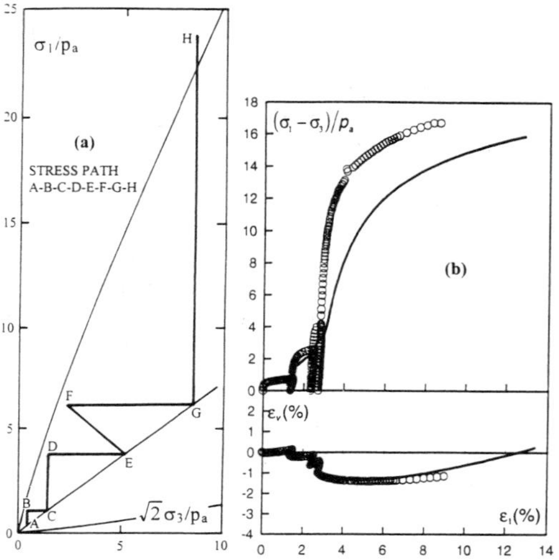

Fig. 9. (a) Complex stress path with generally increasing mean normal stress, and (b) Comparison of measured and predicted stress-strain and volume change relations for triaxial test on loose Santa Monica Beach sand.

perfect, and the prediction of stress-strain behavior deviates only as the soil is primary loaded to failure.

Overall, the predictions from the rotational kinematic hardening model matches the experimental behavior with good accuracy. When deviations occur, they appear to be caused by inconsistencies or scatter in the experimental results. The model seems to capture the stress-strain behavior in principle for stress paths in the triaxial plane.

8 CONCLUSION

A new rotational kinematic hardening model has been proposed which incorporates rotation and intersection of the yield surfaces to achieve a more consistent and physically rational fit with experimentally observed behavior. An existing single hardening constitutive model is used as the basic framework to which the rotational kinematic hardening mechanism has been added. The new combined model preserves the behavior of the isotropic hardening model under monotonic loading conditions, and the extension from isotropic to kinematic hardening is accomplished without introducing new material parameters. The characteristic work hardening law applies to isotropic as well as kinematic behavior of sand. The capability of the new model is examined by comparing predictions with experimental data for stress paths involving large stress reversals in the triaxial plane. Within the scatter of test results, the proposed model is shown to capture the behavior of sand with reasonable accuracy.

Acknowledgment

The research presented here was performed in the Department of Civil Engineering at University of California, Los Angeles, with support from the National Science Foundation under Grant No. MSS 9119272 / MSS 9396271. The junior author was supported partly by the Kenneth L. Lee Memorial Fellowship and partly by the Department of Civil Engineering at UCLA. Grateful appreciation is expressed for this support.

REFERENCES

Kim, M.K. & Lade, P.V. 1988. Single Hardening Constitutive Model for frictional Materials, I. Plastic Potential Function, *Computers & Geotechnics*. 5:307-324.

Lade, P.V. & Boonyachut, S. 1982. Large Stress Reversals in Triaxial Tests on Sand. Proc. 4th Int. Conf. Num. Meth. in Geomech., Edmonton, 171-181.

Lade, P.V. & Kim, M.K. 1988a. Single Hardening Constitutive Model for frictional Materials, II. Yield Criterion and Plastic Work Contours, *Computers & Geotechnics*. 6:13-30.

Lade, P.V. & Kim, M.K. 1988b. Single Hardening Constitutive Model for frictional Materials, III. Comparisons with Experimental Data, *Computers & Geotechnics*. 6:31-48.

Numerical Models in Geomechanics, Pietruszczak & Pande (eds) © 1997 Balkema, Rotterdam, ISBN 90 5410 886 X

Effect of heterogeneity upon stress-strain behavior of particle assemblies

Anil Misra
Department of Civil Engineering, University of Missouri-Kansas City, Mo., USA

ABSTRACT: Due to the irregular, often random, micro-structure of particle neighborhoods, particle assemblies exhibit spatially varying stress and strain fields. These variations have implication upon the prediction of stress-strain behavior of elements of particle assemblies using micro-mechanical modeling. In this paper, we examine the nature of this heterogeneity in the kinematic fields of particle assemblies with the view of obtaining overall kinematic measure in terms of particle-level measures. In the micro-mechanics approach, adopted herein, particle assemblies are considered to be a collection of rigid particles interacting via Hertz-Mindlin type contact interactions. The contact interaction provides a relationship between the forces at the contact and relative motion of the contacting particles. The contact forces and particle motions are related to the overall mean stress and strain fields in the particle packing.

1 INTRODUCTION

The stress-strain behaviors of dense particle assemblies are significantly dependent upon particle contact behavior, and the geometric arrangement of particles, especially, in the particle neighborhoods. In recent times, micro-mechanical considerations of inter-particle interactions have been utilized for devising stress-strain models of particulate materials accounting for the contact stiffness, and the packing structure. A number of these modeling efforts proceed with a mean strain field assumption that, particulate materials deform in accordance with an overall uniform strain field, cf. Digby (1981), Walton (1987), Jenkins (1987), Bathurst and Rothenburg (1988), Chang (1987). However, it is well understood now, that, the mean field models have limited applicability, although, they effectively account for particle contact interactions (Chang and Misra 1989). More recent modeling efforts have focussed upon characterizing fluctuations from uniform strain fields, cf. Koenders (1987), Chang and Liao (1990), Misra and Chang (1993).

In this paper, we examine the nature of the heterogeneity in the kinematic fields of particle assemblies with the view of obtaining overall kinematic measure in terms of particle-level measures. In the following discussion, we first describe the particle interaction model commonly employed in micro-mechanics of particulate media. We then describe the kinematics of granular media with the purpose of defining strain in granular media. Subsequently, a local stress-strain relationship is presented for a micro-element of the granular packing. The local stiffness tensor may be used in conjunction with a homogenization method to obtain the overall stress-strain law for an element of particle assembly.

2 PROBLEM DESCRIPTION

We consider a confined compact assembly of particles that support imposed loads at the boundary through resistance at inter-particle contacts. It is assumed that all the particles have similar stiffness properties. With the intent of keeping the discussions simple, the attention of this paper is focussed upon deformation of particle assemblies under low-level of shear loading. Under such loading conditions, it is reasonable to assume that the particles are bonded together and that there is no loss or gain of contacts.

Under an arbitrary deformation of a particulate assembly, the relative displacement δ_i^{nm} between two particles, m and n, is given by

$$\delta_i^{nm}=u_i^m-u_i^n+e_{ijk}(\omega_j^m r_k^m-\omega_j^n r_k^n) \qquad (1)$$

where u_i = the particle displacement, ω_k = the particle rotation, r_j is the vector joining the centroid of a particle to the contact point, the superscripts, n and m, refer to the particles and e_{ijk} is the permutation symbol. The tensor summation convention is used for the subscripts throughout this paper.

The relative movement between particles in contact leads to development of contact forces and moments which are modeled using the Hertz-Mindlin type contact model (see Misra 1995 for a review). The contact force f_i and the relative displacement δ_j are related via the stiffness K_{ij} as follows:

$$f_i = K_{ij}\delta_j \tag{2}$$

For convenience, the contact stiffness tensor is represented in terms of stiffnesses that describe the contact behavior along the direction normal and tangent to a particle contact, given in the following form

$$K_{ij} = K_n n_i n_j + K_s(s_i s_j + t_i t_j) \tag{3}$$

where K_n and K_s are the contact stiffnesses along the normal and tangential direction of the contact surface respectively. The unit vector **n** is normal to the contact surface and vectors **s** and **t** are arbitrarily chosen such that **nst** forms a local Cartesian coordinate system.

For the m-th particle in the assembly, the equilibrium conditions are written as

$$F_i^m - \sum_\alpha f_j^{m\alpha} = 0 \tag{4}$$

and

$$M_i^m - \sum_\alpha e_{ijk} f_j^{m\alpha} r_k^{m\alpha} = 0 \tag{5}$$

where F_i^m and M_i^m are the externally imposed force and the moment acting at the centroid of the m-th particle, $f_j^{m\alpha}$ is the force acting at the α-th contact of the m-th particle and the summation is performed over the contacts of the m-th particle. Note that contact moment is neglected in this work. For the particles within the granular assembly which are not located at the boundary no external load is acting therefore, F_i^m=0 and M_i^m=0.

3 KINEMATICS OF PARTICLE ASSEMBLY

For the purposes of micro-mechanical modeling, we consider the displacement and rotation of a particle to be decomposed into a part compatible with the overall average strain of a packing and a fluctuation part that varies from particle to particle. Thus, the relative displacement from Eq. (1) may be written as:

$$\begin{aligned}\delta_i^{nm} = {} & \bar{u}_{i,j} l_j^{nm} + e_{ijk}\bar{\omega}_k l_j^{nm} \\ & + \tilde{u}_i^m - \tilde{u}_i^n + e_{ijk}(\tilde{\omega}_j^m r_k^m - \tilde{\omega}_j^n r_k^n)\end{aligned} \tag{6}$$

where, $l_i^{nm} = X_i^m - X_i^n$ is the branch vector joining the centroid of the n-th particle with its m-th neighbor, $\bar{u}_{i,j}$ is the overall or average displacement gradient, and $\bar{\omega}_k$ is the average particle rotation, and the terms with a tilde (~) represent the fluctuations. We introduce, for convenience, a displacement function ϕ_i given for a contact of the n-th particle by

$$\phi_i^n = u_i^n + e_{ijk}\omega_j^n r_k^n \tag{7}$$

such that the relative displacement in Eq. (6) may be rewritten as

$$\delta_i^{nm} = \bar{\epsilon}_{ji} l_j^{nm} - \tilde{\phi}_i^n + \tilde{\phi}_i^m \tag{8}$$

where the average strain tensor, $\bar{\epsilon}_{ji}$ is defined to include the effect of particle rotation as

$$\bar{\epsilon}_{ji} = \bar{\phi}_{i,j} = \bar{u}_{i,j} + e_{ijk}\bar{\omega}_k \tag{9}$$

The strain tensor defined in Eq. (9) is, usually, non-symmetric. The conventional strain tensor may be recovered by taking the symmetric part of the distortion, $\epsilon_{(kl)}$ which is identical to the symmetric part of the displacement gradient $u_{(l,k)}$. The non-symmetric part represents the net particle rotation in excess of rigid body rotation (cf. Chang and Liao 1990).

To characterize the fluctuation terms in Eq. (8), the equilibrium of the particles in the assembly is considered. From force equilibrium of the n-th particle, using Eqs. (2), (4) and (8), we get

$$\begin{aligned}\bar{\epsilon}_{nj} \sum_m K_{ij}^{nm} l_m^{nm} - {} \\ \tilde{\phi}_j^n \sum_m K_{ij}^{nm} + \sum_m K_{ij}^{nm}\tilde{\phi}_j^m = 0\end{aligned} \tag{10a}$$

and for its m-th neighbor, we get

$$\bar{\epsilon}_{mj} \sum_p K_{ij}^{mp} l_m^{mp} - \tilde{\phi}_j^m \sum_p K_{ij}^{mp} + \sum_p K_{ij}^{mp} \tilde{\phi}_j^p = 0 \tag{10b}$$

Similar equilibrium conditions may be written for the neighbors of the m-th neighboring particles and so on. For convenience, the force equilibrium conditions, i.e. Eqs. (10a) and (10b), are rewritten as

$$\tilde{\phi}_j^n - \left(\sum_m K_{ij}^{nm}\right)^{-1} \sum_m K_{il}^{nm} \tilde{\phi}_l^m = \Gamma_{jkl}^n \bar{\epsilon}_{kl} \tag{11a}$$

and

$$\tilde{\phi}_j^m - \left(\sum_p K_{ij}^{mp}\right)^{-1} \sum_p K_{il}^{mp} \tilde{\phi}_l^p = \Gamma_{jkl}^m \bar{\epsilon}_{kl} \tag{11b}$$

where

$$\Gamma_{jkl}^n = \left(\sum_m K_{ij}^{nm}\right)^{-1} \sum_m K_{il}^{nm} l_k^{nm} \quad and$$
$$\Gamma_{jkl}^m = \left(\sum_p K_{ij}^{mp}\right)^{-1} \sum_p K_{il}^{mp} l_k^{mp} \tag{12}$$

Eqs. (11a) and (11b) along with similar equilibrium equations for other particles in the assembly yield a set of 3N linear algebraic equations for N particles in terms of the fluctuations, ϕ_j^n etc.. From the solution of this set of simultaneous equations, the fluctuations, ϕ_j^n and ϕ_j^m are conveniently written in terms of the average strain tensor, $\bar{\epsilon}_{ji}$ as

$$\tilde{\phi}_j^n = \left(\Gamma_{jkl}^n + \hat{\Gamma}_{jkl}^n\right) \bar{\epsilon}_{kl} \tag{13a}$$

and

$$\tilde{\phi}_j^m = \left(\Gamma_{jkl}^m + \hat{\Gamma}_{jkl}^m\right) \bar{\epsilon}_{kl} \tag{13b}$$

From Eq. (12), it is clear that the first term on the right-hand side of Eqs. (13a) and (13b) involves only the nearest neighbor, while the second term is associated with all the other particles in the assembly.

Note that for the present discussion, the particle rotations need not be determined explicitly, hence, only force equilibrium is considered. The fluctuations of particle rotations may be determined explicitly by considering the moment equilibrium of the particles as well. For example, moment equilibrium of the n-th particle yields

$$\bar{\epsilon}_{qp} \sum_m e_{ijk} K_{jp}^{nm} l_q^{nm} r_k^{nm} - \tilde{\phi}_p^n \sum_m e_{ijk} K_{jp}^{nm} r_k^{nm} + \sum_m e_{ijk} K_{jp}^{nm} r_k^{nm} \tilde{\phi}_p^m = (\tag{14}$$

Similar equations may be written for other particles in the assembly. The rotations may now be obtained by simultaneously considering Eqs. (10a) and (10b) along with Eq. (14).

Before we proceed further, it is of interest to examine the structure of coefficient that relate the fluctuations to overall average strains. Consider the coefficients involving nearest neighbors given by Eq. 12. It is simple calculation to show that the sum of all the contact stiffness tensors for a particle with one or more contacts, is non-singular. On the other hand, it may be shown that the sum of the first moments of contact stiffnesses, i.e. $\sum K_{il}\, l_k$, vanishes for centro-symmetric neighborhoods, i.e. if all the contacts are arranged in centrosymmetric fashion. A trivial example of such a neighborhood is a 2-dimensional hexagonal structure or a 3-dimensional body-centered cubic structure. For such particle assemblies, the right hand side of the set of Eqs. 11, vanishes. Hence, for such assemblies, the fluctuations would trivially vanish as well. For non-centro-symmetric neighborhoods, the fluctuations will normally not vanish, and this engenders further discussion

Using results of Eq. (13), the relative displacement, δ_i^{nm} given in Eq. (8), can be written in terms of the average strain tensor, $\bar{\epsilon}_{kl}$ as

$$\delta_j^{nm} = \left(\delta_{jl}\, l_k^{nm} - \Gamma_{jkl}^n - R_{jkl}^n\right) \bar{\epsilon}_{kl} \tag{15}$$

where δ_{ij}(=1 for i=j; =0 for i≠j) is the Kronecker delta. In Eq. (15), Γ_{jkl}^n denotes the term describing the influence of the nearest neighbors on the n-th particle and R_{jkl}^n denotes the remainder terms describing the influence of other particles in the assembly on the n-th particle, where

$$R_{jkl}^n = \hat{\Gamma}_{jkl}^n - \Gamma_{jkl}^m - \hat{\Gamma}_{jkl}^m \tag{16}$$

In order to evaluate the remainder term (Eq. 16), the complete connectivity of the particles in an assembly is required. Consequently, it is desirable to simplify the analysis. To this end, we consider the assembly to be represented by a set of micro-elements. Each micro-element is taken to be composed of a particle and its nearest neighbors. For the n-th micro-element, which is composed of the n-th particle and its nearest neighbor, we define a local average strain denoted by ϵ_{kl}^{n} such that the relative displacement δ_j^{nm} can be written in terms of the nearest neighbors only, as follows:

$$\delta_j^{nm} = (\delta_{jl} l_k^{nm} - \Gamma_{jkl}^{n}) \epsilon_{kl}^{n} \tag{17}$$

where the local average strain ϵ_{kl}^{n} is given by

$$\epsilon_{kl}^{n} = \bar{\epsilon}_{kl} + \Delta\epsilon_{kl}^{n} \tag{18}$$

In Eq. (18), $\Delta\epsilon_{kl}^{n}$ is defined as the fluctuation strain of the n-th micro-element.

Note, for the equivalence of relative displacements given in Eqs. (15) and (17), the following identity holds

$$\left(\delta_{jl} l_k^{nm} - \Gamma_{jkl}^{n}\right) \Delta\epsilon_{kl}^{n} + R_{jkl}^{n} \bar{\epsilon}_{kl} = 0 \tag{19}$$

Interestingly, Eq. (19) represents 3 equations for each contact of the n-th particle while 9 components of the fluctuation strain $\Delta\epsilon_{kl}^{n}$ need to be determined. Thus, for a particle with three contacts explicit solution of local average strain in terms of the overall average strain is possible. For particles with other than three contacts, the system of equations based on Eq. (19) is either over determined or under determined. Therefore, an estimate of the local average strain ϵ_{kl}^{n} in terms of the overall average strain $\bar{\epsilon}_{kl}$, is desired. A general transformation, for such an estimation, may be expressed as

$$\epsilon_{mn}^{n} = H_{mnkl}^{n} \bar{\epsilon}_{kl} \tag{20}$$

or from Eq. (18)

$$\Delta\epsilon_{mn}^{n} = \left(H_{mnkl}^{n} - I_{mnkl}\right) \bar{\epsilon}_{kl} \tag{21}$$

where H_{mnkl}^{n} is an unknown transformation tensor, and I_{mnkl} is a fourth-rank identity tensor defined in terms of Kronecker delta δ_{ij} as

$$I_{ijkl} = \frac{1}{2}\left(\delta_{ik}\delta_{jl} + \delta_{il}\delta_{jk}\right) \tag{22}$$

It is noted that the approximation Eq. (20) resembles the general form of 'self consistent' method used for the determination of strain distributions in an inhomogeneous material with randomly varying local stiffness tensor (Eshelby 1957). In addition to Eshelby's method, a number of schemes could be devised for specifying the transformation tensor in Eq. 20. Quite clearly, Eq. (20) may only yield an approximate relative displacement at a contact. For instance, this approximation may not necessarily guarantee that equilibrium is satisfied for all particles. However, more importantly, it provides considerable simplification in the derivation of overall stress-strain relationship of an element of particle assembly.

4 A LOCAL STRESS-STRAIN RELATIONSHIP

A key requirement of the `self consistent' method is the local stress-strain relationship for a micro-element of the granular media, such that for the n-th micro-element

$$\sigma_{ij}^{n} = C_{ijkl}^{n} \epsilon_{kl}^{n} \tag{23}$$

where the superscript n refers to the micro-element and C_{ijkl}^{n} is the local stiffness tensor. The local stiffness tensor C_{ijkl}^{n} is derived in terms of the contact stiffnesses and the relative position of the neighboring particles. The derivation is facilitated by considering: (a) the relationship between local strain and relative movement of the particle in the micro-element (given in Eq. 17); (b) the interaction of two particles (given in Eq. 2); and (c) the relationship of local stress and contact forces. The local stress σ_{ij}^{n} for the n-th micro-element is given in terms of the contact forces f_j^{nm} as

$$\sigma_{ij}^{n} = \frac{1}{2V^{n}} \sum_{m} l_i^{nm} f_j^{nm} \tag{24}$$

where V^n is the volume associated with the nth particle or micro-element.

Thus, using Eqs. (2), (17) and (24), the local stiffness tensor C_{ijkl}^{n} is obtained to be

$$C_{ijkl}^{n} = \frac{1}{2V^{n}} \sum_{m} l_i^{nm} K_{jl}^{nm} l_k^{nm} - \frac{1}{2V^{n}} \sum_{m} l_i^{nm} K_{jr}^{nm} \Gamma_{rkl}^{n} \tag{25}$$

The stiffness tensor thus obtained is a function of the packing structure measures l_i and V^n and contact stiffnesses K_n and K_s. The local stiffness tensor C_{ijkl}^{n} is, in general, asymmetric with respect to interchange of leading as well as terminal pair of indices. However, due to the symmetry of the contact stiffness tensor K_{ij}, the local stiffness tensor possesses the following symmetry

$$C^n_{ijkl} = C^n_{ilkj} \quad and \quad C^n_{ijkl} = C^n_{klij} \tag{26}$$

5 AVERAGING FOR PARTICLE ASSEMBLY

For an element of particle assembly, an overall stress-strain relationship, that relates the overall mean strain $\bar{\epsilon}_{ij}$, to the overall mean stress $\bar{\sigma}_{ij}$, is desired. To this end, we consider a representative element of particle assembly containing enough number of particles such that the overall stress and strain of the element can be obtained as volume averages of the corresponding local quantities defined for a particle neighborhood. The volume averages are written as

$$\bar{\sigma}_{ij} = \frac{1}{V}\sum_n V^n \sigma^n_{ij} \tag{27}$$

$$\bar{\epsilon}_{ij} = \frac{1}{V}\sum_n V^n \epsilon^n_{ij} \tag{28}$$

where $\bar{\epsilon}_{ij}$ and $\bar{\sigma}_{ij}$, are the overall stress and strain tensors defined for a representative volume, $\sigma_{ij}{}^n$ and $\epsilon_{ij}{}^n$ are the local stress and strain tensors defined for a micro-element, V is the volume of the representative volume given by $\sum_n V^n$, where summation is carried over all the particles in the volume. A volume average stiffness tensor C_{ijkl} may also be defined in terms of the micro-element stiffness $C_{ijkl}{}^n$ as

$$\bar{C}_{ijkl} = \frac{1}{V}\sum_n V^n C^n_{ijkl} \tag{29}$$

Within the representative element, the local field quantities may be written as the summation of the average term uniform everywhere and a fluctuation term for each micro-element such that for stresses

$$\sigma^n_{ij} = \bar{\sigma}_{ij} + \Delta\sigma^n_{ij} \tag{30}$$

and for strains from Eq. (18)

$$\epsilon^n_{ij} = \bar{\epsilon}_{ij} + \Delta\epsilon^n_{ij} \tag{31}$$

Also, the stiffness within the volume may be written as the summation of a quantity uniform everywhere and a fluctuation term

$$C^n_{ijkl} = \bar{C}_{ijkl} + \Delta C^n_{ijkl} \tag{32}$$

By definition the volume averages of all the fluctuation terms vanish, that is

$$\begin{aligned} \frac{1}{V}\sum_n V^n \Delta\sigma^n_{ij} &= \frac{1}{V}\sum_n V^n \Delta\epsilon^n_{ij} \\ &= \frac{1}{V}\sum_n V^n \Delta C^n_{ijkl} = 0 \end{aligned} \tag{33}$$

It is remarked that the volume average stiffness tensor C_{ijkl}, defined in Eq. (29), is the effective stiffness of the representative element if and only if the strain field within the volume is uniform (Chang and Misra 1990). However, it is clear from Eqs. (18) and (19), that the strain field varies within the representative volume. In view of this and Eqs. (20), (23), (27) and (28), the relationship between the overall stress and strain tensors of the representative volume is defined using an effective stiffness tensor C_{ijkl} such that

$$\bar{\sigma}_{ij} = C_{ijkl}\bar{\epsilon}_{kl} \tag{34}$$

where, the effective stiffness tensor C_{ijkl} is written in terms of local stiffness tensor $C_{ijkl}{}^n$ and the transformation tensor $H_{ijkl}{}^n$ of a micro-element as

$$C_{ijkl} = \frac{1}{V}\sum_n V^n C^n_{ijmn} H^n_{mnkl} \tag{35}$$

From Eq. (28), it is seen that volume averaging requires

$$\frac{1}{V}\sum_n V^n H^n_{ijkl} = I_{ijkl} \tag{36}$$

where I_{ijkl} is a fourth rank identity tensor defined in Eq. (22).

6 CONCLUSIONS

The paper outlines a methodology for determining average stress-strain behavior of a particle assembly element. The method is based upon the important micro-mechanical consideration of particle contact behavior. The main result of the paper, contained in Eq. (34), defines the effective stiffness tensor of a particle assembly element. This definition of overall stiffness tensor resembles those given in 'self consistent' and other homogenization theories of heterogeneous media.

There are two key elements of averaging methodology for particle assemblies: (1) the local stiffness tensor, and (2) the transformation tensor. In the methodology described herein, the local stiffness tensor is defined for a micro-element, which is taken to be composed of a particle and its nearest neighbor. The derived local stress-strain relationship is given in Eq. 25. Further, the average local strain, defined for a micro element, is estimated in terms of the overall average strain of a representative element through a fourth rank transformation tensor. A number of methods, including that by Eshelby, may be used for specifying the transformation tensor.

REFERENCES

Bathurst, R.J. and Rothenburg, L. (1988). Micromechanical aspects of isotropic granular assemblies with linear contact interactions. *Journal of Applied Mechanics*, Vol. 55, No. 1, 17-23.

Chang, C.S. (1987). Micromechanical modelling of constitutive relations for granular material. *Micromechanics of Granular Media* (Edited by J.T. Jenkins and M. Satake), 271-278. Elsevier, Amsterdam.

Chang, C.S. and Liao, C.L. (1990). Constitutive relations for particulate medium with the effect of particle rotation. *International Journal of Solids and Structures*, Vol. 26, No. 4, 437-453.

Chang, C.S. and Misra, A. (1990). Application of uniform strain theory to heterogeneous granular solids. *Journal of Engineering Mechanics*, Vol. 116, No. 10, 2310-2328.

Digby, P.J. (1981). The effective elastic moduli of porous granular rocks. *Journal of Applied Mechanics*, Vol. 48, No. 4, 803-908.

Eshelby, J.D. (1957). The determination of the elastic field of an ellipsoidal inclusion and related problems. *Proceedings of Royal Society of London*, Vol. A241, 376-396.

Jenkins, J.T. (1987). Volume change in small strain axisymmetric deformations of a granular material. *Micromechanics of Granular Media* (Edited by J.T. Jenkins and M. Satake), 245-252. Elsevier, Amsterdam.

Koenders, M.A. (1987). The incremental stiffness of an assembly of particles. *Acta Mechanica*, Vol. 70, 31-49.

Misra, A. (1995). Interfaces in particulate materials. *Mechanics of Geomaterial Interfaces* (Edited by A.P.S. Selvadurai and M.J. Boulon), 513-536. Elsevier, Amsterdam.

Misra, A. and Chang, C.S. (1993). Effective elastic moduli of heterogeneous granular solids. *International Journal of Solids and Structures*, Vol. 30, No. 18, 2547-2566.

Walton, K. (1987). The effective elastic moduli of a random packing of spheres. *Journal of Mechanics and Physics of Solids*, Vol. 35, No. 2, 213-226.

Numerical Models in Geomechanics, Pietruszczak & Pande (eds) © *1997 Balkema, Rotterdam, ISBN 90 5410 886 X*

A refined superior sand model

A. Drescher
University of Minnesota, Minneapolis, Minn., USA
Z. Mroz
Institute of Fundamental Technological Research, Warsaw, Poland

ABSTRACT: This paper describes an elasto-plastic constitutive model Superior Sand refined to incorporate the effect of deposition density on the stress/strain response in drained isotropic and undrained triaxial loading. The modified hardening rule contains an additional, reference state-dependent term. For a loose sand the resulting undrained stress path displays two hardening and one softening regimes, and for a dense sand only the hardening regime. The implications for modeling static liquefaction are discussed.

1 INTRODUCTION

There have been numerous constitutive models proposed for sand in order to simulate the inelastic deformation under monotonic and variable loading (Lade & Duncan 1975, Lade 1977, Nova & Wood 1979, Mroz & Pietruszczak 1983, Vermeer 1984). In most cases the sand was treated as an elasto-plastic, hardening material, for which the instantaneous plastic deformation is governed by the flow rule. The non-associative flow rule was used in recent papers by Drescher et.al. (1995), Birgisson & Drescher (1996), and earlier by Nova & Wood (1979), Vermeer (1988), Jarzebowski & Mroz (1988). To specify more accurately the undrained response, the combined, volumetric and distortional hardening models were proposed (Nova and Wood 1979, Mroz & Pietruszczak 1983). The latter paper incorporated the anisotropic hardening rule, with an infinite number of loading surfaces translating and expanding within the domain bounded by the external loading surface. The application to monotonic and cyclic loading phenomena was presented in that paper as well as in Jarzebowski & Mroz (1988).

The aim of this paper is to present a refined isotropic hardening model for sand by starting from the Superior Sand model formulation proposed by Drescher et. al. (1995) which uses the non-associative flow rule and the assumption of volumetric hardening. However, as initial sand deposition affects noticeably the deformation response, both the initial density, or specific volume, and the applied consolidating pressure will be introduced as state variables. Further, a modified hardening rule will be introduced to describe the undrained stress-strain response differing for loose and dense sands. The constitutive equations are formulated in customary triaxial invariants p', q, ε_p, and ε_q, and the specific volume v; compression is taken as positive, and (′) denotes effective stress.

2 SUPERIOR SAND MODEL

The elasto-plastic Superior Sand model bears similarity with the Cam Clay models in that the loading/unloading isotropic compression

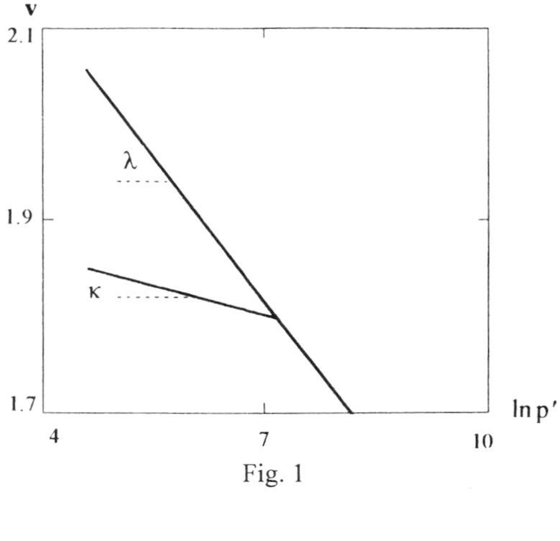

Fig. 1

relationship is represented in the $v:\ln p'$-plane by two straight lines (Fig. 1), and the elastic strain-rates are given by

$$\dot{\varepsilon}_p{}^e = \frac{\kappa}{v}\frac{\dot{p}'}{p'}, \quad \dot{\varepsilon}_q{}^e = \frac{1}{3\mu}\dot{q} \tag{1}$$

where μ is the shear modulus.

The plastic loading surface F = 0 and the plastic potential G = 0 are differing, however (Fig.2)

$$F = |q| - Lp'\sqrt{3}\sqrt{1 - \frac{p'}{p'_c}} = 0$$

$$G = |q| - Mp'\sqrt{3}\sqrt{1 - \frac{p'}{p'_l}} = 0 \tag{2}$$

where the parameter p'_l is determined from the condition that F = 0 and G = 0 correspond to the actual stress point.

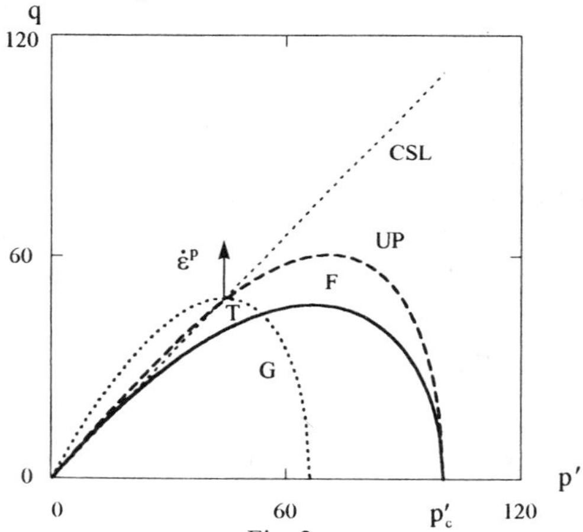

Fig. 2

The plastic strain-rates are specified by the non-associative flow rule

$$\dot{\varepsilon}_p{}^p = \dot{v}\frac{\partial G}{\partial p'}, \quad \dot{\varepsilon}_q{}^p = \dot{v}\frac{\partial G}{\partial q} \tag{3}$$

and $\dot{v}$ is the plastic multiplier. The plastic volumetric strain-rate $\dot{\varepsilon}_p{}^p = 0$ at the critical state line (CSL) given by $q = Mp'$.

Likewise in the Cam Clay models, the parameter p'_c in (2a) is assumed to be only a function of the plastic volumetric strain, with the resulting hardening rule derived from the loading/unloading $v:\ln p'$ diagram (Fig. 1)

$$\dot{p}'_c = \frac{p'_c v}{\lambda - \kappa}\dot{\varepsilon}_p{}^p \tag{4}$$

The assumptions above in the Superior Sand were selected to model the undrained behavior ($\dot{\varepsilon}_p = 0$) of very loose sands. In fact, as shown in Drescher et al. (1995), and Birgisson & Drescher (1996), the model predicts adequately the effective undrained stress path (UP) originating from $p'_c = p'_{ci}$, which is given by

$$q = Lp'\sqrt{3}\sqrt{1 - \left(\frac{p'}{p'_{ci}}\right)^{\frac{\lambda}{\lambda-\kappa}}} \tag{5}$$

and which terminates at the critical state line (point T in Fig. 2). The corresponding $q:\varepsilon_q$ curve upon reaching a peak displays monotonic softening. However, the assumption of volumetric hardening (4) is deficient in neglecting the dependency of the $v:\ln p'$ loading curve on the sand deposition density, as reported by several authors (Sladen & Handford 1987, Jefferies & Been 1987, Sladen & Oswell 1989, Jefferies 1993, Ishihara 1993, Pestana & Whittle 1995). Furthermore, the model does not describe the undrained response of medium dense and dense sands, when upon softening the material hardens again (Ishihara 1993). The following refined model is an attempt to eliminate these deficiencies.

3 REFINED SUPERIOR SAND MODEL

To account for the dependence of the v:lnp loading curve on the initial density, we postulate that the hardening rule (4) contains an additional term related to the distance β between the pressure p'_c and the reference pressure p'^*_c, with the $v:\ln p'^*_c$ relationship taken as linear (Fig. 3). The reference pressure can be regarded as a parameter of a bounding surface $F^* = 0$ similar to F = 0, which was introduced by Mroz & Pietruszczak (1983), and Jarzebowski & Mroz (1988). Alternatively, the distance between the current specific volume v and the reference volume v^* can be introduced as suggested by Wood et. al. (1994) in a simplified model for drained response. We also postulate that the additional term in the hardening rule depends on both the volumetric and weighted deviatoric plastic strain-rates. The resulting hardening rule can be written as

$$\dot{p}'_c = \frac{p'_c v}{\lambda - \kappa}\dot{\varepsilon}_p{}^p + |\dot{\varepsilon}^p| p'_c < f(\beta) > \tag{6}$$

where

$$\dot{\varepsilon}^p = \sqrt{\alpha^2\left(\dot{\varepsilon}_q{}^p\right)^2 + \left(\dot{\varepsilon}_p{}^p\right)^2} \tag{7}$$

and α is a weighting parameter. The function $f(\beta)$ is taken as

$$f(\beta) = m\left(\frac{p'^*_c}{p'_c} - \eta s - 1\right)^n \tag{8}$$

where m and n are constants, and $\eta = q/p'$; s is a parameter which is so adjusted that $f(\beta)=0$ when the undrained stress path terminates at the critical state line. The McCauley bracket denotes: $< f > = f$ for $f \geq 0$ and $< f > = 0$ for $f < 0$.

The elasto-plastic constitutive equation in the matrix form can be written as

$$\begin{bmatrix}\dot{\varepsilon}_q \\ \dot{\varepsilon}_p\end{bmatrix} = \begin{bmatrix}\frac{1}{3\mu} + \frac{1}{h}\frac{\partial G}{\partial q}\frac{\partial F}{\partial q} & \frac{1}{h}\frac{\partial G}{\partial q}\frac{\partial F}{\partial p'} \\ \frac{1}{h}\frac{\partial G}{\partial p'}\frac{\partial F}{\partial q} & \frac{\kappa}{vp}\frac{\partial G}{\partial p'}\frac{\partial F}{\partial p'}\end{bmatrix}\begin{bmatrix}\dot{q} \\ \dot{p}'\end{bmatrix} \tag{9}$$

where the gradients are given by

$$\frac{\partial F}{\partial q} = \pm 1, \quad \frac{\partial F}{\partial p'} = \frac{3}{2}\frac{L^2 - \eta^2}{\eta}$$
$$\frac{\partial G}{\partial q} = \pm 1, \quad \frac{\partial G}{\partial p'} = \frac{3}{2}\frac{M^2 - \eta^2}{\eta} \tag{10}$$
$$\frac{\partial F}{\partial p'_c} = -\frac{1}{6L^2}\frac{(3L^2 - \eta^2)^2}{\eta}$$

Note that for q>0, $\partial F/\partial q = \partial G/\partial q = 1$. Since the expression (7) can be written as

$$\dot{\varepsilon}^p = \dot{v}k(\eta), \quad k(\eta) = \sqrt{\alpha^2 + \frac{9}{4}\left(\frac{M^2 - \eta^2}{\eta}\right)^2} \tag{11}$$

the plastic hardening modulus h in (9) is

$$h = \frac{1}{6L^2}\frac{\left(3L^2 - \eta^2\right)^2}{\eta}\left[\frac{p'_c v}{\lambda - \kappa}\frac{3(M^2 - \eta^2)}{2\eta}\right.$$
$$\left. + k(\eta)p'_c < f(\beta) > \right] \tag{12}$$

4 RESPONSE OT THE REFINED MODEL

To illustrate the response of the refined Superior Sand model, two particular loading histories are considered below: a) isotropic compression, and b) undrained triaxial compression.

4.1 *Response in isotropic compression*

The response to isotropic compression, $\dot{q} = 0$, $\dot{p}' > 0$, is derived from (6) and (7) by setting $\dot{\varepsilon}_q{}^p = 0$; this yields

$$\dot{p}'_c = \dot{\varepsilon}_p{}^p\left[\frac{p'_c v}{\lambda - \kappa} + p'_c < f(\beta) >\right] \tag{13}$$

or

$$\frac{dp'_c}{p'_c} = -\frac{\left[\frac{1}{\lambda - \kappa} + \frac{< f(\beta) >}{v}\right]}{\frac{\lambda}{\lambda - \kappa} + \frac{\kappa}{v} < f(\beta) >}dv \tag{14}$$

where $f(\beta)$ for η=0 is

$$f(\beta) = m\left(\frac{p'^*_c}{p'_c} - 1\right)^n \tag{15}$$

The presence of the terms containing $f(\beta)$ in (14) results in the $v{:}\ln p'$ relationship dependent on the initial specific volume, and no longer linear. This is illustrated in Fig. 3 for three different initial specific volumes v_0 = 1.93, 1.96, 1.99, initial pressure ratio $p'^*_c / p'_c = 2.58$ for $v_0 = 1.99$, and for the constants λ = 0.1, κ = 0.02, m = 30, and n = 2.

4.2 *Response in undrained triaxial tests*

Consider the undrained triaxial test on a fully saturated sand. From the constitutive equation (9) and condition $\dot{\varepsilon}_p = 0$, we can derive the following differential equation defining the resulting undrained stress path in the $p':q$ plane

$$\frac{dq}{dp'}=\frac{\dot{q}}{\dot{p}'}=-\frac{\dfrac{\kappa h}{vp'}+\dfrac{\partial G}{\partial p'}\dfrac{\partial F}{\partial p'}}{\dfrac{\partial G}{\partial p'}\dfrac{\partial F}{\partial q}} \tag{16}$$

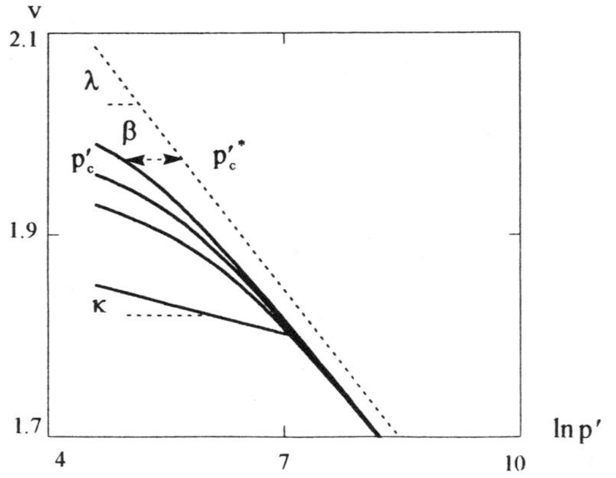

Fig. 3

Using expressions (12), (13), and (14), we can write (16) as

$$\frac{dq}{dp'}=-\frac{1}{2\eta}\left[(3L^2-\eta^2)\frac{\kappa}{\lambda-\kappa}+3(L^2-\eta^2)\right] -\frac{\kappa}{3v}\frac{3L^2-\eta^2}{M^2-\eta^2}k(\eta)<f(\beta)> \tag{17}$$

The differential equation (17) is singular at $\eta = M$, and cannot be integrated analytically. When the second term is neglected, however, that is when only volumetric hardening is assumed, analytic integration is possible and leads to Eq.(5).

The deviatoric response is given by

$$\dot{\varepsilon}_q=\left(\frac{1}{3\mu}+\frac{1}{h}\frac{\partial G}{\partial q}\frac{\partial F}{\partial q}\right)\dot{q}+\left(\frac{1}{h}\frac{\partial G}{\partial q}\frac{\partial F}{\partial p'}\right)\dot{p}' \tag{18}$$

which, upon expressing $\dot{p}'$ in terms of $\dot{q}$, we can write as

$$\dot{\varepsilon}_q=\frac{\dfrac{\kappa}{3\mu vp'}+\dfrac{1}{3\mu h}\dfrac{\partial G}{\partial p'}\dfrac{\partial F}{\partial p'}+\dfrac{\kappa}{vp'h}\dfrac{\partial G}{\partial q}\dfrac{\partial F}{\partial q}}{\dfrac{\kappa}{vp'}+\dfrac{1}{h}\dfrac{\partial G}{\partial p'}\dfrac{\partial F}{\partial p'}}\dot{q} \tag{19}$$

and

$$\frac{dq}{d\varepsilon_q}=\frac{\dfrac{\kappa h}{vp'}+\dfrac{\partial G}{\partial p'}\dfrac{\partial F}{\partial p'}}{\dfrac{\kappa h}{3\mu vp'}+\dfrac{1}{3\mu}\dfrac{\partial G}{\partial p'}\dfrac{\partial F}{\partial p'}+\dfrac{\kappa}{vp'}\dfrac{\partial G}{\partial q}\dfrac{\partial F}{\partial q}} \tag{20}$$

The minimal point on the $p':q$ path, and the limit point on the $q:\varepsilon_q$ curve, are specified by the same condition

$$\frac{\kappa h}{vp'}+\frac{\partial G}{\partial p'}\frac{\partial F}{\partial p'}=0 \tag{21}$$

which, upon utilizing (10) and (12), we can write as

$$\left(\frac{\partial F}{\partial p'}-\frac{p'_c}{p'}\frac{\kappa}{\lambda-\kappa}\frac{\partial F}{\partial p'_c}\right)\frac{\partial G}{\partial p'} -\frac{\partial F}{\partial p'_c}\frac{\kappa p'_c}{vp'}k(\eta)<f(\beta)=0 \tag{22}$$

Figure 4 shows three undrained stress paths originating at one initial consolidation pressure p_{ci} = 100, initial reference pressure p'^*_{ci} = 258 at v_0 = 1.99, and three deposition densities with v_0 = 1.93, 1.96, and 1.99; the following material constants were selected: $\lambda = 0.1$, $\kappa = 0.02$, $\alpha = 5$, m = 30, n = 2, L = 0.7, M = 1.1. The corresponding values of the parameter s are: v_0 = 1.93, s=0.401; v_0 = 1.96, s=0.403; v_0 = 1.99, s=0.408.

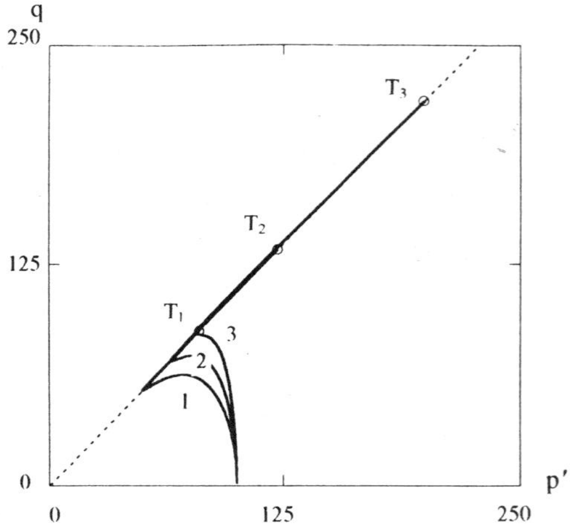

Fig. 4

It is clearly seen that in contrast to the original Superior Sand, the refined model predicts a stress path which crosses the critical state line (CSL) at point C, and subsequently gradually approaches this line from above. Figure 5 is a magnified portion of Fig. 4, and illustrates the rate at which the stress path turns above the CSL. The location of point T at which the stress path terminates at the CSL depends on the initial specific volume; the lower is this volume (higher density) the further up is located this terminal point. The stress paths depicted in Fig. 4 indicate that for low densities (loose sand) the resulting $q{:}\varepsilon_q$ curve will display hardening/softening/ hardening regimes, whereas for a dense sand only the hardening regime will be present (Ishihara 1993). The $q{:}\varepsilon_q$ curves corresponding to the three stress path of Fig. 4 and μ=10000 are shown in Fig. 6.

5 CLOSING REMARKS

We have demonstrated that postulating an additional term in the Superior Sand hardening rule (4) leads to model response that matches qualitatively the experimentally observed behavior of sands of differing initial densities. In particular, the refined model accounts for the influence of deposition density on both the purely isotropic (drained) response, and deviatoric undrained response.

In the new hardening rule (6) the first term is directly related to $\dot{\varepsilon}_p^{\,p}$ which, in turn, is related to the plastic potential G=0, and its contribution vanishes at all p', q points where $\partial G / \partial p' = 0$, i.e. on the critical state line. The contribution of the second term in (6), however, is more complex, for the term consists of two parts which behave differently. The first part is a function of $\dot{\varepsilon}_p^{\,p}$ and $\dot{\varepsilon}_q^{\,p}$ related to G=0, and represents the strength of hardening always greater then zero. The second part depends on the distance β between the loading surface F=0 and the reference surface F^*=0, and the function $f(\beta)$ determines when the second term ceases to contribute to hardening. The form of $f(\beta)$, and the distance β at which $f(\beta)$=0, cannot be deduced from the isotropic response only, for in the refined Superior Sand model the isotropic and the undrained responses derive from the same constitutive equation. In other words, there is no guarantee that function $f(\beta)$ postulated on the basis of isotropic compression experiments will lead to a stress path that is smooth, crosses the CSL only once, and terminates exactly at the CSL. This requirement can be met, for example, by introducing the term ηs in (8) which allows for adjustments unrelated to isotropic response. Obviously, other forms of $f(\beta)$ can also be postulated.

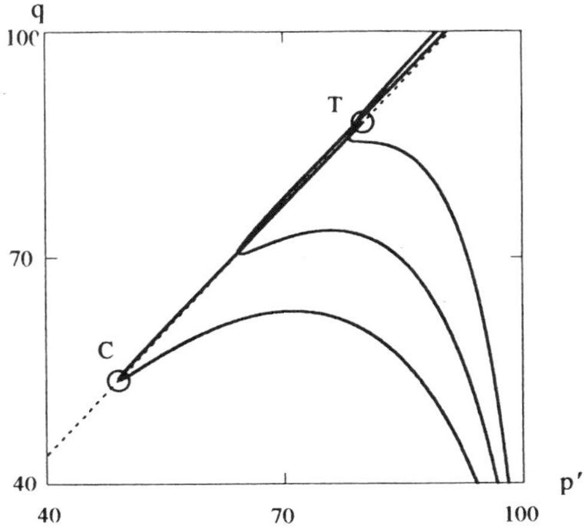

Fig. 5

The appropriate choice of the function $f(\beta)$ is critical for the refined Superior Sand model to predict adequately the asymptotic deviatoric stress q in undrained loading as observed in experiments (Ishihara, 1993). The significance of limiting deviatoric stresses becomes evident when the model is applied for describing static liquefaction of saturated sands considered by several authors (Sladen et al. 1985, Kramer & Seed 1988, Lade

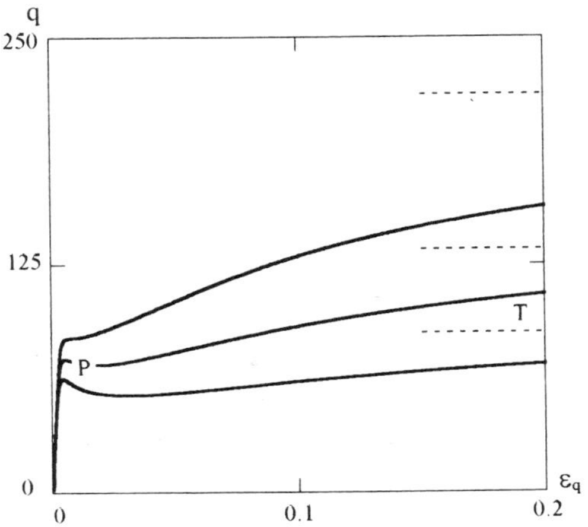

Fig. 6

1992, Sasitharan et al. 1994). Referring to Fig. 6, the static liquefaction may occur if the $q:\varepsilon_q$ curve displays a softening regime, with the area between the curve and a horizontal line through the peak representing the energy released upon liquefaction. If the terminal point T lies above the peak point P, the energy released can be absorbed by the material; if, in addition, the strains at which the energy is absorbed are sufficiently small, the effect of liquefaction may not be disastrous. On the other hand, if point T is located below point P, unmitigated collapse may occur.

REFERENCES

Birgisson, B., & A. Drescher (1996). A model for flow liquefaction in saturated loose sand. In S. Erlingsson & H. Sigursteinsson (eds), *XII Nordic Geotech. Conf. NGM-96*: 1-7.

Drescher, A., B. Birgisson & K. Shah (1995). A model for water saturated loose sand. In G.N. Pande and S. Pietruszczak (eds), *V Int. Symp. Num. Models Geomech. NUMOG V*: 109-112. Rotterdam: Balkema.

Ishihara, K. (1993). Liquefaction and flow failure during earthquakes. *Geotechnique* 43:351-415.

Jarzebowski, A.& Z. Mroz (1988). A constitutive model for sands and its application to monotonic and cyclic loadings. In A. Saada and M. Bianchini (eds), *Constitutive Equations for Granular Non-cohesive Soils*: 307-323. Rotterdam: Balkema.

Jefferies, M.G. (1993). Nor-Sand: a simple critical state model for sand. *Geotechnique* 43:91-103.

Jefferies, M.G. & K. Been (1987). Use of critical state representations of sand in the method of stress characteristics. *Can. Geotech. J.* 24:441-446.

Kramer, S.L. & H.B. Seed (1988). Initiation of soil liquefaction under static loading condition. *J. Geotech. Eng. Div., ASCE* 114:412-430.

Lade, P.V. (1977). Elasto-plastic stress-strain theory for cohesionless soil with curved yield surface. *Int. J. Solids Struct.* 13:1019-1035.

Lade, P.V. (1992). Static instability and liquefaction of loose fine sandy slopes. *J. Geotech. Eng., ASCE* 118:51-71.

Lade, P.V., & J.N. Duncan (1975). Elastoplastic stress-strain theory for cohesionless soil. *J. Geotech. Eng. Div., ASCE* 101:412-430.

Mroz, Z. & S. Pietruszczak (1983). A constitutive model for sand with anisotropic hardening rule. *Int. J. Num. Anal. Meth. Geomech.* 7:305-320.

Nova, R. & D.M. Wood (1979). A constitutive model for sand in triaxial compression. *Int. J. Num. Anal. Meth. Geomech.* 3:255-278.

Pestana, J.M. & A.J. Whittle (1995). Compression model for cohesionless soils. *Geotechnique* 45:611-631.

Sasitharan, S., P.K. Robertson, D.C. Sego & N.R. Morgenstern (1994). State-boundary surface for very loose sand and its practical implications. *Can. Geotech. J.* 31:321-334.

Sladen, J.A. & G. Handford (1987). A potential systematic error in laboratory testing of very loose sands. *Can. Geotech. J.* 24:462-466.

Sladen, J.A. & J.M. Oswell (1989). The behavior of very loose sand in the triaxial compression test. *Can. Geotech. J.* 26:103-113.

Sladen, J.A., R.D. D'Hollander & J. Krahn (1985). The liquefaction of sands, a collapse surface approach. *Can. Geotech. J.* 22:564-578.

Vermeer, P.A. (1984). A five-constant model unifying well established concepts. In G. Gudehus, et.al. (eds), *Constitutive Relations for Soils*: 175-197, Rotterdam: Balkema.

Wood, D.M., K. Belkheir. & D.F. Liu (1994). Strain softening and state parameter for sand modelling. *Geotechnique* 44:235-339.

Numerical Models in Geomechanics, Pietruszczak & Pande (eds) © 1997 Balkema, Rotterdam, ISBN 90 5410 886 X

Behavior and modeling of static liquefaction of silty sands

Jerry A. Yamamuro
Clarkson University, Potsdam, N.Y., USA

Poul V. Lade
The Johns Hopkins University, Baltimore, Md., USA

ABSTRACT: Undrained triaxial compression tests on loosely deposited silty sands exhibit unconventional soil behavior. Complete static liquefaction occurs at low confining pressures, while increases in confining pressure result in increasing stability. This makes predictions of static liquefaction difficult. Modifications to the Single Hardening Model yield surface formulation enables predictions of this behavior pattern.

INTRODUCTION

Most historic cases of static liquefaction have occurred in alluvial deposits of loose silty sands. The effect of fines (particles smaller than the No. 200 sieve size) is currently viewed as negligible or its presence inhibits static liquefaction (Kurerbis et al. 1988, Pitman et al. 1994, Ishihara 1993). However, recent experiments involving loosely deposited silty sands indicates a strong correlation between the fines content and the liquefaction potential of the soil (Yamamuro and Lade 1997a, Lade and Yamamuro 1997). The tests also indicate an unconventional behavior pattern with respect to confining pressure.

1 EXPERIMENTAL OBSERVATIONS

The sand used in the experimental investigation was Nevada sand (D_{50} = 0.2 mm). The fines were obtained from the original Nevada sand and are non-plastic. Specimens were prepared by placing dry sand in a funnel that has a tube attached to the spout. The tube was placed in the bottom of a split-mold. The tube was then slowly raised along the axis of symmetry of the specimen, such that the soil was not allowed any drop height. To increase the density, the mold was gently tapped in a symmetrical pattern around the mold.

Fig. 1 shows effective stress paths of a series of undrained triaxial compression tests performed on Nevada sand containing six percent non-plastic fines. The initial relative density was 12 percent and the initial confining pressures ranged from 25 to 500 kPa. The unconventional nature of the behavior is readily apparent. Complete static liquefaction (zero effective stress) was observed in undrained tests at low initial confining pressures (up to 125 kPa), while at higher initial confining pressures the soil exhibits more dilatant behavior resulting in temporary liquefaction. Thus, the soil exhibits increasing stability as the confining pressure increases. This is opposite the trend observed for most soils. It is hypothesized that the low energy soil deposition allows the formation of a particle structure (fabric) between the silt and sand grains that is highly compressible (Yamamuro and Lade 1997a).

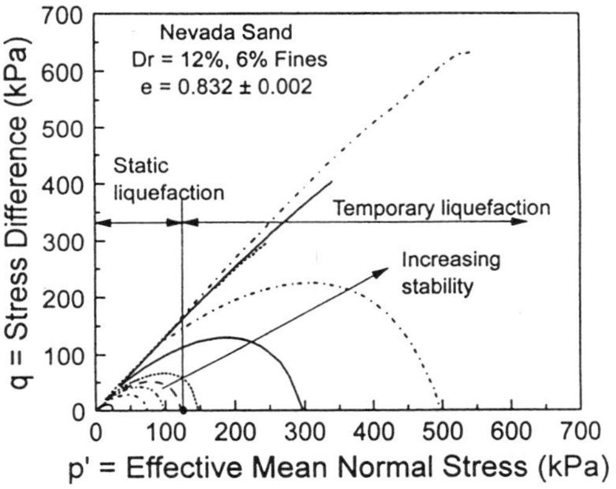

Fig. 1 Undrained stress paths from tests on Nevada sand with 6 percent fines content

2 MODELING STATIC LIQUEFACTION

To predict static liquefaction a constitutive model must be able to predict unstable behavior inside the effective stress failure surface. The model must also be capable of capturing the observed unconventional soil behavior with respect to confining pressure. The Single Hardening Model

(Kim and Lade, 1988, Lade and Kim, 1988a, Lade and Kim, 1988b) was used to make the static liquefaction predictions. This model was developed from extensive analyses of experimental data for all types of frictional materials (sand, clay, rock, concrete). The basis for the model is elasticity combined with work-hardening plasticity. The model has successfully predicted instability of granular materials at high pressures (Yamamuro and Lade 1997b). However, predicting the observed unconventional soil behavior in the undrained tests on Nevada sand proved to be difficult. In its current form the Single Hardening Model is not able to perform accurate predictions. It was realized that small modifications to the yield criterion and work-hardening law were required to enable prediction of the unconventional soil behavior.

For development of the required modifications to the model, drained and undrained tests were performed on Nevada sand with 20 percent fines content. The higher fines content produced more distinct "unconventional" sand behavior, including complete static liquefaction at higher confining pressures and greater initial relative densities. Parameter evaluation was determined on the basis of (1) three drained triaxial compression tests performed at different confining pressures and (2) an isotropic compression test. Undrained tests were then predicted from these parameters.

2.1 *Parameter Determination for Elastic Behavior, Failure Criterion and Plastic Potential*

Since the focus of the model modifications is on the yield surface formulation, the determination of model parameters for other portions of the Single Hardening Model are briefly presented below. These were all evaluated according to existing procedures.

Parameters for the elastic model (Lade and Nelson, 1987) were determined to be M = 371, λ = 0.246 and Poisson's ratio = 0.23.

Failure criterion (Lade 1977) parameters were determined to be m = 0.0713 and η_1 = 20.0.

Model parameters representing the plastic potential surface (Kim and Lade 1988) were evaluated to be Ψ_2 = -4.067 and μ = 2.29.

2.2 *Original Yield Criterion and Work-Hardening Law*

The Single Hardening Model employs a single isotropic yield surface that is expressed as a contour of constant plastic work as measured from the stress origin. The yield surface is shaped as an asymmetric teardrop and is expressed as follows:

$$f_p = f_p'(\sigma) - f_p''(W_p) = 0 \qquad (1)$$

in which

$$f_p' = \left(\Psi_1 \cdot \frac{I_1^3}{I_3} - \frac{I_1^2}{I_2}\right) \cdot \left(\frac{I_1}{p_a}\right)^h \cdot e^q \qquad (2)$$

where I_1, I_2, and I_3 are the stress invariants, p_a is atmospheric pressure, and e is the base of the natural logarithm. ψ_1 is a weighting factor between the triangular shape (from the I_3 term) and the round shape (from the I_2 term). The exponent h is a constant and q varies from zero at the hydrostatic axis to unity at the failure surface.

For hardening the second function f_p'' takes the form:

$$f_p'' = \left(\frac{1}{D}\right)^{\frac{1}{\rho}} \cdot \left(\frac{W_p}{p_a}\right)^{\frac{1}{\rho}} \qquad (3)$$

in which Wp is the plastic work, and ρ and D are constants for a given material.

The value of h is determined on the basis that the plastic work is constant along a yield surface. Thus, for two stress points, A on the hydrostatic axis and B on the failure surface, the following expression is obtained for h:

$$h = \frac{\ln \dfrac{\left(\Psi_1 \cdot \dfrac{I_{1B}^3}{I_{3B}} - \dfrac{I_{1B}^2}{I_{2B}}\right) \cdot e}{27\Psi_1 + 3}}{\ln \dfrac{I_{1A}}{I_{1B}}} \qquad (4)$$

The value of q can be determined from the test data and Eqs. 2 and 3.

$$q = \ln \frac{\left(\dfrac{W_p}{D \cdot p_a}\right)^{\frac{1}{\rho}}}{\left(\Psi_1 \cdot \dfrac{I_1^3}{I_3} - \dfrac{I_1^2}{I_2}\right)\left(\dfrac{I_1}{p_a}\right)^h} \qquad (5)$$

and the variation of q with stress level, S, is expressed as:

$$q = \frac{\alpha \cdot S}{1 - (1 - \alpha) \cdot S} \qquad (6)$$

In which the stress level S is defined as:

$$S = \frac{1}{\eta_1} \cdot \left(\frac{I_1^3}{I_3} - 27\right)\left(\frac{I_1}{p_a}\right)^m \qquad (7)$$

The stress level S varies between zero at the hydrostatic axis and unity at the failure surface. The constant α in Eq. 6 is determined from the S - q variation. The values of ρ and D are given by:

$$\rho = \frac{p}{h} \tag{8}$$

$$D = \frac{C}{(27\Psi_1 + 3)^\rho} \tag{9}$$

in which C and p are used to model the plastic work during isotropic compression:

$$W_p = C \cdot p_a \left(\frac{I_1}{p_a}\right)^p \tag{10}$$

For isotropic compression the plastic work parameters C and p were evaluated using established procedures. They were found to be C = 0.00022 and p = 2.627.

2.3 *Modifications to Yield Criterion and Work-Hardening Law*

When the parameter h was calculated from Eq. 4, for the unconventional sand behavior, three widely different values were obtained from the three drained tests. For most soils the range in the magnitudes of h is narrow, and an average value can be selected without affecting the results of the predictions. However, the unconventional behavior pattern exhibited by loose silty sands does not lend itself to accurate predictions by the current formulation of the yield surface. Therefore, modifications to the yield criterion were sought that would not alter the capabilities of the existing model relative to conventional soil behavior.

For comparison purposes parameters were evaluated for the original formulation of the yield surface using established procedures (Lade and Kim 1988a). They are h = 0.85 and α = 0.583.

After extensive evaluation, a new expression of the yield criterion was found that would effectively capture the unconventional soil behavior exhibited by loose silty sands. The new yield function f_p is expressed as in Eq. 1 but with:

$$f'_p = \left(\Psi_1 \cdot \frac{I_1^3}{I_3} - \frac{I_1^2}{I_2}\right) \cdot \left(\frac{I_1}{p_a} + \beta\right)^h \cdot e^q \tag{11}$$

in which h is a constant and q varies from zero at the hydrostatic axis to unity at the failure surface. The constant β is a new modeling parameter. This formulation was constructed such that it returns to the original formulation (Eq. 2) when β is zero. For hardening:

$$f''_p = (27\Psi_1 + 3)\left[\left(\frac{W_p}{C \cdot p_a}\right)^{\frac{1}{\rho}} + \beta\right] \tag{12}$$

The parameter β in Eq. 11 is determined in a similar manner as h was determined in Eq. 4. Plastic work is constant along a yield surface. Thus, for two stress points, A located on the hydrostatic axis and B located on the failure surface, the following expression is obtained for β as a function of h.

$$\beta = \frac{\left[e \cdot \left(\Psi_1 \frac{I_{1B}^3}{I_{3B}} - \frac{I_{1B}^2}{I_{2B}}\right) - (27\Psi_1 + 3)\right]\left(\frac{I_{1B}}{P_a}\right)^h}{\left[(27\Psi_1 + 3) - e \cdot \left(\Psi_1 \frac{I_{1B}^3}{I_{3B}} - \frac{I_{1B}^2}{I_{2B}}\right)\right]} \tag{13}$$

If β is zero, Eq. 13 returns to its original form, for which h is expressed by Eq. 4. Since both β and h need to be determined for the modified yield criterion, the variation of β with h was computed for each of the three drained triaxial compression tests. The results are shown in Fig. 2. The three curves are widely separated at β = 0, indicating that h is not constant for the original formulation. The three curves cross each other at different points. The ideal selection of β and h is obtained where the values of both variables are approximately constant for all three curves. The best fit for the data in Fig. 2 is obtained when β = 22.5 and h = 1.71.

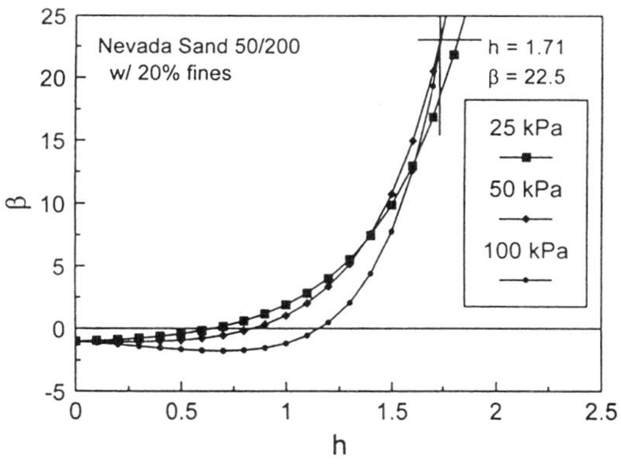

Fig. 2 Variation of β with h

Once β and h are selected, the S - q variation is evaluated to obtain the parameter α. A new expression for q was derived for the new yield criterion:

$$q = \ln\left[\frac{(27\Psi_1 + 3)\left[\left(\frac{W_p}{C \cdot pa}\right)^{\frac{1}{\rho}} + \beta\right]}{\left(\Psi_1\frac{I_1^3}{I_3} - \frac{I_1^2}{I_2}\right)\left[\left(\frac{I_1}{pa}\right)^h + \beta\right]}\right] \quad (14)$$

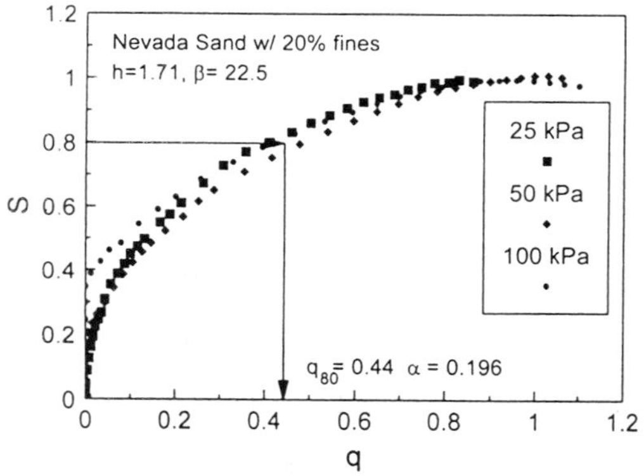

Fig. 3 Variation of q with S

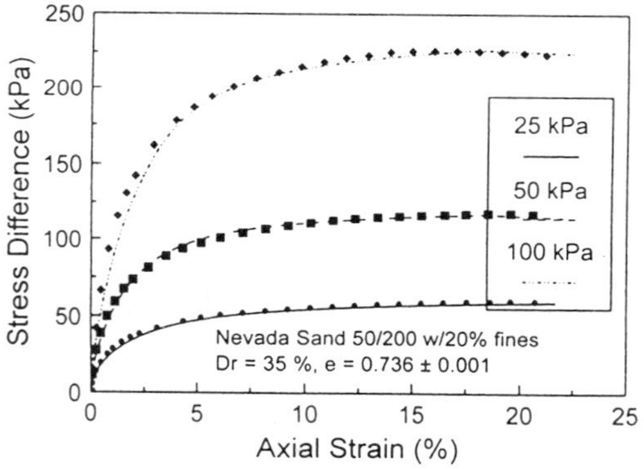

Fig. 4 Stress-strain behavior from predictions and experimental results on Nevada sand

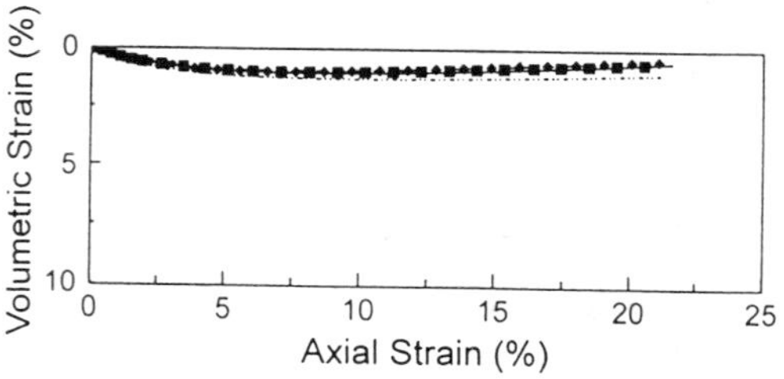

Fig. 5 Volumetric strain behavior from prediction and experimental results on Nevada sand

This expression for q also returns to the original form as expressed by Eq. 5 when β is zero. Using Eq. 14 for q and Eq. 7 for the stress level, the S - q variation can be calculated as shown in Fig. 3. The three different curves representing the three drained triaxial compression tests fall very close to each other as opposed to the wide separation observed using the original formulation of the yield surface. The parameter α is evaluated at S = 0.80. A q_{80} = 0.44 results in a value of α = 0.196.

2.4 *Modeling Drained Behavior*

Back-calculations of the stress-strain and volume change behavior of the three drained triaxial compression tests were performed based upon the parameters derived for the elastic model, failure criterion, plastic potential surface and the new yield criterion/work-hardening law. The results are shown in Figs. 4 and 5. Experimental data is shown as symbols and the model back-calculations are indicated by lines. The drained behavior is effectively captured by the new formulation.

2.5 *Modeling Undrained Behavior*

The original purpose of this modeling effort was to perform predictions of undrained tests on Nevada sand with 20 percent fines content. Using the model parameters obtained from the drained tests, predictions were performed for undrained tests. Effective stress paths of two model predictions and test data from undrained triaxial compression tests performed at an initial confining pressures of 100 and 150 kPa are shown in Fig. 6. Initially, it appears that the model predictions are considerably different from the observed experimental behavior. The

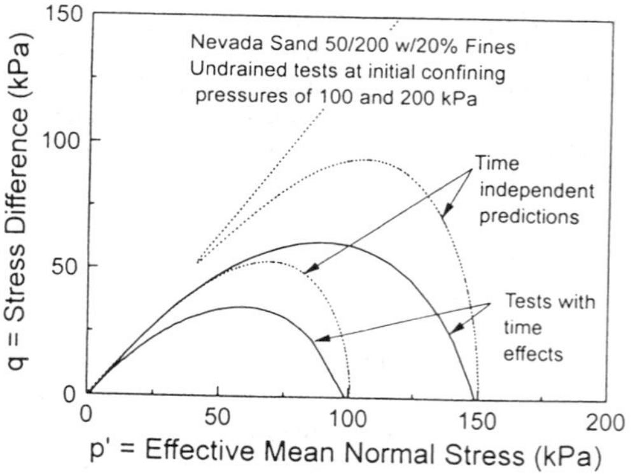

Fig. 6 Undrained stress paths from predictions and experimental results on Nevada sand

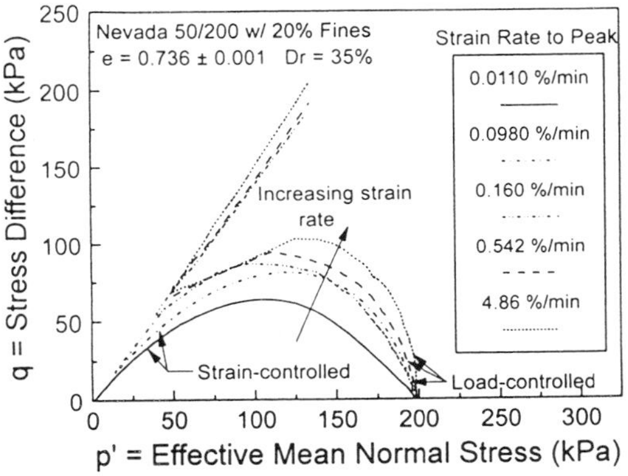

Fig. 7 Effect of strain rate on effective stress path

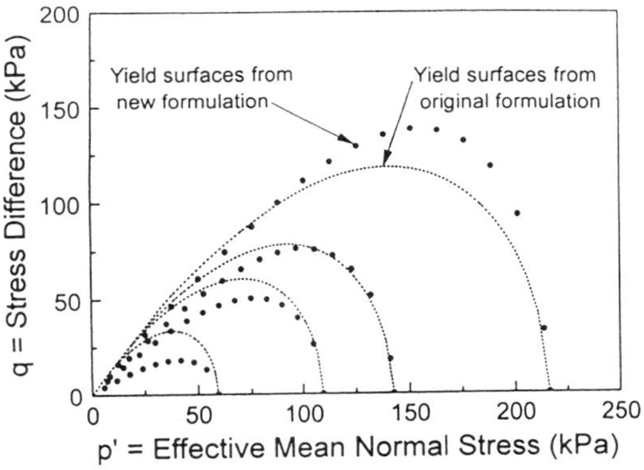

Fig. 8 Shapes of yield surfaces during expansion from new and original formulations

experiments exhibit highly contractive behavior, while the model predictions indicate more dilatant behavior resulting in a higher initial peak and ultimate strength. However, the model predictions are based upon time-independent plasticity. Loose silty sands exhibit considerable time effects. Undrained tests on Nevada sand at different strain rates result in different effective stress paths as shown in Fig. 7. Thus, the differences between the two effective stress paths shown in Fig. 6 can be attributed to time effects.

It is also observed in Fig. 6 that the prediction at the lower initial confining pressure (100 kPa) indicates complete liquefaction, where the prediction for the higher confining pressure (150 kPa) does not. Therefore, the new yield surface formulation effectively captures the unconventional pattern of behavior with respect to confining pressure as discussed and shown in Fig. 1.

2.6 *Shapes of Original and New Yield Surfaces*

To evaluate the new yield surface formulation and its improved capability to predict unconventional soil behavior, parameters for both the original and the new formulations were used to calculate the yield surface shapes and locations in stress space. The amount of plastic work was varied to show how the yield surfaces change shape during expansion. The results for triaxial compression are shown on the Cambridge p' - q diagram in Fig. 8. During expansion, the shape of the original yield surface appears to remain the same, while the size changes. The new yield surface changes both in shape and size during expansion. At low stress magnitudes the shapes of the new yield surfaces appear flatter than the original shapes. This provides the reason that static liquefaction predictions are possible at low pressures. Undrained stress paths tend to follow the yield surface very closely without much expansion until it moves close to the failure surface. At low pressures this results in a much lower deviator stress than obtained from the original yield surface formulation. Static liquefaction can more easily occur because of the close proximity to the stress origin. As they expand, the yield surfaces associated with the new formulation increase in height until their location in stress space actually move outside those of the original formulation. Therefore, the undrained stress paths would exhibit greater deviator stresses resulting in increased stability at higher confining pressures.

3 CONCLUSIONS

Experimental data is presented indicating that loose silty sands exhibit an unconventional pattern of behavior. Static liquefaction occurs at low pressures and increased stability is observed as the confining pressure is increased.

To model the unconventional behavior, the Single Hardening Model is modified with a new yield surface formulation. A new parameter β is introduced. No additional testing is required to evaluate the value of β. When β is zero, the new formulation returns to its original form.

Model parameters are evaluated using experimental data from three drained triaxial compression tests on Nevada sand with 20 percent fines content. The unconventional pattern of behavior observed in undrained tests is predicted. However, time effects caused individual predictions to differ systematically from experimental results.

Changes in yield surface shapes during expansion explain how the unconventional behavior is captured by the new yield surface formulation.

Acknowledgment

The research presented here was sponsored by the Air Force Office of Scientific Research, USAF, under Grant Numbers 910117 and F49620-94-1-0032. Grateful appreciation is expressed for this support.

4 REFERENCES

Ishihara, K. (1993). Liquefaction and flow failure during earthquakes. *Geotechnique.* 43(3):351-415.

Kim, M. K. and Lade, P.V. (1988). Single hardening constitutive model for frictional materials, I. plastic potential function. *Comp. and Geotech.* 5(4):307-324.

Kuerbis, R., Negussey, D., and Vaid, Y.P. (1988). Effect of gradation and fines content on the undrained response of sand. *Hydraulic Fill Structures,* GSP No. 21. ASCE. D.J.A.Van Zyl and S.G. Vick (eds). 330-345.

Lade, P.V. (1977). Elasto-plastic stress-strain theory for cohesionless soil with curved yield surfaces. *Int. J. Solids Struct.* 13:1019-1035.

Lade, P.V. and Kim, M.K. (1988a). Single hardening constitutive model for frictional materials, II. yield criterion and plastic work contours. *Comp. and Geotech.* 6(1):13-29.

Lade, P.V. and Kim, M.K. (1988b). Single hardening constitutive model for frictional materials, III. comparisons with experimental data. *Comp. and Geotech.* 6(1):30-47.

Lade, P.V. and Nelson, R. B. (1987). Modeling the elastic behavior of granular materials. *Int. J. Numer. Anal. Methods Geomech.*, 11:521-542.

Lade, P.V. and Yamamuro, J.A. (1997) Effects of non-plastic fines on static liquefaction of sands. Submitted to the *Can. Geotech. J.*

Pitman, T.D., Robertson, P.K., and Sego, D.C. (1994). Influence of fines on the collapse of loose sands. *Can. Geotech. J.* 32:728-739.

Yamamuro, J.A. and Lade, P.V. (1997a) Static liquefaction of very loose sands. Submitted to the *Can. Geotech. J.*

Yamamuro, J.A. and Lade, P.V. (1997b). Prediction of instability conditions for sand. Accepted for publication in the XIVth Int. Conf. on Soil Mech. and Fnd. Engrg.

Numerical Models in Geomechanics, Pietruszczak & Pande (eds) © 1997 Balkema, Rotterdam, ISBN 90 5410 886 X

A double hardening model based on generalized plasticity and state parameters for cyclic loading of sands

F. Bahda
Delft University of Technology, Department of Civil Engineering, Netherlands

M. Pastor
CETA, Madrid, Spain

A. Saitta
Centre d'Etudes des Tunnels (CETU), Bron, France

ABSTRACT : Cyclic loading of sands is a difficult phenomenon to describe accurately. Especially phenomena like liquefaction and cyclic mobility are hard to capture without resorting to extremely complicated constitutive models which rely on a large number of parameters. Also to predict loose, medium and dense behaviour of sand, one needs to consider the combined influence of the density and the consolidation pressure. This combined influence is related to the state parameter notion.
For this reason, new state parameters are defined and a double hardening model is adapted for use within a generalized plasticity context. The use of the new state parameters makes it possible to properly describe effects that are observed during cyclic loading of loose, medium and dense sands. The present approach allows for a good match with experimental data while using only eleven parameters and preserving a transparency that allow for relatively easy implementation in a finite element code.
In this contribution we shall first present the new state parameters and outline the structure of the proposed generalized plasticity double hardening model. Then, we shall compare model predictions with experimental data for different sand densities in elementary (homogeneous) tests.

1 INTRODUCTION

In the modelling of mechanical behaviour of sand, it is admitted to distinguish between three typical kinds of behaviour. Depending on the initial conditions of deposit, they correspond to initially loose, medium dense and dense particle arrangements. To each of these conditions, a different set of model parameters is used to be attributed. However, despite the fact that the sand particles are the same in all cases, parameters have different values and are thus not intrinsic to the soil. This is not only a theoretical point of view but also a very important internal physical mechanism of granular materials. To overcome this limitation, the influence of the initial conditions has to be included in the constitutive equations. This influence can be represented via the state parameter notion.

Roscoe & Poorooshab (1963) have shown that two soil samples exhibit the same strains if they are sheared during a drained triaxial test, from initial states with the same difference between the initial void ratio and the void ratio at the critical state, for the same mean effective stress. Cole (1967) and Been & Jefferies (1985) proposed the state parameter concept for sands. Almost similar to the above definition and denoted ψ, it corresponds to the vertical distance in (e, lnp') plane between the initial state and the critical state line. To improve this notion, Ishihara (1993) introduced the state index which is a new parameter to characterize the initial state of a sand. The state index, denoted I_s, include more features of sand behaviour such as the influence of the sample fabrication mode and the zero residual strength range of initial state. The state index has been defined in the (e, p') plane, using the quasi-steady state line (Alarcon-Guzman et *al.*, 1988, Ishihara, 1993), as follows :

$$I_s = \frac{e_0 - e}{e_0 - e_s} \qquad (1)$$

e_0 is the smallest void ratio with a zero residual strength, e is the initial void ratio and e_s is the void ratio at the quasi-steady state, for the same effective mean stress.

The introduction of the state parameter notion in constitutive equations formulation has been investigated by different authors such as Sladen et *al.* (1988) with a loose sand model, Jefferies (1993) with a rigid plastic model and Saitta (1994) in the framework of a classical elastoplasticity theory. However these formulations seem incomplete as soon as the initial conditions contain a non zero deviatoric component as in an anisotropic consolidation cases or in cyclic loading conditions.

The objective of this paper is to propose a model based on a new state couple for sand within a generalized plasticity context. The new state couple permit to capture the essential features of sand behaviour in every initial condition. Especially it allows the reproduction of liquefaction and cyclic mobility with the same set of parameters.

2 A STATE COUPLE FOR SAND

To take into account the influence of non zero deviatoric stress, two parameters are considered. Defined in the (e, p') diagram, the first one is the volumetric state parameter I_v and the second one, defined in the (q, p') plane is the deviatoric state parameter I_d. They are expressed as follows :

$$\begin{cases} I_v = \dfrac{p'_s}{p'} \\ I_d = \dfrac{\eta}{\eta_f} \end{cases} \qquad (2)$$

p'_s is the effective mean stress at the steady state corresponding to the current void ratio, p' is the effective mean stress, η_f is the slope of the steady state in (q, p') plane and η is the stress ratio q/p'.
These parameters are no longer related to the initial state only but they are varying during the test. They can then give an information about the stress path and strain at every current loading point.

3 BASIC ASSUMPTIONS

The following assumption are used :

1. infinitesimal strain;
2. elastoplastic theory;
3. generalized plasticity context;
4. non linear elasticity;
5. non associative potential;
6. double loading surface;

4 MODEL DESCRIPTION

4.1 *Elastic part of strain increment*

The elastic part of the total strain increment is modelled through non linear elastic model with constant void ratio ν and bulk modulus K expressed by formula :

$$K = \frac{1+e_0}{\kappa}(\langle p'-p_L\rangle + p_L) \qquad (3)$$

where e_0 is the initial void ratio, κ is the slope of secondary compression line in e-ln(p') system, p' is the effective mean pressure and p_L is some arbitrary pressure limiting K value (Dafalias & Herrmann, 1986b).

4.2 *Plastic part of strain increment*

The plastic strain increment is defined, in a double surface model, as total of plastic strains caused by each mechanism. Depending on the position of the current stress point, they can be expressed as follows :

$$\begin{aligned} d\underline{\varepsilon}_1^p &= \frac{1}{h_1^2}\underline{n}_{G1}(\underline{n}_{F1}^T d\underline{\sigma}) \\ d\underline{\varepsilon}_2^p &= \frac{1}{h_2^1}\underline{n}_{G2}(\underline{n}_{F2}^T d\underline{\sigma}) \end{aligned} \qquad (4)$$

h_1^2 et h_2^1 are the plastic modulus of the first and the second plastic mechanism respectively.
$\underline{n}_{F1}$ et $\underline{n}_{G1}$ denote the plastic loading and the flow vectors related to the first mechanism, $\underline{n}_{F2}$ et $\underline{n}_{G2}$ to the second mechanism.
As only one plastic potential is considered, the flow vector can be written :

$$\underline{n}_{G1} = \underline{n}_{G2} = \underline{n}_G \qquad (5)$$

4.3 *Loading surfaces*

The loading surface F1 is chosen depending of the initial density (its form is given in the figure 1). In generalized plasticity context, only the plastic loading vector $\underline{n}_{F1}$ components are required. Evaluated in stress invariant space, they are given by :

$$n_{F1}^{p} = \frac{n-(n+1)I_d^2}{\sqrt{\left(n-(n+1)I_d^2\right)^2+\left(\frac{I_d}{\eta_f}\right)^2}}$$

$$n_{F1}^{q} = \frac{I_d}{\eta_f\sqrt{\left(n-(n+1)I_d^2\right)^2+\left(\frac{I_d}{\eta_f}\right)^2}} \qquad (6)$$

where I_d is the deviatoric state parameter, η_f is the slope of the steady state in (q, p') space and n is a parameter depending of the density as follows :

$$\begin{cases} n = n_0 \dfrac{e_0}{e_{eff}} & pour \quad e_0 \leq e_{eff} \\ n = n_0 & pour \quad e_0 \geq e_{eff} \end{cases} \qquad (7)$$

e_0 is the initial void ratio, e_{eff} is the smallest void ratio with a zero residual strength in a liquefaction test. (equivalent to Ishihara e_0, eq. 1), n_0 is a model parameter.

The second surface F2 is a Vermeer deviatoric surface type. The components of the plastic loading vector $\underline{n}_{F2}$ are :

$$n_{F2}^{p} = \frac{-I_d}{\sqrt{\frac{1}{\eta_f^2}+I_d^2}}$$

$$n_{F2}^{q} = \frac{1/\eta_f}{\sqrt{\frac{1}{\eta_f^2}+I_d^2}} \qquad (8)$$

4.4 Loading criteria

For a hardening state, the loading criteria can be written :

$$\underline{n}_F^T d\underline{\sigma} > 0 \qquad (9)$$

For the second mechanism, it can also be expressed :

$$\underline{n}_{F2}^T d\underline{\sigma} = \frac{1/\eta_f}{\sqrt{\frac{1}{\eta_f^2}+I_d^2}}\left(-\eta dp' + dq\right) = \frac{p'/\eta_f}{\sqrt{\frac{1}{\eta_f^2}+I_d^2}} d\eta > 0 \qquad (10)$$

As p' is always positive, this leads to :

$$\underline{n}_{F2}^T d\underline{\sigma} > 0 \Leftrightarrow dI_d > 0 \qquad (11)$$

4.5 Flow rule

Following the experimental results obtained by Touati (1982) in drained triaxial tests, it may be observed that dilatancy d can be approached by a function of the stress ratio η. Using η_c as the slope defining zero dilatancy in (q, p') space :

$$d = e\,(\eta_c^2 - \eta^2) \qquad (12)$$

The zero dilatancy in (q, p') plane defines the characteristic state line (Luong, 1978) or the phase transformation line (Ishihara et *al.*, 1975).
The flow vector $\underline{n}_G$ is given as function of the dilatancy by :

$$\begin{cases} n_{Gp} = \dfrac{d}{\sqrt{1+d^2}} \\ n_{Gq} = \dfrac{1}{\sqrt{1+d^2}} \end{cases} \qquad (13)$$

These expressions of the flow vector components lead to a convex plastic potential as shown in figure 1.

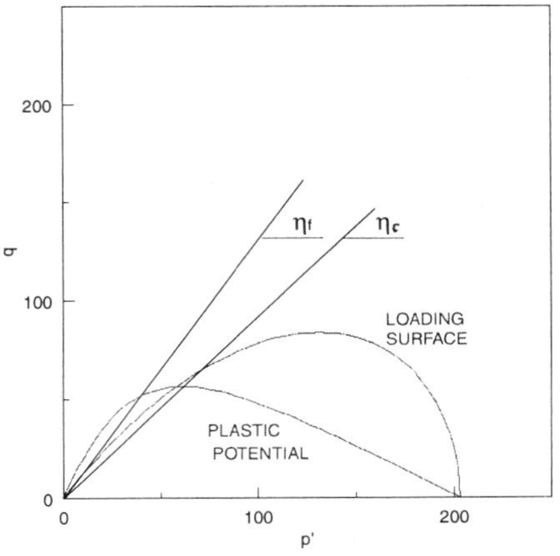

Figure 1 Used yielding surface and plastic potential form

4.6 Plastic modulus expressions

In generalized plasticity context, the plastic modulus expression must be in agreement with a form (Pastor et al. 1990) which respects at least the isotropic virgin compression paths ($h=h_0\ p'$) and the flow deformation at the steady state ($h=0$ at $\eta=\eta_f$).
Following these conditions, an expression of the plastic modulus for each mechanism is proposed :

$$\begin{cases} h_1^2 = h_1 \left\| h_1 \underline{n}_{F2} + h_2 \underline{n}_{F1} \right\| \\ h_2^1 = h_2 \left\| h_1 \underline{n}_{F2} + h_2 \underline{n}_{F1} \right\| \end{cases} \tag{14}$$

This form allows a certain coupling between the two plastic mechanisms. The variables h_1 and h_2 are given as function of the state couple as follows :

$$\begin{cases} h_1 = h_1^0 p' I(t) \\ h_2 = h_2^0 D(I_d^0)(h_v + h_d) \end{cases} \tag{15}$$

$$\begin{cases} h_v = I_v \\ h_d = h_d^0 \left(\dfrac{\eta_c}{\eta_f} - I_d^2 \right) \\ D(I_d^0) = (1-(I_d^0)^2) \end{cases} \tag{16}$$

Following the bounding surface concept (Dafalias & Herrmann, 1982), I(t) is an extrapolation function which informs about the distance run inside the surface F1. Its expression will be defined forward.
I_d^0 is the value of I_d in the beginning of unloading from F2. If $I_d \geq I_d^0$, then I_d^0 is taken equal to I_d.
$D(I_d^0)$ informs about the distance run inside the surface F2. Approaching the steady state, this function becomes close to zero. This allows us to simulate the flow deformation at the steady state.
h_1^0, h_2^0, h_d^0 are model parameters.

4.7 Extrapolation characteristics

The essential feature of the bounding surface concept is that the plastic strains inside the yield surface are obtained by an extrapolation rule. To characterize this rule, one needs to locate a reference point in the (q, p') space and an image point on the yield surface (which becomes a bounding surface) and to define a loading direction available also inside the bounding surface. In our case, this direction is given directly with the second surface F2.

The reference point M_R is taken as the last point of loading (rep. Unloading) before unloading (rep. loading) from the surface F1 or F2. The image point M_I is given by F1 and the vertical line passing through the current point M (figure 2).
The extrapolation rule is given by :

$$t = \frac{\eta - \eta_R}{\eta_I - \eta_R} \tag{17}$$

η, η_R, et η_I are the stress ratios at the points M, M_R et M_I respectively.
To satisfy the conditions on plastic modulus expression, the function I(t) must be equal to 1 when the current point M is at M_R and it must decreases from M_R to M_I. The used function is then chosen of the following form :

$$I(t) = \left(\frac{2}{1+t} \right)^{\gamma} \tag{18}$$

γ is a model parameter which controls the magnitude of plastic strains inside F_1.

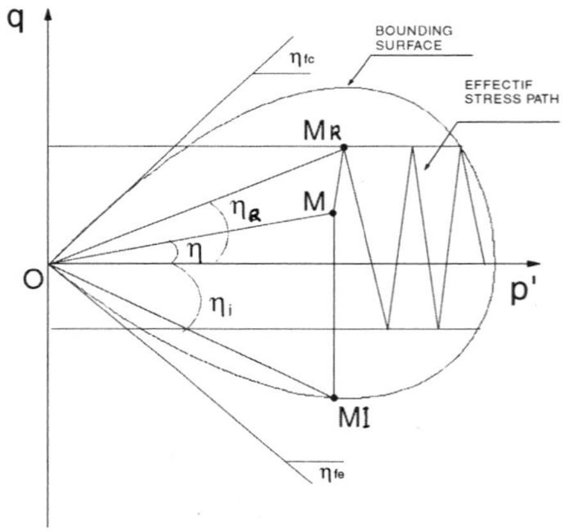

Figure. 2 Example of determination of the image point

4.8 Plastic strains expressions

The value of the total strain depends on type of loading and hardening of each surfaces. Four regions can be considered :

- Zone 1: $\underline{n}_{F1}^T d\underline{\sigma} \geq 0$ et $\underline{n}_{F2}^T d\underline{\sigma} \geq 0$

$$d\underline{\varepsilon}^P = d\underline{\varepsilon}_1^P + d\underline{\varepsilon}_2^P = \frac{1}{h_1 h_2} \underline{n}_G \left(\underline{n}^T d\underline{\sigma} \right) \tag{19}$$

where $\underline{n} = \dfrac{h_2 \underline{n}_{F1} + h_1 \underline{n}_{F2}}{\left\| h_2 \underline{n}_{F1} + h_1 \underline{n}_{F2} \right\|}$ the total unit vector.

- Zone 2: $\underline{n}_{F1}^T d\underline{\sigma} < 0$ et $\underline{n}_{F2}^T d\underline{\sigma} \geq 0$

$$\underline{n} = \underline{n}_{F2} \tag{20}$$

$$d\underline{\varepsilon}^P = d\underline{\varepsilon}_2^P = \frac{1}{h_1 h_2} \underline{n}_G \left(\underline{n}_{F2}^T d\underline{\sigma} \right) \tag{21}$$

- Zone 3: $\underline{n}_{F1}^T d\underline{\sigma} < 0$ et $\underline{n}_{F2}^T d\underline{\sigma} < 0$

$$\underline{n}'_{F2} = -\underline{n}_{F2} \text{ , } \underline{n}'_G = \begin{pmatrix} n_{Gp} \\ -n_{Gq} \end{pmatrix} \tag{22}$$

$$d\underline{\varepsilon}^P = d\varepsilon'^P_2 = \frac{1}{h_1 h_2} \underline{n}'_G \left(\underline{n}'^T_{F2} d\underline{\sigma} \right) \tag{23}$$

- Zone 4: $\underline{n}_{F1}^T d\underline{\sigma} \geq 0$ et $\underline{n}_{F2}^T d\underline{\sigma} < 0$

$$\underline{n}' = \frac{h_2 \underline{n}_{F1} - h_1 \underline{n}_{F2}}{\left\| h_2 \underline{n}_{F1} - h_1 \underline{n}_{F2} \right\|} \tag{24}$$

$$d\underline{\varepsilon}^P = d\underline{\varepsilon}_1^P + d\underline{\varepsilon}'^P_2 = \frac{1}{h_1 h_2} \underline{n}_G \left(\underline{n}'^T d\underline{\sigma} \right) \tag{25}$$

4.9 Material parameters

The proposed model needs definition of the following parameters : κ, ν, λ, e_{eff}, η_c, η_F, n_0, h_1^0, h_2^0 h_d^0 et γ. The first six parameters are issue from the critical state theory. The detailed procedure of the evaluation of all parameters from monotonic and cyclic triaxial compression tests is described by Bahda (1997).

5 APPLICATION

Because of the scope of this article, only two undrained triaxial compression tests results of numerical simulations are presented. Figure 3 shows model predictions and experimental data of a test carried out on RF Hostun sand in a loose state. It concerns liquefaction under cyclic loading which is well reproduced by the model. The assumed values of model parameters are as follows : λ=0.03, κ=0.015, ν=0.3, η_f=1.42, η_c=1.06, e_{eff}=0.96, h_1^0=5, h_2^0=12, h_d^0=0.8, n_0=0.5, γ=10. The second test presents model prediction for cyclic mobility during reversible shear stress controlled cyclic triaxial test performed by Tatsuoka et al. (1974) on Niigata sand. The used values of material parameters are : λ=0.06, κ=0.02, ν=0.3, η_f=1.5, η_c=1.2, e_{eff}=0.9, h_1^0=5, h_2^0=12, h_d^0=0.8, n_0=1, γ=10. The simulation, shown in Figure 4, exhibits a reasonably good agreement with the experimental results.

6 CONCLUDING REMARKS

A simple model able to deal with the essential futures of sand behaviour under monotonic and cyclic loading has been developed within the framework of the generalized plasticity theory. The introduction of state couple in its formulation permits to take into account the influence of the initial conditions on the soil behaviour and to reduce considerably the number of material parameters.

ACKNOWLEDGEMENTS

The present work was carried out while the first author was a PhD student at CERMES, ENPC, LCPC. The author would like to express her gratitude to both institutions. The collaboration between the authors has been conducted in the framework of the activities of the A.L.E.R.T. Geomaterials network.

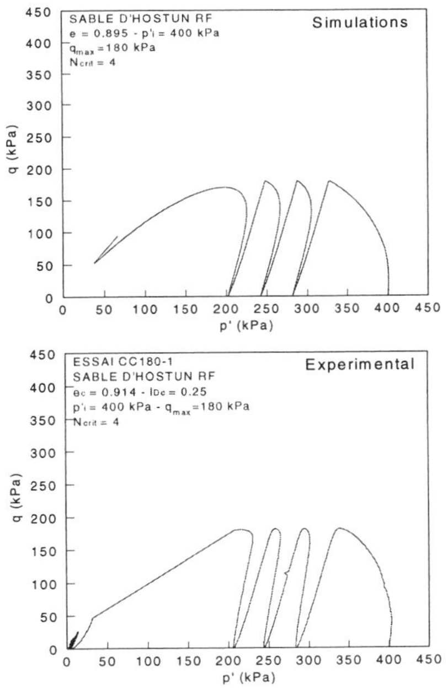

Figure. 3 Simulation and experimental results of liquefaction test

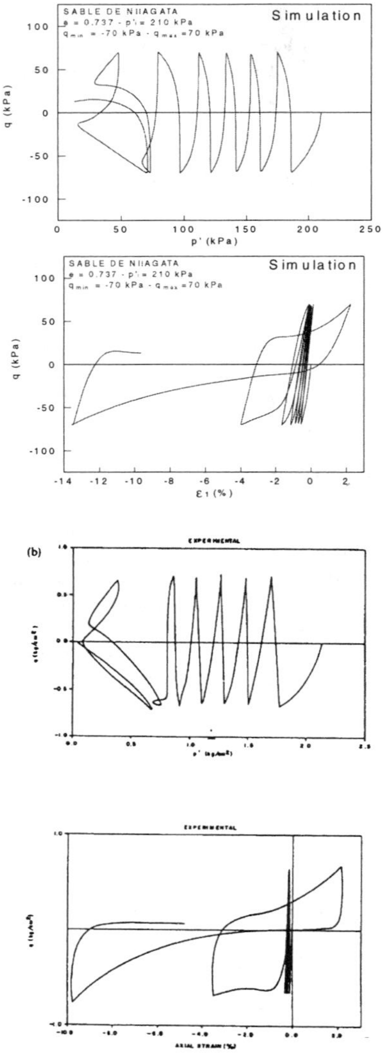

Figure. 4 Simulation and experimental results of cyclic mobility test

REFERENCES

Alarcon-Guzman, A., Leonards, G. A. & Chameau, J. L. 1988. Undrained monotonic and cyclic strength of sands. J. Geotech. Eng. Div. ASCE, 114 (10). 1089-1109.

Bahda, F. 1997. Etude du comportement du sable au triaxial. Expérience et modélisation PhD thesis, ENPC. Paris.

Been, K. & Jefferies, M. G. 1985. A sate parameter for sands. Geotechnique, 35 (2). 99-112.

Cole, E. R. 1967. The behaviour of soils in the simple shear apparatus. PhD. Thesis. Univ. Cambridge.

Dafalias, Y. F. & Herrmann, L. R. 1986b. Bounding Surface Plasticity. II Application to isotropic Cohesive Soils. J. Eng. Mech., 112 (12). 1263-1291.

Dafalias, Y. F. & Herrmann, L. R. 1982. Bounding surface formulation of soil plasticity. In G. N. Pande & O. C. Zienkiewicz (eds), Soil Mech.-Trans. and cyclic loads, Wiley, Chichester. Chap. 10, p. 253-282.

Jefferies, M. G. 1993. Nor-sand: a simple critical state model for sand. Geotechnique, 43 (1). 91-103.

Ishihara, K. 1993. Liquefaction and flow failure during earthquakes. Geotechnique, 43 (3). 315-415.

Ishihara, K., Tatsuoka, F. & Yasuda, S. 1975. Undrained deformation and liquefaction of sand under cyclic stress. Soils & Found., JSSMFE 15 (1): 29-44.

Luong, M. P. 1978. Etat caractéristique du sol. C. R. Académie des Sciences, Paris 287B. 305-307.

Pastor, M., Zienkiewicz, O. C. & Chan, A. H. C. 1990. Generalized plasticity and the modelling of soil behaviour. Int. J. Num. Anal. Meth. Geomech. 14: 151-190.

Roscoe, K. H. & Poorooshab, H. B. 1963. A fundamental principle of similarity in model tests for each pressure problems, Proc. 2nd Conf. Asia. Soils Mec., 1. 134-140.

Saïtta, A. 1994. Modélisation élastoplastique du comportement mécanique des sols. Application à la liquéfaction des sables et à la sollicitation d'expansion de cavité. Ph.D. Thesis, ENPC. Paris.

Sladen, J. A. & Oswell, J. M. 1988. The behaviour of very loose sand in the triaxial compression test, Can. Geotech. J., 26. 103-113.

Tatsuoka, F. & Ishihara, K. 1974. Yielding of sand in triaxial compression. Soils & Found., 14 (2). 63-76.

Touati, A. 1982. Comportement mécanique des sols pulvérulents sous fortes contraintes. PhD Thesis, ENPC, Paris.

Numerical Models in Geomechanics, Pietruszczak & Pande (eds) © 1997 Balkema, Rotterdam, ISBN 90 5410 886 X

Experimental and numerical analysis of a 2-D granular material

Francesco Calvetti
Milan University of Technology (Politecnico), ALERT Geomaterials, Italy

ABSTRACT: The micro-mechanical study of granular materials behaviour can be performed following different techniques and we often distinguish between experimental and numerical approaches. Our aim is to establish some links between such approaches, performing numerical DEM simulations of laboratory tests on a 2-D analogous model of a granular material.
On the experimental side, the results of tests performed with the shear device 1γ2ε, using a specimen composed of wooden rollers stacks, are available. The employed analogous model allows direct measurements of local kinematics, which is derived from specimen pictures taken during testing. On the numerical side, we performed simulations using a commercial DEM based code (PFC-2D). Our main interest is to verify the reliability of the numerical model, comparing its global and local behaviour with that exhibited by the laboratory one. This point is a crucial one, since the numerical model involves direct measurements of local statics also.
In the paper, we first describe the testing devices and the procedures used to directly measure particle properties. Then we detail the steps that lead to the definition of an adequate numerical model, focusing interest on the extent to which direct laboratory measures can be used to assess numerical parameters. Then, the numerical model reliability is evaluated comparing boundary and local results for simple loading paths.

1 AIM OF THE PAPER

The micro-mechanical study of granular materials can be tackled from different viewpoints and many researches have been conducted using numerical or experimental techniques to determine the basic aspects of the mechanical behaviour of highly idealised granular materials. In addition, several attempts were made to find relationships that relate local (contact forces, grain kinematics) to global (stress and strain) variables, with the aim to deduce the global behaviour from the local one.
In this paper, we test the possibility to reproduce in a quantitative way the experimental behaviour of a 2-D specimen composed by a roller stack by performing DEM simulations on disks.
In particular, we refer to laboratory tests performed with the shear device 1γ2ε on a rollers stack, for which a complete series of measurements is available, regarding the boundary behaviour of specimens (stress-strain curves, volumetric behaviour), their initial configuration (position of rollers and contact network), and the mechanical properties of individual particles (stiffness and surface friction). As it will be shown in the following, laboratory measurements are used to build up numerical specimens as similar as possible to the corresponding laboratory ones: in fact, the numerical loading device, the initial specimen configuration and the DEM parameters are directly derived from experiments. It is important to note that no parameter calibration procedure takes place, and all efforts to make the numerical model 'identical' to the experimental device disregard the overall behaviour of the specimen during testing; therefore, a blind comparison between laboratory and numerical results is proposed, to verify to which extent laboratory and numerical experiments can be considered inter-exchangeable.

2 EXPERIMENTAL DEVICE

The reference laboratory tests were performed at Laboratoire 3S in Grenoble, by means of a biaxial shear apparatus named 1γ2ε. This machine allows a specimen composed of a rollers stack to be subjected to the most general 2-D loading conditions, applying vertical and horizontal strains and shear, independently. A detailed description of the mentioned device is given in Joer et al. (1992).

During early nineties, 1γ2ε was used to test under complex loading paths an analogic material composed of small PVC rollers (diameter 1.5 to 3 mm, 6 cm long); a DEM simulation of such tests is presented by Calvetti et al. (1995). In that case, the numerical parameters choice was based on the observed boundary behaviour during simple compression tests (isotropic, oedometric and biaxial). More recently, the device 1γ2ε was used to perform a micro-mechanical investigation of the granular materials behaviour (Calvetti et. al. (1997)), which was possible substituting PVC roller with bigger wooden ones (diameter 13, 18 and 28 mm; 6 cm long). The experimental procedure is detailed in the given reference, where the attention is focused on the micro-kinematics (particle displacements and rotations) and on the relation between the imposed stress or strain path and the specimen fabric evolution.

2.1 Test apparatus

The loading frame, which lays in a vertical x-y plane, consists of four rigid smooth plates that are splitted in five segments and can elongate to apply an almost uniform strain at the specimen boundaries, which reduces the effects of rollers-plate friction.
During the tests, the length of the plates and the forces acting on them are measured, and the state of stress and strain is simply derived, assuming a uniform distribution within the specimen.
The control of the machine is kinematic, with a servo device allowing stress control. In particular, during the compression test presented in the following, the upper plate is given a vertical speed of about 0.01 mm/s. The horizontal speed of the left plate is constantly adjusted by the servo device to maintain the horizontal stress constant. The corner where the right plate and the lower one are connected is fixed to the laboratory frame, which implies non-symmetric loading conditions. A relevant point for the numerical simulations (see §3.3) is that the loading speed is such that the test can be considered 'quasi-static'.

2.2 The analogic material

Specimens are composed by a stack of wooden rollers of three nominal diameters (13, 18, 28 mm), 6 cm long. Rollers cross-section is not exactly circular, and actual diameters slightly differ from the nominal ones. Each roller is numbered and an oriented diameter is traced on its front face, which allow to track rollers by taking pictures at different stages of the tests. After digitalisation of pictures, each roller position and orientation can be deduced, as well as the number and orientation of contacts; if two pictures are compared, rollers displacements and rotations and the evolution of the contact network can be determined.
The concentric growing circular surfaces that characterise the wooden structure of rollers can be considered planar at the roller scale, the roller axis being parallel to the growing planes. As a consequence, the mechanical behaviour of the rollers is not isotropic in the testing plane. The mechanical properties of rollers were measured in laboratory, using the procedures detailed in §3.1.1 (surface friction) and §3.1.2 (roller stiffness).

2.3 Reference Compression test

In order to verify the possibility to reproduce via DEM the laboratory tests, a vertical compression test with constant horizontal confining stress (test GCV1) was considered (Combe (1995)). The boundary results obtained during GCV1 are shown in the following figure.

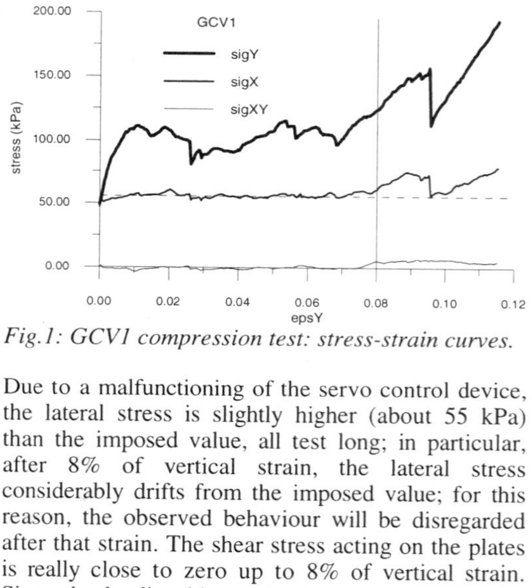

Fig.1: GCV1 compression test: stress-strain curves.

Due to a malfunctioning of the servo control device, the lateral stress is slightly higher (about 55 kPa) than the imposed value, all test long; in particular, after 8% of vertical strain, the lateral stress considerably drifts from the imposed value; for this reason, the observed behaviour will be disregarded after that strain. The shear stress acting on the plates is really close to zero up to 8% of vertical strain. Since the loading kinematics is non-symmetric, this results cannot derive from a self-equilibrated shear stress distribution along the plates. It is concluded that plates can actually be considered smooth.

3 NUMERICAL MODEL

The numerical model employed to reproduce laboratory tests is based on a commercial 2-D DEM code (PFC-2D), that allows to perform simulations on samples composed by rigid disks. The boundary

of the numerical specimen is defined by 'wall' elements, which represent geometrical conditions (see Fig.5). In order to perform a numerical simulation it is necessary to define the initial specimen configuration (size and position of each disk and position of walls), the loading conditions and the mechanical properties of particles and contacts. As usual for DEM based codes, particles are considered rigid, and their deformation is represented by a small overlapping at contact points. The magnitude of normal and tangential contact forces is linked to the relative normal and tangential displacements of particles in contact by a linear elastic relation, the tangential forces being limited by Coulomb's friction.
As it will be detailed in the following, all data regarding the specimen configuration and the properties of particles and contacts are derived from laboratory measurements.

3.1 Evaluation of DEM parameters

3.1.1 Contact friction

The roller to roller friction angle, Φ, was directly measured using the experimental technique described in Fig.2.

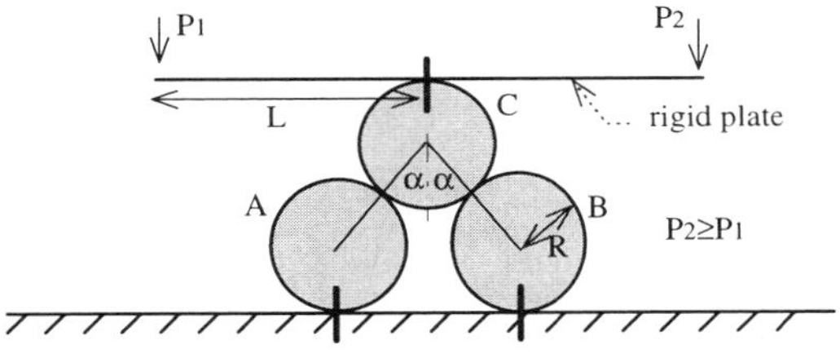

Fig.2: Experimental measurement of roller friction

The two rollers A and B are fixed and support the roller C, which is loaded by two weights, P_1 and P_2. Initially, P_1 and P_2 are equal and the roller C is obviously in equilibrium. By increasing P_2, a limit equilibrium is reached and the roller C rotates sliding at contacts points with A and B. If $\alpha > \Phi$, both contacts are maintained when the sliding condition is reached, and the limit equilibrium equation yields to the following expression:

$$sin2\Phi = 2 \cdot \frac{P_2 - P_1}{P_2 + P_1} \cdot \frac{L}{R} \cdot \cos\alpha \qquad (1)$$

Several tests were performed with different α and P_1 values and the average value, Φ = 28°, of the surface friction angle was estimated.

The average Φ value was used in simulations to characterise disk to disk contacts. On the contrary, wall to disk contacts are given zero friction, for the reasons exposed in §2.3.

3.1.2 Contact stiffness

The stiffness of wooden rollers (contact stiffness in the DEM model) was directly measured using the experimental technique described in Fig.3.

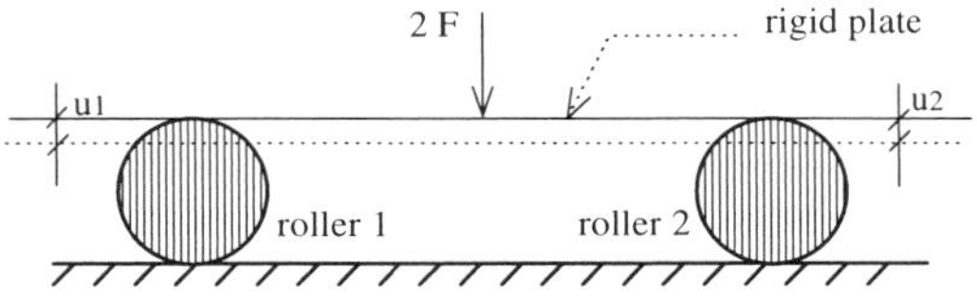

Fig.3: Experimental measurement of roller stiffness

Two equal sized and oriented rollers resting on a rigid support are loaded under a force, F, applied through a rigid plate. Compression tests are performed for the three different roller sizes and, for each roller diameter, horizontal and vertical wood textures are considered. As an example, in the following figure, results obtained using 18 mm diameter rollers with vertical wood textures are presented.

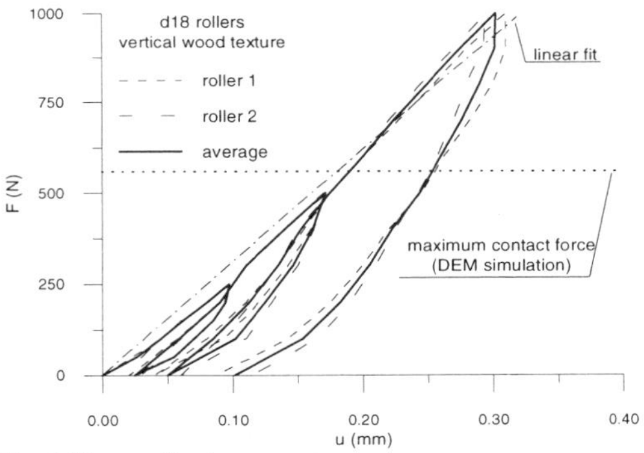

Fig.4:Force-displacements curves

The maximum applied force (1000 N) is higher than the threshold represented by the maximum contact force measured during the DEM simulation that will be presented in the following (Fig.4, dotted line). The obtained force-displacement curves show that rollers undergo non reversible deformation and the incremental stiffness slightly increases with the current load.
These aspects are not incorporated in the numerical model, where linear elastic contacts are used; therefore, a secant value of stiffness (Fig.4, dash-dot

line) is evaluated for each roller size and wood texture orientation (see Tab.1).

Tab. 1: particles stiffness (MN/m)

diameter mm	wood texture: vertical	wood texture: horizontal	mean value
13	2.98	2.47	2.73
18	3.12	2.89	3.01
28	5.55	4.98	5.27

The anisotropic roller structure corresponds to an anisotropic mechanical behaviour, and the global stiffness of particles depends on roller size.
The orientation dependency of particles stiffness cannot be considered in the DEM model: for this reason, the mean stiffness values of Table 1 are used in simulations. In lack of information about the global tangential particle stiffness, the tangential stiffness of contacts is assumed equal to the normal one.

3.2 Numerical specimen

The numerical specimen is generated by directly picking the position and size of each particle from the picture taken at the end of the laboratory isotropic compression. Walls are placed to reproduce the size and geometry of the laboratory frame, with rounded walls at corners where the hinges that connect the laboratory device plates are placed (Fig.5).

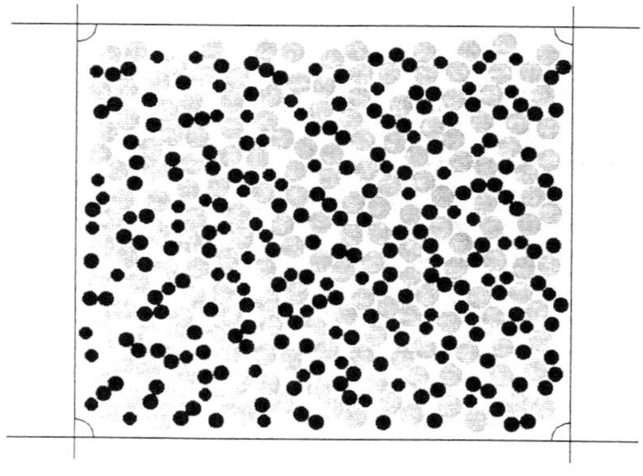

Fig.5: The numerical model: initial configuration

It is necessary to precise that the numerical specimen slightly differs from the experimental one because of the shape imperfection of wooden rollers; moreover, no gravity field is initially applied to the disk assembly.
Anyway, the main discrepancy is that the generated numerical specimen is unstressed, which is linked to the fact that it is not possible to derive any information about statics from digitalised pictures. In fact, rollers are very stiff with respect to the imposed state of stress (see §3.1.2), and their negligible deformation cannot be read on pictures. On the other hand, the analysis of digitalised pictures would be impossible if the rollers would considerably deform during loading.
Before applying a vertical compression, the numerical specimen must be subjected to an isotropic compression up to the value of 55 kPa, corresponding to the measured value of the confining stress during GCV1. The isotropic compression was obtained by increasing (by about 0.1%) the diameter of the numerical rollers, the consequent small overlapping resulting in contact forces in equilibrium with the required boundary confining pressure. The gravity field is introduced after the isotropic compression. The employed numerical technique has the advantage to preserve the initial specimen configuration from the disturbance involved with a compression imposed from the boundaries.

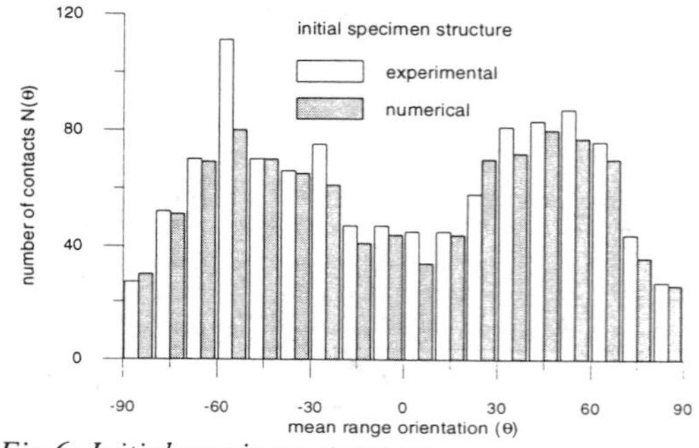

Fig.6: Initial specimen structure

The described procedure leads to a numerical specimen structure which is really close to that of laboratory specimen, characterised by a strong anisotropy (due to the manual specimen construction in the gravity field) (Fig.6). The parameter $N(\theta_0)$ represents the number of contacts with normal orientation belonging to a 10° wide range around θ_0.

3.3 Loading conditions

The control of the numerical simulation is obtained by applying an adequate speed to the walls that define the boundary of the specimen. In particular, during a vertical compression, the upper face is given a constant vertical speed (loading speed), while a servo control applies to the left plate a speed such that the lateral stress is constant; the right and the bottom plates are fixed, which reproduces the non-symmetric loading conditions characterising the laboratory tests.

It is important to underline the fact that it would not be possible to perform numerical simulations applying a loading speed equal to the laboratory one, because simulations would take too much time: it is therefore necessary to increase the loading rate in a relevant way. Of course, an upper limit is fixed by the fact that the simulations, as well as the laboratory tests, must be performed under 'quasi-static' equilibrium conditions. The possibility to perform 'quick' but 'quasi-static' numerical simulations depends on the numerical damping acting on disks accelerations.

The loading speed must be selected in a range over which the results are loading speed independent, and dynamic effects can be neglected.

Since walls are smooth, it is possible to evaluate global dynamic effects by analysing the walls reaction during compression: in fact, the difference between the reactions exerted by two opposite walls increases with the loading speed, which corresponds to increasing inertia forces raising inside the specimen.

As an example, we show results obtained applying a loading speed of 0.01 m/s (Fig.7), that represents a good compromise between the discussed opposite requirements: in this case, the global effect of inertia forces keeps within the 5% of the plates reaction.

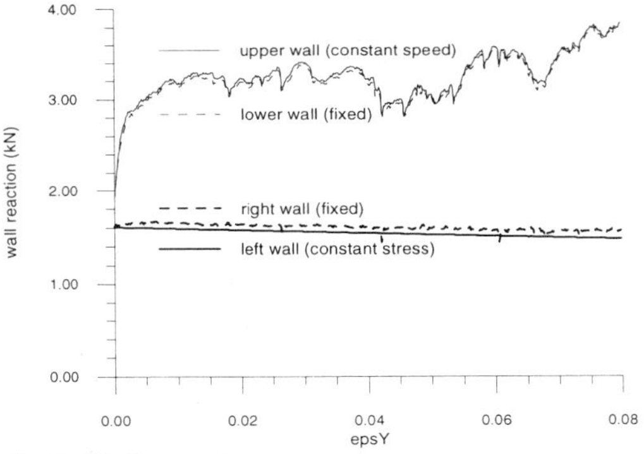

Fig.7: Walls reactions

4 DISCUSSION OF RESULTS

4.1 Boundary results

In this paragraph the numerical and experimental results are compared, focusing interest on the global specimen behaviour, evaluated at the boundaries.

The laboratory stress-strain curves are reproduced in a way that can be considered satisfactory (Fig.8a), taking into account the fact that the numerical results are a prediction of the experimental ones, since no numerical parameter calibration was made, and the numerical model was built without any deliberate attempt to fit the results of laboratory tests.

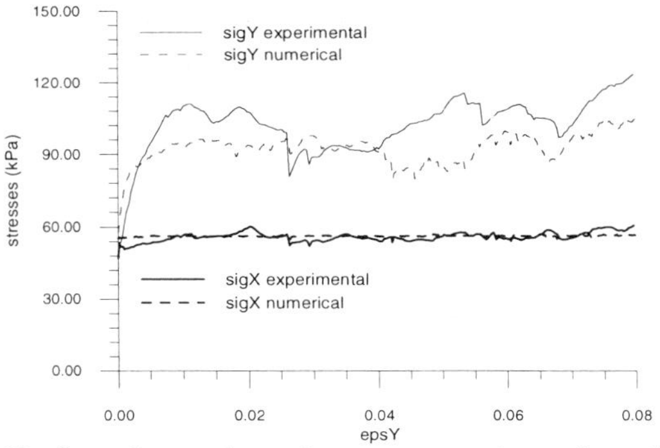

Fig.8a: Comparison between experimental and numerical results: stress-strain curves.

On the other hand, the volumetric behaviour is only qualitatively reproduced: the numerical model is far less compactant than the experimental one in the first part of the test, which influences the observed behaviour all test long (Fig.8b).

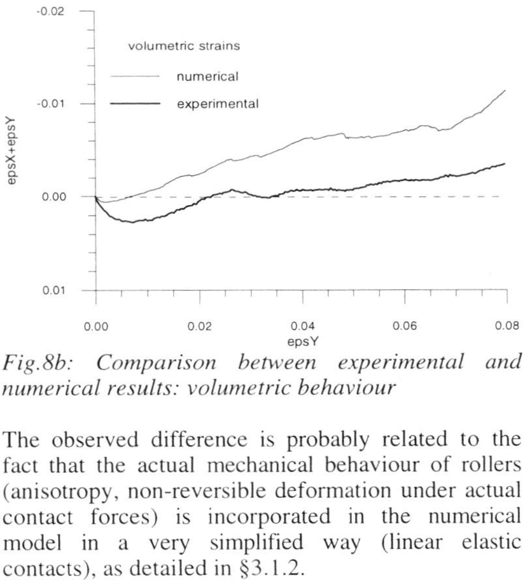

Fig.8b: Comparison between experimental and numerical results: volumetric behaviour

The observed difference is probably related to the fact that the actual mechanical behaviour of rollers (anisotropy, non-reversible deformation under actual contact forces) is incorporated in the numerical model in a very simplified way (linear elastic contacts), as detailed in §3.1.2.

4.2 Micro-mechanical results

A first comparison of the observed micromechanical behaviour regards the evolution of the specimen structure (parameter $S(\theta)$) (Fig.9). For each contact range, $S(\theta_0)$ is defined as the ratio between $N(\theta_0)$ at the end of the compression and $N(\theta_0)$ at the beginning of the test (isotropic state of stress). The specimen structure evolution is well reproduced by the numerical model; it is however important to underline the fact that the parameter $S(\theta)$ takes into account the orientation of contacts, disregarding their spatial distribution within the specimen. In other words, S(θ) is related to the global, or average, specimen behaviour.

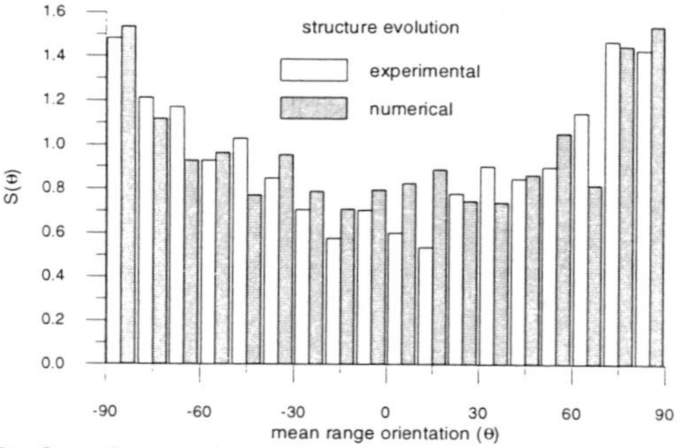

Fig.9: Comparison between experimental and numerical results: structure evolution

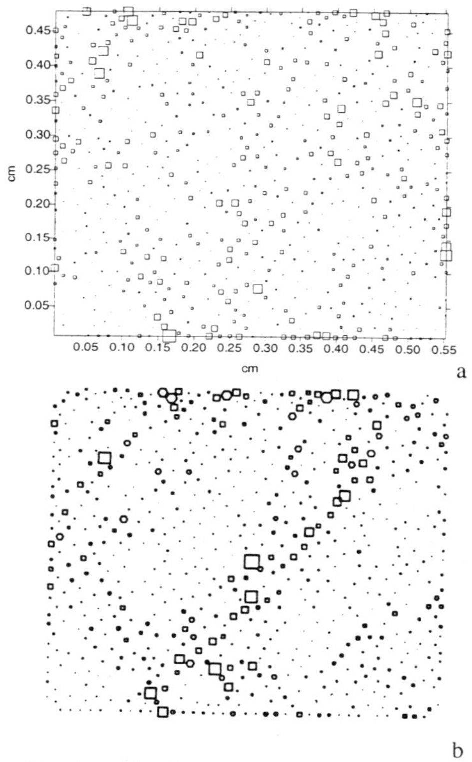

Fig. 10: chart of roller rotations; a): numerical, b): experimental.

The reliability of the numerical model in reproducing the behaviour of individual particles can be verified plotting the chart of rollers rotations (Fig.10a,b). In fact, the numerical model don't reproduce the pattern observed in the experiment, that corresponds to a strain localisation band (Calvetti et al. (1997)). Nevertheless, the average values of the numerical and experimental rotations are really close to zero, which is the rigid rotation of the specimen during compression.

5 FINAL REMARKS

In this paper, the possibility to perform DEM simulations of laboratory tests performed on a rollers stack is evaluated. The numerical model is built to reproduce as close as possible the laboratory conditions, and numerical parameters are directly derived from experiments. The comparison between laboratory and numerical results can be considered satisfactory if the global behaviour of specimens is considered. On the contrary, the numerical model fails to reproduce the actual distribution of local variables.
Further research will be made to understand in which way local conditions can be incorporated into the numerical model.

ACKNOWLEDGEMENTS

This research is part of the activities of the network ALERT-Geomaterials supported by EU in HCM programme.
The author acknowledges Gaël Combe and Jack Lanier (3S-Grenoble) for the help in performing experiments and the kind and careful guidance during his stay in Grenoble, respectively.
Financial contribution from MURST is also gratefully acknowledged.

REFERENCES

Calvetti F., Lugan A., Nova R. (1995). *Numerical analysis of a Schneebeli material specimen* Proceedings NUMOG - V, G.N. Pande and S. Pietruszczak eds., 51-56.

Calvetti F., Combe G., Lanier J. (1997). *Experimental micro-mechanical analysis of a 2-D granular material: relation between structure evolution and loading path.* Mechanics of Cohesive-frictional Materials (in print).

Combe G. (1995). *Etude micromécanique d'un assemblage de rouleaux.* Mémoire de D.E.A. Laboratoire 3-S. Grenoble.

Joer H., Lanier J., Desrues J., Flavigny E. (1992). *1γ2ε: a new shear apparatus to study the behaviour of granular materials.* Geotechnical Testing Journal 15-2, 129-137.

Numerical Models in Geomechanics, Pietruszczak & Pande (eds) © 1997 Balkema, Rotterdam, ISBN 90 5410 886 X

A constitutive model for marine sand

Q.S.Yang
The Hong Kong University of Science and Technology, Hong Kong

H.B.Poorooshasb
Concordia University, Montreal, Que., Canada

ABSTRACT: The seabed response to the natural actions such as earthquake, wind and water wave is one of the challenging problems to marine geotechnical engineering. It is believed that seabed soil liquefaction is an important mechanism causing offshore structure foundation failure. To study the problems associated with seabed instability, a versatile constitutive model capable of describing the engineering properties of marine soils is essential. A non-associated plastic model within the general framework of the bounding surface plasticity is proposed and a computer program is developed for the model validation. The capability of this model is examined. A comparison of model predictions and test results is presented.

1 INTRODUCTION

The phenomenon of seabed liquefaction during ocean storm has been recognized , which may exert damaging influences to the coastal and offshore installations. The load condition caused by earthquakes and sea waves are further complicated due to "rotational shear" (Ishihara, 1983), i.e. the values and directions of the principal stress in seabed soil are continuously changing which has been observed in many in-situ records. Yamada and Ishihara (1979, 1981, 1983, 1985) published a series of test results from a true triaxial apparatus and a torsional device on Fuji River sand and Toyoura sand samples, which emphasized the effect of rotational shear. The test results indicated that the conventionally defined cyclic stress ratio causing liquefaction was reduced by 10-20% under rotational shear loading condition.

The classical plasticity could not adequately describe the very important aspects of soil response to cyclic loading. While some constitutive models developed in the last two decades, performing well in simulating soil behaviour under uni-direction cyclic loading, fail to capture the response of soil under rotational shear.

The generalized plasticity (Zienkiewicz and Mroz, 1984) and the bounding surface concept (Dafalias, 1986) are combined to provide the formulation of a relatively simple constitutive model proposed for marine sand, which is an extension of a two-surface non-associative plasticity model for sand (Poorooshasb and Pietruszczak, 1985). The idea of reflecting plastic potential is retained implicitly by a vanishing yield surface for reverse loading which significantly reduces the numerical complexity of the inner yield surface kinematics. The performance of the model is compared with experimental results for a number of tests. The predictions indicate that the model is able to appropriately describe several fundamental aspects of marine sand behaviour.

2 MODEL FORMULATION

In generalized plasticity formulation, a unit vector $\mathbf{n}$ in the effective stress space is required to determine both loading and unloading. The incremental stress-strain relations could be written in compact matrix notation in which the subscripts l and u refer to loading and unloading respectively,

$$d\sigma' = \mathbf{D}_l \, d\varepsilon \quad \text{if } \mathbf{n}^T d\sigma' > 0 \text{ loading} \quad (1)$$

$$d\sigma' = \mathbf{D}_u \, d\varepsilon \quad \text{if } \mathbf{n}^T d\sigma' < 0 \text{ unloading} \quad (2)$$

where the tangent matrices are related (Zienkiewicz and Mroz, 1984) by

$$(\mathbf{D}_l)^{-1} = (\mathbf{D}^e)^{-1} + \mathbf{n}_{gl} \, \mathbf{n}^T / H_l \quad (3)$$

$$(\mathbf{D}_u)^{-1} = (\mathbf{D}^e)^{-1} + \mathbf{n}_{gu} \, \mathbf{n}^T / H_u \quad (4)$$

in which H_l and H_u are scalar parameters known as loading and unloading plastic moduli. $\mathbf{D}^e$ is the elastic matrix. This relation simply ensures the uniqueness of strain increments when $\mathbf{n}^T d\sigma' = 0$,

$$d\varepsilon = (\mathbf{D}^e)^{-1} d\sigma' \quad (5)$$

$\mathbf{n}_{gl}$ and $\mathbf{n}_{gu}$ are unit vectors which define the flow rule. If they differ from $\mathbf{n}$ then the flow is non-associative and will lead to asymmetric tangent matrices $\mathbf{D}_l$ and $\mathbf{D}_u$. The material behaviour can be fully described if $\mathbf{D}^e$, $\mathbf{n}$, $\mathbf{n}_{gl/u}$ and $H_{l/u}$ could be fully prescribed. By straightforward algebraic manipulation, equation (3) and (4) can be converted into the so-called elasto-plastic matrix,

$$\mathbf{D}_{l/u} = \mathbf{D}^e - \frac{\mathbf{D}^e \, \mathbf{n}_{g\,l/u} \, \mathbf{n}^T \, \mathbf{D}^e}{\mathbf{H}_{l/u} + \mathbf{n}^T \, \mathbf{D}^e \, \mathbf{n}_{g\,l/u}} \quad (6)$$

The constitutive equations for the proposed model in terms of $\mathbf{D}^e$, $\mathbf{n}$, $\mathbf{n}_{gl/u}$, $H_{l/u}$ are determined by a bounding surface F and a plastic potential G in conjunction with a conjugate surface F_c and a potential G_c, as shown in the triaxial configuration in Figure 1. To simulate marine sand behaviour under the general stress conditions, the model is formulated in the full effective stress space. A set of stress invariant (r, θ, Z) is used, which could be related to (I_1, J_2, J_3) as following,

$$r = (2\,J_2)^{1/2}$$
$$\theta = \sin^{-1}(-3\sqrt{3}\ J_3 / (2\,J_2^{3/2})) / 3, \quad -\pi/6 \le \theta \le \pi/6$$
$$Z = I_1 / \sqrt{3} \quad (7)$$

where I_1 is the first invariant of the effective stress tensor σ'_{ij}, while J_2 and J_3 are the second and the third invariant of the stress deviation tensor s_{ij},

$$I_1 = \sigma'_{kk}, \qquad s_{ij} = \sigma'_{ij} - \sigma'_{kk}\,\delta_{ij} / 3$$
$$J_2 = s_{ij}\, s_{ij} / 2, \qquad J_3 = s_{ij}\, s_{jk}\, s_{ki} / 3 \quad (8)$$

in which r and θ give the true value on the octahedral plane ($\theta = -\pi/6$ represents triaxial compression and $\theta = \pi/6$ is for triaxial extension) and z gives the true value along the diagonal of the principal stress space, as shown in Figure 2.

The bounding surface and plastic potential are extended into the full effective stress space in terms of a function g(θ),

$$F = F(r, \theta, Z, e^*) = r - \eta(e^*)\, g(\theta)\, Z = 0 \quad (9)$$
$$G = G(r, \theta, Z, Z_0) = r + M_{cr}\, g(\theta)\, Z \ln (Z / Z_0) = 0$$

where e* is a measure of the total plastic distortion,

$$e^* = \int de^*, \quad de^* = (de^p_{ij}\, de^p_{ij})^{1/2}$$
$$de^p_{ij} = d\varepsilon^p_{ij} - d\varepsilon^p_{kk}\,\delta_{ij} / 3 \quad (10)$$

in which de^p_{ij} is the incremental plastic deviatoric strain. The function η(e*) is given as,

$$\eta(e^*) = \eta_f [1 - \exp(- e^* / A)], \quad (11)$$
$$\eta_f = 2\sqrt{2}\ \sin\phi_f / (3 - \sin\phi_f) \quad (12)$$

by which the size of the bounding surface is determined. When e* increases continuously, the bounding surface approaches the failure surface. As shown in Figure 2, the shape of the bounding surface is a cone having straight meridians and a noncircular cross section with its apex at the origin and its axis along the diagonal of the stress space. The shape of the cross section of the bounding surface on the octahedral plane is describe by the function g(θ). An ellipse formulation first proposed by William and Wande (1975) is adopted to quantify the function g(θ) (see Figure 3), which is symmetric, and convex if $1 < R = g_{max} / g_{min} < 2$. For this model, the convexity is always ensured since $g_{max} = g(-\pi/6) = 1$ is chosen for triaxial compression and $g_{min} = g(\pi/6) = (3 - \sin\phi_f) / (3 + \sin\phi_f)$ for triaxial extension, which is consistent with the Mohr-Coulomb failure criterion.

The line $\eta = \eta(e^*) = r / (g(\theta)\, Z) = M_{cr}$ is synonymous with the 'critical state line' (CSL), on which no plastic volumetric strain occurs, which could be related to the residual angle of the internal friction ϕ_{cr} as

$$M_{cr} = 2\sqrt{2}\ \sin\phi_{cr} / (3 - \sin\phi_{cr}) \quad (13)$$

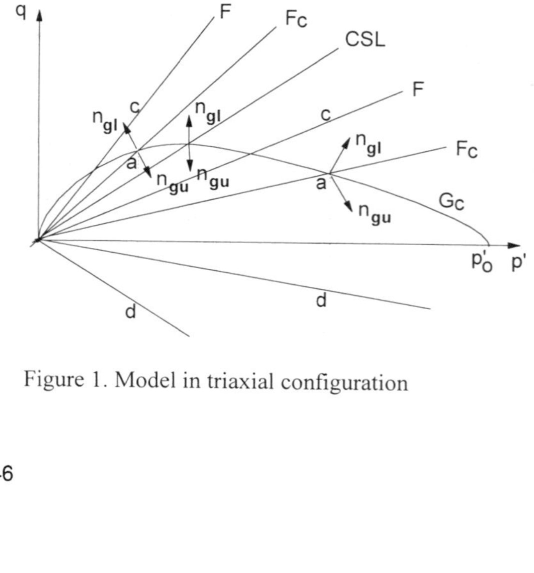

Figure 1. Model in triaxial configuration

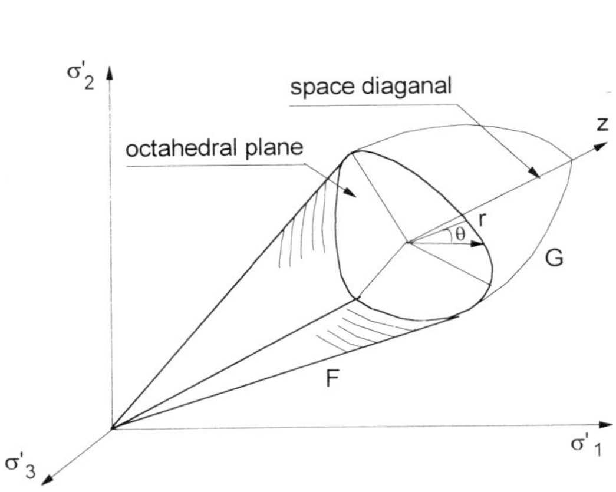

Figure 2. The expression of r, θ, Z in the principal

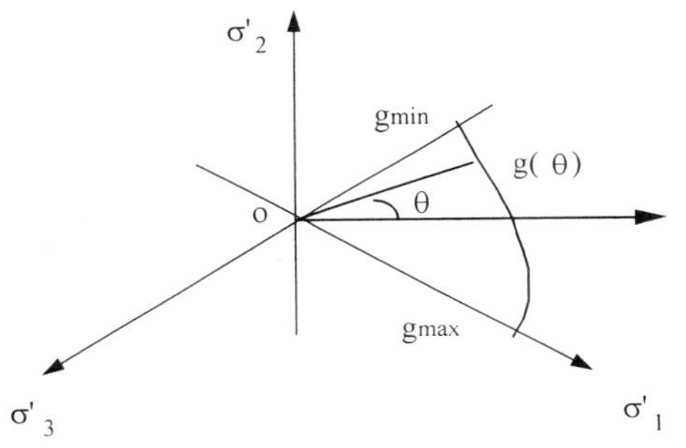

Figure 3. Elliptic trace of the function g(θ)

For virgin loading process, the vector **n** is given by the unit normal vector of the bounding surface, $F(\sigma', e^*) = 0$. If $\mathbf{n}^T d\sigma' > 0$, the loading flow vector $\mathbf{n}_{gl}$ is determined by the plastic potential $G(\sigma')$ and the loading plastic modulus H_l is obtained by imposing the consistency condition $dF = 0$,

$$H_l = \frac{r}{A}\left(\frac{\eta_f}{\eta} - 1\right)\left[\left(\mathrm{dev}\frac{\partial G}{\partial \sigma'}\right)^T\left(\mathrm{dev}\frac{\partial G}{\partial \sigma'}\right)\right]^{1/2} \Big/ \left[\left(\frac{\partial F}{\partial \sigma'}\right)^T\left(\frac{\partial F}{\partial \sigma'}\right)\right]^{1/2}\left[\left(\frac{\partial G}{\partial \sigma'}\right)^T\left(\frac{\partial G}{\partial \sigma'}\right)\right]^{1/2} \quad (14)$$

with

$$\eta = \eta(e^*) = r / (g(\theta) Z) \quad (15)$$

If $\mathbf{n}^T d\sigma' = 0$ or $\mathbf{n}^T d\sigma' < 0$, pure elastic response is assumed, $H_u = \infty$, $\mathbf{D}_u = \mathbf{D}^e$.

For stress reversal process, $F(\sigma', e^*) < 0$. The stress point σ' is located inside the bounding surface, and a vector $\mathbf{n}_c^T$ is determined by the conjugate surface, $F_c(\sigma', e^*_c) = 0$, which is passing through the current stress point and homologous to the bounding surface, $F(\sigma', e^*) = 0$. If $\mathbf{n}_c^T d\sigma' > 0$, it is loading and the loading flow vector $\mathbf{n}_{gl}$ is given by the conjugate potential $G_c(\sigma')$, which passes through the current stress point σ'. If $\mathbf{n}_c^T d\sigma' = 0$, it is pure elastic response. If $\mathbf{n}_c^T d\sigma' < 0$, it is unloading. The unloading flow vector $\mathbf{n}_{gu}$ is related to the loading flow vector $\mathbf{n}_{gl}$ in the same manner as indicated in the triaxial configuration (see Figure 1),

$$\mathbf{n}_{gu}(n^v_{gu}, n^s_{gu}) = (-n^v_{gl}, -n^s_{gl}), \quad \eta > M_{cr} / g(\theta)$$

$$\mathbf{n}_{gu}(n^v_{gu}, n^s_{gu}) = (n^v_{gl}, -n^s_{gl}), \quad \eta < M_{cr} / g(\theta)$$

where superscript s refers to the deviatoric component in the octahedral plane and v stands for the hydrostatic component along the space diagonal.

To specify the loading/unloading plastic moduli $H_{l/u}$, an 'image' point on the bounding surface is defined by a 'deviatoric radial' mapping rule. The image point is the intersection of the bounding surface and a radial on the octahedral plane passing through the current stress point. For loading, the image point and the current point are located on the same side, and for unloading, they are on the opposite sides. This could be expressed by the distance δ as shown in Figure 1,

$$\delta = \underline{ac} \quad \text{if loading}$$
$$\delta = \underline{ad} \quad \text{if unloading} \quad (16)$$
$$\delta_0 = \underline{ac} + \underline{ad} \quad \text{(see Figure 1)}$$

The loading /unloading plastic moduli $H_{l/u}$ is determined by an interpolation rule:

$$\frac{1}{\mathbf{H}_{l/u}} = \frac{1}{\mathbf{H}_B}\left(1 - \frac{\delta}{\delta_0}\right)^{\gamma} \quad (17)$$

where γ is a material parameter and H_B is the plastic loading moduli at the image point. It may be noted that although both H_l and H_u are expressed by equation (17), indeed they are different since the image points, and thus H_B and δ are not the same for loading and unloading.

3 COMPARISONS WITH EXPERIMENTAL RESULTS

In order to identify the model, elastic parameters are required to determine $\mathbf{D}^e$. In this study, Poissonís ratio ν is taken as constant and K, the bulk modulus is assumed to depend on the effective mean stress σ'_m as,

$$K = p_a \frac{1 + e_0}{\kappa} \left(\frac{\sigma'_m}{p_a}\right)^{1/2} \quad (18)$$

where p_a is the atmospheric pressure for the nondimensionalization of the constant κ, and e_0 is the initial void ratio. There are seven material constants (three elastic parameters, e_0 , κ, ν and four plastic parameters, η_f , M_{cr}, A, γ) in this model. All the parameters can be determined from the results of the conventional triaxial compression test except for γ, which can be evaluated from the results of an undrained cyclic test by a trial and error procedure.

Yamada and Ishihara published a series of experimental results. For drained and undrained monotonic tests, the numerical predictions agreed with the experimental results quite well, but will be not presented here due to the space limitation.

Figures 4(a) and 5(a) show the experimental data of Fuji River sand specimens tested on a true triaxial apparatus (Yamada and Ishihara, 1983). Figure 4(b) and 5(b) are the results of corresponding numerical simulations for undrained cyclic loading. The model parameters used for Fuji River sand are listed in Table 1. The specimens were first consolidated to an effective confining pressure of 98 kPa. After consolidation a back pressure of 49 kPa was applied to the specimens to ensure full saturation and the total mean principal stress was kept at a constant value of 147 kPa during the undrained cyclic tests.

The effective stress paths are plotted in terms of $\tau^*_{oct} = \tau_{oct} \cos\theta$, where the angle, θ, indicates the direction of the shear stress in the octahedral plane. The value of τ^*_{oct} means the projection of τ_{oct} in vertical direction in the octahedral plane. As shown in Figure 4 for undrained uni-directional cyclic

Table 1. Parameters for Numerical Simulations

Figure No.	4	5	6	7
Stress Path	UZC-ZE Cyclic	True Triaxial Circular	Triaxial Torsion	Torsion
e_0	0.827	0.822	0.818	0.784
κ	0.025	0.025	0.015	0.015
υ	0.25	0.25	0.25	0.25
η_f	0.73	0.73	0.63	0.63
M_{cr}	0.66	0.66	0.55	0.55
A	0.0055	0.0055	0.0065	0.0065
γ	7.0	7.0	5.0	5.0

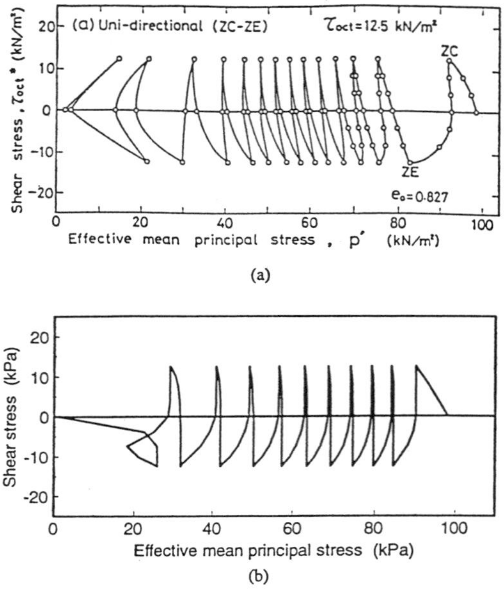

Figure 4. Effective mean principal stress p′ (kPa) vs. octahedral shear stress $\tau^*_{oct} = \tau_{oct} \cos\theta$ (kPa) for undrained uni-directional cyclic loading.

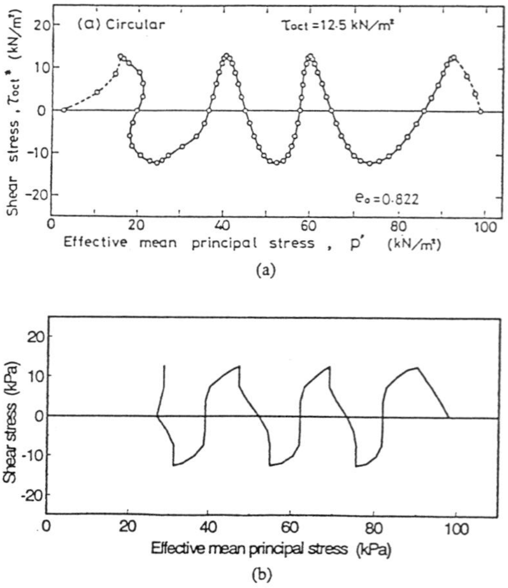

Figure 5. Effective mean principal stress p′ (kPa) vs. octahedral shear stress $\tau^*_{oct} = \tau_{oct} \cos\theta$ (kPa) for undrained circular shear loading.
(a) test result (Yamada and Ishihara, 1983);
(b) numerical simulation.

loading with constant value of τ_{oct} = 12.5 kPa, the pore water pressure increases gradually and the effective stress ratio τ_{oct}/p', eventually reaches a value corresponding to the angle of phase transformation, whereupon a significant increase in pore water pressure and decrease in effective mean principal stress to liquefaction.

For undrained circular rotational shear test with the same constant value of τ_{oct} = 12.5 kPa as shown in Figure 5, the pore water pressure was generated more remarkably and in a similar manner to that of the uni-directional cyclic stress path until the phase transformation line reached. Then the effective stress path showed entirely different from those for the uni-directional cyclic stress path. The pore water pressure never became equal to the initial confining pressure, and the effective stress path moved back and forth along almost the same loop. In the test with a rotational stress path it was not reached the state of initial liquefaction except the shear stress was brought back to zero by monotonic unloading. The monotonic loading at the beginning and monotonic unloading at the end of the stress path are indicated with broken lines in Figure 4(a).

Figures 6(a) and 7(a) show the test data of Japanese standard sand known as Toyoura sand by a triaxial torsion apparatus (Yamada and Ishihara, 1985). Figure 6(b) and 7(b) are the results of corresponding numerical simulations for undrained cyclic loading. The model parameters used for Toyoura sand are listed in Table 1. Specimens were isotropically consolidated under 294 kPa and a back pressure of 29.4 kPa was applied. In cyclic triaxial-torsion shear test (Figure 6), triaxial compression

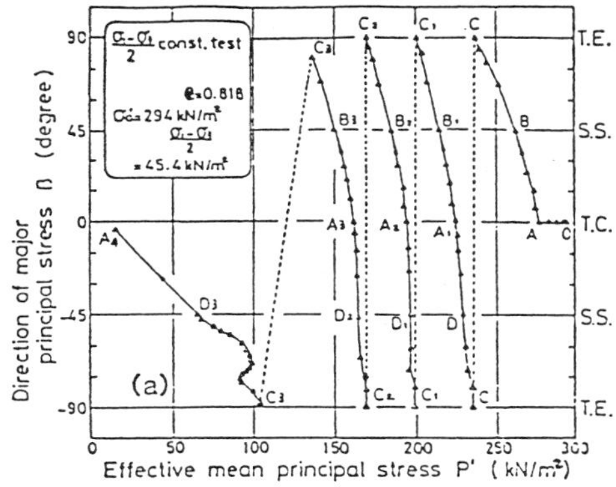

(a)

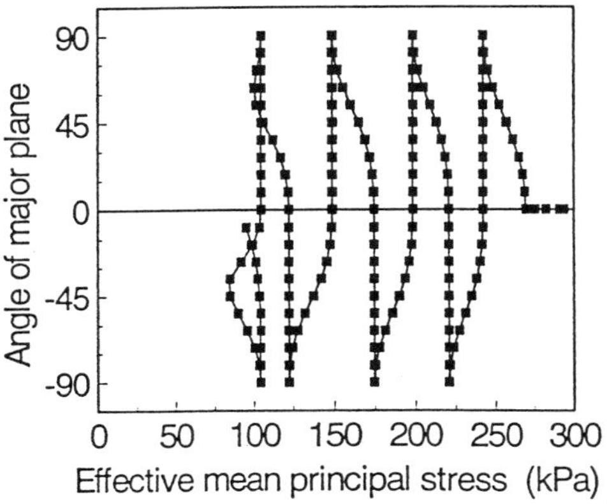

(b)

Figure 6. β (degree) vs. p′ (kPa) relationship in undrained cyclic triaxial-tortion.
(a) test result (Yamada and Ishihara, 1985)
(b) numerical simulation.

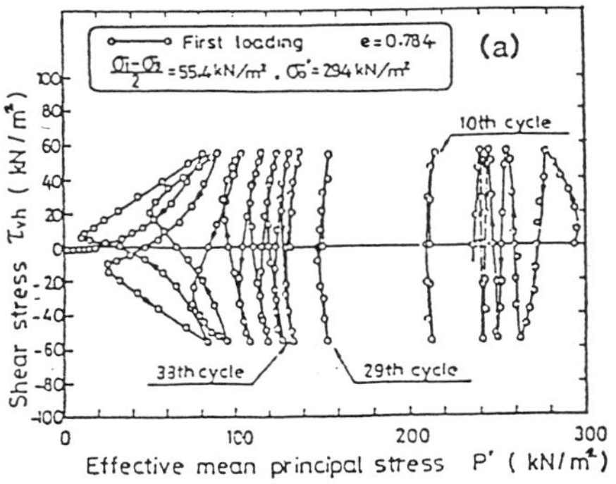

(a)

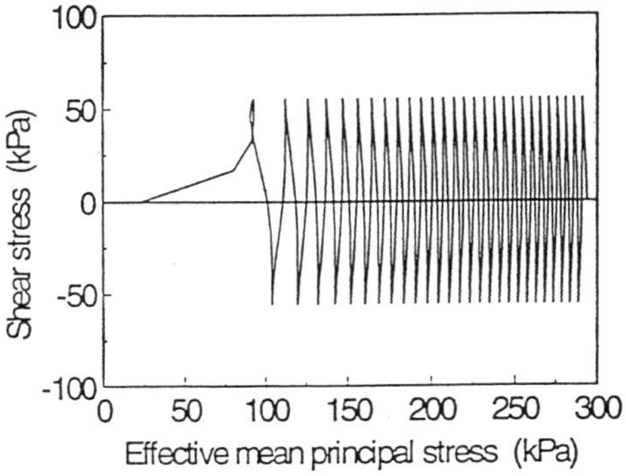

(b)

Figure 7. Stress path, τ_{vh} (kPa) vs. p′ (kPa) in undrained cyclic tortion shear.
(a) test result (Yamada and Ishihara, 1985);
(b) numerical simulation.

was applied to make the deviate stress equal to a specified value. Then the principal stress axes were made to rotate continuously with the deviate stress equal to a specified value of τ_{max} = 45.4 kPa. Test result is shown in Figure 6 in terms of the principal stress direction, β, plotted versus the effective mean principal stress, p'. It can be seen that , while the angle, β, changes, p' is reduced and excess pore water pressure is generated drastically in the first few cycles. Figure 7 shows the result of conventional cyclic torsion shear test with τ_{max} = 55.4 kPa carried out in undrained condition which requires more than 30 cycles to reach liquefaction. Comparison of Figure 6 and Figure 7 clearly indicates that the continuous rotation of principal stress axes reduces the resistance of sand to liquefaction.

4 CONCLUSION

The model is relative simple. A comparison of the model predictions with the experimental results shows that it is capable of simulating several fundamental aspects of marine sand behaviour under various loading conditions either monotonic or cyclic, in particular, the rotational shear which is related to an important class of loading conditions occurred due to ocean wave propagation over the seabed deposits.

REFERENCE

Dafalias, Y.F. (1986). "Bounding surface plasticity. I: Mathematical foundation and hypoplasticity," *J. Engrg. Mech., ASCE*, Vol 112, No 9, pp 966-987.

Ishihara, K. (1983). "Soil response in cyclic loading induced by earthquakes, traffic and waves," *Proc. 7th Asian Reg. Conf. on Soil Mechanics and Foundation Engineering*, Vol 3, pp 42-66.

Poorooshasb, H.B. & S. Pietruszczak 1985. "On yielding and flow of sand; A generalized two-surface model," *Computers and Ceotechnics*, Vol 1, p 33-58.

William, K.T. & E.P. Wande 1975. "Constitutive model for the triaxial behaviour of concrete," *Proc. Int. Ass. Bridge Struct. Engrg.*, Vol 19, pp 1-30.

Yamada, Y. & K. Ishihara 1979. "Anisotropic deformation characteristics of sand under three dimensional stress conditions," *Soils and Foundations*, Vol 19, No 2, pp 79-94.

Yamada, Y. & K. Ishihara 1981. "Undrained deformation characteristics of loose sand under three dimensional stress conditions," *Soils and Foundations*, Vol 21, No 1, pp 97-107.

Yamada, Y. & K. Ishihara 1983. "Undrained deformation characteristics of sand in multi directional shear," *Soils and Foundations*, Vol 23, No 1, pp 61-79.

Yamada, Y. & K. Ishihara 1985. "Undrained strength of sand undergoing cyclic rotation of principal stress axes," *Soils and Foundations*, Vol 25, No 2, pp 135-147.

Zienkiewicz, O.C. & Z. Mroz 1984. "Generalized plasticity formulation and application to geomechanics," Ch 33, *Mechanics of Engineering Materials*, Eds., Desai, C.S.S. and Gallagher, R.H., Wiley.

Numerical Models in Geomechanics, Pietruszczak & Pande (eds) © 1997 Balkema, Rotterdam, ISBN 90 5410 886 X

Constitutive conjectures for response of sand under multiaxial loading

D. Muir Wood
Department of Civil Engineering, University of Bristol, UK

M. I. Alsayed
Department of Civil Engineering, University of Newcastle upon Tyne, UK

W. M. Stewart
Department of Civil Engineering, University of Glasgow, UK

ABSTRACT: A true triaxial apparatus has been used to perform distortional stress probing tests on dry sand. Cycles of unloading and reloading show significant hysteresis indicating plastic response. Comparison of results from tests in which the stress level has been doubled shows that the strains are remarkably similar implying that, in terms of mobilised friction, the behaviour is identical. Friction plays an important role in the stress:strain response. The results indicate significantly greater strains in probes which raise the mobilised angles of friction to progressively higher levels than in probes which reach lower angles of friction and which, from inspection, can be thought of as passing within a kinematic yield locus established by the previous probing. Deviatoric stiffness is not greatly affected by angle of corner in stress path on stress reversal.

1 INTRODUCTION

Stress probe tests have been performed in a rigid boundary true triaxial apparatus in order to explore the distortional response of dry sand. This work follows from earlier work (Alawaji et al, 1990) which showed that some aspects of the distortional response of sand can be described by kinematic hardening models. The present tests have been used to produce a number of constitutive conjectures that can be used to guide the formulation of constitutive models for the distortional response of the sand.

2 EXPERIMENTAL PROCEDURE

Tests have been performed in the True Triaxial Apparatus which was developed and used at Cambridge University from 1969-1987 (Airey and Muir Wood, 1988). This apparatus has a load capacity on each axis of about 10kN which, with a typical sample size of 0.1 × 0.1 × 0.1m gives a maximum stress of about 1MPa.

Tests have been performed on samples of dry Lochaline sand, which is a fine silicaceous uniform sub-rounded sand with d_{50} = 0.25mm.

The medium dense sand samples were prepared by pluviation from a height of about 1.5m through a series of sieves into a cubical membrane stretched over a sample former. Once full the top of the sample was levelled and a flap of latex glued across the top of the membrane. A needle inserted through the flap was used to impose a small vacuum on the sand - a procedure similar to that adopted by Lanier and Zitouni (1989) for preparation of true triaxial sand samples. The sample could then retain its shape without external support and be lifted into the true triaxial apparatus. Once the platens of the apparatus were in contact with the sample the vacuum could be removed and the test could proceed.

Software has been written to control all aspects of the use of the apparatus. The control algorithm that has been devised for applying stress path control introduces some assessment of the current error and the current rate of change of the error in a nonlinear fashion. As has been found previously (Airey and Muir Wood, 1988) it is difficult using stepper motor control of three independent stresses to impose specified rates of loading. The stress state is recorded by a data logger at regular intervals but the stepper motor speeds are capped in a way approximately equivalent to the imposition of a maximum strain rate on a stress controlled process.

3 TEST PROGRAMME

The tests reported here form a group in which multiple stress probes have been applied to single samples of Lochaline sand. The stress states which

Table 1. Stress probe tests

Sample	Test	Path
LA7	T1	oOAB1BAB2BAB3BAB4BAB5BAB6BAB7BAB8BAOo
	T2	oOAD1DAD2DAD3DAD4DAD5DAD6DAD7DAD8DAOo
	T3	oOAC1CAC2CAC3CAC4CAC5CAC6CAC7CAC8CAOo
	T4	oO”o
LA8	T1	oOAB1BAB2BAB3BAB4BAB5BAB6BAB7BAB8BAOo
	T2	oOAB7BAB3BAB6BAB8BAB1BAB4BAB2BAB5BAOo
	T3	oO”o
LA9	T1	oO”o
	T2	oO’AB1BAB2BAB3BAB4BAB5BAB6BAB7BAB8BAO’o
	T3	oO’AB7BAB3BAB6BAB8BAB1BAB4BAB2BAB5BAO’o
	T4	oO”o

have been visited are shown in Fig 1 and the paths followed are indicated in Table 1. The isotropic stress states o, O, O’, O” are given in Table 2. Once the chosen isotropic stress level, O or O’, has been reached the remainder of the stress path is applied at the same constant mean stress. The stress states A, B, C, D all lie on a circle centred on the origin of the deviatoric plane, with stress values and mobilised angles of friction given in Table 3. The numbered probes, 1…8, are always in the same direction in the

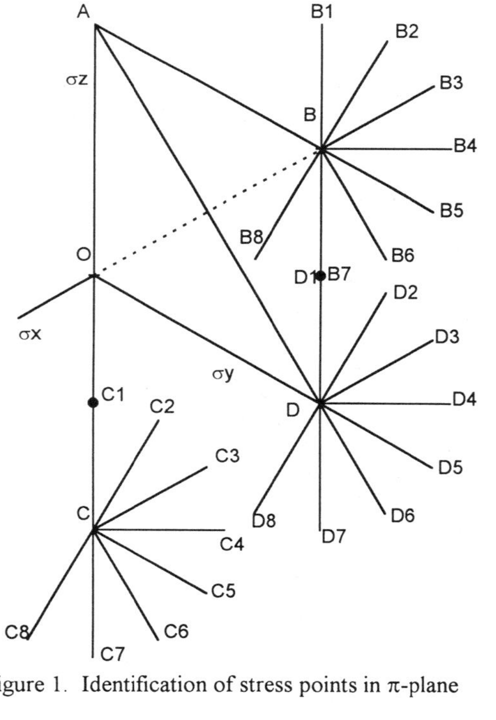

Figure 1. Identification of stress points in π-plane

deviatoric plane relative to the vertical stress axis, σ_z, as shown in Table 4 with angles measured clockwise from this axis. The length of these probes in the deviatoric plane is half the radius of the circle on which A, B, C, D lie: the incremental stresses are also given in Table 4. The mobilised angles of friction on these probes depend on the stress state from which they have been applied (Table 5).

Tests LA7 T1, T2 and T3 apply the same sequence of ultimate stress probes from three different starting points B, D, C. Tests LA8 T1 and T2 apply the same set of eight ultimate stress probes but in a different sequence. Tests LA9 T2 and T3 apply the same paths as tests LA8 T1 and T2 but at an isotropic stress level of 300kPa instead of 150kPa and with the magnitude of all the stress probes correspondingly doubled.

Table 2. Isotropic stress states

Stress state	Isotropic stress kPa
o	30
O	150
O’	300
O”	500

Table 3. Starting points for stress probes

Point	σ_x kPa	σ_y kPa	σ_z kPa	Mobilised friction degrees
A	120	120	210	15.8
B	90	180	180	19.5
C	180	180	90	19.5
D	120	210	120	15.8

Table 4. Orientation of stress probes

Probe	Orientation degrees	$\delta\sigma_x$ kPa	$\delta\sigma_y$ kPa	$\delta\sigma_z$ kPa
1	0	-15	-15	30
2	30	-26	0	26
3	60	-30	15	15
4	90	-26	26	0
5	120	-15	30	-15
6	150	0	26	-26
7	180	15	15	-30
8	210	26	0	-26

Table 5. Mobilised angles of friction on stress probes

Probe	Mobilised friction degrees
B1	28.3
B2	31.7
B3	32.0
B4	31.7
B5	28.3
B6	23.1
B7	17.5
B8	12.5
C1	9.1
C2	12.5
C3	17.5
C4	23.1
C5	28.3
C6	31.7
C7	32.0
C8	31.7
D1	17.5
D2	22.4
D3	25.4
D4	25.5
D5	23.0
D6	25.5
D7	25.4
D8	22.4

4 CONSTITUTIVE CONJECTURES

Some of the conclusions that can be drawn from the results of these stress probe tests will be summarised here in the form of a series of conjectures concerning the deviatoric response of medium dense Lochaline sand. Plots of stress:strain response are presented in terms of length of deviatoric stress path and length of deviatoric strain path, with appropriate factors being applied so that these plots would reduce to plots of deviator stress q and distortional strain ε_q under conditions of axial symmetry, with two equal principal stresses.

Conjecture 1: Isotropic hardening is an inadequate model. Cycles of unloading and reloading show significant hysteresis indicating plastic response which is not reproduced by an isotropic hardening model (Fig 2).

Conjecture 2: Initial loading is significantly less stiff than subsequent loading. After initial loading, cycles of unloading and reloading produce nearly closed hysteresis loops with an average stiffness much higher than that seen on the original loading. Compare the successive stress:strain responses for section OA of tests LA7T1, LA7T2, LA7T3 (Fig 3).

Conjecture 3: Deviatoric stiffness is not greatly affected by angle of corner in stress path on stress reversal. The results obtained from the first applications of paths AB, AD, AC in tests LA7 T1, T2, T3 show very similar stiffnesses over the first sections even though the angle through which the stress path turns are 120, 150 and 180 degrees (Fig 3). Similarly, the response on paths BA after successive stress probes in different directions is not greatly affected by the direction of the preceding

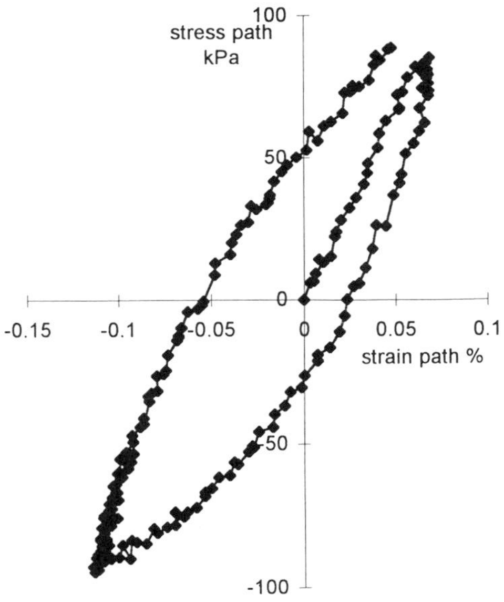

Figure 2. Test LA7T3: path OAC1CA

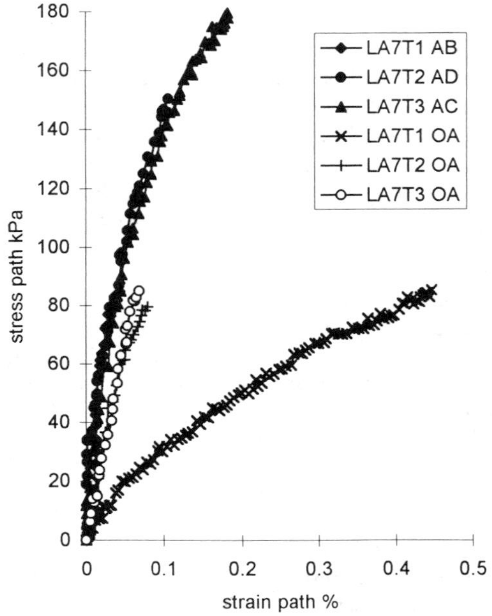

Figure 3. Tests LA7T1, LA7T2, LA7T3: comparison of response on path OA; effect of angle of turn in stress path

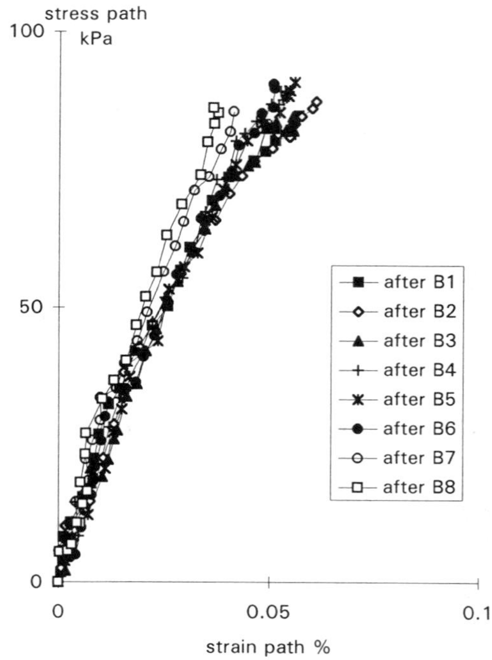

Figure 4. Test LA7T1: effect of angle of turn on response for path BA

probe (Fig 4): there is no obvious trend in these stress:strain responses where the angle of turn is sharpest after B1 (120 degrees) and lowest after B5 (zero) (see stress paths in Fig 1).

Conjecture 4: Friction plays an important role in stress:strain response. The results shown in Fig 5 from test LA8 T1 indicate significantly greater strains in probes B1, B2, B3, which are extending the mobilised angles of friction to progressively higher levels, than in probes B4, B5, B6, B7, B8 which reach lower angles of friction and which, from inspection, might be thought of as passing entirely within a kinematic yield locus established by the

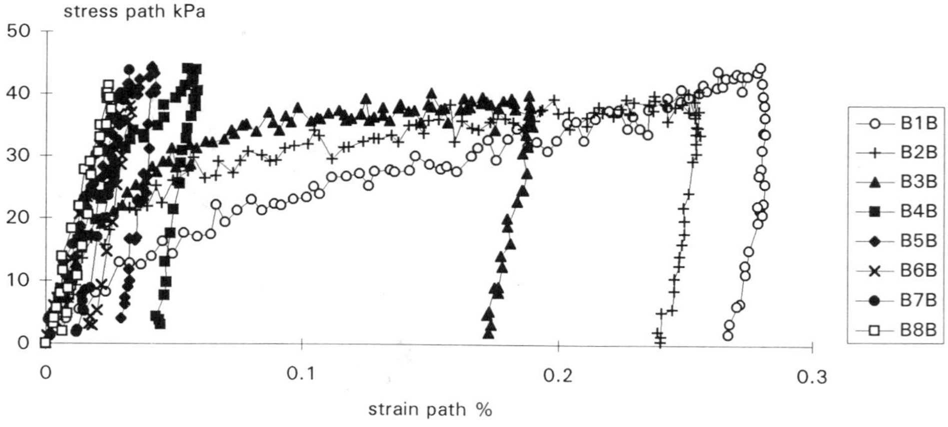

Figure 5. Test LA8T1: stress probes

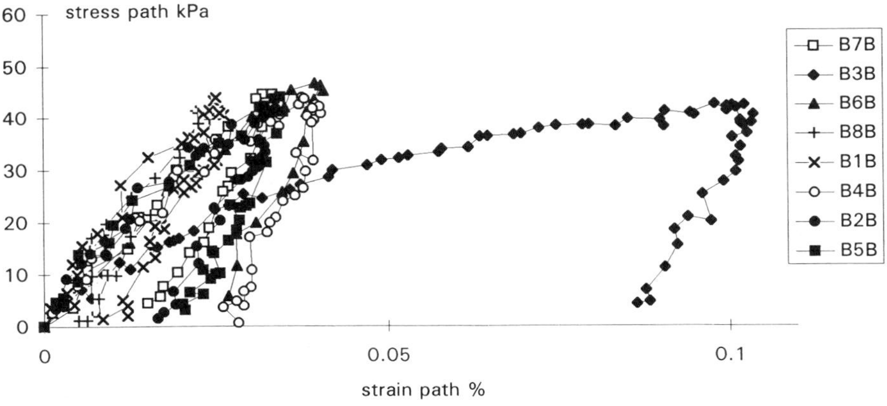

Figure 6. Test LA8T2: stress probes

previous probing. In the random probing of test LA8 T2 (Fig 6), probe B3 is applied before probes B1 and B2 and takes the mobilised friction to the highest level reached in the set of probes. This probe alone generates significant strains.

Conjecture 5: Previously mobilised friction influences kinematic yielding. Comparison of the stress:strain responses observed on path BAB after different probes within test LA7T1 (Fig 7) gives clear indication of yielding after probe B1 but not after probe B8 by which time the sand has been subjected to a higher mobilised friction. The data from these tests can be used to deduce a link between size of kinematic yield locus and previously mobilised friction. It may be noted that the interpretation of low mean stress level stress probe tests on dry sand reported by Alawaji et al (1990) implied deviatoric kinematic yield loci of roughly circular shape. It appears from the present tests that the kinematic yield locus has a diameter which is eventually of the same order as the distance from a stress probe point back to A. It is only on subsequent unloading to O and isotropic unloading and reloading to o that the memory of the deviatoric history starts to be erased.

Conjecture 6: Similar stress ratio: strain response is found when the stress level is doubled. Comparison of results from tests LA8 T2 and LA9 T3 shows that the strains observed in the several stress probes are remarkably similar (Fig 8). The length of the stress probes is proportional to mean stress level implying that, in terms of stress ratio - as plotted in Fig 8 - (or mobilised friction), the behaviour is essentially identical.

Conjecture 7: Volumetric response shows limited hysteresis. The volumetric response of the sand is

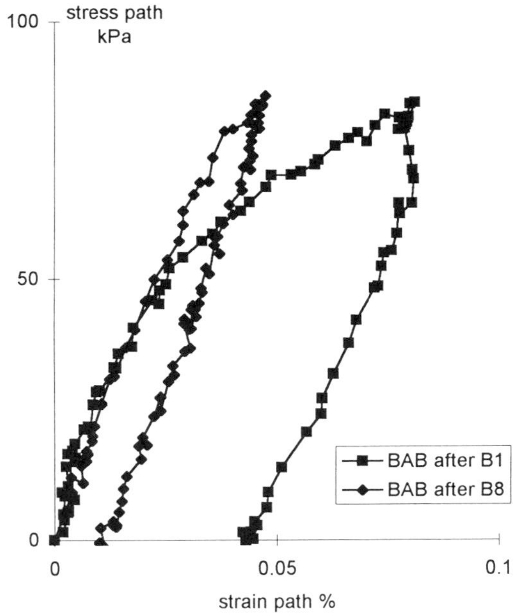

Figure 7. Test LA7T1: cycle BAB after stress probes

nonlinear (results for test LA9 are shown in Fig 9) and combines irrecoverable and recoverable deformations. Each test begins with some isotropic compression to 150kPa (tests 7 and 8) or 500kPa/300kPa (test 9) and each test ends with isotropic compression and unloading from 30kPa to 500kPa. The overall volumetric response is rather similar in all tests and is not greatly influenced by the preceding history of deviatoric stress probes, which

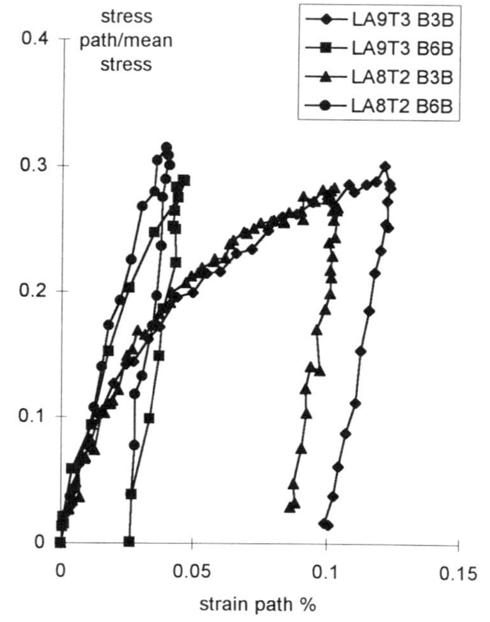

Figure 8. Effect of stress level on response

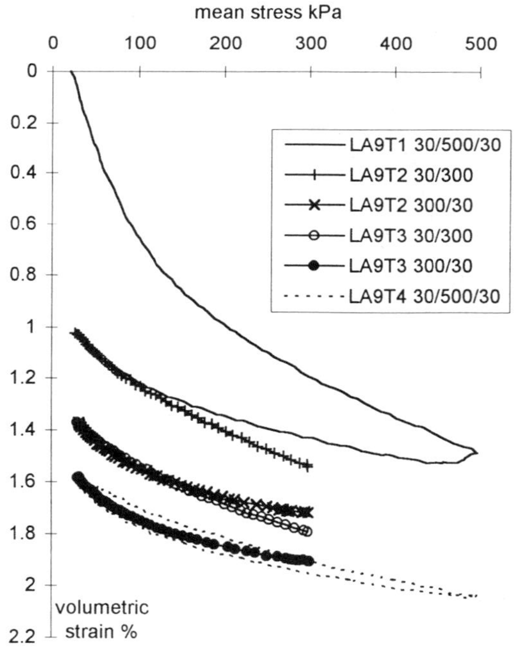

Figure 9. Test LA9: volumetric strain (numbers in table indicate mean stresses on successive paths)

are accompanied by small volumetric deformations. Unloading and reloading produce similar responses with typically slightly more volume strain on reloading than on the previous unloading. Volumetric strains occurring during the stress probing are not insignificant (Fig 9). Much present constitutive modelling assumes that distortional and volumetric responses can be modelled independently: conjectures 4 and 5 concerning the importance of friction make this same assumption.

5. CONCLUSIONS

The tests that have been reported here confirm for rather general distortional stress probes that friction is important in controlling the stress:strain response. It is natural to see these results as supporting the modelling of the deviatoric behaviour of sand using kinematic hardening models.

ACKNOWLEDGEMENTS

The research described here was performed at Glasgow University funded by a grant from the UK Engineering and Physical Sciences Research Council.

REFERENCES

Airey, D.W. and Muir Wood, D. (1988) Cambridge True Triaxial Apparatus. *Advanced triaxial testing of soil and rock* (eds R.T. Donaghe, R.C. Chaney and M.L. Silver) ASTM:STP 977, 796-805.

Alawaji, H., Alawi, M., Ko, H.-Y., Sture, S., Peters, J.F. and Muir Wood, D. (1990) Experimental observations of anisotropy in some stress-controlled tests on dry sand. *Yielding, damage, and failure of anisotropic solids* (ed J.P. Boehler) Mechanical Engineering Publications, London EGF5, 251-264.

Lanier, J. and Zitouni, Z. (1988) Development of a data base using the Grenoble true triaxial apparatus. *Constitutive equations for granular non-cohesive soils* (eds A. Saada and G. Bianchini) Balkema, Rotterdam 47-58.

Numerical Models in Geomechanics, Pietruszczak & Pande (eds) © 1997 Balkema, Rotterdam, ISBN 90 5410 886 X

Static liquefaction: Performances and limitations of two advanced elastoplasticity models

T. Doanh & E. Ibraim
Laboratoire Géomatériaux, Ecole Nationale des Travaux Publics de l'Etat, Vaulx-en-Velin, France

Ph. Dubujet
Ecole Centrale de Lyon, France

R. Matiotti
Milan University of Technology (Politecnico), Italy

ABSTRACT: The objective of this paper is to provide a comprehensive evaluation of two advanced elastoplasticity models to describe the undrained behaviour of very loose sands. This evaluation is based on a complete set of experimental data on very loose Hostun saturated sand, with emphasis on the triaxial extension tests on anisotropically consolidated samples. This study demonstrates the effectiveness of using kinematic hardening mechanism to simulate the phenomena of induced anisotropy, static liquefaction and instability domain. It shows the important role of the non-linearity of the elastic component in the elastoplasticity framework, and finally indicates some limitations imposed by the stress-hardening approach.

1 INTRODUCTION

Over the past decade, many experimental works have been devoted to static liquefaction. This particular phenomenon occurs only in very loose saturated sand and can be observed in classical triaxial undrained compression and extension tests.

Experimental data evidence some main factors affecting this phenomenon, such as the initial void ratio, the initial isotropic or anisotropic consolidation stress ratio, the overconsolidation or the drained axial strain prior to the undrained shearing. Successively, concepts of tstate parameters, Been et al (1985), undrained instability line, Lade et al (1988), collapse surface, Sladen et al (1989), or minimum undrained strength framework, Konrad (1990), have been introduced as attempts to characterize this phenomenon.

At the same time, different theoretical approaches have been proposed to simulate these experimental findings. Non-linear incremental constitutive equations are used to describe the static liquefaction of isotropic samples, Darve (1994). Within the non-associated elastoplasticity framework, the observed behaviour of isotropic sand is simulated by Sladen et al (1989), Saïtta et al (1992). Matiotti (1995) explained partially the behaviour of anisotropic Hostun sand. Furthermore, Molenkamp (1991) formulated the mathematical conditions for the material instability, and Dubujet et al (1997) obtained analytically the undrained instability domain.

This paper proposes a comprehensive evaluation of two recent elastoplasticity models to predict some complex aspects of the triaxial undrained behaviour of very loose saturated Hostun sand. The comparison is based on a complete set of experimental data, including complex stress loading. The effects of some influencing factors are examined. The validity of these models and the advantages of using strain-hardening approach are also discussed.

2 ELASTOPLASTICITY MODELS

Within the non-associated elastoplasticity framework, the two chosen models are very similar. The first model, named Lyon model in this paper, is developed by Cambou et al (1988). Di Prisco et al (1993) conceive the second Milan model, which is largely inspired by the first one, with some new evolution functions of the internal variables depending on the plastic history. These two models have many similar features such as incorporating in its developments the induced anisotropy by preloading. They are characterized by a combination of isotropic and kinematic hardening mechanisms.

Lyon model chooses the stress-hardening approach and incorporates explicitly the characteristic state concept or the phase transformation line, and deduces the plastic potential surface from the yield surface. Milan model prefers the strain-hardening approach, and has explicitly the usual plastic potential surface; thus the phase transformation line can be calculated. A detailed description of these models can be found in Dubujet et al (1997) and di Prisco et al (1993). Table 1 gives some differences between these models.

Table 1. Main differences between two models.

Features	Lyon	Milan
Mixed kinematic hardening	stress	strain
Plastic potential surface	implicit	explicit
Phase transformation surface	explicit	implicit
Parameters	10+2	14
Identification	3CIU +1ISO	1CIU+1CID +1ISO

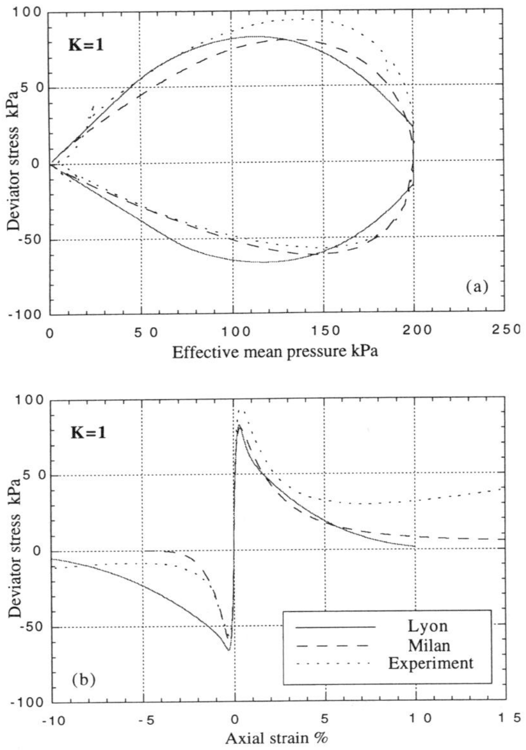

Figure 1. (a) Complete stress paths. (b) Stress-strain behaviour of isotropic samples.

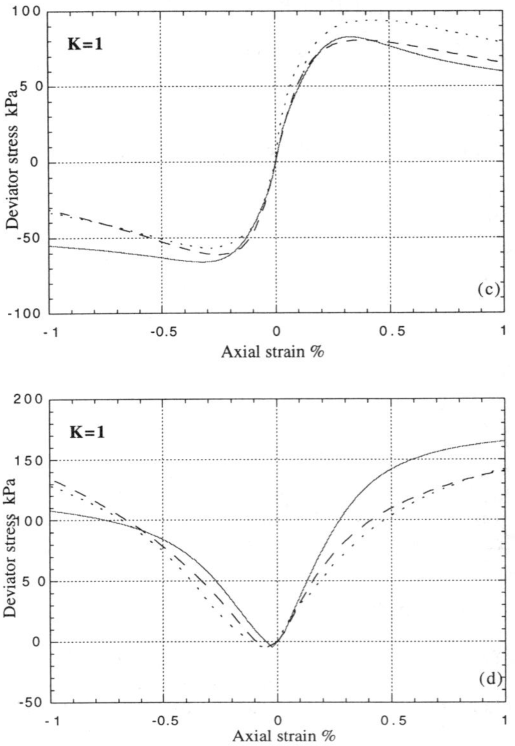

Figure 1. (c) Stress-strain behaviour at small strains. (d) Pore pressure generation of isotropic samples.

Lyon model requires 3 conventional undrained triaxial compression CIU tests to identify the 10 necessary parameters, and an isotropic drained compression test ISO with unloading to obtain the elastic parameters.

In fact, the model has two additional parameters that can be ignored in the case of undrained condition on loose sands, because the mean effective stress of the effective stress path is always below the initial consolidation pressure, thus the third mechanism is not activated.

The 14 parameters of Milan model are identified with an undrained compression test, a drained compression test CID, and an isotropic compression test including unloading. The values of model parameters considered in this comparative study are given in Table 2.

Table 2a. Values of parameters of Lyon model.

G^e	K^e	n	γ	a	ϕ_1
15.10^6	16.10^6	0.6	0.825	17.10^{-5}	0
ϕ_2	A	B	R_e	β	K^p
4.85	0.89	39.66	0.235	-3	---

Table 2b. Values of parameters of Milan model.

Bo	RF	α_c	γ_c	θ_c	θ_e	ζ_c
750	.28	.35	3.5	.253	.118	-.258
ζ_e	B_p	r_c	β_{f0}	β_f	C_p	t_p
-.488	.003	.35	1.1	0.5	18.	10.

3 EXPERIMENTAL OBSERVATIONS

Experimental results were extracted from a complete study of undrained behaviour of very loose saturated Hostun sand in triaxial apparatus, Doanh et al (1997). Samples are subjected to various initial isotropic or anisotropic consolidations along constant effective stress ratio paths $K=\sigma'_r/\sigma'_a$. The bilinear stress paths with one sharp bend in this study fulfill some of the requirements suggested by Gudehus (1985).

It is shown that loose Hostun sand exhibits partial static liquefaction under triaxial compression and extension tests, indicating a large loss of effective mean pressure due to a continuous generation of pore pressure. A stress reversal at large strains is needed to obtain the liquefaction. The stress-strain behaviour is characterized by a sharp deviatoric stress peak at small strains, following by a progressive drop of deviatoric stress before reaching a residual state at large strains.

The deviatoric stress peak is strongly influenced by the monotonic consolidation history. It always increases in compression, independently of the positive anisotropic consolidation, but a reverse trend may occur on the extension side, meaning that the stress ratio increment, $\eta = q/p'$, stabilizes asymptotically. A large positive anisotropic consolidation always produces a mobilized undrained compressive strength increment.

Furthermore, a complex stress history can modify significantly the deviatoric stress peak, depending on the amount of axial strains reached during complex loading.

These relevant points are considered as a qualitative verification in the evaluation of the constitutive models.

4 MODELS EVALUATION

4.1 Isotropic consolidation.

Concerning the isotropic consolidation case, K=1, a very good agreement is obtained in compression as well as in extension between the experimental data and the numerical results of the two models.

Figure 1a gives an example of the effective stress paths in the q-p' plane. In this paper, the final mean effective pressure of all experiments at the end of the consolidation stage is 200 kPa. As expected, the two models fully liquefy the samples without the presence of the dilatancy domain; whereas only partial liquefaction is obtained in experiments. It is worth noting the continuity of the stress paths at the beginning of the undrained shearing revealed by experiments and well simulated by the two models.

The initial vertical stress paths at constant mean effective pressure of Lyon model indicate the size of the elastic domain chosen in order to arrive at realistic simulation of the extension tests. Figure 1c shows the complete stress-strain behaviour with some discrepancies at large strains; and Figure 1d the satisfactory simulation of deviatoric stress peaks at relatively small strains.

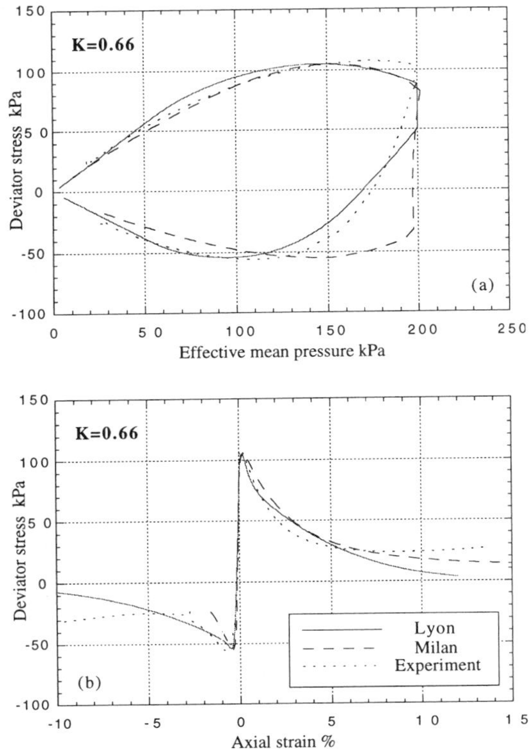

Figure 2. (a) Complete stress paths. (b) Stress-strain behaviour of anisotropic samples.

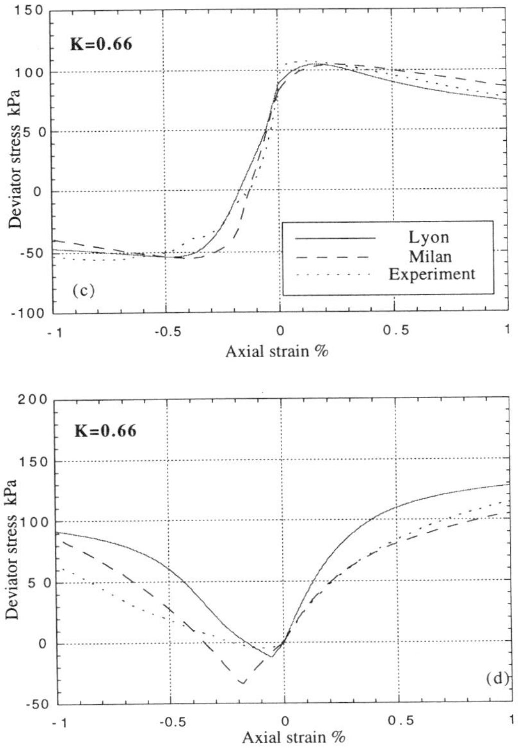

Figure 2. (c) Stress-strain behaviour at small strains. (d) Pore pressure generation of anisotropic samples.

The evolution of pore pressure of the numerical simulations at small strains agrees well with experimental observations, particularly the small drop at the beginning of the extension test. Figure 1b reveals a small initial dilatancy behaviour in extension and indicating a continuous response of the material. Thus, the continuity of the stress paths implies the continuity of the stress-strain behaviour.

4.2 Anisotropic consolidation

When subjected to a positive initial drained consolidation ratio, K=0.66, which is not far from the isotropic axes, the effective stress paths in Figure 2a emphasize the effects of kinematic hardening in the non-associated elastoplasticity framework.

The models reproduce well the effective stress path in compression, but not its continuity at the beginning of the tests. In this figure, a limited elastic nucleus associated with mixed kinematic hardening mechanism, as in Lyon model, describe correctly the initial plastic behaviour observed at the beginning of the extension tests. An oversized isotropic elastic nucleus leads to an unrealistic stress path in extension part with an incorrect curvature, as in Milan model.

Since isotropic elasticity is assumed in both models, correct stress paths, particularly in extension test, may be simulated with the introduction of an anisotropic elasticity component to take into account the effects of the anisotropic consolidation. Again, the models predict inevitably the liquefaction of the sample. This result is expected because the two models ignore completely the void ratio reduction during the drained anisotropic consolidation of the stress history.

Nevertheless, many important aspects of the undrained stress-strain behaviour are correctly simulated; namely the overall stress-strain behaviour in Figure 2c with a sharper drop of deviatoric stress at large strains, the deviatoric stress peaks at small strains, Figure 2d, and the continuity of the soil's response at the beginning of the tests, Figure 2b.

Numerical predictions of other higher positive anisotropic consolidation level, K=0.50, are similar to the results of Figure 2 with same problemes.

Experimentally, the highest positive anisotropic consolidation level in the experimental study, K=0.35, produces a small but noticeable undrained compressive strength increment and stabilizes asymptotically the stress ratio increment. Only Milan model can reproduce qualitatively this difficult test in Figure 3 despite the incorrect aspect of the stress path. At the end of the consolidation stage, Milan model gives 14% of axial strain preshearing, which is twice the amount of experimental data, against an unrealistic 50% of axial strain of Lyon model. Consequently, it is decided to stop the simulations of Lyon model in this case.

4.3 Asymptotic stabilization

Laboratory tests show the well-known drop of deviatoric stress in undrained compression tests. Large reduction of deviatoric stress occurs as the stress consolidation ratio decreases. Experimentally, this deviatoric stress reduction was preceded by a sharp increase, almost vertical, of the stress paths.

Lyon model predicts an immediate drop of deviatoric stress while Milan model gives a modest increase. Nevertheless, the initial increase of stress paths cannot be simulated precisely in both models. This salient feature can be interpreted as an elastic behaviour that may be reproduced by a simulation of an overconsolidation or by the introduction of a visco-plastic component in the constitutive equations.

Figure 4 shows the effective stress ratio increment, $\delta\eta = \eta_{peak} - \eta_{cons}$, against the anisotropic consolidation level η_{cons}. It indicates the dependency of the instability concept on the monotonic consolidation history and confirms the asymptotic stabilization of the effective stress ratio increment.

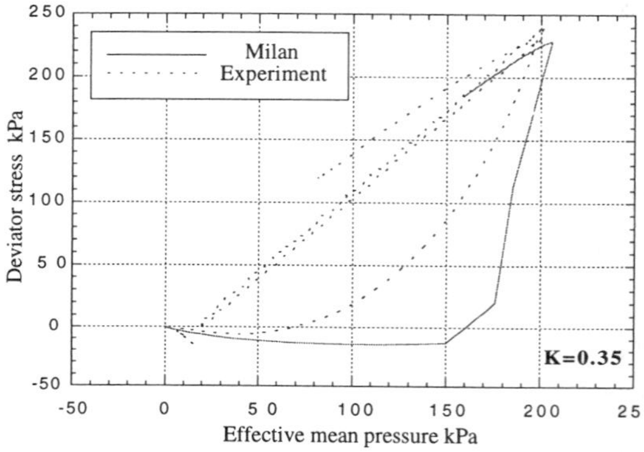

Figure 3. Stress paths with large anisotropic consolidation level.

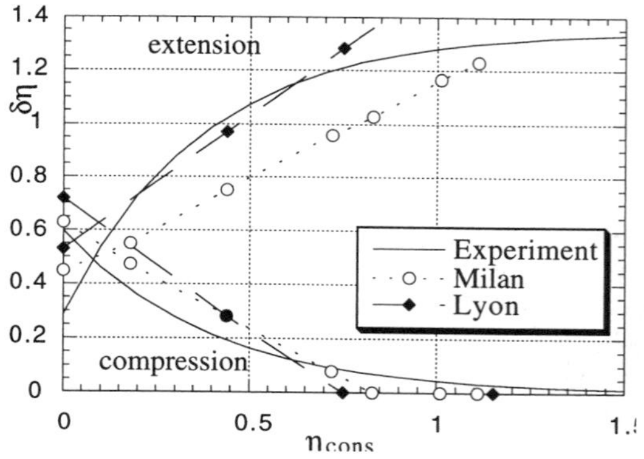

Figure 4. Asymptotic stabilization of the effective stress ratio increment.

In contrast to these experimental evidences, the theoretical analysis of Lyon model gives only two distinct lines in compression side: one line connecting the stress deviator at peak, the other horizontal line indicating an immediate drop of deviator stress peak if the consolidation level K is above the peak line defined by isotropic samples. It also indicates an independence of deviatoric stress peaks on anisotropic consolidation level in extension side. This figure shows the advantages of strain-hardening hypothesis in Milan model, taking into account the accumulated plastic deformations. Qualitatively, the asymptotically stabilization of the stress ratio increments is obtained with bilinear segments in compression and one line with smaller slope in extension.

4.4 Complex stress loading

The case when large irreversible axial strains are accumulated before undrained shearing presents a new challenge for constitutive modelling. In Figure 5a, a special test was conducted to clarify the effects of complex stress history.

The loose sample was consolidated isotropically up to 100 kPa of effective mean pressure, then loaded in compression under drained condition to a maximum axial strain of 15% and unloaded to the initial isotropic state. The sample was sheared monotonically under undrained condition in the

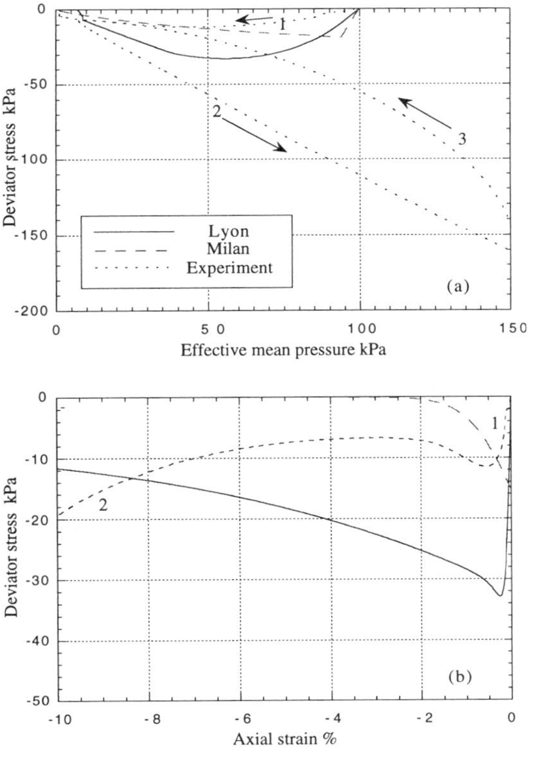

Figure 5. (a) Stress paths. (b) Stress-strain behaviour of preloading in compression samples.

opposite direction. Partial liquefaction is still observed, after a small and gradual stress reduction, arrow n° 1. Then the sample behaved as dense sand and the stress path climbed up along the failure envelope, arrow n° 2. As noticed before, only a stress reversal, arrow n° 3, liquefied the samples.

An undrained compression test after a preloading in drained extension produces a similar drastic change of the usual liquefaction behaviour. Similarly, partial liquefaction in extension on Hostun dense sand after a preloading in compression was reported by Lanier et al (1993).

Concerning the numerical predictions, the initial loading-unloading simply shifts the elastic domain and plastic strains appear at the beginning of the undrained extension tests. As noted before, the stress deviator peak is independent of stress history in the case of Lyon model. This test illustrates the limitations of the stress-hardening approach and shows qualitatively the advantages of strain-hardening plasticity as in the case of Milan model. As usual, the two models liquefy this loose sample, even being preloaded with a complex stress history.

The predicted stress-strain behaviours of both models in Figure 5b gives only the qualitatively results. The experimental observations such as the reduction of the stress deviator peak, or the dilatancy behaviour are not satisfactory simulated.

These preloading tests on loose sand suggest the dependency of the constitutive parameters on the density in a continuous manner. This feature is currently not supported in any proposed model.

Table 3 summarizes the check list when comparing the theoretical simulations of the undrained behaviour of loose sand.

Table 3. Check list of static liquefaction verification.

Features	Lyon	Milan
Isotropic state, Comp&Extension	✓	✓
Anisotropic state, compression	✓	✓
Anisotropic state, extension	✓	–
Partial static liquefaction	–	–
Continuous stress paths	–	–
Continuous response	✓	✓
Large anisotropic state	–	✓
Complex stress history	–	✓
Stress history dependent	–	✓

As indicated in the above check list, many complex aspects of the undrained behaviour of loose sand have already been reproduced by numerical simulations. Nevertheless, these two advanced models fail to recognize the partial static liquefaction and the continuity of stress paths at the beginning of the undrained tests on anisotropic specimens. Predictions of the stress paths of a loose sample in extension from a large anisotropic state or from an isotropic state after a preloading in compression present a real difficulty in constitutive modelling.

Despite these limitations, the theoretical simulations present an overall good agreement in several aspects of the observed behaviour, and demonstrate the interest of their use.

5 CONCLUSIONS

This paper proposes a comprehensive evaluation of two advanced elastoplasticity models within the non-associated elastoplasticity framework.

The effectiveness of the elastoplasticity modelling to predict some complex aspects of the undrained behaviour of very loose saturated Hostun sand in triaxial compression and extension tests is demonstrated against a complete set of experimental data. It emphasizes the advantages of using kinematic hardening mechanism to simulate the static liquefaction phenomenon.

This study indicates the fundamental role of the non-linearity of the elasticity part, and gives some limitations of the stress-hardening approach.

ACKNOWLEDGMENT

Part of the present work was developed within the ALERT Geomaterial program.

REFERENCES

Been, K., Jefferies, M.G. 1985. A state parameter for sands. *Géotechnique* 35(2), 99-112.

Cambou, B., Jafari, K. 1988. Modèle de comportement des sols non cohérents. *Revue française de Géotechnique*. (44), 43-55.

Darve, F. 1994. Stability and uniqueness in geomaterials constitutive modelling. *Localisation and bifurcation theory for soils and rock*, R. Chambon Editor. 43-55.

di Prisco, C., Nova, R., Lanier, J. 1993. A mixed isotropic kinematic hardening constitutive law for sand. *Modern approaches in plasticity*. D. Kolymbas Editor. 83-124.

Doanh, T., Ibraim, E., Matiotti, R. 1997. Undrained instability of very loose Hostun sand in triaxial compression and extension. Part 1: Experimental observations. *J. of Cohesive-Frictional Materials.*, (to appear).

Dubujet, Ph., Doanh, T. 1997. Undrained instability of very loose Hostun sand in triaxial compression and extension. Part 2: Theoretical analysis using an elastoplasticity model. *J. of Cohesive-Frictional Materials.*, (to appear).

Gudehus, G. 1985. Requirements for constitutive relations for soils. *Mechanics of Geomaterials* Z. Bazant Editor. 47-64.

Konrad, J.M. 1990. Minimum undrained strength versus steady state strength of sand. *J. Geotech. Engrg*., ASCE, 116(6): 948-963.

Lade, P.V., Nelson, R.B., Ito, Y.M. 1988. Instability of granular materials with nonassociated flow. *J. Engrg. Mech. ASCE*. 114(12), 2173-2191.

Lanier, J., di Prisco, C., Nova, R. 1993. Etude expérimentale et analyse théorique de l'anisotropie induite du sable Hostun. *Revue française de Géotechnique*. (57), 59-74.

Matiotti, R., Ibraim, E., Doanh, T. 1995. Undrained behaviour of very loose RF sand in compression and extension tests. *NUMOG V*. Pande and Pietruszczak Editor.119-124.

Molenkamp, F. 1991. Material instability for drained and undrained behaviour. Part2: combined uniform deformation and shear-band generation. *Int. J. for Num. Ana. Met. in Geomechanics* 15(2), 73-83.

Saïtta, A., Canou, J., Dupla, J.C, Dormieux, L.. 1992. Application of a generalized elastoplastic model to simulation of sand behaviour. *NUMOG IV*. Pande and Pietruszczak Editor.119-124.

Sladen, J.A., D'Hollander, R.D., Krahn, J. 1985. The liquefaction of sand, a collapse surface approach. *Canadian Geotechnical Journal*, 22(4): 564-578.

Sladen, J.A., Oswell, J. M. 1989. The behaviour of very loose sand in triaxial compression test. *Canadian Geotechnical Journal*, 26(1): 103-113.

Numerical Models in Geomechanics, Pietruszczak & Pande (eds) © *1997 Balkema, Rotterdam, ISBN 90 5410 886 X*

Modelling of granular materials under drained cyclic loading

C.M.Tsang & G.L.England
Department of Civil Engineering, Imperial College of Science, Technology and Medicine, UK

ABSTRACT: The behaviour of drained granular materials under cyclical strain and cyclical stress loading is discussed in relation to soil/structure interaction problems, for which cyclical solar heating and cooling of the structures create stress and strain changes in the soil. Cases investigated include cyclical stress and cyclical strain changes without change of principal stress directions *and* with jump changes of 90°. Cyclical straining leading to a shakedown state is described. A simple soil model is then introduced and used to make numerical predictions under drained monotonic and repeated loading. It incorporates a stress-dilatancy relationship to allow for density changes during straining and makes allowance for stress-induced anisotropy. The model relies on a stress relaxation concept and operates on incremental strain steps. It is similar to an endochronic approach but with significant differences in formulation. The physical meaning of model parameters is discussed and it is concluded that the model contains the major features necessary for the prediction of monotonic and cyclical loading behaviour of granular materials under drained conditions.

1 INTRODUCTION

The paper describes two soil/structure interaction situations for which the cyclic behaviour of granular soil plays an important role in defining long-term structural serviceability. A simple numerical model, based on an incremental visco-elastic formulation, is then presented. Finally, model predictions are compared with test results, and suggestions for further development of the model are discussed.

2 CYCLIC SOIL/STRUCTURE INTERACTION

Granular soils exhibit stiffening and volume change, usually densification, when subjected to cyclic loading. These features can lead to important changes of loading and loss of serviceability for structures designed to remain in contact with granular soil. A detailed understanding of soil behaviour under cyclical stress and cyclical strain loading is therefore necessary, if interaction behaviour is to be fully understood and engineering predictions are to be made with confidence.

One source of interaction stems from the repeating thermal displacements exhibited by structures during daily and seasonal solar heating cycles. The *soil* loadings to silos, soil containing structures, bridge abutments, and slender structures supported on shallow foundations, are all influenced by solar heating over long periods of time.

In the long term one of three final states will normally be approached. (a) A safe shakedown state for which stresses and displacements repeat on a cycle by cycle basis everywhere within the soil/structure system. (b) Instability leading to collapse/failure of the soil or the soil/structure system. (c) Stability after soil densification without significant stress escalation.

The nature of the soil/structure interaction is determined primarily by the form of the repeating soil loading, e.g. cyclical stress or cyclical strain imposition, and whether or not principal stress directions remain unchanged during the cyclical process. Cyclical straining usually leads to greater stability than does cyclical stressing.

Because soil behaviour is strongly dependent on the principal stress ratio, $R=\sigma_1/\sigma_3$, a superimposed fluctuating stress on to a soil element already loaded will lead to different behaviours depending of the peak stress ratios created. Strain ratcheting occurs at all stress ratios, $R\neq1$, and is accompanied by soil densification when $R<3$ and dilation when $R>3$, Figure 1. The development of ratcheting strains is thus faster at the higher stress ratios.

Grains of a soil element subjected to repeating strains in one direction will organise themselves, during volume change, into a shakedown state whereby deformations repeat and there is no residual volume change over any complete strain cycle. In a plane strain situation, with one principal stress imposed, the cyclically imposed strains in the third co-ordinate direction, lead to stress fluctuations about the isotropic stress state. The magnitude of the peak stress ratio is independent of the strain amplitude if the soil is considered to obey a unique energy dissipation equation (England et al. 1997), Figure 2. However, the soil density at shakedown is greater for smaller cyclical strain amplitudes.

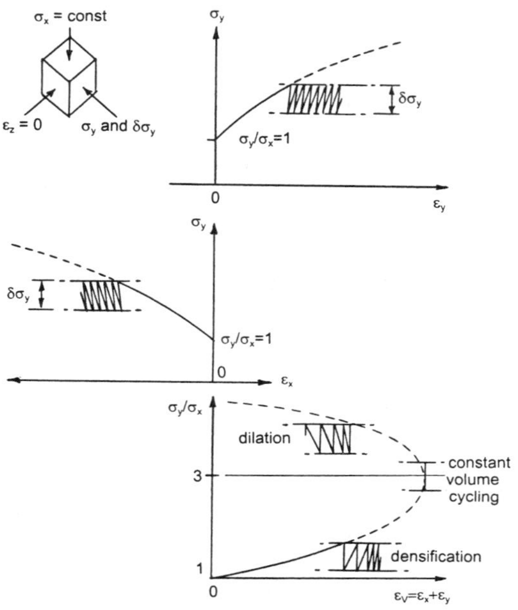

Figure 1. Cyclical stress loading showing strain ratcheting, densification and dilation.

When the repeating soil strains are imposed from an elastic structure the changing soil stresses create elastic deformations in the structure which thus permit ratcheting soil strains to occur simultaneously with stress changes leading to a shakedown state. This behaviour has been demonstrated in experiments and has been observed in circular biological filter containments (England, 1994). Figure 3 illustrates this behaviour when a soil element is loaded through an elastic spring to mimic the effect of an elastic structure.

Stress escalation has also been reported behind the soil retaining abutments of integral bridges (Broms, 1972), although no upper limit has yet been identified. Research in this area is still active.

2.1 *Biological Filters*

These are simple cylindrical containments, typically 2m to 3m in height and from 5m to 50m in diameter. They contain drained granular material and may be of continuous or multi-panel construction. They suffer considerable radial interface stress escalation from cyclical solar heating of the containment wall. Many have failed in circumferential tension (CIRIA, 1976) as the direct result of stress escalation.

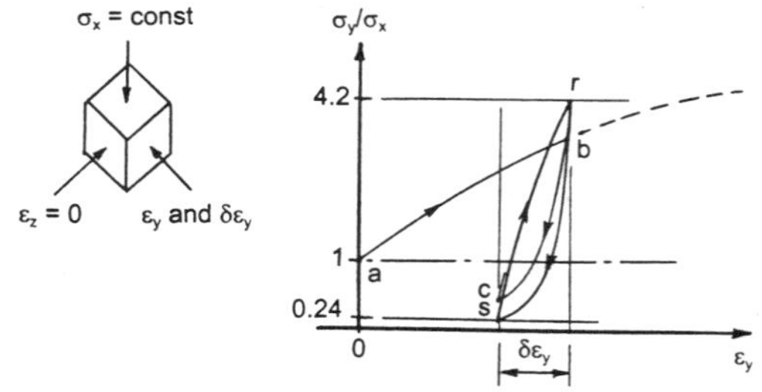

Figure 2. Cyclical strain loading with offset strain, $\delta\varepsilon_y$, leading to shakedown state with peak stress ratios of 4.2, corresponding to the energy dissipation equation of Dunstan et al. 1988. **a-b** is the virgin loading curve; **b-c** is the first half strain cycle; and **r-s** is the shakedown loop for repeating strain cycles.

The soil behaviour adjacent to the containment wall is of interest. Initially, fluctuating radial stresses occur without change of principal stress directions. In this condition the soil exhibits flow properties (towards the wall) with densification and stress escalation. Continued stress escalation leads eventually to 90° changes in the direction of the principal stresses as a shakedown state, centred on the isotropic stress state (determined by the local self-weight vertical soil stress) is approached. The maximum amplitude of the fluctuating stresses at shakedown is governed by the stiffness of the structure and the magnitude of the fluctuating thermal displacements (England & Tsang, 1996).

2.2 *Integral Bridge*

Soil retaining abutments of integral bridges are usually *stiff* structures (England & Tsang, 1996). Thermal elongation and contraction of the bridge deck is responsible for creating the repeating soil loading which in turn leads to, soil densification, stiffening and stress escalation on the face of the abutment. The initial soil behaviour is similar to that of the filter bed containments but because integral bridges behave as stiff structures, stress escalation is bounded theoretically only by the *active* and *passive* soil strengths and not by structural properties. It is believed that soil strength bounds will apply only to cyclical displacements above a certain amplitude. More experimental data are needed to confirm this belief and to identify those soil parameters which govern the shakedown solution at small strain amplitudes.

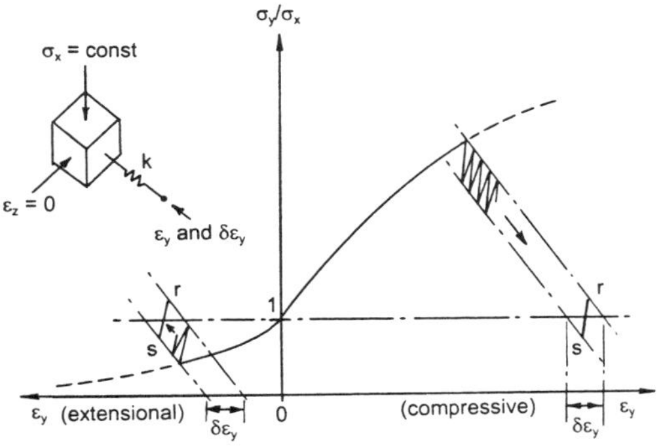

Figure 3. Soil/structure interaction at the element level. Approach to shakedown is accompanied by strain ratcheting *and* stress changes.

2.3 *Slender Structure on Shallow Foundation*

Repeated thermal distortions of slender structures can cause redistribution of the foundation stresses, leading to strain ratcheting and consequential differential settlements. These create further stress redistribution and escalation, leading eventually to structural collapse (England et al. 1995).

3 CYCLIC SOIL BEHAVIOUR

Soil behaviour under plane strain conditions (ε_z=0), leading to strain ratcheting, stress escalation and the stabilisation at a shakedown state, is demonstrated for the following cases.

(a) Figure 1. σ_x=constant and fluctuating $\delta\sigma_y$ imposed. Strain ratcheting occurs in the y and x directions with densification when $\sigma_y/\sigma_x<3$ and with dilation when $\sigma_y/\sigma_x>3$

(b) Figure 2. Constant σ_x; constant ε_y and fluctuating $\delta\varepsilon_y$ imposed. Cyclical stress changes occur and lead to a shakedown state defined by peak stress ratios $\sigma_y/\sigma_x=\sigma_x/\sigma_y=4.2$, when a single energy dissipation equation is used (Dunstan et al. 1988).

(c) Figure 3. The soil element is connected here to a linear spring (structure) to describe simple soil/structure interaction under imposed cyclical strains. Constant σ_x is imposed with an initial ε_y applied through a series-connected spring to the soil element. Then a fluctuating strain $\delta\varepsilon_y$ is imposed. The resulting behaviour shows strain ratcheting and stress changes leading to a shakedown state about the isotropic stress state. When the initial stress ratio $\sigma_y/\sigma_x<1$ stress escalation and extensional strains occur in the y direction. This behaviour is similar to that described earlier for the biological filter bed and integral bridge abutment.

4 UNIAXIAL MODEL

The model relies on a stress relaxation concept and operates on incremental "strain" steps. The total incremental stress is taken to consist of three parts (Figure 4). These are an instantaneous elastic component, a flow component and an internal fabric component as represented in Eq.(1) respectively.

$$\Delta\sigma = \Delta\sigma_e + \Delta\sigma_f + \Delta\sigma_i \quad (1)$$

The one dimensional behaviour can be represented by a "rheological" model of a Burgers unit consisting of a Maxwell element in series with a Kelvin element. The "time" parameter in the model is replaced by "strain" which can be treated as "pseudo-time" in the formulation. Yield surfaces are not required. Both loading and unloading conditions are simulated by the same set of equations. It is similar to an endochronic type (Valanis 1971 and Bazant 1974) of models. The detail, however, is significantly different.

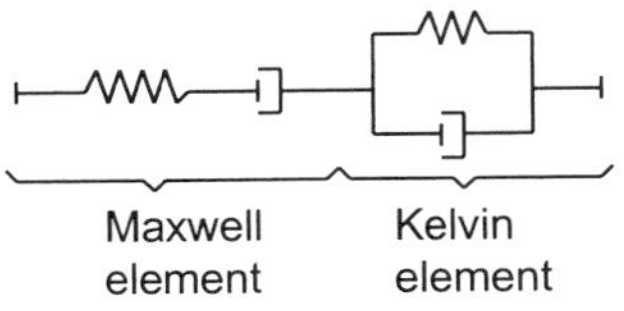

Figure 4. Rheological Model

Due to the non-linear material parameters which are also dependent on stress level and void ratio, a numerical step-by-step method is adopted to evaluate the result of each incremental "strain" step, $\Delta\varepsilon$. The total elastic response is evaluated first. Then, the relaxation due to the dashpot of the Maxwell element is considered. The final stress level is then calculated with the effect due to the Kelvin element in series with the Maxwell spring. Material parameters are updated after each step. The procedures for each incremental step are setout as follows:

4.1 *Elastic stress increment*

The elastic response of the material is determined by the "spring" of the Maxwell element. The elastic modulus of the "spring", E, is dependent on the current void ratio and mean confining pressure. The relationships can be expressed as

$$\Delta\sigma_e = E\,\Delta\varepsilon \qquad (2)$$

where

$$E = E_o F_e p^n \qquad (3)$$

$$F_e = \frac{(A-e)^2}{1+e} \qquad (4)$$

$$A = \sqrt{1+e_{cr}} + e_{cr} \qquad (5)$$

p is the mean confining pressure, e_{cr} is the critical void ratio at p. n and E_o are constant material parameters.

From critical state soil mechanics, e_{cr} is dependent on the mean confining pressure, p, and can be expressed as

$$e_{cr} = e_{cro} - \lambda \log \frac{p}{p_o} \qquad (6)$$

where e_{cro} is the critical void ratio at a reference confining pressure, p_o.

4.2 *Viscous stress relaxation*

The viscous element in Figure 4 is a nonlinear dashpot, govern by the relaxation law

$$\Delta\sigma_f = s_1 \exp\left(-\frac{E\Delta\varepsilon}{\eta}\right) - s_1 \qquad (7)$$

where

$$\eta = \eta_o F_e p \qquad (8)$$

$$\eta_o = \sin\varphi_{cr} \qquad (9)$$

E and F_e are defined as in Eqs.(3) and (4) respectively. η_o is a constant material parameter depending on the friction angle at critical state, φ_{cr}, as in Eq.(9). s_1 is the deviator stress. Under uniaxial condition,

$$s_1 = \sigma_1 - p \qquad (10)$$

$$\sigma_1 = \sigma_o + \Delta\sigma_e \qquad (11)$$

where σ_o is the stress level at the beginning of the incremental step. At the end of this stage, the soil stress is

$$\sigma_2 = \sigma_o + \Delta\sigma_e + \Delta\sigma_f \qquad (12)$$

4.3 *Stress increment due to Kelvin element*

In this stage, the sign of the incremental stress is dependent on the difference of the stress level between the "Kelvin spring", σ_k , and the "Maxwell spring", σ_2 which is also the global soil stress level.

$$\Delta\sigma_i = Q(\sigma_k - \sigma_2)(1-\exp(-h)) \qquad (13)$$

where

$$q = \frac{E_k}{E} \qquad (14)$$

$$Q = \frac{1}{1+q} \qquad (15)$$

$$h = \frac{E_k\,\Delta\varepsilon}{qQ\,\eta_k} \qquad (16)$$

E_k and η_k can be expressed in a similar fashion to E and η , and are:

$$E_k = E_{ko} F_e p^n \qquad (17)$$

$$\eta_k = \eta_{ko} F_e p \qquad (18)$$

The change of σ_k in each step can be found by

$$\Delta\sigma_k = -q\,\Delta\sigma_i \qquad (19)$$

At the end of this stage, the soil stress is

$$\sigma_3 = \sigma_o + \Delta\sigma_e + \Delta\sigma_f + \Delta\sigma_i \qquad (20)$$

This stress is also the final soil stress in the incremental strain step. σ_3 will become σ_o for next incremental step.

4.4 *Stress-Dilatancy and confining pressure*

Volumetric change of the soil during each strain increment is updated according to the starting stress state. Therefore, the void ratio, e, changes as well as other material parameters according to Eqs.(3), (4), (8), (17) and (18). For monotonic loading, the modified Rowe's stress dilatancy theory proposed by Wan and Guo (1997) can be incorporated, while for cyclic loading, the unique stress-dilatancy relationship proposed by Dunstan et at (1989) is more appropriate.

The material parameters are also affected by the change in confining pressure through Eqs. (3), (4), (5), (6) and (8). These parameters are updated based on the stress state at the beginning of each strain increment.

4.5 *Poisson's effect*

For a three dimensional case, the elastic Poisson's ratio, ν_e, and the viscous flow Poisson's ratio, ν_c, should be incorporated into the procedures in sections 4.1 and 4.2. The modification follows the usual elastic and visco-elastic formulations and will not be discussed further. For non-compressible material, ν_c is 0.5. For granular material, ν_c should be adjusted according to the volumetric change evaluated as in section 4.3.

4.6 *Implication of Kelvin element*

The effect of anisotropy to the soil stress-strain response could be significant for cyclic load cases. It has been observed (England et al. 1997) that the build-up of soil structure under stress control biaxial tests caused significant soil stiffening in principle stress directions even though the change in void ratio was small. In the proposed model, the deviation between the Kelvin spring stress and the Maxwell spring stress reflects some degree of anisotropy.

5 MODEL TESTS

The model as set out above requires the specification of 9 parameters. Apart from E_{ko} and η_{ko}, all other parameters can be determined from conventional tests. E_{ko} and η_{ko} are unique for the proposed model and at this stage are determined on a trial-and-error basis to give the best fit to test results.

Since the objective is to use the proposed model in a numerical simulation of granular material behind a retaining wall, the model has been implemented for a plane strain soil element with constant stress (40 kPa) in one direction. The values of material parameters used are given in Table 1.

Three numerical test cases have been considered:

1. Monotonic loading for various initial void ratios.
2. Cyclic loading with 90° jump of major principle stress direction.
3. Unloading and reloading without change of major principle stress direction.

Figures 5, 6 and 7 presents the numerical results of these three cases. There is good agreement between model predictions and the experimental results for cyclically imposed strains (England 1994), with 90° jump changes of the major principle stress direction.

Table 1. Material parameters

Property	Value
E_o	7000 kPa
φ_{cr}	30°
n	0.45
λ	0.0467
e_{cro}	0.79
p_o	100 kPa
ν_e	0.2
E_{ko}	2500 kPa
η_{ko}	0.8

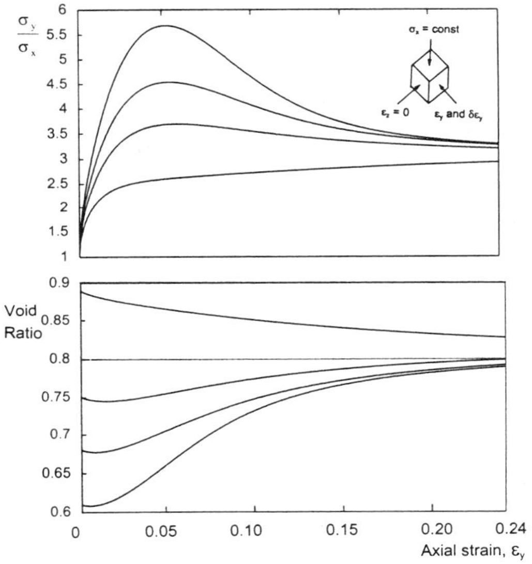

Figure 5. Numerical simulation for various initial void ratios under monotonic loading.

6 DISCUSSION

Although the numerical model has produced good agreement with experimental results under drained monotonic loading (Figure 5) and cyclic loading with 90° jump of principle stress directions (Figure 6), Figure 7 shows that further development is still required for eliminating the excessive ratcheting strains predicted by the model. One of the schemes, currently being investigated by the authors, is to use the Kelvin spring stress as an index to modify the elastic modulus of the soil in the three principle stress directions.

Another improvement which should be made relates to the stress-dilatancy theory under small cyclic strain. Although the stress-dilatancy laws proposed by Rowe et al (1962,1969), Taylor (1948) and Dunstan et al (1988) indicate a unique soil stress-dilatancy behaviour under various void ratios and stress levels, Pradhan et al (1989) suggested that this may not be true for small strains. This belief is supported by recent work of the authors which indicates the existence of a lower transformation stress level under small strain

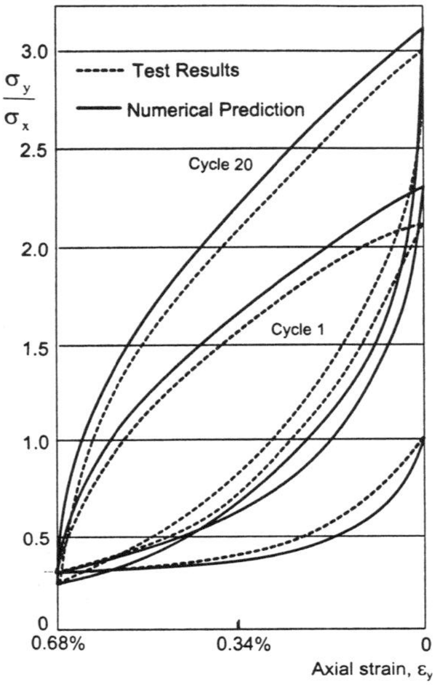

Figure 6. Numerical results compared with test data after England (1994) under cyclic load.

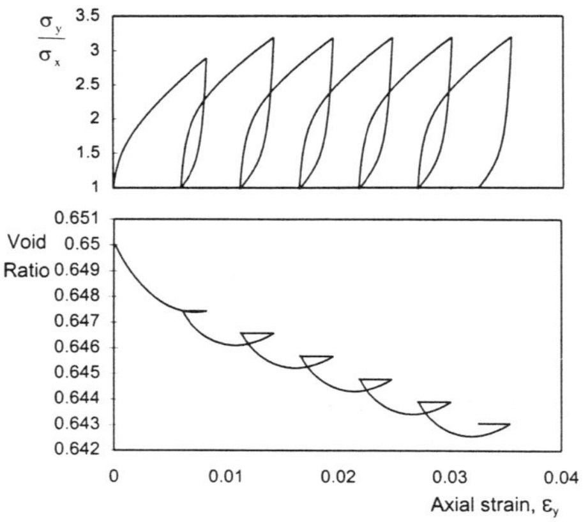

Figure 7. Numerical simulation for unload-reload without principle stress change.

7 CLOSING REMARKS

The existence of a shakedown state for granular soil under cyclical strain loading has been described in relation to two soil/structure interaction cases. Site observation is also discussed. The simple numerical model introduced to describe the behaviour of granular material under drained cyclic loading has produced some encouraging results. However, improvement is still required to overcome the excessive ratcheting strains which it predicts.

ACKNOWLEDGMENTS

The authors acknowledge the partial financial support received from the Highways Agency, and the PhD Scholarship granted to the first author by the Mott MacDonald Charitable Trust.

REFERENCES

Bazant, Z.P. 1974, A new approach to inelasticity and failure of concrete, sand and rock: endochronic theory, *Proc. Soc. of Engng Sci. 11th Annual Meeting*, 158-159, Duke University, Durham.

Broms, B.B., and Inglesson, I, 1972, Lateral earth pressure on a bridge abutment, *Proc. 5th Euro Conf. Soil Mech. and Fdn. Engng*, Madrid, 1, 117-123.

CIRIA, 1976, Design and construction of circular biological filted walls, *CIRIA Rep.No.58*, 26pp.

Dunstan, T., Arthur, J.R.F., Dalili, A., Ogunbekun, O.O. & Wong, R.K.S. 1988, Limiting mechanisms of slow dilatant plastic shear deformation of granular media, *Nature*, 336, 52-54.

England, G.L., 1994, The performance and behaviour of biological filter walls as affected by cyclical temperature changes, *ASCE Geotechnical Special Pub. No.42*, 57-76.

England, G.L., & Tsang, C.M. 1996, Thermally induced problems in civil engineering structures, *Thermal Stresses IV*, Pub.Elsevier Science B.V., Ed. R.B. Hetnarski, 4, 155-275.

England, G.L., Dunstan, T., Tsang, C.M., Mihajlovic, N., & Bolouri-Bazaz, J.,1996, Ratcheting flow of granular materials, *ASCE Special Geotechnical Publication No.56*, 64-76.

England, G.L., Tsang, C.M., Dunstan, T., and Wan, R. 1997, Drained granular material under cyclic loading, *Applied Mechanics Reviews* (in press).

Pradhan, T.B.S., Tatsuoka, F. & Sato, Y. 1989, Experimental stress-dilatancy relations of sand subjected to cyclic loading, *Soils and Foundation*, 29, 45-64.

Valanis, K.C. 1971, A theory of viscoplasticity without a yield surface, *Archiwum Mechaniki Stossowanej* (Archives of Mechanics, Warszaw), 23, 517-551.

Wan, R. G. & Guo, P. J. 1997, Private Communication.

Numerical Models in Geomechanics, Pietruszczak & Pande (eds) © *1997 Balkema, Rotterdam, ISBN 90 5410 886 X*

Modelling of grain breakage influence on mechanical behavior of sands

A. Daouadji & P.Y. Hicher
Laboratoire de Génie Civil Nantes-St Nazaire, Ecole Centrale de Nantes, France

ABSTRACT: Grains as components of granular materials can under certain circumstances be subjected to ruptures. The main consequence of this phenomenon is the increase of the material compressibility. In this paper, we provide a method for integrating this phenomenon in an elastoplastic model by turning the critical state into a function of the plastic work. The modified model was used in order to simulate triaxial tests on a quartzic sand (up to 15 MPa) and a calcareous sand at ordinary stresses (from 0.1 MPa to 1 MPa). Numerical simulations were found to be in accordance with experimental results and hence justified our approach.

INTRODUCTION

Granular material components are usually assumed to have an elastic behavior. This assumption is reasonable for current loads in Civil Engineering. However, structures such as dam embankments, deep foundations, dynamic loads with a high intensity (pile driving) or explosions, can induce significant stresses.
In this case, a particular phenomenon may occur: grain breakage. This phenomenon leads us to observe a modification in the behavior of granular material.
Grain breakage may also occur under usual stresses if the grains have a small strength as in the case of carbonate sands.
In order to model the behavior of granular materials under a wide range of stresses, we should take into account this particular phenomenon.
The purpose of this article, therefore, is to provide a method for its introduction in an elastoplastic model.

1 GRANULAR MATERIALS SUBJECT TO ELEVATED OR HIGH STRESSES

For the ranges of stresses, we adopt the definition of A. S. Vesic et G. W. Clough (1968).
During testing under elevated or high stresses, granular materials are submitted to a modification of the grain (local scale) as well as to a modification of the behavior of the assembly of these grains (global scale).

1.1 *Factors influencing grain breakage*

A number of experimental investigations (Marsal, 1967; Lee and Farhoomand, 1967; Vesic and Clough, 1968; Hardin, 1985; Colliat-Dangus, 1986) highlight the effects of two kinds of parameters throughout particule crushing. Those characterizing the grain: nature parameters and those characterizing both the level and state of stress and the state of strain: mechanical parameters (Biarez and Hicher, to be published).

Nature parameters:

The amount of grain breakage is governed by the mineralogy and by the individual grain strength. The shape of the grain also has a great influence. Indeed, the amount of grain breakage increases with the angularity of the particules. Furthermore, grain size distribution has an important effect on crushing. Tests on two materials different only by their gradation show us more rupture for the poorly graded material than for the well graded one (Hicher and al., 1995).

Mechanical parameters:

All these investigations point out the importance of the stress path during a test. Indeed, the grain crushing phenomenon manifests itself with a different amplitude depending on whether the test is

isotropic, oedometric or triaxal. It is more important for triaxial tests than it is for isotropic or oedometer tests, and this, for the same mean pressure.
It is obvious that the stress value has a particular influence on this phenomenon. The strain value however is also a sensitive parameter. For the same stress level, the amount of grain breakage increases with strain (Hicher and al, 1995).
Let us observe that the initial relative density has an influence during isotropic or oedometer testing whatever the stress value. On the other hand, this factor plays no role beyond a certain value of the mean stress (depending on the material) during triaxial testing.

1.2 *Effects of the grain breakage*

While subjected to crushing, the grain size decreases. This leads to a modification of the grain size distribution as well as to a decrease of the maximal and minimal void ratios. The grains obtained are less brittle and retain the same shape.
As a consequence, the compressibility of the granular material incrases. The explanation may be that the relation of the critical state has been modified. Indeed, grain breakage always leads to an increase of the Uc as defined by:

$$U_c = \frac{d_{60}}{d_{10}} \tag{1}$$

Biarez and Hicher (1994) have shown that for any given grains, we can assume that the increase of U_C leads to the displacement of the critical state's relation towards the lowest void ratios and that the slope may be considered constant. Therefore, in the presence of grain breakage, the diminution of the volume has to be greater to rejoin the critical state.
Moreover, we can observe an increase in the value of the strain corresponding to the maximum strength.

1.3 *Definition of different grain breakage factors*

Several authors have already attempted to quantify grain breakage by defining factors based on the modification of the grain size distribution curves before and after tests.
These empirical factors have been written either as the variation of a particular grain diameter (Lee and Farhoomand, 1967; Lade and Yamamuro, 1996) or as the shift of the whole grain size distribution curve (Marsal, 1967; Hardin, 1985).

2 INTRODUCTION OF GRAIN BREAKAGE PHENOMENON IN AN ELASTOPLASTIC MODEL

2.1 *Presentation of Hujeux's elastoplastic constitutive model*

Hujeux's elastoplastic model is a multimechanism model including three deviatoric plane strain mechanisms with a Mohr-Coulomb criterion and an isotropic mechanism (Aubry and al., 1982; Hujeux, 1985). Each deviatoric mechanism has two hardening variables, one associated to the deviatoric plastic strain of the mechanism, the other one to the volumetric plastic strain. The hardening in density is common to the four mechanisms which are therefore coupled.
We present here only the monotonous model.
We assume the following decomposition of the strain tensor:

$$d\varepsilon = d\varepsilon^e + d\,\varepsilon^p \tag{2}$$

The elastic part is non linear and is written as:

$$K = K_i \left(\frac{P}{P_i}\right)^n \quad \text{and} \quad G = G_i \left(\frac{P}{P_i}\right)^n \tag{3}$$

where K and G are respectively the bulk and the shear modulus.
The equation of the yield surface for each k -plan deviatoric mechanism is:

$$f_k = q_k + p_k \sin\phi \left(1 - b \,\mathrm{Ln}\left(\frac{p^*}{p_c}\right)\right) r_k \tag{4}$$

$$q_k = \frac{\sigma_{ii} - \sigma_{jj}}{2} \tag{5}$$

is a term of shearing and

$p_k \sin\phi \left(1 - b \,\mathrm{Ln}\left(\frac{p^*}{p_c}\right)\right) r_k$ is a term of friction where:

$p^* = p$ ou p_k
with:

$$p = \frac{\sigma_{ii} + \sigma_{jj} + \sigma_{kk}}{3} \quad \text{and} \quad p_k = \frac{\sigma_{ii} + \sigma_{jj}}{2} \tag{6}$$

and σ_{ii} , σ_{jj} and σ_{kk} are the components of the stress tensor σ.

$$p_c = p_{co} \exp\left(\beta\, \varepsilon_v^p\right) \tag{7}$$

p_{co} is the critical stress corresponding to the void

ratio after isotropic consolidation.
ß is the plastic compressibility modulus (slope in the ε_v^p - logp plan).
ε_v^p is the plastic volumetric strain.
r_k is called "the degree of mobilization of the mechanism k" and is given by:

$$r_k = r_k^{el} + \frac{\gamma_k^p}{a + \gamma_k^p} \quad (8)$$

r_k^{el} is the initial elastic domain of the mechanism k.

$$\left(r_k^{el} \leq r_k \leq 1\right) \quad (9)$$

γ_k^p is the plastic distortion, ϕ is the frictional angle and b is a model parameter.
p_c is a density hardening variable and r_k is a deviatoric hardening variable.
The yield surface of the isotropic mechanism is written as:

$$f_{iso} = |p| - d\, p_c\, r_{iso} \quad (10)$$

d is the distance between the normally consolidated line and the critical state line.
For each plane, the constitutive model is non associated. We assume that the total plastic strain is the sum of the plastic strains of the four elementary mechanisms:

$$\frac{\partial \varepsilon^p}{\partial t} = \sum_{k=1}^{4} \partial_t \varepsilon_k^p \quad (11)$$

where

$$\frac{\partial \varepsilon_k^p}{\partial t} = \lambda^p \Psi^k \quad (12)$$

The tensor Ψ^k can be decomposed into a deviatoric tensor Ψ_k^D and an isotropic tensor Ψ_k^v.
For the deviatoric mechanisms:

$$\Psi_k^D = \frac{\partial f_k}{\partial \mathbf{S}_k} = \frac{\mathbf{S}_k}{\|\mathbf{S}_k\|} \quad (13)$$

and

$$\Psi_k^v = \left(\sin\psi - \frac{\mathbf{S}_k : \Psi_k^D}{p_k} \right) \delta_k = \Psi_{vk}\, \delta_k \quad (14)$$

where ψ is the dilatancy angle usually considered to be equal to the angle of friction ϕ.

For the isotropic mechanism:

$$\Psi_{iso}^D = 0 \quad (15)$$

$$\Psi_{iso}^v = \mathrm{sign}(p)\, \frac{1}{3}\, \mathbf{I} \quad (16)$$

where $\mathbf{I}$ is the unit tensor.
The equation of compatibility is given by:

$$\frac{\partial f_k}{\partial t} = \frac{\partial f_k}{\partial \sigma} : \frac{\partial \sigma}{\partial t} + \frac{\partial f_k}{\partial \varepsilon_v^p} \frac{\partial \varepsilon_v^p}{\partial t} + \frac{\partial f_k}{\partial r_k} \frac{\partial r_k}{\partial t} = 0 \quad (17)$$

The evolution law of the hardening variable r_k is, for the deviatoric mechanisms:

$$\frac{\partial r_k}{\partial t} = \frac{(1 - r_k)^2}{a_{cyc} + \xi\left(a_m - a_{cyc}\right)} \quad (18)$$

and for the isotropic mechanism:

$$\frac{\partial r_{iso}}{\partial t} = \frac{(1 - r_{iso})^2}{c} * \frac{p_{ref}}{p_c} \quad (19)$$

where a_m, a_{cyc}, ξ, c are model parameters.
Lastly, the critical state is reached when:

$r_k = 1$ and $p^* = p_c$

Thus, we obtain the Mohr-Coulomb failure criterion:

$$q_k = p_k \sin\phi \quad (20)$$

2.2 *Introduction of the breakage phenomenon*

The strains of the granular materials are due on one hand to the relatives displacements of the grains (by sliding or by rotation) and on the other hand to the existence of particles crushing. The latter, as mentioned above, is influenced by both the nature and the mechanical parameters. A simple method to simultaneously take them into account is to introduce into the model a dependency of the plastic work W^p, which is written as:

$$W^p = \int \sigma : d\,\varepsilon^p \quad (21)$$

Furthermore, the stress path is an important factor. A way to take it into account is to write the plastic work corresponding into each mechanism. Thus, we use W_k^p for the deviatoric mechanisms and W_{iso}^p for the isotropic mechanism.
The grain breakage can be explained as either a breakage of the edges or a failure of the grains

because of their weakness. Under these circumstances, it is better to consider as a breakage indicator the whole surface (S) which is defined by the initial grain size distribution curve and the current grain size distribution curve (obtained after breakage). Since the parameters influencing this breakage are the nature and the mechanical parameters, the evolution of S can be expressed in terms of the plastic work. As a general form:

$$S = \Psi_{ph}(W^p) \tag{22}$$

where Ψ_{ph} is a function depending on the nature of the initial granular material.

The position of the critical state line as defined in the e-logp plan depends on the amount of grain breakage (cf. 1.2), and therefore, on the plastic work. So the value of the critical pressure decreases during the loading. Thus we can write:

$$p_{co} = p_{coi}\left(1 - \chi(S)\right) \tag{23}$$

where p_{coi} is the initial critical stress, and $\chi(S)$ is a rising function of S.

Correlations made by Rahma (1995) have highlighted a relation between p_{co} and W^p. He has clearly shown a decrease of the value of the critical pressure according to the plastic work.

The curve given p_{co}/p_{coi} against W^p can be estimated by a hyperbolic function. The value of this ratio varies between 0 and 1.

The proposed equation allowing us to follow the evolution of p_{co} is:

$$\chi(S) = 1 - \frac{p_{co}}{p_{coi}} = \frac{W^p}{B + W^p} \tag{24}$$

where B is a parameter depending on the initial granular material.

Finally, the critical pressure is given for each mechanism by:

$$p_{ck} = p_{coi}\left(1 - \frac{W_k^p}{B_k + W_k^p}\right)\exp\left(\beta\, \varepsilon_v^p\right) \tag{25}$$

3 RESULTS

The examples we have simulated are taken from the thesis of Colliat-Dangus (1986). They are drained triaxial tests on two kinds of sands: a quartzic poorly graded sand (fine Hostun sand) with angular grains and a calcareous sand with rounded grains better graded than the first sand. This second sand consists of brittle grains, containing a lot of shells and the percentage of calcium carbonate is up to 98%. Information about the two sands are presented in Table 1. All the tests which we simulated involved dense sands.

table 1: parameters characterizing the fine Hostun sand and the calcareous sand.

	e_{max}	e_{min}	D_r	U_c
Hostun	1.00	0.656	0.9-0.95	1.70
Calcar.	1.670	1.014	0.75	2.80

For simulating these drained triaxial tests we first determined the set parameters which could be measured such as the Young's modulus E, the Poisson's coefficient ν and the frictional angle ϕ. Thereafter, we adjusted the parameters of the model as a, b, a_m, a_{cyc}, ξ etc. All these parameters were obtained from tests under low stresses and remained constant for tests under elevated or high confining pressure. For example, calculating the triaxial test on fine Hostun sand with a confining pressure of 15 MPa, we first simulated the phase of isotropic consolidation from 0.5 MPa to 15 MPa followed by shearing up to 30% of axial strain.

Grain breakage is introduced in the model by means of plastic work. More precisely, it is by means of the new factor B mentioned in the modified expression of the critical pressure p_c that the amount of grain breakage can be regulated. Thus, the more B is small, the more the critical pressure decreases so that grain breakage (simulated) will increase. This is clearly highlighted in fig. 1a and 1b.

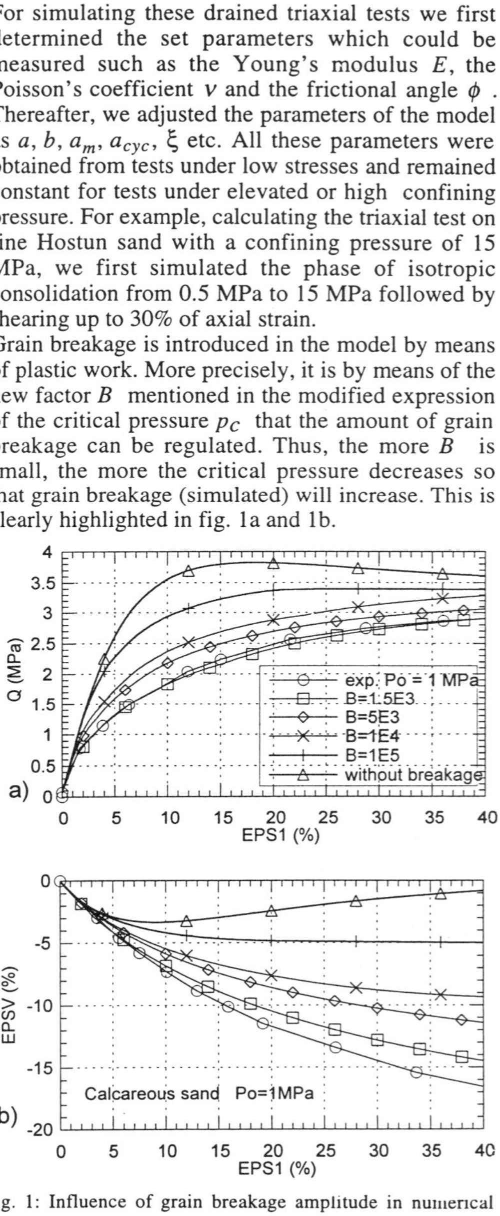

fig. 1: Influence of grain breakage amplitude in numerical simulations of a triaxial test on calcareous sand at Po=1 MPa

This is a simulation of a drained triaxial test on the calcareous sand with a confining pressure of 1 MPa and a final axial strain of 40% with and without breakage.
We can observe in fig. 1a that, when we increase the amount of grain breakage the peak in the stress-strain curve given by the model without breakage vanishes, the material becomes more ductile and the axial strain corresponding to the maximum value of the stress deviator increases. We can note in fig. 1b that the dilatancy phenomenon vanishes with the increase of breakage. The volumetric strain reaches the value noted during the test. Let us point out that at the end of the simulation the value of the volumetric strain increases from 1% (without breakage) to 15% (with breakage).
We present in fig. 2a and 2b simulations under lower confining pressure. We note that we succeeded in simulating both the test under a confining pressure of 0.05 MPa where high dilatancy occures and the test under 1 MPa where a significant grain breakage occured.

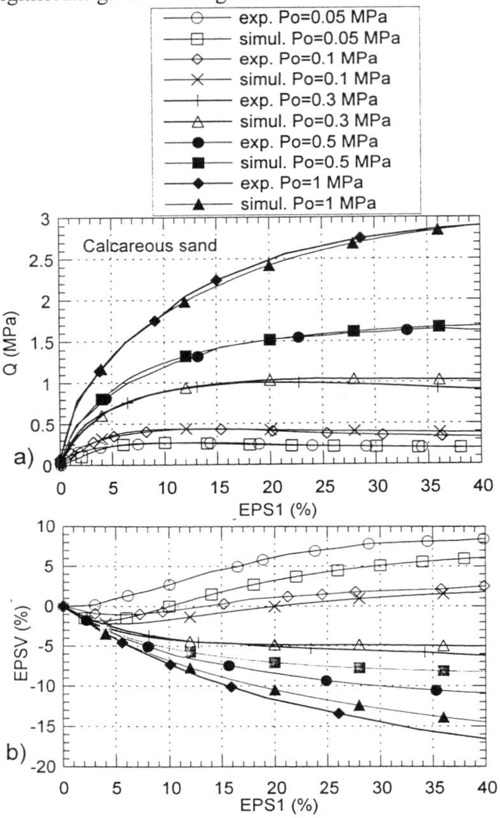

fig.2: Triaxial test on fine calcareous sand. Comparison between numerical and experimental results.

Simulations of fine Hostun sand are shown in fig. 3a and 3b. Simulations without breakage lead to results in the q - ε_1 plan which approximate the experimental curves. On the other hand, values obtained in the $\varepsilon_v - \varepsilon_1$ plan for simulations under confining pressure greater than 3 MPa are much less than values obtained during testing. For a confining pressure under 2 MPa, little grain breakage occurs as we can see from the evolution of the percentage of fines (fig.6). There is still a tendency in dilatancy up to 4 MPa of confining. With grain breakage, we can simulate more correctly triaxial tests up to 15 MPa. This test is presented in fig. 4a and 4b with and without grain breakage. In the case of grain breakage, we observe that for 30% of axial strain, we can reduce the deviator of stress for more than 5 MPa and increase significantly the volumetric strain.

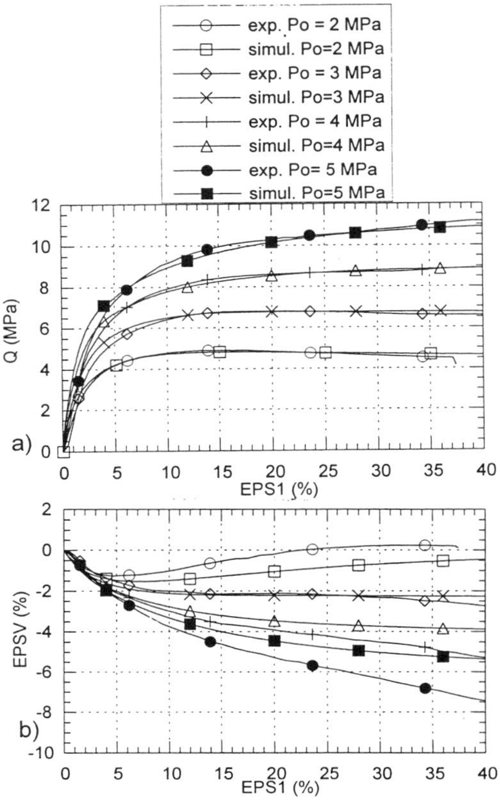

fig.3: Triaxial test on Hostun sand. Comparison between numerical and experimental results.

Contrary to tests on calcareous sand which require little energy in order to induce grain breakage, tests on fine Hostun sand show that several MPa in confinement are needed in order to induce the noticeable influence of grain breakage.

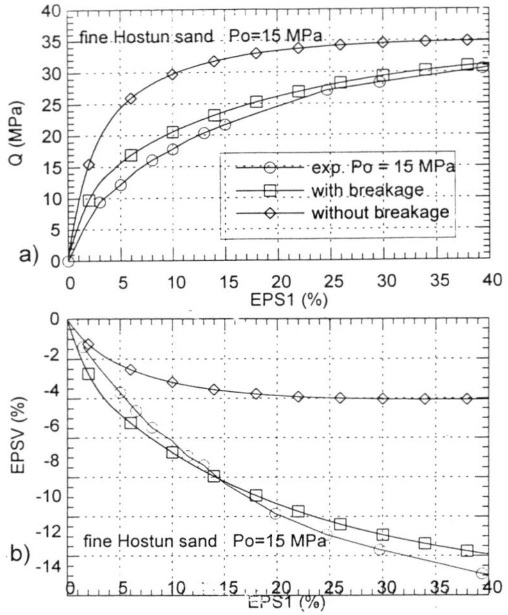

fig. 4: Influence of the grain breakage mechanism in the modelling of a triaxial test on fine Hostun sand at Po=15MPa

Indeed, in fig. 5 which presents a crushing coefficient *Ccr* as a function of the mean stress, we can see that the value of this coefficient is slightly superior in the case of the calcareous sand under pressure of 1 MPa compared to Hostun sand under a confinement of 5 MPa.

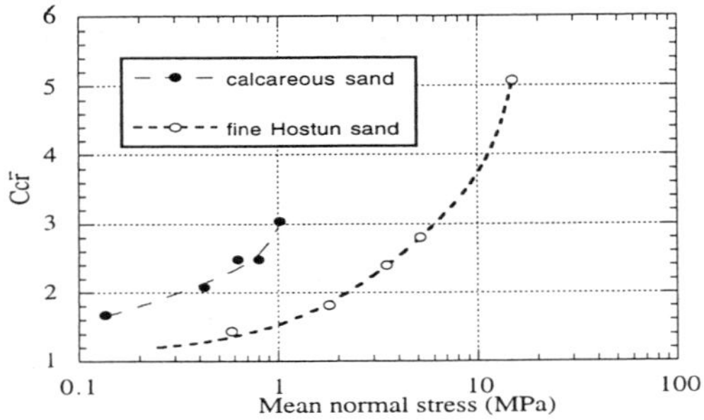

fig. 5: Evolution of crushing coefficient with mean effective stress

CONCLUSION

There are numerous cases where the phenomenon of grain breakage occurs which must be taken into account in simulation. The purpose of this paper has been to introduce a grain breakage mechanism in connection with plastic work. We assumed that the critical state was modified due to grain ruptures by being subjected only to a translation in the e-logp plan. In an elastoplastic model, we introduced the evolution of critical state and by simulating drained triaxial tests, we validated our new model on two different sands: quartzic and calcareous.

The present results are very promising but they require further validation from other granular materials. This work must now be extended in order to link parameter *B* to the physical parameters of the grains.

REFERENCES

Aubry,D., Hujeux, J.C, Lassourdière, F.,Meimon, Y. (1982). *A double memory model with multiple mechanisms for cyclic soil behavior* Int. symp. on Numerical Models in Geomechanics, Zurich, Suisse, 3-13.

Biarez, J., Hicher, P.Y.(1994) *Elementary Mechanics of Soil Behaviour*, Ed. Balkema.

Biarez, J., Hicher, P.Y., *Influence de la granulométrie et de son évolution par ruptures de grains sur le comportement mécanique des matériaux granulaires.* to be published in Revue Française de Génie Civil.

Colliat-Dangus, J.L., *Comportement des matériaux granulaires sous fortes contraintes. Influence d la nature minéralogique du matériau étudié.* , thèse de doctorat, Institut National Polytechnique de Grenoble, France.

Hardin, B.O. (1985) *Crushing of soil particles* J. Geotech. Engrg., ASCE, 111(10), 1177-1192.

Hicher, P.Y., Kim, M.S, Rahma, A. (1995). *Experimental Evidence and Modelling of Grain Breakage. Influence on Mechanical Behaviour of Granular Media* Int. Workshop Homogenization, Theory of Migration and Granular Bodies, Gdansk, Pologne, 125-133.

Hujeux, J.C (1985). *Une loi de comportement pour les chargements cycliques des soils* Génie Parasismique, Ed. Presse des ponts et Chaussées, Paris.

Lade, P.V., Yamamuro, J.A (1996). *Significance of Particule Crushing in Granular Materials* J. of Geotech. Engrg, ASCE, 122(4), 309-316.

Lee, K.L., Farhoomand, I. (1967) *Compressibility and crushing of granular soils in anisotropic triaxial compression.* Can. Geotech. J., Ottawa, Canada, 4(1), 68-86.

Marsal, R.J. (1967) *Large scale testing of rockfill materials* Journal of Soil Mechanics and Foundations Division, ASCE, 93(2), 27-43.

Rahma, A. (1995) *Internal rapport* Ecole Centrale de Paris.

Vesic, A.S., Clough, G.W. (1968) *Behavior of granular materials under high stresses.* J. of Soil Mech. and Found. Div., ASCE, 94(3), 661-688.

Numerical Models in Geomechanics, Pietruszczak & Pande (eds) © *1997 Balkema, Rotterdam, ISBN 90 5410 886 X*

A thermo-poro-elastoplastic constitutive model for geomaterials

G. Xu & M. B. Dusseault
Geomechanics Group, PMRI, Department of Earth Sciences, University of Waterloo, Ont., Canada

ABSTRACT: In this paper, an elastoplastic formulation, coupled with the degradation of stiffness and strength of the material, is proposed to describe the thermo-mechanical behaviour of rocks. Using the effective stress principle, the constitutive relations governing the undrained response of saturated rocks subjected to heating are derived. The build-up of pore fluid pressure due to heating, leading to material failure, can be simulated by the proposed formulation. Thermo-mechanical effects on borehole stability in saturated rocks during drilling are also discussed.

1. INTRODUCTION

Modelling the thermo-mechanical behaviour of fluid-saturated porous media has application to petroleum drilling, injection and production activity (Dusseault et al. 1988, Wang & Dusseault 1995), nuclear waste disposal (Hueckel et al. 1987), geothermal energy production (Bear & Corapcioglu 1980), etc. In general, these problems involve coupling between heat transfer, diffusive pore fluid flow, and rock deformation. Because of the complexity of these coupling processes, some simplifying assumptions, such as no heat transport by fluid flow through pores (Booker & Savvidou 1984), no coupling between the displacement field and the pore pressure and temperature fields (Kurashige 1989), steady-state pore pressure and temperature distribution (Wang & Dusseault 1995), poro-elasticity (Booker & Savvidou 1984, Kurashige 1989), etc., are usually made to obtain solutions for specific problems. It should be emphasized that with the assumption of a steady-state pore pressure distribution, pore pressure build-up resulting from the difference in expansion rates between the pore fluid and solid skeleton can not be considered. Distinguishing between the thermal expansion of the rock skeleton and that of the diffusing fluid is important and has been addressed before (Hueckel et al. 1987, Kurashige 1989, Britto et al. 1989). Experiments (Hueckel et al. 1987) performed by heating a clay sample in undrained conditions demonstrated that this pore pressure build-up can lead to material failure. As pointed out in Palciauskas & Domenico 1982, ideal undrained conditions may be used to simulate pore pressure build-up for low permeability media with relatively high heat diffusion rates.

This paper addresses the issue of thermo-mechanical effects on borehole stability. During circulation for drilling, the drilling fluid becomes heated at depth and heats the rock higher in the borehole (Dusseault 1994). This effect can be substantial in long open-hole sections in regions with high geothermal gradients. The magnitude of expected temperature change is about 10-30°C, a relatively narrow range in which it can be assumed that linear approximations are reasonable. Usually, the thermo-mechanical effect in poro-elastic materials is considered to lead to an increase in tangential stress (normal stress in the circumferential direction) due to the constraint of expansion for a borehole configuration. The present study will show that, for plastic materials, the increase in tangential stress may not be significant because the material around the borehole is already in a stress state close to failure. Because of the difference in expansion rates between pore fluid and solid skeleton, the pore pressure builds up, leading to material failure for low permeability materials, such as shales, the culprit in most borehole problems. This problem is investigated based on a thermo-poro-elastoplasticity formulation. In the following, an elastoplastic formulation coupled with the degradation of stiffness and strength of the material is first proposed to describe the thermo-mechanical behaviour of rocks. Using the effective stress principle, the constitutive relations governing the undrained response of saturated rocks subjected to heating are derived. The build-up of pore fluid pressure due to heating leading to material failure can be simulated by the proposed formulation. The thermo-mechanical effects on borehole stability in saturated rocks during drilling are also discussed.

2. MATHEMATICAL FORMULATION

2.1 *Thermo-elastoplasticity theory*

Assuming additivity of elastic and plastic strain rates, the constitutive relation takes the form

$$\dot{\boldsymbol{\sigma}}' = \boldsymbol{D}^e(\dot{\boldsymbol{\epsilon}}^e - \dot{\boldsymbol{\epsilon}}^p - \frac{1}{3}\dot{\epsilon}_T\boldsymbol{\delta}); \qquad \epsilon_T = k(\boldsymbol{\delta}^T\boldsymbol{\sigma}, T) \tag{1}$$

In eq.(1) $\boldsymbol{\sigma}'$ is effective stress, $\mathbf{D}^e$ is elastic stiffness and may depend on temperature, i.e., $\mathbf{D}^e = \mathbf{D}^e(T)$, $\boldsymbol{\delta}$ is the Kronecker's delta and ϵ_T represents the thermal expansion. The function k may be assumed to depend on the temperature T, and possibly also the effective hydrostatic pressure $\boldsymbol{\delta}^T\boldsymbol{\sigma}'$. In order to define the plastic strain rates, assume that the functional form of the yield surface, f=0, is affected by the temperature T, i.e.

$$f(\boldsymbol{\sigma}', \epsilon^p, T) = 0; \qquad \dot{\epsilon}^p = \dot{\lambda}\frac{\partial Q}{\partial \boldsymbol{\sigma}'} \tag{2}$$

where $Q = Q(\boldsymbol{\sigma}')$=const. is the plastic potential function. Using the consistency condition, the following constitutive relation is obtained after some algebra

$$\begin{gathered}\dot{\boldsymbol{\sigma}}' = \boldsymbol{D}^{ep}\dot{\boldsymbol{\epsilon}} + \boldsymbol{D}^{th}\dot{T}; \quad \boldsymbol{D}^{ep} = \boldsymbol{D}^e - \frac{1}{H}\boldsymbol{D}^e\frac{\partial Q}{\partial \boldsymbol{\sigma}}\left(\frac{\partial f}{\partial \boldsymbol{\sigma}}\right)^T\boldsymbol{D} \\ \boldsymbol{D}^{th} = -\frac{1}{3}k(\boldsymbol{\delta}^T\boldsymbol{\sigma}', T)\boldsymbol{D}^{ep}\boldsymbol{\delta} - \frac{1}{H}\boldsymbol{D}^e\frac{\partial Q}{\partial \boldsymbol{\sigma}'}\frac{\partial f}{\partial T^T}\end{gathered} \tag{3}$$

where $H = H_e + H_p$, and H_e and H_p are defined as

$$H_e = \left(\frac{\partial f}{\partial \boldsymbol{\sigma}'}\right)^T\boldsymbol{D}^e\frac{\partial Q}{\partial \boldsymbol{\sigma}'}; \qquad H_p = -\left(\frac{\partial f}{\partial \epsilon^p}\right)^T\frac{\partial Q}{\partial \boldsymbol{\sigma}'} \tag{4}$$

In order to complete the formulation, consider an arbitrary stress-controlled process. Invoking the additivity postulate again

$$\dot{\boldsymbol{\epsilon}} = \dot{\boldsymbol{\epsilon}}^e + \dot{\boldsymbol{\epsilon}}^p = \boldsymbol{C}^e\dot{\boldsymbol{\sigma}}' + \frac{1}{3}\dot{\epsilon}_T\boldsymbol{\delta} + \dot{\boldsymbol{\epsilon}}^p \tag{5}$$

where $\mathbf{C}^e = \mathbf{C}^e(T)$ is the elastic compliance. Utilizing the consistency condition, the following relation is obtained

$$\begin{gathered}\dot{\boldsymbol{\epsilon}} = \boldsymbol{C}^{ep}\dot{\boldsymbol{\sigma}}' + \boldsymbol{C}^{th}\dot{T} \;; \quad \boldsymbol{C}^{ep} = \boldsymbol{C}^e - \frac{1}{H_p}\frac{\partial Q}{\partial \boldsymbol{\sigma}'}\left(\frac{\partial f}{\partial \boldsymbol{\sigma}'}\right)^T \\ \boldsymbol{C}^{th} = \frac{1}{3}k(\boldsymbol{\delta}^T\boldsymbol{\sigma}, T)\boldsymbol{\delta} + \frac{1}{H_p}\frac{\partial Q}{\partial \boldsymbol{\sigma}'}\frac{\partial f}{\partial T}\end{gathered} \tag{6}$$

Eq.(6) governs the response of the material under a stress-controlled process. For a stationary stress field, $\dot{\boldsymbol{\sigma}} = \mathbf{0}$, the strain rates are given by the last term in eq.(6), which represents the volumetric expansion and degradation of plastic properties associated with temperature change.

2.2 *Undrained formulation*

According to the Terzaghi effective stress principle, the total stress $\boldsymbol{\sigma}$, the effective stress $\boldsymbol{\sigma}'$ and the pore fluid pressure p_f are related as

$$\boldsymbol{\sigma} = \boldsymbol{\sigma}' + p_f\,\boldsymbol{\delta} \tag{7}$$

Eq.(7) should be supplemented by the constitutive relation for the pore fluid(water) and the kinematic constraint of undrained deformation. Assuming that the water is linearly compressible, the volumetric response is defined as

$$\dot{p}^f = K_f\dot{\epsilon}^f; \qquad \dot{\epsilon}^f = \frac{1}{n}[\dot{\epsilon}_{ii} - n\alpha_f\dot{T}] \tag{8}$$

where K_f is the bulk modulus, n is porosity and α_f is the thermal expansion coefficient. Combining eqs.(3), (7) and (8) gives

$$\dot{\boldsymbol{\sigma}} = [\boldsymbol{D}^{ep} - \frac{K_f}{n}\boldsymbol{\delta}\boldsymbol{\delta}^T]\dot{\boldsymbol{\epsilon}} + [\boldsymbol{D}^{th} - K_f\alpha_f\boldsymbol{\delta}]\dot{T} \tag{9}$$

Eq.(9) represents the relation between the total stress and strain rates for a saturated poroelastoplastic material subjected to heating under undrained condition. For a stress-controlled process, the constitutive relations governing the undrained response may be obtained by combining eqs.(6), (7) and (8)

$$\dot{\boldsymbol{\epsilon}} = [\boldsymbol{I} + \frac{K_f}{n}\boldsymbol{C}^{ep}\boldsymbol{\delta}\boldsymbol{\delta}^T]^{-1}[\boldsymbol{C}^{ep}\dot{\boldsymbol{\sigma}} + (\boldsymbol{C}^{th} + \frac{K_f}{n}\boldsymbol{C}^{ep}\boldsymbol{\delta})\dot{T}] \tag{10}$$

3. ROCK BEHAVIOUR DUE TO HEATING IN UNDRAINED CONDITIONS

3.1 *Thermal expansion coefficient*

The volumetric expansion rate is derived from

$$\dot{\epsilon}_T = k_1(\delta^T\sigma')\dot{k}_2(T)$$
$$k_1(\delta^T\sigma') = e^{C_1\frac{\delta^T\sigma'}{f_{c0}}}; \quad k_2(\Delta T) = C_2\Delta T \tag{11}$$

In eq.(11), C_1 and C_2 are material constants, $\Delta T = T - T_g$ and T_g is the temperature in in-situ geostatic state.

3.2 *An elastoplastic model with thermal softening*

Assume that in the elastic range the only degrading parameter with the increase of temperature is the Young's modulus E. For a small range of temperature change, the degradation function may be approximated as linear

$$E = E_0(1 - C_3\Delta T) \tag{12}$$

where C_3 is a material constant and E_0 designates the modulus when $T = T_g$.

Let the properties of the material in the elastoplastic range be described by the constitutive relation, similar to that proposed by Pietruszczak et al. 1988. According to this formulation, the failure surface F=0 is defined in the form

$$F = \sqrt{J_2} - g(\theta)\sqrt{J_{2c}} = 0$$
$$\sqrt{J_{2c}} = \frac{-c_2 + \sqrt{(c_2^{\,2} + 4c_1(c_3 + I/\sigma_c))}}{2c_1}\sigma_c \tag{13}$$

In above equations $I = \sigma_{ii}$, $J_2 = 1/2\, s_{ij}s_{ij}$, $\theta = 1/3 \sin^{-1}(3\surd 3\, J_3 / 2\, J_2^{\,3/2})$ and $J_3 = 1/3 s_{ij}s_{jk}s_{ki}$ are stress invariants. Moreover, the parameters c_1, c_2 and c_3 represent dimensionless material constants and σ_c denotes the uniaxial compressive strength in a dry state. The yield surface is chosen in a functional form similar to that of eq.(13)

$$f = \sqrt{J_2} - \beta(\xi)g(\theta)\sqrt{J_{2c}} \tag{14}$$

where $\beta(\xi)$ represents a hardening function. The internal variable ξ is related to the history of accumulated plastic distortions

$$\beta = \xi/(A + B\xi); \quad \dot{\xi} = (\dot{e}_{ij}^{\,p}\dot{e}_{ij}^{\,p})^{1/2}/\bar{\Phi} \tag{15}$$

where $de_{ij}^{\,p}$ represents the deviatoric part of the plastic strain increment and $\Phi = \Phi(I, \theta)$.

Assume also that the increase of temperature results in the degradation of the uniaxial compressive strength σ_c. For a small range of temperature change, the degradation function may be approximated as linear

$$\sigma_c = \sigma_{c0}(1 - C_4\Delta T) \tag{16}$$

where C_4 is a material constant and σ_{c0} designates the strength when $T = T_g$. It should be noted that the degradation of σ_c is accompanied by a proportional reduction in uniaxial tensile strength, which is ensured by the functional form of eq.(13). For $\sigma_c = \sigma_c(T)$, eqs.(13) and (14) yield

$$\frac{\partial f}{\partial T} = \frac{\partial f}{\partial\sqrt{J_{2c}}}\frac{\partial\sqrt{J_{2c}}}{\partial\sigma_c}\frac{\partial\sigma_c}{\partial T}$$
$$= \beta(\xi)g(\theta)\left(\frac{\sqrt{J_{2c}}}{\sigma_c} - \frac{I}{2c_1\sqrt{J_{2c}} + c_2\sigma_c}\right)\frac{\partial\sigma_c}{\partial T} \tag{17}$$

The direction of plastic flow is governed by a non-associated flow rule. The details concerning the form of the plastic potential function, the function $g(\theta)$, etc. are provided in Xu & Dusseault 1997 or Pietruszczak et al. 1988.

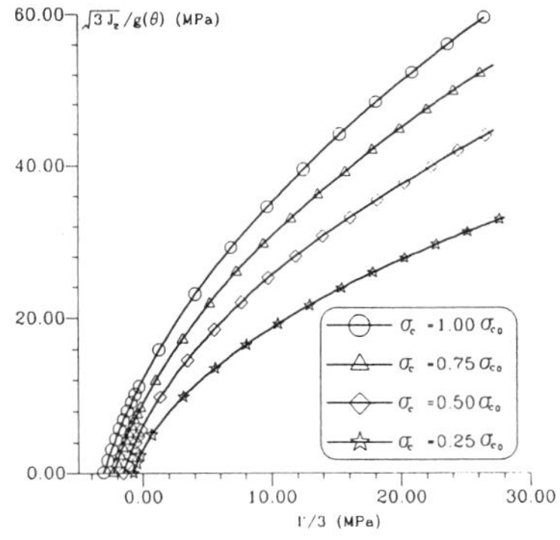

Fig.1 Failure surfaces evolution due to degradation of material properties

Given eqs.(11) through (17), the general formulation of the problem, as outlined in the previous section, is complete. The assumed linear thermal expansion and its pressure dependence, as well as the linear degradation of stiffness and strength, can be replaced by any function available from experimental results. It is interesting to note that under a stationary stress field the degradation in plastic properties is described through the evolution of the failure surface, eq.(13), which undergoes a progressive contraction with the increase of temperature. This is similar to the framework proposed in the article by Pietruszczak 1996. For illustration, the evolution of the failure surface with degradation of strength is provided in Fig.1.

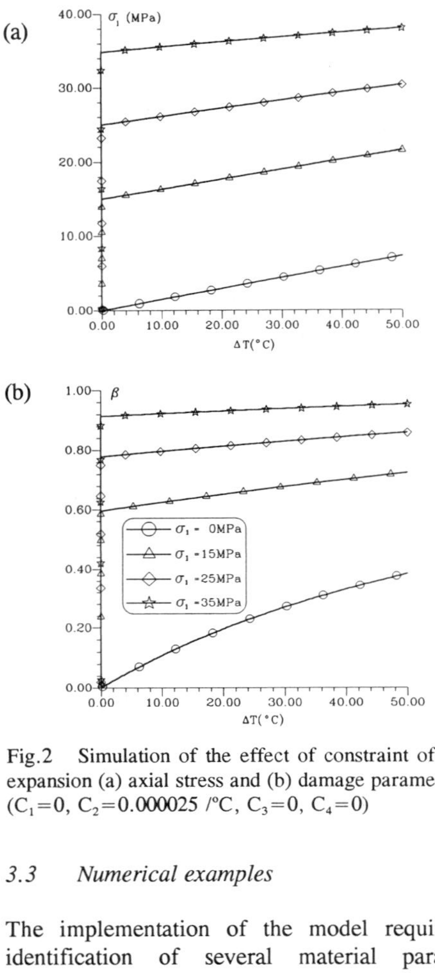

Fig.2 Simulation of the effect of constraint of thermal expansion (a) axial stress and (b) damage parameter ($C_1=0$, $C_2=0.000025$ /°C, $C_3=0$, $C_4=0$)

3.3 Numerical examples

The implementation of the model requires the identification of several material parameters introduced in the previous section. The material properties associated with the elastoplastic response are assumed as: $\sigma_c=42$MPa, $c_1=1.925$, $c_2=1.0$, $A=0.0001$, $B=0.95$, etc.(refer to Xu & Dusseault 1997 for details). For parameters associated with thermal expansion and thermal softening, different values may be adopted. The objective here is to investigate the effect of the constraint of thermal expansion and the effect of thermal degradation in both drained and undrained conditions. To demonstrate this clearly, a cylindrical specimen involving the constraint of the axial thermal expansion at sustained axial load is adopted.

In drained conditions, the excess pore pressure is assumed to be dissipated and thus the difference of expansion rates between the fluid and solid skeleton will have no effect. The numerical simulations pertain to a material subjected to temperature increase at different intensities of axial load. The sample is first subjected to uniaxial compression and subsequently subjected to the constraint of the axial thermal expansion. To achieve the maximum increase of axial stress, no degradation is assumed. Fig.2a show the evolution of axial stress (σ_1) for the sample. The damage evolution $\beta(\xi)$ due to the constraint of expansion , as defined in eqs.(14) and (15), is illustrated in Fig.2b. It is evident that at higher stress intensity the increase of axial stress and damage become smaller. It is expected that with thermal degradation the increase of axial stress becomes less. In fact, the degradation of material properties may result in axial contraction which counteracts the thermal expansion, as shown in Fig.3a. The thermal degradation can also lead to

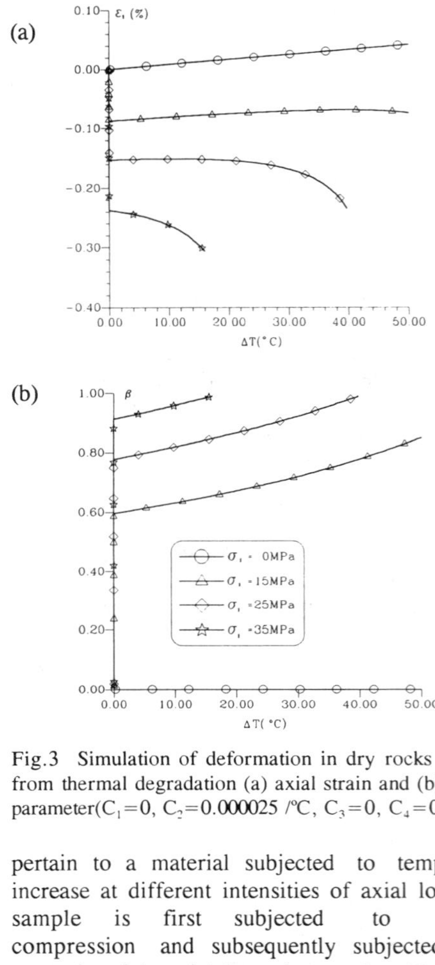

Fig.3 Simulation of deformation in dry rocks resulting from thermal degradation (a) axial strain and (b) damage parameter($C_1=0$, $C_2=0.000025$ /°C, $C_3=0$, $C_4=0.01$ / °C)

material failure, as illustrated in Fig.3b.

For undrained conditions (n=0.2, K_f=2200MPa), the effect of the constraint of the axial thermal expansion and the effect of thermal degradation are demonstrated in Figs. 4 and 5, respectively. It is evident that the dominant failure mechanism is pore pressure build-up resulting from the different expansion rates between solid skeletion and pore fluid, which leads to material failure (Figs.4a and 5a). The pore pressure increase with temperature is shown in Figs.4b and 5b. The effect of the constraint of the axial thermal expansion may promote failure at lower stresses; At higher stresses, the difference becomes less (Fig.4c). The effect of thermal degradation always promotes failure, as shown in Fig.5c.

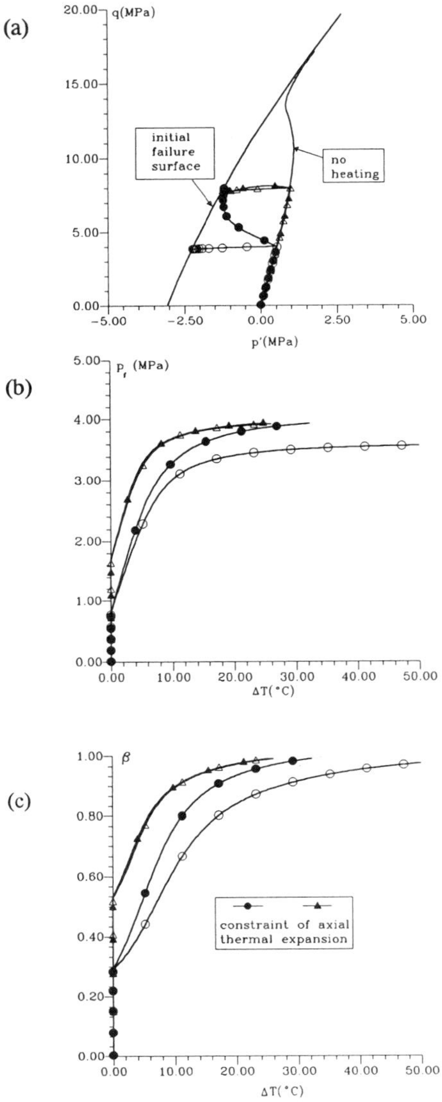

Fig.4 Simulation of the effect of constraint of thermal expansion in undrained conditions (a) stress path, (b) pore fluid pressure and (c) damage parameter
(C_1=0, C_2=0.000025 /°C, α_f=0.00004 / °C, C_3=0, C_4=0)

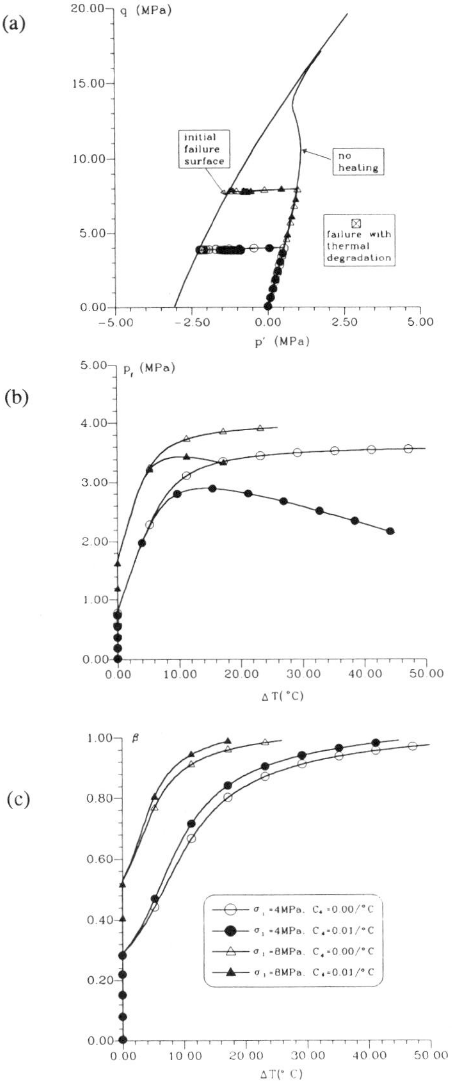

Fig.5 Simulation of deformation in undrained conditions resulting from thermal degradation (a) stress path, (b) pore fluid pressure and (c) damage parameter
(C_1=0, C_2=0.000025 /°C, α_f=0.00004 / °C, C_3=0)

4. FINAL REMARKS

4.1 *Discussion*

In the elastic borehole case, a temperature increase inevitably leads to an increase in tangential stress. However, if the material is elastoplastic or experiences progressive damage, there may be little or no increase in tangential stress at the borehole wall. In this case, a yield (damaged) zone propagates into borehole wall, and there may be stress relaxation, combined with stress distribution farther into the formation, as originally pointed out for an ideal elastoplastic drained case (Wang & Dusseault 1995). Because of the different expansion rates of the solid skeleton and the pore fluid, pore pressure build-up in an undrained condition may exacerbate failure, especially for material with a low heat diffusion rate or the case of slow heating (i.e. slow drilling rates). In both cases, the fully coupled formulation needs to be pursued. The process rate will be governed by the thermal conductivity, the permeability, the thermal expansion coefficient, and the magnitude of the temperature and pressure changes(Hojka, 1991).

4.2 *Conclusion*

A relatively general thermo-poro-elastoplasticity formulation is developed by combining the effective stress principle and thermo-elastoplasticity theory, coupled with the degradation of stiffness and strength of the material. Based on numerical simulations, the thermo-mechanical effect on material damage is demonstrated. The increase of stress due to expansion may not be significant because the material around the borehole is already in a very high stress state close to failure, especially when there is thermal degradation causing stress relaxation. However, because of the difference of expansion rates between the solid skeleton and pore fluid, the build-up of pore fluid pressure in undrained conditions can lead to material failure, which can be simulated by the proposed formulation. Further research is under way to implement the proposed formulation in a finite element algorithm for borehole instability analysis.

ACKNOWLEDGMENTS

The support for Waterloo Shale Project from the National Sciences and Engineering Research Council of Canada and the oil industry is appreciated. G. Xu is thankful to S. Pietruszczak for his critical review.

REFERENCES

Bear, J. & Corapcioglu 1980. A mathematical model for consolidation in a thermoplastic aquifer due to hot water injection or pumping. *Water Resources Res.* **17:** 723-736.

Booker, J.R. & Savvidou, C. 1984. Consolidation around a spherical heat source. *Int. J. Solids Struct.* **20**: 1079-1090.

Britto, A.M., Savvidou, Maddocks, D.V., Gunn, M.J. & Booker, J.R. 1989. Numerical and centrifuge modelling of coupled heat flow and consolidation around hot cylinders buried in clay. *Geotechique* **39**: 13-25.

Dusseault, M.B. 1994. Analysis of borehole stability, *Comp.Meth.& Advances Geomech.* Siriwardane & Zaman(eds), Balkema:125-137.

Dusseault, M.B., Wang, Y & Simmons, J.V. 1988. Induced stresses near a fireflood front, *AOSTRA J Res.* **4**: 153-170.

Hojka, K. 1991. Temperature and stress solutions for a plane-strain borehole in saturated non-isothermal thermoporoelastic media, MSc Thesis, Earth Sciences, University of Waterloo.

Hueckel, T., Borsetto, M. & Peano, A. 1987. Modelling of coupled thermo-elastoplastic-hydraulic response of clays subjected to nuclear waste heat. *Numerical Methods for Transient and Coupled problems*,Lewis,Hinton,Bettess,& Schrefler(eds), John Wiley &Sons Ltd.: 213-233.

Kurashige, M. 1989. A thermoplastic theory of fluid-filled porous materials. *Int. J. Solids Struct.***25**: 1039-1052.

Palciauskas,V.V. & Domenico, P.A. 1982. Characterization of drained and undrained response of thermally loaded repository rocks. *Water Resources Res.***18:** 281-290.

Pietruszczak, S. 1996. On the mechanical behaviour of concrete subjected to alkali-aggregate reaction, *Int.J.Computers Struct.* **58**: 1093-1099.

Pietruszczak, S, Jiang, J & Mirza, F.A. 1988. An elastoplastic constitutive model for concrete. *Int. J. Solid Struct.* **24**: 705-722.

Wang, Y & Dusseault, M.B. 1995. Response of a circular opening in a friable low-permeability medium to temperature & pore pressure changes. *Int.J.Num.Anal.Meth.Geomech.***19**: 157-179.

Xu, G. & Dusseault, M.B. 1997. The influence of anisotropies on deformation and damage around underground openings, submitted to *Canadian Geotech. Journal.*

Numerical Models in Geomechanics, Pietruszczak & Pande (eds) © 1997 Balkema, Rotterdam, ISBN 90 5410 886 X

An elasto-viscoplastic constitutive model with strain softening

T. Adachi
Civil Engineering Department, Kyoto University, Japan

F. Oka
Civil Engineering Department, Gifu University, Japan

F. Zhang
Chuo Fukken Consultants Co. Ltd, Japan

ABSTRACT: In this paper, an elasto-viscoplastic constitutive model with strain softening is proposed based on Adachi-Oka's model. In the model, a time measure, instead of strain measure adopted in previous model, is introduced as an internal variable to account for not only stress history but also time change. The yielding function is defined in the way that the structural deterioration of geologic materials with time is considered. The application of the model to the experimental results of soft sedimental rock indicates that the model can not only describe the time dependency, such as strain rate dependency, creep and stress relaxation, but also the strain softening behavior of geologic materials.

1 INTRODUCTION

Generally speaking, the mechanical behavior of soft sedimentary rock is elasto-plastic, dilatant, strain hardening-strain softening and time dependent. The softening behavior of soft sedimentary rock plays an important rule in the long-term stability of geomechanical engineering. As for the time dependent behavior, it involves three aspects, namely, the strain rate effect, creep and stress relaxation. Many models have been proposed to describe the time dependency of geologic materials. They are found in the studies by Andersland and Al-Nouri (1970), Ting (1983). Singh and Mitchell (1968) clarified the mechanism of creep for soil under undrained condition. Oka (1985) proposed a viscoplastic model with memory and an internal variable based on the generalized theory of Wang (1969) and Perzyna (1980). Adachi et al. (1990) proposed a viscoplastic model which can describe both the strain rate effect and strain softening of frozen sand.

Adachi et al(1991). published in their research that Adachi-Oka's model with strain softening is applicable to the numerical analyses of the strain-hardening and the strain-softening behaviors of geological materials. Based on the model, the finite element analysis can lead to a unique solution for initial value problems in Valanis's sense(1985) and the analytical results have only a very small dependency on the mesh size. In this paper, a new type of elasto-viscoplastic model is proposed to describe not only strain softening but also strain rate dependency, creep and stress relaxation.

2 ELASTO-VISCOPLASTIC MODEL WITH STRAIN SOFTENING

Adachi et al.(1990) proposed an elasto-viscoplastic model for frozen sand exhibiting strain softening, using a time measure which is similar to that proposed by Valanis (1971). In present paper, however, the definition of a new incremental time measure is introduced in following form:

$$dz = C \exp(-(z - z_1)/C)\, dt \qquad (1)$$

where dz is an incremental time measure, t is the real time and C is the parameter of time dependency which will be discussed later in detail. z_1 is the initial time measure. In the present study, the stress history tensor is expressed by introducing a single exponential type of kernel function similar to that in a previous work by Adachi and Oka (1990), namely,

$$\sigma^*_{ij} = \frac{1}{\tau}\int_0^z \exp(-(z-z')/\tau)\sigma_{ij}(z')dz' \qquad (2)$$

where τ is a material parameter which expresses the retardation of stress with respect to the time measure. The total strain increment tensor is composed of the elastic and plastic components:

$$d\varepsilon_{ij} = d\varepsilon^e_{ij} + d\varepsilon^p_{ij} \qquad (3)$$

The viscoplastic strain increment is assumed to be given by the non-associated flow rule,

$$d\varepsilon_{ij}^{\,p} = H(\partial f_p / \partial \sigma_{ij})df_y\ , \qquad (4)$$

where f_p is the plastic potential function, f_y is the yield function and H is the loading index describing

the hardening-softening characteristics.
The subsequent yield function is defined by

$$f_y = \eta^* - \kappa = 0 \qquad (5)$$

$$\eta^* = \sqrt{S^*_{ij}\, S^*_{ij}} \,/\, (\sigma^*_m + b) \qquad (6)$$

where S^*_{ij} is the deviatoric stress history tensor, b is the plastic potential parameter which can be determined by the assumption that the stress ratio η is kept constant in the over-consolidated region along the M_{rr} line. σ^*_m is the mean stress history and κ is the strain hardening-softening parameter. The strain hardening-softening parameter is assumed to be given by the following evolution equation for strain hardening and softening:

$$\dot{\kappa} = \dot{\gamma}^{p}\, G'(M^*_f + \kappa)^2 / M^{*2}_f \qquad (7)$$

where $\dot{\gamma}^p = (\dot{e}^p_{ij}\, \dot{e}^p_{ij})^{1/2}$. In the case of proportional loading, it can be integrated as

$$\kappa = M^*_f G' \gamma^{p} / (M^*_f + G' \gamma^{p}) \qquad (8)$$

where γ^p is the second invariant of the deviatoric plastic strain, that is, $\gamma^p = (e_{ij}{}^p e_{ij}{}^p)^{1/2}$. G' and M^*_f are strain hardening-softening parameters. M^*_f was supposed to be constant in the previous study by Adachi et al.(1992). While in this paper, it is assumed to change with time on account of the structural deterioration of geologic materials. M^*_f takes the form of

$$M^*_f = M^*_{f\infty} + (M^*_{f0} - M^*_{f\infty})\, e^{-A(z - z_1)} \qquad (9)$$

where $M^*_{f\infty}$ is the residual value of $\eta^* = \sqrt{2J^*_2}/(\sigma^*_m + b)$. A is a material parameter and can be determined by triaxial drained creep tests. M^*_{f0} is dependent on the initial stress and is defined as

$$M^*_{f0} = \begin{cases} (2J_2)^{1/2}/\sigma_m|_{at\ t=0} & if\ (2J_2)^{1/2}/\sigma_m|_{at\ t=0} > M^*_{f\infty} \\ M_{f\infty} & if\ (2J_2)^{1/2}/\sigma_m|_{at\ t=0} \le M^*_{f\infty} \end{cases} \qquad (10)$$

Eq.9 illustrates the deterioration of the shear strength of a given material with the time measure. Fig.1 shows schematically how the failure line changes in accordance with time measure in stress history space.

The loading conditions are given by the following relations:

$$d\varepsilon^p_{ij} \begin{cases} \neq 0 \ if\ f_y = 0,\ df_y > 0 & loading \\ = 0 \ if\ f_y = 0,\ df_y = 0 & neutral \\ = 0 \ if\ f_y = 0,\ df_y < 0 & unloading \end{cases} \qquad (11)$$

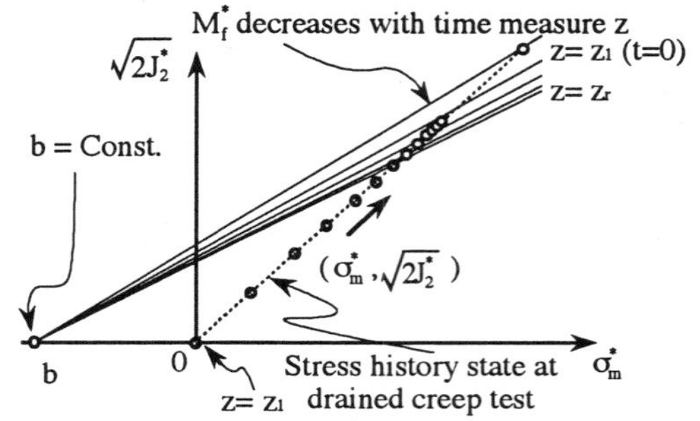

Fig.1 Variation of M^*_f with time measure

It is assumed that plastic potential function is expressed by the relation as

$$f_p = \bar{\eta} + \bar{M} \ln [(\sigma_m + b)/(\sigma_{mb} + b)] = 0 \qquad (12)$$

where $\bar{\eta} = (S_{ij} S_{ij} / (\sigma_m + b)^2)^{1/2}$, in which S_{ij} is the deviatoric stress tensor, σ_m is the mean stress and $\bar{M}$ is the parameter that controls the development of the volumetric strain. σ_{mb}, the plastic potential parameter, is determined by isotropic consolidation tests and takes the value of the pre-consolidated stress.

The concept of boundary surface, introduced in previous studies, is expressed by the following relation:

$$f_b = \bar{\eta} + \bar{M}_m \ln [(\sigma_m + b)/(\sigma_{mb} + b)] = 0 \qquad (13)$$

Based on this relation, the value of $\bar{M}$ can be determined by considering the following condition:

$$\begin{cases} \bar{M} = -\bar{\eta} / \ln [(\sigma_m + b)/(\sigma_{mb} + b)] & f_b \le 0 \\ \bar{M} = \bar{M}_m & f_b > 0 \end{cases} \qquad (14)$$

Combining Eqs.4, 5, 12, 13, 14, we can derive the following equation for the viscoplastic strain increment tensor:

$$d\varepsilon^p_{ij} = \Lambda\, [\bar{\eta}_{ij}/\bar{\eta} + (\bar{M} - \bar{\eta})\, \delta_{ij}/3]\, d\eta^*$$

or

$$d\varepsilon^p_{ij} = \Lambda \left[\frac{\bar{\eta}_{ij}}{\bar{\eta}} + (\bar{M} - \bar{\eta})\frac{\delta_{ij}}{3}\right] \left[\frac{\eta^*_{kl}}{\eta^*} - \eta^* \frac{\delta_{kl}}{3}\right] \frac{d\sigma^*_{kl}}{\sigma^*_m + b} \qquad (15)$$

where

$$\Lambda = \frac{M^{*2}_f}{G(M^*_f - \eta^*)^2}, \quad \eta_{ij} = S_{ij}/(\sigma_m + b),$$

$$\eta^*_{kl} = S^*_{kl} / (\sigma^*_m + b). \qquad (16)$$

Finally, the elastic strain increment is evaluated by

$$d\varepsilon^e_{ij} = ds_{ij}/2G + d\sigma_m\delta_{ij}/3K \ . \tag{17}$$

3 APPLICATION TO DRAINED CREEP TESTS ON SOFT ROCK

In drained triaxial creep tests, the stress is kept constant, that is, $\Delta\sigma_{ij} = 0$, $\sqrt{J_2} = \mathrm{const.}$ and $\sigma_m = \mathrm{const.}$ By integrating Eq.1, the expression for the time measure can be given in Eq.18 if we take into consideration the fact that the initial time measure, $z\,|_{t=0} = z_1$, is usually not equal to zero, namely,

$$z - z_1 = C\ln(1+t) \ . \tag{18}$$

Here z_1 is the time measure for the time at which the creep tests begin. In drained creep tests, the stress is not applied all at once to the specimens. The loading process from zero to a certain stress state usually takes some time. It is reasonable, therefore, to conclude that the initial time measure is not equal to zero, which means that the stress history tensor will also not be equal to zero. Its value is dependent on the loading rate and the loading path. Here, we can assume that the load increases proportionally to time measure z during the interval $[0, z_1]$. The stress history tensor can then be calculated in the following manner:

$$\sigma^*_{ij} = \sigma^*_{ij1} + \sigma^*_{ij2} \tag{19}$$

where

$$\sigma^*_{ij1} = \frac{1}{\tau}\int_0^{z_1} \exp(-(z-z')/\tau)\,\sigma_{ij}(z')\,dz' \tag{20}$$

$$= \sigma_{ij}\frac{e^{-z/\tau}}{z_1}\int_0^{z_1} e^{z'/\tau}\frac{z'}{\tau}dz' = \sigma_{ij}e^{-(z-z_1)/\tau}\,\frac{\tau}{z_1}\left(\frac{z_1}{\tau} - 1 + e^{-z_1/\tau}\right)$$

is the initial stress history at t=0 resulted in the loading process before creep. While in creep tests, stress is kept constant, namely,

$$\sigma^*_{ij2} = \frac{1}{\tau}\int_{z_1}^{z} \exp(-(z-z')/\tau)\,\sigma_{ij}(z')\,dz' = \sigma_{ij}(1 - e^{-(z-z_1)/\tau}) \tag{21}$$

$$\sigma^*_{ij} = \sigma^*_{ij1} + \sigma^*_{ij2} = \sigma_{ij}[1 - e^{-(z-z_1)/\tau}g(z_1)] = \sigma_{ij}f(z) \tag{22}$$

$$f(z) = 1 - e^{-(z-z_1)/\tau}g(z_1),\ g = g(z_1) = (1 - e^{-z_1/\tau})\tau/z_1 \tag{23}$$

In axisymmetric condition, the detailed expression for the deviatoric and the volumetric components of viscoplastic strain can be given by some algebraic as

$$de^p_{11} = \frac{2}{3G}\frac{M_f^{*2}}{(M_f^* - \eta^*)^2}\,g(z_1)\,\frac{bq\,e^{-(z-z_1)/\tau}}{(\sigma^*_m + b)^2}\,\frac{dz}{\tau} \tag{24}$$

$$dv^p = \Lambda\left(\bar{M} - \frac{\sqrt{2/3}\,q}{\sigma_m + b}\right)d\eta^* = \left(\bar{M} - \frac{\sqrt{2/3}\,q}{\sigma_m + b}\right)\sqrt{3/2}\,de^p_{11} \tag{25}$$

where, $q = (\sigma_1 - \sigma_3)$. Eq.24 can be rewritten as

$$\dot e_{11} = \dot e^p_{11} = \frac{2bq}{3G'\tau}\,g(z_1)\,\frac{M_f^{*2}}{(M_f^* - \eta^*)^2}\,\frac{e^{-(z-z_1)/\tau}}{(\sigma^*_m + b)^2}\,\dot z(t) \ . \tag{26}$$

Substituting Eq.18 into Eq.26, we obtain

$$\ln\dot e_{11} = \ln\frac{2qC\,g(z_1)}{3G'b\tau} - \frac{C}{\tau}\ln(1+t) + 2\ln\left(\frac{M_f^*}{M_f^* - \eta^*}\,\frac{b}{\sigma^*_m + b}\,\frac{1}{\sqrt{(1+t)}}\right) \ . \tag{27}$$

In Eq.27, the first term of the right-hand side is constant with respect to time and can be regarded as the initial strain rate. The second and third terms correspond to steady creep and accelerating creep, respectively.

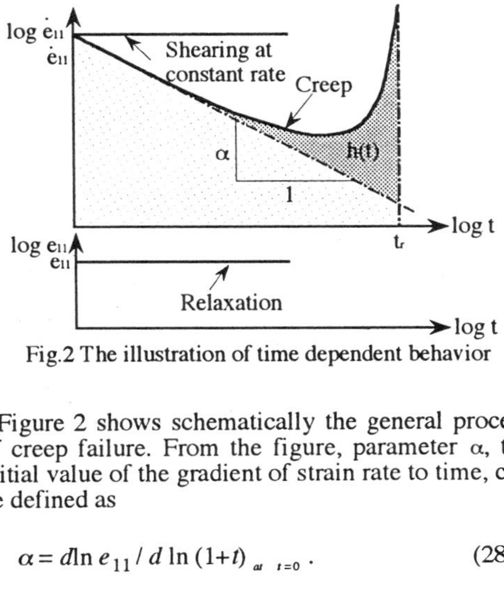

Fig.2 The illustration of time dependent behavior

Figure 2 shows schematically the general process of creep failure. From the figure, parameter α, the initial value of the gradient of strain rate to time, can be defined as

$$\alpha = d\ln\dot e_{11}/d\ln(1+t)_{\ at\ t=0} \ . \tag{28}$$

The value of C is defined here as $C = \alpha\tau$ (29)
so that,

$$\dot e^0_{11} = \frac{2q\alpha}{3G'b}\,g\,\frac{M_f^{*2}}{(M_f^* - \eta_0^*)^2}\,\frac{b^2}{(\sigma^*_{m0} + b)^2} \ . \tag{30}$$

In Eq.31, $\dot e^0_{11} = \dot e_{11}\,|_{t=0}$, and

$$\eta_0^* = \frac{\sqrt{2J_2}\,(1-g)}{\sigma_m(1-g)+b}, \tag{31}$$

$$\sigma_{m0}^* = \sigma_m(1-g) \tag{32}$$

$$\frac{M_f^{*2}}{(M_f^*-\eta_0^*)^2}\,\frac{b^2}{(\sigma_{m0}^*+b)^2} = \left\{\frac{M_f^* b}{M_f^* b + (\sigma_m M_f^* - \sqrt{2J_2})(1-g)}\right\}^2 . \tag{33}$$

Because $M^*_{f\,at\,t=0} = M^*_{f0} = \sqrt{2J_2}/\sigma_m$, Eq.33 will become

$$\frac{M_f^{*2}}{(M_f^*-\eta_0^*)^2}\,\frac{b^2}{(\sigma_{m0}^*+b)^2} = 1 \tag{34}$$

Based on Eqs.10, 30, 31, 32, 33, 34, the expression for the initial creep rate is obtained as

$$\dot{e}_{11}^{0} = \frac{2q\alpha\, g(z_1)}{3G'b} \quad \Rightarrow g(z_1) = \frac{3G'b}{2q\alpha}\,\dot{e}_{11}^{0} \tag{35}$$

Based on Eq.35, parameter $g(z_1)$ can be determined by triaxial drained creep tests. Eq.27 can be rewritten as

$$\ln \dot{e}_{11} = \ln \dot{e}_{11}^{0} - \alpha \ln(1+t) + \ln h(t) \tag{36}$$

In Eq.36, the first term at the right-hand side of the equation represents the initial deviatoric strain creep rate. The second term stands for the steady creep strain in the decreasing stage, which can also be called the secondary consolidation. The third term can be regarded as the acceleration of creep strain at rupture. The expression for the accelerating creep strain is

$$h(t) = \left(\frac{M_f^*}{M_f^*-\eta^*}\,\frac{b}{\sigma_m^*+b}\,\frac{1}{\sqrt{1+t}}\right)^2 \tag{37}$$

It is known from Eq.10 that if $\sqrt{2J_2}/\sigma_m|_{at\,t=0} < M^*_{f\infty}$, $M^*_{f\infty} = M^*_j$ is valid. Furthermore, in drained creep test, $\sqrt{2J_2^*}/\sigma_m^* = \sqrt{2J_2}/\sigma_m$ is valid. Therefore, when the initial stress ratio is $\sqrt{2J_2}/\sigma_m|_{at\,t=0} < M^*_{f\infty}$,

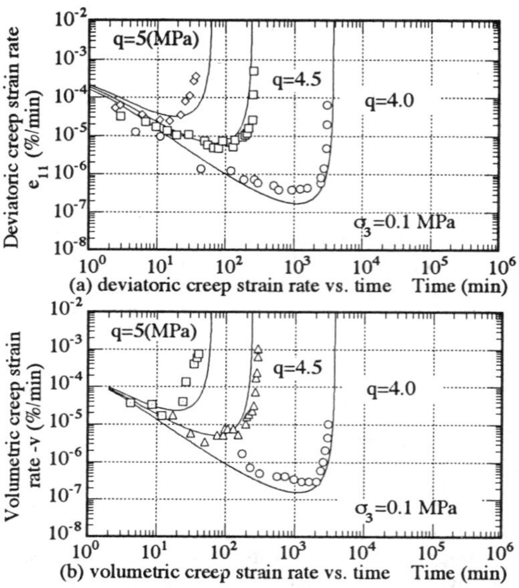

Fig.3 Comparison between theoretic and experimental results of creep strain rates

then

$$\eta^* = \frac{\sqrt{2J_2^*}}{\sigma_m^*+b} < \frac{\sqrt{2J_2^*}}{\sigma_m^*} = \frac{\sqrt{2J_2}}{\sigma_m} < M^*_{f\infty} = M_f^* \tag{38}$$

which means that the right-hand side of Eq.26 will never become infinite. In other words, creep failure never occur when the stress ratio $\sqrt{2J_2}/\sigma_m|_{at\,t=0}$ is less than $M^*_{f\infty}$.

In application, the drained creep test results for a saturated sample of Ohya stone (porous tuff) deposited in the Miocene Epoch of Tertiary Period is analyzed with the present model. The material parameters of the model are listed in Table 1.

In the present study, the results of drained creep tests under confining pressure of 0.1 MPa are considered. Fig.3 and Fig.4 show the comparison between the theoretical and the experimental results of deviatoric and volumetric creep strain rates at a confining pressure of 0.1 MPa. It is found that the

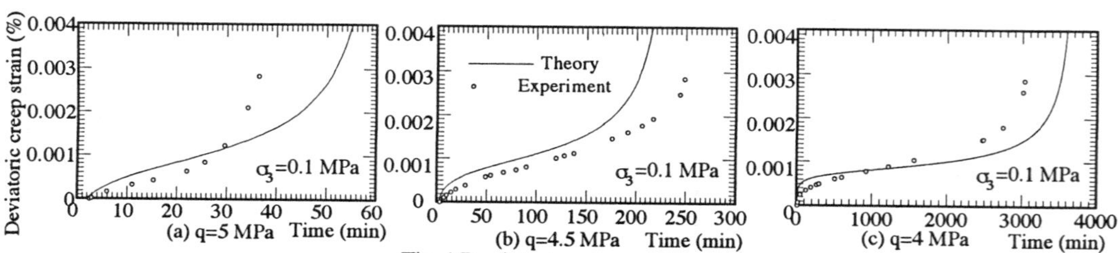

Fig.4 Deviatoric creep strain vs. time

Table 1 Material parameters of Ohya stone

K(kgf/cm^2)	4403.0	a	-0.101
G(kgf/cm^2)	5029.0	α	0.614
b(kgf/cm^2)	20.0	g	0.924
σ_{mb}(kgf/cm^2)	110.	G'	1563.
$\bar{M}_m$	1.694	t	14.15
		$M_{f\infty}^*$	0.392

Table 2 Comparison of theoretical and tested results

q (kgf/cm^2)	Creep failure time t_r(min)		Minimum strain rate (10^{-6} % / min)	
	experiment	theory	experiment	theory
40	2923	3765	0.410	0.17
45	253	242	6.0	9.33
50	40	64	26.0	31.5

general characteristics of creep behavior, such as the initial creep rate, the steady creep stage and the creep rupture, can be simulated well.

Table 2 lists the results of the theoretical and the experimental results of creep failure times and minimum strain rates.

4 APPLICATION TO TRIAXIAL SHEAR TESTS ON SOFT ROCK

In drained triaxial shear tests at a constant strain rate the initial time measure is equal to zero. Therefore Eq.18 becomes $z=Cln(1+t)$ In strain-controlled conventional triaxial tests at a constant strain rate, the following relations exist:

$$\dot{\varepsilon}_{ij} = \dot{\varepsilon}^e_{ij} + \dot{\varepsilon}^p_{ij} = const \tag{39}$$

$$\dot{e}^e_{ij} = \frac{s_{ij}}{2G}, \quad \dot{\varepsilon}^e_v = \frac{\sigma_m}{K}, \quad \dot{\varepsilon}^e_{ij} = \dot{e}^e_{ij} + \delta_{ij}\,\dot{\varepsilon}^e_v \tag{40}$$

By considering the relation of $d\sigma^*_{ij}=(\sigma_{ij}-\sigma^*_{ij})dz/\tau$ and Eq.15, we obtain

$$\dot{\varepsilon}^p_{ij} = \Lambda[\frac{\bar{\eta}_{ij}}{\bar{\eta}} + (\bar{M}-\bar{\eta})\frac{\delta_{ij}}{3}][\frac{\eta^*_{kl}}{\eta^*} - \eta^*\frac{\delta_{kl}}{3}]\frac{1}{\sigma^*_m+b}(\sigma_{kl}-\sigma^*_{kl})\frac{C}{1+t} \tag{41}$$

In the present case, unlike the creep, the stress history tensor cannot be integrated explicitly and should be solved by the associated differential equations found in Eq.39~41.

The parameter of time dependency C in the present case is determined in the following way. By differentiating the Eq.18, the following relation is found

$$dz = C\,dt/(1+t). \tag{42}$$

It is known from experimental results that parameter C here is dependent on the shear strain rate. As a reference, the stress history parameter τ is firstly determined under the conditions of $\dot{\varepsilon}_{11} = \dot{\varepsilon}_0 = 1\% / \min$, assuming that $C=1$. The value of τ here is denoted by τ_0. Secondly, by using the curve fitting method, we evaluate the values of τ at an arbitrary shearing strain rates $\dot{\varepsilon}_{11}$ with the same assumption of $C=1$. τ is evaluated in such a way that the peak strength in the theoretical results is the most suitable to the experimental results. Through the above analysis, we find, for an arbitrary shearing strain rates $\dot{\varepsilon}_{11}$ and the correspondent τ, there exists such a relation as,

$$\tau = (\dot{\varepsilon}_{11}/\dot{\varepsilon}_0)^a\,\tau_0 \;. \tag{43}$$

where a is called the parameter of strain-rate-dependency. By omparing the following two relations,

$$dz/\tau_0 = dt/\tau_0/(1+t)\;;\; dz/\tau = C\,dt/\tau/(1+t) \tag{44}$$

it can be find that if the parameter for time dependency C takes the form expressed in Eq.45, the values of dz/τ will become independent of the strain rate.

$$C = (\dot{\varepsilon}_{11}/\dot{\varepsilon}_0)^a \tag{45}$$

The proposed model is used to analyzed the behavior of Ohya stone under different constant axial strain rate. The strain-controlled conventional triaxial tests is supposed to be carried out at different strain rates.

Figure 5 shows the relations of deviatoric stress ($q=\sigma_{11}-\sigma_{33}$) vs. deviatoric strain. It is found that the rate dependency can be simulated well and the theoretical result is in agreement with the experimental result at a constant strain rate of 0.0025%/min.

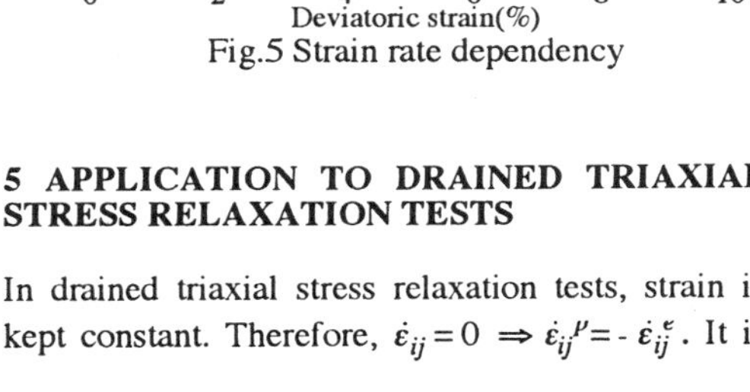

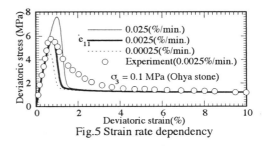

Fig.5 Strain rate dependency

5 APPLICATION TO DRAINED TRIAXIAL STRESS RELAXATION TESTS

In drained triaxial stress relaxation tests, strain is kept constant. Therefore, $\dot{\varepsilon}_{ij} = 0 \Rightarrow \dot{\varepsilon}^p_{ij} = -\dot{\varepsilon}^e_{ij}$. It is

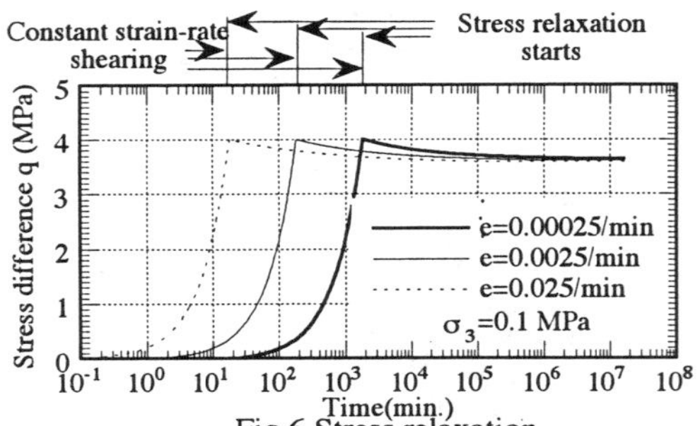

Fig.6 Stress relaxation

found that parameter C is dependent on the strain rate in the process where the specimen is loaded to the prescribed stress state. C can be determined by drained triaxial stress tests.

The proposed model is used to simulate drained triaxial stress relaxation tests. As drained triaxial stress relaxation tests were not conducted on Ohya stone, only a theoretical estimation of the stress relaxation is given. Fig.6 shows the estimated stress relaxation behavior of Ohya stone using the present model. The sample is first loaded to a deviatoric stress state of 4 MPa with a confining pressure of 0.1 MPa, the strain ε_{11} at the moment being 0.415%. Parameter C here takes the values of 2.29, 2.88 and 3.63, which correspond to the strain rates of 0.025%/min., 0.0025%/min. and 0.00025%/min. at constant strain-rate shearing respectively. It is found from the figure that in the process of stress relaxation, the stresses decreases almost linearly at the beginning in stress-logarithmic time axes. After decreased to a certain value, the stress almost remain unchanged. The stress relaxation will almost be completed within 10^4 minutes after the relaxation starts and the stress will decrease from 4.0 MPa to 3.6 MPa after 10^7 minutes.

6 DETERMINATION OF THE PARAMETERS IN THE MODEL

Eleven material parameters are involved in the model, that is, G, K: Young's modules; b, σ_{mb}: plastic potential parameters; G', A, $M^*_{f\infty}$ strain hardening-softening parameters; τ: stress history parameter; M_m: parameter of over-consolidated boundary; C : parameters of time dependency and g: parameter of the initial time measure.

Young's modulus and the bulk modulus are determined directly by measuring the tangent to the deviatoric stress-strain curve and the tangent to the volumetric stress-strain curve at the initial stage of loading. The parameter of the over-consolidated boundary, M_m, takes the value of stress ratio $\sqrt{S_{ij}S_{ij}}/(\sigma_m+b)$ at the point where the maximum contraction occurs in the shearing process.

Stress history parameter τ is determined by a process of trial and error using the peak shearing strength. Strain hardening-softening parameter $M^*_{f\infty}$ is determined as the value of η^* at the residual state. G' can be determined as the initial gradient of the stress-strain curve in the unloading-reloading process at the residual state. As for parameter A, it is found from creep tests that it increases linearly with stress ratio $\bar{\eta}$ and can be given as $A = A_0 + \bar{\eta}A_1$, where, the value of A_0 is evaluated to be -0.664 and A_1 to be 0.854 based on creep tests at a confining pressure of 0.1 MPa.

7 CONCLUSION

In this paper, an elasto-viscoplastic model with memory is proposed for geologic materials. A comparison of the analytical results and the experimental results indicates that it is not only able to describe all the aspects of time dependency, such as strain rate dependency, creep and stress relaxation, but also the strain softening behavior of geologic materials. Due to the fact that this model is based on a previous work by Oka and Adachi (1985), most of the parameters involved in the model can be determined in same way as those in the previous work. The new parameters introduced in this model can be determined by triaxial and drained creep tests.

REFERENCES

Adachi, T., Oka, F., and Poorooshasb, H. B., 1990, "A Constitutive Model for Frozen Sand," Trans. of ASME, Vol. 112, pp. 208-212.

Adachi, T., Zhang, F., Oka, F. and Yashima, A., 1991, "A FEM analysis of strain localization using a non-local strain-softening plasticity," Proc. 3th Int. Conf. on Constitutive Laws for Engineering Materials, Theory and Application, Tucson, USA, 83-94.

Andersland, O. B., and Al-Nouri, I. : 1970, "Time Dependent Strength Behavior of Frozen Soils," Journal of Mechanics and Foundation Division, ASCE, 96SM(4), pp. 1249-1265.

Oka, F., 1985, "Elasto/viscoplastic constitutive equation with memory and internal variables," Computer and Geomechanics, 1, pp. 59-69.

Perzyna, P., 1980, "Memory Effects and Internal Changes of a Material," International Journal of Nonlinear mechanics, Vol. 6, pp. 707-716.

Singh, A., and Mitchell, J. K., 1968, "General stress-strain-time function for soils, Proc. ASCE, Vol. 94, No.SM1, pp. 21-46.

Ting, J. M., 1983, "Tertiary Creep Model for Frozen Sands," ASCE Journal of Geotechnical Engineering, 109(7), pp. 932-945.

Valanis, K. C., 1971, " A Theory of Viscoplasticity without a Yielding Surface," Arch. Mech. Stos. Vol. 23, No. 4, pp. 517-533.

Valanis, K. C., 1985, "On the uniqueness of solution of the initial value problem in softening materials," J. Appl. Mech. 52, 649-653

Wang, C. C., 1969, "Generalized Simple Bodies," Arch. Rational Mech. Analysis, Vol. 32, pp. 1-30.

Numerical Models in Geomechanics, Pietruszczak & Pande (eds) © 1997 Balkema, Rotterdam, ISBN 90 5410 886 X

On a general flow rule

K. Hashiguchi & T. Okayasu
Kyushu University, Fukuoka, Japan

S. Sakajo
Kiso-jiban Consultants Co., Ltd, Tokyo, Japan

ABSTRACT: Keeping a single and smooth (regular) yield surface for the steady development of elastoplasticity, an extended flow rule describing the dependencies of the magnitude and the direction of plastic stretching on those of stress rate is proposed by incorporating an additional term of the stretching tensor in degree zero into the associated flow rule. The elastoplastic constitutive equation with this flow rule is thought to be a pertinent one which fulfills the mechanical requirements, i.e. the rate-nonlinearity, the continuity condition and the work rate-stiffness relaxation and which is applicable to an arbitrary loading process including unloading, reloading and reverse loading.

1 INTRODUCTION

The direction of the plastic stretching does not depend on the stress rate in the traditional elastoplastic constitutive equation with a single and smooth plastic potential surface. In addition, the magnitude of the plastic stretching does not depend on the stress rate component tangential to the yield surface but depends only on the component normal to that surface in the traditional one. The extension of the constitutive equation so as to describe these dependencies pertinently is one of the most fundamental but unsolved problems in elastoplasticity at present.

In order to describe the irreversible deformation of materials pertinently, one would first have to ascertain the mechanical requirements which have to be fulfilled in constitutive equations concerning irreversible deformation and then one would need to formulate a pertinent flow rule by taking account of these requirements. An elastoplastic constitutive equation with an arbitrary flow rule is formulated, which fulfills the *continuity condition* (Hashiguchi 1993a, b) as the mechanical requirement, while a single and smooth (regular) yield surface is kept as a steady development of the elastoplasticity in physical and mathematical aspects. An extended flow rule is then proposed by incorporating an additional term of the stretching tensor in degree zero into the associated flow rule, which describes the dependencies of the magnitude and the direction of plastic stretching on those of stress rate. It fulfills the *work rate-stiffness relaxation* (Hashiguchi 1993a) which is one of the mechanical requirements for elastoplastic constitutive equations. The constitutive equation with this flow rule is thought to be a pertinent one applicable to an arbitrary loading process including unloading, reloading and reverse loading processes for plastically compressible/incompressible and pressure-dependent/independent materials by selecting material functions appropriately. Based on this equation, the constitutive equation of metals with the von Mises yield condition obeying the isotropic-kinematic hardening is formulated and its mechanical response is examined by a numerical calculation of the stress rate response to the stretching the direction of which rotates on the principal deviatoric stretching plane.

2 MECHANICAL REQUIREMENTS

2.1 *Reversible/irreversible response: Rate-linearity/nonlinearity*

Let stress rates induced by stretchings (rates of deformation, i.e. a symmetric parts of velocity gradient) $\boldsymbol{D}$ and $\boldsymbol{D}'$ which have the same magnitude but opposite directions to each other be denoted as $\overset{\circ}{\boldsymbol{\sigma}}$ and $\overset{\circ}{\boldsymbol{\sigma}}'$, respectively. It is defined that $\boldsymbol{D}$ and $\boldsymbol{D}'$ are reversible if $\overset{\circ}{\boldsymbol{\sigma}}$ and $\overset{\circ}{\boldsymbol{\sigma}}'$ have the same magnitude and opposite directions to each other and that $\boldsymbol{D}$ or $\boldsymbol{D}'$ is irreversible if not so. That is,

$$\left.\begin{array}{l} \boldsymbol{D} \text{ and } \boldsymbol{D}' \text{ are reversible if } \overset{\circ}{\boldsymbol{\sigma}} = -\overset{\circ}{\boldsymbol{\sigma}}' \text{ for } \boldsymbol{D} = -\boldsymbol{D}', \\ \boldsymbol{D} \text{ or } \boldsymbol{D}' \text{ is irreversible if } \overset{\circ}{\boldsymbol{\sigma}} \neq -\overset{\circ}{\boldsymbol{\sigma}}' \text{ for } \boldsymbol{D} = -\boldsymbol{D}', \end{array}\right\} \quad (1)$$

where $\boldsymbol{\sigma}$ is the stress and $\overset{\circ}{(\ )}$ denotes the proper corotational rate.

While a constitutive equation of time-independent materials is described by a homogeneous function of rate variables, the relation between the stress rate $\overset{\circ}{\boldsymbol{\sigma}}$

and the stretching $\boldsymbol{D}$ has to be *linear* and *nonlinear* in constitutive equations for *reversible* and *irreversible* deformation, respectively, because of eqn (1). That is to say, the *reversible* and the *irreversible* constitutive equation have the mathematical structures of *rate-linearity* and *rate-nonlinearity*, respectively. The traditional elastoplastic constitutive equation with a smooth plastic potential and yield surface has the lowest rate-nonlinearity, i.e. the *rate-bilinearity* of the stress rate-stretching relation by a loading criterion which will be described in 2.4.

2.2 *Continuity and smoothness conditions*

In real materials the stress rate $\overset{\circ}{\boldsymbol{\sigma}}$ will change continuously for a continuous change of the stress state $\boldsymbol{\sigma}$ and the stretching $\boldsymbol{D}$. Then, the following relation, called the *condition of continuity* or *continuity condition* of mechanical response, is required to hold (Hashiguchi 1993a, b).

$$\lim_{\delta \boldsymbol{D} \to \boldsymbol{O}} \overset{\circ}{\boldsymbol{\sigma}}(\boldsymbol{\sigma}, S_i, \boldsymbol{D}+\delta\boldsymbol{D}) = \overset{\circ}{\boldsymbol{\sigma}}(\boldsymbol{\sigma}, S_i, \boldsymbol{D}) \qquad (2)$$

when the response of the stress rate to the stretching in a current state of the stress and plastic internal-state variables is designated as $\overset{\circ}{\boldsymbol{\sigma}}(\boldsymbol{\sigma}, S_i, \boldsymbol{D})$, where S_i ($i = 1, 2, \cdots, m$) denotes collectively scalar- or tensor-valued internal-state variables describing the change of mechanical response due to the irreversible deformation.

Further, the ratio of a stress rate to a stretching will change continuously for a continuous change of a stress state. Then, the following relation, called the *condition of smoothness* or *smoothness condition* of mechanical response, is required to hold.

$$\lim_{\delta\boldsymbol{\sigma} \to \boldsymbol{O}} \frac{\partial \overset{\circ}{\boldsymbol{\sigma}}(\boldsymbol{\sigma}+\delta\boldsymbol{\sigma}, S_i, \boldsymbol{D})}{\partial \boldsymbol{D}} = \frac{\partial \overset{\circ}{\boldsymbol{\sigma}}(\boldsymbol{\sigma}, S_i, \boldsymbol{D})}{\partial \boldsymbol{D}} \qquad (3)$$

A violation of smoothness condition leads to the following disadvantages:

1) A smooth stress-strain curve is not predicted even for proportional-monotonic loading, since the stress rate-stretching ratio, i.e. the inclination of the stress-strain curve changes discontinuously due to a continuous change of the stress state. This is an imperfection for the description of the mechanical behavior of materials.

2) The judgment of "which side of the boundary violating the condition (3) a stress lies on" is required, since the stress rate-stretching ratio changes abruptly at this boundary in the stress space. This creates inconvenience in numerical calculations. For instance, in the conventional elastoplastic constitutive equation assuming that the interior of the yield surface is a purely elastic domain, the stress rate-stretching relation changes abruptly when a stress reaches the yield surface and thus it is necessary to incorporate a special method, e.g. the Euler method , the radial return method or the mean normal method.

In what follows, discussions are limited to the *elastoplasticity* which premises on the additive decomposition of the stretching $\boldsymbol{D}$ into the elastic stretching $\boldsymbol{D}^e$ and the plastic stretching $\boldsymbol{D}^p$, i.e.

$$\boldsymbol{D} = \boldsymbol{D}^e + \boldsymbol{D}^p, \qquad (4)$$

provided that the elastic stretching $\boldsymbol{D}^e$ is linearly related to the stress rate as

$$\boldsymbol{D}^e = \boldsymbol{E}^{-1} \overset{\circ}{\boldsymbol{\sigma}}, \qquad (5)$$

where $\boldsymbol{E}$ is the elastic modulus, a function of the stress and plastic internal-state variables in general, and $(\ )^{-1}$ designates the inverse, while the plastic stretching $\boldsymbol{D}^p$ has to be nonlinearly related to the stress rate as was described in 2.1.

2.3 *Work rate-stiffness relaxation*

It holds for the second-order *work rate* w that

$$w = w^{es} - w^{pr}, \qquad (6)$$

where

$$w \equiv \mathrm{tr}(\overset{\circ}{\boldsymbol{\sigma}}\boldsymbol{D}), \qquad (7)$$

$$w^{es} \equiv \mathrm{tr}(\overset{\circ}{\boldsymbol{\sigma}}{}^e\boldsymbol{D}), \qquad (8)$$

$$w^{pr} \equiv \mathrm{tr}(\overset{\circ}{\boldsymbol{\sigma}}{}^p\boldsymbol{D}), \qquad (9)$$

$$\overset{\circ}{\boldsymbol{\sigma}} = \boldsymbol{E}\boldsymbol{D}^e, \quad \overset{\circ}{\boldsymbol{\sigma}}{}^e \equiv \boldsymbol{E}\boldsymbol{D}, \quad \overset{\circ}{\boldsymbol{\sigma}}{}^p \equiv \boldsymbol{E}\boldsymbol{D}^p. \qquad (10)$$

$\mathrm{tr}(\ )$ stands for the trace. $\overset{\circ}{\boldsymbol{\sigma}}{}^e$ and $\overset{\circ}{\boldsymbol{\sigma}}{}^p$ are called the *elastic stress rate* and the *plastic relaxation stress rate*, respectively. w^{es} and w^{pr} are called the *elastic stress work rate* and the *plastic relaxation work rate*, respectively, in second-order (Hashiguchi 1991, 1993a), while w^{es} and w^{pr} are different from the *elastic work rate* $w^e \equiv \mathrm{tr}(\overset{\circ}{\boldsymbol{\sigma}}\boldsymbol{D}^e)$ and the *plastic work rate* $w^p \equiv \mathrm{tr}(\overset{\circ}{\boldsymbol{\sigma}}\boldsymbol{D}^p)$ in second-order.

Further, it holds for the *stiffness modulus* D_d that

$$D_d = D_d^e - D_d^{pr}, \qquad (11)$$

where

$$D_d \equiv \mathrm{tr}\left(\overset{\circ}{\boldsymbol{\sigma}}\frac{\boldsymbol{D}}{\|\boldsymbol{D}\|}\right) / \|\boldsymbol{D}\| = \frac{w}{\|\boldsymbol{D}\|^2}, \qquad (12)$$

$$D_d^e \equiv \mathrm{tr}\left(\overset{\circ}{\boldsymbol{\sigma}}{}^e\frac{\boldsymbol{D}}{\|\boldsymbol{D}\|}\right) / \|\boldsymbol{D}\| = \frac{w^{es}}{\|\boldsymbol{D}\|^2}, \qquad (13)$$

$$D_d^{pr} \equiv \mathrm{tr}\left(\overset{\circ}{\boldsymbol{\sigma}}{}^p\frac{\boldsymbol{D}}{\|\boldsymbol{D}\|}\right) / \|\boldsymbol{D}\| = \frac{w^{pr}}{\|\boldsymbol{D}\|^2}, \qquad (14)$$

where D_d^e and D_d^{pr} are called the *elastic stiffness modulus* and the *plastic relaxation stiffness modulus*, respectively (Runesson and Mroz 1989, Hashiguchi 1993a), and $\|\ \|$ designates the magnitude.

There may not exist a material in which the work rate w is greater than the elastic stress work rate w^{es} which is calculated supposing that the induced deformation is elastic, nor any material that exhibits a stiffness modulus D_d which is greater than the elastic stiffness modulus D_d^e. Accordingly, it has to hold that

$$w^{pr} \geq 0, \quad D_d^{pr} \geq 0 \qquad (15)$$

which can be reduced to the unique inequality

$$\mathrm{tr}(\boldsymbol{D}^p\boldsymbol{E}\,\boldsymbol{D}) \geq 0\,. \tag{16}$$

The inequality (16) is called the *work rate-stiffness relaxation* (Hashiguchi 1993a). The deformation process in the *stability in the small* $\mathrm{tr}(\overset{\circ}{\boldsymbol{\sigma}}\boldsymbol{D}^p) \geq 0$ fulfills the inequality (17) because of the relation

$$\mathrm{tr}(\boldsymbol{D}^p\boldsymbol{E}\,\boldsymbol{D}) = \mathrm{tr}(\boldsymbol{D}^p\boldsymbol{E}\boldsymbol{D}^p) + \mathrm{tr}(\overset{\circ}{\boldsymbol{\sigma}}\boldsymbol{D}^p), \tag{17}$$

provided that the elastic stiffness modulus $\boldsymbol{E}$ is a positive-definite tensor.

Thus, let it be assumed that the work rate-stiffness relaxation has to be fulfilled in elastoplastic constitutive equations.

2.4. *Loading criterion*

The decomposition (4) of the stretching requires the judgment as to "whether or not a plastic stretching will be produced". The rule for the judgment is called the *loading criterion*.

The loading criterion for the material which has an elastic modulus $\boldsymbol{E}$ determined uniquely by the stress and plastic internal state variables and a single smooth yield surface is given as follows (Hill 1967):

$$\left.\begin{aligned} \boldsymbol{D}^p \neq \boldsymbol{O} &: f = 0 \text{ and } \mathrm{tr}(\boldsymbol{NED}) > 0, \\ \boldsymbol{D}^p = \boldsymbol{O} &: f < 0 \text{ or } \mathrm{tr}(\boldsymbol{NED}) \leq 0, \end{aligned}\right\} \tag{18}$$

where

$$f(\boldsymbol{\sigma}, \boldsymbol{H}_i) = 0 \tag{19}$$

represents a yield condition, $\boldsymbol{H}_i$ ($i = 1, 2, \cdots, n$) denoting collectively scalar- or tensor-valued plastic internal state variables, and $\boldsymbol{N}$ is the normalized outward-normal of the yield surface, i.e.

$$\boldsymbol{N} \equiv \frac{\partial f}{\partial \boldsymbol{\sigma}} \Big/ \left\| \frac{\partial f}{\partial \boldsymbol{\sigma}} \right\|. \tag{20}$$

A mechanical background of the loading criterion (18) was considered by the author (Hashiguchi 1994). An elastoplastic constitutive equation obeying the loading criterion (18) has a stress rate-stretching relation with a bilinearity for a single smooth plastic potential flow rule or a higher-order nonlinearity for a more general flow rule.

A constitutive equation fulfilling the smoothness condition (3), e.g. the subloading surface model (Hashiguchi 1980, 1989), in which the interior of the yield surface is assumed not to be an elastic domain but to be an elastoplastic one, describing always a smooth transition from an elastic to an elastoplastic state, i.e. the *elastic-plastic transition*, does not require the judgment as to whether or not the yield condition is satisfied , i.e. $f = 0$ or $f < 0$ in eqn (18) as was described in 2.2. The loading criterion is simply given for such a constitutive equation as

$$\left.\begin{aligned} \boldsymbol{D}^p \neq \boldsymbol{O} &: \mathrm{tr}(\boldsymbol{NED}) > 0, \\ \boldsymbol{D}^p = \boldsymbol{O} &: \mathrm{tr}(\boldsymbol{NED}) \leq 0. \end{aligned}\right\} \tag{21}$$

The continuity condition (2) can be written as

$$\lim_{\mathrm{tr}(\boldsymbol{NED})\to 0} \boldsymbol{D}^p = \boldsymbol{O}, \text{ i.e. } \lim_{\mathrm{tr}(\boldsymbol{NED})\to 0} \overset{\circ}{\boldsymbol{\sigma}}{}^p = \boldsymbol{O} \tag{22}$$

for the loading-unloading boundary for elastoplastic constitutive equations with a single smooth yield surface (Hashiguchi 1993a, b). The continuity condition is fulfilled at the boundary between the plastic loading and unloading processes in the elastoplastic constitutive equations obeying the consistency condition of a yield surface as will be described in section 3.

3 EXTENDED FLOW RULE

Assume the generalized flow rule

$$\boldsymbol{D}^p = \ll\lambda\gg \boldsymbol{P}\,, \tag{23}$$

where λ is a proportionality factor, a function of the stress, plastic internal state variables and the stress rate or the stretching in degree one, and $\boldsymbol{P}$ is a function of the stress, plastic internal state variables and the stretching in degree zero, i.e.

$$\lambda = \lambda(\boldsymbol{\sigma}, \boldsymbol{H}_i, \overset{\circ}{\boldsymbol{\sigma}}) \text{ or } \lambda(\boldsymbol{\sigma}, \boldsymbol{H}_i, \boldsymbol{D})\,, \tag{24}$$

$$\boldsymbol{P} = \boldsymbol{P}(\boldsymbol{\sigma}, \boldsymbol{H}_i, \boldsymbol{D})\,. \tag{25}$$

The bracket $\ll\ \gg$ designates $\ll a \gg = a$ for $f(\boldsymbol{\sigma}, \boldsymbol{H}_i) = 0$ and $\mathrm{tr}(\boldsymbol{NED}) > 0$ and $\ll a \gg = 0$ for $f(\boldsymbol{\sigma}, \boldsymbol{H}_i) < 0$ or $\mathrm{tr}(\boldsymbol{NED}) \leq 0$, where a is an arbitrary scalar variable. Moreover, $\ll\ \gg$ is replaced by $<\ >$ for constitutive equations fulfilling the smoothness condition, where $<\ >$ designates $<a> = a$ for $\mathrm{tr}(\boldsymbol{NED}) > 0$ and $<a> = 0$ for $\mathrm{tr}(\boldsymbol{NED}) \leq 0$, respectively, in accordance with the loading criterion (18).

The substitution of the flow rule (23) into the consistency condition of the yield condition (19) leads to

$$\boldsymbol{D}^p = \ll \frac{\mathrm{tr}(\boldsymbol{N}\overset{\circ}{\boldsymbol{\sigma}})}{D} \gg \boldsymbol{P}\,, \tag{26}$$

$$D \equiv -\sum_{i=1}^{n} \mathrm{tr}\left(\frac{\partial f}{\partial \boldsymbol{H}_i}\boldsymbol{h}_i\right) \Big/ \left\| \frac{\partial f}{\partial \boldsymbol{\sigma}} \right\|, \tag{27}$$

where $\boldsymbol{h}_i$ is a function of the stress, plastic internal state variables and $\boldsymbol{P}$ in degree one, which is related to $\overset{\circ}{\boldsymbol{H}}_i$ as

$$\overset{\circ}{\boldsymbol{H}}_i = \ll\lambda\gg \boldsymbol{h}_i \tag{28}$$

since $\overset{\circ}{\boldsymbol{H}}_i$ includes $\ll\lambda\gg$ in degree one.

Eqns (4), (5) and (26) lead to

$$\boldsymbol{D} = \boldsymbol{E}^{-1}\overset{\circ}{\boldsymbol{\sigma}} + \ll \frac{\mathrm{tr}(\boldsymbol{N}\overset{\circ}{\boldsymbol{\sigma}})}{D} \gg \boldsymbol{P}\,. \tag{29}$$

Equation (26) is rewritten using (4) and (5) as

$$\boldsymbol{D}^p = \ll \frac{\mathrm{tr}(\boldsymbol{NED})}{D + \mathrm{tr}(\boldsymbol{NEP})} \gg \boldsymbol{P}\,. \tag{30}$$

Obviously, eqn (30) fulfills the continuity condition (22) at the boundary between the plastic loading and unloading processes at which $\mathrm{tr}(\boldsymbol{NED}) = 0$ holds.

The quantity in the bracket $\ll\ \gg$ of eqn (30) has to be positive for the plastic loading process. Thus, $\boldsymbol{P}$ cannot be arbitrary but is required to fulfill the inequality

$$D + \mathrm{tr}(\boldsymbol{NEP}) \geq 0 \quad \text{for } \mathrm{tr}(\boldsymbol{NED}) \geq 0. \quad (31)$$

Furthermore, $\boldsymbol{P}$ has to fulfill the inequality

$$\mathrm{tr}(\boldsymbol{PED}) \geq 0 \quad \text{for } \mathrm{tr}(\boldsymbol{NED}) \geq 0 \quad (32)$$

according to the work rate-stiffness relaxation (16).

Eqns (4), (5) and (30) lead to

$$\overset{\circ}{\boldsymbol{\sigma}} = \boldsymbol{ED} - \boldsymbol{E} \ll \frac{\mathrm{tr}(\boldsymbol{NED})}{D + \mathrm{tr}(\boldsymbol{NEP})} \gg \boldsymbol{P} \quad (33)$$

Now, consider the function $\boldsymbol{P}$ in the flow rule (23).

The simple flow rule

$$\boldsymbol{P} = \boldsymbol{N} + P_t \frac{\boldsymbol{D}}{\|\boldsymbol{D}\|}, \quad (34)$$

where P_t is a material parameter, fulfills obviously the work rate-stiffness relaxation (16) because of

$$\mathrm{tr}(\boldsymbol{PED}) = \mathrm{tr}(\boldsymbol{NED}) + P_t \,\mathrm{tr}(\boldsymbol{DED})/\|\boldsymbol{D}\| \geq 0 \quad \text{for } \mathrm{tr}(\boldsymbol{NED}) \geq 0 \quad (35)$$

provided that $\boldsymbol{E}$ is the positive-definite tensor. It is, however, incapable of describing the plastic incompressibility observed in metals. Then, let $\boldsymbol{P}$ be assumed as follows:

$$\boldsymbol{P} = \boldsymbol{N} + P_t^v \frac{D_v}{\|\boldsymbol{D}^*\|}\boldsymbol{I} + P_t^* \frac{\boldsymbol{D}^*}{\|\boldsymbol{D}^*\|}, \quad (36)$$

or

$$\boldsymbol{P} = \boldsymbol{N}\|\boldsymbol{D}^*\| + P_t^v D_v \boldsymbol{I} + P_t^* \boldsymbol{D}^*, \quad (36)'$$

where

$$D_v \equiv \mathrm{tr}\boldsymbol{D}, \quad \boldsymbol{D}^* \equiv \boldsymbol{D} - \tfrac{1}{3} D_v \boldsymbol{I}. \quad (37)$$

P_t^v and P_t^* are material parameters, and $\boldsymbol{I}$ stands for the unit tensor. The stretching term in eqn (36) is decomposed into the part concerning the ratio of the volumetric stretching to the magnitude of the deviatoric stretching and the part concerning the direction of the deviatoric stretching. Thus, the flow rule (36) with $\mathrm{tr}\boldsymbol{N} = 0$ and $P_t^v = 0$ fulfills $\mathrm{tr}\boldsymbol{P} = 0$, i.e. $\mathrm{tr}\boldsymbol{D}^p = 0$, resulting in plastic incompressibility. Although these parts involve $\|\boldsymbol{D}^*\|$ in their denominators, the plastic stretching $\boldsymbol{D}^p$ is not singular in the state $\boldsymbol{D}^* = \boldsymbol{O}$ since the tensor $\boldsymbol{P}/(D + \mathrm{tr}(\boldsymbol{NEP}))$ in eqn (30) is regular in this state. In fact, the direction of plastic stretching is given by $\boldsymbol{I}$ in the state $\boldsymbol{D}^* = \boldsymbol{O}$ and $D_v \neq 0$.

The relation between the plastic stretching and the stress rate is obtained by substituting $\boldsymbol{D} = \boldsymbol{E}^{-1}\overset{\circ}{\boldsymbol{\sigma}} + \boldsymbol{D}^p$ into eqn (30) incorporating eqn (36). Obviously, not only the magnitude but also the direction of plastic stretching depend on those of stress rate.

The plastic modulus D for the flow rule (36) includes the stretching in degree zero but does not include a stress rate so that the stress rate is expressed in terms of the stretching as in eqn (33). However, the inverse relation of eqn (33), i.e. the expression of the stretching in terms of the stress rate cannot be obtained in general, while the right-hand side in eqn (29) includes the stretching.

In what follows, let the elastic property be given by an equation of the HOOKE's type

$$E_{ijkl} = L\delta_{ij}\delta_{kl} + G(\delta_{ik}\delta_{jl} + \delta_{il}\delta_{jk}), \quad (37)$$

where L and G are the material parameters corresponding to the Lame's constant and the elastic shear modulus, respectively, for the elastic body, and δ_{ij} is the KRONECKER's delta.

It holds for eqns (36) and (37) that

$$\mathrm{tr}(\boldsymbol{PED}) = \mathrm{tr}(\boldsymbol{NED}) + (3L + 2G)P_t^v \frac{D_v^2}{\|\boldsymbol{D}^*\|} + 2G P_t^* \|\boldsymbol{D}^*\| \geq 0 \quad \text{for } \mathrm{tr}(\boldsymbol{NED}) \geq 0 \quad (38)$$

Therefore, the constitutive equation (33) with the flow rule (36) and the Hooke's type of elastic property (37) fulfills the continuity condition (22) and the work rate-stiffness relaxation (16). It is desirable that eqn (33) is jointed with a constitutive model fulfilling the smoothness condition, e.g. the subloading surface model (Hashiguchi 1989), especially for cyclic loading behavior as was described by the author (Hashiguchi 1993b).

In the case of elastic property (37) in addition to $\mathrm{tr}\,\boldsymbol{N} = 0$ and $P_t^v = 0$ it holds that

$$\left.\begin{aligned} \mathrm{tr}(\boldsymbol{NED}) &= 2G\,\mathrm{tr}(\boldsymbol{ND}) = 2G\,\mathrm{tr}(\boldsymbol{ND}^*), \\ \mathrm{tr}(\boldsymbol{NEP}) &= 2G\{1 + P_t^*\mathrm{tr}(\boldsymbol{ND}^*)/\|\boldsymbol{D}^*\|\} \\ &= 2G + P_t^*\mathrm{tr}(\boldsymbol{NED})/\|\boldsymbol{D}^*\|, \end{aligned}\right\} \quad (39)$$

and the plastic modulus D depends only on the deviatoric stretching $\boldsymbol{D}^*$ as a rate variable. Therefore, the plastic stretching (30) with the flow rule (36) depends only on $\boldsymbol{D}^*$ as a rate variable and thus it becomes independent of the mean stress rate by the relation $\boldsymbol{D}^* = \overset{\circ}{\boldsymbol{\sigma}}{}^*/(2G) + \boldsymbol{D}^p$ ($\overset{\circ}{\boldsymbol{\sigma}}{}^*$: deviatoric stress rate), resulting in plastic pressure-independence in addition to the plastic incompressibility mentioned earlier. Furthermore, it holds that

$$\mathrm{tr}(\boldsymbol{NEP}) > 0 \quad \text{for } \mathrm{tr}(\boldsymbol{NED}) > 0 \quad (40)$$

because of eqn $(39)_2$.

4 CONSTITUTIVE EQUATION OF METALS

Based on eqn (33) with the flow rule (36) in section 3, let a constitutive equation of metals be formulated in this section.

Now, adopt the von Mises yield condition with the isotropic-kinematic hardening:

$$f(\hat{\boldsymbol{\sigma}}) - F(H) = 0, \quad (41)$$

where

$$f(\hat{\boldsymbol{\sigma}}) = \sqrt{\tfrac{3}{2}}\|\hat{\boldsymbol{\sigma}}^*\|, \quad (42)$$

$$\overset{\circ}{\boldsymbol{\alpha}} = K_1 \hat{\boldsymbol{\sigma}} - K_2 \boldsymbol{\alpha}, \quad (43)$$

$$K_1 = k_1 \|\boldsymbol{D}^p\|, \quad K_2 = k_2 \|\boldsymbol{D}^p\|, \quad (44)$$

$$F = F_0[1 + h_1\{1 - \exp(-h_2 H)\}], \quad (45)$$

$$\dot{H} = \sqrt{\tfrac{2}{3}}|\boldsymbol{D}^p| \,. \tag{46}$$

$\hat{\boldsymbol{\sigma}}$ and $\hat{\boldsymbol{\sigma}}^*$ stand for

$$\hat{\boldsymbol{\sigma}} \equiv \boldsymbol{\sigma} - \boldsymbol{\alpha} \,, \tag{47}$$

$$\hat{\sigma}_m \equiv \tfrac{1}{3}\mathrm{tr}\,\hat{\boldsymbol{\sigma}}, \;\; \hat{\boldsymbol{\sigma}}^* \equiv \hat{\boldsymbol{\sigma}} - \hat{\sigma}_m \boldsymbol{I} \,. \tag{48}$$

k_1 and k_2 are material constants, while K_1 and K_2 are generally scalar functions of the plastic stretching in degree one, the stress and plastic internal-state variables. F_0 is the initial value of F, and h_1 and h_2 are material constants.

Substituting the flow rule (23) into the consistency condition of yield condition (41), the plastic modulus D is given as

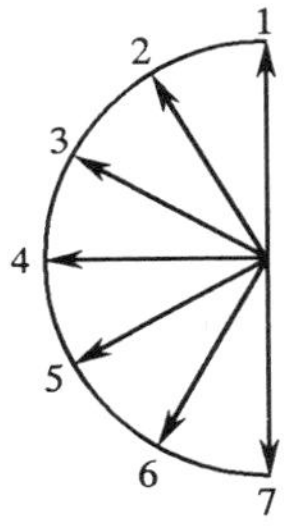

(a) Input of stretching

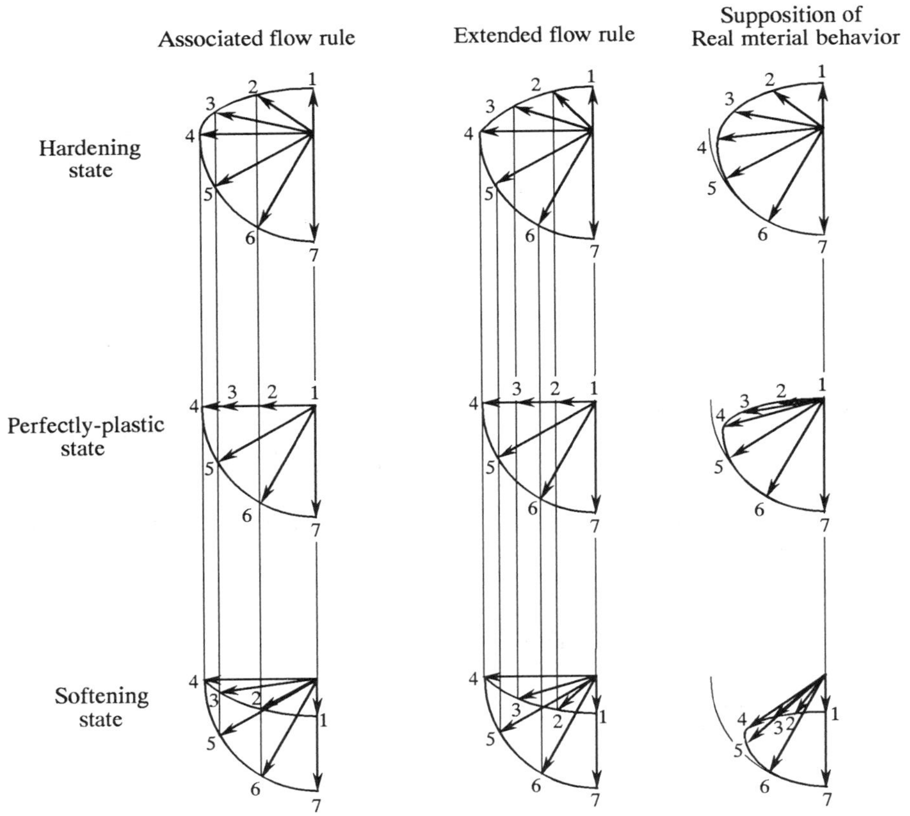

(b) Response of stress rate

Fig. 1 Mechanical response of flow rules and a supposed real material behavior

$$D = \mathrm{tr}(\boldsymbol{N}\boldsymbol{a}) + F'h \Big/ \left\| \frac{\partial f}{\partial \boldsymbol{\sigma}} \right\| \quad (>0)\,, \tag{49}$$

where

$$\boldsymbol{N} = \frac{\hat{\boldsymbol{\sigma}}^*}{\|\hat{\boldsymbol{\sigma}}^*\|}\,, \tag{50}$$

$$\boldsymbol{a} \equiv \mathring{\boldsymbol{\alpha}} \,/ \ll\lambda\gg = (k_1\hat{\boldsymbol{\sigma}} - k_2\boldsymbol{\alpha})\|\boldsymbol{P}\|\,, \tag{51}$$

$$F' \equiv \frac{dF}{dH} = F_0\, h_1\, h_2 \exp(-h_2 H) \quad (>0)\,, \tag{52}$$

$$h \equiv \dot{H} \,/ \ll\lambda\gg = \sqrt{\tfrac{2}{3}}\|\boldsymbol{P}\|\,, \tag{53}$$

$$\left\| \frac{\partial f}{\partial \boldsymbol{\sigma}} \right\| = \sqrt{\tfrac{2}{3}}\,. \tag{54}$$

It holds that $D > 0$ (hardening $\mathrm{tr}(\boldsymbol{N}\mathring{\boldsymbol{\sigma}}) > 0$) because of $\mathrm{tr}(\boldsymbol{N}\boldsymbol{a}) \equiv \mathrm{tr}(\boldsymbol{N}\mathring{\boldsymbol{\alpha}})/\ll\lambda\gg > 0$, $F' > 0$ and $h > 0$. This fact plus eqn (40) lead to the fulfillment of the inequality (31). The magnitude of the plastic stretching is given from eqn (26) for eqn (27) with eqns (51) and (53) as

$$\boldsymbol{D}^p = \ll \frac{\mathrm{tr}(\boldsymbol{N}\mathring{\boldsymbol{\sigma}})}{\mathrm{tr}\{\boldsymbol{N}(k_1\hat{\boldsymbol{\sigma}} - k_2\boldsymbol{\alpha})\} + F'} \gg \frac{\boldsymbol{P}}{\|\boldsymbol{P}\|}\,. \tag{55}$$

Therefore, the magnitude of the plastic stretching is same as that in the case of the associated flow rule $\boldsymbol{P} = \boldsymbol{N}$, i.e. the J_2-flow rule since the evolution rules of the hardening variables $\boldsymbol{\alpha}$ and H depend only on $\|\boldsymbol{D}^p\|$.

Furthermore, putting $P_t^v = 0$ in eqn (36), let $\boldsymbol{P}$ be given as

$$\boldsymbol{P} = \boldsymbol{N} + P_t^* \frac{\boldsymbol{D}^*}{\|\boldsymbol{D}^*\|}\,, \tag{56}$$

and let the elastic property be given by eqn (37) of the HOOKE's type.

5 MECHANICAL RESPONSE

The responses of the stress rates on the principal deviatoric stress rate plane to the stretchings on the principal deviatoric stretching plane are depicted in Fig. 1. The response for the associated flow rule is shown in the left of Fig.1(b). The response for the elastoplastic constitutive equation (33) with the extended flow rule (36) is shown in the middle of Fig.1(b), which describes the dependencies of the magnitude and the direction of plastic stretching on those of stress rate The extended flow rule keeps a single smooth (regular) yield surface without incorporating plural yield surfaces or a corner of the yield surface for the steady development of elastoplasticity. Furthermore, the constitutive equation may be applicable to the prediction of the deformation for the general loading process including cyclic loading for plastically compressible/incompressible and pressure-dependent/independent materials by incorporating the subloading surface model (Hashiguchi 1989) which fulfills the smoothness condition. In addition, it may be the simplest flow rule to describe the dependencies, fulfilling the mechanical requirements, i.e. the rate-nonlinearity, the continuity condition and the work rate-stiffness relaxation. Needless to say, it results in a high-order nonlinearity of the stress rate-stretching relation. Therefore, the inverse expression between the stress rate and the stretching does not hold. Whereas, the stress rate is expressed in terms of the stretching. This feature is convenient for ordinary finite element programming based on the displacement method.

It should be, however, noted that a plastic stretching cannot be induced by the stress rate tangential to the yield surface in order to fulfill the continuity condition as far as a single and smooth yield surface is assumed. Further, the peculiar response envelope of the stress rate in the crescent -shape is predicted for the softening state by the associated flow rule. The response envelope for real materials is supposed to be more smooth and have a convex shape as shown in the right of Fig. 1 (b). It is desirable to find the fact by experiments and formulate a more general flow rule reflecting the fact.

REFERENCES

Hashiguchi, K. 1980. Constitutive equations of elastoplastic materials with elastic-plastic transition. *J. Appl. Mech.* (ASME) 47: 266-272.

Hashiguchi, K. 1989. Subloading surface model in unconventional plasticity. *Int. J. Solids Struct.* 25: 917-945.

Hashiguchi, K. 1993a. Fundamental requirements and formulation of elastoplastic constitutive equations with tangential plasticity. *Int. J. Plasticity* 9: 525-549.

Hashiguchi, K. 1993b. Mechanical requirements and and structures of cyclic plasticity models. *Int. J. Plasticity*. 9: 721-748.

Hashiguchi, K. 1994. On the loading criterion. *Int. J. Plasticity*. 10: 871-878.

Hill, R. 1967. On the classical constitutive relations for elastic/plastic solids. *Recent Progress in Appl. Mech.* (*The Folke Odqvist Volume*), John Wiley Sons. 241-249.

Runesson, K. and Mroz, Z. 1989. A note on non-associated flow rules. *Int. J. Plasticity*. 5: 639-658.

Numerical Models in Geomechanics, Pietruszczak & Pande (eds) © 1997 Balkema, Rotterdam, ISBN 90 5410 886 X

Cyclic elastoplastic constitutive equation of cohesive and noncohesive soils

K. Hashiguchi, Z.-P. Chen & S. Tsutsumi
Kyushu University, Fukuoka, Japan

S. Sakajo
Kiso-jiban Consultants Co., Ltd, Tokyo, Japan

ABSTRACT: The subloading surface model fulfills the mechanical requirements for constitutive equations, i.e. the continuity condition, the smoothness condition and the work rate stiffness relaxation and describes pertinently the Masing effect. The constitutive equation of soils is formulated by introducing the subloading surface model and formulating the evolutional rule of rotational hardening for the description of anisotropy. The applicability of the constitutive equation to the prediction of real soil deformation behavior is verified by predicting monotonic and cyclic loading behavior of sands under drained and undrained conditions and comparing them with test data.

1 INTRODUCTION

Among the existing plasticity models the subloading surface model (Hashiguchi, 1989) fulfills the fundamental requirements for constitutive equations of materials, i.e. the *continuity condition*, the *smoothness condition* and the *work rate-stiffness relaxation*, obeying the associated flow rule, and describes pertinently the *Masing effect* which is required to predict cyclic loading behavior. On the other hand, Sekiguchi and Ohta (1977) proposed a simple description of an inherent anisotropy of K_0 consolidated soils by assuming a yield surface rotated around the origin of stress space. Hashiguchi (1977) called it a *rotational hardening* and studied its evolutional rule. In this article, elastoplastic constitutive equation of soils ranging from clays to sands is formulated by incorporating the concept of the subloading surface and formulating the evolutional rule of the rotational hardening for the description of inherent and induced anisotropy. Its adequacy is examined by predicting monotonic and cyclic loading behavior of sands under drained and undrained conditions and comparing them with experimental data. The signs of a stress (rate) and a stretching (a symmetric part of velocity gradient) are positive for tension, and the stress stands for the effective stress, i.e. the stress excluded a pore pressure from a total stress throughout this article.

2 SUBLOADING SURFACE MODEL WITH ROTATIONAL HARDENING

The stretching $\boldsymbol{D}$ (symmetric part of velocity gradient) is additively decomposed into the elastic stretching $\boldsymbol{D}^e$ and the plastic stretching $\boldsymbol{D}^p$ as usual, i.e.

$$\boldsymbol{D} = \boldsymbol{D}^e + \boldsymbol{D}^p, \tag{1}$$

where the elastic stretching is given by

$$\boldsymbol{D}^e = \boldsymbol{E}^{-1}\overset{\circ}{\boldsymbol{\sigma}}. \tag{2}$$

$\boldsymbol{\sigma}$ is a stress and $\overset{\circ}{(\ )}$ indicates the corotational rate. $\boldsymbol{E}$ is the elastic modulus given in the Hooke's type as

$$E_{ijkl} = (K - \tfrac{2}{3}G)\delta_{ij}\delta_{kl} + G\,(\delta_{ik}\delta_{jl} + \delta_{il}\delta_{jk}), \tag{3}$$

where K and G are the bulk modulus and the shear modulus, respectively.

Let the normal-yield surface which passes through the origin of stress space and obeys isotropic and rotational hardening be described as

$$f(\hat{p}, \hat{\chi}) = F(H), \tag{4}$$

where, letting the stress on the normal-yield surface be denoted as $\hat{\boldsymbol{\sigma}}$,

$$\hat{\sigma}_m \equiv \tfrac{1}{3}\mathrm{tr}\,\hat{\boldsymbol{\sigma}},\quad \hat{p} \equiv -\hat{\sigma}_m,\quad \hat{\boldsymbol{\sigma}}^* \equiv \hat{\boldsymbol{\sigma}} + \hat{p}\boldsymbol{I}, \tag{5}$$

$$\hat{\boldsymbol{\eta}} \equiv \hat{\boldsymbol{Q}} - \boldsymbol{\beta},\quad \hat{\boldsymbol{Q}} \equiv \frac{\hat{\boldsymbol{\sigma}}^*}{\hat{p}}, \tag{6}$$

$$\hat{\chi} \equiv \frac{\|\hat{\boldsymbol{\eta}}\|}{\hat{m}}. \tag{7}$$

$\hat{m}$ is a function of

$$\sin 3\hat{\theta}_\sigma \equiv -\sqrt{6}\frac{\mathrm{tr}\,\hat{\boldsymbol{\eta}}^3}{\|\hat{\boldsymbol{\eta}}\|^3}, \tag{8}$$

including the material constant ϕ, i.e.

$$\hat{m} = f_m(\sin 3\hat{\theta}_\sigma; \phi), \tag{9}$$

$\| \|$ denoting a magnitude. H is the isotropic hardening variable. The central axis of the normal-yield surface is described as $\boldsymbol{\sigma}^*/p = \boldsymbol{\beta}$, where

$$\sigma_m \equiv \frac{1}{3}\mathrm{tr}\,\boldsymbol{\sigma},\ \ p \equiv -\sigma_m,\ \ \boldsymbol{\sigma}^* \equiv \boldsymbol{\sigma} + p\boldsymbol{I}. \quad (10)$$

Let $\boldsymbol{\beta}$ be called the *rotational hardening variable*. The equation $\|\hat{\boldsymbol{\eta}}\| = \hat{m}$ describes a conical surface whose tip exists at the origin and central axis coincides with the central axis $\boldsymbol{\sigma}^*/p = \boldsymbol{\beta}$ of the normal-yield surface, the radius varying with the variable $\hat{\theta}_\sigma$, in the principal stress space.

Hereinafter, it is assumed that the normal-yield surface keeps a similarity to itself. Then, f is a homogeneous function of $\hat{\boldsymbol{\sigma}}$, satisfying Euler's theorem for a homogeneous function. Therefore, by selecting the function f to be homogeneous degree one, the following expression holds, while $\hat{\chi}$ is a dimensionless variable.

$$f(\hat{p}, \hat{\chi}) = \hat{p}g(\hat{\chi}). \quad (11)$$

The parameter $\hat{\boldsymbol{\eta}}$ was introduced by Sekiguchi and Ohta (1977) in order to describe the rotation of yield surface concisely.

Then, let the evolutional rule of $\boldsymbol{\beta}$ be assumed as

$$\overset{\circ}{\boldsymbol{\beta}} = b_r \|\boldsymbol{D}^{p*}\| \|\hat{\boldsymbol{\eta}}\| \hat{\boldsymbol{\eta}}_b, \quad (12)$$

where b_r is the material constant and

$$D_v^p \equiv \mathrm{tr}\,\boldsymbol{D}^p,\ \ \boldsymbol{D}^{p*} \equiv \boldsymbol{D}^p - \frac{1}{3}D_v^p\boldsymbol{I}, \quad (13)$$

$$\hat{\boldsymbol{\eta}}_b \equiv \hat{m}_b\hat{\boldsymbol{t}} - \boldsymbol{\beta}, \quad (14)$$

$$\hat{\boldsymbol{t}} \equiv \frac{\hat{\boldsymbol{\eta}}}{\|\hat{\boldsymbol{\eta}}\|}. \quad (15)$$

Now, the subloading surface is given by the similarity to the normal-yield surface (4) as

$$f(\bar{p}, \bar{\chi}) = RF(H), \quad (16)$$

where

$$\bar{\boldsymbol{\sigma}} \equiv \boldsymbol{\sigma} - \bar{\boldsymbol{\alpha}}, \quad (17)$$

$$\bar{p} \equiv -\frac{1}{3}\mathrm{tr}\,\bar{\boldsymbol{\sigma}},\ \ \bar{\boldsymbol{\sigma}}^* \equiv \bar{\boldsymbol{\sigma}} + \bar{p}\boldsymbol{I}, \quad (18)$$

$$\bar{\boldsymbol{\eta}} \equiv \bar{\boldsymbol{Q}} - \boldsymbol{\beta},\ \ \bar{\boldsymbol{Q}} \equiv \frac{\bar{\boldsymbol{\sigma}}^*}{\bar{p}}, \quad (19)$$

$$\bar{\chi} \equiv \frac{\|\bar{\boldsymbol{\eta}}\|}{\bar{m}}. \quad (20)$$

$\bar{m}$ is a function f_m of

$$\sin 3\bar{\theta}_\sigma \equiv -\sqrt{6}\frac{\mathrm{tr}\,\bar{\boldsymbol{\eta}}^3}{\|\bar{\boldsymbol{\eta}}\|^3}, \quad (21)$$

including the material constant ϕ, i.e.

$$\bar{m} = f_m(\sin 3\bar{\theta}_\sigma;\ \phi). \quad (22)$$

Needless to say, $\boldsymbol{\sigma}$ stands for a current stress which exits always on the subloading surface. $\bar{\boldsymbol{\alpha}}$ on the subloading surface is the conjugate point of the null stress on the normal-yield surface. R $(0 \le R \le 1)$ is the ratio of the size of the subloading surface to that of the normal-yield surface. Hereinafter, let $\hat{\boldsymbol{\sigma}}$ be regarded as the conjugate stress on the normal-yield surface for the current stress on the subloading surface. $\boldsymbol{s}$ is the *center of similarity* (or *similarity-center*) of the normal-yield and the subloading surfaces. Besides, the following expression similar to (11) holds.

$$f(\bar{p}, \bar{\chi}) = \bar{p}g(\bar{\chi}). \quad (23)$$

The variable R is calculated from (16) with the substitution of

$$\bar{\boldsymbol{\sigma}} = \boldsymbol{\sigma} - (1 - R)\boldsymbol{s} \quad (24)$$

which holds on the similarity of the normal and subloading surfaces, and then $\bar{\boldsymbol{\alpha}}$ is calculated by the equation $\bar{\boldsymbol{\alpha}} = (1 - R)\boldsymbol{s}$.

The evolutional rule of rotational hardening for the general state is given from equations (12)-(15) as

$$\overset{\circ}{\boldsymbol{\beta}} = b_r \|\boldsymbol{D}^{p*}\| \|\bar{\boldsymbol{\eta}}\| \bar{\boldsymbol{\eta}}_b, \quad (25)$$

where

$$\bar{\boldsymbol{\eta}}_b = \bar{m}_b\bar{\boldsymbol{t}} - \boldsymbol{\beta}, \quad (26)$$

$$\bar{\boldsymbol{t}} \equiv \frac{\bar{\boldsymbol{\eta}}}{\|\bar{\boldsymbol{\eta}}\|}, \quad (27)$$

$$\bar{m}_b = f_m(\sin 3\bar{\theta}_\sigma;\ \phi_b). \quad (28)$$

Let the evolutional rule of the similarity-center $\boldsymbol{s}$ of the normal-yield and the subloading surfaces be formulated below.

The surface which passes through the similarity-center $\boldsymbol{s}$ and is similar to the normal-yield surface with respect to the origin of stress space, called the *similarity-center surface*, is described as

$$f(p_s, \chi_s) = R_s F(H), \quad (29)$$

where R_s $(0 \le R_s \le 1)$ is the ratio of the size of similarity-center surface to that of normal-yield surface. The following expression which is similar to (11) and (23) holds.

$$f(p_s, \chi_s) = p_s g(\chi_s), \quad (30)$$

where

$$p_s \equiv -\frac{1}{3}\mathrm{tr}\,\boldsymbol{s},\ \ \boldsymbol{s}^* \equiv \boldsymbol{s} + p_s\boldsymbol{I}, \quad (31)$$

$$\boldsymbol{\eta}_s \equiv \boldsymbol{Q}_s - \boldsymbol{\beta},\ \ \boldsymbol{Q}_s \equiv \frac{\boldsymbol{s}^*}{p_s}, \quad (32)$$

$$\chi_s \equiv \frac{\|\boldsymbol{\eta}_s\|}{m_s}, \quad (33)$$

m_s is a function f_m of

$$\sin 3\theta_s \equiv -\sqrt{6}\frac{\mathrm{tr}\,\boldsymbol{\eta}_s^3}{\|\boldsymbol{\eta}_s\|^3}, \quad (34)$$

including the material constant ϕ, i.e.

$$m_s = f_m(\sin 3\theta_s;\ \phi). \quad (35)$$

The evolutional rule of the similarity-center is given as follows:

$$\overset{\circ}{\boldsymbol{s}} = c\|\boldsymbol{D}^p\|\tilde{\boldsymbol{\sigma}} + \frac{1}{F}\left\{\dot{F} - \mathrm{tr}\left(\frac{\partial f(p_s, \chi_s)}{\partial \boldsymbol{\beta}}\overset{\circ}{\boldsymbol{\beta}}\right)\right\}\boldsymbol{s}, \quad (36)$$

where c is a material constant and

$$\tilde{\boldsymbol{\sigma}} \equiv \boldsymbol{\sigma} - \boldsymbol{s}. \quad (37)$$

The evolutional rule of R in the plastic loading process is given as

$$\dot{R} = U\,\|\boldsymbol{D}^p\| \quad \text{for } \boldsymbol{D}^p \neq \boldsymbol{O}, \tag{38}$$

where U is the monotonically decreasing function of R satisfying $U = 0$ for $R = 1$, while the following function is used in the later calculation.

$$U = u_1(1/R^{m_1} - 1), \tag{39}$$

where u_1 and m_1 are material constants.

Substituting (38) into (36), the following extended consistency condition for the subloading surface is obtained.

$$\mathrm{tr}\left\{\frac{\partial f(\bar{p}, \bar{\chi})}{\partial \bar{\boldsymbol{\sigma}}}\mathring{\bar{\boldsymbol{\sigma}}}\right\} + \mathrm{tr}\left\{\frac{\partial f(\bar{p}, \bar{\chi})}{\partial \boldsymbol{\beta}}\mathring{\boldsymbol{\beta}}\right\} = R\dot{F} + U\,\|\boldsymbol{D}^p\|F, \tag{40}$$

where $\mathring{\bar{\boldsymbol{\sigma}}}$ is given from (24) as

$$\mathring{\bar{\boldsymbol{\sigma}}} = \mathring{\boldsymbol{\sigma}} - (1-R)\mathring{\boldsymbol{s}} + \dot{R}\boldsymbol{s}. \tag{41}$$

The associated flow rule is adopted:

$$\boldsymbol{D}^p = \lambda\bar{\boldsymbol{N}} \quad (\lambda > 0), \tag{42}$$

where λ is the positive proportionality factor, and the second-order tensor $\bar{\boldsymbol{N}}$ is the normalized outward-normal of the subloading surface.

Substituting (25), (41) and (42) into (40) and further the result into (1) and (2), λ is obtained as

$$\lambda = \frac{\mathrm{tr}(\bar{\boldsymbol{N}}\mathring{\boldsymbol{\sigma}})}{D_p} = \frac{\mathrm{tr}(\bar{\boldsymbol{N}}\boldsymbol{E}\boldsymbol{D})}{D_p + \mathrm{tr}(\bar{\boldsymbol{N}}\boldsymbol{E}\bar{\boldsymbol{N}})}, \tag{43}$$

where

$$D_p \equiv \mathrm{tr}(\bar{\boldsymbol{N}}\bar{\boldsymbol{a}}) + \mathrm{tr}(\bar{\boldsymbol{N}}\bar{\boldsymbol{\sigma}})\left\{\frac{F'}{F}h - \frac{1}{RF}\mathrm{tr}\left(\frac{\partial f(\bar{p}, \bar{\chi})}{\partial \boldsymbol{\beta}}\boldsymbol{b}\right) + \frac{U}{R}\right\}, \tag{44}$$

$$\bar{\boldsymbol{a}} \equiv \frac{\mathring{\bar{\boldsymbol{\alpha}}}}{\lambda} = (1-R)\boldsymbol{z} - U\boldsymbol{s}, \tag{45}$$

$$F' \equiv \frac{dF}{dH}, \quad h \equiv \frac{\dot{H}}{\lambda} \quad (\dot{F} = F'\lambda h), \tag{46}$$

$$\boldsymbol{b} \equiv \frac{\mathring{\boldsymbol{\beta}}}{\lambda} = b_r\|\bar{\boldsymbol{N}}^*\|\|\boldsymbol{\eta}\|\bar{\boldsymbol{\eta}}_b, \tag{47}$$

$$\boldsymbol{z} \equiv \frac{\mathring{\boldsymbol{s}}}{\lambda} = c\tilde{\boldsymbol{\sigma}} + \frac{1}{F}\left\{F'h - \mathrm{tr}\left(\frac{\partial f(p_s, \chi_s)}{\partial \boldsymbol{\beta}}\boldsymbol{b}\right)\right\}\boldsymbol{s}, \tag{48}$$

$$\bar{\boldsymbol{N}}^* \equiv \bar{\boldsymbol{N}} - \tfrac{1}{3}(\mathrm{tr}\bar{\boldsymbol{N}})\boldsymbol{I}. \tag{49}$$

The loading criterion is given by

$$\left.\begin{aligned} \boldsymbol{D}^p &\neq \boldsymbol{O} : \mathrm{tr}(\bar{\boldsymbol{N}}\boldsymbol{E}\boldsymbol{D}) > 0, \\ \boldsymbol{D}^p &= \boldsymbol{O} : \mathrm{tr}(\bar{\boldsymbol{N}}\boldsymbol{E}\boldsymbol{D}) \leq 0. \end{aligned}\right\} \tag{50}$$

3 MATERIAL FUNCTIONS

Let the following material functions be adopted.

$$g(\bar{\chi}) = 1 + \bar{\chi}^2, \quad g(\chi_s) = 1 + \chi_s^2, \tag{51}$$

Further, let the following equations be also assumed.

$$\left.\begin{aligned} \bar{m} &= \frac{2\sqrt{6}\sin\phi}{3(1.1 - 0.1\sin^2 3\bar{\theta}_\sigma) - \sin\phi\sin 3\bar{\theta}_\sigma}, \\ m_s &= \frac{2\sqrt{6}\sin\phi}{3(1.1 - 0.1\sin^2 3\theta_s) - \sin\phi\sin 3\theta_s}, \\ \bar{m}_b &= \frac{2\sqrt{6}\sin\phi_b}{3(1.1 - 0.1\sin^2 3\bar{\theta}_\sigma) - \sin\phi_b\sin 3\bar{\theta}_\sigma}. \end{aligned}\right\} \tag{52}$$

Let the isotropic hardening/softening function be given as

$$F = (F_0 + P_i)\exp\left(\frac{H}{\rho - \gamma}\right) - P_i, \tag{53}$$

$$\dot{H} = -D_v^p + D_s^p, \tag{54}$$

where

$$D_s^p \equiv \mu\|\boldsymbol{D}^{p*}\|\left(\frac{\|\boldsymbol{\sigma}^*\|}{p} - m_d\right) \tag{55}$$

and μ is a material constant. F_0 is the initial value of F. P_i is the material constant, while $-P_i$ is a negative pressure for which a volume becomes infinite. ρ and γ are the material constants describing the slopes of the normal consolidation curve and the swelling curve, respectively, in the ($\ln p$, $\ln v$) space.

$$m_d = f_m(\theta_\sigma;\phi_d) = \frac{2\sqrt{6}\sin\phi_d}{3\{1 + a(1 - \sin^2 3\theta_\sigma)\} - \sin\phi_d\sin 3\theta_\sigma}, \tag{56}$$

where ϕ_d is the material constant and

$$\sin 3\theta_\sigma \equiv -\sqrt{6}\frac{\mathrm{tr}\,\boldsymbol{\sigma}^{*3}}{\|\boldsymbol{\sigma}^*\|^3}. \tag{57}$$

From (53)-(55) it holds for (46) that

$$F' = \frac{F + P_i}{\rho - \gamma}, \quad h = -\mathrm{tr}\bar{\boldsymbol{N}} + \mu\|\bar{\boldsymbol{N}}^*\|\left(\frac{\|\boldsymbol{\sigma}^*\|}{p} - m_d\right) \tag{58}$$

Let the elastic bulk modulus K be given as

$$K = \frac{p + P_i}{\gamma}. \tag{59}$$

4 COMPARISONS WITH EXPERIMENTS

4.1 *Drained tests*

The material used for prediction is the Hostun sand in the dense state with the initial void ratio e_0 = 0.616 under the initial pressure p_0 = 100 kPa, using the true triaxial (cubical) test apparatus (CTA) and the hollow cylinder test apparatus (HCTA) (Saada and Bianchi, 1989).

The following material constants and initial values are used in all calculations and all the calculations are done from the same initial isotropic stress state $\boldsymbol{\sigma}_0 = -100\boldsymbol{I}$ kPa.

material constants:

yield surface (ellipsoid) shape $\phi = 27°$

hardening/softening:

isotropic { volumetric $\rho = 0.008$, $p_i = 10$ kPa; deviatoric $\mu = 0.6$, $\phi_d = 25°$ }

rotational $b_r = 110$, $\phi_b = 26°$

evolution of R $u_1 = 1.5$, $m_1 = 3.8$

movement of similarity-center $c = 20$

elastic constants $\gamma = 0.003$, $G = 200{,}000$ kPa

initial values:

$F_0 = 400$ kPa, $\boldsymbol{\beta}_0 = \boldsymbol{O}$, $s_0 = -50\boldsymbol{I}$ kPa,

$\boldsymbol{\sigma}_0 = -100\boldsymbol{I}$ kPa

where the subscript $_0$ stands for initial values.

The test data and the calculated results are depicted by the dashed and the solid curves, respectively.

(a) Cyclic isotropic loading (CTA)

The cyclic isotropic loading is shown in Fig. 1.

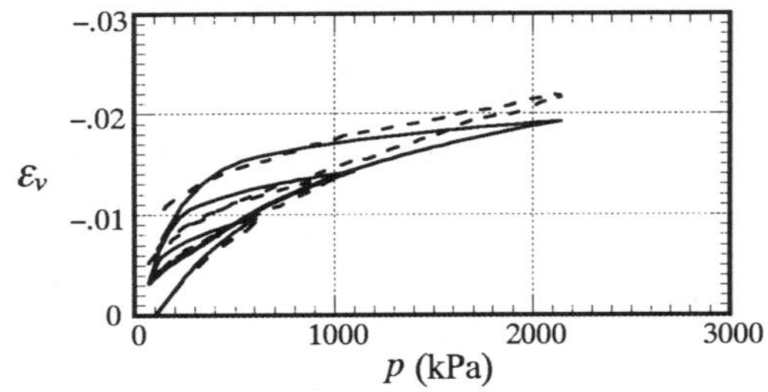

Fig. 1 Cyclic isotropic loading.

(b) Axisymmetric compression with a constant lateral stress (CTA)

The compression with a constant lateral stress ($-\sigma_x = -\sigma_y = 200$ kPa $\le -\sigma_z$) from $\boldsymbol{\sigma} = -200\boldsymbol{I}$ kPa is shown in Fig. 2. The response in unloading and reloading is predicted to be stiffer than the test data and thus the hysteresis loop is predicted to be smaller than that. The volumetric strain is predicted well, while the associated flow rule is adopted.

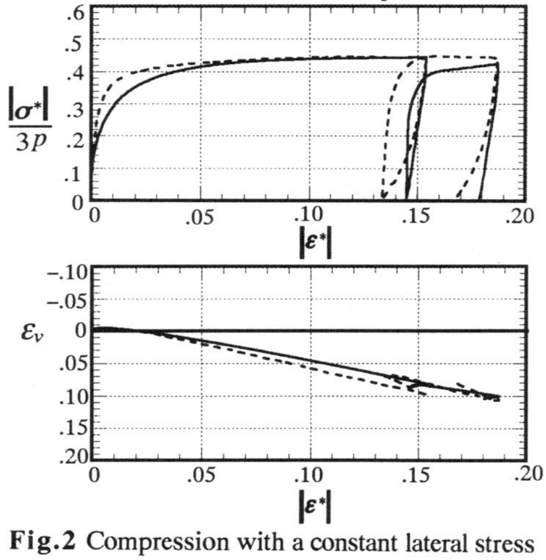

Fig.2 Compression with a constant lateral stress

(c) Axisymmetric extension with a constant lateral stress (CTA)

The Axisymmetric extension with a constant lateral stress ($-\sigma_x = -\sigma_y = 200$ kPa $\ge -\sigma_z$) from $\boldsymbol{\sigma} = -200\boldsymbol{I}$ kPa is shown in Fig. 3.

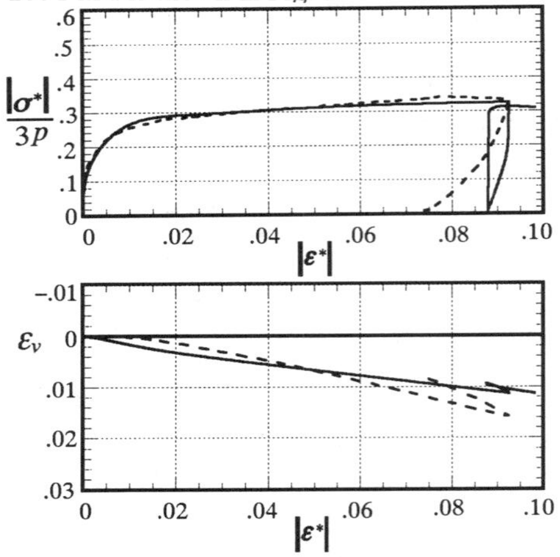

Fig. 3 Extension with a constant lateral stress.

(d) Proportional loading with $b = 0.666$ (CTA)

The proportional loading with $b = 0.666$ ($\theta_\sigma = -10°51'$) from $\boldsymbol{\sigma} = -500\boldsymbol{I}$ kPa is shown in Fig. 3. That is, from the initial isotropic stress state of -500 kPa, the axial stress σ_z was increased by 330 kPa to -170 kPa, while the lateral stress σ_x was decreased by -165.5 kPa to -665.5 kPa proportionately keeping $d\sigma_x/d\sigma_z = -0.5015$. The other stress σ_y was kept constant.

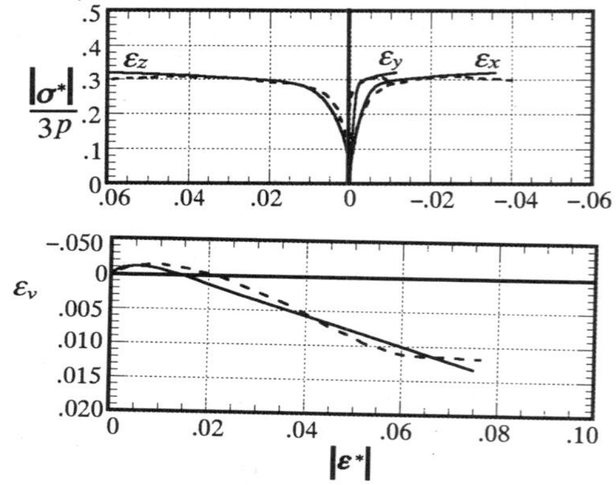

Fig. 4 Proportional loading with $b = 0.666$.

(e) Circular stress path in the deviatoric stress plane (CTA)

This test is split into two stages:

i) The axial stress σ_z was decreased by -343 kPa to -843 kPa from the isotropic stress state of -500 kPa, while σ_x and σ_y were both increased proportionately by 171 kPa to -329 kPa, keeping the mean stress σ_m at -500 kPa. The magnitude of deviatoric stress $\|\boldsymbol{\sigma}^*\|$ finally became 420 kPa.

ii) A circular stress path $(\theta_\sigma = 30° \rightarrow 750°)$ was depicted in a deviatoric stress plane by varying three principal stresses σ_x, σ_y, σ_z in sinusoidal forms, σ_m and $\|\boldsymbol{\sigma}^*\|$ being kept to be constant. The variations of the principal strains ε_x, ε_y, ε_z and the volumetric strain ε_v vs. the angle θ_σ for the circular stress path are shown in Fig.5.

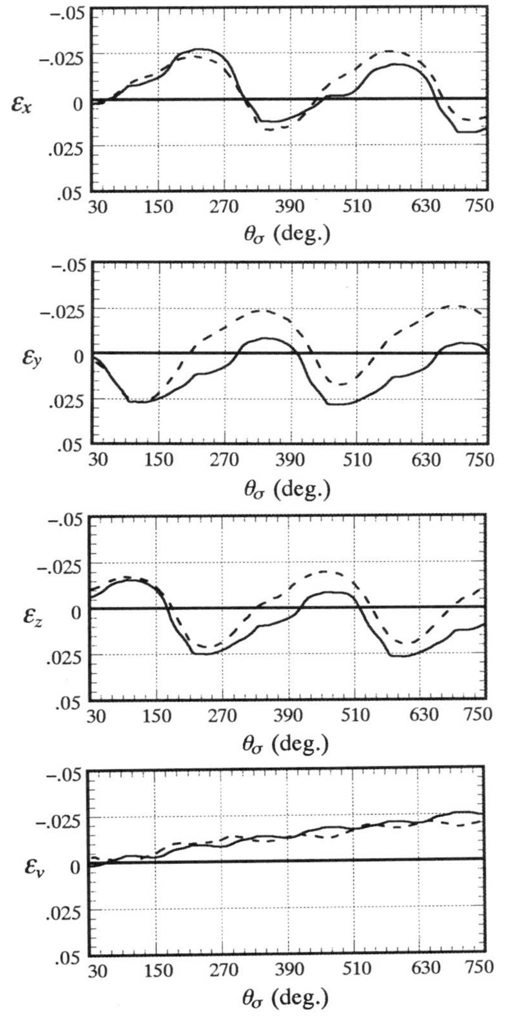

Fig. 5 circular stress path in the deviatoric stress plane

(f) Cyclic torsional loading (HCTA)

The cyclic torsional loading (5 cycles) between the shear stress limits $\tau_{\theta z} = \pm 135$ kPa under $\sigma_r = \sigma_\theta = -500$ kPa and $\sigma_z = -1020$ kPa loaded from the isotropic stress state $\boldsymbol{\sigma} = -500\,\boldsymbol{I}$ kPa is shown in Fig.6.

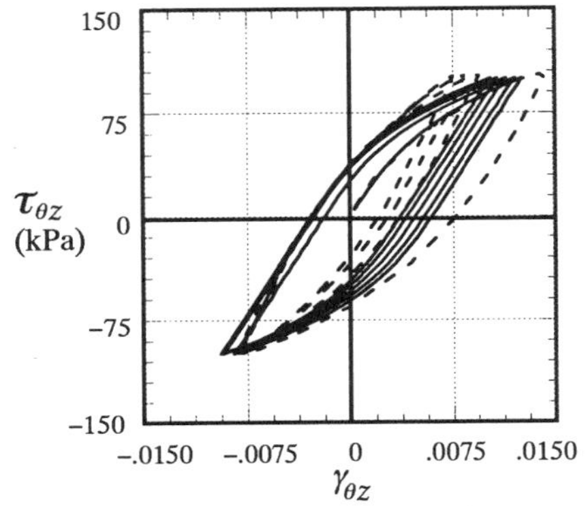

Fig. 6 Cyclic torsional loading.

4.2 *Undrained test*

(a) Monotonic loading

The test data (Castro, 1969) and the calculated results of the stress paths in the (p, q) plane and the relations of q vs. the axial strain ε_a for the monotonic loading in the axisymmetric compression under the undrained condition are shown in Fig.14. The material is Banding sand with the initial relative densities D_r = 0.27, 0.44, 0.47, 0.64 under the initial pressure p_0 = 400 kPa. Material constants and initial values are selected as follows:

material constants:

yield surface (ellipsoid) shape $\phi = 26, 30, 31, 32°$

hardening/softening

- isotropic
  - volumetric
    - $\rho = 0.025, 0.018, 0.014, 0.010$
    - $p_i = 0, 10, 30, 80$ kPa
  - deviatoric
    - $\mu = 1.00, 0.65, 0.30, 0.10$
    - $\phi_d = 40, 33, 30, 20°$
- rotational $b_r = 10$, $\phi_b = 20°$

evolution of R

$u_1 = 0.1, 0.3, 0.5, 1.0$, $m_1 = 0.1, 0.4, 0.5, 0.7$

movement of similarity-center $c = 20, 18, 14, 8$

elastic constants

$\gamma = 0.0067, 0.0065, 0.0060, 0.0058$,

$G = 18{,}000, 23{,}000, 25{,}000, 35{,}000$ kPa

initial values:

$F_0 = 410, 480, 520, 580$ kPa,

$\boldsymbol{\beta}_0 = \boldsymbol{O}$, $\boldsymbol{s}_0 = -200, -110, -100, -80\,\boldsymbol{I}$ kPa

$\boldsymbol{\sigma}_0 = -400\,\boldsymbol{I}$ kPa,

where the four numbers written for constants or an initial value correspond to the case of D_r = 0.27, 0.44, 0.47, 0.64, respectively, in this order.

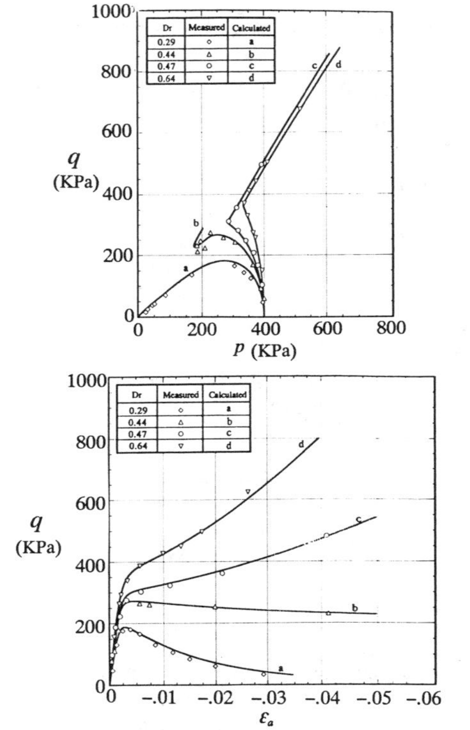

Fig. 7 Undrained behavior of Banding sand. (Calculated results are shown by solid lines)

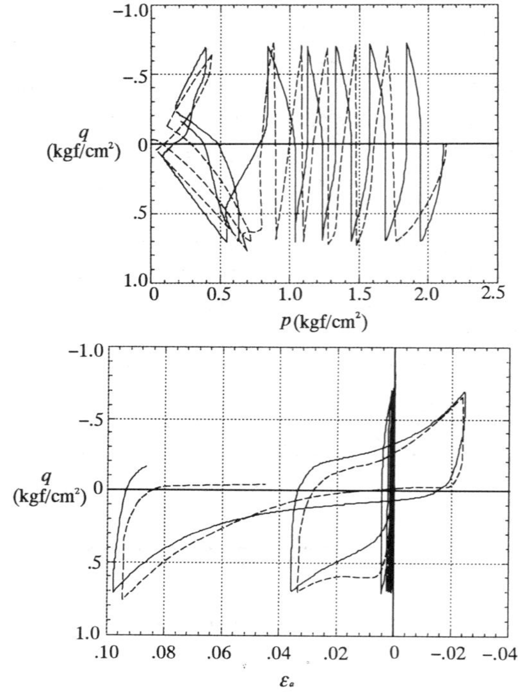

Fig. 8 Cyclic mobility of loose Niigata sand. (Test and calculated results are shown by dashed and solid line, respectively)

(b) Cyclic mobility

The test data (Isshihara, Tatsuoka and Yasuda, 1975) and the calculated results of the stress path and the stress-strain relation for the cyclic mobility with the constant stress amplitude $q = \pm 0.71$ kgf/cm² in the axisymmetric compression under the undrained condition are shown in Figs. 8, respectively. The material is Niigata sand with the initial void ratio $e_0 = 0.737$ under the initial pressure $p_0 = 2.1$ kgf/cm². The material constants and initial values are uselected as follows:

material constants:

yield surface (ellipsoid) shape $\phi = 28^\circ$

hardening/softening

isotropic: volumetric $\rho = 0.01$, $p_i = 0.05$ kgf/cm²; deviatoric $\mu = 1.0$, $\phi_d = 35^\circ$

rotational $b_r = 120$ $\phi_b = 35^\circ$

evolution of R $u_1 = 8.0$, $m_1 = 1.3$

movement of similarity-center $c = 34$

elastic constants $\gamma = 0.0065$, $G = 1{,}800$ kgf/cm²

initial values:

$F_0 = 5.5$ kgf/cm², $\boldsymbol{\beta}_0 = \boldsymbol{O}$, $\mathbf{s}_0 = -0.2\,\boldsymbol{I}$ kgf/cm², $\boldsymbol{\sigma}_0 = -2.1\,\boldsymbol{I}$ kgf/cm²

REFERENCES

Castro, G. 1969. Lquifaction of sands, *Ph.D. Thesis, Harvard Soil Mech. Series No.81.*

Hashiguchi, K. 1977. An expression of anisotropy in a plastic constitutive equation of soils, *Constitutive Equations of Soils (Proc. 9th Int. Conf. Soil Mech. Found. Eng., Spec. Session 9)*: 302-305.

Hashiguchi, K. 1989. Subloading surface model in unconventional plasticity. *J. Solids & Struct.* 25: 917-945.

Ishihara, K., Tatsuoka, F. and Yasuda, S. 1975. Undrained deformation and liquefaction of sand under cyclic stresses. *Soils & Found.* 15: 29-44.

Saada, A.S. and Bianchini, G. (ed.) 1989. *Proc. Int. Workshop on Constitutive Equations for Granular Non-Cohesive Soils*, Cleveland.

Sekiguchi, H. and Ohta, H. 1977. Induced anisotropy and time dependency in clays, *Constitutive Equations of Soils (Proc. 9th Int. Conf. Soil Mech. Found. Eng., Spec. Session 9)*:, 229-238.

Numerical Models in Geomechanics, Pietruszczak & Pande (eds) © 1997 Balkema, Rotterdam, ISBN 90 5410 886 X

A constitutive model for expansive and non expansive clays

C. Pothier & J.C. Robinet
Euro-Géomat Consulting, Orleans, France

A. Jullien
ESEM, University of Orleans, France

ABSTRACT: An elastoplastic model for saturated expansive and non expansive clays was developed. The original features of this model are that two independent plastic mechanisms were introduced considering the microscopic phenomena inside the material. At the normally consolidated state, an increase of the external stresses is balanced by an increase of the contact stresses : a contact plastic yield surface of the modified Cam-clay type is used. For overconsolidated states the variations of the external stresses are balanced by an increase of the internal stresses : a swelling plastic yield surface with non associated flow rules is then used. A unique set of model parameters allows to simulate oedometric and triaxial tests. This model was implemented in a finite element code to simulate the tunnelling of a circular gallery in an expansive clay.

I INTRODUCTION

In the field of waste disposal for radioactive containers, compacted expansive clays are studied as possible barrier materials. So an elastoplastoplastic constitutive model for expansive and non expansive deep clays has been developed. Both the microscopic swelling mechanisms and the macroscopic behaviour of several clays were considered Israelachvili (1978), pointed out that for highly consolidated expansive clays, an excess of internal repulsive forces is created at the scale of the micropores when the water molecules get into or out of the interlamellar space. These forces produce irreversible microscopic strains and reversible macroscopic swelling strains of the material. However, for non expansive clays these repulsive forces are balanced by attractive forces. In order to describe the strain-stress behaviour of expansive clays, Mitchell (1976) and Sridharan (1982) have extended the Terzaghi's concept of effective stresses. They assumed that the strains of expansive clay were controlled by three stress states : the contact stresses, the interstitial pressure and the internal repulsive stresses : $\sigma = \sigma'_c + u_w + \sigma_{R-A}$.

The model hypothesis were based on the above analysis. The first part of this paper presents the constitutive model and its parameters determined on laboratory paths for three saturated clays (kaolin, smectite and Boom clay). Then the model is evaluated through comparisons with other experimental results using the same set of parameters. Finally the model has been implemented on a finit element code to simulate the tunnelling of a circular gallery in Boom Clay.

2 PRESENTATION OF THE MODEL

2.1 *Main features of the model*

Experimental oedometer results on expansive and non expansive clays showed an hysteresis loop along an unloading reloading path. The more expansive the clay is, the wider the hysteresis loop is. Further, the unloading slope increases with the consolidation pressure. Thus, the main model hypothesis developed by Pakzad and Robinet (1994) were the following. Along a loading path the increase of the external stresses is balanced by the increase of the contact stresses between the clay particules. Along an unloading or a reloading path the variations of the external stresses are balanced by those of the internal repulsive stresses. The stress variations were associated with two independent plastic mechanisms : i) For the effective contact stress variations, a contact yield surface of the modified Cam Clay model (F_c) was used ; ii) For the internal stress variations an elliptic swelling yield surface (F_{R-A}) was considered. Strain hardening was obtained with non associated flow rules for the swelling yield surface. Further, cyclic oedometer tests with measurements of the horizontal stress ($\sigma 3$) were recently realised. The experimental variations of K0 versus q/P' are plotted on figure 1. These curves exhibit total reversible variations of $K0$ with q/P' for the loading-unloading-reloading paths. More, quite no differences between the measured values of $K0$ for the smectite and the kaolinite are pointed out at given q/P' values.Then , calculating $K0$:

$$K0 = \frac{\sigma 3}{\sigma 1} \quad (1)$$

and also considering that :

$$q = \sigma 1 - \sigma 3 \text{ and } p' = \frac{\sigma 1 + 2.\sigma 3}{3} \quad (2)$$

it comes :

$$K0 = 1 - \frac{3 * q / P'}{2 * q / P' + 3} \quad (3)$$

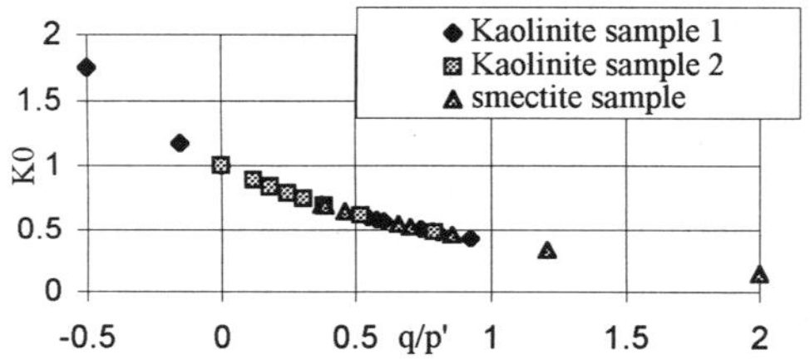

Figure 1 : Experimental variation of K0 versus q/P' on oedometer paths.

Theses experimental results show the effects of the horizontal stresses in an oedometer test. So the model was modified using non associated plasticity flow rules based on the K0 variations.

2.2 *Constitutive equations*

The elastic incremental behaviour is classically described by :

$$d\varepsilon^{e}_{ij} = \frac{dq}{3G} + \frac{dP'}{3K}.\delta_{ij} \quad (4)$$

$$\text{with } K = K_a P_a \left(\frac{P'}{P_a}\right)^n, G = G_a P_a \left(\frac{P'}{P_a}\right)^n \quad (5)$$

where Ka and Ga, are respectively the volumetric and the deviatoric initial elastic modulus.

The contact yield surface equation is given by :

$$F_C = q^2 + M^2.P'.(P' - P_{Cf}).R^2(\theta) = 0 \quad (6)$$

$$P'_{Cf} = P'_{C0}.\exp(\beta_1.\varepsilon^p_v) \quad (7)$$

P'_{cf} is the preconsolidation pressure at the critical state, P'_{c0} the mean effective preconsolidation and β_1 the plastic compressibility modulus. ε^p_v is the strain hardening variable, R(θ) is a shape function which allows for defining a failure criterion close to the Mohr-Coulomb's one. According to Ohnaki (1982), the plastic strains are given by :

$$d\varepsilon^p_v = \lambda\frac{\partial F_c}{\partial P'} \text{ and } d\varepsilon^p_v = \varpi\lambda\frac{\partial F_c}{\partial q} \quad (8)$$

where the parameter ϖ depends on K0.Thus the volumetric and the deviatoric plastic strains for an oedometer path are given by :

$$d\varepsilon^p_v = \frac{\lambda_{1-k}}{\lambda_1}d\varepsilon_1 \quad (9)$$

$$d\varepsilon^p_d = \frac{2}{3}d\varepsilon_1 - d\varepsilon^e_q = d\varepsilon_1\left[\frac{2}{3} - \frac{d\varepsilon^e_q}{d\varepsilon_1}\right] \quad (10)$$

considering that :

$$\frac{d\varepsilon^e_d}{d\varepsilon_1} \langle\langle 1 \quad (11)$$

it comes :

$$\frac{d\varepsilon^e_d}{d\varepsilon^p_v} = \frac{2}{3}\frac{\lambda_1}{\lambda_1 - k} \quad (12)$$

Considering the following relationships :

$$K0 = \frac{6 - 2M}{6 + M} \quad (13)$$

$$M_{k0} = \frac{q}{P'} = \frac{3(1 - K0)}{1 + 2K0} \quad (14)$$

with k0 given by the equation (3), we obtained

$$\varpi = \frac{M^2 - M_{k0}{}^2}{3M_{k0}}\left[\frac{\lambda_1}{\lambda_1 - k}\right] \quad (15)$$

The consistency condition is given by:

$$\lambda = \frac{\frac{\partial F_c}{\partial \sigma'}\underline{\underline{C}}.d\varepsilon}{\frac{\partial F_c}{\partial \sigma'}\underline{\underline{C}}\frac{\partial F_c}{\partial \sigma'} - \frac{\partial F_c}{\partial P_{cf}}\frac{\partial F_c}{\partial \varepsilon^p_v}\frac{\partial F_c}{\partial P'}} \quad (16)$$

The second plastic mechanism is associated with an elliptic swelling yield surface called $F_{R\text{-}A}$ based on non-associated plasticity and combined isotropic and kinematics strain hardening. The equation of the yield surface $F_{R\text{-}A}$ is :

$$F_{R\text{-}A} = (q - \alpha_q)^2 + M^2[P' - \alpha_p + a_i][P' - \alpha_p - a_i] = 0 \quad (17)$$

α_p and α_q, the coordinates of the F_{R-A} surface centre, depend on the fix point between the two surfaces and the stress state along the unloading or the reloading path.

$$\alpha_p = \frac{P'+P'_1}{2};\ \alpha_q = \frac{q+q_1}{2} \qquad (18)$$

a_i, the strain hardening parameter, is the major radius of F_{R-A} and represents the variables (ε_v^p and ε_d^p) associated with the yield of the surface. Thus :

$$da_i = B_1 . d\varepsilon_v^p + B_2 . d\varepsilon_d^p \qquad (19)$$

$$B_1 = P'.\beta_2 . \frac{\partial F_{R-A} / \partial P'}{-2\partial F_{R-A} / \partial a_i} \qquad (20)$$

$$B_2 = M^2 . P'.\beta_2 . \frac{\partial F_{R-A} / \partial q}{-2\partial F_{R-A} / \partial a_i} \qquad (21)$$

At the initial state a_i=a'_0 with

$$a'_0 = a_0 \exp\left(\frac{2P'_0}{P'_{c0}}\right) \qquad (22)$$

with P'_0 the initial mean effective pressure.

β_2 is the plastic compressibility modulus at the overconsolidated state :

$$\beta_2 = \beta_1 . \left(1 + \frac{P_{C(ref)}}{(P_C)_{dech.}}\right) . \frac{P_C}{P'} \text{, unloading} \qquad (23)$$

$$\beta_2 = \beta_1 . \left(1 + \frac{P_{C(ref).}}{P_C}\right) . \frac{2P_{Cf}}{3\ P_{C dech.}} . \frac{P'}{P_C} \text{ reloading} \qquad (24)$$

$P_{C dech.}$ is the critical unloading pressure. $P_{C(ref).}$ is a reference value of pressure in the model for swelling during unloading. Beyond this pressure, it is possible to determine β_2 for any other unloading pressure. It has to be noticed that this pressure corresponds to the pressure from which unloading in an oedometer test is done such that : $\beta_d = 2\beta_1$.

The usual criterion allowing for separation between unloading and reloading is obtained from the sign of the increment of the strain hardening parameter da_i. The latter has to be positive, so that a negative value of da_i during the loading path imposes to change the direction of loading and to initialise a new yield surface F_{R-A}. The non associated plastic strains are given by :

$$d\varepsilon_v^p = d\lambda_{R-A} . \frac{\partial F_{R-A}}{\partial P'} \qquad (25)$$

$$d\varepsilon_d^p = d\lambda_{R-A} . \left(\frac{\partial F_{R-A}}{\partial q} - \alpha . \frac{\partial F_{R-A}}{\partial P'}\right) \qquad (26)$$

α is a parameter which permits to define non associated plasticity :

$$\alpha = \left(1{,}0 - \frac{|P'_1 - \alpha_{p1}|}{a'_0}\right)\left(\frac{1{,}5P_C}{P_{C dech.}} - 0{,}5\right) \qquad (27)$$

when unloading

$$\alpha = \left(1{,}0 - \frac{|P'_1 - \alpha_{p1}|}{a'_0}\right)\left(1{,}0 - 1{,}5\frac{P_C}{P_{Cf}}\right) \qquad (28)$$

when reloading

$d\lambda_{R-A}$ is the plasticity multiplier, determined from the consistency condition :

$$\begin{cases} d\lambda_{R-A} = \dfrac{\partial F_{R-A}}{\partial \underline{\sigma}'} \underline{\underline{C}} . d\underline{\varepsilon} / (A+B) \\ A = \left(\dfrac{\partial F_{R-A}}{\partial P'}\right)^2 (K + P'\beta_2) \\ B = \dfrac{\partial F_{R-A}}{\partial q} (3G + M^2 P'\beta_2)\left(\dfrac{\partial F_{R-A}}{\partial q} - \alpha . \dfrac{\partial F_{R-A}}{\partial P'}\right) \end{cases} \qquad (29)$$

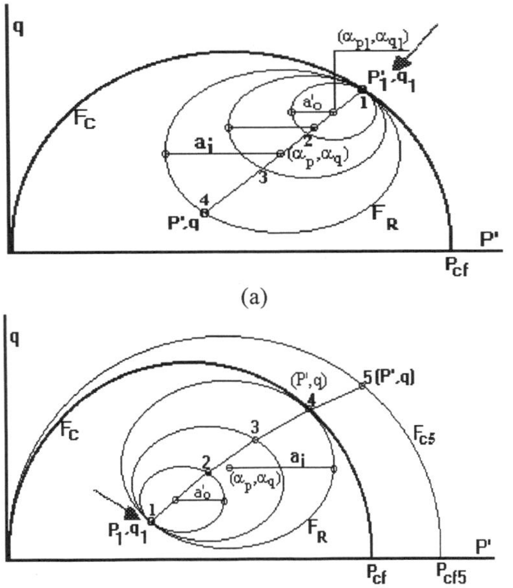

Figure 2 : Schematic view of the activation of F_C and F_{R-A} surfaces on a oedometer path a) unloading b) reloading .

Figure 2 presents the yield surface activation along an oedometric path during a loading-unloading cycle. Along a loading path, at the normally consolidated state, the contact yield surface Fc is the only plastic mechanism activated. Along an unloading path the swelling yield surface is initialised and swells until the final unloading stress is reached (point 4), while the contact yield surface is constant and the contact point between the two surfaces is fix. When reloading, the swelling yield surface is initialised again and swells until the normally consolidated state is reached : $F_{R\text{-}A}$ always passes by a fix point. After, if the external stresses are increased the contact surface is activated again.

2.3 *Model parameters*

The following rheological parameters need to be determined: i) the parameters associated with the elastic mechanism (K_a, G_a, n), ii) these which correspond to the F_C surface (P_{C0}, M, β_1), iii) these which correspond to the $F_{R\text{-}A}$ surface (a_0, $P_{C(ref)}$). This parameters can be determined on a cyclic oedometer test (high pressure) and on a triaxial test, both on a normally consolidated material. The parameters used for the materials of the study, kaolin, smectite and boom clay, are listed in table 1, and n=1 .

Table 1. Model parameters for kaolin(k), smectite (s) and boom clay (b).

	K_a	G_a	M	β_1	a_0 (MPA)	$P_{C(ref)}$ (MPa)
k	83	37	0.59	16	0.1	30
s	33	15	0.8	11.5	0.05	2
b	33	15	0.6	18	0.25	15

3 RESULTS

3.1 Modelling oedometer and triaxial tests

The comparison between the previously experimental results used to determinate model parameter and the numerical simulations is presented on figure 3 for the oedometer test and on figure 4 for the triaxial tests. The hysteresis loop is well described by the model. Other laboratory tests simulated with the same set of parameters, are compared with numerical simulations to complete the validation of the model.

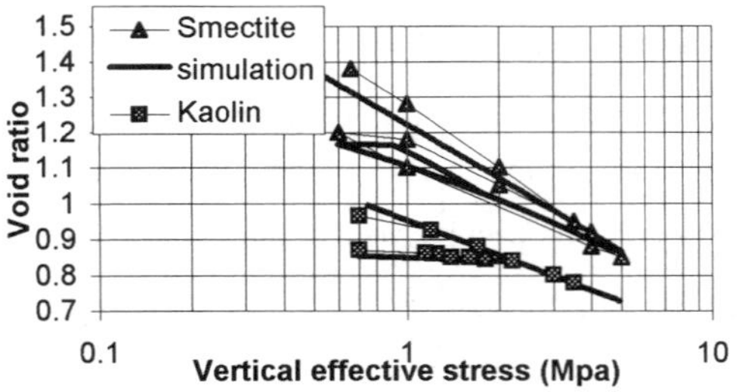

Figure 3 : Oedometer test on clay samples initially at the state of mud.

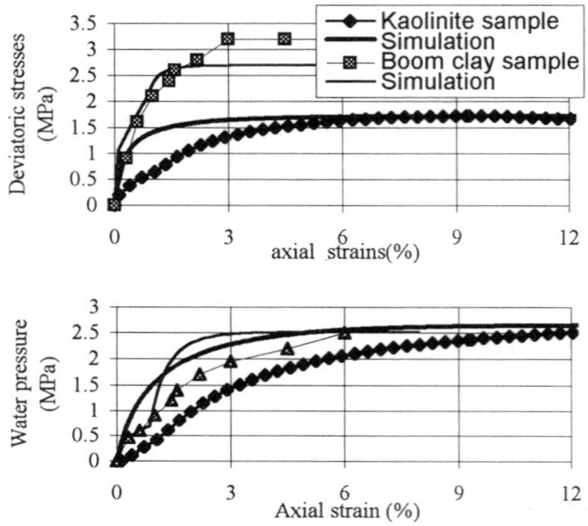

Figure 4 : Undrained triaxial test on a kaolin sample normally consolidated at 5 MPa and on a Boom clay overconsolidated OCR=1.25, σ_3=4.75 MPa (Baldi 1987).

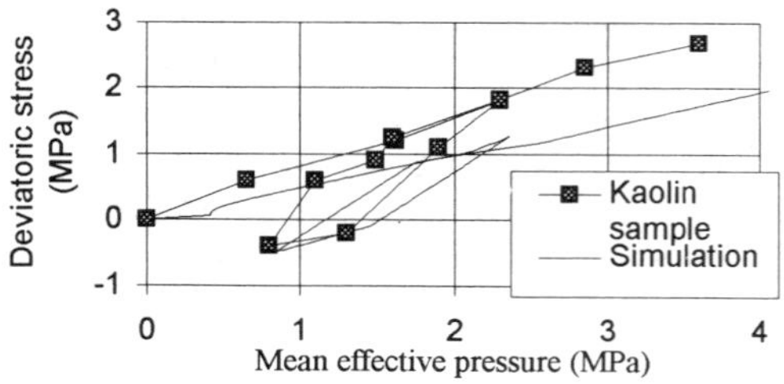

Figure 5 : Oedometer test on a kaolin sample

Oedometer results are presented in the (q-P') plane figure 5 and 6 thanks to K0 measurements. The unloading and reloading cycles are well described with the non associated flows rules.

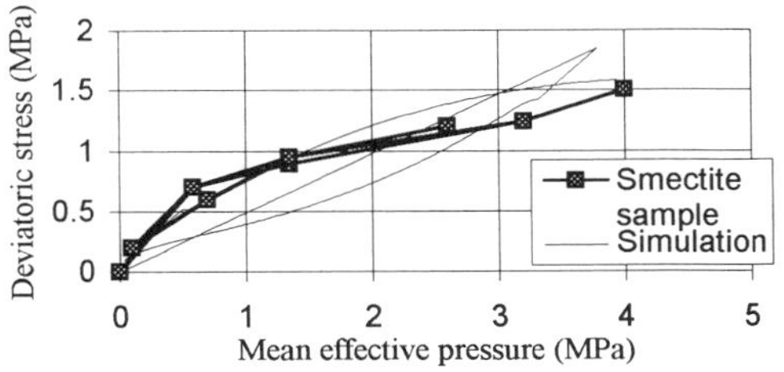

Figure 6 : Oedometer test on a smectite sample

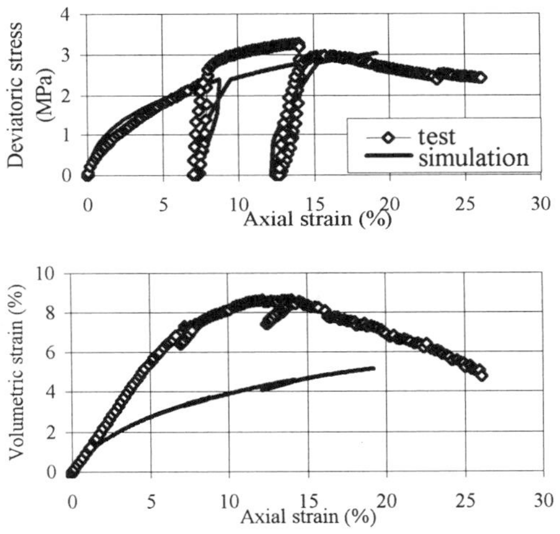

Figure 7 : Drained cyclic triaxial test on a kaolin sample normally consolidated at 5 MPa.

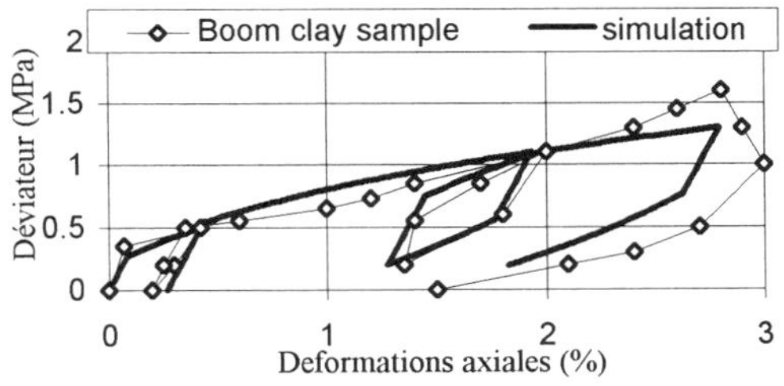

Figure 8 : Drained cyclic triaxial test on Boom clay sample overconsolidated, OCR=7.5, σ_3=0.8MPa (Baldi 1987).

A comparison between a cyclicrained trixial test on a kaolin sample, a cyclicrained triaxial test for an overconsolidated boom clay sample and the numerical simulation is respectively presented on figures7 and 8. The correlation between the experimental results and the simulation is good. The unloading-relaoding loops are well described both for the normally consildated and overconsolidated cases. However, the volumetric strain is underestimated by the numerical simulation..

3.2 Modelling of the tunnelling of a circular gallery

A circular gallery with a diameter of 6 m is realised at a depth of 250 m in Boom clay. The numerical model considers under plane strains a vertical section of the groundwork. Considering symmetry, only half of the gallery is studied ; the medium is simply supported along its perimeter. At the initial state, normal stresses are applied to the contour of the gallery at the wall. The excavation is simulated by an incremental unloading until a zero normal stress at the wall is matched. The normal vector of each finite element at the wall of the gallery is inclined with an angle δ from the horizontal axis. Thus ,the normal stress applied on a face inclined with an angle δ is given by :

$$|\sigma'_n| = \sigma'_1 \sqrt{\frac{1}{2}(1+K_0^2)+\frac{1}{2}.(K_0^2-1).\cos 2\delta} \quad (30)$$

with $\sigma_1 = \sigma'_1 + u_w, \sigma'_3 = K_0\, \sigma'_1, u_w = \gamma_\omega . z$=2.5 MPa

and $\sigma'_{nx} = \sigma'_n \cos\delta; \sigma'_{ny} = \sigma'_n \sin\delta$

Then, the tunnelling is simulated with the assumption that the initial total vertical and horizontal stresses values are: σ_1=4.98 MPa σ_3= 4.48 MPa ; K0 = 0.8 at a depth of 250 m for a density of 19.91 kN/m^3. The horizontal and vertical components of the total stresses applied on the contour of the galleryare summarised in table 2 Simulations were performed with different rheological models. Figure 9 presents the vertical displacement obtained at x=0 along the vertical axis of the medium (fig. 10). The elastic calculations underestimate the displacements and give a maximum displacement value of 4 cm at the wall of the gallery (at x=0; y=3 m). The results of the Mohr-Coulomb criterion show that important displacements arise in the elastic domain of the medium, the maximum displacement is equal to 6.7 cm (at x=0; y=3m). Numerical simulations including strain hardening models lead to a higher and more regular convergence. Furthermore, the value of the convergence and the extent of the plastic domain are greater with the swelling model than with other ones.

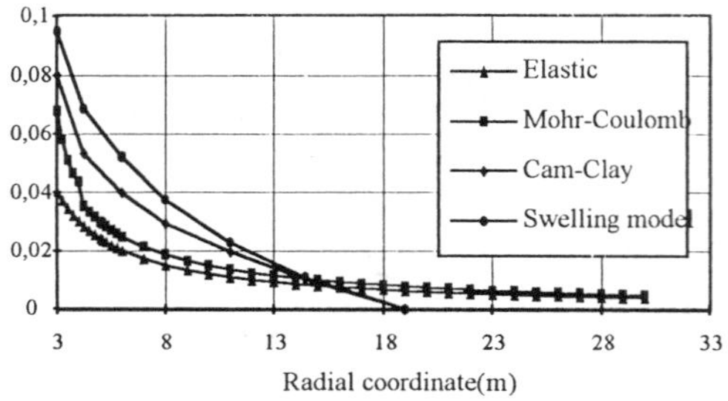

Figure 9 : Vertical displacement obtained at x=0 along the vertical axis of the medium.

Table 2 : Horizontal and vertical total stress (in MPa) components applied on the nodes of the gallerywall

δ	82.51°	67.5°	52.47°	37.53°	22.51°	7.49°
σ_n	4.97	4.90	4.79	4.67	4.55	4.49
σ_{nx}	0.65	1.88	2.92	3.70	4.21	4.45
σ_{ny}	4.93	4.53	3.80	2.84	1.74	0.58

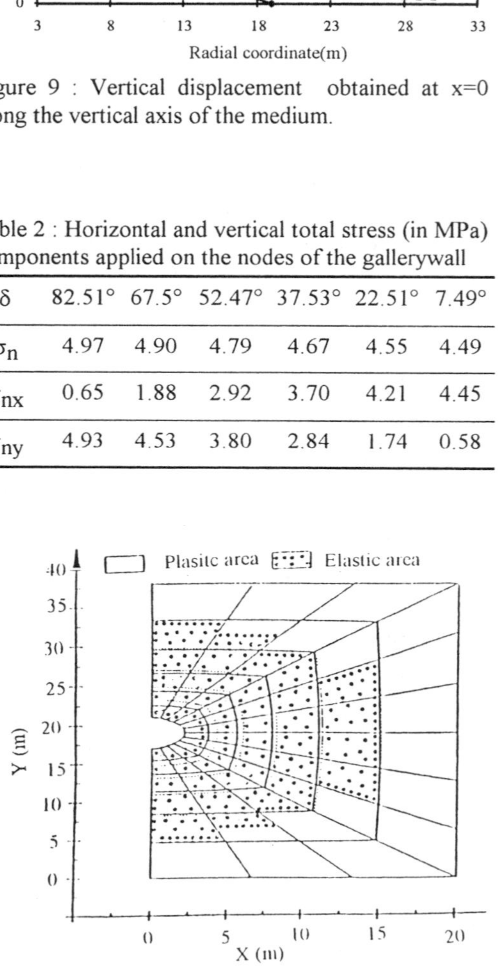

Figure 10 : Plastic domain inside the medium with the swelling model

CONCLUSION

The analysis of the microstructural mechanisms which govern the macroscopic behaviour of expansive clays was used to propose an elastoplastic rheological model. Thus, irreversible strains during unloading are produced because of unbalanced repulsive stresses at the scale of the microporosities. This plastic phenomenon is taken into account by a second yield surface using nonassociated plasticity F_{R-A} together with a classical contact yield surface. This model well simulates the phenomena observed on laboratory paths with a unique set of parameters especially the hysteresis loops where unloading - reloading. A simulation of tunnelling a circular gallery at great depth in an expansive clay was performed . The results pointed out an increase of the convergence of the gallery and of the plastic areas in comparison with the results obtained with an elastoplastic model of the Cam-Clay type.

REFERENCES :

Baldi G., Borsetto M & Hueckel T; 1987 Calibration of mathematical model for the simulation of thermal, seepage and mechanical behaviour of boom clay, CEC Publishers, Luxembourg, EUR 10924 EN

Dormieux L., Barboux P. et Coussy O., 1994.Une modélisation macroscopique du gonflement des argiles saturées *C.R. Acad. Sci.* Séri II, 256

Israelachvili, J.N. & Adams, G.E. 1978. Measurement of forces between two mica surfaces in aqueous electrolyte solutions in the range 0-100 nm *Journal of the chemical society*, Faraday Transactions I, 74: 975-1001

Ohmaki, S. 1982 Stress-strain behaviour of anisotropically, normally consolidated cohesive soil. Proc 1 st Int. Symp. Num. Mod Geomech, Zurich 250-269.

Mitchell J.K.,1976 *Fundamentals of soil behaviour*, John Wiley, New York

Robinet, J.C., Pakzad, M. & Plas, F.1994 Un modèle rhéologique pour les argiles gonflantes. Revue Française de Géotechnique, 67: 57-67.

Sridharan A. & Joyadava M.S.1982. Double layer theory and compressibility of clays *Géotechnique*, 32, 133-144

Tacherifet S.1995 Modélisation du comportement mécanique des argiles profondes gonflantes, application aux ouvrages de stockages',.Ph.D. Thesis. University of. Orléans,France

Numerical Models in Geomechanics, Pietruszczak & Pande (eds) © 1997 Balkema, Rotterdam, ISBN 90 5410 886 X

An elasto-thermo-viscoplastic model for natural clay and its application

F. Oka & A. Yashima
Department of Civil Engineering, Gifu University, Japan

S. Leroueil
Department of Civil Engineering, Laval University, Que., Canada

ABSTRACT: The behavior of natural clays during one dimensional consolidation is strongly influenced by strain rate as well as temperature. The preconsolidation pressure is a function of both strain rate and temperature. An elasto-thermo-viscoplastic constitutive model is developed to reproduce the one dimensional behavior of natural clays at different strain rates and temperatures.

1 INTRODUCTION

Natural clays are subjected to the action of heat in many different circumstances. Studies on the effect of temperature on clays have been thoroughly studied in the sixties. More recently, a renewal interest for this topic has come, mainly in relation with new concerns such as nuclear waste isolation and the use of soil deposits for heat energy storage.

Boudali et al. (1994) performed constant rate of strain (CRS) oedometer tests on the soft silty clay from Berthierville, Canada, at different strain rates and temperatures. The specimens were 20 mm high and 71 mm in diameter. Fig.1 shows typical results obtained at strain rates of 1×10^{-5} and 1.6×10^{-7} s^{-1}, and at temperatures of 5°C and 35°C. It can be seen on Fig.1 that, at a given temperature, the higher is the strain rate, the higher the vertical effective stress at a given strain and that, at a given strain rate, the higher the temperature, the smaller is the vertical effective stress at a given strain. In fact, strain rate and temperature effects are two aspects of the viscous nature of clays, and can thus be combined. In particular, for the clay considered in Fig.1, the compression curves obtained at (T = 5°C and $\dot{\varepsilon}_1 = 1.6\times10^{-7}$ s^{-1}) and at (T = 35°C and $\dot{\varepsilon}_1 = 1\times10^{-5}$ s^{-1}) almost coincide.

The first conceptional model for the behavior of clay to heating was proposed by Campanella and Mitchell (1968). Thermal consolidation has been also studied by Schiffmann (1971), Derski and Kowalski (1979), and Bear and Corapcioglu (1981). In these works, no plastic strain was considered explicitly. Hueckel and Borsetto (1990), Hueckel and Baldi (1990) and Lingnau et al. (1995) developed a thermoplasticity model by generalizing the critical state theory (Schofield and Wroth, 1968) by incorporating thermal effect.

To take into account the effect of strain rate, Oka (1981) and Adachi and Oka (1982) proposed an elasto-viscoplastic constitutive model based on Perzyna's theory of an elasto-viscoplastic continuum (1963) to describe the rate-sensitive and plastic behavior of normally consolidated clay. They assumed that the normally consolidated clay never reaches a static equilibrium state, even at the end of primary consolidation, and took a volumetric viscoplastic strain as a hardening parameter.

As viscosity means effect of strain rate as well as temperature, it appears very interesting to examine how the model proposed by Adachi and Oka (1982) can be extended to include the influence of temperature on clay behavior. For this purpose, an elasto-thermo-viscoplastic constitutive model extended from the original Adachi and Oka's model is proposed for one dimensional consolidation on natural clay.

2 ELASTO-THERMO-VISCOPLASTIC CONSTITUTIVE MODEL FOR CLAY

2.1 *Elasto-thermo-viscoplastic constitutive model*

An elasto-viscoplastic constitutive model, originally developed by Adachi and Oka (1982), is extended to include the influence of temperature on the viscous behavior of natural clays. As for numerical simulation by a finite element method, a three dimensional constitutive equation is needed. The general formulation of the Adachi and Oka's model

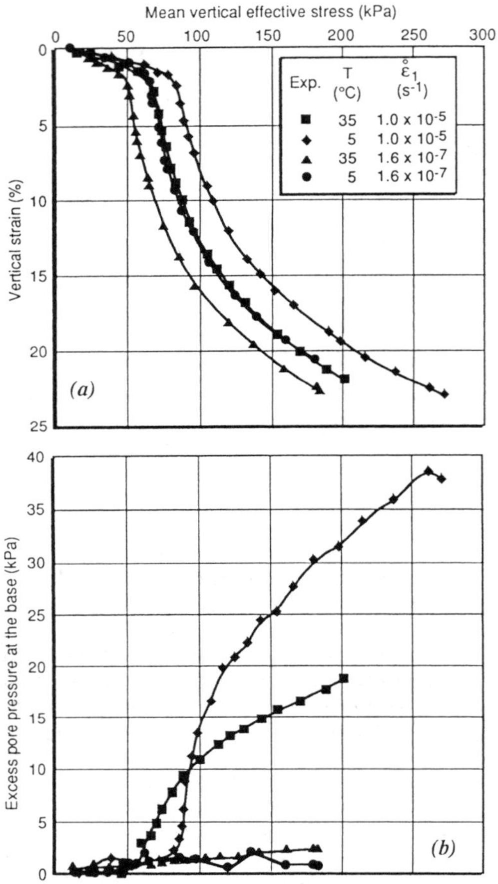

Fig.1 Effects of strain rate and temperature on the one-dimensional consolidation behavior of Berthierville clay (Boudali et al. 1994).

including temperature effect is described under three dimensional conditions. However, because the experimental informations available were obtained in one dimensional conditions, the temperature dependent viscoplastic parameter introduced in this study is described in one dimensional conditions.

Adachi and Oka (1982) proposed a general constitutive equation for a normally consolidated clay based on the principle of the Cam-clay model(Roscoe et al. 1958;Schofield and Wroth 1968) and the Perzyna's overstress type viscoplastic theory (1963). The viscoplastic flow rule is given by

$$\dot{\varepsilon}_{ij}^{vp} = C < \Phi(F) > \frac{\partial f}{\partial \sigma'_{ij}} \tag{1}$$

$$< \Phi(F) >= \begin{cases} 0, & F \leq 0 \\ \Phi(F), & F > 0 \end{cases} \tag{2}$$

$$F = \frac{f - k_s}{k_s} \tag{3}$$

where $\dot{\varepsilon}_{ij}^{vp}$ is the viscoplastic strain rate tensor, σ'_{ij} is the effective stress tensor, f is the dynamic yield function, C is the viscoplastic parameter, Φ is the material function for strain rate effect and $F = 0$ denotes the static yield function. The dynamic yield function extended for anisotropically consolidated clay (Oka et al. 1986) is expressed as follows:

$$f = \frac{\bar{\eta}^*}{M^*} + ln\frac{\sigma'_m}{\sigma'_{m0}} = ln\frac{\sigma'_{my}}{\sigma'_{m0}} = k_s \tag{4}$$

in which M^* is the stress ratio at critical state, σ'_m is the effective mean stress, σ'_{my} is the hardening parameter, σ'_{m0} is the unit of σ'_m and $\bar{\eta}^*$ is the stress ratio invariant expressed by

$$\bar{\eta}^* = \sqrt{(\eta_{ij} - \eta_{ij(0)})(\eta_{ij} - \eta_{ij(0)})} \tag{5}$$

$$\eta_{ij} = s_{ij}/\sigma'_m \tag{6}$$

where s_{ij} is the deviatoric stress tensor, and $\eta_{ij(0)}$ is the value of η_{ij} at the initial stress state. The hardening parameter σ'_{my} is calculated by the following evolutional equation.

$$\frac{d\sigma'_{my}}{\sigma'_{my}} = \frac{(1+e)}{\lambda - \kappa} dv^{vp} \, . \tag{7}$$

Integrating Eq.(7) under the condition that the initial value of σ'_{my} is σ'_{myi} and the initial value of v^{vp} is zero, we obtain

$$v^{vp} = \frac{\lambda - \kappa}{(1+e)} ln\frac{\sigma'_{my}}{\sigma'_{myi}} \, . \tag{8}$$

Based on experimental studies, material function $\Phi(F)$ is derived. Using Eq.(8), we get

$$C\Phi(F) = C\Phi(k_s \cdot F)$$

$$= C_0(T)exp\{m'(-ln[\frac{\sigma'_{myi}}{\sigma'_{me}}])\}$$

$$\times \; exp\{m'(\frac{\bar{\eta}^*}{M^*} + ln\frac{\sigma'_m}{\sigma'_{me}} - \frac{1+e}{\lambda - \kappa}v^{vp}\}$$

$$= C(T) \cdot exp\{m'(\frac{\bar{\eta}^*}{M^*} + ln\frac{\sigma'_m}{\sigma'_{me}} - \frac{1+e}{\lambda - \kappa}v^{vp}\} \tag{9}$$

$$C(T) = C_0(T)exp\{m'(-ln[\frac{\sigma'_{myi}}{\sigma'_{me}}])\} \tag{10}$$

where m' and is the viscoplastic parameter, σ'_{me} is the initial consolidation (effective confining) pressure, λ is the compression index, κ is the swelling index, e is the void ratio and v^{vp} is the volumetric plastic strain. It is worth noting that $C(T)$ depends on the initial hardening parameter σ'_{myi} and may also depend on the temperature.

The total strain rate, $\dot{\varepsilon}_{ij}$, is calculated by the summation of the elastic strain rate and the viscoplastic strain rate as

$$\dot{\varepsilon}_{ij} = \frac{1}{2G}\dot{s}_{ij} + \frac{\kappa}{3(1+e)\sigma'_m}\dot{\sigma}'_m\delta_{ij}$$

$$+\frac{C}{M^*\sigma'_m} < \Phi(F) > \frac{\eta_{ij} - \eta_{ij(0)}}{\bar{\eta}^*}$$

$$+\frac{C}{3M^*\sigma'_m} < \Phi(F) > [M^* - \frac{S_{kl}(\eta_{kl} - \eta_{kl(0)})}{\bar{\eta}^*}]\delta_{ij} \tag{11}$$

where G is the elastic shear modulus and δ_{ij} is the Kronecker's delta.

Equations summarized above are proposed in three dimensional stress space. In the case of one dimensional consolidation process, if the stress ratio during one dimensional compression is assumed to be constant and the initial hardening parameter corresponds to the preconsolidation pressure σ'_p, the viscoplastic part in Eq.(11) takes a simple form as

$$\dot{\varepsilon}_1^{vp} = C(T)\cdot exp\{m'(ln\frac{\sigma'_v}{\sigma'_0} - \frac{1+e}{\lambda-\kappa}\varepsilon_1^{vp})\} \tag{12}$$

$$C(T) = C_0 exp\{m'(-ln[\frac{\sigma'_p}{\sigma'_0}])\} \tag{13}$$

where $\dot{\varepsilon}_1^{vp}$ is the viscoplastic vertical strain rate, ε_1^{vp} is the viscoplastic vertical strain and σ'_0 is the initial consolidation pressure.

From Fig.2 which is prepared from the experimenral results by Boudali et al. (1994), it can be considered as a first approximation that the preconsolidation pressure is a function of temperature as,

$$\frac{\sigma'_p}{\sigma'_{pr}} = [\frac{T_r}{T}]^\alpha, \qquad \alpha = 0.15 \tag{14}$$

where σ'_{pr} is the value of σ'_p at the referential temperature, T is the temperature and T_r is the referential temperature. Both T and T_r are expressed in °C. If the viscoplastic parameter C depends on temperature only through the preconsolidation pressure, the following relation can be derived.

$$C(T) = C_0 exp\{m'(-ln[\frac{\sigma'_{pr}}{\sigma'_0}\cdot(\frac{T_r}{T})^\alpha])\}$$

$$= C_0 exp\{m'(-ln[\frac{\sigma'_{pr}}{\sigma'_0}])exp\{m'(-ln(\frac{T_r}{T})^\alpha)\}$$

$$= C(T_r)exp\{m'(-ln(\frac{T_r}{T})^\alpha)\} \tag{15}$$

In this form, the viscoplastic parameter $C(T)$ is expressed by $C(T_r)$, the value of $C(T)$ at the referential temperature T_r as:

$$\frac{C(T)}{C(T_r)} = [\frac{T}{T_r}]^{\alpha m'} \tag{16}$$

In order to verify the assumption that the viscoplastic parameter C depends on the temperature T only through the preconsolidation pressure, the relation between $C(T)$ and temperature must be examined. By substituting the viscoplastic strain rate (approximately equal to total strain rate), vertical effective stress and viscoplastic strain (ap-

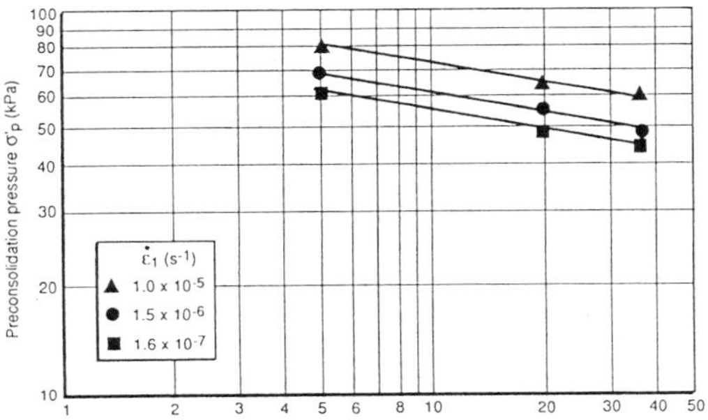

Fig.2 Relationship between log σ'_p and log T from oedometer tests.

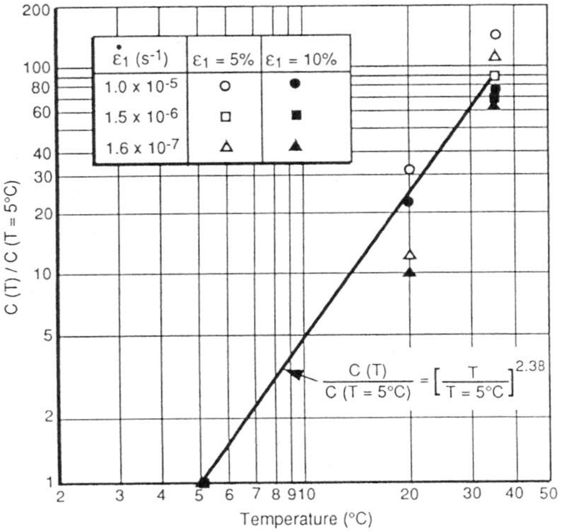

Fig.3 Relationship between viscoplastic parameter, C and temperature.

proximately equal to total strain) into Eq.(12), the viscoplastic parameter C can be calculated. Fig.3 is obtained with the assumption that the viscoplastic strain rate and viscoplastic strain are approximately equal to total strain rate and total strain respectively. The viscoplastic parameters C at total strains of 5 % and 10 % are plotted as a function of temperature for 3 different strain rates. From Fig.3, the following relation can be obtained.

$$\frac{C(T)}{C(T_r = 5°C)} = [\frac{T}{T_r = 5°C}]^{\beta} \tag{17}$$

where $\beta = 2.38$ and β corresponds to $\alpha m'$ in Eq.(16). Since m' is independently determined as 17 from the experimental results, α is obtained as 0.14. The α value obtained from Fig.2 (α = 0.15) is very similar to the value of α in Eq.(16) (α = 0.14). It thus seems reasonable to assume that the viscoplastic parameter $C(T)$ depends on the temperature T only through the temperature dependence of the preconsolidation pressure. An elasto-thermo-viscoplastic constitutive model for one dimensional consolidation behavior on clay under different temperature conditions is defined by Eqs. (12) and (17).

2.2 *Determination of material parameters*

In order to analyze the one dimensional consolidation on Berthierville clay by an effective stress based finite element method together with Eq.(12), the material parameters for the soil skeleton as well as the characteristics of the clay permeability have to be determined.

Compression index and swelling index, λ and κ

Based on the experimental results shown in Fig.1, the compression and swelling indices are independent on strain rate and temperature. The swelling index, κ, was easily obtained as 0.025. The compression index has been determined considering its dependence on the magnitude of the vertical strain. It has been considered constant for vertical strains between the strain at yielding (2 %) and 5 %, and variable as a parabola with respect to the vertical strain between 5 % and 20 %.

Poisson's ratio and initial void ratio

Poisson's ratio is assumed to be 0.33 for this soft clay. The initial void ratio at a vertical effective stress of 10 kPa is 1.70. The stress ratio σ'_h/σ'_v is assumed to be constant and equal to 0.5.

Viscoplastic parameter, m' and α

The viscoplastic parameter, m' in Eq.(12) is determined from the relationship between preconsolidation pressure and strain rate. Assuming the stress ratio during one dimensional consolidation constant, the constitutive equation for viscoplastic vertical strain rate becomes

$$ln\frac{\dot{\varepsilon}^{vp}_{1(1)}}{\dot{\varepsilon}^{vp}_{1(2)}} = m' ln\frac{\sigma'_{p(1)}}{\sigma'_{p(2)}}. \tag{18}$$

From the test results, m' is equal to 17 and is independent of temperature.

Viscoplastic parameter, C

If the reference strain rate and temperature are specified, the value of the viscoplastic parameter C can be calculated. In this study, the condition at (T = 5°C and $\dot{\varepsilon}_1$ = 1×10^{-5} s^{-1}) was selected as the reference state. Under this condition, if the preconsolidation pressure, σ'_p = 81 kPa, is substituted into Eq.(12), the viscoplastic parameter, C, becomes equal to 1.40×10^{-5} s^{-1}. From Eq.(17) and Fig.3 in which $\beta = 2.38$ and $\alpha = 0.14$, it becomes equal to 1.43×10^{-3} s^{-1} at T = 35°C.

Initial permeability, k_0

The initial coefficient of permeability at a void ratio = 1.7 can be determined from the oedometer test. Assuming a linear relationship between the logarithm of permeability and void ratio between 1.0 and 1.7, and considering back-calculated k values reliable for void ratios smaller than 1.4, k_0 at T = 5°C is 3.0×10^{-9} m/s. The permeability at T = 35°C is 6.33×10^{-9} m/s, i.e., 2.1 times larger than the value at T = 5°C.

3. SIMULATION OF THE ONE DIMENSIONAL BEHAVIOR OF NATURAL CLAY

Finite element analyses were carried out to simu-

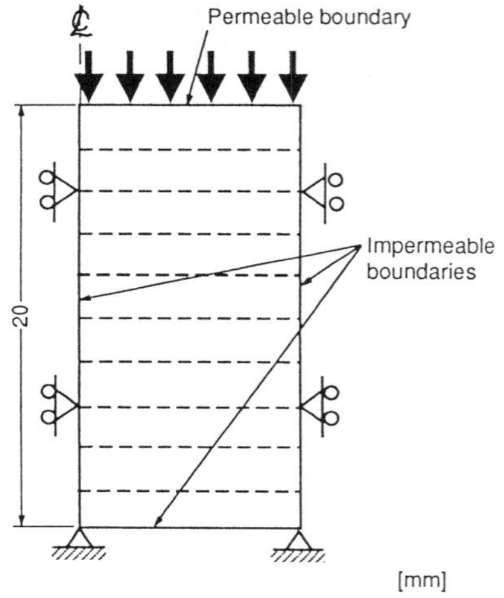

Fig.4 Finite element mesh and boundary conditions.

late the CRS oedometer tests performed on Berthierville clay at different strain rates and temperatures. The finite element mesh used in the analysis is shown in Fig.4, with the 2 cm high specimen divided into 10 elements. The boundary conditions for the displacements as well as for the water flow are also shown in Fig.4. The analysis was carried out under axis-symmetric conditions. The strain increment used is 0.01 %/step and a total of 2000 numerical steps were necessary to reproduce the vertical strain of 20 %. Before applying the constant rate of strain, the specimen was subjected to a uniform vertical effective stress of 10 kPa.

Fig.5 shows the simulated results together with the experimental results at strain rates of 1×10^{-5} and 1.6×10^{-7} s^{-1}, and temperatures of 5°C and 35°C. From Fig.5a, it is clear that the simulated results with material parameters obtained at the reference condition (T = 5°C and $\dot{\varepsilon}_1 = 1\times10^{-5}$ s^{-1}) reproduce all compression curves very well. In particular, the compression curves obtained at (T = 5°C and $\dot{\varepsilon}_1 = 1.6\times10^{-7}$ s^{-1}) and at (T = 35°C and $\dot{\varepsilon}_1 = 1\times10^{-5}$ s^{-1}) are found to almost coincide, as in Fig.1. In Fig.5b, the predicted pore pressure build-up at the bottom of the specimen is compared with the pore pressure measured in experiments. The pore pressure behavior in the normally consolidated region is well reproduced by the numerical analysis in all cases. On the other hand, in the overconsolidated region, the simulations predict at high strain rate pore pressures larger than the experiments. This is certainly due to an underestimation of the permeability because of the assumption of a linear relationship between the logarithm of the permeability and the void ratio around the initial void ratio. Except for the prediction of the pore pressure behavior in the overconsolidated region, it can be concluded that the proposed elasto-thermo-viscoplastic constitutive model satisfactorily reproduce the one dimensional viscous behavior of the natural Berthierville clay at different strain rates and temperatures.

The homogeneous behavior of the clay, as calculated from Eq.(12) at T = 5°C, is shown with dashed lines in Fig.6 for different strain rates. The simulated results for (T = 5°C and $\dot{\varepsilon}_1 = 1\times10^{-5}$ s^{-1}) and (T = 35°C and $\dot{\varepsilon}_1 = 1\times10^{-5}$ s^{-1}) obtained by the finite element method are also plotted. For the elements close to the impervious boundary (bottom element in Fig.4) and close to the drainage boundary (top element in Fig.4) the mean vertical effective stress - strain relations are plotted on this diagram, with a shaded area in between. Different elements in the specimen undergo different paths. For example, the top element in the specimen at (T

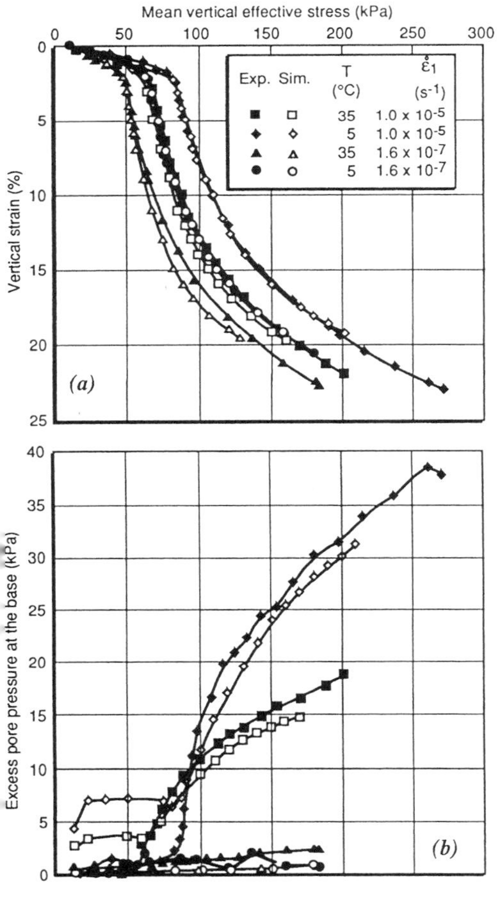

Fig.5 Simulated and experimental results at strain rates of 1×10^{-5} and $1.6 \times 10^{-7} s^{-1}$.

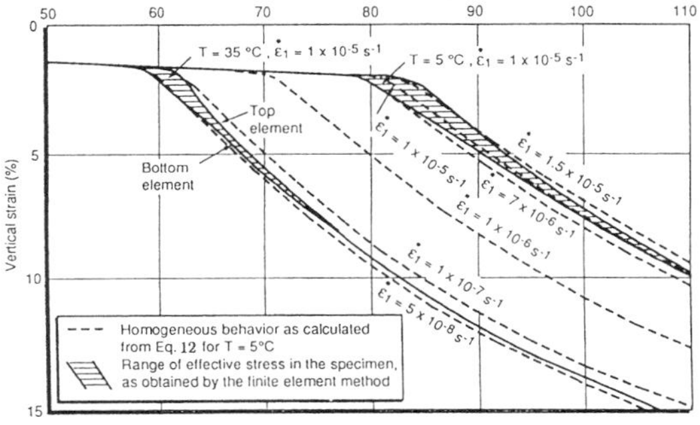

Fig.6 Homogeneous behavior of clay calculated by Eq.(12) and simulated results obtained by the finite element method.

$= 5°C$ and $\dot{\varepsilon}_1 = 1\times10^{-5}$ s^{-1}) experiences a vertical effective stress - strain path which corresponds to that for a higher strain rate (approximately $\dot{\varepsilon}_1 = 1.5\times10^{-5}$ s^{-1}) at strains smaller than 4 %. On the other hand, the bottom element in the specimen follows a path near the passage of the preconsolidation pressure corresponding to a lower strain rate (about 7×10^{-6} s^{-1}). These different paths, however, finally come into a unique compression line corresponding to the line ($\dot{\varepsilon}_1 = 1\times10^{-5}$ s^{-1}). A similar tendency can be observed for the case at $T = 35°C$.

This large inhomogeneous strain distribution in the specimen can be understood by the distribution of the vertical effective stress in the specimen. In Fig.7, the distributions of the vertical effective stress in the specimen are plotted at several stages. From Fig.1, the preconsolidation pressure is 81 kPa under the strain rate and temperature considered. Even when the vertical effective stress is slightly larger than the preconsolidation pressure, i.e., 82 kPa, the lower part of the specimen is still in the overconsolidated region while the rest of the specimen is already in the normally consolidated region. That explains why the strain distribution becomes more inhomogeneous at the passage of the preconsolidation pressure.

4. CONCLUSIONS

The conclusions from this study are as follows:

1) An elasto-thermo-viscoplastic constitutive model has been developed by extending the Adachi and Oka's model to reproduce the behavior of natural clays at different strain rates and temperatures. The parameter $C(T)$ is found to depend on the temperature only through the temperature dependence of the preconsolidation pressure.

2) It is found that the proposed elasto-thermo-viscoplastic constitutive model satisfactorily reproduce the one dimensional viscous behavior of the natural Berthierville clay at different strain rates and temperatures.

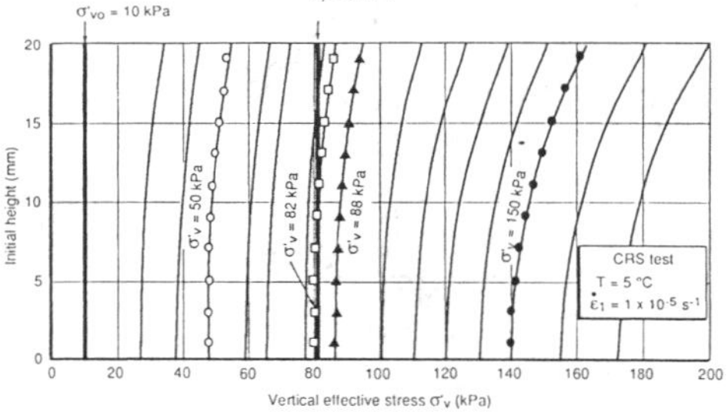

Fig.7 Calculated distributions of the vertical effective stress in the specimen at several stages of the CRS oedometer test.

REFERENCES

Adachi,T. and Oka,F. 1982. Constitutive equations for normally consolidated clay based on elasto-viscoplasticity. *Soils and Foundations.* 22(4):57-70.

Bear,J. and Corapcioglu,C.M.Y. 1981. A mathematical model for consolidation in a thermoelastic aquifer due to hot water injection or pumping. *Water Resour. Res.* 17(3):723-736.

Boudali,M., Leroueil,S. and Srinivasa Murthy,B.R. 1994. Viscous behaviour of natural clays. *Proc. 13th ICSMFE.* New Delhi. 1:411-416.

Campanella,R.G. and Mitchell,J.K. 1968. Influence of temperature variations on soil behavior. *ASCE Journal of Soil Mechanics and Foundation Division.* 94(3):709-734.

Derski,W. and Kowalski,S.T. 1979. Equations of linear thermoconsolidation. *Archives of Mech.* 31(3):303-316.

Hueckel,T. and Baldi,G. 1990. Thermoplasticity of saturated clays: Experimental constitutive study. *ASCE Journal of Geotechnical Engineering.* 116(12):1778-1796.

Hueckel,T. and Borsetto,M. 1990. Thermoplasticity of saturated soils and shales: Constitutive equations. *ASCE Journal of Geotechnical Engineering.* 116(12):1765-1777.

Lingnau,B.E., Graham,J. and Tanaka,N. 1995. Isothermal modeling of sand-bentonite mixtures at elevated temperatures. *Canadian Geotechnical Journal.* 32(1):78-88.

Oka,F. 1981. Prediction of time dependent behavior of clay. *Proc. 10th ICSMFE.* Stockholm. 1:215-218.

Oka,F., Adachi,T. and Okano,Y. 1986. Two-dimensional consolidation analysis using an elasto- viscoplastic constitutive equation. *Int. J. Numerical and Analytical Methods in Geomechanics.* 10:1-16.

Perzyna,P. 1963. The constitutive equation for work-hardening and rate sensitive plastic materials. *Proc. Vibrational Problems.* Warsaw, 4(3):281-290.

Roscoe,K.H., Schofield,A.H. and Wroth,C.P. 1958. On the yielding of soils. *Géotechnique.* 8:22-53.

Schiffmann,R.L. 1971. A thermoelastic theory of consolidation. *Environmental and Geophysical Heat Transfer.* ASME. New York.

Schofield,A.N. and Wroth,C.P. 1968. Critical state soil mechanics. *McGraw-Hill.* London. United Kingdom.

Numerical Models in Geomechanics, Pietruszczak & Pande (eds) © 1997 Balkema, Rotterdam, ISBN 90 5410 886 X

A generalized model for clay under monotonic and cyclic loading conditions

E.Q.Chowdhury & T.Nakai
Nagoya Institute of Technology, Japan

ABSTRACT: In this paper a kinematic hardening model for clay has been proposed to replace a previously proposed kinematic hardening model which could not predict the observed stabilization characteristics of strains and change of stress-dilatancy relation of clay during cyclic loading. Also, the present model can account for both positive and negative dilatancy as well as the increase of strength (friction angle) due to over consolidation or cyclic loading. All these features have been achieved by modifying hardening function and plastic potential function taking over consolidation ratio as the controlling factor. This model is equally applicable to the normally consolidated clay and closely predicts the strength and dilatancy characteristics. Apart from the conventional clay models (e.g. Cam clay model), this model uses a modified stress tensor (t_{ij}), which has made it possible to consider the effect of intermediate principal stress on the strength and dilatancy characteristics. It has been assumed that the plastic strain increment vector is composed of two components, one satisfies associated flow rule in t_{ij}-space and the other is linked with the increase of normal stress. Finally, the new kinematic clay model has been verified by experimental results.

1 INTRODUCTION

Numerous constitutive equations for clay have been developed during the last three decades of the history of elastoplastic constitutive equations for soils. The primary objective of those models were to predict the behavior of clay under triaxial compression conditions due to simplicity of laboratory testing and abundance of test data, for example Cam-clay models (Roscoe, Schofield and Thurairajah 1963 and Roscoe and Burland 1968). Most of those models including Cam-clay model have failed to predict realistically the strength and dilatancy characteristics of normally consolidated clay under triaxial extension, true triaxial or under rotation of principal stress conditions.

One of the authors proposed a clay model named t_{ij}-clay model (Nakai and Matsuoka 1986) which can incorporate the strength and dilatancy behavior of normally consolidated clay not only under triaxial compression but also under generalized three dimensional stress conditions. The prime objective of the present paper is to introduce a model for clay which is equally applicable to normally and over consolidated clay under monotonic or cyclic loading conditions. In doing this we strictly adhere to the objective to keep the number of soil parameters to the minimum and to easy determination of these parameters from simple triaxial tests. Since, t_{ij}-concept uses unconventional stress and stress ratio variables these are listed in table 1.

Table 1. Comparison of ordinary with t_{ij} concept

Ordinary Concept	t_{ij}-Concept
σ_{ij}	$t_{ij} = \sigma_{ik} a_{kj}$
δ_{ij} (Unit tensor)	a_{ij}
$p = \sigma_{ij}\delta_{ij}/3$	$t_N = t_{ij} a_{ij}$
$s_{ij} = \sigma_{ij} - p\delta_{ij}$	$t'_{ij} = t_{ij} - t_N \delta_{ij}$
$q = \sqrt{(3/2) s_{ij} s_{ij}}$	$t_S = \sqrt{t'_{ij} t'_{ij}}$
$\eta_{ij} = s_{ij}/p$	$x_{ij} = t'_{ij}/t_N$
$\eta = q/p = \sqrt{(3/2)\eta_{ij}\eta_{ij}}$	$X = t_S/t_N = \sqrt{x_{ij} x_{ij}}$
$\eta^*_{ij} = (\eta_{ij} - \eta_{ij0})$	$x^*_{ij} = (x_{ij} - n_{ij})$
$\eta^* = \sqrt{\eta^*_{ij}\eta^*_{ij}}$	$X^* = \sqrt{x^*_{ij} x^*_{ij}}$

2 MODEL FORMULATION

2.1 *Yield Function*

The yield function of the t_{ij}-clay model (Nakai-Matsuoka 1986) has been given by assuming associated flow rule and observing the experimental evidence that a linear stress-dilatancy relation exists for normally consolidated clay. But, critical investigation reveals that at low stress ratios, stress-dilatancy asymptotically approaches to very large values and hence indicate an elliptic shape of the yield surface near the tip. This features can easily be reflected by assuming an yield function as in Eq. 1.

$$f = \ln\left(\frac{t_N}{t_{N0}}\right) + AX^{\beta} - \frac{\varepsilon_v^p}{C_p} = 0 \quad (1)$$

Internal state variables to describe normal stress and stress ratio are t_N and X respectively, are the same as those used in the t_{ij}-clay model. β and $C_p = C_t - C_e$ are the soil parameters. Yield function of Eq. 1 has many advantages, for instance a smooth transition of stiffness is possible when stress reversals occur. Stiff response at low stress ratio can be obtained. Yield function used in the original t_{ij}-clay model had a singularity point at the tip but no such point exists if function defined in Eq. 1 is used. Yield function given by Eq. 1 and that of the original t_{ij}-clay model are shown schematically in Figure 1.

The stress-dilatancy relation corresponding to Eq. 1 can be given by Eq. 2 and schematically shown and compared with that of t_{ij}-clay model in figure 2. From figure 2 we can see that there is no significant change in stress-dilatancy at relatively high stress ratio but at low stress ratio shearing component of strain increment is much smaller than that of the original model, thus a stiffer response at low stress ratio can be obtained. It is observed that original Cam-clay model gives relatively flexible and modified Cam-clay model gives stiffer response than those observed under triaxial compression and extension conditions at low stress ratio. A similar yield function as in Eq. 1 for Cam-clay model would give better predictions.

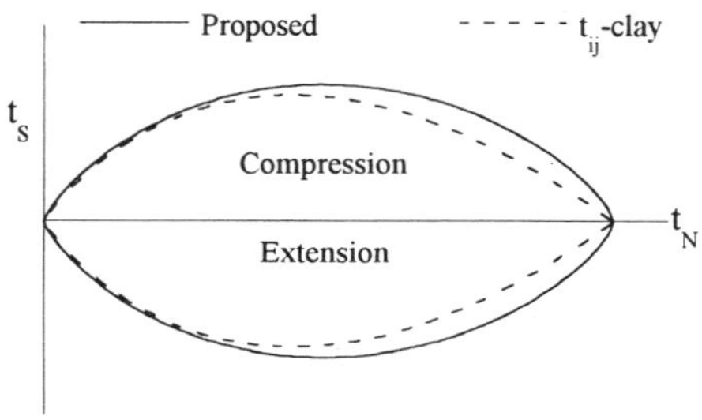

Figure 1. Yield functions in t_N- t_s plane

$$Y = d\varepsilon^*_{SMP} / d\gamma^*_{SMP} = 1/\left(\beta A X^{\beta-1}\right) - X \quad (2)$$

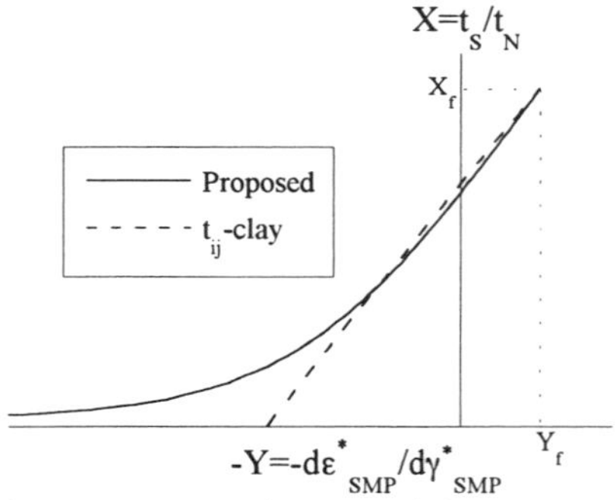

Figure 2. Stress ratio vs. Strain increment ratio.

Yield function given in Eq. 1 is for an isotropic model for normally consolidated clay which is extended to a kinematic one by assuming $X \cong X^* + n$ (which holds for monotonic loading paths. Nakai and Hoshikawa 1991), also paying attention to Hashiguchi's subloading surface model (Hashiguchi, 1993) Eq. 1 has been modified for over consolidated clay and is given by Eq. 3. Where G is the inverse of over consolidation ratio (ratio of current to normal yield surface) in accordance with the t_{ij}-concept. Physical significance of the last term of Eq. 3 is that magnitude of hardening for over consolidated clay will be less than that of normally consolidated clay (since $0 < G \leq 1$). For normally consolidated clay last term of Eq. 3 disappears.

$$f = \ln\left(\frac{t_N}{t_{N0}}\right) + A\left(X^* + n\right)^{\beta} - \left(\frac{\varepsilon_v^p}{C_p} + \ln(G)\right) = 0 \quad (3)$$

Where $n = \sqrt{n_{ij} n_{ij}}$ and n_{ij} is the so called back stress ratio which is assumed to follow the following evolution rule:

$$dn_{ij} = k\left(x_{ij} - n_{ij}\right) \quad (4)$$

$$dn = K_n \left\| dx_{ij} \right\| \quad (5)$$

K_n in above equation is a soil parameter and k can be evaluated from Eq. 5.

Unlike conventional models we assumed that current stress point always remains on the current yield surface and plastic strains are produced when loading occurs as Hashiguchi's subloading surface model (Hashiguchi 1993). When stress point moves inside i.e. when unloading occurs, current yield surface shrinks. Normal yield surface which is similar in shape as current surface but its size is determined by the current void ratio will coincide with the current yield surface for normally consolidated clay while for over consolidated clay normal yield surface will be larger than current yield surface. Based on the foregoing discussion following loading condition may be adopted (Hashiguchi, 1994).

$$\left.\begin{array}{l}\text{Loading: } d\varepsilon^p D^e d\varepsilon > 0 \text{ (hardening / softening)} \\ \text{Nutral or unloading: } d\varepsilon^p D^e d\varepsilon \le 0\end{array}\right\} \quad (6)$$

2.2 *Plastic potential*

Since it is customary to assume associated flow rule for normally consolidated clay, plastic potential can equivalently be given by Eq. 7. But it is observed

$$g = \ln(t_N) + A\left(X^* + n\right)^\beta \quad (7)$$

for over consolidated clay that it does not follow the same stress dilatancy as normally consolidated one, to reflect changing stress-dilatancy with over consolidation, Eq. 7 has been modified as Eq. 8. For NC clay Eq. 7 and 8 are the same.

$$g = \ln(t_N) + \frac{1}{G^l} A\left(X^* + n\right)^\beta \quad (8)$$

2.3 *Strain increments*

Total strain increments are composed of elastic and plastic components, again plastic strains are composed of two separate components as Eq. 9.

$$d\varepsilon_{ij} = d\varepsilon_{ij}^e + d\varepsilon_{ij}^p = d\varepsilon_{ij}^e + d\varepsilon_{ij}^{p(FR)} + d\varepsilon_{ij}^{p(IC)} \quad (9)$$

Elastic part is assumed to follow hook's law:

$$d\varepsilon_{ij}^e = \frac{1+\nu_e}{E_e} d\sigma_{ij} - \frac{\nu_e}{E_e} d\sigma_{kk}\delta_{ij} \quad (10)$$

Plastic component which satisfies flow rule in t_{ij}-space $\left(d\varepsilon_{ij}^{p(FR)}\right)$ is as follows:

$$d\varepsilon_{ij}^{p(FR)} = \Lambda \frac{\partial g}{\partial t_{ij}} \quad (11)$$

Other component is only produced if $dt_N > 0$.

$$d\varepsilon_{ij}^{p(IC)} = \frac{\delta_{ij}}{3} K\langle dt_N \rangle \quad (12)$$

The proportionality constant Λ of Eq. 11 can be evaluated from the consistency condition $df = 0$.

$$df = \frac{\partial f}{\partial \sigma_{ij}} d\sigma_{ij} - \left\{ \frac{1}{C_p} \Lambda \frac{\partial g}{\partial t_{kk}} + \frac{dG}{G} \right\} = 0 \quad (13)$$

Since the current yield surface can never go beyond the normal yield surface, the evaluation equation for G can be assumed as follows:

$$dG = U \left\| d\varepsilon_{ij}^p \right\| \quad (14)$$

where the scalar U is a monotonically decreasing function of G, satisfying following patch tests. (Hashiguchi, 1993)

$$\left.\begin{array}{ll} U = +\infty & \text{for } G = 0 \\ U = 0 & \text{for } G = 1 \\ U < 0 & \text{for } G > 1 \end{array}\right\} \quad (15)$$

There are many other functions of U which satisfy patch tests of Eq. 15 but we considered following function:

$$U = -a\ln(G) \quad (16)$$

Consistency condition of Eq. 13 yields

$$df = \frac{\partial f}{\partial \sigma_{ij}} d\sigma_{ij} - \left\{ \frac{1}{C_p} \Lambda \frac{\partial g}{\partial t_{kk}} - \frac{a\ln(G)}{G} \Lambda \left\| \frac{\partial g}{\partial t_{st}} \right\| \right\} = 0$$

$$\Lambda = \frac{\frac{\partial f}{\partial \sigma_{ij}} d\sigma_{ij}}{\frac{1}{C_p}\frac{\partial g}{\partial t_{kk}} - \frac{a\ln(G)}{G}\left\|\frac{\partial g}{\partial t_{st}}\right\|} \tag{17}$$

Now, formulation of the model is complete. In section 4 we shall see the simulations by the present model.

3 DETERMINATION OF SOIL PARAMETERS

Table 2. Soil parameters for Fujinomori clay

$C_t = \lambda/(1+e_0)$	4.44×10^{-2}	ν_e	0.00
$C_e = \kappa/(1+e_0)$	0.47×10^{-2}	K_n	0.52
φ'_{comp}	33.7°	l	0.50
β	1.40	a	0.10

A total of three triaxial tests are to be performed for the complete determination of the soil parameters listed in table 2. λ and κ are the slopes of loading and unloading-reloading lines of a consolidation test (or an oedometer test) plotted on e vs. $\ln(t_N)$ plane and e_0 is the void ratio at reference state (1 kgf/cm² or 98 kpa in our case). φ'_{comp} can be calculated from a conventional triaxial compression test. ν_e for clay can be assumed close to zero without significant error because plastic strains are predominant in case of clay. K_n has been calculated assuming that stress induced anisotropy builds up linearly with increasing stress ratio and is shown on figure 3. At this stage parameter β should be determined by trial analysis of the triaxial compression test. β value lies between 1 and 2, close to 1.5 gives the best result. Finally l and a can be determined by trial analyses of the test of figure 3 or analyzing triaxial test on over consolidated clay.

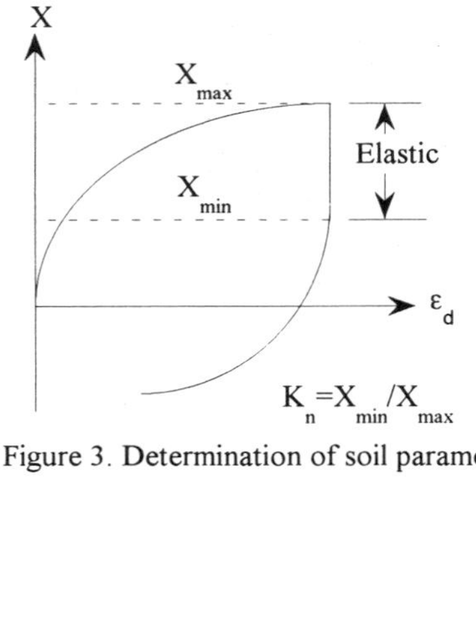

Figure 3. Determination of soil parameter K_n

4 TEST RESULTS AND SIMULATIONS

Strain controlled triaxial compression and extension tests at constant mean principal stress on normally consolidated Fujinomori clay have been performed and corresponding analyses are shown in figure 4 and 5 respectively. Though the analyses slightly over

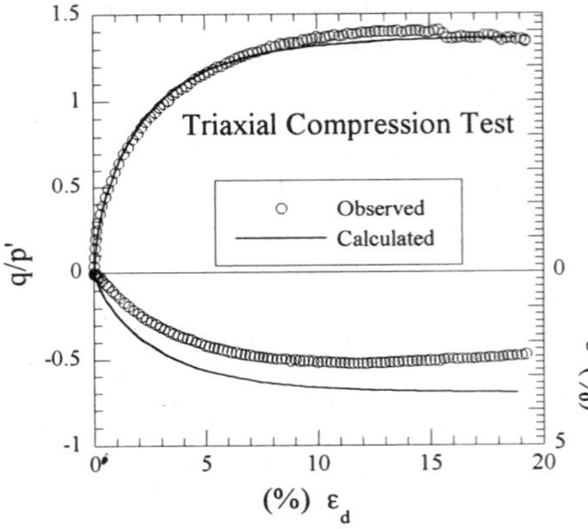

Figure 4. Stress-strain relation of triaxial comp. Test

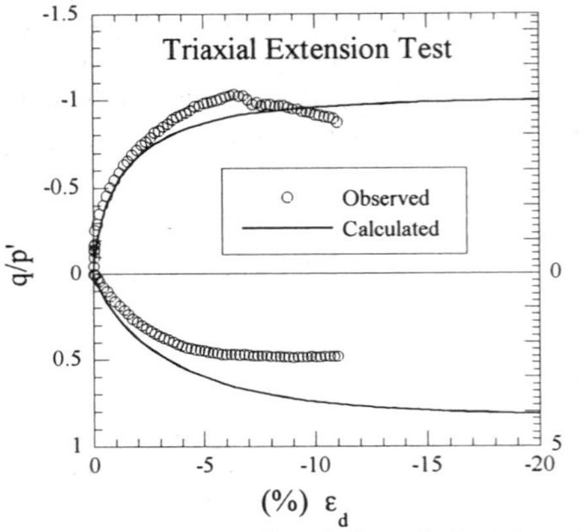

Figure 5. Stress-strain relation of triaxial ext. Test

predict volumetric strains otherwise these well match the observed responses.

Simple cyclic triaxial tests at constant mean principal stress (as shown in figure 6) are also performed (Nakai, Hoshikawa and Chowdhury 1995) and presented in figures 7 through 10. In test Cy-1 soil sample was first sheared along compression side to a principal stress ratio (σ_1/σ_3) of three then sheared along extension side up to failure. Test Cy-2 has been performed in the reverse order. In both the tests calculated ones are close to observed responses. Present model also gives higher strength than NC clay for the above mentioned tests. At the final stage of shearing it shows positive dilatancy as observed in the tests.

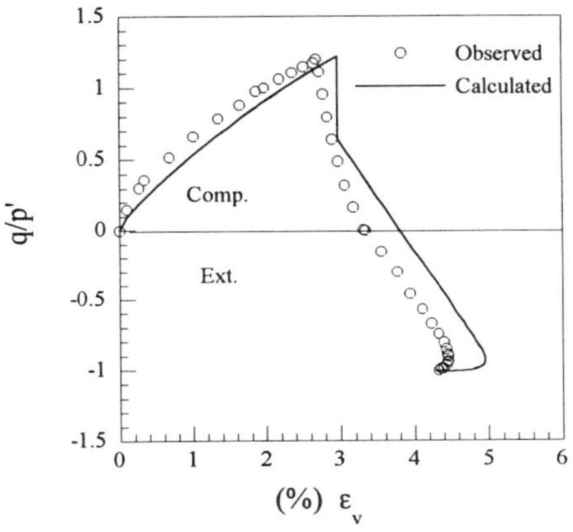

Figure 8. Stress ratio vs. Volumetric strain for Cy-1

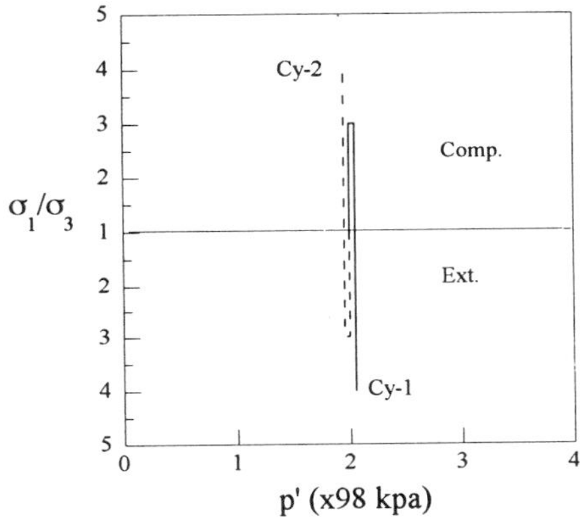

Figure 6. Stress paths of simple cyclic tests

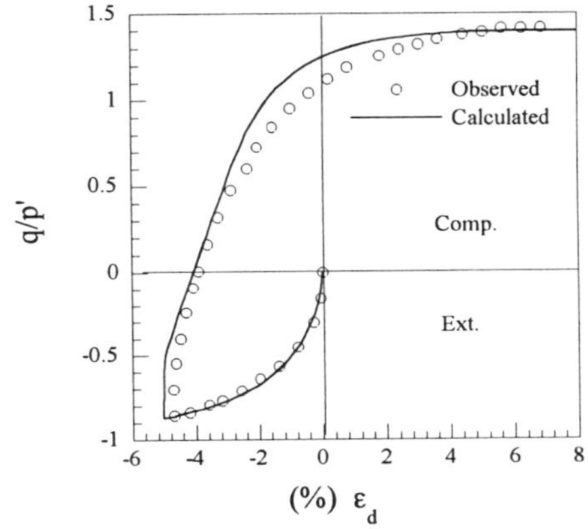

Figure 9. Stress ratio vs. Shearing strain for Cy-2

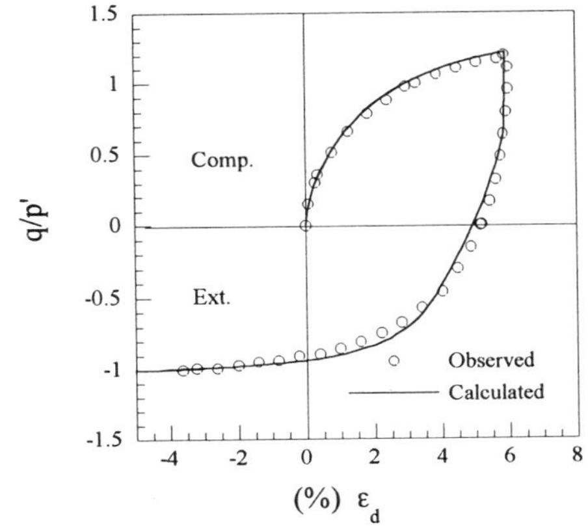

Figure 7. Stress ratio vs. Shearing strain for Cy-1

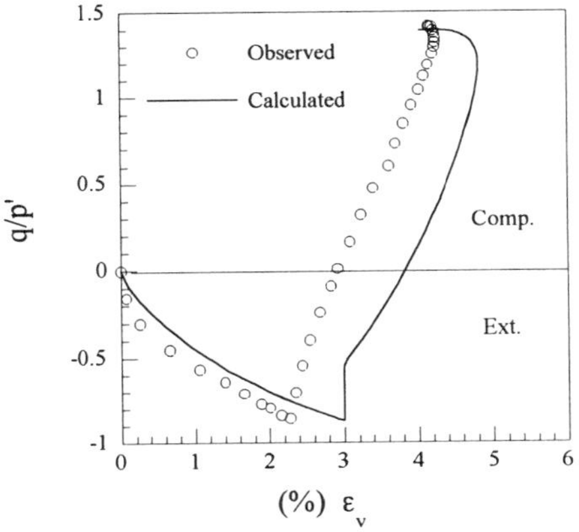

Figure 10. Stress ratio vs. Volumetric strain for Cy-2

Conventional triaxial compression tests at constant mean principal stress on over consolidated clay with various over consolidation ratios (2, 4 and 8) are performed and results are plotted in figure 11. In these tests strengths are closely predicted with little deviation in volumetric strains.

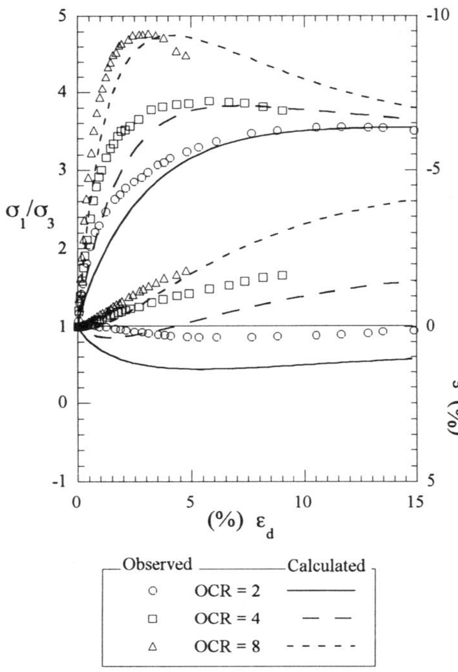

Figure 11. Stress ratio vs. Axial strain vs. Volumetric strain for OC clay with various OCR.

5 CONCLUSIONS

Thus making some numerical treatments as mentioned in section 2 to the conventional approach of modeling, over consolidated and cyclically loaded clay behavior (higher strength, positive dilatancy etc.) can be predicted.

Proposed model uses non-associated flow rule by dividing plastic strain increment and by assuming plastic potential of Eq. 8 but for normally consolidated clay at constant mean principal or other stress paths where $dt_N \leq 0$ associated flow holds.

Number of soil parameters is fewer than many other models which can predict over consolidated clay behavior, hence it can be applicable to boundary value problems.

6 REFERENCES

Hashiguchi, K. 1993. Mechanical requirements and structures of cyclic plasticity models. *Int. J. of Plasticity* 9:721-748.

Hashiguchi, K. 1994. On the loading criterion. *Int. J. of Plasticity* 10(8):871-878.

Nakai, T. and H. Matsuoka 1986. A generalized elastoplastic constitutive model for clay in three dimensional stresses. *Soils and Foundations* 26(3):81-98.

Nakai, T. and T. Hoshikawa 1991. Kinematic hardening model for clay in three dimensional stresses. *Proc. 7th Int. Conf. Comp. Meth. Adv. Geom.* 1:655-660.

Nakai, T., T. Hoshikawa and E.Q. Chowdhury 1995. Stress-strain behavior of clay under cyclic loading and its modeling. *Proc. IS-TOKYO '95* (1):655-660

Roscoe, K.H., A.N. Schofield and A. Thurairajah 1963. Yielding of clays in states wetter than critical. *Geotechnique* 13(3):211-240.

Roscoe, K.H. and J.B. Burland 1968.On the generalized stress-strain behavior of 'wet' clay. In J. Heyman and F.A. Leckie (eds.), *Engineering plasticity,* Cambridge university press: 211-240.

Numerical Models in Geomechanics, Pietruszczak & Pande (eds) © *1997 Balkema, Rotterdam, ISBN 90 5410 886 X*

A three-scale model for consolidation of swelling clay soils

M.A. Murad
LNCC/CNPq, Rio de Janeiro, Brazil

John H. Cushman
Mathematics Department, Purdue University, W. Lafayette, Ind., USA

Abstract: A three scale theory of swelling clay soils is developed which incorporates physico-chemical effects and delayed adsorbed water flow during secondary consolidation. At the microscale the clay platelets and adsorbed water (water between the platelets) are considered as distinct nonoverlaying continua. At the intermediate (meso) scale the clay platelets and the adsorbed water are homogenized in the spirit of hybrid mixture theory, so that, at the mesoscale they may be thought of as two overlaying continua, each having a well defined mass density. Within this framework the swelling pressure is defined thermodynamically and it is shown to govern the effect of physico-chemical forces in a modified Terzaghi's effective stress principle. A homogenization procedure can used to upscale the mesoscale mixture of clay particles and bulk water (water next to the swelling mesoscale particles) to the macroscale. The two-scale governing equations are discretized using a finite element procedure and numerical simulation of a heat source in a compacted bentonitic clay are presented.

1 Introduction

Swelling clay soils consisting of an assemblage of clusters of hydrous alluminium and magnesium silicates with an expanding layer lattice are widely distributed in the earth's crust. Their behavior is of paramount importance in almost all aspects of life, where they are responsible for many reactions and processes. For example, compacted clays such as bentonite have been extensively used to impede the movement of water through cracks and fissures. They play a critical role in various high waste isolation scenerios such as barriers for commercial land fills. Swelling clays also play a critical role in the consolidation and failure of foundations, highways and runways.

Since Terzaghi [16] and Biot [3] first proposed linear poroelastic models for deformable media, the criterion for rupture and failure of soils has been based on the concept of effective stress. It has been experimentally verified that the classical Terzaghi effective stress principle accurately describes coarse-grained soils such as sands, silts and low and medium plastic clays such as kaolinite or illite. However, in its classical form it has been found to be inadequate for explaining deformation of swelling clays, in particular active plastic clays such as bentonite and montmorillonite [10],[15]. The reason is that the classical effective stress principle assumes no other forces except the effective stress and pore pressure are present. The existence of physico-chemical forces within and between the clay particles are excluded.

It has been shown that interparticle forces arising from physico-chemical mechanisms are of paramount importance for active clays. Researchers have heuristically modified Terzaghi's effective stress principle to account for the physico-chemical forces and consequently different mechanistic pictures of the various stresses have been derived [14],[10]).

The nature of the physico-chemical forces remains controversial. In contrast to the effective stress, net attractive(A)-repulsive(R) forces between the clay particles do not depend upon direct contact. They have at least three components: the Van der Walls attraction, electrostatic (or osmotic) repulsion and surface hydration (a structural component). The electrostatic component arises from the electro negativity of natural smectites. The hydrophilic structure of the platelets manifest short range hydration forces between the minerals and water. In the case of clay and many other hydrophilic colloidal particles, hydration forces are believed to arise from the hydrophilic character of the mineral surfaces [9]. It has been argued by Low and co-workers [11], Derjaguin and co-workers [6] and Israelachvili and co-workers [9] that for intersticies smaller than 30Å, the structural hydration forces play a crucial rule in swelling. More specifically, these authors have advocated that for small interlamellar spaces the diffuse double layer forces play a negligible role in swelling and are too weak to explain the anomalous behavior of the adsorbed water.

Hybrid mixture theory, HMT [8] consists of classical mixture theory applied to a multiphase system with volume averaged balance equations. An average value for each phase property is established at every point in the mixture, forming M coexisting continua at each point. Constitutive equations are developed on the averaged scale and are subject to constraints placed by the entropy inequality [5]. In earlier papers [2, 12, 13] the authors have extensively applied HMT to improve the understanding of flow and deformation in swelling systems such as smectitic clays. Within the framework of the HMT, our goal is to provide a natural thermodynamical definition for physico-chemical forces within the clay particles. This is accomplished by adopting a proper theory of constitution which includes appropriate internal variables needed to capture the swelling character of the system. In particular, the approach developed herein provides a thermodynamical basis for the role hydration forces play on the consolidation of a swelling clay soil and also explains some modified effective stress principles which account for hydration forces discussed in [14],[10]. In addition by treating the adsorbed water as a phase different from the clay minerals we get a novel form of Darcy's law, which governs the averaged adsorbed water flow. This form involves an additional interaction potential gradient accounting for the adsorptive character of the clay platelets. Within the current framework, we can overcome some limitations of the works of Lambe [10] where the adsorbed water is considered part of the solid phase with only one total particle stress assumed equal in the platelets and adsorbed water

2 Constitutive Relations for the Swelling Clay Particles

We present the constitutive assumptions and constitutive theory for a two-phase system composed of clay-platelets and adsorbed water (clay particles). The average balance laws can be found in [8]. At this scale the clay particles are viewed as two liquid-solid coexisting continua. The clay systems we have in mind are smectite swelling clays such as montmorillonite. This system may swell under hydration and shrink under desiccation. We assume the macroscopic medium is non-heat conducting and the macroscopic fluid is non-viscous. The behavior of the system is then dictated by the following independent variables:

$$T, \ \rho_l, \ \rho_s, \ \boldsymbol{E}_s, \ \phi_l, \ \boldsymbol{\nabla}\phi_l, \ \boldsymbol{v}_{l,s} \tag{1}$$

where T is the temperature, ρ_l and ρ_s are the averaged densities of the adsorbed water and clay minerals, ϕ_l the volume fraction, $\boldsymbol{v}_{l,s}$ is the mass-average velocity of the adsorbed water relative to the solid phase and $\boldsymbol{E}_s$ is the averaged strain tensor of the solid phase given by $2\boldsymbol{E}_s = \boldsymbol{F}_s^T\boldsymbol{F}_s - \boldsymbol{I}$ with $\boldsymbol{F}_s$ denoting the deformation gradient (Eringen [7]). The novelty in the above set of independent variables is the inclusion of ϕ_l and $\boldsymbol{\nabla}\phi_l$ which allows for the medium to swell at the averaged scale [13],[12]).

It is usually postulated that the Helmholtz free energies of the phases A_α ($\alpha = l, s$ denotes the liquid and solid phase, respectively) depend only on a subset of the set of independent variables. For the system under consideration we postulate the following dependence of the Helmholtz free energies as

$$\begin{aligned} A_s &= A_s(T, \rho_s, \boldsymbol{E}_s) && (2)\\ A_l &= A_l(T, \rho_l, \phi_l). && (3)\end{aligned}$$

Note that A_l depends on ϕ_l in (3). This dependence is motivated by the experimental observations of Low [11] relating the behavior of the adsorbed water to the platelet separation.

For simplicity, consider the isothermal case where temperature gradients and heat fluxes are absent. Let $\boldsymbol{d}_\alpha$, $\boldsymbol{t}_a$ and η_α ($\alpha = l, s$) denote, respectively the symmetric part of the gradient of velocity, averaged stress tensor and entropy of the α-phase. To obtain a constitutive theory for the swelling clay particles we exploit the entropy inequality [8],[12]

$$\begin{aligned} T\Lambda = \sum_{\alpha=l,s} -\phi_\alpha\rho_\alpha\left(\frac{D_\alpha A_\alpha}{Dt} + \eta_\alpha\frac{D_\alpha T}{Dt}\right)\\ +\phi_\alpha \mathrm{tr}(\boldsymbol{t}_\alpha\boldsymbol{d}_\alpha) - \boldsymbol{v}_{l,s}\cdot\widehat{\boldsymbol{T}}_l \geq 0\end{aligned}$$

where Λ is the rate of net entropy production, D_α/Dt denotes the material time derivative following the α-phase and $\widehat{\boldsymbol{T}}_l$ represents the net gain of momentum of the adsorbed water from the solid phase arising in the momentum equation,

$$\phi_\alpha\rho_\alpha\frac{D_\alpha\boldsymbol{v}_\alpha}{Dt} - \mathrm{div}(\phi_\alpha\boldsymbol{t}_\alpha) = \widehat{\boldsymbol{T}}_\alpha \quad (\alpha = l, s) \tag{4}$$

where gravity effects have been neglected. To exploit the restrictions placed by the entropy inequality on the constitutive theory we apply the Coleman and Noll method [5]. Within this framework the total derivatives of the free energies are rewritten in terms of partial derivatives using the chain rule and the functional forms postulated in (2)-(3). Denoting the volume fraction of the solid phase by $\phi_s = 1 - \phi_l$ and the thermodynamic pressure of the α-phase by $p_\alpha = \rho_\alpha^2(\partial A_\alpha/\partial\rho_\alpha)$ and using the relation $D_s\boldsymbol{E}_s/Dt = \boldsymbol{F}_s^T\boldsymbol{d}_s\boldsymbol{F}_s$ (Eringen [7]), the entropy inequality can be rewritten as [12]

$$\begin{aligned} T\Lambda &= \sum_{\alpha=l,s} -\phi_\alpha\rho_\alpha\left(\frac{\partial A_\alpha}{\partial T} + \eta_\alpha\right)\frac{D_sT}{Dt}\\ &+ \phi_l \mathrm{tr}\left((\boldsymbol{t}_l + p_l\boldsymbol{I})\boldsymbol{d}_l\right)\\ &+ \phi_s \mathrm{tr}\left((\boldsymbol{t}_s + p_s\boldsymbol{I} - \boldsymbol{t}_s^e)\boldsymbol{d}_s\right)\\ &- \boldsymbol{v}_{l,s}\cdot(\phi_l p_*\boldsymbol{\nabla}\phi_l - p_l\boldsymbol{\nabla}\phi_l + \widehat{\boldsymbol{T}}_l)\\ &- \frac{D_s\phi_l}{Dt}(\phi_l p_* - p_l + p_s)\\ &\geq 0\end{aligned}$$

where

$$\boldsymbol{t}_s^e = \rho_s\boldsymbol{F}_s\frac{\partial A_s}{\partial\boldsymbol{E}_s}\boldsymbol{F}_s^T, \qquad p_* = \rho_l\frac{\partial A_l}{\partial\phi_l} \tag{5}$$

denote respectively the effective stress tensor and the hydration pressure [12]. The above relation provides an alternative thermodynamical definition for the effective stress tensor $\boldsymbol{t}_s^e$. For example, for a linear isotropic elastic solid with a given pair of Lame constants $\{\lambda_s, \mu_s\}$ the Helmholtz free energy of the solid phase has the quadratic form

$$\rho_s A_s = \frac{\lambda_s}{2}(\text{tr}\boldsymbol{E}_s)^2 + \mu_s \text{tr}\boldsymbol{E}_s^2, \qquad (6)$$

which when combined with (5) implies that $\boldsymbol{t}_s^e$ is given as in the classical Biot theory [3]

$$\boldsymbol{t}_s^e = \lambda_s \text{tr}\boldsymbol{E}_s + 2\mu_s \text{tr}\boldsymbol{E}_s\,. \qquad (7)$$

The new quantity inherent to swelling porous media is the hydration pressure p_* [12],[13]. It appears because we have postulated the dependency of A_l on ϕ_l. This dependency is the key point of this work since it distinguishes our theory from other thermodynamical theories for granular non-swelling porous media [8].

We now linearize the entropy inequality about equilibrium to derive near equilibrium results. For example, if z is a variable of the set $\{\boldsymbol{v}_{l,s}, D_s\phi_l/Dt\}$ which vanishes at equilibrium and f is the coefficient of z within the entropy inequality, the linearization procedure gives an approximation for the near-equilibrium value of f as, $f_{\text{neq}} \approx f_{\text{eq}} + Cz$, where C is the linearization constant. In addition, Λ is a linear function of $\{D_sT/Dt,\ \boldsymbol{d}_l$ and $\boldsymbol{d}_s\}$ which are neither dependent (constitutive) nor independent. Hence to satisfy the entropy inequality for all possible processes, the coefficients of these variables must be identically zero. Note that $D_s\phi_l/Dt$ is not included in this latter set but assumed to be a constitutive variable. In order to have the same number of equations as unknowns an additional equation must be added to the system. This is a typical closure issue discussed in [4]. We postulate that $D_s\phi_l/Dt$ is a dependent variable with dependence given in terms of a volume fraction topological law [4]. In applying the above procedure, the entropy inequality yields the following relations

$$\sum_{\alpha=l,s} \varepsilon_\alpha \rho_\alpha \left(\frac{\partial A_\alpha}{\partial T} + \eta_\alpha\right) = 0 \qquad (8)$$

$$\phi_l \boldsymbol{t}_l = -\phi_l p_l \boldsymbol{I} \qquad (9)$$

$$\phi_s \boldsymbol{t}_s = -\phi_s p_s \boldsymbol{I} + \boldsymbol{t}_s^e \qquad (10)$$

$$\phi_l p_* \boldsymbol{\nabla}\phi_l - p_l \boldsymbol{\nabla}\phi_l + \widehat{\boldsymbol{T}}_l = -\boldsymbol{R}_l \boldsymbol{v}_{l,s} \qquad (11)$$

$$p_l - p_s = \phi_l p_* + \mu_* \frac{D_s\phi_l}{Dt} \qquad (12)$$

where R_l and μ_* are material coefficients arising from the linearization procedure.

3 Modified Effective Stress Principle

Equation (8) is a classical result stating that entropy and temperature are dual variables. Equation (12) is crucial in the present formulation. To exploit its physical significance let us introduce the total particle stress tensor $\boldsymbol{t} = \phi_s \boldsymbol{t}_s + \phi_l \boldsymbol{t}_l$. By adding (9) and (10) and using (12) we obtain

$$\boldsymbol{t} + p_l + \mu_* \phi_s \frac{D_s\phi_l}{Dt}\boldsymbol{I} = \boldsymbol{t}_s^e + p_* \phi_l \phi_s \boldsymbol{I}\,. \qquad (13)$$

The above result provides important information related to the stress analysis of the swelling particles. For simplicity consider the equilibrium case ($D_s\phi_l/Dt = 0$). First note that if we postulate that the free energy of the adsorbed water A_l is independent of ϕ_l, as in the case of a non-swelling granular medium, then $p_* = 0$, and at equilibrium (13) reduces to the form

$$\boldsymbol{t} + p_l \boldsymbol{I} = \boldsymbol{t}_s^e\,. \qquad (14)$$

In classical soil mechanics (14) is recognized as the Terzaghi's effective stress principle for non-swelling granular media with p_l and $\boldsymbol{t}_s^e$ normally being referred to as pore pressure (or bulk phase pressure) and the effective stress tensor. The modified effective stress principle (13) for swelling media has the additional term, $p_* \phi_l \phi_s \boldsymbol{I}$. In contrast with coarse-grained soils where stress mechanisms are primarily controlled by the contact stresses $\boldsymbol{t}_s^e$, for swelling clays such as montmorillonite the additional stress component $p_* \phi_l \phi_s \boldsymbol{I}$ governs the deformation of swelling particles. Clearly this additional intra-particle stress resulting from the presence of adsorbed water within the particles is of physico-chemical nature and can be viewed as a stress structural component arising from surface hydration. Eq (13) resembles in form some modified effective stress principles for clays [14] or [10]. Historically, physico-chemical forces have heuristically been modeled at the macroscale through the addition of a term in the Terzaghi's principle which incorporates the net repulsive ($R\boldsymbol{I}$) and attractive ($A\boldsymbol{I}$) forces between particles. This stress is commonly denoted by $(R - A)\boldsymbol{I}$ (see [14],[10]). Equation (13) is the first rational attempt to extend the modified Terzaghi's principle and to the partition of particle stress into its adsorbed water, effective matrix, and physico-chemical stress components. The modified effective stress principle (13) also provides important information regarding the constitutive behavior of the swelling particles near equilibrium. Note that though the solid is considered elastic, the appearance of the retardation term $\mu_* \phi_s D_s\phi_l/Dt$ in (13) leads to viscoelastic behavior for the volumetric stresses. The coefficient μ_* may be thought of as a retardation factor which among other effects, accounts for the reordering of the adsorbed water molecules as they are disturbed, i.e. an entropic effect. If this is the only source of retardation, then it follows that for a granular media, $\mu_* \approx 0$, since there is very little ordering of the liquid phase in such a medium.

4 Modified Darcy's Law for the Adsorbed Water Flow

Defining the permeability tensor of the clay particles as $\boldsymbol{K}_l = \phi_l^2 \boldsymbol{R}_l^{-1}$ then eliminating $\widehat{\boldsymbol{T}}_l$ in (11) using the momentum equation (4), using (9) and neglecting inertial effects we obtain Darcy's law for the adsorbed water

$$\phi_l \boldsymbol{v}_{l,s} = -\boldsymbol{K}_l \left(\boldsymbol{\nabla} p_l + p_* \boldsymbol{\nabla} \phi_l \right) , \tag{15}$$

In addition to a pressure gradient, the above form of Darcy's law contains a gradient of a generalized interaction potential which accounts for swelling. The appearance of this additional term is consistent with the fact that volume fraction gradients provide a potential for adsorbed water flow in a swelling medium. Note that if we postulate that A_l is independent of ϕ_l, then $p_* = 0$ and (15) reduces to the classical form of Darcy's law.

5 Two-Scale Linear Model

The infinitesimal theory for the clay particles is obtained following the standard linearization procedure: Assume that particles are initially homogeneous, isotropic and at equilibrium. Let us consider that the clay particles are initially at an equilibrium state given by $\boldsymbol{E}_s = 0$, $\phi_l = \overline{\phi}_l$ and $\phi_s = \overline{\phi}_s$. Let $\overline{A}_l = A_l(\overline{\phi}_l)$ denote the free energy of the adsorbed fluid at the reference configuration. Further let $\overline{p}_* = p_*(\overline{\phi}_l)$, $\overline{K}_l = K_l(\overline{\phi}_l)$ and $\overline{\mu}_* = \overline{\phi}_s \mu_*(\overline{\phi}_l)$ and let the strain tensor be identified with its linearized form $\boldsymbol{E}_s = 1/2(\boldsymbol{\nabla} \boldsymbol{u}_s + \boldsymbol{\nabla} \boldsymbol{u}_s^T)$ where $\boldsymbol{u}_s$ denotes the diaplacement of the solid phase. Expand A_s and A_l in a Taylor series about equilibrium and retain quadratic terms. In addition to (6) postulate

$$\rho_l A_l = \rho_l \overline{A}_l + \overline{p}_* (\phi_l - \overline{\phi}_l) + \frac{\gamma}{2} (\phi_l - \overline{\phi}_l)^2$$

Using the above expansions in (5) the linearized form of $\boldsymbol{t}_s^e$ is given as in (7. In addtion the linearized relation for the hydration pressure is

$$p_* = \overline{p}_* + \gamma (\phi_l - \overline{\phi}_l) \,. \tag{16}$$

We are now ready for our linearized governing equation in the clay particle domain. By neglecting all inertial and convective effects, for $\boldsymbol{q}_l \equiv \phi_l \boldsymbol{v}_{l,s}$, our system of linearized equations governing the swelling clay particles written in terms of the unknowns $\{\boldsymbol{u}_s, \boldsymbol{q}_l, \boldsymbol{t}_s^e, \phi_l, p_*, p_l, \boldsymbol{t}\}$ is

Mass of the Adsorbed Water

$$\frac{\partial \phi_l}{\partial t} + \overline{\phi}_s \mathrm{div} \boldsymbol{q}_l = 0. \tag{17}$$

Total Mass

$$\mathrm{div} \boldsymbol{q}_l + \mathrm{div} \frac{\partial \boldsymbol{u}_s}{\partial t} = 0. \tag{18}$$

Total Momentum

$$\mathrm{div} \boldsymbol{t} = 0.$$

Total Particle Stress Constitutive Equation

$$\boldsymbol{t} = -p_l \boldsymbol{I} + \boldsymbol{t}_s^e + \left(p_* \phi_l \phi_s + \overline{\mu}_* \frac{\partial \phi_l}{\partial t} \right) \boldsymbol{I}. \tag{19}$$

Linearized Effective Stress Constitutive Relation

$$\boldsymbol{t}_s^e = \lambda_s \mathrm{div} \boldsymbol{u}_s \boldsymbol{I} + 2 \mu_s \boldsymbol{\nabla}^s \boldsymbol{u}_s. \tag{20}$$

Linearized Hydration Stress Constitutive Relation

$$p_* \phi_l \phi_s = \overline{p}_* \overline{\phi}_l \overline{\phi}_s + \overline{g} (\phi_l - \overline{\phi}_l).$$

Modified Darcy's Law for the Adsorbed Water

$$\boldsymbol{q}_l = -\overline{\boldsymbol{K}}_l (\boldsymbol{\nabla} p_l + \overline{p}_* \boldsymbol{\nabla} \phi_l).$$

where

$$\overline{g}(\overline{\phi}_l) = \overline{\phi}_l \overline{\phi}_s \frac{dp_*}{d\phi_l} \bigg|_{\phi_l = \overline{\phi}_l} + \overline{p}_* (\overline{\phi}_s - \overline{\phi}_l)$$

The above system can be rewritten in terms of a single equation for ϕ_l of viscoelastic type. In terms of $\{\boldsymbol{u}_s, p_l, \phi_l\}$ this system is given by

$$\mu_s \Delta \boldsymbol{u}_s + (\lambda_s + \mu_s) \boldsymbol{\nabla} \mathrm{div} \boldsymbol{u}_s - \boldsymbol{\nabla} p_l + \overline{g} \boldsymbol{\nabla} \phi_l + \overline{\mu}_* \boldsymbol{\nabla} \frac{\partial \phi_l}{\partial t} = 0, \tag{21}$$

$$\frac{\partial \phi_l}{\partial t} - \overline{\phi}_s \overline{K}_l (\Delta p_l + \overline{p}_* \Delta \phi_l) = 0, \tag{22}$$

$$\phi_l - \overline{\phi}_l = \overline{\phi}_s \mathrm{div} \boldsymbol{u}_s. \tag{23}$$

In the above system the effects of hydration effects appear through the coefficients $\overline{p}_*$ and $\overline{\mu}_*$. Note that if we set $\overline{p}_* = \overline{\mu}_* = 0$ as in the case of a granular non-swelling medium, the aboce system reproduces the classical Biot's model of consolidation of linear elastic media.

6 Remarks on the Three-Scale Model

A three-scale model (micro, meso and macro) of a porous matrix consisting of porous swelling particles is depicted in Fig 1. The particles are in contact with one another and bulk water. Each particle consists of clay colloids and vicinal water. At the microscale the model has two phases, the disjoint clay minerals and the vicinal water. At the mesoscale (the homogenized microscale) the model consists of the clay particles (where our two-scale governing equations hold) and the bulk water. The macroscale consists of the bulk water homogenized with the swelling particles. In

Murad et al. [12, 13] the two-scale model was coupled with Stokesian slow bulk-water movement and the homogenization procedure was used to upscale the mesoscopic governing equations to the macroscale. In the latter case the homogenized equations give rise to a dual porosity model in which at the macroscale the vicinal water is represented by sources/sinks to the bulk phase flow (see [13]). The homogenization technique for deriving three-scale models has been successfully used to model naturally fractured reservoirs in which the system of fractures play the role of the bulk phase s ystem (where the macroscopic flow takes place) and the matrix blocks behave as t he analogue of the clay particles (see e.g. Arbogast [1])

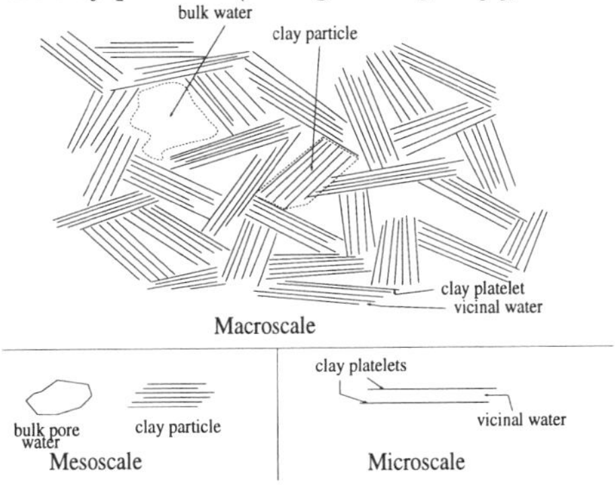

Fig 1: Three-Scale Model for Clay

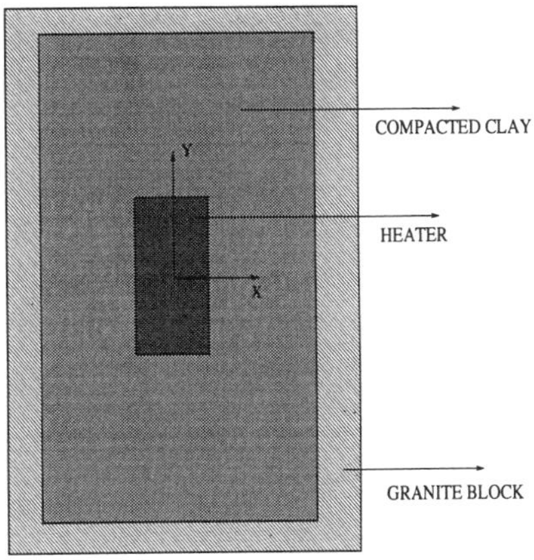

Fig 2: Heat Source in compacted clay

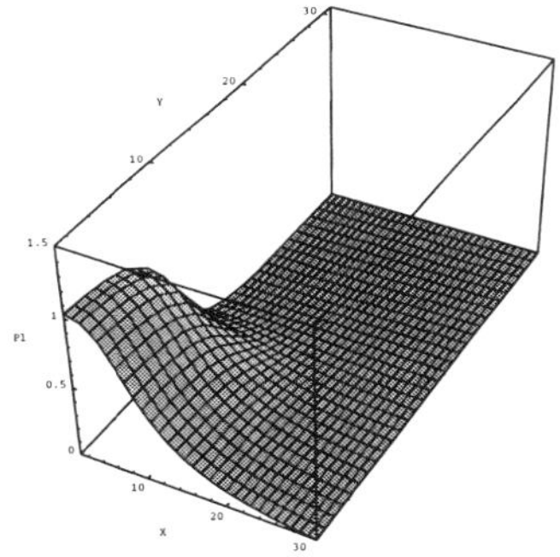

(a)

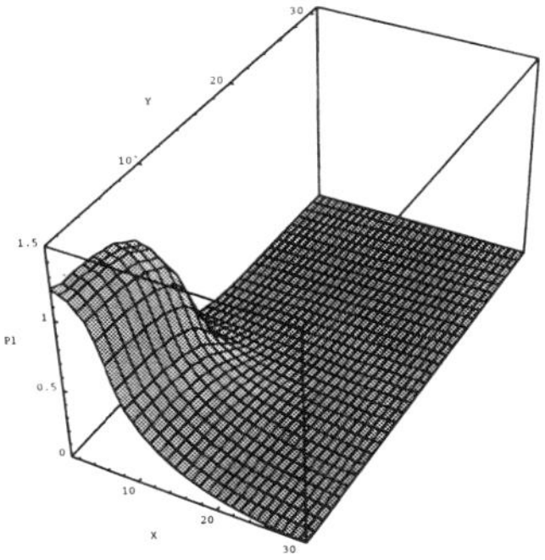

(b)

Fig 3: Adsorbed and bulk water pressures

7 Numerical Simulation

The governing equations arising from the two-scale model describe highly compacted clays with a monomodal distribution with most of the water essentially adsorbed. In thise section our purpose is to illustrate the effect of physico-chemical forces on the adsorbed water. Since in the proposed model physico-chemical forces are governed by the hydration pressure p_*, our aim is to show the influence of p_* on the adsorbed water presure p_l. The selected problem consists of a heat source in a compacted clay (fig. 2). The temperature field is obtained by solving the heat equation and is gradient treated as a source term in the momentum balance (21). Two-scale governing equations (21),(22),(23) are discretized by a finite element procedure, Taylor-Hood elements (biquadratic elements for displacements and bilinear for pore pressure) were adopted and the time domain is discretized by the backward Euler method. Fig. 3(a)-(b) depict respectively the adsorbed water pressure for two values of the hydration pressure p_*. The former exhibits physico-chemical effects ($p_* = 2$), where the modified Terzaghi's principle (13) holds and the latter behaves as a bulk fluid ($p_* = 0$ satisfying the classical principle (14). Note that physico-chemical effects amplify the magnitude of p_l, especially close to the heater.

8 Conclusions

A hybrid mixture theory for swelling clay soils was presented. Clay minerals and adsorbed water were treated as distinct phases in order to capture swelling at the particle level and physico-chemical effects. This physico-chemical component has a thermodynamical definition in terms of change of the free energy of the adsorbed fluid with respect to the volume fraction. Some notable consequences of the theory developed herein are that it provides a rational basis for some heuristically modified effective stress principles for smectite clays. In addition a modified form of Darcy's law was derived for the adsorbed water flow. Numerical simulation of a heat source in betonitic clay is presented illustrating the effect oh hydration stresses.

References

[1] T. Arbogast. Gravitational forces in dual-porosity systems. *Transport in Porous Media*, 13:179–220, 1993.

[2] L. S. Bennethum and J. H. Cushman. Multiscale hybrid mixture theory for swelling systems: Part II: Constitutive theory. *Int. J. Engrg Sci*, 34(2):147–169, 1996.

[3] M. Biot. General theory of three-dimensional consolidation. *J. Appl. Phys.*, 12:155–164, 1941.

[4] J. A. Bouré. Two-phase flow models: The closure issue. In G. F. Hewitt, J. M. Delhaye and N. Zuber, editor, *Multiphase Science and Technology*, volume 3, pages 3–30. Marcel Dekker, New York, 1987.

[5] B. D. Coleman and W. Noll. The thermodynamics of elastic materials with heat conduction and viscosity. *Arch. Rat. Mech. Anal.*, 13:167–178, 1963.

[6] B. V. Derjaguin, N. Churaev, and V. M. Muller. *Surface Forces.* Plenum press, New York, 1987 .

[7] A. C. Eringen. *Mechanics of Continua.* John Wiley and Sons, New York, 1967.

[8] S. M. Hassanizadeh and W. G. Gray. General conservation equations for multiphase systems: 3. Constitutive theory for porous media. *Adv. Water Resour.*, 3:25–40, 1980.

[9] J. Israelachvili. *Intermolecular and Surface Forces.* Academic Press, New York, 1991.

[10] T. W. Lambe. A mechanistic picture of shear strength in clay. In *Proceedings of The ASCE research conference on shear strength of cohesive soils*, pages 503–532, Boulder Colorado, 1960.

[11] P. F. Low. Structural component of the swelling pressure of clays. *Langmuir*, 3:18–25, 1987.

[12] M. A. Murad, L. S. Bennethum, and J. H. Cushman. A Multiscale theory of swelling porous media: I Application to one-dimensional consolidation. *Transport in Porous Media*, 19:93–122, 1995.

[13] M. A. Murad and J. H. Cushman. Multiscale flow and deformation in hydrophilic swelling porous media . *Int. J. Engrg Sci*, 34(3):313–336, 1996.

[14] A. Sridharan and G. V. Rao. Mechanisms controlling volume change of saturated clays and the role of the effective stress concept. *Geotechnique*, 23(3):359–382, 1973.

[15] A. Sridharan and G. V. Rao. Mechanisms controlling the secondary compression of clays. *Geotechnique*, 32(3):249–260, 1982.

[16] K. Terzaghi. *Theoretical soil mechanics.* John Wiley and Sons , 1942.

Numerical Models in Geomechanics, Pietruszczak & Pande (eds) © 1997 Balkema, Rotterdam, ISBN 90 5410 886 X

A soft soil model and experiences with two integration schemes

D.F.E.Stolle
Department of Civil Engineering, McMaster University, Hamilton, Ont., Canada

P.G.Bonnier
Plaxis B.V., Rhoon, Netherlands

P.A.Vermeer
Institut für Geotechnik, Universität Stuttgart, Germany

ABSTRACT: A constitutive model is developed within the framework of volumetric hardening elasto/viscoplasticity for predicting the stress-strain-time behaviour of soft soils. The paper also presents two time-stepping algorithms for integrating the sensitive creep law. Examples are given to demonstrate the appropriateness of the constitutive description and the performance of the time-stepping schemes.

1 INTRODUCTION

The prediction of creep movements associated with the excavation of tunnels and cuts and the long term settlement of structures is often an important design consideration. Use of conventional time-independent procedures for the interpretation of laboratory results and the analysis of geotechnical boundary-valued problems involving rate sensitive soils may result in solutions and interpretations which do not properly capture the actual in situ soil response; see e.g., Skempton (1964) and Bishop(1966).

The objectives of this paper are to: (a) present a general soft soil model which is capable of taking into account the rate sensitivity of undrained shear strength and creep rupture, as well as apparent preconsolidation pressures; and (b) discuss the use of numerical procedures for time-stepping using large time increments. Examples are given to demonstrate the appropriateness of the model. Given the space limitations, only the literature which is most relevant to the paper is cited.

2 CONCEPTUAL MODEL

The model described in this section is constructed within an elasto/viscoplastic framework. The time-dependent, isothermal deformation of an isotropic clay is assumed to be given by an equation of state that relates the volumetric creep strain rate $\dot{\varepsilon}_v^c$ to an equivalent pressure p_a, that depends on the effective confining pressure p and deviatoric stress q, and the volumetric creep strain ε_v^c that acts as a hardening parameter; i.e.

$$\dot{\varepsilon}_v^c = \frac{C}{\tau}\left(\frac{p_a}{p_c}\right)^{B/C} \tag{1}$$

where p_c is a reference pressure defined as

$$p_c = p_{co} e^{\varepsilon_v^c / B} \tag{2}$$

with p_{co} being the reference pressure which at the beginning of an analysis is related to the overconsolidation ratio OCR ; i.e. p_{co} = OCR · p_a. The parameter p_{co} may be viewed as an apparent isotropic preconsolidation pressure. When using eq. (2) it is understood that $\varepsilon_v^c = 0$ for time $t = 0$ and that compressive stresses and strains are positive.

Given the initial void ratio e_o, critical state theory paramters λ and κ, or better known compression C_c and recompression C_r indices,

$$B = \frac{(\lambda - \kappa)}{(1+e_o)} \approx \frac{(C_c - C_r)}{\ln 10 \cdot (1+e_o)} \tag{3}$$

The parameter τ is introduced to provide a time scale, often given the value one day for secondary compression analysis, and $C = C_\alpha/[\ln 10 \cdot (1+e_o)]$ with C_α being the secondary compression index.

If we were to restrict ourselves for the moment to fully drained conditions corresponding to zero deviatoric stress q and constant effective confining pressure of p_a as shown in Figure 1, eq. (1) indicates that the volumetric creep strain rate depends on the ratio p_a/p_c. As a soil creeps, the value of p_c increases according to eq (2), thereby causing a continuous reduction in creep rate over time. As long as the state of stress remains on the hydostatic axis, p_a represents the actual state of effective stress. For constant p_a it is possible to integrate eq. (1) subject to (2) over time Δt yielding an expression for creep strain increment

$$\Delta\varepsilon_v^c = C\ln[1+\frac{\dot{\varepsilon}_v^c\Delta t}{C}] \tag{4}$$

in which $\dot{\varepsilon}_v^c$ is the creep rate at the beginning of the time step.

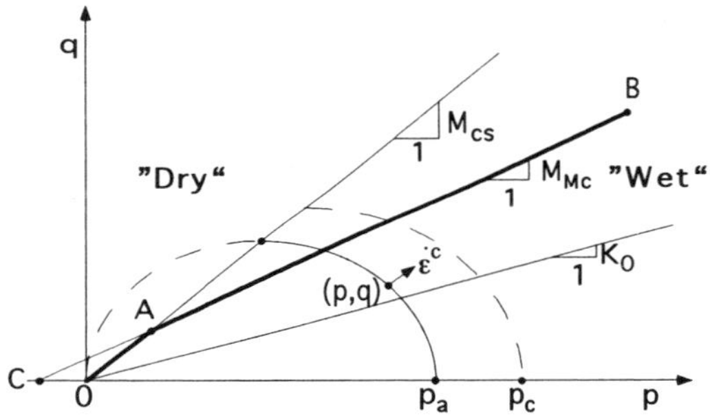

Figure 1. Stress space description of model.

For more general stress conditions (p,q), the creep rates are still given by eq.(1). The parameter p_a however now represents an equivalent pressure which is determined by some functional relation to the actual state of stress. It is assumed that the function defining the creep potential Q provides an appropriate relation. This relation may also be viewed as representing a curve of constant volumetric creep strain rate.

When deviatoric stresses are present, shear failure can develop if the state of effective stress makes contact with the *'critical state'* line 0A or the Mohr-Coulomb failure envelope defined by AB in Figure 1; i.e., the feasible stress space is restricted to the area between the p-axis and the locus of points along 0AB. Under constant deviatoric stress, failure can only occur under undrained conditions where an increase in volumetric creep strain is accompanied by an increase in excess pore pressure thus a decrease in effective confining stress p. Although the creep rate continuously decreases due to volumetric hardening, a continuous reduction in p, albeit slow for small deviatoric stresses, will eventually lead to creep failure. For many clays $M_{mc} \approx M_{cs}$ and $c \approx 0$.

3 MATHEMATICAL MODEL

3.1 *Invariant Definitions*

For materials which are isotropic, a soil's constitutive response to loading can be conveniently described in terms of invariant stresses and strains that are independent of the frame of reference. Given that σ_{oct} and τ_{oct} are the octahedral normal and shear stresses, respectively, the failure and creep potential functions can be conveniently defined for general states of stress in terms of: pressure $p = \sigma_{oct}$; deviatoric stress $q = 3\tau_{oct}/\sqrt{2}$; and Lode angle θ. The conjugate strain invariants for p and q are volumetric ε_v and deviatoric $\varepsilon_q = \gamma_{oct}/\sqrt{2}$ strain, respectively, with γ_{oct} representing the octahedral shear strain.

The usual soil mechanics convention of compression being positive is adhered to in the following development. For the most part, the notation used by Zienkiewicz and Taylor (1991) is adopted.

3.2 *Elasto/viscoplastic Constitutive Law*

For the moment it is assumed that the stress levels are well below failure conditions, thereby allowing us to relate stress rates $\dot{\underline{\sigma}}$ to strain rates $\dot{\underline{\varepsilon}}$ via

$$\dot{\underline{\sigma}} = \underline{\underline{D}}(\dot{\underline{\varepsilon}} - \dot{\underline{\varepsilon}}^c) \tag{5}$$

where $\underline{\underline{D}}$ is the elastic constitutive matrix which is a function of Poisson's ratio ν and a pressure sensitive bulk modulus $K = (1+e_o)p/\kappa$. The creep strain rate vector $\dot{\underline{\varepsilon}}^c$ is defined according to

$$\dot{\underline{\varepsilon}}^c = \frac{\dot{\varepsilon}_v^c}{Q,_p}\frac{\partial Q}{\partial \underline{\sigma}} \tag{6}$$

where $Q,_p$ represents the derivative of Q with respect to p and Q the *Modified Cam-Clay* plastic potential defined as

$$Q = \sqrt{\left(p - \frac{p_a}{2}\right)^2 + \left(\frac{q}{M_{cs}g(\theta)}\right)^2} - \frac{p_a}{2} = 0 \tag{7}$$

with $g(\theta)$ being a function that describes the shape of the potential function in the deviatoric plane and M_{cs} representing the slope of the *'critical state'* line which is related to the friction angle for triaxial states of stress via $M_{cs} = (6\sin\phi_{cs})/(3-\sin\phi_{cs})$.

4 NUMERICAL ALGORITHMS

4.1 *Integration for Creep Response (Scheme A)*

A close examination of eq. (1) indicates that we have a power law with exponents typically varying between 15 and 25. Although various procedures exist for integrating eq. (5) over various stress/strain histories, severe restrictions on the size of admissible time step exist to maintain numerical stability unless an implicit scheme is adopted. The semi-implicit procedure developed by Stolle (1991) was the first scheme implemented for this study; i.e. for $F = \Delta t\dot{\varepsilon}_v^c$, time step $\Delta t = t_{n+1} - t_n$ and degree of implicitness factor θ, usually equal to 0.5, we have the following set of equations, for relating stress increment $\Delta\underline{\sigma}$ to strain increment $\Delta\underline{\varepsilon}$ subject to creep loading $\Delta\underline{\sigma}^c$,

$$\Delta\underline{\sigma} = \underline{\underline{D}}^{ec}\Delta\underline{\varepsilon} - \Delta\underline{\sigma}^c \tag{8}$$

where

$$\underline{\underline{D}}^{ec} = \underline{\underline{D}} - \frac{\theta\underline{\underline{D}}\frac{\partial Q^T}{\partial\underline{\sigma}}\frac{\partial F}{\partial\underline{\sigma}}\underline{\underline{D}}}{Q,_p + \theta(H_e + H_c)} \tag{9}$$

$$\Delta\underline{\sigma}^c = \frac{F_o}{Q,_p + \theta(H_e + H_c)}\underline{\underline{D}}\frac{\partial Q}{\partial\underline{\sigma}} \tag{10}$$

$$H_e = \frac{\partial F}{\partial\underline{\sigma}}\underline{\underline{D}}\frac{\partial Q}{\partial\underline{\sigma}}, \quad H_c = -\frac{\partial F}{\partial\varepsilon_v^c}\frac{\partial Q}{\partial p} \tag{11}$$

with the subscript 'o' denoting a value at the beginning of a time step. A close examination of these equations indicates that all matrices and vectors remain bounded as $\Delta t \to \infty$. Furthermore for states of stress lying on the *'critical state'* line we find that $\dot{\varepsilon}_v^c = 0$ and the tangent matrix $\underline{\underline{D}}^{ec} \to \underline{\underline{D}}^{ep}$ which corresponds to the elastoplastic constitutive matrix. In the absence of a separate failure envelope, this would imply that the constitutive description accomodates failure in a natural manner.

Although creep analyses are often performed using initial strain algorithms, a tangential, initial stress procedure was found to be more efficient in which $\underline{\underline{D}}^{ec}$ is used when assembling the stiffness matrix, yet the effect of $\Delta\underline{\sigma}^c$ is introduced at the stress update level. To improve accuracy a sub-increment scheme had to be implemented at the local level whereby the strain increment is subdivided and the stress update takes place over n steps, where n is approximately given by 5-10 times $\Delta t/\Delta t_{max}$ with Δt_{max} representing the critical time step for non-oscillating decay; see Stolle (1991).

4.2 *Alternative Integration Procedure (Scheme B)*

A problem with Scheme A is that n can become quite large; particularly for $p_a/p_c > 1$. To overcome this problem an efficient fully implicit scheme can be developed by defining $F = \Delta\varepsilon_v^c$ using eq. (4). The resulting constitutive equation for the formation of the global equations is the same as that given by eq.'s (8) to (11) with $\theta = 1$ and $H_c = 0$ It can be shown that this alternative scheme reduces to the fully implicit form of that given in Section 4.1 for small $\Delta t\dot{\varepsilon}_v^c$.

On the local stress update level eq.'s (4) and (5) are used directly together with an iterative procedure similar to that developed by Borja and Lee (1990) for plasticity. A stress update is first performed in the *(p,q)* space using a radial return procedure before the Cartesian components are modified. The algorithm starts by updating stresses assuming elastic behaviour; i.e., given initial stresses (p_o, q_o) then $p^e = p_o + K\Delta\varepsilon_v$ and $q^e = q_o + 3G\Delta\varepsilon_q$ with K and G being the pressure dependent bulk and shear moduli, respectively.. The state of stress at the end of the time step can now be calculated iteratively via

$$F = C\ln\left[1 + \left(\frac{p_a}{p_{co}}\right)^{B/C}\frac{\Delta t}{\tau}\right] \tag{12}$$

$$p = p^e - KF \qquad (13)$$

$$q = q^e - 3GF\frac{2pq}{M_{cs}^2 p^2 - q^2} \qquad (14)$$

A stepwise uncoupled procedure was implemented to solve these equations holding p_{co} constant; i.e., (a) given estimated stresses $(\bar{p},\bar{q})$, $\bar{F}$ is calculated using eq. (12); (b) given $\bar{F}$ then stresses are updated analytically using eq.'s (13) and (14); and (c) with F evaluated using eq.(12) and the new estimates for (p,q), $\bar{F}$ is updated according to

$$\bar{F} \to \bar{F} + \frac{F - \bar{F}}{1 - F_{,\bar{F}}} \qquad (15)$$

Steps (b) and (c) are repeated until F $\approx \bar{F}$ at which time p_c can also be updated.

After state (p,q) has been determined, the Cartesian stresses are obtained according to

$$\underline{\sigma} = p\underline{I} + \frac{q}{q^e}\underline{s}^e \qquad (16)$$

where $\underline{s}^e$ is the deviatoric stress vector corresponding to state (p^e, q^e) and $\underline{I}$ ensures that p is only added to the normal stresses. The algorithm presented in this section was found to converge extremely well for all the stress paths tested when compared to predictor-corrector schemes which tended to diverge.

At this point it is important to realize that, since the global system of equations are nonlinear the updated stresses are not necessarily the same as the equilibrium stresses thus an iteration loop is required within each time step for both Schemes A and B; see, for example Vermeer and van Langen (1989).

4.3 *Treatment of Failure*

Although the creep formulation already accomodates shear failure, a separate rate independent failure criterion is introduced to accommodate failure under instantaneous loading and to avoid numerical difficulties that can develop due to a negative or semi-definite constitutive matrix that appears when the state of stress is on or above the critical state. Furthermore, failure may not always correspond to critical state conditions.

After updating stresses a check is made to ensure that the stress point lies within the feasible stress space; i.e., $f < 0$ where f represents the yield function. For cases where yielding takes place, a stress correction is calculated via

$$\Delta\underline{\sigma}^{corr} = -\underline{\underline{D}}\frac{\langle f \rangle}{h}\frac{\partial g}{\partial \underline{\sigma}} \qquad (17)$$

where h is the usual hardening parameter for plasticity, g is a plastic potential and the brackets $\langle\rangle$ are introduced as an operator where $\langle x \rangle = x$ for $x > 0$ and $\langle x \rangle = 0$ for $x \leq 0$. It was assumed for the purpose of this study that failure can be defined using the Mohr-Coulomb criterion and that the plastic potential takes a similar form, with a dilatency angle $\psi = 0$ to ensure that no negative pore pressure can develop under undrained conditions.

5 MODEL VALIDATION

To carry out this phase of the study, some of the undrained triaxial tests performed by Vaid and Campanella (1977) on Haney clay were simulated using the material parameters recommended by Matsui and Abe (1988) and summarized in Table 1. The Poisson's ratio of 0.25 was assumed, along with a *'critical state'* angle slightly larger than the failure friction angle of 32°. All triaxial tests had been completed by initially consolidating the samples under an effective confining pressure of 525 kPa for 36 hours and then allowing them to stand for 12 hours under undrained conditions before starting a test.

Table 1. Material Properties for Haney Clay

$\kappa = 0.031$	$\phi_{cs} = 32.1°$	$\phi_{mc} = 32.0°$
$\nu = 0.25$	$\tau = 1$ day	$\psi = 0°$
$e_o = 0.896$	$\lambda = 0.20$	$c = 0$ kPa
	$C = 0.004$	

By examining eq.'s (1) and (2) one may appreciate that the values of p_{co} and τ are not independent. The value of p_{co} was determined by simulating the creep test corresponding to a deviatoric stress of $q = 300.3$ kPa, assuming a τ of 1 day and then matching the creep rupture life. The p_{co} value of 373 kPa was found to give a reasonable result. As one can see, based on the definition $OCR = p_{co}/p_a$, there was insufficient time for the reference pressure to attain the value of 525, which would have been required for an overconsolidation ratio of 1.0. Assuming that the model accurately describes the constitutive behaviour, it is clear that the 'preconsolidation pressure' not only depends on the applied maximum stress, but also on time as suggested by Bjerrum (1967).

Up to this point of the model validation process, only one set of results has been used to determine missing information. Figure 2 compares the experimental and predicted creep rupture lives at the various deviatoric stress levels. Based on this comparison, it is clear that the constitutive model is capable of identifying the creep rupture characteristic of the material under consideration . Figure 3 confirms that the strain histories predicted by the model are also reasonable.

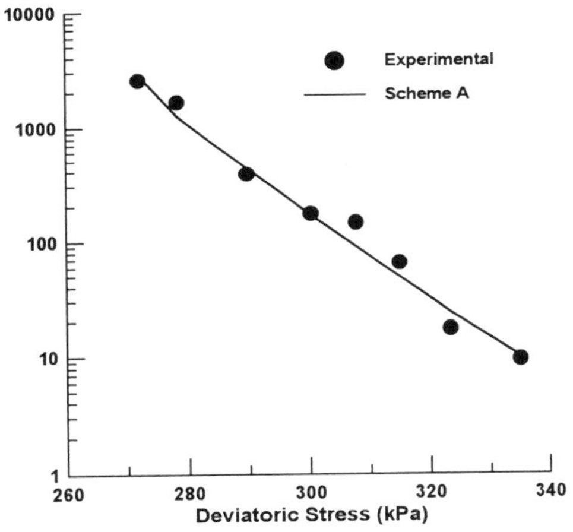

Figure 2. Prediction of creep rupture life.

If the creep behaviour follows an equation of state, then it should also be possible to properly account for the rate effects associated with CU triaxial tests. Figure 4 verifies that Haney clay appears to obey the equation of state selected for the soft soil model. A sensitivity analysis further revealed that the undrained shear strength c_u may be approximated by

$$c_u / c_{u1\%/hr} \approx 1.02 + 0.09 \log \dot{\varepsilon} \tag{18}$$

which agrees fairly well with the experimental data summarized by Kulhawy and Mayne (1990). The c_u in eq. (18) is normalized with respect to the strength corresponding to a strain rate of 1 percent per hour. It was found that eq. (18) appeared to be independent of the choice of creep law but appeared to depend on the form of creep potential.

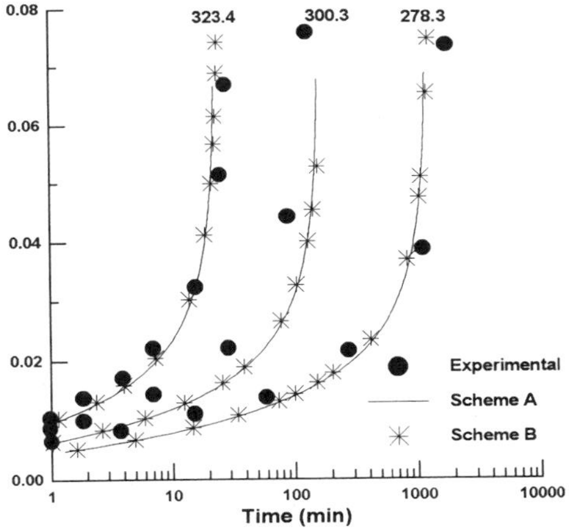

Figure 3. Prediction of strain history for creep test.

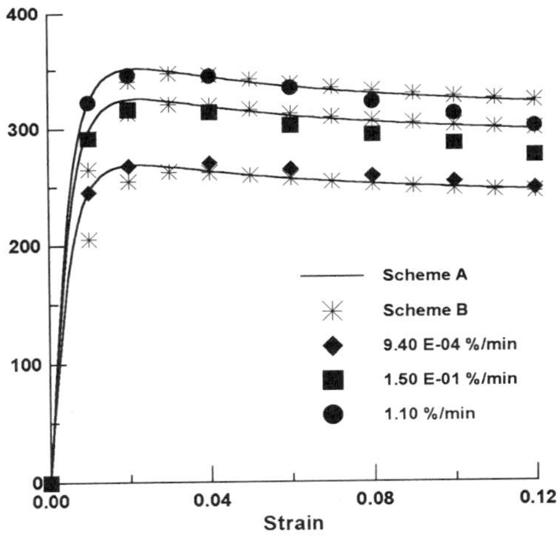

Figure 4. Rate effects on stress-strain behaviour.

6 NUMERICAL PERFORMANCE

The creep simulations with Scheme A for model validation were performed using 2000 steps. While this may be satisfactory for one point integation, finite element simulations are usually carried out using fewer steps. A sensitivity analysis indicated that it was possible to reduce the number of steps to 30, typically the number required when using Scheme B, however 20 to 40 substeps were necessary to properly capture the strain histories. When using Scheme B, stress updates could be achieved typically within 3 to 4 local iterations; i.e. a factor of 10 fewer substeps/iterations. Whereas the number of substeps for Scheme A must be defined, a priori, the number of iterations used in Scheme B depends on the rate of convergence which may change from step to step.

The stress-strain predictions using Scheme B, shown in Figure 4, were obtained using 12 steps. Similar results were obtained with Scheme A in which 12 steps were selected with n = 16. An important observation is that the efficiency with respect to modelling the stress-strain-time behaviour of the consolidated undrained triaxial tests did not provide a good measure for evaluating the numerical schemes. The performance associated with modeling the undrained creep behaviour of the Haney clay proved to be a much better test for comparing the efficiency of the algorithms.

The question arises why many substeps are required for accurate results when using Scheme A whereas rapid convergence is possible using the alternative, fully implicit algorithm? If one closely examines eq.'s (8) to (11), one can recognize that for $\Delta t >> \Delta t_{max}$, $\underline{\underline{D}}^{ec} \rightarrow \underline{\underline{D}}^{ep}$ and that $\Delta\underline{\sigma}^c$ remains bounded; i.e. there is a limit to the degree of nonlinearity that can be modelled. Consequently, the linearized, implicit equations used for stress updating in Scheme A can underestimate the rapid changes in stress that can occur during a creep test. To capture these changes, substeps are required. For Scheme B, on the other hand, the rapid changes are accounted for by iterating using the integrated creep relation given by eq. (12) that more realistically takes into account the accumulation of creep strain, thus stress change for large Δt. Although solutions from fully implicit schemes often tend to suffer poor accuracy when using large steps, this is not the case for the scheme presented here since the response during time step Δt is dictated by the stress conditions at the end of the increment, particularly for $p_d/p_c > 1$.

7 CONCLUDING REMARKS

A model has been presented which is capable of accurately describing the stress-strain-time behaviour of soft soils. The two integration algorithms, which were described, were found to be robust and capable of providing accurate solutions in a few steps. Although Scheme A performed well, the experience gained during this study leads to the conclusion that the updating of stresses can, from a computational point of view, be handled more efficiently by Scheme B. Nevertheless, given the nature of the constitutive law described here, it is important from a numerical stability viewpoint, that the stiffness matrix be assembled using a tangential constitutive matrix, which for both schemes involves a linearization.

REFERENCES

Bishop, A.W. 1966. The strength of soils as engineering materials. Sixth Rankine Lecture. *Geotechnique* 16: 91-130.

Bjerrum, L. 1967. Engineering geology of Norwegian normally-consolidated marine clays as related to settlements of buildings. Seventh Rankine Lecture. *Geotechnique* 17: 81-118.

Borja, R.I. & S.R. Lee 1990. Cam-clay plasticity, part 1: implicit itegration of elasto-plastic constitutive relations. *Computer Methods in Applied Mechanics and Engineering* 78: 48-72.

Kulhawy, F.H. & P.W. Mayne 1990. *Manual of Estimating Soil Properties for Foundation Design, Geotechnical Engineering Group*, Cornell University, Ithica.

Matsui, T. & N. Abe 1988. Verification of elasto-viscoplastic model of normally consolidated clays in undrained creep. *Proceedings, 6th International Conference Numerical Methods in Geomechanics, Innsbruck*: 453-459.

Skempton, A.W. 1964. Long term stability of slopes. Fourth Rankine Lecture, *Geotechnique* 14: 77-101.

Stolle, D.F.E. 1991. An interpretation of initial stress and strain methods, and numerical stability. *International Journal for Numerical and Analytical Methods in Geomechanics* 15: 399-416.

Vaid, Y. & R.G. Campanella 1977. Time dependent behaviour of undisturbed clay. *ASCE Journal of the Geotechnical Engineering Division*, 103(GT7), pp. 693-709.

Vermeer, P.A. & H. van Langen 1989. Soil collapse computations with finite elements. *Ingenieur-Archiv* 59: 221-236.

Zienkiewicz, O.C. & R.L. Taylor 1991. *The Finite Element Method, Volume 2*, London: McGraw-Hill.

Numerical Models in Geomechanics, Pietruszczak & Pande (eds) © *1997 Balkema, Rotterdam, ISBN 90 5410 886 X*

Disturbed state concept for partially saturated soils

F. Geiser, L. Laloui & L. Vulliet
Soil Mechanics Laboratory, EPFL, Lausanne, Switzerland

C. S. Desai
Department of Civil Engineering and Engineering Mechanics, University of Arizona, Ariz., USA

ABSTRACT: A general elasto-plastic model for partially saturated soils is proposed, based on the disturbed state concept (DSC). This model is unified and simplified, includes both saturated and unsaturated states and is able to capture some typical features of partially saturated soils. The saturated state can be considered as a special case. Experimental results on a remolded sandy silt are used to calibrate the model.

1. INTRODUCTION

Geological materials such as soils are often partially saturated. However, the majority of theories and models concern fully saturated soils. With the importance of topics such as geo-environmental engineering, dams, tunnels and landslides, considerable attention is now given to the testing and modeling of partially saturated soils.

The Soil Mechanics Laboratory of Lausanne has been active for many years on this subject, and recently focusing on an unsaturated sandy silt, which involves rheological characterization by extensive experimental tests (Laloui *et al.*, 1995) and numerical modeling (Geiser *et al.*, 1997).

The first part of this paper presents a constitutive model for unsaturated soils based on the hierarchical single surface law (HISS models from Desai and coworkers (see Desai, 1994)). The HISS-model initially developed for saturated soils is extended to partially saturated soils by taking into account the evolution of material parameters with the suction. The second part of this paper presents an extension of this approach to allow for the stress softening behavior typically observed in case of low degree of saturation and low net mean pressure (see Cui *et al.*, 1996 & Laloui *et al.*, 1997). The disturbed state concept (DSC) is used for this purpose. It includes the HISS plasticity model as a special case and allows for microcracking, damage and softening, and stiffening (Desai, 1995). The theory of the DSC presented here, for partially saturated soils, was developped by Desai, and is described in details by Desai *et al.* (1996).

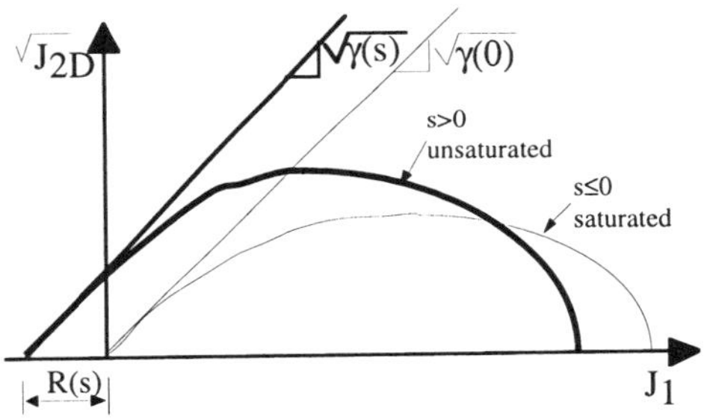

Figure 1: Yield function F

2. MODIFIED $\delta_{1\text{-}m}$

To represent the unsaturated behavior of a porous medium, the non-associative elastoplastic hierarchical single surface model HISS-δ_1, developed initially for saturated soils (Desai, 1994), is modified.

In the following the suction s will be defined as the excess of pore air-pressure u_a to water-pressure u_w(compression positive):

$$s = u_a - u_w \tag{1}$$

The modified HISS-δ1 includes the suction as a state parameter governing the evolution of the yield surface. The expanding yield surface F is defined as:

$$F \equiv J^*_{2D} - \left[-\alpha(J^*_1)^n + \gamma(J^*_1)^2\right](1 - \beta\ S_r)^{-0.5} \tag{2}$$

where material parameters are explicit functions of the suction. Figure 1 illustrates the effect of the suction on the shape of F in the $J_1 - \sqrt{J_{2D}}$ plane.

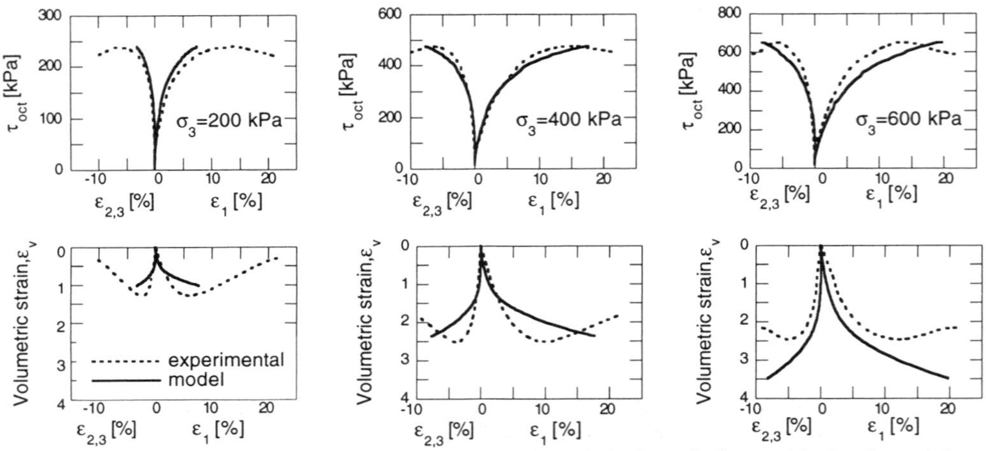

Figure 2: Comparison between experimental results and numerical predictions with the δ_1-model, at three different lateral pressures: σ_3 = 200 kPa, σ_3 = 400 kPa and σ_3 = 600 kPa.

Expressions in Eq. (2) are the following:

$J^*_{2D} = J_{2D}/p_a^2$, with J_{2D} is the second invariant of the deviatoric stress tensor, t_{ij};

$J^*_1 = (J_1 + R)/p_a$, with J_1 is the first invariant of the net stress tensor J_1=3(p-u_a) and R is a bonding stress;

p is the mean pressure: p=(σ_1+2σ_3)/3;

p_a is a constant atmospheric pressure;

γ and β are ultimate parameters;

S_r is the stress ratio with $S_r = \sqrt{27/2}\ J_{3D} \cdot J_{2D}^{-3/2}$,

J_{3D} is the third invariant of the deviatoric stress tensor t_{ij};

α is the hardening function defined as:

$$\alpha = \frac{a_1}{\xi^{\eta_1}} \tag{3}$$

where a_1 and η_1 are the hardening parameters and ξ is the trajectory of total plastic strains given by

$$\xi = \int (d\varepsilon^p_{ij} d\varepsilon^p_{ij})^{1/2}; \tag{4}$$

n is the phase change parameter related to the state of stress at which transition from compaction to dilation occurs or at which the change in the volume vanishes.

The plastic non-associative potential function is defined as:

$$Q \equiv J^*_{2D} - \left[-\alpha_Q (J^*_1)^n + \gamma (J^*_1)^2\right](1 - \beta\ S_r)^{-0.5} \tag{5}$$

where

$$\alpha_Q = \alpha + \kappa(\alpha_0 - \alpha) + (1 - r_v)$$

$$r_v = \xi_v / \xi$$

ξ_v is the volumetric part of ξ

α_0 is the α at the beginning of shear loading and requires a nonassociative parameter κ.

Inside the yield surface a linear elastic behavior is assumed, defined by two more parameters: the Young's modulus E and the Poisson's ratio ν.

As such the model contains 9 material parameters.

3. NUMERICAL SIMULATIONS WITH THE MODIFIED δ_1-MODEL

Experimental results obtained on a sandy silt (Laloui *et al.*, 1997) are used to analyze the performance of the modified δ_1-model. Figure 2 shows stress-strain curves obtained from triaxial compression tests on saturated samples and from the modified δ_1-model (material parameter see Table 1). τ_{oct} is defined as:

$$\tau_{oct} = \frac{\sqrt{2}}{3} q \tag{6}$$

with $q = \sigma_1 - \sigma_3$ the deviatoric stress.

ε_1, $\varepsilon_{2,3}$ represent the strain values in the principal directions 1 and 2 or 3; ε_v is the volumetric strain.

Table 1: Material parameters (saturated state)

Parameter	Symbol	Value
Young Modulus	E	145000 kPa
Poisson's ratio	ν	0.4
Shape of F	β	0.58
Ultimate parameter	γ	0.08
Bonding stress	R	0 kPa
Hardening parameter	a_1	0.005
Hardening parameter	η_1	0.36
Phase change parameter	n	2.45
Non-assoc. parameter	κ	0.5 .

To simulate the unsaturated tests, the following four parameters are taken as non linear functions of the suction: γ, R, a_1 and n. The corresponding evolution functions are determined experimentally (Figure3).

The experimental results show that the bonding

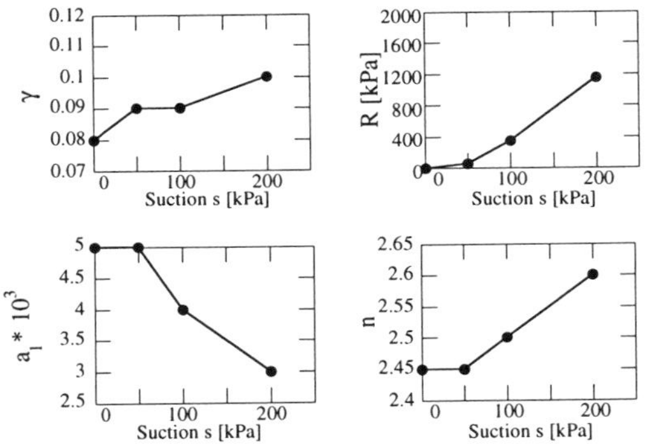

Figure 3: Experimental determination of γ, R, n and a_1 versus suction s

stress R increases with the suction (what corresponds in more conventional terms to an increase of the cohesion). The parameter γ (which defines the slope of the ultimate envelope) only slightly increases with increasing suction while the hardening parameter a_1 decreases. This denotes a stiffening of the soil with increasing suction. The phase change parameter n seems also to be slightly affected by the suction. The parameters β, η_1, κ, E and ν are assumed for the time being to be independent of the suction.

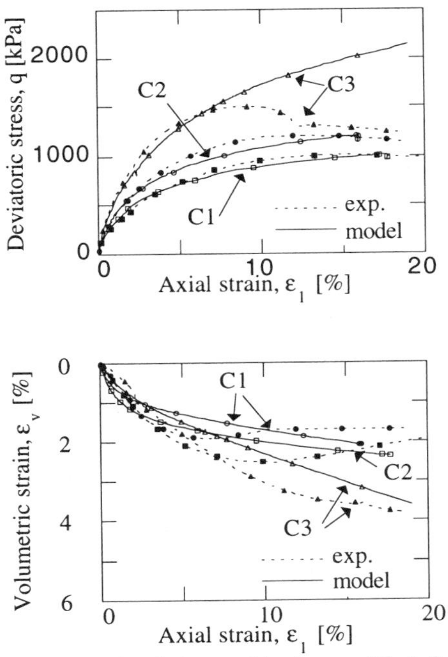

Figure 4: Simulations with the modified δ_1-model

Three drained triaxial compression tests at the same initial net mean pressure $p-u_a$=400 kPa are simulated with the modified δ_1-model, namely:

- C1, saturated case: σ_3=400 kPa, s=0kPa
- C2, small suction: σ_3=450 kPa, s=50kPa
- C3, high suction: σ_3=600 kPa, s=200kPa

Two other tests will be used later, namely
- C4, average suction: σ_3=600 kPa, s=100kPa
- C5, average suction: σ_3=400 kPa, s=100kPa

The comparison between experiments and computation for cases C1 to C3 are presented in Figure 4. The saturated case C1 corresponds to a calibration path. The numerical simulation of the test C2 (small suction, s=50 kPa) is close to the experimental results. In the case C3 (high suction) the experimental stress-strain behavior shows an increase of strength followed by a loss of resistance like a brittle failure; this effect appears more pronounced for higher suctions. Due to its intrinsic feature of continuous yielding, the modified δ_1-model is not able to predict such behavior. A possible approach is to extend the DSC theory to partially saturated soils, as presented below.

4. NEW APPROACH BASED ON THE DSC

4.1 *Introduction*

The use of a general concept in the framework of the disturbed state concept (DSC) (Desai, 1995) is proposed to model the particular behavior of unsaturated soils, namely the loss of strength in the stress-strain relationship for low degrees of saturation and low net mean pressures in the post-peak phase.

The DSC is based on the idea that a deforming material element can be treated as a mixture of two constituent parts in the relative intact (RI) and fully adjusted (FA) states, referred to as reference state. During external loading, the material experiences internal changes in its microstructure due to a self-adjustment process and, as a consequence, the initial RI state transforms continuously to the FA state. The observed stress σ^a_{ij} is defined as:

$$\sigma^a_{ij} = (1-D)\sigma^i_{ij} + D\,\sigma^c_{ij} \qquad (7)$$

where σ^i_{ij} is the RI stress, σ^c_{ij} the FA stress and D the disturbance function ($0 \le D \le 1$).

As a first approach the FA state is considered as the stress state with no disturbance (D=1). This case represents here the saturated state. The RI state is obtained with the modified δ_1-model presented in the previous section. Figure 5 shows those different states.

The disturbance function D is expressed as a scalar, which represents the microstructural changes leading

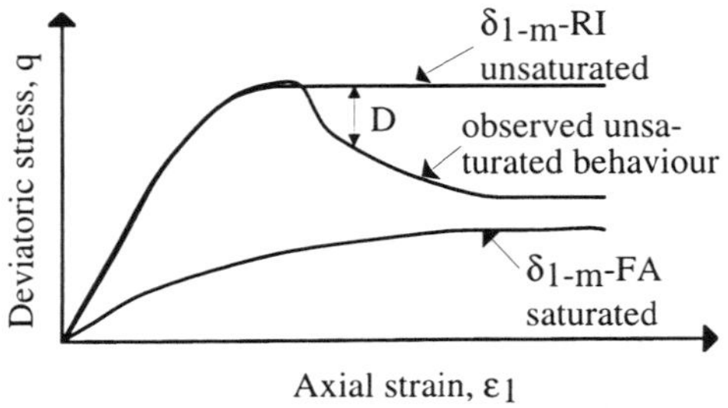

Figure 5: DSC concept for unsaturated soil

to microcraking, damage and softening. It is proposed as a first approximation to express D as a function of the deviatoric stress invariant.
The disturbance function is then:

$$D = \frac{\sqrt{J_{2D}}^{i} - \sqrt{J_{2D}}^{a}}{\sqrt{J_{2D}}^{i} - \sqrt{J_{2D}}^{c}} \qquad (8)$$

with the exponent "i" corresponding to the RI-state, "a" corresponding to the observed feature and "c" corresponding to the FA-state.

The disturbance is assumed to be a function of the trajectory of the deviatoric plastic strains ξ_D and the suction s. It is expressed as:

$$D = D_u(1 - e^{-A\xi_D^Z}) \qquad (9)$$

where D_u is the ultimate value of D (a material parameter), which may be assumed to be independent of the suction;
A and Z are material parameters function of the suction (see later Eq. 11);
The trajectory of the deviatoric plastic strains is defined as:

$$\xi_D = \int \left(dE_{ij}^p \cdot dE_{ij}^p\right)^{1/2} \qquad (10)$$

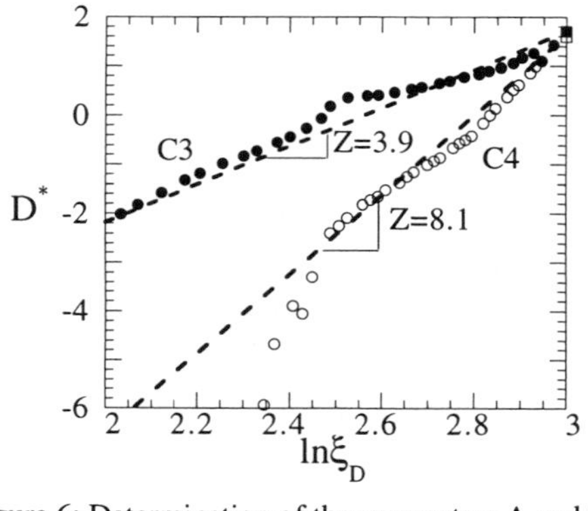

Figure 6: Determination of the parameters A and Z

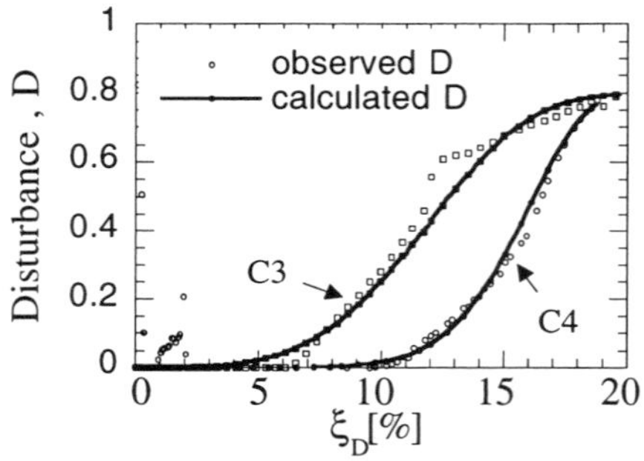

Figure 7: Comparison between the observed and calculated disturbance

where $dE_{ij}^p = d\varepsilon_{ij}^p - \frac{1}{3} d\varepsilon_v^p \, \delta_{ij}$

with $d\varepsilon_{ij}^p$ the plastic strain rate and $d\varepsilon_v^p = tr\left(d\varepsilon_{ij}^p\right)$

4.2 *Determination of the disturbance function D*

The determination of D versus ξ_D and s is done experimentally based on two unsaturated drained triaxial compression tests showing a clear "softening" behavior:
- test at a net mean pressure $p\text{-}u_a$=400 kPa with suction s=200 kPa (C3)
- test at a net mean pressure $p\text{-}u_a$=500 kPa with suction s=100 kPa (C4)

Applying Equation 8 through the entire stress path together with the computing of the plastic strain increments (from the elasto-plastic decomposition of strain) gives the evolution of D with ξ_D at a given suction.

From Equation 9, the parameters Z and A are determined as follow:

$$\ln A + Z \ln(\xi_D) = \ln\left[-\ln\left(\frac{D_u - D}{D_u}\right)\right] \equiv D^* \qquad (11)$$

Finally a plot of D^* vs $\ln(\xi_D)$ (see Figure 6) gives the parameter Z as the slope and the parameter A as the intercept of the regression line. The values of the parameters are given in Table 2.

Table2: Parameters of the disturbance function (Eq.6)

Suction s	Parameter Z	Parameter A	Du
100 kPa	8.1	1.4e-10	0.85
200 kPa	3.9	4.5e-5	0.85

The comparison between the observed disturbance function D (Eq. 8) and the calculated one (Eq. 9) is given on Figure 7. The correlation is very good.

4.3 *Numerical simulations with the DSC concept*

Figures 8 & 9 show the numerical modeling of the

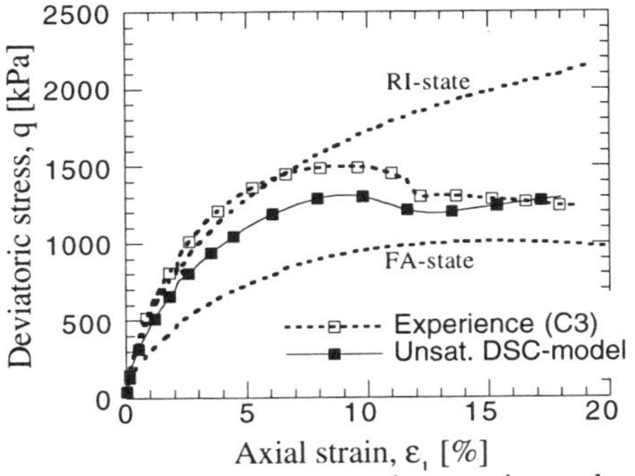

Figure 8: Comparison between the experimental result and the numerical simulation: p-ua=400 kPa, s=200 kPa (C3)

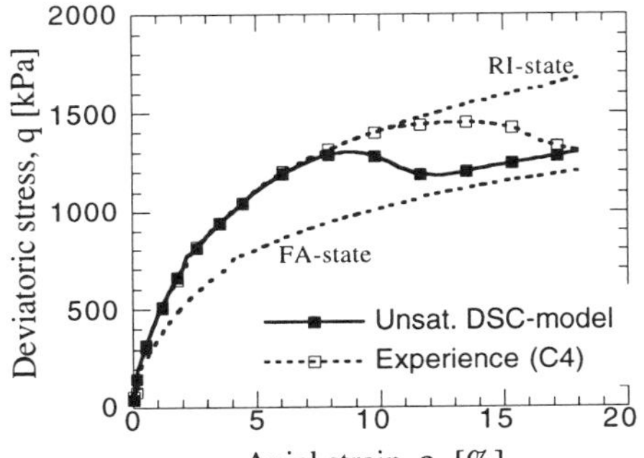

Figure 9: Comparison between the experimental result and the numerical simulation: p-ua=500 kPa, s=200 kPa (C4)

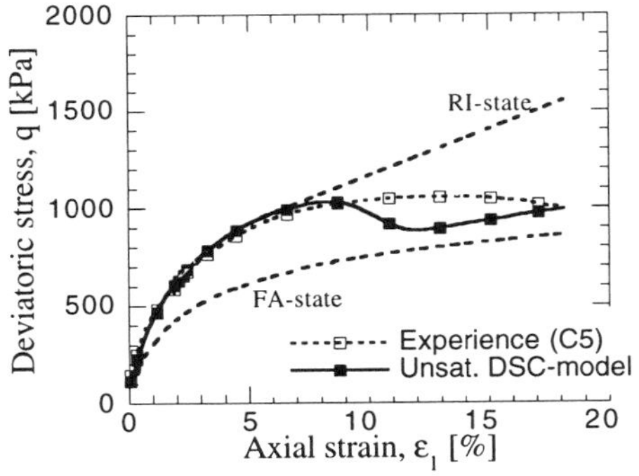

Figure 10: Comparison between the experimental result and the numerical prediction: p-ua=300 kPa, s=100 kPa (C5)

experimental tests used for the calibration of the material parameters (backanalysis). Using Equation 7, the unsaturated DSC gives better results than the modified δ_1-model, especially for the test C3.

Figure 10 represents the numerical prediction of a drained triaxial compression test (C5) at an initial net mean pressure of 300 kPa and a suction of 100 kPa. The comparison with the experimental results constitutes a good validation of the proposed formulation.

5. CONCLUSION

Two different models have been proposed in this paper. It is shown first that a modified δ_1-model (from the HISS elastoplatic family of constitutive model) constitutes a satisfactory model for partially saturated soil. Material parameters are simply non-linear function of the suction. However, the post peak behavior cannot be captured by this continuously yielding model. As a further improvement, the second model, unsaturated DSC (Disturbed State Concept), is capable of predicting the post peak behavior. The disturbance function can be obtained from the laboratory tests and the first results presented here are encouraging. It is believed that this unsaturated DSC has the potential to encompass most of the specific features of partially saturated soil.

6. REFERENCES

Cui, Y.J. & P. Delage 1996. Yielding and plastic behaviour of an unsaturated compacted silt. *Géotechnique* 46(2): 291-311.

Desai, C.S. 1994. Hierarchical single surface and the disturbed state constitutive models with emphasis on geotechnical application. *Geotechnical Engineering: emerging trend in design and practice*. K. R. Saxeng. New Delhi, Oxford IBH Pub. Co.: 115-154.

Desai, C.S. 1995. Constitutive modelling using the disturbed state as a microstructure self-adjustement concept. Chapter 8 in *Continuum Models for material with Microstructure*. H.B. Mühlhaus, John Wiley & sons, U. K.

Desai, C.S., L. Vulliet, L. Laloui & F. Geiser 1996. *Disturb state concept for constitutive modeling of partially saturated porous materials*, Internal report, EPFL.

Geiser, F., L. Laloui & L. Vulliet 1997. Constitutive modelling of unsaturated sandy silt. *Proc. 9th. Int. Conf. of the Int. Assoc. for Computer Meth. and Advances in Geomechanics (IACMAG)*: Wuhan'97: Balkema.

Laloui, L., F. Geiser, L. Vulliet, X.L. Li, A. Bolle & R. Charlier 1997. Characterisation of the mechanical behaviour of an unsaturated sandy silt. *Proc. XIV Int. Conf. on Soil Mech. & Found. Eng.*: Hambourg'97: Balkema.

Laloui, L., L. Vulliet & G. Gruaz 1995. Influence de la succion sur le comportement mécanique d'un limon sableux. in "Unsaturated Soils". Ed. E.E. Alonso & P. Delage, Balkema: 133-138.

Numerical Models in Geomechanics, Pietruszczak & Pande (eds) © 1997 Balkema, Rotterdam, ISBN 90 5410 886 X

Calibration and numerical validation of a micromechanical damage model

V. Renaud & D. Kondo
Laboratoire de Mécanique de Lille, URA CNRS, Université de Lille, Villeneuve d'Ascq, France

ABSTRACT : The mechanical behaviour of brittle geomaterials is strongly determined by the growth and nucleation of microcracks. In the present work, a micromechanical damage model is numerically implemented and calibrated on a sandstone. First we present the general framework and the physical features assumed for the anisotropic damage. Then, the model is used to simulate the mechanical behaviour of the material under triaxial and proportional compression. The results show good agreements between the experience and the model predictions, and allow to give more insight on the dominant mechanisms of deformation.

1 INTRODUCTION

Since two decades, Continuum Damage Mechanics (CDM) has been the subject of tremendous researches (Krajcinovic, 1989 ; Lemaître, 1990). The conventional phenomenological approach of damage modelling is based on the use of internal variables (scalar, vectorial or tensorial). This variable is supposed to represent the salient features of the microcracking which generates the brittle damage. In Kondo et al. (1992) such modelling approach was used for a britlle sandstone. In spite of its efficiency in some situations, the phenomenological approach lacks to describe precisely some specific features of brittle damage : stress - induced anisotropy, unilateral effects due to microcracks closure (see Chaboche, 1992). On the other hand, micromechanical models, by using informations at mesoscopic level, try to give more insight on the damage phenomenon. The main purpose of the present study is to calibrate and validate a three-dimensional micromechanical constitutive theory inspired from Fanella and Krajinovic (1988) and Ju and Lee (1991). An outline of the paper is as following. First, we summarize the mechanical behaviour of the material characterized by a stress-induced anisotropy and the dilatant deformations. Then, the general framework of the micromechanical model is presented. Calibration on triaxial tests and numerical simulations allow to demonstrate the strong capabilities of the model to reproduce the experimental data. More particularly, the comparisons of the numerical predictions with the damaged moduli (experimentally determined by cyclic tests) show good agreements.

2 SUMMARY OF EXPERIMENTAL RESULTS

Experiments have been conducted on a brittle sandstone (Fontainebleau sandstone). This material is constituted mainly by quartz grains (98%), the rest being clay minerals. The average quartz grains size is about 250 μm and the initial relative porosity is low (about 10%). The average specific density is 23.7 (±0.2) $kN.m^{-3}$. The specimens (cylinders) measured 37.5 mm in diameter by 75 mm in length. They are tested under the same boundary conditions and the current laboratory environment conditions. Great care has been taken in the design and sequences of the experimental frame to warrant uniform loading of samples.

Hydrostatic and Triaxial compressive tests

Hydrostatic compressive tests are first performed. Stress-strains (axial and lateral) curves indicate that Fontainebleau sandstone is initially isotropic. The experimental results for triaxial tests are presented in terms of deviatoric stress $(\sigma_1 - \sigma_3)$. σ_1 is the axial stress and σ_3 is the confining pressure. Figure 1 shows the stress-strain curves for triaxial tests with different confining pressures. These curves correspond to a typical brittle behaviour with stress induced anisotropy and a large dilatancy in deformation (related to the strong non linearity of the lateral strains). Proportional compression tests are also performed and will serve as validation of the modelling (see section 4).

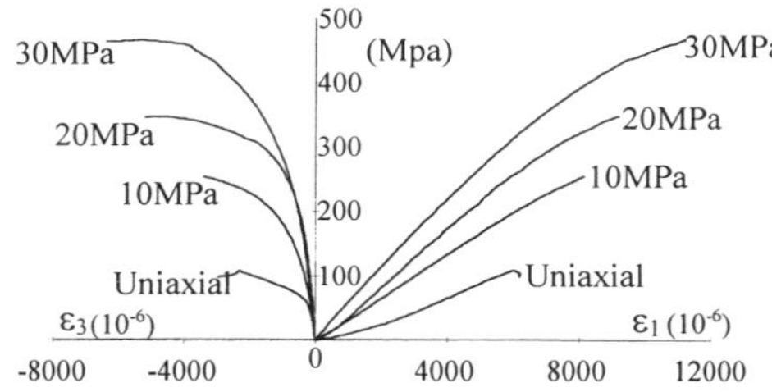

Figure 1 : Stress-strain curves for monotonous triaxial compression tests

3 BRIEF PRESENTATION OF THE 3-D MICROMECHANICAL DAMAGE MODEL

3.1 General framework

Consider a representative elementary volume (REV) containing a great number of microcracks (see for example Nemat-Nasser and Horii (1993 for conditions on REV). It is assumed that the solid matrix (brittle) is homogeneous and has a linear elastic behaviour, with a initial compliance denoted $\bar{S}^0$. The constitutive relations of the microcracks-weakened solid, linking the macroscopic stress $\bar{\sigma}$ and the macroscopic strain $\bar{\varepsilon}$ is described through the overall compliance matrix $\bar{S} = \bar{S}^0 + \bar{S}^d$. $\bar{S}^d$ is the inelastic part of the compliance due to microcracks present in the REV. Such constitutive relations can be summarised as follow (Ju, 1991) :

complementary free energy :

$$\psi^*(\bar{\sigma}, \bar{S}) = \frac{1}{2}\bar{\sigma}:\bar{S}:\bar{\sigma} = \frac{1}{2}\bar{\sigma}\ (\bar{S}^0 + \bar{S}^d)\bar{\sigma} \quad (1)$$

state laws : $$\bar{\varepsilon} = \frac{\partial \psi^*}{\partial \bar{\sigma}} = \bar{S}:\bar{\sigma} = (\bar{S}^0 + \bar{S}^d)\bar{\sigma} \quad (2)$$

The damage dissipation inequality which is $\frac{1}{2}\bar{\sigma}\dot{\bar{S}}\,\bar{\sigma} \geq 0$ depends strongly on the rate of the overall compliance.

Voigt's notations are used in all the study.

$$\begin{bmatrix} e_1 \\ e_2 \\ e_3 \\ e_4 \\ e_5 \\ e_6 \end{bmatrix} = \begin{bmatrix} \varepsilon_{11} \\ \varepsilon_{22} \\ \varepsilon_{33} \\ 2\varepsilon_{23} \\ 2\varepsilon_{13} \\ 2\varepsilon_{12} \end{bmatrix} \text{ et } \begin{bmatrix} \tau_1 \\ \tau_2 \\ \tau_3 \\ \tau_4 \\ \tau_5 \\ \tau_6 \end{bmatrix} = \begin{bmatrix} \sigma_{11} \\ \sigma_{22} \\ \sigma_{33} \\ \sigma_{23} \\ \sigma_{13} \\ \sigma_{12} \end{bmatrix} \text{ with } e_i = S_{ij}\tau_j .$$

The development of the damage micromechanical model requires :

i) evaluation of the overall compliance for the elastic brittle material weakened by numerous interactive microcracks
ii) use of precise kinetic equations for microcracks growth.

We examine now these two points

3.2 Overall compliance of the microcracked solid

The inelastic part of the compliance $\bar{S}^d$ is estimated from the displacement discontinuities field (cracks opening displacements, COD) b'_i. The general form of the solution for penny-shapped cracks in anisotropic medium loaded in compression (figure 2) is as given by Lee and Ju, (1991) and Ju and Lee (1991) (see see Horii and Nemat-Nasser, 1983 for 2-D solutions) :

$$b'_i(x', y') = 2a\sqrt{1 - x'^2 - y'^2}\, C''^{-1}_{ik}\, \bar{\sigma}'_{1k} \quad (3)$$

Matrix C'^{-1} is a tensor which relates COD to the applied (macroscopic) stress tensor. It depends on material properties (anisotropic elastic parameters). The prime indicates that COD are calculated in the local frame of each microcracks. Contribution of all microcracks to inelastic compliance is given by :

$$\bar{\varepsilon}^d = \bar{S}^d \bar{\sigma} = \frac{1}{2V} \sum_k \left[\int_{A_k} (b \otimes n + n \otimes b) dA \right]^{(k)} \quad (4)$$

V is the volume occupied by the RVE. A_k is the area of k-th microcrack. n is the normal to the microcrack surface.

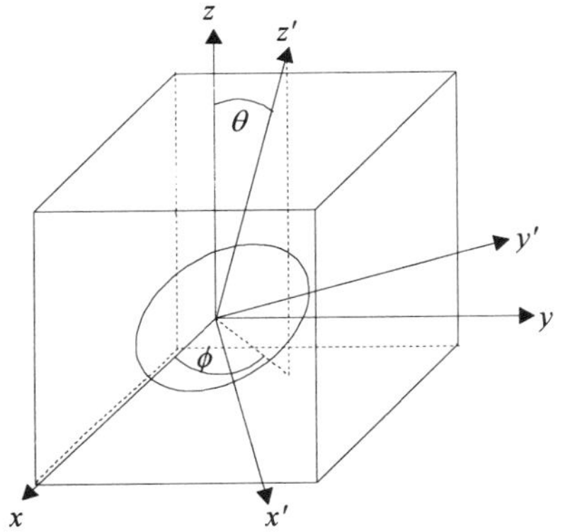

Figure. 2 : 3-D penny shaped microcrack geometry and definition of axes

In fact, the summation in expression (4) is semi analytically evaluated by average techniques. For this, microcracks distribution is supposed to be spatially continued and their location and orientation are random : the initial radius and orientation of defects are randomly distributed on $\left[a_{0_{\min}}, a_{0_{\max}}\right]$ for radius, on $\left[0, \frac{\pi}{2}\right]$ for θ , and on $[0, 2\pi]$ for ϕ . Microcracks interactions are taken into account by use of the Self Consistent Method (SCM). In this method each defect is assumed to be already in the unknown effective medium. Since the SCM is used, an iterative algorithm is needed for the calculation of the overall compliance.

3.3 Microcracks growth mechanisms and Kinetic equations

Based on the previous statistical assumptions, the initial microstructure is determined by the minimum and maximum grain size and by the microcrack density. Preexisting microcracks are supposed to be at grains-matrix interfaces (the material can be viewed as a composite aggregate). The progressive damage is the result of microcracks growth when the solid is loaded. Under tensile load, cracks may propagate in mode I (opening mode), whereas the propagation is more complex under compression. Based on works of Nemat-Nasser and Horii, (1982) and Zaitsev (1983) the kinetic equations of microcrack growth under compressive loading is summarised as follow (figure 3).

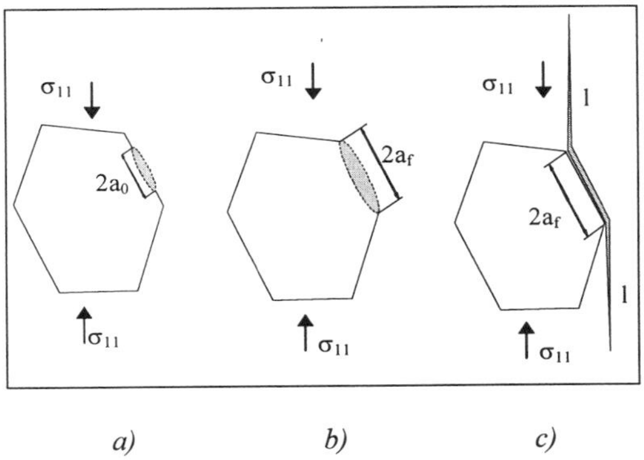

Figure 3 : Mechanisms of microcracks growth under compressive load - a) sliding without propagation ; b) propagation in unstable mode II ; c) Kinking in mixed mode

Case b) corresponds to instantaneous crack growth at the interface ; such growth is temporarily stopped by the matrix energy barrier. The crack begins to kink (case c) when the stress intensity factor attains the critical mode I value K_{Ic}^m in the matrix. Propagation conditions are obtained from classical fracture mechanics analysis. Details of such analysis can be found in Fanella and Krajcinovic (1988) or in Ju and Lee (1991).

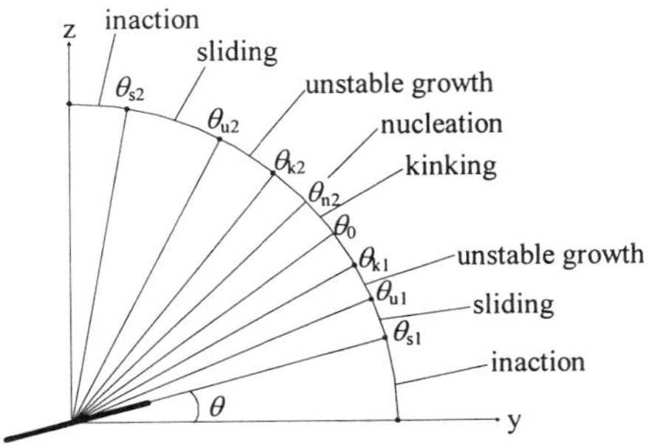

Figure 4 : 3-D compressive active damage domain

At each load step a domain of active damage (sliding, unstable growth, kinking and opening) depending on microcracks orientation is determined (see figure 4).

For completeness, note that microcracks nucleation mechanism is integrated in the analysis via the classical Zener-Stroh model based on a critical debonding shear stress.

4 CALIBRATION OF THE MODEL AND NUMERICAL PREDICTIONS

4.1 Constitutive parameters and calibration of the model

The 3-D micromechanical model requires 9 parameters which have physical meanings and then are easy to be identified. These parameters can be divided in two distinct classes :

Macroscopic parameters :

These parameters are the Young modulus E, the Poisson's ration ν, the friction coefficient, the interfacial fracture toughness K_{Ic}^{if} and the matrix fracture toughness K_{Ic}^{m}

Mesoscopic parameters :

The precise description of the initial microstructure requires parameters such as the minimum and maximum grain size $D_{\min}$ and $D_{\max}$, the initial microcracks density w_0 and the critical debonding stress τ_c^0 (for nucleation mechanism).

We intend here to show the capabilities of the micromechanical damage model. The calibration is done using the set of triaxial compression tests presented at figure 1. Note that, except for E and ν, a single set of parameters is used for all the triaxial tests. Figure 5 shows two examples of comparisons of the numerical results and the experimental data. The computed results correlate well these data and confirm the efficiency of the model.

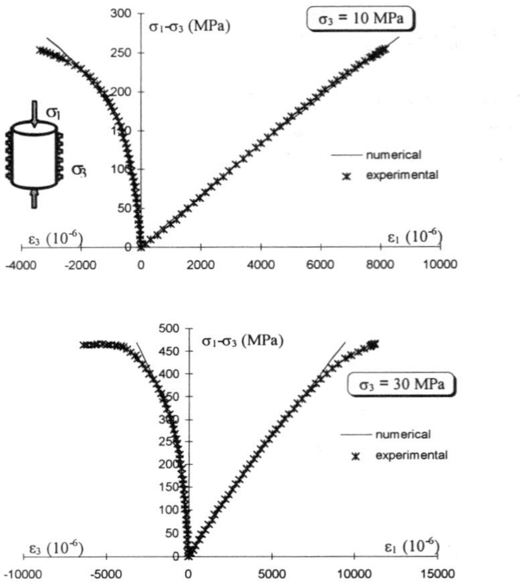

Figure 5 : Triaxial compressive tests - Comparisons between experiments data and numerical results

4.2 Numerical predictions of damaged moduli : comparisons with experimental data

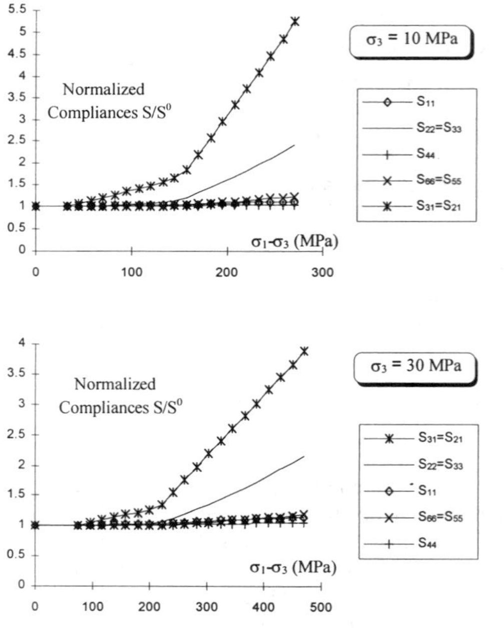

Figure 6 : Normalised compliances variation in triaxial compression

The detailed analysis of the results shows that the dilatancy is essentially due to microcracks kinking (axial opening mode). In order to show the importance of damage anisotropy, we have plotted at figure 6 the variation of the normalised components of the compliance tensor. Variation of the lateral compliance $\overline{S}_{22}$ is observed to be much larger than for the axial one $\overline{S}_{11}$. The more important increase of $\overline{S}_{21} = \overline{S}_{31}$ explains the strong non-linearity of lateral strain and then the dilatancy.

These results are interpreted in term of the moduli and Poisson's ratio variation (see figures 7 for test with $\sigma_3 = 10\text{MPa}$). Comparisons of predicted values with experimental data (obtained from cyclic compression tests) give good agreements. Note however that the model underestimates slightly the decrease of axial modulus.

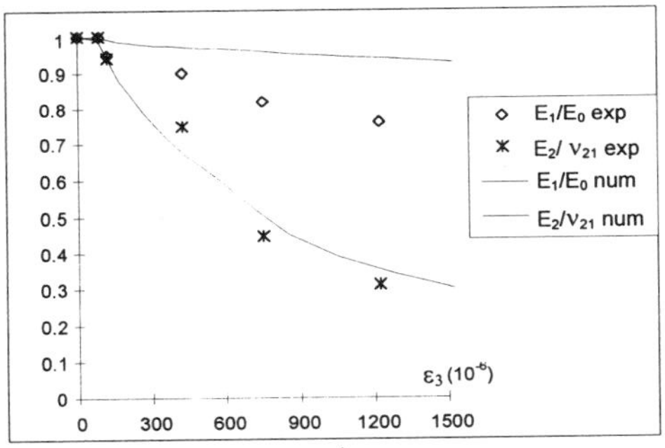

a)

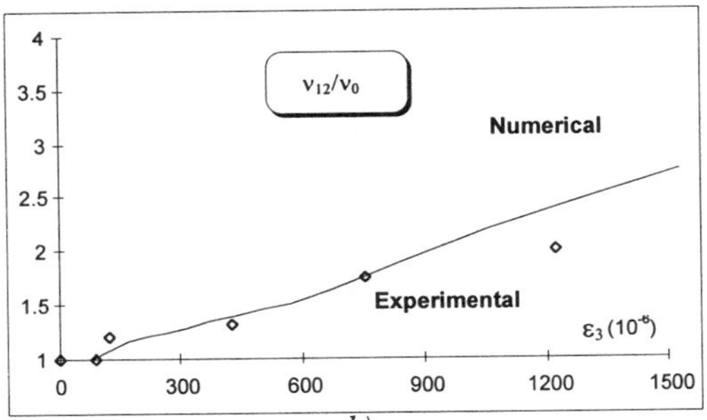

b)

Figure 7 : Predictions of the moduli decrease
a) Axial and lateral moduli b) Poisson's ratio

4.3 Proportional stress path test :

On the basis of the preceding calibration, numerical simulations are conducted on proportional stress path tests (experimental results are available for this test). We present on figure 8a the predictions for a test performed at a ratio $k = \dfrac{\sigma_1}{\sigma_3} = 30$. It is observed that the numerical predictions are in accordance with the tests data. In comparison with the triaxial data $(\sigma_3 = 10\text{MPa})$, damage in the present case is more pronounced for similar deviatoric stress (see figure 8b).

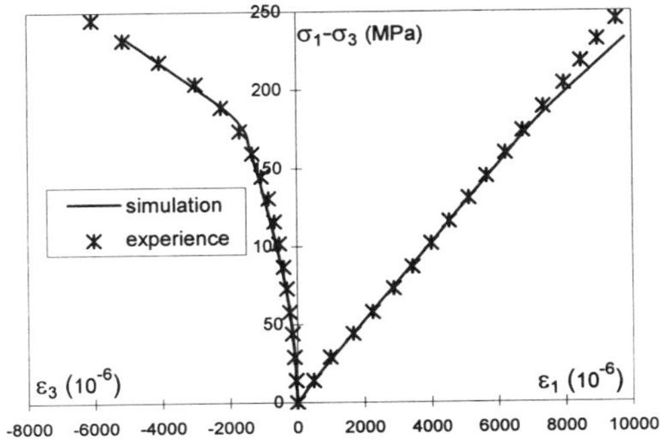

a)

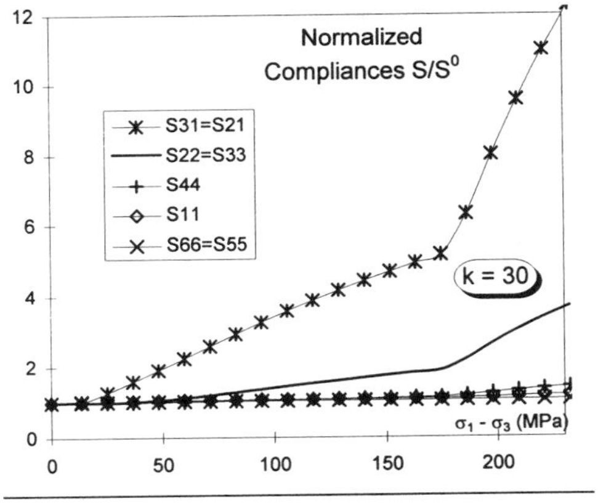

b)

a) Comparisons between experiments data and numerical results b) Normalised compliances variation under proportional compression

Figure 8 : Proportional compressive tests ($k = 30$) -

4.4 Influence of parameters :

A study of the influence of some parameters has been done. We present here some illustrations for the initial microcracks density w_0 (figure 9) and for nucleation mechanism (figure 10) in a triaxial test ($\sigma_3 = 10\text{MPa}$). From figure 9 we note that the material response depends strongly on its initial damage state. The material is more dilatant when the initial density w_0 is more important. Finally the nucleation mechanism appears to be important in the anisotropic damage process. We observe that the simulation without generation of new microcracks (0 pn on figure 10) leads to non negligible differences with experimental data.

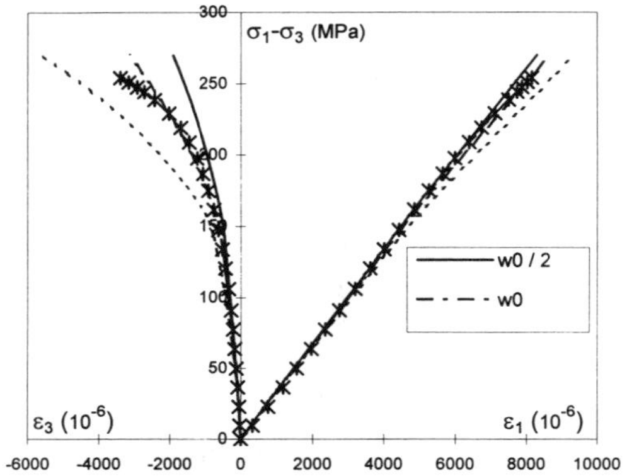

Figure 9 : Uniaxial compression - Effect of initial microcracks density (ω_0 is the reference density)

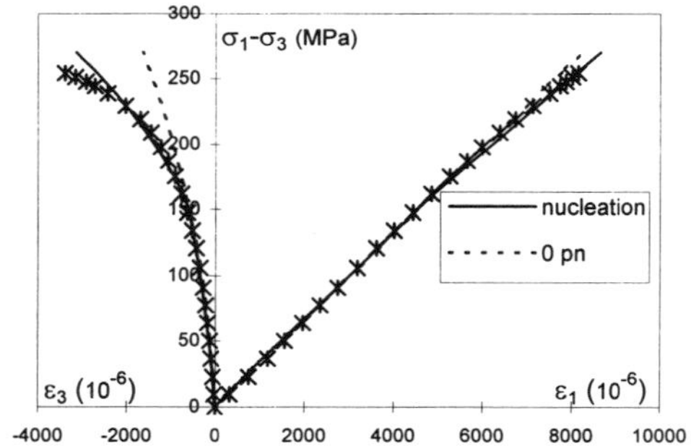

Figure 10 : Influence of nucleation mechanism

5 CONCLUSIONS

A 3-D micromechanical damage model is implemented and used for simulate the behaviour of a sandstone under triaxial and proportional loadings. The numerical results confirm the capabilities of the modelling approach and the salient features of the brittle damage are well reproduced by the model. Furthermore, the numerical simulations give some information on the importance of the various mesoscopic mechanisms involved in the brittle damage process. More complex stress paths are under investigation in order to check cracks closure effects on damage (damage desactivation). Finally, in order to study the effect of damage on stability of structures, we plan to introduce the model in a Finite Element code.

References

J.L. Chaboche (1992), Damage induced anisotropy : on difficulties associated with the active/passive unilateral condition. *Int. J. of Damage Mechanics*, Vol. 1(2), pp. 148-171.

D. Fanella & D. Krajcinovic (1988), A micromechanical model for concrete in compression. *Eng. Fract. Mech.*, vol. 29, pp. 59-66.

H. Horii & S. Nemat-Nasser (1983), Overall moduli of solids with microcracks load-induced anisotropy. *J. Mech. Phys. Solids*, vol. 31, n° 2, pp. 155-171.

J. W. Ju (1991), On two-dimensional self-consistent micromechanical damage models for brittle solids. *Int. J. Solids Structures,* Vol. 27, n° 2, pp. 227-258.

J. W. Ju & X. Lee (1991), Micromechanical damage models for brittle solids, part I : tensile loadings. *J. Eng. Mech.* vol. 117, pp. 1495-1514.

D. Kondo, J.F. Shao & J.P. Henry (1992), Numerical modelling of induced anisotropic damage in a sandstone. *NUMOG IV, Swansea, U.K. Balkema,* pp. 191-199.

X. Lee & J. W. Ju (1991), Micromechanical damage models for brittle solids, part II : compressive loadings. *J. Eng. Mech.* vol. 117, pp. 1515-1536.

J. Lemaitre (1990), A course on Damage Mechanics, *2nd edition Cambridge University Press*

S. Nemat-Nasser & H. Horii (1982), Compression-induced noplanar crack extension with application to splitting, exfoliation and rockburst. *J. Geophys. Res.*, vol. 87, pp. 6805-6821.

S. Nemat-Nasser & H. Horii (1993), Micromechanics : overall properties of heterogeneous materials, North - Holland, Amsterdam.

Y. B. Zaitsev (1983), Crack propagation in a composite material. *in F. H. Wittmann,* ed. *Fracture mechanics of concrete*, Elsevier, Amsterdam.

Numerical Models in Geomechanics, Pietruszczak & Pande (eds) © 1997 Balkema, Rotterdam, ISBN 90 5410 886 X

Plasticity and imbibition-drainage curves for unsaturated soils: A unified approach

P. Dangla, L. Malinsky & O. Coussy
Laboratoire Central des Ponts et Chaussées, Paris, France

ABSTRACT: The basic concepts of plasticity are extended to partially saturated porous materials in the so called poroplasticity. In addition to plastic strains, a new internal variable emerges: the irrecoverable volumetric water content. It captures hysteresis phenomena observed during drying-wetting cycles. The poroplasticity is formulated in a consistent framework provided by the thermodynamics of open systems. The energies involved in the formulation are discussed. Starting from the model proposed by Alonso et al for unsaturated soils it is shown how to render it consistent with hardening poroplasticity.

1. INTRODUCTION

To formulate the constitutive equations for unsaturated soils, the primary idea was to use an effective stress by referring to the saturated situation for which the concept is firmly founded. This met with some difficulty as reported by many investigators (Bishop et al 1961, Burland 1965). Now most of them have accepted the idea of two effective stresses as a necessary condition to describe the observed features of unsaturated soils. A review of the literature on the mechanical behaviour of partially saturated soils is presented in the comprehensive report of Alonso et al 1987.

To account for the main features of soil behaviour, Alonso et al (1990) formulated a model in the framework of hardening plasticity. These features can be summarized as follows. An increase in capillary pressure stiffens the soil. Changes in capillary pressure may induce irrecoverable volumetric deformations. A reduction in capillary pressure (wetting) exhibits a transition from swelling to collapse. An increase in capillary pressure (drying) results in an increase in shear strength. The volumetric response in highly stress path dependent (due to irreversibilities).

However this model only deals with stress-strain relationship which is only part of the complete set of the constitutive equations since there is the capillary pressure versus water content relationship as well. During drying-wetting process the capillary pressure-water content curves generally exhibit hysteresis. This particular pattern is not taken into account in the model proposed by Alonso et al. The purpose of this paper is to describe these two aspects of the behaviour into a unique model. To this aim the basic concepts of plasticity are extended to partially saturated porous media in the so called poroplasticity. This gives rise to a new internal state variable: the irrecoverable volumetric water content. This variable captures the hysteresis phenomena in the drying-wetting processes. The constitutive equations of the poroplasticity are derived in a consistent framework provided by the thermodynamics of open systems. To focus on the hysteresis of the capillary pressure-water content curves, the case of non deforming skeleton is firstly considered. Then the poroplasticity of deforming porous media is presented. Finally starting from the model proposed by Alonso et al, it is shown how to render it consistent with hardening poroplasticity.

2. ELEMENT OF THERMODYNAMICS

An element of porous medium, the deformation of which is identified by that of the skeleton, can gain or lose fluid masses due to external actions. In this sense any element of porous medium can be considered as an open thermodynamic system. In many situations two fluids can be distinguished: the liquid water (w) and the air (a). The air is assumed to behave as an ideal gas and the liquid water as a standard compressible fluid. Then it can be shown that the laws of thermodynamics applied to this open system entail the following (Clausius-Duhem) inequality (Coussy 1995, Dangla et al 1996)

$$\sigma_{ij}d\varepsilon_{ij} + p_a d\phi_a + p_w d\phi_w - S_s dT - d\Psi_s \geq 0 \quad (1)$$

In (1) σ_{ij} and ε_{ij} are, respectively, the stress and strain tensors, p_α is the pressure of fluid phase α, T the temperature while ϕ_α , S_s and Ψ_s are the volumetric fluid content of fluid phase α, entropy and free energy of the "skeleton" per unit initial volume of the bulk material. The "skeleton" considered here is the system composed of both the solid grains and the phase interfaces having their own thermomechanical properties. The left hand side of (1) represents the intrinsic dissipation of an element of skeleton.

Let $\phi = \phi_w + \phi_a$ be the porosity referred to the initial volume of the skeleton. Assuming that the volume change of solid grains is negligible, then the volumetric strain ε ($=\varepsilon_{ii}$) is only due to the variation of porosity, yielding

$$\phi = \phi_0 + \varepsilon \quad (2)$$

Due to (2) equation (1) can be written as

$$(\sigma_{ij} + p_a\delta_{ij})d\varepsilon_{ij} - p_c d\phi_w - S_s dT - d\Psi_s \geq 0 \quad (3)$$

where $p_c = p_a - p_w$ is the capillary pressure. As the modelling of dissipative phenomena may be achieved by the use of internal state variables, denoted with χ_n, the following class of free energies are considered:

$$\Psi_s = \Psi_s(\varepsilon_{ij}, \phi_w, T, \chi_n) \quad (4)$$

In case of reversible transformations, (3) turns out to be an equality, the internal state variables not varying. As a result the conjugate state variables associated with strain ε_{ij}, water content ϕ_w and temperature T are respectively $\sigma_{ij} + p_a\delta_{ij}$, p_c and S_s according to

$$\sigma_{ij} + p_a\delta_{ij} = \frac{\partial \Psi_s}{\partial \varepsilon_{ij}} \quad p_c = -\frac{\partial \Psi_s}{\partial \phi_w} \quad S_s = -\frac{\partial \Psi_s}{\partial T} \quad (5)_{a,b,c}$$

Then the irreversible processes, involving the evolution of internal variables χ_n, necessarily have to satisfy the fundamental inequality (3):

$$-\frac{\partial \Psi_s}{\partial \chi_n} d\chi_n \geq 0 \quad (6)$$

3. POROPLASTICITY FOR NON DEFORMING SKELETON

3.1 Formulation of the model

Let us consider, in this section, non deforming skeleton (ε_{ij}=0) and isothermal transformations (T=constant). The figure 1 shows an example of the capillary pressure versus water degree of saturation relationship as it is measured in a pack of smooth glass beads. The water degree of saturation is defined as follows:

$$S_w = \frac{\phi_w}{\phi} \quad (7)$$

It is custom to start the measurement of the capillary pressure curves with the sample 100% saturated with the water. Then the air is injected by very slowly increasing the capillary pressure applied to it. The response (denoted with R_0 in the figure 1) is measured until the water cannot be displaced by further increase of capillary pressure. This curve is called the primary drainage curve. When decreasing the capillary pressure from a high value to zero the obtained curve (denoted with A in the figure 1) is called the secondary imbibition curve. Then the secondary drainage curve (denoted R in the figure 1) constitute with the secondary imbibition curve a closed and reproducible hysteresis loop (RA). As indicated in the figure 1 scanning curves within the main hysteresis loop RA can be obtained by reversing the direction of pressure change at some intermediate point along either the drainage curve or the imbibition curve.

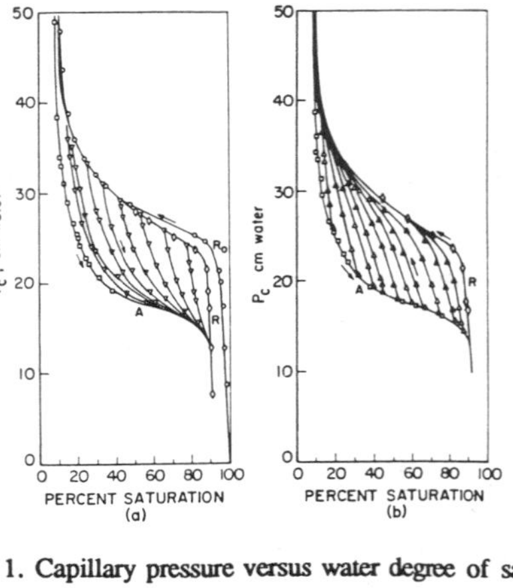

Figure 1. Capillary pressure versus water degree of saturation relationship (after Morrow and Harris, 1965)

This observed behaviour is simplified by referring to the plasticity of solid materials. The ensuing idealized behaviour is presented in the figure 2. The scanning curves are assumed to be single-valued, each of them defining a domain in which the response is reversible (elastic response). The drainage and imbibition curves result from the hardening response of the model.

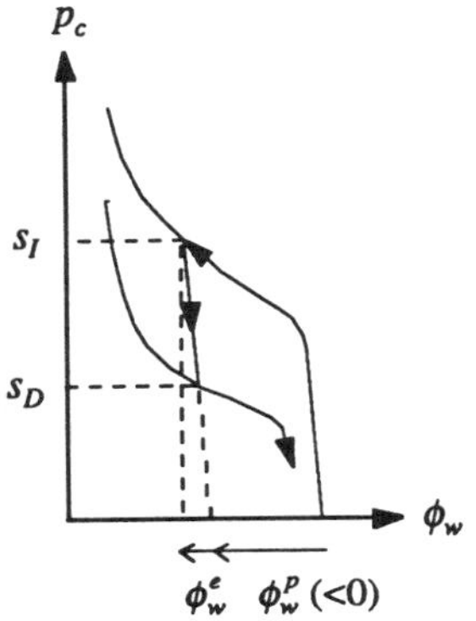

Figure 2. Idealization of the capillary pressure versus water content relationship.

The starting point of the model is the following decomposition:

$$\phi_w - \phi_{w_0} = \phi_w^e + \phi_w^p \tag{8}$$

where ϕ_w^e and ϕ_w^p are the reversible and the irrecoverable variations of the volumetric water content respectively (figure 2). The free energy of the skeleton is assumed to have the form:

$$\Psi_s(\phi_w^e, \phi_w^p) = W(\phi_w^e) + U(\phi_w^p) \tag{9}$$

Then the state equation $(5)_b$ may be written incrementally:

$$d\phi_w^e = -N(p_c)dp_c \tag{10}$$

On the basis of the measurements performed on a pack of smooth glass beads (figure 1), the function $N(p_c)$ may be approximated by the constant $2\ 10^{-5}$ Pa^{-1}.

The reversible domain, in the p_c space, is the range defined, at any time, as

$$s_D \le p_c \le s_I \tag{11}$$

where s_D and s_I are the boundaries of the present reversible domain (figure 2). By referring to the formulation used in plasticity, the reversible domain can also be defined with the help of a scalar function $f(p_c, s_M, \pi)$:

$$f(p_c, s_M, \pi) = |p_c - s_M| - \pi \le 0 \tag{12}$$

where $s_M = (s_I + s_D)/2$ and $\pi = (s_I - s_D)/2$. Then the flow rules are defined as

$$d\phi_w^p = -\delta\lambda \frac{\partial f}{\partial p_c} \tag{13}$$

where $\delta\lambda$ is the plastic multiplier obtained through plastic consistency condition (i.e. $df=0$). The evolution of the reversible domain is controlled by the hardening parameters s_D and s_I. They depend on the irrecoverable volumetric water content ϕ_w^p as follows:

$$d\phi_w^p = -H_I(s_I)ds_I \qquad d\phi_w^p = -H_D(s_D)ds_D \qquad (14)_{a,b}$$

Thanks to (10) the functions $H_I(s_I)$ and $H_D(s_D)$ can be identified on the basis of the drainage and imbibition experimental curves.

3.2 *Physical interpretation*

The hysteresis, discussed here, is caused by pore structure effects. Its physical interpretation is illustrated in the figure 3. For $p_c=s_I$ (figure 3a) the corresponding meniscus is located in a neck, namely an unstable equilibrium. If the capillary pressure increases by dp_c then decreases by $-dp_c$, describing an infinitesimal cycle, then pores will empty, until the meniscus finds a stable equilibrium. An irrecoverable water content $d\phi_w = d\phi_w^p$ is then recorded. Similarly for $p_c=s_D$ (figure 3b) the meniscus has a maximum radius of curvature corresponding to another unstable equilibrium and irrecoverable water content is recorded on decreasing the capillary pressure.

The free energy Ψ_s is that of the surfaces which separate the three phases: solid, liquid, gas (here the solid grains are not considered since they don't deform). These surfaces are characterized by surface tensions denoted with γ_{wa}, γ_{sw} and γ_{sa}, where the subscripts s, w and a stand for solid, water and air respectively. By design the free energy Ψ_s of these surfaces is

$$\Psi_s = \gamma_{wa}\Sigma_{wa} + \gamma_{sw}\Sigma_{sw} + \gamma_{sa}\Sigma_{sa} \tag{15}$$

where Σ_{wa}, Σ_{sw} and Σ_{sa} are the corresponding surface areas. Due to the irrecoverable water content recorded after an infinitesimal cycle, $dp_c=0$, the wetted surface Σ_{sw} varies by $d\Sigma_{sw}$. Consequently the surface free energy $\gamma_{sw} d\Sigma_{sw}$ is substituted for $\gamma_{sa} d\Sigma_{sw}$. Thus part of the free energy is « frozen » under the form:

$$dU = (\gamma_{sw} - \gamma_{sa}) d\Sigma_{sw} \quad (16)$$

where the magnitude of γ_{sa}- γ_{sw} is about 0.05 N.m^{-1}.

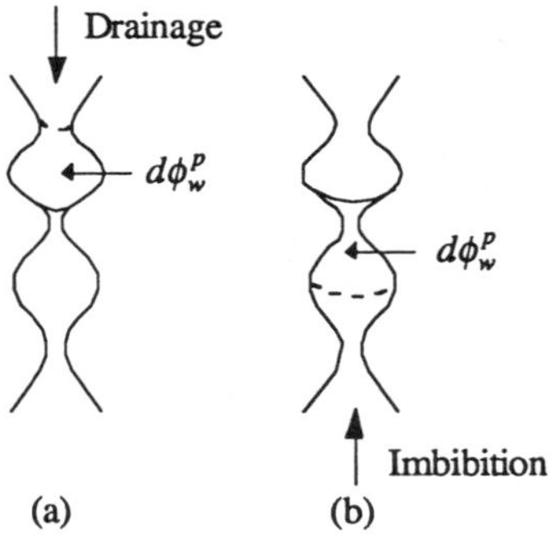

Figure 3 Hysteresis due to pore structure effects. Irrecoverable volumetric water content is recorded during drainage (a) or imbibition (b).

Equation (16) gives the interpretation of the « frozen » energy $U(\phi_w^p)$, responsible for hardening. The fundamental inequality, reading,

$$-p_c d\phi_w^p - dU \geq 0 \quad (17)$$

stipulates that the work element $-p_c d\phi_w^p$ supplied to the skeleton, during an infinitesimal cycle, $dp_c=0$, is part frozen in mechanical form and part dissipated in heat.

4. POROPLASTICITY FOR DEFORMING SKELETON

Only isothermal transformations is considered. The following decomposition is then assumed

$$\varepsilon_{ij} = \varepsilon_{ij}^e + \varepsilon_{ij}^p \quad \text{and} \quad \phi_w - \phi_{w_0} = \phi_w^e + \phi_w^p \quad (18)_{a,b}$$

Similarly to (12) the free energy of the skeleton is assumed to have the form:

$$\Psi_s(\varepsilon_{ij}, \phi_w, \varepsilon_{ij}^p, \phi_w^p) = W(\varepsilon_{ij}^e, \phi_w^e) + U(\varepsilon_{ij}^p, \phi_w^p) \quad (19)$$

In addition let us assume that the deviatoric behaviour is uncoupled from the volume change behaviour (precisely $W(\varepsilon_{ij}^e, \phi_w^e) = W_v(\varepsilon^e, \phi_w^e) + W_d(e_{ij}^e)$). Then differentiating (5) yields the incremental relations

$$d\sigma + dp_a = K(\varepsilon^e, p_c) d\varepsilon^e + b(\varepsilon^e, p_c) dp_c \quad (20)_a$$

$$d\phi_w^e = b(\varepsilon^e, p_c) d\varepsilon^e - N(\varepsilon^e, p_c) dp_c \quad (20)_b$$

where $\sigma = \frac{1}{3}\sigma_{ii}$. Finally, assuming an isotropic linear deviatoric behaviour ($W_d(e_{ij}^e) = 2\mu e_{ij}^e e_{ij}^e$) yields

$$s_{ij} = 2\mu e_{ij}^e \quad (21)$$

where s_{ij} and e_{ij}^e are the stress and elastic strain deviatoric tensors respectively.

In equations (20)-(21) K and μ are the bulk and shear modulus, while b and N are generalized Biot's coefficients (Lassabatère 1994, Coussy 1995).

The poroelastic (reversible) domain in $\{(\sigma_{ij} + p_a\delta_{ij}) \times p_c \times \zeta_n\}$ space is defined by the criterion (see, for example, Chateau 1995, for a study of this domain by homogenization techniques):

$$f(\sigma_{ij} + p_a\delta_{ij}, p_c, \zeta_n) \leq 0 \quad (22)$$

where ζ_n are the hardening parameters. The flow rules have to specify the evolution of $(\varepsilon_{ij}^p, -\phi_w^p)$ as

$$d\varepsilon_{ij}^p = \delta\lambda h_{ij}(\sigma_{kl} + p_a\delta_{kl}, p_c, \zeta_n) \quad (23)_a$$
$$d\phi_w^p = -\delta\lambda h_\phi(\sigma_{kl} + p_a\delta_{kl}, p_c, \zeta_n) \quad (23)_b$$

In (23) h_{ij} and h_ϕ are prescribed functions which defined the direction of plastic flow. $\delta\lambda$ is the non negative plastic multiplier. Finally additional laws, the hardening laws, must specify the evolution of the parameters ζ_n as functions of ε_{ij}^p and ϕ_w^p.

5. APPLICATION TO UNSATURATED SOILS

In 1990 Alonso et al presented a model for unsaturated soils. This model was formulated in the framework of hardening plasticity and is able to represent many of the fundamental features of the stress-strain behaviour. With little changes this model may enter in the framework of poroplasticity. Then not only the stress-strain behaviour but the capillary pressure versus water content behaviour, including hysteresis, can be described into a unique model. The resulting model is described in the following.

.1 Poroelastic behaviour

Alonso et al stated that, during elastic unloading, the equation $(20)_a$ takes the form

$$d\varepsilon^e = -\kappa\frac{d(\sigma+p_a)}{\sigma+p_a} - \kappa_s\frac{dp_c}{p_c+p_{atm}} \tag{24}$$

where κ and κ_s are soil parameters and p_{atm} is the atmospheric pressure. Then (24) implies that $(20)_b$ necessarily reads

$$d\phi_w^e = \kappa_s d(\frac{\sigma+p_a}{p_c+p_{atm}}) - N(p_c)dp_c \tag{25}$$

where the function $N(p_c)$ is the slope of the scanning curves (obtained for $\sigma+p_a=0$).

.2 Yield surfaces

Originally Alonso et al defined the reversible domain with the help of two yield functions f_1 and f_2. To account for the imbibition curve a third yield surface f_3 has been added. The resulting poroelastic domain, in $\{p\times q\times p_c\}$ space with p=-σ-p_a and $q=\sqrt{(3/2 s_{ij}s_{ji})}$, is shown in the figure 4.

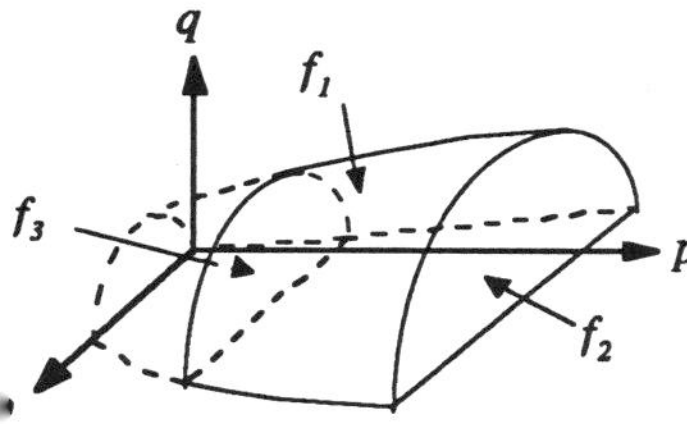

Figure 4 The poroelastic domain (solid lines), in $\{p\times q\times p_c\}$ space, as defined from the original Alonso et al's yield surfaces (solid+dotted lines).

The three yield functions f_1, f_2, f_3 are defined in terms of the stress variables: p ; q; p_c :

$$f_1(p,q,p_c,p_0) = q^2 + M^2(p+kp_c)(p-p_{LC}) \tag{26}$$

$$f_2(p_c,s_I) = p_c - s_I \tag{27}$$

$$f_3(p_c,s_D) = s_D - p_c \tag{28}$$

where $p_{LC}/p_r = (p_0/p_r)^{[\lambda(0)-\kappa]/[\lambda(p_c)-\kappa]}$ and $\lambda(p_c) = \lambda(0)[(1-r)\exp(-\beta p_c)+r]$. The constants $M, k, p_r, \lambda(0), \beta$ and r are soil parameters (Alonso et al 1991).

In (26)-(28) p_0, s_I and s_D are the hardening parameters. They are governed by hardening laws given hereafter.

5.3 Flow rules

Concerning the flow rules associated with the yield surface f_1, an associated law can be proposed for the plastic strains even though Alonso et al's model was non associated to better fit the observed behaviour in saturated situation. For lack of knowledge the plastic volumetric water content increment $d\phi_w^p$ is prescribed to zero:

$$d\varepsilon^p = -\delta\lambda\frac{\partial f_1}{\partial p}\ ;\ de_{ij}^p = \delta\lambda\frac{\partial f_1}{\partial s_{ij}}\ ;\ d\phi_w^p = 0 \tag{29}$$

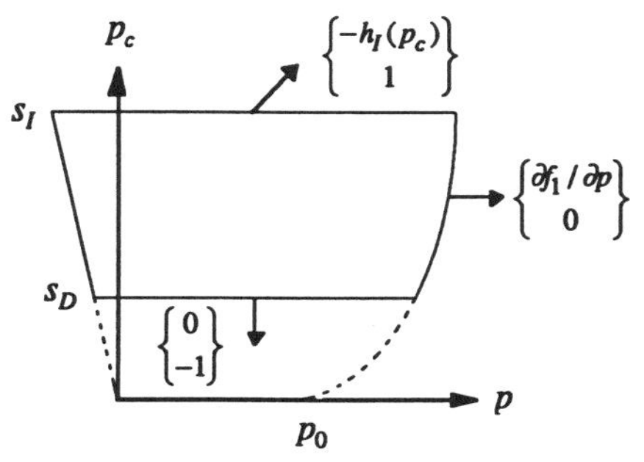

Figure 5 The yield surfaces in $\{p\times p_c\}$ space and the direction of the plastic flow vector $(-d\varepsilon^p, -d\phi_w^p)$.

On the yield surface f_2 the deviatoric plastic strain increments are prescribed to zero according to Alonso et al's model. Concerning the plastic volumetric strain and plastic volumetric water content, the flow rules can be defined consistent with the Alonso et al's model and the drainage curve:

$$d\varepsilon^p = \delta\lambda\, h_I(p_c)\ ;\ de_{ij}^p = 0\ ;\ d\phi_w^p = -\delta\lambda \tag{30}$$

where $h_I(p_c) = -\dfrac{\lambda_s - \kappa_s}{(p_c + p_{atm})H_I(p_c)}$. The constant λ_s is a soil parameter of the Alonso et al's model. The function $H_I(p_c)$ is related to the drainage curve as described hereafter.

On the yield surface f_3 the plastic strain increments are prescribed to zero by lack of knowledge. Thus the flow rules only accounts for the imbibition curve:

$$d\varepsilon^p = 0\,;\, de^p_{ij} = 0\,;\, d\phi^p_w = \delta\lambda \tag{31}$$

A general view of the plastic flows, in $\{p \times p_c\}$ space, is given in figure 5.

5.4 Hardening laws

The evolution of the yield surfaces are controlled by the parameters p_0, s_I and s_D which depend on plastic variables ε^p and ϕ^p_w. According to Alonso et al's model, the motion of the yield surface f_1 is governed by the relation

$$\frac{dp_0}{p_0} = -\frac{d\varepsilon^p}{\lambda(0) - \kappa} \tag{32}$$

The motion of f_2 and f_3 are governed by

$$H_I(s_I)ds_I = -d\phi^p_w \tag{33}$$
$$H_D(s_D)ds_D = -d\phi^p_w \tag{34}$$

The functions $H_I(p_c)$ and $H_D(p_c)$ are such that the sums $H_I(p_c)+N(p_c)$ and $H_D(p_c)+N(p_c)$ correspond to the slope of the drainage and imbibition curves respectively. Combining (30) and (33) yields

$$\frac{ds_I}{s_I + p_{atm}} = -\frac{d\varepsilon^p}{\lambda_s - \kappa_s} \tag{35}$$

which is the hardening law originally proposed par Alonso et al. It should be noted that the motion of the yield surface f_2 is not coupled to that of the yield surface f_1 as it is in the Alonso et al's model.

6 CONCLUSION

The concepts of plasticity are extended to partially saturated porous media in the so called poroplasticity. In addition to irreversible strains, an internal variable, the irrecoverable volumetric water content, captures the hysteresis phenomena exhibited in drying-wetting cycles. The ensuing model involves a fully coupled hydromechanical behaviour. This model is then applied to unsaturated soils on the basis of the model presented by Alonso et al. With little changes the Alonso et al's model may be considered as a poroplastic model. However some consequences of this coupled model should be confirmed experimentally as, for example, the simultaneous occurrence of the plastic strains and the plastic volumetric water content associated with the drainage curve.

REFERENCES

Alonso E. E., Gens A. and Hight D. W. 1987, Special problem soils: General report, in *Proceedings, 9th European Conference on Soil Mechanics and Foundation Engineering, vol. 3*, pp. 1087-1146, Balkema, Rotterdam.

Alonso E. E., Gens A. and Josa A. 1990, A constitutive model for partially saturated soils, *Géotechnique*, 40 (3): 405-430.

Bishop A. W. and Donald I. B. 1961, The experimental study of partly saturated soils in triaxial apparatus, *Proc. Int. Conf. Soil Mech. Found. Eng. 5th, vol. 1*, 13-21.

Burland J. B. 1965, Some aspects of the mechanical behaviour of partially saturated soils. *Moisture Equilibria and Moisture Changes Beneath Covered Areas*: 270-278. Butterworths. Sydney.

Chateau X. and Dormieux L. 1995, Homogenization of a non-saturated porous medium: Hill's lemma and applications, C. R. A. S. Paris, t. 320, série II b, p. 627-634, (in french with abridged english version).

Coussy O. 1995, *Mechanics of porous continua*, J. Wiley & Sons.

Dangla P.and Coussy O. 1996, Drainage and drying of deformable porous materials: one dimensional case study, *IUTAM Symposium on Mechanics of Granular and Porous Materials*, Cambridge.

Lassabatère T. 1994, Couplages hydromécaniques en milieu poreux non saturé avec changement de phase: application au retrait de dessication. Thèse de doctorat de l'École Nationale des Ponts et Chausées, Paris.

Morrow N. R., and Harris C. C. 1965, SPE Jour. 5, 15.

Numerical Models in Geomechanics, Pietruszczak & Pande (eds) © *1997 Balkema, Rotterdam, ISBN 90 5410 886 X*

Stress and strain non-uniformities in triaxial testing of structured soils

Rocco Lagioia
Dipartimento di Ingegneria Strutturale, Milan University of Technology, Italy (Formerly: Imperial College, London, UK)

David M. Potts
Department of Civil Engineering, Imperial College of Science, Technology and Medicine, London, UK

ABSTRACT: This study presents the results of a numerical investigation into the effects of end restraint in triaxial testing of structured soils. Finite element analyses were performed in which soil behaviour was reproduced by means of a constitutive model which takes into account the effects of the initial undisturbed structure and of its progressive structural degradation as plastic straining occurs. The paper analyses the uniformity of stress and strain fields within soil specimen and also the physical uniformity of the material as loading progresses. Comparison is also made between the deviatoric stress-axial strain curve given by the integration of the constitutive law and that obtained by the finite element analysis of the boundary value problem.

1 INTRODUCTION

The use of numerical methods to analyse geotechnical engineering problems is becoming wide spread. However the ability of these methods to accurately predict real soil behaviour depends on the constitutive model employed. Such models are usually based on the results of laboratory tests. Ideally these tests should apply uniform stresses and/or strains to a sample such that the results reflect the behaviour of an infinitesimal soil element. Unfortunately this rarely occurs as most testing devices involve loading platens which apply unwanted frictional forces to the soil sample. This results in non-uniform stress and strain fields. Consequently soil behaviour computed on the basis of the tests results may not be representative of the real constituent material behaviour.

The discrepancy between the behaviour of an infinitesimal soil element and a specimen of finite size tested in the laboratory depends on soil type and it is likely to be particularly large for structured soils. As such soils occur naturally on a large scale this could have important practical consequences. For such soils it is usually possible to identify an initial stress domain characterised by stiff behaviour and high strength. When the applied stress exceeds the limit of this domain a sharp reduction in stiffness is observed and the soil structure is progressively destroyed. When dealing with structured soils the complete elimination of structure produces a physical change of the soil, from a "rock-like" to a "soil-like", appearance, as described by Pellegrino (1970).

Better knowledge of the behaviour of structured soils has been achieved in the last twenty years as a result of extensive laboratory investigations performed mainly using the conventional triaxial apparatus. More recently several authors have proposed constitutive models to reproduce the complex behaviour of natural structured soils.

This paper presents the results of a study to investigate the effects of non-uniformities involved in the triaxial testing of structured soil specimens. Finite element analyses have been performed using a constitutive model specifically developed to model these soils.

2 FINITE ELEMENT ANALYSES

The numerical analyses were performed with the Imperial College Finite Element Program (ICFEP).

Due to the symmetry of the problem only the top right handside quarter of a triaxial specimen was analysed. The specimen was assumed to be 38mm in diameter and 76mm high and was discretised into 200 (10 by 20) 8 node isoparameteric axi-symmetric

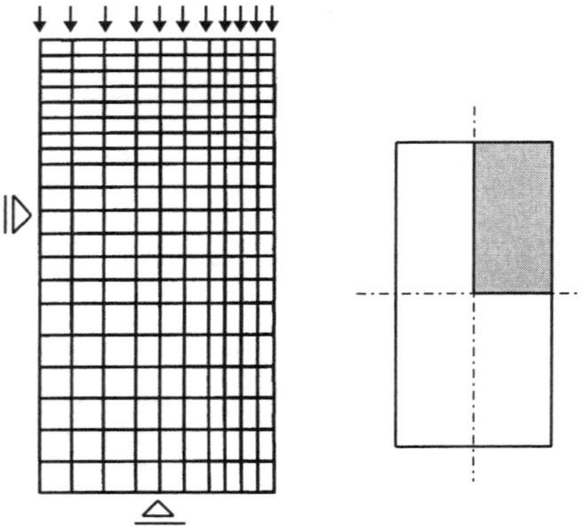

Figure 1. Finite element mesh and boundary conditions.

quadrilateral elements. Fig.1 shows the mesh and boundary conditions applied. Drained triaxial compression tests on isotropically consolidated samples were simulated by first imposing initial stresses in the sample and then applying increments of vertical displacement to the top of the sample (typically 400 increments were used) while maintaining the radial stress on the curved outer surface constant. To model the rough nature of the top cap no horizontal movements along the top boundary of the mesh were permitted. The sample was assumed to behave in a drained manner throughout.

A modified Newton-Rapson approach was used for the non-linear solver together with a modified Euler scheme for the stress point algorithm. This combination was shown to be very efficient both in terms of accuracy and speed of computation by Potts and Ganendra (1992).

3 CONSTITUTIVE MODEL

The material behaviour was reproduced by means of the constitutive model proposed by Lagioia and Nova (1995). This model is based on the classical theory of plasticity and allows for strain hardening/softening as well as strain degradation of soil structure. The basic assumption is that the presence of soil structure causes an expansion of the initial yield locus as compared to the non-structured material. This expansion occurs both in the positive and the negative direction of the mean effective stress axis, the latter being necessary to account for the tensile strength of the material. The size of the current yield surface is controlled by three hardening parameters (hidden variables). The first of these parameters, p_s controls the softening/hardening behaviour of the soil due to plastic straining and is similar to that used in conventional critical state models. The other two parameters, p_t and p_m, deal with the effects of soil structure and account for the softening behaviour due to the contraction of the yield surface on the negative (p_t) and positive (p_m) side of the mean effective stress axis, as structure is progressively removed by plastic straining.

The current values of model parameters p_t and p_m are therefore a measure of the amount of structure still present in the soil specimen. When these parameters are zero the effects of soil structure has been completely removed, i.e. in the case of a cemented sandstone the behaviour of a granular material is to be expected thereafter. For further details of the model see Lagioia and Nova (1995).

4 SOIL BEHAVIOUR

Finite element analyses have been performed using two sets of model parameters which differ only in the rate of structure degradation (Table 1). In set one the degradation behaviour is abrupt while in set two a more gradual structure degradation is considered.

The first set of parameters were selected to accurately reproduce the behaviour of the Gravina calcarenite described by Lagioia and Nova (1995). Figures 2 and 3 show theoretical curves given by the Lagioia and Nova's model with the two sets of parameters of table 1 for an isotropic compression test and a drained triaxial compression test with a cell pressure of 1100kPa. These curves have been obtained by numerical integration of the constitutive equations. They can also be reproduced from Finite Element analysis if a single element is subject to uniform (no end effects) stress and strain fields.

Table 1. Parameters for constitutive model.

	γ	p_{s0} kPa	p_{m0} kPa	p_{t0} kPa	β	B_p	ρ_m	ρ_t	ξ	M_g	μ_g	E kPa	ν
set 1	4.4	800	1600	142	1.5	0.06	8333	500	0	2.167	1.075	171101	0.129
set 2	4.4	800	1600	142	1.5	0.06	2500	20	0	2.167	1.075	171101	0.129

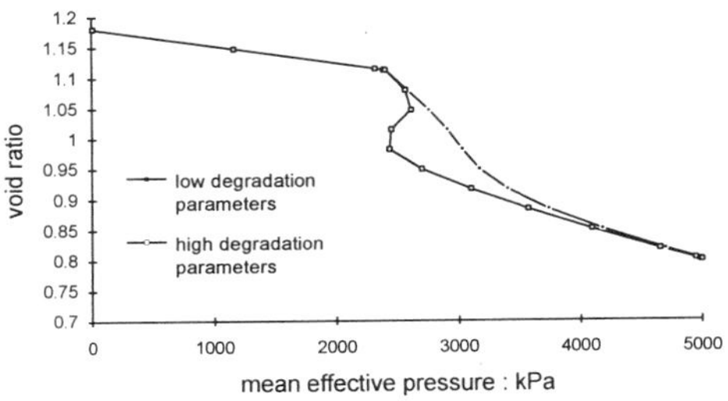

Figure 2. Isotropic compression curve from Lagioia and Nova's model for the two sets of parameters of table 1.

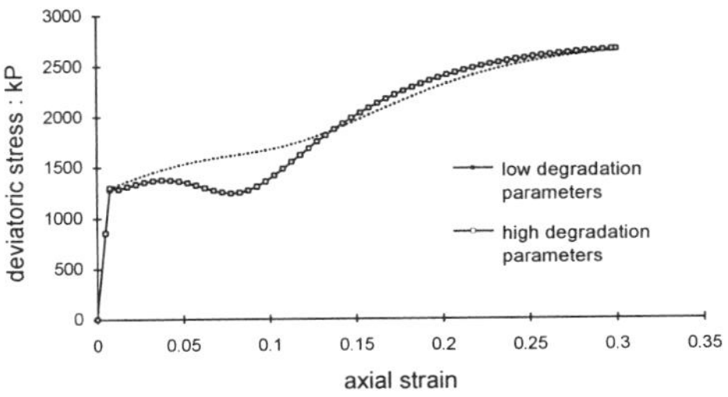

Figure 3. Drained shear test from Lagioia and Nova's model for the two sets of parameters of table 1.

It is apparent that the calcarenite needs high degradation parameters to reproduce the abrupt degradation of inter-particles cement, whilst most structured soils show a more gradual structure degradation. The second set of parameters gives curves which are typical of many natural soils, such as those shown for example by Coop and Atkinson (1993), Leroueil and Vaughan (1990),...

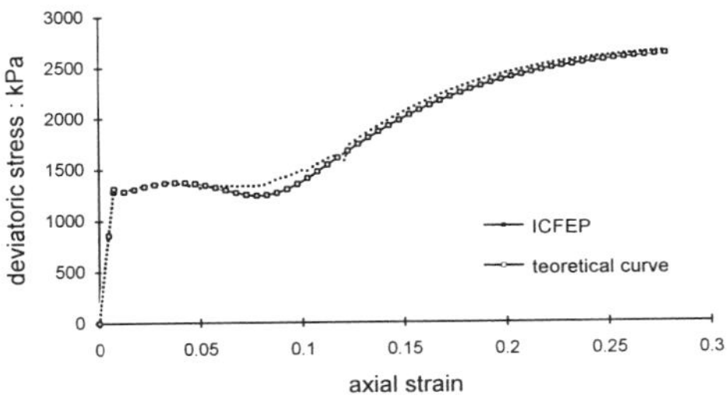

Figure 4. Comparison between stress-strain curves from FE analyses and constitutive model.

5 RESULTS

Numerical analyses of drained triaxial compression tests were performed using both sets of model parameters. The specimens being initially isotropically consolidated prior to shearing to 1100kPa mean effective stress. Despite the different shapes of the theoretical stress-strain curves due to the different rates of structural degradation, the results from the finite element analyses indicate qualitatively the same effects of end restraint. Consequently only the results of the Gravina calcarenite (set 1 parameters) will be described in detail. All plots of FE results refer to the top right handside quarter of the triaxial specimen.

The predicted deviatoric stress versus axial strain curve is compared with the theoretical curve in Fig. 4. The finite element curve has been obtained by using the reactions on the top boundary of the sample to calculate $\Delta\sigma_v$ and therefore the deviatoric stress and by taking the ratio of the change in sample height to its original height to give the axial strain. This procedure is similar to that followed when interpreting the results from laboratory tests. It can be seen that reasonable agreement is obtained between the two curves. Therefore, at first sight, it might appear that there is little effect of end restraint. However, as will now be shown, such an interpretation is misleading.

Lagioia (1996) showed the existence of an initial elastic domain for the calcarenite. This is also produced in the finite element analysis were for a sufficiently small vertical displacement of the top boundary the specimen remains elastic. Figure 5 shows results from the last elastic displacement increment (ε_a=0.58%). Even at this small strain the stress and strain fields are non uniform due to end effects. Figure 5a shows that the deviatoric stress, J, is high under the top right hand corner of the mesh, i.e. at the edge of the rough contact between the loading platen and specimen. Despite this, the principal stress and strain directions only deviate slightly from the vertical in the immediate vicinity of the top cap. Further investigation has shown that such behaviour appears to be independent of the Poisson's ratio.

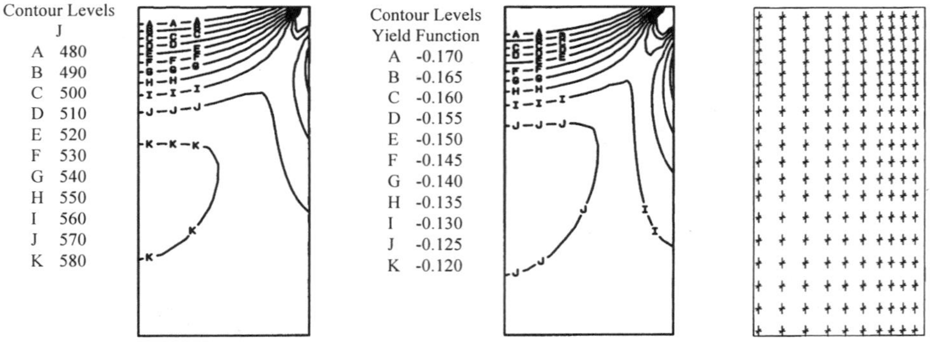

deviatoric stress (a) yield function (b) principal stress (c)

Figure 5. Last elastic increment (ε_a=0.58%): a) contours of deviatoric stress, b) yield function and c) principal stress directions.

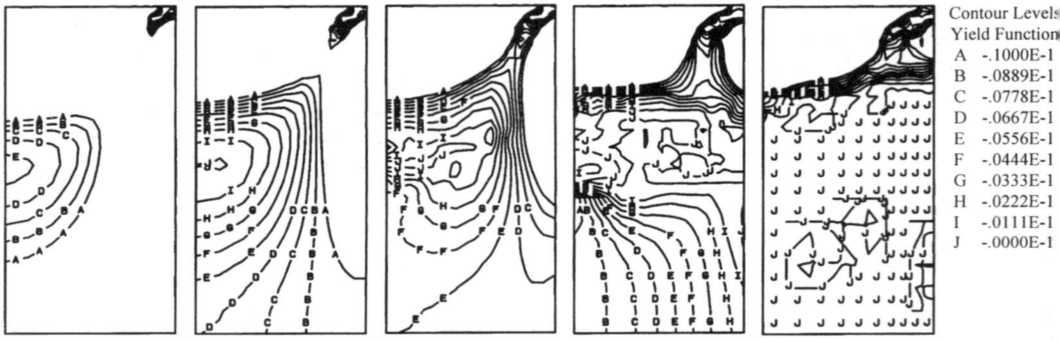

ε_a=0.73% (a) ε_a=0.74% (b) ε_a=0.75% (c) ε_a=0.79% (d) ε_a=1.18% (e)

Figure 6. Contour levels of yield function for increasing accumulated vertical displacement of top boundary.

Figure 5b shows contours of the yield function, F. The value of F is -ve if the soil is elastic and zero if the soil is plastic. Due to the end effects the stresses vary non uniformly throughout the sample and this causes some parts of the specimen to be nearer to yield than others. Two areas in which the sample is close to yield can be identified. The first is close to the edge of the top boundary which is suffering high deviatoric stresses, see Fig. 5a. The second area is located on the vertical axis of symmetry of the sample at 1/4 specimen height. On further straining yield is first reached at the top edge of the specimen. This is followed by yielding at 1/4 height. While the yield zone near the top edge remains comparatively small that at the 1/4 height expands both toward the top and the outer edge of the specimen. This process can be clearly seen in the contour plots of yield function given in Fig. 6. Each plot in this figure corresponds to a different value of axial strain.

As discussed by Lagioia and Nova (1995) plastic straining causes progressive structural degradation. In the case of the calcarenite this means that the material progressively deteriorates towards a calcareous sand. In the constitutive model the hidden variables p_t and p_m measure the amount of structural degradation. Figure 7 shows contour levels of one of these variables, p_t for the same series of axial strain values as used in Fig. 6. The contour level labelled "A" indicates intact material (i.e. no degradation) whilst "K" indicates that the cohesion has been completely removed by plastic straining.

At the onset of yielding degradation starts under the top edge and then at 1/4 specimen height (figure 7a and 7b). As loading proceeds degradation

Contour Levels	
	p_t
A	-142
B	-128
C	-114
D	-99.4
E	-85.2
F	-71.0
G	-56.8
H	-42.6
I	-28.4
J	-14.2
K	-0.00

$\varepsilon_a=0.73\%$	$\varepsilon_a=0.74\%$	$\varepsilon_a=0.75\%$	$\varepsilon_a=0.79\%$	$\varepsilon_a=1.18\%$
(a)	(b)	(c)	(d)	(e)

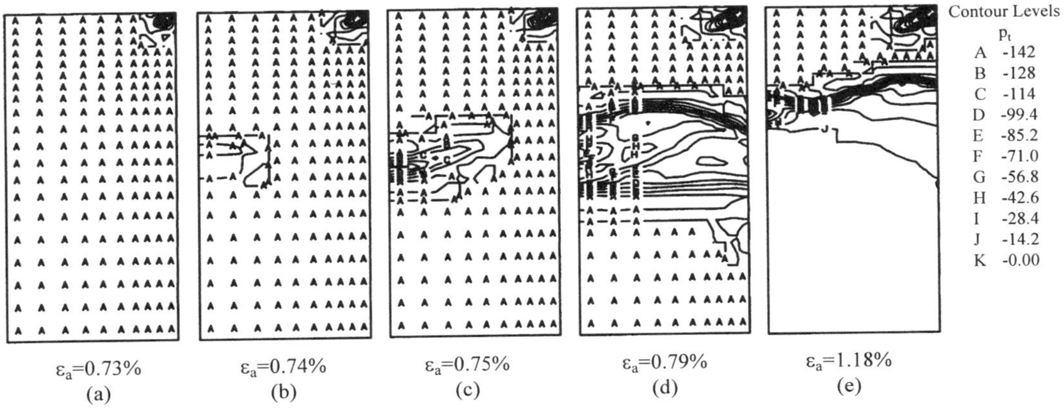

Figure 7. Contour levels of hidden variables p_t for increasing accumulated vertical displacement of top boundary.

$\varepsilon_a=0.73\%$	$\varepsilon_a=0.74\%$	$\varepsilon_a=0.75\%$	$\varepsilon_a=0.79\%$	$\varepsilon_a=1.18\%$
(a)	(b)	(c)	(d)	(e)

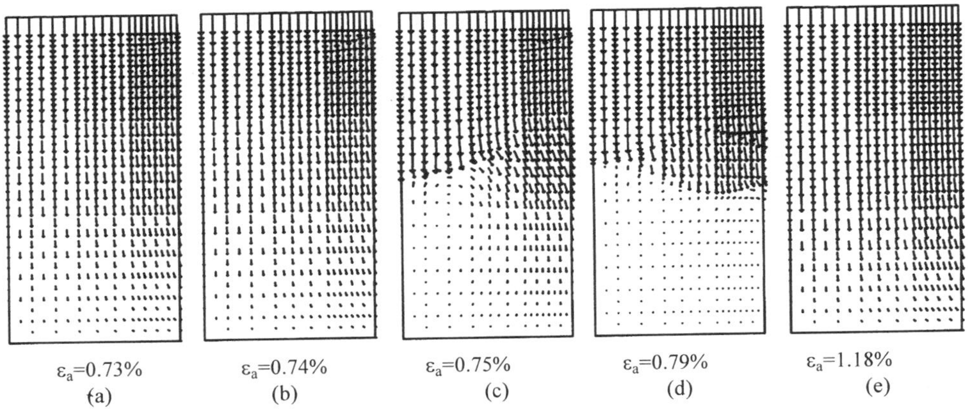

Figure 8. Incremental displacement vectors for increasing accumulated vertical displacement of top boundary.

expands within the soil specimen. Figure 7d shows that a localised zone of completely degraded material develops at roughly 1/4 height. Figure 7e shows a soil specimen which is completely degraded except for the top central part, which is characterised by low shear stresses.

Structure degradation takes place in a non uniform manner within the specimen during shearing. Once yielding begins the specimen is not any more a uniform "rock-like" material, but part of it is gradually becoming a non cemented sand. In this condition it seems difficult to consider the specimen equivalent to an infinitesimal material element.

The localised degradation which develops at 1/4 height produces a very non uniform deformation of the soil specimen. Figure 8 shows incremental displacement vectors for the same load increments of figures 6 and 7. Each vector plot shows the magnitude and direction of movement for a single increment of the analysis. While the absolute magnitude of the vectors is not important the relative magnitude and the directions indicate how the behaviour of the specimen changes with loading. As would be expected by a material which progressively transforms from a soft rock to a loose sand, an abrupt increase in radial displacement takes place at 1/4 height.

Fig. 9 shows contours of deviatoric stress. As loading progresses and structural degradation takes place, the initially non uniform stress level within the specimen tends to become more uniform.

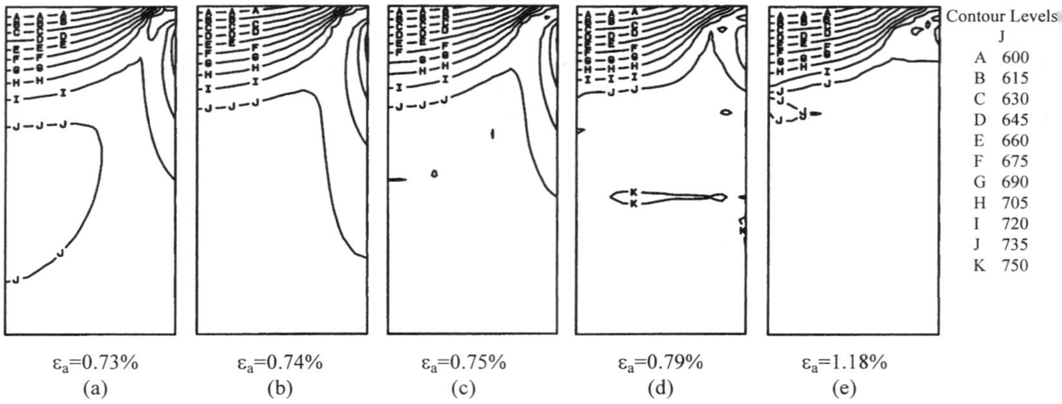

Figure 9. Contour levels of deviatoric stress for increasing accumulated vertical displacement of top boundary.

6 CONCLUSIONS

This study presents the results of a numerical investigation into the end effects in triaxial testing of structured soils. The analyses were performed assuming that friction between specimen ends and loading platens restricts the radial movement of the top and bottom boundaries of the specimen. This is particularly appropriate to tests in which capping is adopted as a technique to reduce bedding at the contact between porous stones and specimen ends.

Structured soils were considered in this study as improved sampling and testing techniques and better understanding of soil behaviour have shown that most natural soils are structured (Leroueil and Vaughan (1990)). Soil behaviour was then modelled using a constitutive model proposed by Lagioia and Nova (1995) to reproduce the stress-strain behaviour of structured soils.

The analysis of the stress and strain fields within the specimen show that the triaxial specimen cannot be considered as a "perfect" infinitesimal element. The stress and strain fields are infact highly non-uniform. As plastic straining takes place the material becomes physically non uniform and localised zones of non structured soil develop within the specimen. These localised zones make the specimen bulge at approximately 1/4 specimen height.

The results of the analyses show that despite the complex soil behaviour, and the clearly observed non-uniformities, the deviatoric stress-axial strain curve derived from the forces and movements imposed at the specimen ends is practically identical to that for an ideal sample with no end effects. Therefore the stress-strain characteristics as measured from an "imperfect" triaxial test seems to be representative of the behaviour of an "infinitesimal" constituent soil element.

AKNOWLEDGMENT

The first author wishes to thank the support received from the European Commission, under the Human Capital and Mobility Project.

REFERENCES

Coop, M.R. & Atkinson, J.H. 1993. The mechanics of cemented carbonate sands. *Géotechnique* 43(1): 53-67.

Lagioia, R. & Nova, R. 1995. An experimental and theoretical study of the behaviour of a calcarenite in triaxial compression. *Géotechnique* 45(4): 633-648.

Lagioia, R. 1996. Comportamento meccanico di una calcarenite di Gravina di Puglia. *Rivista Italiana di Geotecnica* (in Italian), in print.

Leroueil, S. & Vaughan, P.R. 1990. The general and congruent effects of structure in natural soils and weak rocks. *Géotechnique* 40(3): 467-488.

Pellegrino, A. 1970. Mechanical behaviour of soft rocks under high stresses. *Proc. 2nd Int. Cong. Rock Mechanics, Belgrad 2*: 173-180.

Potts, D.M & Ganendra, D. 1992. A critical assessment of solution strategies for nonlinear finite element analysis of geotechnical problems. *Proc III Portuguese Conference on Computational Mechanics, Coimbra:* C3.1-C3.21

Numerical Models in Geomechanics, Pietruszczak & Pande (eds) © 1997 Balkema, Rotterdam, ISBN 90 5410 886 X

Thermo-mechanical analysis of concrete structures subjected to alkali-aggregate reaction

S. Pietruszczak & M. Huang
McMaster University, Hamilton, Ont., Canada

V. Gocevski
Hydro-Quebec, Montreal, Que., Canada

ABSTRACT: In this paper, a non-linear continuum theory is presented for modelling the thermo-mechanical behaviour of concrete subjected to alkali-aggregate reaction (AAR). The formation of silica gel around the aggregate particles leads to a progressive expansion of the material. The expansion rate is assumed to be controlled by the alkali content, the magnitude of confining stress as well as the time history of the temperature. The progress in the reaction is assumed to be coupled with degradation of the mechanical properties of concrete. First, the constitutive model developed previously in the context of isothermal deformation is extended to describe the influence of temperature on the mechanical effects of the reaction. Subsequently, a numerical example is provided to assess the performance of the proposed formulation. In particular, the model is applied to analyse the right wing dam of the Beauharnois powerhouse, situated in Quebec, Canada.

1. INTRODUCTION

An overview of the physical nature of AAR and its structural implications can be found, for example, in the article by Leger et al. (1995). The reaction was identified more than 50 years ago, prompting extensive interdisciplinary research. Most of the research has been directed at physicochemical aspects of the reaction, and a few studies have also been published on the effects of AAR on the degradation of material properties, i.e. compressive/tensile strength, elastic properties, etc. (c.f., Nixon and Bollinghaus, 1985; Swamy and Al-Asali, 1987; Curtil and Habita, 1994). However, relatively little work has been done in the development of numerical models that are able to follow the time history of the reaction and predict the structural damage resulting from AAR.

Recently, the first author has proposed a non-linear continuum theory to describe the mechanical behaviour of concrete subjected to AAR (Pietruszczak, 1996). The formulation is restricted to isothermal conditions and invokes the assumption that the formation of expansive phases results in a progressive degradation of mechanical properties of the material. The latter have been described within the framework of plasticity. This formulation has been recently applied in the context of 3D finite element analysis and the details on the numerical implementation, including an implicit integration scheme, are provided in the article by Huang and Pietruszczak (1996).

In general, concrete structures built in northern regions, are subjected to significant seasonal temperature fluctuations. This affects the progress in the reaction and results in a more severe damage owing to non-uniform volumetric expansion. Therefore, in the first part of this paper the original formulation is extended to non-isothermal conditions. Both the rate of volumetric expansion and the degradation of mechanical properties, such as compressive/tensile strength and the elastic modulus, are assumed to be affected by the temperature of the concrete mass. The evolution laws are fomulated by invoking the notion of 'reaction time', which is defined in a local sense. Subsequently, the paper focuses on the implementation of the model in the numerical analysis of concrete structures. In particular, the model is applied to analyse the right wing dam of the Beauharnois powerhouse, situated in Quebec, Canada. Both isothermal and non-isothermal finite element analyses are carried out and the results are compared.

2. A CONSTITUTIVE MODEL FOR CONCRETE SUBJECTED TO AAR

2.1 General formulation; extension to non-isothermal case

In the elastic range, the constitutive relation can be written in a general form

$$\boldsymbol{\varepsilon}^e = \boldsymbol{C}^e \boldsymbol{\sigma} + \frac{1}{3}\varepsilon_A \boldsymbol{m} + \boldsymbol{\varepsilon}_T \qquad (1)$$

where C^e is the elastic compliance matrix, $\boldsymbol{m}$ is the operator analogous to Kronecker's delta, ε_A is the volumetric strain due to expansion of the silica gel, and $\boldsymbol{\varepsilon}_T$ is the thermal strain. The latter is defined as

$$\boldsymbol{\varepsilon}_T = \frac{1}{3}\beta_T(T - T_0)\boldsymbol{m} \qquad (2)$$

where β_T is the coefficient of thermal expansion, T and T_0 represent the absolute and the reference temperature, respectively. Differentiation of eq. (1) with respect to time gives

$$\dot{\boldsymbol{\varepsilon}}^e = \boldsymbol{C}^e\dot{\boldsymbol{\sigma}} + \dot{\boldsymbol{C}}^e\boldsymbol{\sigma} + \frac{1}{3}\dot{\varepsilon}_A\boldsymbol{m} + \frac{1}{3}\beta_T\dot{T}\boldsymbol{m} \qquad (3)$$

where the second term, on the right-hand side, describes the degradation of elastic properties due to continuing reaction. Assuming now the additivity of elastic and plastic strain rates, the constitutive relation (3) can be generalized to

$$\dot{\boldsymbol{\varepsilon}} = \boldsymbol{C}^e\dot{\boldsymbol{\sigma}} + \dot{\boldsymbol{C}}^e\boldsymbol{\sigma} + \frac{1}{3}\dot{\varepsilon}_A\boldsymbol{m} + \frac{1}{3}\beta_T\dot{T}\boldsymbol{m} + \dot{\boldsymbol{\varepsilon}}^p \qquad (4)$$

or

$$\dot{\boldsymbol{\sigma}} = \boldsymbol{D}^e(\dot{\boldsymbol{\varepsilon}} - \dot{\boldsymbol{\varepsilon}}^p - \dot{\boldsymbol{C}}^e\boldsymbol{\sigma} - \frac{1}{3}\dot{\varepsilon}_A\boldsymbol{m} - \frac{1}{3}\beta_T\dot{T}\boldsymbol{m}) \qquad (5)$$

where $\boldsymbol{D}^e$ is the elastic constitutive matrix. In order to define the plastic strain rates, the functional form of the yield surface $f = 0$ is assumed to be affected by the progress in the reaction, i.e.

$$f(\boldsymbol{\sigma}, \boldsymbol{\varepsilon}^p, t) = 0\,; \qquad \dot{\boldsymbol{\varepsilon}}^p = \dot{\lambda}\frac{\partial Q}{\partial \boldsymbol{\sigma}} \qquad (6)$$

with $Q(\sigma)$ = const. representing the plastic potential. Writing the consistency condition as

$$\dot{f} = (\frac{\partial f}{\partial \boldsymbol{\sigma}})^T\dot{\boldsymbol{\sigma}} + (\frac{\partial f}{\partial \boldsymbol{\sigma}})^T\dot{\lambda}\frac{\partial Q}{\partial \boldsymbol{\sigma}} + \frac{\partial f}{\partial t} = 0 \qquad (7)$$

and following the standard plasticity procedure, the following constitutive relation is obtained

$$\dot{\boldsymbol{\sigma}} = \boldsymbol{D}^{ep}(\dot{\boldsymbol{\varepsilon}} - \frac{1}{3}\dot{\varepsilon}_v^A\boldsymbol{m} - \frac{1}{3}\beta_T\dot{T}\boldsymbol{m}) - \left(\frac{1}{H}\boldsymbol{D}^e\frac{\partial Q}{\partial \boldsymbol{\sigma}}\left(\frac{\partial f}{\partial \boldsymbol{\sigma}}\right)^T - \boldsymbol{I}\right)\dot{\boldsymbol{D}}^e\boldsymbol{C}^e\boldsymbol{\sigma} - \frac{1}{H}\frac{\partial f}{\partial t}\boldsymbol{D}^e\frac{\partial Q}{\partial \boldsymbol{\sigma}}$$

where $\boldsymbol{I}$ is the identity matrix, and

$$\boldsymbol{D}^{ep} = \boldsymbol{D}^e - \frac{1}{H}\boldsymbol{D}^e\frac{\partial Q}{\partial \boldsymbol{\sigma}}\left(\frac{\partial f}{\partial \boldsymbol{\sigma}}\right)^T\boldsymbol{D}^e\,;$$

$$H = \left(\frac{\partial f}{\partial \boldsymbol{\sigma}}\right)^T\boldsymbol{D}^e\frac{\partial Q}{\partial \boldsymbol{\sigma}} - \left(\frac{\partial f}{\partial \boldsymbol{\varepsilon}^p}\right)^T\frac{\partial Q}{\partial \boldsymbol{\sigma}} \qquad (8)$$

2.2 Evolution/degradation laws

In order to formulate the evolution law for the AAR-induced expansion, it may be convenient to introduce the notion of 'reaction time', t', which is considered as a local property influenced by the temperature history. With this notion, the evolution law may be assumed in the form

$$\dot{\varepsilon}_v^A = g_1(\boldsymbol{m}^T\boldsymbol{\sigma})\,\dot{g}_3(t')\,; \qquad dt' = g_2(T)\,dt \qquad (9)$$

Here g_3 represents the free expansion for a constant alkali content; whereas g_1 and g_2 specify the constraining effect of hydrostatic pressure and the temperature. Based on the exisiting experimental information (Kennerlay et al., 1981; Hobbs, 1988), the functions g_1 and g_3 may be chosen as

$$g_3(t) = \frac{\epsilon\, t'}{A_3 + t'}\,; \qquad t' \to \infty \Rightarrow g_3 \to \epsilon$$

$$g_1(\boldsymbol{m}^T\boldsymbol{\sigma}) = \exp(A_1\boldsymbol{m}^T\boldsymbol{\sigma}/f_{co})\,; \qquad 0 \le g_1 \le 1 \qquad (10)$$

where A_3, A_1 are the material constants, f_{co} represents the initial uniaxial compressive strength and ϵ defines the maximum expansion rate at the pessimum temperature $T = T_p$. The functional form of g_2 may be established based on the available experimental information. In general, the rate of alkali-silica reaction, which directly affects the expansion, is proportional to soluble alkali content. The lower the temperature, the lower the solubility of alkali. Therefore, at a low temperature the reaction is slowed down. As the temperature increases the alkali solubility and concentration are increased, so that the reaction speeds up. As the temperature exceeds the pessimum temperature, T_p, the solubility is decreased due to the over saturation of alkali contents. As a result, the reaction slows down again. Given the above information, the function $g_2(T)$ may be chosen in the following hyperbolic form

$$g_2(T) = 2\,sech(\frac{T - T_p}{A_2}) \qquad (11)$$

where A_2 is a material constant. The value of A_2 employed later in the numerical simulations has been identified from the experimental results reported by Pleau et al. (1989).

The degradation of mechanical properties of concrete, particularly the elastic modulus and uniaxial compressive/tensile strengths, comes mainly from the mechanical damage to the material. During the continuing AAR, the expansive silica gels destroy the concrete matrix bonds and cause fracturing between aggregates and the adjacent cement paste. The degradation functions for the Young's modulus and the uniaxial compressive /tensile strength may be chosen in a simple form

$$E = E_0[1-(1-B_1)\frac{g_3}{\epsilon}];\ f_c = f_{co}[1-(1-B_2)\frac{g_3}{\epsilon}] \qquad (12)$$

where B's are material constants. Apparently as $g_3 \to \epsilon$, there is $E \to B_1 E_o$; $f_c \to B_2 f_{co}$.

3. NUMERICAL ANALYSIS

The mathematical formulation outlined in the previous section has been applied to analyse the deformation history in the right wing dam of the Beauharnois complex, situated in Quebec, Canada. The dam is a gravity structure, 115.44m long and 22.10m high, founded on a firm rock (Potsdam sandstone).

The concrete used for the construction was prepared with Portland cement which contained alkali from 0.8% to a maximum of 1.12%. The aggregate was taken from the excavated rock. The only experimental information on AAR-induced expansion of concrete, available at this stage, is that pertaining to the rate of free expansion, viz. $g_3(t)$. Based on this information, ϵ and A_3 were identified as ϵ= 0.05 and A_3= 190 years. The constants appearing in the function $g_2(T)$, eq.(11), were selected based on experimental results provided by Pleau et al. (1989), i.e. T_p=40°C and A_2=11.04°C. The remaining material parameters were chosen, on a rather intuitive basis, as A_1=0.1 , B_1=0.7 and B_2=0.9. The values of B_1 and B_2 correspond to a maximum of 30% reduction in elasticity modulus and 10% reduction in compressive/tensile strength (at $T=T_p$). It is noted that an experimental program is currently under way at University of Sherbrooke aimed at quantifying the influence of confining pressure on the rate of expansion, as well as the aspects related to degradation of material parameters.

The simulations were carried out based on a constitutive model for concrete proposed by Pietruszczak et al. (1988). The formulation invokes a non-associated flow rule and the yield surface is expressed in a functional form

$$f = \bar{\sigma} - \beta(\xi)\, k(\theta)\, \bar{\sigma}_c = 0\,; \qquad \dot{\xi} \sim (2J_{2\dot{\varepsilon}})^{1/2}$$

where

$$\bar{\sigma}_c = \frac{-a_1 + \sqrt{a_1^2 + 4a_2(a_3 + I/f_c)}}{2a_2} f_c$$

In the above equations $I=-I_1$, $\bar{\sigma}=(J_2)^{1/2}$, $\theta=1/3\sin^{-1}(3\sqrt{3}J_3/2\bar{\sigma}^3)$, where I_1 and (J_2, J_3) are the basic invariants of the stress tensor and the stress deviator, respectively. Moreover, the parameters a_1, a_2 and a_3 represent dimensionless material constants and f_c denotes the uniaxial compressive strength of concrete. The function $\beta(\xi)$ is designated as the hardening function and its evolution depends on $J_{2\varepsilon}$, i.e. the second invariant of the deviatoric part of plastic strain. This function is defined in such a way that, in a stable regime, $\xi\to\infty \Rightarrow \beta\to 1$, which describes a homogeneous deformation mode associated with formation of microcracks. At low confining pressures, the material characteristics are said to become unstable (strain softening). The expression defining the rate of strain softening incorporates the 'size effect' by attributing a unique softening rate to the characteristic volume associated with the integration domain. The details on specification of this function are given in the original article. In general, the unstable response, which corresponds to formation of macrocracks, is characterized by a progressive decrease in the value of β, such that $\beta\to 1-\phi_r$ for $\xi\to\infty$, where ϕ_r defines the residual strength of the material. It has to be emphasized that the formulation pertaining to brittle response, although not rigorous, can still render the solution which is only weakly dependent on discretization. Apparently, other more meticulous descriptions can be incorporated. In general, the distribution of β, in the context of any boundary-valued problem, will provide an indication of the extend of the structural damage in the system.

The numerical simulations discussed here, have been carried out assuming E_o=15,000MPa, ν=0.2 and f_{co}=27MPa. The material parameters associated with elastoplastic response were the same as those cited by Pietruszczak et al. (1988). The coefficient of

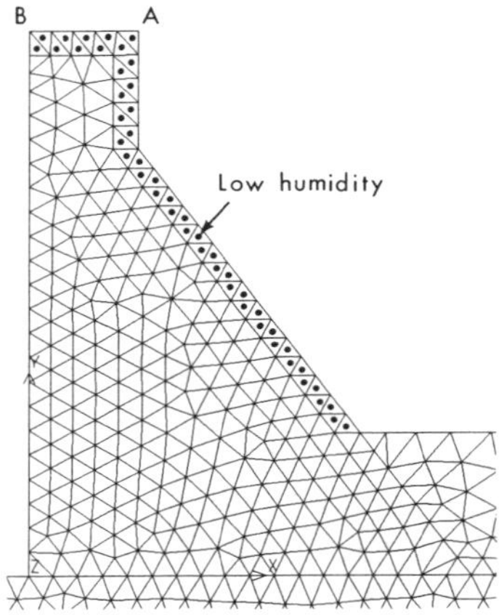

Fig.1 Finite element mesh

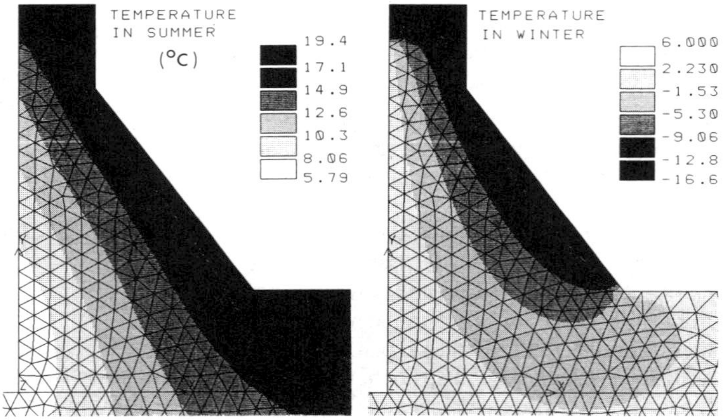

Fig.2 Temperature distribution (a) in the summer ; (b) in the winter

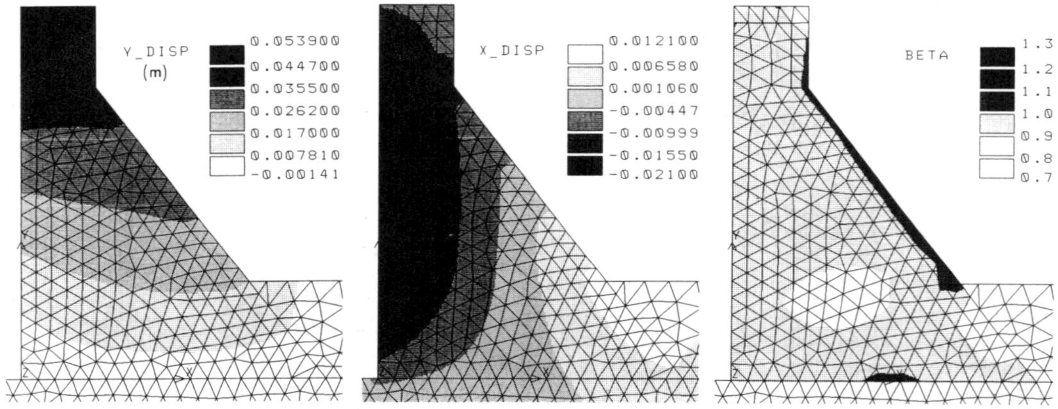

Fig.3 Isothermal case: (a) contours of vertical displacements; (b) contours of horizontal displacements; (c) distribution of the stress intensity factor β after 25 years of reaction

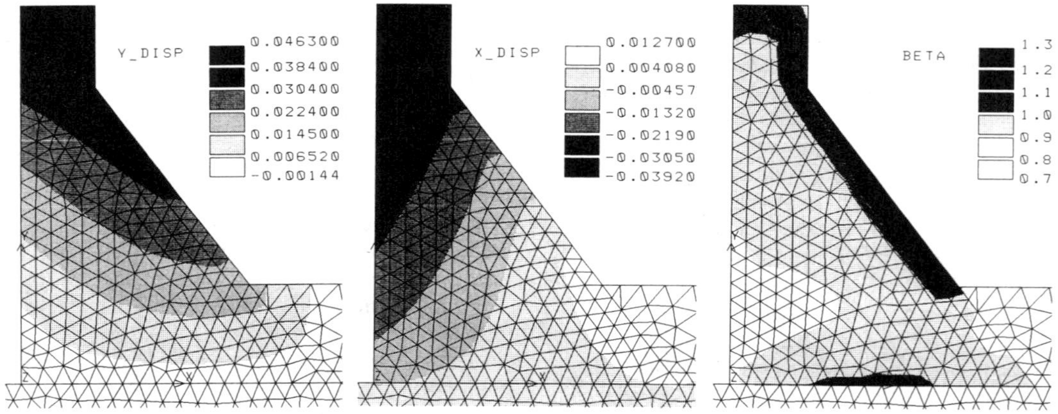

Fig.4 Non-isothermal case: (a) contours of vertical displacements; (b) contours of horizontal displacements; (c) distribution of the stress intensity factor β after 25 years of reaction

thermal expansion of concrete was taken as β_T = 1.5x10^{-5} /°C. The finite element discretization of the problem is shown in Fig.1. The mesh employs 1107 triangular elements. Only a weakly coupled analysis has been performed. Thus, a separate heat transfer analysis was carried out first and the resulting temperature distribution was then used for the mechanical analysis. The thermal boundary conditions were selected as follows:

Dam-Air Interface
For simplicity, only the convection and radiation boundary conditions were considered while the externally supplied heat flux from the sun, i.e. the non-reflected solar energy, was neglected. The evolution of the air temperature was described by a sine function. The period of this function was taken as P=365days. The minimum average daily air temperature was taken as -14.93°C (at the beginning of the year), while the maximum average daily air temperature was assumed as 18.13°C (at $P/2$ days). The temperature record is similar to that used by Leger et al. (1993).

Ground Surface Boundary
The temperature history at the ground surface was described by a sine curve from $P/4$ to $3P/4$ with the maximum temperature of 19.4°C. During the remaining time interval, the temperature was said to remain at -1.2°C. This is due to snow insulation of the ground, which constraints the heat transfer during the period of cold months.

Dam-Reservoir Interface
The maximum temperature was assumed to occur at $P/2$ days. The ice period was assumed to be approximately from $P/2$ to next year's $P/4$. During this period, no temperature fluctuations were allowed since the reservoir remains insulated by the ice. During the ice-free period, the time history of temperature was again described by a sine curve. The maximum and minimum temperatures were assumed to vary with the reservoir depth. On the water surface, they were assumed as 14°C and 0°C, respectively. At the depth of 15m, they were taken as 6°C and 2°C, respectively. Below the depth of 60m, the temperature was said to remain constant at 4°C.

Foundation Bottom Boundary
In general, the penetration depth of annual temperature oscillations beneath the ground surface is about 10m. Therefore, the temperature at the base of the foundation may be considered as constant. Based on the interpolation of the available data, the temperature along the bottom boundary of the foundation was fixed at 6.2°C.

The heat transfer analysis was carried out assuming the thermal conductivity of 2.62 N/s°C, the specific heat capacity of 912.0 m^2/s^{2}°C, the convective coefficient of 23.2 N/ms°C and the radiation coefficient of 4.2 N/ms°C. Fig.2a shows the predicted temperature field in the summer. The temperature on the exposed surfaces reaches approximately 19°C, which enables AAR expansion to develop rapidly; whereas parts of the structure near the reservoir side remain at a lower temperature, 5.8°C. Fig.2b shows the distribution in the winter. The temperature for most part of the dam is below 0°C, indicating that the reaction is substantially slowed down and virtually no expansion occurs.

In the second stage, an isothermal analysis was performed simulating the mechanical response of the structure over a period of 25 years of continuing reaction. The analysis corresponds to a constant temperature of $T = 5.46°C$, which was selected based on the temperature distribution from the heat transfer analysis. At this temperature, the rate of free expansion in the structure is close to the average rate under non-isothermal condition. Figs.3a and 3b show the horizontal and vertical displacement fields after 25 years of continuing reaction. In general, the predicted horizontal movement is substantially smaller than that recorded from in-situ measurements. Fig.3c shows the distribution of the stress intensity factor β corresponding to 25 years of AAR expansion. As explained earlier, $\beta \to 1$ signifies the local failure through formation of microcracks (strain hardening regime). The values of stress intensity factor β exceeding one correspond to $\beta = \int |\dot{\beta}| dt$ and indicate an unstable response associated with formation of macrocracks. The extend of structural damage is quite limited; the macrocracks form only near the exposed surface, which is primarily due to a lower humidity (and thus a lower expansion rate) of concrete in this region.

Finally, a non-isothermal analysis was performed, which was based on the temperature distribution from the heat transfer analysis. Figs. 4a and 4b present the horizontal and vertical displacement fields after 25 years of the reaction. The average vertical displacement at the crest of the dam is close to that under isothermal condition. However, the displacements near the exposed surfaces are higher than those on the reservoir side, which is triggered by a non-uniform summer temperature distribution. The maximum horizontal displacement at the crest of the

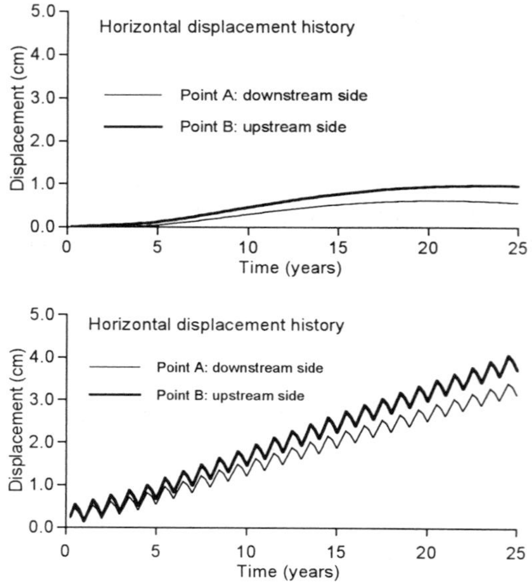

Fig.5 Comparison of horizontal displacements on the crest of the dam from isothermal and non-isothermal analyses

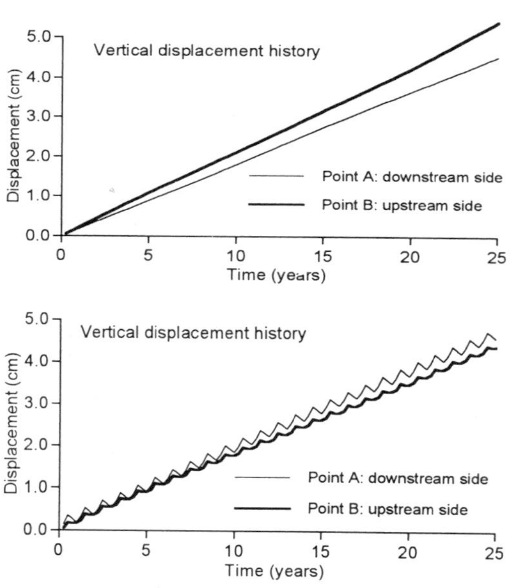

Fig.6 Comparison of vertical displacements on the crest of the dam from isothermal and non-isothermal analyses

dam reaches 38.8mm; this is much higher than that corresponding to the isothermal condition. Figs.5 and 6 compare the horizontal and vertical displacement histories at the crest of the dam obtained from isothermal and non-isothermal analyses. The distribution of β corresponding to 25 years of reaction is shown in Fig.4c. In general, the distribution differs from that corresponding to the isothermal condition. The values of β are relatively high near the exposed surfaces and lower near the dam-reservoir interface. On the exposed surfaces, the macrocracking is more severe than that predicted by the isothermal formulation.

REFERENCES

Curtil, L. and M.F. Habita. 1994. Study of the alkali-aggregate reaction on concrete prisms', *Cement and Concrete Research*, 24, 473-478.

Hobbs, D.W. 1988. *Alkali - silica reaction in concrete*. London: Thomas Telford.

Huang, M. & S. Pietruszczak. 1996. Numerical analysis of concrete structures subjected to alkali-aggregate reaction. *Mechanics of cohesive-frictional materials*. 1: 305-319.

Kennerley, R.A., D. St John & L.M. Smith. 1981. A review of thirty years of investigation of alkali-aggregate reaction in New Zeland. *Proc. 5th International Conference on alkali-aggregate reaction in concrete*. Cape Town. 1-10.

Leger, P., R. Tinawi and N. Mounzer. 1995. Numerical simulation of concrete expansion in concrete dams affected by alkali-aggregate reaction: state-of-the-art. *Can. J. Civ. Eng.*. 22: 692-713 (1995).

Nixon, P.J. and R. Bollinghaus. 1985.The effect of alkali aggregate reaction on the tensile and compressive strength of concrete. *Durability of Building Materials*. 2: 243-248.

Pietruszczak, S. 1996. On the mechanical behaviour of concrete subjected to alkali-aggregate reaction. *Int. J. Computers & Struct.*58: 1093-1099.

Pietruszczak, S., J. Jiang & F.A. Mirza. 1988. An elastoplastic constitutive model for concrete. *Int. J. Solids Structures*. 24: 705-722.

Swamy R.N. and M. M. Al-Asali. 1987. Engineering properties of concrete affected by alkali-silica reaction. *ACI Materials Journal*, 85: 367-369.

Pleau, R., M.A. Berube, M. Pigeon, B. Fournier & S. Raphael. 1989. Mechanical behaviour of concrete affected by ASR. *Proc. 8th International Conference on alkali-aggregate reaction*, Kyoto. 721-726.

Numerical Models in Geomechanics, Pietruszczak & Pande (eds) © 1997 Balkema, Rotterdam, ISBN 90 5410 886 X

Equivalent continuum anisotropic microplane model for rock: Theory and applications to finite element analysis

P.C. Prat, F. Sánchez & A. Gens
Department of Geotechnical Engineering, Technical University of Catalunya, Barcelona, Spain

ABSTRACT: The paper presents the formulation of a simplified micromechanics model suitable for the numerical analysis of anisotropic media such as rock, hard soils, masonry constructions, composite materials, and discontinuous media in general. The model is based on the microplane models developed in recent years, where the macroscopic stress/strain laws are obtained by superimposing the stress/strain behavior on planes of arbitrary orientation distributed uniformly in space. The formulation permits developing anisotropic models simply by giving different values to the model parameters as a function of the orientation of the different planes. The model has been developed to be used in finite element analysis, and is well suited for that purpose. A preliminary application example is presented in the paper.

1 INTRODUCTION

Numerical analysis of many engineering materials, natural or manufactured, necessitates of models accounting for anisotropy. Anisotropic behavior is inherent to materials exhibiting jointed structure such as rock masses, masonry constructions, composite materials and others. The same type of anisotropy is also apparent in the behavior of materials containing cracks or micro-cracks like concrete, rock, stiff clay, ceramics and other brittle materials. The effects of anisotropy are further enhanced by the fact that during loading the material develops additional directional properties that depend on the stress/strain history of the material. It is customary to refer to this latter anisotropy component as *induced* anisotropy whereas *inherent* anisotropy denotes the anisotropy initially present in the material. Inherent anisotropy includes not only the anisotropy due to discontinuities but that corresponding to the material matrix (i.e. the intact material between discontinuities) as well.

The use of numerical analysis techniques for the analysis of structures made up of anisotropic materials requires the adoption of constitutive laws capable of describing this type of behavior. Although it would be possible in principle to reproduce the anisotropy due to discontinuities (joint, interfaces, fissures) using special finite elements (such as joint elements), this technique can lead to very costly analysis in terms of computer and manpower resources. On the other hand, information is seldom complete enough to be able to describe numerically the real situation in sufficient detail. It appears obvious, therefore, that performance of numerical analysis of engineering problems involving this class of materials requires models capable of reproducing the different types of anisotropic behavior by means of an equivalent continuum medium.

This paper describes an approach for modelling anisotropy based on the use of the microplane technique. This technique uses micromechanical considerations, albeit simplified, to obtain a macroscopic stress/strain relationship based on the superposition of behavior laws valid for planes of arbitrary orientation that cover all possible space orientations. The original idea on which microplane models are based can be traced back to the *'slip theory'* of metal plasticity and, more recently, to the *'multilaminate models'* developed in Swansea by Pande and co-workers (Zienkiewicz and Pande, 1977b; Pande and Sharma, 1983; Sadrnejad and Pande, 1989). The term *'microplane model'* was coined by Bažant (Bažant and Oh, 1983), who has developed the model together with his co-workers and other authors (Bažant and Prat, 1987; Bažant and Prat, 1988a; Bažant and Prat, 1988b; Bažant and Ožbolt, 1990; Carol et al., 1991; Prat and Bažant, 1991; Carol et al., 1992; Ožbolt and Bažant, 1992; Cofer and Kohut, 1994; Yamada et al., 1994; Prat and Gens, 1995; Bažant et al., 1996a; Bažant et al., 1996b; Carol and Bažant, 1997). As pointed out above, in these models the behavior laws are defined on a plane arbitrarily orientated in order to add up afterwards the contributions of all planes to obtain the macroscopic stress/strain relationship. On each plane, the constitutive law is expressed in terms of normal and shear stress/strain relations, which makes for an easier formulation than the conventional tensorial constitutive law formulation using six components. Moreover, in these models, tensorial invariance is automatically achieved through the appropriate superposition process.

Because of the nature of this model, it is relatively simple to reproduce the anisotropy of the material without restricting it to a special configuration as is the case in classical models. Since the constitutive laws are specified independently for each microplane, anisotropy easily follows from choosing different values of the material parameters on each plane as a function of its orientation. Besides anisotropy due to discontinuities, the model also allows for anisotropy of the material matrix, i.e., the continuum medium between discontinuities. The model can reproduce, as particular cases, isotropy, transverse anisotropy and orthotropy (Prat and Gens, 1994). It can also reproduce induced anisotropy, or anisotropy due to the loading process, if a suitable parameter is chosen to control the variation of the material parameters with the loading state. In the formulation presented, a classical work- hardening plasticity formulation has been adopted.

2 MICRO/MACRO EQUIVALENCE EQUATIONS

For the development of the model presented herein, a strain-driven formulation using a micro-macro kinematic constraint has been selected. Thus, the basic hypothesis is that the components of the *strain* vector ($\boldsymbol{\epsilon}_n$) acting on a plane of arbitrary orientation are equal to the components of the macroscopic *strain* tensor ($\boldsymbol{\epsilon} \equiv \epsilon_{ij}$) on this plane (Fig. 1). In addition, as the deformation process in soils and other materials usually involves significant volume changes, often irreversible, the volumetric ($\epsilon_V = \frac{1}{3}\epsilon_{kk}$) and deviatoric ($e_{ij} = \epsilon_{ij} - \epsilon_V \delta_{ij}$) components of the macroscopic strain tensor are treated separately.

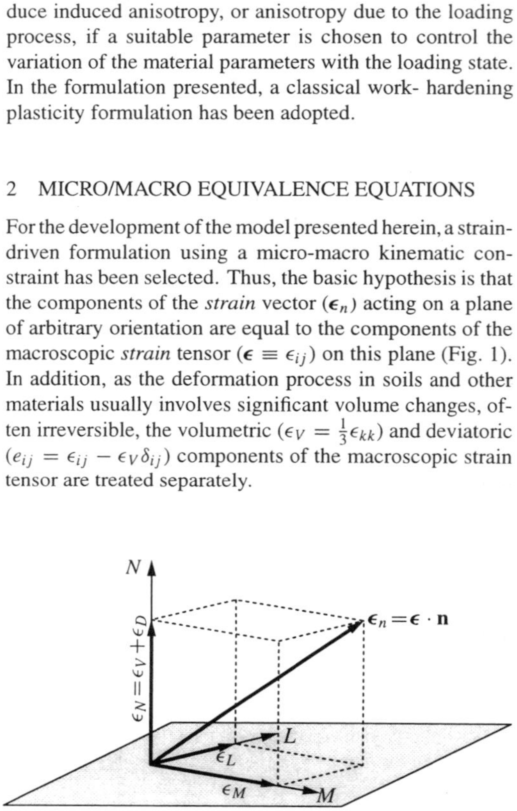

Figure 1. Strain components on a microplane.

The volumetric component of the strain tensor leads to a spherical state of strain yielding the same value of volumetric strain (ϵ_V) for all microplanes. In addition the strain vector for each plane due to the deviatoric part of the strain tensor is computed. This three-component vector is referred to an orthonormal base composed by the unit vector normal to the plane (**n**) and two vectors contained in it (**m** and **ℓ**). These latter vectors are chosen arbitrarily from all possible linear combinations that constitute, together with (**n**), an orthonormal base. Denoting the direction normal to the plane as N and the in-plane directions defined by vectors **m** and **ℓ** as M and L, the strains at a particular point are defined by the four quantities $\{\epsilon_V, \epsilon_D, \epsilon_M, \epsilon_L\}$:

$$\begin{aligned} \epsilon_V &= V_{ij}\epsilon_{ij}, & \epsilon_D &= D_{ij}\epsilon_{ij} \\ \epsilon_M &= M_{ij}\epsilon_{ij}, & \epsilon_L &= L_{ij}\epsilon_{ij} \end{aligned} \tag{1}$$

where V_{ij}, D_{ij}, M_{ij} and L_{ij} are second rank tensors defined in terms of the components n_i, m_i and ℓ_i of the unit vectors in the direction of N, M and L respectively:

$$\begin{aligned} V_{ij} &= \frac{1}{3}\delta_{ij}, & D_{ij} &= n_i n_j - V_{ij} \\ M_{ij} &= \frac{1}{2}(n_i m_j + n_j m_i), & L_{ij} &= \frac{1}{2}(n_i \ell_j + n_j \ell_i) \end{aligned} \tag{2}$$

Calling $\{\sigma_V, \sigma_D, \sigma_M, \sigma_L\}$ the stresses associated with the strains $\{\epsilon_V, \epsilon_D, \epsilon_M, \epsilon_L\}$ on a microplane, the following relationships between stress and strain components are assumed:

$$\begin{aligned} \sigma_V &= \sigma_V(\epsilon_V), & \sigma_D &= \sigma_D(\epsilon_D) \\ \sigma_M &= \sigma_M(\epsilon_M, \sigma_N), & \sigma_L &= \sigma_L(\epsilon_L, \sigma_N) \end{aligned} \tag{3}$$

Note that the total strain on each microplane, $\epsilon_N = \epsilon_V + \epsilon_D$, is the resolved component of the macroscopic strain tensor, whereas the quantity $\sigma_N = \sigma_V + \sigma_D$ can be interpreted as the total normal stress acting on the plane under consideration, but *is not in general* the resolved component of the macroscopic stress tensor. The four functions appearing in (3) are the material-dependent core of the model and will be defined in the following section (Bažant and Prat, 1987; Bažant and Prat, 1988a; Prat and Bažant, 1991; Carol et al., 1992).

The macroscopic stress/strain law can be obtained, after defining the microscopic relationships on each plane, from the principle of virtual work that in this case establishes the equilibrium between the stresses σ_{ij} inside a sphere of unit radius and the three components of stress (σ_N, σ_M and σ_L) acting on the surface of the same sphere:

$$\frac{4\pi}{3}\sigma_{ij}\delta\epsilon_{ij} = 2\int_{\Omega}\Big(\sigma_N\delta\epsilon_N + \sigma_M\delta\epsilon_M + \sigma_L\delta\epsilon_L\Big)\mathrm{d}\Omega \tag{4}$$

where Ω denotes the surface of a half-sphere of unit radius. Replacing $\sigma_N = \sigma_V + \sigma_D$, $\delta\epsilon_N = N_{ij}\delta\epsilon_{ij}$ ($N_{ij} = V_{ij} + D_{ij}$), $\delta\epsilon_M = M_{ij}\delta\epsilon_{ij}$ and $\delta\epsilon_L = L_{ij}\delta\epsilon_{ij}$ in Eqn. (4) the following expression is obtained (Bažant and Prat, 1988a; Carol et al., 1992):

$$\sigma_{ij} = \sigma_V\delta_{ij} + \frac{3}{2\pi}\int_{\Omega}\Big(\sigma_D D_{ij} + \sigma_M M_{ij} + \sigma_L L_{ij}\Big)\,\mathrm{d}\Omega \tag{5}$$

Equation (5) yields the macroscopic stress tensor as a function of the stresses on each of the planes considered in the formulation. Note that for the volumetric stress to come out of the integral in (5), we have assumed that the parameters controlling this component do not have directional properties. Otherwise σ_V would not be the same for all microplanes as expected.

3 IN-PLANE CONSTITUTIVE LAWS

According to Eqns. (3) one needs to define four constitutive equations on one given plane, relating the volumetric, normal deviatoric and (two) shear components of stress and strain. The present formulation is written with total strain constitutive laws for the volumetric and deviatoric part, whereas the shear components are given a classical elastoplastic formulation.

3.1 *Volumetric law*

For simplicity, an elastic constitutive law is assumed under hydrostatic compression conditions. For hydrostatic tension, an exponentially decaying strain-softening law with parameters k_t and p_t is used (Fig. 2):

$$\begin{aligned} \sigma_V &= E_V^\circ \epsilon_V && \text{compression} \\ \sigma_V &= E_V^\circ \epsilon_V \exp(-k_t |\epsilon_V|^{p_t}) && \text{tension} \end{aligned} \qquad (6)$$

where $E_V^\circ = E/(1-2\nu)$ is the initial volumetric elastic modulus.

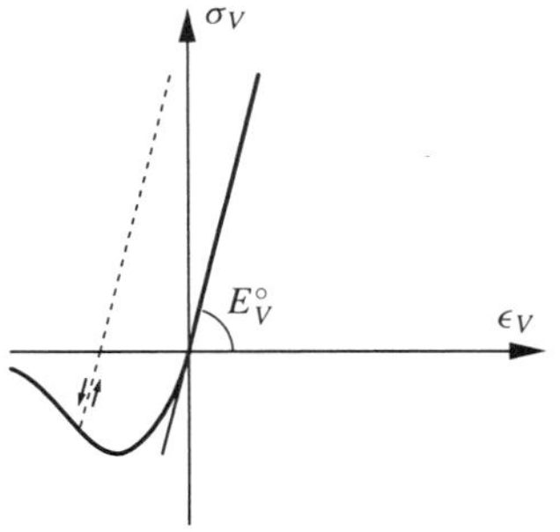

Figure 2. Volumetric law.

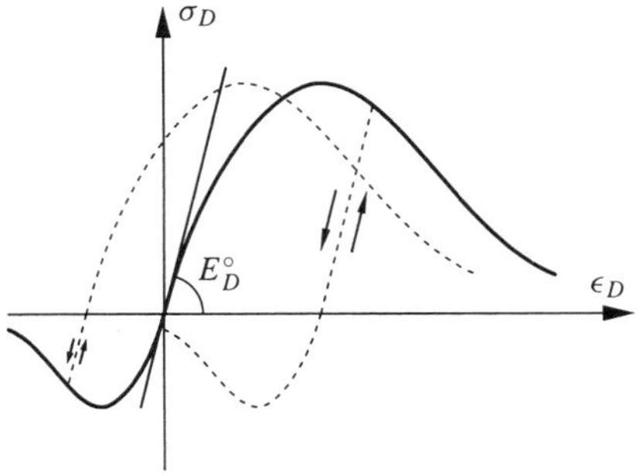

Figure 3. Deviatoric law.

3.2 *Deviatoric law*

The deviatoric constitutive law is given a strain-softening behavior both in tension and compression (Fig. 3), with parameters k_t and p_t in tension, and k_c and p_c in compression. Parameter k_c is about one order of magnitude larger than k_t to model the difference between tension and compression normal to a plane:

$$\begin{aligned} \sigma_D &= E_D^\circ \epsilon_D \exp(-k_c |\epsilon_D|^{p_c}) && \text{compression} \\ \sigma_D &= E_D^\circ \epsilon_D \exp(-k_t |\epsilon_D|^{p_t}) && \text{tension} \end{aligned} \qquad (7)$$

where $E_D^\circ = E/(1-2\nu)$ is the initial deviatoric elastic modulus.

3.3 *Shear law*

The formulation of the shear law is done within the framework of classical elastoplasticity. For simplicity, an associated law with no dilatancy is chosen in this paper. Further refinements will be introduced in the future.

The yield function adopted for this model is an hyperbola (Zienkiewicz and Pande, 1977a; Gens et al., 1989) in the $\tau - \sigma_N$ plane, given by (see Fig. 4)

$$\begin{aligned} F &\equiv F(\sigma_N, \sigma_M, \sigma_L) = \\ &= \tau^2 - \left[(c + \sigma_N \tan\phi)^2 - (c - q_t \tan\phi)^2\right] = 0 \end{aligned} \qquad (8)$$

where $\tau^2 = \sigma_M^2 + \sigma_L^2$, c is the cohesion, ϕ the friction angle and q_t the tensile strength of the material. The cross-section of the yield surface perpendicular to the σ_N axis is a circle (see Fig. 5) with its radius function of the current σ_N.

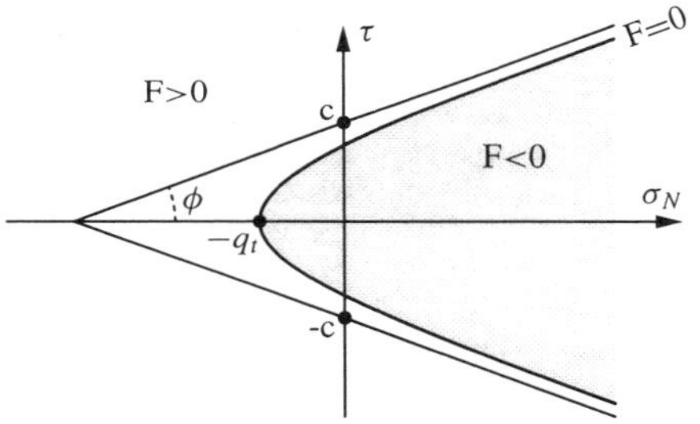

Figure 4. Yield surface in the $\tau - \sigma_N$ space.

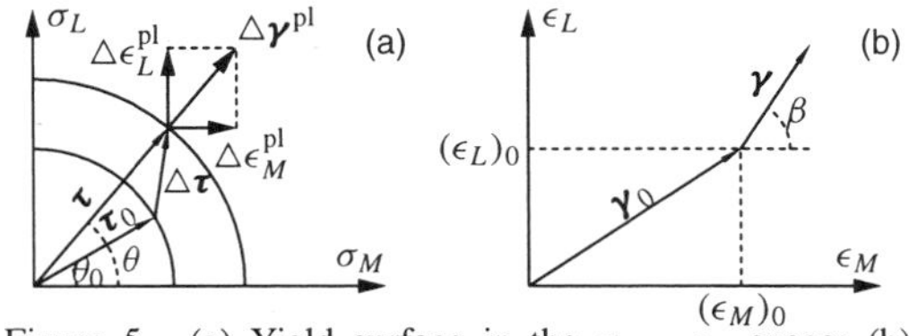

Figure 5. (a) Yield surface in the $\sigma_M - \sigma_L$ space; (b) variables in the deformation space.

These parameters change during loading according to the following softening rules (Fig. 6):

$$\begin{aligned} c &= c_0 \exp\left(-\frac{\chi}{\chi_c}\right) \\ q_t &= q_{t0} \exp\left(-\frac{\chi}{\chi_q}\right) \\ m &= m_r + (m_0 - m_r) \exp\left(-\frac{\chi}{\chi_m}\right) \end{aligned} \qquad (9)$$

where $m = \tan\phi$; c_0, q_{t0} and m_0 are the initial values of the corresponding parameters; m_r is the residual value of the friction angle for a purely frictional behavior; χ_c, χ_q and χ_m are model parameters controlling the decaying of the softening laws, and χ is the softening parameter taken as the shear plastic work

$$d\chi = \boldsymbol{\tau}^T \cdot d\boldsymbol{\gamma}^{pl} = \sigma_M d\epsilon_M^{pl} + \sigma_L \epsilon_L^{pl} \tag{10}$$

where

$$\boldsymbol{\tau} = \begin{pmatrix} \sigma_M \\ \sigma_L \end{pmatrix}; \qquad d\boldsymbol{\gamma}^{pl} = \begin{pmatrix} d\epsilon_M^{pl} \\ d\epsilon_L^{pl} \end{pmatrix} \tag{11}$$

Because the evolution of parameters c, ϕ and q_t is different in general for each microplane orientation, anisotropy induced by the loading process is achieved.

The shear stress increments on each plane can be obtained using a classical plasticity formulation, with the modification introduced by the fact that in this model the normal stress increment $d\sigma_N = d\sigma_V + d\sigma_D$ is already known from the preceding equations (6) and (7):

$$d\boldsymbol{\tau} = \mathbf{D}^e(d\boldsymbol{\gamma} - d\boldsymbol{\gamma}^{pl}) \qquad d\boldsymbol{\gamma}^{pl} = d\lambda \frac{\partial F}{\partial \boldsymbol{\tau}} \tag{12}$$

where

$$d\boldsymbol{\tau} = \begin{pmatrix} d\sigma_M \\ d\sigma_L \end{pmatrix}; \quad d\boldsymbol{\gamma} = \begin{pmatrix} d\epsilon_M \\ d\epsilon_L \end{pmatrix}; \quad \mathbf{D}^e = \begin{pmatrix} E_T^\circ & 0 \\ 0 & E_T^\circ \end{pmatrix} \tag{13}$$

and $E_T^\circ = E/(1+\nu)$ is the shear elastic modulus. From the consistency condition, $dF = 0$, the plastic multiplier $d\lambda$ is obtained as

$$d\lambda = \frac{\left(\frac{\partial F}{\partial \boldsymbol{\tau}}\right)^T \mathbf{D}^e d\boldsymbol{\gamma} + \frac{\partial F}{\partial \sigma_N} d\sigma_N}{H + \left(\frac{\partial F}{\partial \boldsymbol{\tau}}\right)^T \mathbf{D}^e \frac{\partial F}{\partial \boldsymbol{\tau}}} \tag{14}$$

where H is the plastic modulus

$$H = -\frac{\partial F}{\partial \chi} \boldsymbol{\tau}^T \frac{\partial F}{\partial \boldsymbol{\tau}} = -\frac{\partial F}{\partial \chi}\left[\sigma_M \frac{\partial F}{\partial \sigma_M} + \sigma_L \frac{\partial F}{\partial \sigma_L}\right] \tag{15}$$

and

$$\frac{\partial F}{\partial \chi} = \frac{\partial F}{\partial c}\frac{\partial c}{\partial \chi} + \frac{\partial F}{\partial q_t}\frac{\partial q_t}{\partial \chi} + \frac{\partial F}{\partial m}\frac{\partial m}{\partial \chi} \tag{16}$$

Then the shear stress increments are finally calculated

$$d\boldsymbol{\tau} = \mathbf{D}^e \left[\mathbf{I} - \frac{\frac{\partial F}{\partial \boldsymbol{\tau}}\left(\frac{\partial F}{\partial \boldsymbol{\tau}}\right)^T \mathbf{D}^e}{H + \left(\frac{\partial F}{\partial \boldsymbol{\tau}}\right)^T \mathbf{D}^e \frac{\partial F}{\partial \boldsymbol{\tau}}}\right] d\boldsymbol{\gamma} - \frac{\mathbf{D}^e \frac{\partial F}{\partial \boldsymbol{\tau}} \frac{\partial F}{\partial \sigma_N} d\sigma_N}{H + \left(\frac{\partial F}{\partial \boldsymbol{\tau}}\right)^T \mathbf{D}^e \frac{\partial F}{\partial \boldsymbol{\tau}}} \tag{17}$$

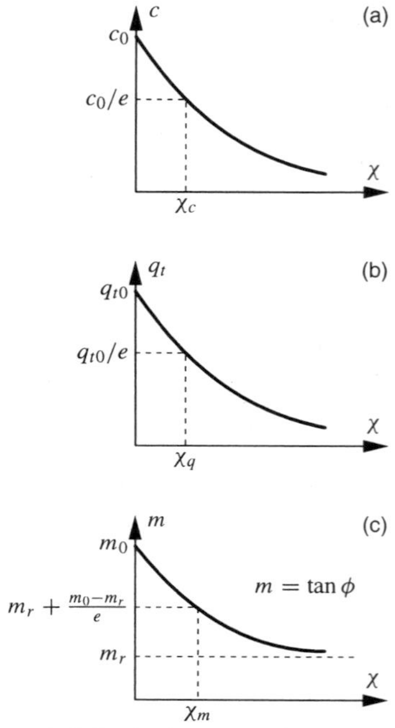

Figure 6. Softening laws: (a) cohesion; (b) tensile strength; (c) friction angle.

4 IMPLEMENTATION AND APPLICATIONS

The constitutive law described in the previous sections has been implemented for use in a finite element program as constitutive model. Although the implementation has been made according to the design of a particular finite element code developed by the authors, DRAC (Prat et al., 1993), it can also be used as a standalone subroutine for the purpose of constitutive model verification and for use in other finite element codes.

It transpires from the theoretical derivations in Section 2 that a kinematic constraint between the micro and macro variables has been used for the model, as described elsewhere (Bažant and Prat, 1988a). A simplified flowchart of the microplane subroutines is as follows:

1) Initial data consists of the initial stress state $\boldsymbol{\sigma}_0$, initial strain state $\boldsymbol{\epsilon}_0$, given strain increment $\Delta\boldsymbol{\epsilon}$ and the initial state variables.
2) Calculate the current strain $\boldsymbol{\epsilon} = \boldsymbol{\epsilon}_0 + \Delta\boldsymbol{\epsilon}$ and the associated in-plane strains $\{\epsilon_V, \epsilon_D, \epsilon_M, \epsilon_L\}$ (for all microplanes) according to the formulas in Eqns. (1).
3) With the current value of ϵ_V obtain the value of the volumetric stress component σ_V from Eqn. (6).
4) With the current value of ϵ_D obtain the value of the deviatoric stress component σ_D, for all microplanes, from Eqn. (7).
5) Estimate the current normal stress $\sigma_N = \sigma_V + \sigma_D$ for

all microplanes.

6) For all microplanes: with the current value of the shear strain vector $\boldsymbol{\gamma} = [\epsilon_M, \epsilon_L]$ and of the current normal stress σ_N, evaluate the current shear stress vector $\boldsymbol{\tau} = [\sigma_M, \sigma_L]$ according to the procedure shown in Section 3.3 and Eqn. (17). Because of the nonlinearity of the formulation, iterations are needed at this level to evaluate the shear stress increments.
7) After all stress components have been obtained for all microplanes, evaluate the current macroscopic stress $\boldsymbol{\sigma}$ from Eqn. (5), and update all the variables. Proceed to the next loading step.

The model requires the following 13 parameters:

- Elastic constants, E and ν
- Parameters of the normal stress/strain law, Eqn. (6): k_c, k_t, p_c, p_t. These parameters are related to the tensile and compressive strength of the material, and to the rate of decaying of the corresponding softening branches both in tension and compression. Work is currently under way (Sánchez, 1997) to rewrite these equations in terms of the tensile and compressive strengths of the material, thus giving the parameters a more physical meaning.
- Parameters of the shear stress/strain law, Sec. 3.3: cohesion (c_0), tensile strength (q_{t0}), initial and residual friction angle (m_0, m_r), and softening parameters (χ_c, χ_q, χ_m) that control that rate of change of the yield surface.

Figure 7 shows a comparison of the results obtained using the present formulation and the experimental results reported by Kawahara et al. (1981), using the data obtained from the uniaxial cyclic compression tests made at the Nakayama tunnel (Japan) construction site. The basic material parameters obtained from the test results are indicated in the figure. The remaining model parameters are $k_c = 3550$, $k_t = 70$, $p_c = 1.5$, $p_t = 0.5$, $c_0 = 144$ kg/cm^2, $q_{t0} = 45$ kg/cm^2, $m_0 = 0.577$, $m_r = 0.2$ and $\chi_c = \chi_q = \chi_m = 0.85$. It can be seen from this figure that the model presented is capable of reproducing the experimental results with an acceptable accuracy, including unloading/reloading behavior. The slight hysteresis loops that the analytical curve shows are probably due to numerical error accumulation because the present formulation does not currently account for such behavior.

5 ANISOTROPY

Although in the example shown in the previous section the sample is supposed to be initially isotropic, inherent anisotropy (existing before the loading starts) can be easily modeled using the present type of formulation. Because the constitutive laws are defined on a plane, it is very easy to make the material properties dependent on the orientation of the planes, thus giving different stiffness, cohesion, friction, etc. for different orientations.

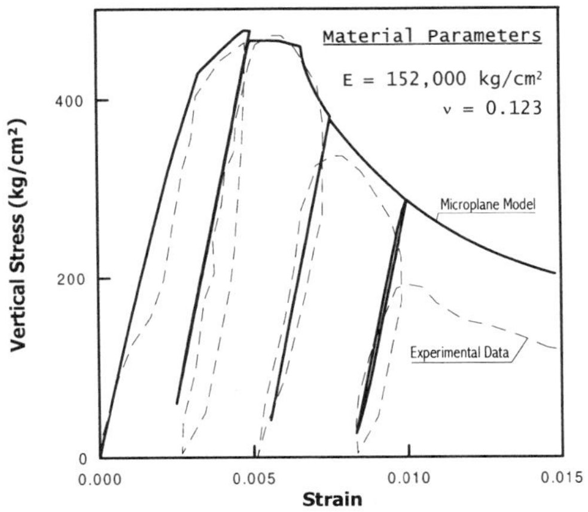

Figure 7. Comparison of analytical and experimental results from the Nakayama tunnel uniaxial compression test (Kawahara et al., 1981).

To accomplish this, one can easily define the material properties not as a single scalar value, but as a set of values in a vector the size of the maximum number of plane orientations for the particular integration formula used. Another way is to prescribe some values for the material parameter for some specific directions, and then interpolate the values for the remaining directions. In that way, the directions for which the material properties are prescribed do not need to coincide with the directions of the particular set of microplanes selected.

In the latter case, let's assume that the number of prescribed directions is N_p, and that θ_i^* and ϕ_i^* $(i = 1, \ldots, N_p)$ are the spherical coordinate angles defining these prescribed directions. Then a generic material parameter Q can be obtained as:

$$
\begin{aligned}
Q_\mu &= Q(\theta_\mu, \phi_\mu) \\
&= \sum_{i=1}^{N_p} Q_i^* \exp\left\{-k_i\left[(\theta_\mu - \theta_i^*)^2 + (\phi_\mu - \phi_i^*)^2\right]\right\}
\end{aligned} \tag{18}
$$

where Q_μ are the material properties for the direction μ of the numerical integration formula used, θ_μ and ϕ_μ the spherical coordinate angles defining this direction, and Q_i^* the prescribed values of parameter Q. For a more detailed explanation of this procedure, see Prat and Bažant, 1997. Figure 8 shows a sketch of the function defined by Eqn. (18) for the case of two-dimensional analysis.

ACKNOWLEDGEMENTS

Partial finantial support from grant PB94-1204, Ministerio de Educación y Cultura (Madrid, Spain) is gratefully acknowledged. The second author acknowledges the financial support received from Consultec, S.C. (México DF).

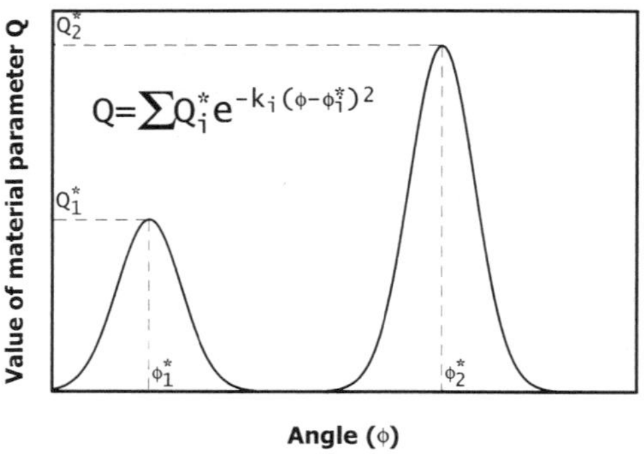

Figure 8. Example of the initial anisotropy distribution function (Eqn. 18) for two-dimensional analysis.

REFERENCES

Bažant, Z.P. and Oh, B.H. (1983). Microplane model for fracture analysis of concrete structures. In U.S. Air Force Academy (ed.), *Proc. Symp. on the Interaction of Non-Nuclear Munitions with Structures*, pp. 49–55, Colorado Springs.

Bažant, Z.P. and Ožbolt, J. (1990). Non-local microplane model for fracture, damage and size effect in structures. *ASCE J. Engng. Mech.*, 116(11):2485–2504.

Bažant, Z.P. and Prat, P.C. (1987). Creep of anisotropic clay: new microplane model. *ASCE J. Engng. Mech.*, 103(7):1050–1064.

Bažant, Z.P. and Prat, P.C. (1988a). Microplane model for brittle-plastic material: I. Theory. *ASCE J. Engng. Mech.*, 114(10):1672–1688.

Bažant, Z.P. and Prat, P.C. (1988b). Microplane model for brittle-plastic material: II. Verification. *ASCE J. Engng. Mech.*, 114(10):1689–1702.

Bažant, Z.P., Xiang, Y., and Prat, P.C. (1996). Microplane model for concrete: I. Stress-strain boundaries and finite strain. *ASCE J. Engng. Mech.*, 122(3):245–262.

Bažant, Z.P., Xiang, Y., Adley, M.D., Prat, P.C., and Akers, S.A. (1996). Microplane model for concrete: II. Data delocalization and verification. *ASCE J. Engng. Mech.*, 122(3):263–268.

Carol, I. and Bažant, Z.P. (1997). Damage and plasticity in microplane theory. *Int. J. Solids and Structures*. In press.

Carol, I., Bažant, Z.P., and Prat, P.C. (1991). Geometric damage tensor based on microplane model. *ASCE J. Engng. Mech.*, 117:2429–2448.

Carol, I., Prat, P.C., and Bažant, Z.P. (1992). New explicit microplane model for concrete: theoretical aspects and numerical implementation. *Int. J. Solids and Structures*, 29(9):1173–1191.

Cofer, W.F. and Kohut, S.W. (1994). A general nonlocal microplane concrete material model for dynamic finite element analysis. *Computers and Structures*, 53(1):189–199.

Gens, A., Carol, I., and Alonso, E.E. (1989). Elasto-plastic model for joints and interfaces. In Owen, D.R.J., Hinton, E., and Oñate, E. (eds.), *Computational Plasticity (COMPLAS II)*, pp. 1251–1264, Barcelona. Pineridge Press.

Kawahara, M., Kanoh, Y., Kaneko, N., and Yada, K. (1981). Strain-softening finite element analysis of rock applied to tunnel excavation. In ISRM (ed.), *International Symposium on Weak Rock. Soft, Fractured and Weathered Rock*, pp. 108–113, Tokyo, Japan. ISRM. Paper III-2-20.

Ožbolt, J. and Bažant, Z.P. (1992). Microplane model for cyclic triaxial behavior of concrete. *ASCE J. Engng. Mech.*, 118(7):1365–1386.

Pande, G.N. and Sharma, K.G. (1983). Multilaminate model of clays — A numerical evaluation of the influence of rotation of principal axes. *ASCE J. Engng. Mech.*, 109(7):397–418.

Prat, P.C. and Bažant, Z.P. (1991). Microplane model for triaxial deformation of saturated cohesive soils. *ASCE J. Geotech. Engng.*, 117(6):891–912.

Prat, P.C. and Bažant, Z.P. (1997). Tangential stiffness of elastic materials with systems of growing or closing cracks. *J. Mech. Phys. Solids*. In press.

Prat, P.C. and Gens, A. (1994). Microplane formulation for quasibrittle materials with anisotropy and damage. In Bažant, Z.P., Bittnar, Z., Jirásek, M., and Mazars, J. (eds.), *Fracture and damage in quasi-brittle structures*, pp. 67–74. E & FN SPON, London.

Prat, P.C. and Gens, A. (1995). Panelist discussion: Microplane approach to modelling deformation of soils. In Shibuya, S., Mitachi, T., and Miura, S. (eds.), *Pre-failure Deformation of Geomaterials*, Vol. 2, pp. 1203–1208. Balkema.

Prat, P.C., Gens, A., Carol, I., Ledesma, A., and Gili, J.A. (1993). DRAC: A computer software for the analysis of rock mechanics problems. In Liu, H. (ed.), *Application of computer methods in rock mechanics*, Vol. 2, pp. 1361–1368, Xian, China. Shaanxi Science and Technology Press.

Sadrnejad, S.A. and Pande, G.N. (1989). A multilaminate model for sands. In Pietruszczak, S. and Pande, G.N. (eds.), *Numerical Models in Geomechanics (NUMOG III)*, pp. 17–27, Niagara Falls, Canada. Elsevier.

Sánchez, F. (1997). Anisotropic microplane model for rock: theory and application to finite element analysis. Master's thesis, ETSECCPB-UPC, E-08034 Barcelona (Spain). In spanish.

Yamada, K., Kawamaki, M., and Tanabe, T.-A. (1994). Experimental verification of the microplane model. *Transactions of the Japan Concrete Institute*, 16:139–146.

Zienkiewicz, O.C. and Pande, G.N. (1977a). Some useful forms of isotropic yield surfaces for soil and rock mechanics. In Gudehus, G. (ed.), *Finite Elements in Geomechanics*, pp. 179–190. John Wiley.

Zienkiewicz, O.C. and Pande, G.N. (1977b). Time-dependent multi-laminate model of rocks — A numerical study of deformation and failure of rock masses. *Int. J. Num. Anal. Methods in Geomechanics*, 1:219–247.

2 Instability and strain localization in geomaterials

Numerical Models in Geomechanics, Pietruszczak & Pande (eds) © 1997 Balkema, Rotterdam, ISBN 90 5410 886 X

Material instabilities and bifurcations in granular media

F. Darve, O. Pal & X. Roguiez
Laboratoire Sols Solides Structures, INPG/UJF/CNRS, ALERT Geomaterials, Grenoble, France

ABSTRACT : It is now well recognised that loose granular materials are subjected to different instability mechanisms. The aim of this paper is essentially to generalise these results to the field of dense materials. For this purpose we consider first of all, well established experimental features of soils mechanics and analyse them by means of the Lyapunov's definition of stability. Then we present numerical results issued from constitutive relations and analysed by the Hill's sufficient condition of stability. Finally a new incrementally non-linear relation with 5 material constants is used for a post-bifurcation analysis which exhibits two possible bifurcated branches. These last results show the existence of potential homogeneous bifurcations, which differ basically from the discontinuous bifurcations by plastic strains localisation into shear bands.

1. INTRODUCTION

For associated elasto-plastic materials it is classically established that loss of material stability and loss of material uniqueness coincide with the plastic limit condition essentially because the constitutive matrix is symmetric. Geomaterials are strongly non-associated materials and their constitutive relations are considered as non-symmetric. From a rough point of view we can estimate that the difference between friction angle and dilatancy angle is about equal to 30°, while in the case of associativity both these angles should be equal.

The first class of bifurcations which has been exhibited developing strictly inside the plastic limit condition is constituted by the phenomenon of strains localisation. In fact the bifurcation of the strain mode from a diffuse one into a strictly localised one by formation of shear bands appears experimentally in an indisputable manner before the stress peak for dense sands. Thus it could be conjectured that other classes of instabilities and bifurcations could also develop strictly inside the plastic limit condition for non-associated materials as geomaterials.

This point is the main objective of this paper : one shows that it is possible to derive for sands at different densities, different types of instabilities and bifurcations before the plastic failure.

This result has several consequences either from a practical point of view or from a numerical one. In fact that implies that in practice one will obtain failures which do not obey to a Mohr-Colomb criterion with the usual material friction angle. With respect to numerical aspects that could explain difficulties which are encountered in relation with convergency algorithms near to the failure condition when the material is strongly non-associated.

In a first paragraph we define in a proper manner « material stability », while the following paragraphs analyse possible losses of stability and uniqueness by some criteria applied to incrementally piece-wise linear and non-linear constitutive relations.

Finally one considers a simple incrementally non-linear model based on 5 material constants, which allows to exhibit, for a given set of constants, two different bifurcated branches, one stable and the other unstable.

2. EXPERIMENTAL EVIDENCES

Before considering from a theoretical/numerical viewpoint possible material instabilities, it is necessary to set clearly the existence of such instabilities from experimental bases.

In this perspective let us consider undrained axisymmetric loading on loose and dense granular media. The results presented on fig.1 are issued from our constitutive relation but are typical of experiments (the plotted stresses are all effective) :

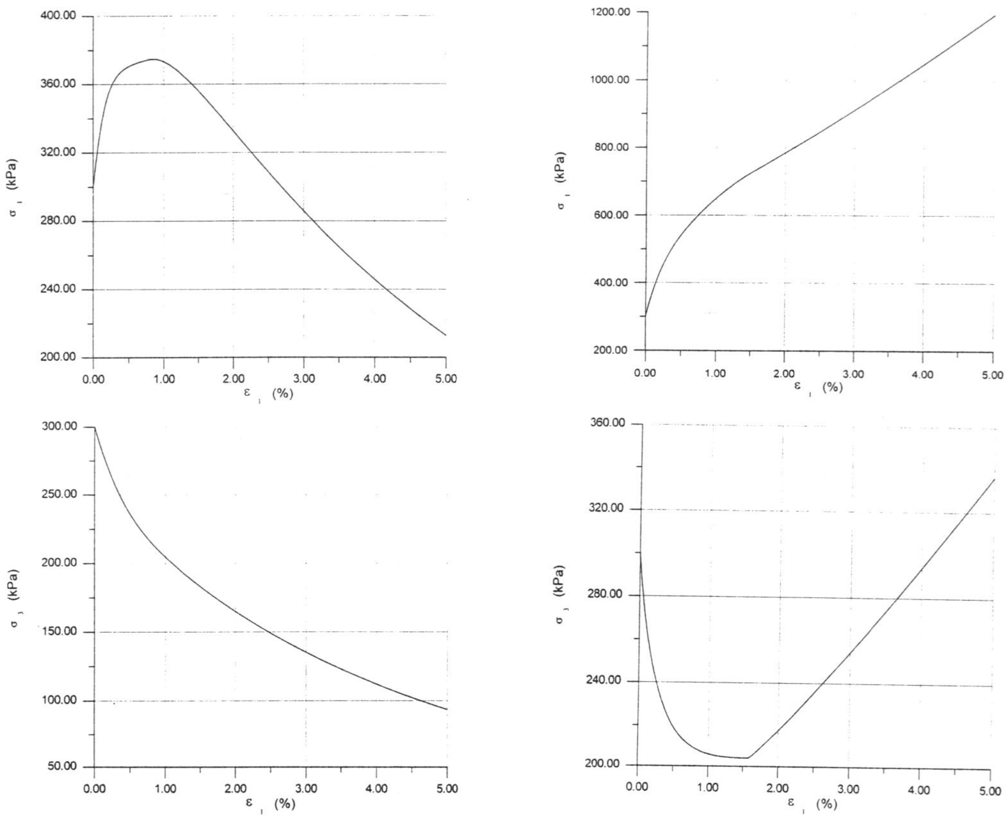

Fig. 1 Typical behaviour of loose Hostun sand on the left and of dense Hostun sand on the right for axisymmetric isochoric (undrained) loading. The plotted stresses are all effective. The maximum of σ_1 and the minimum of σ_3 respectively are unstable states according to the Lyapunov's definition.

• for a loose sand the axial stress passes through a maximum value while the lateral stress is monotonously decreasing,
• for a dense sand the lateral stress passes through a minimum value while the axial stress is monotonously increasing.

The undrained loading path is classically defined by :

$$\begin{cases} d\varepsilon_1 = \text{positive constant} \\ d\varepsilon_2 = d\varepsilon_3 = -d\varepsilon_1/2 \text{ (isochoric axisymmetry)} \end{cases}$$

However with a stress-strain controlled triaxial apparatus and a dry sand sample it is possible to apply this path by the following incremental loading :

• for a loose sand :

$$\begin{cases} d\sigma_1 = \text{positive constant} \\ d\varepsilon_2 = d\varepsilon_3 = -\, d\varepsilon_1/2 \end{cases}$$

• for a dense sand :

$$\begin{cases} d\sigma_3 = d\sigma_2 = \text{negative constant} \\ d\varepsilon_1 = -2\, d\varepsilon_3 \end{cases}$$

In such cases the maximum of σ_1 and respectively the minimum of σ_3 appear to be unstable states, since at these extrema we would obtain a sudden strains failure for infinitesimal variations respectively of σ_1 and σ_3.
This intuitive point of view can be corroborated by a general definition of stability, as for example the one which has been proposed by Lyapunov (1907) at the beginning of the century. This definition, applied to the field of the machanics of continuum media, states that :
a stress-strain state for a given material after a given strain history is called <u>stable</u> if for every positive scalar ε there exists a positive number $\eta = \eta(\varepsilon)$ such that for all incremental loading bounded by : $\| \underset{\sim}{dl} \| < \eta$,

the corresponding responses remain bounded :

$\| \underset{\sim}{dr} \| < \varepsilon$.

According to this definition the previous extrema are clearly unstable states.

3. STABILITY ANALYSIS

The previous definition of stability is not a convenient tool to analyse analytically/numerically the stability of a constitutive relation. Drucker's postulate (Drucker 1951) or the more general criterion constituted by the Hill's sufficient condition of stability (Hill 1958) is more appropriated.

This criterion states that, considering a stress-strain state for a given material after a given strain history, a sufficient condition of stability is the positiveness of the work of second order for any pair ($\underset{\sim}{d\sigma}$, $\underset{\sim}{d\varepsilon}$) related by the constitutive relation :

$$\forall\ \underset{\sim}{d\sigma}, \underset{\sim}{d\varepsilon} : d^2w = \underset{\sim}{d\sigma} : \underset{\sim}{d\varepsilon} > 0$$

This condition can be conveniently checked on polar diagrams of the normalised work of second order :

$$t = d^2w / \| \underset{\sim}{d\sigma} \| / \| \underset{\sim}{d\varepsilon} \|$$

following $\underset{\sim}{d\sigma}$ (or $\underset{\sim}{d\varepsilon}$) direction.

Fig. 2 shows such results for a dense and loose Hostun sand derived from an incrementally non-linear constitutive relation, while fig. 3 presents results issued from an incrementally piece-wise linear model. The general expression of the non-linear model is the following relationship :

$$d\varepsilon_{ij} = M_{ijkl}\, d\sigma_{kl} + \frac{1}{\| \underset{\sim}{d\sigma} \|} N_{ijklpq}\, d\sigma_{kl}\, d\sigma_{pq}$$

In stress-strain principal axes (as considered for the applications presented in this paper) the previous expression becomes :

• for the non-linear model :

$$\begin{bmatrix} d\varepsilon_1 \\ d\varepsilon_2 \\ d\varepsilon_3 \end{bmatrix} = \frac{1}{2}\left(\underset{\sim}{N}^+ + \underset{\sim}{N}^-\right) \begin{bmatrix} d\sigma_1 \\ d\sigma_2 \\ d\sigma_3 \end{bmatrix} + \frac{1}{2\| \underset{\sim}{d\sigma} \|}\left(\underset{\sim}{N}^+ - \underset{\sim}{N}^-\right) \begin{bmatrix} (d\sigma_1)^2 \\ (d\sigma_2)^2 \\ (d\sigma_3)^2 \end{bmatrix}$$

• for the piece-wise linear model :

$$\begin{bmatrix} d\varepsilon_1 \\ d\varepsilon_2 \\ d\varepsilon_3 \end{bmatrix} = \frac{1}{2}\left(\underset{\sim}{N}^+ + \underset{\sim}{N}^-\right) \begin{bmatrix} d\sigma_1 \\ d\sigma_2 \\ d\sigma_3 \end{bmatrix} + \frac{1}{2}\left(\underset{\sim}{N}^+ - \underset{\sim}{N}^-\right) \begin{bmatrix} |d\sigma_1| \\ |d\sigma_2| \\ |d\sigma_3| \end{bmatrix}$$

Fig. 2 and 3 present polar diagrams of t for different values of :

$\eta = q/p = 3(\sigma_1 - \sigma_3)/(\sigma_1 + 2\sigma_3)$

along a triaxial compression path. The figures exhibit the « first » stress state for which the work of second order is vanishing first in a unique direction. Then a cone of unstable directions will be developing.

It can be concluded from fig. 2 and 3 that a wide domain of the stress space corresponds to unstable stress-strain states for dense as for loose sands. Obviously this domain includes the unstable states which have been detected in paragraph 2. The first stress states for which the work of second order is vanishing are different for loose and dense sands : q/p = 0.455 and 0.830 respectively in the case of the incrementally non-linear model, 0.327 and 0.659 respectively for the piece-wise linear relation.

However the stress vector direction that sets the work of second order to zero is about the same independently of the sand density.

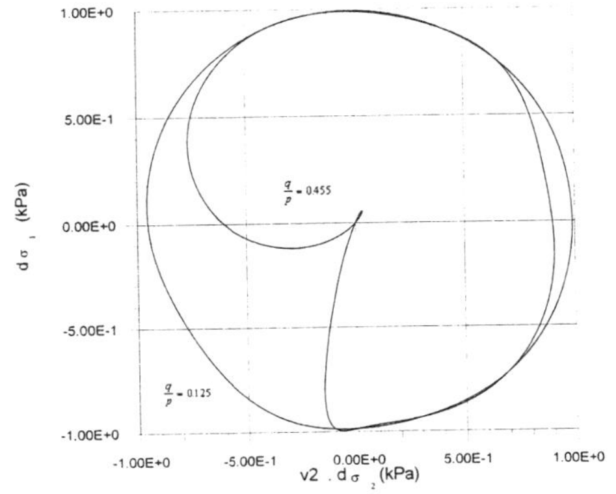

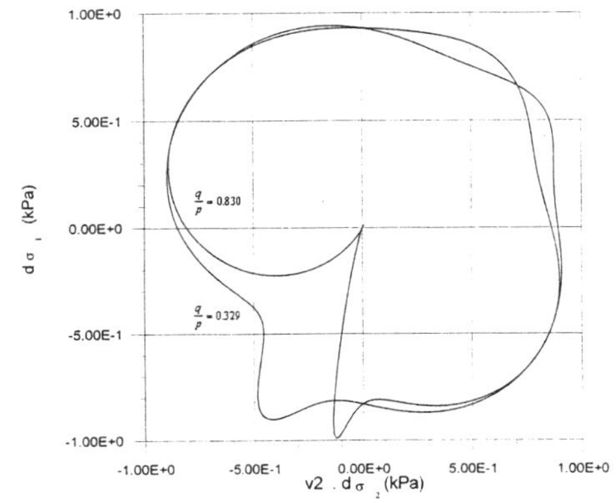

Fig. 2 : Polar diagrams of the normalised work of second order : $t = d^2W/\| \underset{\sim}{d\sigma} \| / \| \underset{\sim}{d\varepsilon} \|$ for loose Hostun sand at the top and dense Hostun sand at the bottom. The diagrams are plotted for 2 different values of $\eta = q/p$. The incrementally non-linear relation is utilised.

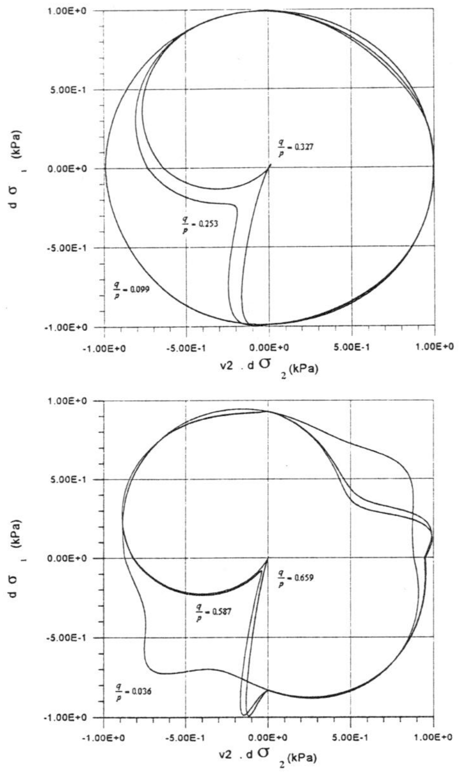

Fig. 3 : Polar diagrams of the normalised work of second order : $t = d^2W/\|d\sigma\|/\|d\varepsilon\|$ for loose Hostun sand at the top and dense Hostun sand at the bottom. The diagrams are plotted for 2 different values of $\eta = q/p$. The incrementally piece-wise linear relation is utilised.

4. BIFURCATION ANALYSIS

A softening material presents along its softening regime a loss of constitutive uniqueness since for a given $d\sigma$ two possible solutions exist in the incremental strains space : the first corresponds to the elastic unloading, while the second is related to a continued plastic loading. According to the Hill's sufficient condition of stability, the elastic unloading branch is stable while the plastic loading branch is unstable. These bifurcation states are situated on the plastic limit condition and have been extensively studied as material failure points.

These results can be derived from the previous constitutive relations by considering their respective inverse forms :

$$d\sigma = M^{-1}(u)\ d\varepsilon$$

with : $u = d\sigma / \|d\sigma\|$

The vanishing constitutive determinant :

$$\det M^{-1}(u) = 0$$

corresponds to the plastic criterion. Its interpretation in the framework of the incrementally piece-wise linear model corresponds to vanishing tangent moduli.

The condition : $M^{-1}(u)\ d\varepsilon = 0$ constitutes a generalised flow rule and can be related in simple cases, to the notion of dilatancy/contractancy angle.

Let us consider now $d\varepsilon$ as a function of $d\sigma$:

$$d\varepsilon = M(u)\ d\sigma.$$

It is also possible to analyse the variations of det M. The study is restricted here to the axisymmetric case in fixed stress-strain principal axes. In order to simplify the analysis as much as possible, the incrementally piece-wise linear model is utilised. Thus for loading paths defined by :

$$\begin{cases} d\sigma_1 > 0 \\ d\sigma_2 = d\sigma_3 < 0, \end{cases}$$

it comes :

$$\det M = (1 + V_3^{3-})(1 - V_3^{3-} - 2\,V_1^{3+} V_3^{1-}) \,/\, E_1^{+}/(E_3^{-})^2,$$

where E_i are tangent moduli and V_i^j tangent Poisson's ratio (see their rigorous definitions in Darve et al. 1982, for example).

This relation shows obviously that det M^{-1} vanishes for E_1^{+} or E_3^{-} equal to zero (plastic criterion). What is more interesting is to note that det M can vanish for particular values of the Poisson's ratio, what result is to be related to volume variations characteristics.

In order to investigate such possibilities, particular radial stress paths are considered as follows :

$$\begin{cases} d\varepsilon_1 = \text{positive constant} \\ dq/dp = -3.33 \text{ or } d\sigma_1/d\sigma_3 = -0.58 \end{cases}$$

These stress paths are presented at fig.4. This

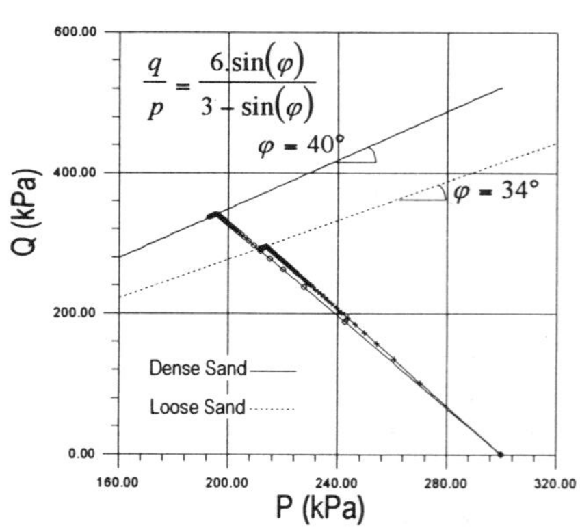

Fig. 4 : Loading paths chosen to analyse the variations of the constitutive determinant.

particular class of loading paths has been chosen according to the first stress vector direction where the constitutive determinant becomes nil.

Fig. 5 shows the variations of :
V_1^{3+}, $1 - V_3^{3-} - V_3^{1-}$ and finally of :
$1 - V_3^{3-} - 2\ V_1^{3+}\ V_3^{1-}$ for loose and dense Hostun sand with the incrementally piece-wise linear constitutive relation (the so-called « octolinear » model). It is clear that the last expression is vanishing before reaching the plastic limit condition.

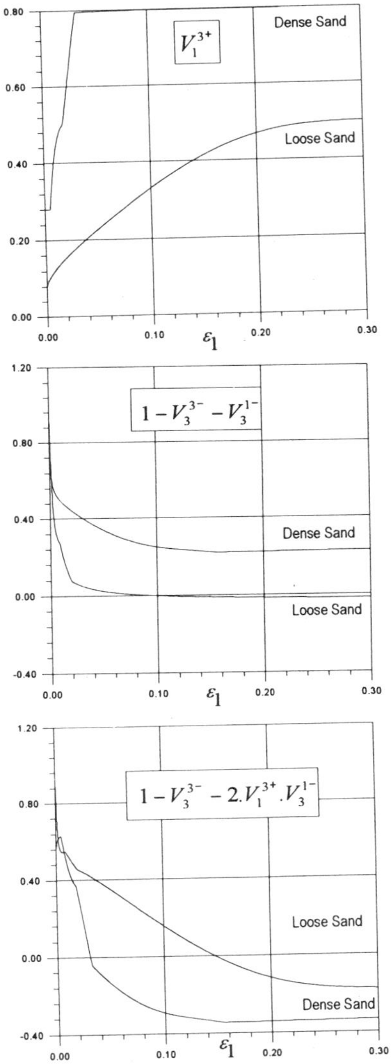

Fig. 5 : Variations of some quantities along the loading paths of fig. 4 in order to explain the change of sign of the constitutive determinant.

Basically it results from the following characteristics :
- the dilatant behaviour in compression ($V_1^{3+} > 0.5$), for a dense sand
- the contractant behaviour in extension ($1 - V_3^{3-} - V_3^{1-} < 0$) for a loose sand.

Fig. 6 presents the variations of the constitutive determinant for both the initial densities. It is possible to check on these figures that the constitutive determinant becomes nil at the same stress-strain point as for : $1 - V_3^{3-} - 2V_1^{3+}V_3^{1-}$.

It is then necessary to characterise the type of degeneration which has been obtained at these points of vanishing constitutive determinants. For this purpose an adequate tool is constituted by the « response-envelopes » as proposed by Gudehus (1979). For a given stress-strain state after a given strain history, one considers the particular class of loading constituted by all the axisymmetric incremental stress vectors of same magnitude ($\|d\underset{\sim}{\sigma}\| = 1$). The extremities of the axisymmetric

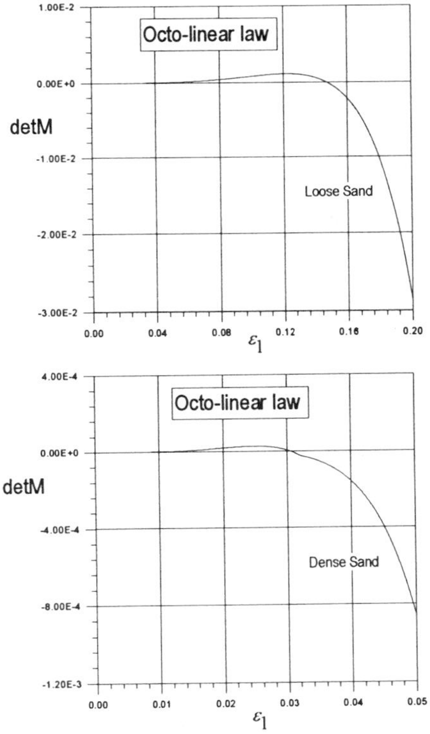

Fig. 6 : Variations of the constitutive determinant along the loading paths of fig. 4 for loose Hostun sand at the top and dense Hostun sand at the bottom with the « octo-linear » constitutive relation.

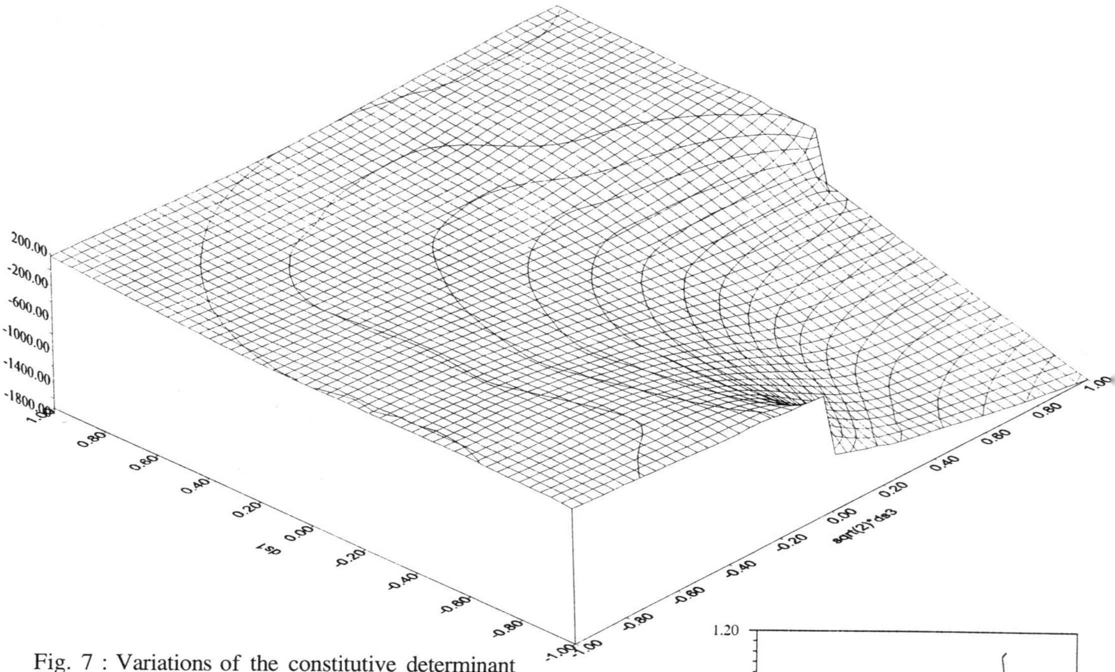

Fig. 7 : Variations of the constitutive determinant with the incremental stress vector directions on the plastic limit condition for the incrementally non-linear model based on 5 parameters.

incremental strain response vectors constitute the « response-envelope ». One can demonstrate analytically and verify numerically that, when $1 - V_3^{3-} - 2V_1^{3+}V_3^{1-}$ is vanishing, the response-envelope is a straight line defined by the following equation :

$$d\varepsilon_3/d\varepsilon_1 = - V_1^{3+}.$$

That means that for any incremental stress vector direction the strain vector keeps the same direction : it is a kind of flow rule. These results generalise previous ones obtained in the case of : V_1^{3+} equal to 0.5 (incrementally incompressible state) (Darve et al. 1995).

A vanishing constitutive determinant is a sufficient condition for a loss of material uniqueness (Darve 1994). Thus it can be conjectured that at this state there are several bifurcated branches, as we have seen before in the softening regime for the well known case of a negative tangent modulus.

Such a post-bifurcation analysis has been carried out by the simplest version of the incrementally non-linear model, which depends only on 5 material parameters. In such a case the unstable

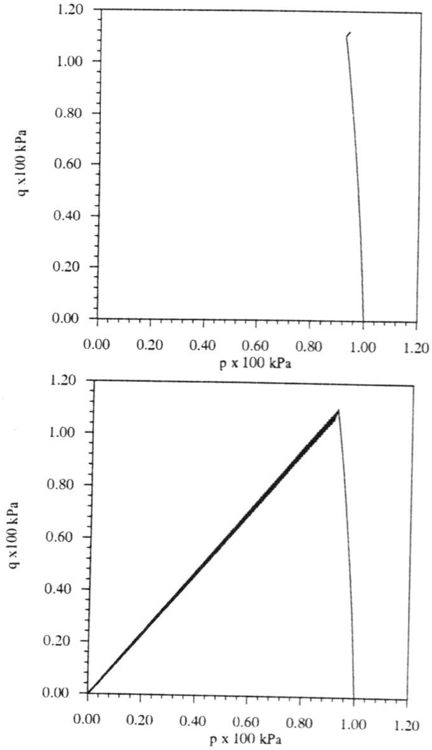

Fig. 8 : Post-bifurcation analysis in the case of fig. 7. Both the bifurcated branches are plotted, the ascending branch is stable while the descending one is unstable.

points and the bifurcation states coincide with the plastic limit surface. Thus the constitutive determinant is strictly positive inside the plastic limit condition and becomes negative when the plastic condition is reached. Fig. 7 presents the variations of the constitutive determinant on the plastic limit for loose Hostun sand and axisymmetric conditions. For some stress vector directions this determinant is positive while it becomes negative for others. For the same strain loading path defined by :

$$\begin{cases} d\varepsilon_1 = \text{positive constant} \\ d\varepsilon_2 = d\varepsilon_3 = -d\varepsilon_1/2 \text{ (isochoric axisymmetry)} \end{cases}$$

two possible stress responses have been found, one corresponding to a negative value of the constitutive determinant and an other to a positive value. As the incremental stress response is proportional to the determinant, both these stress responses are opposite. It means that one corresponds to drecreasing values

$$(d\sigma_1 < 0,\ d\sigma_3 < 0)$$

and the other to increasing values

$$(d\sigma_1 > 0,\ d\sigma_3 > 0).$$

Fig. 8 illustrates both these stress responses in q-p plane.

Now let us apply to these bifurcated branches the Hill's sufficient condition of stability :

$$d^2W = \underset{\sim}{d\sigma} : \underset{\sim}{d\varepsilon} = d\sigma_1\, d\varepsilon_1 + 2d\sigma_3\, d\varepsilon_3 = dq\, d\varepsilon_1.$$

Thus the ascending branch is stable according to Hill's condition, while the descending branch is unstable.

In practice it implies that the imperfections which are always effective in reality will make the stress point to jump from the unstable descendant branch to the ascendant stable branch. Such jumps are frequently observed in experimental measurements performed during undrained loading on loose sands. These jumps are seen as sudden reversals of the stress path in q-p plane.

5 - CONCLUSIONS

The first part of this paper has been devoted to a clarification of the question of material stability in the field of the mechanical behaviour of sands. The use of the Lyapunov's definition of stability allows to exhibit unstable states strictly inside the plastic limit condition of Mohr-Coulomb. Thus the question is open, at least for non-associated materials : how to predict this kind of material instabilities ?

For this purpose we have applied the Hill's sufficient condition of stability to our constitutive relations. For the incrementally piece-wise linear model as for the incrementally non-linear one a wide domain of the stress space appears as potentially unstable for loose as for dense granular materials. The analysis of stability is related to the sign of the determinant of the symmetric part of the constitutive tensor. As for non-associated materials the constitutive tensor is non-symmetric, the question of constitutive uniqueness which is related to the sign of the constitutive determinant must be analysed by itself.

A novel class of bifurcation points has been exhibited according to the vanishing constitutive determinant. These bifurcations are not related, as for classical analyses, to negative tangent moduli but to high values of tangent Poisson's ratios (in triaxial compression for dense sands and in triaxial extension for loose sands). Both the bifurcated branches have been exhibited. It is conjectured that these results could explain some experimental results during undrained loading.

REFERENCES

Lyapunov, A.M. 1907. Problème général de la stabilité de mouvement. Annales de la Faculté des Sciences de Toulouse, 9 : 203-474.

Drucker, D.C. 1951. A more fundamental approach to stress-strain relations. First U.S. Nat. Cong. of Applied Mechanics, ASME : 487-491.

Hill, R. 1958. A general theory of uniqueness and stability in elastic-plastic solids. J. of Mech. and Phys. of Solids, 6 : 236-249.

Darve, F., Labanieh, S. 1982. Incremental constitutive law for sands and clays. Simulation of monotonic and cyclic tests. Int. J. for Num. and Anal. Meth. in Geomechanics, 6 : 243-275.

Gudehus, G. 1979. A comparison for some constitutive laws for soils under radially symmetric loading and unloading. 3rd Int. Conf. on Num. Meth. in Geomechanics, W. Wittke (ed.), Balkema : 1309-1323.

Darve, F., Flavigny, E. Meghachou, M. 1995. Constitutive modelling and instabilities of soil behaviour. Computers and Geotechnics, 17 : 203-224.

Darve, F. 1994. Stability and uniqueness in Geomaterials constitutive modelling. In : Localisation and Bifurcation Theory for Soils and Rocks. Chambon, Desrues and Vardoulakis (eds.). Balkema : 73-88.

Numerical Models in Geomechanics, Pietruszczak & Pande (eds) © *1997 Balkema, Rotterdam, ISBN 90 5410 886 X*

Numerical modelling of bifurcation and post-bifurcation in soils

R.de Borst, A.Groen & H.van der Veen
Koiter Institute Delft and Faculty of Civil Engineering, Delft University of Technology, Netherlands

ABSTRACT: The numerical treatment of bifurcation phenomena is discussed including the continuation onto different possible post-bifurcation paths. Two techniques are discussed for branch switching in elasto-plastic soils, namely the more traditional vector perturbation method and a newly developed deflation-like technique. Examples are given for both methods. The results suggest a better performance for the deflation-like technique. Finally, an enhanced continuum, the so-called Cosserat continuum, is used to trace the entire post-bifurcation behaviour. A comparison is made with recent tests by Tatsuoka *et al.* (1994).

1 INTRODUCTION

A fundamental issue in the behaviour of many mechanical systems is their sensitivity to imperfections, either from a geometrical nature or from a material nature. Geometrical imperfections are due to the impossibility to manufacture specimens that perfectly satisfy the design. This type of imperfection prevails in mechanical or aerospace engineering, where buckling of thin-walled structures plays a dominant role. The relation between imperfection sensitivity and bifurcation of the underlying perfect structure was established by Koiter (1945) in his landmark dissertation. Important progress in terms of simulating these phenomena numerically could be made because of the introduction of finite element methods and specifically of branch-switching and path-following techniques to trace the entire post-bifurcation path (Riks 1979).

In soil mechanics, material imperfections. which arise from locally varying strengths or stiffnesses of the soil are much more important. The result with respect to the mechanical behaviour of the structure is essentially the same: the imperfect structure tends to have a significantly lower load-carrying capacity and a lower ductility than the underlying perfect structure. However, their theoretical treatment and numerical simulation is more complicated. This is because we no longer have an injective relation between stress and strain. Indeed, plasticity, which is commonly used to describe the non-linear behaviour of soils, implies a multi-valued constitutive relation, where, from a given point, a different behaviour is obtained depending whether the local strain rate indicates loading or unloading. This observation not only implies that Koiter's imperfection sensitivity theory can no longer be applied, it also provides additional difficulties in numerical computations, where densely clustered bifurcation points arise and branch-switching techniques have met with limited success owing to local loading-unloading possibilities (de Borst 1986, 1988, 1989).

In this contribution we shall first recall the basic issues in bifurcation analyses. Then, we shall outline two different techniques for branch switching, namely perturbing the trivial solution by the eigenvector related to the lowest (negative) eigenvalue, and applying a deflation-like technique (van der Veen *et al.* 1997). After a concise description of the constitutive model, some examples are offered, showing a preliminary comparison between both approaches. In the future, this comparison should be extended to non-standard continuum models which are needed for a realistic description of the post-bifurcation branch, in particular the localisation behaviour that is normally encountered in this regime. Now, only a computation is shown in which an enhanced (Cosserat) continuum is used to trace the entire post-bifurcation behaviour of a biaxial test on sand (Tatsuoka *et al.* 1994).

2 ON BIFURCATION STATES

A discussion of the uniqueness of response of a body to a further load increment is best started from the strong form of the equilibrium rate equations:

$$\mathbf{L}^{\mathrm{T}}\dot{\boldsymbol{\sigma}} + \rho\dot{\mathbf{g}} = \mathbf{0} \,, \qquad (1)$$

with $\mathbf{L}$ a differential operator matrix, $\boldsymbol{\sigma}$ the stress tensor, ρ the specific density and $\mathbf{g}$ the gravity acceleration vector. The superimposed dots in eq. (1) denote time differentiation and the superscript T is the transpose symbol.

Loss of uniqueness of the differential rate problem requires the existence of at least *two* stress rate fields which both satisfy equilibrium. Labelling the solutions by $\dot{\boldsymbol{\sigma}}_A$ and $\dot{\boldsymbol{\sigma}}_B$ respectively, we find upon subtraction of both solutions that the identity

$$\mathbf{L}^{\mathrm{T}}\Delta\dot{\boldsymbol{\sigma}} = \mathbf{0} \,, \qquad (2)$$

with $\Delta\dot{\boldsymbol{\sigma}} = \dot{\boldsymbol{\sigma}}_A - \dot{\boldsymbol{\sigma}}_B$, has to be satisfied non-trivially for bifurcation to occur. Finite element applications are usually rooted in the weak form of the equilibrium equation (1). In this spirit we will also rephrase the condition for loss of uniqueness (2) in a weak form:

$$\int_V (\delta\Delta\dot{\mathbf{u}})^{\mathrm{T}}\mathbf{L}^{\mathrm{T}}\Delta\dot{\boldsymbol{\sigma}}\,\mathrm{d}V = 0 \,, \qquad (3)$$

where the spatial integration extends over the entire domain of the body. The δ-symbol represents the first variation of a quantity and $\Delta\dot{\mathbf{u}}$ is the difference of the two kinematically admissible velocity fields $\dot{\mathbf{u}}_A$ and $\dot{\mathbf{u}}_B$ which yield the stress rate fields $\dot{\boldsymbol{\sigma}}_A$ and $\dot{\boldsymbol{\sigma}}_B$ respectively.

Eq. (3) can be used directly to assess uniqueness of discretised systems that are composed of materials which have the same stiffness for loading and unloading, e.g., non-linear elastic materials, so that a unique, one-to-one relation exists between stress rate and strain rate at each generic strain level. For material models which have a different behaviour in loading and unloading eq. (3) is better modified in order that an assessment can be made whether eq. (3) is a sharp indicator for bifurcation states. Employing Gauss' divergence theorem and using the identity

$$\Delta\dot{\boldsymbol{\varepsilon}} = \mathbf{L}\Delta\dot{\mathbf{u}} \,, \qquad (4)$$

for the difference between both considered strain fields $\dot{\boldsymbol{\varepsilon}}_A - \dot{\boldsymbol{\varepsilon}}_B$, we can rewrite eq. (3) as

$$\int_V (\delta\Delta\dot{\boldsymbol{\varepsilon}})^{\mathrm{T}}\Delta\dot{\boldsymbol{\sigma}}\,\mathrm{d}V = 0 \,, \qquad (5)$$

since either the difference between both velocity fields or the difference between the tractions must vanish along the boundary.

In its general form the elasto-plastic stiffness tensor $\mathbf{D}^{ep}$ which sets the relation between stress rate and strain rate, reads

$$\dot{\boldsymbol{\sigma}} = \mathbf{D}^{ep}\dot{\boldsymbol{\varepsilon}} \,. \qquad (6)$$

Substituting eq. (6) into the bifurcation condition (5), we obtain

$$\int_V (\delta\Delta\dot{\boldsymbol{\varepsilon}})^{\mathrm{T}}\mathbf{D}^{ep}\Delta\dot{\boldsymbol{\varepsilon}}\,\mathrm{d}V = 0 \,, \qquad (7)$$

where it has been assumed that both strain rates are related to the stress rates by the same stiffness modulus $\mathbf{D}^{ep}$. Put differently, it has been assumed that both stress fields show loading for at least an infinitesimally small time increment after bifurcation. Strictly speaking this possibility is but one of infinitely many possibility in which different parts of the body show unloading, i.e. in these parts the stress rate is related to the strain rate by the elastic operator $\mathbf{D}^e$. While such a search should be undertaken in the general case it appears that for the important subclass of associative plasticity, the combination for which both stress fields show loading for the entire part of the body which has already been plastified, is always the most critical.

In the remainder of this paper we shall assume that the case in which the entire plastified part of the body shows further plastic loading for at least an infinitesimally small increment after bifurcation is the most critical case also for non-associative plasticity. Inserting eq. (4) into eq. (7) yields

$$\int_V (\delta\Delta\dot{\mathbf{u}})^{\mathrm{T}}\mathbf{L}^{\mathrm{T}}\mathbf{D}^{ep}\mathbf{L}\Delta\dot{\mathbf{u}}\,\mathrm{d}V = 0 \,. \qquad (8)$$

We next discretise the continuum into an arbitrary number of finite elements:

$$\mathbf{u} = \mathbf{H}\mathbf{a} \qquad (9)$$

where the vector $\mathbf{a}$ contains the nodal displacements and $\mathbf{H}$ contains the interpolation polynomials. With eq. (9) the following relation ensues:

$$\delta\Delta\dot{\mathbf{a}}^{\mathrm{T}} \int_V \mathbf{H}^{\mathrm{T}}\mathbf{L}^{\mathrm{T}}\mathbf{D}^{ep}\mathbf{L}\mathbf{H}\,\mathrm{d}V\,\Delta\dot{\mathbf{a}} = 0 \,, \qquad (10)$$

since the differences between both velocity distributions at the nodes $\Delta\dot{\mathbf{a}}$ are now independent quantities and can be brought outside of the integral. We next define the tangential stiffness matrix

$$\mathbf{K} = \int_V \mathbf{H}^{\mathrm{T}}\mathbf{L}^{\mathrm{T}}\mathbf{D}^{ep}\mathbf{L}\,\mathbf{H}\,\mathrm{d}V \qquad (11)$$

and require that eq. (10) holds for any virtual velocity vector $\delta\Delta\dot{\mathbf{a}}$. This results in:

$$\mathbf{K}\,\Delta\dot{\mathbf{a}} = \mathbf{0} \,. \qquad (12)$$

We next write $\Delta\dot{\mathbf{a}}$ as a linear combination of the n right eigenvectors $\mathbf{v}_i$ and the n left eigenvectors $\mathbf{w}_i$ of the matrix $\mathbf{K}$,

$$\Delta\dot{\mathbf{a}} = \sum_{i=1}^{n}(\mathbf{w}_i^{\mathrm{T}}\Delta\dot{\mathbf{a}})\mathbf{v}_i \ , \tag{13}$$

where it has been assumed that $\mathbf{w}_i$ and $\mathbf{v}_i$ have been normalised such that $\mathbf{w}_i^{\mathrm{T}}\mathbf{v}_i$. Eq. (12) can then be recast in the form

$$\sum_{i=1}^{n} \lambda_i(\mathbf{w}_i^{\mathrm{T}}\Delta\dot{\mathbf{a}})\ \mathbf{v}_i = \mathbf{0} \tag{14}$$

since $\mathbf{K}\mathbf{v}_i = \lambda_i\mathbf{v}_i$ (no summation implied). Assuming that $\mathbf{K}$ is not defect, the right eigenvectors $\mathbf{v}_i$ and the left eigenvectors $\mathbf{w}_i$ each constitute a set of n linearly independent eigenvectors. Consequently, $(\mathbf{w}_i^{\mathrm{T}}\Delta\dot{\mathbf{a}})\lambda_i$ must vanish for each i. Since $\Delta\dot{\mathbf{a}}$ can not be orthogonal to each left eigenvector $\mathbf{w}_i$, this means that at least one eigenvalue, say λ_1, must vanish at a bifurcation point. In practical numerical analyses a point where the tangential stiffness has exactly one or more vanishing eigenvalues will never be encountered. Instead it is assumed that a bifurcation point has been passed when at least one of the eigenvalues has changed sign within a loading step.

3 BRANCH SWITCHING

When a bifurcation point has been detected according to the procedure outlined in the preceding sections, the incremental solution $\Delta\mathbf{a}$ can be perturbed by adding a part of the right eigenvector $\mathbf{v}_1$ which corresponds to the lowest eigenvalue λ_1. Note that the Δ-symbol now stands for a finite increment and no longer for the difference between two quantities. In consideration of eq. (13) we obtain for the perturbed displacement increment $\Delta\mathbf{a}^*$:

$$\Delta\mathbf{a}^* = \Delta\mathbf{a} + \omega(\mathbf{w}_1^{\mathrm{T}}\Delta\mathbf{a})\ \mathbf{v}_1 \ , \tag{15}$$

with ω a damping/amplification factor, which can be determined from the requirement that (Riks 1979, de Borst 1986, 1988, 1989)

$$\Delta\mathbf{a}^{\mathrm{T}}\Delta\mathbf{a}^* = 0 \ . \tag{16}$$

Combination of eqs. (15) and (16) gives for the perturbed displacement increment

$$\Delta\mathbf{a}^* = \Delta\mathbf{a} - \frac{\Delta\mathbf{a}^{\mathrm{T}}\Delta\mathbf{a}}{\mathbf{v}_1^{\mathrm{T}}\Delta\mathbf{a}}\ \mathbf{v}_1 \ . \tag{17}$$

Eq. (16) states that the search direction for the non-trivial solution is orthogonal to the trivial or basic solution path. In general, the non-trivial solution will not be in the search direction as defined by eq. (17), and equilibrium iterations are necessary to converge on a non-trivial (localised) solution. This solution is not necessarily the lowest bifurcation branch when more equilibrium branches emanate from a bifurcation point. When we have converged on another than the lowest post-bifurcation path, this will be revealed by negative eigenvalues of the obtained solution. The above described procedure can be repeated until we ultimately arrive at the lowest bifurcation path.

Although the procedure outlined above is straightforward, its practical use in complicated mechanical systems with thousands degrees-of-freedom may entail difficulties. Sometimes the perturbation as defined in eq. (17) is not 'strong' enough and the equilibrium iterations that are added bring the system back on the basic equilibrium path. If, on the other hand, the perturbation is 'too strong' and the prediction according to eq. (17) deviates too much from the bifurcated equilibrium path, the additional equilibrium equations will not be able to restore equilibrium.

An alternative techique would be to substitute eq. (15) into the linearised and discretised equilibrium condition

$$\mathbf{K}\,\Delta\mathbf{a}^* = \mathbf{f}_{\mathrm{ext}} - \mathbf{f}_{\mathrm{int}} \ , \tag{18}$$

where $\mathbf{f}_{\mathrm{ext}}$ and $\mathbf{f}_{int}$ denote the external and internal force vector, respectively. With $\Delta\mathbf{a}^*$ the perturbed displacement increment, we obtain for the first iteration of the loading step after detection of bifurcation:

$$\mathbf{K}\,\Delta\mathbf{a} + \omega(\mathbf{w}_1^{\mathrm{T}}\Delta\mathbf{a})\ \mathbf{K}\mathbf{v}_1 = \mathbf{f}_{\mathrm{ext}} - \mathbf{f}_{\mathrm{int}} \ . \tag{19}$$

Rearranging and noting that by definition $\mathbf{K}\mathbf{v}_1 = \lambda_1\mathbf{v}_1$ one arrives at

$$(\mathbf{K} + \omega\lambda_1\mathbf{v}_1\mathbf{w}_1^{\mathrm{T}})\Delta\mathbf{a} = \mathbf{f}_{\mathrm{ext}} - \mathbf{f}_{\mathrm{int}} \ , \tag{20}$$

or

$$\mathbf{K}^*\Delta\mathbf{a} = \mathbf{f}_{\mathrm{ext}} - \mathbf{f}_{\mathrm{int}} \ , \tag{21}$$

where $\mathbf{K}^*$ is a perturbed tangential stiffness matrix, on which the perturbation is applied in a deflation-like manner:

$$\mathbf{K}^* = \mathbf{K} + \omega\lambda_1\mathbf{v}_1\mathbf{w}_1^{\mathrm{T}} \ . \tag{22}$$

Post-multiplication of $\mathbf{K}^*$ with the right eigenvectors $\mathbf{v}_i$ or pre-multiplication of $\mathbf{K}^*$ with the left eigenvectors $\mathbf{w}_i$ shows that the eigenvectorspace is unaltered. Also, the eigenvalues are the same, except for the lowest eigenvalue λ_1, which has now changed into $\lambda_1^* = (1+\omega)\lambda_1$. The original tangential stiffness matrix $\mathbf{K}$ is indefinite, i.e. it has one or more negative eigenvalues λ_i. A stable path is associated with a tangential stiffness matrix that is positive definite, i.e. it has only positive eigenvalues. This idea is used to steer the perturbed solution such that $\mathbf{K}^*$ is positive definite. This can be achieved by choosing ω such that $\omega < -1$, since then $\lambda_1^* > 0$. In the examples that follow ω has been chosen equal to -100. A problem related to the computation of $\mathbf{K}^*$ is that a singular perturbation of the type $\mathbf{w}_1\mathbf{v}_1^{\mathrm{T}}$ destroys the bandedness of the original tangential stiffness matrix **K**. To preserve numerical efficiency, it has been decided to perturb **K** only within the original band.

4 EXAMPLES

In order not to obscure the essential characteristics of both branch-switching techniques a plasticity model has been chosen that is as simple as possible, while of loss of uniqueness and the possible ensuing strain localisation is still retained. A cohesionless Drucker-Prager model has been selected with a yield function

$$f = q + \alpha p \ , \tag{23}$$

with p and q the hydrostatic pressure and the second invariant of the deviatoric stress tensor respectively. The friction parameter α is related to the more common angle of internal friction ϕ via

$$\alpha = \frac{6 \sin\phi}{3 - \sin\phi} \ . \tag{24}$$

A non-associative flow rule is obtained by defining a separate plastic potential function g

$$g = q + \beta p \ , \tag{25}$$

from which the plastic strain rates are derived according to

$$\dot{\varepsilon}^p = \dot{\lambda} \frac{\partial g}{\partial \sigma} \tag{26}$$

where $\dot{\lambda}$ now denotes the plastic flow rate. The dilatancy parameter β is related to the more familiar dilatancy angle ψ in a fashion similar to the relation between α and the friction angle ϕ. Hardening/softening is introduced via

$$\sin\phi = \sin\phi(\gamma^p) \quad , \quad \gamma^p = \int \dot{\gamma}^p \mathrm{d}t \ , \tag{27}$$

with $\dot{\gamma}^p$ a function of the plastic strain rate tensor.

In the analyses which have been carried out to assess the performance of the different perturbation techniques a plane-strain biaxial specimen has been considered with dimensions 120 × 240 mm. The bottom of the specimen has been assumed smooth and loading is by a vertical force at the top of the specimen, while dependence relations have been applied to ensure equal displacements of the top of the specimen. A rather coarse 6×12 mesh has been used in which each quadrilateral is composed of four crossed triangles with a quadratic interpolation field. Isotropic elasticity has been assumed with a Young's modulus equal to $E = 0.96$ MPa and a Poisson's ratio $\nu = 0.2$. The angle of internal friction was assigned $\sin\phi = 0.34$ and two different values of the dilatancy angle were used:

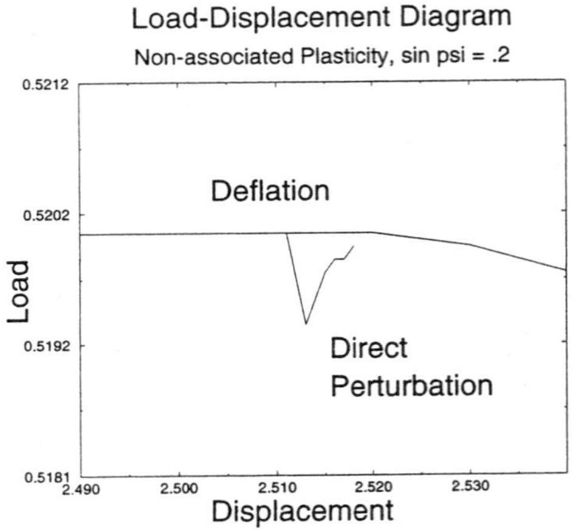

Figure 1. Load-displacement diagram around bifurcation point for a moderate degree of non-associativity.

$\sin\psi = 0.2$ and $\sin\psi = 0$.

When the degree of non-associativity is moderate ($\sin\psi = 0.2$) two negative pivots are found at a generic stage in the loading process. In an eigenvalue analysis using Lanczos' method two negative eigenvalues were extracted. Perturbing the trivial solution with the eigenvector related to the most negative eigenvalue resulted in a smooth continuation on a post-bifurcation path which features a localised (shear-band) solution when the deflation-like technique was employed. However, a rather irregular behaviour was obtained when the perturbation was applied directly on the incremental displacement vector, and adding iterations could not drive back the solution on a global equilibrium path. Figure 1 illustrates the smooth continuation of the equilibrium path after bifurcation when the deflation-like method is used, while the irregular behaviour before final divergence is also clearly brought out when the perturbation is applied on the incremental displacement vector. When the degree of non-associativity is increased, i.e. for $\sin\psi = 0$, a similar behaviour is obtained. However, for this case the particularity was observed that two negative pivots occurred prior to the moment that the lowest eigenvalue turned negative. Perturbation was carried out at the latter instant, and resulted again in a smooth continuation of the deflation-like technique and in divergence of the equilibrium-finding iterative procedure if the perturbation was applied directly on the incremental displacement vector.

To trace the equilibrium path well beyond primary bifurcation enhanced continuum models are necessary to prevent loss of ellipticity of the governing differential equations when strain localisation develops (e.g., de Borst *et al* 1993). Herein, the Cosserat continuum has been used. In particular, the Drucker-Prager yield function has been embedded in a Cosserat framework (de Borst 1993, Groen 1997). It would have been desirable to perform the above bifurcation analyses with a Cosserat-based Drucker-Prager plasticity model, but at the time of the preparation of this contribution these analyses had not yet been completed. For this reason, analyses on a biaxial specimen have been carried out with a Cosserat model where strain localisation is triggered by inserting a small imperfection in the right upper corner of the specimen, so that the bifurcation problem is transferred into a limit problem. The dimensions of the specimen that was tested by Tatsuoka *et al.* (1994) were 80×200 mm and smooth boundary conditions were again adopted at the top and the bottom of the specimen. Six-noded triangular elements in a cross-diagonal patch were again adopted with a six-point Gaussian integration. The total number of elements varied from 600 for the coarsest mesh to 9600 for the finest mesh. To characterise the elastic behaviour a Young's modulus E = 50 MPa and a Poisson's ratio ν = 0.3 have been adopted. A constant cell pressure σ_3 = 78.4 kPa has been assumed. To match the experimental results a friction-hardening relation has been used according to Figure 2.

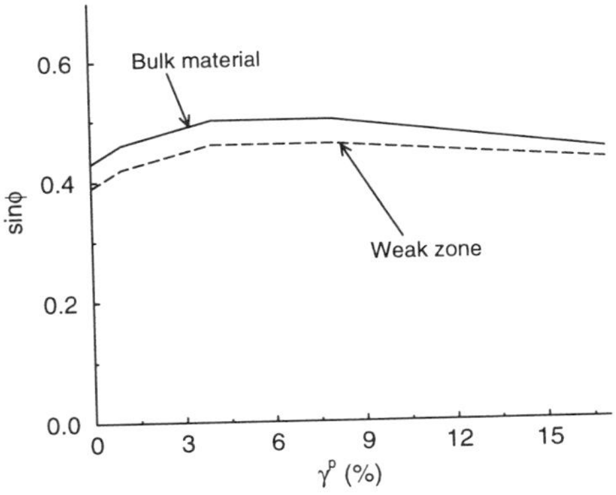

Figure 2. Multi-linear friction-hardening diagram.

The assumed stress-strain relation leads to the load-displacement curves depicted in Figure 3 for the homogeneous solutions of the bulk material and of the weak zone, respectively. Figure 4 shows the computed load-deformation curves as well as the experimental results of Tatsuoka *et al.* (1994). Apparently, the maximum and the residual strengths could be retro-fitted within a tolerance of 5%. An interesting observation is that shear banding and the ensuing structural softening occur already when both the bulk material and the weak

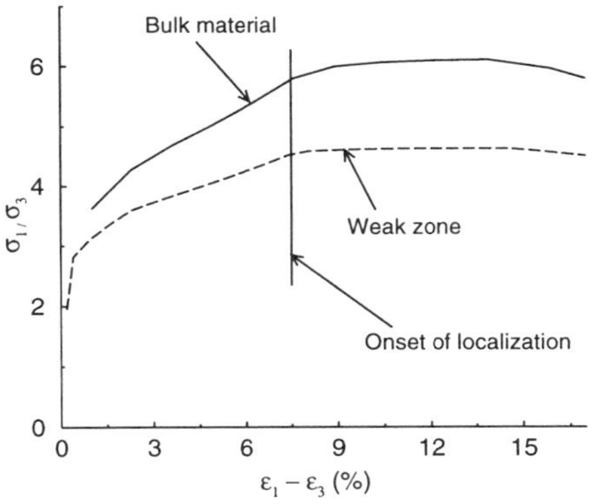

Figure 3. Stress ratio vs shear deformation for the homogeneous solutions of the bulk material and the weak zone.

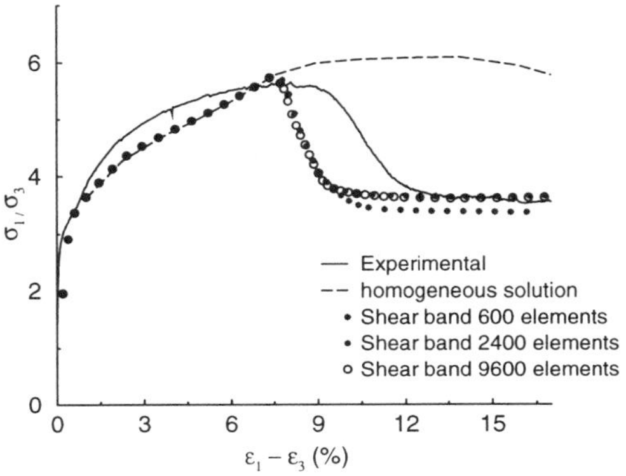

Figure 4. Stress ratio vs shear deformation for a biaxial test modelled using a Cosserat continuum.

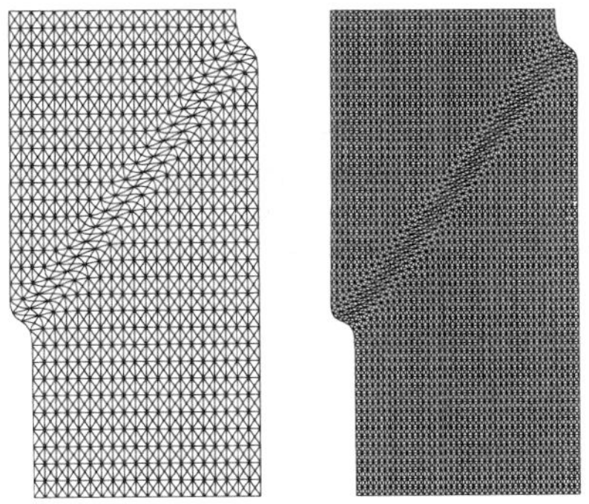

Figure 5. Deformations at termination of loading for the two finer meshes (2400 and 9600 elements).

zone are still in the hardening regime. This confirms that structural softening can be caused solely by non-associative plastic flow.

The deformations at termination of loading are shown in Figure 5 for the two finer meshes. We observe that the predicted shear band thickness and the inclination converge to a unique solution. The coarse mesh of 600 elements does not have enough regularising effect as the shear band inclination is still affected by the direction of the mesh lines. However, the medium mesh (2400 elements) and the fine mesh (9600 elements) predict a shear band inclination that is not influenced by the mesh lines. Its value, about 54°, is smaller than the experimental value of 60°, which is probably attributable to the application of the Drucker-Prager yield contour, which does not involve the third stress invariant. On the other hand, the thickness of the band is of the order of 6 mm, which is in reasonable agreement with the experimental observations by Tatsuoka *et al.* (1994).

5 FURTHER COMMENTS

Branch switching after a bifurcation point has been located is a delicate issue in elasto-plastic solids, in particular if non-associative flow rules are used. Two such techniques have been discussed, one based on perturbing the tangential stiffness matrix in a deflation-like manner, and another which directly perturbs the incremental displacement vector. From preliminary analyses the former method seems to be more robust, although more numerical experimentation needs to be done in order to confirm this finding. Also, extension must be done to enhanced continuum models such as the Cosserat model, since they must be utilised in the post-bifurcation regime, in particular when strain localisation develops.

ACKNOWLEDGEMENTS

Financial support from the Dutch Technology Foundation STW under grant DCT22.2930 and from the Dutch Centre for Civil Engineering, Codes and Specifications is gratefully acknowledged.

REFERENCES

de Borst, R. 1986. *Non-Linear Analysis of Frictional Materials.* Dissertation, Delft University of Technology: Delft.

de Borst, R. 1988. Bifurcations in finite element models with a nonassociated flow law, *Int. J. Num. Anal. Meth. Geomech.* 12: 99-116.

de Borst, R. 1989. Numerical methods for bifurcation analysis in geomechanics, *Ing.-Arch.* 59: 160-174.

de Borst, R. 1993. A generalisation of J_2-flow theory for polar continua, *Comp. Meth. Appl. Mech. Eng.* 103: 347-362.

de Borst, R., L.J. Sluys, J. Pamin & H.-B. Mühlhaus 1993. Fundamental issues in finite element analysis of localisation of deformation, *Eng. Comp.* 10: 99-122.

Groen, A.E. 1997. *Three-Dimensional Elasto-Plastic Analysis of Soils.* Dissertation, Delft University of Technology: Delft.

Koiter, E. 1945. *Over de Stabiliteit van het Elastisch Evenwicht* (in Dutch). Dissertation, Delft University of Technology: Delft.

Riks, E. 1979. An incremental approach to the solution of snapping and buckling problems, *Int. J. Solids Struct.* 15: 529-551.

Tatsuoka, F., M.S.A. Siddique, T. Yoshida, C.S. Park, Y. Kamegai, S. Goto & Y. Kohata 1994. *Testing methods and results of elementary tests and testing conditions of plane strain model bearing capacity tests using air-dried dense Leighton Buzzard sand.* Internal Report, ISS, University of Tokyo: Tokyo.

van der Veen, H., K. Vuik & R. de Borst 1997. Post-bifurcation behaviour in soil plasticity: eigenvector perturbation compared to deflation. *Proc. Int. Conf. Computational Plasticity V.* Pineridge Press: Swansea.

Numerical Models in Geomechanics, Pietruszczak & Pande (eds) © *1997 Balkema, Rotterdam, ISBN 90 5410 886 X*

Modelling strain localization in anisotropically consolidated samples

H.F. Schweiger
Institute for Soil Mechanics and Foundation Engineering, Technical University Graz, Austria

M. Karstunen
Department of Civil Engineering, University of Glasgow, UK

G.N. Pande
Department of Civil Engineering, University of Wales, Swansea, UK

ABSTRACT: An important aspect in the analysis of strain localization is addressed in this paper, namely the influence of anisotropic consolidation for biaxial tests. A multilaminate model is used for capturing the onset of localization and by monitoring the mobilized plastic shear strains on integration planes the effect of the stress history can be visualized in a simple manner. After localization, homogenisation technique is applied to describe the material response. The results of this study clearly indicate that the effect of anisotropic consolidation has to be included in the analysis. Employing the proposed framework a change of direction of the minimum principal stress after the consolidation phase is also readily accounted for. This latter aspect is of importance when analysing practical boundary value problems where complex construction stages and support measures may lead to significant and abrupt changes in orientation of principal stresses.

1 INTRODUCTION

The problem of strain localization has been addressed by a number of research groups and several methods have been developed to overcome the shortcomings of standard finite element formulations, which are not suitable for describing the deformation pattern after the onset of localization as the problem becomes ill-posed. Classical constitutive models with material softening cannot be used because they do not have an internal length scale, which leads to severe mesh-sensitivity and non-converging iteration procedures.

The most commonly adopted techniques are the enhanced finite element models (e.g. Belytschko & al. 1988, Pietruszczak & Mróz 1981), the non-local models (e.g. Bazant & Pijaudier-Cabot 1988), the viscoplastic models (e.g. Loret & Prevost 1990), the gradient plasticity models (e.g. Vardoulakis & Aifantis 1991) and the polar continua (e.g. de Borst 1991). Mesh adaptivity techniques have also been proposed (e.g. Hicks 1995). Most analyses presented so far have considered plane strain conditions only.

An approach, which seems to have some potential for practical applications, because it is readily extended to three dimensions, has been proposed by Pietruszczak & Niu (1993) namely, the application of homogenisation technique. This approach is adopted here together with multilaminate framework (Pande & Sharma 1983).

2 MULTILAMINATE FRAMEWORK AND HOMOGENISATION TECHNIQUE

It is beyond the scope of this contribution to describe details of the employed method. The multilaminate framework (Pande & Sharma 1983) has been extended by Karstunen & Pande (1997) for defining the onset of localization by monitoring the plastic shear strains on integration planes. As will be shown later, this concept provides a convenient and physically appealing way of capturing the transition from homogeneous to localized deformation. After the shear band has formed, the material can be treated as a composite material consisting of materials inside and outside the shear band respectively (Pietruszczak & Niu 1993). The averaging rule is based on volume averages occupied by both constituents. A slight modification of this concept has been presented by Schweiger et al. (1997) by introducing the volume fraction occupied by the shear band as additional model parameter, and this has been employed for the present study.

The yield function, used here to describe the non-linear material behaviour before onset of localization, is an extension of the well-known Mohr-Coulomb criterion to account for deviatoric hardening (Pietruszczak & Niu 1993), and can be defined for the local co-ordinate system (assuming tension positive) for each sampling plane as

$$F = |\tau_{xy}| + \tan \varphi_m (\sigma_y' - c \cot \varphi_m) = 0 \qquad (1)$$

where

$$\tan \varphi_m = \tan \varphi_0 + (\tan \varphi - \tan \varphi_0) \frac{\gamma_{xy}^{\,p}}{A + \gamma_{xy}^{\,p}} \tag{2}$$

τ_{xy} and σ'_y are the shear and normal (effective) stresses on the plane respectively, φ_m is the mobilised friction angle, c is the cohesion and φ_0 defines the elastic region. The material parameter A controls the rate of hardening and γ_{xy}^p is the strain hardening parameter, which is different for each plane. Assuming non-associated plasticity, the plastic potential function can be selected as (Pietruszczak & Niu 1993)

$$Q = |\tau_{xy}| - \eta_c\,(\sigma'_y - c\,\cot\varphi_m)\,\ln\left(\frac{(\sigma'_y - c\,\cot\varphi_m)}{\sigma_0}\right) = 0 \tag{3}$$

where η_c specifies the slope of the zero dilatancy line and σ_0 is a scalar depending on the current stress state.

For the formulation of the 'equivalent matrix' after formation of the shear band, constitutive relations for both constituents need to be established (Pietruszczak & Niu 1993). For the shear band a modified Mohr-Coulomb model (Eqns. 4 and 5) to account for deviatoric softening (in the local coordinate system of the shear band) is used

$$F = |\tau_{xy}| + \tan \varphi_m\, \sigma'_y - c = 0 \tag{4}$$

where

$$\tan \varphi_m = \tan \varphi_r - (\tan \varphi_r - \tan \varphi_i)\, \exp\,[-B_\varphi\, g_{xy}^p] \tag{5}$$

B_φ is a material constant governing the rate of strain softening, φ_m is the mobilised friction angle, c is the cohesion, φ_r is the residual value of the friction angle, φ_i is the mobilized friction angle of the material at the onset of localization and g_{xy}^p is the tangential component of the irreversible part of the velocity discontinuity vector. Similarly, by adopting a non-associated flow rule we can express the plastic potential function as

$$Q = |\tau_{xy}| + \tan \psi_m \sigma'_y = 0 \tag{6}$$

$$\tan \psi_m = \tan \psi_r - (\tan \psi_r - \tan \psi_i)\, \exp[-B_\psi\, g_{xy}^p] \tag{7}$$

Again, B_ψ is the parameter controlling the rate of change of the angle of dilatancy, ψ_m is the mobilised dilatancy angle and ψ_r the residual value. For simplicity, the behaviour of the material outside the shear band is assumed to be elastic.

The parameters given in Tables 1 and 2 have been used for the numerical study presented here. The additional parameter μ representing a measure for the volume fraction occupied by the shear band has been assumed to be $\mu = 7.0$ (Schweiger et al. 1997).

Table 1. Parameters used for the deviatoric hardening model

Parameter	Value
Young's modulus, E	30 000 kN/m^2
Poisson's ratio, ν	0.3
Peak friction angle, φ	35°
Initial friction angle, φ_o	1°
Cohesion, c	1 kN/m^2
Parameter A, (Eq. 2)	0.0015
Parameter η_c, (Eq. 3)	0.30

Table 2. Parameters used for the deviatoric softening model for the shear band

Parameter	Value
Normal stiffness, K_N	40000 kN/m^3
Tangential stiffness, K_T	8000 kN/m^3
Residual friction angle, φ_r	0.35φ
Residual dilatancy angle, ψ_r	0°
Parameter B_φ, (Eq. 5)	30
Parameter B_ψ, (Eq. 7)	20
Parameter μ	7.0

3 INFLUENCE OF THE ANISOTROPIC CONSOLIDATION OF SAMPLES

Using the technique described above a parametric study investigating the influence of the consolidation history on the onset of localization has been performed. A computer program integrating the constitutive equations has been developed and employed to predict the stress strain response of a strain controlled biaxial test. It should be mentioned that no attempts were made to match experimental results in detail. The purpose of this study is merely to show the potential advantages of the proposed approach and to highlight the influence of the consolidation history on the onset of localization. Particular emphasis has been placed on the development of plastic shear strains on integration planes which will be shown for different consolidation stresses and various loading stages.

3.1 *Reference calculation (isotropic consolidation)*

In order to establish a reference for subsequent comparisons the parameters given in section 2 have been used and an isotropic consolidation stress of $\sigma_x = \sigma_y = -100$ kN/m^2 has been applied (the out of plane stress σ_z, which has no influence in the present formulation was also set to -100 kN/m^2). In this case no shear strains develop on planes and strains due to the consolidation can be ignored.. Figure 1 depicts deviatoric stress q (= σ_y - σ_x) vs axial strain ε_1.

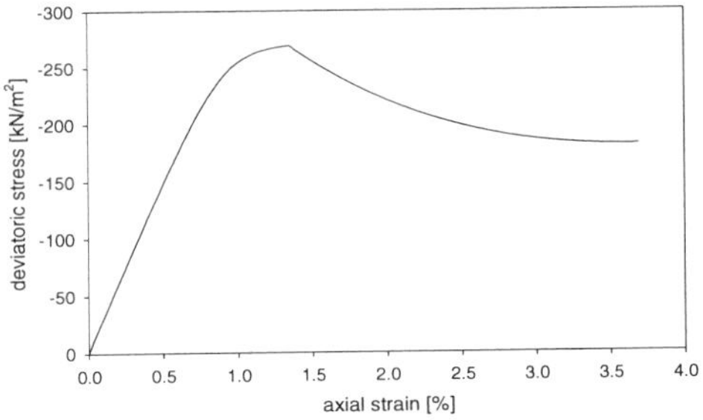

Figure 1. Deviatoric stress vs axial strain - isotropic consolidation

In Figures 2 to 4 plastic shear strains on integration planes are shown for stress levels q = -75, -150 and -250 kN/m^2 respectively (note that the scale is different). These Figures provide a clear physical interpretation of the proposed concept, the concentration of plastic strains on a small number of integration planes with increasing deviatoric stress is evident. The criterion for defining the point of localization is discussed in detail by Karstunen & Pande (1997). Figure 5 shows the plastic shear strains at the onset of localization and the concentration of shear strains into a narrow band is apparently described very well with this formulation. The obtained bifurcation angle of 62^0 is also in the range what would be expected for the assumed material parameters.

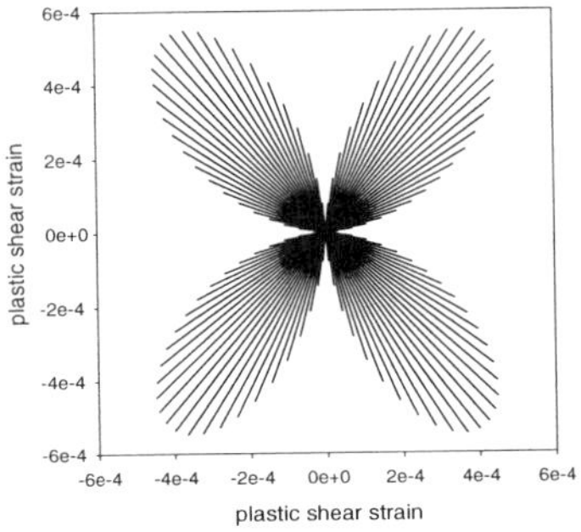

Figure 2. Plastic shear strains on integration planes - isotropic consolidation, q = -75 kN/m^2

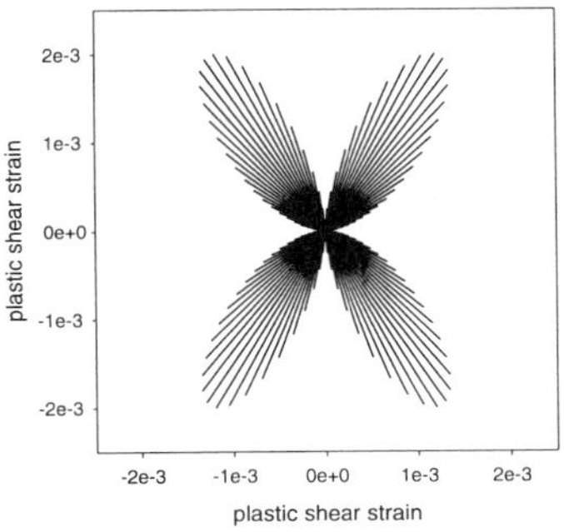

Figure 3. Plastic shear strains on integration planes - isotropic consolidation, q = -150 kN/m^2

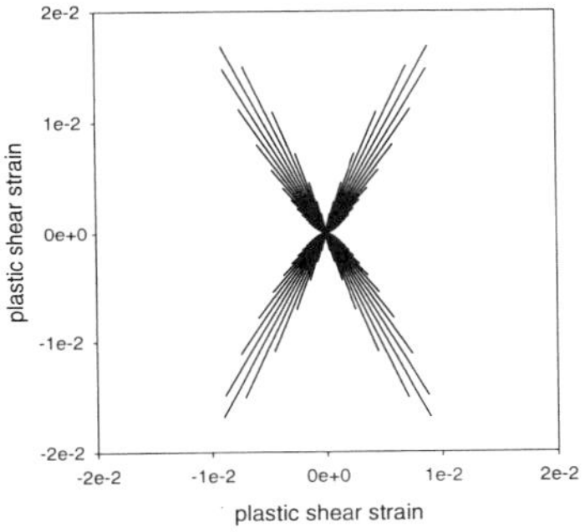

Figure 4. Plastic shear strains on integration planes - isotropic consolidation, q = -250 kN/m^2

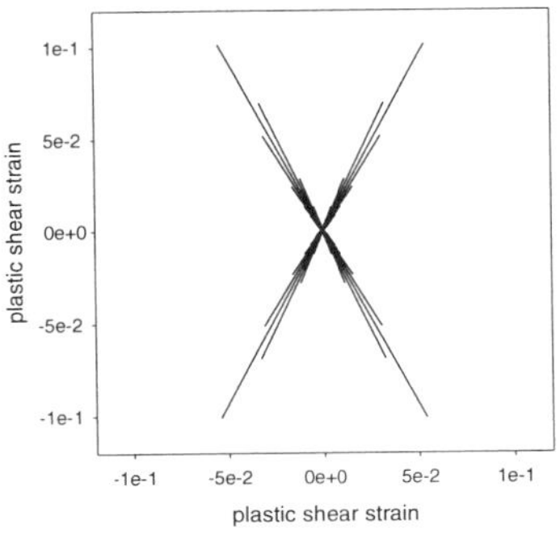

Figure 5. Plastic shear strains on integration planes at onset of localization - isotropic consolidation

3.2 *Influence of confining pressure and anisotropic consolidation*

In this section the influence of consolidating the

sample anisotropically before the strain controlled biaxial test is investigated. Starting from a stress state close to zero, anisotropic consolidation was simulated by a stepwise increase of the vertical and horizontal stress σ_y and σ_x, keeping the ratio σ_x/σ_y constant. The cases with K (σ_x/σ_y) = 0.35, 0.75 and 1.25 respectively have been simulated with the final value of σ_y being fixed as -100 kN/m^2. The latter case is particularily interesting because it simulates $K_0 > 1$ conditions and a change of orientation of the minimum principal stress which may be of paramount importance for some cases in practical engineering. A K_0-consolidation has also been simulated. Figure 6 shows deviatoric stress vs axial strain for all cases analysed.

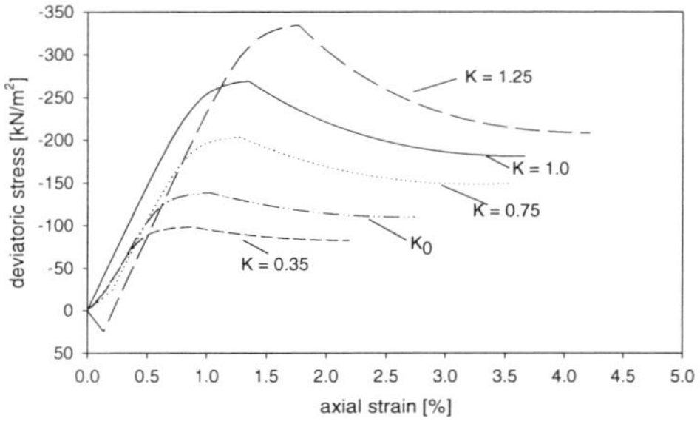

Figure 6. Deviatoric stress vs axial strain for different consolidation stresses

At first sight Figure 6 seems to represent only the influence of the confining pressure on the peak strength of the sample. However this is only one aspect which becomes apparent when the plastic strains mobilized on the different integration planes are compared. This is done in Figures 7 to 11 for a deviatoric stress level of q = -75 kN/m^2. The scale is the same on all plots to allow an easy comparison and shear strains for the case with K = 0.35 are visibly higher. Most of these strains are arising from the consolidation phase and ignoring them would clearly lead to a different result. It is also evident that mobilized plastic strains for the case with K = 1.25 are very small at this stage. The latter case demonstrates that the multilaminate framework easily captures the change of the orientation of the minimum principal stress (σ_x for consolidation, σ_y for strain controlled biaxial test) which is reflected by the shear strains accumulated on certain planes. Similarly, the model is capable of capturing the plastic flow induced by the rotation of the principal stress axes. This feature is especially important when solving practical problems involving excavation sequences with subsequent support installations where an abrupt change in the direction of principal stresses may be encountered.

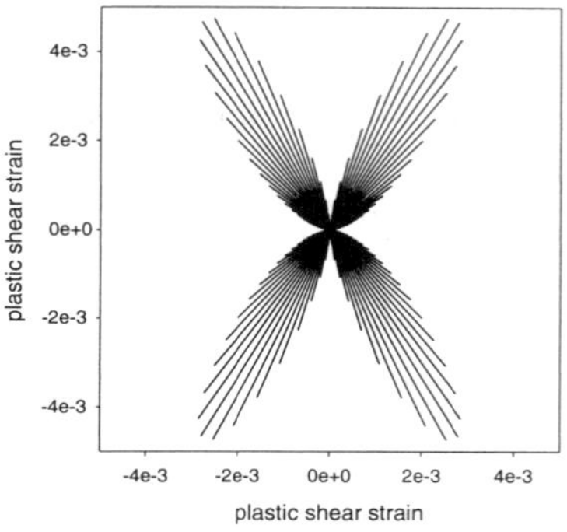

Figure 7. Plastic shear strains on integration planes for K = 0.35 at stress level q = -75 kN/m^2

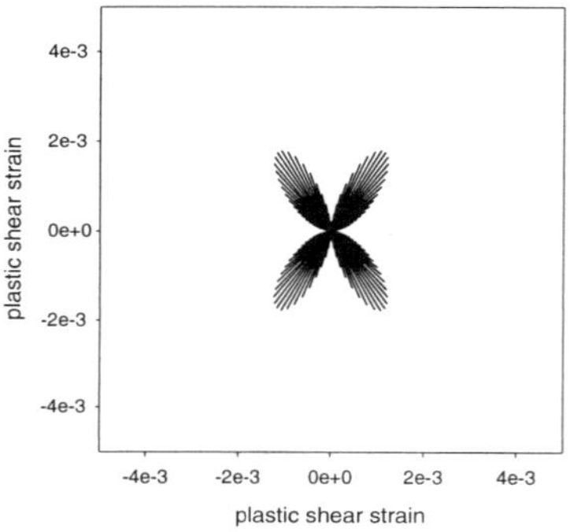

Figure 8. Plastic shear strains on integration planes for K_0 - consolidation at stress level q = -75 kN/m^2

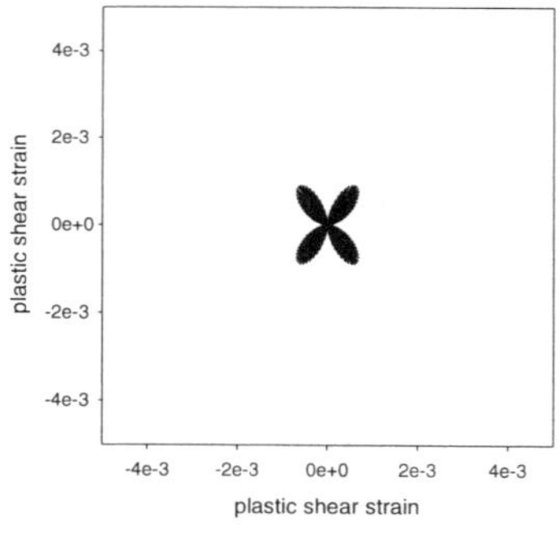

Figure 9. Plastic shear strains on integration planes for K = 0.75 at stress level q = -75 kN/m^2

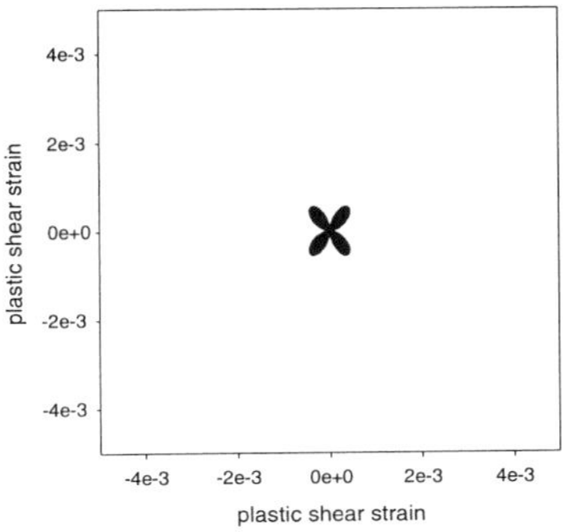

Figure 10. Plastic shear strains on integration planes for K = 1.0 (reference case) at stress level q = -75 kN/m^2

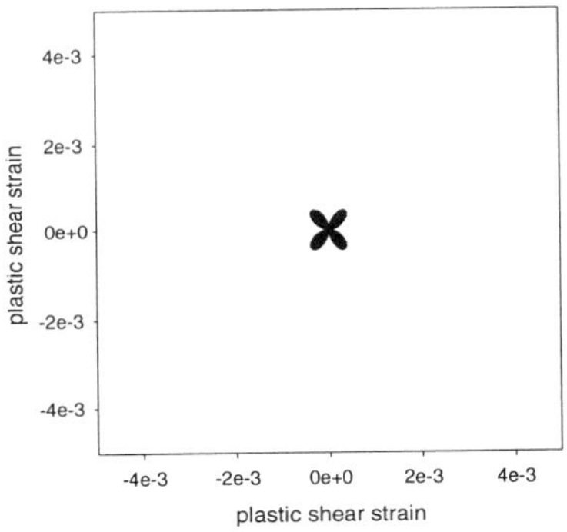

Figure 11. Plastic shear strains on integration planes for K = 1.25 at stress level q = -75 kN/m^2

3.3 *Influence of shear strains on integration planes during consolidation*

It has been mentioned earlier that the results shown in Figure 6 do not necessarily indicate, that the onset of localization depends on whether or not plastic shear strains on integration planes during anisotropic consolidation have been taken into account or not. In order to emphasize this aspect in more detail an additional comparison is discussed in the following.

Figures 12 and 13 show deviatoric stress and shear stress in the band vs axial strain assuming an anisotropic "initial" stress (ignoring strains due to consolidation) and taking into account strains developed on planes during consolidation respectively. The confining pressure σ_x was taken as -50 kN/m^2 in both cases, i.e. K = 0.5. The importance of including the history of the consolidation process is clearly indicated because the peak stress is much lower in this case. The maximimum shear strain at the onset of localization differs by about 15%, with the higher value for the case when consolidation is ignored.

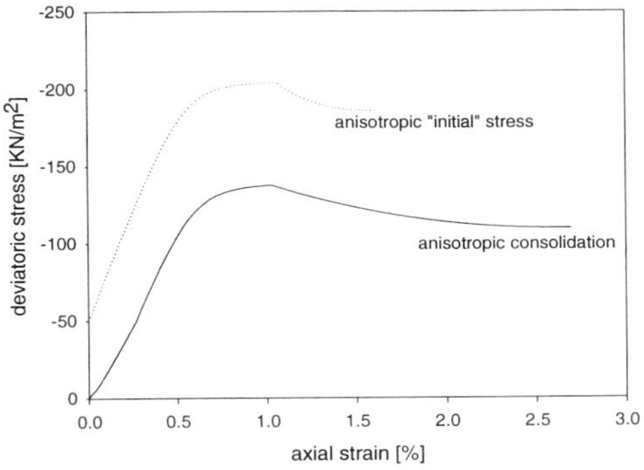

Figure 12. Deviatoric stress vs axial strain for K = 0.5

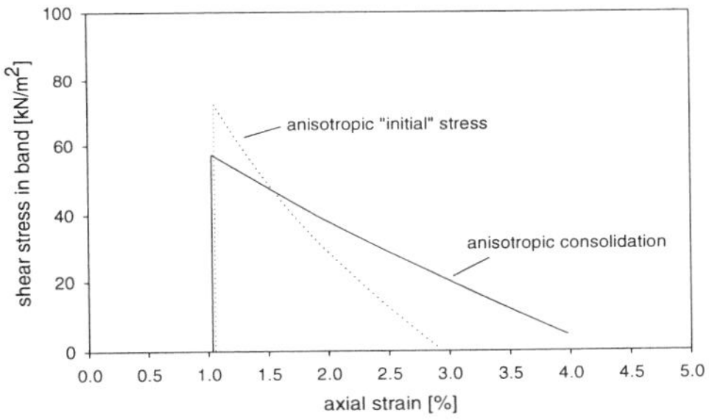

Figure 13. Shear stress in band vs axial strain for K = 0.5

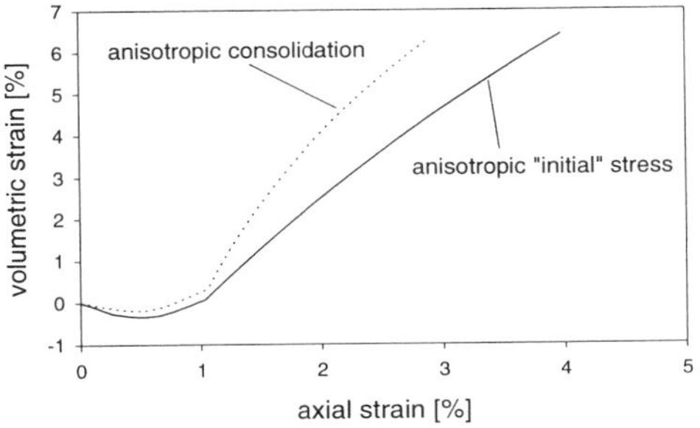

Figure 14. Volumetric strain vs axial strain for K = 0.5

The bifurcation angle obtained also changes from 62^0 (consolidation considered) to 67^0 (consolidation ignored). In the multilaminate model the bifurcation angle, naturally, has to correspond with an integration plane and thus the accuracy which can be achieved depends on the integration rule. For the present study a 64-plane integration has been adopted i.e. the accuracy is 1.4^0. This is an important result with far reaching consequences and further investigation are currently under progress to study the effect of anisotropic consolidation on the bifurcation angle. After localization the shear stress in the band decreases much faster when consolidation has not been considered (the jump in the shear stress in Figure 13 at the onset of localization at approx. 1 % axial strain is due to the change from the global to a local coordinate system). Figure 14 shows the corresponding volumetric strains.

Figure 15 shows a similar result. In this case the mean consolidation stress $(\sigma_x + \sigma_y)/2$ was fixed, as in the reference case, at -100 kN/m^2 but instead of consolidating the sample to $\sigma_x = \sigma_y = -100$ kN/m^2, a stress state of $\sigma_x = -75$ kN/m^2 and $\sigma_y = -125$ kN/m^2 was assumed (as mentioned before the out of plane stress σ_z has no influence in the present formulation). The differences in the results are crucial, in the case of an anisotropic starting stress the peak depends only on the total value of the deviatoric stress, as expected, but in the latter case plastic shear strains developed in the consolidation phase lead to lower peak stress. Again a different bifurcation angle (65^0) is obtained for the case where consolidation has not been taken into account thus the reference solution and the solution for the anisotropic "initial" stress show a similar peak stress but lead to different bifurcation angles.

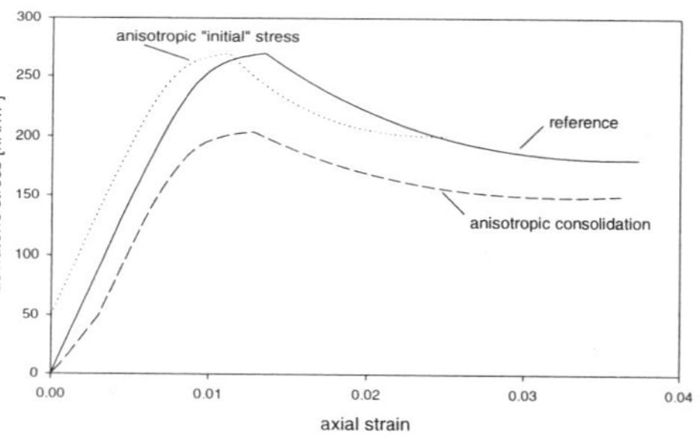

Figure 15. Deviatoric stress vs axial strain for constant mean consolidation stress $(\sigma_x + \sigma_y)/2= -100$ kN/m^2

CONCLUSION

A numerical study investigating the influence of anisotropic consolidation for subsequent analysis of strain localization in a biaxial test has been presented.

It is shown that the multilaminate framework employed to capture the onset of localization together with a homogenisation technique is well suited to describe the influence of anisotropic consolidation.

By monitoring the mobilized plastic shear strains on integration planes a physically appealing way of demonstrating the effects of anisotropic consolidation has been presented. Deviatoric peak stress is significantly reduced if the anisotropic consolidation process is taken into account. The effect of change of orientation of the minimum principal stress after consolidation has also been considered. The results presented are of significance from a practical point of view, because in practical problems a change of direction of principal stresses is often encountered. The proposed framework offers the tools for handling these problems in an efficient way.

REFERENCES

Bazant, Z. P. & G. Pijaudier-Cabot (1988). Non-local continuum damage, localization instability and convergence. *Journal of Applied Mechanics* 55: 287-293.

Belytschko, T., J. Fish & B.E. Engelmann (1988). A finite element with embedded localization zones. *Comp. Meth. Appl. Mech. Eng.* 70: 59-89.

de Borst, R. (1991). Simulation of strain localization: A reappraisal of the Cosserat continuum. *Eng. Comp.* 83: 17-332.

Hicks, M.A. (1995). Computation of localisation in undrained soil using adaptive mesh refinement. In: G.N.Pande & S.Pietruszczak (eds.), *Proc. 5th Int. Symp. Num. Models Geomech.*: 203-208. Rotterdam: Balkema.

Karstunen, M. & G.N. Pande (1997). Criteria for strain localization based on multilaminate framework.In: G.N.Pande & S.Pietruszczak (eds.), *Proc. Int. Symp. Num. Models in Geomech.*, Montreal (to be published).

Loret, B. & J.H. Prevost (1990). Dynamic strain-localization in elasto-(visco-)plastic solids -Part 1: General formulation and one-dimensional example. *Comp. Meth. Appl. Mech. Eng.* 83: 247-273.

Pande, G.N. & K.G. Sharma (1983). Multi-laminate Model of clays - a numerical evaluation of the influence of rotation of the principal stress axes. *Int. J. Num. Analyt. Meth. Geomech.* 7: 397-418.

Pietruszczak, S. & Z. Mróz (1981). Finite element analysis of deformation of strain-softening materials. *Int. J. Num. Meth. Eng.* 17: 327-334.

Pietruszczak, St. & X. Niu (1993). On the description of localized deformation. *Int. J. Num. Anal. Meth. Geomech.* 17, 791-805.

Schweiger, H.F., M. Karstunen & G.N. Pande (1997). Modelling strain lokalization in soils using multilaminate model and homogenisation technique. *Proc. Int. Conf. Symp. Deformation and Progressive Failure in Geomechanics*, Nagoya (submitted for publication).

Vardoulakis, I. & E.C. Aifantis (1991). A gradient flow theory of plasticity for granular materials. *Acta Mechanica* 87: 197-217.

Numerical Models in Geomechanics, Pietruszczak & Pande (eds) © 1997 Balkema, Rotterdam, ISBN 90 5410 886 X

Daphnis: A model, consistent with CLoE, for the description of post-localization behaviour

R.Chambon & S.Crochepeyre
Laboratoire Sols Solides Structures, UJF INPG CNRS, Grenoble, France

ABSTRACT: This paper addresses some recent developments in a constitutive model able to describe the post-localized behaviour of structures composed of geomaterials. The behaviour of the shear band is defined by means of a specific nonlinear constitutive relation in the framework of large strains. Particular attention has been given to the 'consistency' of this model with the CLoE model. In the post-localization regime the behaviour of the material is modelled by introducing the concept of critical void ratio. After a short description of CLoE model and the Rice criterion, basic concepts of Daphins model are introduced. The last part of the paper is devoted to the numerical analysis of biaxial tests on initially homogenous samples.

1 INTRODUCTION

Though the analysis of the localization of deformation has received considerable attention is the last two decades, the modelling of post-localization behaviour of the structures is still subject to keen debate. At the onset of localization, high strain gradients are involved so that the classical Continuum Mechanics is inefficient to describe this phenomenon, especially because of the lack of an internal length scale. In order to model the post-localization, many approaches have already been considered :

1. on one hand, the Continuum Mechanics approach is still considered but additional terms are introduced, leading to special constitutive equations. Different types of enhanced models have been developed so far.

2. on the another hand, the behaviour of the shear band is modelled by means of a specific constitutive equation expressing the evolution of the traction vector (or its time derivative) as a function of the jump of displacement between the boundaries of the band. Such an approach has been considered by Larsson et al. (1993). A kind of interfacial equation is then obtained using a kinematical assumption involving discontinuous displacements. For the sake of effectiveness in numerical computations, a pertinent relationship can be defined by simply deriving the constitutive equation describing the behaviour of the continuously deforming material. In that way, the response envelop (Gudehus 1979) in terms of time derivative of the traction vector of the global volumetric model and of the specific model for the band are the same at the onset of localization. In the framework of this paper, this concept will be denoted as a 'consistency condition'. In the post-localization regime the behaviour of the shear band is defined by this constitutive equation and the remaining part of the structure is described by the classical stress-strain equation.

Such an approach is considered in the present paper but significant differences with previous works about that topic are introduced :

1. the model is defined in the large strain framework.

2. Up to now, few experimental results about the intrinsic behaviour of the shear band are available. Nevertheless, experimental results (Desrues 1996) have highlighted a specific evolution of the material inside the shear band. Indeed it is the locus of high dilation and it evolves to a critical void ratio different from the one observed in the remaining part of the structure. On the basis of these observations, it was the author's choice to introduce the void ratio only inside the band as a specific state variable in order to govern the evolution of the interfacial constitutive equation in the post localization regime.

Thus, after some brief recalls about CLoE models, basic concepts of the interface model called Daphnis will be introduced. Basic features of the implementation of this model will then be discussed and the last part of this paper will be devoted to the numerical simulation of a frictionless biaxial test on an initially homogeneous sample.

2 THE CLoE MODEL FRAMEWORK

2.1 *Introduction*

The CLoE model framework defines a class of non linear rate type constitutive laws developed by Chambon et al. (1994) with the objective to be reliable for two classes of problems :

1. the effectiveness of the equations in finite element computations. Thus CLoE models are based on a set of basic mathematical requirements such as the inversibility, the concept of flow rule, the consistency at the limit surface (Chambon et al. 1994). These concepts represent the foundations of any CLoE model and they ensure robustness in numerical computations.

2. the strain localization. Despite its non linearity, CLoE generic equation is simple enough so that a tractable shear band analysis can be performed following the developments introduced by Rice (1976). From this analysis, a bifurcation criterion is obtained and some parameter of the model can be back identified as shown in section 2.3.

2.2 *Basic concepts*

This class of generic constitutive equations is defined by the non linear generic form :

$$\tilde{\sigma}_{ij} = A_{ijkl}\left(\dot{\varepsilon}_{kl} + b'_{kl}\|\dot{\varepsilon}\|\right) \tag{1}$$

where $\tilde{\sigma}_{ij}$ = Jaummann derivative of the Cauchy stress rate; $\dot{\varepsilon}_{kl}$ = strain rate defined as the symmetric part of the velocity gradient; A_{ijkl} and b'_{kl} = fourth order tensor and second order tensor (respectively) depending on the state which is reduced here to the Cauchy stress tensor.

CLoE models have been developed in order to clearly control inversibility. As one postulates the existence of a limit surface coinciding with the failure surface (defined in the stress space as the locus where the material can flow under steady stress state), inversibility is ensured for any stress state located inside the limit surface. Mathematically, insersibility is ensured while :

$$\|\underline{b}'\| < 1 \tag{2}$$

Consequently, the law is no more a one to one correspondence as soon as :

$$\|\underline{b}'\| = 1 \tag{3}$$

The limit surface is then defined in the CLoE framework as the locus in the stress space such as equation (3) is met. It should be noticed that for such stress states, $-\underline{b}'$ can be regarded as a flow direction.

It should be noted that $\|\underline{b}'\| > 1$ involves softening in the constitutive equations. In that case, the solution of an elementary problem can be no more unique.

Finally, when the stress state lies on the limit surface, no outer stress rate has to be generated by the constitutive equation. This requirement is the consistency at the limit surface which ensures the effectiveness of the constitutive equations for ultimate stress states. Mathematically, the consistency condition reads :

$$A_{klmn}\frac{\partial f(\sigma)}{\partial \sigma_{kl}} = -\alpha b'_{mn} \tag{4}$$

where $f(\underline{\sigma}) = 0$ is the limit surface equation, α is a positive number.

2.3 *Shear band analysis involving CLoE models*

CLoE model have been developed in order to be reliable for problems involving strain localization. In the present section, the main results about a shear band analysis involving CLoE models and following an extension of the developments introduced by Rice (1976) are recalled. More details are available in Chambon et al. (1994, 1997).

A shear band analysis involving CLoE models leads to the following non linear criterion

$$\left\|\frac{1}{2}\left(\Lambda^{-1}_{ki}b_{ij}n_j n_l + \Lambda^{-1}_{li}b_{ij}n_j n_k\right)\right\| = 1 \tag{5}$$

where $b_{ij} = A_{ijkl}b'_{kl}$;
n = normal unit vector to the shear band ;
$\Lambda_{ik} = D_{ijkl}n_j n_l$;

$$D_{ijkl} = \frac{1}{2}\left(A_{ijkl} + A_{ijlk} - \sigma_{ik}\delta_{jl} + \sigma_{il}\delta_{jk} + \delta_{ik}\sigma_{jl} - \delta_{il}\ldots\right.$$

Meeting equation (5) implies that the ratio between the strain rate norm inside and outside the band becomes infinite as localization occurs. In other words, at the onset of localization, the whole strain rate of the specimen is concentrated is a part which can be seen as the inside of the band. The outer part of the band behaves then as a rigid body. Consequently, the static condition involved in the shear band analysis becomes :

$$\dot{T}^b_i = 0 \tag{6}$$

where $\dot{T}^b_i = \dot{\sigma}^b_{ij}n_j$.

3.THE PLANE MOHR-COULOMB CLoE MODEL

As this analysis is aimed to validate an interface model which will be presented in the next section

The behaviour of sand will be described in the numerical simulations by means of a very simple model. This model defines in the CLoE framework a plane Mohr-Coulomb behaviour.

Assuming no cohesion, the classical Mohr-Coulomb limit surface equation reads :

$$\frac{\sigma_{22}-\sigma_{11}}{2}+\frac{\sigma_{22}+\sigma_{11}}{2}\sin\phi=0 \qquad (7)$$

where σ_{11} = minor principal stress ; σ_{22} = major principal stress ; ϕ = friction angle.

The equation fulfilled by the strain rates when the material flows under constant stress is :

$$\dot{\varepsilon}_{22}+\dot{\varepsilon}_{11}+(\dot{\varepsilon}_{22}-\dot{\varepsilon}_{11})\sin\psi=0 \qquad (8)$$

where ψ = dilatancy angle.

Considering equations (3) and (8), the expression of tensor b' is defined for limit stress states by :

$$b'^{\lim}=\left\{\frac{1}{\sqrt{1+\left(\frac{1+\sin\psi}{1-\sin\psi}\right)^2}} \quad -\frac{1}{\sqrt{1+\left(\frac{1-\sin\psi}{1+\sin\psi}\right)^2}} \quad 0\right\}^T \qquad (9)$$

Assuming a reversible behaviour for isotropic stress states, b' is chosen equal to zero for such stress states. For any stress state, a linear function of the mobilized friction angle ϕ_{mob} sine is used to describe the evolution of b' between its isotropic and limit value such as :

$$b'_{kl}=b'^{\lim}_{kl}\frac{\sin\phi_{mob}}{\sin\phi} \qquad (10)$$

For any stress state, tensor A is defined as :

$$\underline{\underline{A}}=\begin{bmatrix} c & s & 0\\ -s & c & 0\\ 0 & 0 & 1\end{bmatrix}\begin{bmatrix}\frac{E(1-\nu)}{(1+\nu)(1-2\nu)} & \frac{E\nu}{(1+\nu)(1-2\nu)} & 0\\ \frac{E\nu}{(1+\nu)(1-2\nu)} & \frac{E(1-\nu)}{(1+\nu)(1-2\nu)} & 0\\ 0 & 0 & 2g\end{bmatrix} \qquad (11)$$

where $c=\cos\theta$; $s=\sin\theta$.

θ evolves from 0 (for isotropic stress states) to $\theta^{\lim}$ (for limit stress states). $\theta^{\lim}$ is defined such as consistency condition (4) is met and reads :

$$\theta^{\lim}=\frac{\sin\phi-\sin\psi(1-2\nu)}{1+(1-2\nu)\sin\phi\sin\psi} \qquad (12)$$

where ν = Poisson ratio.
For any stress state, the evolution of θ reads :

$$\theta=\theta^{\lim}\frac{\sin\phi_{mob}}{\sin\phi} \qquad (13)$$

Moreover, slight changes have been introduced on the out of axis shear modulus in order to fit the onset of localization to experimental data. Thus for any stress state, the out of axis shear modulus g is defined as :

$$g=G(1-\omega\frac{\sin\phi_{mob}}{\sin\phi}) \qquad (14)$$

G = shear modulus ; ω = parameter to be fitted owing to experiments exhibiting strain localization.

Tensor A is yet completely defined. It is analogue to the fourth order isotropic elastic tensor for isotropic states and meets the consistency condition (4) for limit stress states.

4 DAPHNIS : BASIC CONCEPTS

4.1 *Introduction*

Numerical simulations have shown that CLoE models are reliable to predict the onset of localization. As mentioned in the introduction it was our choice to develop a specific model for the shear band behaviour in order to perform some analysis in the post-localization regime. This model has been built up following two essential requirements :

1. At the onset of localization, it has to be 'consistent' with its associate CLoE model. In other words, the response envelops of both CLoE and Daphnis in terms of time derivative of the traction vector are the same. In that sense, at the onset of localization, Daphnis can be regarded as the exact 'cutting off' of CLoE.

2. In the post-localization regime, this model evolves in a specific manner. On the basis of experimental observations, the shear band is the locus of high dilation till a limit value of the void ratio is reached (which is the so-called critical void ratio). Thus, the void ratio inside the band has been chosen as a state variable for this model

4.2 *The generic mathematical form*

The generic form for Daphnis is directly derived from the CLoE one. It expresses, in a non linear fashion the evolution of a traction rate $\hat{\dot{T}}$ ($\hat{\dot{T}}$ is a time derivative of the traction vector with respect to a frame bounded to the boundaries of the band) with respect to the relative velocity $[\dot{u}]$ as shown in figure 1.

The generic form of Daphnis at the onset of localization is derived from the one of CLoE model in the specific case of a strain rate defined by $\vec{g}\otimes\vec{n}$. It should be noted that all the developments are performed in the framework of large strain. Let us define the relative velocity $[\dot{u}]$ such as :

$$\frac{[\dot{u}_k]}{d} = \frac{1}{2}\left(g_k + g_l n_k n_l\right) \qquad (15)$$

where d = shear band thickness.

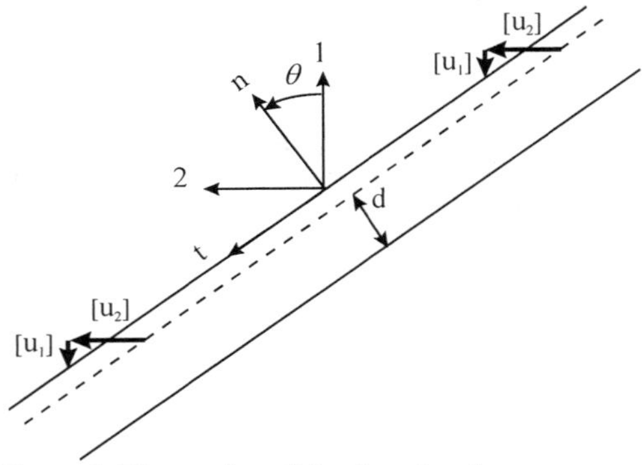

Figure 1. Kinematics of the shear band

After a long but quite straightforward calculation (details are given in Chambon at al. 1997), the form of Daphnis at the onset of localization reads :

$$\hat{T}_i = \frac{1}{d} A_{ik}\left([\dot{u}_k] + B'_k \sqrt{2[\dot{u}_k][\dot{u}_k] - [\dot{u}_k][\dot{u}_l] n_k n_l}\right) \qquad (16)$$

where A_{ik} = second rank tensor ; B'_k = vector.

This form will be used in the latter as the Daphnis generic form. Both A_{ik} and B'_k depend on the tensors involved in the CLoE generic form, on the stress state at the onset of localization and on the normal to the shear band. In that sense large deformation terms are introduced in tensor A_{ik} and vector B'_k since the time derivatives involved in CLoE and Daphnis models are performed with respect to different frames.

It should be noted that the norm involved in the non linear term is not a Euclidean one. In order to perform with Daphnis the same kinds of reasoning as for CLoE models about concepts of inversibility or softening, let us split $[\dot{u}]$ up into its normal part $[\dot{u}^n]$ and its tangential part $[\dot{u}^t]$ and introduce a vector $[\dot{U}]$ so that :

$$[\dot{U}_k] = [\dot{u}_k^n] + \sqrt{2}[\dot{u}_k^t] \qquad (17)$$

The generic form of the Daphnis model then becomes :

$$\hat{T}_i = \frac{1}{d} a_{ik}\left([\dot{U}_k] + {}^b b'_k \left\|[\dot{U}]\right\|\right) \qquad (18)$$

where $a_{ik}[\dot{U}_k] = A_{ik}[\dot{u}_k]$ and $A_{ik}B'_k = a_{ik}\,{}^b b'_k$.

By means of equation (18) and on the basis of basic CLoE concepts introduced in section 2.2, the behaviour of Daphnis model is completely dependent on the value of $\|{}^b\vec{b}'\|$. Thus :

1. when $\|{}^b\vec{b}'\| < 1$, inversibility of Daphnis is ensured.
2. when $\|{}^b\vec{b}'\| = 1$, the traction rate vector vanishes for a particular direction of $[\dot{U}]$. In that case and in the same way as for CLoE models, $-{}^b b'$ can be regarded as a kind of flow direction since, for a vanishing traction rate vector :

$$ {}^b b'_k = -\frac{[\dot{U}_k]}{\left\|[\dot{U}]\right\|} \qquad (19)$$

3. when $\|{}^b\vec{b}'\| > 1$, the Daphnis model exhibit softening.

4.3 *Condition on Daphnis at the onset of localization.*

Daphnis is built up on the basis of two main requirements as mentioned in section 4.1. At the onset of localization 'consistency' between Daphnis and its associate CLoE model requires that the traction rate vector vanishes for a non vanishing relative velocity. Thus :

$$\left\|{}^b\vec{b}'^{loc}\right\| = 1 \qquad (20)$$

Let us then denote by e_b^{loc} the void ratio inside the band at the onset of localization.

4.4 *Daphnis at the end of the post-localization regime.*

In the post-localization regime, Daphnis evolves in a specific way. In order to assess dilation inside the band, the void ratio is introduced as a state variable. As the void ratio reaches its critical value, no more volumetric deformation occurs inside the band and the traction vector remains constant. The sample keeps then on deforming in pure shear. The normal part of the relative velocity vanishes. Thus, at the end of post-localization regime.

$$\left\|{}^b\vec{b}'^{ecr}\right\| = 1 \qquad (21)$$

On the basis of the flow rule considerations (equation 19), the corresponding components of ${}^b\vec{b}'$ expressed in a frame bounded to the shear band read :

$$\begin{cases} {}^b b_n'^{ecr} = 0 \\ {}^b b_t'^{ecr} = -\dfrac{[\dot{U}_t]}{\left\|[\dot{U}_t]\right\|} \end{cases} \qquad (22)$$

Let us also denote by e_b^{ecr} the void ratio inside the band at the onset of localization.

4.5 *Evolution of the Daphnis model*

For intermediate states, softening is introduced in Daphnis. Thus, for a void ratio contained between e_b^{loc} and e_b^{ecr},

$$\left\| {}^{b}\vec{b}' \right\| > 1 \qquad (23)$$

Basic principles of Daphnis are yet completely defined. The model depends only on a single state variable and on an internal parameter.

4.6 *Daphnis model in a plane case*

All basic concepts introduced in previous subsections are available in a general case. In order to validate Daphnis and for the sake of simplicity, the analysis will be reduced to a plane case in the following. Thus the explicit form of Daphnis expressed in a frame bounded to the band reads :

$$\begin{Bmatrix} \hat{T}_n \\ \hat{T}_t \end{Bmatrix} = \begin{bmatrix} a_{nn} & a_{nt} \\ a_{tn} & a_{tt} \end{bmatrix} \left(\begin{Bmatrix} [\dot{U}_n] \\ [\dot{U}_t] \end{Bmatrix} + \begin{Bmatrix} {}^{b}b'_n \\ {}^{b}b'_t \end{Bmatrix} \left\| [\dot{\vec{U}}] \right\| \right) \qquad (25)$$

As soon as the underlying CLoE model is specified, the components of a and of ${}^{b}\vec{b}'$ can be computed at the onset of localization. As far as the evolution of Daphnis is concerned and for the sake of simplicity, a remains constant and ${}^{b}\vec{b}'$ evolves with respect to the void ratio such as :

$$\begin{cases} {}^{b}b'_n = {}^{b}b'^{loc}_n \dfrac{e^{ecr} - e}{e^{ecr} - e^{loc}} \\ {}^{b}b'_t = \left({}^{b}b'^{loc}_t - {}^{b}b'^{ecr}_t \right) \left(\dfrac{e^{ecr} - e}{e^{ecr} - e^{loc}} \right)^4 + {}^{b}b'^{ecr}_t \end{cases} \qquad (26)$$

5 APPLICATION TO PLANE STRAIN PROBLEMS

In order to check Daphnis behaviour, a frictionless biaxial test on an initially homogeneous, initially isotropic sample is computed. The sample has the shape of a parallelepiped. The sample is assumed to remain homogeneous till the shear band criterion introduced in section 2.3 is met. As soon as localization occurs, the structure is divided into three parts. Homogeneity is assumed in each part. The initial direction of the shear band is defined by the shear band analysis. The behaviour of the shear band is computed by means of Daphnis whereas the behaviour of the two remaining parts is computed by means of the CLoE Mohr-Coulomb model as shown in figure 2.

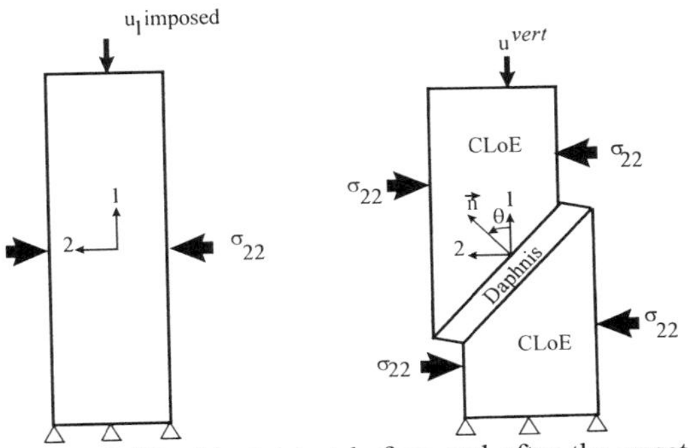

Figure 2. The biaxial test before and after the onset of localization

The sample is 338 mm high and 102.5 mm wide. The confining pressure is taken equal to 200kPa.

The material parameters used to describe in a very simple fashion the behaviour of a dense Hostun sand for CLoE Mohr-Coulomb are :
E=30000kPa $\nu = 0.32$ $\phi = 45°$ $\psi = 25°$ $\omega = 0.6$
ω has been chosen so that the angle of the normal to the band with respect to direction 1 obtained from a shear band analysis and that the onset of localization are fitted to experimental data. The orientation of the angle of the shear band is 71°. As far as Daphnis is concerned, $e_b^{loc} = 0.67$, $e_b^{ecr} = 0.82$ and the shear band thickness d is taken equal to 7 mm. In that way, these simulations aim to fit the experiment shf41 performed by Hammad (1991). Simulations are performed in a large strain framework.
Three global curves are presented.

1. the evolution of the stress ratio (figure 3)
2. the global volumetric deformation (figure 4)
3. a zoom of the stress ratio curve for several loading step sizes (figure 5)

6 RESULTS ANALYSIS

The results obtained seem quite satisfying since they fit the experimental results in a quantitative fashion. Moreover, at the onset of localization, theoretical result expressed by equation (6) is retrieved by numerical simulations involving small steps as shown by figure 5. However, at the beginning of post-localization regime, the slope of the volumetric curve much higher than the experimental one. In fact, the CLoE Mohr-Coulomb model exhibits too much dilation, as shown by figure 4. It is the author's thought that such a drawback will be circumvented as soon as a more accurate CLoE model will be used.

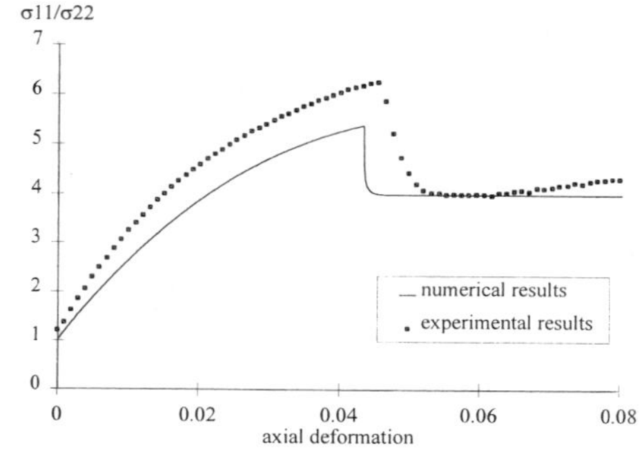

Figure 3. Stress ratio curve. Comparison with experimental data

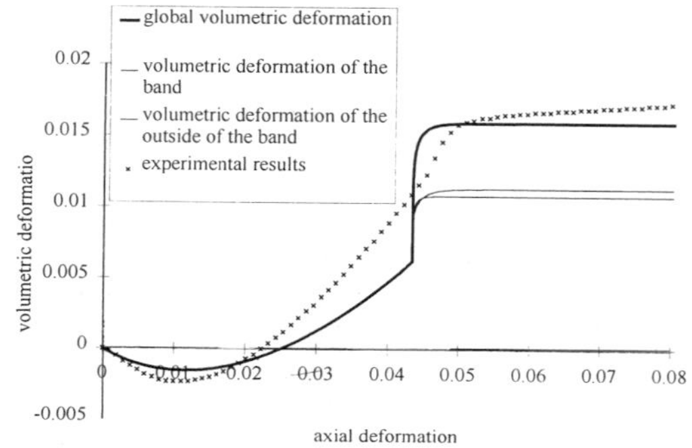

Figure 4. Volumetric curve. Comparison with experimental data

7 CONCLUSION

The basic concepts of a model able to describe the post-localized behaviour of structures composed by geomaterials have been presented in that paper. The first results obtained seem quite promising. It should be recalled and emphasized that Daphnis is the exact 'cutting-off' of CLoE at the onset of localization and that it evolves in a specific way in post localization regime. During post-localization regime, Daphnis and CLoE are only connected by means of the equilibrium equation at the boundaries of the band. It should also be recalled that softening is only introduced in the shear band zone whereas the outside of the band never exhibits softening. Finite element implementation of Daphnis has been performed and the first results seem promising since results presented in that paper are retrieved.

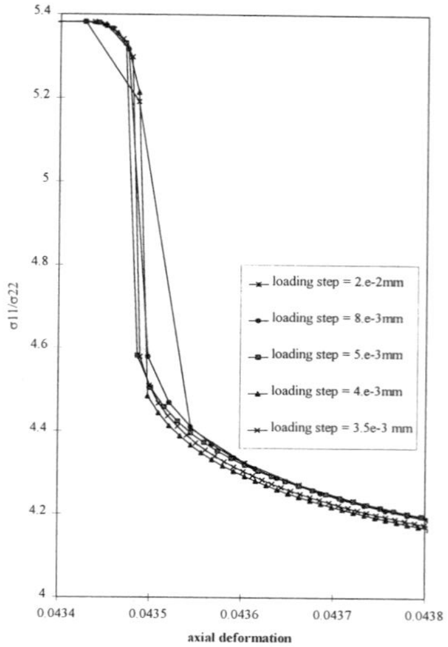

Figure 5 : Influence of the loading step size at the beginning of the post-localization regime

REFERENCES

Chambon R., Desrues J., Charlier R. & Hammad W. 1994. CLoE, a new rate type constitutive model for geomaterials : theoretical basis and implementation. *Int. J. Num. Anal. Meth. Geom.*, Vol 18-4, 253-278.

Chambon R. Desrues J. & Tillard D. 1994. Shear moduli identification versus experimental data. *Localization and bifurcation theory for soils and rocks*. R. Chambon, J . Desrues and I. Vardoulakis ed. A.A. Balkema, Rotterdam.

Chambon R. & Crochepeyre S., 1997. Daphnis : a new model for the description of post-localization behaviour : application to sands. *Mech. Of Coh. Frict. Mate.*, Accepted.

Desrues J., Chambon R., Mokni M. & Mazerolle F., 1996. Void ratio evolution inside shear bands in triaxial sand specimens studied by computed tomography. *Geotechnique*, Vol 46-3, 529-546.

Gudehus G.. 1979. A comparison of some constitutive laws under radially symmetric loading and unloading, *Num. Meth. in Geomech*, Vol 4, 1309-1323, W Wittke ed. A.A. Balkema, Rotterdam.

Hammad W. 1991. *Modelisation non lineaire et etude experimentale des bandes de cisaillement dans les sables*, Ph.D. Thesis.

Larsson R., Runesson K. & Ottosen N.S., 1993. Discontinous displacement approximation for capturing plastic localization, *Int. Jour. for Num. Meth in Engng.*, Vol 36, 2087-2105.

Rice J., 1976. The localization of plastic deformation, *Int. Conf. of Theor. and Appl. Mech.*, W.D. Koiter ed., North Holland Publ. Comp.

Numerical Models in Geomechanics, Pietruszczak & Pande (eds) © 1997 Balkema, Rotterdam, ISBN 90 5410 886 X

Non-uniqueness of the incremental response of soil specimens under true-triaxial stress-paths

R. Nova & S. Imposimato
Milan University of Technology (Politecnico), ALERT Geomaterials, Italy

ABSTRACT: Piecewise linear constitutive relationships characterised by non-symmetric stiffness matrices. Such as elastoplastic laws with non-associated flow rule, allow the existence of bifurcated homogeneous solutions under particular loading programmes. Such bifurcations are possible when the stiffness matrix is no more positive definite, what occurs for a strain-hardening material in the hardening regime, well before the limit state (conventionally assumed as failure state) is reached.
The paper is concerned with the occurrence of homogeneous bifurcations which may arise in true-triaxial tests when appropriate stress and/or strain perturbations are imposed on homogeneously loaded specimens. Reference will be made to samples of loose sand described by a strain-hardening model with mixed isotropic/kinematic hardening. It is shown that, for stress states in the region between the loci corresponding to loss of the positive definiteness of the stiffness matrix and the limit state, many types of different bifurcations may take place under appropriate stress/strain perturbations.

1. INTRODUCTION

The loss of the one-to-one relationship between stress and strain rates at yield for an elastic-perfectly plastic material is one of the basic facts which the theory of plasticity is built on. For an elastoplastic hardening material such a loss-of-uniqueness under load control occurs when the hardening modulus H is zero, what is usually associated with the failure condition, otherwise called limit state. If the same stress path is followed by controlling displacements instead of loads, the stress-strain relationship may exhibit a peak for H = 0 followed by a strain-softening branch. These results are summarised in Fig.1.
The point at which yield occurs for an elastic perfectly plastic material or at which limit state is reached for an elastic hardening material (in both cases H = 0) is associated to the loss of control of the test, if this is load controlled. Loss of control corresponds then to loss of uniqueness of the incremental response. On the other hand under displacement control the stress point for which H = 0 is associated either to an asymptotic state or to a peak in the stress-strain relationship. Indeed, this is the reason why failure or limit state were associated to the H = 0 condition.

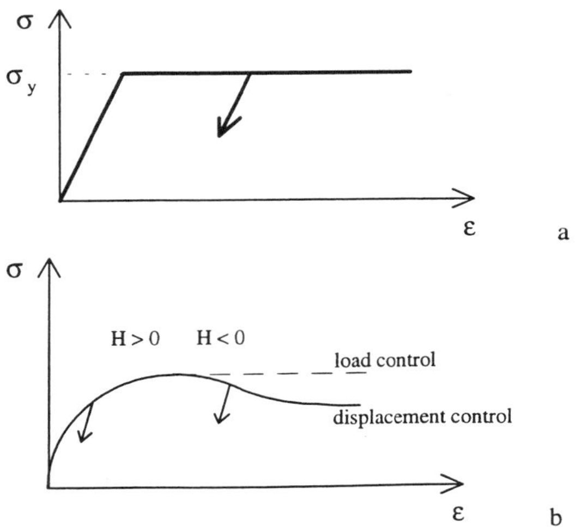

Fig.1: Non uniqueness of the incremental response at limit state a) for an elastic-perfectly plastic material; b) for a strain-hardening material

Fig1a,b are representative of the constitutive law of a material, which can be derived from experimental tests on homogeneously strained specimens. In actual specimens the strains tend to localise in a narrow band, for load levels large enough, so that uniformity of the stress-strain states within the specimen is lost. The level at which this occurs depends on the constitutive law of the material and the relative rôle played by the constitutive parameters (for instance elastic shear modulus and hardening modulus). In general, the localisation condition (nullity of the determinant of the acoustic tensor A) is reached at a stress level for which $H = H_b \neq 0$. H_b may be negative (for associated flow rules) or even positive (for non-associated flow rules and special loading conditions, such as plane strain). In the latter case, therefore, the failure of the specimen (which *ipso facto* coincides with the occurrence of the strain localisation) does not coincide with the limit state of the material, Fig.2.

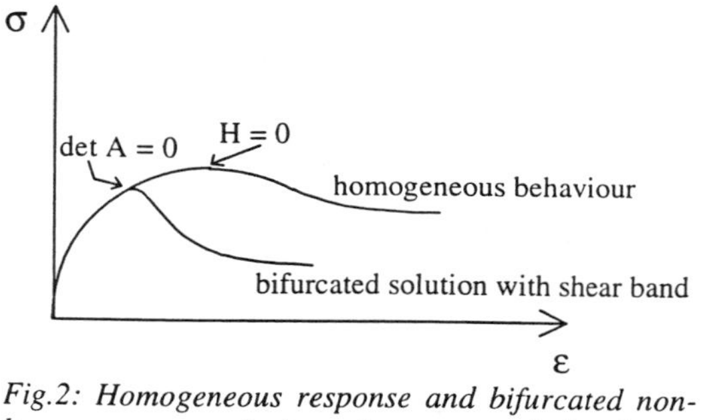

Fig.2: Homogeneous response and bifurcated non-homogeneous solution for a strain-hardening material (with non-associated flow rule)

The nullity of the determinant of the acoustic tensor is associated to the possibility of a multiplicity of solutions under the same displacement perturbation. A jump in the stress-strain state across the band is possible, in fact, violating neither equilibrium nor strain compatibility. Thus, the failure of the localised specimen can be associated to the loss of the uniqueness of the material response.
When saturated soils are tested in undrained condition, moreover, the loss-of-uniqueness (i.e. of control) of the incremental response under load controlled conditions, may occur well before the limit condition is reached, without the occurrence of any, apparent, shear band. Fig.3 shows for instance the effective stress paths of a loose and a medium dense sand specimens in an undrained triaxial tests. Such paths can be followed only under displacement control. Under load control, points A and B would be associated to large strain increments, unlimited in the former case, while in the latter the specimen able to regain stiffness, approximately at point B'. is worth-noting that when points A and B a reached the material is in the hardening regime. Th hardening modulus, which in the elastoplast framework does not depend on the stress increme direction but only on the stress state, is positive, fact, as can be easily demonstrated experimentally b allowing drainage to occur after peak is reached. A shown by di Prisco et al. (1995), the material ha positive stiffness under drained loading, what impli a positive hardening modulus. The path after peak by no means a softening branch, even thoug unstable under load control.

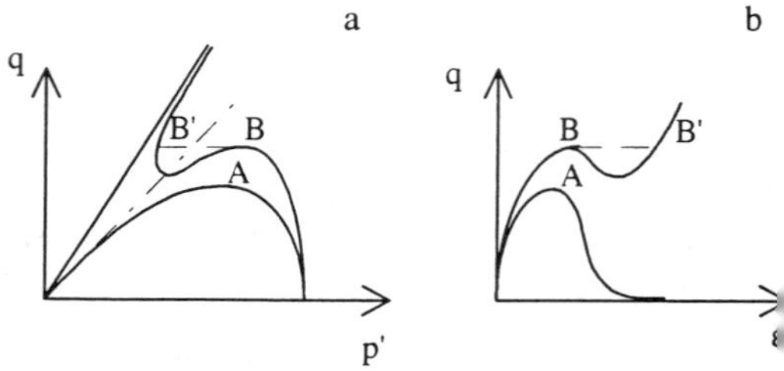

Fig.3: Experimental behaviour of a loose an medium dense sand specimens in undrained triaxia test: a) stress paths, b) stress-strain relationship Points A and B are associated to loss of control (i.e non-uniqueness) in load controlled tests.

The key point is in fact that materials obeying a non associated flow rule lose the one-to-one relationshi between stress and strain increments in the hardenin regime well before the limit state, so that special tes paths become uncontrollable (Nova(1994)). Darv (1994) has discussed the problem of stability an uniqueness of the incremental response for material obeying an incrementally non linear constitutive law appropriate for describing the behaviour of sand reaching essentially the same conclusions.
In this paper we shall examine various stress paths that can be followed by means of an ideal true triaxial apparatus, where either principal strains o principal stresses can be controlled. It will be shown that there exists a wide region in the stress space, within the limit locus, where a peak in the generalised stress-strain relationship occurs. Such a peak is associated to loss-of-uniqueness of the incremental response, which in turn corresponds to an unstable behaviour under proper controlling variables. Soils, and sand in particular, are therefore highly unstable materials. Such instabilities are not apparent, however, since they are put in evidence only by special load paths at particular stress points.

BIFURCATIONS UNDER MIXED STRESS ND STRAIN CONTROL

is well known that for an elastic plastic hardening aterials loss-of-uniqueness of the incremental sponse is possible when the hardening modulus is ro (limit state) which is associated to the nullity of e determinant of the stiffness matrix D:

$$\det D = 0. \qquad (1)$$

aier (1966) showed that loss-of-uniqueness of the cremental response under displacement control (i.e. ain, if homogeneity of the stress-strain state is sumed) occurs when:

$$H = -H_C = \tilde{n}\, D^e\, m, \qquad (2)$$

here a tilde means transposed, n and m being the adients of the loading function and of the plastic tential, respectively, and D^e being the elastic ffness matrix. Equation (2) is fulfilled when the terminant of the compliance matrix C ($=D^{-1}$) is ro.

ore recently, Nova (1989) has considered mixed ntrol of stress and strains. He proved that loss-of-iqueness of the incremental response in a loading ogramme in which we control some stress rate mponents grouped in a vector $\dot{\sigma}_\alpha$ and some other ain rate components grouped in a vector $\dot{\varepsilon}_\beta$

$$\left\{\begin{matrix} \dot{\sigma}_\alpha \\ \dot{\varepsilon}_\beta \end{matrix}\right\} = \begin{bmatrix} D_{\alpha\alpha} - D_{\alpha\beta}D_{\beta\beta}^{-1}D_{\beta\alpha} & D_{\alpha\beta}D_{\beta\beta}^{-1} \\ -D_{\beta\beta}^{-1}D_{\beta\alpha} & D_{\beta\beta}^{-1} \end{bmatrix} \left\{\begin{matrix} \dot{\varepsilon}_\alpha \\ \dot{\sigma}_\beta \end{matrix}\right\} \qquad (3)$$

ere D_{ij} are the four submatrices in which the ffness matrix is partitioned, is possible when

$$\det D_{\alpha\alpha} = 0. \qquad (4)$$

this case in fact no solution exists under arbitrary lues of $\dot{\sigma}_\alpha$, $\dot{\varepsilon}_\beta$, but for

$$\dot{\sigma}_\alpha = D_{\alpha\beta}\dot{\varepsilon}_\beta \qquad (5)$$

re exist infinite solutions of the type

$$\begin{cases} \dot{\varepsilon}_\alpha = \varphi\dot{\varepsilon}^*_\alpha \\ \dot{\sigma}_\beta = \varphi D_{\beta\alpha}\dot{\varepsilon}^*_\alpha + D_{\beta\beta}\dot{\varepsilon}_\beta \end{cases} \qquad (6)$$

ere φ is an arbitrary scalar and $\dot{\varepsilon}^*_\alpha$ is the rmalised eigenvector of matrix $D_{\alpha\alpha}$.

The nullity of the determinant of a submatrix cannot occur before the nullity of the determinant of the matrix itself, if this is positive definite and the stress state varies continuously. If the flow rule is associated, the stiffness matrix is symmetric and the loss of positive definiteness occurs when

$$\det D_S = 0 \qquad (7)$$

D_S being the symmetric part of the stiffness matrix. which coincides in this case with equation (1). Bifurcations of the type predicted by Eq.(6) are possible only when the determinant of the stiffness matrix is negative, i.e. in the softening regime.

If the flow rule is non-associated, however, a more interesting situation arise, since the conditions of loss of positive definiteness and limit state do not coincide and by virtute of the Ostrowski and Taussky (1951) theorem, loss of positive definiteness occurs when the determinant of the stiffness matrix is non negative (i.e. in the hardening regime). There is therefore a region in the stress space between the locus given by Eq.(7) and the limit state, where bifurcations of the type given by Eq.(6) are possible under special perturbations of the stress-strain state. For instances, by using an elastoplastic model with mixed isotropic/kinematic hardening (di Prisco et al. (1993)) and parameters appropriate for loose Hostun sand R.F. as in di Prisco et al. (1995), we can obtain the results shown in Fig.4, where the limit locus, the locus of loss of positive definiteness of the stiffness matrix and the loci for which Eq.(4) is fulfilled when controlling two principal stress rates ($\dot{\sigma}_2$ and $\dot{\sigma}_3$) and one strain rate ($\dot{\varepsilon}_1$) are plotted.

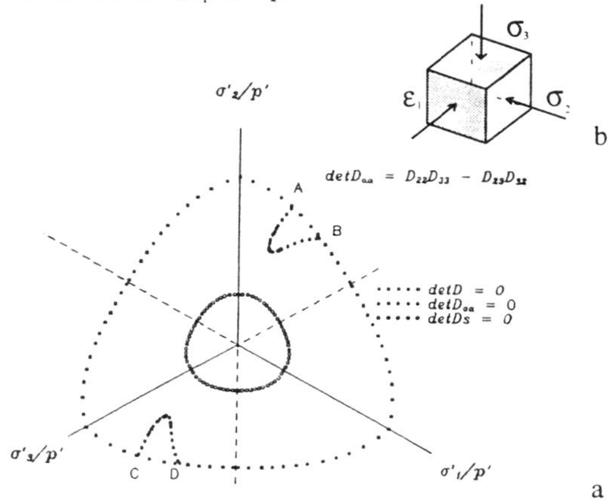

Fig4: a) Limit locus of loss of positive definiteness of the stiffness matrix loci for which multiplicity of solutions occur under control of $\dot{\varepsilon}_1, \dot{\sigma}_2, \dot{\sigma}_3$.
b) principal stress-strain axes.

Eq.(6) states that by imposing plane strain conditions (in the plane orthogonal to x_1) and giving no increment to stresses in directions x_2, x_3, if the stress state is on the locus AB of Fig.4a, spontaneous variations of the stress-strain state of the type

$$\begin{cases} \dot{\sigma}_1 = \left(D_{12} - D_{13}\dfrac{D_{32}}{D_{33}} \right)\dot{\varepsilon}_2 \\ \dot{\varepsilon}_2 > 0 \\ \dot{\varepsilon}_3 = -\dfrac{D_{32}}{D_{33}}\dot{\varepsilon}_2 \end{cases} \tag{8}$$

where here D_{ij} are the (scalar) components of the stiffness matrix (in the principal stress space), are possible.

3. BIFURCATIONS IN PROGRAMMES WHERE LINEAR COMBINATIONS OF STRAINS ARE CONTROLLED

In certain geotechnical testing programmes we may decides to control instead of stress or strain directly, linear combinations of them. For instance, in an undrained test we control the sum of the principal strain rates or we can control strain rates in such a way they are proportional to each other. The control variables are in this case generalised stress or strain variables. For instance, in an undrained test in axisymmetric conditions (standard triaxial test) under load control we have as generalised stress (ξ) or strain (η) the following control variables:

$$\begin{cases} \dot{\xi}_1 = \dot{\sigma}'_2 - \dot{\sigma}'_3 = 0 \\ \dot{\xi}_2 = \dot{\sigma}'_1 - \dot{\sigma}'_2 > 0 \\ \dot{\eta}_3 = \dot{\varepsilon}_1 + \dot{\varepsilon}_2 + \dot{\varepsilon}_3 = 0 \end{cases} \tag{9}$$

In order to determine the corresponding response variables, we must enforce the requirement that the generalised stress-strain variables produce the same work density as the principal stress-strain ones:

$$\tilde{\dot{\sigma}}\dot{\varepsilon} = \tilde{\dot{\xi}}\dot{\eta} \tag{10}$$

If we call T_σ and T_ε the transformation matrices linking stresses to generalised stress and strains to generalised strains, respectively, so that

$$\dot{\xi} = T_\sigma \dot{\sigma} \tag{11}$$

and

$$\dot{\eta} = T_\varepsilon \dot{\varepsilon} \tag{12}$$

Equation (10) implies

$$T_\varepsilon = (\tilde{T}_\sigma)^{-1} \tag{13}$$

In this case, the dual response variables can derived as:

$$\begin{cases} \dot{\eta}_1 = -\dfrac{1}{3}(\dot{\varepsilon}_1 + \dot{\varepsilon}_2) + \dfrac{2}{3}\dot{\varepsilon}_3 \\ \dot{\eta}_2 = \dfrac{2}{3}\dot{\varepsilon}_1 - \dfrac{1}{3}(\dot{\varepsilon}_2 + \dot{\varepsilon}_3) \\ \dot{\xi}_3 = \dfrac{1}{3}(\dot{\sigma}'_1 + \dot{\sigma}'_2 + \dot{\sigma}'_3) = \dot{p}' \end{cases} \tag{14}$$

Note that the axial strain corresponds to the devia stress and not to the axial stress directly.
Let Γ be generalised compliance matrix

$$\dot{\eta} = \Gamma\dot{\xi} \tag{15}$$

It can be proven (Imposimato and Nova (1998)) t when

$$\Gamma_{33} = 0 \tag{16}$$

the determinant of the matrix linking variables (9) (14) is zero.
Increments (9) cannot be assigned arbitrar therefore, and infinite eigensolutions exist. T corresponds to the sudden failure of a load controll test in undrained condition (onset of sta liquefaction of loose sand specimens, Castro (1969 The locus for which $\Gamma_{33} = 0$ in the p', q ($\equiv \xi_2$) pla was called instability line by Lade (1991). It can shown that when Eq.(16) is fulfilled, the theo predicts that under an isotropic total stress increa unlimited pore pressure increase can be generate Imposimato and Nova (1998) have generalised su a locus to conditions other than triaxial. Fig.5 sho a section of it in the deviatoric (Π) plane.

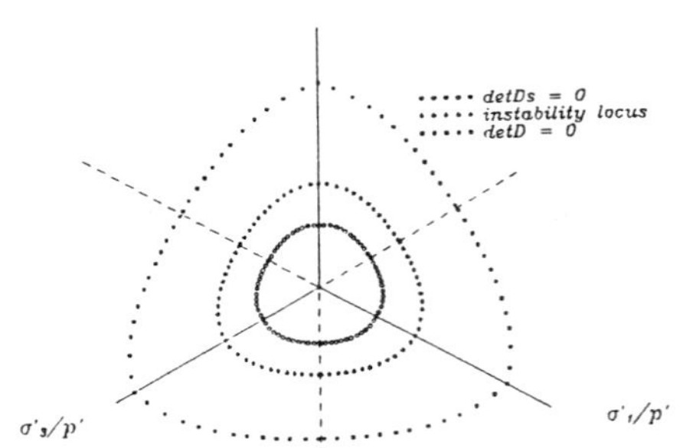

Fig.5: Instability locus in undrained conditions the deviatoric plane

nsider now the tests performed by Megachou 993) on Hostun R.F. sand and recently reported by rve (1994). It is about a series of axisymmetric axial tests in which horizontal strains decrease portionally to axial ones showing that instabilities the specimens occur leading to static liquefaction. the light of what stated above this is equivalent to form tests in which the control variables are:

$$\begin{cases} \dot{\xi}_1 = \dot{\sigma}'_2 - \dot{\sigma}'_3 = 0 \\ \dot{\eta}_2 = \dot{\varepsilon}_1 > 0 \\ \dot{\eta}_3 = (1-\alpha)\dot{\varepsilon}_1 + \dot{\varepsilon}_2 + \dot{\varepsilon}_3 = 0 \end{cases} \tag{17}$$

coefficient α being the ratio between the umetric strain increment and the axial strain rate:

$$\alpha \equiv \frac{\dot{v}}{\dot{\varepsilon}_1} = 1 + 2\frac{\dot{\varepsilon}_3}{\dot{\varepsilon}_1} . \tag{18}$$

ring this test α is kept constant.

a

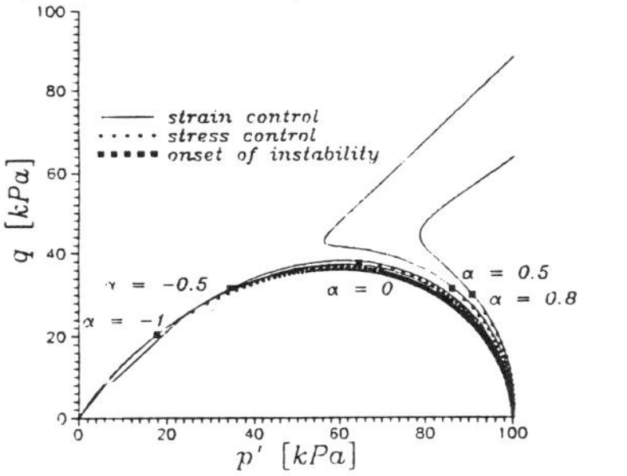

b

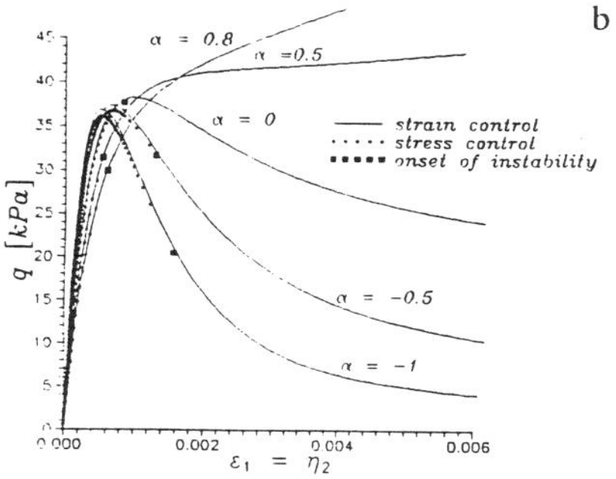

.6: *Rectilinear axisymmetric strain tests lumetric strain rate proportional to the axial). a) stress paths in the p'-q plane; b) stress ain relationship.*

.6 shows the stress paths followed in the p'-q ne and the corresponding stress strain ationships (q,ε_1). It is apparent that for some ues of α, total loss of strength is achieved ('liquefaction'). It is worth-noting, however, that the onset of instability (corresponding to the nullity of second order work) is not the peak in q but it occurs at the points marked by black squares, either after or before the peak, depending on the value of α. In fact, with the load programme given by (17) the appropriate response variable corresponding to η_2, via Eq.(13), is:

$$\xi_2 \equiv \sigma_1 - \alpha\sigma_2 + (2\alpha - 1)\sigma_3 \tag{19}$$

We can see from Fig.7 that the squares correspond in fact to peaks in ξ_2. Indeed stress (ξ_2) controlled tests will give rise to unlimited strains when the peak is reached. Note that when $\sigma_2 = \sigma_3$ and $\alpha = 0$ (undrained conditions) $\xi_2 \equiv q$ and the peak in q observed in undrained triaxial tests on loose sand specimens is therefore effectively the onset of instability for this test.

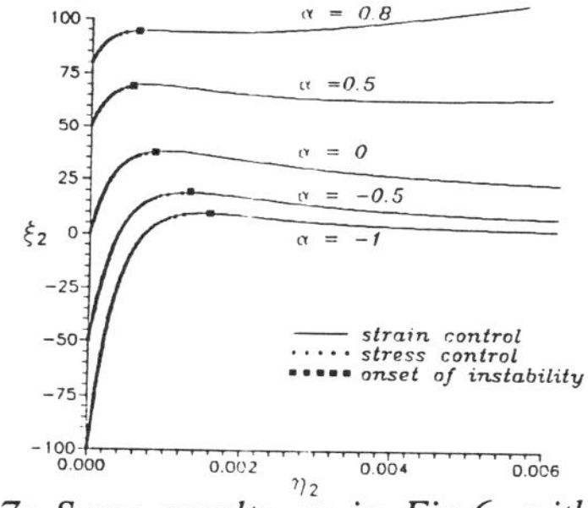

Fig.7: Same results as in Fig.6, with appropriate generalised stress variables showing onset of instability at peak.

Programme (17) can be also considered a perturbations of stress-strain states other than axisymmetric. Fig.8 show the loci for which Eq.(16) is fulfilled for positive and negative α values.
Fig.9 show the same loci for particular α values with permutation of indices of stress and strain variables. It is interesting to note that the locus for which the stiffness matrix ceases to be positive definite can be seen as the envelope of such loci.

CONCLUSIONS

It was shown in this paper that the familiar concept of loss-of-uniqueness of the incremental response of a geomaterial, characterised by a non-symmetric stiffness matrix, can be generalised. In particular it is apparent that the region of the stress space between the locus of positive definiteness and the limit locus (Fig.10) is a region in which at each point one or more possibilities of instability (non-uniqueness of incremental responses) following appropriate loading programmes are possible.

a

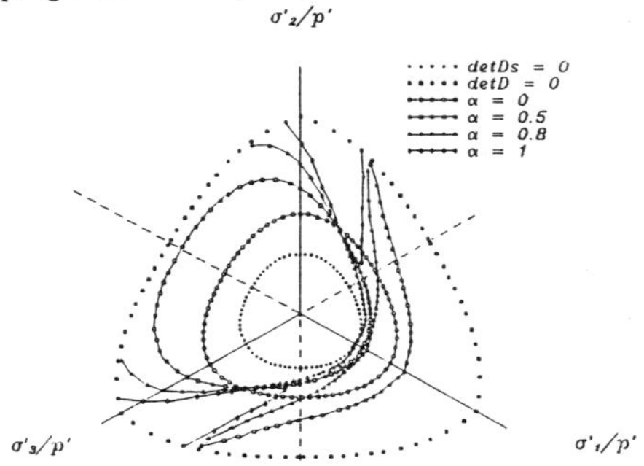

b

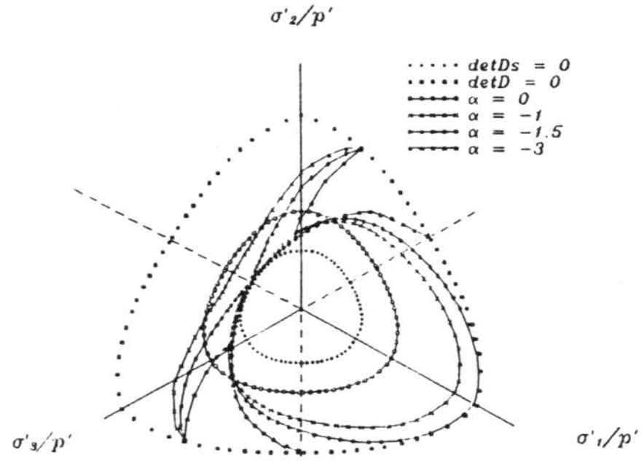

Fig.8: Instability loci for perturbations of the type given by programme Eq.(17) for stress states other than axisymmetric. a) positive α; b) negative α.

a

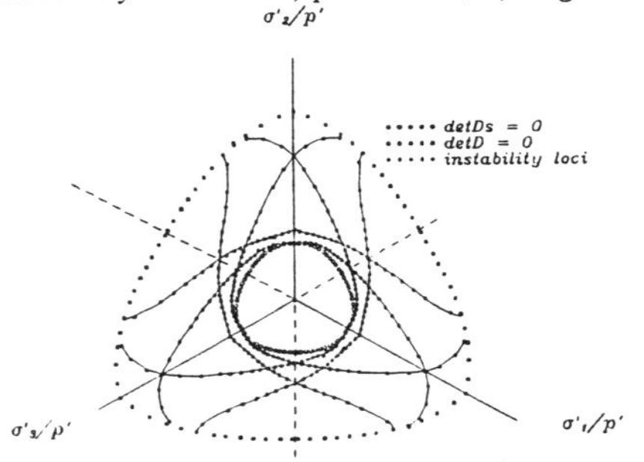

b

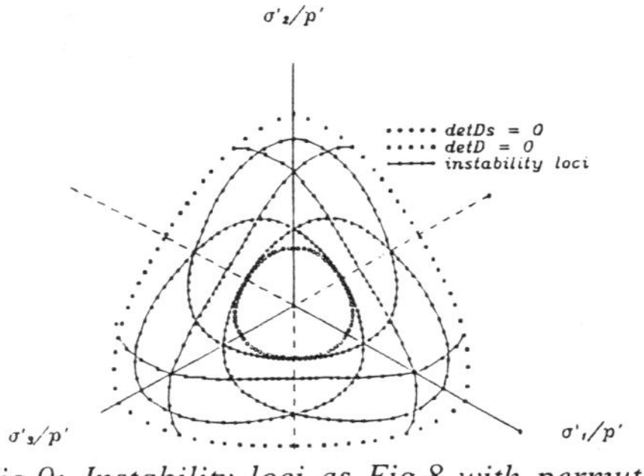

Fig.9: Instability loci as Fig.8 with permutation of indices of principal stresses and strains: a) α = 0.8 b) α = -1

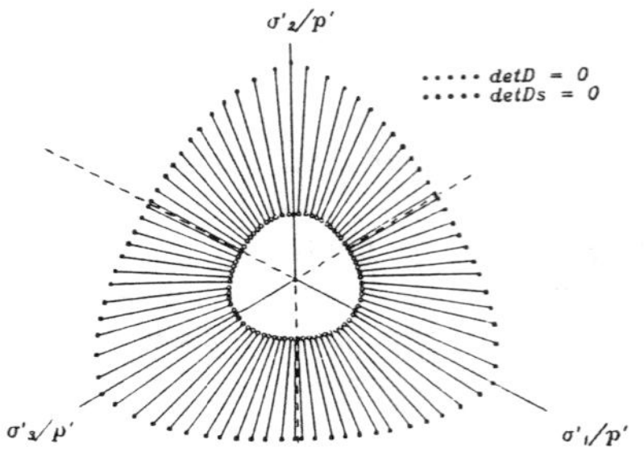

Fig.10: Region of potential instability

REFERENCES

Castro G. (1969) 'Liquefaction of sand' Ph.D. Thes Harvard Soil Mechanics Series, N° 81, Harva University, Cambridge, MA

Darve F. (1994) 'Liquefaction phenomeno Modelling, stability and uniqueness', in *Verificati of numerical procedures for the analysis of s liquefaction problems*, Arulanandan & Scott Ed 1305-1319

di Prisco C., Nova R. and Lanier J. (1993) 'A mix isotropic-kinematic hardening constitutive law f sand' *Modern approach to plasticity*, D.Kolymb (ed.), Balkema, 83-124

di Prisco C.,Matiotti R.,Nova R. (1995) 'Theoretic investigation of the undrained stability of shallo submerged slopes', *Géotechnique*, 45, 479-496

Imposimato S., Nova R. (1998) 'An investigation existence and uniqueness of the incremental respon of elastoplastic models for virgin sand' to appear

Lade P. V. (1991) 'Nonassociated flow and instabili of slopes', *Computer Methods and Advances Geomechanics*, Balkema, 487-492

Maier G. (1966) 'Sui legami associati tra sforzi deformazioni incrementali in elastoplasticità' *Rendiconti dell'Istituto Lombardo di Scienze Lettere*, 100, 809-838

Megachou M. (1993) 'Stabilité de sables lache Essais et modélisastions' Thése de doctorat INP Grenoble

Nova R. (1989) 'Liquefaction, stability, bifurcatio of soil via strainhardening plasticity' *Numeric methods for localization and bifurcation of granul bodies*, International Workshop, Gdans Sobieszewo, September, 25-30

Nova R. (1994) 'Controllability of the increment response of soil specimens subjected to arbitra loading programmes', *Journal of Mechanic Behaviour of Materials*, 5, No 2, 193-201

Ostrowsky A. and Taussky O. (1951), 'On t variation of the determinant of a positive defini matrix' *Nederl. Akad. Wet. Proc.* (A)54, 383-351

Numerical Models in Geomechanics, Pietruszczak & Pande (eds) © 1997 Balkema, Rotterdam, ISBN 90 5410 886 X

Criterion for strain localization based on multilaminate framework

M. Karstunen
Department of Civil Engineering, University of Glasgow, UK

G. N. Pande
Department of Civil Engineering, University of Wales, Swansea, UK

ABSTRACT: The phenomena of strain localization in biaxial test conditions has been studied by means of a numerical model which utilises multilaminate framework. The material response is described via a simple deviatoric hardening constitutive model with non-associated flow rule embedded in the multilaminate framework. As a result, the evolution of plastic shear strains on randomly oriented planes can be traced throughout the loading history. The criterion for localization has been defined in a practical and physical way in terms of relative shear strain concentration instead of the more commonly used bifurcation criterion.

1 INTRODUCTION

Strain localization in geomaterials is a physical phenomenon in which the nearly homogenous deformation of the body is changed into a highly concentrated deformation mode, usually in the form of a single or multiple narrow shear bands or cracks. This mode of deformation is likely to correspond to the actual types of failure in geotechnical engineering practice. Problems such as the failures of earth-dams, retaining wall structures, foundations, embankments and slopes are examples of the manifestation of strain localization.

Because of this practical significance, several researchers have been working on the constitutive and numerical aspects of the problems associated with strain localization during the last decade. An accurate numerical modelling of the phenomenon would lead to better understanding of the behaviour of the structures and would improve the current design practise into a more cost effective and safer direction.

In practice, localization can arise from initial local inhomogeneities within the material, stress concentrations imposed by external loading, or from the onset of some physical or geochemical mechanism which degrades abruptly the strength of the material leading to local inhomogenities. Alternatively, strain localization can be considered as a bifurcation from a smoothly varying pattern of deformation as a result of an instability in the inelastic behaviour of the material considered due to the plastic flow rule being non-associated and/or strain softening.

2 EXPERIMENTAL EVIDENCE

Before constitutive and theoretical framework for shear band formation and localization of deformation can be established, the material and the conditions leading to strain localization needs to studied in controlled laboratory tests. Strain localization in soils have been experimentally studied mainly by Vardoulakis and his co-workers (Vardoulakis & Graf 1985, Han & Vardoulakis 1991), by Finno and his co-workers (Finno & Rhee 1993, Viggiani et al., 1994) and Desrues and his co-workers (Desrues et al. 1985, Desrues 1990). Evidence of strain localization has been found in all types of elementary tests (i.e. tests which are conducted on homogeneous samples in homogeneous stress conditions) with different types of materials (dense sand, loose sand, overconsolidated clay).

The most convenient test for studying the strain localization is a purpose-built biaxial apparatus which allows free shear band formation and the visual inspection of the sample throughout the test. Thus special techniques, like false relief stereophotogrammetry can be used and it is possible to get quantitative information on global and local deformation fields of the sample throughout the experiment. A vast amount of data has been published by Desrues and his co-workers especially for sands (Hammad 1991, Mokni 1992).

The experimental studies have shown that the shear band formation is influenced by many factors, the most obvious ones being test conditions (drained or undrained), porosity of the sample, initial anisotropy, grain shape and size together with confining pressure.

In this study drained biaxial tests have been simulated numerically. The data used as a reference for the numerical studies are from biaxial test conducted in Laboratoire 3S, Grenoble by Hammad (1991) and Mokni (1992) on dense Hostun sand.

3 MATERIAL INSTABILITY AND CRITERION FOR LOCALIZATION

From theoretical point of view, localisation phenomena can be considered to be related to material instability. Unfortunately, the definitions of stability and uniqueness do still not have generally accepted forms. Drucker's stability postulate (Drucker 1950) assumes that for any load increment the plastic work of second order is strictly positive. For associated plasticity the second order plastic work becomes negative in the softening regime. In the case of non-associated plasticity, the postulate may be violated even within the hardening regime of the material behaviour.

The more general criterion by Hill (1958) considers the positiveness of the work of second order for all possible stress and strain increments. This product becomes typically negative on strain softening regime, but there are also another class of material instabilities which may cause the inner product to be negative (Rudnicki & Rice 1975). In non-associative elastoplasticity the non-symmetric relation between the stress-rate tensor and strain-rate tensor in itself is sufficient to cause the loss of material stability.

The crucial consequence of the loss of material stability is that it may result in loss of ellipticity of the local rate equilibrium (Bigoni & Hueckel 1991). Typically, the necessary condition for bifurcation is derived by employing the kinematic constraint and the equilibrium condition across the singular surface (Rudnicki & Rice 1975, Rice 1976):

$$\det(n_i D_{ijkl} n_l) = 0 \tag{1}$$

where D_{ijkl} is the tangential stiffness tensor and n_i is the unit normal vector acting normal to the shear band (which reduces to $n_i = \{\cos\theta \ \sin\theta\}^T$ in two-dimensional case). The summation with respect of repeated indexes has been adopted. The angle θ is generally known as the bifurcation angle.

Historically, two classical solutions for the bifurcation angle exist, namely the Coulomb solution, θ_C (2), and the Roscoe solution, θ_R (3).

$$\theta_C = 45^\circ \pm \varphi_m / 2 \tag{2}$$

$$\theta_R = 45^\circ \pm \psi_m / 2 \tag{3}$$

where φ_m and ψ_m are the mobilised friction and dilatancy angles respectively. The difference between Coulomb orientation and Roscoe orientation is quite significant, as the dilatancy angle is usually much smaller than the friction angle. Typically, the difference is of order of 30° (Vermeer 1990). There is experimental evidence for both the Coulomb solution and the Roscoe solution.

Based on experiments Arthur et al. (1977) suggested an intermediate orientation, namely

$$\theta_A = 45^\circ \pm \tfrac{1}{4}(\varphi_m + \psi_m) \tag{4}$$

Later on, Vardoulakis (1980) presented theoretical background for this intermediate orientation; for realistic values of φ and ψ, Arthur's equation gives θ values which are in agreement with those solved from the Rice criterion (1).

In this study a different approach has been adapted: the onset of localization and the bifurcation angle are determined by using the multilaminate framework. The idea of multilaminate framework is to describe the behaviour of the material through the response of an infinite number of randomly oriented planes in space. The mathematical description involved in accurately describing the behaviour of any one contact plane is assumed to hold for all the planes. The global, or macroscopic, behaviour can be obtained by integrating the contributions made by each of the contact planes.

In practice, a limited number of planes is chosen and the integration is performed numerically over the surface of a unit sphere. In plane strain a further simplification arises if the yield/plastic potential functions are such that no plastic strains take place in the direction normal to the plane of analysis (Pande & Yamada 1994): the integration can be carried out on the circumference of a circle and the number of planes can then be chosen according to desired accuracy.

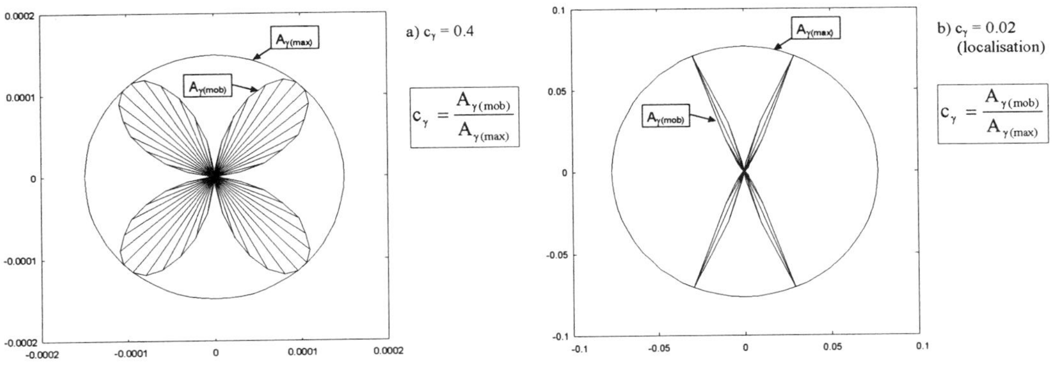

Figure 1. Mobilisation of plastic shear strains on planes. Plastic shear strain concentration.
a) No bifurcation. b) Bifurcation criteria is met.

The model has links with the slip theory of plasticity for metals (Taylor 1938, Batdorf & Budiansky 1949). Similarly, it can be considered as a generalisation of Coulomb's idea of rupture planes to include slip effects on failure of planes at different inclinations. This analogy has been expanded to the idea of strain localization and used to determine the bifurcation angle (Karstunen et al. 1997).

The non-linear behaviour of granular materials can be described by extending the well-known Mohr-Coulomb criterion to account for deviatoric hardening. The yield function (Pietruszczak & Niu 1993) can be defined in local co-ordinate system (assuming tension positive) on each sampling plane as

$$F = \left|\tau_{xy}\right| + \tan\varphi_m(\sigma'_y - c\cot\varphi_m) = 0 \quad (5)$$

where

$$\tan\varphi_m = \tan\varphi_0 + (\tan\varphi - \tan\varphi_0)\frac{\gamma_{xy}^{p}}{A + \gamma_{xy}^{p}}$$

and τ_{xy} and σ'_y are the shear and normal (effective) stresses on the plane respectively, φ_m is the mobilised friction angle, c is the cohesion, φ_0 defines the elastic region, A is the material parameter controlling the rate of hardening and γ_{xy}^{p} is the strain hardening parameter (unique for each plane).

Similarly, by assuming a non-associated flow rule, the plastic potential function can be selected as (Pietruszczak & Niu 1993)

$$Q = \left|\tau_{xy}\right| - \eta_c(\sigma'_y - c\cot\varphi_m)\ln\left(\frac{(\sigma'_y - c\cot\varphi_m)}{\sigma_0}\right) = 0 \quad (6)$$

where η_c specifies the slope of the line in stress space at which the minimum volume is obtained and σ_0 is a scalar defined by current stress state.

By adopting appropriate transformation matrices, the increment of plastic strain in global co-ordinate system can be evaluated:

$$\dot{\underset{\sim}{\varepsilon}}^{p} = \frac{1}{n}\left(\sum_{i=1}^{n}[T_\varepsilon]_i^{-1}[C^p]_i[T_\sigma]_i\,\dot{\underset{\sim}{\sigma}}'\right) \quad (7)$$

where $\dot{\underset{\sim}{\varepsilon}}^{p}$ is the incremental plastic strain vector and $\dot{\underset{\sim}{\sigma}}'$ is the incremental stress vector (both now in global co-ordinate system), $[T_\varepsilon]_i$ and $[T_\sigma]_i$ are the corresponding transformation matrices for strains and stresses respectively, $[C^P]_i$ is the plastic compliance matrix (on a plane), and *n* is the number of planes.

The development of plastic shear strain on different planes are monitored, and after a certain amount of plastic flow the strains start to concentrate on certain symmetric planes. In Figure 1, the mobilised plastic shear on the planes have been plotted for two load increments (Fig. 1a & 1b). The orientation of the lines in a fan represent the orientation of the planes and the length of the lines represents the magnitude of the plastic shear strain on that plane. As illustrated in Figure 1, the plastic shear strain concentration, c_γ, can be calculated (at any stage of loading) as the area of the fan of mobilised shear strains on the planes ($A_{\gamma(mob)}$) divided by the area of a circle with its radius equal to the maximum shear strain ($A_{\gamma(max)}$) at that increment. When the plastic shear strain concentration exceeds a pre-defined value, sample is assumed to have bifurcated and the bifurcation angle is defined by the inclination of the plane consistent with the global mode of deformation.

The method proposed has some features which makes it superior to other approaches. Firstly, the criterion for strain localization has been defined in a practical and physically meaningful way. Secondly, the effects of anisotropy of the plastic flow and the

Table 1. Parameters for the deviatoric hardening model for dense Hostun sand.

Parameter	Value
Young's modulus, E	30 000 kPa
Poisson's ratio, ν	0.42
Peak friction angle, φ	47 °
Initial friction angle, φ_0	0.1 °
Cohesion, c	5 kPa
Parameter A	0.007
Parameter η_c	0.550

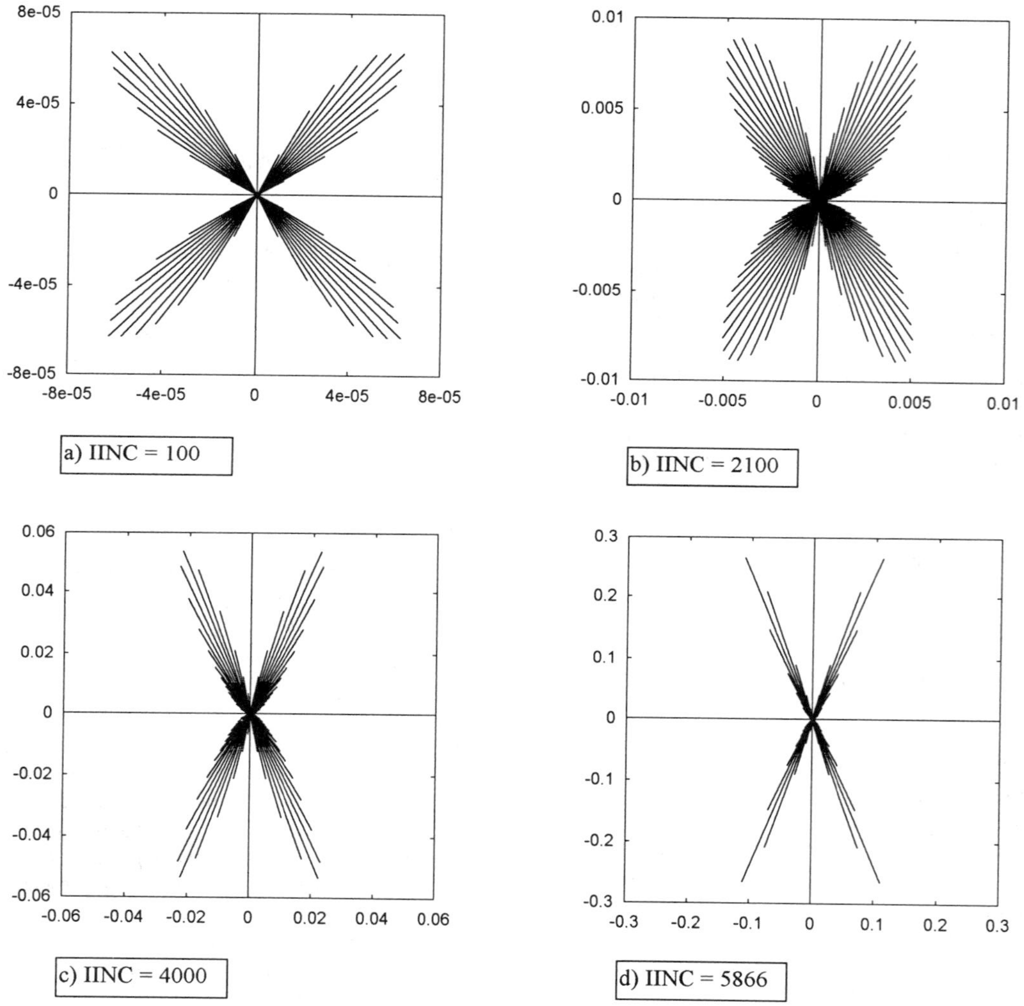

Figure 2. Mobilised plastic shear strains on planes. Increment number: a) IINC=100, b) IINC=2100, c) IINC= 4000, d) IINC=5866 (localization with 3% criterion).

influence of the rotation of the principal stress axes are implicitly embedded in the multilaminate formulation. Therefore, the history of loading influences the bifurcation angle (Schweiger et al. 1997a). Furthermore, the idea can easily be extended to 3D -space.

4 NUMERICAL EXAMPLES

The constitutive models have been implemented into an incremental elasto-plastic program with the boundary conditions corresponding to an 'ideal' strain controlled biaxial test ('ideal' means no friction between the top and bottom plates and the sample).

The parameters adopted for the multilaminate deviatoric hardening model defined based on biaxial and triaxial tests on dense Hostun sand (Hammad 1991, Mokni 1992) are presented in Table 1. For isotropic sample, the same parameters can be used for each plane. If the preceding loading had been anisotropic, it can be taken into account (Schweiger et al. 1997b) or alternatively, different values of φ_0 for different planes could be input.

When multilaminate model is used for the criteria of localization, the accuracy, like the accuracy with multilaminate model in general, depends on the number of planes. More planes means stiffer response. For these simulations a 64-plane rule has been adopted, which means that the bifurcation angle obtained numerically can be in error by ± 1.4 °.

Firstly, the physical interpretation of the strain localization will be demonstrated. In Fig. 2, the mobilised plastic shear strains on the planes have been plotted for different load increments. At the early stages of loading, the mobilised friction angle is close to zero, hence the major plastic shear strains occur on symmetric planes with 45° angle with the co-ordinate axis. In subsequent loading the strains start to rotate as the friction mobilises and finally, as displayed in Fig. 2d, the strain have visibly localized. The relative shear strain concentration, as defined in Fig. 1, is at this stage less than 3%.

The bifurcation angle predicted with multilaminate model corresponding this stage is 67.5°, slightly less than the Coulomb angle (68.75°). The same calculation was repeated by increasing the number of the planes from 64 to 360. With 360-plane integration rule the bifurcation angle would be 68.5°. The corresponding experiments with the same confining pressure on dense Hostun sand yielded a bifurcation angle of 68.5°. For practical purposes the 64 plane-rule is relatively accurate, and it is to be expected that for FE -simulations it would be advisable to decrease the number further. For the same set of parameters, the Rice criterion would have predicted an angle of 63.7°.

Next the value of the relative shear strain concentration will be studied. Firstly, for a confining pressure of 100 kPa the results are compared with the experimental results (Hammad (1991)) in Figure 3. The solid line represents the homogeneous deformation and the arrows mark when the different values of plastic shear strain concentration have been reached (2%, 3%, 4% and Rice criterion). Similar results for a confining pressure of 200 kPa are represented in Figure 4 (experiments by Hammad (1991)). Based on these simulations, a criterion for localization by choosing a limiting value of c_γ = 3% for the multilaminate model seems realistic.

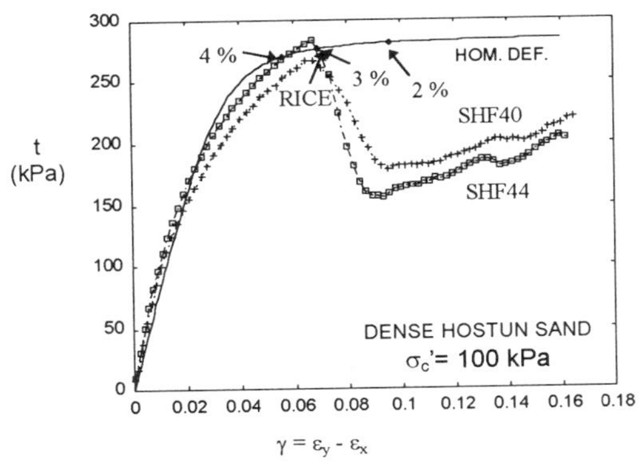

Figure 3. Different criteria for onset of localization. Confining pressure σ_c = 100 kPa. Experiments by Hammad (1991).

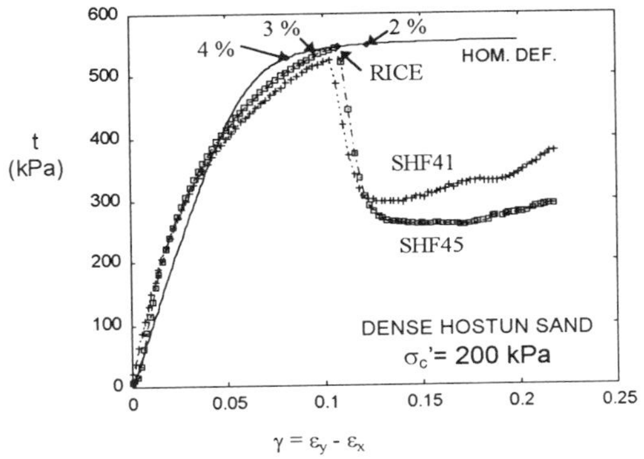

Figure 4. Different criteria for onset of localization. Confining pressure σ_c = 200 kPa. Experiments by Hammad (1991).

Complementary studies on the model prediction for different confining pressures have been presented by Karstunen & al. (1997) and a separate study on the effect of an anisotropic loading prior to the biaxial loading has been presented by Schweiger & al. (1997b). Those studies involve also modelling the post-peak behaviour. Further studies on the influence of the rotation of the principal stress axes on the bifurcation angle will be presented in the near future.

REFERENCES

Arthur, J. R. F., Dunstan, T., Al-Ani, Q. A. J. L. & Assadi, A. (1977). Plastic deformation and failure in granular media. *Géotechnique 27(1)*: 53-74.

Batdorf, S. B. & Budiansky, B. (1949). A mathematical theory of plasticity based on the concept of slip. *National Advisory Committee for Aeronautics TN 1871*.

Bigoni, D. & Hueckel, T. (1991). Uniqueness and localization-I. Associative and non-associative elastoplasticity. *Int. J. Solids Structures 28(2)*: 197-213.

Desrues, J. (1990). Shear band initiation in granular materials: experimentation and theory. In: *Geomaterials: Constitutive Equations and Modelling*, Elsevier, Rotterdam. pp 283-310.

Desrues, J., Lanier, J. & Stutz, P. (1985). Localization of the deformation in tests on sand sample. *Eng. Fract. Mech. 21(4)*: 909-921.

Drucker, D. C. (1950). Some implications of work hardening and ideal plasticity. *Q. Appl. Math. 7*: 411-418.

Finno, R. J. & Rhee, Y. (1993). Consolidation, pre- and post peak shearing from internally instrumented biaxial compression device. *Geotechnical Testing Journal 16(4)*: 496-509.

Hammad, W. (1991). *Modélisation non linéaire et étude expérimentale de la localisation dans les sables*. Thèse de doctorat. UJF-INPG, Grenoble.

Han, C. & Vardoulakis, I. G. (1991). Plane-strain compression experiments on water-saturated fine-grained sand. *Géotechnique 41(1)*: 49-78.

Hill, R. (1958). A general theory of uniqueness and stability in elastic-plastic solids. *Journal of Mechanics and Physics of Solids 6*: 236-249.

Karstunen, M., Pande, G. N. & Desrues, J. (1997). Strain localization and rotation of principal stress axis in biaxial tests. *Proc. IACMAG 97, Wuhan, China,* A.A. Balkema, Rotterdam.

Mokni, M. (1992). *Relations entre déformations en masse et déformations localisées dans de matériaux granulaires*. Thèse de doctorat. UJF-INPG, Grenoble.

Pande, G. N. & Yamada, M. (1994). The multilaminate framework of models for rock and soil masses. *Applications of Computational Mechanics in Geotechnical Engineering,* A. A. Balkema, Rotterdam. pp. 105-123.

Pietruszczak, S. & Niu, X. (1993). On the description of localized deformation. *Int. J. Num. Anal. Meth. Geomech. 17*: 791-805.

Rice, J. R. (1976). The localization of plastic deformation. *Proc 14 th IUTAM Congress: Theo. and Appl. Mech.*, pp. 207-220

Rudnicki, J. W. & Rice, J. R. (1975). Conditions for the localization of deformation in pressure-sensitive dilatant materials. *Journal of Mechanics and Physics of Solids 23*: 371-394.

Schweiger, H. F., Karstunen, M. & Pande, G. N. (1997a). Modelling strain localization in soils using multilaminate model and homogenisation technique. *Proc. IS-Nagoya'97, Nagoya, Japan,* Elsevier, Rotterdam.

Schweiger, H. F., Karstunen, M. & Pande, G. N. (1997b). Modelling strain localization of anisotropic consolidated samples. *Proc. NUMOG VI, Montreal, Canada,* A.A. Balkema, Rotterdam.

Taylor, G. I. (1938). Plastic strain in metals. *J. Inst. Metals 62*: 307-324.

Vardoulakis, I. (1980). Shear band inclination and shear modulus of sand in biaxial tests. *Int. J. Num. Anal. Meth. Geomech. 4*: 103-119.

Vardoulakis, I. & Graf, B. (1985). Calibration of constitutive models for granular materials using data from biaxial experiments. *Géotechnique 35(3)*: 299-317.

Vermeer, P. A. (1990). The orientation of shear bands in biaxial tests. *Géotechnique 40(2)*: 223-236.

Viggiani, G., Finno, R. J. & Harris, W. W. (1994). Experimental observations of strain localisation in plane strain compression of a stiff clay. *Localisation and Bifurcation Theory for Soils and Rocks,* A.A. Balkema, Rotterdam. pp. 189-198.

Numerical Models in Geomechanics, Pietruszczak & Pande (eds) © *1997 Balkema, Rotterdam, ISBN 90 5410 886 X*

On the localized deformation in fluid-infiltrated soils

S. Pietruszczak & M. Huang
McMaster University, Hamilton, Ont., Canada

ABSTRACT: This paper addresses the issue of the mechanical behaviour of fluid-infiltrated soils in the presence of localized deformation. The cases of full as well as partial saturation are considered. In the latter case, it is assumed that the soil contains discrete gas inclusions entrapped in the liquid phase. The formulation of the problem is based on a homogenization procedure. The case when both constituents (i.e., the intact and shear band material) respond in undrained manner is considered first. Later, the problem is reformulated by allowing for an exchange of fluids (partial drainage) in constituents due to dissipation of the excess pore pressure gradient. The formulation is applied in the context of finite element analyses of plane strain drained/undrained deformation process. The issues related to the mesh size and mesh alignment sensitivity are addressed.

1. INTRODUCTION

Over the last few decades, the phenomenon of localized deformation in saturated/dry porous media has been the object of intense experimental and theoretical research. The experimental evidence on deformation instabilities and the failure modes in saturated granular materials comes from conventional triaxial (Lade 1988; Lade 1994) as well as plane strain biaxial tests (Vardoulakis and Graf 1985; Han and Vardoulakis 1991). The results indicate that the uniform response is often followed by the onset of a diffused, non-homogeneous deformation mode, after which distinct shear bands form. The onset of localized deformation is typically observed during the stage of progressive decrease in pore pressure, i.e. when the specimen is dense enough to behave in a dilatant manner.

The mathematical formulation of the problem involves two aspects. The fist one is related to the establishment of an appropriate criterion governing the inception of the localized mode. The second concerns the development of a constitutive relation for the description of the inhomogeneous deformation process. In the last decade, several conceptually different approaches for modelling of strain localization have emerged. A detailed review of these can be found, for example, in De Borst *et al.* (1993). The modelling of the localized deformation in the context of a boundary-value problem also presents significant difficulties associated primarily with the issues of mesh-size and mesh-alignment dependencies.

The present paper is a direct extension of the research reported by Pietruszczak (1995). The formulation of the problem is based on a simple homogenization procedure, which is used to estimate the average macroscopic response of a continuum intercepted by a shear band. The mathematical description outlined here is different from that in the original reference, which is primarily due to a different choice of the referential coordinate system. This formulation is believed to be computationally more efficient in the context of a numerical analysis. In the next section, a general framework for describing the localized deformation in dry, saturated and partially saturated soils is outlined. Subsequently, the formulation is applied to study the response of a class of geomaterials subjected to drained/undrained uniaxial compression under plane strain regime.

2. FORMULATION OF THE PROBLEM

Consider a soil sample, of a given geometry, which is initially under a homogeneous state of stress. At the onset of localization, which may be considered as a bifurcation problem, the material is intercepted by a distinct shear band. In order to formulate the problem, attach the frame of reference to the sample and denote by $\dot{\sigma}^{(i)}$, $\dot{\varepsilon}^{(i)}$ (i=1,2) the stress/strain rates in both constituents involved, i.e. the intact and the shear band material. Since the shear band penetrates through the entire sample and its thickness is negligible compared to all other dimensions, both of these fields may be considered as homogeneous within themselves.

(i) *Homogenization procedure*

In order to establish the average macroscopic response of the sample a homogenization procedure, similar to that proposed by Pietruszczak and Niu (1993), has been employed. The formulation presented here is different from that in the original reference, which stems primarily from a different choice of the referential coordinate system. The formulation employs the stress/strain rate decomposition based on volume averaging

$$\dot{\boldsymbol{\sigma}} = \mu_1\,\dot{\boldsymbol{\sigma}}^{(1)} + \mu_2\,\dot{\boldsymbol{\sigma}}^{(2)}\,; \qquad \dot{\boldsymbol{\varepsilon}} = \mu_1\,\dot{\boldsymbol{\varepsilon}}^{(1)} + \mu_2\,\dot{\boldsymbol{\varepsilon}}^{(2)} \tag{1}$$

Here i=1,2 refers to the intact and the shear band material, respectively and μ's are the volume fractions of constituents. The strain rates in the shear band material can be conveniently expressed in terms of velocity discontinuity $\dot{\boldsymbol{g}}$, as a symmetric part of a dyadic product

$$[\dot{\varepsilon}^{(2)}] = \frac{1}{2t}(\boldsymbol{n}\,\dot{\boldsymbol{g}}^T + \dot{\boldsymbol{g}}\,\boldsymbol{n}^T)\,; \rightharpoonup \dot{\boldsymbol{\varepsilon}}^{(2)} = \frac{1}{t}[N]\dot{\boldsymbol{g}} \tag{2}$$

where $\boldsymbol{n}$ is the unit vector normal to the shear band, and t is the thickness. The equilibrium requires that the tractions along the interface remain continuos. Thus,

$$\dot{\boldsymbol{T}} = [N]^T\,\dot{\boldsymbol{\sigma}}^{(1)} \tag{3}$$

Assuming now the constitutive relations in the general form

$$\dot{\boldsymbol{\sigma}}^{(1)} = [D]\,\dot{\boldsymbol{\varepsilon}}^{(1)}\,; \qquad \dot{\boldsymbol{T}} = [K]\dot{\boldsymbol{g}} \tag{4}$$

and substituting in eqn.(3) one obtains, after some transformations

$$\dot{\boldsymbol{g}} = [S]\dot{\boldsymbol{\varepsilon}}\,; \;\; [S] = \left([K] + \mu\,[N]^T[D]\,[N]\right)^{-1}[N]^T[D] \tag{5}$$

where $\mu=\mu_2/t$ is defined as the ratio of the shear band surface area to the volume of the sample. Given the fact that $\mu_2 \ll \mu_1$ and μ is independent of the thickness of the shear band, the latter can be formally eliminated from the macroscopic considerations. The global constitutive relation can now be obtained by invoking the decomposition (1). Noting that $\dot{\boldsymbol{\sigma}} \approx \dot{\boldsymbol{\sigma}}^{(1)}$ and substituting eq.(5) into eq.(1) yields

$$\dot{\boldsymbol{\sigma}} = [D]\left([I] - \mu\,[N]\,[S]\right)\dot{\boldsymbol{\varepsilon}} \tag{6}$$

where $[I]$ is the identity matrix.

(ii) *Drained behaviour*

Assume first that both constituents respond in a drained manner. The behaviour of the intact material is described by the first relation in eq.(4), where $[D]$ specifies the drained properties. In order to establish the functional form of the constitutive relation for the shear band material, it is convenient to choose the local base {**s**, **n**, **t**} such that

$$\boldsymbol{s} = \boldsymbol{T}_s / T_s\,; \quad \boldsymbol{T}_s = \boldsymbol{T} - (\boldsymbol{n}^T\boldsymbol{T})\boldsymbol{n} = ([I] - \boldsymbol{n}\boldsymbol{n}^T)\boldsymbol{T} \tag{7}$$

and $\boldsymbol{t} = \boldsymbol{s} \times \boldsymbol{n}$. In this case T_t=**0**, so that the yield and plastic potential functions can be assumed in a general form

$$f = f(T_n, T_s, g^p) = 0\,; \quad \Psi = \Psi(T_n, T_s) = const. \tag{8}$$

Following now the standard plasticity procedure, the constitutive relation can be expressed as

$$\dot{\bar{\boldsymbol{T}}} = [\bar{K}]\,\dot{\bar{\boldsymbol{g}}} \tag{9}$$

where the bar signifies that the quantities are referred to the local frame of reference {**s**, **n**, **t** }. Transforming eq.(9) to the global frame, attached to the sample, one obtains

$$\dot{\boldsymbol{T}} = [K]\,\dot{\boldsymbol{g}}\,; \quad [K] = [Q]^T\,[\bar{K}]\,[Q] \tag{10a}$$

where

$$[Q] = \begin{bmatrix} \boldsymbol{s}^T \\ \boldsymbol{n}^T \\ \boldsymbol{t}^T \end{bmatrix} \tag{10b}$$

Eq.(10a) defines the operator $[K]$ required to complete the formulation of the problem, viz. eq.(6).

(ii) *Saturated material; undrained conditions*

Assume first that both constituents respond in an undrained manner and establish the functional form of the constitutive relations (4). The behaviour of the intact material is defined by

$$(\dot{\boldsymbol{\sigma}}^{(1)})' = [D']\,\dot{\boldsymbol{\varepsilon}}^{(1)}\,; \quad \dot{p}^{(1)} = \frac{K_f}{\eta^{(1)}}\,\boldsymbol{\delta}^T\,\dot{\boldsymbol{\varepsilon}}^{(1)} \tag{11}$$

where η is the porosity, p represents the excess of pore pressure and $\boldsymbol{\sigma}'$ designates the effective stress. Invoking the Terzaghi's decomposition, yields

$$\dot{\boldsymbol{\sigma}}^{(1)} = [D]\,\dot{\boldsymbol{\varepsilon}}^{(1)}; \quad [D] = [D'] + \boldsymbol{\delta}\,\frac{K_f}{\eta^{(1)}}\,\boldsymbol{\delta}^T \tag{12}$$

For the shear band material,

$$\dot{\bar{\boldsymbol{T}}}' = [\bar{K}']\,\dot{\bar{\boldsymbol{g}}}; \quad \dot{\bar{\boldsymbol{T}}} = \dot{\bar{\boldsymbol{T}}}' + \dot{p}^{(2)}\,\bar{\boldsymbol{n}} \tag{13}$$

where

$$\dot{p}^{(2)} = \frac{K_f}{\eta^{(2)}\,t}\,\boldsymbol{n}^T\,\dot{\boldsymbol{g}} \tag{14}$$

Substituting now eq.(14) in eqs.(13), the following constitutive relation is obtained

$$\dot{\bar{\boldsymbol{T}}} = [\bar{K}']\,\dot{\bar{\boldsymbol{g}}} + \bar{\boldsymbol{n}}\,\frac{K_f}{\eta^{(2)}\,t}\,\boldsymbol{n}^T\,\dot{\boldsymbol{g}} \tag{15}$$

Finally, transforming eq.(15) to the global frame of reference

$$\dot{\boldsymbol{T}} = [K]\dot{\boldsymbol{g}}; \; [K] = [Q]^T[\bar{K}'][Q] + \boldsymbol{n}\frac{K_f}{\eta^{(2)}\,t}\,\boldsymbol{n}^T \tag{16}$$

Eqs.(12) and (16) specify the functional forms of $[D]$ and $[K]$ required to define the global macroscopic response, viz. eq.(6).

It should be noted that the above formulation is rather restrictive as it assumes that both constituents are subjected to undrained deformation. In many practical situations involving localization in cohesionless granular media, the time required for the dissipation of the pore pressure gradient in both constituents may be short compared to the time interval associated with an increase in the external load. This is due to the fact that $t \to 0$ and the permeability of the adjacent intact material is relatively high. Thus, it may be desirable to reformulate the problem by accounting for a partial drainage in both constituents under the global constraint of undrained deformation.

Express the change in pore pressure in both constituents as

$$\dot{p}^{(1)} = \frac{K_f}{\eta^{(1)}}\boldsymbol{\delta}^T(\dot{\boldsymbol{\varepsilon}}^{(1)} - \dot{\boldsymbol{\varepsilon}}_d^{(1)}); \; \dot{p}^{(2)} = \frac{K_f}{\eta^{(2)}t}\boldsymbol{n}^T(\dot{\boldsymbol{g}} - \dot{\boldsymbol{g}}_d) \tag{17}$$

where the subscript d designates the strain rates/velocity discontinuities resulting from partial drainage. It is noted that, since K_f is very large compared to drained moduli of both constituent materials, the mechanical response is virtually insensitive to the values of η's, so that one can assume, without a loss in accuracy, $\eta^{(1)}=\eta^{(2)}=\eta$.

In order to define the macroscopic response of the system, it may be sufficiently accurate to describe the deformation process in terms of undrained behaviour constraint by $\dot{p}^{(1)}=\dot{p}^{(2)}$. Thus,

$$\frac{K_f}{\eta}\,\boldsymbol{\delta}^T(\dot{\boldsymbol{\varepsilon}}^{(1)} - \dot{\boldsymbol{\varepsilon}}_d^{(1)}) = \frac{K_f}{\eta\,t}\,\boldsymbol{n}^T(\dot{\boldsymbol{g}} - \dot{\boldsymbol{g}}_d) \tag{18}$$

Combining eq. (18) with the mass balance equation

$$\boldsymbol{\delta}^T\,\dot{\boldsymbol{\varepsilon}}_d = \mu_1\,\boldsymbol{\delta}^T\,\dot{\boldsymbol{\varepsilon}}_g^{(1)} + \mu\,\boldsymbol{n}^T\,\dot{\boldsymbol{g}}_d \tag{19}$$

one obtains, after some algebraic transformations,

$$\dot{p}^{(1)} = \frac{K_f}{\eta}\,\boldsymbol{\delta}^T\dot{\boldsymbol{\varepsilon}} = \dot{p}^{(2)} \tag{20}$$

In order to specify the global constitutive relation, it is convenient now to formulate the static constraint of traction continuity, eq.(3), in terms of effective components

$$\dot{\boldsymbol{T}}' = [N]^T(\dot{\boldsymbol{\sigma}}^{(1)})' = [Q]^T[\bar{K}'][Q]\,\dot{\boldsymbol{g}} = [K']\,\dot{\boldsymbol{g}} \tag{21}$$

which leads to

$$[N]^T[D'](\dot{\boldsymbol{\varepsilon}}^{(1)}) = [K']\,\dot{\boldsymbol{g}} \tag{22}$$

Substituting in eq.(22) the strain decomposition (1), together with representation (2), the following relations are obtained

$$\dot{\boldsymbol{g}} = [S']\,\dot{\boldsymbol{\varepsilon}}; \quad \dot{\boldsymbol{\varepsilon}}^{(1)} = ([I] - \mu[N][S'])\,\dot{\boldsymbol{\varepsilon}} \tag{23a}$$

where

$$[S'] = \left([K'] + \mu[N]^T[D'][N]\right)^{-1}[N]^T[D'] \tag{23b}$$

Writing now the constitutive relation for the intact material in the form

$$\dot{\boldsymbol{\sigma}}^{(1)} = [D']\,\dot{\boldsymbol{\varepsilon}}^{(1)} + \boldsymbol{\delta}\,\frac{K_f}{\eta}\,\boldsymbol{\delta}^T\,\dot{\boldsymbol{\varepsilon}} \tag{24}$$

and substituting the second equation in (23a), the following global relation can be established

$$\dot{\boldsymbol{\sigma}} \approx \dot{\boldsymbol{\sigma}}^{(1)} = \left([D] - \mu[D'][N][S']\right)\dot{\boldsymbol{\varepsilon}} \tag{25}$$

(iii) *Partially saturated material*

Consider now the case when both constituent materials are partially saturated. Assume that the liquid phase is continuous while the gas phase is in the form of discrete inclusions (bubbles) embedded in the liquid. This case corresponds to a high degree of saturation and the mathematical formulation of the problem has been presented by Pietruszczak and Pande (1996). Assume again that both constituents respond in an undrained manner. The constitutive relation governing the response of the intact material can be derived from

$$(\dot{\boldsymbol{\sigma}}^{(1)})' = [D'] \dot{\boldsymbol{\varepsilon}}^{(1)}; \quad \dot{p}^{(1)} = \frac{K_m^{(1)}}{\eta^{(1)}} \boldsymbol{\delta}^T \dot{\boldsymbol{\varepsilon}}^{(1)} \tag{26}$$

where K_m represents the bulk modulus of the air-water mixture. The latter is defined as

$$\frac{K_m^{(1)}}{\eta^{(1)}} = B_w^{(1)} K_f \tag{27}$$

where

$$B_w^{(1)} = \left[\eta^{(1)} \{ S_r^{(1)} + (1 - S_r^{(1)}) \frac{K_f}{K_a^{(1)} - \beta^{(1)}} \} \right]^{-1}$$
$$\beta^{(1)} = \frac{T}{3\rho_V} \frac{S_r^{(1)}}{\sqrt{1 - S_r^{(1)}}} \tag{28}$$

In the above equation, T is the surface tension force (per unit length of the air-water meniscus) and K_a is the bulk modulus of air. Moreover, S_r is the degree of saturation and ρ_v is the 'average pore size' defined as $\rho_v = V_v / S_s$, where V_v is the volume of voids and S_s is the internal solid surface area. The average pore size has the dimension of length and is physically analogous to the concept of hydraulic radius. The evolution law for the degree of saturation takes the form

$$\dot{S}_r^{(1)} = S_r^{(1)}(1 - S_r^{(1)}) B_a^{(1)} \boldsymbol{\delta}^T \dot{\boldsymbol{\varepsilon}}^{(1)}; \quad B_a^{(1)} = \frac{B_w^{(1)} K_f}{(K_a^{(1)} - \beta^{(1)})} \tag{29}$$

Invoking now Terzaghi's decomposition, together with eqs. (26) and (27), the following constitutive relation is obtained

$$\dot{\boldsymbol{\sigma}} = [D] \dot{\boldsymbol{\varepsilon}}; \quad [D] = [D'] + \boldsymbol{\delta} B_w^{(1)} K_f \boldsymbol{\delta}^T \tag{30}$$

For the shear band material, the constitutive relation can be derived from

$$\dot{\boldsymbol{T}}' = [\bar{K}'] \dot{\boldsymbol{g}}; \quad \dot{p}^{(2)} = \frac{K_m^{(2)}}{\eta^{(2)}} \boldsymbol{n}^T \dot{\boldsymbol{g}} = \frac{K_f}{t} B_w^{(2)} \boldsymbol{n}^T \dot{\boldsymbol{g}} \tag{31}$$

where the definition of K_m and B_w is analogous to that in eqs.(26) and (27). Substituting eq.(31) in Terzaghi's decomposition (13) and transforming the resulting expression to the global frame of reference, yields

$$\dot{\boldsymbol{T}} = [K] \dot{\boldsymbol{g}}; \quad [K] = [Q]^T [\bar{K}'] [Q] + \boldsymbol{n} \frac{K_f}{t} B_w^{(2)} \boldsymbol{n}^T \tag{32}$$

Eqs.(30) and (32) specify the functional forms of $[D]$ and $[K]$ required to define the global macroscopic response, viz. eq.(6).

Finally, the problem can again be reformulated by accounting for partial drainage in both constituents under the global constraint of undrained deformation. The simplest approach (which is rather restrictive in the context of a partial saturation) is that based on the procedure adopted for saturated soils. In this case, the response is governed by eqs.(17)-(25), in which K_f is replaced by $B_w^{(i)} K_f$ (for i=1,2).

3. NUMERICAL EXAMPLES

(i) *Drained behaviour*

The first problem considered here involves the specimen of a cemented geomaterial (e.g., rock) subjected to drained uniaxial compression under plane strain conditions. The sample is assumed to contain an inhomogeneity in the form of a weak inclusion (compressive strength reduced by 16.7%), Fig.1a. The loading process consists of applying uniform vertical displacements along the top surface under the condition of no friction at the end platens.

The response of the material in the homogeneous range has been described using a plasticity framework outlined in the article by Pietruszczak et al. (1988). The following material parameters have been chosen: E=50,000MPa, ν=0.167, f_c=30MPa, f_t=0.1f_c, A=0.000085, B=0.95, where E and ν are the elastic constants, f_c and f_t represent the uniaxial compressive and tensile strength, and A,B are constants employed in the hardening function. The transition to localized mode has been considered as a classical bifurcation problem. The properties of the interface have been described based on a simple formulation provided in the article by Pietruszczak (1995), in which k_n=100,000 MN/m, k_s=50,000 MN/m, B=50 m^{-1}, m_c=0 and m_r=0.6m_o. Here, k_n , k_s are the elastic moduli, m_r , m_c and B are the material parameters employed in the definition of the plastic potential and the degradation functions, whereas m_o represents the stress ratio at the onset of localization.

The numerical analysis has been carried out using three different meshes consisting of 32, 128 and 512 four-node isoparametric elements with full integration. The results of the simulations are presented in Fig.1. Fig.1b provides the mechanical characteristics, load against displacement, obtained for different types of discretization, whereas Fig.1c shows the fracture zone inside the specimen at the maximum load intensity. The load-displacement characteristics, obtained based on homogenization procedure, eq.(6), exhibit a transition from stable to unstable regime and are only marginally affected by the discretization. For comparison, Fig.1b also presents the results of a similar analysis based on conventional formulation, which ignores the inhomogeneous nature of deformation after the onset of localization. These simulations have been completed using the strain softening parameters provided in the original reference (Pietruszczak et al., 1988). As expected, the prediction of the ultimate load is quite unreliable as the results display a strong dependence on the details of discretization.

ii) *Saturated material; undrained conditions*

The next problem addressed is that of a saturated sample of a medium dense sand subjected to undrained compression under plane strain regime. The specimen, which is assumed to be initially homogeneous, is compressed under the constraint of perfect bonding at the end platens. The test is said to be conducted at the initial confining pressure of 450 kPa. The material properties have been described based on the formulation provided by Pietruszczak and Stolle (1987). In the range of homogeneous deformation, the following material parameters have been selected:
E=62,000kPa, ν=0.24, A=0.0015, η_f=0.407, η_c=0.4.
Here, η_f and η_c define the slope of the failure and zero dilatancy lines in the meridional plane corresponding to 'triaxial' compression and A is a constant employed in the hardening function. The properties of the shear band material have been described by invoking, once again, the formulation given in Pietruszczak (1996), in which:
k_n=150,000 kN/m, k_s= 75,000 kN/m, B=50 m^{-1}
m_c= 0, m_r= 0.6m_o.
The analysis has been conducted assuming undrained conditions in both constituents, i.e. employing the representation (6), together with (12) and (16).

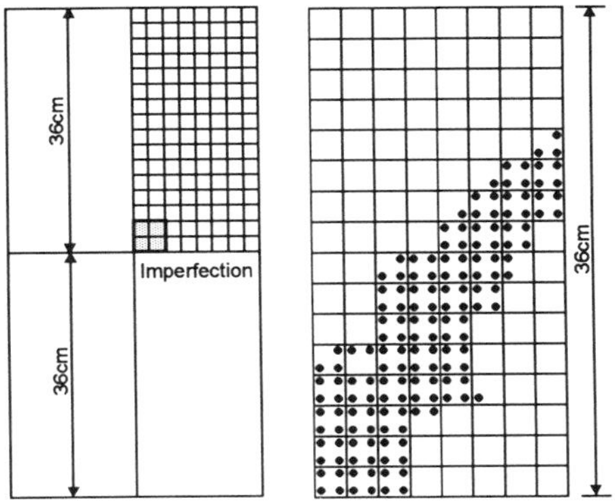

(a) (c)

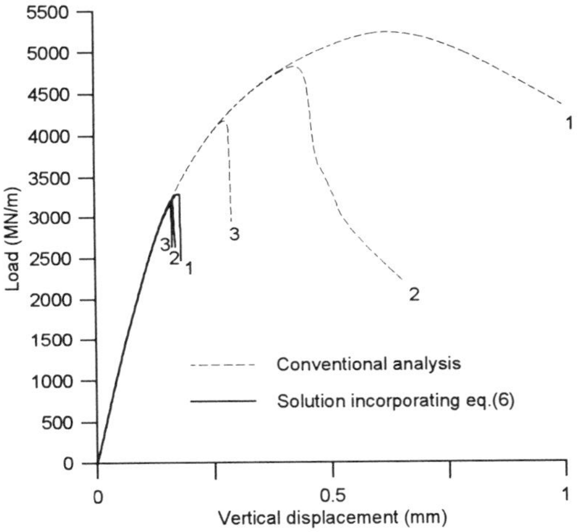

(b)

(1) - 4x8 elements; (2) - 8x16 elements
(3) - 16x32 elements

Fig.1. Rock specimen, with an initial imperfection, subjected to plane strain uniaxial compression
(a) geometry of the problem; (b) load-displacement characteristics; (c) fracture zone at the ultimate load

The simulations have been carried out for two differently aligned meshes, as shown in Fig.2a. Each mesh consisted of 128 six-node triangular elements. The main results are presented in Fig.2b, which shows the load-displacement characteristics together with the location of the fracture zone inside the sample. Clearly, the mechanical characteristics are virtually the same for both grids. It is noted however, that for this particular example the initiation of localized deformation takes place at the load intensity which is close to the ultimate load. Thus, although the results are mesh-independent, no general conclusions on mesh alignment sensitivity can be drawn here. Further studies are required examining a broader class of boundary-value problems.

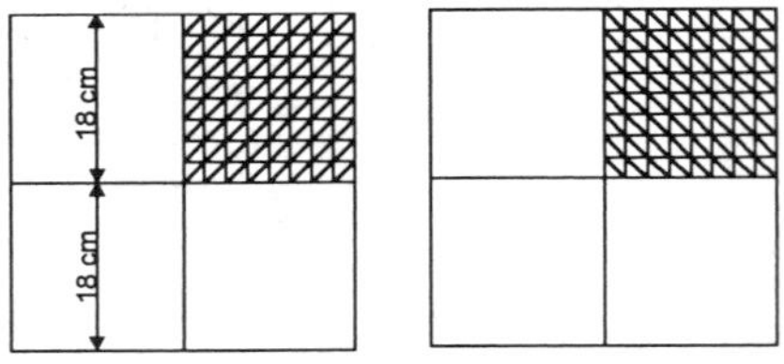

(a)

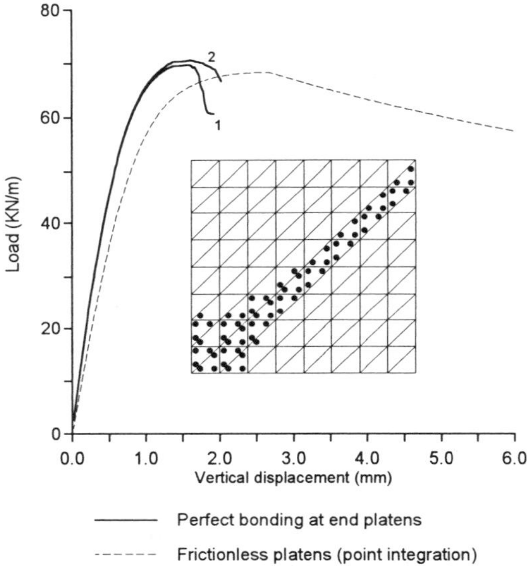

(b)

Fig.2. Sand specimen subjected to undrained plane strain uniaxial compression; (a) geometry of the problem; (b) load-displacement characteristics

(iii) *Partially saturated sample*

In order to examine the performance of the proposed formulation, viz. eqs. (30) and (32), consider the behaviour of a dense sand subjected to plane strain deformation. In particular, choose an example analogous to that discussed by Pietruszczak (1995). While the emphasis in that paper was on simulation of the experimental results corresponding to saturated sample, the main objective here is to study the influence of partial saturation. The example is based on experimental data provided by Han and Vardoulakis (1991). A fully saturated specimen (of height 0.135m and porosity of 0.374) was tested under the initial effective confining pressure of 206kPa and the back-pressure of 300kPa. The test was conducted under undrained conditions. The numerical simulations presented here, which are based on direct integration of the governing constitutive equations, have been completed using the formulation and material parameters identical to those employed in the original reference. The only additional variable was the degree of saturation of the sample.

The main results of the simulations are shown in Fig.3. The results are presented in terms of the stress/strain measures usually adopted for plane strain analysis, i.e. $t=(\sigma_{22}-\sigma_{11})/2$, $\gamma=\varepsilon_{22}-\varepsilon_{11}$. Fig.3a shows the results corresponding to drained and fully saturated conditions. For the drained conditions, the classical bifurcation analysis predicts the onset of localized deformation at γ=3.7% (the shear band inclination angle β, with respect to the horizontal axis, is β=54°), followed by an unstable response. For fully saturated sample, the homogeneous deformation process is associated with a build up of negative excess of pore pressure. After the onset of localization the deviatoric characteristic remains stable. Fig.3b shows the results for a partially saturated sample at the degree of saturation of 0.9. In general, the mechanical characteristics of partially saturated soils are sensitive to the initial back-pressure, p_{wo}, which influences the compressibility of the air-phase. This is clearly evident in Fig.3b, which provides the results corresponding to the actual back-pressure of 300kPa and that of 100 kPa. While the sample at p_{wo}=300 kPa remains homogeneous throughout the deformation process, at p_{wo}=100 kPa a transition to localized deformation takes place. In the case when both constituents respond in an undrained manner, the resulting deviatoric characteristic is only marginally affected by localization and it remains stable. The solution based on coupled formulation (i.e., eqs.(17)-(25) with K_f replaced by $B_w^{(i)} K_f$), which permits the exchange of fluids between constituents, predicts an unstable response.

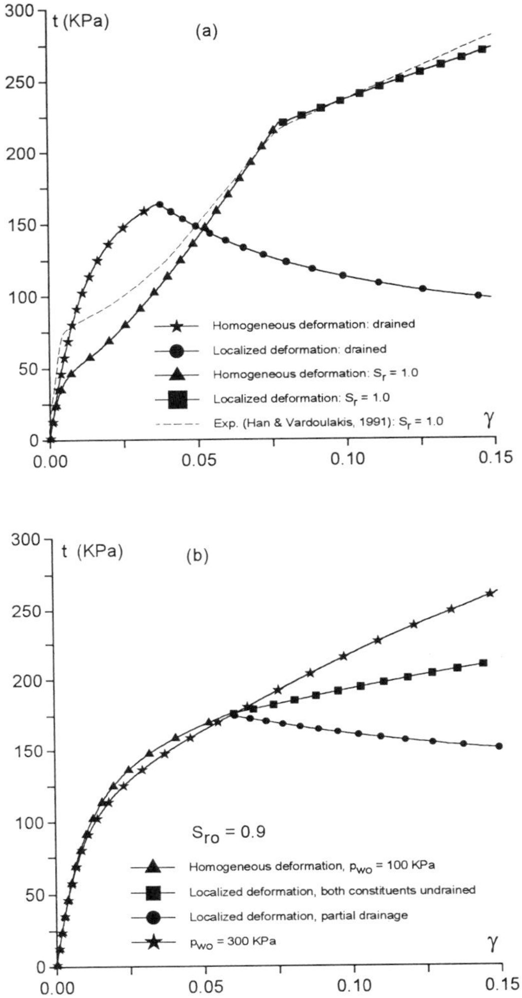

Fig.3 Influence of the degree of saturation on the deviatoric characteristics in plane strain deformation of a dense sand

4. FINAL REMARKS

The paper outlines the methodology for describing the mechanical response of fluid-infiltrated soils in the presence of localized deformation. A general formulation has been provided, which includes the case of drained as well as undrained deformation of saturated and partially saturated soils. The framework has been illustrated by a series of numerical examples pertaining to each of these cases. The transition to localized deformation has been based on classical bifurcation analysis. It should be pointed out that this criterion gives quite reasonable predictions for drained analysis, under undrained conditions however its performance is rather questionable (Pietruszczak, 1995). In general, the finite element solutions display little sensitivity to the mesh size. It is important to note however, that the 'weak' dependance, which is pertinent to the finite element method itself, is still present. The issue of mesh alignment sensitivity requires further investigation in a broader context of different boundary-value problems. Finally, the solutions incorporating the homogenization procedure will, in general, tend to overestimate the ultimate (collapse) load, which is primarily due to kinematic simplifications embedded in the averaging procedure.

REFERENCES

De Borst., R., L.J. Sluys, H.B. Muhlhaus & J. Pamin. 1993. Fundamental issues in finite element analyses of localization of deformation. *Engrg.Comput.* 10: 99-121.

Han, C. & I. Vardoulakis. 1991. Plane strain compression experiments on water-saturated fine-grained sand. *Geotechnique*. 41: 49-78.

Lade, P.V. 1994. Instability and liquefaction of granular materials. *Comp.&Geotech.* 16: 123-151.

Lade, P.V., R.B. Nelson & Y.M. Ito. 1988. Instability of granular materials with nonassociated flow. *J.Engrg.Mech.*, ASCE, 114: 2173-2191.

Pietruszczak, S. 1995. Undrained response of granular soil involving localized deformation. *J.Engrg.Mech.*, ASCE, 121: 1292-1297.

Pietruszczak, S., J. Jiang & F.A. Mirza. 1988. An elastoplastic constitutive model for concrete. *Int. J. Solids Structures.* 24: 705-722.

Pietruszczak, S. & X. Niu. 1993. On the description of localized deformation. *Int.J.Num.Anal.Meth.Geomech.* 17: 791-805.

Pietruszczak, S. & G.N. Pande. 1996. Constitutive relations for partially saturated soils containing gas inclusions. *J.Geotechn.Engrg.* ASCE. 122: 50-60.

Pietruszczak, S. & D.F.E. Stolle. 1987. Modelling of sand behaviour under earthquake excitation. *Int.J.Num.Anal.Meth.Geomech.* 11: 221-240.

Vardoulakis, I. & B. Graf. 1985. Calibration of constitutive models for granular materials using data from biaxial experiments. *Geotechnique*. 35: 299-317.

Numerical Models in Geomechanics, Pietruszczak & Pande (eds) © *1997 Balkema, Rotterdam, ISBN 90 5410 886 X*

Continuum finite element analysis of strain localization in slopes

R.A. Regueiro & R.I. Borja
Department of Civil Engineering, Stanford University, Calif., USA

ABSTRACT: This paper presents a continuum finite element analysis of strain localization in slopes. The stability problem is viewed from the standpoint of strong discontinuity (jump in displacement field) as opposed to weak discontinuity (jump in strain field). For this type of problem, an enhanced finite element solution has previously been shown to be independent of mesh refinement and insensitive to mesh alignment. Numerical simulations of a load-driven slope stability problem in infinitesimal plane strain will demonstrate these characteristics, which are necessary for a finite element solution to be meaningful.

1 INTRODUCTION

Analytical limit equilibrium methods are currently used to analyze slope stability problems where certain simplifying assumptions are made with regard to soil material properties, problem geometry, and shape of slip surface. For complex slope geometries and soil constitutive models, however, analytical limit equilibrium solutions are unwieldy, and the assumption that limit is reached at the load at which a standard finite element (FE) solution fails to converge is unsound. Thus, the need for a more sophisticated numerical tool is in order.

For a FE analysis to be meaningful, it must be objective with respect to mesh refinement and insensitive with respect to mesh alignment (i.e. element sides need not be aligned with the slip surface for the numerical solution to proceed). Both of these characteristics have previously been demonstrated in the context of strong discontinuities (Simo et al. 1993, Simo & Oliver 1994, Armero & Garikipati 1995, Garikipati 1996). A numerical example presented in this paper will demonstrate these attractive qualities of the model.

From the perspective of the practicing geotechnical engineer, post-limit behavior of geotechnical structures is considered useless because the soil has already failed, which is assumed manifest in the formation of a slip surface and in a softening behavior of the soil. In reality, however, the geotechnical structure could contain a slip surface and still be in a pre-limit (pre-collapse) state. Thus, from an analysis standpoint, post-localization behavior should be represented in the model, as long as it is pre-limit. This paper will demonstrate the use of the FE method with strain enhancements to accurately capture slip surfaces in slopes and to predict post-localization behavior, which, it turns out, occurs *before* the standard FE solution fails to converge for a load-driven problem. The critical height of slope predicted by the standard and enhanced FE analyses agrees with that predicted by a simple analytical limit equilibrium analysis. With respect to settlement under a surcharge load at the slope crest, the standard FE solution is shown to be *unconservative* because it predicts a smaller settlement than the enhanced FE solution. In addition, results of the enhanced FE solution motivate the use of the geometrically nonlinear theory. As mentioned above, the usefulness of the FE approximation can be fully realized when conducting numerical slope stability analyses of complex slope geometries and soil constitutive models for which analytical limit equilibrium solutions are unmanageable.

The FE strain enhancements arise naturally from a treatment of classical continuum plasticity with strong discontinuity. It is noteworthy to recognize that the assumption of a discontinuous displacement field leading to a singular strain field is valid for soils because as a shear band forms between two rigid soil masses sliding along one another, the thickness of the band is negligible (the band reduces to a surface), and the strain across the band approaches infinity (see pictures in Vardoulakis et al. 1978). Important results of the formulation with strong discontinuity are the distributional form of the hardening/softening modulus H, the model-specific stress-displacement relation along the slip surface, and the model-specific localization condition (by "model-specific" it is meant results specific to the classical continuum plasticity model formulated in the context of strong discontinuities). Here, a $J2$ flow plasticity model is formulated in the context of strong discontinuities.

2 $J2$ PLASTICITY WITH STRONG DISCONTINUITY

The following theory is essentially a re-working of that developed by the late Professor Juan C. Simo and his co-workers (Simo et al. 1993, Simo & Oliver 1994, Armero & Garikipati 1995, Garikipati 1996). Also, to keep redundancy to a minimum, reference will be made to the development already given in Borja & Regueiro (1997).

2.1 *Review of classical $J2$ plasticity*

Before showing the results of the formulation of $J2$ plasticity with strong discontinuities, it is appropriate to review the classical theory. The yield function $\mathcal{F}$ may be written as

$$\mathcal{F}(\boldsymbol{\sigma}) := J_{2D}^2 - \kappa^2 = 0 \qquad (1)$$

where $J_{2D} = \sqrt{\frac{1}{2}\boldsymbol{\sigma}' : \boldsymbol{\sigma}'}$ is the second invariant of the deviatoric Cauchy stress tensor $\boldsymbol{\sigma}'$, and κ is a value constant in stress space used to express the radius $R = \sqrt{2}\kappa = \|\boldsymbol{\sigma}'\|$ of the cylindrical yield surface formed in stress space. An associative flow rule is assumed as

$$\dot{\boldsymbol{\epsilon}}^p = \lambda \hat{\boldsymbol{n}} \qquad (2)$$

where λ is the plastic consistency parameter and $\hat{\boldsymbol{n}} = \boldsymbol{\sigma}'/\|\boldsymbol{\sigma}'\|$ is the unit normal to the yield surface in stress space. The implication of the assumed associative form of the flow rule (i.e. the direction of the plastic flow normal to the yield surface in stress space) is that in order for there to be loss of strong ellipticity of the acoustic tensor, the hardening/softening modulus must be less than or equal to zero (i.e. perfect of softening plasticity); otherwise, for a nonassociative flow rule, this condition is not necessary (Ortiz et al. 1987).

2.2 *Model-specific stress-displacement relation along discontinuity, and localization condition*

From the form of the singular part of the plastic consistency parameter, λ_δ, resulting from required regularity (i.e. non-distributional or not singular) of the yield function (Borja & Regueiro 1997), the stress-displacement relation may be written as

$$\begin{aligned} \zeta(t) &= \psi^{-1}\boldsymbol{\sigma}' : \boldsymbol{c}^e : \operatorname{symm}(\boldsymbol{\nabla}\dot{\bar{\boldsymbol{u}}}) \ ; \\ \psi &= \frac{2\sqrt{2}}{3}\kappa H_\delta \left[\frac{\boldsymbol{\sigma}' : \boldsymbol{c}^e : \operatorname{symm}(\boldsymbol{m}\otimes\boldsymbol{n})}{\boldsymbol{\sigma}' : \boldsymbol{c}^e : \hat{\boldsymbol{n}}}\right] \end{aligned} \qquad (3)$$

where $\zeta(t)$ is the magnitude of the jump displacement rate across the discontinuity as a function of time t; $\boldsymbol{c}^e$ is the fourth-order elastic modulus tensor; $\bar{\boldsymbol{u}}$ is the regular part of the displacement vector $\boldsymbol{u}(\boldsymbol{x}, t)$; H_δ is the singular part of the hardening/softening modulus; $\boldsymbol{m}$ is the unit vector denoting the direction of the jump displacement; and $\boldsymbol{n}$ is the unit normal to the discontinuity surface. This expression (3) may be further simplified using results from the localization condition.

From a basic consideration of Newton's third law of motion , we see that the traction across the discontinuity must be continuous (Malvern 1969); also see the discussion by Simo & Oliver (1994) for traction continuity demonstrated in the context of the weak form (variational equation of equilibrium). The model-specific localization condition arises from the satisfaction of this condition that the traction rate be regular on the discontinuity (Armero & Garikipati 1995, Borja & Regueiro 1997). For the case of plane strain this condition may be written as

$$\frac{\|\boldsymbol{\sigma}'\|}{\sqrt{2}|r|} = 1 \quad ; \qquad \theta = \pm 45^\circ \qquad (4)$$

where $r = \frac{1}{2}(\sigma_1 - \sigma_2) = \boldsymbol{n}\cdot\boldsymbol{\sigma}\cdot\boldsymbol{m}$ is the *resolved* stress along the discontinuity; σ_1 and σ_2 are the major and minor principal stresses ($\boldsymbol{\sigma}$ written in principal stress space); and θ is the angle to the discontinuity normal $\boldsymbol{n}$ from the major principal stress axis. Note that there are two possible directions of $\boldsymbol{n}$ ($\theta = \pm 45^\circ$), a result accounted for in the numerical implementation by choosing the one which aligns the slip surface with the element deformation.

Now, with the results of the localization condition, the model-specific stress-displacement relation (3) representing the softening along the discontinuity surface may be rewritten as

$$r(t) = r(0) + \frac{1}{3} H_\delta\, \alpha(t) \quad ; \quad H_\delta \leq 0 \qquad (5)$$

where $r(0)$ is the resolved stress on the discontinuity at the inception of localization (i.e. satisfaction of (4)), and $\alpha(t)$ is the magnitude of the jump displacement ($\dot{\alpha}(t) = \zeta(t)$, where $\zeta(t)$ is defined in (3)).

3 NUMERICAL IMPLEMENTATION

The finite element formulation begins with a treatment of the standard weak form with strong discontinuities in the context of the assumed enhanced strain method (Simo & Rifai 1990). The discretized form is as follows (Simo & Oliver 1994)

$$\begin{aligned} \int_\Omega \boldsymbol{\nabla}\bar{\boldsymbol{\eta}}^h : \boldsymbol{\sigma}^h \mathrm{d}\Omega &= \int_\Omega \bar{\boldsymbol{\eta}}^h \cdot \boldsymbol{f}\, \mathrm{d}\Omega + \int_{\Gamma_t} \bar{\boldsymbol{\eta}}^h \cdot \boldsymbol{t}\, \mathrm{d}\Gamma \\ \int_{\Omega_{e,loc}} \tilde{\gamma}_e^h : \boldsymbol{\sigma}^h \mathrm{d}\Omega &= 0 \end{aligned} \qquad (6)$$

where $\bar{\boldsymbol{\eta}}^h \in \mathcal{V}^h$ is the regular part of the discretized displacement variation in the space of admissible discretized test functions $\mathcal{V}^h$; $\boldsymbol{f}$ and $\boldsymbol{t}$ are the prescribed body and traction forces, respectively; and $\tilde{\gamma}_e^h \in \tilde{\mathcal{E}}^h$ is the enhanced strain variation in the space of admissible strain variations $\tilde{\mathcal{E}}^h$ over the localized element e with a form chosen to satisfy the patch test (Simo & Rifai 1990). The superscript h denotes finite element discretization. From (6), with the choice of appropriate standard and enhanced shape functions (Borja & Regueiro 1997), the following discretized governing finite element equations result

$$\begin{aligned} \boldsymbol{R} &:= \boldsymbol{f}^{\mathrm{ext}} - \mathbf{A}_{e=1}^{n_{el}} \int_{\Omega_e} \boldsymbol{B}_e^T \boldsymbol{\sigma}^h \mathrm{d}\Omega = \boldsymbol{0} \\ b_e &:= -\int_{\Omega_e} \boldsymbol{G}_{\mathrm{patch}}^T : \boldsymbol{\sigma}^h \mathrm{d}\Omega + \int_{S_e} r_S\, \mathrm{d}S = 0 \end{aligned} \qquad (7)$$

where $\boldsymbol{f}^{\mathrm{ext}}$ is the standard external force vector, $\mathbf{A}_{e=1}^{n_{el}}$ is the finite element assembly operator, $\boldsymbol{B}_e$ is the standard strain-displacement matrix for element e, $\boldsymbol{G}_{\mathrm{patch}}$ is the coupling matrix chosen to ensure satisfaction of the patch test, $r_S = (\int_{\Omega_e} \boldsymbol{n} \cdot \boldsymbol{\sigma}^h \cdot \boldsymbol{m} \mathrm{d}\Omega)/A_e$ is the average resolved stress on the discontinuity, and $\boldsymbol{R}$ and b_e are the corresponding residuals. Note that the governing equation of resolved stress on the discontinuity $(7)_2$ is discontinuous from element to element, which is consistent with the standard C^0 finite element approximation (i.e. the displacements at most must be continuous across element boundaries leading to discontinuous strains). This characteristic of the enhanced strain method is an attractive feature of the model from a numerical implementation standpoint: static condensation of the jump displacement for element e, α_e, may occur at the element level. Equation (7) is consistently linearized to make the equations amenable to solution by the Newton-Raphson method.

4 NUMERICAL EXAMPLE: LOAD-DRIVEN SLOPE STABILITY PROBLEM

To demonstrate the model, a simple load-driven slope stability analysis in infinitesimal plane strain will be performed with the enhanced finite element method discussed in Section 3 and the $J2$ plasticity model with strong discontinuities discussed in Section 2. With such a constitutive model (i.e. deviatoric plastic flow) it is appropriate to model an overconsolidated (significant elastic region) cohesive soil in undrained condition (friction angle $\phi = 0$, zero volumetric plastic flow). Material properties are: modulus of elasticity $E = 10^3$ kPa, Poisson's ratio $\nu = 0.4$, uniaxial yield strength $\bar{\sigma}_o = 40$ kPa, standard hardening/softening modulus $\bar{H} = 0$, hardening/softening modulus along slip surface, $H_\delta = 0$, and saturated unit weight of soil $\gamma = 20$ kN/m^3. The assumed value of the yield strength $\bar{\sigma}_o$ may be interpreted as the failure strength of a cohesive soil with undrained shear strength $c_u = \bar{\sigma}_o/2 = 20$ kPa. For a slope at an angle of 63.43° with c_u and γ as given, an analytical limit equilibrium analysis (Atkinson 1993) predicts a critical height of $H_{cr} \approx 5$ m. Thus, it may be assumed that such a slope with height of 5 m is nearly unstable.

Consider the coarse and fine mesh discretizations of a 5 m high slope at angle 63.43° shown in Figures 1 and 2. The coarse mesh depicted in Figure 1 is composed of 151 nodes and 125 bilinear quadrilateral elements, and the fine mesh consists of 551 nodes and 500 elements. Each mesh is 20 m wide by 10 m high with the crest of the slope at the horizontal center of each mesh. Roller supports are assumed on the two vertical boundaries of each mesh, while the bottom boundary is assumed to be pinned to represent a firm base. The elements were numerically integrated with the $\bar{B}$-

method (2×2 deviatoric, 1-point volumetric) to alleviate mesh locking in the incompressible plastic regime. Points A and B shown in Figure 1 are used on each mesh to calculate the angle of rotation of the slope as it deforms under the surcharge q.

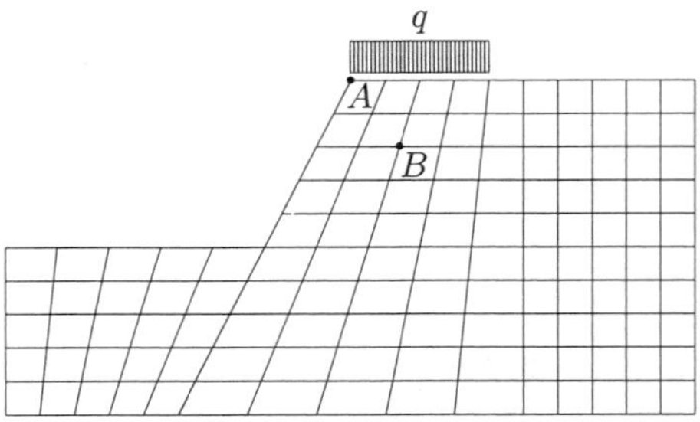

Figure 1. Coarse mesh showing surcharge load and points A and B used to calculate angle of rotation.

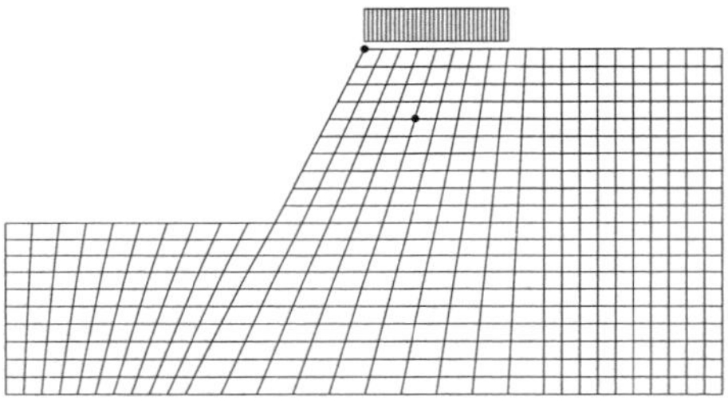

Figure 2. Fine mesh showing surcharge load and points A and B used to calculate angle of rotation (labels left off here).

After application of the gravity load, a surcharge of $q = 14$ kPa at the crest of the slope acting over 4 m is applied to perturb the problem; the small value of q was chosen because at 15 kPa the standard FE solution failed to converge (a uniformly distributed load of this magnitude is equivalent to a 70 cm-thick layer of soil). Along with the result that elements at the slope base begin to localize at the start of surcharge loading, this non-convergence of the standard FE problem at a small surcharge confirms the critical height of 5 m predicted by the analytical limit equilibrium analysis. In addition, when attempting to excavate the first soil layer 1m deep to the left of the continuous slope line extending through the mesh from the slope crest to the base, both the standard and enhanced FE solutions failed to converge.

Figures 3 and 4 show the deformed coarse mesh at end of surcharge loading for the standard and enhanced FE solutions, respectively, while Figures 5 and 6 show the same for the fine mesh (magnification factor for displacements = 1.0). The slip surfaces in Figures 4 and 6 are nearly toe circles, which is the shape assumed by the analytical limit equilibrium analysis, and have essentially the same shape for both the coarse and fine meshes, which demonstrates objectivity of the model with respect to mesh refinement. The other characteristic of the finite element model with strain enhancement needed for a numerical analysis to be meaningful—namely, insenstivity to mesh alignment—is easily observed as the slip surface traces across elements not in line with the elements' sides. Note that the elements that have been traced by the slip surface in Figures 4 and 6 have localized according to (4) and thus contain strain enhancement via the enhanced shape functions (Borja & Regueiro 1997) and softening due to plastic flow localized to the discontinuity (see (5) with $H_\delta = 0$). The elements not traced by the slip surface behave according to the standard FE formulation and standard $J2$ plasticity model.

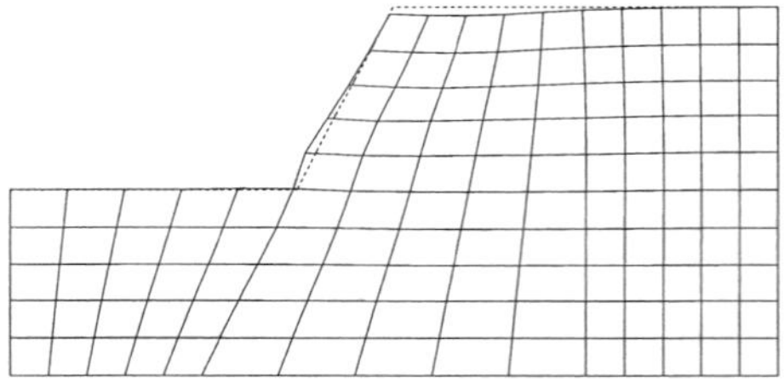

Figure 3. Standard FE solution: coarse mesh at end of surcharge loading.

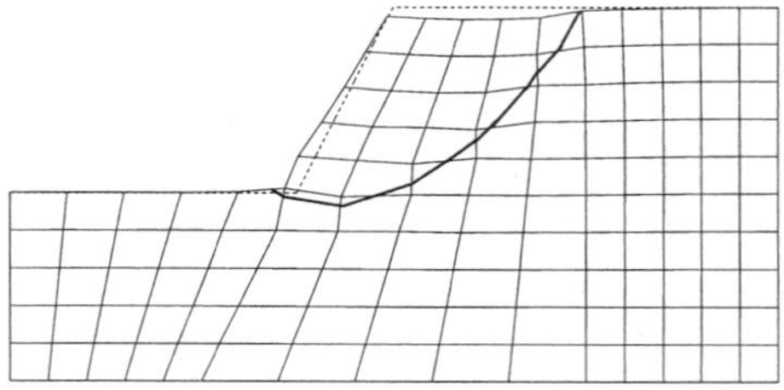

Figure 4. Enhanced FE solution: coarse mesh at end of surcharge loading showing slip surface.

To further prove objectivity of the model besides demonstrating similar slip surfaces, a plot of applied surcharge load versus resulting average downward displacement of the top surface of the

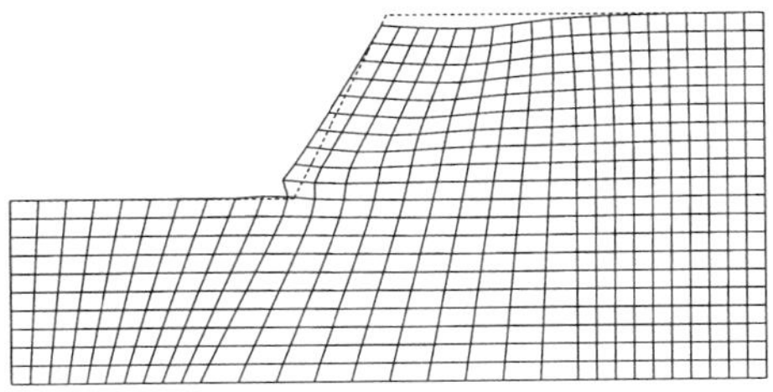

Figure 5. Standard FE solution: fine mesh at end of surcharge loading.

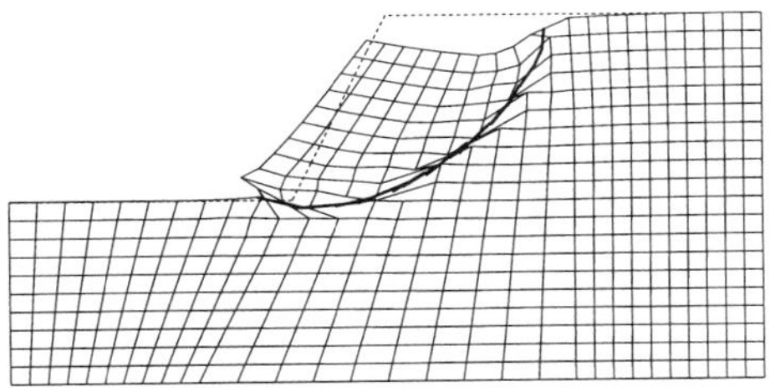

Figure 6. Enhanced FE solution: fine mesh at end of surcharge loading showing slip surface.

slope is shown in Figure 7 (Figure 8 is the same plot with abscissa range reduced); and a plot of applied surcharge load versus resulting angle of rotation of point B relative to point A (see Figure 1) is shown in Figure 9 (Figure 10 is the same plot with abscissa range reduced). Note that the displacements and rotations due to the enhanced FE solution for each mesh are in excess of those due to the standard FE solution by approximately the same amount for each load step, demonstrating objectivity with respect to mesh refinement. An already well-known result of the standard FE approximation, the fine mesh shows a softer response than the coarse mesh. As the load steps progress, the enhanced FE solution for both the coarse and fine meshes shows displacements and rotations increasing over those of the standard FE solution, but especially for the fine mesh during the last load step. Such large displacement and rotation at 56 kN/m for the fine mesh shown in Figure 6 motivates the geometrically nonlinear theory (Armero & Garikipati 1996b).

Since this is a load-driven problem, as the load approaches an apparent limit close to 60 kN/m, the FE solutions will eventually not be able to converge, needing a numerical solution method such as an arc-length method to advance the solution. A displacement-driven problem is easier to perform numerically for this reason, but for problems of interest to the geotechnical engineer, the load-driven problem is more appropriate. On this note, this analysis differs from the one conducted by Armero & Garikipati (1996a).

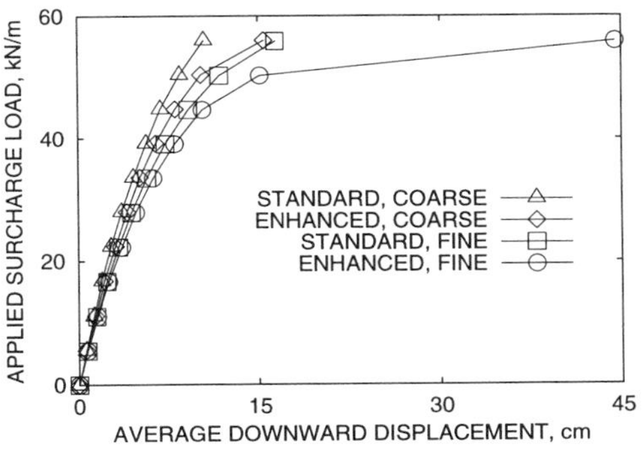

Figure 7. Variation of average downward displacement of top surface due to applied surcharge load.

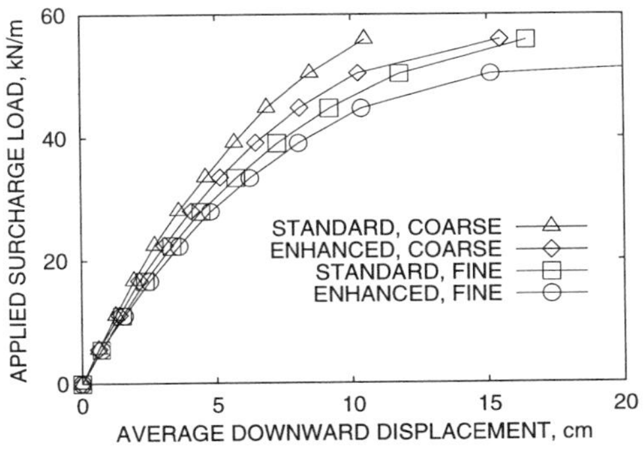

Figure 8. Variation of average downward displacement of top surface due to applied surcharge load (reduced abscissa range).

5 SUMMARY

This paper has demonstrated the usefulness of an enhanced FE method which meets two essential criteria in order for a finite element analysis to be meaningful: objectivity with respect to mesh refinement and insensitivity to mesh alignment. The enhanced FE method has been used to analyze a simple load-driven slope stability problem in order to make a comparison with an analytical limit equilibrium method. The usefulness of the

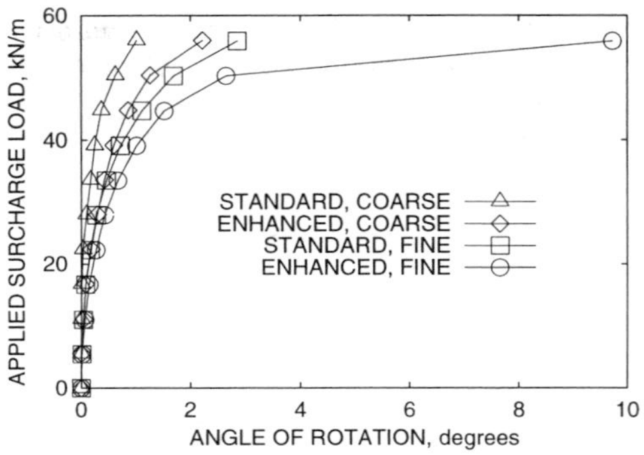

Figure 9. Variation of angle of rotation of point B relative to point A (see Figure 1) due to applied surcharge load.

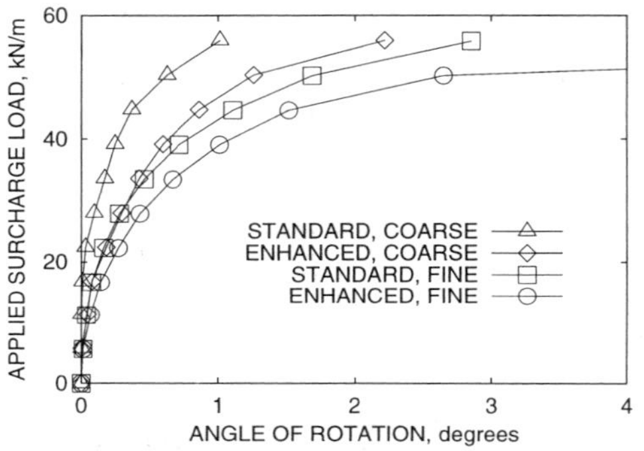

Figure 10. Variation of angle of rotation of point B relative to point A (see Figure 1) due to applied surcharge load (reduced abscissa range).

model to the practicing geotechnical engineer can be fully utilized when performing numerical slope stability analyses of complex slope geometries involving more complicated soil constitutive models. Work is now in progress to test the model with more appropriate soil constitutive models such as Drucker-Prager and Modified Cam-Clay.

6 ACKNOWLEDGEMENT

The authors would like to thank Dr. Krishna Garikipati of Stanford University for providing invaluable technical support.

REFERENCES

Armero, F. & K. Garikipati 1995. Recent advances in the analysis and numerical simulation of strain localization in inelastic solids. In D.R.J. Owen, E. Onate, & E. Hinton (eds), CIMNE, *Proceedings of Computational Plasticity IV, Barcelona, Spain, April 1995:* 547–561.

Armero, F. & K. Garikipati 1996a. An analysis of strong-discontinuities in inelastic solids with applications to the finite simulation of strain localization problems. In Y.K. Lin & T.C. Su (eds), *Proceedings of 11th Conference on Engineering Mechanics, Fort Lauderdale, FL, May 1996:* 136-139.

Armero, F. & K. Garikipati 1996b. An analysis of strong-discontinuities in multiplicative finite strain plasticity and their relation with the numerical simulation of strain localization in solids. *Int. J. Solids Structures.* 33:2863–2885.

Atkinson, J. 1993. *An introduction to the mechanics of soils and foundations through critical state soil mechanics.* McGraw-Hill.

Borja, R.I. & R.A. Regueiro 1997. Finite element analysis of strain localization in excavations. In F. Oka (ed.), *Proceedings of the International Symposium on Deformation and Progressive Failure in Geomechanics, Nagoya, Japan, October 1997:* in press.

Garikipati, K.R. 1996. *On strong discontinuities in inelastic solids and their numerical simulation.* Ph.D. Dissertation, Stanford University, Stanford, California.

Malvern, L.E. 1969. *Introduction to the mechanics of a continuous medium.* Prentice-Hall: 242.

Ortiz, M., Y. Leroy, & A. Needleman 1987. A finite element method for localized failure analysis. *Comput. Methods Applied Mech. Engrg.*, 61:189–214.

Simo, J.C., J. Oliver, & F. Armero 1993. An analysis of strong discontinuities induced by strain- softening in rate-independent inelastic solids. *Computational Mechanics.* 12:277–296.

Simo, J.C., & J. Oliver 1994. A new approach to the analysis and simulation of strain softening in solids. In Z.P. Bazant, Z. Bittnar, M. Jirasek, & J. Mazars (eds), *Fracture and Damage in Quasibrittle Structures.*

Simo, J.C., & M.S. Rifai 1990. A class of mixed assumed strain methods and the method of incompatible modes. *Int. J. Num. Methods in Engrg* 29:1595–1638.

Vardoulakis, I., M. Goldschieder, & G. Gudehus 1978. Formation of shear bands in sand bodies as a bifurcation problem. *Int. J. Num. Anal. Methods Geomech.* 2:99–128.

Numerical Models in Geomechanics, Pietruszczak & Pande (eds) © 1997 Balkema, Rotterdam, ISBN 90 5410 886 X

Water movement effect on the strain localization during a biaxial compression

R.Charlier, J.P.Radu & J.D.Barnichon
MSM Department, Université de Liège, Belgium & ALERT Geomaterials

ABSTRACT : This paper is devoted to the numerical modelling of the strain localisation in a water saturated sample of soil, using a large strain finite element code. First an internal friction constitutive law is used. It includes a Lode angle dependency. Then the coupling with a pore fluid is considered, and the linkages between the seepage and the soil strain and stress evolution is taken in account through an effective stress postulate and an adaptation of the storage law. Coupled monolythical finite elements are developed. Unsaturated media are considered using the Bishop formulation and an adaptation of the seepage model. Finally the developed finite element code is applied to the modelling of the plane strain compression (including a strain localisation) of samples with different initial pore pressures and different drainage conditions. Drained case, undrained fully saturated and undrained partly saturated cases are considered.

1. INTRODUCTION

Strain localisation has been investigated in soils and in rocks for about two decades. Mainly drained behaviour has been studied in soils and rocks. However practically most of these materials are fully or partially saturated by water, oil, gas,... The question of the bifurcation to a localised strain mode in a biphasic soil remains quite open. Now it is particularly important for example in geotechnics (analysis of landslides, of foundation stability,...) or in tectonophysics (sedimentary basin evolution,...).

Only very few authors have proposed solution to the strain localisation problems for saturated soils. Desrues and Mokni (1992) have conducted experiments on localisation in undrained saturated sand. They performed a series of biaxial compression tests (in plane strain state) on Hostun sand in order to characterise the localisation appearance and the shear band mode. Vardoulakis (1995a, 1995b) and Han (1991) have conducted similar experiments on a clay. Loret and Prevost (1991) first have proposed some theoretical and numerical analysis of such problems. Schrefler (1996) has more recently proposed a finite element modelling of a multiphase localisation problem. But this analysis is limited to small strains problems and is based on dynamic and seepage coupled model.

The present paper is devoted to a finite element modelling of the strain localisation in a (partly) saturated soil sample during a biaxial compression. A Van Eekelen - Drücker Prager constitutive law is used. The hydromechanical coupling is based on the Terzaghi's postulate and on the storage law. Unsaturated behaviour is then derived as an extension of the previous equations. The developed finite element are monolythical : they are associating 2 displacements degrees of freedom and 1 water pressure one. The developed code is used to model some biaxial compressions in various states : drained, undrained saturated, undrained unsaturated and under different permeabilities.

2. SOIL CONSTITUTIVE LAWS

2.1 Stress invariants and stress space

I_σ, $II_{\hat{\sigma}}$, $III_{\hat{\sigma}}$ and β are defined as the first stress tensor invariant, the second deviatoric stress tensor invariant, the third deviatoric stress tensor invariant and the Lode angle, respectively

$$I_\sigma = \sigma_{ii} \qquad (1)$$

$$II_{\hat{\sigma}} = \sqrt{\frac{1}{2}\hat{\sigma}_{ij}\hat{\sigma}_{ij}} \qquad (2)$$

$$\hat{\sigma}_{ij} = \sigma_{ij} - \frac{I_\sigma}{3}\delta_{ij} \qquad (3)$$

$$\beta = -\frac{1}{3}\sin^{-1}\left(\frac{3\sqrt{3}}{2}\frac{III_{\hat{\sigma}}}{II_{\hat{\sigma}}^{3}}\right), \text{ with } III_{\hat{\sigma}} = \frac{1}{3}\hat{\sigma}_{ij}\hat{\sigma}_{jk}\hat{\sigma}_{ki} \quad (4)$$

2.2 Mohr Coulomb criterion (MC)

The Mohr-Coulomb failure criterion is an intrinsic curve criterion. It expresses a linear relationship between the shear stress τ and the normal stress σ_N acting on a failure plane

$$\tau = c + \sigma_{\mathrm{N}} \tan\phi \quad (5)$$

where c is the cohesion and ϕ the friction angle.

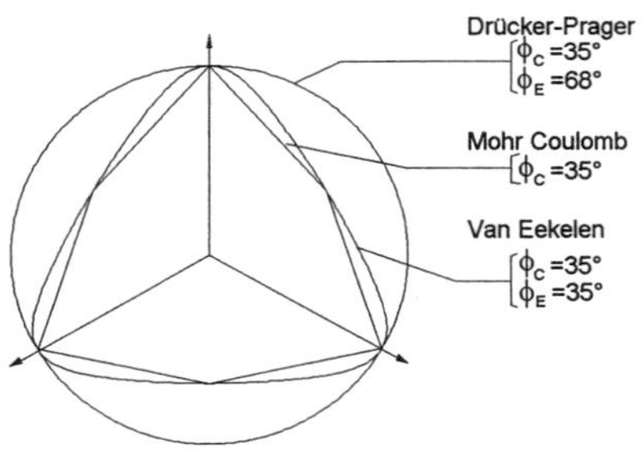

Figure 1. Limit surface for Mohr-Coulomb, Drücker-Prager and Van Eekelen criteria in the deviatoric plane.

This criterion predicts identical friction angles under triaxial compression paths (referred as ϕ_C) and triaxial extension paths (ϕ_E). Geometric representation of this criterion in the deviatoric plane gives an irregular hexagon (see figure 1). This model is not convenient to implement in the classical plasticity framework as the gradient of this yield surface is undefined on the hexagon corners. Therefore more continuously derivable yield surfaces are preferred.

2.3 Drücker Prager criterion (DP)

A simple approximation of the Mohr Coulomb surface has been proposed by Drücker and Prager, who defined the yield function f using a linear relationship between the first stress tensor invariant and the second deviatoric stress tensor invariant

$$f = II_{\hat{\sigma}} + \frac{2\sin\phi_C}{\sqrt{3}(3-\sin\phi_C)}\left(I_\sigma - \frac{3c}{\tan\phi_C}\right) = 0 \quad (6)$$

In the principal stress space, the plasticity surface becomes a cone which is much easier to use in numerical algorithms. The trace of this plasticity surface on the deviatoric plane is then a circle (see figure 1). Although this simple criterion is widely used in geomechanics to represent frictional behaviour, it does not depend on the third stress invariant and thus on the Lode angle β. This is a main drawback of this model.

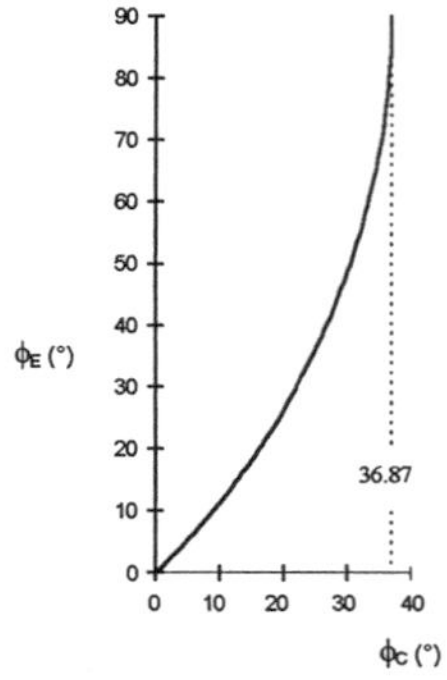

Figure 2. ϕ_E versus ϕ_C for Drücker-Prager criterion.

The radius is constant in the Drücker Prager model, this yields

$$\sin\phi_{\mathrm{E}} = \frac{3\sin\phi_{\mathrm{C}}}{3-\sin\phi_{\mathrm{C}}} \Bigg/ 1 - \left(\frac{\sin\phi_{\mathrm{C}}}{3-\sin\phi_{\mathrm{C}}}\right) \quad (7)$$

A plot of ϕ_C versus ϕ_E computed from equation (7) is given on figure 2 and shows that ϕ_E does not increase linearly with ϕ_C. There is a limit value of ϕ_E=90° for $\phi_C \approx 36.87° = \phi_C^{lim}$.

If non-associated plasticity is considered, a plastic potential g can be defined in a similar fashion than the yield surface f replacing ϕ by ψ in equations (6).

2.4 Van Eekelen criterion (VE)

A more sophisticated model can be built from the Drücker-Prager cone by introduction of a dependence on the Lode angle β in order to match more closely the Mohr Coulomb criterion. It consists in smoothing the Mohr Coulomb plasticity surface. The formulation proposed by Van Eekelen (1980) is adopted

$$f = II_{\hat{\sigma}} + m\left(I_\sigma - \frac{3c}{\tan\phi_C}\right) = 0 \quad (8)$$

with the coefficient m defined by

$$m = a(1 + b\sin 3\beta)^n \qquad (9)$$

The only difference between DP and VE criteria comes from the point that the coefficient m is constant for DP whereas it is a function of the Lode angle for VE. Coefficients a and b allow an independent choice for ϕ_C and ϕ_E. The exponent n actually controls the convexity of the yield surface. Following the conclusion of Van Eekelen (1980), the default value n=-0.229 has been chosen. The trace of this plasticity surface in the deviatoric plane is displayed on figure 2. If non-associated plasticity is considered, a flow potential g can also be defined in a similar fashion than the plastic potential f.

2.5 Comparison between Mohr Coulomb, Drücker Prager and Van Eekelen criteria

At very low friction angles, the three criteria are pretty much similar. The difference between the DP criterion on the one hand and the MC or VE criteria on the other hand increases as friction angle gets larger. This is directly linked to the relation between ϕ_C and ϕ_E : from equation (7) it is found that for DP ϕ_E=26° for ϕ_C=20°. However as friction angle ϕ_C gets closer to the limit value 36.89°, the corresponding angle ϕ_E approaches 90°. Therefore if low friction angles are considered (let say below 20°), the three criteria will give approximately the same results. However, above this value of 20°, some significant differences can be expected between the DP criterion on the one hand and the MC or VE criteria on the other hand.

3. LARGE STRAINS IN SOLID MECHANICS

Strain localisation is generally associated to large strains and large rotations. The strain amount observed inside the hereafter modelled shear bands can as high as 50% to 100%. In the following the mechanical equilibrium is formulated in the current configuration using the Cauchy stresses. The Jaumann correction is used in order to give an objective stress rate.

When a small strain formulation is assumed, the strain decomposition between its elastic and plastic parts is assumed. In a large strain formulation one often postulates an unloaded configuration, which is leading to a multiplicative decomposition of the elastic and plastic Jacobian matrix :

$$\underline{F} = \underline{F}^p \underline{F}^e \qquad (10)$$

We assume here that the elastic part of the deformation is small compared to the unity. This yields an approximately additive decomposition of the strain rate :

$$\underline{D} = \underline{D}^E + \underline{D}^P \qquad (11)$$

4. FLOW IN POROUS MEDIA

In a saturated porous medium, flow is assumed to follow the Darcy's law :

$$\underline{v} = -K\,\underline{\nabla}\left(\frac{p}{\gamma} + z\right) \qquad (12)$$

where K is the permeability, p the pore pressure, z the level, γ the fluid specific weight and $\underline{v}$ the Darcy's velocity. If partly saturation is to be considered, the Darcy's law (12) can be used assuming that the permeability K is varying with the saturation degree :

$$K = K(S_r) \qquad (13)$$

We will suppose hereafter that this law is a linear one

$$K = K_0\, S_r \qquad (14)$$

The storage law is giving the evolution of the amount of fluid mass per unit of soil volume. When the fluid saturation degree S_r is varying, two storage terms are to be taken in account :

$$\dot{S} = n\frac{\dot{p}}{\chi_w} + n\dot{S}_r \qquad (15)$$

When two fluids are partly saturating the pores, two sets of equations (12-15) should be written. However if one fluid is a gas, its behaviour can be neglected, considering that its pressure remains quite constant, or equivalently that it is highly compressible.

On the other hand it is necessary to formulate a retention curve associating the saturation degree to the suction s (which is the difference between the non-wetting and the wetting fluid pressures). We here are using the following relation :

$$S_r = \frac{1}{\pi}\,arctg\frac{s-\beta}{\alpha} + \frac{1}{2} \qquad (16)$$

$$s = p_{gas} - p_{liquid}$$

When considering a water saturated and dilatant soil under shear loading (e.g. in the hereafter considered biaxial compression) the material tends to dilate but the pore fluid (which is quite incompressible) does not allow that. The fluid pressure therefore decreases from its initial value (generally a back-pressure imposed to ensure the good saturation by dissolving air in water). When the water pressure tends to the atmospheric pressure, the dissolved air appears as bubbles. The fluid is now biphasic, what means that the soil is partly saturated in air and water. If the initial air content is low enough, a similar desaturation will appear when cavitation will occur. In that case the gas is composed of water vapour bubbles. Whatever the way the soil becomes partly saturated, the fluid compressibility increases quickly at that time. This phenomenon will be simply modelled by using a retention curve postulating a desaturation around the atmospheric pressure.

5. HYDROMECHANICAL COUPLING

Stresses are affected by the seepage. This is modelled for saturated media by the Terzaghi's postulate. It is here written in an incremental form:

$$\underline{\dot{\sigma}} = \underline{\dot{\sigma}'} - \dot{p}\underline{I} \tag{17}$$

If the soil is partly saturated by two fluids, this postulate is not more valid. Bishop has proposed a modified form :

$$\underline{\dot{\sigma}} = \underline{\dot{\sigma}'} - \chi(S_r)\dot{p}\underline{I} \tag{18}$$

where $\chi(S_r)$ is a new function to be defined from experimental results. This relation can be analysed as a mixture law (Schrefler, 1990). It becomes then :

$$\underline{\dot{\sigma}} = \underline{\dot{\sigma}'} - S_r\dot{p}_{liquid}\underline{I} - (1 - S_r)\dot{p}_{gas}\underline{I} \tag{19}$$

On the other hand, the fluid flow is affected by the soil mechanics through the storage law in which the pores volume is modified according to the volumetric strain rate :

$$\dot{S} = n\frac{\dot{p}}{\chi_w} + n\dot{S}_r + \dot{\varepsilon}_{ii} \tag{20}$$

In this equation, it is supposed that the soil grains (or the soil skeleton) volume does not change with respect to the mean effective stress variation.

The soil mechanics is involving large strains. This does not affect the fluid flow in the sense that the balance equation is formulated at one instant per time step (generally at the step end). Moreover the storage and Darcy's flow laws are not depending on the history. Therefore no objective correction is needed. On the other hand, the additive strain decomposition (11) is here assumed to remain valid (Bourgeois et al (1995)).

The hydromechanical coupling is implemented in the finite element code LAGAMINE trough a monolythical scheme, which has proved to be efficient for the considered problems.

6. APPLICATION

We consider in the following the biaxial compression which is a classical test for the study of the strain localisation in soils (Mokni and Desrues (1992), Vardoulakis (1991, 1995 a & b). A parallelepiped sample of soil is under plane strain conditions. Two sample sizes are considered : 175 x 350 mm^2 (rectangular) and 164 x 173 mm^2 (square). The initial stress state is isotropic. A pressure is applied to the lateral boundaries. It remains constant during the axial loading. The lower base lies on a frictionless platen ; the upper one is compressed by a second frictionless platen.

6.1 Rectangular sample

The initial stress state is σ'_0 = 100 kPa. The sample is discretised by 10 x 20 8-nodes finite elements. The soil is modelled by an elastic - perfectly plastic Drücker-Prager model whose parameters are indicated in table 1. The seepage parameters are given in the same table. The compression velocity (upper plateau) is $\dot{H} = 1{,}8\ 10^{-3}\ mm/s$. Some different cases have been modelled. Most results are not illustrated here, due to the lack of space.

table 1 : Rectangular sample - mechanical and seepage parameters.

parameter	symbol	value
Young modulus	E	2.6 GPa
Poisson's ratio	ν	0.3
compression friction angle	ϕ_c	25°
dilatancy angle	ψ	10°
water compressibility	χ_w	3. GPa
porosity	n	0.45

Case 1 : It corresponds to a drained sample. A shear band clearly appears, as generally under such kind of condition. Small imperfections have been introduced in order to help this bifurcation. However it should be noted that numerical imperfection are sufficient to induce the strain localisation.

Case 2 : The sample is fully saturated by water and the initial back-pressure is high enough to avoid any cavitation or other de-saturation. The permeability is low : $K=10^{-20}\ ms^{-1}$. The mesh remains rectangular and the stress and pore pressure state is homogeneous during the whole compression process. The global volume remains constant. The volumetric strains are zero. The pore pressure variation is Δp = 500 kPa at the end of the simulation.

Case 3 : The permeability is much more larger than in the preceding example: $K=10^{-7}\ ms^{-1}$. Two shear bands now appear. The global volume remains always constant. However the volumetric strains are not homogeneous (figure 3) and varies from about 4% dilatation inside the band to about 1.5% contraction outside it. This is allowing the necessary dilation inside the shear bands, and explains why they are appearing as in the drained case.

According to these variations of the void ratio the water storage and the pore pressure are also varying. It results in pressure gradients and in water flows. These Darcy's velocities are showing that the two shear bands are not always active together.

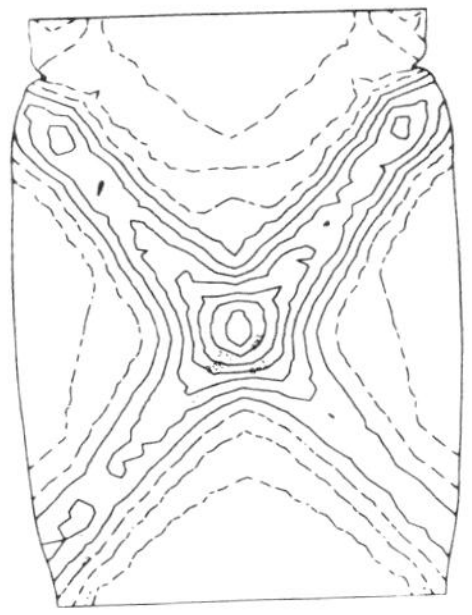

Figure 3 : Case 3 : volumetric strains in the deformed mesh after 15 % of axial strain.

6.2 Square sample

The initial stress state is $\sigma'_0 = 100$ kPa. The sample is discretised by 30 x 30 8-nodes finite elements. The soil is modelled by an elastic - perfectly plastic Van Eekelen model whose parameters are given in table 2. The seepage parameters are indicated in the same table. Globally undrained states are considered. Therefore the total pore fluid mass remains constant. An initial back-pressure $p_0 = 30$ kPa is applied. The rate of displacement of the upper boundary is 10^{-3} $\%.s^{-1}$ (of the initial height).

table 2 : Square sample: mechanical and seepage parameters.

parameter	symbol	value
Young modulus	E	35 MPa
Poisson's ratio	ν	0.4
compression friction angle	ϕ_c	41°
extension friction angle	ϕ_e	41°
dilatancy angle	ψ	10°
permeability	K	10^{-16}m/s
water compressibility	χ_w	3. GPa
porosity	n	0.3

No strain localisation is apparent from the deformed mesh. However the water pore pressure map shows clearly a localisation band where the pore pressure has highly decreased (from 130 kPa initially to -120 kPa after 9% of axial deformation -figure 4). The water pore pressure has decreased because of the band dilatancy. A little later (10% compression) the localisation scheme has changed (figure 5) and two localised band have appeared. The water pressure is a little lower (-160 kPa). These negative pore pressures are associated with a partial saturation (cf. retention curve). At the band centre the saturation degree has decreased to 95%.

The sample shearing concentrates in the band and it is associated with a local volume increase. Water accommodates this by decreasing its pressure. Due to the resulting pressure gradient water flow from outside the band to inside it (figure 6).

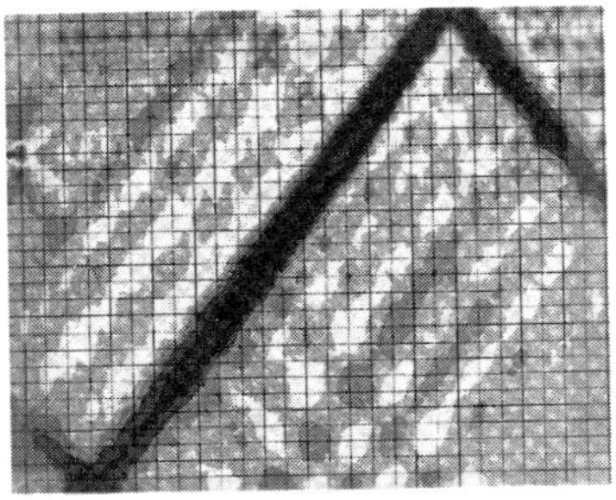

Figure 4. Contours of water pressure after 9 % compression.

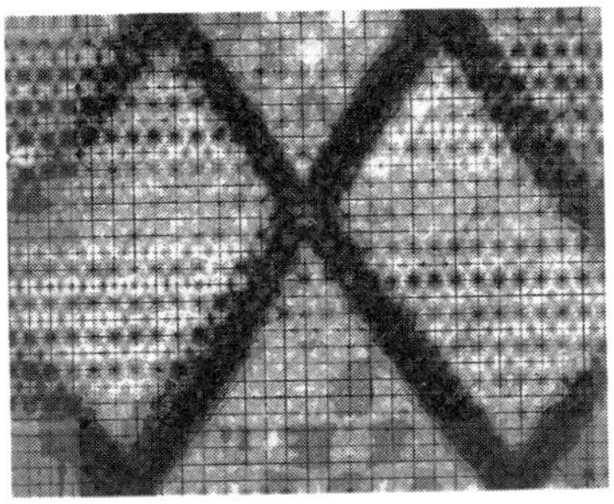

Figure 5. Contours of water pressure after 10 % compression.

7. CONCLUSIONS

A hydromechanical large strains finite element code has been developed for soils and rocks modelling. It has been applied to the modelling of biaxial compressions on various soil samples.

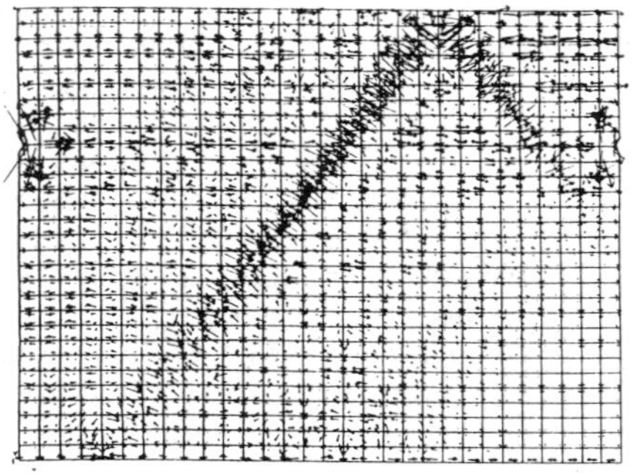

Figure 6. Water flow - Darcy's velocities after 10% compression.

Pure mechanical analysis (on drained samples) shows clearly a strain localisation. The modelling is possible up to very large strains.

In a saturated soil sample, if the permeability is small enough compared to the loading velocity, water has not enough time to move during any strain localisation process. Then the unsaturated state is a global and a local one. No strain localisation can appear.

In a dilatant sample the pore pressure is continuously decreasing during the stress deviator increase (uniaxial compression phase). When that pressure becomes small enough, gas bubbles are appearing due to cavitation in pure water or to the end of air dissolution in other cases. Then the pore fluid becomes a biphasic one with a high compressibility, and a drained like localisation is possible.

If the permeability is larger, then water moves and allows a strain localisation in a dilatant medium. The undrained condition is a global one. This means that the total volume remains constant due to the low fluid compressibility. However at a local level, dilatancy as well as contractancy are appearing. The volumetric strain integral over the whole sample volume is quite null (with respect to the fluid compressibility).

ACKNOWLEDGEMENTS

The authors are grateful to the *Groupement de Recherche en géomécanique des Roches Profondes,* to the EC through the financial support to the network *ALERT Geomaterials,* and to the FNRS.

REFERENCES

E. BOURGEOIS, P. de BUHAN et L. DORMIEUX. Formulation d'une loi élastoplastique pour un milieu poreux saturé en transformation finie. C.R. Acad.Sci, Paris, t.321, Série IIb, pp. 175-182, 1995.

M. MOKNI Relations entre déformations en masse et déformations localiseés dans les matériaux granulaires. Thesis UJF-INPG, Grenoble, 1992.

B. LORET, J.H. PREVOST Dynamic strain localization in fluid-saturated porous media. Jl. of Eng. Mech. Vol. 117, N° 4, pp. 907-922, 1991.

B.A. SCHREFLER, L. SIMONI, X.K. LI, O.C. ZIENKIEWICZ. Mechanics of partially saturated porous media. Numerical Methods and Constitutive Modelling in Geomechanics. C.S. Desai and G. Gioda, ed., CISM Courses and Lectures, N° 311, Springer Verlag, 169-209, 1990.

B.A. SCHREFLER, L. SANAVIA and C.E. MAJORANA. A multiphase medium model for localisation and postlocalisation simulation in geomaterials. Mechanics of Cohesive Frictional Materials, Vol. 1, 95-114 (1996).

C. HAN and I.G. VARDOULAKIS. Plane strain compression experiments on water-saturated fine-grained sand. Géotechnique 41, N° 1, pp. 49-78, 1991.

H.A.M. van EEKELEN. Isotropic yield surfaces in three dimensions for use in soil mechanics. Int. Jl. for Numerical and Analytical Methods in Geomech., Vol.4, 89-101, 1980.

I.G. VARDOULAKIS. Deformation of water-saturated sand: I. Uniform undrained deformation and shear banding. Géotechnique, 46, n° 3, 441-456, 1995.

I.G. VARDOULAKIS. Deformation of water-saturated sand: II. The effect of pore water flow and shear banding. Géotechnique, 46, n° 3, 457-472, 1995.

Numerical Models in Geomechanics, Pietruszczak & Pande (eds) © 1997 Balkema, Rotterdam, ISBN 90 5410 886 X

Coupling of shear bands in modeling the localization problem in geomechanics

H. Hazarika, Y. Terado & T. Nasu
Japan Foundation Engineering Co., Ltd, Tokyo, Japan

ABSTRACT: A numerical formulation, for analyzing the localized deformation in geomaterials, is presented. In contrast to the conventional shear band formulation, it considers two shear bands inside a localized element. The constitutive law for the localized element is formulated by coupling the concentrated strains inside the two bands. The developed formulation was applied to analyze the passive earth pressure against retaining wall under seismic loading. It was found to be very efficient in capturing the progressive failure of the backfill. The resulting stress non-uniformity, at the point of rotation (toe) of the wall due to the wall movement, could be well captured by the method. The results show that the friction angle mobilizes progressively within an element. The obtained results are in good agreement with the established experimental trend.

1 INTRODUCTION

The vulnerability of granular materials to the fascinating phenomenon of shear band localization is now well established. In shearing deformation of such materials, the microstructural damage results in concentrated deformation. Failure of many structures is preceded by the formation and propagation of shear bands. Thus, the modeling of localization is of paramount importance.

Researchers, in the past three decades, have been putting commendable efforts in the intricate field of localization. Their findings revolutionized the field of constitutive behavior of geomaterials. All these researches considered only one shear band inside a localized element. Theoretical and experimental investigations, however, established that there exists two bands along which the plastic strain is concentrated. Mohr's diagram too gives two directions of the failure planes along which the failure develops and progresses. The concentrated strains in both directions contribute to the softening. Therefore, coupling of the effect of the bi-directional bands is indispensable in deriving the constitutive law. In this paper, a numerical model is described that incorporates the effects of the incepted shear bands inside an element in the two preferred directions. The constitutive relation, which is based on the "smeared shear band technique" of Pietruszczak and Stolle (1987), is formulated by coupling the contributions of the two bands to the total strain.

A brief review of the model is presented in the following section. The model is validated by applying the concept to analyze passive earth pressure problem.

2 A REVIEW AND ESSENTIALS OF THE MODEL

2.1 *Smeared shear band technique*

An element undergoing shear band bifurcation can be schematically represented as shown in Figure 1. The element can be divided into two sub-elements, one with fundamental (elastic) response, and the other with shear band response engulfing the shear band, where the strain is localized.

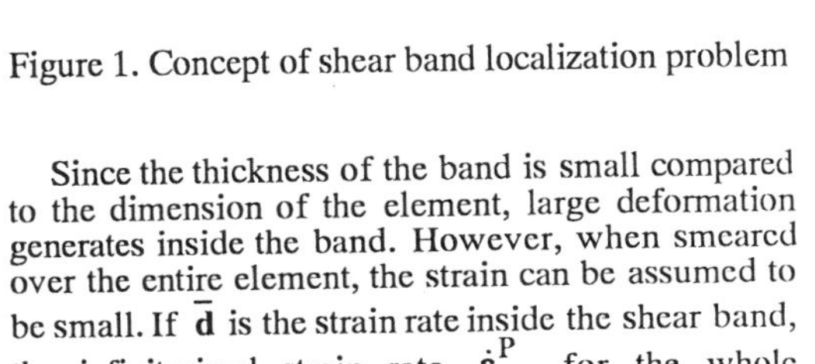

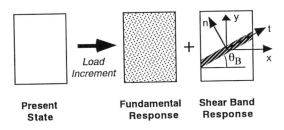

Figure 1. Concept of shear band localization problem

Since the thickness of the band is small compared to the dimension of the element, large deformation generates inside the band. However, when smeared over the entire element, the strain can be assumed to be small. If $\bar{\mathbf{d}}$ is the strain rate inside the shear band, the infinitesimal strain rate, $\dot{\boldsymbol{\varepsilon}}^P$, for the whole element is given by the following relation.

$$\dot{\boldsymbol{\varepsilon}}^P = \zeta\bar{\mathbf{d}} \quad (1)$$

Here ζ is the smearing factor defined to be:

$$\zeta = \frac{\text{Area of Shear Band } (A_b)}{\text{Area of Element } (A_e)} \tag{2}$$

The total strain in the localized element is given by:

$$\dot{\varepsilon} = \dot{\varepsilon}^e + \dot{\varepsilon}^P \tag{3}$$

where $\dot{\varepsilon}^e$ is the elastic strain rate. This is the so called "Smeared Shear Band Approach" (will be referred as SSBA hereafter) initially proposed by Pietruszczak and Mroz (1981).

2.2 *Band orientation and coupling*

The post-localization behavior was idealized as pseudo-uniform deformation using SSBA with single band by Pietruszczak and Stolle (1987). However, theoretical evidence demonstrates that there exists two bands along which the strain is localized. Experimentally too it is confirmed that if co-axiality of the ends of the sample is imposed by the apparatus generally two shear bands will be observed crossing each other (Desrues 1990). Therefore, for the proper formulation of the constitutive behavior, the consideration of both the bands is essential. The schematic diagram of this concept is shown in Figure 2. Sub-element 1 behaves as an elastic material under the applied load. Sub-elements 2 and 3 entrap the shear bands inclined at θ_{BI} and θ_{B2} respectively. In this research, the constitutive laws for the post-localization behavior are developed using two shear bands based on the concept of SSBA explained in the previous sub-section. The constitutive model, named *Coupled Shear Band Method* (CSB Method), is formulated by coupling the effect of the two shear bands, the description of which follows.

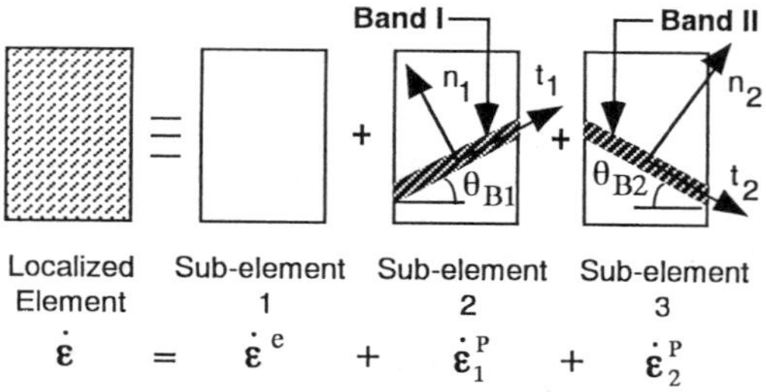

Figure 2. Schematic diagram showing the state of localization

A simplifying assumption is made here regarding the onset of localization -- it initiates at the failure. Therefore, when the yield surface reaches the failure surface, the strain in the element can be assumed to have localized inside the two shear bands. At that instant, the constitutive relation for the elements needs to be modified taking into the account the developed stress discontinuity. Assuming that the stress and strain rate tensors are co-axial, the orientation of the two bands can be obtained as:

$$\theta_{BI} = (\theta_m - \frac{\pi}{4}) - \frac{\phi_P}{2} \tag{4}$$

$$\theta_{B2} = (\theta_m - \frac{3\pi}{4}) + \frac{\phi_P}{2} \tag{5}$$

Here θ_m is the angle from the X axis to the major principal axis in the Mohr diagram, and ϕ_P is the angle of internal friction at peak load.

2.3 *Constitutive relation*

The Jaumann rate of stress tensor, $\hat{\sigma}$, is used as the stress rate inside the bands. Relative to the $\{n_1, t_1\}$ co-ordinate system the constitutive relation for band I takes the form given below.

$$\begin{Bmatrix} \bar{d}_{n_1} \\ \bar{d}_{t_1} \\ \bar{d}_{nt_1} \end{Bmatrix} = \frac{1}{S} \begin{bmatrix} \frac{\partial g_b}{\partial \sigma_{n_1}} \frac{\partial f_b}{\partial \sigma_{n_1}} & 0 & \frac{\partial g_b}{\partial \sigma_{n_1}} \frac{\partial f_b}{\partial \sigma_{nt_1}} \\ 0 & 0 & 0 \\ \frac{\partial g_b}{\partial \sigma_{nt_1}} \frac{\partial f_b}{\partial \sigma_{n_1}} & 0 & \frac{\partial g_b}{\partial \sigma_{nt_1}} \frac{\partial f_b}{\partial \sigma_{nt_1}} \end{bmatrix} \begin{Bmatrix} \hat{\sigma}_{n_1} \\ \hat{\sigma}_{t_1} \\ \hat{\sigma}_{nt_1} \end{Bmatrix} \tag{6}$$

Here, f_b and g_b refers to the yield function and the plastic potential function respectively, and S is the softening modulus. Similarly, the constitutive relation for band II is given by,

$$\begin{Bmatrix} \bar{d}_{n2} \\ \bar{d}_{t2} \\ \bar{d}_{nt2} \end{Bmatrix} = \frac{1}{S} \begin{bmatrix} \frac{\partial g_b}{\partial \sigma_{n2}} \frac{\partial f_b}{\partial \sigma_{n2}} & 0 & \frac{\partial g_b}{\partial \sigma_{n2}} \frac{\partial f_b}{\partial \sigma_{nt2}} \\ 0 & 0 & 0 \\ \frac{\partial g_b}{\partial \sigma_{nt2}} \frac{\partial f_b}{\partial \sigma_{n2}} & 0 & \frac{\partial g_b}{\partial \sigma_{nt2}} \frac{\partial f_b}{\partial \sigma_{nt2}} \end{bmatrix} \begin{Bmatrix} \hat{\sigma}_{n2} \\ \hat{\sigma}_{t2} \\ \hat{\sigma}_{nt2} \end{Bmatrix} \tag{7}$$

Eqs. (6) and (7) indicate the diminishing rate of deformations across the bands.

The Jaumann stress rate can be expressed in terms of Cauchy's stress rate by the following transformation equation.

$$\hat{\boldsymbol{\sigma}} = \dot{\boldsymbol{\sigma}} + \boldsymbol{\delta}^T \bar{\mathbf{d}}\, \boldsymbol{\beta} \tag{8}$$

Here the vector $\boldsymbol{\beta}$ and the transformation matrix δ^T are given by the following equations.

$$\boldsymbol{\beta} = \{-2\sigma_{xy}, 2\sigma_{xy}, (\sigma_x - \sigma_y)\}^T \tag{9}$$

$$\delta^T = \{-sc,\ sc,\ \frac{1}{2}(c^2 - s^2)\} \tag{10}$$

Where, $s = \sin(-\theta)$ and $c = \cos(-\theta)$. Using the transformation (Eq. 8) yields the following two equations in the global co-ordinate system.

$$\bar{\mathbf{d}}_1 = \frac{1}{S}([\mathbf{I}] - \frac{1}{S}[\bar{C}_1]\boldsymbol{\beta}\delta_1{}^T)^{-1}[\bar{C}_1]\dot{\boldsymbol{\sigma}} \tag{11}$$

$$\bar{\mathbf{d}}_2 = \frac{1}{S}([\mathbf{I}] - \frac{1}{S}[\bar{C}_2]\boldsymbol{\beta}\delta_2{}^T)^{-1}[\bar{C}_2]\dot{\boldsymbol{\sigma}} \tag{12}$$

Here, δ_1 and δ_2 are the transformation matrices defined by the angles θ_{B1} and θ_{B2}. $[\bar{C}_1]$ and $[\bar{C}_2]$ are the compliance matrices (Vermeer 1982) in the global co-ordinate system, and defined by the following equations:

$$[\bar{C}_1] = [T_1]^T[\bar{C}_1^*][T_1] \tag{13}$$

$$[\bar{C}_2] = [T_2]^T[\bar{C}_2^*][T_2] \tag{14}$$

In which $[\bar{C}_1^*]$ and $[\bar{C}_2^*]$ represent the compliance matrices in the local co-ordinate systems $\{n_1, t_1\}$ and $\{n_2, t_2\}$ respectively, and can be evaluated by the bracketed terms in the Eqs. (6) and (7). $[T_1]$ and $[T_2]$ are the transformation matrices.

The infinitesimal plastic deformation in the two sub-elements (sub-elements 2 & 3) can be expressed as follows.

$$\dot{\boldsymbol{\varepsilon}}_1^P = \zeta\bar{\mathbf{d}}_1 \tag{15}$$

$$\dot{\boldsymbol{\varepsilon}}_2^P = \zeta\bar{\mathbf{d}}_2 \tag{16}$$

Therefore, the total strain in the "localized element" can be written as,

$$\dot{\varepsilon} = \dot{\varepsilon}^e + \zeta(\bar{\mathbf{d}}_1 + \bar{\mathbf{d}}_2) \tag{17}$$

On substituting Eqs. (11) and (12) into Eq. (17), yields the following constitutive law.

$$\dot{\boldsymbol{\varepsilon}} = \{([\mathbf{I}] - \frac{1}{S}[\bar{C}_1]\boldsymbol{\beta}\delta_1{}^T)^{-1}[\bar{C}_1] + ([\mathbf{I}] - \frac{1}{S}[\bar{C}_2]\boldsymbol{\beta}\delta_2{}^T)^{-1}[\bar{C}_2] + \frac{S}{\zeta}[C^e]\}\frac{\zeta}{S}\dot{\boldsymbol{\sigma}} \tag{18}$$

In which C^e is given by the following matrix.

$$[C^e] = \frac{1}{E}\begin{bmatrix} (1-\nu^2) & -\nu(1+\nu) & 0 \\ -\nu(1+\nu) & (1-\nu^2) & 0 \\ 0 & 0 & 2(1+\nu) \end{bmatrix} \tag{19}$$

E and ν are the Young's modulus and Poisson's ratio respectively. Eq. (18), containing the contributions from both the shear bands, forms the foundation of the so called *Coupled Shear Band (CSB) Method*.

3 MODEL APPLICATION

It is important that the method of analysis employed for a boundary value problem is appropriate both to the soil model which is being employed and to the problem itself. Only by this means, reliable, accurate and economical results can be obtained. For instance, progressive failure mechanism invariably leads to localized deformation into shear bands. Therefore, in the analysis of the progressive failure of granular materials, it is of vital importance to consider the localized strains. A classical example of progressive failure is the backfill deformation during movement of a retaining wall. Analysis on the sensitivity of soils under different stress condition showed that the plane strain condition is more prone to localization (Wan 1990). Since analysis of earth pressure involves plane strain condition, consideration of localized deformation in such analysis yields added advantages.

The capability of *CSB* model in analyzing the static active earth pressure has already been established (Hazarika and Matsuzawa, 1996). Application of the model to the dynamic active earth pressure analysis can be found in Hazarika and Matsuzawa (1997). A good prediction of the progressive failure could be made using the model. In this paper, a passive earth pressure problem is analyzed using the developed model.

3.1 *Finite element model*

Figure 3 shows the plane strain finite element model for a 10 m high rigid retaining wall supporting dry cohesionless backfill. The interface between the wall and the backfill was modeled using a simplified interface model (Hazarika 1996).

As shown in the figure, the reflected boundaries (roller support) are restrained in the horizontal directions, while the fixed boundary is restrained against both horizontal as well as vertical movements.

3.2 *Analysis*

The nodal points along the wall were all given the same horizontal displacements. The analyses were performed by applying horizontal forces, pushing the wall toward the backfill. A pure sinusoidal motion with a frequency of 3 Hz was applied as dynamic excitation. The amplitudes of the accelerations were increased from 0 gals (static case) to a maximum of 400 gals. The Wilson θ method was used in the time domain analysis to calculate the applied dynamic increment on the retaining wall.

The angle of internal friction at peak was taken as 40^0. The unit weight of the soil was assumed to be 16.0 kN/m^3. The coefficient of earth pressure at-rest was taken as 0.5. Young's modulus, E and Poisson's ratio, ν (Eq. 19) was taken to be 26000 kPa and 0.3 respectively.

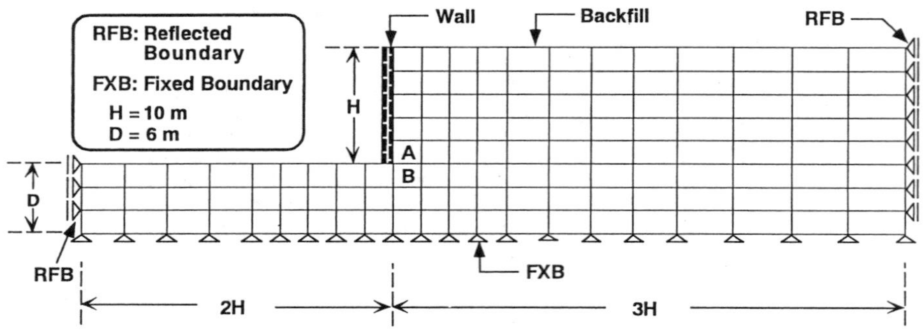

Figure 3. FE grid for retaining wall-backfill system

3.3 *Constitutive model*

It is recognized that the concept of mixed hardening is more appropriate for the problems involving dynamic loading. The key features of the phenomenology of shear band development under the dynamic loading conditions are the same as under quasi-static loading conditions, and a delay in shear band development arises due to the inertial effects (Needleman 1989). The visco-plastic approach was used by Prevost and Loret (1990) for dynamic localization problems by introducing the artificial viscosity into the rate-independent materials.

In contrast to the free field vibration, the stress-strain response in the course of seismic earth pressure development takes the form shown in Figure 4. The envelope of the response is somewhat similar to the static stress-strain relation. Hence, the same constitutive relation based on isotropic hardening can be applied for the dynamic earth pressure calculation as well. One of the main advantages of the model is that it can predict the total earth pressure (static + dynamic) not the dynamic component of earth pressure alone.

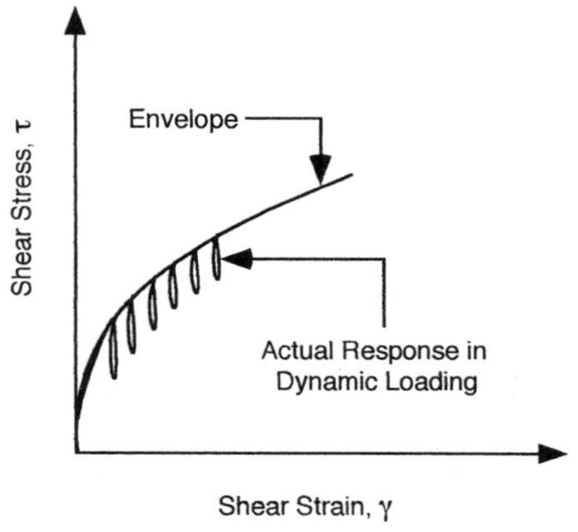

Figure 4. Stress-strain curve in seismic earth pressure generation

4 NUMERICAL RESULTS

The behavior of the backfill under static loading conditions (acceleration = 0) is discussed first, followed by the discussions under dynamic loading.

4.1 *Effect of stress concentration and dilatancy*

Figure 5 shows the calculated stress-paths (using CSB method) of two representative elements A and B located immediately above and immediately below the toe of the wall (see Figure 3). It can be observed that the stress paths for the two elements are different. The difference can be partly attributed to the stress concentration occurring at the toe of the wall. The stress paths show a marked difference in the case of a wall rotating about its base or top which are beyond the discussion of this paper. Only by means of the localized deformation analysis such discontinuity can be accurately captured.

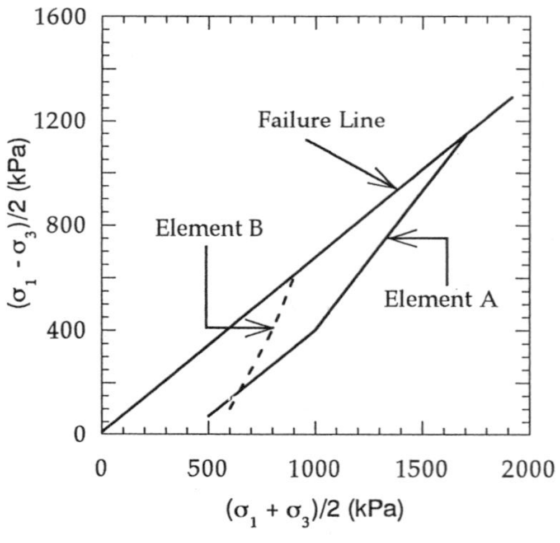

Figure 5. Stress-paths for representative elements (static case)

Figure 6 shows the stress-paths for the element B calculated by using the conventional analysis and the proposed formulation. The accurate prediction of dilatancy using the presented formulation accounts for the difference in the behavior as shown.

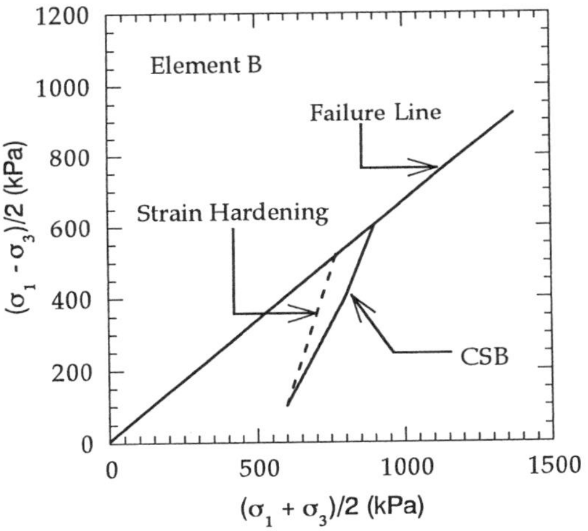

Figure 6. Comparison of the stress-paths (static case)

4.2 *Development of passive thrust*

The progressive development of passive thrust against the wall can be seen in Figure 7. A pronounced peak is observed for both the static (acceleration = 0) and the sinusoidal loading. An interesting observation was that the displacement required to reach the critical state is independent of the loading conditions (static or dynamic) and the acceleration levels. This observation is in conformity with the experimental results (Matsuzawa and Matsumura 1981).

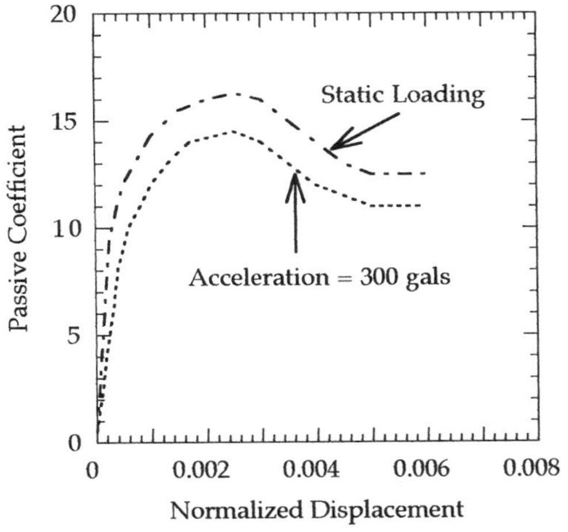

Figure 7. Passive thrust vs. wall displacement

4.3 *Friction angle mobilization*

The mobilized friction angle at the element B, when a complete rupture surface develops in the backfill, is shown in Figure 8 at the acceleration of 300 gals. The figure reveals a typical trend of the mobilization (increases to reach the peak and then decreases).

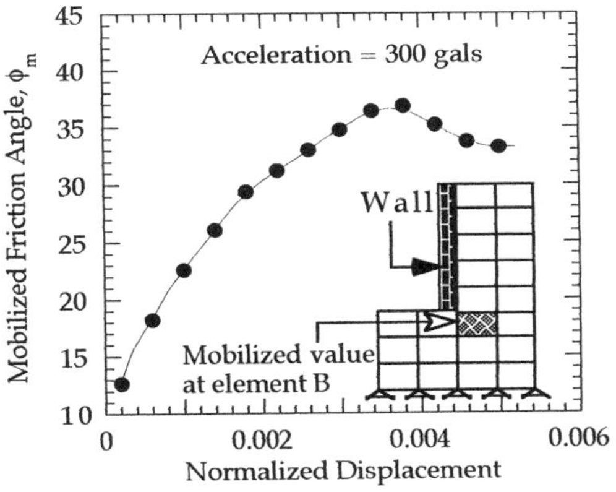

Figure 8. Friction angle mobilized at the critical state

4.4 *Failure state*

The failure zones in the backfill at the acceleration of 300 gals, which were obtained from the three methods of analysis (Strain Hardening formulation, Single Shear Band (SSB) and the CSB formulation), are shown in Figure 9. The figures present a clear picture of the merit of the CSB formulation (9c) over the other formulations. A wide failure zone was obtained using the conventional formulation (9a). The failure zone domain obtained using the CSB method is in close agreement with the observation of other researchers (Bakeer and Bhatia 1985).

5 CLOSURE

The numerical method based on the coupling of the bi-directional localized strain will be very effective in analyzing such progressive deformation phenomena in which local discontinuity of stress poses special problems. Earth pressure was selected, in this research, as a particular application of the model.

The consideration of the shear band localization, in general, and the coupling technique, in particular, assumes special importance in analyzing earth pressure against a rigid wall when the wall rotates about its base or top. The failure characteristics of the backfill influence the progressive mobilization of the friction angle, which in turn influence the magnitude of the resultant earth pressure. In order to capture such phenomenon numerically, consideration of the bi-directional shear bands is a viable alternative to the analysis method conventionally adopted.

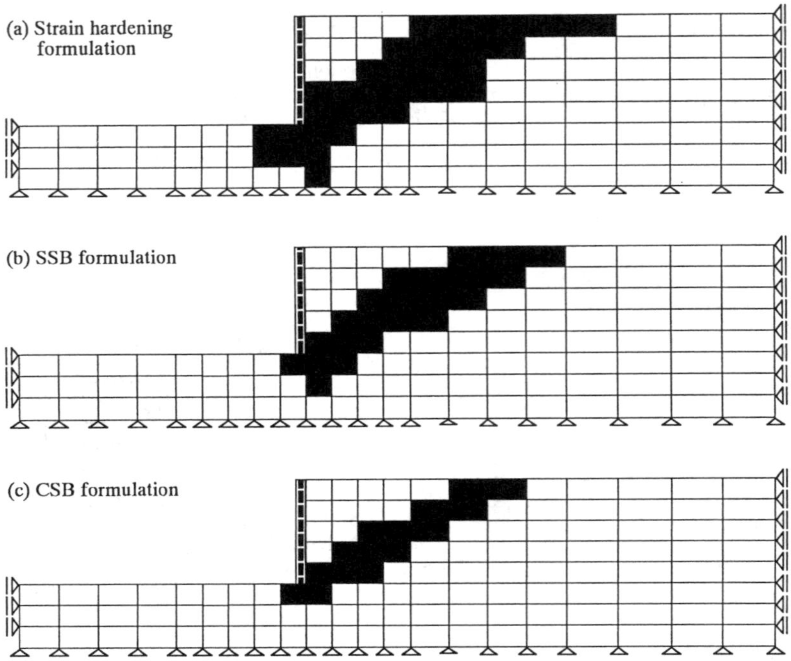

Figure 9. Comparison of the state of failure in the backfill (Acceleration = 300 gals)

ACKNOWLEDGMENT

The first author would like to thank Prof. Akira Asaoka of Nagoya University, Nagoya, Japan for his encouraging suggestions which opened up author's interest in the application of the developed model to analyze passive earth pressure problems.

REFERENCES

Bakeer, R.M. and S.K. Bhatia 1985. Dynamic earth pressure behind gravity walls subjected to sinusoidal motion. *Proceeding of the 2nd International Conference on Soil Dynamics and Earthquake Engineering*: 3-12.

Desrues, J. 1990. Shear band initiation in granular material: experimentation and theory. In F. Darve (ed), *Geomaterials: Constitutive Equations and Modelling*: 283-310. Elsevier Applied Science: London.

Hazarika, H. 1996. Computation modeling of static and dynamic earth pressure against retaining wall based on localized deformation. *Ph. D. Thesis.* Nagoya University, Nagoya: Japan.

Hazarika, H. and H. Matsuzawa 1996. Wall displacement modes dependent active earth pressure analyses using smeared shear band method with two bands. *Computers and Geotechnics.* 19(3): 193-219.

Hazarika, H. and H. Matsuzawa 1997. Coupled shear band method and its application to seismic earth pressure problems. *Soils and Foundations, JGS.* (To appear).

Loret, B. and J.H. Prevost 1990. Dynamic strain localization in elasto-(visco)plastic solids, part 1. General formulation and one-dimensional examples. *Computer Methods in Applied Mechanics and Engineering.* 83: 247-273.

Matsuzawa, H. and A. Matsumura 1981. Passive earth pressure during earthquakes. *Proceedings of the International Conference on Recent Advances in Geotechnical Earthquake Engineering and Soil Dynamics.* Missouri, USA: 2: 715-720.

Needleman, A. 1989. Dynamic shear band development in plane strain. *Journal of Applied Mechanics, ASME*. 56: 1-9.

Pietruszczak, S. and Z. Mroz 1981. Finite element analysis of deformation of strain softening materials. *International Journal of Numerical Method in Engineering.* 17: 327-334.

Pietruszczak, S. and D.F.E. Stolle 1987. Deformation of strain softening materials, part II. *Computers and Geotechnics.* 4: 109-123.

Vermeer, P.A. 1982. A simple shear-band analysis using compliances. *Proceedings of the IUTAM Conference on Deformation and Failure of GranularMaterials.* Delft, Holland: 493-499.

Wan, R.G. 1990. The numerical modeling of shear bands in geological materials. *Ph. D. Thesis.* University of Alberta.

3 Modelling of reinforced soil

Numerical Models in Geomechanics, Pietruszczak & Pande (eds) © 1997 Balkema, Rotterdam, ISBN 90 5410 886 X

Anisotropic strength of fiber-reinforced soils

R.L. Michalowski & J. Čermák
The Johns Hopkins University, Baltimore, Md., USA

ABSTRACT: A homogenization approach is used to determine failure stresses for soils reinforced with fibers. The method leads to a closed-form solution for the failure criterion of an isotropic composite, and a numerical solution for an anisotropic fiber-reinforced soil. Predictions of the failure stress for isotropic composites compare favorably with the experimental tests.

1 INTRODUCTION

Fiber reinforcement of soils has recently been tried as an alternative method to a traditionally engineered reinforced soil mass. Design with reinforced soil is based predominantly on limit state techniques and the failure criterion for reinforced soil is an essential material (constitutive) function in the design process. This paper indicates how an energy-based homogenization method can be used to obtain the macroscopic limit stress criterion for fiber-reinforced soils. Attention is paid to both isotropic and anisotropic composites. Anisotropy of fiber-reinforced soils occurs because of the preferred plane of the fiber deposition.

Earlier attempts at describing the strength of fiber-reinforced soils were based on considerations of soil-fiber interaction in a localized shear band (Waldron 1977, Wu *et al.* 1979, Gray and Ohashi 1983). More recent efforts toward description of macroscopic properties of fiber-reinforced or continuous filament-reinforced soils include homogenization (di Prisco and Nova 1993, Michalowski and Zhao 1996), which has been also used to describe the behavior of soils reinforced with plane or strip inclusions (Sawicki 1983, Kulczykowski 1989, de Buhan *et al.* 1989, Jommi *et al.* 1995, Michalowski and Zhao 1995).

Triaxial tests on fiber-reinforced sand reveal a surprising effect where, at first, the specimens deform with decreasing stiffness, and, later in the process, the specimens harden. A clear point of inflection can be identified on the stress-strain curves. This hardening effect can be attributed to the change in configuration of the fibers - a kinematic hardening effect leading to anisotropic strength of the deformed specimen.

Examples of failure surfaces for fiber-reinforced soil are presented for both isotropic and anisotropic cases.

2 SOIL-FIBER INTERACTION

The mechanism of load transfer from the soil (matrix) to the fibers (reinforcement) is an important part of modeling of soils reinforced with short fibers or thin continuous filaments. This mechanism depends greatly on the relative size of the matrix grains and the reinforcement. With relatively large-diameter fibers, Fig. 1(a), the matrix-reinforcement interface can be considered frictional, whereas for a very thin filament, Fig. 1(b), the load transfer mechanism is of a very different type.

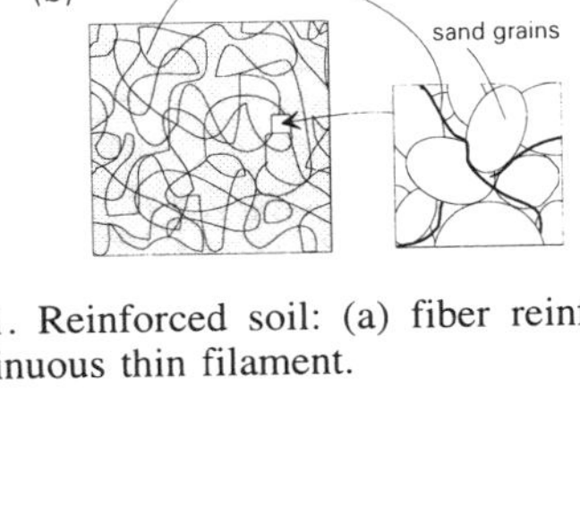

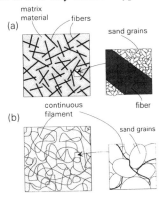

Figure 1. Reinforced soil: (a) fiber reinforcement, (b) continuous thin filament.

Considering rigid-plastic fibers, the (axial) stress in fibers during the composite deformation increases from zero at the fiber ends to a maximum (not exceeding fiber yield stress σ_0) at the center part, Fig. 2(a). Fiber ends will then slip in the process of composite deformation whereas the mid-part of the fiber will yield plastically. A pure slip case is, of course, conceivable for short fibers or small isotropic pressure in the composite.

If the diameter of the fiber is an order of magnitude smaller than the grain size, the flexible fiber may be accommodated (in a 3-D grain assembly) entirely by the pore space even if the fiber aspect ratio is large. Such a reinforcement will become effective only if it is a continuous and flexible filament so that the force in the filament can be induced due to the "belt friction effect", since the filament will be "wrapped" around the grains or around clusters of grains.

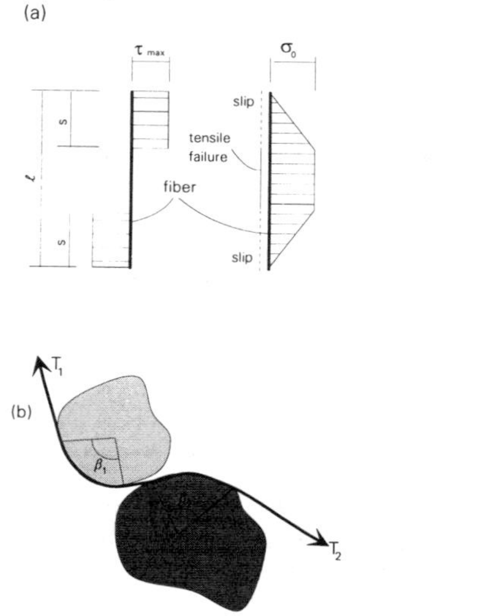

Figure 2. Soil-reinforcement interaction: (a) short fibers, (b) continuous filament.

The belt friction effect is illustrated in Fig. 2(b). If the filament moves opposite to the direction of force T_1, then force T_2 is equal to

$$T_2 = T_1 e^{\mu\beta} \tag{1}$$

where μ is the coefficient of friction between the grains and the filament, and β is the envelope (wrap) angle, equal here to $\beta_1 + \beta_2$. The force in the filaments is induced because of the deformation of the matrix. A tensile force so induced will not relax due to slippage because of the "serpentine" deposition of the filament in the matrix. During a deformation process, however, only a portion of the filament is subjected to extension (stretching), whereas the remaining part is likely to kink because of the filament's inability to carry compression (also, portions of filaments at transition from the extension to the compression regime may slip). It is the portion of the filament subjected to extension that will contribute foremost to the composite strength.

3 MACROSCOPIC FAILURE STRESS

An average stress in the composite is represented here as a *macroscopic stress* calculated in a homogenization procedure. The stiffness of the material at failure is zero, and all internal energy is dissipated during the deformation process.

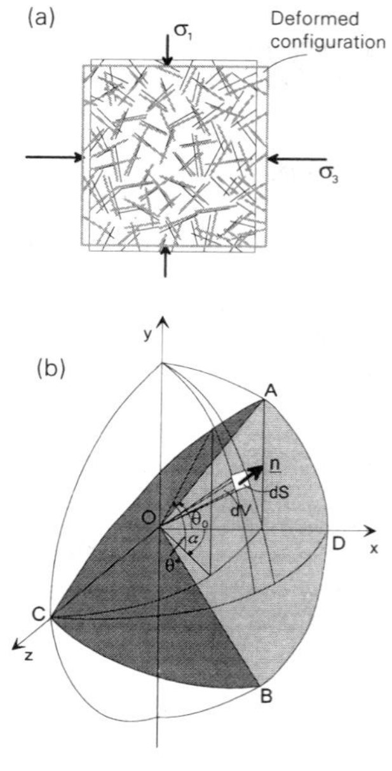

Figure 3. Homogenization procedure:
(a) deformation pattern, (b) integration space.

A deformation increment of a representative element of the composite is considered, Fig. 3(a), and the work increment of the macroscopic stress $\overline{\sigma}_{ij}$ is required to be equal to the increment of the work dissipation rate in the composite constituents

$$\overline{\sigma}_{ij}\dot{\overline{\varepsilon}}_{ij} = \frac{1}{V}\int_V \dot{D}(\dot{\varepsilon}_{ij})\, dV \tag{2}$$

The formula in (2) will be used to find the components of the macroscopic stress tensor at failure. A plane-stain deformation increment will be considered, though the composite structure is three-dimensional. This procedure is straightforward, particularly for isotropic composites. When response of the composite is in the hardening regime, a similar procedure can be used to determine the stress-strain behavior, but with the second-order work in eq. (2) (Jommi *et al.* 1995).

Macroscopic failure stress is expected to be dependent on the content of the fibers and their aspect ratio. The average fiber concentration is defined as

$$\bar{\rho} = \frac{V_r}{V} \tag{3}$$

where V_r is the volume of the fibers and V is the volume of the entire representative composite element.

The fibers are considered to be cylindrical in shape, and their slenderness is described here by the aspect ratio

$$\eta = \frac{l}{2r} \tag{4}$$

where l is the length of the fiber and r is its radius.

4 FAILURE SURFACE FOR ISOTROPIC COMPOSITES

Fiber-reinforced soil is expected to have isotropic strength if the orientation of the fiber reinforcement is distributed uniformly in all directions. During plastic deformation all fibers working in the tensile regime are expected to contribute to the composite strength. Fibers in the compressive regime are expected to kink (or buckle), and they should be ignored when calculating the work dissipation rate in eq. (2). Whether the fiber is in compression or tension depends on its inclination. Deformation of the composite is assumed here to conform to the normality rule for the matrix. With the matrix failure described by the Mohr-Coulomb function, the rate of the two nonzero (plane strain) principal strain rates is

$$\frac{\dot{\varepsilon}_1}{\dot{\varepsilon}_3} = -\tan^2\left(\frac{\pi}{4} - \frac{\varphi}{2}\right) \tag{5}$$

where φ is the internal friction angle of the matrix material. The strain rate in the direction inclined at θ to the x-axis (Fig. 3(a)) is

$$\dot{\varepsilon}_\theta = \dot{\varepsilon}_{ij} n_i n_j = \dot{\varepsilon}_1 \sin^2\theta + \dot{\varepsilon}_3 \cos^2\theta \tag{6}$$

Using eqs. (5) and (6) one can find the inclination angles of planes which contain all fibers with a zero strain rate ($\dot{\varepsilon}_\theta = 0$)

$$\theta_0 = \pm\left(\frac{\pi}{4} + \frac{\varphi}{2}\right) \tag{7}$$

The work dissipation rate of all fibers in the tensile regime can now be integrated in the volume contained between planes AOC and BOC (Fig. 3(b)).

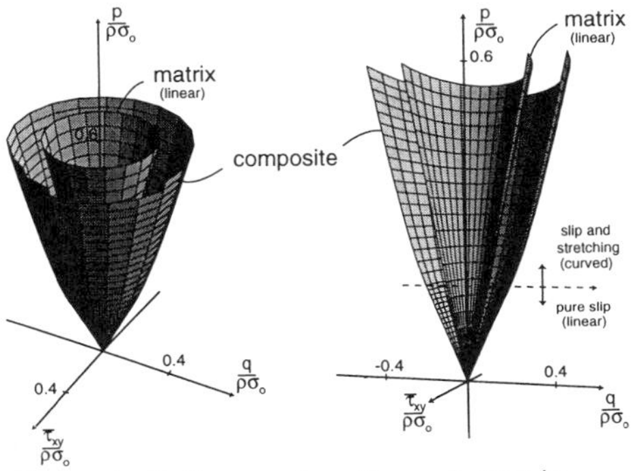

Figure 4. Failure surface for an isotropic soil-fiber composite ($\varphi = 30^0$, $\varphi_w = 20^0$).

These planes are inclined at angles θ_0 to the horizontal plane. A closed-form solution to such an integral was found, and the limit condition was derived in terms of macroscopic stresses. It was convenient to represent this function as

$$R = F(p, \psi) \tag{8}$$

where R is the maximum shear stress ($R = \sqrt{(\bar{\tau}_{xy}^2 + q^2)}$, $q = (\bar{\sigma}_x - \bar{\sigma}_y)/2$), p is the in-plane mean stress ($p = (\bar{\sigma}_x + \bar{\sigma}_y)/2$; both R and p are the in-plane invariants), and ψ is the angle of inclination of the major principal stress to the x-axis. Function (8) was found in a closed form for the isotropic fiber-reinforced soil. It is a piece-wise function

$$R = p\sin\varphi + \frac{1}{3}\bar{\rho}\sigma_0 N\left(1 - \frac{1}{4\eta\rho}\frac{\cot\varphi_w}{\frac{p}{\rho\sigma_0}}\right) \tag{9}$$

and

$$R = p\bar{\rho}\sigma_0\left(\sin\varphi + \frac{1}{3}N\rho\eta\tan\varphi_w\right) \tag{10}$$

where

$$N = \frac{1}{\pi}\cos\varphi + \left(\frac{1}{2} + \frac{\varphi}{\pi}\right)\sin\varphi \qquad (11)$$

with segments in eqs. (9) and (10) corresponding to the case where the yield stress is reached in the fibers, and the case of a pure slip of the fibers in the matrix soil, respectively (for details see Michalowski and Zhao, 1996). The failure surface is shown in Fig. 4.

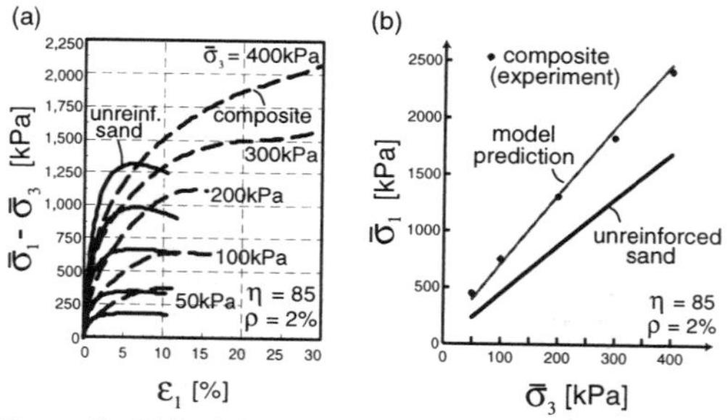

Figure 5. Triaxial tests: (a) stress-strain curves, (b) experimental and predicted failure stresses.

Examples of drained triaxial compression test results on initially isotropic fiber-reinforced soil (polyamide fibers) are presented in Fig. 5(a). Comparison of the predicted failure stress and that from triaxial tests is shown in Fig. 5(b).

5 STRENGTH OF ANISOTROPIC COMPOSITES

5.1 *Distribution of Fibers or Filaments*

Strength of fiber-reinforced soils is likely to be anisotropic because of the mixing and compaction (rolling) process. The same is true for continuous thread reinforcement because of the technique of placement of continuous filament reinforced soils. In either case there is a plane of preferred orientation, and therefore the distribution of fiber/filament orientation should be expected to be axisymmetrical. The following function was found convenient to describe such distributions

$$\rho(\theta) = \rho_{min} + a\,|\cos^n\theta| \qquad (12)$$

where ρ_{min} is the minimum density (volumetric concentration), n is a parameter indicating anisotropy of distribution (zero if isotropic), and a is the difference between the maximum and minimum $\rho(\theta)$. A more convenient set of parameters may include a ratio b of minimum to the maximum of $\rho(\theta)$ ($b = \rho_{min}/\rho_{max}$), and the average filament concentration $\bar{\rho}$, in which case the distribution of orientation in (12) becomes

$$\rho(\theta) = \bar{\rho}\,B\left(\frac{b}{1-b} + |\cos^n\theta|\right) \qquad (13)$$

Coefficient B must be derived from the requirement that the integrated average fiber concentration must be equal to $\bar{\rho}$. Restricting exponent n in (13) to positive even integers, and $b = 0$ (a realistic case where no fibers are vertical, or, in the case of continuous threads, no filament has a tangent perpendicular to the bedding plane), a convenient form of the concentration function results

$$\rho(\theta) = \bar{\rho}\,\frac{(1+n)!!}{n!!}\cos^n\theta = B\,\bar{\rho}\,\cos^n\theta \qquad (14)$$

where !! denotes the double factorial [$n!! = 2{\cdot}4{\cdot}6...n$ and $(n+1)!! = 3{\cdot}5{\cdot}7...(n+1)$].

5.2 *Limit Condition for Anisotropic Reinforced Soil*

For the plane strain deformation of the element in Fig. 6, eq. (2) can be written as

$$\bar{\sigma}_{x'}\dot{\varepsilon}_{x'} + \bar{\sigma}_{y'}\dot{\varepsilon}_{y'} + 2\bar{\tau}_{x'y'}\dot{\varepsilon}_{x'y'} = \dot{D}(\dot{\varepsilon}_{ij}) \qquad (15)$$

where bars denote the macroscopic stresses. The work dissipation rate due to fibers or filaments was integrated at failure, introducing an integration space as presented in Fig. 3(b). The fiber orientation distribution was mapped on the surface in Fig. 3(b), and the internal work rate during plastic deformation was calculated as a surface integral over the part of the surface associated with the fibers in the tensile regime. Two cases need to be considered: one when composite failure is associated with pure slip of fibers in the matrix, and the second one when the yield stress (σ_0) is reached in the fibers.

Calculations of the work dissipation rate are considerably more elaborate now compared to those in the isotropic case. When the yield stress is reached in the fibers, the dissipation integral is

$$\dot{D}(\dot{\varepsilon}_{ij}) = \frac{1}{\pi}\bar{\rho}\sigma_0\left(1 - \frac{1}{4\eta}\frac{\sigma_0}{p\tan\varphi_w}\right)\dot{\varepsilon}_1 B \cdot \int_0^{\frac{\pi}{2}}\int_{-\theta^*}^{\theta^*}\left(\frac{b}{1-b} + |\cos^n(\theta-\omega^*)|\right) \cdot (K_p\cos^2\alpha\cos^2\theta - \sin^2\theta)\cos\theta\, d\theta\, d\alpha \qquad (16)$$

where angle ω^* is a function of α and the deviation angle ω in the plane of deformation ($\tan\omega^* = \cos\alpha \tan\omega$, see Figs. 3(b) and 6), and $\dot{D}$ is the dissipation rate per unit volume of the composite.

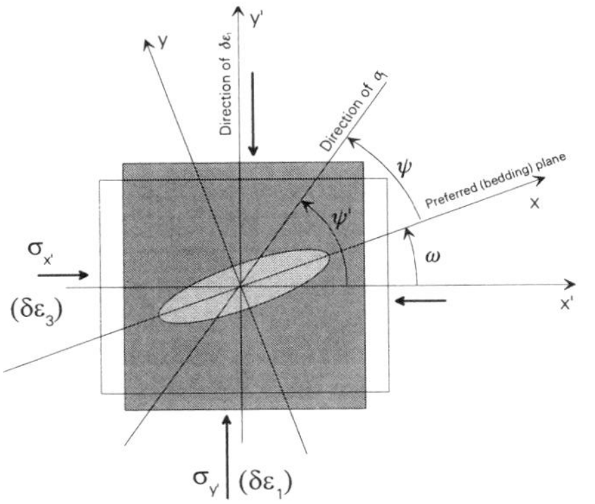

Figure 6. Plane deformation used in homogenization.

Considering that the Mohr-Coulomb failure criterion used to describe the granular matrix is independent of the intermediate principal stress, the failure criterion for the composite is also expected to be independent of σ_z. The failure criterion then will depend on three independent stress components $\bar{\sigma}_x$, $\bar{\sigma}_y$ and $\bar{\tau}_{xy}$, with direction x uniquely related to the anisotropy plane (plane of preferred fiber orientation). Consequently, the equation in (15) can be used to calculate one of the stress components, while the other two are given. It was found more convenient, however, to introduce parameters R, p, and ψ instead (see eq. (8)), and calculate the maximum shear stress R for given in-plane average stress p and angle of inclination of the major principal stress ψ. Because of the anisotropy of the composite, this scheme is not straightforward.

In order to find the failure criterion in stress space R, p, $\bar{\tau}_{xy}$, calculations of R were performed for a constant p and varied inclination angle ψ of the major principal stress to the x-axis. This makes it possible to trace the contour of the failure criterion in plane p = constant. The direction of x describes the anisotropy of the composite (x is the trace of the bedding plane on the plane of deformation).

The principal directions of the strain rate field are kept constant (x', y', Fig. 6), and they do not coincide with the principal stress directions, since the composite is anisotropic. The problem to be solved is formulated as follows (Fig. 6): given the principal directions of the macroscopic stress tensor (ψ), what would the preferred plane of fiber orientation (ω) need to be in order for the principal directions of strain to be x' and y'? Angle ω is a "deviation" angle defined here as an angle between axis x' and axis x (axis x coincides with the preferred plane of filament orientation).

The approach in eq. (2) (or (15)) is analogous to the application of the upper bound theorem of limit analysis. The deviation angle (ω), therefore, needs to satisfy the requirement that the calculated magnitude of R is a minimum. An optimization procedure was used to find such ω for a single p and ψ, and the calculations were repeated for a large set of p and ψ, to trace the entire failure surface in the macroscopic stress space. This failure surface, for a specific case of fiber distribution, is shown in Fig. 7.

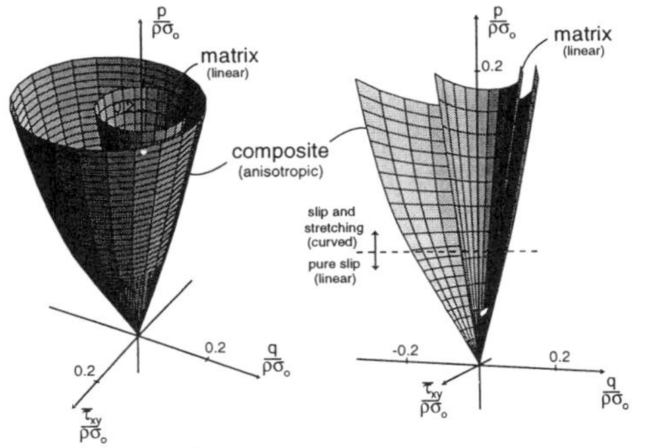

Figure 7. Failure surface for an anisotropic fiber-soil composite ($\varphi = 36^0$, $\varphi_w = 27^0$).

Numerical details are omitted here because of limited space (for details see Michalowski, 1997, where a similar problem is considered with an anisotropic distribution of continuous filament).

5.3 *Anisotropic hardening of fiber-reinforced soil*

Triaxial tests on initially isotropic fiber-reinforced soil exhibit a kinematic hardening effect. An initially isotropic distribution of fiber orientation becomes anisotropic because of fiber reorientation due to progressive strain of the specimen. For a relatively large confining pressure and large fiber content, this effect occurs prior to the specimen reaching instability, as illustrated in Fig. 5(a) for confining pressures of 300 and 400 kPa.

Preliminary results of simulating the process of kinematic hardening are shown in Fig. 8. The initial distribution of fiber orientation is isotropic, and it can be illustrated by a spherical surface. The change of fiber orientation then occurs due to specimen deformation, and it was assumed that this evolving distribution can be represented as an ellipsoid. The principal fiber concentrations (axes of the ellipsoid) were calculated from the requirement that the mean fiber content ($\bar{\rho}$) must be constant for an incompressible composite, and must reduce accordingly for a dilatant process.

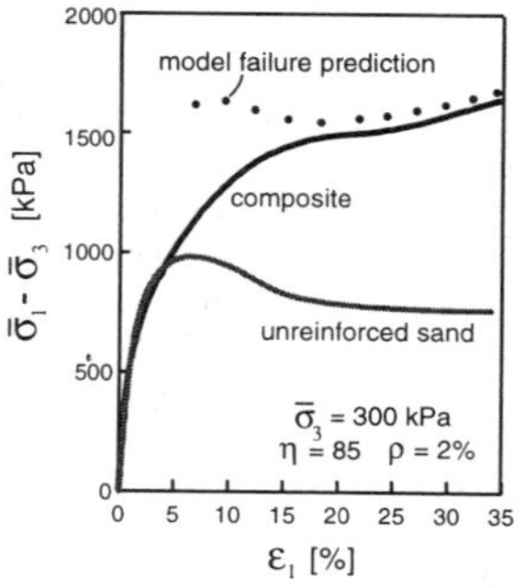

Figure 8. Kinematic hardening of fiber-reinforced soil.

The change in the fiber orientation distribution then becomes the evolution rule for the kinematic hardening process. The work dissipation rate was calculated for a sequence of deformation stages of a specimen in a triaxial compression test, and the failure stress was calculated from eq. (2). The bullets in Fig. 8 correspond to the calculated stresses. It is not surprising that the calculations overestimate the failure stress at the beginning of the process when the fiber stress is not fully mobilized. In the advanced phase, however, the method suggested seems to be quite effective.

6 CONCLUDING REMARKS

The method presented for calculating the limit stress for fiber-reinforced soils is based on the homogenization technique in which the work dissipation rate at failure is equated to the rate of macroscopic composite stress. Experimental tests for soils reinforced with fibers distributed uniformly in all directions (isotropic composites) appear to confirm the effectiveness of the method. Application of the homogenization method to anisotropic fiber reinforcement is more elaborate, though straightforward in principle. The method may be equally effective in estimating strength of soils reinforced with continuous threads.

Acknowledgments
The work presented in this paper was sponsored by the Air Force Office of Scientific Research, grant No. F49620-97-1-0109, and the National Science Foundation, grant No. CMS-9634193. This support is greatly appreciated.

REFERENCES

de Buhan, P., Mangiavacchi, R., Nova, R., Pellegrini, G. and Salençon, J. (1989). "Yield design of reinforced earth walls by homogenization method." *Géotechnique*, vol. 39(2), 189-201.
Gray, D.H. and Ohashi, H. (1983). "Mechanics of fiber reinforcement in sand." *ASCE J. Geot. Eng.*, vol. 109, 335-353.
Jommi, C., Nova. R. and Gomis, F. (1995). "Numerical analysis of reinforced earth walls via a homogenization method." *5th Int. Symp. Num. Meth. Geomech.* (NUMOG V), G.N. Pande and S. Pietruszczak, eds., Balkema, 231-236.
Kulczykowski, M. (1989). "Bearing capacity of foundation situated on reinforced earth." (in Polish) *Arch. Hydrotech.*, vol. 36(1-2), 121-164.
Michalowski, R.L. (1997). "Limit stress for granular composites reinforced with continuous filament." *ASCE J. Eng. Mech.*, vol. 123 (to appear).
Michalowski, R.L. and Zhao, A. (1996). "Failure of fiber-reinforced granular soils." *ASCE J. Geot. Eng.*, vol. 122(3), 226-234.
Michalowski, R.L. and Zhao, A. (1995). "Continuum versus structural approach to stability of reinforced soil." *ASCE J. Geot. Eng.*, vol. 121(2), 152-162.
di Prisco, C. and Nova, R. (1993). "A constitutive model for soil reinforced by continuous threads." *Geotextiles and Geomembranes*, vol.12, 161-178.
Sawicki, A. (1983). "Plastic limit behavior of reinforced earth." *ASCE J. Geot. Eng.*, vol. 109(7), 1000-1005.
Waldron, L.J. (1977). "The shear resistance of root-permeated homogeneous and stratified soil." *Soil Sci. Soc. Am.*, vol. 41, 843-849.
Wu, T.H., McKinnell III, W.P. and Swanston, D.N. (1979). "Strength of tree roots and landslides on Prince of Wales Island, Alaska." *Can. Geot. J.*, vol. 16(1), 19-33.

Numerical Models in Geomechanics, Pietruszczak & Pande (eds) © *1997 Balkema, Rotterdam, ISBN 90 5410 886 X*

An upper bound analysis of stone columns

A. Noorzad & H. B. Poorooshasb
Department of Civil Engineering, Concordia University, Montreal, Que., Canada
N. Miura
Institute for Lowland Technology, Saga University, Japan

ABSTRACT: An upper bound analysis of the performance of foundations with stone column inclusions is presented. It is shown that the dilatation properties of the column material plays some role in the over all performance of the system but the determining factor is the column spacing. The results of the present analysis are compared with an earlier study by Poorooshasb and Meyerhof (1997) which appeared in the journal *Computers and Geotechnics.*

1 INTRODUCTION

Stone columns provide a simple and economically feasible solution to improve the properties of the foundations for low rise buildings and structures that can tolerate appreciable settlements. In the present paper an analysis is presented which is based on an earlier study by Poorooshasb and Meyerhof (1997). It also utilizes the concept of the *infinitely flexible membrane* to obtain an upper bound to estimate the magnitude of settlement of the foundation.

2 FLEXIBLE SMOOTH MEMBRANE

A flexible smooth membrane is defined as a hypothetical *frictionless* membrane of *infinitesimal thickness.* Being of infinitesimal thickness it can offer no resistance to bending or tensile or compressive forces. All it does it provides admissible stress discontinuities at the location where it is inserted. A judicial choice of the position of such membranes can facilitate the analysis immensely.

A hypothesis central to the present study may now be stated: "*insertion of one (or more) smooth membranes will not reduce the work done by the external forces on the system*". Consider the system shown in Figure 1,a which represents the actual system and that shown in Figure 1,b which shows the same system with the membrane inserted.

Let the external forces increase from some predetermined value (say zero) to their present value of F and denote the work done on the system by W_a. Next consider the system with the membrane inserted subjected to the same loading process and denote the work done by W_b. Then according to the hypothesis;

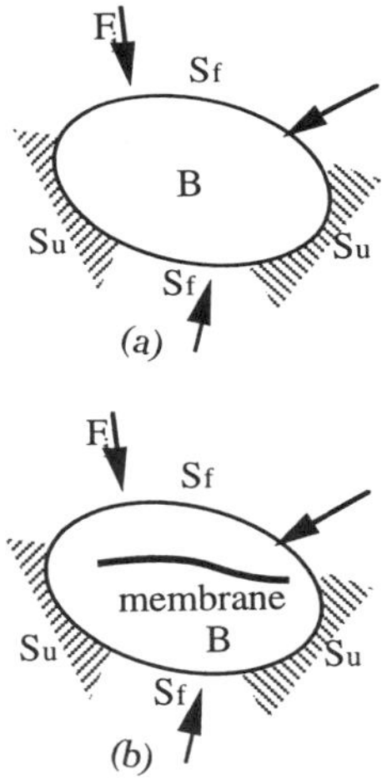

Figure 1. Actual system (a) and system with membrane inserted (b).

$$W_b \geq W_a \qquad (1)$$

For certain types of material this hypothesis can be stated as a theorem. For example for an elastic solid it may be shown that [see, for example Hill (1960) for a concise proof];

$$\int_V \sigma_{ij}^* \varepsilon_{ij}^* dV \geq \int_S F_i u_i dS \qquad (2)$$

where σ_{ij}^* and ε_{ij}^* are the stress and strain fields in the *system with the membrane*, F_i are the components of the external forces and u_i are the components of the surface displacements. Since σ_{ij}^* is statically admissible then it follows;

$$\int_V \sigma_{ij}^* \varepsilon_{ij}^* dV = \int_S F_i u_i^* dS \quad (3)$$

Equation (3) in combination with expression (2) leads to;

$$\int_S F_i u_i^* dS \geq \int_S F_i u_i \, dS \quad (4)$$

which establishes the truth of the theorem as the left hand side of the expression (4) is equal to $2W_b$ and that on the right hand side equal to $2W_a$. Cohesionless granular do not qualify as either elastic or elastic-plastic materials with an associated flow rule. Thus the above theorem is at this stage proposed as a hypothesis.

3 SETTLEMENT OF STONE COLUMNED FOUNDATIONS

Stone columned foundations consist of a large number of columns in a regular (usually triangular or square) pattern and spaced at equal distances from one another. Thus an individual column acts within a cylindrical cell (sometime referred to as the tributary region) of radius b, say, Figure 3,b.

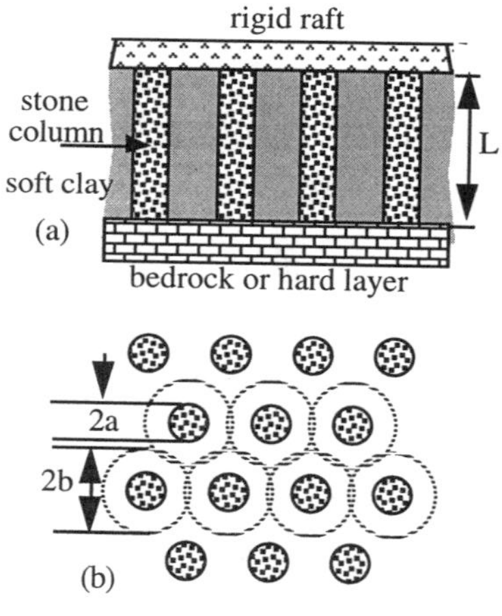

Figure 2- Section (a) and plan view of a stone columned foundation (b).

In view of the symmetry of the pattern of installation it is assumed that the soil particles in the neighborhood of the cell walls 3,a shows a stone column within its cell. The same system is modeled in Figure 3,b with a large number of horizontal membranes and one vertical membrane surrounding the column. Note that the horizontal membrane pass through the clay layer as well as the column.

The work done for the system shown in Figure 3,a (the actual system) is;

$$W_a = \frac{\pi b^2}{8} UDL \cdot \delta_{actual}$$

and the work done on the system with membranes is;

$$W_b = \frac{\pi b^2}{8} UDL \cdot \delta_{membraned}$$

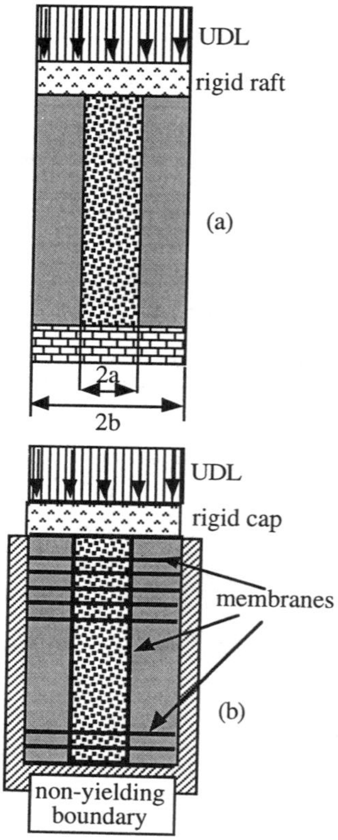

Figure 3- Stone column within tributary cell (a) and model with membrane (b).

Using expression 4 results in;

$$\frac{\pi b^2}{8} UDL \cdot (\delta_{membraned} - \delta_{actual}) \geq 0;$$

or;

$$\delta_{membraned} \geq \delta_{actual}.$$

Thus the settlement results obtained from the present analysis are upper bound values.

Before advancing further it is appropriate to point out the way in which introduction of the iembranes facilitate the calculations. Under this :heme the soil sandwiched between two adjacent iembranes behaves as thick cylinder subjected to a niform internal pressure (exerted by the stone olumn plus some residual pressures), a uniformly istributed vertical load p (due to application of the DL) and is constrained from lateral movement at e boundary of the cell where r=b. The column ection between the same membranes is subjected to n axial compression and an ambient stress (exerted y surrounding clay.) Thus the column section nder consideration is experiencing a loading process milar to that experienced in a large triaxial ompression test. Figure 4 shows the situation;

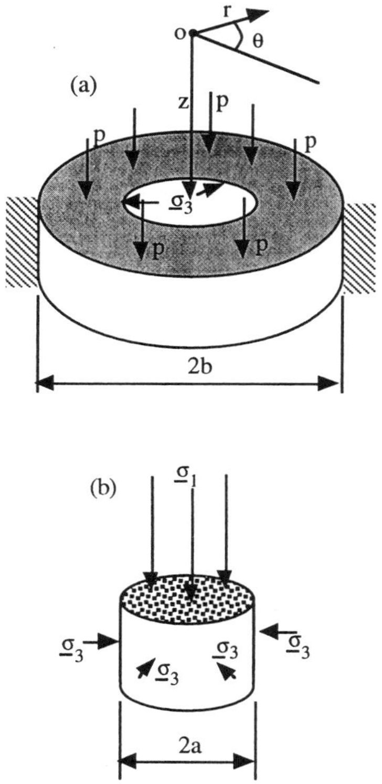

Figure 4- Boundary forces and constraints acting on a section of soft soil (a) and a section of column (b).

1 *Analysis of column section*

et auxiliary parameters c_1 and c_2 be defined by uations;

$$c_1 = \frac{v}{1-v};$$

$$c_2 = E_s^{(e-p)} \frac{(1+v)a^2 + (1-v)b^2}{(1-v^2)(b^2-a^2)}$$

then it can be shown [Poorooshasb and Meyerhof (1997)] that the contact pressure σ_3 *caused by the surcharge UDL* is given by;

$$\sigma_3 = c_1 p - c_2 \varepsilon_3$$

In the last equation $E_s^{(e-p)}$ is the elastic-plastic Young modulus for the soil and v is its Poisson's ratio. Thus the total lateral stress $\underline{\sigma}_3$ is;

$$\underline{\sigma}_3 = c_1 p - c_2 \varepsilon_3 + \sigma_{res} \qquad (5)$$

The ratio of the vertical stress to the ambient stress at depth z is;

$$\eta = \frac{\sigma_1 + \gamma'_g z}{\underline{\sigma}_3} \qquad (6)$$

where σ_1 is the vertical stress imposed by the raft on the column and γ'_g is the effective unit weight of the material in the column. It may be shown that if the material behaves as the CANAsand model then one may obtain the axial strain ε_1 from the equation;

$$\varepsilon_1 = \frac{a(\eta - 1)}{\eta_f - \eta} - \varepsilon_0 \qquad (7)$$

where a is a material constant η_f is the stress ratio corresponding to the failure state of the stones and ε_0 is the axial strain the sample had experience under residual stresses (i.e. before the loading of the mat had proceeded). In deriving Eq. (7) the hyperbolic relation is assumed.

Let the volumetric strain of the material in the column be denoted by v. Then to a very close approximation one may write;

$$v = \varepsilon_1 + 2\varepsilon_3 = m\varepsilon_1$$

where m is dependent on the degree of interlocking of the particles (and in an indirect way on the angle of friction of the material). From the last equation one may get;

$$\varepsilon_1 = \frac{2\varepsilon_3}{m-1} = \varepsilon_1^{column} \qquad (8)$$

which when combined with Eq. (7) leads to;

$$\eta = \frac{K_1 + 2\varepsilon_3 \eta_f}{K_2 + 2\varepsilon_3} \qquad (9)$$

where;

$$K_1 = a(m-1) + (m-1)\varepsilon_0 \eta_f;$$

$$K_2 = a(m-1) + (m-1)\varepsilon_0.$$

Combining Eq. (9) with Eq. (6) and rearranging terms results in the equation;

$$A\varepsilon_3^2 + B\varepsilon_3 + C = 0 \tag{10}$$

where;

$$A = 2c_2\eta_f;$$

$$B = 2(\sigma_1 + \gamma'_g z) - 2(\sigma_{res} + c_1 p)\eta_f + K_1 c_2$$

$$C = (\sigma_1 + \gamma'_g z)K_2 - (\sigma_{res} + c_1 p)K_1$$

which determines the value of ε_3 [and hence ε_1^{column} from Eq. (8)] *had the value of* σ_1 *and p been known..* The settlement of the column is obviously given by;

$$\delta_{col} = \int_0^L \varepsilon_1^{column} dz \tag{11}$$

The vertical strain in the soil section is given by the equation;

$$\varepsilon_1^{soil} = c_3 p + c_4 \varepsilon_3 \tag{12}$$

where;

$$c_3 = \frac{1 - 2v^2 - v}{(1 - v)E_s^{(e-p)}};$$

$$c_4 = \frac{2va^2}{(1 - v)(b^2 - a^2)}$$

and hence the settlement of the ground surface is;

$$\delta_{soil} = \int_0^L \varepsilon_1^{soil} dz \tag{13}$$

If one defines the parameter ξ such that

$$\xi = \delta_{soil} - \delta_{col} \tag{14}$$

then obviously the solution to the problem obtains when ξ=0. With the aid of the equation;

$$p = \frac{b^2 \cdot UDL - a^2\sigma_1}{b^2 - a^2} \tag{15}$$

and an iterative technique this is possible. To start with the value of σ_1 is assumed to be equal to the UDL. This yields a value for the soil settlement much larger than the column settlement. The process is repeated increasing σ_1 successively until ξ assumes a very small negative value. The answer is obtained by interpolation between the old values (when ξ was small but positive and now that ξ is small but negative). The flow chart for this operation is shown in Figure (5)

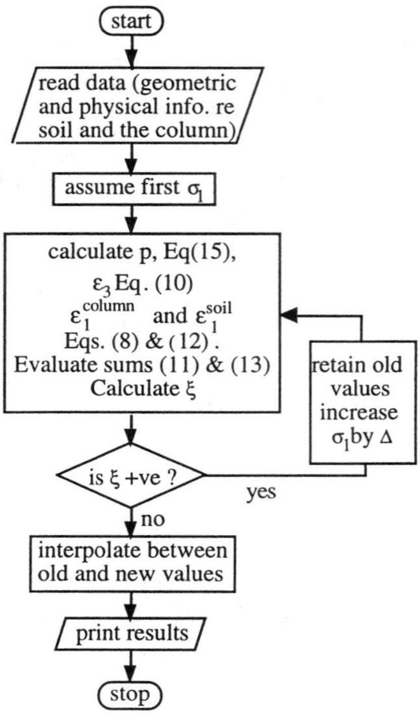

Figure 5- Flow chart for the evaluation of the performance of the system

3.2 *Typical results*

Figures 6 and 7 show some typical results from th above analysis. The material in the column shown i Figure 6 is assumed to be medium dense with a angle of friction at failure equal to 38 degrees. Th spacing between the columns is assumed to be 3, spacing beyond which stone columns will cease to b effective in reducing the settlement of the foundatio scheme. In this case as well as that shown in Figur 7 the soil layer is assumed to have a "hard crust" du to overconsolidation by desiccation. The depth o the hard crust is assumed to be 1.5 meters. Al columns are assumed to be 10 meters in length with diameter of 100 cm.

The performance of the system is determine by its performance ratio which is defined as the rati of the settlement of the treated ground (foundatio with stone column inclusion) to that of the natura ground. Obviously a high performance ratio, P.R. indicates very poor performance. A P.R.=1 indicate that inclusion of columns has had no effect at al The P.R. value associated with the case shown i Figure 6 is 0.78.

Figure 7 shows the performance of a syste with the same physical features. Here the material in

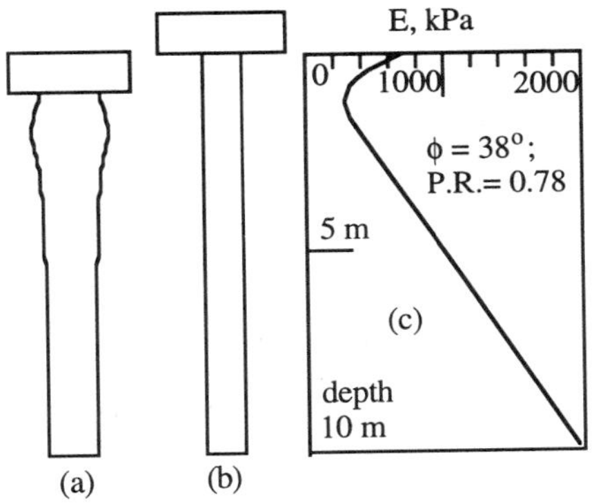

Figure 6- Deformed (a) and undeformed column, material medium dense ϕ=38 degrees, Spacing= 3.dia..

the column is assumed to have an angle of friction equal to 44 degrees. There is an improvement in the P.R. value, .61 compared to .78 as in the previous case, representing an improvement of about 20%.

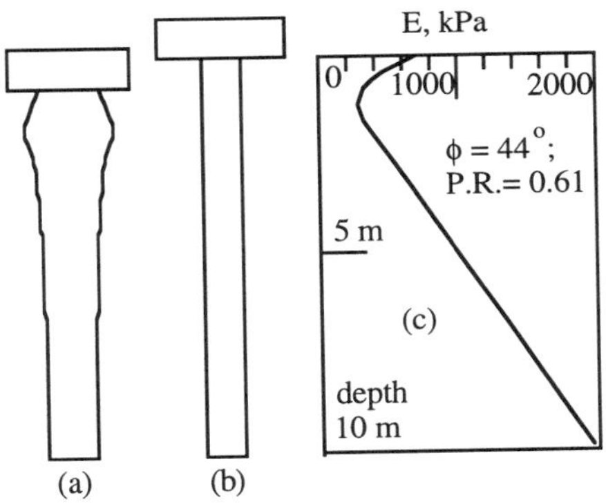

Figure 7- Deformed (a) and undeformed column, material medium dense ϕ=44 degrees, Spacing=3.dia.

When column spacing is smaller the dilatation properties play a slightly more important role. For example if a spacing of two is used in the design then the resulting P.R. value for a medium dense column is 0.56, Figure 8.

If the columns from the previous case were filled with very dense material then the corresponding P.R. value would be 0.35 which represents an improvement of about 38 %.

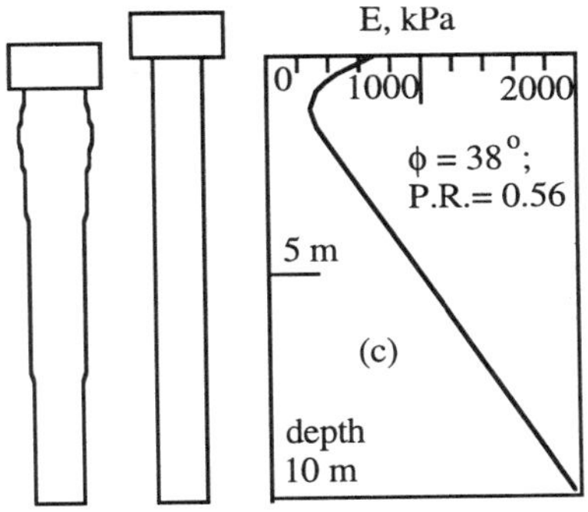

Figure 8- Deformed (a) and undeformed column, material medium dense ϕ=38 degrees, Spacing=2.dia.

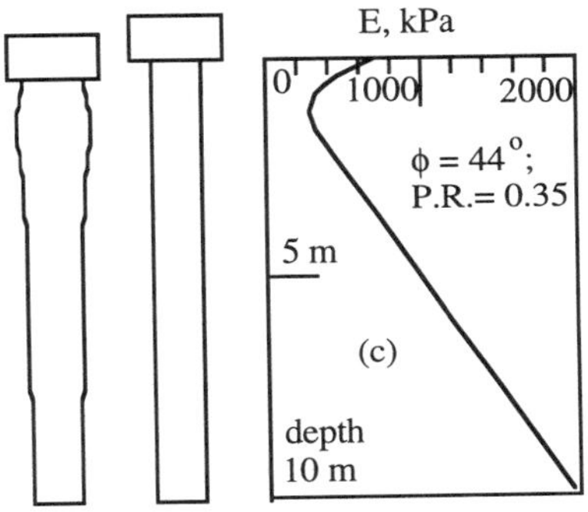

Figure 9- Deformed (a) and undeformed column, material medium dense ϕ=44 degrees, Spacing=2.dia.

3.3 *Number of sections required*

At this stage it is necessary to discuss the number of sections required to obtain an "acceptable solution". The answer is very few indeed. This section also serves as check on some previously published results by Poorooshasb and Meyerhof (1997). The analysis by these authors is, in fact, equivalent to assuming only two horizontal membranes (underneath the raft and at the elevation of the bedrock) and one vertical membrane wrapped around the column. The table shown below provides a comparison between results obtained by these authors and those obtained in this study which uses 51 membranes. The symbol P&M are the results obtained Poorooshasb and Meyerhof which assumes a constant Young modulus, (1) indicates a variation of E with depth as shown in curve (1) of Figure 10 and (2) assumes a variation of E as shown by curve (2) of the same Figure.

In all cases the value of Poisson's ratio was assumed to be 0.2.

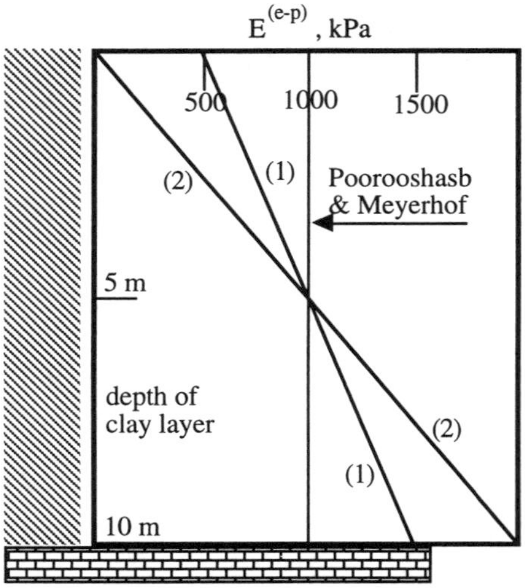

Figure 10- Variation of E with depth in the soft clay layer

P.R. for ϕ =44 degrees.

Spacing	P&M	(1)	(2)
1.5	.15	.17	.14
2	.31	.34	.33
3	.56	.61	.60
10	.94	.96	.95

P.R. for ϕ =41 degrees.

Spacing	P&M	(1)	(2)
1.5	.2	.22	.2
2	.39	.45	.44
3	.64	.71	.70
10	.97	.97	.97

P.R. for ϕ =38 degrees.

Spacing	P&M	(1)	(2)
1.5	.27	.31	.28
2	.48	.54	.53
3	.72	.78	.77
10	.97	.98	.98

From the above table it may be concluded that the results of the evaluations are by far more sensitive to the physical dimensions and material properties as they are to the number of membranes inserted.

ACKNOWLEDGMENT

The financial support provided by the Natural Science and Engineering Council (NSERC) of Canada and the Ministry of Energy , Islamic Republic of Iran in support of this study is gratefully acknowledged.

REFERENCES

Hill, R. 1960. *The mathematical theory of plasticity*, Oxford at the Clarendon Press.

Poorooshasb, H.B. and Meyerhof, G.G. 1997. Analysis of Behavior of stone columns, Computers and Geotechnics, Vol. 20, No.1, pp. 47-70.

Poorooshasb, H.B., Miura. N. And Noorzad, A. 1997. An upper bound analysis of the behavior of stone columns, Lecture to be delivered at the opening ceremony of the new facilities of the Institute for Lowland Technology, Saga University, Japan.

Numerical Models in Geomechanics, Pietruszczak & Pande (eds) © 1997 Balkema, Rotterdam, ISBN 90 5410 886 X

Finite element modeling of a soil nailed wall: Earth pressures and arching effects

B. Benhamida
Ecole Nationale des Ponts et Chaussées, Paris & CERMES (Soil Mechanics Research Centre), Noisy-le-Grand, France

F. Schlosser
Terrasol, Montreuil, France

P. Unterreiner
Direction Départementale de l'Equipement, Martinique, France

ABSTRACT: An acceptable engineering design relies on the successful identification of the different mechanisms which may take place during and after the construction of a structure. One aspect of such design in soil nailing deals with the magnitude and distribution of lateral stresses acting on the facing. With this objective in mind, a finite element analysis of a full scale experimental soil nailed wall was performed. This paper presents numerical results of the distribution of earth pressures acting on the facing. The influences of the soil movements, the wall flexibility and the construction method of soil nailing on the development of these are identified and both local and global arching effects are well illustrated.

1 INTRODUCTION

Soil nailing, an in-situ earth support system, has been widely used in recent years to stabilize excavated walls and slopes. In this technique, whilst the excavation proceeds, a series of steel bars are placed into the excavated native soil to artificially improve and strengthen its mechanical characteristics. In the 1980s, this new technique of soil reinforcement was the subject of an important development (Schlosser, 1982, 1983), particularly in France where a National Research Project was launched with the objective to better understand the behavior of soil nailed walls during construction, in service and at failure. Five years of research, studies and tests on soil nailing concluded with the publication of French Specifications entitled "*CLOUTERRE Recommendations 1991*" (French version, 1991 and English version, 1993). The design recommended at the present time is based on classical limit equilibrium methods (method of slices) generalized to reinforced soils using a multicriteria approach (Schlosser, et al., 1983). This method allows to analyze and check the stability of a structure but does not allow to estimate displacements.

Research are going on within the frame of a second French National Research Project entitled CLOUTERRE II whose principal objective is to develop a serviceability limit state design based on displacements calculation methods. This article present a finite element study of the first full scale experimental soil nailed wall constructed during the CLOUTERRE project and describes the significant results concerning the distribution of lateral stresses.

2 CHARACTERISTICS OF THE EXPERIMENTAL SOIL NAILED WALL

The nailed wall is 7 m high, 7.5 m wide and constrained between two lateral walls covered with a double layer of polyethylene sheet greased in between to ensure plane strains conditions. The wall was built by alternating 1 m high excavations with the placing of the nails according to an horizontal spacing of 1.15 m. The nails were inclined 10° with respect to the horizontal direction and their lengths range from 6 to 8 m (Figure 1).

2.1 Soil

The soil used in the experimentation is the Fontainebleau sand type and was placed and compacted to obtain a relative density of 0.6 and an average unit weight of 16.6 kN/m^3. The mechanical properties were evaluated based on data published on the sand Fontainebleau type, the pressuremeter test results conducted at the site and from triaxial test results performed at the CERMES, Table 1.

Table 1:

I_D	σ_c [kPa]	z [m]	E_t [MPa]	E_U [MPa]	ϕ [°]	ψ^* [°]	ν	E_M ** [MPa]
0.65	100	10.35	59	153	36.4	22.3	0.37	20.7
0.60	75	7.76	46	121	37	21.2	0.42	17.0
0.65	50	5.18	35	108	39.1	32.6	0.40	13.2
0.60	25	2.59	21	60	40.3	33.2	0.41	9.55

(*) : $\sin\psi = \sqrt{(2/3)\cdot s/(1-s)}$, where s is the slope of the (ε_v, ε_1) curve.

(**) : E_M is the average Ménard pressumeter modulus at a depth such that $\sigma_c = \dfrac{\sigma_v + 2K_0\sigma_v}{3}$ with $K_0 = 1 - \sin\phi$.

2.2 Reinforcement

Soil reinforcement consists of hollow aluminum tubes grouted in the sand. Figure 1 summarizes the lengths, diameters and thicknesses regarding each nail row. Laboratory tests conducted on a nail constructed in the same sand and with the same technique as for the full scale structure show a yield stress of 245 MPa for a 40 mm diameter tube and 235 MPa for a 30 mm diameter tube. The elastic limit, measured at 0.2% is 127 MPa and 94 MPa, respectively. The equivalent 2D parameters needed for the Finite Element analysis have been estimated from the values obtained on the real nails taking into account the horizontal spacing between the nails.

2.3 Facing

The facing is made of a mesh reinforced shotcrete. Its Young's modulus is taken equal to 25 10^6 kPa with a unit weight, γ, of 24 kN/m^3 and a Poisson coefficient, ν, of 0.2.

2.4 Soil - nail interface

Both the elastic coefficient of mobilization of lateral friction, k_τ, and the unit limit lateral friction, q_s, have been estimated from CLOUTERRE correlations. Those correlations are based on the pressuremeter modulus, E_M, for k_τ and on the limit pressuremeter pressure, p_l, for q_s.

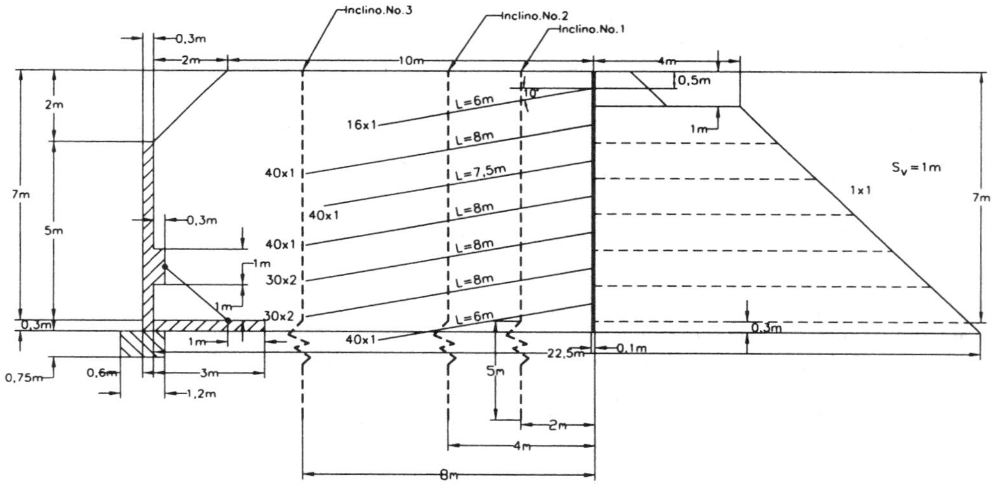

Figure 1. Cross section of the full scale experimental soil nailed wall

3 MODELING OF THE DIFFERENT COMPONENTS OF THE SOIL NAILED WALL AND THEIR INTERACTION

Both the foundation and the backfill soils are modeled using a linear elastic perfectly plastic model with a Mohr-Coulomb criterion and a non associated flow rule. The behavior of the nails is represented by a linear elastic model for the range of tensile forces which was measured during the first five phases of construction. The facing is approximated with a linear elastic model. The constitutive behavior of the soil-nail interface is modeled with a linear elastic perfectly plastic model. Tables 2, 3 and 4 summarize the different values of the parameters retained for the calculation

Despite the complexity of the structure, a simple mesh using a relatively small number of rectangular 8 nodes elements is used. The foundation soil elements are about 1 m by 1 m, while the backfill ones are 0.25 m by 1 m. The soil-nail interface elements have 6 nodes. Boundary conditions are located so that they have negligible effects on the structure. (Benhamida et al. 1997).

Table 2: Mechanical properties of the backfill and foundation soils used in the calculation

Backfill soil		
Young's modulus constant with depth ($\langle E_M \rangle_{z=0 \text{ to } 7.3\text{ m}} = 10.3$ MPa)	$E = 3\,E_M$	$31\ 10^3$ kPa
Unit weight	γ	16.6 kN/m^3
Poisson coefficient	ν	0.39
Angle of internal friction	φ	38°
Angle of dilatancy	ψ	27°
Cohesion	c	3 kPa
Foundation soil		
Young's modulus constant with depth ($\langle E_M \rangle_{z=7.3 \text{ to } 14.3\text{ m}} = 28$ MPa)	$E = 3\,E_M$	$84\ 10^3$ kPa
Unit weight	γ	17 kN/m^3
Poisson coefficient	ν	0.37
Angle of internal friction	φ	36°
Angle of dilatancy	ψ	22°
Cohesion	c	0 kPa

Table 3: Summary of the nails parameters

Young's modulus (real nails)	E	$67\ 10^6$ kPa
Limit elastic stress (real nails)	σ_e	$110\ 10^6$ kPa
Young's modulus (model nails)	$\tilde{E} = E/S_h$	$58\ 10^6$ kPa
Limit elastic stress (model nails)	$\tilde{\sigma}_e = \sigma_e/S_h$	$95.65\ 10^3$ kPa
Model nail section (equal to real nail section in order to have the relationship $\tilde{E}\tilde{S} = ES/S_h$)	$\tilde{S} = S$	varies with each nail layer

Table 4: Summary of the soil-nail interface parameters

Coefficient of mobilization of lateral friction : $K_\tau = 2\langle E_M \rangle / m D_c$, with $\langle E_M \rangle_{z=0 \text{ to } 7.3\text{ m}} = 10.3$ MPa and $m = 4.6$	K_τ	$71\ 10^3$ kPa/m
Unit skin friction (real nails) : $\langle p_l \rangle_{z=0 \text{ to } 7.3\text{ m}} = 1.05$ MPa	q_s	80.0 kPa
Boring diameter (real nails)	D_c	$63\ 10^{-3}$ m
Horizontal spacing (real nails)	S_h	1.15 m
Tangential rigidity of interface $K_s = \dfrac{\pi D_c}{2 S_h} K_\tau$	K_s	$6.117\ 10^3$ kPa/m
Cohesion of interface elements : $c_i = \dfrac{\pi D_c}{2 S_h} q_s$	c_i	6.88 kPa
Friction angle of interface elements	φ_i	0°
Dilatancy angle of interface elements	ψ_i	0°

4 COMPARISON OF CALCULATIONS WITH MEASURES

The finite element modeling of the construction of the first full scale experimental soil nailed wall of the CLOUTERRE PROJECT gives a reasonable agreement between predicted and measured tensile forces in the nails as well as for the displacements of the facing and the soil (Unterreiner et al, 1997; Benhamida et al., 1997).

5 EARTH PRESSURES IN RETAINING STRUCTURES

5.1 Classical earth pressures theories

Classical earth pressure theories and the behavior of earth-retaining structures are based on considerations of shear strength. These theories use the concept of the earth pressure coefficient to describe the development of lateral stresses behind a retaining structure. The general approach proposed by Terzaghi, in the early 1930s, is to assume that the wall is held in a fixed position and the soil behind is "at rest" condition and the earth pressure distribution increases linearly with depth ($\sigma_h' = K_0\sigma_v'$). Then, depending on the wall movement two conditions may take place. The wall is allowed to move, by rotation about its base, under the action of the in-situ pressure, then it moves away from the soil and the pressure decreases until it reaches a minimum value called active earth pressure, or the wall is moved towards the soil which causes the earth pressure to rise and continues to do so until it attains a constant value once again, known as the passive earth pressure. The reason for change in the earth pressure due to wall movement is that, as the wall moves away from its initial "at rest" condition, shear stresses are applied to the soil and mobilize its full shear strength. In the active case the shear stress, τ, acts to support the weight of the soil, whereas in the passive case the shear stress, τ, acts against the force pushing the wall into the soil. Terzaghi (1936)* concluded his observations by recognizing that other modes of wall movement could yield very different pressure distributions. Also, Meem (1908)* and Moulton (1920)*, based on observations of excavations, reported that pressure distributions did not increase linearly with depth. Nowadays, the multitude of the different earth support systems along with the variety of construction techniques have made the classical solutions of calculations of earth pressures in retaining structures completely not satisfactory. Accurate calculations have to take into account not only the way how the wall moves, but to consider the construction technique used and the wall flexibility.

Most retaining structures are not rigid but bend under the applied loads. Often they do not rotate about the base but may well translate or rotate about the top or other part of the structure. Also, depending on the technique used to construct a wall, it is clear that the in-situ earth pressures will be affected. In the case of excavations walls, such a soil nailed wall, it is certain that, during a construction phase, leaving an excavation of 1 to 2 m height of native soil unsupported for a period of time before placing the nails and the confining skin (facing) results in significant total stress relief. The total horizontal stress on the boundary of the newly excavated soil will reduce from the initial in-situ horizontal total stress in the undisturbed soil to approximately zero.

5.2 Earth pressures in soil nailed walls

The development of the earth pressures in soil nailed walls is complex due to the different mechanisms which take place, during construction and at service, between the different components, namely the soil-nail interaction, the soil-facing interaction, the nail-facing interaction and the behavior of the newly reinforced soil mass. During successive excavations, the soil that will form the soil nailed wall is subject simultaneously to lateral decompression and to settlements. As a result, at the end of construction a slight tilting of the facing occurs where horizontal and vertical displacements are at their maximum at the head. Figure 2 illustrates the way the facing moves at the end of construction of phase 3 and 5 and also indicates the evolution of vertical and horizontal displacements at the head during the construction.

Besides the particular mode of movement of the facing, the influence of the construction technique pertaining to soil nailing is of fundamental importance to the magnitude and the distribution of earth pressures. The impact of installation effects is even more pronounced by the presence of nails at the top of the wall. In their finite element analysis, Gunn* et al. (1992) have demonstrated that the greater the restraint imposed at the top of the wall (by propping or anchoring), the larger will be the effects of installation.

---

* cited by Clayton et al., (1993)

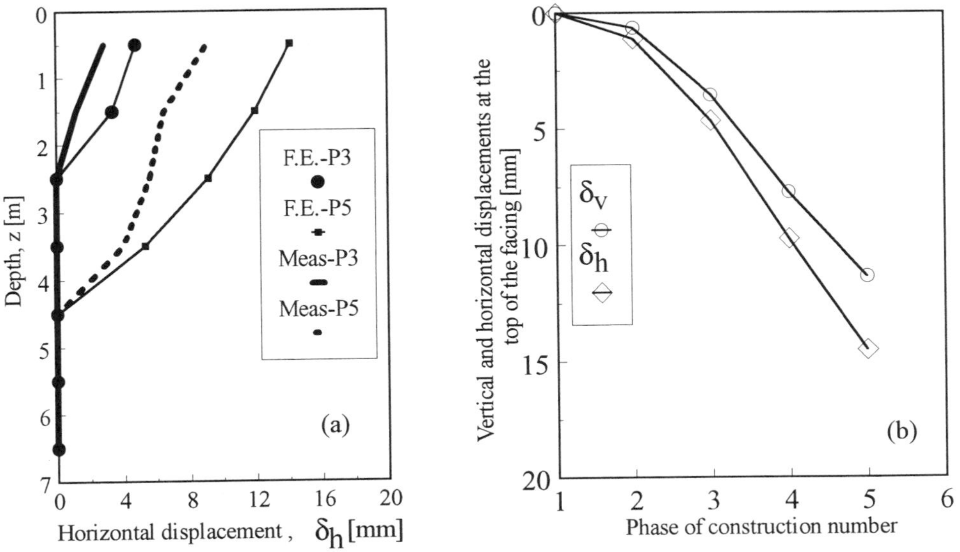

Figure 2 : Movements of the wall facing: (a) Horizontal displacements of the facing : comparison F.E. calculation and measures at end of phase 3 and phase 5 - (b) Evolution of vertical and horizontal displacements at the head of the wall facing during the successive stages of construction.

The primary role of the flexible facing in soil nailed wall is to be associated to the soil newly excavated in order to bring a slight confinement and to neutralize the drop of stress at the boundary.

Considering the above, it is clear that, in the case of soil nailed walls the magnitude and the distribution of earth pressures will deviate from what would predict the classical solutions. To better visualize this difference, two vertical cross-sections within the soil nailed wall have been considered. The numerical results obtained for the distribution of the vertical stress, σ_V , and the horizontal stress, σ_h , at 0.5 m and 1.0 m of the facing are compared respectively to the overburden geostatic vertical stress, γz , and to the corresponding passive, $K_p\gamma z$, active, $K_a\gamma z$, and at rest, $K_0\gamma z$ earth pressures where the cohesion has been neglected and K_0 has been chosen equal to $1-\sin\phi$ (Figures 3 and 4).

Next to the facing at 0.5 m, (Figure 3), for the upper part of the wall, the pressures generated are close to those predicted for "at rest" condition. At the bottom, the pressures are close to active state. At the top of the wall the nails are mobilized sufficiently to restore "at rest" condition in the soil. This mobilization of the nails along with the restoration of earth pressures to "at rest" condition result in the development of local arching effect between the nails rows with a concentration of σ_h around the nails heads. At the base of wall, earth pressures are still in active state due to the last phases of excavation and the nails are not mobilized yet. The only resistance available at the base comes from the friction of the foundation soil which increases both horizontal and vertical stresses (as indicated at 5 m depth). The soil over the whole height of the wall is held in its "at rest" condition by the mobilization of the nails above and is subject to the mobilization of shear strength of the soil foundation below, and will attempt to follow the tilting of the facing wall. Under these conditions, the soil will develops a "global arching effect" between the top and the bottom of the wall.

At 1.0 m from the facing (Figure 4), the vertical stress σ_V is larger than the overburden geostatic pressure γz especially in the bottom of the wall. This excessive vertical stress corresponds to the previously observed global arching effect which induces near the facing very low vertical stresses. The horizontal stresses distribution is more uniform and closer to "at rest" condition for most of the wall height compared to the ones in figure 3 since local arching effects between the nails rows are less encountered.

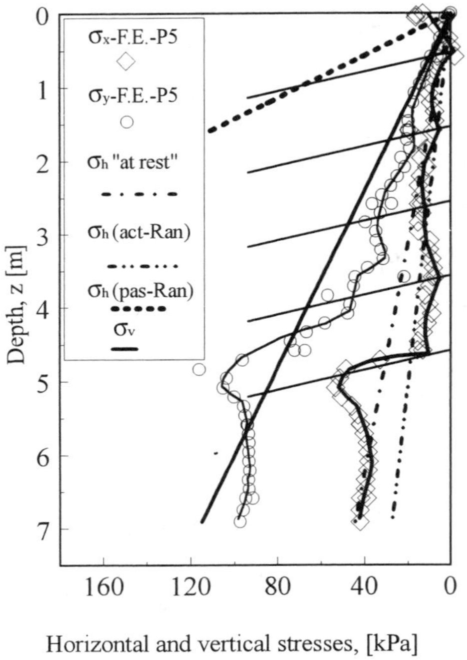

Figure 3 : Earth pressures at 0.5 m from the facing

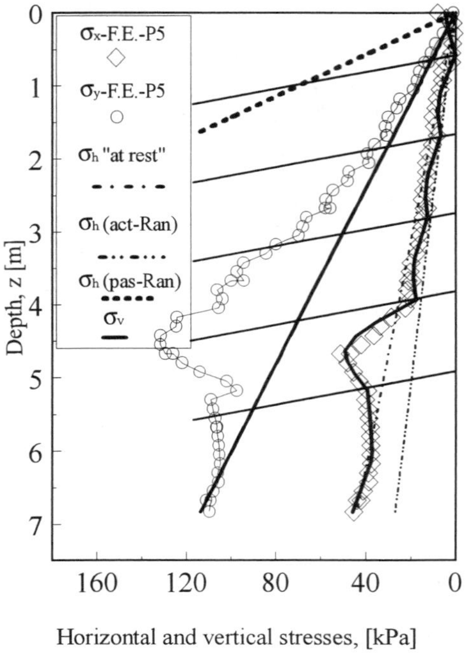

Figure 4 : Earth pressures at 1.0 m from the facing

6. CONCLUSIONS

The finite element analysis performed demonstrates clearly that classical theories are not completely satisfactory in predicting earth pressures acting on the soil nailed wall facing. Type of wall movement, wall flexibility as well as the construction technique have to be taken into account to achieve a good design of such earth retaining structure. The numerical results indicated that sufficiently mobilized, the upper rows of nails are able to restore soil condition from active to "at rest" state condition by taking a part of the load imposed upon the soil during the previous stages of excavation. This, results in the development of a local arching effect along with stress concentration around the nail heads which has to be accounted for in the facing design. In its movement attempting to follow facing deflection, the nailed soil will develop a global arching between the top held by the mobilization of the upper rows of nails and the bottom where the friction of soil foundation is acting.

These results give an idea how complex is a real behavior of soil nailing structures. It is clear that further F.E. studies are needed to better understand the mechanisms taking place within soil nailed walls in order to ameliorate existing design methods.

REFERENCES

Benhamida, B., P. Unterreiner, F. Schlosser 1997. Numerical analysis of a full scale experimental soi nailed wall. *3rd Int. Conf. on Grnd. Imp., London*

Clayton, C.R.I., Milititsky, J., Woods, R.I. 1993. *Earth pressure and earth retaining strucrures.* Chapman & Hall.

Recommandations CLOUTERRE 1991 pour la conception, le calcul, l'exécution et le contrôle de soutènements réalisés par clouage des sols. Presses de l'ENPC, Paris.

Schlosser, F. (1982). "Behavior and design of soil nailing." *Proc. Soil and Rock Impr. Tech.: Geotextile, Reinf. Earth and Modern Piling Methods, Bangkok, pp. 319 - 413.*

Schlosser, F. (1983). "Analogies et différences dans le comportement et le calcul des ouvrages de soutènement en Terre Armée et par clouage des sols." *Annales de l'ITBTP, No. 418, pp. 8 - 23.*

Schlosser, F., Jacobsen, H. M., Juran, I. (1983). Le renforcement des sols. *C. R. VIIIème Conf. Europ. Méc. des Sols et Trav. de Fond., Helsinki.*

Unterreiner P., Benhamida B., Schlosser F., 1997. Finite Element modeling of the construction of a full scale experimental soil nailed wall. Int. Journa Ground Improvement. N° 1, V. 1, pp. 1-8.

Numerical Models in Geomechanics, Pietruszczak & Pande (eds) © 1997 Balkema, Rotterdam, ISBN 90 5410 886 X

lastic-plastic modelling of geosynthetic reinforced soft soils

I.Yin
partment of Civil and Structural Engineering, The Hong Kong Polytechnic University, Hong Kong

Zhan
nie Consultants Limited, Hong Kong

STRACT: This paper presents the results of a parametric study of geosynthetic reinforced granular base over t soil. A finite difference program is used for numerical modelling. The soil and interface are considered as soplastic materials. Varying parameters are (a) the thickness of the sand base (b) analysis with and without osynthetic membrane (c) smooth rigid footing (uniform displacement boundary) and flexible footing (uniform ssure boundary). The paper examines the effects of these parameters on the ultimate bearing capacity, tlement and tensile force in the geomembrane as well as the displacement characteristics. The results obtained useful for understanding the mechanism of load transfer in reinforced structures.

INTRODUCTION

ft foundation soils suffer problems such as cessive settlement and low bearing capacity. One techniques suggested for improving soft indation soils is to place a high-strength osynthetic layer (geotextile, geomembrane or ogrid) over soft foundation soils, then a granular se (sand or gravel fills) on the geosynthetic layer. ues facing civil engineers are: Can this technique uce the settlement and increase the bearing pacity of soft foundation soils? How to calculate settlement and bearing capacity of the osynthetic reinforced soft foundation soils?

is paper suggests a numerical model based on a ite difference (FD) program - FLAC (1995) for delling the behaviour of a geosynthetic reinforced nular base over soft ground. This FD model is sed on continuum mechanics. The reinforcement stem studied in this paper is a thin high-strength osynthetic sheet (geomembrane) which is buried rizontally between a granular base and soft ground g.1).

ried parameters in the modelling study are (a) the ckness of the sand base, (b) with or without osynthetic membrane, (c) smooth rigid footing nstant displacement boundary) and flexible footing (pressure boundary). This paper studies the effects of these parameter changes on (a) the ultimate bearing capacity, (b) settlement and tension force of the geomembrane, and (c) displacement characteristics. The FD model and main results are presented and discussed in following sections.

2 FINITE DIFFERENCE MODEL

Fig.1 shows the dimensions of a finite difference (FD) model for a half foundation and soil profile due to symmetry (left vertical line is symmetric line). Elements of FD grid are uniform and are not shown here. Each element is 0.125m (horizontal length) by 0.0917m (vertical length). The top left is the footing boundary - constant downward displacement for a smooth rigid footing or constant pressure for a flexible footing, The right vertical side in Fig.1 is constrained horizontally only. The bottom horizontal boundary is constrained vertically only.

The FD model is based on a commercial program - FLAC (1995). The behaviour of granular base (sand) and clay is modelled using an isotropic elastic-perfectly plastic model with Mohr-Coulomb strength criterion. The dilation angle ψ is assumed to be zero which is a lower bound value. Properties of clay and sand are listed in Table 1.

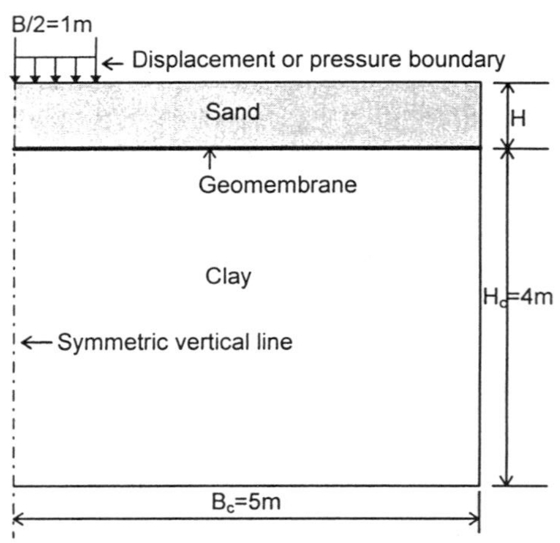

Fig.1 Dimensions of FD model with H/B changes

Table 1 Properties of clay and sand

Soils	E (MPa)	ν	c (kPa)	φ (deg.)	γ (kN/m^3)
Clay	10	0.45	10	0	20
Sand	20	0.3	0	40	20

The properties of the geomembrane are (a) tension modulus E_g=40000kN/m, (or E=2000MPa, thickness H_g=0.02m and completely flexible in FD model), (b) shear stiffness between soil and geomembrane =5000MPa/m (almost rigid plastic interface), and (c) between sand (or clay) and geomembrane, friction=40°, cohesion=0.

The interface properties between sand and geomembrane and between clay and geomembrane may be different. Since FLAC cannot specify different properties on both sides of the geomembrane, the interface properties are assumed to be the same. The friction angle value chosen is a higher bound value. Since the behaviour of sand or clay is elastic-plastic, the plastic flow in the interface area can be modelled. The geomembrane cannot take any bending moment in the FD model.

3 FINITE DIFFERENCE RESULTS

3.1 *Clay only or sand only*

Two reference cases of clay only and sand only are considered. Fig.2(a) and (b) shows the displacement vectors for a uniform vertical downward

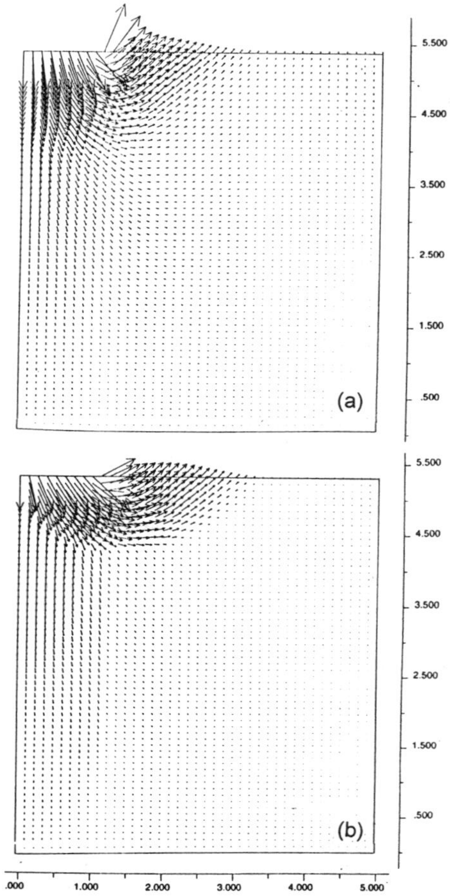

Fig.2 Displacement vectors, (a) clay only and (b) sand only

displacement of 0.2m for clay only and sand only without geosynthetics. It is seen in Fig.2(a) and (b) that large displacements occur in a shallow depth, less than 0.6B for clay only and 0.5B for sand only. The slip line angle at the exit side is approximately 45° for clay only case and 60° for sand only case. The ultimate bearing capacity pressure is 54 kPa for clay only case and 700kPa for sand only case. The predicted bearing capacity of 700kPa is lower that the value from classic theory for a rough foundation. It is noted that in the FD model, the footing is assumed smooth and the dilation angle ψ is assumed to be zero. If ψ >0, the bearing capacity computed is higher.

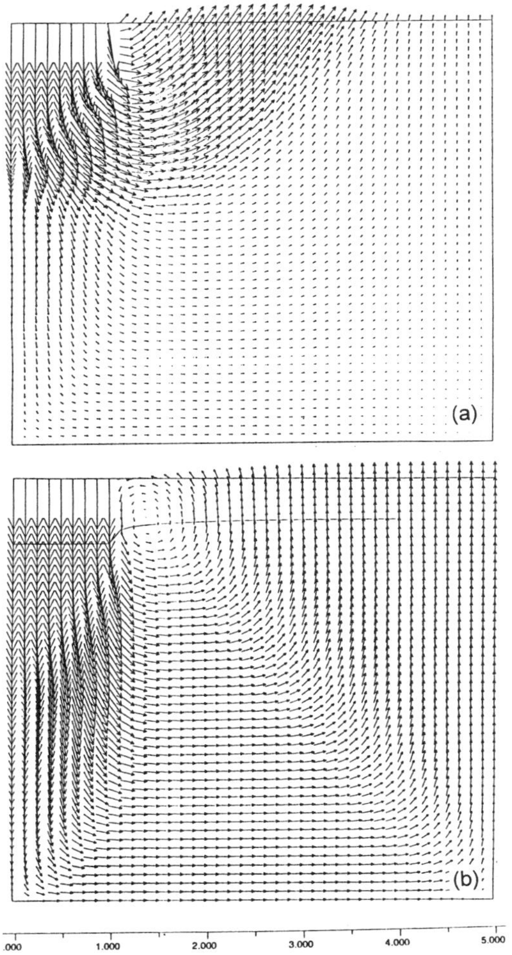

Fig.3 Displacement vectors for H/B=0.25, (a) without geomembrane and (b) with geomembrane

3.2 *Changes in sand thickness with rigid footing*

Four different thickness values of the sand base are studied, that is, H/B=0.25, 0.5, 0.75 and 1 (B=2m) with a smooth rigid footing (uniform vertical downward displacement up to at least 0.2m).

Fig.3(a) and (b) show the displacement vectors for H/B=0.25 without and with a geomembrane. Without geomembrane as in Fig,3(a), the slip line in the clay and the sand looks like the slip line predicted from the classic theory. But the influence depth is increased to approximately 1B, larger than 0.6B in Fig.2(a). When geomembrane is used as shown in Fig.3(b), the influence depth is deeper, approximately 2B, larger than 0.5B in Fig.2(b). The displacement of the sand layer within footing width is almost uniformly downward. The failure mode is almost a "punching" failure mode. The displacement vectors in the geomembrane interface area on the right footing side show an anti-clockwise soil movement. The ultimate bearing values, geomembrane settlements and tension forces in all cases in this section are shown in Fig.7 and Fig.8 and to be discussed in Section 4.

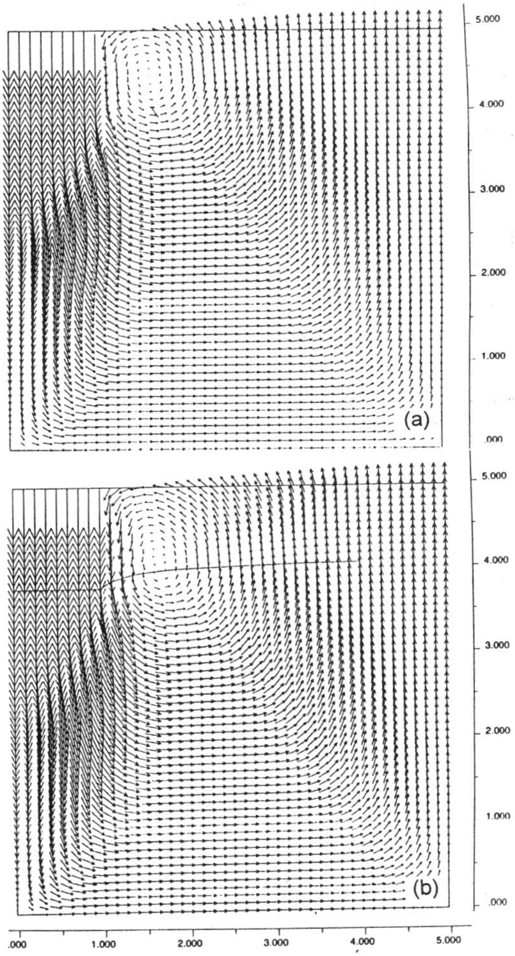

Fig.4 Displacement vectors for H/B=0.5, (a) without geomembrane and (b) with geomembrane

Fig.4(a) and (b) show the displacement vectors for H/B=0.5 without and with a geomembrane. Without geomembrane as in Fig,4(a), the failure mode is similar to the model in Fig.3(b) with a geomembrane and H/B=0.25. There is no clear classic slip line in the clay in Fig.4(a) as in Fig.2(a) and Fig.3(a). The failure mode looks like a "punching" failure mode. The failure mode in Fig.4(b) with a geomembrane is similar to the mode in Fig.4(a). However the degree "punching" (downward influence depth) in Fig.4(b) with geomembrane is bigger than that in Fig.4(a) without geomembrane.

Fig.5(a) and (b) show the displacement vectors for H/B=0.75 without and with a geomembrane. The failure modes are very similar the modes in Fig.4(a) and (b). All failure modes look like a "punching" failure mode. The degree "punching" (downward influence depth) in Fig.5(b) with geomembrane is bigger than that in Fig.5(a) without geomembrane.

Fig.6(a) and (b) show the displacement vectors for H/B=1 without and with a geomembrane. The failure modes are very similar the modes in Fig.5(a) and (b). All failure modes look like a "punching" failure mode. The degree "punching" (downward influence depth) in Fig.6(b) with geomembrane is bigger than that in Fig.6(a) without geomembrane.

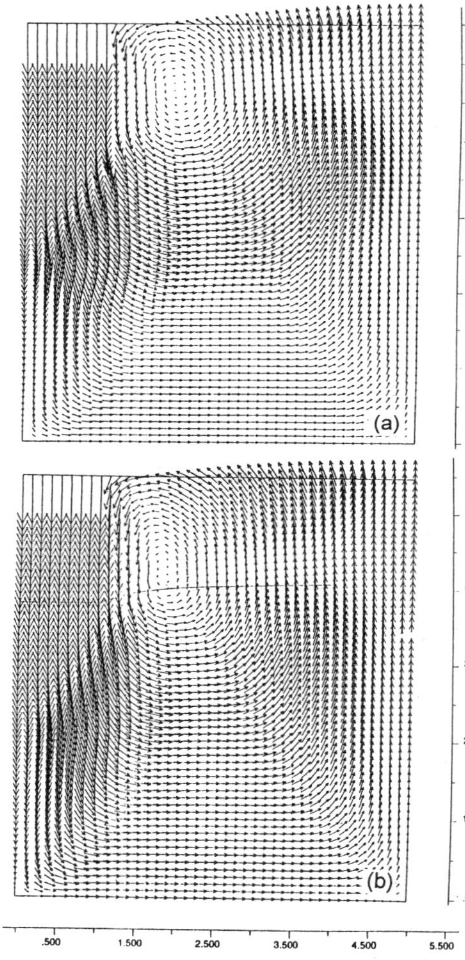

Fig.5 Displacement vectors for H/B=0.75, (a) without geomembrane and (b) with geomembrane

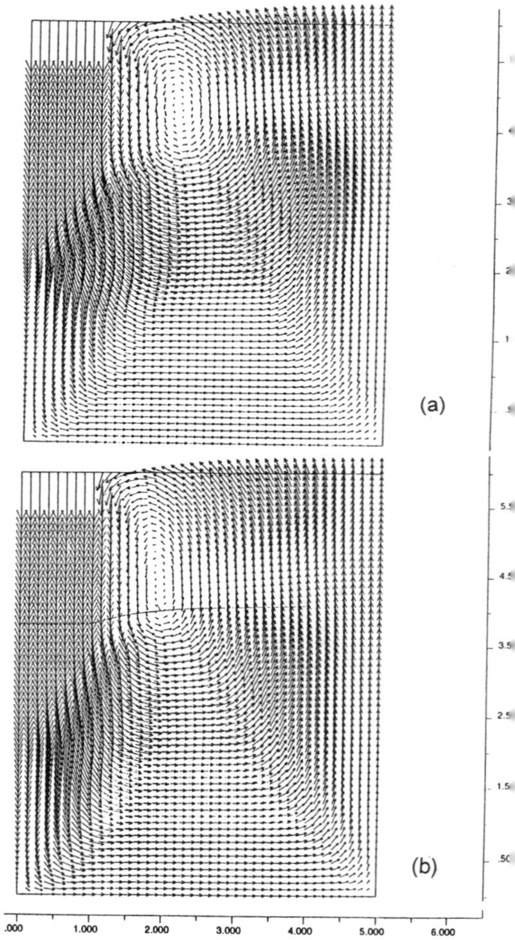

Fig.6 Displacement vectors for H/B=1, (a) without geomembrane and (b) with geomembrane

4 EFFECTS OF SAND LAYER THICKNESS AND GEOMEMBRANE

4.1 *Increase in bearing capacity*

Fig.7 shows the curves of settlement vs. average bearing pressure for a smooth rigid footing. It is seen that the bearing capacity increases with the thickness of sand base (H or H/B ratio). For the same sand base thickness, the bearing capacity value with geomembrane is only slightly larger that the value without geomembrane. The maximum increase is 5% for H/B=0.75. The thickness of sand layer has significant influence on the bearing capacity value. When H/B=1, the bearing capacity pressure is 2.3 times of the value of clay only as shown in Fig.7.

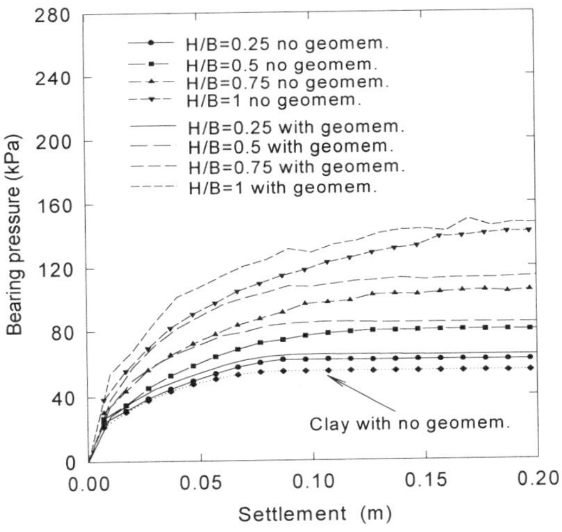

Fig.7 Settlement vs. pressure

4.2 *Geomembrane settlement and tension force*

Fig.8(a) and (b) show the settlement and mobilised tension force (negative for tension) of the geomembrane. The vertical downward settlement of the smooth rigid footing is 0.2m, the same for all cases. It is seen from Fig.7(a) that the settlement on the right side of the footing (distance from 1m to 2.75m) increases with H/B, but decreases with H/B for far right side (distance from 2.75m to 4m). It is seen in Fig.8(b) that the tension force increase with H/B. All these are an indication of the effects of geomembrane reinforcement and sand layer strengthening.

Fig.9 shows the settlement vs. average bearing pressure for smooth rigid footing and for a flexible footing (using the settlement in the middle of the footing). The ultimate bearing capacity for a flexible footing is very close to that of a rigid footing. The baring capacity with geomembrane for both rigid and flexible footings is slightly larger than that without geomembrane.

5 SUMMARY AND CONCLUSIONS

This paper used a finite difference (FD) model for studying the deformation and bearing capacity behaviour of geomembrane reinforced sand layer over soft foundation soils. An elastic-plastic model was used for the soil and the interface behaviour. The effects of sand layer thickness and geomembrane reinforcement were studied. The following conclusions can made from the results obtained:

(a) When H/B=0.25, the failure mode is almost a general shear failure in the clay without geomembrane, but is a "punching" failure with geomembrane. When H/B=0.5 or higher, the failure modes with or without geomembrane are like a "punching" failure.

(b) The bearing capacity increases with H/B. The inclusion of a geomembrane increases slightly (≤5%) the bearing capacity of a smooth rigid footing for the same H/B in the cases studied in this paper.

(c) The geomembrane tension force increases with H/B under the ultimate load.

The results presented in this paper are preliminary. Further study is needed in order to have a better understanding on the reinforcement mechanism and to establish a simple 1-D model for the calculation of both settlement and bearing capacity of the geomembrane reinforced granular base over soft foundation soils.

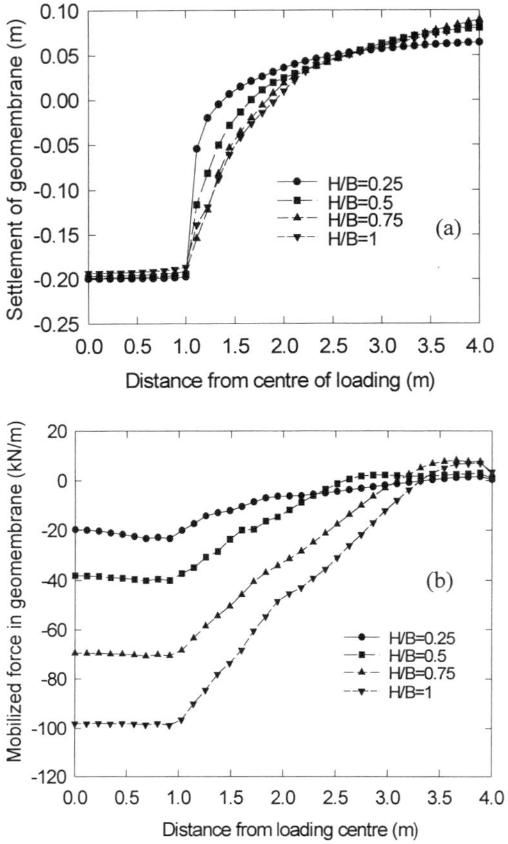

Fig.8 (a) Geomembrane settlement and (b) mobilised tension force

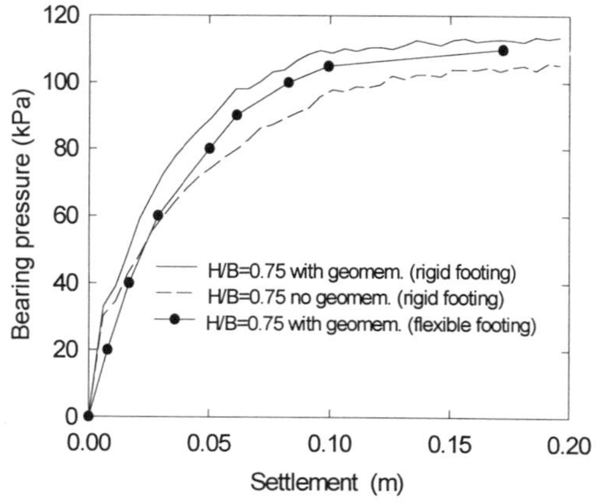

Fig.9 Settlement vs. pressure for rigid and flexible footings.

ACKNOWLEDGEMENTS

Financial support from the Hong Kong Polytechnic University is acknowledged.

REFERENCE

FLAC Manual, Version 3.3, Itasca Consulting Group Inc., Minneapolis, Minnesota, U.S.A., 1995

Numerical Models in Geomechanics, Pietruszczak & Pande (eds) © *1997 Balkema, Rotterdam, ISBN 90 5410 886 X*

Finite element analysis of centrifuge model tests

A. Porbaha
Technical Research Institute, Toa Corporation, Yokohama, Japan

M. Kobayashi
Port and Harbour Research Institute, Yokosuka, Japan

ABSTRACT: A numerical technique was applied to study the failure and prefailure behavior of geotextile reinforced steeply sloped walls of 80.5^0 (1H:6V) that brought to failure under self-weight using a geotechnical centrifuge. In the finite element analysis solid elements and bar elements were used with elasto-plastic material properties and adoption of Mohr-Coulomb failure criterion. The predicted centrifugal accelerations and the prototype equivalent heights, at development of tension crack and when failure occurred, were compared with relevant experimental results. In addition, the effects of maximum shear strain contours and plastic yield zones on the positions and traces of slip surfaces were investigated for unreinforced and reinforced steeply sloped walls.

1 INTRODUCTION

Reinforced soil retaining structures are getting more popular since they came to civil engineering practice in 1960's. The main reason is that these structures are more cost-effective than conventional concrete retaining structures, particularly as the height of the structure increases. Additionally, steepend slopes and battered walls are usually more economical than vertical retaining wall alternatives, and are often more cost-effective than flatter slopes. Savings are expected both from material, construction time reductions, and reduced right of way requirements. In addition, significant savings can be realized by use of local on-site fills, rather than importing granular soil to the site.

Several researchers have applied small scale tests to model geotechnical problems (see, for example, Mitchell et al. ,1988; and Bolton et al., 1978). The literature contains a large number of cases in which finite element analysis was applied for predicting the performance of field or laboratory tests (see, for example, Rowe and Soderman, 1985; Wu et al.,1992).

This paper describes the numerical simulation of scaled models of geotextile reinforced cohesive backfill sloping walls of 1H:6V that brought to failure under increasing self-weight using the geotechnical centrifuge.

2 LABORATORY INVESTIGATION

2.1 *Scaled model construction*

Hydrite kaolin was used for the backfill and for the foundation of all models. The liquid limit of the kaolin is 49% and the plastic limit is 33%. The model sloped walls were constructed in a rigid aluminum container with inside dimensions of 400 mm by 300 mm in area, by 300 mm in depth. For these models the foundation was firm in which the soil was mixed at optimum moisture content and then compressed, increasing stress slowly using a single loading plate to reach a maximum vertical stress of 337 kN/m^2. When that stress was reached, the load was immediately removed. This produced a clay foundation with a dry unit weight of 13.5 kN/m^3. The result was a foundation layer that was firmer than the same kaolin prepared for the retained fill and the backfill of the model walls.

After foundation preparation, an aluminum block was laid on the foundation at the toe of the wall to be constructed, to provide lateral support during model construction. The inside vertical side boundaries of the container were sprayed with silicon, and overlain with a thin plastic film to reduce boundary friction effects. The first layer of reinforcement was then placed on the exposed portion of the foundation, a layer of soil placed on it, and the geotextile folded back into the soil to provide a flexible facing for the wall. A compressive stress was then applied increasing slowly to a maximum of 175 kN/m^2. The result was a lift of backfill and retained fill with dry unit weight of 12.3 kN/m^3. This process was repeated for successive layers, each of which had finished thicknesses of 19 mm, until the model wall reached the desired height. A profile of a model is shown in Figure 1. The lateral support blocks were then removed before the centrifuge test. A full height lateral support during construction is not desirable in the field, since it is beneficial to develop gradual tensioning of the reinforcement as a wall is constructed. This gradual tensioning is achieved in the centrifuge models during the steadily increasing self-weight loading. The top of the model was sprayed with dark paint to highlight the development of tension cracks on the surface of the white clay. On the vertical side of the model, which would be visible during the test through a Plexiglas window, a grid of dots was painted to highlight cross-sectional deformation.

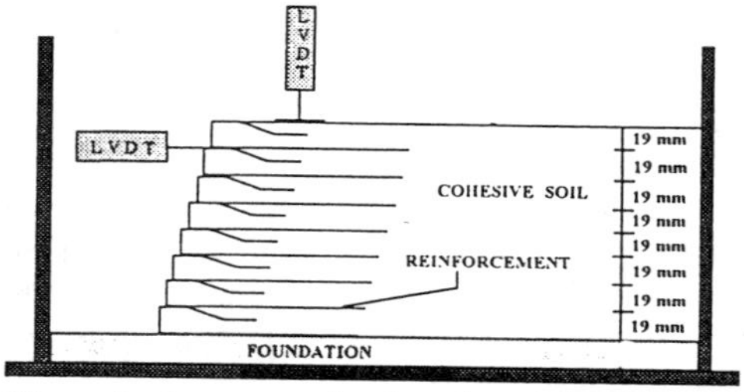

Figure 1: Geometry of a sloped model wall

2.2 *Centrifuge tests*

Models were exposed to an artificial gravity induced by geotechnical centrifuge by increasing gradually the self-weight of the model until failure occurred. The rate of these increases was 2g/minute until cracking was observed, at which point the rate was decreased, allowing any wall movement to cease before further increases were made. After a test, the model was disassembled to examine the deformations of the reinforcements at different elevations. The coordinates of the failure surfaces were recorded using a profilometer, measuring the vertical profile at 10 mm horizontal intervals through various model cross-sections. Direct shear tests were performed on specimens retrieved at various depths in the unfailed rear portion of the model after failure, subjecting each to normal stresses equal to the maximum experienced by the specimen during a test due to overburden load.

Table 1 presents the model geometry of the model tests built on firm foundations. Reinforcement configurations varied from no reinforcement, to a maximum reinforcement length of 114 mm or 0.75 times model wall height, with eight layers of reinforcement in every model. Further discussion on the behaviors of these models in terms of development of tension crack, foundation rigidity, and stability analysis is presented by Porbaha and Goodings (1996).

Table 1. Geometries and properties of model reinforced slopes

Model No.	Length (mm)	L/H ratio	Cohesion (kN/m^2)	Friction angle (deg.)
M-29	0	0	17.8	21.7
M-37	0.76	0.50	18.1	20.2
M-33	100	0.66	19.3	21.4
M-32	114	0.75	23.8	20.6

3 NUMERICAL SIMULATION

In this study a finite element program developed by Kobayashi (1984) was applied to investigate the behavior of geotextile reinforced slopes of 80.50^0 (1H:6V). The details of modeling technique are discussed in the following section.

3.1 *FEM mesh*

The two dimensional finite element mesh used to model the typical reinforced slope is shown in Figure 2. The geometry of the mesh is selected to ensure proper modeling of the slope identical to the physical model. The finite element mesh consists of 336 nodes and 141 elements to simulate the backfill, the foundation, and the reinforcement.

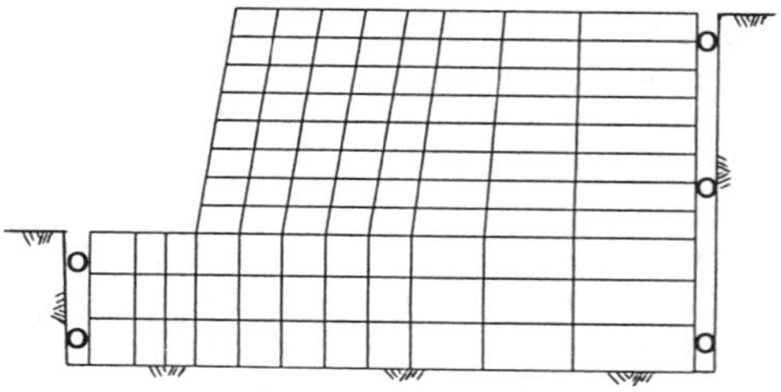

Fig. 2 Finite element mesh and boundary conditions

3.2 *Soil and reinforcement modeling*

Different types of elements are used to model the backfill, the foundation, and the reinforcement. The simulation comprises of eight-noded quadrilateral isoparametric solid elements to model the soil in the backfill and the foundation, and three-noded bar elements for the reinforcement.

The six-noded joint elements is commonly used to simulate the soil-interface interaction and to apply the properties based on the results obtained from laboratory pullout tests.However, the implementation of such a process seems obscure in this study. The reason is that the backfill material is a partially saturated soil and the hydraulic characteristics of nonwoven geotextile allows drainage in both sides of the reinforcement. Accordingly, the moisture content of the soil which tends to be peak value at the center of the soil layer reaches a minimum value at the interface in the vicinity of the reinforcement, and thus creating a complex soil profile that is difficult to model in ordinary pullout tests. For this reason to resolve this issue the interface was modeled the same as the soil model, despite some deficiencies with the complex real situations.

3.3 *Input parameters*

The constitutive models commonly used to study the behavior of reinforced retaining structures at failure are based on Mohr-Coulomb yield criterion. The elasto-plastic soil properties based on Mohr-Coulomb failure criterion were derived from laboratory tests for each model slope, as listed in Table 1. Poisson ratio of 0.30 was adopted for the analysis. The reinforcement was modeled as beam element with axial stiffness of 0.1 kN/m and no flexural rigidity (due to the extensibility of the reinforcement). The tensile strength of the geotextile from zero-span laboratory test was 0.117 kN/m that was input to the program. The fixities at the boundaries allow deformations in vertical direction (roller hinge in y-direction). Unit weights of the backfill and of the foundation soil are 17.3 and 18.2 kN/m^3, respectively, based on the laboratory results.

3.4 *Analysis*

Each analysis was carried out by increasing the gravitational acceleration at a very small load increment to reach the point where no convergence was acquired. Then, the analyses continued with steps back by oscillating back and forth around the smaller increments to ensure convergence with an accuracy of 0.0001g. A large number of iterations was applied based on the fictitious viscoplasticity algorithm (Zienkiewicz et al., 1975; Kobayashi, 1984 and 1988; Zienkiewicz and Taylor, 1990) to calculate the load and consequently the collapse height of the slopes. During the analysis the vertical deformation of the crest is measured, as the gravitational acceleration increases (see Figure 3).

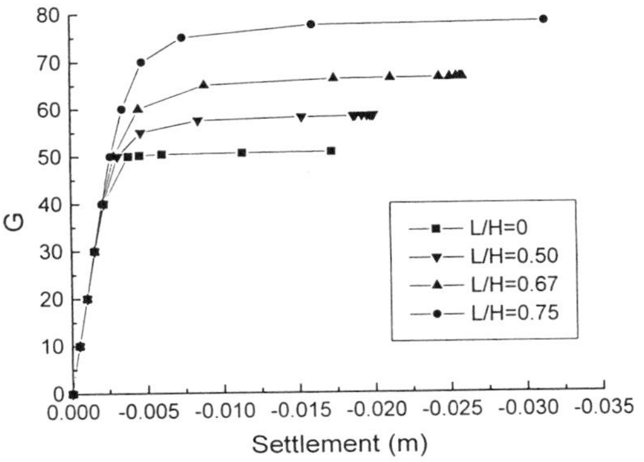

Fig. 3 Gravitational acceleration versus crest settlement

4 DISUSSION OF RESULTS

The results of physical and numerical modeling of different sloped wall models in terms of centrifugal accelerations and prototype equivalent heights are summarized in Tables 2 and 3. The characterization of tension crack is not possible during the analysis due to material discontinuity which results from

Table 2. Comparison of centrifugal accelerations from physical and numerical approaches

Model No.	$(N_c)_{EXP}$	$(N_f)_{EXP}$	$(N_c)_{FEM}$	$(N_f)_{FEM}$
M-29	43	47	-	50.5
M-37	44	56	40	59.0
M-33	49	70	48	66.8
M-32	51	75	58	77.5

$(N_c)_{EXP}$=gravitational acceleration at tension crack obtained from centrifuge tests
$(N_f)_{EXP}$=gravitational acceleration at failure obtained from centrifuge tests
$(N_c)_{FEM}$=estimated gravitational acceleration at tension crack obtained from FEM analyses
$(N_c)_{FEM}$=gravitational acceleration at failure obtained from FEM analyses

cracking. However, the centrifugal acceleration at the time when tension crack occurs is approximately estimated from the load-settlement curves for each model. The criteria is to determine the gravitational acceleration at which the initial tangent in the elastic region intersects the tangent to the curve at a small deformation that results from development of tension crack. The failure, however, is clearly defined when large deformation occurs at constant stress levels, as shown in Figure 3.

Table 3. Comparison of prototype equivalent heights from physical and numerical approaches

Model No.	$(H_c)_{EXP}$	$(H_f)_{EXP}$	$(H_c)_{FEM}$	$(H_f)_{FEM}$
M-29	6.54	7.14	-	7.68
M-37	6.69	8.51	6.08	8.97
M-33	7.49	10.64	7.30	10.15
M-32	7.75	11.40	8.82	11.78

$(H_c)_{EXP}$= prototype equivalent height at tension crack obtained from centrifuge tests
$(H_f)_{EXP}$= prototype equivalent height at failure obtained from centrifuge tests
$(H_c)_{FEM}$= estimated prototype equivalent height at tension crack obtained from FEM analyses
$(H_c)_{FEM}$= prototype equivalent height at failure obtained from FEM analyses

4.1 *Gravitational acceleration*

In terms of gravitational acceleration at tension crack the prediction by this approximate technique is within 10 to 14% of the experimental values. In terms of gravitational acceleration at failure, which represents the collapse prototype equivalent height, the best prediction is for the reinforced model (M-32). The relative difference in this case is 2.5 g that accounts for the prototype equivalent height of 0.38 m, which is practically insignificant. However, the differences in prediction of failure accelerations are higher for the cases of walls with shorter reinforcements (i.e., M-33 and M-37). This is not surprising when the failure characteristics of these models are taken into consideration. Experimental evidences based on the post-test observations revealed that the failure surfaces of these models occurred either far beyond the back of the reinforcement, as in model M-37, or it occurred just behind the reinforced zone, as in the case of M-33. Unlike model M-32 in which the failure was entirely internal, in both these cases the slip surfaces occurred outside the reinforced zone. These differences in the failure behaviors may justify the disparity. One source of discrepancy may also be

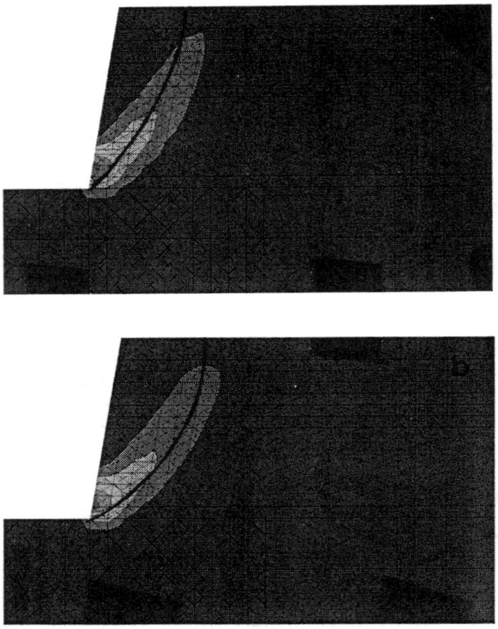

Fig. 4 Contours of maximum shear strain (a) unreinforced model (M-29), (b) reinforced model (M-32)

attributed to the initial state of the stresses in the model which was not modeled in this study due to the complexities in stress path during model construction and the partially saturated soil of the backfill. Overall, despite the limitations, the predictions are considered to be reasonable.

4.2 *Shear strain contours*

Figures 4 through 7 show the contours of maximum shear strains, yield zones with plastic deformations, displacement vectors, and the deformed meshes after failure for both unreinforced and reinforced cases. The traces of slip surfaces obtained from centrifuge model tests are also plotted in those Figures.

In the case of unreinforced model (M-29) the slip surface coincides with the peaks of shear strain contours at the bottom of the wall. However, it diverges as the height of the slope increases. On the other hand, for the case of reinforced model (M-32), in which the failure surface is contained within the reinforced zone, the prediction of slip surface is quite good. In this case the slip surface fits the peak of shear strain contours and better prediction is achieved as the slip surface reaches the top of the sloped wall.

4.3 *Deformed mesh and displacement vector*

The deformed meshes for both unreinforced and reinforced cases are shown in Figure 5. The deformed mesh at the toe for the case of unreinforced model will be improved as a result of selecting a finer mesh size. Figure 6 shows the displacement vector for reinforced model M-32. The vectors are more enhanced in the active zone of the reinforced area, particularly at the toe of the wall.

4.4 *Yield zones*

Figure 7 shows the contours of yield zones for these models that represents the regions with plastic deformations. The slip surfaces from physical modeling are located within these yield zones, even though the identification of clear-cut slip surfaces at the crest of the slopes are not as explicit as the body of the slopes. In the case of unreinforced model, however, the slip surface from physical test is closer to the face of th sloped wall.

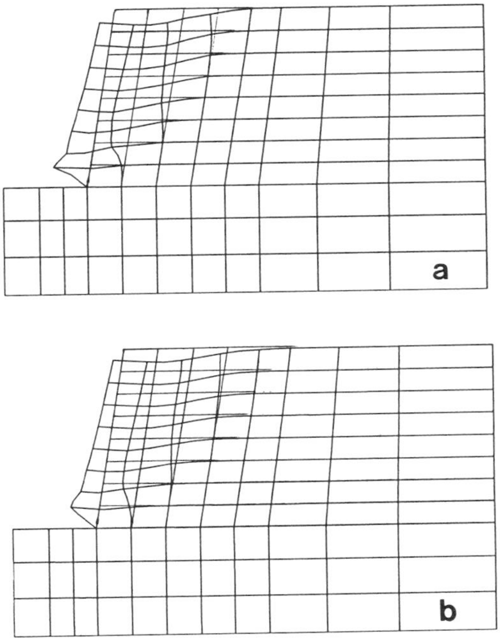

Fig. 5 Deformed mesh (a) unreinforced model (M-29), (b) reinforced model (M-32).

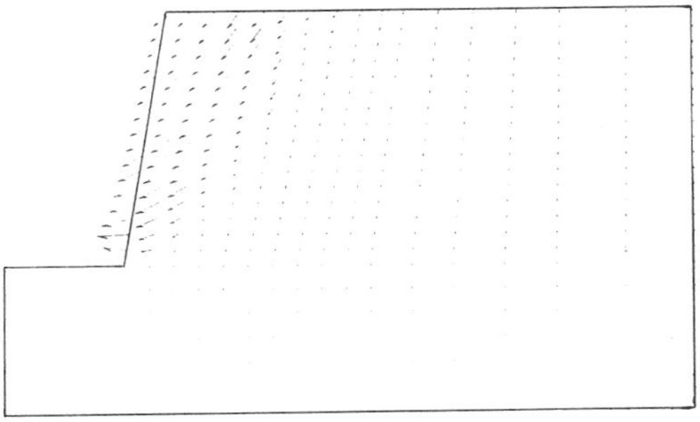

Fig. 6 Displacement vector for model M-32

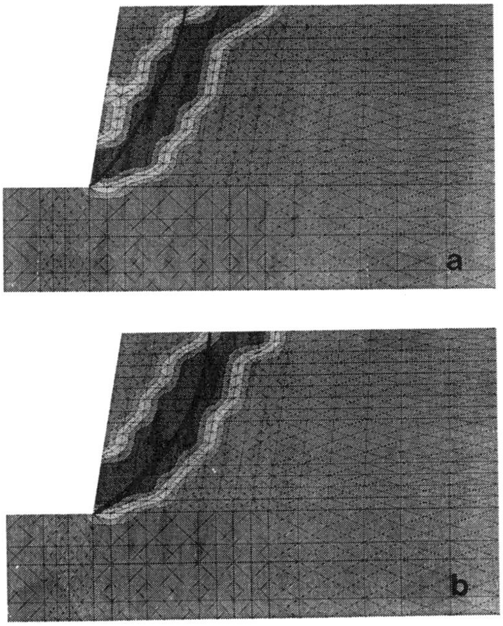

Fig. 7 Contours of yield zones (a) unreinforced model (M-29), (b) reinforced model (M-32).

5. CONCLUSIONS

The combined physical and numerical techniques can contribute to better understanding of a complex engineering system. Along this line this investigation presents the results of numerical simulation to predict the behavior of geotextile reinforced cohesive sloped walls with different reinforcement lengths that failed under under self-weight in a geotechnical centrifuge. The eight-noded solid elements with elasto-plastic material properties and Mohr-Coulomb failure criterion was adopted to model the backfill and the foundation. The reinforcements were modeled by bar elements without flexural rigidities. The results of numerical simulation were compared with physical models and discussed in terms of tension crack development,gravitational acceleration and prototype equivalent height at failure. In general, the simulation is more successful to predict the behavior of the reinforced case in which failure is internal, i.e., the slip surface contains in the reinforced zone. The results show some deviations for the case where the failure is external, however, the differences are reasonable from a practical viewpoint. The implications of this study have practical significance in cost-effective parametric studies of soil retaining structures.

ACKNOWLEDGMENTS

The first author is grateful to the Science and Technology Agency of Japan for support during his stay in Port and Harbour Research Institute.

REFERENCES

Almeida, M.S.S., Britto, A.M., and Parry, R.H.G. 1986. Numerical modeling of a centrifuged embankment on soft clay, *Canadian Geotechnical Journal*, 23, 103-114.

Bolton, M.D., Sun, H.W., and Britto, A.M. 1993. Finite element analyses of bridge abutments on firm clay, *Computers and Geotechnics*, 15, 221-245.

Kobayashi, M. 1984. Stability analysis of geotechnical structures by finite elements, *Report of Port and Harbour Research Institute*, Vol. 23, No.1: 83-110.

Kobayashi, M. 1988. Stability analysis of geotechnical structures by adaptive finite element procedure, *Report of Port and Harbour Research Institute*, Vol. 27, No.2, June: 3-22.

Porbaha, A. and Goodings, D.J. 1996. Centrifuge modeling of geotextile reinforced walls, *Journal of Geotechnical Eng.,* ASCE, Vol.122, No.10, October 1996.

Row, R.K., and Soderman, K.L. 1985. An approximate method for estimating the stability of geotextile reinforced embankments, *Canadian Geotechnical Journal*, 22, No.3: 392-398.

Wu, T.H.J., Christopher, B. Siel, D., Nelson, N.S., Chou, H., and Helwany, B. 1992. The effectiveness of geosynthetic reinforced embankments constructed over weak foundation, *Geotextiles and Geomembranes*, 11: 133-150.

Zienkiewicz, O.C., and Humpheson, C., and Lewis, R.W. 1975. Associated and non-associated viscoplasticity and plasticity in soil mechanics, Geotechnique, Vol. 25: 671-689.

Zienkiewicz, O.C., and Taylor, R.L. 1990. The finite element method, McGraw Hill, fourth edition, Vol.2: 256-260.

Numerical Models in Geomechanics, Pietruszczak & Pande (eds) © *1997 Balkema, Rotterdam, ISBN 90 5410 886 X*

Numerical modelling of an in-situ pull-out test on soil reinforcing bars

T. Ochiai
Civil Engineering Planning Department, Yahagi Construction Co. Ltd, Nagoya, Japan

K. Hattori & G. Pokharel
Technical Development Division, Civil Engineering Planning Department, Yahagi Construction Co. Ltd, Nagoya, Japan

ABSTRACT: In this paper, the current state of the conventional design and analysis of the reinforced soil structures is discussed. The need for in-situ pull out tests on reinforcing bars is first identified. Later, results from a set of in-situ pull out test carried out at a full scale test site for a newly developed soil nailing technology called PAN Wall Method, are presented. A new proposal is made to identify the effective length of reinforcing members behind a critical potential slip surface which seems appropriate and exhibits promising features for its future application. Finally, a numerical model is introduced to simulate the in situ test results by employing the RPFEM extended by Asaoka and co-workers to the reinforced soil structures. The modeling procedures presented through this paper seems powerful if the strength parameters for the soil elements in tension and compression are correctly defined.

1 INTRODUCTION

Attention is growing up for the minimization of damage to the existing environments, e.g. the protection of trees, vegetation, top soil layer, etc., during construction and/or operation of even small geotechnical structures like excavations of natural slopes to pass road, highway, railway tracks through the mountains. To satisfy this requirement, cut slopes should sometime be made very steep. Conventional retaining walls may not achieve such steep slopes, and therefore, new slope stabilization techniques like soil nailing methods or earth anchors have to be used. The designed length of nails becomes very long for such steep slopes and the existing design methods are considered insufficient to guarantee the performance of long nails because of several reasons, e.g., the effects of increasing confining pressure with depth and orientation of principal stresses around reinforcing bars are usually neglected, the contribution of the rigid facing is also neglected, etc. Therefore, to confirm the performance of nails, 3% nails should be pulled out up to the mobilization of design axial force in the reinforcements.

This paper presents results from a set of in-situ pull out test on the newly developed PAN Wall® (Panel And Nail) method (Figs.1,2&3) and subsequent numerical simulations. The PAN Wall method is a light and economical soil nailing technology with prefabricated rigid concrete panels(1.8mX1.2m) to stabilize natural slopes and excavations developed at Yahagi Constructions Co. Ltd. in collaboration with the Nagoya University's Geo-mechanics Laboratory. The applicability of the PAN Wall was initially demonstrated through a series of proto type model tests and a full scale test construction. Then, the real applications of PAN Wall technology was started at the beginning of 1993(e.g. Fig. 3).

The PAN Wall components are illustrated in Fig. 2. So far, the PAN Wall has stabilized more than 10,000 sq.m. surface area of slopes. The slope height varied from one panel height to as much as 16.5m. The length of nails varied from 3m to 10m. All these structures have performed very well. Still, aforementioned 3% nails are pulled out till the mobilization of design axial force. Similarly, a set of pull out test was carried out on the instrumented test construction site to verify that the PAN Wall is performing well even after 3 years of construction. These results are simulated using Rigid Plastic FEM (hereinafter called as RPFEM) developed by Asaoka and his co-workers for the analysis of reinforced soil structures. Few new concepts are discussed on the computation of effective length of the reinforcing bars behind the most critical slip surface and effective length of sleeve for pull out tests. The results show that the PAN Wall exhibits challenging and interesting features. The part of the reinforcement length beyond the effective length behind the most critical slip surface can not develop enough strain in soil mass in order to mobilize effective axial force with respect to the total cost of the reinforcement. Thus, the axial force transfer per unit price is too low for very long bars. The maximum length of the reinforcing bars in the PAN Wall method does not exceed 10m, which seems quite logical based on the economical and the real behavior of nails in soil mass.

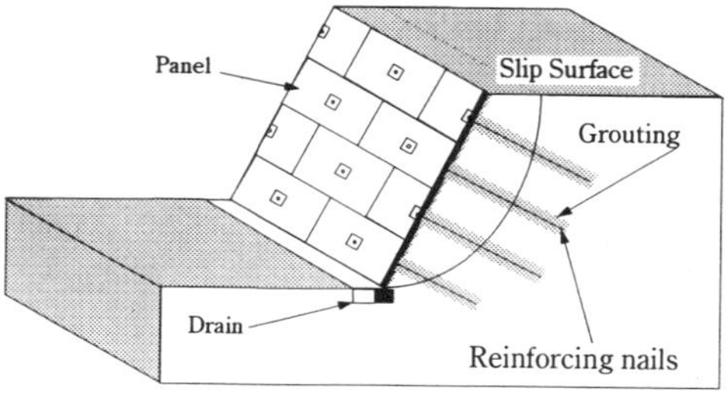

Figure 1: Schematic view of the PAN Wall Technology

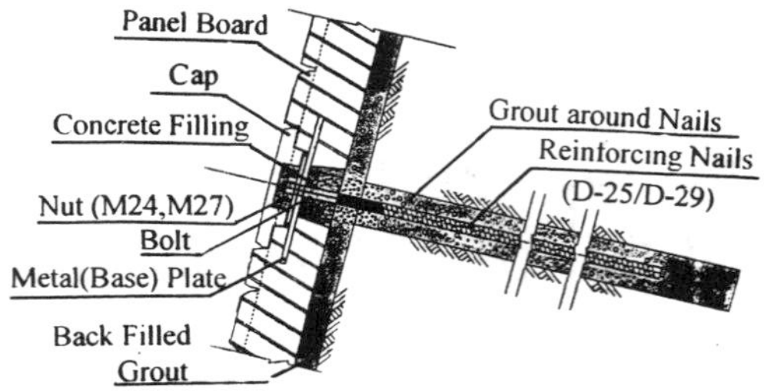

Figure 2 Basic components of the PAN Wall Technology

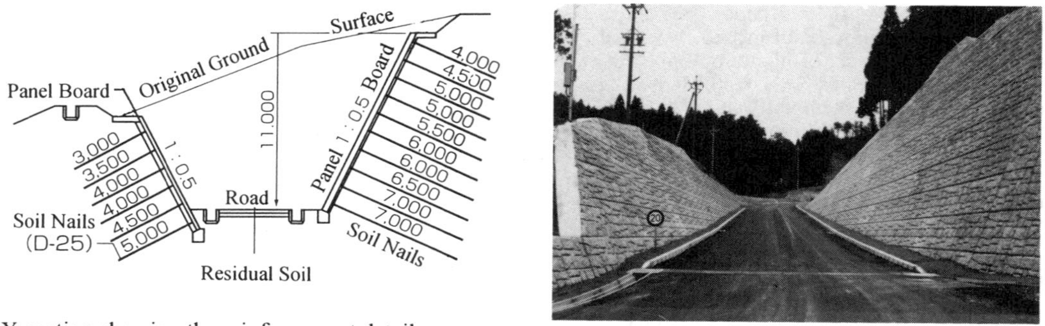

X-section showing the reinforcement details

Figure 3 A typical application of the PAN Wall method to stabilize slopes and excavations.

2 NUMERICAL MODEL (*RPFEM*)

2.1 *General*

The tensile force acting on a reinforcing member should be considered as a typical internal force because the force develops due to the interaction between soil and the reinforcing member. Such force developed during deformation or plastic flow at limit state of soil mass should therefore be estimated with an existence of a deformation and/or a velocity constrained mechanism in the soil system(Asaoka et al., 1994). The reinforced soil system with such internal force at limit equilibrium state was originally formulated by Asaoka and his coworkers (Asaoka et al., 1994) based on the rigid plastic FEM by imposing a new linear constraint condition called "*no length change*" upon the plastic flow of soil mass at limit state. The rate of plastic energy dissipation of the soil is minimized under the constraint condition and mathematically it was shown that the Lagrange multiplier for the constraint condition is interpreted as an axial force developed along a reinforcing member when the equilibrium of forces of the system is considered. The principal advantage of the method is said to lie in the simultaneous computation of a load factor, axial force along the reinforcing members, and the velocity flow of soil mass at limit state.

The analysis method is used in the present study to simulate the pull-out test presented in this paper. The essence of the methodology is summarized here for the sake of completeness. The details may be referred to the aforementioned research papers.

2.2 *Formulation of reinforced soil system*

The "*no-length change*" condition is first illustrated in terms of a velocity field assuming that the soil mass exists at the limit equilibrium state. This is for the sake of incorporating these conditions into the rigid plastic FEM.

No-length change condition:

The reinforcing members stiff in axial tension suppresses the plastic flow of adjoining soil elements around the reinforcements. Thus, the length between two consecutive soil nodes in FE mesh along a reinforcing member, is kept constant throughout the plastic flow at limit state.

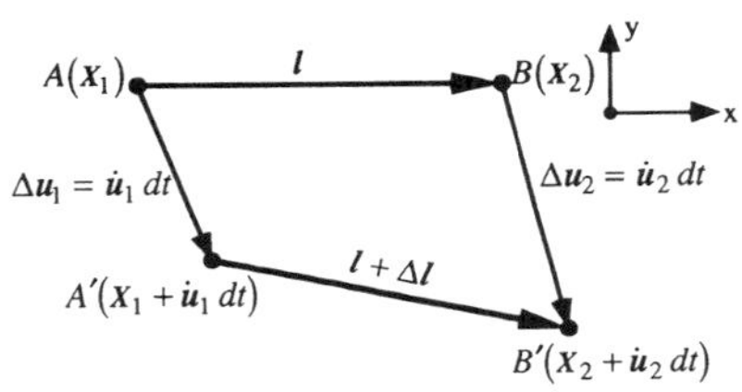

Fig. 4 Concept of "*no-length change*" condition. (*Asaoka et al.,1994*)

Mathematically, plastic flow of a reinforcing member AB to A' B' in Fig. 4 for a small time increment *dt* can be expressed as $|l| = |l+\Delta l|$. Finally, it reduces to the following linear equation:

$$(X_2 - X_1)^T(\dot{u}_2 - \dot{u}_1) = 0 \quad ...(1)$$

All reinforcing members can be integrated in one global matrix as follows:

$$C_t \ \dot{u}=0 \quad ...(2)$$

where $\dot{u}$ is the vector of all nodal velocities. The matrix C_t contains constrained nodes in $\dot{u}$ and gives Eq. 1 only to those respective reinforcing nodes. For details please refer Asaoka et al. (1994).

2.3 *Formulation of the rigid plastic FEM*

The rigid plastic FEM (*abv. RPFEM*) is obtained through minimizing the rate of internal plastic energy dissipation at the limit state of soil mass, with respect to the velocity field under several linear constraint conditions (Tamura et al., 1984) including the no-length change and no-bending conditions. Various constraint conditions explained below are incorporated in an energy function by introducing a set of Lagrange multipliers λ, υ, and μ as follows:

$$\varphi(\dot{u},\lambda,\upsilon,\mu) = \int_V D(\dot{u})dV + \lambda^T(L\dot{u}-0) + \upsilon^T(C_t\dot{u}-0) + \mu(F^T\dot{u}-1) \quad ...(3)$$

in which $D(\dot{u})$ is the rate of internal plastic energy dissipation. L is the matrix defined such as $\dot{v} = L\dot{u}$ where $\dot{v}$ is the rate of volume changes in all elements. Therefore the first constraint condition, $L\dot{u} = 0$ indicates that rate of plastic volume change is always zero for all elements at the limit state (Mises material). As mentioned before, the second condition, $C_t\dot{u}=0$ is the "*no length change*" condition. The third constraint condition, $F^T\dot{u}=1$ defines the provisional norms of velocity vectors at every node. Minimizing the energy function, φ, with respect to the velocity field and the Lagrange multipliers, the following set of algebraic equations can be derived as a solution to the reinforced soil structures at limit state of soil mass.

$$\int_V B^T s\, dV + L^T\lambda + C_t^T\upsilon + \mu F = 0 \quad ...(4)$$

$$L\dot{u} = 0 \quad ...(5)$$

$$C_t\dot{u} = 0 \quad ...(6)$$

$$F^T\dot{u} = 1 \quad ...(7)$$

in which s denotes the deviator stress vector while Lagrange multipliers λ and μ are interpreted as interpreted as the indeterminate isotropic stress and the load factor, respectively (Tamura et al., 1984). The Lagrange multiplier, υ, is interpreted as the unit nodal force acting on the constrained nodes along the reinforcement direction (Asaoka et al. 1994).

2.4 *Stress-Strain rate relations at limit state*

Eqs. (4)-(7) define statically indeterminate limiting equilibrium equation problems and these equations are solved with the aid of a constitutive relation of soils at limit equilibrium state. As the no rate of volume change is assumed in Eq. (5), the following the Mises type plastic flow is employed (Asaoka et al., 1994).

$$\dot{s}_{ij} = \frac{\sigma_0}{\bar{\dot{e}}}\dot{\varepsilon}_{ij} \quad \text{where} \quad \bar{\dot{e}} = \sqrt{\dot{\varepsilon}_{ij}.\dot{\varepsilon}_{ij}} \quad ...(8)$$

where ε_{ij} denotes a plastic strain rate. Asaoka et al. (1994) modeled the frictional material as a cascaded Mises yield criteria where the Mises constant depends on the mean confining pressure (Fig. 5). The Mises constant is mathematically expressed in the form, $\sigma_0 = \sqrt{2}(c\, cos\phi + p\, sin\phi)$. Details may be referred to Asaoka et al.(1994). The flow chart of the methodology is shown in Fig. 6.

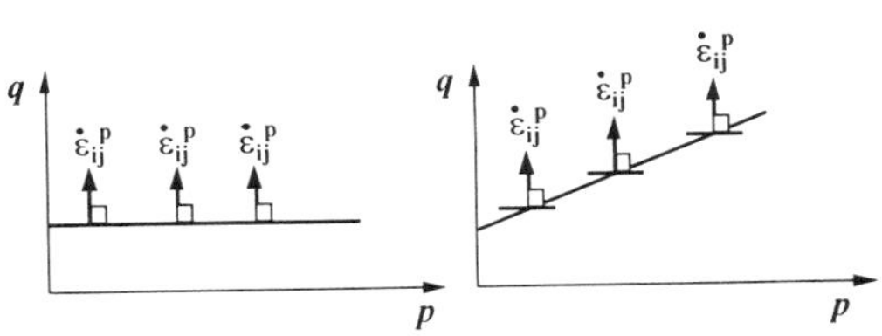

a. purely cohesive clay b. frictional(c-ϕ) soil

Fig. 5 Idealization of the frictional material (c-ϕ) as an assembly of inhomogeneous Mises materials of varying σ_o

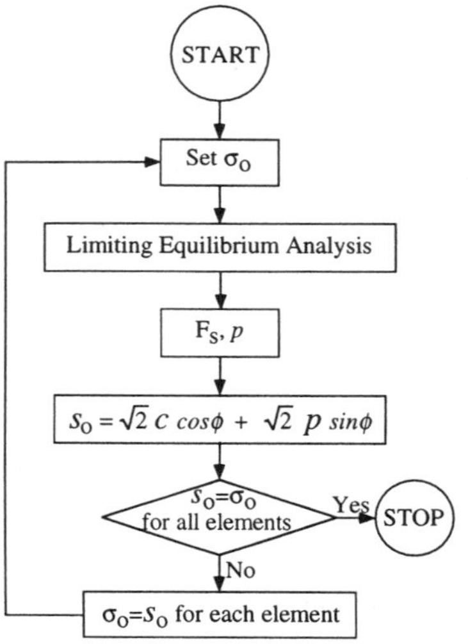

Fig. 6 Flow chart adopted for the numerical simulations based on the RPFEM

3 PULLOUT TESTS ON SOIL NAILS

3.1 *Outline of pullout tests*

After completing the model tests on 10 reinforced soil slope in March 1993, a full scale test construction of the newly developed PAN Wall method was initiated at the warehouse campus of Yahagi Constructions Co. Ltd. in Nagakutte-cho. The completed view of the first full scale PAN Wall method is shown in Fig. 7. Prefabricated R.C. Panels and the standard bars designed for rock bolting or for tunneling (NATM) are used as facing panels and reinforcing bars, respectively.

a. Full scale instrumented test construction

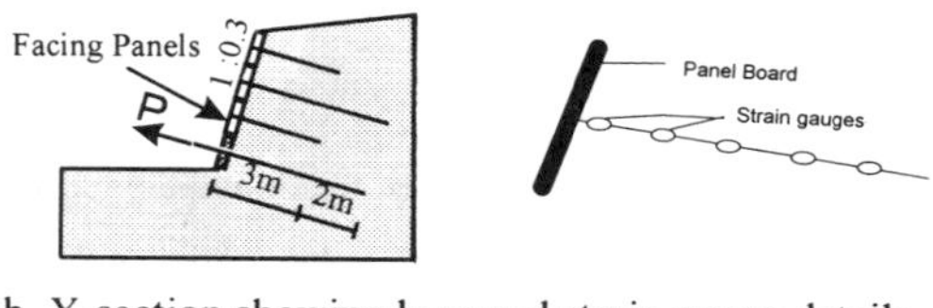

b. X-section showing bars and strain gauge details.
Fig. 7 Details of the pull out test site for the PAN Wall

The nails are either 3m or 5m long. Strain gauges were installed at every 1m length of the bars (Fig.7b). The nails are periodically checked to verify their performance by pull out tests (Fig. 8). Similarly, on May 1996, some nails were pulled out up to 10 or 15 ton mobilization of axial tension depending on the length of the nails. Out of the 6 nails tested, one nail failed at 13.6 ton of pull out load. Numerical simulation of the failure of the nail is presented in the next chapter using the RPFEM.

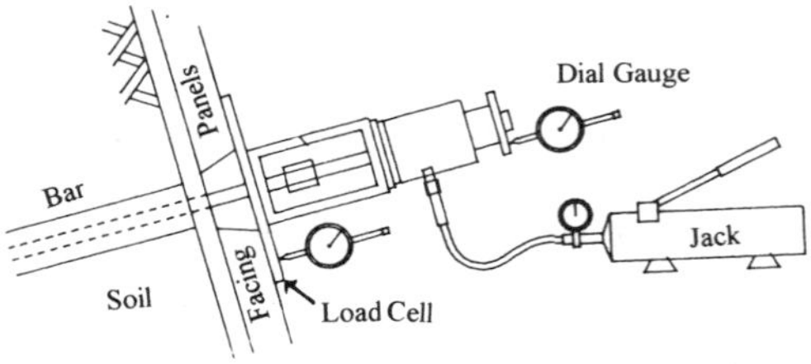

Fig. 8 Schematic view of the pull out tests

3.2 *Test results and discussions*

The tests are actually conducted for two full loading-unloading cycles. At different load steps, various observations, e.g. dial and strain gauge readings corresponding to the reinforcing bar, SWP 4-2, were taken as illustrated in Fig. 9.

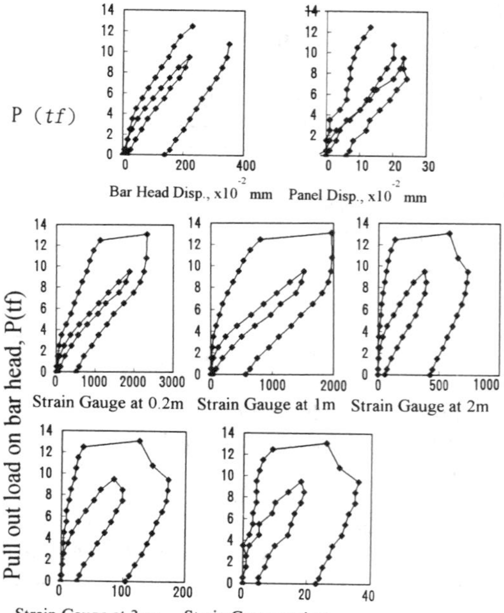

Fig. 9 Typical observations taken at different pull out load steps for the SWP 4-2 reinforcing bar.

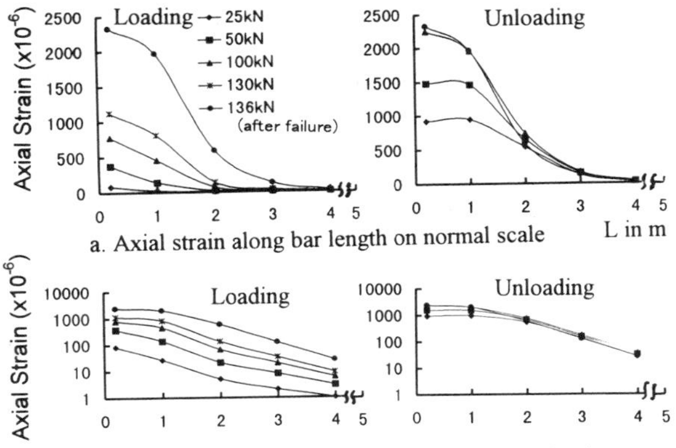

a. Axial strain along bar length on normal scale

b. Axial Strain on logarithmic scale (Bar Length in m)

Fig. 10 Observed axial strain distribution at different loads.

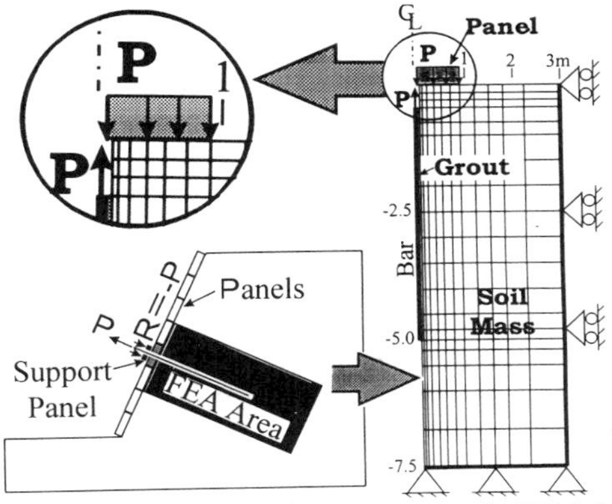

a. Schematic diagram of pull out test area b. FE Mesh

Fig. 11 FE Modelling of the Pull-out Test

The first tangent lines on the loading-unloading curves in Fig. 9 are almost similar. On the first cycle of loading, slip of bar was observed at 136kN of pull out load. The axial strain along the bar length before and after slip failure was decaying exponentially (*see* Fig.10) and it should be attributed to the loss of confining pressure on the rear side of the bar. Assumption of constant shear stress mobilization on soil-grout-bar interface in conventional design methods should be revised considering the effect of confining pressure. In standard Japanese design guidelines, the interface shear strength should be decided w.r.t. SPT N (or soil type) and is independent of the bar position (i.e. confining pressure). This is the reason why the conventional design approaches are conservative over FE Analysis.

4 NUMERICAL SIMULATIONS VIA RPFEM

4.1 *Outline of FE Simulations of the test results*

Fig. 10 shown below illustrates the FE Mesh and the boundary conditions considered in the numerical simulation using the Rigid Plastic FEM summarized in the Chapter 2. The pullout load was applied on the nail head and the reaction of the pull out load was transferred to the soil mass through the Facing Panels. The facing panel which directly supports the load cell is considered vary similar to a rigid footing (i.e. b=1.8m). The effect of other panels and bars is considered to be negligible throughout this present paper for the sake of simplicity and practicability.

A hole of 45cm depth called sleeve in pullout test terminology is assumed in FE Mesh to model the slip of the bar-grout block w.r.t. soil element. This means no axial force is transferred to the soil until this depth.

The material parameters for soil mass were γ =20kN/m3, cohesion, c=20kN/m^2 and angle of internal friction, ϕ=30°. Though the frictional angle was fixed for the soil type, friction angle was increased from 0° up to 30° degrees at 5° increments in order to understand the effect of friction angle. Meanwhile, the grouting has c=1600 kN/m^2 and ϕ=0°. The reinforcing bar was modeled by imposing the constrained condition of "*no length change*" for nodal lengths along the bar axis as illustrated in the earlier chapter.

The area around the pull out bar (3m either side) was considered enough for FEA assuming that the existing bar force is negligible compared to the pullout load applied at failure. The total depth of soil mass was 1.5times the total bar length (1.5x5.0=7.5m) to avoid the depth effect. The types of boundaries considered in FE Modeling are shown in Fig. 11(b). Total number of isoparametric finite elements are 225 and total nodes are 257 considering symmetry along the bar axis. The FE mesh is considered fine enough to understand the mechanism during pull out tests.

4.2 *FE Simulated results and discussions*

The safety factor, velocity field, mean confining pressure and axial force distribution in bar were simultaneously computed and are presented in the following Figs. 12-14. The results were very interesting and numerical methodology seems very much promising.

Table 1 shows that the computed failure load increases from 149kN (*c.f.* P_{test}=136kN) for cohesive soil mass(ϕ=0°) to about 250kN for ϕ=30°. The failure load for soils ϕ>15° was almost remained constant. The reason for this can be

Table 1 Computed Pullout load at limit state of soil mass (P_{test}=136kN)

Friction angle, ϕ(deg.)	0	5	10	15	20	25	30
Failure Load, P (kN)	150	188	237	245	250	250	250

illustrated through the Fig. 12. Very high confining pressure prevails close to the footing (i.e. facing panels), and the confining pressure decays in a very short distance. Thus, the frictional part of soil strength is negligible around the inner end of the reinforcement. Therefore, the failure load remained constant for even very high frictional material. The exponential decay of axial force in pull out test should be attributed to this fact.

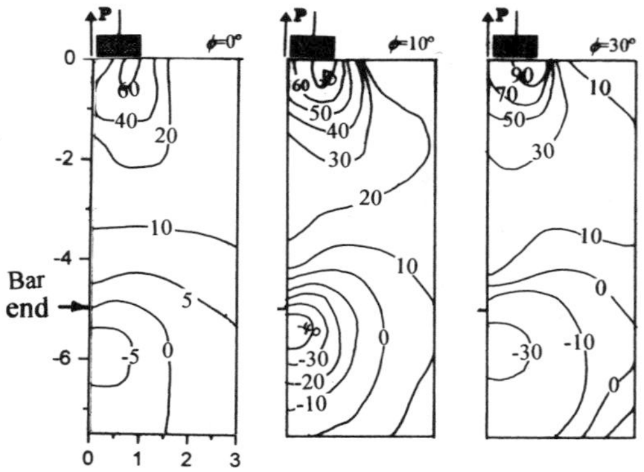

Fig. 12 Computed mean confining pressure contours.

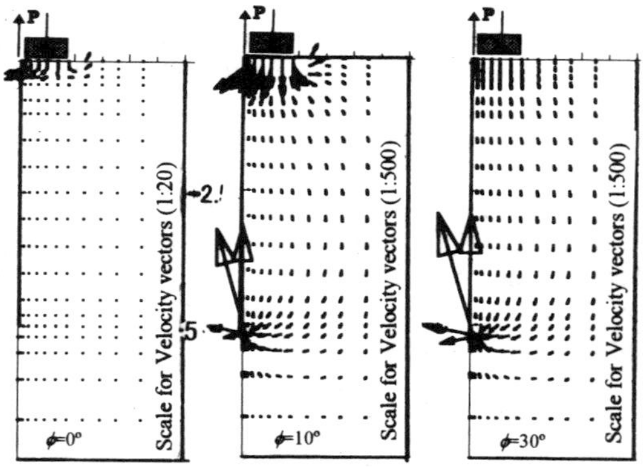

Fig. 13 Velocity field at limit state of soil mass

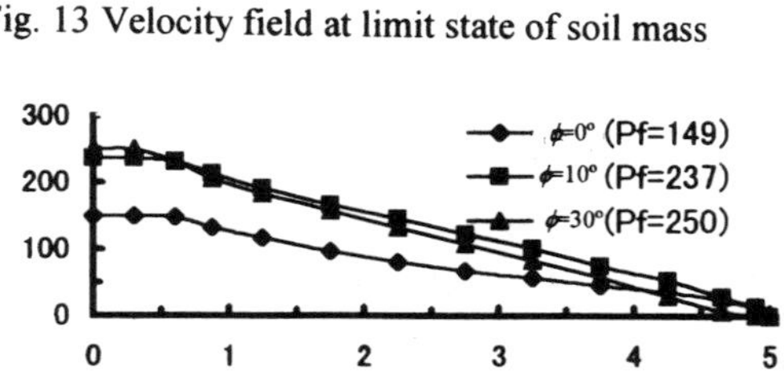

Fig. 14 Axial force distribution at limit state

The computed axial force distribution along the bar (Fig.14) is decreasing almost linearly which is quite different from the pull out test. It may be explained through the velocity field (Fig. 13) around the rear end of the bar. The direction of velocity vectors in this rear zone should mean that the very high tensile stresses prevail in the zone and the soil may not bear such high tensile stress in the real world while in FE analysis the same strength was entered for compression and tension. So, the tensile and compressive strengths should be correctly described in FEA to get real pattern.

5 CONCLUSIONS AND RECOMMENDATIONS

The following conclusions and recommendations are made based on the present study on a pull out test on soil nails and subsequent numerical simulations using RPFEM at limiting state of soil mass:

1. Pull out tests indicate that the bar force decays in a very short length (exponentially) compared to the total bar length because of very low confining pressure on the rear end of bars. This is quite different from the present design concepts of uniform shear strength mobilization along the bar length.
2. To avoid the effect of the very high confining pressure just behind the pull-out load supporting panels, a minimum sleeve length is required. Most appropriate depth would be 1/4 of the footing width.
3. The efficiency of the long bars can be improved by increasing the cohesion of the soil around the rear end of the bar e.g. by grouting.
4. The appropriate strength parameters should be described (i.e. for tension and compression) in the FEA to get real patterns from numerical computation procedures.

Overall, the proposed pull-out test simulation model and recommendations presented through this paper are promising and useful to the practicing engineers in correctly carrying out the pull-out tests and subsequent interpretations for design purposes.

REFERENCES

Asaoka, A., T. Kodaka and G. Pokharel(1995), Stability Analysis of Reinforced Soil Structures, *Soils and Foundations*, Vol. 34(1), pp.107-118

Tamura, T., Kobayashi, S. and Sumi T.(1984): Limit Analysis of Soil Structure by Rigid Plastic Finite Element Method, *Soils and Foundations*, Vol. 24(1), p. 34-42.

Numerical Models in Geomechanics, Pietruszczak & Pande (eds) © 1997 Balkema, Rotterdam, ISBN 90 5410 886 X

Finite element analysis of reinforced geocell-soil covers over large span conduits

M.A. Knight & R.J. Bathurst
Royal Military College of Canada, Kingston, Ont., Canada

ABSTRACT: A novel technique to improve the load-deformation performance of thin soil cover layers over flexible long span soil-steel bridge conduits is proposed. The soil cover is reinforced by a composite layer of geocell-soil whose properties have greater strength and stiffness than the aggregate soil infill in an unconfined condition. The results of numerical simulations using a large strain non-linear finite element model are used to demonstrate the improved load-deformation response of centrally loaded steel conduits with geocell-soil covers compared to conventional unreinforced soil covers. The simulation results show that thinner depths of soil cover are possible using this reinforcement technique.

1 INTRODUCTION

Soil-steel bridges are corrugated structural steel plate conduits that are assembled on site in circular, elliptical or arch shapes and backfilled with granular soils. They are typically used to support highway pavements and railway tracks. The soil-steel bridge carries load through interaction between the conduit shell and the confining soil.

The economic viability of using soil-steel structures for long span structures (i.e. spans in excess of 3 m) is often controlled by the depth of soil cover required over the conduit crest (Ontario Highway Bridge Design Code (OHBDC) 1992). Field experience has shown that shear or tension failure of the cover soil is the typical failure mechanism. Relatively large long span conduits up to 16.8 m have also been built in recent years (Abdel-Sayed and Bakht 1982). However, some large long span soil-steel structures with shallow covers have failed due to shell buckling (Mohammed and Kennedy 1996).

Current design codes such as the American Association of State and Highway Transportation Officials (AASHTO 1992), and the OHBDC (1992) restrict cover soil thickness to a minimum of about $D_h/6$ in order to prevent soil cover failure, where D_h is the horizontal conduit diameter. Thus, these design codes may require significant soil cover thickness for large span conduits and in turn render the structures uneconomical or unable to satisfy project geometric constraints.

In this paper the authors propose that the strength and stiffness of the soil cover can be artificially increased by replacing all or a portion of the conduit cover depth with a composite geocell-soil reinforcement layer. The improved mechanical properties of the composite geocell-soil layer allow a reduced cover soil depth to be used over flexible steel conduits.

Non-linear large strain finite element modelling is used to demonstrate that the proposed reinforcement technique leads to enhanced load and deflection response of the cover soil-flexible conduit system. Alternatively, thinner reinforced cover soil layers may give the same performance as a greater thickness of the same soil placed in an unreinforced condition.

2 GEOCELL-SOIL REINFORCEMENT

The term *geocell* is a generic term describing a class of geosynthetic products manufactured from thin strips of polymeric material (usually high density polyethylene) bonded or welded together to form a three-dimensional cellular network that can be filled with compacted soil (**Figure 1**). The effect of cellular confinement on the infill soil is to increase the stiffness and shear strength of the confined soil. Bathurst and Karpurapu (1993b) have demonstrated from large triaxial compression tests taken to collapse that cellu-

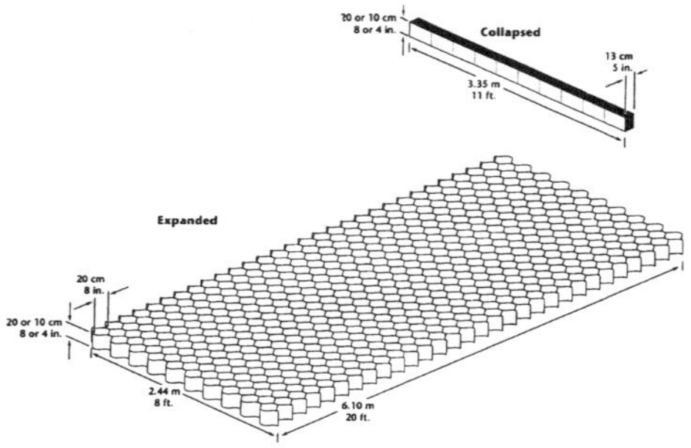

Figure 1. Single layer of polymeric geocell material.

lar confinement imparts additional apparent cohesion to aggregate infill soils while leaving the peak friction angle of the aggregate essentially unchanged. The composite geocell-soil layer in road base applications has been demonstrated to act as a stiffened mat that provides greater load bearing capacity and stiffness than the same granular base constructed without cellular confinement (Bathurst and Jarrett 1988). Layers of geocell-soil reinforcement can be stacked to create a composite material zone of any thickness (Bathurst and Crowe 1992).

3 PROBLEM CONFIGURATIONS

The plane strain geometry used to investigate the influence of unreinforced and composite geocell-soil reinforcement on system load-deformation performance is illustrated in **Figure 2**. A circular steel conduit of diameter (span) D located below a soil cover thickness H is loaded at mid-span by a line load P applied over a centrally located strip of width B placed at the soil surface. The contact between the bearing area B and the soil surface is assumed to be perfectly rigid and fully bonded. A total of 11 problem geometries grouped in 5 test series was investigated (**Table 1**). The conduit diameter was kept constant at 6 m.

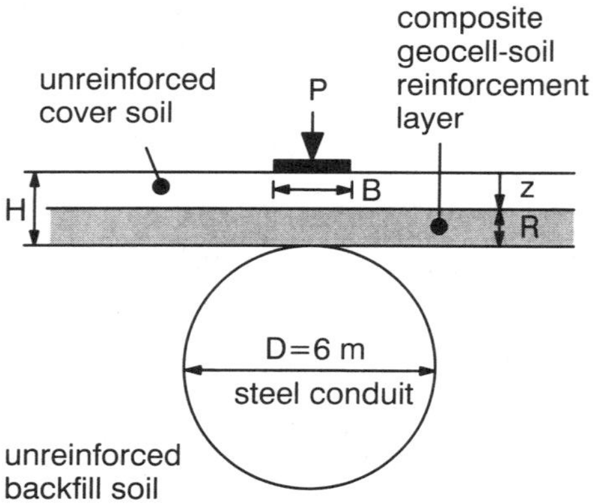

Figure 2. Problem geometry: B = loading width; P = applied load; H = soil cover depth; z = depth to the top of composite geocell-soil layer; R = geocell-soil thickness; and D = conduit diameter.

Table 1. Test configurations.

Test series	H (m)	R (m)	z/B
1 (unreinforced)	1.0	–	–
2	1.0	0.2, 0.4, 0.6, 0.8	0, 1, 2, 3
3	0.8	0.2, 0.4, 0.6	0, 1, 2
4	0.6	0.2, 0.4	0, 1
5	0.4	0.2	0

4 NUMERICAL SIMULATION

4.1 *Program GEOFEM*

The non-linear finite element program GEOFEM (Geotechnical Finite Element Modelling) was used to carry out the numerical simulations in the current study. GEOFEM is a general purpose finite element program for the analysis of structural and geotechnical problems. The program was specifically developed to investigate soil-structure interaction problems and uses a linearized updated Lagrangian method to accommodate large deformation behaviour (Bathurst and Karpurapu 1993a; Karpurapu and Bathurst 1992, 1995).

Prior to the current investigation, the results of program GEOFEM were validated against the closed form arch solution (Spofford 1937), reduced-scale physical model experiments reported by Hafez and Abdel-Sayed (1983a) and full-scale field tests (Bakht 1981). These validation exercises showed that the program gives accurate predictions of conduit wall axial thrust and failure loads for shallow cover large span circular flexible steel conduits under centrally applied surface loads.

4.2 *Details*

The finite element mesh that was used in the current study is shown in **Figure 3**. For the FEM analysis the following element types were used:

- 60 two-noded beam elements to represent the circular steel conduit constructed out of 5 gauge steel with 51 × 152 mm corrugations.
- a maximum of 392 eight-noded quadrilateral continuum elements to represent the composite geocell-soil reinforcement layer, backfill and cover soil.
- 30 six-noded interface elements to connect the beam elements to the continuum elements.

Linear elastic beam element material properties were obtained from McVay and Selig (1982) and are presented in **Table 2**.

Table 2. Corrugated plate properties.

Parameter	Value
unreduced area (m^2/m)	6.8×10^{-3}
modulus of elasticity (MPa)	2.0×10^{5}
Poisson's ratio	0.33
moment of inertia (m^4/m)	2.08×10^{-6}
plate thickness (m)	5.5×10^{-3}

McVay and Selig note that bolted seams in a flexible steel conduit can cause the circumferential stiffness of the conduit wall under compressive loads to be lower than that of a continuous corrugated-steel plate. Thus,

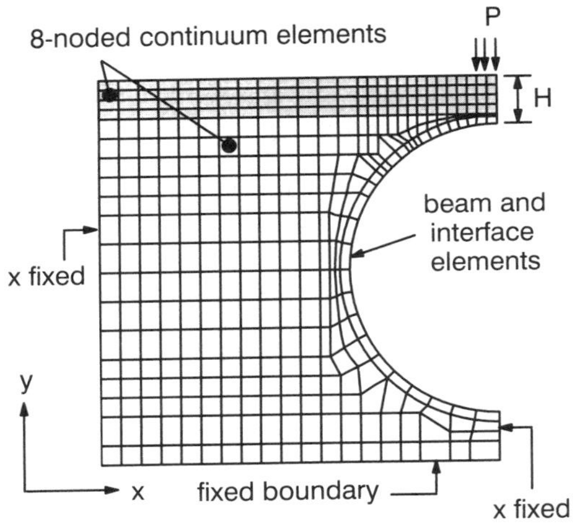

Figure 3. Finite element mesh.

to model the bolted corrugated-steel plate used for large span conduits a reduction in thrust stiffness (EA) is required to maintain the same bending stiffness (EI) of the shell. McVay and Selig also found that reducing the cross-sectional area of beam elements by six times their original area provided a better model response. This same reduction factor has been used in this study.

The composite geocell-soil material and unreinforced soils were assigned hyperbolic material properties (Duncan et al. 1980) interpreted from the results of single unit geocell-soil triaxial compression tests reported by Bathurst and Karpurapu (1993b)(**Table 3**).

Table 3. Hyperbolic parameters.

Parameter	Composite Geocell-soil	Unreinforced soils
Young's modulus component, K_s	1033	507
exponent for Young's modulus, n	0.77	0.76
Bulk modulus, K_b	1755	570
exponent for bulk modulus, m	0.41	0.5
minimum Poisson's ratio, υ	0.25	0.25
failure ratio, R_f	0.85	0.64
cohesive strength, c (kPa)	190	1
friction angle at 1 atm. pressure, ϕ_o (degrees)	45	45
change in friction due to ten fold increase in confining pressure, $\Delta\phi$	0	0
dilation angle, ψ	0	0
unit weight, γ (kN/m^3)	15.7	15.7
lateral earth pressure coefficient, K_o	0.35	0.35

Non-linear (hyperbolic) soil-beam interface element material properties were obtained from Hafez and Abdel-Sayed (1983b) and are presented in **Table 4**.

Table 4. Non-linear interface element properties.

Parameter	Value
Initial tangent stiffness	43070
Initial normal stiffness	2.7 x 10^{-8}
Residual normal stiffness	2.7 x 10^{-8}
Exponent	0.6
Failure Ratio	0.834
Interface cohesive strength (kPa)	1.0
Interface friction angle (degrees)	23

The finite element mesh shown in **Figure 3** was designed so that the continuum elements above the conduit are 0.2 m high matching the 0.2 m thick geocell thickness that is typical for a single layer of this material in the field. For each test series the following construction procedure was adopted:

- The finite element mesh was constructed in 12 increments to simulate a typical construction sequence and to allow conduit stresses during construction to develop. Prior to the application of the next construction increment the gravity force for each element was applied in 10 equal load steps.
- Using a plane strain displacement controlled boundary, a rigid footing of width B = 0.2 m was pushed into the cover soil in increments of 0.005 m until the total surface deflection was 0.5 m. This loading sequence simulates a standard OHBDC test vehicle.

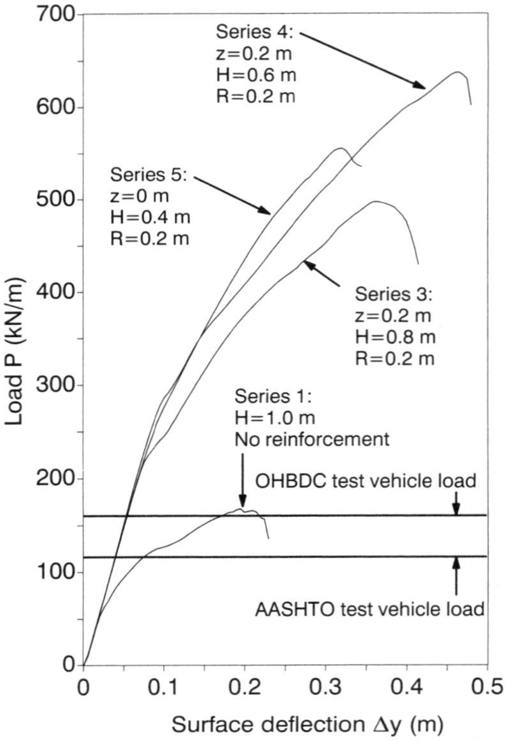

Figure 4. Applied load versus surface deflection for selected cases.

5 RESULTS

Typical simulation results are presented in **Figure 4**. The numerical results for the reference unreinforced system (Series 1) shows that the maximum applied surface load is approximately 167 kN/m at approximately 0.19 m of surface deflection (**Figure 4**). The maximum conduit axial thrust, at peak surface load was found to occur at approximately halfway between the conduit crown and haunch and to be approximately 210 kN/m. The magnitude and location of the maximum conduit thrust was consistent with the results of field measurements reported by Bakht (1981). Since the maximum conduit axial thrust for the unreinforced system is well below the OHBDC conduit yield value of 436.5 kN/m, shear failure of the cover soil would be the expected collapse mechanism in the field.

The effect of reducing the minimum OHBDC cover thickness and incorporating a 0.2 m thick layer of geocell-soil reinforcement is also clear from the figure. Based on the data shown in **Figure 4** the following observations can be made:

- Cover soils with a total thickness H = 0.4, 0.6 or 0.8 m and reinforced with a composite geocell-soil layer 0.2 m thick have an ultimate load capacity that is about 3 to 4 times greater than the unreinforced system with H = 1.0 m.
- The initial stiffness of the load-deflection curves (i.e. $\Delta y < 0.025$ m) of the thinner reinforced cover soil sections is unchanged from the reference unreinforced section with H = 1.0 m.

The significant increase in system load capacity is attributed to the apparent cohesion and increased stiffness that is imparted to the granular cover soil due to soil confinement (Bathurst and Karpurapu 1993b).

Results of parametric analysis for all reinforced cases are shown in **Table 5** through **8**. The collapse load and load recorded after 0.025 m of surface deflection are shown for each case. The measured loads for the reinforced cases have been normalized with respect to the standard unreinforced case (H=1.0 m) at collapse and after 25 mm of deflection. In some cases the cover soil thickness is reduced to a value that leads to overstressing of the conduit in which case the system capacity is assumed to be governed by conduit failure (i.e. axial thrust in any beam element exceeding 436.5 kN/m based on OHBDC guidelines for the conduit type and properties assumed in this study).

Table 5. Normalized load capacity for H = 1.0 m.

	P_f/P_{unr} (system collapse)			
	Reinforcement layer thickness, R (m)			
z/B	0.2	0.4	0.6	0.8
0	1.4	2.4	3.0	3.5
1	2.3	2.4	3.2	–
2	2.4	2.5	–	–
3	1.1	–	–	–

Note: underline denotes conduit failure

	P_f/P_{unr} (surface deflection = 0.025 m)			
	Reinforcement layer thickness, R (m)			
z/B	0.2	0.4	0.6	0.8
0	1.2	1.5	1.8	1.4
1	1.2	1.4	1.7	–
2	1.2	1.3	–	–
3	1.1	–	–	–

Table 6. Normalized load capacity for H = 0.8 m.

	P_f/P_{unr} (system collapse)			
	Reinforcement layer thickness, R (m)			
z/B	0.2	0.4	0.6	0.8
0	1.3	2.5	3.0	–
1	1.6	2.6	–	–
2	1.7	–	–	–

Note: underline denotes conduit failure

	P_f/P_{unr} (surface deflection = 0.025 m)			
	Reinforcement layer thickness, R (m)			
z/B	0.2	0.4	0.6	0.8
0	1.1	1.4	1.7	–
1	1.2	1.4	–	–
2	1.1	–	–	–

Table 7. Normalized load capacity for H = 0.6 m.

	P_f/P_{unr} (system collapse)			
	Reinforcement layer thickness, R (m)			
z/B	0.2	0.4	0.6	0.8
0	1.8	2.1	–	–
1	1.7	–	–	–

Note: underline denotes conduit failure

	P_f/P_{unr} (surface deflection = 0.025 m)			
	Reinforcement layer thickness, R (m)			
z/B	0.2	0.4	0.6	0.8
0	1.1	1.4	–	–
1	1.2	–	–	–

Table 8. Normalized load capacity for H = 0.4 m.

	P_f/P_{unr} (system collapse)			
	Reinforcement layer thickness, R (m)			
z/B	0.2	0.4	0.6	0.8
0	2.7	–	–	–

Note: underline denotes conduit failure

	P_f/P_{unr} (surface deflection = 0.025 m)			
	Reinforcement layer thickness, R (m)			
z/B	0.2	0.4	0.6	0.8
0	1.2	–	–	–

Based on **Tables 5** through **8** the following conclusions can be made:

- In all reinforced cases the collapse loads and loads at 0.025 m deflection were greater than the reference unreinforced system with a cover soil thickness of H = 1.0 m.
- The initial stiffness of the system was observed to generally increase with thickness, R, of the reinforcing layer (based on loads at a surface deflection $\Delta y = 0.025$ m). The increase in initial stiffness of the reinforced systems varied from 1.1 to 1.8 times the stiffness of the reference unreinforced system.
- The magnitude of load capacity benefit increases with increasing magnitude of surface deflection.
- As a general observation, the optimum placement depth of a layer of composite geocell-soil reinforcement with thickness R = 0.2 m is z/B = 1 to 2.
- For R $\geq$ 0.4 m and H $>$ 0.4 m, the ultimate collapse load of the system was controlled by conduit

axial thrust capacity. For thinner reinforcement layers (R = 0.2 m) shear failure of the cover soil controlled system capacity with the exception of the thinnest reinforced soil cover case (H = 0.4 m).

6 FINAL REMARKS

Plane strain finite element model simulations of a centrally loaded circular flexible steel conduit with and without a layer of composite geocell-soil reinforcement in the granular soil cover layer were carried out. The results of the parametric analysis show that the ultimate load capacity and load required to achieve a surface deflection of 0.025 m were increased by factors of 1.1 to 3.5 and 1.1 to 1.8, respectively, for the reinforced configurations when compared to a standard 1 m cover soil unreinforced configuration. Similarly, the results of the analyses show that the OHBDC minimum specified cover soil thickness of 1.0 m may be reduced to 0.4 m with no reduction in initial surface stiffness while significantly increasing the ultimate load capacity of the system. The novel technique of improving the mechanical properties of a granular soil cover through the use of polymeric cellular confinement holds promise to reduce the cost and increase the range of soil-steel bridge applications.

REFERENCES

American Association of State Highway and Transport Officials (AASHTO) 15th Edition. Standard Specifications for Highway Bridges, Washington, D.C., 1992.

Abdel-Sayed, G., and Bakht, B., 1982. Analysis of live-load effects in soil-steel structures. Transportation Research Record 878, pp. 49-55.

Bakht, B., 1981. Soil-steel structure response to live loads. Proceeding of the ASCE Vol. 107, No. GT6, pp. 779-798.

Bathurst, R.J. and Crowe, E.R., 1992. Recent case histories of flexible geocell retaining walls in North America, Recent Case Histories of Permanent Geosynthetic-Reinforced Soil Retaining Walls, (editors Tatsuoka and Leshchinsky), Tokyo, Japan, pp. 3-20, 6-7 November 1992 (published by A.A. Balkema, Rotterdam).

Bathurst, R.J. and Karpurapu, R.G., 1993a. User manual for Finite Element Modelling GEOFEM Vol. 1, 2 & 3. Department of Civil Engineering Royal Military College, Kingston, Ontario, Canada.

Bathurst, R.J. and Karpurapu, R.G., 1993b. Large scale triaxial compression testing of geocell-reinforced granular soils. Geotechnical Testing Journal, Vol. 16 No. 3, pp. 296-303.

Bathurst, R.J. and Jarrett, P.M., 1988. Large-Scal Model Tests of Geocomposite Mattresses Over Pea Subgrades, Transportation Research Record 1188 pp. 28-36.

Duncan, J.M., Byrne, P., Wong, K.S. and Mabry, P 1980. Strength, stress-strain and bulk modulu parameters for finite element analyses of stresse and movements in soil masses. Geotechnical Engineering Report UCB/GT/80-01, University c California, Berkeley, CA.

Hafez, H. and Abdel-Sayed, G., 1983a. Soil failure i shallow covers above flexible conduits. Can. J. Civ Eng., Vol. 10, pp. 654-661.

Hafez, H. and Abdel-Sayed, G., 1983b. Finite elemer analysis of soil-steel structures. Can. J. Civ. Eng Vol. 10, pp. 287-294.

Karpurapu, R.G. and Bathurst, R.J., 1992. Numerica investigation of controlled yielding of soil-retaining wall structures. Geotextiles and Geomembranes, Vol. 11, No.2, pp.115-131.

Karpurapu, R.G. and Bathurst, R.J., 1995. Behaviou of geosynthetic reinforced soil retaining walls using the finite element method, Computers and Geotechnics, Vol. 17, No. 3, pp. 279-299, 1995

McVay, M.C. and Selig, E.T., 1982. Performance an Analysis of a Long-Span Culvert. Transportatio Research Record 878, pp. 23-29.

Mohammed, H. and Kennedy, J.B., 1996. Economica design for long-span soil-metal structures. Can. J Civ. Eng., Vol. 23, pp. 838-849.

Ontario Highway Bridge Design Code (OHBDC) 3r Edition, 1992, Ministry of Transportation o Ontario, Downsview, Ontario.

Spofford, C.M., 1937. The theory of continuous structures and arches. McGraw Hill Inc.

4 Modelling of transient/coupled problems

Numerical Models in Geomechanics, Pietruszczak & Pande (eds) © 1997 Balkema, Rotterdam, ISBN 90 5410 886 X

ome considerations in numerical modelling of leachate collection system logging

.K.Rowe & A.J.Cooke
eotechnical Research Centre, University of Western Ontario, London, Ont., Canada

.E.Rittmann
epartment of Civil Engineering, Northwestern University, Evanston, Ill., USA

Fleming
ow Consulting Engineers, Brampton, Ont., Canada

BSTRACT: Landfill leachate collection systems may fail due to the accumulation of material ithin the voids of the granular drainage blanket such that this "clogging" reduces the hydaulic conductivity to the point that the collection system can no longer control the leachte levels to the design value. This paper outlines a numerical model for predicting the ate of clogging of granular drainage layers caused by biological growth and biochemically riven precipitation. By representing the porous media as a collection of elements, and ssuming each element acts as a separate, fixed film reactor, the model predicts the severity f the clogging by combining biological process (wastewater treatment) and geotechnical ngineering concepts. The model may be used to predict the transient substrate utilization, rowth and decay of biofilm, and accumulation of inert biomass and calcium carbonate on the ranular media at any time, or position, in the drainage layer. An example of the pplication of the model to typical laboratory column test conditions is presented.

. INTRODUCTION

odern municipal solid waste (MSW) landfills ypically include a barrier system consisting of a leachate collection system (or sysems) and either a natural or engineered iner (or liners). The leachate collection ystem (LCS) is intended to control the eight of the leachate mound on the base of he landfill and collect most, if not all, f the leachate generated by the landfill, hereby removing contaminant from the landill and minimizing the risk of contaminatng either surface water or groundwater.

Leachate collection systems typically onsist of a series of perforated pipes connected to manholes to allow cleaning) in granular layer (sand, gravel or crushed tone). The granular material around the ipes can not be cleaned and will be exected to experience a build up of organic nd inorganic material with time (with the ssociated decrease in pore space available or flow and a decrease in hydraulic conducivity). When the occlusion of pore space eaches a critical level, the hydraulic onductivity of the granular layer is no onger sufficient to transport the leachate o the pipes without an increase in driving orce (head), and hence a leachate mound egins to develop.

Once a leachate mound exceeds the design evel, the collection system is said to have logged (i.e. it is no longer performing its rimary function of controlling the leachate evel to the design level). However, it will generally still conduct some leachate towards the collection system (i.e. it has not become impermeable). Further clogging will result in further decrease in hydraulic conductivity and consequently further increase in the height of the leachate mound.

The build up in leachate levels associated with clogging will usually increase advective contaminant transport through the liner and into the underlying groundwater system. The time at which the collection system clogs and the magnitude of the leachate mound at subsequent times may be critical in terms of assessing the potential impact of the landfill on groundwater. If the leachate system operates without failure for long enough, the concentration of contaminant in the landfill leachate will reduce to levels that will have negligible impact if released to the environment (see Rowe, 1991, 1995; Rowe et al., 1995). However, if the collection system fails prematurely, the concentration of contaminant in the leachate that is now transported through the liner system may cause unacceptable impact.

Rowe and Booker (1995, 1997) have developed techniques that can be used to model the time history of a landfill operation including the effect of the operation and failure of leachate collection systems (with associated changes in advection due to leachate mounding) on contaminant migration to adjacent aquifers. However, the time at which collection system clogging occurs is an essential input to these models. The objective of the present paper is to discuss some considerations associated with the modelling of the clogging processes in the granular layer around pipes.

2. CLOGGING PROCESSES

The clogging of leachate collection systems may be the result of sedimentation, chemical precipitation, biological growth and biochemically driven precipitation (Rowe et al., 1995a,b; McBean et al., 1993). Sedimentation can be controlled by appropriate filter design and by ensuring adequate slopes on the leachate collection system. This paper focuses on the more difficult problem related to clogging due to the presence of microorganisms causing biological growth and biochemically driven precipitation. As indicated by Ramke (1989), biological clogging involves slime growth, filamentous growth, biomass formation and ferric incrustations. Also attributed to biological processes were the bacterial production of sulphide and carbonate precipitations.

The passage of water through MSW causes the generation of leachate that has significant concentrations of organic compounds (especially volatile fatty acids) and dissolved inorganic elements and compounds (calcium, iron, carbonate etc.). The organic compounds provide a substrate for biological growth and the formation of a biofilm on the particles in the porous media. The constant supply of leachate (substrate) flowing past the biofilm provides an environment for the biofilm to grow (increase in thickness); this alone will reduce the effective porosity (i.e. pore space available for fluid flow). However, the environment around the biofilm also encourages precipitation of a mineral film, which typically consists of calcium carbonate and iron sulphide under anaerobic conditions (which are typically established relatively early in the life cycle of a landfill). The mineral film thickness will also grow with time, further decreasing the effective porosity. Eventually, the porosity (and permeability) is reduced sufficiently in critical areas (e.g. close to the leachate collection pipes) such that a leachate mound develops in order to maintain the flow required for continuity of flow.

3. PREDICTING THE RATE OF CLOGGING

This paper outlines the characteristics of a numerical model to simulate the effects of biofilm growth due to leachate flow through a column test or landfill drainage layer. The model combines concepts associated with transient anaerobic, fixed film biological processes (wastewater treatment) with concepts of geotechnical engineering. The model uses a time marching algorithm to model the evolution of the influent and effluent organic concentration, biofilm thickness, inert biofilm plus mineral film thickness and porosity at any position or time.

3.1 *Fluid flow*

Flow is assumed to occur through a porous media that consists of a volume of packed spheres that is discretized into subregions. Each subregion or "element" is assumed to act as a separate, fixed film reactor. The fluid flow route through the porous media is controlled by the type of flow field being examined, either a column test or a LCS drainage layer.

The column test involves "one dimensional" flow, where the fixed film reactors are in series as illustrated schematically in Figure 1a. Flow through this system is controlled by the element with the greatest clogging; once the pore space in the element is fully occluded, the flow stops.

The leachate collection system flow field involves "two dimensional" or "bi-directional" flow consisting of unsaturated downward flow zones that feed horizontally flowing layers. The flow field in this case maps a drainage path from the top to bottom of a slope, with the length of the field being discretized into sub-horizontal segments and the segments being discretized into layers (see Figure 1b) with each layer in a segment being an "element" that is a fixed film reactor. The LCS flow field allows each segment to have a number of saturated horizontal flow layers, a number of unsaturated horizontal flow layers (a "splash zone") and a downward flow zone (similar in behaviour to a trickling filter). When a horizontal flow layer is deemed clogged in this field, it is assumed that the flow rises to the layer directly above. The type of flow through each layer in this segment is moved to the layer above, and a new layer is added in the downward flow zone.

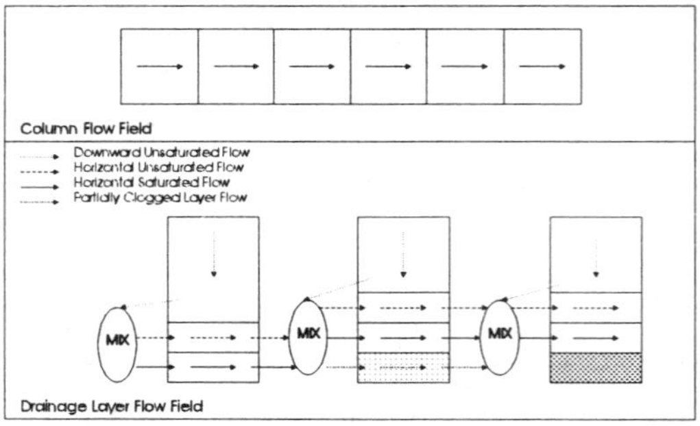

Figure 1. (a) Column test flow field, (b) LCS drainage layer flow field.

2 *Flow distribution*

2.1 Unsaturated downward flow zone

e unsaturated downward flow zone of each gment of the drainage path receives an ual amount of substrate at flow Q_{unsat}. It ould be noted that generally the granular terial ranges from gravel to cobbles (or ushed stone), and the vertical flow is ntrolled by infiltration through the landll cover rather than consideration of unturated soil mechanics. Following treatnt within the downward unsaturated zone, e flow is combined with flow exiting the stream segment, Q_{prev}, which has experienc a different level of biological treatment d is, therefore, at a different substrate ncentration. The mixed flow is distrited among the horizontal flow layers. is flow pattern is shown in Figure 1b.

2.2 Partially and discontinuous clogged zones

critical total film thickness, Lft_{crit} ists at which pore discontinuity occurs e to the total build-up of inert biomass, ecipitation (e.g. of $CaCO_3$), and the acve biofilm layer. If the critical porosty is reached while including the active ofilm the layer becomes partially clogged. e layer is considered only partially ogged because flow must occur through the yer to supply nutrients to the active biolm. Once a layer is found to be partially ogged, a minimum flow rate is calculated ch that the amount of substrate available the amount required to arrive at the cical porosity exactly. The total amount of ow through the partially clogged layers, Q_{clogs}, must be subtracted from the total rizontal flow to get the flow in the unogged layers. If the critical porosity is ached due to the total build-up of inert omass and precipitation of $CaCO_3$ alone, e layer is considered a discontinuous ogged layer. A partially clogged layer y become a discontinuous clogged layer er time as the inactive film builds up low the active biofilm and reaches the itical porosity. A discontinuous clogged yer transmits no flow.

2.3 Unsaturated and saturated horizontal flow layers

e flow through the unsaturated and satuted layers is distributed as follows: the ow through each saturated layer is equal, d the flow through each unsaturated layer calculated as a linearly distributed ighting of the flow through the saturated yers. Equating the flows into and out of e horizontal flow layers results in

$$Q_{unsat} + Q_{prev} = \sum_{i=1}^{n} W_i Q_{sat} + mQ_{sat} + \sum_{j=1}^{p} Q_{clogsj} \qquad (1)$$

where Q_{unsat} is the total flow through the unsaturated downward flow zone, Q_{prev} is the horizontal flow from the previous segment, W_1 is the weight given to the topmost unsaturated layer, W_2 is given to the next unsaturated layer down, etcetera, $\sum_j Q_{clogsj}$ is the sum of the flows through partially clogged layers, m is the number of unsaturated (splash zone) layers, n is the number of saturated layers and p is the number of partially clogged layers in a given segment.

In order to simulate trickling or splashing, the flow is partitioned into an untreated portion and a portion in direct contact with the stone or biomass surface and is thus available for treatment. This treated portion of flow, Q', is assumed to be a portion of a total flow, Q

$$Q' = \chi Q \qquad (2)$$

where χ must be established based on experimental evidence but in the following is assumed to be given by $\chi = \theta/n$ where θ is the volumetric water content and n is the total porosity.

The concentration of substrate of the treated portion of flow exiting the layer, Se', can be calculated as discussed in Section 3.3, and this is then added proportionally to the concentration of the untreated flow (which bypasses significant contact with the biofilm during its passage through the unsaturated downward flow zone) at the influent concentration, S_0, to give the concentration of the layer's effluent using:

$$S_e = S_0(1-\chi) + S_e'\chi \qquad (3)$$

In recognition of the short residence time of the unsaturated zones, an adjustment is made to the time available for treatment in these zones.

3.3 *Non-steady state biofilm growth model*

A non-steady state biofilm has a thickness, which may change with time. Since substrate flux into the biofilm is a function of this thickness and the substrate concentration, the substrate flux may also change with time. Growth of cell matter is proportional to substrate flux into the biofilm, while losses occur due to bacterial detachment and respiration decay. The substrate is considered to be the single required nutrient which is not available in excess, and therefore limits the rate of biofilm growth. The primary substrate can be represented as COD. The rate of growth will depend on the bio-kinetic parameters, the influent concentrations, flow rates and mass density, the physical dimensions of the flow field, the average diameter of the granular media, the choice of packing arrangement, the system temperature, and the film properties such as densities and mineral accumulation yield rate.

3.3.1 Model description

Non-steady state biofilm growth and loss can be modelled using an algorithm modified from Rittmann and McCarty (1981) and Rittmann and Brunner (1984). The non-steady state growth and loss algorithm is combined with equations to model the change in the porosity and specific surface as the biofilm and mineral film grow, mass balance equations and shearing equations. Due to non-linearity, two coupled iterative procedures are required to derive each layer's effluent COD concentration, film thicknesses, porosity, and specific surface area at each time increment.

3.3.2 Biofilm idealization

The biofilm is assumed to have a uniform thickness L_{fa} and uniform density X_{fa}. The biofilm is supplied with substrate by diffusion from the bulk fluid through a stagnant liquid layer adjacent to the biofilm and thence by diffusion through the biofilm. This mass transport is modelled using Fick's first law of diffusion.

The biofilm growth and loss modelling must deal with five processes occurring simultaneously: mass transport of the substrate across the diffusion layer, diffusion of the substrate in the biofilm, the utilization of the substrate for cell growth, the growth of active cells due to this utilization, and the loss of biomass due to detachment and decay of cell matter.

3.3.3 Substrate flux into the biofilm

An iterative procedure similar to that outlined in Rittmann and McCarty (1981) is used to calculate the substrate flux into the biofilm, J. The procedure of Rittmann and McCarty utilizes the work of Atkinson and Davies (1974) which assumes substrate utilization follows a Monod kinetic relationship, and modifies it to include mass transport resistance. By accounting for the mass resistance of the stagnant diffusion layer, the substrate flux can be calculated as a function not of the surface substrate concentration, but of the bulk (measurable) substrate concentration. The procedure is applicable to biofilms of any thickness.

3.3.4 Non-steady growth and loss of biofilm

The expression of non-steady growth and loss of biofilm (developed by Rittmann and Brunner (1984)) gives the biomass growth rate, r_{gr} as:

$$r_{gr} = \frac{Y\hat{q}X_{fa}S_f}{K_S + S_f} \tag{4}$$

where Y, $\hat{q}$, S_f and K_s are defined by

- Y is the true yield coefficient,
- $\hat{q}$ is the maximum specific rate of substrate utilization,
- S_f is the substrate concentration within the biofilm, and
- K_s is the Monod half-maximum velocity concentration.

The biomass loss rate, r_{loss} was expressed by

$$r_{loss} = -b'X_{fa} \tag{5}$$

where b' is the first-order biofilm loss coefficient which includes decay losses, b, and detachment losses, b_s. The transient growth and loss expression may be derived by first combining Eqs. 4 and 5 and simplifying to

$$\frac{d(X_fL_f)_a}{dt} = YJ - b'X_{fa}L_{fa} \tag{6}$$

(Rittmann and McCarty, 1980; Rittmann and Brunner, 1984), which may be integrated with respect to time. Assuming the biofilm density, X_{fa} to be constant with respect to time, Eq. 6 can be used to calculate changes in biofilm thickness for small, finite time steps.

3.3.5 Mineral precipitation and inert biomass accumulation

The composition of the inorganic portion of clog material has been found (Brune et al. (1991), Rowe et al. (1995)) to be predominantly calcium carbonate, $CaCO_3$. Rittmann et al. (1996), reported that there is a relationship between the microbial oxidation of COD to inorganic carbon to the mass of calcium carbonate precipitated out of landfill leachate. Laboratory experiments were performed in which leachate collection was simulated using landfill materials infiltrated with actual landfill leachate and water quality monitoring was performed on the influent and effluent. The experiments found that the removal of COD, mostly acetic acid, and its partial substitution with H_2CO_3, resulted in major increases in pH and total carbonate, which together caused a large increase in CO_3^{2-}

oncentration, allowing or accelerating aCO_3 precipitation. From data obtained rom Rowe et al. (1995), a linear elationship between COD and $CaCO_3$ recipitation may be derived and a yield oefficient, Y_c, may be calculated such that ne rate of mineral precipitation of $CaCO_3$ s

$$\frac{d(X_f L_f)_m}{dt} = Y_c J \qquad (7)$$

ne accumulation of inert biomass may be ound by assuming a fraction of the active iomass, f_d, is degradable due to decay. ne remaining fraction, $(1 - f_d)$, is not egradable and thus becomes inert. The rate f accumulation of inert biomass is

$$\frac{d(X_f L_f)_i}{dt} = (1 - f_d)\, b X_{fa} L_{fa} \qquad (8)$$

.3.6 Effluent substrate concentration by mass balance

ne calculation of the substrate flux and hus rate of growth or loss of biofilm escribed in the previous sections requires ne bulk substrate concentration as input. t is assumed that the bulk substrate concentration is equal to the geometric mean of ne influent and effluent substrate concentrations. By considering the non-steady rowth and loss of biofilm, the change in ne porosity and specific surface, and mass alance, the substrate concentration can be alculated using an iterative procedure.

. APPLICATION OF MODEL

he following examples illustrate the use of he model applied to a typical column test ssuming the fluid has bio-kinetic parameters similar to acetic acid. The variation n effluent substrate concentration and orosity the first layer in line is shown in igure 2. The variation in substrate concentration and porosity with position along he column (after 60, 70 and 80 days) is hown in Figure 3. The effect of changing he stone size from 1.4 cm to 2.8 cm is vident from a comparison of Figures 3 and . It can be seen that the model predicts hat an increase in the stone size causes a ecrease in the substrate removal and biological matter accumulation through the olumn. This is a direct result of the ecrease in surface area corresponding to he increase in stone size.

An increase in flow through the column is redicted to cause a decrease in the removal f substrate, but an increase in biofilm ccumulation. In this case, the decrease in emoval is due to the shorter residence time n each layer of the column and the increase n accumulation is due to the increase in he amount of fresh substrate available.

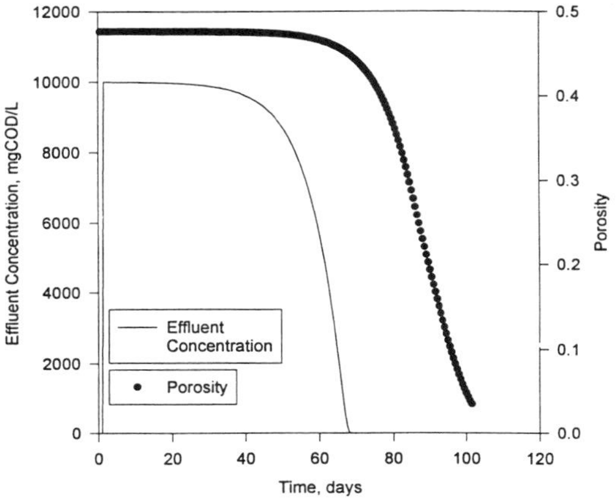

Figure 2. Predicted variation in column effluent concentration and porosity of first-in-line layer.

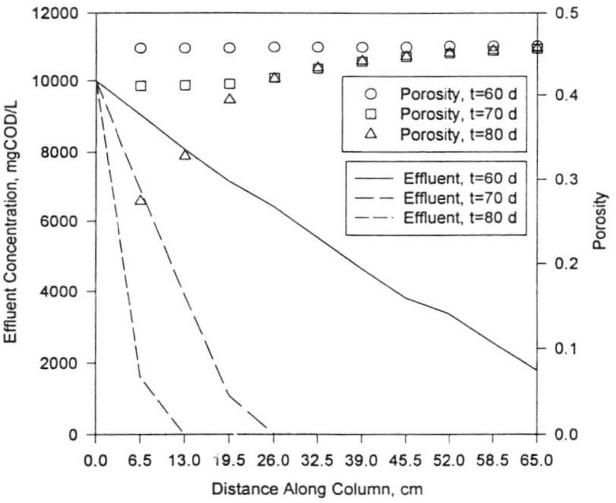

Figure 3. Predicted variation of layer effluent concentration and porosity with time and position in column (particle diameter 1.4 cm).

5. CONCLUSIONS

A numerical model has been developed to predict the transient growth, decay, and substrate utilization of biofilm, the accumulation of inert biomass and calcium carbonate, the corresponding changes in media porosity and subsequent time to clogging for landfill drainage applications. The model has been developed to simulate both one dimensional column tests and drainage layers where there is two-dimensional flow.

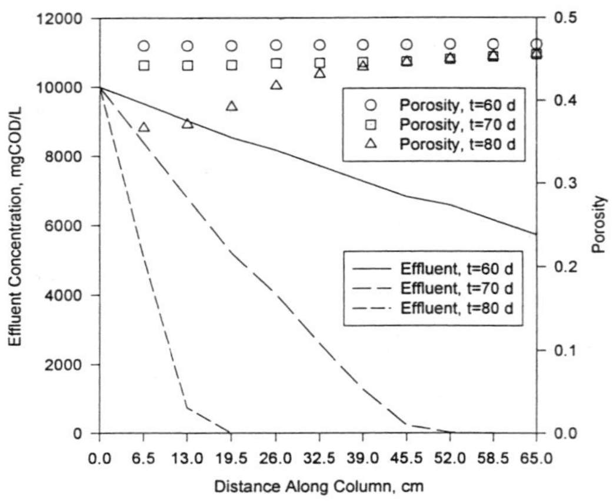

Figure 4. Predicted variation of layer effluent concentration and porosity (particle diameter 2.8 cm).

6. ACKNOWLEDGEMENT

Funding of the program of research into the clogging of leachate collection systems came from Collaborative Grant CPG0163097 provided by the Natural Sciences and Engineering Research Council of Canada.

7. REFERENCES

Atkinson, B. and I.J. Davies 1974. The overall rate of substrate uptake (reaction) by microbial films. Part I.a. Biological rate equation. *Transactions, Institution of Chemical Engineers*, London, England, 52: 248-259.

Brune, M., H.G. Ramke, H.J. Collins and H.H. Hanert 1991. Incrustation processes in drainage systems of sanitary landfills. *Proc. Third Inter. Landfill Symp.*, Cagliari, Italy, 999-1035.

McBean, E.A., F.R. Mosher and F.A. Rovers 1993. Reliability-based design for leachate collection systems. *Proc. Sardinia 93, Fourth Inter. Landfill Symp.*, Cagliari, Italy: 433-441.

Ramke, H.G. 1989. Leachate collection systems. *Sanitary Landfilling: Process, Technology and Environmental Impact*. Christensen, Th.H., R. Cossu, R. Stegmann (Eds.), Academic Press, London.

Rittmann, B.E. and C.W. Brunner 1984. The nonsteady-state-biofilm process for advanced organics removal. *Journal WPCF*, 56(7): 874-880.

Rittmann, B.E. and P.L. McCarty 1980. A model of steady-state-biofilm kinetics. *Biotechnol. Bioeng.*, 22: 2343-2357.

Rittmann, B.E. and P.L. McCarty 1981 Substrate flux into biofilms of an thickness. *J. Environ. Eng. Div.* ASCE, 107(EE4): 831-849.

Rittmann, B.E., I. Fleming and R.K. Row 1996. Leachate chemistry: It implications for clogging. *Nort American Water and Environmen Congress '96*, Anaheim, CA, June, pape 4 (CD Rom) 6 p, Session GW-1, Biological Processes in Groundwater Quality

Rowe, R.K. 1991. Contaminant impac assessment and the contaminant life span of landfills. *Can. J. Civi Engrg.*, 18(2): 244-253.

Rowe, R.K. 1995. Leachate characterizatio for MSW landfills. *Proc. Fifth Int Landfill Symp.*, Sardinia, Italy, 2 327-344.

Rowe, R.K. and J.R. Booker 1995. A finit layer technique for modelling comple landfill history. *Can. Geotech. J.* 32(4): 660-676.

Rowe, R.K. and J.R. Booker 1997. Modellin impacts due to multiple landfill cell and clogging of leachate collectio systems. *Can. Geotech. J.* (accepted)

Rowe, R.K., I. Fleming, D.R. Cullimore, N Kosaric and R.M. Quigley 1995. research study of clogging and encrustation in leachate collection system at municipal solid waste landfills Submitted to Interim Waste Authorit (131 pages).

Rowe, R.K., R.M. Quigley and J.R. Booke 1995. *Clayey Barrier Systems for Wast Disposal Facilities*. E &FN Spo (Chapman & Hall), London, 390 pp.

Numerical Models in Geomechanics, Pietruszczak & Pande (eds) © 1997 Balkema, Rotterdam, ISBN 90 5410 886 X

On the validation of a model of coupled heat and moisture transfer in unsaturated soil

H.R.Thomas
Cardiff School of Engineering, University of Wales Cardiff, UK

ABSTRACT: An examination of the validity of a theoretical formulation of coupled heat and moisture in non-deforming unsaturated soil is presented. In particular, results from a theoretical formulation were compared with those obtained from controlled laboratory heating experiments. The work considered one material, a highly compacted medium sand, for which the thermo/physical parameters had been established independently of the heating experiments. A solution of the theoretical formulation was achieved by means of a numerical analysis, using the finite difference method. Very good correlations were achieved. Confidence in the ability of the theoretical formulation to describe the physical processes involved is thus enhanced.

1 INTRODUCTION

Coupled heat and mass transfer in unsaturated soil is a phenomenon of importance in a number of engineering applications. Two major approaches to the development of a theoretical formulation for the problem can generally be seen to have been attempted. These are methods which evolve from the application of the principles of irreversible thermodynamics (Luikov, 1966; Taylor and Cary, 1964) or models developed from a physical mechanistic approach (Philip and De Vries, 1957; De Vries, 1958; De Vries, 1987).

The Philip and De Vries/De Vries approach has gained much favour with research workers and practitioners in the field. However, because of the very nature of a formulation developed from a physical mechanistic approach, questions of the validity of the theoretical model are particularly important. In a recent paper De Vries reviewed these issues (De Vries, 1987) and concluded that for the range of problems considered to date, the model had proved useful in describing and analysing experimental results. However limitations, in terms of the predictive capability of the model, were acknowledged to still exist. In a recent paper, (Thomas and Li, 1997(a)) an attempt was made to continue the process of validation, with an examination of the performance of the complete theoretical formulation proposed. This paper briefly reviews the work performed and presents some of the salient highlights, in terms of the results achieved.

The approach adopted consists of an assessment of the validity of the model, by comparison of results from the theoretical formulation with those obtained from controlled laboratory experiments.

The formulation under investigation is essentially the Philip and De Vries/De Vries approach, but developed to take into account more fully, vapour transfer characteristics in unsaturated soil (Ewen and Thomas, 1989). The theoretical formulation is strictly applicable only to materials that do not deform as a result of heat and mass transfer. Hysteresis effects are not included at this stage. Continuity of both liquid water flow and water vapour flow are taken into account via a conservation of mass equation. The energy conservation equation is based on heat transfer by means of conduction, convection and latent heat of vaporisation.

A numerical solution of the governing equations was achieved by the use of the finite difference method. In particular, a self implicit scheme was adopted (Howells and Marshall, 1983) for the evaluation of the time derivatives. Six laboratory heating experiments were performed (Ewen, 1987), covering a range of power inputs and initial moisture contents. The geometrical configuration and test layout is essentially as reported previously (Ewen and Thomas, 1989). The material selected for investigation is a medium sand (Ewen and Thomas, 1987). A sample of the sand was compacted into a glass cylinder and heated by an embedded circular heater passing through the centre of the cylinder.

2 THEORETICAL FORMULATION

The theoretical formulation employed has been presented previously (Ewen and Thomas, 1989) and therefore, for reasons of conciseness, will not be repeated in full here. The governing differential moisture transfer equation can be written as:-

$$G1\frac{\partial\theta}{\partial t}+G2\frac{\partial T}{\partial t}=-\left[\frac{1}{r}\frac{\partial}{\partial r}(rU_r)+\frac{1}{r}\frac{\partial U_\phi}{\partial\phi}\right] -\left[\frac{1}{r}\frac{\partial}{\partial r}(rV_r)+\frac{1}{r}\frac{\partial V_\phi}{\partial\phi}\right] \tag{1}$$

where

$$G1=1+\frac{(\eta-\theta)}{v\eta}\frac{D_{\theta v}}{D_{atm}}-\frac{h\rho_o}{\rho_l} \tag{2}$$

$$G2=\frac{(\eta-\theta)h}{\rho_l}\frac{d\rho_o}{dT} \tag{3}$$

$$U_r=\frac{\rho_l}{\rho_{lr}}u_r=-D_{Tl}\frac{\partial T}{\partial r}-D_{\theta l}\frac{\partial\theta}{\partial r}-D_g\sin\phi \tag{4}$$

$$U_\phi=\frac{\rho_l}{\rho_{lr}}u_\phi=\frac{1}{r}\left(-D_{Tl}\frac{\partial T}{\partial\phi}-D_{\theta l}\frac{\partial\theta}{\partial\phi}-rD_g\cos\phi\right) \tag{5}$$

$$V_r=\frac{\rho_v}{\rho_{lr}}v_r=-D_{Tv}\frac{\partial T}{\partial r}-D_{\theta v}\frac{\partial\theta}{\partial r} \tag{6}$$

and

$$V_\phi=\frac{\rho_v}{\rho_{lr}}v_\phi=\frac{1}{r}\left(-D_{Tv}\frac{\partial T}{\partial\phi}-D_{\theta v}\frac{\partial\theta}{\partial\phi}\right) \tag{7}$$

The diffusivity terms in equations (4) and (6) are defined according to

$$D_{Tl}=\frac{1}{\sigma_r}\frac{d\sigma}{dT}\frac{\mu_r}{\mu}K_r\varphi_r \tag{8}$$

$$D_{\theta l}=\frac{\sigma}{\sigma_r}\frac{\mu_r}{\mu}\frac{d\varphi_r}{d\theta}K_r \tag{9}$$

$$D_g=\frac{\mu_r}{\mu}K_r \tag{10}$$

$$D_{Tv}=\begin{cases}\dfrac{D_{atm}v\eta h}{\rho_{lr}}\dfrac{d\rho_o}{dT}\dfrac{(\nabla T)_a}{\nabla T} & \theta<\eta\\ O & \theta=\eta\end{cases} \tag{11}$$

and

$$D_{\theta v}=\begin{cases}\dfrac{D_{atm}v\eta gh}{\rho_{lr}}\dfrac{\rho_o}{RT}\dfrac{\sigma}{\sigma_r}\dfrac{d\varphi_r}{d\theta} & \theta<\eta\\ O & \theta=\eta\end{cases} \tag{12}$$

where σ is the surface energy per unit area at any given absolute temperature, σ_r is the surface energy per unit area at the absolute reference temperature, μ is the dynamic viscosity of liquid water at any given absolute temperature, μ_r is the dynamic viscosity of liquid water at the absolute reference temperature, K_r is the unsaturated hydraulic conductivity at the absolute reference temperature, φ_r is the capillary potential at the absolute reference temperature, $(\nabla T)_a/\nabla T$ is the ratio of the microscopic temperature gradient in the pore space to the macroscopic temperature gradient, g is the gravitational constant and R is the specific gas constant.

Similarly, the governing differential energy equation is given by

$$G3\frac{\partial\theta}{\partial t}+G4\frac{\partial T}{\partial t}=\frac{1}{\rho_l}\left[\frac{1}{r}\frac{\partial}{\partial r}\left(rk\frac{\partial T}{\partial r}\right)+\frac{1}{r^2}\frac{\partial}{\partial\phi}\left(k\frac{\partial T}{\partial\phi}\right)\right] -\frac{C_{pl}}{r}\left\{\frac{\partial}{\partial r}[r(TU_r)]+\frac{\partial}{\partial\phi}(TU_\phi)-T_r\left[\frac{\partial}{\partial r}(rU_r)+\frac{\partial U_\phi}{\partial\theta}\right]\right\} -\frac{C_{pv}}{r}\left\{\frac{\partial}{\partial r}[r(TV_r)]+\frac{\partial}{\partial\phi}(TV_\phi)-T_r\left[\frac{\partial}{\partial r}(rV_r)+\frac{\partial V_\phi}{\partial\phi}\right]\right\} -\frac{L_w}{r}\left[\frac{\partial}{\partial r}(rV_r)+\frac{\partial V_\phi}{\partial\phi}\right] \tag{13}$$

where

$$G3=C_{pl}(T-T_r)+\left[\frac{(\eta-\theta)}{v\eta}\frac{D_{\theta v}}{D_{atm}}-\frac{h\rho_o}{\rho_l}\right]\left[C_{pv}(T-T_r)+L_w\right] \tag{14}$$

and

$$G4=\frac{H}{\rho_l}+\frac{(\eta-\theta)h}{\rho_l}\frac{d\rho_o}{dT}\left[C_{pv}(T-T_r)+L_w\right] \tag{15}$$

$G3$ and $G4$ are the generalised storage terms, T_r is the absolute reference temperature, C_{pl} and C_{pv} are respectively the specific heat capacity of liquid water and water vapour, k is the thermal conductivity of moist soil, L_w is the latent heat of vaporisation of liquid water and H is the bulk specific heat capacity which comprises the capacity of the soil solid, liquid soil water and soil water vapour.

3 NUMERICAL ALGORITHM

Adopting a self implicit method, employing a forward difference approximation to the time derivative, upwind differencing of the second king (Roache, 1972) to the convective terms in equation (13) and the central difference approximation to other space analogues, the finite difference equation for equation (1) applicable at any node (i,j) in the finite difference grid may be written as

$$\left(G1\Big|_{i,j}^{n} - \varsigma\Delta t \frac{\partial S_\theta}{\partial \theta}\Big|_{i,j}^{n} \right) \omega_\theta + \left(G2\Big|_{i,j}^{n} - \varsigma\Delta t \frac{\partial S_\theta}{\partial T}\Big|_{i,j}^{n} \right) \omega_t = S_\theta \quad (16)$$

where the superscript n denotes the time index, the subscripts i and j indicate the spatial indices respectively in the r and ϕ directions, Δt is the time step size, S_θ is the source term of the finite difference approximation representing the analogues of the right-hand side of equations (1), $\partial S_\theta / \partial\theta|_{i,j}^{n}$ and $\partial S_\theta / \partial T|_{i,j}^{n}$ are the rate of change of the source term respectively with the volumetric liquid content and the temperature at the node (i,j), ζ is the time weighting factor taking on values between zero and unity, ω_θ and ω_T are respectively the finite difference approximations to the time derivative of the volumetric liquid content and the temperature. A similar expression may be derived for the finite difference equation representation of equation (13), in terms of S_E rather than S_θ.

The two derivative terms $\partial S_\theta / \partial\theta|_{i,j}^{n}$ and $\partial S_\theta / \partial T|_{i,j}^{n}$ are introduced from the source term which are approximated at time $t+\zeta\nabla t$ by using a Taylor's series expansion in time. The introduction of these terms, which involve the rate of change of the source term with the volumetric liquid content and the temperature at the node (i,j), adds a stabilising influence to the solution which in turn enables a larger time step size to be used (Thomas and Li, 1997(b)). This implicitness only in the dependent variables at the node (i,j) gives rise to the name, the self implicit method.

The time weighting factor in this paper is set to 0.5, i.e., the mean of the finite difference representations in the spatial derivatives on the nth and (n+1)th time levels.

4 MATERIAL PARAMETER DETERMINATION

Full details of the determination of the range of thermo/physical parameters required for the modelling work have been given previously (Ewen and Thomas, 1987; Ewen and Thomas, 1989). Only the salient highlights will therefore be repeated here.

The soil used for the laboratory heating experiments was Garside grade 21 medium sand. This type of sand has a quartz content of over 96%. Therefore typical values for quartz were adopted for the density (2700 kg m^{-3}) and the specific heat capacity (800 J kg^{-1} K^{-1}) of the soil particles.

The capillary potential was determined experimentally using standard volumetric pressure plate extraction equipment (Ewen and Thomas, 1989). A mathematical fit to within ±6% error was found to the capillary potential yielding the expressing

$$\varphi_r = \left(-2.41 - 0.002\theta^{-1.75}\right) / g\,[m] \quad (17)$$

The unsaturated hydraulic was determined using the horizontal infiltration scaling method and the short column-large increment method (Ewen and Thomas, 1989). The combined results, yielded the following expressions

$$K_r \begin{cases} 1.91x10^{-13} \exp\left(76.565\dfrac{\theta}{\eta}\right) \\ 1.5x10^{-10} \exp\left[28.061\dfrac{\theta}{\eta} - 12.235\left(\dfrac{\theta}{\eta}\right)^2\right] \end{cases} \quad (18)$$

in units of ms^{-1} for the ranges $\theta \le 0.05174$ and $0.05174 < \theta < \eta$ respectively.

The experimental determination of the thermal conductivity was achieved using the thermal probe method (Ewen and Thomas, 1987). A curve was fitted to the data obtained yielding the expression

$$k = k_{dry} + \left(k_{sat} - k_{dry}\right)\left[1 - \exp\left(-8.9\frac{\theta}{\eta}\right)\right] \quad (19)$$

in units of Wm^{-1}K^{-1}. k_{dry} is the thermal conductivity of the dry medium sand and k_{sat} is the thermal conductivity of the saturated medium sand.

5 HEATING EXPERIMENTS

Six experimental tests were performed, designated as Tests T1-S1, T2-S2, S3, S4, S5 and S6. Tests T1-S1 and T2-S2 included both transient and steady state experimental data while Tests S3, S4, S5 and S6 only recorded steady state data.

The experimental configuration for the heating tests performed is essentially the same as reported previously (Ewen and Thomas, 1989). For completeness, the layout is repeated in this paper, as Figure 1. The internal diameter, length and thickness of the glass cylinder are 219.6 mm, 522 mm and 5.3 mm respectively. The heater placed at the centre of the sample is 12.7 mm in diameter and 525 mm in length.

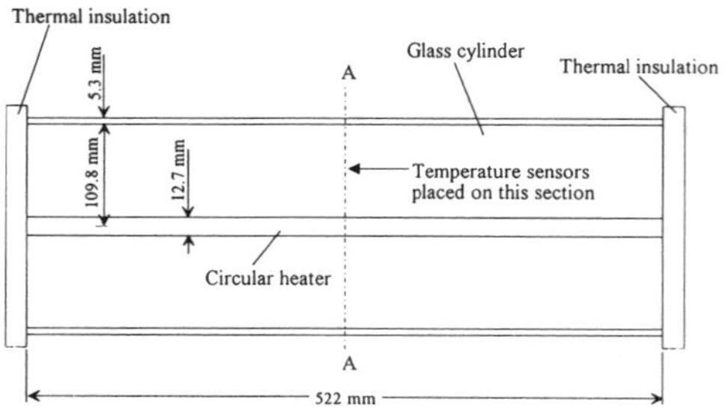

Fig. 1 The glass cylinder with circular heater in position.

Temperature measurements were taken by placing the temperature sensors on section A-A. For Tests T1-S1, T2-S2, S3, S4 and S5, three groups of nine temperature sensors were installed. Two groups were placed along the upper and lower vertical radial axis while one group was placed along the horizontal radial axis. Only one group of temperature sensors was used in Test S6, placed along the horizontal radial axis. In each of the experiments a temperature sensor recorded the room air temperature and a sensor was used to record the temperature of the outside wall of the glass cylinder.

Once the sample, prepared to a prescribed volumetric initial liquid content, was compacted into the glass cylinder, the entire assembly was laid horizontally in an environmentally controlled laboratory at 20 ± 1.5°C for several days to come to thermal and moisture equilibrium. During this time a non-uniform moisture content distribution would have been created due to gravitational effects. The circular heater was then switched on and the temperature of each temperature sensor recorded at selected intervals. The experiment was terminated when no further temperature rise was recorded. Hence, it was assumed that steady state conditions had been achieved.

The experiments were designed to cover a range of initial moisture contents, temperature changes and constant heating powers. Table 1 summarises the power input, the bulk dry density and the porosity for all the tests. Tests S3, S4 and S5 were the continuation of Test T2-S2 with different heating powers. The initial temperature and volumetric liquid contents for Tests S3, S4 and S5 were therefore the steady state conditions from Tests S2, S3 and S4 respectively.

Table 1 Summary of Test Configurations for the Laboratory Heating Experiments.

Test	Heating Power (Q_h) [W m^{-1}]	Bulk Dry Density (ρ_{dry}) [kg m^{-3}]	Porosity (η)
T1-S1	53.8	1,580	0.415
T2-S2	20.0	1,580	0.415
S3	30.0	1,580	0.415
S4	40.0	1,580	0.415
S5	53.8	1,580	0.415
S6	96.0	1,740	0.356

Experimental temperature results showed that the thermal response was radially one-dimensional for all the tests. It was found that, at any two temperature sensors at the same radius from the heat source, the worst instantaneous temperature difference was 0.7 K in a temperature rise of 10 K and 1.3 K in a temperature rise of 60 K.

The moisture distribution at steady state conditions for Tests T1-S1, S5 and S6 was visually examined. Two distinct regions, a dry region and a wet region, were found in the three tests. In addition, a detailed examination of the steady state moisture distribution in Test S5 was performed. Sub-samples were removed from several radii and their moisture content determined.

6 NUMERICAL MODELLING

The region of interest was the annulus of sand between the circular heater at $r = r_1$ and the inside wall of the sample glass cylinder at $r = r_2$. Since the solution in this region is symmetrical about the vertical line passing through the centre of the circular heater, only one half of the overall domain was modelled. The modelled region was divided into a 15x15 nodal array, i.e., 15 nodes in the radial direction and 15 nodes in the circumferential direction with each node placed at the centre of the mesh.

The numerical model was first used to simulate the behaviour of the soil as it reached thermal and moisture equilibrium. The results obtained at the end of this period were then taken to be the initial conditions for that heating test. The initial and boundary conditions for simulation are defined by

$$T = 293K, \theta = \theta_{ini}, when\ t = 0, r_1 \le r \le r_2, \quad -\frac{\pi}{2} \le \phi \le \frac{\pi}{2}$$
$$T = 293K, U = 0, V = 0, for\ t > 0, r = r_1\ and\ r_2, \quad \phi = -\frac{\pi}{2}\ and\ \frac{\pi}{2} \tag{20}$$

For the simulation of the heating experiments, as all the boundaries are impermeable to moisture flow, the moisture boundary conditions are defined by

$$U = V = 0\ \ for\ t > 0,\ r = r_1\ and\ r_2,\ \phi = -\frac{\pi}{2}\ and\ \frac{\pi}{2} \tag{21}$$

For the temperature boundary conditions at $\phi = -\pi/2$ and $\phi = \pi/2$ there can be no heat transfer across the line of symmetry. The thermal gradient at these boundaries can therefore be set to zero, i.e.,

$$\frac{\partial T}{\partial \phi} = 0\ \ for\ t > 0,\ \phi = -\frac{\pi}{2}\ and\ \frac{\pi}{2} \tag{22}$$

At the inner boundary where $r = r_1$ the circular heater is in contact with the soil sample. A subsidiary equation which describes the flow of heat from the heater into the soil is employed, as described previously (Ewen and Thomas, 1989). Thermal boundary conditions at the outer radius where $r = r_2$ are also as proposed previously (Ewen and Thomas, 1989). The rate of heat transfer through the glass cylinder wall to the room is set equal to the rate of thermal conduction in the layer of sand adjacent to the glass wall. An overall heat transfer coefficient (U_o) for thermal conduction through the glass wall and for the surface convection and radiation heat loss is thus defined.

7 COMPARISON OF EXPERIMENTAL AND SIMULATED RESULTS

A comparison of the simulated and the experimental temperature results for tests T2-S2, S3, S4 and S5 are presented. The values at 10 hours, 25 hours and steady state conditions of 313 hours for test T2-S2 are shown in Figure 2(a). A very good match between the numerical simulation and the experimental results can be seen. Assuming a Student's t-distribution of absolute errors the statistical figures for the profiles of 10 hours, 25 hours and 313 hours were computed. It was found that there is a 95% probability that the numerical simulation, at any radius, is within 0.56 K, 0.63 K and 0.28 K of the experimental measurement respectively for the profiles of 10 hours, 25 hours and 313 hours. Alternatively, the results of the numerical simulation are within an error of 4.92%, 4.58% and 1.54% of the experimental measured temperature range for the profiles of 10 hours, 25 hours and 313 hours respectively.

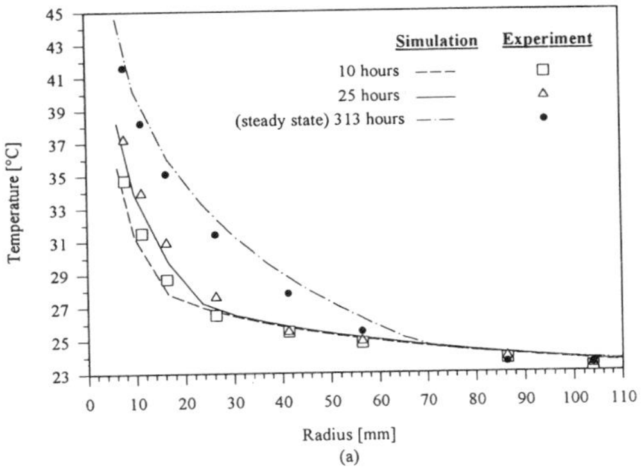

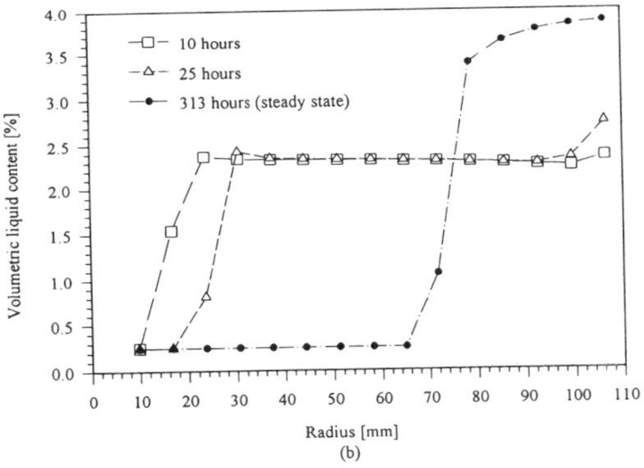

Fig 2: Test S2: (a) measured and simulated temperatures; (b) simulated volumetric liquid content distributions.

Figure 2(b) shows the simulated nodal volumetric liquid contents respectively at 10 hours, 25 hours and 313 hours. It can be seen that a dry/moist front moves away from the heat source. This is due to the fact that vapour transfer occurs from the hotter region to the colder region. The position of the dry/moist front for the steady state profile occurs at a radius between 65.0 mm and 78.8 mm.

The simulated temperature profile and the experimental temperature profile at steady state of 240 hours for test S3, are presented in Figure 3(a). Again, the simulated results can be seen to match the experimental data very well in both the general shape and magnitude. Assuming a Student's t-distribution of absolute errors, it has been calculated that there is a 95% confidence level that the numerical simulation, at any radius, is within 0.34 K of the experimental measurement or is within error of 1.13% of the experimental measured temperature range.

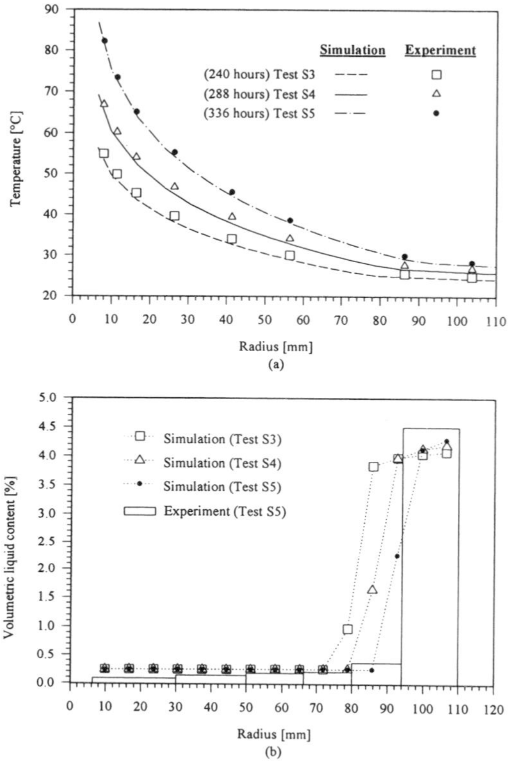

Fig 3: Tests S3, S4 and S5: (a) measured and simulated temperatures; (b) measured and simulated volumetric liquid content distributions.

The simulated nodal volumetric liquid contents at steady state conditions are shown in Figure 3(b). Comparing the results with those for Test T2-S2, it can be seen that the dry/moist front of the volumetric liquid content moves a radial distance of 6.9 mm. It now occurs at a radius between 71.9 mm and 85.7 mm.

Figure 3(a) also illustrates the steady state temperature results for both the simulation and the experiment for Test S4. An excellent agreement can be seen. Assuming a Student's t-distribution of absolute errors, it has been calculated that there is a 95% probability that, at any radius, the simulated results are within 0.35 K of the experimental measurement. Alternatively, there is a 95% probability that the numerical simulation is within an error of 0.87% of the experimental measured temperature range.

The simulated nodal volumetric liquid contents at steady state conditions are again included in Figure 3(b). The dry/moist front of the volumetric liquid content moves a radial distance of 6.9 mm further from the position of the simulation for Test S3. The position of the dry/moist front is now at a radius between 78.8 mm and 92.6 mm from the centre of the heater.

The simulated and experimental steady state temperature profiles for Test S5 are also given in Figure 3(a). It may be seen that the comparison between the results is again excellent. Assuming a Student's t-distribution of absolute errors at 95% confidence level, it has been calculated that the numerical simulation, at any radius, is within 0.25 K of the experimental measurement or is within an error of 0.47% of the experimental measured temperature range.

By increasing the heating power from Test S4, the dry zone of the sample is further increased and the dry/moist front moves a radial distance of 6.9 mm. It can be seen that most of the liquid has accumulated around the glass wall. The experimental profile displays a region of virtually dry soil between the circular heater and a radius of 94 mm. At this point, the volumetric liquid content of the soil changes from virtually dry (0.37%) to 4.5%. The dry/moist front for the simulation occurs at a radius between 85.7 mm and 99.5 mm or at an average radius of 92.6 mm. The positions of the two dry/moist front are therefore within 1.4 mm of each other. Expressing this value as a percentage of the full radius range, this represents an error of less than 1.35%. Such a small discrepancy indicates that the numerical simulation and the experimental results are in excellent agreement.

8 CONCLUSIONS

The work presented in this paper examines the validity of a theoretical formulation of coupled heat and mass transfer in non-deforming unsaturated soil. In particular results from the theoretical formulation were compared with those obtained from controlled laboratory heating experiments. The work considered one material, a highly compacted medium sand, for which the thermo/physical parameters had been determined independently of the heating experiments. A solution of the theoretical formulation was achieved by means of a numerical analysis, using the finite difference method.

It can be concluded that overall very good correlations have been achieved. Confidence in the ability of the theoretical formulation to describe the physical processes involved is thus enhanced.

9 ACKNOWLEDGEMENTS

The financial support of the former SERC, via research grant GR/F/60403, from the Geotechnics steering group of the civil engineering sub-committeee is gratefully acknowledged. Dr John Ewen and Dr Welkin Li's contribution to the work, whilst colleagues in the Cardiff School of Engineering, are also gratefully acknowledged.

10 REFERENCES

De Vries, D A (1958). Simultaneous transfer of heat and moisture in porous media. *Trans. Amer. Geophysical Union 39(5), 909-916.*

De Vries, D A (1987). The theory of heat and moisture transfer in porous media re-visited. *Int. J. of Heat Mass Transfer* 30(7), 1343-1350.

Ewen, J (1987). *Combined heat and mass transfer in unsaturated sand surrounding a heated cylinder.* Ph.D Thesis, University of Wales Cardiff, UK.

Ewen, J and Thomas, H R (1987). The thermal probe - A new method and its use on an unsaturated sand. *Géotechnique* 37(1),91-105.

Ewen, J and Thomas, H R (1989). Heating unsaturated medium sand. *Géotechnique* 39(3), 455-470.

Howells, P B and Marshall, R H (1983). An improved computer code for the simulation of solar heating systems. *Solar Energy* 30(2), 99-108.

Luikov, A V (1966*). Heat and mass transfer in capillary porous bodies.* Oxford: Pergamon Press.

Philip, J R and De Vries, D A (1957). Moisture movement in porous materials under temperature gradients. *Transactions, American Geophysical Union* 38(2), 222-232.

Roache, P J (1972*). Computation fluid dynamics.* Albuquerque, New Mexico: Hermosa Publishers.

Taylor, S A and Cary, J W (1964). Linear equations for the simultaneous flow of water and energy in a continuous soil system. *Soil Sci. Soc. Amer. Proc.* 28, 167-172.

Thomas, H R and Li, C L W (1997(a)). An assessment of a model of heat and moisture transfer in unsturated soil. *Geotechnique* 47(1), 113-131.

Thomas, H R and Li, C L W (1997(b)). Modelling heat and moisutre tranfer in unsaturated soil using a finite difference self-implicit method on parallel computers. *Int. J. Num. and Anal. Methods in Geomechanics,* 21, In press..

Numerical Models in Geomechanics, Pietruszczak & Pande (eds) © 1997 Balkema, Rotterdam, ISBN 90 5410 886 X

A new suction-based mathematical model for thermo-hydro-mechanical behavior of unsaturated porous media

B.Gatmiri
Ecole Nationale des Ponts et Chaussées, Paris & CERMES (Soil Mechanics Research Centre), Noisy-le-Grand, France & University of Tehran, Iran

M.Seyedi
University of Tehran, Iran

P.Delage
Ecole Nationale des Ponts et Chaussées, Paris & CERMES (Soil Mechanics Research Centre), Noisy-le-Grand, France

J.J.Fry
Centre National d'Equipement Hydraulique, Electricité de France

ABSTRACT: A set of fully coupled thermo-hydro-mechanical equations of unsaturated porous media behaviour are given. In this formulation, heat and moisture transfer equations in an alternative form based on water and air pore pressures are presented. The effects of deformations on the temperature and suction distribution in soil skeleton and the inverse effects are included in the formulation via a new temperature-dependent formulation of state surfaces of void ratio and degree of saturation. The nonlinear (hyperbolic) constitutive law is assumed. The mechanical and hydraulic properties of porous media are temperature-dependent. The effect of soil nonhomogeneity and phase changes are introduced. To the author knowledge, this alternative framework is the most complete formulation of non-isothermal behavior of unsaturated porous media under the temperature, loading and suction changes.

1. INTRODUCTION

In a multi-barrier waste disposal system, the canister and the metal containers form the first barrier. The surrounding sand annulus and bentonitic clay barrier as a part of near field and initial conditions for the best performance of host rocks which are the long term barrier against contamination should be received a considerable attention.

The first step in a theoretical development of a fully coupled thermohydromechanical model for an unsaturated soil, is choosing the adequate and independent variables which would be able to present all significant interaction effects among the different components involved in a coupled process in a deformable unsaturated porous medium with three phases (skeleton, water and air)under heating.

The phase changes between liquid and gas, evaporation, condensation, induced moisture transfer under thermal and pore pressure gradients and the effects of moisture distribution on the heat flow are important aspects in undeformable unsaturated porous media. If the deformation of porous media, which is significant in engineered clay barriers, is considered, the coupling effects among deformation, moisture, and heat should be also regarded in addition of all above aspects.

The theory of Philip and de Vries (1957) and de Vries (1958) is known as a basic frame and more comprehensive and representative theory of moisture and heat movement in an unsaturated and incompressible porous medium. In this theory the moisture and heat transfer equations are formulated in term of temperature (T) and volumetric moisture content (θ). In this paper this kind of formulation is named θ-based formulation.

The modeling of infinitesimal deformation of unsaturated porous media is another important part of establishment of a theoretical framework of non-isothermal behavior of such material. The previous works of authors in modeling of isothermal behavior of unsaturated soils (Gatmiri et al (1992) , Gatmiri (1993, 1994) have consisted the basis of development of the non-isothermal theory. This model has been developed along similar lines to that of Alonso et al. (1988) with difference on the relationship between stain increments and net stress increments and the assumption of hyperbolic variations for both K and E, bulk modulus and Young's modulus which has led to a new form of equation of void ratio state surface. This model is developed by using the two widely used independent variables, net stress and suction as the state variables in order to describe the water and air pore pressures distribution and deformation of skeleton. The coupling effects among suction-stress-deformation are taken into account by introducing the concept of void ratio and degree of saturation state surfaces.

Extending such isothermal theory of deformation of unsaturated soil to non-isothermal theory needs the combination of two above mentioned theories with a lot of important modifications in the basic concepts and formulation of both theories. Gatmiri(1997) has attempted to form a framework of non-isothermal behavior of unsaturated porous media. In this paper a summary of equations are given.

2. THE PROPOSED MODEL

In this model two basic theories are modified and combined in order to describe a fully coupled behavior of unsaturated porous medium under heating. In one hand the nonlinear theory of isothermal behavior of unsaturated soil under coupled effects of net stress and suction via the concept of state surfaces of void ratio and degree of saturation based on previous work of senior author (1992-1995) is modified and extended to non-isothermal behavior. In the other hand the Philip and de Vries' theory of heat and moisture transfer is modified in order to consider the deformation of skeleton and to be presented in a new form, suction-based formulation, which is more suitable for combination with deformation theory of unsaturated soils. The basic assumptions considered in this development are the followings :

-The medium is consisted of superposition of three continuos media

-The poroelastic medium of skeleton is isotropic and nonlinear.

-Quasi-static conditions and small transformation are considered.

-Fluids and solid grains are compressible.

-Energy transfer by all phases, and phase changes between liquid and gas are considered.

-Generalized Darcy's law is valid for motion of water and dry air.

-Fourier's law is considered for conductive heat flow.

-Solid and pore water densities are pressure and temperature-dependent.

-Air and water permeabilities depend on matric suction, strain level and temperature.

-Void ratio and degree of saturation state surfaces are temperature-dependent.

-Thermal expansion coefficient of mixture depends on suction, stress and temperature.

-Dissolving of air in water is considered

-Vapor pressure and thermal gradient are considered.

-The state variables are the net total stress σ-P_a, matric suction P_a-P_w and temperature T which leads, in numerical formulation, to five degree of liberty.

-Creep phenomenon is neglected.

The field equations can be formulated in following manner :

2.1 Total moisture transfer

The total moisture movement in unsaturated soil due to temperature gradient and its resulting moisture content gradient is equal to the sum of the flows which take place in both phases, vapor and liquid. Thus based on theory of Philip and de Vries(1957), the θ-based equation of total moisture transfer can be written as:

$$\frac{q}{\rho_w}=\frac{q_{vap}}{\rho_w}+\frac{q_{liq}}{\rho_w}=V+U=-\left(D_{Tv}+D_{Tw}\right)\nabla T \qquad (1)$$
$$-\left(D_{\theta v}+D_{\theta w}\right)\nabla\theta-D_w\nabla Z$$

which can be presented in following global form :

$$\frac{q}{\rho_w}=-D_T\nabla T-D_\theta\nabla\theta-D_w\nabla Z \qquad (2)$$

where D_T is thermal moisture diffusivity and is equal to D_{Tv}+D_{Tw},

D_θ is isothermal moisture diffusivity and is equal to $D_{\theta v}$+ $D_{\theta w}$.

D_{Tw} is thermal liquid diffusivity

$$D_{Tw}=K(\theta,T)\frac{\psi_r(\theta)}{\sigma_r}\frac{d\sigma(T)}{dT}$$

$D_{\theta w}$ is isothermal liquid diffusivity

$$D_{\theta w}=K(\theta,T)\frac{\sigma(T)}{\sigma_r}\frac{d\psi_r}{d\theta}$$

D_w is gravitational diffusivity, $D_w=K(\theta,T)$.

D_{Tv} is thermal vapor diffusivity,

$$D_{TV}=\frac{D_0}{\rho_w}.v.n.h\frac{(\nabla T)a}{\nabla T}.\frac{d\rho_0}{dT} \qquad \theta\langle n$$

$D_{\theta v}$ is isothermal vapor diffusivity

$$D_{\theta v}=\frac{D_0}{\rho_w}.v.n.\frac{\rho_0 hg}{RT}\frac{\partial\psi}{\partial\theta} \qquad \theta\langle n$$

and

$D_{TV}=0, \qquad D_{\theta v}=0 \qquad$ for $\qquad \theta=n$

in which

D_0 is the molecular diffusivity of water vapor in air, m^2 sec

ν is the « mass flow factor » introduced to allow for the mass flow of vapor arising from the difference in boundary conditions governing the air and vapor components of the diffusion system (cited from Philip and de Vries(1957)), it is given by the expression : $\nu=P_g/(P_g-p_v)$ where P_v is partial pressure vapor. ν is quite close to unity at normal soil temperature and its deviation from unity can be calculated by above formula.

α is a tortuosity factor

a is volumetric air content which can be expressed by $a=n(1-S_r)$ with n : porosity and S_r : degree of water saturation.

ρ_v is the density of water vapor, kgm / m^3

$$\rho_v = \rho_0 . h \qquad (3)$$

$$h = \exp\left[\psi . g / R.T\right] \qquad (4)$$

where ρ_0 is the density of saturated water vapor, kgm / m^3

h is relative humidity, and R is the gas constant

ψ is thermodynamic potential of water in soil.

$((\nabla T)_a/\nabla T)$ is the ratio of microscopic temperature gradient in pore space to macroscopic temperature gradient (Philip and de Vries (1957)).

2.2 Moisture mass conservation

Considering the moisture as the sum of liquid and vapor phases, the differential equation of moisture transfer can be written as :

$$\frac{\partial \rho_m}{\partial t} = -\mathrm{div}\left(\rho_w (U + V)\right) \qquad (5)$$

in which

$$\rho_m = \theta \rho_w + (n-\theta)\rho_v = nS_r\rho_w + n(1-S_r)\rho_v \qquad (6)$$

After development of the left-hand side and right-hand side terms of general equation Gatmiri(1997) has given a new suction-based formulation of moisture movement as :

$$nS_r\beta_T \frac{\partial T}{\partial t} + nS_r\beta_P \frac{\partial p_w}{\partial t} + (\rho_w - \rho_v) n \frac{\partial S_r}{\partial t} + \left(S_r\rho_w + \rho_v(1-S_r)\right)\frac{\partial n}{\partial t} + n(1-S_r)\frac{\partial \rho_v}{\partial t} =$$

$$\mathrm{div}(\rho_w D_w \nabla z) + \mathrm{div}(\rho_w D_T \nabla T) + \mathrm{div}\left(\rho_w D_p \nabla (P_w - P_g)\right) + Q_m \qquad (7)$$

where

$$D_{Pw} = K(\theta, T)\frac{\sigma(T)}{\sigma_r . \gamma_w} \qquad (8)$$

$$D_{Pv} = \frac{D_0}{\rho_w} \nu . n \frac{\rho_v g}{RT.\gamma_w} . \frac{\sigma(T)}{\sigma_r} \qquad (9)$$

$$D_P = D_{Pw} + D_{Pv} \qquad (10)$$

in which surface energy can be written as :

$$\sigma(T) = -75.882 + 0.165T \qquad (11)$$

where T is temperature in degree Celcius and σ is surface energy in dyne/cm^2.

It is important to note that in derivation of this recast form of Philip and de Vries theory, soil heterogeneities are considered and this form is more suitable for combination with the theory of deformation of unsaturated media.

2.3 gas flow equation

Considering that P_g is a function of temperature ($P_g = P_g(T)$), Darcy's law is valid, the gas flow equation can be written as

$$V_g = \frac{-K_g}{\gamma_g}\frac{\partial P_g}{\partial T}\nabla T - K_g\left(\nabla\left(\frac{P_g}{\gamma_g}\right) + \nabla Z\right), \qquad (12)$$

where

V_g is vector of gas velocity, q_g is vector of gas flow, ρ_g is density of gas, K_g is gas permeability, P_g is gas pressure, γ_g is specific weight of gas ∇Z is elevation.

with

$$K_g = \frac{b\gamma_g}{\mu_g}\left[e(1-S_r)\right]^C \qquad (13)$$

a, b are the constants. In equation 13 , gas permeability is assumed to be a function of water content. It does not depend on temperature.

2.4 Mass Conservation of gas

The governing differential equation of mass conservation of air in a control volume of an unsaturated porous media can be given as

$$\frac{\partial}{\partial t}\left[n\rho_g\left(1-S_r+HS_r\right)\right]=-\mathrm{div}\left(\rho_g V_g\right)-\mathrm{div}\left(\rho_g HU\right)+\rho_w \mathrm{div}V \qquad (14)$$

where H is Henry constant which is equal to 0.02.

The general differential equation of gas movemoent in unsaturated base in suction-based form has bben found by Gatmiri(1997) as:

$$\rho_g\left(1-S_r(1-H)\right)\frac{\partial n}{\partial t}+n\left(1-S_r(1-H)\right)\frac{\partial \rho_g}{\partial t}-(1-H)n\rho_g\frac{\partial S_r}{\partial t}=$$
$$\mathrm{div}\left[\left(K_g\rho_g\beta_{Pg}+H\rho_g D_{Tw}-\rho_w D_{TV}\right)\nabla T\right]+\mathrm{div}\left[\left(\frac{K_g\rho_g}{\gamma_g}-H\rho_g D_{Pw}+\rho_w D_{PV}\right)\nabla P_g\right]$$
$$+\mathrm{div}\left[\left(H\rho_g D_{Pw}-\rho_w D_{Pv}\right)\nabla P_w\right]+\mathrm{div}\left[\left(K_g\rho_g+H\rho_g D_w\right)\nabla Z\right]$$

2.5 Energy conservation equation:

Energy conservation equation in a porous medium can be expressed by

$$\frac{\partial\varphi}{\partial t}+\mathrm{div}Q=0 \qquad (16)$$

in which Q is heat flow and φ is the volumetric bulk heat content of medium which can be defined by

$$\varphi=C_T\left(T-T_0\right)+(n-\theta)\rho_v h_{fg} \qquad (17)$$

where C_T is the specific heat capacity of unsaturated mixture and can be written as

$$C_T=(1-n)\rho_s C_{PS}+\theta\rho_w C_{Pw}+(n-\theta)\rho_v C_{PV}+(n-\theta)\rho_g C_{Pg} \qquad (18)$$

Total flow of latent and sensible heat in an unsaturated porous medium is given based on Philip and de Vries theory as :

$$Q=-\lambda \mathrm{grad}T+\rho_w h_{fg}V+\rho_v V_g h_{fg}+\left[C_{Pw}\rho_w U+C_{pv}\rho_w V+C_{pg}\rho_g V_g\right]\left(T-T_0\right) \qquad (19)$$

where λ accounts for Fourier heat diffusion coefficient and can be evaluated by following proposition :

$$\lambda=(1-n)\lambda_s+\theta\lambda_w+(n-\theta)\lambda_v \qquad (20)$$

Through this equation which gives the upper limit of heat conductivity in unsaturated porous media, the continuity between saturated and unsaturated case is ensured.

ρ_s is density of solid grain, ρ_w is density of liquid, ρ_v is density of vapor, ρ_g is density of gas

C_{PS} is specific heat capacity of solid, C_{PW} is specific heat capacity of liquid, C_{PV} is specific heat capacity of vapor, C_{Pg} is specific heat capacity of gas, T_0 is an arbitrary reference temperature and, h_{fg} is latent heat of vaporization.

The final differential form of heat flow equation in unsaturated porous media has been given by Gatmiri(1997) as :

$$C_T\frac{\partial T}{\partial t}+\left(T-T_0\right)\frac{\partial C_T}{\partial t}+\left(1-S_r\right)\rho_v h_{fg}\frac{\partial n}{\partial t}-n\rho_v h_{fg}\frac{\partial S_r}{\partial t}+n\left(1-S_r\right)h_{fg}\frac{\partial \rho_v}{\partial t} \quad (15$$
$$-\mathrm{div}[\lambda(\theta)\nabla T]+C_{Pw}\rho_w\mathrm{div}\left[\left(-D_{Tv}\nabla T-D_{Pv}\nabla\left(P_w-P_g\right)\right)\left(T-T_0\right)\right]+$$
$$C_{Pv}\rho_w\mathrm{div}\left[\left(-D_{Tv}\nabla T-D_{Pv}\nabla\left(P_w-P_g\right)\right)\left(T-T_0\right)\right]+$$
$$C_{Pg}\mathrm{div}\left[\left(-\rho_g K_g\beta_{Pg}\nabla T-\rho_g K_g\left(\frac{\nabla\rho_g}{\gamma_g}+\nabla Z\right)\right)\left(T-T_0\right)\right]$$
$$+h_{fg}\mathrm{div}\left[-\rho_v K_g\beta_{Pg}\nabla T-\rho_v K_g\left(\frac{\nabla P_g}{\gamma_g}+\nabla Z\right)\right]+$$
$$\rho_w h_{fg}\mathrm{div}\left[-D_{Tv}\nabla T-D_{Pv}\nabla\left(P_w-P_g\right)\right]=0 \qquad (21)$$

2.6 Solid skeleton behavior

Considering the two stress state variables as suction and net stress, equilibrium equation and constitutive law for a non-isothermal isotropic and non-linear case based on isothermal equations can be written as follows :

2.7 Equilibrium equation :

$$(\sigma_{ij}-\delta_{ij}p_g)_{,j}+p_{g,j}+b_i=0 \qquad (22)$$

2.8 *Incremental constitutive law*

The constitutive law for solid skeleton in saturated soil which is under suction and thermal effect assuming the small deformation can be written as :

$$d(\sigma_{ij} - \delta_{ij} p_g) = D d\varepsilon - F d(p_g - p_w) - C dT \quad (23)$$

$$F = DD_s^{-1} \text{ with } D_s^{-1} = \beta_s m$$

$$\text{in which } \beta_s = \frac{1}{1+e}\frac{\partial e}{\partial (p_g - p_w)}$$

$$C = DD_t^{-1} \text{ with } D_t^{-1} = \beta_t m \quad (24)$$

$$\text{in which } \beta_t = \frac{1}{1+e}\frac{\partial e}{\partial T}$$

$$\text{and} \quad m = \begin{bmatrix} 1 & 1 & 0 \end{bmatrix}$$

D is non-linear and temperature-dependent.

2.9 *Thermal void ratio state surface :*

In order to calculate the bulk modulus, the volumetric strain can be taken into account via void ratio state surface which depends on stress, suction and temperature. Using the same approach for determination of state surface of void ratio which is used by Gatmiri (1994), and Gatmiri and Delage (1995), a new formulation of void ratio state hyper surface is proposed as following:

$$e = \frac{(1+e_0)\exp[-c_e(T-T_0)]}{\exp\left[\dfrac{[a(\frac{\sigma - p_g}{p_{atm}}) + b(1 - \frac{\sigma - p_g}{\sigma_c})(\frac{p_g - p_w}{p_{atm}})]^{1-m}}{K_b(1-m)}\right]} - \quad (25)$$

Through this equation the compatibility with non-linear behavior of soil is ensured.

The bulk modulus B can be defined by :

$$B = (1+e_0)\left(\frac{\partial e}{\partial(\sigma - P_g)}\right)^{-1} \quad (26)$$

thus

$$D = D\,(B, E) = D\,(\sigma - P_g,\ P_g - P_w, T) \quad (27)$$

2.10 *Thermal degree of saturation state surface :*

Besides stress-strain behavior coupled with temperature, the description of coupled state of volumetric moisture content with temperature of unsaturated soil is also necessary under the stress and suction effects. On the basis of experimental data the following state surface of degree of saturation is proposed :

$$sr = 1 - \frac{[a_s + b_s(\sigma - p_g)][1 - \exp(c_s(p_g - p_w))]}{\exp(-d_s(T - T_0))} \quad (28)$$

a_s, b_s, c_s and d_s are constants.

3. SOLUTION APPROACH AND FINITE ELEMENT DISCRETIZATION

Application of weighted residual method and the Galerkin choice of weighted functions to the equations represented in the terms of nodal point values of the field variables for the total spatial discrete form of domain Ω, and the single-step integration in time has resulted in a global matrix form which is encoded in a particular purpose finite element package.

4. APPLICATION AND RESULTS

The laboratory experiment reported by Villar et al (1993) has been considered, but all of parameters involved in this model have not been mesured. These parameters are chosen in a logic range for a clay. Thus the results can be qualitatively representative. In the upper part of this cell, a heat source at 100°C is located. The initial degree of saturation was 0.5. The mesh, details of parameters and boundary conditions are given by Gatmiri (1997). The contours of degree of saturation and temperature distributions after 20 hours are presented in figure 1 and 2.

These figures show that with heat increasing, water content near the heat source deacreases and suction increases. The degree of saturation in the lower part of the cell increases. This results agrees quantitatively very well with the experimental results. This result has not been found by other models. A good agreement is found for temperature distibution within the cell.

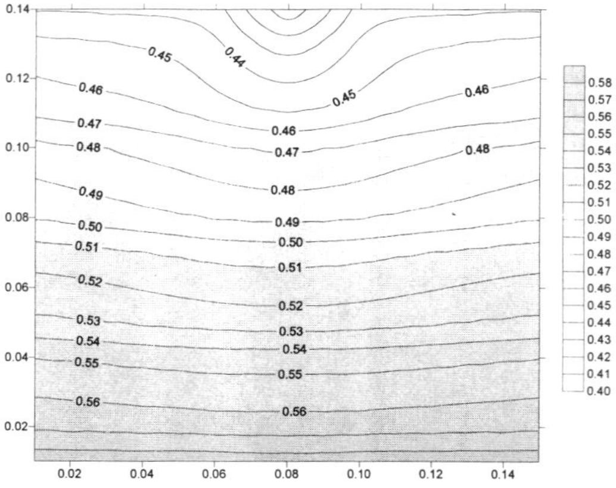

FIG. 1 : DEGREE OF SATURATION CONTOURS

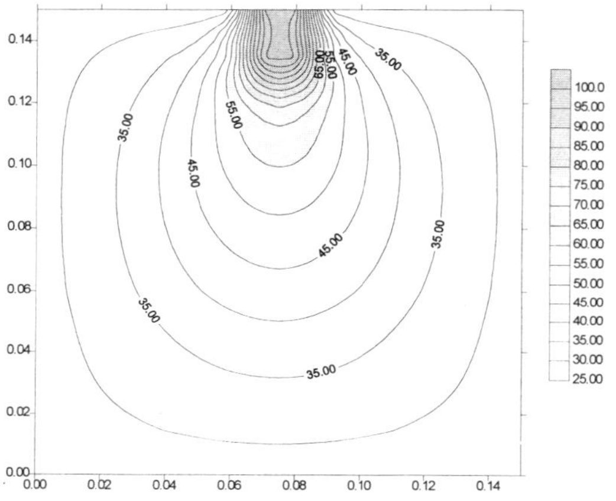

FIG. 2 : TEMPERATURE CONTOURS

5. CONCLUSIONS

In unsaturated field, according to author's knowledge, a new theoretical framework for analysis of fully coupled moisture, heat, gas and skeleton deformation is established. The new theoretical formulation is a combination of two extended theories, the first part is an extension of moisture transfer theory of Philip and de Vries. The second part is the extension of the isothermal deformation theory of unsaturated soil to thermal effects. A complete set of equations is presented in which the suction-based equations of moisture, heat and air transport are combined with the equilibrium equation of solid skeleton and constitutive law relation with void ratio and degree of saturation hyper state surfaces. These state surfaces are temperature-dependent.

A good agreement between the predictions of this new model and experimental results reported by others has shown the strong ability of this new model which is encouraging.

6. REFERENCES

Alonso. E.E, Battle F., Gens A. and Lloret A. 1988 Consolidation analysis of partially saturated soils. Application to earth dam construction. *Proc. Int. Conf. on Numerical Methods in Geotechnics. Innsbruck*, 1303-1308.

de Vries D.A. 1958 `Simultaneous transfer of heat and moisture in porous media, Trans. Am. Geophys. Un. 39, No. 5, 909-916.

Gatmiri B., Delage P.and A. Nanda 1992 'Consolidation des sols non saturés: simulation des essais au laboratoire-application aux barrages en remblai', Rapport du CERMES-ENPC.

Gatmiri B. 1993 'Validation du code U-DAM'. Rapport du CERMES, ENPC.

Gatmiri B. 1994 'Surfaces d'états et déformation des barrages en remblai avec drains sous pression négative', Rapport du CERMES, ENPC.

Gatmiri B. and P. Delage, 1995 'Nouvelle formulation de la surface d'état en indice des vides pour un modèle non linéaire élastique des sols non saturés', *Proc. 1st Int. Con. Unsaturated Soils,* **2,** 1049-1056.

Gatmiri B.and P. Delage, 1997 'A formulation of fully coupled thermal-hydraulic-mechanical behavior of saturated porous media - numerical approach', *Int. j. numer. anal. methods geomech* , **21(3),** 199-225.

Gatmiri B. 1997 'Analysis of fully coupled behaviour of unsaturated porous media under stress, suction and temperature gradient', Final report of CERMES-EDF.

Philip J.R & D.A. de Vries, 1957 'Moisture Movement in porous materials under temperature gradents', Trans. Am. Geophys. Un. 38, 222-232

Villar M. V., J. Cuevas, A.M. Fernandez and P.L. Martin, 1993 'Effects of the interaction of heat and waler flow in compacted bentonite', *Proceedings of International workshop thermomechanics of clays and clay barriers, Bergamo : ISMES.*

Numerical Models in Geomechanics, Pietruszczak & Pande (eds) © 1997 Balkema, Rotterdam, ISBN 90 5410 886 X

A strategy for numerical analysis of the transition between saturated and unsaturated flow conditions

J.Vaunat & A.Gens
Escuela Técnica Superior de Ingenieros de Caminos, Canales y Puertos de Barcelona, Spain

C.Jommi
Politecnico di Milano, Italy

ABSTRACT: A strategy to cope with saturation/desaturation problems in porous media is presented. The strategy consists of introducing in the governing equations terms related to the dissolution process of gas into liquid. The numerical formulation is presented and two cases involving saturation and desaturation are simulated. Validity and limitation of the method are then discussed and the extension to the limit case of immiscible species is studied. In the latter case, an equation which expresses the memory of the saturation condition is proposed to replace mass balance of the vanishing phase in the saturated zone.

1. INTRODUCTION

Numerical models of multiphase flow in porous media generally focus either on saturated conditions, where the pores are filled by a single phase (see for example Advani *et al.*, 1993), or on unsaturated conditions where two phases are present in the medium (see for example Thomas & King, 1992). Implementation of multiphase flow when local saturated and unsaturated conditions coexist in the mesh is still a challenging point (Gawin & Schrefler, 1996). In the following text, this type of situation will be referred to as the *transition problem.*

Two major difficulties arise when dealing with transition problem. On the one hand, the mechanical behaviour of saturated media is controlled by one stress dimension variable —the effective stress— whether two independent stress dimension variables, combination of total stress, liquid pressure and gas pressure are responsible of the deformational behaviour of unsaturated media. A switch between both sets of variable has then to be introduced, potentially leading to oscillations and convergence problems. From the point of view of classical F.E. implementation, this problem requires an adequate treatment of secondary variables. Various solutions exist and are discussed by Vaunat *et al.* (1997).

On the other hand, the gas mass balance equation degenerates into a trivial equality (0 = 0) when gas phase vanishes. The global matrix of the discrete equation system may then become singular. This problem involves primary variables, generally harder to manipulate. In this paper, a strategy is presented to cope with this difficulty. It consists in considering the process of dissolution of gas into liquid as a way to regularize the system of equations.

2. GOVERNING EQUATIONS

Governing equations are written using a compositional approach (Panday & Corapcioglu, 1989, Olivella *et al.*, 1994). This approach consists in writing mass balance, not for each phase, but for each species present at a given time t in the porous medium. Since a species i can exist in various phases ϕ, a generic form of the mass balance equation is:

$$\frac{\partial \sum_\phi m^{i\phi}}{\partial t} + div\left(\sum_\phi \mathbf{q}_m^{i\phi}\right) + \sum_\phi Q_m^{i\phi} = 0 \quad (1)$$

where $m^{i\phi}$ is the mass of species i in phase ϕ *per* unit volume of porous medium, $\mathbf{q}_m^{i\phi}$ is the mass flux of species i in phase ϕ *per* unit area of porous medium and $Q_m^{i\phi}$ is a source/sink term of mass of species i in phase ϕ. $m^{i\phi}$ can be written as $\omega^{i\phi}\rho^\phi S_v^\phi$ where $\omega^{i\phi}$ is the mass fraction of species i in phase ϕ, ρ^ϕ is the density of phase ϕ and S_v^ϕ the volume of phase ϕ *per* unit volume of porous medium. With this approach, the mass flux $\mathbf{q}_m^{i\phi}$ has to be split generally into an advective term $\mathbf{q}_m^{i\phi} = \omega^{i\phi}\rho^{\phi}S_v^{\phi}\mathbf{q}^{*\phi}_v$ and a non advective term $\mathbf{i}_m^{\phi}$ which expresses the mass movement of specie i within phase ϕ. $\mathbf{q}^{*\phi}_v$ is the real velocity of fluid ϕ and is related to Darcy's flux $\mathbf{q}_v^{\phi}$ by $\mathbf{q}_v^{\phi} = S_v^{\phi}(\mathbf{q}^{*\phi}_v - \mathbf{q}^{*s}_v)$, where $\mathbf{q}^{*s}_v$ is the real velocity of solid particles. Generalized Darcy's law (Bear, 1972):

$$\mathbf{q}_v^{\phi} = \frac{K_i K_r^{\phi}}{\mu^{\phi}}\left(\nabla p^{\phi} + \rho^{\phi}\mathbf{g}\right) \tag{2}$$

is used to relate advective fluxes $\mathbf{q}_v^{\phi}$ to phase pressure p^{ϕ}.

For the sake of clarity, a simple case will be considered in which balance equations are reduced to the terms relevant for the solution of transition problem. In this case, two phases ($\phi = g$ for gas and $\phi = l$ for liquid) and two species ($i = a$ for air and $i = w$ for water) are present in the pores. Gas contains only dry air while liquid phase is an ideal dilute solution of air into water. No source/sink terms $Q_m^{i\phi}$ are considered. Non advective fluxes $\mathbf{i}_m^{\phi}$ are not taken into account.

2.1 *Air mass balance*

In the simplified case, S_v^g and S_v^l are respectively equal to $n(1-S_r)$ and nS_r, where n is the porosity and S_r the degree of saturation. Air mass balance then becomes:

$$\frac{\partial\left(n\rho^g\omega^{ag}(1-S_r) + n\rho^l\omega^{al}S_r\right)}{\partial t} + div\left(\rho^g\omega^{ag}\mathbf{q}_v^g + \rho^l\omega^{al}\mathbf{q}_v^l\right) = 0 \tag{3}$$

ω^{ag} is equal to 1 and gas phase variables p^g, ρ^g, μ^g and K_r^g are identical to air phase variables p^a, ρ^a, μ^a and K_r^a. ω^{al} is assumed governed by Henry's law:

$$\omega^{al} = \frac{M^a}{K_d^a M^w}p^a \tag{4}$$

where M^a and M^w are respectively the molecular mass of air and water, p^a is the air pressure above the solution and K_d^a is a constant. Combining Eq. 4 with perfect gas law, the classical relationship:

$$\frac{\rho^w\omega^{al}}{\rho^a} = \frac{\rho^w M^a}{K_d^a M^w}\frac{RT}{M^a} = H\left(\rho^w, T\right) \tag{5}$$

can be recovered, where R is the perfect gas constant and T the temperature. For isothermal processes and incompressible water $\rho^w\omega^{al}/\rho^a$ is a constant called Henry's constant. Introducing Eq. 5 into Eq. 3, a simple form of air mass balance results:

$$\frac{\partial\left(n\rho^a(1-S_r+HS_r)\right)}{\partial t} + div\left(\rho^a\left(\mathbf{q}_v^g + H\mathbf{q}_v^l\right)\right) = 0 \tag{6}$$

2.2 *Water mass balance*

In the simplified case, $\omega^{wg} = 0$, $\omega^{al} \ll 1$ and $\omega^{wl} \simeq 1$. As a consequence, liquid phase variables p^l, ρ^l, μ^l and K_r^l can be approximated by water phase variables p^w, ρ^w, μ^w and K_r^w. Water mass balance can then be written as:

$$\frac{\partial\left(n\rho^w S_r\right)}{\partial t} + div\left(\rho^w\mathbf{q}_v^l\right) = 0 \tag{7}$$

2.3 *Extension to saturated conditions*

In saturated conditions, water mass balance (Eq. 7) remains unchanged. Air mass balance can not *a priori* be expressed by Eq. 6, since Henry's law is not valid when no free air exists above the solution. Going back to Eq. 3 with $\omega^{ag} = 0$, air mass balance can be rewritten in saturated conditions as:

$$\frac{\partial\left(n\rho^l\omega^{al}S_r\right)}{\partial t} + \rho^l\omega^{al}div\left(\mathbf{q}_v^l\right) = 0 \tag{8}$$

An artifice consists in defining a dummy air pressure having the same linear relationships with respect to ω^{al} and ρ^a as in unsaturated conditions. The air mass balance recovers then the same form as Eq. 6. However, in this case, air pressure has to be related to dissolved air density and not to gas pressure. The criterion that defines the transition between the two conditions is $p^l = p^g$ on the unsaturated side and $p^l = \omega^{al}K_d^a M^w/M^a$ on the saturated side, which reduces in both cases to $p^w = p^a$. With this formulation, air pressure will evolve independently of water pressure either in saturated and unsaturated conditions.

2.4 *Stress equilibrium*

The stress equilibrium equation is:

$$\frac{\partial\sigma_{ij}}{\partial x_j} + b_i = 0 \tag{3}$$

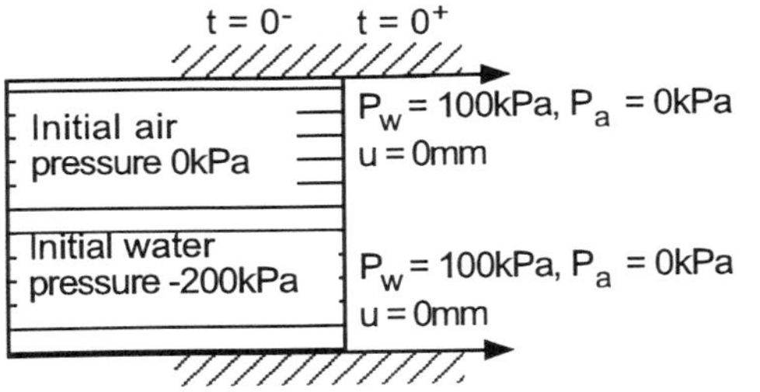

Figure 1: Swelling pressure test: problem definition

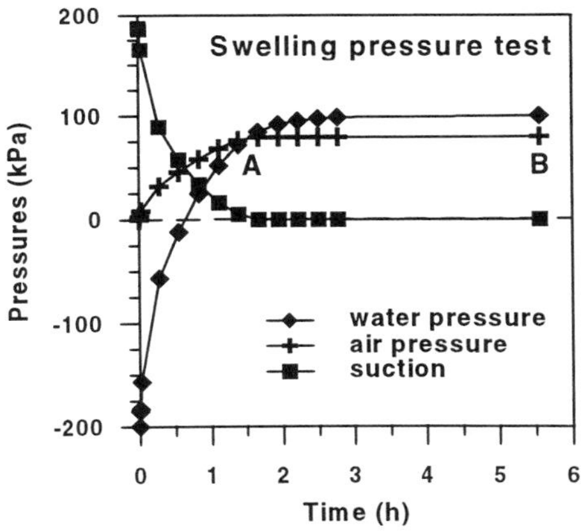

Figure 2: Swelling pressure test: time evolution of water pressure, air pressure and suction

where $\boldsymbol{\sigma}$ is Cauchy's stress tensor and $\mathbf{b}$ the vector of body forces. In unsaturated conditions, net stress concept $\sigma_{ij}^{n} = \sigma_{ij} - \delta_{ij} p^{g}$ is used, while in saturated conditions effective stress concept $\sigma_{ij}^{\prime} = \sigma_{ij} - \delta_{ij} p^{l}$ is implemented. The strategy to perform the switch between the adopted stress variables consists in defining the net stress as $\sigma_{ij} - \delta_{ij} p^{l} - \delta_{ij} s$, where suction $s = p^{g} - p^{l}$, when $p^{g} - p^{l} > 0$ and $s = 0$, when $p^{g} - p^{l} \leq 0$. A discontinuity in the derivatives of suction occurs then at $p^{g} = p^{l}$. To avoid convergence problems, a smoothing of these derivatives at $p^{g} = p^{l}$ is performed when computing the Jacobian matrix (Vaunat *et al.*, 1997).

2.5 *Numerical implementation*

From this set of partial differential equations, a F.E. model is built. Space is discretized using Standard Galerkin's method. Main unknown are air pressure, water pressure and displacements. Linear and quadratic interpolations are implemented for pressures and displacements respectively. Time integration is performed by an implicit Finite Difference scheme. Newton-Raphson procedure copes with non linearities. A mass conservative approach is used for time integration of storage terms. The velocities of solid particles and the convective terms $\mathbf{q}_{v}^{\phi} \nabla \left(\omega^{i\phi} \rho^{\phi}\right)$ are assumed negligible.

3. SIMULATION RESULTS

3.1 *Case A: Swelling pressure test*

An initially unsaturated sample at a suction of 200kPa is wetted on both sides at a positive water pressure (100kPa). The initial degree of saturation is very high (98%) to allow dissolution of all the air mass initially present in the medium. Null vertical displacements are prescribed at the top and at the bottom of the sample. Oedometer conditions are prescribed on the lateral sides. A linear elastic soil skeletton is assumed. Initial and boundary conditions are summarized in Fig. 1.

Time evolution of air and water pressure at the middle of the sample is shown in Fig. 2. During a first stage, gas pressure increases within the sample as a results of the increasing pressure exerted by the liquid phase. During this process, more and more free air dissolves into water. At point A, all air has dissolved and only the liquid phase is present in the medium: full saturation takes place. At B, water flow reaches steady state conditions. From point A to point B, air pressure remains constant, since the density of dissolved air in liquid phase remains constant. Air pressure value is then smaller than water pressure value in the saturated zone. If the process is reversed from point B, the sample remains locally saturated up to point A. At this point, water pressure passes below air pressure and desaturation is triggered by air release. It is worth noting that no arbitrary assumption is made on the value of air pressure in the saturated zone. Saturation as well as desaturation processes are naturally reproduced.

Let us consider now the case of a desaturation process starting from initially saturated conditions (point B in Fig. 2). In this case, conditions at which saturation occurred (point A in Fig. 2) are not known. Initial values prescribed for air pressure, which give the initial density of dissolved air and consequently the initial mass of dissolved air available for release,

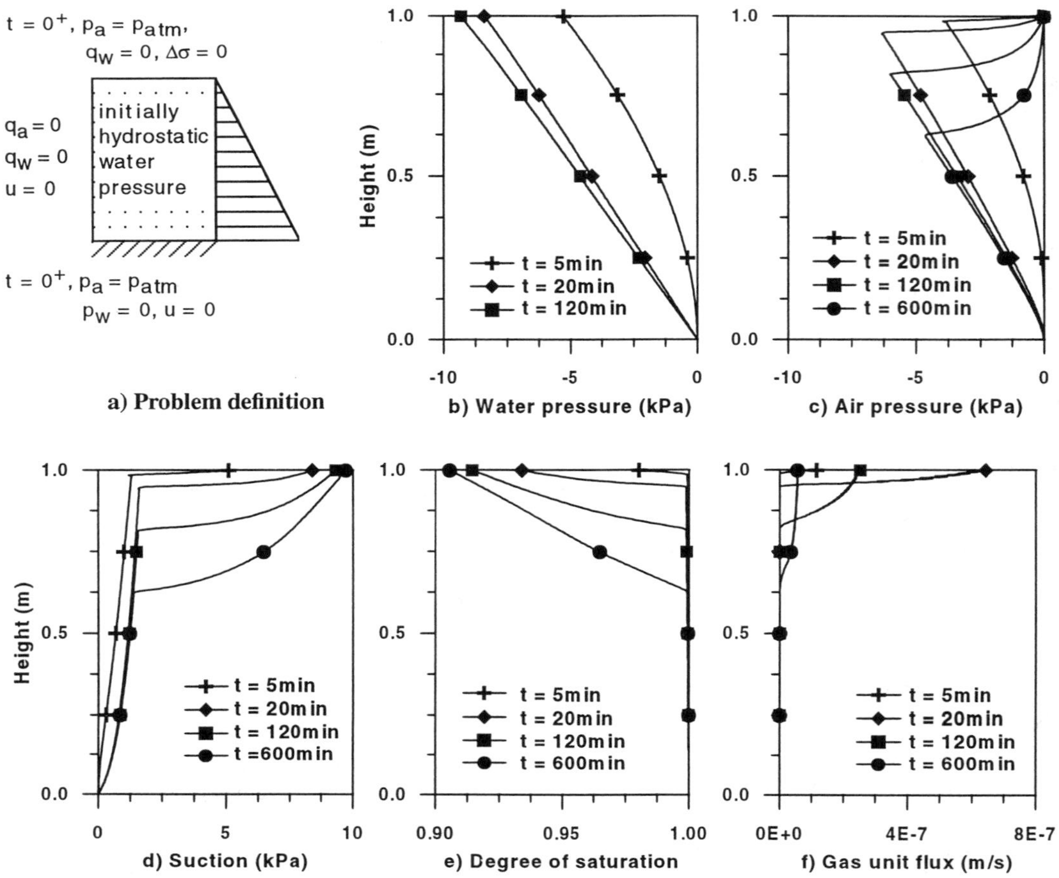

Figure 3: Desaturation of a sand column: problem definition and solution obtained.

will govern the moment of desaturation. Since this initial condition depends on the history of saturation, it is difficult to assess with physical confidence. Some consequences of this aspect are discussed with reference to the results of the second test presented here.

3.2 *Case B: Desaturation of a sand column*

A benchmark exercise proposed within program 4 of the European Network "ALERT Geomaterials" is reproduced. It consists in the drainage by gravity of a 1m high sand column. At $t = 0^-$, hydrostatic water pressure is imposed; at $t = 0^+$, water pressure at the bottom is put equal to atmospheric pressure. Air pressure is equal to atmospheric pressure at both sides of the sample. At the top, a condition of null vertical load is imposed. At the bottom and on the lateral sides of the column, no normal displacements are allowed. Benchmark specifications are depicted in Fig. 3.a. Details of constitutive laws and parameters used can be found in Jommi *et al.* (1997).

Within the framework of the benchmark exercise, Dangla *et al.* (1996) give a mathematical justification of the existence of a saturation front during drainage, which splits the column into a saturated part and an unsaturated part at each time t. They show that the propagation of the front is controlled by the behaviour of the relative air permeability K_a and of the derivative of retention curve dS_r/ds when S_r tends to zero.

Results are depicted in Fig. 3. Water pressure isochrones (Fig. 3.b) show that steady state conditions are reached very quickly for the liquid phase. Suction (Fig. 3.c) equilibration is much slower. In fact, at high degrees of saturation, air permeability may be of various orders of magnitude lower than water permeability and suction evolution is controlled by air pressure evolution. In Fig. 3.c, the propagation of a discontinuity in air pressure gradient can be observed, suggesting that the saturation front evidenced by Dangla *et al.* (1996) is well-reproduced. However, isochrones of suction and degree of saturation (Fig. 3.d and Fig. 3.e) show that the discontinuity does not delimitate a boundary between saturated and unsaturated zone, but occurs at a degree of saturation smaller than one. This can be explained looking at the isochrones of gas (or free air) flux presented in Fig 3.f. They indicate that the discontinuity is located at the point where gas flux becomes equal to 0. This means that the air mass present below the discontinuity is in fact immobile. Two processes appear then to be responsible of the desaturation of the column. On the one hand, a flux of free air is generated from the top of the sample to compensate the outflow of water at the bottom of the sample. On the other hand, the decrease of water pressure triggers a simultaneous release of air in all the saturated part, leading to a general desaturation of the sample. However, since dissolved air is released at the same pressure at all the points, no gradient of gas pressure is built up during this phenomenon, preventing an inflow of gas flux from the upper part of the sample and generating a discontinuity in the gradient of air pressure, suction and degree of saturation.

As a conclusion, results of case B suggest that if suction and degree of saturation are sensitive to the initial content of dissolved air, gas flow pattern is not affected by this condition. As a consequence, the overall pattern of desaturation is respected, particularly as concerns the presence or not of a "saturation" front. As a general rule, it is therefore important to analyse the results not in terms of total air flow, which does not correspond to gas flow, but in terms of free air flow.

4. EXTENSION TO IMMISCIBLE SPECIES

The expression of the air mass balance equation in the saturated conditions (Eq. 8) suggests that the case of immiscible species can be handled as the limit case of miscible species when $H \rightarrow 0$. In fact, considering the simplications defined in paragraph 2 and H constant, Eq. 8 becomes in saturated conditions:

$$H\frac{\partial\left(n\rho^{a}\right)}{\partial t}+Hdiv\left(\rho^{a}\mathbf{q}_{v}^{l}\right)=0 \tag{10}$$

Dividing the left hand term by H, Eq. 10 leads to:

$$\frac{\partial\left(n\rho^{a}\right)}{\partial t}+div\left(\rho^{a}\mathbf{q}_{v}^{l}\right)=0 \tag{11}$$

It is then sufficient to perform a switch between Eq. 10 and Eq. 11 at the nodes where full saturation is detected to extend the numerical model to the case of immiscible species.

The physical meaning of Eq. 11 can be derived in the case of a rigid medium. Multiplying water mass balance (Eq. 7) by $-\rho^{a}$, Eq. 11 by ρ^{w} and adding both equations, the following equality is obtained:

$$\begin{aligned} &n\left(\rho^{w}\frac{\partial\rho^{a}}{\partial t}-\rho^{a}\frac{\partial\rho^{w}}{\partial t}\right)+\\ &\mathbf{q}_{v}^{l}\left(\rho^{w}\nabla\rho^{a}-\rho^{a}\nabla\rho^{w}\right)=0 \end{aligned} \tag{12}$$

Using material derivative of liquid phase $D\xi/Dt^{l}=\partial\xi/\partial t+\mathbf{q^{*}}_{v}^{l}.\nabla\xi$, where by definition $\mathbf{q^{*}}_{v}^{l}$ is the real velocity of liquid phase, Eq. 12 becomes:

$$\frac{D\left(\rho^{a}/\rho^{w}\right)}{Dt^{l}}=0 \tag{13}$$

Since only isothermal processes are considered, constant compressibility for water $\rho^{w}=\beta^{w}p^{w}$ and for air $\rho^{a}=\beta^{a}p^{a}$ can be assumed. Eq. 13 can be rewritten as:

$$\frac{D\left(p^{a}/p^{w}\right)}{Dt^{l}}=0 \tag{14}$$

Eq. 14 expresses the transport by liquid phase of the condition $p^{a}/p^{w}=K$, where K is a constant. The value of K is given by the criterion of saturation, that is $p^{a}/p^{w}=1$. In other words, through Eq. 11, each liquid particle keeps the memory of the pressure condition at which saturation occurred.

From the point of view of numerical implementation, a first consequence of this result is that no degrees of freedom have to be introduced or removed during the analysis, since it exists a equation to replace the balance equation of the vanished phase in saturated condition. A second consequence is that the same formulation can be used for miscible or immiscible species, if a switch between Eq. 10 and Eq. 11 is implemented in saturated conditions.

5. CONCLUSION

In a first part, a strategy is proposed to solve the problem posed by the numerical simulation of saturation/desaturation processes in porous media. It consists in considering the physical process of dissolution of gas into liquid as a way to regularize the system of equations when local saturated conditions occur. In fact, with this strategy, flow of species will always exist independently of the existence or not of gas phase. It is a valuable alternative to considering a residual flow of gas in the saturated part, which is not physicaly justified, since, in most of cases, gas is occluded even before saturation. Simulations performed show that this strategy works well. However, the results are sensitive to the initial values prescribed for the mass of dissolved species into the liquid in the saturated part. When monotonic saturation path is imposed or when history of saturation can be reproduced, knowledge of these values is straightforward. The problem arises when desaturation of an initially saturated sample with an unknown history of saturation has to be simulated. In this case, conclusions from case B indicate that if suction and degree of saturation depend on initial conditions, gas flow pattern is not affected by the latter. As a consequence, the overall pattern of desaturation is quite insensitive to initial conditions.

In a second part, the extension of this strategy to the case of immiscible species is presented in the case of a rigid medium. It is shown that the continuity equation of dissolved species into liquid becomes, in the limit case of immiscible species, the transport equation of the saturation criterion by the liquid phase. As a consequence, the extension of the proposed strategy for immiscible species is straightforward from the point of view of numerical implementation.

ACKOWLEDGEMENT

This work has been carried out within the network "ALERT Geomaterials", financed by the European Commission throught Human Capital and Mobility Programme.

REFERENCES

Advani, S.H., Lee, T.S., Lee, J.K. & Kim, C.S. 1993. Hygrothermomechanical evaluation of porous media under finite deformations. Part I - Finite element formulations. *Int. J. Num. Meth. in Eng.*, 36:147-160.

Bear, J. 1972. Dynamics of fluids in porous media. New York: American Elsevier Pub. Co.

Dangla, P., Coussy, O. & Eymard, R. 1996. Contribution to the benchmark, *ALERT Programme 4*, Aussois.

Gawin, D. & Schrefler, B. 1996. Thermo-hydro-mechanical anaysis of partially saturated porous materials. *Engineering Computations*, 13:113-143.

Jommi, C., Vaunat, J., Gens, G., Gawin, D. & Schrefler, B. 1997. Multiphase flow in porous media: a numerical benchmark. *Proc. NAFEMS World Congress*, Stuttgart.

Olivella, S., Carrera, J., Gens, A. & Alonso, E.E. 1994. Nonisothermal multiphase flow of brine and gas through saline media. *Transport in Porous Media*, 15:271-293.

Panday, S. & Corapcioglu, M.Y. 1989. Reservoir Transport by Compositional Approach. *Transport in Porous Media*, 4:369-393.

Thomas, H.R. & King, S.D. 1992. Coupled heat and mass transfer in unsaturated soil — A potential based solution. *Int. Jnl. for Num. and Anal. Meth. in Geomechanics*, 16:757-773.

Vaunat, J., Jommi, C. & Gens, A. 1997. Constitutive formulations for saturation and desaturation processes in porous media, *to be published.*

Numerical Models in Geomechanics, Pietruszczak & Pande (eds) © *1997 Balkema, Rotterdam, ISBN 90 5410 886 X*

Numerical simulation of the transition from fully saturated to partially saturated state: Application to a drained column

P. Jouanna & M.A. Abellan
DTMC, UMR UM2-CNRS, Université Montpellier II, France

ABSTRACT: This paper proposes a method for tackling the delicate numerical problem of the transition from fully saturated to partially saturated state in a soil column subjected to desaturation either by water flow at the bottom or water evaporation at the top. The method is based on the application of a general thermo-hydro-mechanical and physico-chemical approach for detailed description of the phenomena during the following of the moving desaturation front.

INTRODUCTION

This paper presents a numerical approach for solving the delicate problem of the evolution of the transition front between the fully saturated zone and the partially saturated zone in a soil, this change depending on water evaporation at the top of the unsaturated soil and water flow under gravity.

The application of the general approach to heterogeneous media (Jouanna and Abellan, 1995) enables detailed description of the phenomena that take place following the movement of the desaturation front.

The model derived from the above approach is briefly summarized using Lagrange formalism with five constituents: air, water vapour, capillary water, free water and sand.

Modelling of the phenomena occuring on the axis of a vertical sand column is performed using a 1-D Finite Difference Method. At a given time step and at each level, an iterative process gives access to the different variables, up to 24 different variables being available per constituent. An updated spatial refined mesh is used in the vicinity of the desaturation front.

Among all possible results given by this detailed simulation, attention is paid to the most noticeable phenomena related to the moving desaturation front.

1 CASE STUDY

Generalizing an experiment reported by Liakopoulos (1965), the drainage of an initially saturated 1.0 m-thick sand column is considered, mass exchanges being allowed at the top and liquid water being free to flow at the bottom.

Three varying domains are considered :

- a non-saturated domain with dry air, water vapour, capillary water and sand, in the upper zone.
- a moving intermediate domain with dry air, water vapour, capillary water, free liquid water and sand; in this domain, the refined mesh covers both sides of the desaturation front where transfers occur.
- a saturated domain with free liquid water and sand.

2 MODELLING

The reference velocity field $\mathbf{v}^*$ to be considered for the relative movements of the different constituents is chosen here as being equal to the velocity field of the sand v_s. Modelling covers 3 chemical species, 5 constituents and 3 phases: π=a refers to dry air, π=v to water vapour, g to the gas phase, π=c to capillary water, π=w to free liquid water, π=s to the sand.

The following assumptions are used:

[H0]: Stress ratio $n_{\sigma\pi}$ is equal to the volume ratio n_π

[H1]: Temperatures of constituents are identical, $\hat{E}_\pi{}^* = 0$.

[H2]: For all constituents, $m_\pi = 0$.

[H3]: Air and solid are inert, $\hat{C}_a{}^* = \hat{C}_s{}^* = 0$.

[H4]: The external force considered is gravity (- g).

[H5]: External specific heat sources $r_\pi = 0$.

[H6]: Air and water vapour are perfect gases

[H7]: Fluids are ideal.

[H8]: The sand is assumed to have elastic behaviour.

The exhaustive list of the Euler, Lagrange and physical variables relative to a standard constituent is given in Table 1, notations being defined in the Appendix. Transformation formulas between the three types of variables are given in Abellan (1994, t. 1, Table 2.2, p. 146).

Table 1. Physical, Eulerian and Lagrangian variables considered

		Mass				Momentum					Heat				
Type of variable		V. R.	Pilot var.	Dual rev.	Source terms	Stress ratio	Pilot var.	Dual rev.	Dual irrev.	Source terms	V. R.	Pilot var.	Dual rev.	Dual irrev.	Source terms
Physical	π=a,v,c,w	n_π	μ'_π	ρ'_π	/	n_π	v'_π	σ'^{rev}_π	σ'^{irr}_π	/	n_π	T'	s'_π	q'_π	/
	π=s	n_s	♦	ρ'_s	/	n_s	σ'_s	u'_s rev	u'_s irr	/	n_s	T'	s'_s	q'_s	/
Euler	π=a,v,c,w	/	μ_π	ρ_π	$\hat{c}_\pi$	/	w_π^*	σ_π^{rev}	σ_π^{irr}	$\hat{p}_\pi$	/	T	s_π	q_π	0
	π=s	/	♦	ρ_s	0	/	σ_s	u_s rev	u_s^{irr}	$\hat{p}_s$	/	T	s_s	q_s	0
Lagrange	π=a,v,c,w	/	μ_π^*	ρ_π^*	$\hat{C}_\pi^*$	/	W_π^*	Σ_π^{*rev}	Σ_π^{*irr}	$\hat{P}_\pi^*$	/	$\mathcal{T}^*$	S_π^*	Q_π^*	0
	π=s	/	♦	ρ_s^*	0	/	Σ_s^*	U_s rev	U_s^{irr}	$\hat{P}_s^*$	/	$\mathcal{T}^*$	S_s^*	Q_s^*	0

In this case study, the implicit arguments of all vectors or tensors are the vertical Lagrange coordinate Z* in the reference domain and time t. On this basis, the fundamental relations given below include for any constituent π the mass, momentum and entropy balance relations following **v***, the material state relations, the non-equilibrium relations and the relations linking phenomenological to physical variables. Some complementary definition or constraint relations on the total set of constituents are recalled in the Appendix.

• Generalized mass balance relations (π=a,v,c,w,s)

$$\frac{\partial}{\partial t}\rho_\pi^* = -\frac{\partial}{\partial Z^*}\left[\rho_\pi^* W_\pi Z^* \frac{1}{\frac{\partial}{\partial Z^*}U_s Z^* + 1}\right] + \hat{C}_\pi^* \tag{1}$$

• Generalized momentum balance relations (π=a,v,c,w,s)

$$\frac{\partial}{\partial t}[\rho_\pi^*(W_\pi Z^* + V_s Z^*)] = -\frac{\partial}{\partial Z^*}\left[(\rho_\pi^*(W_\pi Z^* + V_s Z^*)W_\pi Z^* + \Sigma_\pi 1^*)\frac{1}{\frac{\partial}{\partial Z^*}U_s Z^* + 1}\right]$$
$$-\rho_\pi^* g + (W_\pi Z^* + V_s Z^*)\hat{C}_\pi^* + \hat{P}_\pi Z^* \tag{2}$$

• Generalized entropy balance relations (π=a,v,c,w,s)

$$\frac{\partial}{\partial t}(\rho_\pi^* S_\pi^*)$$
$$= -\frac{\partial}{\partial Z^*}\left\{\left[\frac{1}{\mathcal{T}^*}Q_\pi^* + \rho_\pi^* S_\pi^* W_\pi Z^*\right]\frac{1}{\frac{\partial}{\partial Z^*}U_s Z^* + 1}\right\}$$
$$-\frac{1}{\mathcal{T}^*}\left\{\Sigma_\pi^*\frac{\partial}{\partial Z^*}(W_\pi^* + V_s^*)\right\}^{irr}\frac{1}{\frac{\partial}{\partial Z^*}U_s Z^* + 1}$$
$$+\frac{1}{\mathcal{T}^*}\mu_\pi^*\hat{C}_\pi^* - \frac{1}{\mathcal{T}^*}[W_\pi^*\hat{P}_\pi^*]$$
$$-\frac{1}{(\mathcal{T}^*)^2}Q_\pi^*\frac{1}{\frac{\partial}{\partial Z^*}U_s Z^* + 1}\frac{\partial}{\partial Z^*}\mathcal{T}^* \tag{3}$$

• Material state relations relative to mass (π = v, c,w)

$$\mu_\pi^* = \mu_{o\pi}^* + \frac{1}{\mathcal{M}_\pi}R\mathcal{T}^* \ln\frac{\Sigma_{\pi'}^* \text{ rev}}{n_{\pi'} J^*} \quad (\text{with } \pi'=v) \tag{4}$$

• Material momentum state relations

- For air and water vapour (π = a, v)

$$\Sigma_\pi^* \text{ rev} = \frac{\rho_\pi^* R\mathcal{T}^*}{\mathcal{M}_\pi} \tag{5}$$

- For capillary water (π = c) the state relation relative to momentum is replaced by relations taking into account hydraulic hysteresis phenomena.

for S > 0.91

$$\frac{n_\pi}{1 - n_s} = 1 - 1.9722\text{E-}11\left[\frac{\Sigma_a^* \text{ rev}}{n_g J^*} + \frac{\Sigma_v^* \text{ rev}}{n_g J^*} - \frac{\Sigma_\pi^* \text{ rev}}{n_\pi J^*}\right]^{2.4279} \tag{6}$$

for S ≤ 0.91

$$-\delta\left[\left(\frac{\Sigma_a^* \text{ rev}}{n_g} + \frac{\Sigma_v^* \text{ rev}}{n_g} - \frac{\Sigma_\pi^* \text{ rev}}{n_\pi}\right)\frac{n_\pi}{\rho_\pi^* g}\right]$$
$$= F_\pi^{+-}\,\delta\left(\frac{\rho_\pi^*}{J^*}\right)^{+-} + F_{T^*}^{+-}\,\delta\mathcal{T}^{*+-} \tag{7}$$

- For the free water (π = w)

$$d\Sigma_\pi^* \text{ rev}$$
$$= B_{\rho\pi}\,d\rho_\pi^* + B_{\sigma\pi}\,d\left[\frac{1}{\frac{\partial}{\partial Z^*}U_s Z^* + 1}\frac{\partial}{\partial Z^*}U_\pi Z^*\right] + B_{T\pi}\,d\mathcal{T}^* \tag{8}$$

- For the sand (π = s) a simple elastic behaviour law is considered.

• Material state relations relative to heat (π=a,v,c,w,s)

$$d\left(\frac{\rho^*_\pi S_\pi^*}{J^*}\right) = -\mathcal{B}_\pi\,d\left(\frac{\rho^*_\pi}{J^*}\right) - \alpha_\pi\,d\left(\frac{\Sigma_\pi^* \text{ rev}}{J^*}\right)$$
$$+\frac{1}{\mathcal{T}^*}c^v{}_{s|\rho,\sigma}\,d\mathcal{T}^* \tag{9}$$

• Material non-equilibrium relations giving $\hat{C}_\pi{}^*$

- in the non-saturated domain (π = v, c):

$$\hat{C}_v{}^* = J^* L_{rr} \frac{1}{T^*} [\mu_c{}^* - \mu_v{}^*]$$

$$\hat{C}_c{}^* = -\hat{C}_v{}^* \qquad (10)$$

- at the desaturation front (π = v, c, w):

$$\hat{C}_v{}^* = J^* L_{rr} \frac{1}{T^*} [\mu_w{}^* - \mu_v{}^*]$$

$$\hat{C}_c{}^* = J^* L_{rr} \frac{1}{T^*} [\mu_w{}^* - \mu_c{}^*]$$

$$\hat{C}_w{}^* = -\hat{C}_v{}^* - \hat{C}_c{}^* \qquad (11)$$

• Material non-equilibrium relations giving $\hat{P}_\pi{}^*$

$$W_\pi{}^* = \frac{k_\pi T^*}{n_\pi} \frac{1}{J^*} P^{*\wedge}{}_\pi \qquad (\pi = a, v, c, w, s) \quad (12)$$

• Material non-equilibrium relations giving $Q_\pi{}^*$

$$Q_\pi Z^* = -n_\pi \lambda'_\pi \frac{\partial}{\partial Z^*} T^* \qquad (\pi = a, v, c, w, s) \quad (13)$$

• Material non-equilibrium tensorial relations for constituents π = a, v, c, w

$$\Sigma_\pi 1^{*\,irr} = n_\pi [2\tilde{\mu}_\pi + \tilde{\eta}_\pi] [\frac{\partial W_\pi Z^*}{\partial Z^*} + \frac{\partial V_s Z^*}{\partial Z^*}]$$

$$\Sigma_\pi 3^{*\,irr} = n_\pi \tilde{\eta}_\pi [\frac{\partial W_\pi Z}{\partial Z^*} + \frac{\partial V_s Z^*}{\partial Z^*}] \qquad (14)$$

• Volume ratio relations (π = a, v, c, w, s)

$$d_v{*}n_\pi(\mathbf{x},t)=[d_v{*}\rho_\pi(\mathbf{x},t) - n_\pi(\mathbf{x},t)\, d_v{*}(\frac{\rho_\pi(\mathbf{x},t)}{n_\pi(\mathbf{x},t)})] \frac{n_\pi(\mathbf{x},t)}{\rho_\pi(\mathbf{x},t)} \qquad (15)$$

3 NUMERICAL STRATEGY

3.1 General strategy

Equations (1) to (15) are considered only for the constituents present in a given sub-domain.

Initial conditions correspond to a steady-state saturated flow in the column.

For boundary conditions, at the top (Z^*=1m) gas mass exchanges are allowed with the atmosphere and at the bottom (Z^*=0) flow of liquid water is free.

To solve these equations, the strategy consists of associating one variable with one equation by a bijection (Jouanna & Abellan, 1997). In an iterative process, constituent after constituent, equation after equation, each variable is computed by its associated equation in which values of all other necessary variables are estimated at the previous iteration step.

3.2 Special considerations concerning the case study

1. Space-time mesh: all relations are discretized using a 1-Dim Finite Difference Method. The space mesh consists of 40 points located on the vertical axis of the column. A refined mesh (with 20 of the above 40 points) is used in an intermediate domain for following the desaturation front. As the desaturation front moves down the column, due to evaporation at the top or to water flow at the bottom, the intermediate domain also moves. Hence the refined mesh requires updating throughout the process in order to remain linked to the intermediate domain.

The criterion presented below shows when the mesh needs to be updated. Two space interpolations are then made for computing the values of all variables:

• at the new points of the standard mesh replacing the previous 20 points of the refined part of the mesh.

• at the new 20 points of the refined part of the mesh replacing the previous points of the standard mesh.

Two space intervals are necessary for the mesh depending on the domain and two time intervals depending on the type of equation. Hence four pairs of time and space intervals were used and found sufficient to ensure the convergence of the whole numerical iterative process.

2. Criterion for following the desaturation front during the iterative process: the most simple strategy consists of following the variation of the volume ratio of free water. In the intermediate domain, the boundary between the saturated sand and the non-saturated sand is represented by a refined layer. Two points of the mesh are situated on this layer. One point i1 is linked to the non saturated zone and one i2 is linked to the saturated zone.

At the beginning of a given time iteration n, the volume ratio of free water is greater than zero at point i2 : $n_w(i2,t)>0$.

At the end of the iteration the value of $n_w(i2,t)$ is checked:

if $n_w(i2,t)>0$ then:

- the sand remains saturated;
- an other iteration can be computed.

else:

- the desaturation front has moved down about one space refined interval;
- air, water vapour and capillary water must be taken into account. The subroutines linked to the different constituents and the different variables are enlightened at this point of the refined mesh by the main programme.
- inversely, free liquid water disappears. Its subroutines are no longer considered at this point of the refined mesh by the main programme.
- Moreover if point i2 represents the last point of the refined part of the mesh, the mesh must be updated.

4 ILLUSTRATION OF THE RESULTS

Results are available for all variables of all constituents at each time step. Hereafter a zoom is given for phenomena taking place in the vicinity of the desaturation front, in the time interval 95s to 575s after the beginning of the experiment. A special attention is paid to:

• the mass supply to water vapour (Fig. 1) and the mass supply to capillary water (Fig. 2).
• the velocities of air (Fig. 3), water vapour (Fig. 4), free water (Fig. 5) and the vertical displacement of the sand (Fig. 6).
• the volume ratios of the gas phase (Fig. 7), the capillary water (Fig. 8), the free water (Fig. 9) and the sand (Fig. 10).

The legend of these figures is given in Table 2.

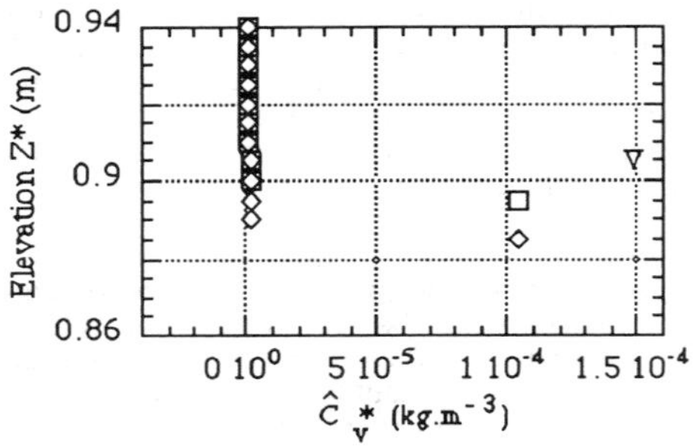

Fig. 1: Profiles of the mass supply $\hat{C}_v$* to water vapour

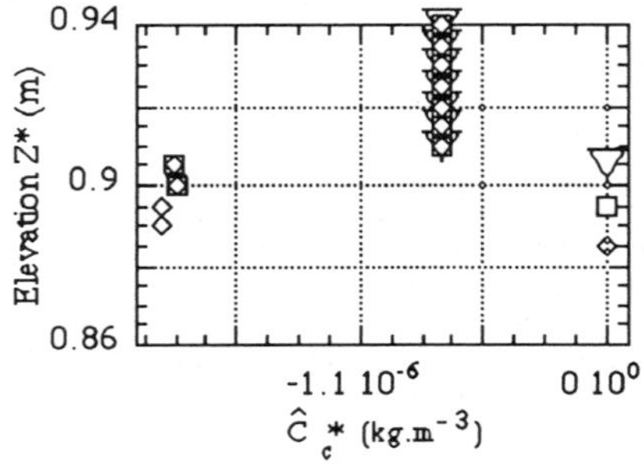

Fig. 2: Profiles of the mass supply $\hat{C}_c$* to capillary water

Table 2. Legend of figures 1 to 10

▽	95 s
□	335 s
◇	575 s

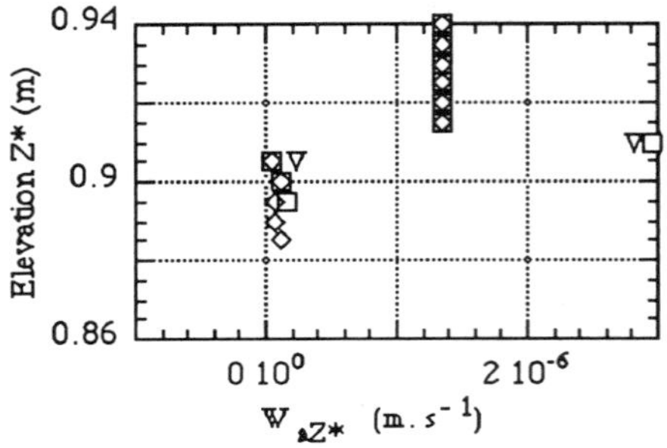

Fig. 3: Profiles of the relative velocity W_{aZ}* of air

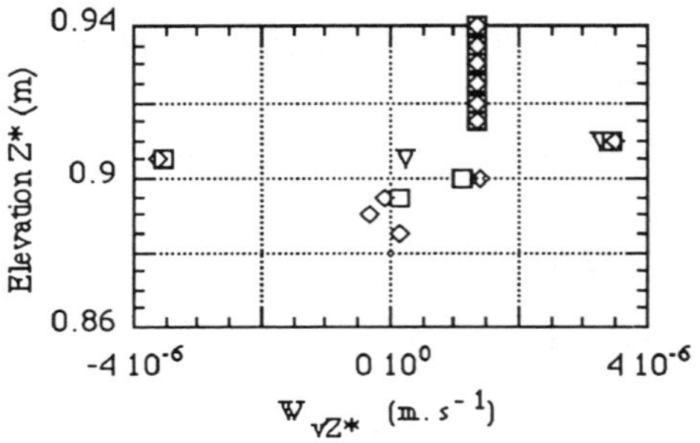

Fig. 4: Profiles of the relative velocity W_{vZ}* of water vapour

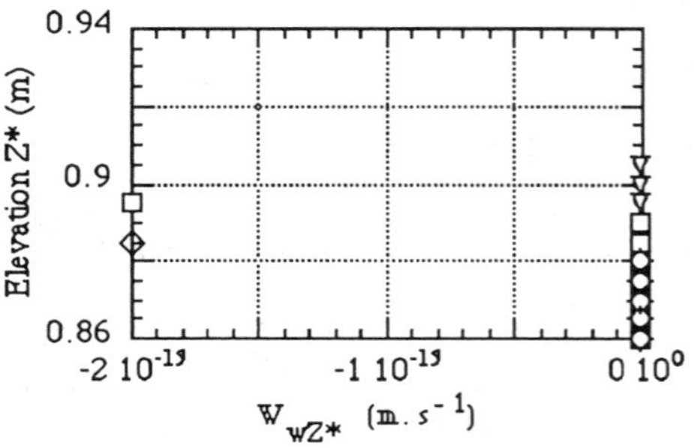

Fig. 5: Profiles of the relative velocity W_{wZ}* of free water

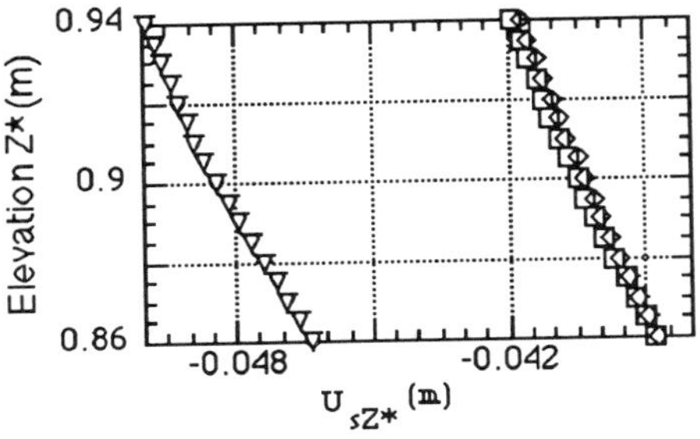

Fig. 6: Profiles of the vertical displacement U_{sZ^*} of the sand

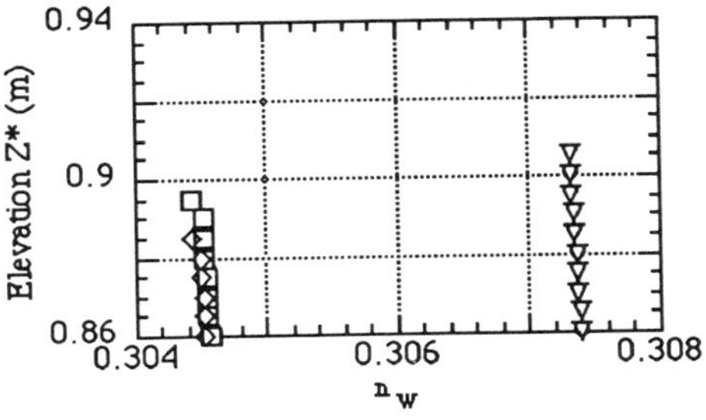

Fig. 9: Profiles of the volume ratio n_w of the free water

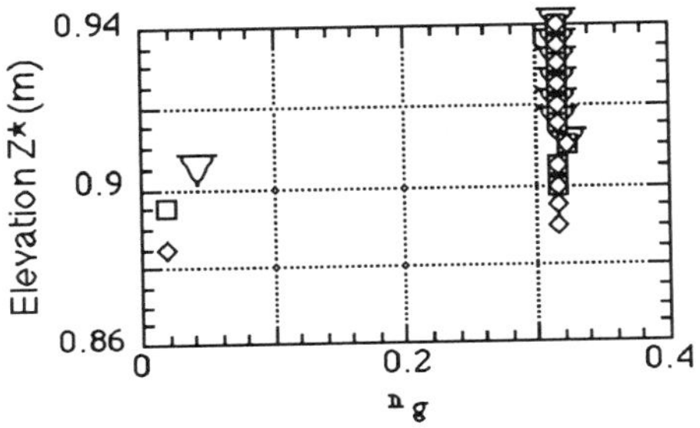

Fig. 7: Profiles of the volume ratio n_g of the gas phase

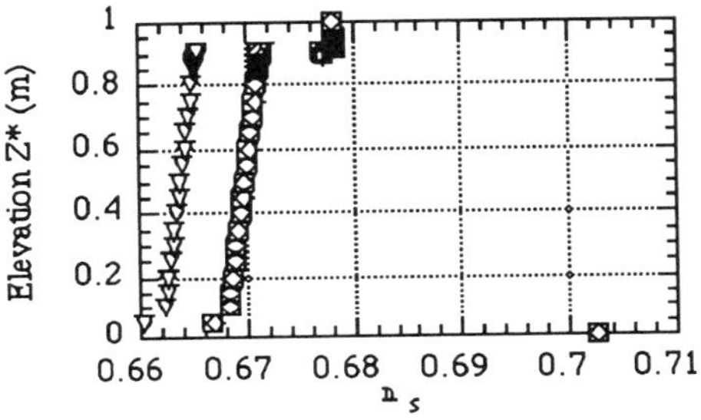

Fig. 10: Profiles of the volume ratio n_s of the sand

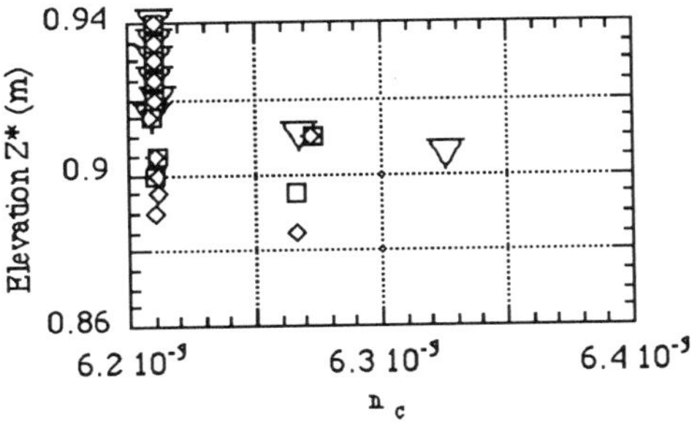

Fig. 8: Profiles of the volume ratio n_c of the capillary water

5 DISCUSSION AND CONCLUSION

(i) Phenomena linked to variables relative to mass:

mass exchanges are observed

- in the non-saturated domain where capillary water evaporates as water vapour,
- and in the intermediate domain where both mass exchanges occur between free water and its vapour and between free and capillary water.

These mass exchanges are linked to the difference of chemical potentials between water vapour, capillary water and free water.

(ii) Phenomena linked to variables relative to momentum:

- in the non-saturated domain, upward movements of both air and water vapour are observed;
- in the intermediate domain, opposite movements of water vapour were observed, i.e. an upward movement in the layers situated in the vicinity of the non-saturated domain and a downward movement in the layers near the saturated domain boundary. The latter movement is also associated with a downward movement of free water (i.e. of the desaturation front) due both to gravity and to evaporation. Settlement of the sand is observed as the result of mechanical couplings with all the other variables.

(iii) More complete results could show heat exchanges and temperature variations in the vicinity of the desaturation front.

(iv) Changes in the volume ratios of all phases are strongly connected to all the phenomena above. In particular, the desaturation front can be easily followed observing the progression of zero values for the volume ratio of the free water and the apparition of non-zero values for the volume ratios of the gas phase and the capillary water.

REFERENCES

Abellan, M-A. 1994. Approche phénoménologique généralisée et modélisation systématique de milieux hétérogènes. Illustration sur un sol non-saturé en évolution thermo-hydro-mécanique et physico-chimique. Tome 1 (Concepts) et Tome 2 (Illustration), *Thèse de Doctorat*, Université Montpellier II, France.

Jouanna, P., Abellan, M-A. 1995. Generalized approach to heterogeneous media, in "*Modern Issues in Non-Saturated Soils*", Chapter 1, C.I.S.M.courses and lectures No. 357, edited by A. Gens/ P. Jouanna/ B.A. Schrefler, Springer-Verlag,Wien, New York, 1-127.

Liakopoulos, A.C. 1965. Transient Flow through Unsaturated Porous Media, *Ph. D. thesis*, Univ. of California, Berkeley.

APPENDIX:

• Main notations non explicitly defined in the text

$\hat{c}_\pi$, $\hat{C}_\pi^*$ mass supply to π due to other constituents per volume and time unit in kg m^{-3} s^{-1}
d differential symbol
d_{v^*} differential following the virtual movement
$\hat{e}_\pi$, $\hat{E}_\pi^*$ energy supply to π due to other constituents per volume and time unit in W m^{-3}
$\mathbf{f}_\pi$ external force acting on the mass unit of π
irr irreversible part of a quantity
J* Jacobian of the transformation
$\mathcal{M}_\pi$ molar mass of π in kg $mole^{-1}$
m_π external mass source of π in s^{-1}
n_π volume ratio of π defined by $\rho_\pi = n_\pi \rho'_\pi$
$n_{\sigma\pi}$ stress ratio of π defined by $\underline{\sigma}_\pi = n_{\sigma\pi}\underline{\sigma}'_\pi$
$\hat{\mathbf{p}}_\pi$, $\hat{\mathbf{P}}_\pi^*$ momentum supply vector to π due to other constituents in kg m^{-2} s^{-2}
$\mathbf{q}'_\pi$, $\mathbf{q}_\pi$, $\mathbf{Q}_\pi^*$ heat influx vector relative to π
R universal gas constant in J K^{-1} $mole^{-1}$
rev reversible part of a quantity
s'_π, s_π, S_π^* specific entropy of π in J kg^{-1} $°C^{-1}$
T', T, $\mathcal{T}^*$ temperature in °C or °K
$\mathbf{u}'_\pi$, $\mathbf{u}_\pi$, $\mathbf{U}_\pi$ displacement of π in m
var abbreviation for variable
$\mathbf{v}'_\pi$, $\mathbf{v}_\pi$ velocity vector of π
V.R. abbreviation for volume ratio
$\mathbf{w}'_\pi, \mathbf{w}_\pi^*$, $\mathbf{W}_\pi^*$ relative velocity vector with $\mathbf{w}_\pi^* = \mathbf{v}_\pi - \mathbf{v}^*$
δ variation symbol
μ'_π, μ_π, μ_π^* specific chemical potential of π
$\mu_{o\pi}^*$ standard chemical potential of π in J kg^{-1}
π index of any constituent
ρ'_π, ρ_π, ρ_π^* mass density of π in kg m^{-3}
$\underline{\sigma}'_\pi$, $\underline{\sigma}_\pi$, $\underline{\Sigma}_\pi^*$ stress tensor relative to π in its movement
ψ', ψ, ψ^* capillary suction ;

$$\psi = \left(\frac{\sigma_g{}^{rev}}{n_g} - \frac{\sigma_w{}^{rev}}{n_w}\right)\frac{n_w}{\rho_w\, g}$$

F_w^{+-}, F_T^{+-}, $B_{\rho\pi}$, $B_{\sigma\pi}$, $B_{T\pi}$, $\mathcal{B}_\pi$, $c^v{}_{\pi\rho\sigma}$, $\underline{\alpha}_\pi$, L_{rr}, k_π, k, λ'_π ,

$\tilde{\mu}_\pi$, $\tilde{\eta}_\pi$ phenomenological coefficients

• Definition relations

- the Lagrange displacement vector of π is :

$$\mathbf{U}_\pi(Z^*,t) = \mathbf{U}_\pi{}^{rev}(Z^*,t) + \mathbf{U}_\pi{}^{irr}(Z^*,t)$$

- the Lagrange velocity vector of π is given by:

$$\mathbf{V}(\mathbf{X}^*,t) \equiv \frac{d}{dt}\,\mathbf{U}(\mathbf{X}^*,t)$$

• Constraint relations on the total set of N constituents are given below:

$$\sum_{\pi=1}^{\pi=N} \hat{C}_\pi^*(Z^*,t) = 0$$

$$\sum_{\pi=1}^{\pi=N} \hat{\mathbf{P}}_\pi^*(Z^*,t) = \mathbf{0}$$

$$\sum_{\pi=1}^{\pi=N} \hat{E}_\pi^*(Z^*,t) = 0$$

$$\sum_{\pi=1}^{\pi=N} n_\pi(\mathbf{x},t) = 1$$

Numerical Models in Geomechanics, Pietruszczak & Pande (eds) © 1997 Balkema, Rotterdam, ISBN 90 5410 886 X

Modelling nonlinear mass transfer problems using the line searches technique

L. Lemoine, J.C. Robinet & A. Pasquiou
EuroGéomat Consulting, Orléans, France

A. Jullien
ESEM Université d'Orléans, France

ABSTRACT : A one dimensional mass transfer model for rigid porous media was developed. The permeability and capillary matrices were determined from experimental results at various hydraulic states. The equations were solved using the Piccard's method associated with the « line searches » technique to avoid convergence problems. Simulations of soaking tests on Boom clay and on a deep marl are presented. Then, the ability to predict desaturation of Boom Clay is discussed from laboratory experimental results.

1 INTRODUCTION

The actual trend in soil mechanics is to take into account desaturation of the material as one component of the stress history. Narasimhan and Witherspoon (1978), Shrefler and Xiaoyong (1993) have developed coupled models for water and air flows in deformable porous media.

In the field of waste disposal for radioactive containers, deep clays are studied as possible barrier materials for storage host rocks. A low hydraulic conductivity is required to limit fluid transfer with time. Further, the hydraulic properties of clays strongly depend on their hydromechanical history. Thus, the hydraulic properties of the materials have to be determined for low saturation degrees up to saturation under known stress states. Further, numerical models should take into account the material properties over a wide range of hydraulic states.

In this study we aimed at analyzing the assumptions of rigid porous medium when simulating mass transfer in a clay barrier material and in a clayey rock. A set of experimental results was simulated with a 1D finite element model. The constitutive equations are first detailed. Then, the unsaturated hydraulic properties of the materials are presented. Finally the ability to predict the experimental results is discussed.

2 THE CONSTITUTIVE EQUATIONS

The equations of conservation of mass are given by :

$$\frac{\partial(nS)}{\partial t} + \mathrm{div}(\vec{q}) = 0 \qquad (1)$$

where :

S is the saturation degree,

n is the porosity,

$\vec{q}$ is the water speed (m/s).

The water motion is governed by Darcy's law :

$$\vec{q} = \frac{K_{rl}k}{\mu_w}(\vec{\mathrm{grad}}(p_w) + \rho_w g\, \vec{\mathrm{grad}}(z)) \qquad (2)$$

with :

K_{rl}, the relative permeability of the medium,

k, the intrinsic permeability of the medium (m^2),

μ_w, the dynamic viscosity of water (1.10^{-3} Pa.s).

ρ_w, the fluid density (1000 kg/m^3)

g, the gravitational acceleration

The general one-dimensional boundary conditions are expressed by :

$$p_w(r,0) = p_{w0}$$
$$p_w(R,t) = p_{wR} \qquad (3)$$
$$q(R_\alpha,t) = q_{re}$$

Equation (2) used in equation (1) gives:

$$\frac{\partial(nS)}{\partial t} + \mathrm{div}\left(\frac{kK_{rl}}{\mu_w}\left(\vec{\mathrm{grad}}(p_w) + \rho_w g\, \vec{\mathrm{grad}}(z)\right)\right) = 0 \qquad (4)$$

Assuming that the medium is rigid, equation (4) gives :

$$n\frac{\partial S}{\partial p_c}\frac{\partial p_w}{\partial t}+\mathrm{div}(X)+\mathrm{div}\left(\frac{kK_{rl}}{\mu_w}\rho_w g\,\mathrm{gr\vec{a}d}(z)\right)=0 \quad (5)$$

and $$X=\frac{kK_{rl}}{\mu_w}\overrightarrow{\mathrm{grad}}(p_w) \quad (6)$$

Considering U^{ad} the admissible pressure space and u such that : $u\in U^{ad}=\left\{u/u\in H^1(\Omega), u(R,t)=0\right\}$

Multiplying (5) by u a test function and integrating on (Ω) gives :

$$\int_\Omega n\frac{\partial S}{\partial p_c}\frac{\partial p_w}{\partial t}u\,d\Omega+$$

$$\int_\Omega u\,\mathrm{div}(X+\frac{kK_{rl}}{\mu_w}\rho_w g\,\mathrm{gr\vec{a}d}(z))d\Omega=0 \quad (7)$$

In the 1D case, the equation (7) becomes :

$$\int_r -n\frac{\partial S}{\partial p_c}\frac{\partial p_w}{\partial t}u\,dr-\int_r u\frac{\partial}{\partial r}(X)dr=0 \quad (8)$$

Integrating (8) by parts, and with u(r,t)=0 and (3) :

$$\int_r -n\frac{\partial S}{\partial p_c}\frac{\partial p_w}{\partial t}u\,dr+\int_r\frac{\partial u}{\partial r}.X.dr-u.q_{re}=0 \quad (9)$$

The variational formulation for an infinite space U^{ad} can be defined as :

$$(Q)\Leftrightarrow\begin{cases}\text{find } p_w\in U^{ad} \text{ such that } \forall\, u\in U^{ad}\\ p_w(R,t)=-p_{wR} \quad \text{with } t\geq 0\\ -n\int_r\frac{\partial S}{\partial p_c}\frac{\partial p_w}{\partial t}u\,dr+\int_r\frac{kK_{rl}}{\mu_w}\frac{\partial u}{\partial r}\frac{\partial p_w}{\partial r}dr=u.q_{re}\end{cases}$$

For the finite element solution, we defined the space $U^h\in U^{ad}$ which has a finite dimension. The finite element formulation of the problem is then :

$(Q_h)\Leftrightarrow$

$$\begin{cases}\text{find } p_w\in U_h^{ad} \text{ such that } \forall\, u\in U_h^{ad}\\ p_{wh}(R,0)=-p_{wR} \quad \text{avec } t>0\\ -n\int_r\frac{\partial S}{\partial p_c}\frac{\partial p_{wh}}{\partial t}u_h\,dr+\int_r.\frac{kK_{rl}}{\mu_w}\frac{\partial u_h}{\partial r}\frac{\partial p_{wh}}{\partial r}dr=u_h.q_{re}\end{cases}$$

U^h admits the following basis functions : $\left\{w_j\right\}_{j=1}^{N_h}$.

Thus :

$$p_{wh}(r,t)=\sum_{i=1}^{N_h}p_{wi}(t)w_i(r) \quad (10)$$

Hence, the differential system to be solved is :

$(Q_h)\Leftrightarrow$

$$\begin{cases}\sum_{i=1}^{Nh}\frac{dp_{wi}(t)}{dt}.C_{ij}(p_w(t))+\sum_{i=1}^{Nh}p_{wi}(t).K_{ij}(p_w(t))=Q_j\\ \text{with } p_{wi}(R,t)=-p_{wr}\end{cases}$$

$$\widetilde{p}_w=\left\{p_{w1},\ldots,P_{wN_h}\right\}$$

where :

$K=[K_{ij}]$ is the permeability matrix

$$K_{ij}=\int_r\frac{K_{rl}k}{\mu_w}\frac{\partial w_i}{\partial r}\frac{\partial w_j}{\partial r}dr$$

$C=[C_{ij}]$ is the capillary matrix

$$C_{ij}=-n\int_r\frac{\partial S}{\partial p_c}w_iw_jdr$$

$$Q=\left\{Q_1,\ldots,Q_{N_h}\right\}^T \quad Q_j=w_j(r).q_{re}$$

The element load vector, stiffness matrix and damping matrix are calculated using the Newton-Cotes integration method with 5 integration points. The higher order of the polynomial which can be exactly calculated is 9.

Time discretization is obtained using a linear interpolation :

$$p_w(t+\alpha\Delta t)=\alpha.p_w(t+\Delta t)+p_w(t).[1-\alpha]$$

$$\frac{dp_w(t+\alpha\Delta t)}{dt}=\alpha\frac{p_w(t+\Delta t)-p_w(t)}{\Delta t} \quad (11)$$

At each time step a set of linear equations is solved

$$A=\alpha.\sum_{i=1}^{Nh}\frac{p_{wi}(t+\Delta t)-p_{wi}(t)}{\Delta t}.C_{ij}$$

$$B=\sum_{i=1}^{Nh}(\alpha.p_{wi}(t+\Delta t)+(1-\alpha)p_{wi}(t)).K_{ij} \quad (12)$$

$$C=\alpha.Q_j(t+\Delta t)+(1-\alpha).Q_j(t)$$

$$A+B=C$$

The time dependent problem to be solved is to find $p_{wh}\in U_h$ such that :

$$\sum_{i=1}^{Nh}p_{wi}(t+\Delta t).\widetilde{K}_{ij}=\widetilde{R}_j(t+\Delta t)$$

$$\widetilde{K}_{ij}=\alpha.C_{ij}+\Delta t.\alpha.K_{ij} \quad (13)$$

$$\widetilde{R}_j=\sum_{i=1}^{Nh}p_{wi}(t).\left[\alpha.C_{ij}-\Delta t(1-\alpha)K_{ij}\right]+C\Delta t$$

Significant non linearities in the equations (13) led to use the Euler implicit method ($\alpha=1$).

The line searches method was used by Schrefler (1991) to actualise the vector p_{wn}^{l+1} such that :

$$p_{n+1}^{l+1}=p_{n+1}^{l}+\Delta p_{n+1}^{l+1} \quad (14)$$

The line searches technique consists in keeping the direction of the modification determined by the vector Δp_{n+1}^{l+1}, whereas the modulus is modified thanks to the scalar η_{k+1} called the "*advance length*".

$$p_{n+1}^{l+1} = p_{n+1}^{l} + \eta_{l+1}.\Delta p_{n+1}^{l+1} \quad (15)$$

The optimal advance length has to be determined. So, let's assume that the pressure field at the first iteration and its increment are known. Hence, the problem is to find η_{l+1} such that :

$$R(\eta_{lk+1}) = R(p_{n+1}^{l} + \eta_{l+1}\Delta p_{n+1}^{l}) = 0 \quad (16)$$

However, as the solution cannot be reached, the value of the residual is minimised. So, the new problem is to find the absolute minimum of the following scalar function :

$$\psi(\eta_{l+1}) = (\Delta p_{n+1}^{l+1})^{T}.R(p_{n+1}^{l} + \eta_{l+1}\Delta p_{n+1}^{l+1}) \quad (17)$$

ψ is the linearized total potential energy associated with point($p_{n+1}^{l} + \eta_{k+1}\Delta p_{n+1}^{l}$). In order to limit the calculation cost, the function ψ found is such that :

$$|\psi(\eta_{k+1})| < \varepsilon|\psi(0)| \quad (18)$$

In this case, ε is equal to 1 and $\eta_{l+1}(\psi(1))$ is obtained from :

$$\psi(0) = (\Delta p_{wn+1}^{l+1})^{T} R^{l} \quad (19)$$

$$\psi(1) = (\Delta p_{wn+1}^{l+1})^{T} R^{l+1} \quad (20)$$

Hence, the scalar η_{l+1} allows to accelerate or to reach the convergence of the Picard's method, whatever the time step is.

3 HYDRAULIC PROPERTIES OF THE CLAYS

The permeability matrix is obtained from the unsaturated hydraulic conductivity K_{wt} (m/s), determined after soaking tests and using drying-wetting isotherms. Thus, the relative and intrinsic permeabilities are given by :

$$K_{rl}(S) = \frac{K_{wt}(S)}{K_{wt}^{sat}} \qquad k = \frac{K_{wt}^{sat}.\mu}{\rho_w.g} \quad (21)$$

The hydraulic conductivity was determined by Perrin (1983) :

$$K_{wt} = C_{wt} \times D_{wt} \quad (22)$$

with :

C_{wt} the capillary capacity (m^{-1})

D_{wt} the coefficient of diffusion (m^2/s).

K_{wt}^{sat} the saturated hydraulic conductivity (m/s)

Cwt is determined from drying-wetting isotherms obtained with the oversaturated salt solution technique described by Robinet and Rhattas (1995). The solution is placed in the bottom of a small tight enclosure and the sample is put on an inert plastic net above it. Mass exchanges are quantified by successive weightings(+/- 1mg) until the equilibrium is reached. The volumetric water content (θ) versus the relative humidity of air (Hr) is considered. Introducing the water potential h (m) it comes :

$$h = 10^{6}\frac{\psi_w}{\rho g}\ ;\ \psi_w = -\frac{\rho_w RT}{M}\ln\ (Hr) \quad (23)$$

with : ψ_w, the suction (MPa),

R the constant of the perfect gases,

T the absolute temperature

M the molar water mass (18g/mol).

The retention curves (θ-h) are fitted by an hyperbolic function :

$$\theta = \frac{\theta_{sat}}{1+\beta h} \quad (24)$$

The capillary capacity is then :

$$C_{wt} = -\left(\frac{d\theta}{dh}\right) = \frac{\beta\theta_{sat}}{(1+\beta h)^2} = \frac{\beta}{\theta_{sat}}\theta^2 \quad (25)$$

For the soaking test the sample is initially unsaturated : θi is the initial volumetric water content at the bottom of the sample. After stopping the test, the sample is cut in several 2 cm thick disks. The middle of each disk is taken as the vertical co-ordinate (x) ; their volumetric water content are determined to obtain the hydraulic profile and then Dwt.

The hydraulic profile at time t (θ-x) is plotted using the Boltzmann's variable (χ ($m.s^{-1/2}$)) :

$$\chi = \frac{x}{\sqrt{t}} \quad (26)$$

Then the coefficient of diffusion is given by :

$$D_{wt} = \frac{1}{2}\left(\frac{d\chi}{d\theta}\right)_{\theta}\int_{\theta}^{\theta_i}\chi\, d\theta \quad (27)$$

The curve (θ-χ) is fitted using an hyperbolic function :

$$\chi = \frac{a+b(\theta-\theta_i)}{b+c(\theta-\theta_i)} \quad (28)$$

Integrating (27) gives the coefficient of diffusion :

$$D_{wt} = \left(\frac{b-ca}{\sqrt{2}(b+c(\theta-\theta_i))c}\right)^2 \times \left[\frac{(\theta-\theta_i)c}{ca-b} + \ln\left(1+\frac{c}{b}(\theta-\theta_i)\right)\right] \quad (29)$$

The hydraulic properties of a natural marl and of remoulded Boom clay consolidated up to 5.8 MPa were investigated. The intrinsic permeability, the hydraulic conductivity and the water content at the saturated state are given in table 1.

The retention curves of both materials are presented on figure 1, the hydraulic profiles are given in Boltzmann's variable in figure 2.

Table 1: Saturated hydrodynamic properties.

Material	θsat	Kwtsat (m/s)	k (m^2)
marl	0,144	5.5610^{-16}	$5.7\ 10^{-23}$
Boom clay	0,51	8.5310^{-13}	$8.7\ 10^{-20}$

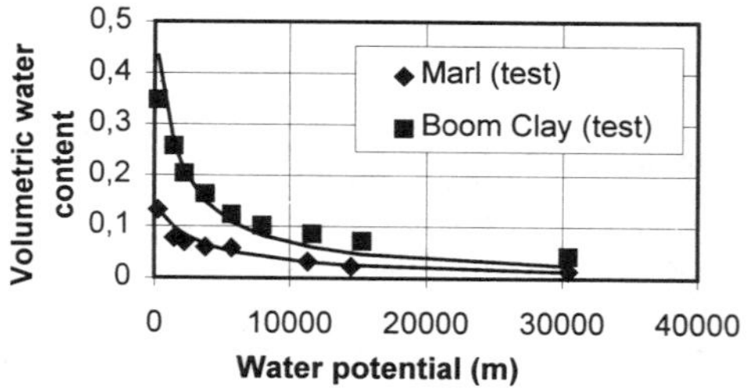

Figure 1 : Fitted retention curve for both materials (straight lines)

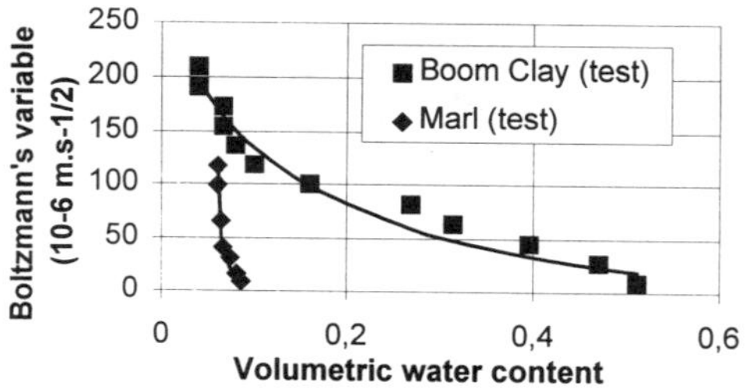

Figure 2 : Fitted hydraulic profiles for both materials (straight lines)

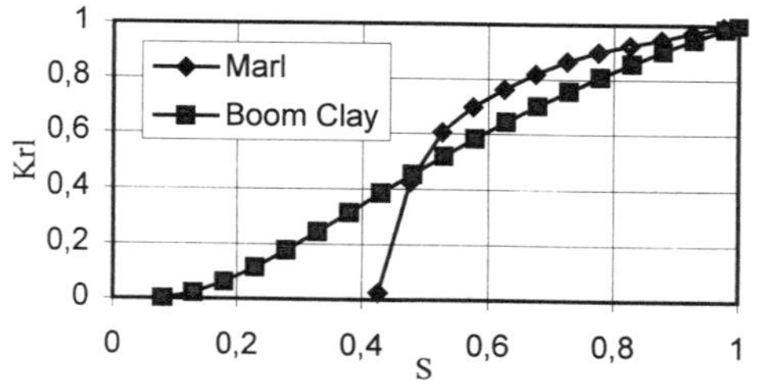

Figure 3 : Krl variations versus S of the materials.

The variations of the relative permeability of Boom clay and of the marl with the saturation degree obtained from the experiments are plotted on figure 3.

Finally, the relationships between the saturation degree and the capillary pressure (Pc) determined from the drying wetting isotherms are given in table 2 for both materials.

Table 2 : saturation degree versus the capillary pressure in MPa.

Material	S(Pc)
marl	S=1/(1+0.0326 Pc)
Boom clay	S=1/(1+0.0613 Pc)

4 SIMULATIONS

The soaking tests presented in section 3 performed on remoulded Boom clay and on the natural marl were simulated. Five elements were used for the mesh. The total time for soaking was of 14 days for Boom clay and of 17 days for the marl associated respectively with initial saturation degree of 0,131 and 0,425. The initial water pressure at the bottom of the column was the atmospheric pressure (Pc=0) ; no feedback flow was considered. The porosity was of 51% for Boom clay and 14.4 % for the marl. Figure 4 presents a comparison between the numerical and experimental water contents for both materials at the end of the test. The model well simulates the experiments for the marl. As for Boom clay the calculated water content underestimates the experimental one. But the finite element model simulate the fitted curves instead of the experimental points and the hyperbolic function doesn't fit very well the experimental boundary conditions.

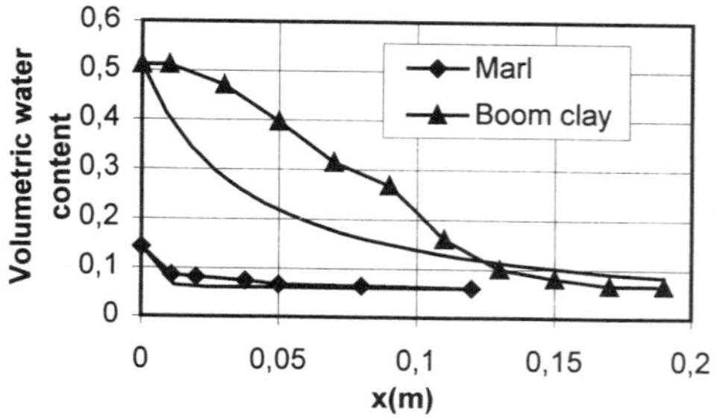

Figure 4 : results of the soaking tests.

'urther, no significant influence of the mesh (10 ·lements) and of the time step (10 steps up to 100 teps) on the results was noticed. Then, the lifference between the model and the experiment for 3oom clay can be explained by the fact that although the material is initially unsaturated, some swelling effects could have occurred during the soaking test .

A drying path was then simulated. Some experimental results were obtained for Boom clay oedometrically compacted up to 5 MPa and dried in the middle under ventilated conditions using the salt solutions technique (Figure 5). After 162 days of ventilation, the clay cylindrical sample (ϕin=50 mm, ϕext= 370 mm) was drilled through a radial direction. Thus, the hydraulic profile along a radial direction was determined.

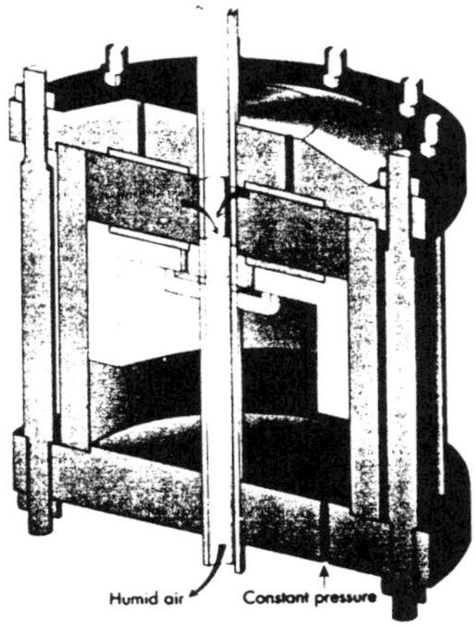

Figure 5 : schematic view of the experimental apparatus.

The experiment was simulated using a 1D mesh (10 elements) along the radius of the sample. An initial boundary capillary pressure of 60 MPa was imposed at the inner diameter, 100 time steps were used. The hydraulic properties obtained from the experiments presented in section 3 were used for simulations.

Assuming the medium to be rigid the first simulation was done with a constant intrinsic permeability. Figure 6 presents a comparison between the numerical and the experimental results after 162 days of ventilation. The simulation underestimates the experimental volumetric water content in the medium. This could be explained by the settlement strains of the clay on a drying path imposed at a constant vertical stress (oedometer test). Figure 7 shows the void ratio variations obtained on a drying-wetting path on highly compacted Boom clay. Thus, the void ratio significantly decreases when suction increases along the first drying path whereas it is quite constant along the wetting path. These variations imply that both the porosity and the hydraulic conductivity decrease when the material is dried.

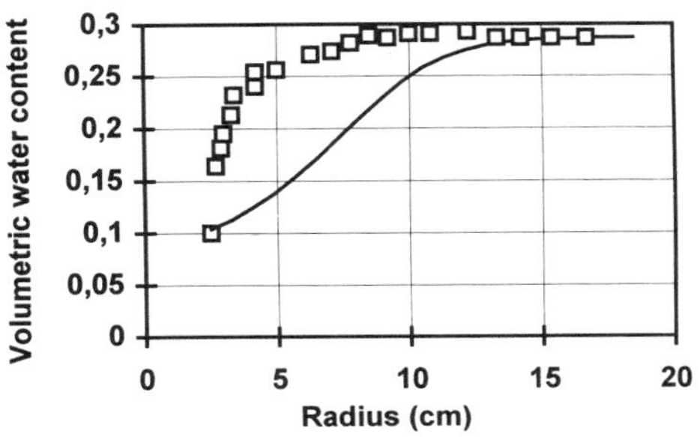

Figure 6 : Simulation of a drying-wetting path with k constant

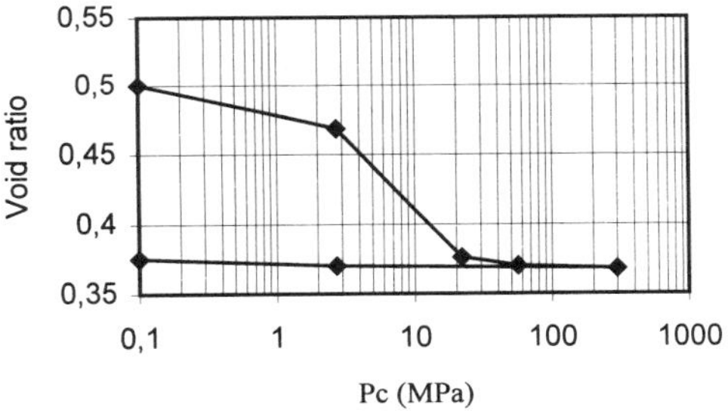

Figure 7 : Variation of the void ratio on a drying-wetting path

A second simulation was then performed considering the decrease with time of the porosity and of the intrinsic permeability. Linear relationships were successively considered. The porosity decreased from 0,33 down to 0,259 to have the same variations as in the drying-wetting test. The intrinsic permeability was arbitrarily divided by 10 along the drying path. The predicted water content (Figure 8) is then closer to the experimental results than before.

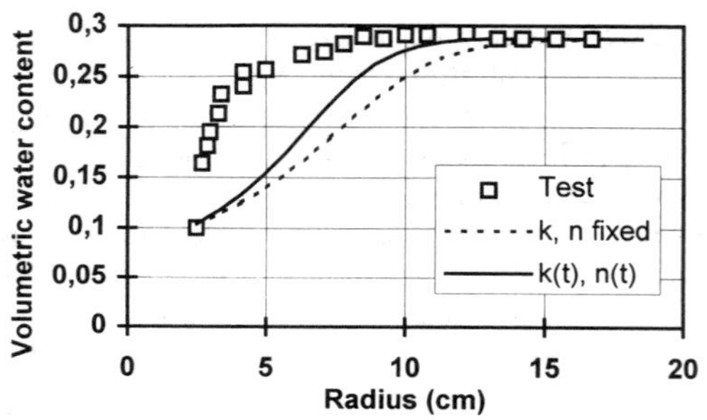

Figure 8 : drying path on Boom clay using a varying porosity and intrinsic permeability.

5 CONCLUSIONS

A one-dimensional finite element mass transfer model was developed with the assumption of a rigid porous medium. Simulations of soaking and drying tests were performed. The results showed that the predictions are better for the marl than for Boom clay. Further, concerning Boom clay, the results are better along the wetting path than along the first drying path. This could be explained by the fact that, contrarily to the marl, the Boom clay is a swelling clay and that the coupling between the porosity and the saturation degree is much more stronger during the first drying path. Besides, this study pointed out that the simulation results depend a lot on the quality of the fits of the experimental curves.

REFERENCES

Narasimhan, T.N., & Witherspoon P.A. 1978. Numerical model for saturated-unsaturated flow in deformable porous media. 3. Applications, *Water Resour. Res.* 14 : 1017-1034.

Perrin, B., Foures, J.C., & Javelas, R.1983. détermination du coefficient de diffusion isotherme de l'humidité dans des matériaux de construction. *Matériaux et construction,* Paris, 16.

Robinet J.C. & Rhattas M. 1995. Détermination de la perméabilité non saturée des matériaux argileux à faible porosité. *Revue Canad. de Géotech.* 32: 6, 1035-1043.

Schrefler, B.A., & Xiaoyong L. 1991. Comparison between different finite elements solutions for immiscible two phase flow in deforming porous media, *Computer methods and advances in Geomechanics, Balkema.* 1215-1220.

Schrefler, B.A., & Xiaoyong L. 1993. A fully coupled model for water flow and airflow in deformable porous media, *Water Resour. Res.* 29 : 155-167.

Numerical Models in Geomechanics, Pietruszczak & Pande (eds) © 1997 Balkema, Rotterdam, ISBN 90 5410 886 X

Coupled thermal-hydraulic-mechanical processes in a cylindrical cavity in rock

T.S. Nguyen
Atomic Energy Control Board, Ottawa, Ont., Canada

A.P.S. Selvadurai
McGill University, Montreal, Que., Canada

ABSTRACT: The concept of geological disposal of heat emitting nuclear fuel wastes being considered by many countries consists of burying the wastes in a geological formations at depths ranging from hundreds to thousands of metres. The only credible agent which could transport contaminants from the stored wastes to the ground surface is the groundwater. It is generally recognized that the groundwater regime might be significantly influenced by thermal and mechanical processes occurring in the host geological medium. In order to further our understanding of the coupled thermal-hydraulic-mechanical processes occuring in both intact and jointed rock, a research program consisting of computational and experimental modelling is being conducted by the Atomic Energy Control Board and McGill University. In this paper we present the results of experimental and computational simulations of pressure pulse and heating tests performed with cylindrical samples of intact saturated granite. The samples contain a central water-filled cavity, where the water pressure transients are continually monitored during a test. The time-varying water pressure in the cavity bears the signature of the coupled T-H-M behaviour of the rock.

1 INTRODUCTION

Currently, many countries are actively seeking solutions for the permanent disposal of heat emitting nuclear fuel wastes (NFW) generated from their nuclear reactors. The most promising method to date consists of burying the wastes in a system of rooms excavated at depths ranging from hundreds to thousands of metres in a geological formation. In Canada, Atomic Energy of Canada Ltd (AECL) has proposed the concept of NFW disposal in a plutonic rock formation of the Canadian Shield (AECL, 1994 a, b,c). The primary agent that could transport the contaminants from the wastes to the surface is groundwater. In order to minimize the rate of contaminant transport, a system of engineered barriers will be provided, consisting of copper or titanium canisters containing the NFW; low permeability, highly sorptive clay/sand/gravel mixtures emplaced around the canisters (the buffer and the backfill). The host rock formation serves as the final barrier between the wastes and the surface biosphere. The rate of contaminant transport in the geological medium will largely be governed by the in-situ permeability, the natural hydraulic gradients and other transport characteristics of the medium. It seems that large regions of competent sparsely fractured rock could be found in the Canadian Shield (AECL, 1994 a,b, c). For these types of rocks, flow will occur both in the microcracks of the intact rocks, and the sparsely distributed joints. Due to the decay of the radionuclides in the wastes, a thermal pulse will be created in the geological medium, with estimated peak temperatures of the order of 100°C. These high temperatures are expected to perturb the mechanical and hydraulic regimes of the geological formation. Scoping calculations (Selvadurai and Nguyen, 1997) taking into account the coupling of thermal-hydraulic-mechanical (T-H-M) processes have shown that under certain adverse conditions the thermal pulse might generate sufficiently high porewater pressures and hydraulic gradients to induce tensile cracks in the rock mass around the NFW rooms. The increases in porewater pressure occur as a result of the thermal expansion of the water and the time lag associated with its dissipation due to the low permeability of the rock mass. Under such conditions, the rate of groundwater flow, and consequently the rate of contaminant transport, could be accelerated in the vicinity of the wastes for the duration of the thermal pulse.

In order to further our understanding of coupled T-H-M processes, a research program consisting of computational and experimental modelling is being conducted at McGill University and the Atomic Energy Control Board (AECB) on both intact and fractured granite. In this paper, we will present some results for the study of intact granitic rock.

2 THEORETICAL FRAMEWORK

Coupled T-H-M processes in saturated geological media could be interpreted with Biot's (1941) theory of consolidation. By extending Biot's theory of consolidation to include thermal effects, the following governing equations could be obtained (Booker and Savvidou, 1985; Selvadurai and Nguyen, 1995 ; Nguyen and Selvadurai, 1995) :

$$\frac{\partial}{\partial x_i}\left(\kappa_{ij}\frac{\partial T}{\partial x_j}\right) = \rho C\frac{\partial T}{\partial t} \qquad (1)$$

$$G\frac{\partial^2 u_i}{\partial x_j \partial x_j} + (G+\lambda)\frac{\partial^2 u_j}{\partial x_i \partial x_j} +$$

$$\alpha\frac{\partial p}{\partial x_i} - \beta K_D\frac{\partial T}{\partial x_i} + F_i = 0 \qquad (2)$$

$$\frac{\partial}{\partial x_i}\left(\frac{k_{ij}}{\mu}\left(\frac{\partial p}{\partial x_j} + \rho_f g_j\right)\right) - \left(\frac{n}{K_f} - \frac{n-\alpha}{K_s}\right)\frac{\partial p}{\partial t} +$$

$$\alpha\frac{\partial}{\partial t}\frac{\partial u_i}{\partial x_i} + \left((1-\alpha)\beta - (1-n)\,\beta_s - n\beta_f\right)\frac{\partial T}{\partial t} = 0 \qquad (3)$$

where the unknowns are the displacement u_i [m], temperature T [°C] and pore pressure p [Pa]

κ_{ij} is the thermal conductivity tensor of the bulk medium [W/m/°C]

k_{ij} is the intrinsic permeability tensor [m^2]

ρ is the density of the bulk medium [kg/m^3]

ρ_f is the density of the fluid [kg/m^3]

C is the specific heat per unit mass of the bulk medium [J/kg/°C]

G and λ are the Lame's constants of the porous skeleton [Pa]

$\alpha = 1 - K_D/K_s$

K_D, K_s and K_f are respectively the bulk modulus of the drained material, the solid phase and the fluid phase [Pa]

F_i is the volumetric body force [N/m^3]

β, β_s and β_f are respectively the coefficient of thermal expansion of the drained material, the solid phase and the fluid phase [$°C^{-1}$]

n : is the porosity of the medium [dimensionless]

μ is the viscosity of the fluid [kg/m/s]

g_i is the ith component corresponding to the acceleration due to gravity [m/s^2]

(In the above parameter definitions the bracketed designations are in typical SI units)

In the above equations, the Cartesian tensor notation, with Einstein's summation convention on repeated indices is adopted. For the present, the sign convention for stresses and the fluid pressure adopt tension fields as positive. In developing these equations, we invoke the basic principles of continuum mechanics (namely conservation of mass, momentum and energy). These principles are universally applicable, independent of the nature of the medium being considered. In order to complete the formulation (i.e. to develop a set of equations in which the number of unknowns equals the number of equations), the following additional assumptions are necessary:

i) *Representation of granitic rock mass as a single porosity continuum.* Granitic rock masses are in general characterized by the presence of fractures. In order to take this characteristic into account, many researchers utilize the concept of a dual-porosity/dual permeability medium (see e.g. Huyakorn and Pinder 1983 and Berryman and Wang, 1995 for a detailed description of the concept). According to that concept the rock mass is idealized into a porous matrix partitioned into blocks by a fracture network. For a saturated rock mass, the fluid would be present in two types of void space: the void in the porous blocks due to the pores and microcracks in the intact rock; and the voids between the walls of the fractures. A representative elementary volume (REV) is then defined in that dual-porosity medium. A REV is a finite volume in the medium that contains a number of porous blocks and fractures and that surrounds a point mathematically defined in that medium. The average properties and also the average parameters in the REV are assigned to the mathematical point. The REV should be large enough to contain a sufficient number of porous blocks and fractures so that the average value of a given property or parameter has a statistical significance. On the other hand, the REV should be sufficiently small so that the variations of these properties and parameters from one domain to the next may be approximated by continuous functions so that the use of infinitesimal calculus is still appropriate. Two types of fluid pressures are defined for a dual-porosity medium: one fluid pressure in the pore and microcracks of the porous blocks, and one fluid pressure in the fractures.

The dual porosity representation is appropriate for

highly fractured rock masses. Rocks at depths beyond a few hundred metres of the Canadian Shield are very sparsely fractured (AECL, 1994 a,b and c) with typical fracture frequency of one every 300-400 metres. The FRACON code has recently been developed with this particular application in mind. Consequently, only one type of porosity, namely that due to the pores and microcracks of the rock matrix, is considered. Discrete fractures that intersect the rock mass are explicitly represented by special joint elements described later in this paper.

ii) *Darcy's law governing pore fluid flow*. Darcy's law is applicable with reasonable accuracy to a variety of geological materials (Freeze and Cherry, 1979) including soils and rocks, provided that the hydraulic gradients are within the laminar flow range, and above a threshold gradient within which the pore fluid is virtually immobile.

In most conventional geotechnical applications, the pore fluid is water at a constant temperature, and the original form of Darcy's law is appropriate. Where thermal effects are important, and/or when the pore fluid is not water, Darcy's law has to be modified (Huyakorn and Pinder, 1983) , i.e.

$$V_{if} - V_{is} = \frac{k_{ij}}{n\,\mu}\left(\frac{\partial p}{\partial x_j} + \rho_f g_j\right) \qquad (4)$$

V_{if}, V_{is} are the velocities, respectively of the fluid and the solid components [m/s]

The use of the generalized Darcy's law (4) is essential when one deals with fluids other than water. It is thus necessary to separate the expression for the hydraulic conductivity K_{ij} (loosely referred to as the permeability in most geotechnical applications) into a fluid independent component k_{ij}, and a fluid-dependent component characterized by its viscosity and density, i.e. :

The viscosity and density of a particular fluid are

$$K_{ij} = \frac{\rho_f g}{\mu} k_{ij} \qquad (5)$$

also temperature-dependent. When thermal effects are considered, appropriate experimentally derived functions of temperature should be used for these two properties.

iii) *The principle of effective stress*. A generalized principle of effective stress is adopted. Several forms of this principle exist (Terzaghi, 1923; Biot, 1941; Zienkiewicz et al., 1977; Rice and Cleary, 1976). We adopt the form of generalized principle of effective stress formulated by Zienkiewicz et al. (1977). In contrast to Terzaghi's (1923) principle of effective stress, this generalized principle takes into consideration the compressibility of the pore fluid and the solid phases. Omission of the compressibility of the pore fluid and the solid phase could lead to an overprediction of pore pressures in competent rocks (Selvadurai and Nguyen, 1995).

iv) *Stress strain relationship for the solid matrix*. Hooke's law for linear isotropic elastic behaviour of the porous skeleton is adopted. This is a useful first approximation for the study of intact competent rocks which are subjected to stress states lower than those which could initiate fracture, failure or damage. Discrete joints on the other hand could behave in a linear elastic or an elasto-plastic fashion (Nguyen and Selvadurai, 1997).

v) *Heat transfer mechanism*. Heat conduction is assumed to be the predominant mechanism of heat transfer. In the current context, there are two dominant mechanisms of heat transfer in a geological medium: heat conduction and heat convection. Heat conduction is the transfer of heat energy by the activation of solid and fluid particles, without their bulk movement. The conduction of heat is governed by Fourier's law, which states that the rate of heat flow is proportional to the temperature gradient. Convective heat transfer on the other hand is due to the bulk motion of the particles. In a poro-elastic medium, the movement or displacement of the solid particles could be neglected; thus it is the fluid flow which is primarily responsible for the convective heat transfer. The rate of heat transfer by convection is proportional to the rate of fluid flow. It can be shown that for geological media with low permeability, such as granitic rock masses, heat convection is negligible in most situations. Even for fracture zones with relatively high permeability of the order of 10^{-14} m^2, Nguyen (1995) has shown that the convective heat transfer component becomes important only when the fracture zones intersect a NFW repository, where heat induced hydraulic gradients of the order of 100% could be found. For a fracture zone situated, say 100 m from the repository, at the time the thermal pulse reaches the fracture, hydraulic gradients are much smaller as a result of pore pressure dissipation. Consequently, the groundwater velocity is still low and the convective heat component could also be neglected in the latter case.

We assume that conduction is the main mechanism of heat transfer, and that Fourier's law applies. We also assume that a state of thermal equilibrium always exists between the fluid and the solid (i.e. at any point, the temperature of the solid equals the temperature of the fluid).

3 FINITE ELEMENT APPROXIMATION OF THE GOVERNING EQUATIONS

A finite element code, FRACON, has been developed in order to numerically solve the governing equations (1)-(3). Two types of elements are considered in the code:

i) *Plane isoparametric elements*: This element (Figure 1) is used to represent the intact rock mass. Displacements within the element are interpolated as functions of the displacements at all 8 nodes, whereas the pore pressure and temperature are interpolated as functions of the values at only the four corner nodes 1, 3, 5 and 7 only. A detailed description of this element is given by Smith and Griffiths (1988).

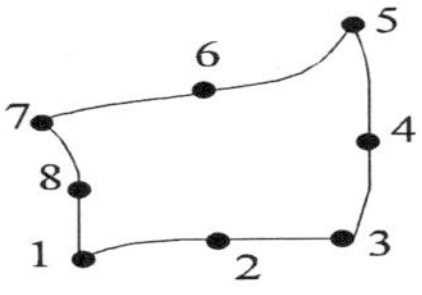

Figure 1 Isoparametric plane element

ii) *Joint Element*: This element (Figure 2) is used to simulate discontinuities in the rock mass such as joints, fracture zones and fault zones. In finite element terminology (see e.g. Goodman et al., 1968; Zienkiewicz et al., 1970; Noorishad et al., 1971; Ghaboussi et al., 1973; Desai and Nagaraj, 1986; and Selvadurai and Boulon, 1995), it is a very thin element, characterized by a thickness b and length L. Nodal displacements are obtained at all six nodes (Figure 2) while nodal pore pressures and temperatures are obtained only at the corner nodes 1, 3, 4 and 6. The mechanical behaviour of the element is dictated by its shear and normal stiffnesses D_s and D_n respectively and its hydraulic and thermal behaviour are governed by the tranverse and longitudinal permeabilities $k_{y'y'}$ and $k_{x'x'}$, and the tranverse and longitudinal thermal conductivities $\kappa_{y'y'}$ and $\kappa_{x'x'}$, respectively .

With the above two types of elements being fully defined, a Galerkin procedure is applied to the differential equations (1)-(3) of non-isothermal consolidation. The resulting matrix equations have the forms:

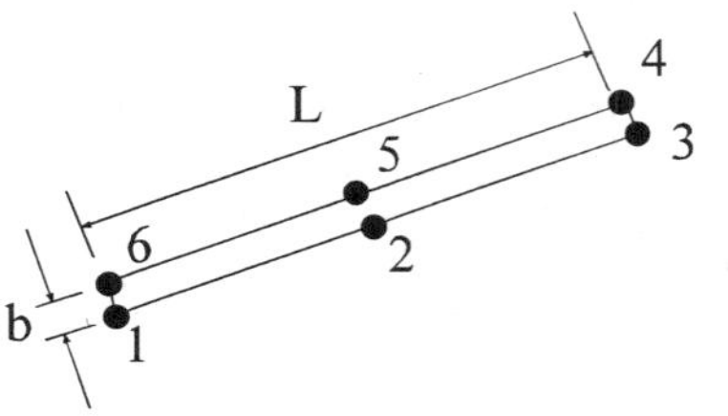

Figure 2 Joint element

$$\left[\theta[\boldsymbol{KH}] + \rho C/\Delta t[\boldsymbol{CM}]\right]\{\boldsymbol{T}\}^1 = \{\boldsymbol{FH}\}+\{\boldsymbol{FQ}\} + \left[(\theta-1)[\boldsymbol{KH}] + \rho C/\Delta t[\boldsymbol{CM}]\right]\{\boldsymbol{T}\}^0 \qquad (6)$$

$$\begin{bmatrix} [\boldsymbol{K}] & \alpha[\boldsymbol{CP}] \\ \alpha[\boldsymbol{CP}]^T & -\theta\Delta t[\boldsymbol{KP}] - c_e[\boldsymbol{CM}] \end{bmatrix} \begin{Bmatrix} \{\boldsymbol{d}\}^1 \\ \{\boldsymbol{p}\}^1 \end{Bmatrix} = \{\boldsymbol{f}\} + \begin{bmatrix} [\boldsymbol{K}] & \alpha[\boldsymbol{CP}] \\ \alpha[\boldsymbol{CP}]^T & (1-\theta)\Delta t[\boldsymbol{KP}] - c_e[\boldsymbol{CM}] \end{bmatrix} \begin{Bmatrix} \{\boldsymbol{d}\}^0 \\ \{\boldsymbol{p}\}^0 \end{Bmatrix} + \begin{bmatrix} \beta K_D[\boldsymbol{CP}] & [\boldsymbol{0}] \\ [\boldsymbol{0}] & -\beta_e[\boldsymbol{CM}] \end{bmatrix} \begin{Bmatrix} \{\boldsymbol{T}\}^1 - \{\boldsymbol{T}\}^0 \\ \{\boldsymbol{T}\}^1 - \{\boldsymbol{T}\}^0 \end{Bmatrix} \qquad (7)$$

w h
ere:

the unknowns are the nodal displacements $\{\boldsymbol{d}\}^1$, the nodal temperatures $\{\boldsymbol{T}\}^1$ and the nodal pore pressures $\{\boldsymbol{p}\}^1$ at the current time step

$\{\boldsymbol{d}\}^0$, $\{\boldsymbol{T}\}^0$ and $\{\boldsymbol{p}\}^0$ are the nodal displacements , nodal temperatures, and nodal pore pressures at the previous time step

$\{\boldsymbol{f}\}$ is the "force" vector ,$\{\boldsymbol{FQ}\}$ and $\{\boldsymbol{FH}\}$ are heat flux vectors, θ is a time integration constant

all the other matrices , $[\boldsymbol{K}]$, $[\boldsymbol{CP}]$, etc. are assembled from element matrices , which are dependent on thermal, mechanical and hydrological properties of the individual elements and the interpolation functions used. Also,

$c_e = n/K_f - n/K_s + \alpha/K_s$

$\beta_e = (1-\alpha)\beta - (1-n)\beta_s - n\beta_f$

Also Δt is the time increment and the time integration constant θ varies between 0 and 1. Using a value of θ close to 1, we observe that after the first few three or four time steps, stability of the solution is generally reliably achieved.

The FRACON code was extensively verified against analytical solutions for both isothermal and thermal consolidation of porous media (Nguyen and Selvadurai, 1995). The code was successfully used to interpret a laboratory experiment involving the heating of a partially saturated cementitious material (Nguyen and Selvadurai, 1995) and in the interpretation of laboratory and field experiments involving the mechanical and coupled hydraulic-mechanical behaviour of rock joints (Nguyen and Selvadurai, 1997).

4 EXPERIMENTAL SETUP

An experimental facility has been developed at McGill University for the study of the T-H-M behaviour of both intact and jointed rock samples. The general arrangement of the facility is shown in figure 3. The maximum dimensions of the cylindrical rock samples were specified at 460 mm in diameter and 508 mm in height. These rock samples are so far the largest ever used in laboratory tests for coupled T-H-M processes. The samples are provided with a central cylindrical cavity filled with water. A pressure transducer continually monitors the water pressure in the central cavity; thermistors and strain gauges could also be introduced at salient points of the samples. A computerized data acquisition system record all the measurements. Immediately prior to the tests, the whole sample is saturated by imposing a constant outward flow rate until outflow is observed at the outer radius of the cylindrical sample. In this paper, we report the results of two tests on a Barre granite sample, from Vermont:

i) *the pulse test*: for this test, the water pump injects water into the central cavity to instantaneously raise the pressure to a given value, and the pump is then deactivated. Due to outward radial flow, the pressure will gradually decay to zero. The pressure decay curve depends on the permeability as well as the elastic properties of the rock sample, according to Biot's theory of consolidation (cf. Equations (2) and (3) where the temperature term is dropped).

ii) *the heating test*: for this test, heat is applied at the exterior surface of the cylinder and the temperature at all points of the sample will gradually increase. In the cavity, following a time lag, the water temperature will also increase and induce an increase in the pressure in the cavity. This water pressure will subsequently dissipate due to outward radial flow through the granite sample.

The Barre granite used in the above experiments is a blue graniodorite composed of approximately 23% quartz, 61% feldspar and 11% mica. The following T-H-M properties of the granite were determined by independent tests:

Young's modulus: 59 GPA
Poisson's ratio: 0.138
Porosity: 0.011
Coefficient of linear expansion: 9.1×10^{-6} m/m/°C
Thermal conductivity: 2.5 W/m/°C
Heat capacity: 905 J/kg/°C
Bulk density: 2630 kg/m^3

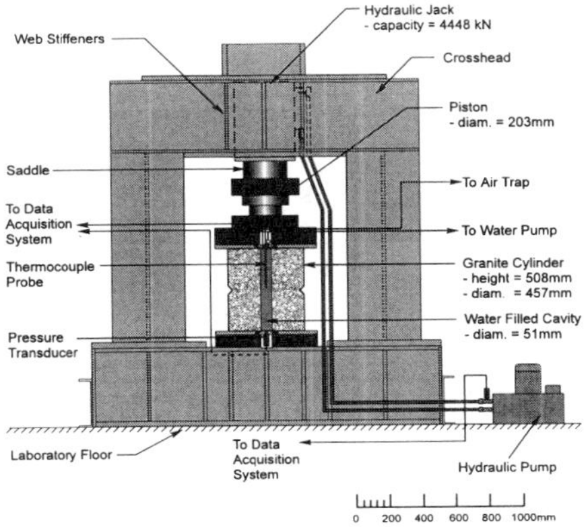

Figure 3 Experimental configuration for T-H-M tests

5 EXPERIMENTAL AND COMPUTATIONAL RESULTS

5.1 Pulse tests

The cylindrical cavity for the tests we consider here has a diameter of 51 mm. The pressure inside the cavity was intantaneously raised to 600 kPa. The test was simulated with the FRACON code using the axisymmetric finite element mesh shown in figure 4. The intact rock was represented with isoparametric solid elements, with a Young modulus of 59 GPa and a Poisson's ratio of 0.138. The water inside the cavity was represented by fluid elements having essentially negligible shear resistance. The compressibility of water is 4.5×10^{-10} Pa^{-1}.

The experimental and calculated decay curves for the pressure inside the cavity is shown in figure 5. Six pressure pulse tests were performed. It could be seen that after the first two tests, all subsequent tests essentially resulted in the same decay curve. For a value of permeability 2.5×10^{-19} m^2, the FRACON results compare well with the experimental results for tests 3 to 6. The decay curve for the same value of

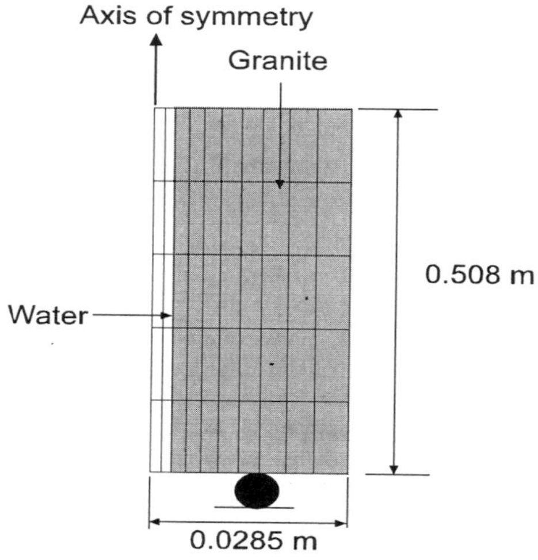

Figure 4 Finite element model for the pulse test

permeability derived from an analytical solution is also shown in figure 5. The analytical solution considers the radial flow from an infinitely long borehole into an infinite porous medium (Selvadurai and Carnaffan, 1997). The analytical solution considers a purely diffusive phenomenon with accommodation for bulk compressibility of the porous skeleton. It could be seen that the FRACON results are also close to the analytical solution. Thus for this particular example, the

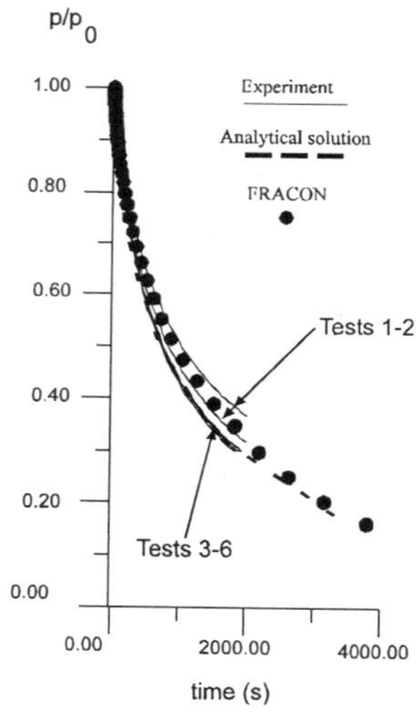

Figure 5 Decay curve for pressure in cavity

approximation used in the analytical solution (infinite length and infinite radius of the test cylinder) is a reasonable one.

5.2 Heating test

At the end of the previous series of pulse test, band heaters were provided around the exterior surface of the same cylinder (figure 6). The heaters were turned on and the temperatures at three points were monitored. The first point is located at mid-height of the outer surface, the second point is located at mid-distance between the outer surface and the cavity on top of the sample, and the third is located in the centre of the cavity. Ceramic blankets at the top and bottom faces of the cylinders provide some insulation. The heating test was simulated with the FRACON code using the axisymmetric finite element mesh shown in figure 7. Assuming symmetry around the horizontal plane at mid-height of the cylinder, only the half-top is represented. The finite element mesh with the boundary conditions are shown in figure 7. We assumed symmetry around the plane intersecting the mid-height of the specimen. The band heaters were turned on for 7.5 hours (450 minutes) and were suddenly turned off. The temperature at monitoring point 1 reached a maximum of 180 °C at the instant when the heaters were turned off (figure 8). To simulate the effects of the band heaters, we prescribed temperature variations which correspond to the recorded temperatures at monitoring point 1 (figure 8). A calibration was performed to fit the calculated and experimental temperature values at monitoring points 2 and 3. In order to obtain a good match, the following assumptions were adopted:

1. All points on the outer radius are assumed to vary with time in a similar fashion to monitoring point 1, except that the peak temperature at the ends are assumed to be 50 °C smaller (the peak temperature attained at 450 min is 130 °C).
2. Heat losses occur through the ceramic blanket according to the equation:

$$Q = h\,(T - T_a) \tag{8}$$

where T is the temperature at the top boundary, T_a is the outside ambient temperature (T_a=23 °C) and h is the heat loss coefficient. A value of h= 0.0005 J/s/m²/°C was used in the analysis.

The hydraulic and mechanical properties for the granite are the same as the ones used in the pulse test, in particular, the permeability was assumed to be 2.5×10^{-19} m². The thermal properties, the coefficient of thermal expansion, the density and porosity of the granite used in the FRACON simulation are the ones determined from separate laboratory tests (cf. Section 4). In addition, the viscosity, density and coefficient of thermal expansion of the water are temperature dependent and their values are tabulated in most textbooks (see e.g. de Marsily, 1986). The following best fit functions for the thermal expansion coefficient, viscosity, and density of water were used in the FRACON simulation:

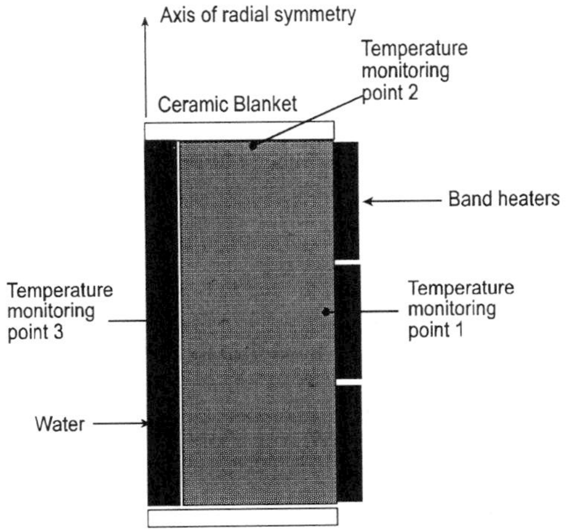

Figure 6 Experimental configuration for heating test

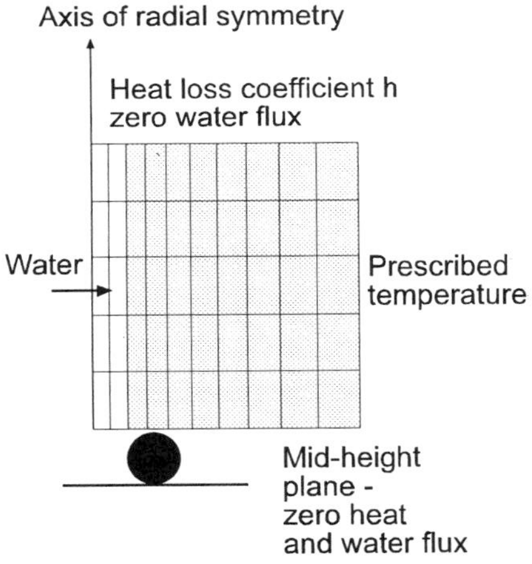

Figure 7 Finite element model for heating test

$$\beta_f = 4.684\text{x}10^{-5} + 7.486\text{x}10^{-6}T \quad (9)$$
$$\mu = 2.285\text{x}10^{-3} + 1.01\text{x}10^{-3}\log_{10}(T) \quad (10)$$
$$\rho = 4.684\text{x}10^{-5} + 7.486\text{x}10^{-6}T \quad (11)$$

The calculated and measured temperatures at monitoring points 2 and 3 are compared in figure 9. There is good agreement in the trends between experimental and calculated temperatures. The calculated and measured water pressure in the cavity are compared in figure 10. It could be seen that although the trends are well simulated by the FRACON code, the absolute values of the peak pressure are overpredicted by a factor of three compared to the experimental results. In particular, very high negative pressures in the cool-down phase are predicted with the FRACON code.

The variations of the water pressure in the central cavity could be divided in three distinct phases, and could be interpreted as follows:

i) In the first phase (time 0 to approximately 50 minutes), due to heating from the outside, there was an instantaneous outward expansion of the cavity. This expansion should induce some negative pressures in the cavity, as calculated by the FRACON code. Experimentally, however, no early negative pressures were detected by the pressure transducer.

ii) In the second phase (time approximately 50 minutes to approximately 250 minutes), the heat pulse started to reach the cavity, after a time lag of approximately less than one hour. The temperature in the cavity started to increase rapidly (point 2 in figure 9). Due to the temperature rise, the water experienced expansion; however, this thermal expansion was inhibited by the surrounding rock, and consequently, the water pressure inside the cavity increased.

iii) In the third phase (after approximately 250 minutes), the water temperature in the cavity increased at a slower rate and subsequently decreased. The pressure inside the cavity gradually dissipated as outward flow through the rock took place . Both the experimental data and the FRACON predictions show that the pressure would eventually become negative (after approximately 500 minutes), because of the expanded volume of the cavity.

In order to further assess the discrepancy between the FRACON results and the experimental data, a second analysis was performed with two adjustments in the FRACON analysis:

i) the radial permeability of the rock was increased to $9\text{x}10^{-19}$ m^2. This is a plausible assumption, since due to thermal effects, tensile stresses are induced in the hoop direction of the cylinder. These tensile stresses

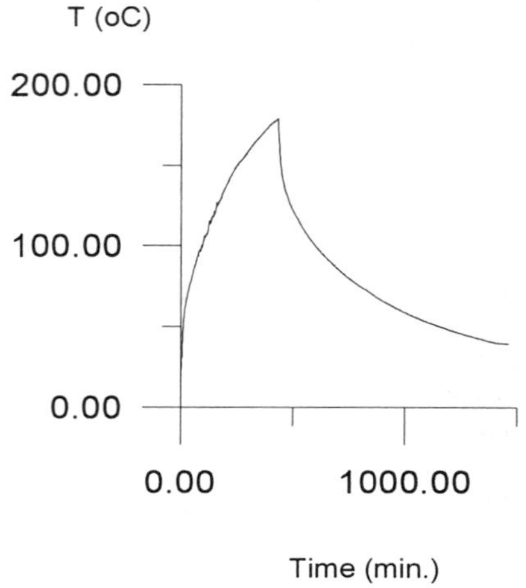

Figure 8 Prescribed temperature at monitoring point 1

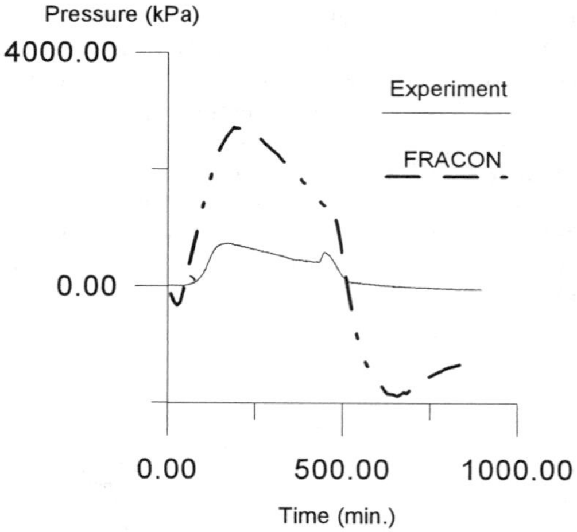

Figure 10 Pressure in central cavity - rock permeability is 2.5×10^{-19} m^2 in FRACON analysis

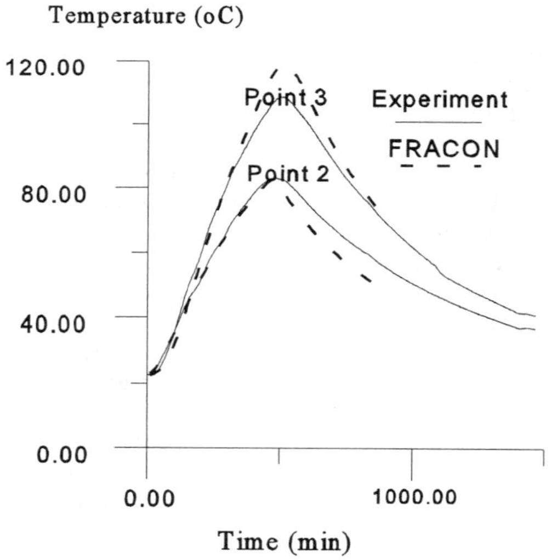

Figure 9 Temperature variations at points 2 and 3

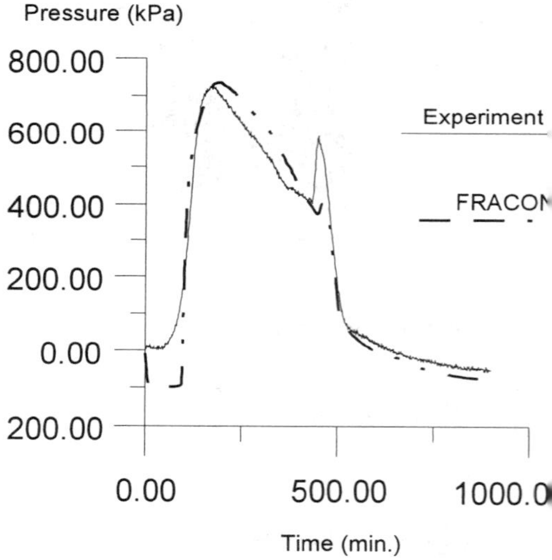

Figure 11 Pressure in central cavity - rock permeability increased to 9×10^{-19} m^2 in FRACON analysis

will open the pores and microcracks of the rock and the permeability should be increased.

ii) At a given temperature, the pressure inside the cavity should be higher than the vapour pressure. For example at 20°C, the vapour pressure is approximately -98 kPa, at which point, the water will cavitate (formation of air bubble), and thus the pressure could not be lower than -98 kPa. We modified the FRACON code so that at any temperature, the vapour pressure is calculated according to steam tables available in most textbooks (see e.g. de Marsily, 1986).When the value of calculated pressure decreases and approaches the vapour pressure, the compressibility of the water is arbitrarily increased to a high value, in order to simulate the effects of occluded air bubbles being formed.

The results of the second FRACON analysis compare well with the experimental results, as shown in figure 11.

6 CONCLUSIONS

This paper essentially presented experimental and computational results for two types of tests performed on intact granite samples, with a central, water-filled cavity. In these tests, the water pressure in the cavity is continuously monitored; this pressure transient reflects the coupled T-H-M characteristics of the granite. The pulse test proved to be a reliable and rapid method of determining the initial permeability of intact granite. The heating tests showed the characteristic increase in water pressure due to thermal expansion, followed by dissipation when radial drainage took place. The experimental and computational results also suggest that during heating, due to the formation of tensile stresses, the initial permeability could be increased by a factor of four.

ACKNOWLEDGEMENT

We sincerely thank the AECB for its financial support, our colleagues at the AECB, Dr. M. Benbelfadhel, Mr. K. Bragg, Mr. P. Flavelle for their peer review of this paper, and Mr. N. Vannelli for providing the experimental data.

REFERENCES

Atomic Energy of Canada Ltd., 1994-a, 'Environmental Impact Statement on the Concept for Disposal of Canada's Nuclear Fuel Waste', AECL-10711,COG-93-1.

Atomic Energy of Canada Ltd., 1994-b, 'The Disposal of Canada's Nuclear Fuel Waste: engineering for a disposal facility', AECL-10715, COG-93-5.

Atomic Energy of Canada Ltd., 1994-c, 'The Disposal of Canada's Nuclear Fuel Waste: the Geosphere Model for Postclosure Assessment', AECL-10795, COG-93-9.

Berryman, J.G. and Wang, H.F., 1995, 'The elastic coefficients of double-porosity models for fluid transport in jointed rock', J. of Geoph. Research, 100, B12, 24 611-24 627.

Biot, M.A., 1941, 'General theory in three dimensional consolidation', J. Appl. Physics, 12, 155-164.

Booker, J.R. and Savvidou, C., 1985, 'Consolidation around a point heat source', Int. J. Numerical and Analytical Geomech., 9, 173-184.

Desai, C.S. and Nagaraj, B.K., 1986, 'Constitutive modelling for interfaces under cyclic loading', in Mechanics of Material Interfaces (Selvadurai, A.P.S. and G.Z. Voyiadjis, Eds.), II, 'Studies in Applied Mechanics', Elsevier, 97-108.

Freeze, R.A. and Cherry, J.A., 1979,' Groundwater', Prentice Hall.

Ghaboussi, J., Wilson, E.L. and Isenberg, J., 1973, 'Finite elements for rock Joints and interfaces', J. Soil Mech. Fdn. Div. Proc. ASCE, 99, 833-848.

Goodman, R.E., Taylor, R.L. and Brekke, T.L., 1968, 'A model for the mechanics of jointed rock', J. Soil Mech. Fdn. Div. Proc. ASCE, 94, 637-659.

Huyakorn, P.S. and Pinder, G.F., 1983, 'Computational Methods in Subsurface Flow', Academic Press.

Noorishad, J., Witherspoon, P.A. and Brekke, T.L., 1971, 'A method for coupled stress and flow analysis of fractured rock masses', Geotechnical Engineering Publication No. 71-6, University of California, Berkeley.

Nguyen, T.S., 1995, 'Computational Modelling of Thermal-Hydrological-Mechanical Processes in geological Media', Ph.D thesis, App. A, Dept. of Civil Engineering and Applied Mechanics, McGill University, Montreal, Canada.

Nguyen, T.S. and Selvadurai, A.P.S., 1995, 'Coupled thermal-hydrological-mechanical processes in sparsely fractured rock', International Journal of Rock Mechanics and Mining Sciences & Geomechanics Abstracts, 32 (5), 465-480.

Nguyen, T.S. and Selvadurai, A.P.S., 1997, 'A model for coupled mechanical and hydraulic behaviour of a rock joint', Int. J. of Numerical and Analytical Meth. Geomech. (In press).

Rice, J.R. and Cleary, M.P., 1976, 'Some basic stress

diffusion solutions for fluid-saturated elastic porous media with compressible constituent', Rev. Geophys. Space Phys., 14, 227-241.
Selvadurai, A.P.S. and Boulon M. (Eds.), 1995, 'Mechanics of geomaterial interfaces', Elsevier.
Selvadurai, A.P.S. and Carnaffan, P., 1997, 'A transient pulse pressure method for the measurement of the permeability of a cement grout', Canadian J. of Civil Eng. (in press)
Selvadurai, A.P.S. and Nguyen, T.S., 1997, 'Scoping analyses of the coupled thermal-hydrological-mechanical behaviour of the rock mass around a nuclear fuel waste repository', Eng. Geology (in press).
Selvadurai, A.P.S. and Nguyen, T.S., 1995, 'Computational modelling of isothermal consolidation of fractured porous media', Computers and Geotechnics , 17 (1), 39-73
Smith, I.M. and Griffiths, D.V., 1988, 'Programming the Finite Element Method', John Wiley & Sons.
Terzaghi, K., 1923, 'Die Berechnung der Durchlassigkeitsziffer des Tones aus dem Verlauf der hydrodynamischen Spannungserscheinungen', Ak. der Wissenschaften in Wien, Sitzungsberichte mathematisch-naturwissenschaftliche Klasse, part Iia, 132(3/4), 125-38.
Zienkiewicz, O.C., Humpheson, C. and Lewis, R.W., 1977, 'A unified approach to soil mechanics problems (including plasticity and viscoplasticity)', in Finite Element in Geomechanics, edited by G. Gudehus, 151-178, John Wiley & Sons Ltd .
Zienkiewicz, O.C., Best, B., Dullage, C. and Stagg, K., 1970, 'Analysis of non-linear problems in rock mechanics with particular reference to jointed rock systems', Proc. 2nd Congress Int. Soc. Rock Mech., Belgrade, Yugoslavia, 3, 501-509 (1970) .

Numerical Models in Geomechanics, Pietruszczak & Pande (eds) © 1997 Balkema, Rotterdam, ISBN 90 5410 886 X

Chemically and biologically induced consolidation problems in geomechanics – An overview

T. Hueckel, Q. Zhang & Y. P. Chang
Department of Civil & Environmental Engineering, Duke University, Durham, N.C., USA

M. Kaczmarek
Polish Academy of Sciences, Laboratory of Mechanics and Acoustics of Porous Media, Poznan, Poland
(Presently: Bydgoska Wyzsza Szkola Inzynierska, Poland)

Abstract: Physico-chemical process resulting from the permeation of clay with an organic contaminant or a salt may result in consolidation of clay at a constant load. In the same line, an organic component of landfill mass undergoes the process of decomposition which eventually leads to consolidation and landfill settlement. Preliminary experimental results for both phenomena are presented and discussed in terms of constitutive relationships. A solution for a linear case of chemically induced consolidation is then obtained.

1 INTRODUCTION

Classical meaning of consolidation established by Terzaghi (1943) defines it as "every process involving a decrease of the water content of saturated soil without replacement of the water by air". "Because of the low permeability of soil does not permit a rapid transfer of the water from one part of the mass of soil to another, ..[consolidation]... produces a time lag between a change of the external forces which act on a feeble permeable, compressible stratum and the corresponding change of the water content of the soil." Terzaghi (1943). Most of the considerations on consolidation in soil mechanics adress mechanically induced consolidation. However, geologic materials of many types are subject to consolidation due to loads other than mechanical. For example, Terzaghi (1943) have addressed consolidation due to surface evaporation or dessication, which was understood as induced by capillary pressure in soil.

This paper focuses on consolidation induced by chemical or biological processes affecting the skeleton of the geological materials. In the first part of this paper some experimental evidence will be presented and constitutive equations for such phenomena will be discussed. Two particular processes will be addressed: the effect of changes in the chemical composition of pore liquid on compressibility and permeability of clays; and effect of biological decomposition of landfill material on its settlement. The first problem relates to stability of landfill and impoundment liners, the clay component of which may increase in permeability up to 1000 times as a result of contamination, depending on the "healing" role of consolidation. The second problem results from aerobic and/or anerobic decomposition of organic parts of landfill mass leading to an increase in porosity and permeability and weakening of the landfill skeleton. However, the waste weight can induce consolidation of the mass, thus reducing porosity, permeability and producing visible settlements. Such settlements are known to reach up to 30% of the landfill height. The first part of the paper discusses the basic formulation of consolidation in the presence of degradation processes, phenomenology involved, and some implications for the constitutive modeling. In the second part an analytical solution will be outlined for one-dimensional consolidation process for a simplified linearized case.

2. GENERAL FORMULATION FOR CHEMICAL AND BIOLOGICAL CONSOLIDATION

In this Section we shall discuss the coupled role of chemical or biological processes and stress in 1-D consolidation following as close as possible the original Terzaghi's formulation of consolidation. There are three mechanisms which will be addressed, through which chemical or biological degradation may be thought to affect effective stress: two linked to porosity changes and one connected to permeability. The processes which will be discussed occur in fluid saturated medium and effectively result in loss of mass of the solid skeleton of the medium. At the same time these processes produce changes in the properties of the fluid. To model these mechanisms in the context of consolidation the mass balance law for fluid should be invoked. The change of fluid mass can be represented as the change in total porosity and a change in density of fluid due to the physico-chemical processes or biological reactions. The latter one may result from either the simple mixing of contaminant and water, and the change of density of water during the transition from adsorbed to free state in the case of contamination, or due to formation of micro-organisms in water as a result of biological decomposition of organic solids. The mass exchange between adsorbed and free water does not affect total fluid mass balance, as internal. Thus, the total fluid mass balance reads

$$-\frac{\partial n}{\partial t}+n_0\frac{1}{\rho}\xi\frac{\partial \rho}{\partial t}=\frac{\partial Q}{\partial z} \tag{1}$$

where ξ is the coefficient of the resulting change of total mass density of pore liquid, ρ, due to degradation, n_0 is the initial porosity, Q is the specific discharge of fluid.

Two of the above mentioned mechanisms refer to changes in porosity. The first mechanism, results from the porosity change due to degradation of the solid. The second one, relates to strain, or porosity change resulting from the weakening of the skeleton at constant load due to chemical or biological process. These two changes in porosity correspond to what is traditionally referred to as change in effective porosity, and to the difference between total and effective porosity. To obtain semi-analytical solution, porosity or 1-D strain need to be employed to describe the deformation. To adhere as close as possible to Terzaghi's formulation, the total porosity change, is expressed in terms of differences, rather than rates of effective stress σ, and degree of reaction $\mu = 1-m/m_0$ (m and m_0 being initial and remaining mass available for degradation reaction)

$$\Delta n = m_{v0}(\overline{\sigma}-\overline{\sigma}_0)+m_{bm}(\overline{\sigma}-\overline{\sigma}_0)\mu+m_b(\mu-\mu_0), \tag{2}$$

where m_{v0} is the initial loading modulus (1-D), m_{bm} is the change in the modulus due to degradation process, and m_b is the coefficient of volume fraction loss due to reaction described by its degree μ. Note that the above relationship is written for the mechanical loading process, in which $\overline{\sigma}\dot{\overline{\sigma}} > 0$.

The degree of reaction is process specific. For the biological decomposition process it is defined through the rate of the mass loss described by the empirical Michaelis-Menten model or the classical Arrhenius rate of reaction as follows (Probstein, 1994)

$$\dot{\mu}=\frac{-[\dot{\mu}]_{\max}c}{c_{1/2}+c} \quad or \quad \dot{\mu}=(1-\mu)^n k\exp\left(\frac{-e}{RT}\right) \tag{3}$$

where c is mass concentration of the species, $c_{1/2}$ is the concentration at half of the volume reacted, $[\mu]_{max}$ is the maximum reaction rate, n is the reaction order, k is the rate constant, e is activation energy, R is the universal gas constant, T is absolute temperature. For the degeneration of the clay the degree of reaction may be considered as proportional to concentration of the contaminant (Fernandez and Quigley, 1988, 1991, Hueckel, 1992b).

Note that as for the biological decomposition problem within this classical formulation we can consider only the consolidation of the submerged part of the landfill, where most of the reaction is anerobic, and when gas pressure buildup is not significant. Processes in non-satuated part of landfill offer interesting challenges. In the following Sections we shall discuss some phenomenological aspects of the processes under consideration.

Finally, the last mechanism to be discussed affecting consolidation concerns the variability of permeability with the degradation process described in what follows by the asssumption that intrinsic permeability and thus hydraulic conductivity depend on concentration of contaminant $k(c)$. Thus the discharge velocity

$$Q = -\frac{k(c)}{\gamma_f}\frac{\partial u}{\partial z} + k_c\frac{\partial c}{\partial z} \qquad (4)$$

where γ_f is the average density of fluid mass, whereas k_c is the osmotic permeability. The above assumptions give rise two to types of biologically or chemically originated consolidation: one depending directly on the current value of degree of degradation through eq.(2), and another depending on the degradation (or chemical concentration) gradient, see also Barbour and Fredlund (1989). The latter one results from the coupling of porosity through the discharge with the concentration gradient. Thus, the gradient of concentration induces the gradient of pore pressure which in turn generates an increase of effective stress (osmotic consolidation).

3 CHEMO-CONSOLIDATION OF CLAY

3. 1 *Experimental*

Consolidation is known to occur in clays subjected to a change in soil salinity (Mitchell et al., 1973) or to a permeation by concentrated organic contaminant (Sridharan and Ventkatappa Rao, 1973 and Fernandez and Quigley, 1991) under sufficient constant effective stress. The technical significance in the later case is related to the role of stress in a possible preventing of an otherwise substantial chemically driven increase in permeability of clay barriers under landfills and impoundments. In fact, Fernandez and Quigley, 1991) have shown that for a natural Sarnia clay virtually no increase in permeability occured at constant vertical effective stress of 40 kPa and 136 kPa respectively during 100% ethanol and 100% dioxane permeation. Without the stress imposed the increase in permeability could be 100-fold and 1000-fold, respectively.

However, it appears that this is not a universal rule and the response depends on the clay content. For economical reasons soils with low clay content (LCC) are preferred as barrier material over more costly high clay content (HCC) mixtures. Usually 5 to 10% clay content provides a sufficiently low hydraulic conductivity (Daniel, 1993). Zhang (1997) has performed experiments similar to those by Fernandez and Quigley (1991) on effect of stress on hydraulic conductivity of organic contaminated, controlled clay content sand-clay mixtures. It was found that the beneficial role of stress may be defeated in LCC barriers. Fig. 1 shows a response to permeation of two water compacted soils, with respectively 20 % (LCC) and 50% (HCC) dry clay fraction by weight, to a sequential permeation with water, 100% ethanol and again with water in a consolidometer with the lateral stress measurement. The consolidometer have two separate concentric fluid collection rings and additionally minuscule grooves in the lateral wall to control and prevent side wall flow (Zhang, 1996).

The experimental results show that:

- for both HCC and LCC soils at the low effective stress (28 kPa or 43 kPa, including seepage stress), a 100% ethanol contamination resulted in an increase in hydraulic conductivity at steady state over two hundred times, Fig. 1a.
- a high pre-applied stress of 152.58 kPa can suppress the hydraulic conductivity hike almost entirely (to 2.5 times) in the HCC mix for the 100% ethanol permeation, while this was not the case of the LCC mix, for which the permeability increase was still about fifty times, Fig. 1b.
- transport in all cases is clearly showing a diffusive/advective character, Fig. 2a, while changes in apparent hydraulic conductivity occur in an abrupt manner when the concentration of ethanol in effluent would exceed 60-70%. Time at which such change would take place is strongly dependent on stress (Fig. 2b).
- at low stress the initial vertical strain of swelling due to permeation with water, is during permeation with ethanol supressed at ab. 30-40% of ethanol in the effluent, in coincidence with a very abrupt decrease in the latteral effective stress, which after this event remains practically constant with no correlation to permeability changes. At ab. 60-70% of ethanol in the effluent, vertical strain suffers a visible drop (consolidation) coincident with the onset of the increase in apparent hydraulic conductivity and both reach steady state concurrently, Fig. 3.
- the changes in hydraulic conductivity are reversible by a post-permeation mechanical

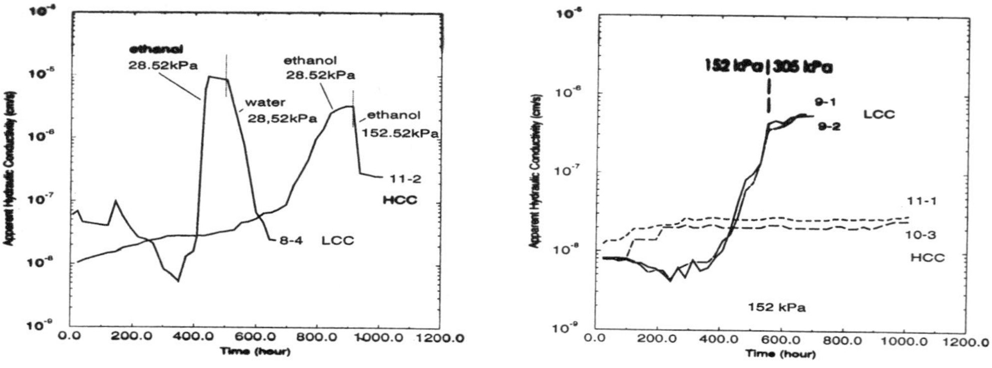

Fig. 1 a and b. Apparent hydraulic conductivity (fluid velocity/fluid gradient) vs time during permeation with water and ethanol for low (left) and high (right) effective vertical stress in LCC and HCC mixtures

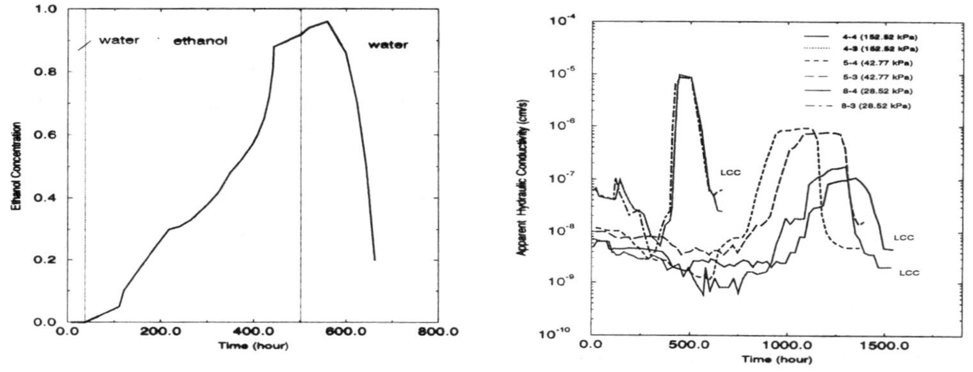

Fig. 2 (a) ethanol concentration in effluent; (b) vertical effective stress dependence of apparent hydraulic conductivity in LCC soil

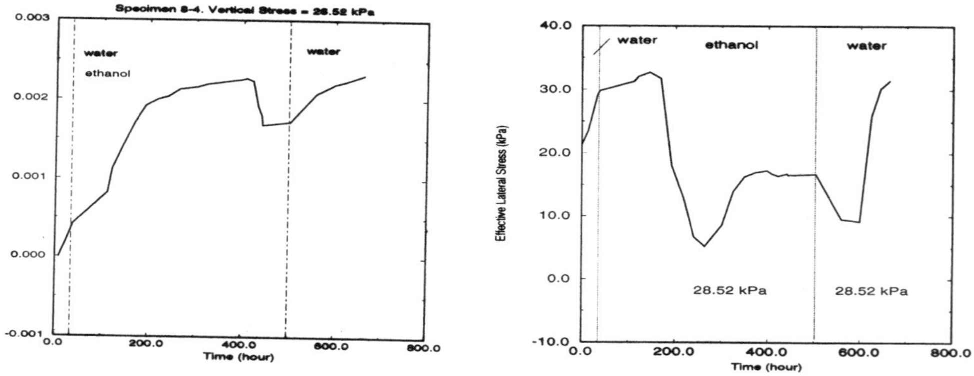

Fig. 3 (a) vertical strain (swelling positive) and (b) effective lateral stress at low stress in LCC mix

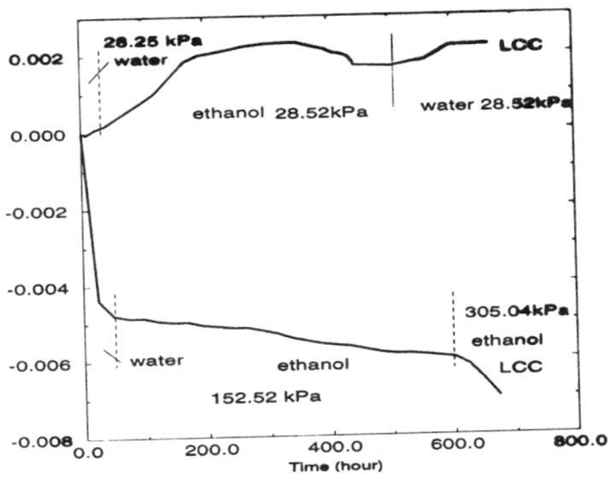

Fig. 4 Vertical strain for LCC mix at low and high effective stress during permeation with ethanol

- loading (at least partially) in HCC soils (Fig. 1a) and irreversible in LCC soils, (see Fig. 1b). However, they are reversible when after ethanol a reverse water permeation occurs at the same stress, in HCC and LCC soils, Fig. 1a & 2b.
- at high stress, the initial swelling is completely supressed in LCC (Fig. 4) soils, and the specimens consolidate with an almost constant strain rate, only slightly higher that the mechanical secondary compression rate. Apparent hydraulic conductivity still increases, even if less than at low stress, despite the consolidation (Fig. 1a).
- in HCC soil, there is also a continuous consolidation, but with a very small permeability increase, and no changes in lateral stress.

3.2 *Constitutive considerations*

To analyze the link between stress and in chemically active conditions a chemically sensitive failure locus has been proposed, Hueckel (1992, 1997) by modifying the Cam clay model (Roscoe and Burland, 1968). The size of yield surface, defined by the apparent maximum preconsolidation pressure, p'_c decreases with the contaminant concentration,

$$f = \left(\frac{2p'}{p_c'(\varepsilon_v^{pl}, c)} - 1\right)^2 + \left(\frac{2q}{M(c)p_c'(\varepsilon_v^{pl}, c)}\right)^2 - 1 = 0 \tag{5}$$

where p' is the mean principal effective stress, q is the second stress deviator invariant, M is the critical state coefficient postulated to be a monotonically increasing function of concentration, c, following indications of Olson and Mesri (1970). The apparent maximum preconsolidation isotropic stress p_c' chosen as the hardening parameter decreases with the increasing concentration. In the case of a continuous and smooth change of p'_c, c, and ε_v^{pl} starting at p'_{c0}, $0,0$, the evolution equations simplifies to

$$p_c' = p_{co}' \exp\left(\frac{1+e_0}{\lambda(c) - \kappa(c)}\varepsilon_v^{pl}\right) S(c) \tag{6}$$

where $S(c)$ *is* chemical softening function, while λ and κ are elasto-plastic and elastic bulk moduli, respectively; e_0 is the initial void ratio.

Analyzing the effective stress principle in chemically affected clays it was also found (Hueckel, 1992) that the total strain in a chemically active clay is composed of the mechanical strain and of the strain caused by changes in physico-chemical properties of the interlayer liquid at constant stress. Both strain parts may either be entirely reversible or in part irreversible. A macroscopic plastic strain rate at constant stress, during flow of contaminants with variable mass concentration c, has been proposed through the usual normality rule (Hueckel, 1997)

$$\dot{\varepsilon}_{ij}^{p} = \frac{1}{H}\frac{\partial f}{\partial c}\frac{\partial f}{\partial \sigma_{ij}'}\dot{c}, \tag{7}$$

where H is the usual plastic modulus to be calculated from the consistency condition of Prager. The above strain is activated, when at constant stress at yielding, the yield surface shrinking due to growing concentration activates plastic strain hardening to counterbalance the chemical softening. For this to occur however, a sufficiently high stress is required to initialize yielding. Plastic strain has been seen to prevent changes in intrinsic permeability due to contamination in HCC soils (Fernandez and Quigley, 1991, Hueckel, 1992).

However, it is clear from the above experiments that the above mechanism does not hold true in LCC soils. While plastic chemo-consolidation strain does occur at high stress, hydraulic conductivity still increases at high concentrations. It appears that stress plays a different role in changes of permeability, depending on the clay content: high stress at a HCC can, and at a LCC cannot, prevent permeability increase. It is postulated that this difference can be explained by a different role of clay fraction and solid grain play in the two types of soils. To start with, it should be realized that clay fraction is the only one that conducts water, and that solid grains are impermeable.

In the LCC soil, the soil structure is made of solid grains nearly in contact with all the clay entirely filling the space left void between the grains. That means that there are no free water channels between the grains, and whole water flows through clay fraction. In the HCC soil the structure is that of clay matrix with solid grains sparcely placed within the matrix.It is subsequently surmised that while in the HCC soil, clay mass carries both the stress and conducts fluids, solid grains being just unconnected inclusions, in the low clay systems stress is carried by the chains of grains in contact, leaving the clay between the grains practically unloaded. The formation of the stress carrying chains were photo-elastically observed in well graded granular assemblies (Konishi et al. 1982). Thus the stress carrying grain chains shield the water carrying clay from the influence of stress, leaving its permeability stress independent. Fig. 5 presents an elementary representative volume of an idealized LCC soil based on simple cubic packing granular system and a mechanical model for the corresponding volumetric behavior of the clay solid. In the idealization the stress carrying chain includes certain limited fraction of clay in accordance with microscopic observations. It is inevitable that during compaction process certain amount of clay will be left outside of the intergranular space contributing to the deformability of the stress carrying chain. It is thus assumed that the total effective stress σ' is split between the stress in the loaded chain, σ'_1, and in the clay fraction, σ'_2:

$$\sigma' = \sigma - u = \sigma'_1 + \sigma'_2 \qquad (8)$$

Following previous discussions on the role of the stress in adsorbed water in clay (Hueckel, 1992) it is postulated that isotropic stress in clay solids is equal to specific forces of electro-chemical repulsion in adsorbed wate and that the strain due to changes in chemical variables is independent from and adds to the stress induced strain in the clays solids, as visualized by the series connection in Fig. 5.

$$\varepsilon_2 = \varepsilon_{sk2}(\sigma'_2) + \delta_2(c,v) \qquad (9)$$

$$\varepsilon_1 = \varepsilon_{sk1}(\sigma'_1) + \delta_1(c,v), \text{ while } \varepsilon_1 = \varepsilon_2$$

where ε_{sk1} and ε_{sk2} are volumetric strain in grain chain and in clay resulting from mechanical stress, while δ_1 and δ_2 are strains in clays resulting from changes in chemical properties of pore liquid, primarily: ionic concentration in water, v and mass concentration of contaminant, c, responsible for the water swelling and contamination induced stress, respectively. It should be noted that while clay fraction throughout the soil is the same, the volume fraction of clay involved in stress carrying chain is much smaller than of clay transporting liquids.Thus functions δ_1 and δ_2 may be substantially different. Note also that the material functions for the model described by Fig.5 and eq. (9) may be identified within certain approximation by analyzing separately the behavior of pure clay and low clay content soil.

4 CONSOLIDATION OF BIO-DEGRADING SOLID

The experiments (Chang, 1996) were performed on uncoated cellophane by Flexel Inc. cut into small poieces and mixed with biologically activated soil (Compost BioActivator, by Necessary Organics Inc., Virginia). Cellophane is known to decompose fastest among various cellulose products. Moisture content was controlled at a fixed value of 80%. The effect of bio-degradation was tested by comparison to the control behavior of the specimen which was autoclaved to eliminate the biological activity. The typical results are presented in Fig. 6. It is clear that consolidation of the degrading cellophane had much higher rate than the autoclaved one. The estimated associated mass loss in the bio-active specimen over the period of one week was significant, but analytical methods failed because of the involvement of the uncontrolled growth of fungi and bacterial colonies and some liquid production. The general conclusion from these experiments is that the biologically induced consolidation is of the order lower than the initial mechanical one, but on a more significant time scale is visibly larger. The bio-consolidation can be quantified on small specimens, but the experimental techniques in this area require further refinements, especially the control of the rate and conditions of the biological process.

5 NUMERICAL SOLUTIONS

An analytical solution is developed for a 1-D chemical consolidation. To achieve an analytical solution a number of limitation has to be made thus restricting the problem's its validity range. First of all, we shall assume that all the non-linear phenomena such as fluid density change, change of porosity with degradation reaction are all linearly dependent on concentration of contaminant, and additionally that hydraulic conductivity is constant. With these assumptions, together with the usual condition of constant total mechanical load the two principal constitutive equations of deformation and flow, eqs. (1 and 3) become

$$-\frac{\partial n}{\partial t} = -m_v\frac{\partial u}{\partial t} + m_c\frac{\partial c}{\partial t} \tag{10}$$

$$-\frac{\partial n}{\partial t} = -\frac{k}{\gamma_f}\frac{\partial^2 u}{\partial z^2} - \left(n_0\xi\frac{\partial^2 c}{\partial t^2} + k_c\frac{\partial c}{\partial t}\right)$$

where m_v and m_c are mechanical and chemical rate compliances. This set of equations can be reduced to one for two variables, c and u, by eliminating porosity rate and completed by a linearized equation for advective-diffusive flux with additional term proportional to pore pressure gradient leading to a system of two equations for pore pressure and concentration

$$c_o m_v\frac{\partial u}{\partial t} - \left(c_0\frac{k}{\gamma_f} - D_u\right)\frac{\partial^2 u}{\partial z^2} - (c_0 m_c - n_0)\frac{\partial c}{\partial t} + (c_0 k_c - D)\frac{\partial^2 c}{\partial z^2} = 0 \tag{11}$$

where c_0 is the initial concentration, and D and D_u are respectively effective coefficient of diffusion and coefficient of ultrafiltration.

By simple substitutions between the above equations a pair of governing equations of chemo-mechanical consolidation can be obtained as

$$\frac{\partial u}{\partial t} - E\frac{\partial^2 u}{\partial z^2} - F\frac{\partial^2 c}{\partial z^2} = 0, \quad \frac{\partial c}{\partial t} - G\frac{\partial^2 u}{\partial z^2} - H\frac{\partial^2 c}{\partial z^2} = 0 \tag{12}$$

where E, F, G and H are combination of the constants involved. The two coupled equations (5.48) can be solved (see Kaczmarek and Hueckel, 1997) using a method developed for coupled problems of diffusion of heat and moisture into solids (Crank, 1956).

The solution of the above system was obtained for the boundary conditions with draining top boundary and non-draining bottom boundary, consisting of an initial mechanical consolidation due to a constant stress equivalent to 10m of water, and subjected to water solution of NaCl at concentration of 23 kg/l. Ultrafiltration and fluid density effects are ignored. The resulting settlements are shown in Fig. 6 for mechanical, chemical (due to concentration of salt), osmotic (due to concentration gradient), and jointly for chemical and osmotic consolidation for a set of realistic material constants Barbour and Fredlund (1989). Time scale, each consolidation stage starts at zero. It may be seen that joint chemo-osmotic consolidation is nearly half of the mechanical

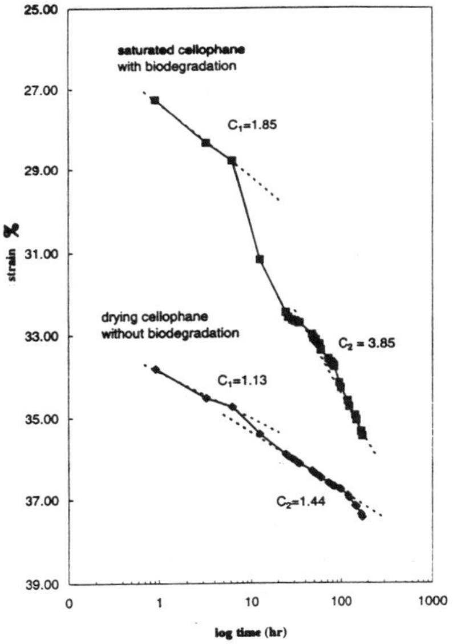

Fig. 5. Strain in 1-D consolidation of saturated mixture of shredded cellophane and compost bio-activator

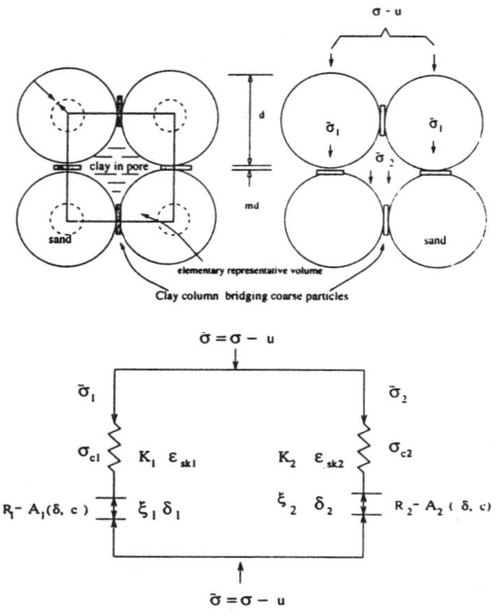

Fig. 6. Representative elementary volume of clay-sand mix with low clay content (a) intergranular force transmission (b)

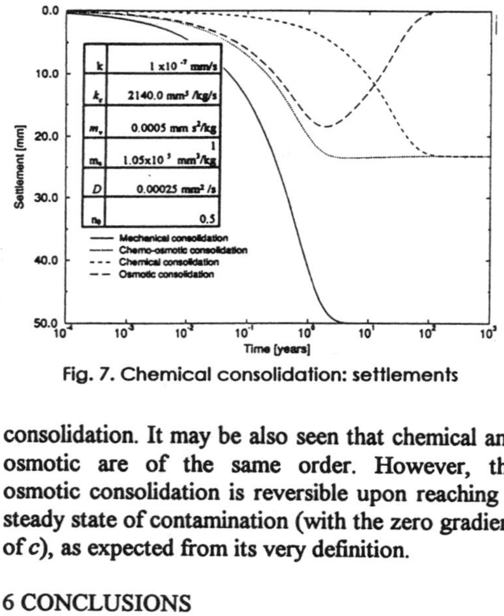

Fig. 7. Chemical consolidation: settlements

consolidation. It may be also seen that chemical and osmotic are of the same order. However, the osmotic consolidation is reversible upon reaching a steady state of contamination (with the zero gradient of *c*), as expected from its very definition.

6 CONCLUSIONS

Two different problems in consolidation theory of porous materials with degrading solid skeleton are formulated: chemo-consolidation of clay-sand barriers and bio-consolidation of organic landfilled material. Consolidation is triggered via three mechanisms: degradation of mechanical and permeability properties of the solid, exchange of mass between solid and liquid phase. Experiments indicate that for low clay content soil, liquid conducting clay is shielded from stress carried by a non-clay fraction. Experiments in bio-degradation of organic show that the related strain rate is comparable with the mechanical one. Analytical solution obtained for 1-D linear consolidation due to clay contamination may be obtained and shows that porosity change due to dehydration may produce the mechanical settlement increase over one third.

7 REFERENCES

Barbour S. L., and Fredlund D. G., 1989, Can. Geotech. J. **26**, 551-562,

Chang, Y.P. 1996, Consolidation of bio-degradable cellulose materials, MS thesis, Duke University

Crank, J., 1956, *The mathematics of diffusion*, Oxford

Daniel, D.E., 1993, in *Geotechnical practice for waste disposal*, Chapman and Hall, London

Fernandez F., and Quigley R. M., 1988, Can. Geotech. J. **25**, 582-589,

Fernandez, F., & R.M. Quigley, 1991, Can. Geotechn. J., **28**, 388-398.

Hueckel T.,1992a, Canadian Geotech. J., **29**, 1120-25,

Hueckel T.,1992b, Canadian Geotech. J., 29, 1071-86

Hueckel T. M. Kaczmarek & P. Caramuscio, 1997, Theoretical assessment of fabric and permeability changes in clays affected by organic contaminants, *Can. Geotech. J.*, 4, 34, in print,

Hueckel, T., M. Kaczmarek and Q. Zhang, 1996, in: *Environmental Geotechnology, v.1*, eds: H.Y. Fang and H.I. Inyang, Technomic Publ. Co, 302-11

Kaczmarek M. and Hueckel T, 1997, Linear 1-D chemo-mechanical consolidation, Duke University, Report for ISMES Spa, Bergamo, Italy

Konishi J., Oda M., and Nemat-Naser S., 1982, in *"Deformation and Failure of Granular Materials"*, ed. H. J. Luger, A. A. BALKEMA,

Mitchell J. K., Greenberg J. A., Witherspoon P., 1973, J. Soil Mech. Found., **99**, 307-321,

Mitchell J. K. and Jaber, M., 1990, *"Waste Containment Systems"*, GSP No. 26, ed. R. Bonaparte, pp. 84-105, ASCE.

Probstein, R.F., 1994, *Physicochemical hydrodynamics*, J.Wiley, New York

Sridharan A. and Ventakatappa Rao G., 1973, Géotechnique, **23**, 359-382,

Terzaghi, K. 1943, *Theoretical Soil Mechanics*, J. Wiley, New York

Zhang, Q, 1996, Experimental study of coupling beteween stress and chemically induced changes in hydraulic conductivity of clay-sand mixtures, MS thesis, Duke University

Numerical Models in Geomechanics, Pietruszczak & Pande (eds) © *1997 Balkema, Rotterdam, ISBN 90 5410 886 X*

Poroelastic behaviour of a rigid anchor plate embedded in a cracked geomaterial

A.T. Mahyari & A.P.S. Selvadurai
Department of Civil Engineering and Applied Mechanics, McGill University, Montreal, Que., Canada

ABSTRACT: The paper examines the poroelastic problem related to axisymmetric indentation of the single surface of a penny-shaped crack embedded in a fluid saturated medium by a smooth rigid disc. The finite element based computational modelling accounts for the singular behaviour of the stress state at the crack tip located in the poroelastic medium. The results of the computational modelling investigates the time dependent consolidation behaviour of the flat disc shaped anchor and the time dependent evolution of stress intensity factors at the crack tip for either prescribed displacements or prescribed total force with a variation in the form of a Heaviside step function. The analysis also focuses on the evaluation of time-dependent quasi-static planar crack extension that results from the loading of penny-shaped crack.

1. INTRODUCTION

The theory of a rigid disc embedded in a geological medium has been applied to the study of a variety of problems in geomechanics including the study of flat earth anchors, response of *in situ* testing devices and in the examination of the mechanical behaviour of deep foundations (Selvadurai 1976, 1980, 1994a; Rowe and Booker, 1979; Selvadurai and Nicholas, 1979). The modelling of the geomaterial has largely been restricted to elastic behaviour although certain problems related to viscoelastic (Selvadurai, 1978), plastic (Rowe and Davis, 1982) and poroelastic (Yue and Selvadurai, 1995) response of the geomaterial has also been investigated. In a majority of these studies, the disc shaped region is assumed to be in bonded contact with the surrounding geomaterial. The consideration of the influences of cracking and delamination between the disc and the geological medium has recently been investigated by Selvadurai (1989, 1994a, 1994b) in the context of the elastic behaviour of the geomaterial. The development of such cracks can be due to use of expansive grout material employed to create the injection anchor regions. Similar crack development is also feasible either during penetration of single helix anchors or during a plate loading test conducted at the base of a borehole. The objective of this paper is to extend these studies to the examination of the poroelastic response of the geomaterial.

In particular we examine the axisymmetric smooth indentation of the single surface of a penny-shaped crack located in a poroelastic medium, by a rigid circular anchor plate (Figure 1). It is assumed that the anchor plate experiences debonding at the lower plane face due to application of the axial anchor load P. The classical theory of poroelasticity proposed by Biot (1941, 1955) for an elastic soil skeleton saturated with a compressible pore fluid is used in the computational formulations. The computational developments also consider the *moving boundary* effects arising from the loading of the penny-shaped crack in a poroelastic medium. The finite element procedure used in the treatment accounts for the singularity in the *effective stresses* at the crack tip. The application of the anchor load is assumed to initiate the quasi-static growth of the crack when the stress intensity factor attains a critical value. During this crack extension, the displacement, traction and pore fluid pressure boundary conditions change. The extension of the crack is governed by the attainment of an effective critical stress intensity factor at the crack tip. The crack extension initiates the moving of the crack tip and the alteration of boundary conditions governing the tractions, displacements

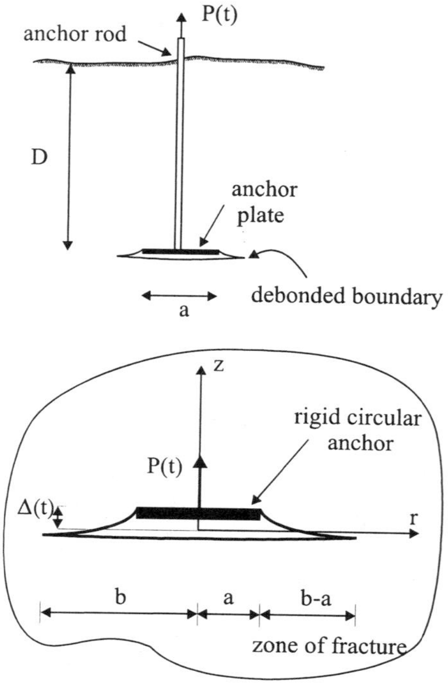

Figure 1. The detached anchor plate problem.

and pore fluid pressures in the incremental crack extension zone. The analysis focuses on the evaluation of time-dependent stress intensity factors and calculation of the time-dependent evolution of crack extension when the anchor plate is subjected to a step function of either prescribed total load or a prescribed displacement.

2. GOVERNING EQUATIONS

The basic equations governing Biot's theory of poroelasticity are summarized for completeness. The constitutive equations governing the quasi-static response of a poroelastic medium, which consists of a porous isotropic elastic soil skeleton saturated with a *compressible pore fluid* take the forms

$$\sigma_{ij} = \frac{2\mu\nu}{1-2\nu}\varepsilon_{kk}\delta_{ij} + 2\mu\varepsilon_{ij} - \frac{3(\nu_u - \nu)}{B(1-2\nu)(1+\nu_u)}p\delta_{ij} \quad (1a)$$

$$p = \frac{2\mu B^2(1-2\nu)(1+\nu_u)^2}{9(\nu_u - \nu)(1-2\nu_u)}\zeta_\upsilon - \frac{2\mu B(1+\nu_u)}{3(1-2\nu_u)}\varepsilon_{kk} \quad (1b)$$

where σ_{ij} is the total stress tensor; p is the pore fluid pressure; ζ_υ = volumetric strain in the compressible pore fluid; ν is Poisson's ratio of the porous fabric; ν_u is the undrained Poisson's ratio; μ is the shear modulus; B is Skempton's pore pressure coefficient; $\kappa = k/\gamma_w$ where k is the hydraulic conductivity and γ_w is the unit weight of pore fluid. Also ε_{ij} is the soil skeleton strain tensor defined by

$$\varepsilon_{ij} = \frac{1}{2}(u_{i,j} + u_{j,i}) \quad (2)$$

where u_i are the displacement components, and a comma denotes a partial derivative with respect to a spatial variable. The effective stresses in the soil skeleton are given by

$$\sigma'_{ij} = \sigma_{ij} - \frac{3(\nu_u - \nu)}{B(1-2\nu)(1+\nu_u)}p\delta_{ij} \quad (3)$$

In the absence of body forces, the quasi-static equations of equilibrium take the form

$$\sigma_{ij,j} = 0 \quad (4)$$

The fluid transport within the pores of the medium is governed by Darcy's law which can be written as

$$\upsilon_i = -\kappa p_{,i} \quad (5)$$

where υ_i is the specific discharge vector in the pore fluid. The equation of continuity associated with quasi-static fluid flow is

$$\frac{\partial \zeta_\upsilon}{\partial t} + \upsilon_{i,i} = 0 \quad (6)$$

Considering requirements for a positive definite strain energy potential (Rice and Cleary, 1976), it can be shown that the material parameters should satisfy the following thermodynamic constraints:

$$\mu>0;\ 0 \le B \le 1;\ -1 < \nu < \nu_u \le 0.5;\ \kappa>0. \quad (7)$$

The resulting equations of equilibrium for a poroelastic medium as introduced by Biot (1941, 1955) and reformulated in more physically relevant variables by Rice and Cleary (1976), can be written in terms of the displacements and pore pressure as

$$\mu\nabla^2 u_i + \frac{\mu}{(1-2\nu)}\varepsilon_{kk,i} - \frac{3(\nu_u - \nu)}{B(1-2\nu)(1+\nu_u)}p_{,i} = 0 \quad (8a)$$

$$\frac{\partial p}{\partial t} - \frac{2\kappa\mu B^2(1-2\nu)(1+\nu_u)^2}{9(\nu_u - \nu)(1-2\nu_u)}\nabla^2 p = - \frac{2\mu B(1+\nu_u)}{3(1-2\nu_u)}\frac{\partial \varepsilon_{kk}}{\partial t} \quad (8b)$$

For a well posed problem, boundary conditions and initial conditions on the variables u_i, σ_{ij} and p also need to be prescribed.

3. FINITE ELEMENT FORMULATION

Finite element method offers an efficient procedure for solution of problems associated with the consolidation of saturated poroelastic media. Sandhu and Wilson (1969) were the first to apply finite element methods to the study of problems associated with consolidating geomaterials. Ghaboussi and Wilson (1973), and Booker and Small (1975) have developed finite element procedures for the analysis of problems associated with surface loading of semi-infinite media. Reviews of both analytical and numerical approaches to the study of soil consolidation related to poroelastic media are given by Lewis and Schrefler (1987) and Selvadurai (1996).

In this section we present a brief review of the finite element formulation of the poroelasticity problem with special reference to the incorporation of singular elements to model the singular stress field at the crack tip.

The governing equations (8) can be approximated by a matrix equation by adopting a standard Galerkin finite element procedure. The details of these procedures are well documented by Sandhu and Wilson (1969), Lewis and Schrefler (1987) and more recently by Selvadurai and Nguyen (1995) in connection with the finite element modelling of isothermal consolidation of sparsely jointed porous media. The governing equations (8) are discretized in the spatial domain using a standard finite element procedure. The element chosen to represent the intact region of the poroelastic medium is the eight-noded axisymmetric isoparametric element where the displacements within the element are interpolated as functions of the 8 nodes, whereas the pore pressures are interpolated as a function of only the four corner nodes i, k, m, and o (Figure 2).

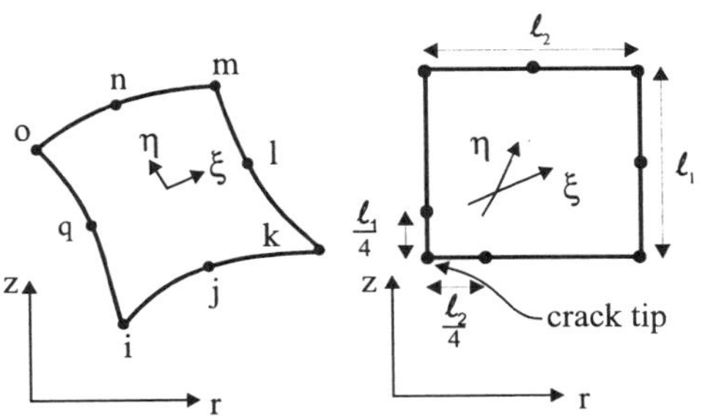

Figure 2. Plane and quarter point isoparametric elements.

The rationale for this procedure is now well documented (see e.g. Lewis and Schrefler, 1987; and Smith and Griffiths, 1988).

The quarter point singularity element introduced independently by Henshell and Shaw (1975) and Barsoum (1976) has been adopted to model the singular behaviour at the crack tip. It has been successfully utilized to model both two dimensional (plane stress and plane strain) and axisymmetric crack problems in classical elasticity. The order of the singularity in the effective stresses at the crack tip, for the porous skeleton, modelled by this approach corresponds to $r^{-1/2}$. The pore pressure field around crack tip is modeled by conventional isoparametric elements.

The application of a Galerkin procedure to the governing equations gives rise to the following discretized forms of the equations governing poroelastic media:

$$\begin{bmatrix} K & C \\ C^T & \{-\alpha\Delta t H + CM\} \end{bmatrix}\begin{Bmatrix} u_{t+\Delta t} \\ p_{t+\Delta t} \end{Bmatrix} = \begin{bmatrix} K & C \\ C^T & \{(1-\alpha)\Delta t H + CM\} \end{bmatrix}\begin{Bmatrix} u_t \\ p_t \end{Bmatrix} + \{F\} \quad (9)$$

where
K = stiffness matrix of the soil skeleton;
C = stiffness matrix due to interaction between soil and the pore fluid;
CM = compressibility matrix for the fluid;
H = permeability matrix;
F = force vectors due to external tractions, body forces and flow fields;
u_t, p_t= nodal displacements and pore pressures at time t;
Δt = time increment;
and α = time integration constant.

The time integration constant α varies between 0 and 1. The criteria for the stability of the integration scheme given by Booker and Small (1975) require that $\alpha \geq 1/2$. According to Lewis and Schrefler (1987) and Selvadurai and Nguyen (1995), the stability of solution can generally be achieved by selecting values of α close to 1.

4. MODELLING OF CRACK EXTENSION

In this paper the $r^{-1/2}$ type stress singularity at the crack tip is modelled by incorporation of singular stress field crack tip element to examine the crack extension problem in poroelastic media. The result of particular interest is the evaluation of the time-dependent stress intensity factors at the crack tip. For axisymmetric problems the crack opening and shearing mode stress intensity factors K_I and K_{II} can be evaluated by the displacement correlation method incorporating the nodes A, B, C , D and the crack tip C (Figure 3). i.e.,

$$K_I = \frac{\mu}{(1+k_\alpha)}\sqrt{\frac{2\pi}{\ell_o}}\begin{Bmatrix} 4[u_z(B)-u_z(D)] \\ +u_z(E)-u_z(A) \end{Bmatrix} \quad (10a)$$

$$K_{II} = \frac{\mu}{(1+k_\alpha)}\sqrt{\frac{2\pi}{\ell_o}}\begin{Bmatrix} 4[u_r(B)-u_r(D)] \\ +u_r(E)-u_r(A) \end{Bmatrix} \quad (10b)$$

where $k_\alpha = (3\text{-}4\nu)$ and ℓ_o is the length of the crack-tip element.

It is postulated that the crack in the poroelastic medium will extend when the mode I stress intensity factor applicable to the singular effective stresses at the crack tip attains a critical value of K_{IC}. The crack extension results in the relocation of the crack tip in an incremental fashion. The boundary conditions associated with the tractions, displacements and pore fluid pressures change with the time-dependent extension of crack. The pore pressures at the boundary of the newly opened crack region are assumed to be zero consistent with permeable crack faces.

5. NUMERICAL RESULTS

The numerical approach described above is applied to examine the following two poroelastic problems.

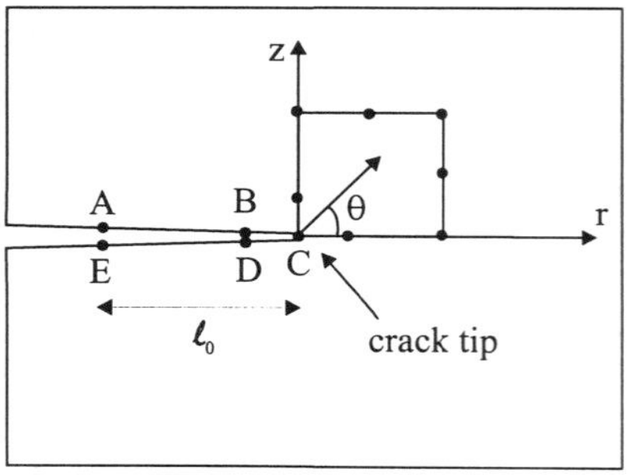

Figure 3. Node arrangement for computation of the stress intensity factors.

5.1 Deep rigid anchor indentation

We consider the axisymmetric problem related to the indentation of a single surface of a penny-shaped crack, of radius *b*, located in an infinite poroelastic medium by a smoothly embedded porous circular rigid anchor plate of radius *a* (Figure 1). The rigid anchor is subjected to either a prescribed load or a prescribed displacement with a time variation in form of a Heaviside step function. The soil skeleton and pore fluid of the porous medium are assumed to be incompressible.

The computational scheme has been verified with analytical results of associated elastic problem given by Selvadurai (1994a) for the axial stiffness

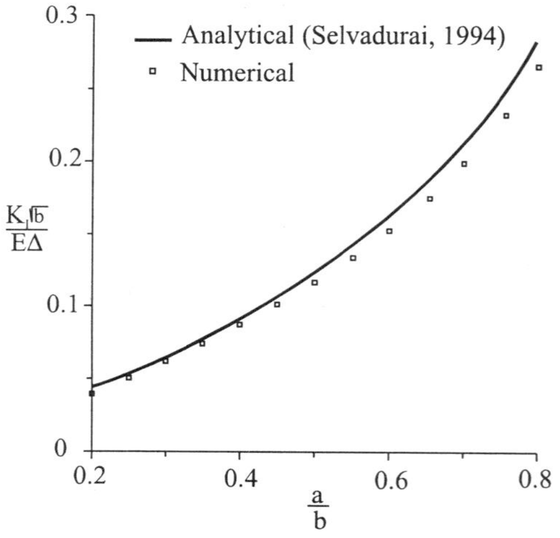

Figure 4. Crack opening mode stress intensity factor.

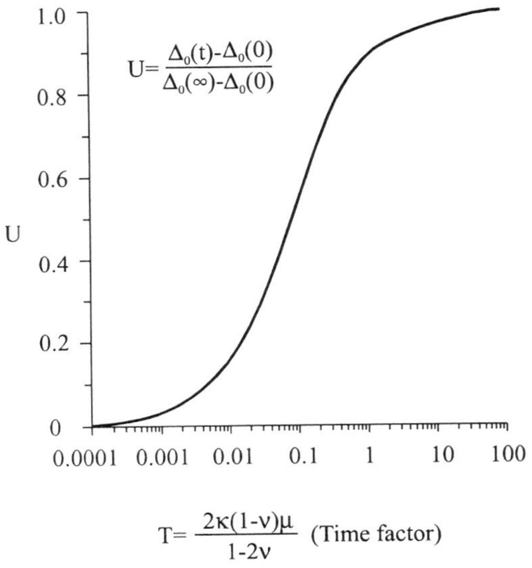

(a)

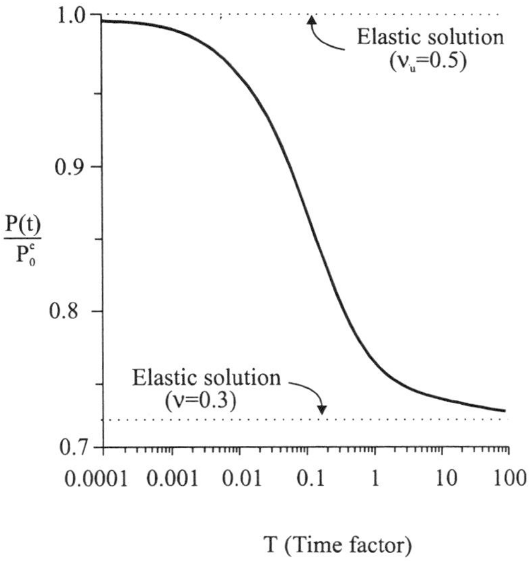

(a)

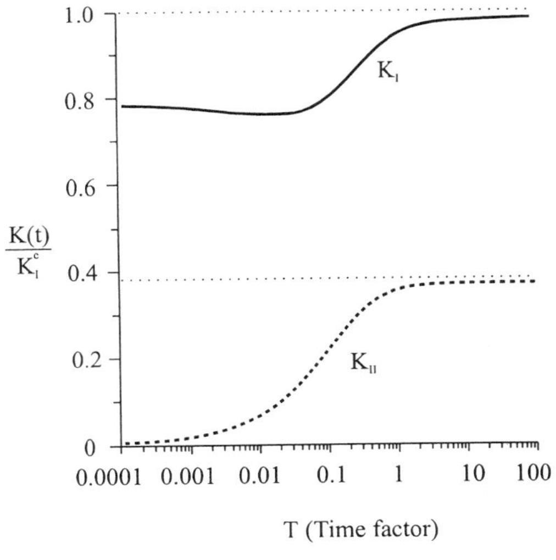

(b)

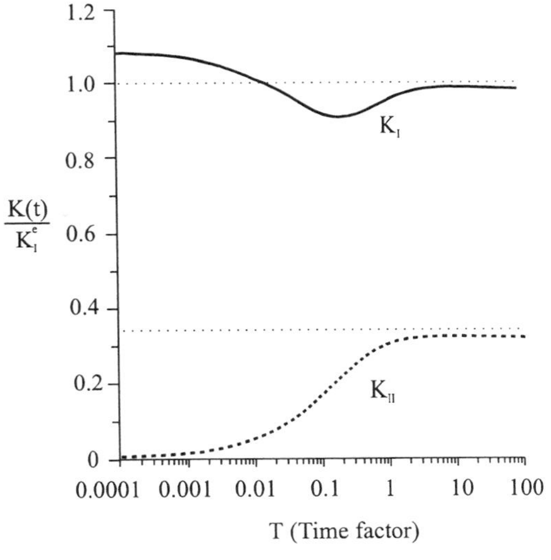

(b)

Figure 5. Variations in (a) the degree of consolidation and (b) the stress intensity factor K_I for the prescribed total load P.

Figure 6. Variations in (a) the load relaxation and (b) the stress intensity factor K_I for the prescribed total load P.

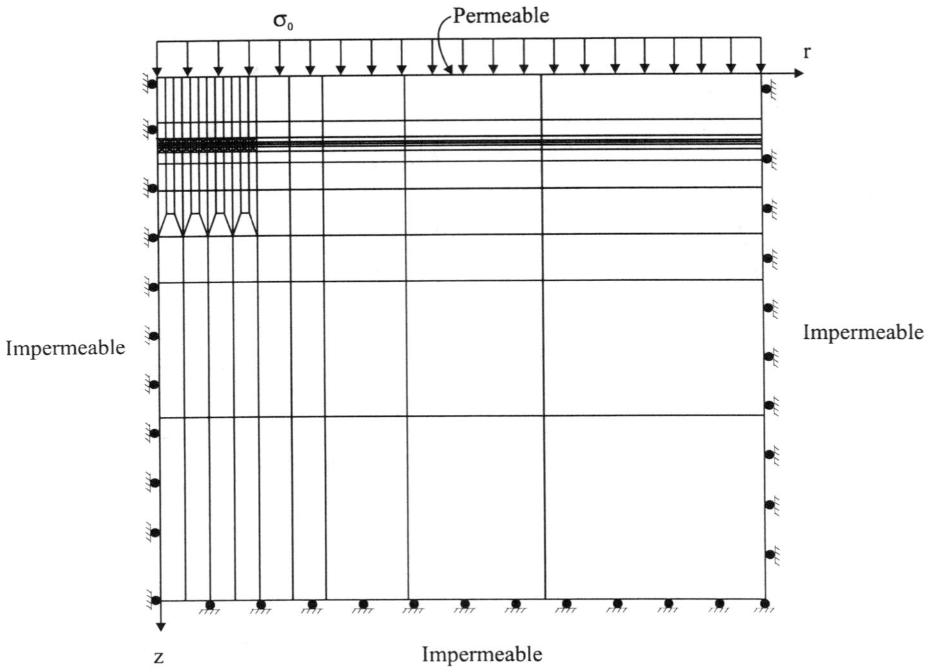

Figure 7. Finite element discretization of the shallow anchor plate problem.

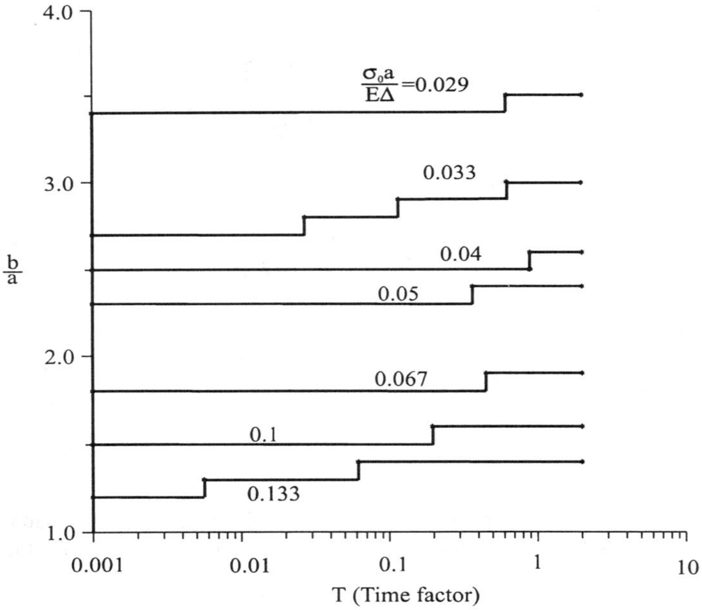

Figure 8. Variation of the crack evolution length for different values of the *in situ* stress σ_0.

characteristics of anchor disc and the stress intensity factors at the tip of the penny-shaped crack where the associated coupled integral equations are solved to develop these results. Figure 4 illustrates the comparison of analytical and numerical solutions for the flaw opening mode crack intensity factor K_I as a function of anchor plate-crack aspect ratio a/b, for drained elastic behaviour.

Results of primary interest relate to the illustration of the time-dependent degree of consolidation, the load relaxation of anchor disc and the evaluation of the time-dependent variation in the stress intensity factors at the crack tip. Figure 5 illustrates these results for the case where the total load is prescribed. Figure 6 presents the case where the displacements are prescribed. The asymptotic values corresponding to the drained elastic behaviour shown are the analytical results given by Selvadurai (1994a). The results indicate that the mode II shearing stress intensity factor is considerably smaller than the mode I stress intensity factor (i.e. $K_{II}/K_I \ll 1$), particularly at early times when the behaviour of material is largely undrained. It is therefore reasonable to assume that a possible mode of crack extension would relate to self similar expansion of the crack. The fluid transport towards the crack tip leads to a volume expansion of material which tends to initially close the crack in the vicinity of crack tip. This results in a decrease in the stress intensity factor. However, as the pore pressure diffusion takes place into the permeable faces of the crack, the crack begins to open up leading to increase the mode I stress intensity factor.

5.2 Shallow rigid anchor indentation

In this section, we examine the problem where the anchor is embedded at a finite depth of D/a=3 (Figure 1) in a poroelastic half-space, and subjected to a *in situ* stress σ_0 at its surface. The finite element discretization of the problem is shown in Figure 7. To examine the quasi-static extension of the crack, it is allowed to extend radially in the plane of the anchor plate. Figure 8 illustrates the time-dependent evolution of crack length for different values of the *in situ* stress σ_0 for a prescribed loading of the total force type. The crack extension tends to be unstable at low values of σ_0.

6. CONCLUDING REMARKS

The computational concepts that have been proposed in the literature for the study of poroelastic media have been adopted and extended to include the analysis of both stable and quasi-statically extending cracks embedded in such media. It is shown that the results for the poroelastic behaviour of the disc anchor indenting a single surface of a crack agree with the limiting solutions for the associated elasticity problem, derived via the solution of complex integral equations. In particular, it is observed that the transition from $t \to 0$ to $t \to \infty$ need not be monotonic, the influence of pore fluid migration can result in the alteration of the stress intensity factors at the crack tip in a poroelastic medium. The paper also presents computational results for situations where the crack is allowed to experience self similar extension. In these circumstances, it is observed that crack extension in to stable states (i.e. locations where $K_I/K_{IC}<1$) occurs relatively rapidly and further pore pressure diffusion phenomena do not result in any appreciable alterations in the crack geometry.

REFERENCES

Barsoum, R.S. 1976. On the use of isoparametric finite elements in linear fracture mechanics. *Int. J. Numer. Meth. Eng.* 10:25-37.

Biot, M.A. 1941. General theory of three-dimensional consolidation. *J. Appl. Physc.* 12:155-164.

Biot, M.A. 1955. Theory of elasticity and consolidation for a porous anisotropic solid. *J. Appl. Physc.* 26:182-185.

Booker, J.R. & J.C. Small 1975. An investigation of the stability of numerical solution of Biot's equations of consolidation. *Int. J. Solids Struct.* 11:907-917.

Ghaboussi, J. & E.L. Wilson 1973. Flow of compressible fluid in porous elastic media. *Int. J. Numer. Meth. Eng.* 5:419-442.

Henshell, R.D. & K.G. Shaw 1975. Crack tip finite elements are unnecessary. *Int. J. Numer. Meth. Eng.* 9:495-507.

Lewis, R.W. & B.A. Schrefler 1987. *The finite element method in the deformation and consolidation of porous media.* John Wiley & Sons.

Rice, J.R. & M.P. Cleary 1976. Some basic stress diffusion solutions for fluid-saturated elastic porous media with compressible constituents. *R. Geoph. S. Physc.* 14:227-241.

Rowe, R.K. & J.R. Booker 1979. A method of analysis of horizontally embedded anchors in an elastic soil. *Int. J. Numer. Anal. Meth. Geomech.* 3:187-203.

Rowe, R.K. & E.H. Davis 1982. The behaviour of anchor plates in clay. *Geotechnique*. 32:9-23.

Sandhu, R.S. & E.L. Wilson 1969. Finite element analysis of flow in saturated porous media. *J. Eng. Mech., ASCE*. 95(EM3):641-652.

Selvadurai, A.P.S. 1976. The load-deflexion characteristics of a deep rigid anchor in an elastic medium. *Geotechnique*. 26:603-612.

Selvadurai, A.P.S. 1978. The time-dependent response of deep rigid anchor in a viscoelastic medium. *Int. J. Rock Min. Sci. & Geomech. Absrt.* 15:11-19.

Selvadurai, A.P.S. 1980. The eccentric loading of a rigid circular foundation embedded in an isotropic elastic medium. *Int. J. Numer. Anal. Meth. Geomech.* 4:121-129.

Selvadurai, A.P.S. 1989. The influence of boundary fracture on the elastic stiffness of deeply embedded anchor plate. *Int. J. Numer. Anal. Meth. Geomech.* 13:159-170.

Selvadurai, A.P.S. 1994a. On the problem of a detached anchor plate embedded in a crack. *Int. J. Solids Struct.* 31(9):1279-1290.

Selvadurai, A.P.S. 1994b. Analytical methods for embedded flat anchor problems in geomechanics. In H.J. Siriwardane & M.M. Zaman (eds.), *Proc. 8th Int. Conf. Comp. Meth. Advan. Geomech., West Virginia*. 305-321.

Selvadurai, A.P.S. (ed.) 1996. *Mechanics of poroelastic media.* AH Dordrecht: Kluwer Academic Publishers.

Selvadurai, A.P.S. & T.J. Nicholas 1979. A theoretical assessment of the screw plate test. *Proc. 3rd Int. Conf. on Numer. Meth. in Geomech.* 3:1245-1252. Aachen.

Selvadurai, A.P.S. & T.S. Nguyen 1995. Computational modelling of isothermal consolidation of fractured porous media. *Computers and Geotechnics*. 17:39-73.

Smith, I.M. & D.V. Griffiths 1988. *Programming the finite element method.* John Wiley & Sons.

Yue, Z.Q. & A.P.S. Selvadurai 1995. On the mechanics of a rigid disc inclusion embedded in a fluid saturated poroelastic medium. *Int. J. Engng. Sci.* 33(11):1633-1662.

Numerical Models in Geomechanics, Pietruszczak & Pande (eds) © 1997 Balkema, Rotterdam, ISBN 90 5410 886 X

Accounting for flow in structured, unsaturated soils

J.R. McDougall & I.C. Pyrah
Department of Civil & Transportation Engineering, Napier University, Edinburgh, UK

ABSTRACT

In fully saturated soils, zones of relatively low hydraulic conductivity constitute impediments to flow. Under unsaturated conditions such impediments continue to modify flow but the adjustment of moisture contents within higher permeability pathways means the impact on overall flux may be muted. This paper examines how numerical simulations, using Richards equation, account for the hydraulic regime in structured porous media under various states of saturation.

In a numerical simulation, two independent impacts on overall flux can be recognised. Firstly, media structure redirects flow through high permeability pathways and produces a reduction in overall flux at all states of saturation. Secondly, depending on the prevailing suctions and state of saturation, a redistribution of moisture and associated increase in hydraulic conductivity in the high permeability pathways can moderate the structural reduction. For the particular configuration examined here, media structure reduces flux from the corresponding homogeneous case by about 35% but with moisture redistribution the overall flux reduction may be only 11%.

1. INTRODUCTION

At the elemental level, the simulation of moisture movement in soils assumes uniformity of media structure and hydraulic conductivity. Spatial variations can be defined using numerical techniques such as the finite element method by modifying individual element properties. Subsequent flow simulations may be run to illustrate the impact of non-homogeneity on the hydraulic regime.

In a non-homogeneous, fully saturated porous domain, zones of relatively low hydraulic conductivity interfere with the flow of moisture so flux is less than that which would occur in a uniform domain of the higher conductivity. The exact magnitude of the reduction will be dependent on the specific structural configuration. For example, where the connectivity between higher conductivity zones produces pathways or 'conduits' which are oriented parallel to the main direction of flow, the reduction in flux will be less than where such zones are occluded or lie in other orientations.

In unsaturated media, there is still interference with flow but redistribution of moisture within the air phase may modify the hydraulic regime such that the impact on overall flux is muted. The following simulations reveal how permeation in a non-homogeneous media under various states of saturation is accounted for by a numerical model.

2. UNSATURATED FLOW MODEL

2.1 Model characteristics

The numerical simulations of flow performed here use a finite element formulation of Richards equation as described by Neuman (1973) and Thomas & Rees (1990) and implemented via linear quadrilateral elements. Fuller details of this particular coding and its verification are contained in McDougall (1996). The fundamental soil water characteristics follow the van Genuchten (1980) functional forms and are dealt with in more detail in section 4.1.

2.2 Geometry & material properties

The tests simulate flow through an 8m high by 2m wide vertical soil column discretised as shown in Fig.1a. Dirichlet boundary conditions are applied to all upper and lower surface nodes so in the homogeneous case, a uniform one dimensional downward flow pattern occurs throughout. The saturated hydraulic conductivity in this case is

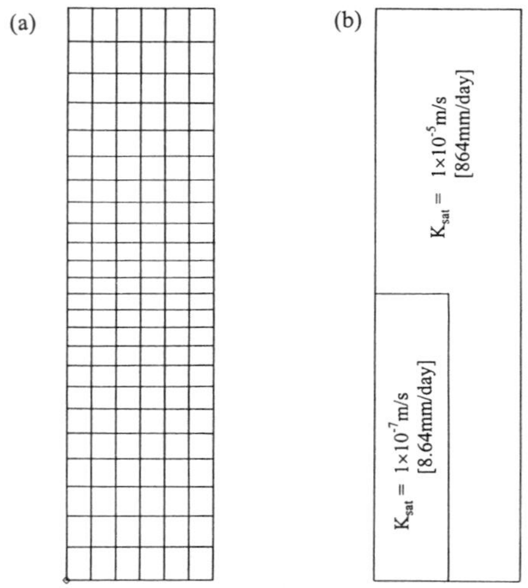

Fig.1. Infiltration column geometry; (a) finite element discretisation, (b) structured hydraulic conductivity

10^{-5}m/s [864mm/day]. For the non-homogeneous case, the lower left quadrant of the column is assigned a hydraulic conductivity of 10^{-7} m/s [8.64mm/day] (Fig.1b). This simple conductivity structure has been adopted to ensure redirection and concentration is confined to the central region and essentially one dimensional flow is preserved in the upper column and lower conduit sections. Boundary flux measurements are thus reliably obtained.

3. SATURATED MEDIA TESTS

3.1 Homogeneous column

The upper and lower boundary conditions ar prescribed pressures of 1m and 5m equivalent hea of water respectively, i.e. hydraulic heads are thu 9m and 5m respectively. For a homogeneou medium, the energy lost by liquid permeating th column is given up evenly. Hydraulic head therefore fall linearly (Fig 2a) and there is a constar hydraulic gradient throughout. In this case, th boundary conditions establish a hydraulic gradier over the column of 0.5 and with a one dimensiona flow pattern, flux volumes can be directly checke using Darcys law - 864.0 l/day. Flow vector obtained from the numerical simulation (see Fig.2b report overall volume influx and efflux of 864.0 and 863.98 l/day respectively so match the manua calculations closely.

3.2 Structured column

Under the same boundary water pressures, th redirection in flux necessary to accommodate th low conductivity zone is shown in Figs.3a & b. Thi zone acts as a constriction on flow through th column and has an impact on overall flux which i dependent on the particular configuration. Fo example, if the low conductivity zone extende uniformly over the full height of the column then th resulting flux could be realistically interpreted a two separate vertical columns. In such a case th geometry and conductivities of Fig.1 would produc a reduction in flux of approximately 50%.

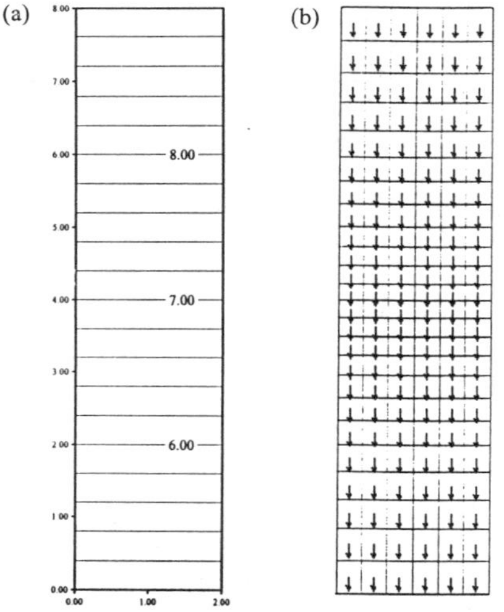

Fig.2. Infiltration in saturated, homogeneous column; (a) hydraulic heads, (b) vector flow

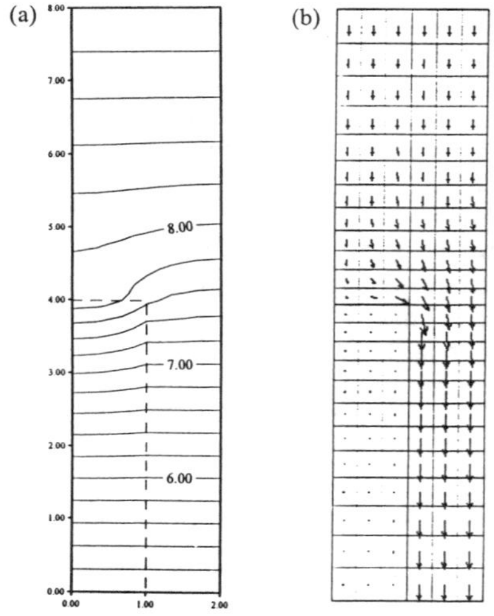

Fig.3. Infiltration in saturated, structured column; (a hydraulic heads, (b) vector flow

For intermediate zone lengths, a quantitative approximation of the resulting flux can be obtained by assuming a bi-linear variation in hydraulic heads along the main flow path, i.e. hydraulic gradients in the upper column and lower conduit are constant. In a steady flux state, the boundary influx and efflux are equal so from Darcy's Law we can write,

$$k\frac{A_u}{A_b} = k\frac{i_b}{i_u} \qquad (1)$$

where k is the hydraulic conductivity, A is the column area normal to the flow direction, i is the hydraulic gradient and subscripts u and b refer to the upper column and lower conduit respectively. Additionally, the assumed piecewise linear variation in hydraulic heads must comply with the hydraulic gradient established by the boundary suctions, i.e.,

$$I = \frac{1}{L}\left(i_u l_u + i_b l_b\right) \qquad (2)$$

where I is the boundary hydraulic gradient, l refers to conduit zone lengths and L the overall column length. Combining equations (1) & (2) and solving for the hydraulic gradients in a column with a low conductivity zone extending to mid-height of the column gives influx and discharge gradients of 0.3333 and 0.6667 respectively. For the respective flux boundary surfaces these gradients correspond to an overall flux of 576 l/day, i.e., a reduction in flux of 33% from the homogeneous case.

Figure 4 gives an alternative presentation of the variation in hydraulic heads. The plots relate to

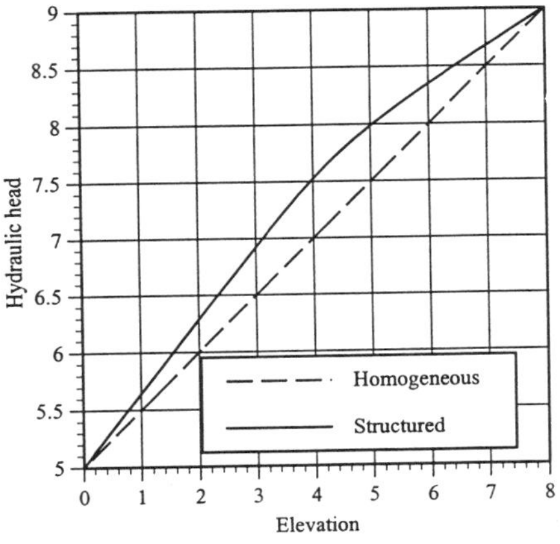

Fig.4. Plot of hydraulic heads with elevation along main flow path for homogeneous & structured columns under full saturation.

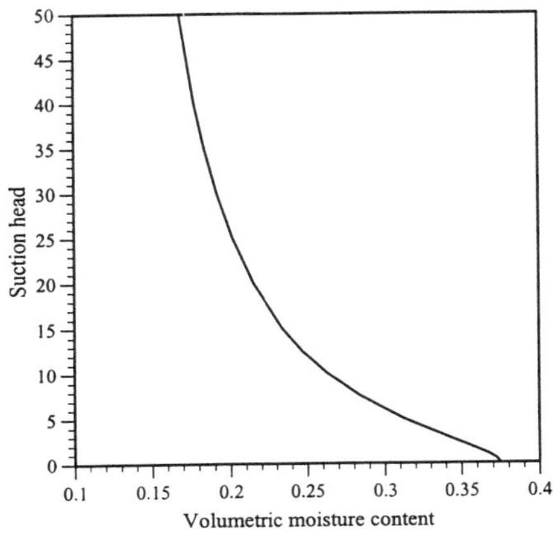

Fig.5. Soil moisture characteristic

the right hand vertical boundary of the column, i.e., the main flow path, and enable results from both homogeneous and structured columns to be more usefully compared. The assumption of a bi-linear head distribution implies an instantaneous hydraulic adjustment at the entrance to the constriction but the numerical tests reveal a smoother and probably more realistic transition. Nevertheless, the transition occurs mainly between 4m and 5m elevations and outwith, a bi-linear approximation fits very well. Since the gradient of this plot is the hydraulic gradient, it is evident that by comparison to the homogeneous case, there is a lower gradient in the upper column and higher gradient in the lower conduit. Monitoring of boundary fluxes in the numerical tests revealed values of 559.2 l/day which represent a reduction in flux of 35% from the homogeneous case and compares well with the 33% reduction predicted analytically.

4. UNSATURATED MEDIA TESTS UNDER CONSTANT GRADIENT

4.1 Homogeneous column, near full saturation

In the unsaturated case, solution of Darcy's Law becomes more difficult since the soil water relations (the soil moisture characteristic and relative permeability function), introduce a non-linearity into the governing equations. The particular soil moisture characteristic used here is defined using the functional form proposed by van Genuchten (1980) and shown in Fig 5. The control parameters are detailed in Table 1. For the homogeneous domain a one dimensional flow pattern would be expected and, for moderate flux rates, would reveal a variation of moisture contents in which the hydraulic gradient

Table 1. Soil water relation parameters
(van Genuchten, 1980)

α	n	θ_s	θ_r
0.2	1.6	0.375	0.1

and relative permeability change so as to jointly maintain the steady flux state.

Figures 6a & b show the contours of equipotentials and moisture contents for a column subject to boundary conditions of 1m and 5m negative pressure heads on the base and upper surface respectively. For an elevation datum at the column base, hydraulic heads are -1m at the base and +3m at the upper surface and the overall hydraulic gradient is 0.5, as in the fully saturated case. Volumetric moisture contents decrease from a near-fully saturated value of 0.367 at the base to about 0.312 at the upper surface of the column. The relative permeability, which is dependent on moisture content, is 0.391 at the base but falls to about 0.046 at the upper surface. Compared with the fully saturated case shown in Fig.2a, the distribution of equipotentials indicates relatively high hydraulic gradients are necessary to maintain the flux against a lower unsaturated permeability in the upper regions whereas higher conductivities close to the phreatic surface mean gradients are lower (Fig.6a). Vector calculations report an overall flux of 68.9 l/day.

4.2 Structured column, near full saturation

In a partially saturated medium, the zone of low conductivity distorts equipotentials (Fig.7(a)) in a manner similar to the saturated case but the impact on overall flux is rather more complex. Inspection of moisture content distributions in Figs.6(b) and 7(b) shows that at the entrance to the constriction an increase of about 0.008 occurs. For the prevailing moisture contents, such an increase, e.g. from say 0.327 to 0.335, produces an increase in relative permeability from 0.077 to 0.102 and will serve to offset any reduction in flux due to structural impediment. In fact the reported flux for this particular run falls to 59.5 l/day which is a reduction of only 13.6% from the corresponding homogeneous flux. However, the non-linear form of the soil water relations means that the potential for moisture redistribution will depend on the state of saturation and, through the soil moisture characteristic, the prescribed boundary suctions.

4.3 Homogeneous & structured columns, lower saturation states

For tests with boundary suction heads of 10m and 14m [hydraulic heads of -10m and -6m] on the base and upper surfaces, fluxes of 3.606 & 2.735 l/day are reported for the homogeneous and structured columns respectively, i.e., a reduction of 24.2% in overall flux. With suction heads of 18m and 22m [hydraulic heads of -18m and -14m] fluxes of 0.7704 & 0.5540 l/day are predicted - a reduction of 28.1%. Tests at different suction heads and saturation states will report overall fluxes whose absolute magnitude is determined by the prevailing relative permeability but the relative impact of a non-homogeneous media is clearly variable.

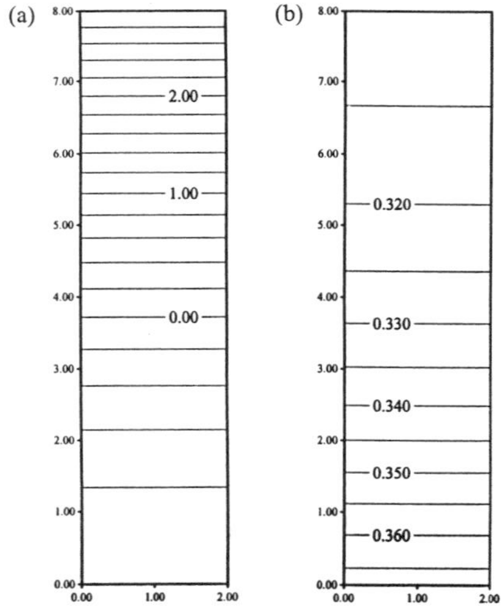

Fig.6. Infiltration in unsaturated, homogeneous column; (a) hydraulic heads, (b) moisture contents.

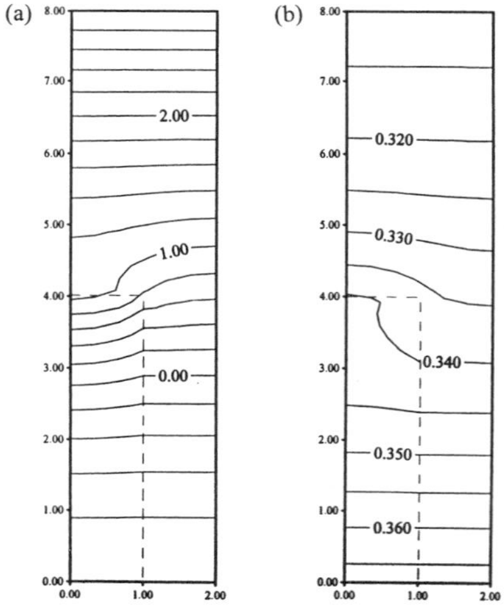

Fig.7. Infiltration in unsaturated, structured column; (a) hydraulic heads, (b) moisture contents.

If the reduction in flux is viewed as a reduction due to structural impediment which is moderated by moisture redistribution and an increase in hydraulic conductivity, then it is the potential for moisture adjustment which controls the moderating influence. With a monotonic variation in hydraulic heads, surface moisture contents, which are fixed by the prescribed suctions, represent upper and lower bounds to the potential moisture adjustment.

The range of moisture contents associated with unsaturated media can be readily illustrated using the soil moisture characteristic. Figure 8 presents the suction boundary conditions and corresponding moisture ranges for the three unsaturated flow tests reported above. The associated percentages denote the reductions in overall flux from the homogeneous case. The greatest range of moisture contents shown occurs between suction heads of 1m and 5m and the flux reduction is 13.6%. Tests at higher boundary suctions reveal progressively smaller moisture ranges until at extreme suctions the corresponding range of moisture content is negligible. This latter condition is analogous to the fully saturated state in which moisture contents are constant so no change in conductivity can occur. Reduction in flux is thus solely structural and might be expected to tend to that observed under full saturation.

For example, under surface boundary conditions of 100m and 104m suction heads, the overall hydraulic gradient remains at 0.5 but the corresponding moisture contents are now 0.1444 and 0.1454. Overall fluxes are 3.12×10^{-2} and 2.06×10^{-2} l/day, i.e., overall flux reduction is 34% and as shown in Fig.9 the variation in hydraulic heads closely resembles the fully saturated case (Fig.4).

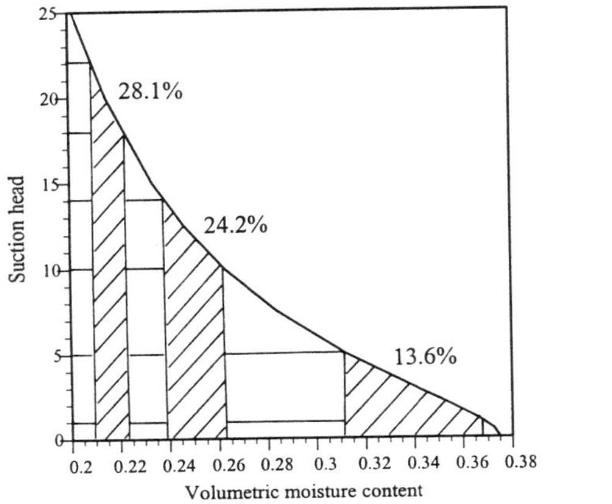

Fig.8. Percentage reduction in overall flux predicted for given boundary suctions and corresponding range of moisture contents

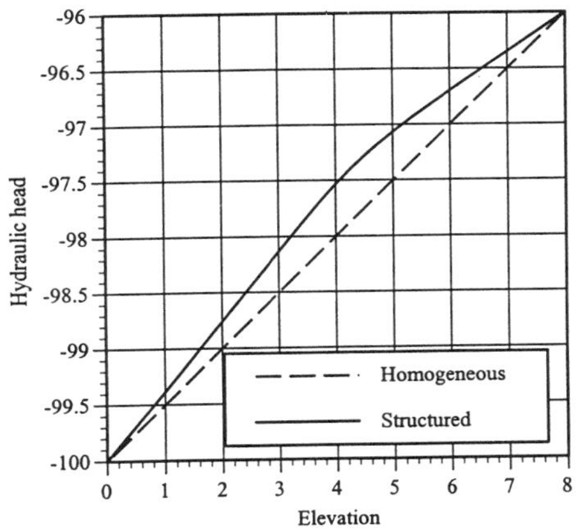

Fig.9. Plot of hydraulic heads with elevation along main flow path for homogeneous & structured columns under low saturation states.

5. UNSATURATED MEDIA TESTS UNDER VARIOUS HYDRAULIC GRADIENTS

5.1 Moisture variation measurements adjusted for hydraulic gradient

The existence of a relationship between flux reduction and boundary moisture contents has so far only been indicated under a constant hydraulic gradient. Reductions in flux are thus solely due to conductivity effects. However, evidence of this relationship at other boundary suctions/hydraulic gradients requires boundary moisture variation measures to be adjusted for the prevailing hydraulic gradient. For example, the variation of moisture contents over a column length z, may be measured by $\frac{\Delta\theta}{\Delta z}$, where θ is the moisture content. This may be expanded to give $\frac{\Delta\theta}{\Delta\psi}\frac{\Delta\psi}{\Delta z}$, where ψ is the suction head. The right hand component of this expansion is the hydraulic gradient and if this is removed, the term $\frac{\Delta\theta}{\Delta\psi}$ reflects the variation in moisture content between given boundary suctions normalised for the prevailing hydraulic gradient. In limit form this term is the specific water capacity which is well known in the study of unsaturated flow and is simply the gradient of the soil moisture characteristic.

5.2 Flux reduction and boundary specific water capacity

Calculation of a 'specific water capacity' over the full column length, provides a simple measure of the

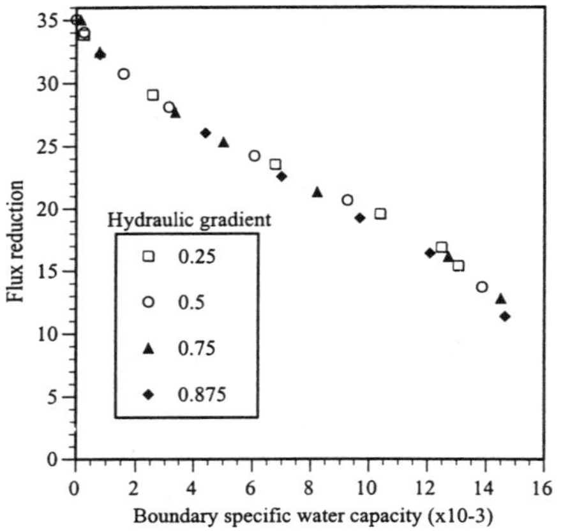

Fig.10. Variation of overall flux reduction with specific water capacity measured between boundary suctions for various hydraulic gradients.

potential moisture variation with respect to a given set of boundary suctions. Flux reductions under various boundary specific water capacities are shown in Fig.10 and for convenience are identified according to hydraulic gradient. The two limiting conditions of full saturation and extreme suctions have already been described. In these cases the specific water capacity is, or tends to, zero and for the given structural configuration, the reduction in flux is approximately 35%. There is clearly a well defined relationship between flux reduction and the specific water capacity which falls to a minimum of about 11%. Here, boundary suction heads are 1m and 3m; they embrace the shallowest part of the soil moisture characteristic and give rise to a hydraulic gradient of 0.875. It is interesting to note that based on the foregoing results, soil water relations of the type usually associated with coarse soils, i.e., with flatter desaturation curves and hence greater specific water capacities, provide the potential for even smaller flux reductions.

6. CONCLUSIONS

Non-homogeneity in media structure is usually associated with variations in hydraulic conductivity. In fully saturated media, low conductivity zones impede the flow of moisture and reduce the effective flux. When unsaturated, the same configuration of media structure similarly reduces flux but a hydraulic adjustment can modify moisture contents in the main flow path and offset the flux reduction. In the tests reported herein, flux reductions under certain partially saturated conditions may be only one third of that occurring under full saturation.

Numerical tests indicate that the attenuation in flux reduction is closely related to the specific water capacity measured over the column length. Where moisture contents are invariant with soil water pressures, i.e. under full saturation and at extreme suctions, the specific water capacity is zero or negligible and flux reduction is due solely to structural impediment. The smallest flux reduction occurs where the specific water capacity is greatest, i.e. when boundary suctions embrace the shallowest part of the soil moisture characteristic.

7. REFERENCES

McDougall, J.R. (1996) Application of variably saturated flow theory to the hydraulics of landfilled waste: A finite element solution. Ph.D. thesis, University of Manchester/Bolton Institute, U.K.

Neuman, S.P. (1973) Saturated-unsaturated seepage by finite elements. Journal of Hydraulics Division, American Society of Civil Engineers, Vol.99, HY12, pp 2233-2250.

Thomas, H.R. & Rees, S.W. (1990) Modelling field infiltration into unsaturated clays. Journal of Geotechnical Engineering, American Society of Civil Engineers, Vol.116, No.10, pp 1483-1501.

van Genuchten, M.T. (1980) A closed form equation for predicting the hydraulic conductivity of unsaturated soils. Soil Science Society of America Journal, Vol.44, pp 892-898.

Numerical Models in Geomechanics, Pietruszczak & Pande (eds) © *1997 Balkema, Rotterdam, ISBN 90 5410 886 X*

A numerical model for flow of particulate material

R. Kitamura, G. Kisanuki, M. Yamada & K. Uemura
Kagoshima University, Japan

S. Fukuhara
Kagoshima Prefecture, Japan

ABSTRACT: An approach to model the flow of multi-phase material such as soil is proposed based on some mechanical and probabilistic consideration of the particle motion. The seepage behavior of rain water into particulate material is numerically simulated by the model for voids. Then the failure and flow of particulate material is numerically simulated by some simple equations for the free falling of a particle and the collision of two particles. It is found out from the numerical results that the proposed approach can simulate the flow.

1. INTRODUCTION

The particulate material such as soil is composed of particles and fluid among them, i.e., the multi-phase material. Therefore the slope failure, debris and mud flow due to heavy rain, and pyroclastic flow are regarded as the multi-phase flow of particulate material. Kagoshima Prefecture is located in a volcanic zone in which the volcanic products are widely sedimented. In the rainy season of almost every year there are heavy rainfalls followed by the occurrences of slope failure and debris flow. In the geotechnical engineering it is one of the important and urgent problems to be solved to establish the effective countermeasure to such natural disasters.

In this paper an approach for modeling of multi-phase flow of particulate material such as progressive slope failure is proposed based on some consideration of the motion of solid and fluid phases on the particle scale. Then the numerical experiment is carried out to simulate the seepage of water into particulate material followed by the occurrence of failure and flow of particulate material by decreasing the cohesive force at contact points of particles. The rate of decrease in cohesive force and increase in deadweight is calculated in accordance with the change in suction and water content in particulate material which are calculated by the model for voids. Then the flow of particulate material is simulated based on only the free falling of a particle and the inelastic collision of two particles.

2. MODEL FOR VOIDS IN PARTICULATE MATERIAL

Figure 1(a) is a typical picture showing the arrangement of several soil particles which are contained in the parallelepiped. The state shown in Fig. 1(a) can be modeled by Fig. 1(b), where voids are represented by a straight pipe whose diameter and inclination angle are D and θ respectively and soil particles are represented by the shaded part which is impermeable. The shape and size of soil particles are random and consequently the distribution of voids is too complicated to be presented deterministically. Therefore, the diameter D and inclination angle θ of pipe in Fig. 1(b) are regarded as random variables and the probability density functions, Pd(D) and Pc(θ), are introduced in the model. Using Poisseulle's law and doing some probabilistic consideration, we can derive the permeability coefficient, void ratio, moisture content by volume and pF-value as follows[1),2)].

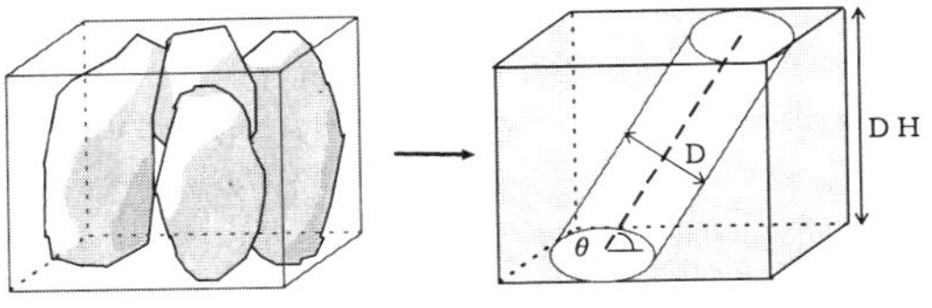

(a) Soil particles and voids

(b) Modeling by pipe for voids

Figure 1 Modeling for voids

$$k = \int_0^d \int_{-\pi/2}^{\pi/2} \frac{\pi \cdot \gamma_W \cdot D^3 \sin\theta}{128\mu(D/\sin\theta + DH/\tan\theta)} P_d(D) P_c(\theta) d\theta dD \quad (1)$$

where
k: permeability coefficient
for $d \to \infty$, k : saturated permeability coefficient
for $d \neq \infty$, k: unsaturated permeability coefficient,
μ : viscous coefficient of pore water,
$\gamma\omega$: unit weight of pore water,
D: diameter of pipe in Fig.1(b),
θ : inclination angle of pipe in Fig.1(b),
$Pd(D)$:probability density function of diameter D,
$Pc(\theta)$: probability density function of inclination θ angle .

$$e = \int_0^\infty \int_{-\pi/2}^{\pi/2} \frac{V_p}{V_e - V_p} P_d(D) P_c(\theta) d\theta dD \quad (2)$$

where
e: void ratio,
Ve: volume of pipe (permeable part) in Fig.1(b),
Vp: volume of container in Fig.1(b).

$$W_v = \frac{\int_0^d \int_{-\pi/2}^{\pi/2} V_p P_d(D) P_c(\theta) d\theta dD}{\int_0^\infty \int_{-\pi/2}^{\pi/2} V_e P_d(D) Pc(\theta) d\theta dD} \quad (3)$$

where
Wv: moisture content by volume.

$$S_r = \frac{\int_0^d \int_{-\pi/2}^{\pi/2} V_p P_d(D) P_c(\theta) d\theta dD}{\int_0^\infty \int_{-\pi/2}^{\pi/2} V_p P_d(D) Pc(\theta) d\theta dD} \quad (4)$$

where
Sr : degree of saturation.

$$pF = \log_{10} h_c = \log_{10}\left(\frac{4 \cdot T_s \cdot \cos\alpha}{\gamma_w \cdot d}\right) \quad (5)$$

where
pF: suction
Ts : capillary tension
α: contact angle

Figure 2 shows two contact particle due to the surface tension of water membrane on particles. Based on the geometric consideration the inter-particle force is derived as follows.[3)]

$$Fi = 2\pi r' Ts + \pi r'^2 h_c \quad (6)$$
$$(r'+a)^2 = (a+\beta)^2 + \alpha^2$$
$$\alpha = r + \Delta r/2 \quad , \beta = r + \Delta r$$

where

hc:height of water head (suction),
Fi:iter-particle force.

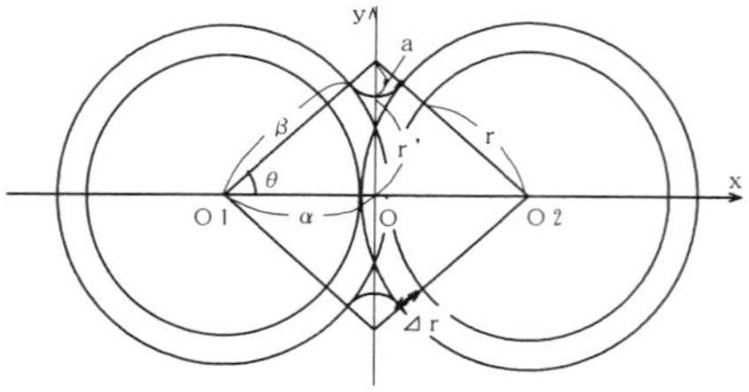

Figure.2 Inter-particle force due to surface tension between two contact particles

3. MODEL FOR FAILURE AND FLOW OF PARTICULATE MATERIAL[4),5)]

Figure 3 shows the loosest packing for uniform spherical particles on a slope with the inclination angle . Using the computer, it is easy to pack the particles in the structure shown in the figure by giving cohesion at each contact point of the particles. When the inter-particle force produced by cohesion is decreased, the particle becomes unstable, and its motion is described by Newton's second law of motion.

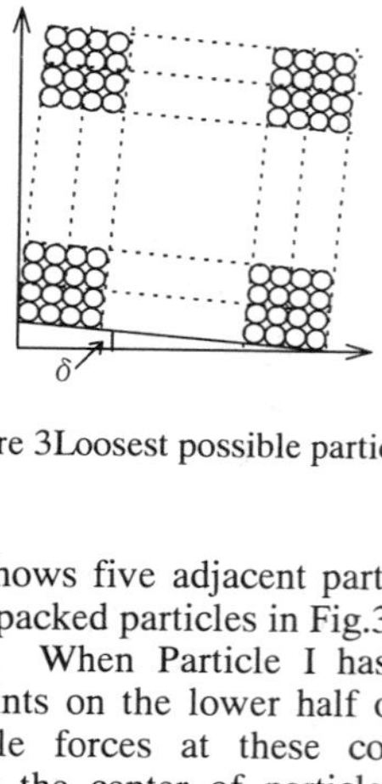

Figure 3Loosest possible particle packing

Figure 4 shows five adjacent particles which are a part of the packed particles in Fig.3, Particles I, J, K, L, and H. When Particle I has more than two contact points on the lower half of its surface, the inter-particle forces at these contact points are directed to the center of particle to be balanced. Under this condition, the inter-particle force at contact point, C(I,J) is expressed by

$$\vec{F}\{I,J\} = \vec{T}\{I,J\} + \vec{M}\{I,J\} \quad (7)$$

where,
$\vec{F}\{I,J\}$: the total inter-particle force at the contact point, C(I,J);
$\vec{T}\{I,J\}$: the cohesion at the contact point, C(I,J)
$\vec{M}\{I,J\}$: the inter-particle force caused by the upper particles' weight at contact point, C(I,J).

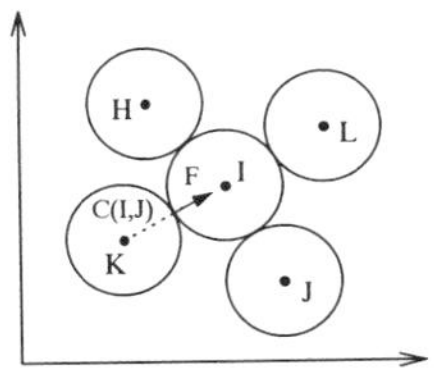

Figure 4

Arrangement of five adjacent particles

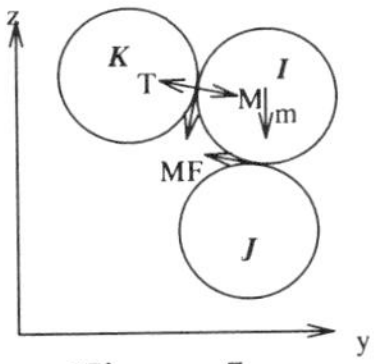

Figure 5

Equilibrium of force at a contact point, C(I,J)

Figure 5 shows the arrangement of particles in which Particle I has only one contact point on the lower half of its surface. In this case the frictional force at the contact point, C(I,J) is estimated by

$$\overrightarrow{MF}\{I,J\} = \kappa \cdot \vec{F}\{I,J\} \tag{8}$$

where

$\overrightarrow{MF}\{I,J\}$: the frictional force

k : the frictional coefficient.

The unstable condition of Particle I is accounted for by the equilibriums of force and moment and is estimated by the following equations:

For the direction of the y-axis

$$\left|\vec{P}\{I,J\}\right| \cdot \cos\delta > \left|\overrightarrow{MF}\{I,K\}\right| \cdot \sin\delta + \left|\overrightarrow{MF}\{I,J\}\right| \cdot \cos\delta \tag{9}$$

for the direction of the z-axis

$$\left|\vec{P}\{I,J\}\right| \cdot \sin\delta > \left|\overrightarrow{MF}\{I,K\}\right| \cdot \cos\delta + \left|\overrightarrow{MF}\{I,J\}\right| \cdot \sin\delta \tag{10}$$

and for the moment about the contact point, C(I,J)

$$m \cdot g \cdot r \cdot \sin\delta + \left|\vec{M}\{I,K\}\right| \cdot r > \left|\vec{T}\{I,K\}\right| \cdot r + \left|\overrightarrow{MF}\{I,K\}\right| \cdot r \tag{11}$$

where

m: the mass of the particle

g: the gravitational acceleration

$\left|\vec{P}\{I,J\}\right|$: the resultant force that causes the relative displacement between Particles I and J at contact the point, C{I,J}

δ : the inclination angle defined in Fig. 3

r: the radius of the particle.

Figure 6 shows the particle motion in the gravity field, as expressed by the following equations:

For free falling

$$m\frac{d^2}{dt^2}(y/\cos\delta) = 0 \tag{12}$$

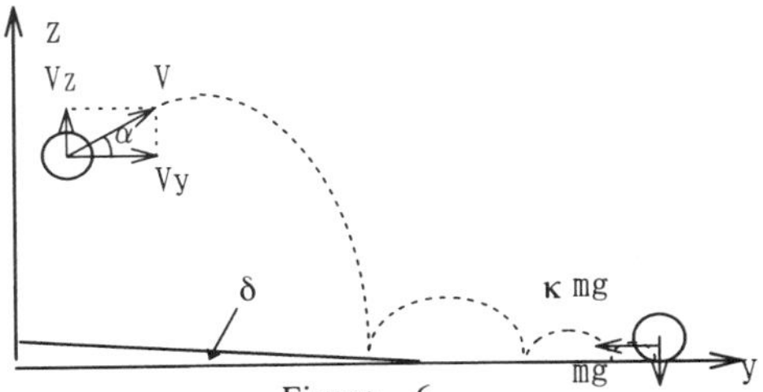

Figure 6

Free falling of a particle under the force of gravity

$$m\frac{d^2}{dt^2}(y/\sin\delta) = -mg \tag{13}$$

When the friction of the particle acts on the slope, the following equation is used instead of Eq.(7);

$$m\frac{d^2}{dt^2}(y/\cos\delta) = mg \cdot \sin\delta - \kappa \cdot mg \cdot \cos\delta \tag{14}$$

where

k : the frictional coefficient between particles, or between the particles and slope.

The conservation of momentum of a two-particle collision is expressed as

$$m_1 \cdot \vec{u_1} + m_2 \cdot \vec{u_2} = m_1 \cdot \vec{u_1}' + m_2 \cdot \vec{u_2}' \tag{15}$$

When the law of elastic collision is assumed, Equation (11) can be used:

$$e = -\frac{u_{1\eta}' - u_{2\eta}'}{u_{1\eta} - u_{2\eta}} \tag{16}$$

where

e : the coefficient of restitution of the particle

The contact angle , defined by the tangential plane at the contact point, is given by

$$\tan\gamma = -\frac{z_1 - z_2}{y_1 - y_2} \tag{17}$$

The many-body problem has been investigated by researchers in various fields of physics, but has not yet been solved analytically. In this model the collision of the particles is assumed to occur at one point at the same time in order to avoid the many-body problem. To satisfy this assumption, the increment time of calculation, Δt , should be as small as possible although it may not be small enough in the following numerical experiment.

4. NUMERICAL EXPERIMENT

(1) Moisture Characteristic Curves

The numerical experiment was carried out to obtain the moisture characteristic curves for Shirasu.

Shirasu. is defined as a non-welded part of pyroclastic flow deposits and distributed in the southern part of Kyushu Island. The values of input parameters of model are listed in Table1, where the height *DH* in Fig.1(b) for each soil is determined to be the same order of D50 which is the diameter of 50 % finer by weight in grain size accumulation curve. The mean value of diameter D of pipe in Fig.1(b) for each soil is determined to be approximately 0.1~0.5 of D50. The standard deviation of logD is determined so that the coefficient of variation of logD is same as the one of the grain size distribution. Here the probability density function is assumed to be a concave pentagonal shape [6]. Figure 7 shows the moisture characteristic curves of Shirasu obtained by the numerical experiments in which the results of pF-tests on these soils are also plotted. The input values used in the numerical experiment are listed in Table 1. It is found out from these results that the proposed model can express the suction and permeability coefficient of unsaturated soil.

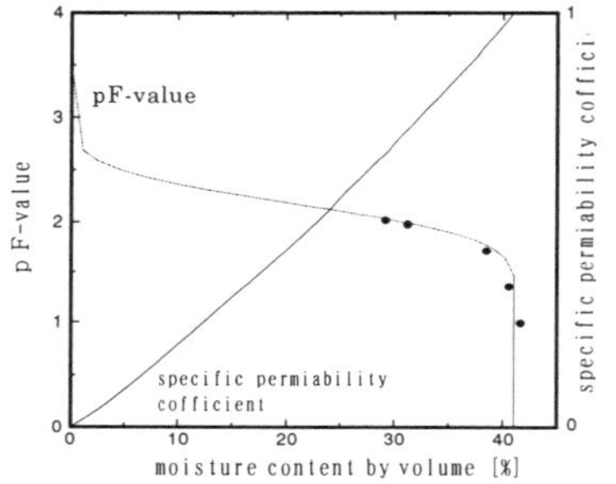

Figure 7
Moisture characteristic curves of Shirasu obtained by numerical experiment and pF-test

(2) Failure and flow of particulate material

Figure 8(a) is the imaginary slope divided into elements and Fig.8(b) is the corresponding particulate mass which is piled by adding the cohesion at contact points of particle which is divided into elements. The hydraulic gradient between adjacent element in Fig.9 is calculated by the following equations.

The hydraulic gradient of adjacent meshes as shown in Fig.6 can be obtained by Eq.(18).

$$i(I,J) = \frac{(z_j + h_j) - (z_i + h_i)}{\sqrt{(y_i - y_j)^2 + (zi - zj)^2}} \tag{18}$$

where

i(I,J): hydraulic gradient between the meshes I and J,

(yi,zi): position of mesh I,

hi: water head of mesh I.

Then the volume of inflow pore water from mesh J to mesh I is obtained by using Darcy's law as follows.

$$\Delta Q(I,J) = k\,i(I,J) A\,\Delta t \tag{19}$$

where

$\Delta Q(I,J)$: volume of inflow pore water from mesh J to mesh I,

k: permeability coefficient,

A: cross section area of mesh,

Δt : time increment.

The water content of mesh I is obtained by the following equation.

$$Q(I,t+\Delta t) = Q(I,t) + \sum_{j-1}^{n} \Delta Q(I,J) \tag{20}$$

where

$Q(I,t)$: water content of mesh I at time t.

n: number of adjacent meshes to mesh I.

Figure 10 shows the flow chart for calculation procedure. A numerical experiments was carried out, i.e., the water is supplied from the bottom of particulate material at the rate of 0.0003 $l/(cm^2*min)$. The results of numerical experiment are shown in Figs.11 . The input values used are listed in Table 2.

Table 1 Input values of parameters for seepage

height of DH in Fig.1(b)(cm)	0.002	number of mesh	100
coefficient of viscocity of water ($gf/s/cm^2$)	$1.16*10^{-5}$	length of mesh (cm)	4*4
surface tension of water (gf/cm)	0.075	inital water content	0.35
height of DH in Fig.1(b) (cm)	0.0012	rate of supplied water (mm/10min)	20.0
mean of pipe D in Fig.1(b) (cm)	0.0008	time increment (sec)	1.0
standard deviation of pipe D in Fig.1(b) (cm)	0.159	inclination angle of slope (°)	3.0

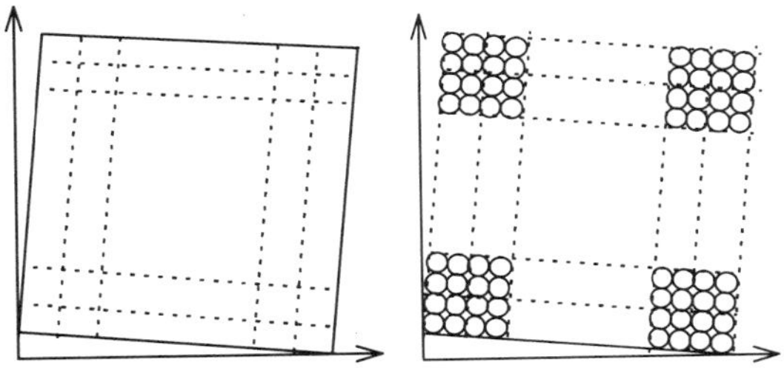

(a) Imaginary slope divided into elements (b)Corresponding particulate mass

Fig.8 Imaginary slope

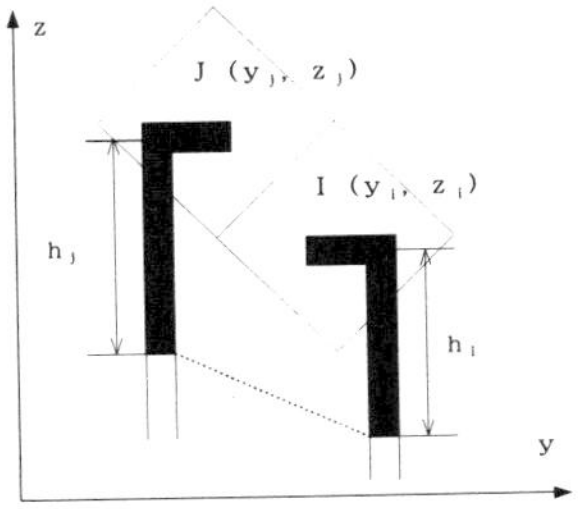

Fig.9 Relation of water head in adjacent elements

Table 2
Input values of parameters for failure and flow of particulate material

number of mesh	400
dadius of particle(cm)	1.0
coefficient of restitution of particles	0.2
density of particle(g/cm^3)	2.7
frictional coefficient	0.577
time increment (sec)	0.001
inclination angle of slope(°)	3.0

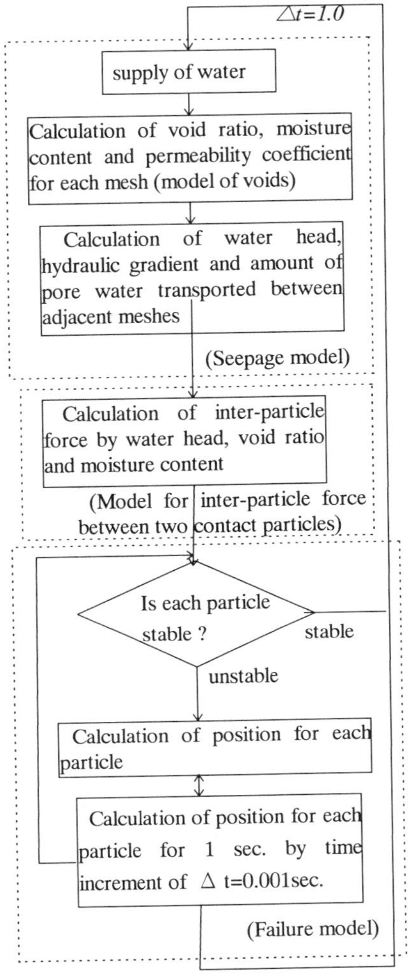

Fig.10 Calculation procedure

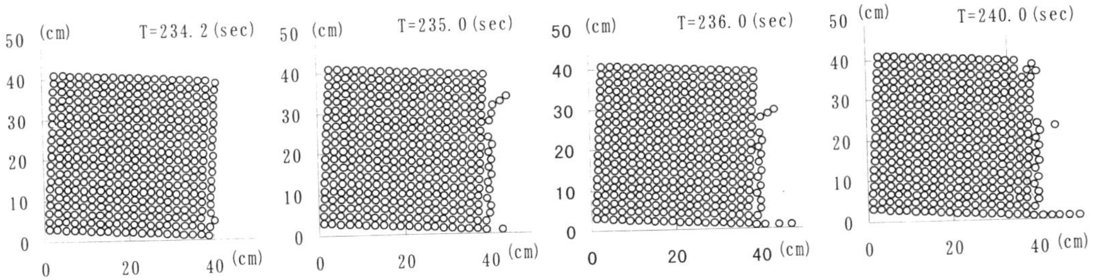

Fig.11 Failure mode of particulate material with time

5. CONCLUSIONS

The numerical models are proposed to simulate the progressive slope failure of particulate material which are caused by the decrease in cohesion at contact points of particles. In geotechnical engineering such phenomena are called slope failure, debris flow etc. which means that the proposed model could be applied to these problems. It is found out that various phenomena of particulate material related to the geomechanics could be simulated by the computer based on the simple equations in the Newtonian mechanics.

REFERENCES

1) Kitamura, R. and Uto, Y. (1991). Unsaturated seepage model based on micro-structure of soil. Proc. Int. Conf. for Computer Methods and Advances in Geomechanics, Vol.2, 1635-1640.

2) Kitamura, R., Fukuhara, S. and Kisanuki, G. (1996). A numerical model for infiltration. Proc.2nd Int. Conf. on Environmental Geotechnics, Vol.1, 239-244.

3) Karube, D. , Namura, K. , Morita, N. and Iwasaki, T. (1978). Fundamental study of the stress-strain behavior of an unsaturated soil . Proc. JSCE , No.269 , 105-119.(in japanese)

4) Kitamura, R. and Kisanuki, G. (1991). Simulation Model for Flowage and Deposition of Particulate Material. Natural Disaster Science, Vol.13, No.2 pp.21-33.

5) Kitamura, R., Fukuhara, S. and Kisanuki, G. (1995). Simulation model for flowage of particulate material. Natural Disaster Science, Vol.17, No.1, 29-52.

6) Kitamura , R. , Fukuhara , S. and Kisanuki , G .(1996) A numerical model for infiltration . Proc. 2nd Int. Congress on Environmental geotechnics (IS-Kyushu'96), 239-244.

Numerical Models in Geomechanics, Pietruszczak & Pande (eds) © 1997 Balkema, Rotterdam, ISBN 90 5410 886 X

A numerical model for the analysis of piezocone dissipation curves

P.U. Kurup & M.T. Tumay
Louisiana Transportation Research Center, Louisiana State University, La., USA

ABSTRACT: Dissipation tests conducted during piezocone penetration tests (PCPT) give good indication of the consolidation characteristics of fine grained soils. This however requires proper interpretation of excess pore pressure dissipation results. A soil constitutive model based on critical state soil mechanics is presented for the interpretation of piezocone dissipation results. The dissipation phase is analyzed by an elastoplastic, coupled stress consolidation analysis. The dissipation results are also analyzed using a simplified extension of the spherical cavity expansion model to two dimensions combined with an uncoupled consolidation analysis. The models are evaluated by experimental PCPT results obtained from calibration chamber studies using a miniature piezocone penetrometer. The finite element analysis under-predicted the excess pore pressure measured above the cone base. The extended cavity expansion model showed very good agreement with pore pressure dissipation monitored above the cone base and qualitative agreement with the radial dissipation results.

1 INTRODUCTION

Many problems in geotechnical engineering especially those involving settlement and seepage, require reliable estimation of consolidation characteristics of fine grained soils. The conventional methods of estimating consolidation parameters are from laboratory oedometer tests conducted on small soil samples or from back analysis using piezometric and settlement records during and after construction. Small soil samples obtained for laboratory tests may not be able to capture the in situ macro fabric and are prone to disturbance during sampling, transport and handling operations. Oedometer tests are performed under simplified boundary conditions and in back analysis variations in soil stratigraphy makes it difficult to estimate the correct drainage path during consolidation. Conventional methods are generally more time consuming and expensive.

In recent years there has been an increasing trend to use in situ techniques for determining the consolidation characteristics of cohesive soils. In situ tests are performed under existing in situ stresses, boundary and environmental conditions and may be more appropriate for the purpose of design. The piezocone penetrometer is a popular in situ investigation tool of choice for the assessment of engineering properties and in situ states of geomaterials. The excess pore pressure dissipation data at the end of steady penetration during a PCPT can provide fairly good estimates of consolidation characteristics of fine grained soils.

This paper describes a soil constitutive model based on critical soil mechanics for the interpretation of excess pore pressure dissipation results from PCPTs. The dissipation phase is analyzed by an elastoplastic, coupled stress-consolidation analysis. The dissipation results are also analyzed using a simplified extension of the spherical cavity expansion model to two dimensions combined with an uncoupled consolidation analysis. The models are evaluated by experimental PCPT dissipation results obtained from calibration chamber studies using a miniature piezocone penetrometer.

2 EXCESS PORE PRESSURE DISTRIBUTION

The initial excess pore pressure distribution due to piezocone penetration in clays is an important factor affecting the interpretation of the coefficient of consolidation (c_r). Detailed parametric studies using different initial excess pore pressure distribution (constant, linear and logarithmic distribution) involving cylindrical and spherical cavities have been performed by Levadoux and Baligh (1980). Different methods have been proposed by investigators in the past based on cavity expansion theories, strain path method, semi-empirical methods and finite element analysis. However prediction of the exact initial excess pore pressure distribution in clays of various plasticity index, rigidity index and overconsolidation ratios (OCR) is still elusive. Material and geometric non linearity, principal stress rotation and soil remolding during penetration further adds to the complexity of the problem. Calibration chamber studies on instrumented clay specimens prepared and tested under controlled conditions help in evaluating theoretical models.

3 EXPERIMENTAL STUDY

In order to verify the efficacy of any theoretical model it is essential to validate their prediction/interpretation capabilities with experimental results. For the research described in this paper controlled calibration chamber testing of clays specimens were conducted. Homogeneous cohesive soil specimens 525 mm in diameter and 812 mm in height were prepared by a two stage slurry consolidation technique. In the first stage soil slurry prepared by mixing Kaolin and Edgar fine sand with de-ionized water at twice the liquid limit was consolidated in an automated rigid wall slurry consolidometer (Kurup, 1993; Voyiadjis et al., 1993). After the first stage of slurry consolidation the soil specimens were transferred into a double wall flexible calibration chamber system and subjected to known stress histories and controlled boundary conditions. The servo-controlled calibration chamber system developed by Tumay and de Lima (1992) allows independent control of vertical and lateral stresses and strains. Four soil specimens were prepared and tested under a back pressure (u_o), of 138 kPa. The K50 specimens were prepared by mixing 50% kaolin and 50% fine sand whereas the K33 specimen was prepared by mixing 33% kaolin and 67% fine sand. The specimens were instrumented with pore pressure access ducts (prior to slurry consolidation) to monitor the spatial pore pressure distribution. The tips of the ducts were located at varying radial distances from the center and at two different depths (level 1 and level 2). Level 1 was located at a depth of 570 mm (measured from the top of the specimen) for specimens 1 and 2 and at a depth of at a depth of 600 mm for specimens 3 and 4. Level 2 was located at a depth of 390 mm for specimens 1 and 2 and at a depth of at a depth of 420 mm for specimens 3 and 4. Piezocone penetration tests were performed using a 1 cm^2 cross-sectional area (radius, r_o = 5.64 mm) miniature piezocone fabricated by Fugro McClelland Engineers B.V., The Netherlands. The piezocone has a choice for the filter element to be located either in the lowest 1/4 of the cone at the very tip (u1 configuration); or starting 0.5 mm above the cone base and 2 mm in vertical height (u2 configuration). Details of the specimen preparation and PCPTs are given by Kurup 1993 and Kurup et al. 1994. A summary of the specimen soil properties, stress histories and PCPT results are given in Table 1.

Table 1. Soil properties and PCPT results

Specimen no.	1	2	3	4
Soil sample	K50	K33	K50	K50
Plasticity index, I_p	14%	6%	14%	14%
$K_o = \sigma'_{ho} / \sigma'_{vo}$	1.0	1.0	1.0	0.52
σ'_{vo} (kPa)	207	207	41.4	207
OCR	1	1	5	1
Undrained shear strength, s_u(kPa)	60	80	40	65
Rigidity index, $I_r=G_{50}/s_u$	267	100	150	567
$c_r x10^{-3}$ cm^2/s. Virgin Reload	 14.1 78.8	 28.3 141	 14.1 78.8	 26.4 105
Cone resistance, q_T-u_o (kPa)	1183	1249	661	644
Δu_1 at tip (kPa)	562	632	528	490
Δu_2 above cone base (kPa)	624	591	406	368

Dissipation tests were performed at the end of all PCPTs at level 1. The excess pore pressure dissipation monitored by the piezocone (in the u2 configuration) and by the four pore pressure ducts located in level 1 were interpreted using various models (Kurup, 1993). In the following sections the dissipation results will be evaluated by a critical state soil constitutive model and also by a simplified model which is an extension of the one-dimensional spherical cavity expansion to two-dimensional situations.

4 CRITICAL STATE SOIL MODEL AND FINITE ELEMENT ANALYSIS

An elastoplastic work hardening constitutive model proposed by Roscoe and Burland (1968) known as the modified cam clay (MCC) is used for the finite element analysis. This model is based on the critical state concepts (Schofield and Wroth, 1968). In the MCC model the yield locus takes the form:

$$f = M^2 p'^2 - M^2 p' p'_c + q^2 = 0 \quad (1)$$

and describes an ellipse in the q-p′ space. The strain hardening parameter is p'_c (defining the subsequent location of the yield surface) and is the intersection of the yield surface and the p′ axis. In the above definition p′ and q (stress invariants) are the mean effective stress and the deviatoric stress, respectively, expressed in terms of principal effective stresses as:

$$p' = \frac{1}{3}(\sigma'_1 + \sigma'_2 + \sigma'_3) \quad (2)$$

$$q = \frac{1}{\sqrt{2}}[(\sigma'_1 - \sigma'_2) + (\sigma'_2 - \sigma'_3)^2 + (\sigma'_3 - \sigma'_1)^2]^{1/2} \quad (3)$$

In addition to the above, a knowledge of the in situ stresses and the preconsolidation pressure (p'_c) is required. The incremental effective stress-strain relationship is given by:

$$\dot{\sigma}'_{ij} = D^{ep}_{ijkl}\, \dot{\epsilon}_{kl} \quad (4)$$

where D^{ep}_{ijkl} is the elastoplastic constitutive tensor given by:

$$D^{ep}_{ijkl} = \left[D^{e}_{ijkl} - \frac{D^{e}_{ijrs} A_{rs} A_{mn} C^{e}_{mnkl}}{H^e + H^p} \right] \quad (5)$$

where D^e_{ijkl} is the elastic constitutive tensor, H^e and H^p are the elastic and plastic moduli, respectively, and defined as:

$$H^e = \left(\frac{\partial f}{\partial \sigma'_{ij}}\right) D^{e}_{ijkl} \left(\frac{\partial f}{\partial \sigma'_{kl}}\right) \quad (6)$$

$$H^p = \left(\frac{1 + e_0}{\lambda - \kappa}\right) p'\, p'_c \left(\frac{\partial f}{\partial \sigma'_{ii}}\right) \quad (7)$$

$$A_{ij} = \frac{\partial f}{\partial \sigma'_{ij}} \quad (8)$$

The material parameters of soil specimen no.1 for the MCC model are: slope of the virgin consolidation line in the e-ln p′ space, λ = 0.11; slope of the unloading-reloading line in the e-ln p′ space, κ = 0.024; slope of the critical state line, M = 1.2; void ratio at unit p′ on the critical state line in the e-ln p′ space, e_{cs} = 1.162; Poisson's ratio, ν = 0.3. The permeability of the soil, k = 0.5×10^{-9} m/s.

A small strain axisymmetric finite element analysis was performed using a mesh consisting of 140 eight noded isoparametric quadrilateral elements and a total of 165 vertex nodes. A mesh of height, 28 r_o and radius, 30 r_0 was selected based on the criteria of minimum influence from boundary effects. The analysis was performed under conditions of constant vertical and lateral stresses (to simulate the calibration chamber boundary condition). The piezocone penetration was simulated up to a depth of 15 r_o, in 10 stages, by imposing radial displacements to the nodes incrementally. The excess pore pressure dissipation was conducted using a coupled elastoplastic consolidation analysis. The finite element method predicted the normalized dissipation profile measured above the cone base reasonably well (figure 1). The radial pore pressure distribution predicted by the finite element analysis is shown in figure 2. The analysis was found to under-predict the excess pore pressure measured above the cone base (at r/r_o = 1 in figure 2). The radial pore pressure distribution and its subsequent dissipation was significantly different from those monitored by the pore pressure access ducts and are hence not shown.

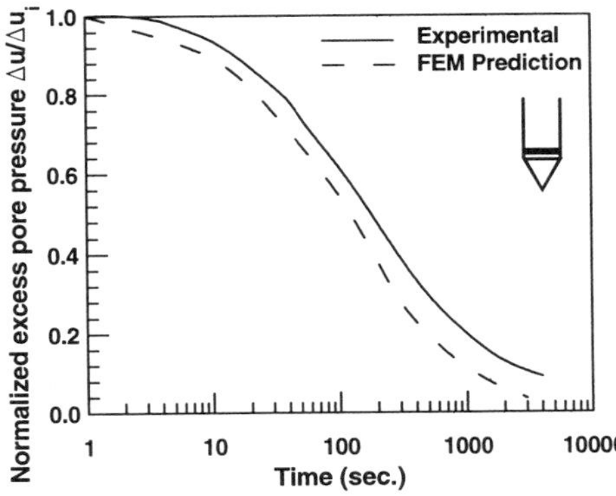

Figure 1. Dissipation profile above the cone base in specimen 1 compared with finite element prediction

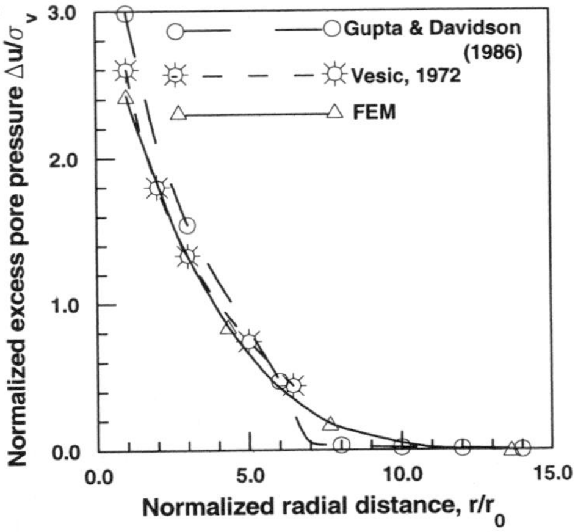

Figure 2. Radial excess pore pressure distribution at level 1 for specimen 1 predicted by various methods

5 AN EXTENDED CAVITY EXPANSION MODEL

Both cylindrical and spherical cavity expansion models are one-dimensional models where as the piezocone penetration problem is actually a two-dimensional process. The one dimensional spherical cavity expansion model may be extended to two dimensional situations by modeling the piezocone advance approximately as a series of successive spherical cavity expansions (Gupta and Davidson, 1986). This method assumes that during the advance of the piezocone, it produces a deformation pattern in the vicinity of the tip approximately similar to that of a spherical cavity expansion. The cylindrical symmetry of pore pressure distribution above the cone base is a result of successive summation of spherical cavity expansions.

The excess pore pressure distribution predicted by the spherical cavity expansion theory (Vesic, 1972) is given by:

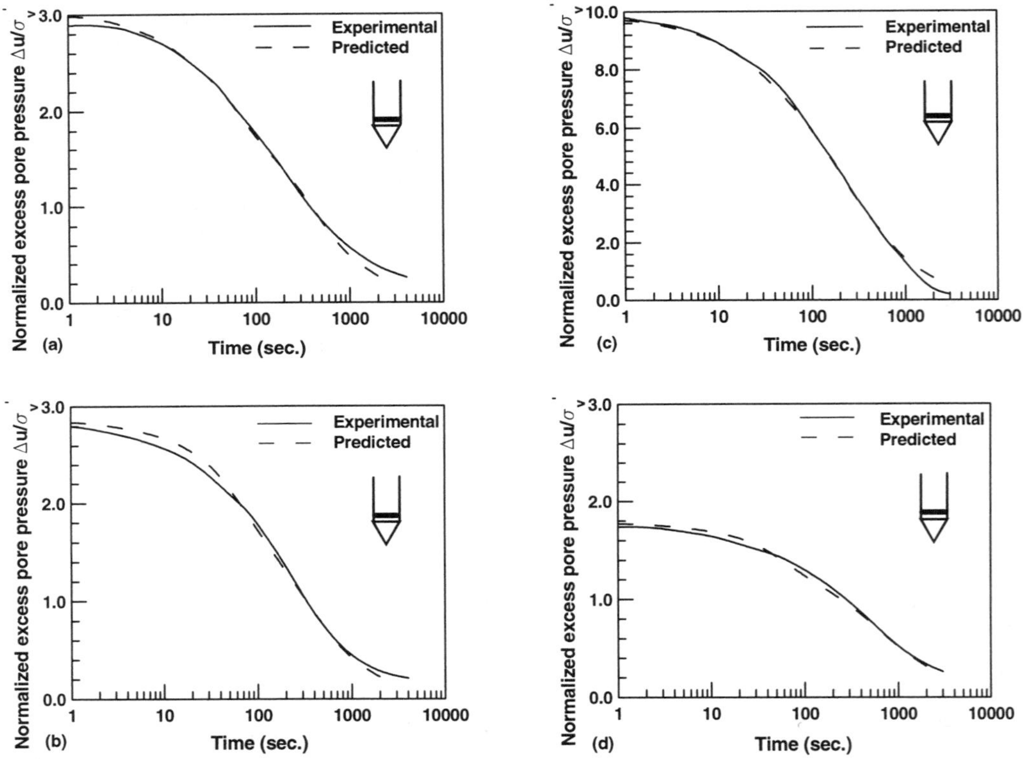

Figure 3. Experimental dissipation profiles at the cone base compared with that predicted by the method of successive spherical cavity expansion: (a) Specimen 1 (b) Specimen 2 (c) Specimen 3 (d) Specimen 4

$$\Delta u_i = s_u \left[0.943 \; \alpha_f + 4 \ln \left(\frac{r_p}{r} \right) \right] \tag{9}$$

where s_u is as defined earlier, r is the radial distance and r_p is the radius of the plastic zone for spherical cavity expansion given by:

$$r_p = r_0 \sqrt[3]{\frac{G}{s_u}} \tag{10}$$

where r_0 is the equivalent penetrometer radius, and G is the shear modulus..

The Henkel's pore pressure parameter, α_f is related to the Skempton's pore pressure parameter at failure, A_f, by:

$$\alpha_f = 0.707\,(3\,A_f - 1) \tag{11}$$

It was found that the magnitude of the excess pore pressure predicted by equation 9 was lower than the excess pore pressure measured by the piezocone in the u2 configuration. An empirical correction procedure was hence performed (Gupta and Davidson, 1986) by modifying equation 9 to exactly predict the excess pore pressure measured above the cone base. The modified initial excess pore pressure distribution is given by (equation 12):

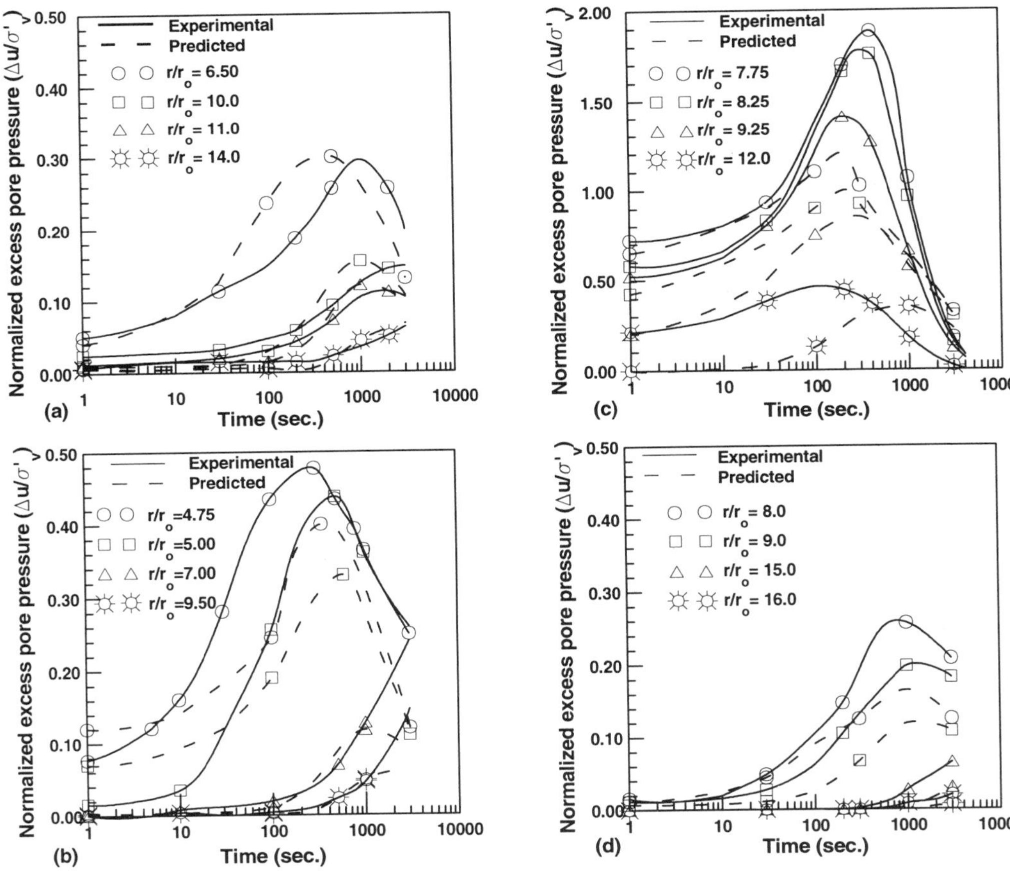

Figure 4. Radial excess pore pressure dissipation at level 1 compared with that predicted by the method of successive spherical cavity expansion: (a) Specimen 1 (b) Specimen 2 (c) Specimen 3 (d) Specimen 4

$$\Delta u_{rc} = \frac{\Delta u_b \left[0.943\ \alpha_f + 4\ \ln\left(\frac{r_p}{r}\right)\right]}{\left[0.943\alpha_f + 4\ \ln\left(\frac{r_p}{r_o}\right)\right]} \tag{12}$$

where Δu_{rc} is the corrected spatial excess pore pressure distribution, and Δu_b is the actual measured excess pore pressure at the base of the piezocone. The method proposed by Gupta and Davidson (1986) for determining the initial excess pore pressure distribution by successive spherical cavity expansion was used to simulate the piezocone penetration mechanism. The pore pressure dissipation that takes place even during piezocone penetration was also taken into account. The corrected pore pressure distribution was used in a dissipation analysis based on the Terzaghi-Rendulic uncoupled consolidation theory. The radial pore pressure distribution predicted by equations 9 and 12 for specimen 1 (at level 1) are shown in figure 2. Comparison between the predicted and measured excess pore pressure dissipation at the cone base (figure 3) for the four specimens showed very good agreement. The predicted and measured radial excess pores pressure dissipation at level 1 are shown in figure 4. The radial distance (r) of the tip of the pore pressure access ducts from the center of the specimen is normalized with respect to the radius (r_o) of the piezocone penetrometer. It can be seen there is qualitative agreement between the measured and predicted dissipation curves.

6 CONCLUSIONS

Excess pore pressure dissipation results from miniature piezocone penetration tests in four clay specimens inside a calibration chamber system are evaluated. The small strain finite element analysis using the modified cam clay constitutive model under-predicted the excess pore pressure above the cone base. The normalized dissipation profile above the cone base was predicted reasonably well. The extended cavity expansion model using successive spherical cavity expansions to simulate piezocone advance showed very good agreement between the experimental and predicted excess pore pressure dissipation above the cone base. The radial excess pore pressure dissipation predicted by the extended cavity expansion model showed qualitative agreement with experimental results.

7 ACKNOWLEDGMENTS

The financial support from the National Science Foundation under Grant CMS 9531782 is gratefully acknowledged. The support of Louisiana Transportation Research Center is also appreciated.

8 REFERENCES

Gupta, R. C. and Davidson, J. L. 1986. Piezoprobe Determined Coefficient of Consolidation. *Soils and Foundation*, 26(3): 12-22.

Kurup, P. U. 1993. *Calibration Chamber Studies of Miniature Piezocone Penetration Tests in Cohesive Soil Specimen, Ph.D. Dissertation*. Louisiana State University, Baton Rouge, LA.

Kurup, P. U., Voyiadjis, G. Z. and Tumay, M.T. 1994. Calibration Chamber Studies of Piezocone Tests in Cohesive Soils. *ASCE, Journal of Geotechnical Engineering Division, 120(1): 81-107.*

Levadoux, J. N. and Baligh, M. M. 1980. *Pore Pressures during Cone Penetration*. MIT, Dept. of Civil Engineering, Report No. MITSG 8--23.

Roscoe, K. H. and Burland, J. B. 1968. On the Generalized Behavior of 'Wet' Clay. In J. Heyman & F.A. Leckie (eds), *Engineering Plasticity*, Cambridge University Press, 535-609.

Schofield, A.N. and Wroth, C.P. 1968. *Critical State Soil Mechanics*. New York: McGraw-Hill.

Tumay, M.T. and de Lima, D.C. 1992. *Calibration and Implementation of Miniature Electric Cone Penetrometer and Development, Fabrication and Verification of the LSU In-situ Testing Calibration Chamber (LSU/CALCHAS)*. LTRC/FHWA Report No. GE-92/08.

Vesic, A. S. 1972. Expansion of Cavities in Infinite Soil Mass. *Journal of Soil Mechanics, Foundation Div., ASCE*, 98, 265-290,

Voyiadjis, G. Z., Kurup, P. U. and Tumay, M. T. 1993. Preparation of Large Size Cohesive Specimens for Calibration Chamber Testing. *ASTM, Geotechnical Testing Journal*, 16(3), 339-349.

Numerical Models in Geomechanics, Pietruszczak & Pande (eds) © 1997 Balkema, Rotterdam, ISBN 90 5410 886 X

Minimizing numerical oscillation in the consolidation analysis

M. Bai & J.-C. Roegiers
Rock Mechanics Institute, The University of Oklahoma, Okla., USA

M. Zaman
School of Civil Engineering and Environmental Science, The University of Oklahoma, Okla., USA

D. Elsworth
Department of Mineral Engineering, Pennsylvania State University, Pa., USA

ABSTRACT: To minimize the numerical instability and numerical oscillation in analyzing the consolidation scenarios, this paper presents a number of effective methodologies which include: (a) maintaining compatibility in magnitude for all coefficients in the system of equations through matrix normalization, (b) reducing the numerical oscillation due to local effect such as in the areas of load application through global finite element mesh refinement, and (c) maintaining the fully implicit time stepping at large times.

1. INTRODUCTION

In analyzing transient consolidation scenarios using numerical approximation, two types of difficulties may be encountered. The first type is numerical instability, which occurs as the difference between the true solution and the numerical solution grows unusually large in a few time steps. This behavior is illustrated in Figure 1. In general, numerical instability can be avoided by using a stable time discretization method such as a backward time marching scheme (fully implicit scheme) or a higher order time discretization scheme, i.e., Crank Nicolson method (half implicit scheme). For solving the transport problems in which the advective flow may be dominant, the instability can be minimized through controlling the Courant number (Huyakorn and Pinder, 1983).

The second type of difficulty is numerical oscillation. This occurs when the computed values fluctuate about the true solution as illustrated in Figure 2. For coupled problems where the systems of equation is solved simultaneously, such as the interactive flow-deformation consolidation, numerical oscillations have been reported despite use of a stable time stepping scheme (Sandhu et al., 1985). The condition of the system of equations is a major source of numerical oscillation (Ghaboussi & Wilson, 1973; Brady, 1979). Numerical oscillation may appear in the modeling of solute transport for the higher Peclet numbers when using a Galerkin's finite element method (Huyakorn and Pinder, 1983). The numerical oscillation may be reduced by transforming partial differential equations into a better conditioned forms (Bai et al., 1993), or using the method of Laplace transform before using the finite element discretization to minimize the adverse impact due to time stepping (Sudicky, 1989).

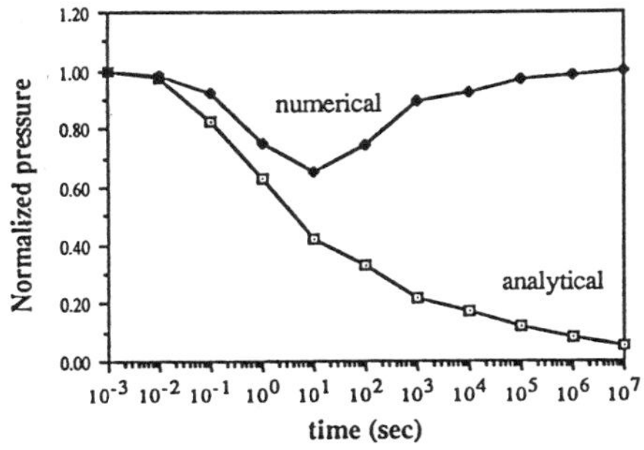

Figure 1. Example of numerical instability.

More effective approaches in reducing the numerical oscillation may remain to be further explored. The objective of this paper, however, is to discuss some simple alternatives (e.g., using intuition) which may provide certain simplified solutions in minimizing numerical oscillation for the analysis of consolidation scenarios.

2. CONDITIONING OF SYSTEM OF EQUATIONS

The conditions of a system of equations perhaps are the first and important issues to be examined. For an ill-conditioned system, additional remedies may have to be taken. The standard system of equations can be written in a matrix form as:

$$\mathbf{Ax} = \mathbf{b} \tag{1}$$

where **A** is the system matrix, **x** is a vector of unknowns and **b** is a vector of known boundary conditions. The system is robust if all the elements of **A** are approximately the same size or if the matrix is diagonally dominant. A simple sensitivity analysis can be performed using slightly different data to determine how sensitive the solution is to changes in the data. A more sophisticated method is to apply a perturbation technique (Rice, 1981) on one, or both sides of the system of equations.

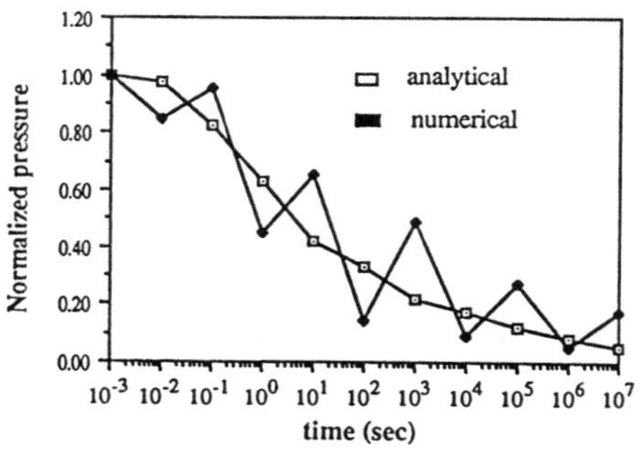

Figure 2. Example of numerical oscillation.

A posterior estimate may be used to evaluate the accuracy and condition of a system of equations. The following estimate for equation (1) can be derived as

$$\frac{\|\mathbf{x}-\hat{\mathbf{x}}\|}{\|\mathbf{x}\|} \leq K(\mathbf{A})\frac{\|\hat{\mathbf{r}}\|}{\|\mathbf{b}\|} \tag{2}$$

where $K(\mathbf{A})$ is the condition number and is equal to $\|\mathbf{A}\|\|\mathbf{A}^{-1}\|$, $\mathbf{x}$ is the exact solution, $\hat{\mathbf{x}}$ is the approximate solution, and $\hat{\mathbf{r}}$ is the residual calculated by the equation

$$\hat{\mathbf{r}} = \mathbf{b} - \mathbf{A}\hat{\mathbf{x}} \tag{3}$$

A more careful examination of equation (2) reveals that the left hand side of equation (2) is a relative error estimate for the solution, while $\frac{\|\hat{\mathbf{r}}\|}{\|\mathbf{b}\|}$ on the right hand side is the residual. The condition number, $K(\mathbf{A})$, is merely an estimate of how much the uncertainty in **b** may be magnified. In actuality, if the computed condition number is large, a good approximation to the true solution may not be achievable, even though the residual is small.

For inherently ill-conditioned systems of equations, great caution has to be exercised to prevent arriving at only a conditionally stable solution. There is, apparently, no unique way to deal with an ill-conditioned system of equations. Frequently, trial and error methods are used to obtain a stable solution.

3. CONDITION OF SYSTEM MATRIX FOR CONSOLIDATION

From equation (2) it is clear that the features of the system matrix, **A**, exert a controlling influence over the condition number, and hence the condition of the total system. For the case of transient deformation-dependent flow behavior in consolidation, the symmetric system matrix, **A**, can be symbolically written as

$$\mathbf{A} = \begin{pmatrix} \mathbf{E} & \mathbf{R} \\ \mathbf{R}^T & \Delta t\mathbf{C}(k) + \mathbf{S} \end{pmatrix} \tag{4}$$

where **E** is the elastic stiffness matrix, **R** is a matrix coupling solid deformation and fluid pressure, $\mathbf{R}^T$ is the transpose of **R**, Δt is the time step, $\mathbf{C}(k)$ is a conductance matrix that is a linear function of formation permeability k, and **S** is the storage matrix representing the capacity of fluid mass released from a storage domain. The matrix will be well conditioned if the order of magnitude of the diagonal terms are comparable and the matrix remains diagonally dominant. These requirements may not be satisfied for the period immediately after initial loading such as pumping or injection, because, resembling an undrained case at the initial loading, the formation may be stiff and induced strains and pore pressure may not be in the compatible magnitudes. As a result, the system of equation can be ill-conditioned, and the numerical oscillation would occur at this stage. It should be noted that this stage is temporary. The system may not be inherently ill-conditioned.

4. RELATION TO SOLUTION TECHNIQUE

There are many methods available for the solution of linear systems of equations. The relation of these methods to the numerical oscillation may be represented by the dynamic changes made to the system matrix during the solution process. Substantial modifications can lead to the ill-conditioned system at worst. Among the numerous techniques, the Gaussian elimination method is perhaps the simplest. One advantage of using this method is that the original system matrix is preserved during the solution process. However, the Gaussian elimination method may fail to produce a correct solution due to excessive computer round-off error if the diagonal terms in the system matrix, **A** in equation (4), are near-zero. The configuration of the system of equations may be arranged in such a way that the influence of the near-zero diagonal terms is reduced through pivoting. Full pivoting (interchanging both rows and columns) provides greater accuracy than partial pivoting (interchanging rows only), at the expense of doubling the computational cost. Table 1 shows a comparison of the CPU times consumed in solving a simple 3 by 3 matrix using Gaussian elimination, Gaussian elimination with partial pivoting, and Gaussian elimination with full pivoting, respectively.

Table 1. CPU time vs various solution techniques.

Method:	Gauss	Gauss(P)	Gauss(F)
CPU (0.001 s):	0.35	0.49	0.66

where P: partial pivot, F: full pivot.

In order to determine the influential factors for the numerical oscillation in the analysis of consolidation, a variety of matrix solution techniques [LU factorization, previous three Gaussian methods, Successive Over Relaxation (SOR) method] are implemented to study a consolidation scenario where the fluid flows vertically downwards across a formation which is subject to a sudden tectonic load on the top boundary. Very little difference has been identified in terms of solution accuracy. In other words, the numerical oscillation appears not to be associated with the variation in the solution techniques, at least for this case. However, the CPU times consumed by various methods show the differences which are compared in Table 2. Clearly, LU factorization, which does not use pivoting, consumes the least amount of CPU time, as opposed to full pivoting which requires the largest CPU time.

Table 2. CPU time vs various solution techniques.

Number	Method	CPU (sec/1000)
1	LU	0.10
2	Gauss	0.13
3	Gauss-P	0.13
4	SOR	0.20
5	Gauss-F	0.24

5. DATA MODIFICATION TECHNIQUE

The pivoting technique, described previously, is in essence a data normalization method. Brady (1979) reported a successful application of the normalization technique to boundary element method. However, for a coupled deformation-dependent flow system such as the consolidation scenario described by equation (4), it is not practical to normalize both solid displacement and fluid pressure concurrently, since, interchanging rows may result in a non-symmetric new system of equations, which could be a source of further numerical errors such as oscillation. As an example, formulation of flow in homogeneous porous media always leads to a symmetric system matrix, in which the parabolic partial differential equations preserve the numerical stable and oscillation-free environment. In contrast, modeling of transport scenario can result in a non-symmetric system of equations with the advection dominant features where the hyperbolic partial differential equations invoke numerical difficulties such as instability and oscillation.

A stable and oscillation-free solution is likely to be achieved if all the elements of the system of equations are of similar magnitude and are diagonally dominant. As a specific example, in the consolidation matrix of equation (4), the magnitude of diagonal terms in $\mathbf{E}$ is in general much larger than those in $\Delta t\mathbf{C}(k) + \mathbf{S}$, assuming that a small time step is used for a slightly compressible system. In order to retain compatibility in magnitude between displacement, $\mathbf{u}$, and pressure, $\mathbf{p}$, it may be advisable to normalize both displacement and pressure before solving the system of equations. If the absolute values are desired whilst the magnitudes between $\mathbf{u}$ and $\mathbf{p}$ are not compatible, the condition number $K(\mathbf{A})$ must be checked to ensure that the system is well-conditioned. Frequently, a decoupled approach may become the only choice if the system matrix becomes ill-conditioned.

6. MESH REFINEMENT

For the deformation-dependent flow consolidation problem as discussed previously, the numerical instability or oscillation tends to occur immediately following application of the load, and at the location of loading. In view of modeling, increasing element density in this region may substantially improve computational accuracy (Sandhu, et al., 1985). The appearance of numerical oscillation in solution may result from an abrupt change in mesh density. A mesh of equally spaced, instead of unequally spaced, elements may be designed to minimize the oscillation, at the expense of additional computational cost. Examining a one-dimensional consolidation scenario (column problem) using a 5 element configuration, Figure 3 shows the spatial pressure variation along the depth of the column. The result indicates that numerical oscillation occurs at the location on the first top half of the column with a time step Δt, of five seconds.

The problem can be avoided by using a larger time step (Δt = 10 seconds). This manipulation, however, does not provide the result for the time at 5 seconds, frequently required for the small time calculation. Alternatively, the numerical oscillation can be eliminated for the time at 5 seconds by increasing the number of elements to 10 (Figure 4). In this configuration, numerical oscillation may reappear for a time step of 1 second. However, the numerical oscillation for this time step magnitude may be suppressed by further increasing mesh density to 40 elements (Figure 5). The numerical oscillation will not reappear until the time step reduces to 0.1 second.

Similarly, this numerical difficulty may be overcome using 100 elements in the designated domain for even smaller time interval ($\Delta t = 0.01$ second), as shown in Figure 6. In general, using a denser mesh at small time levels and using a sparser mesh at large time may help to obtain an oscillation-free solution, again, at the cost of consuming more computational time. This global mesh refinement scheme is becoming increasingly practical and possible due to the rapid advance of current computer technology (e.g., larger RAM and greater speed).

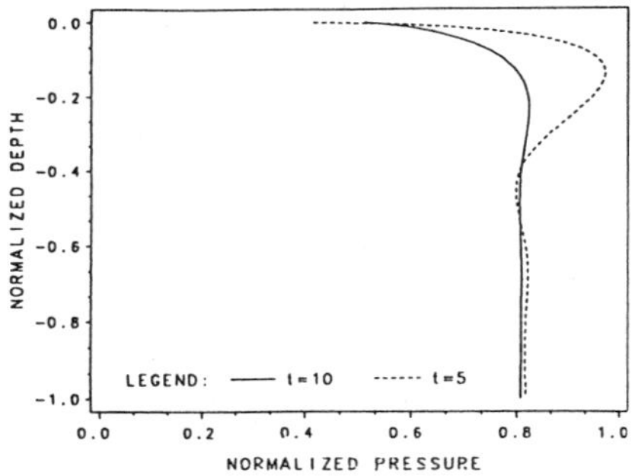

Figure 3. Spatial pressure using 5 elements.

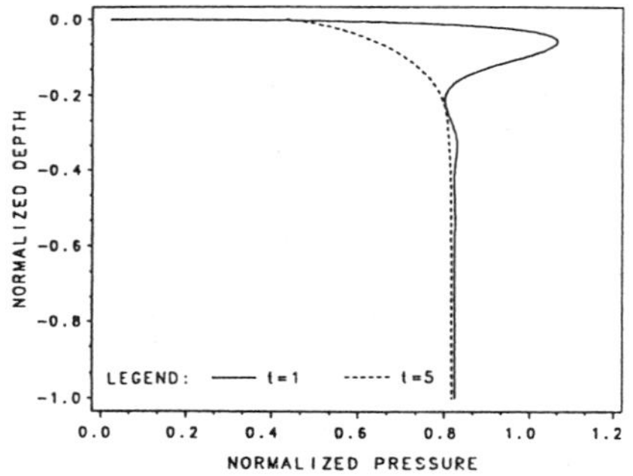

Figure 4. Spatial pressure using 10 elements.

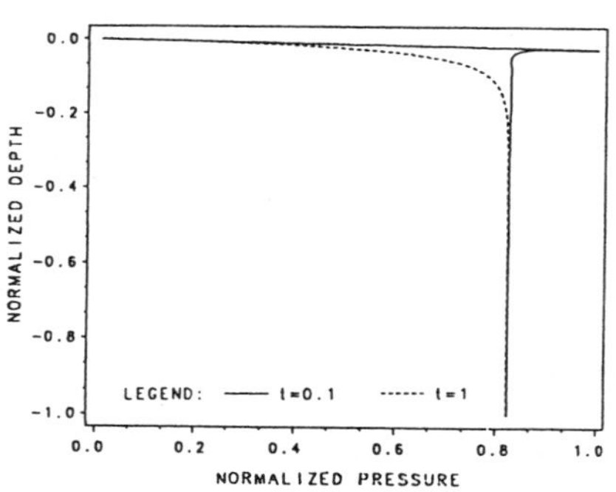

Figure 5. Spatial pressure using 40 elements.

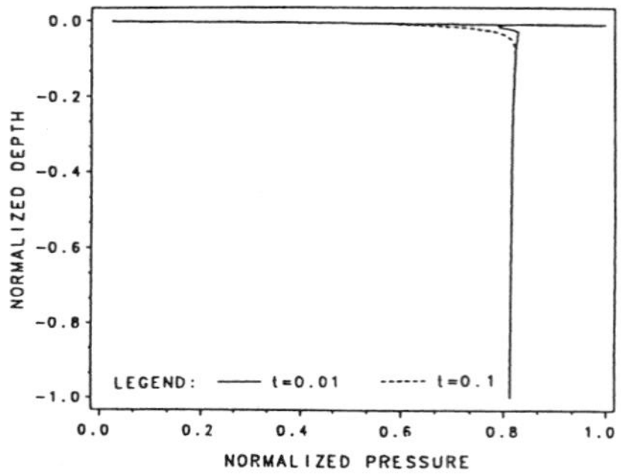

Figure 6. Spatial pressure using 100 elements.

7. MULTIPLE COUPLING AND ALTERNATIVE TIME DISCRETIZATION

In the previous examination of a consolidation scenario, the coupled process is represented by the interactive behavior of fluid flow and solid deformation. If the porous media are also naturally fractured, additional coupling, in view of interporosity flow between the fractures and the matrix blocks, needs to be incorporated. This multiple coupling leads to the aggravated situation in minimizing numerical oscillation, as a result of expanded system matrix and multiplicities in the compatibility issues among the increased primary variables.

On another related issue, the time of evaluation within the calculated domain appears to be a strong factor of influence, as shown in the previous example. The finite difference method becomes a traditional approach in the finite element scheme for the time discretization. For analyzing a coupled scenario, explicit finite difference method is undesirable due to stability constraints. Instead, an unconditionally stable method, such as the fully implicit method used in the previous case, should be adopted. For more accurate calculation, a higher order half implicit method, i.e., Crank-Nicolson's method, can be attempted. The relationship between the methods of time discretization and the impact on the numerical oscillation seems to be of interest for the investigation.

The study of the multiple coupling and the alternative time discretization scheme is performed in the analysis of a column consolidation scenario. The fractured porous column is conceptualized as a dual-porosity medium (Elsworth and Bai, 1992). For a limited number of finite elements (i.e., 10) the numerical oscillation becomes a dominant feature near the loading area, even at large time. Figure 7 indicates the normalized spatial pressures in matrix blocks at time of 1000 seconds. Even though more severe numerical oscillation is depicted in the pressure profile by Crank-Nicolson's method (CN), the difference between this method and the fully implicit method (FI) is not significant.

At a larger time (t=10000 seconds), the numerical oscillation becomes less severe and closer to the loading zone, as expected (Figure 8).

However, the difference between the CN and the FI methods seems to be more substantial. Similar results with slightly different magnitudes are obtained for the spatial pressures in fractures, as illustrated in Figures 9 and 10.

In the dual-porosity approach, pressures in the fractures and in the matrix are derived separately. The common link is made between the two continua through a leakage function. In the calculation of displacements, however, the lumped treatment is used with regard to the strains generated in the fractures and in the matrix, respectively. As a result, the numerical oscillation is reduced, as shown in Figure 11 for t=1000 s, where the oscillation is limited to the small area, and very small discrepancy is observed between the CN and the

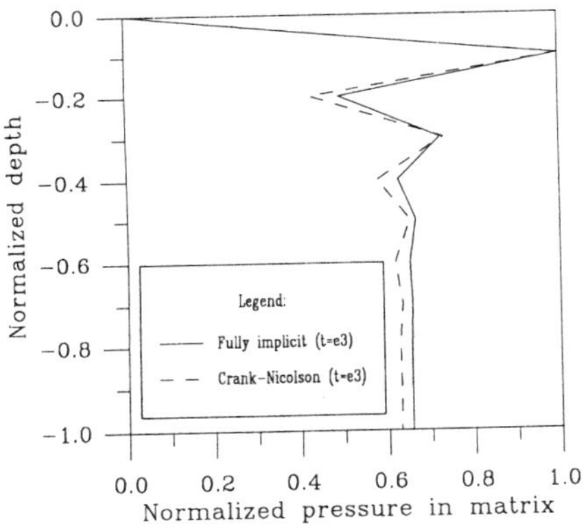

Figure 7. Spatial pressure in matrix (t=1000 s).

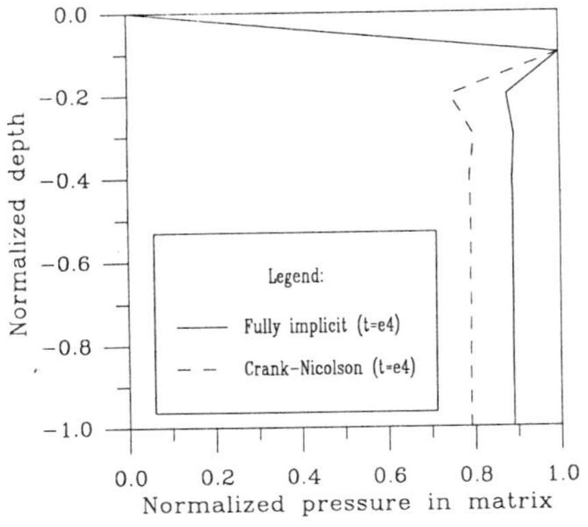

Figure 8. Spatial pressure in matrix (t=10000 s).

FI methods. The oscillation becomes less severe for larger time (t=10000 s), as shown in Figure 12.

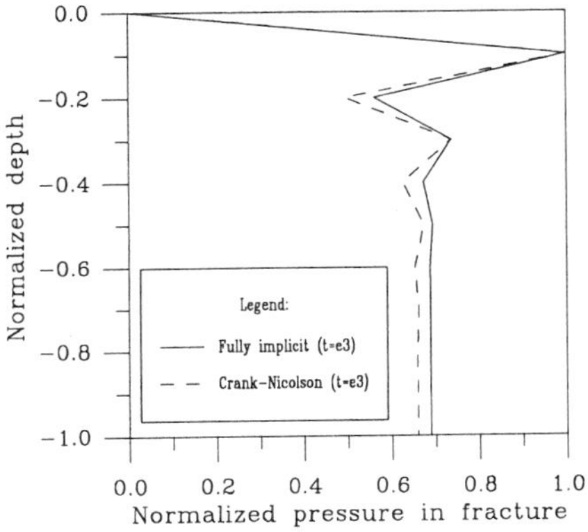

Figure 9. Spatial pressure in fractures (t=1000 s).

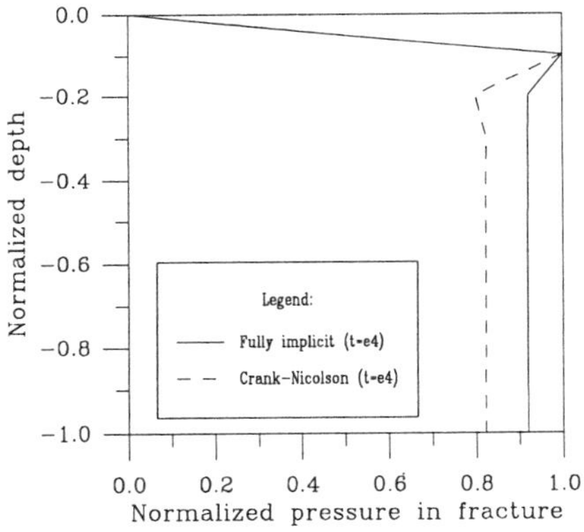

Figure 10. Spatial pressure in fractures (t=10000 s).

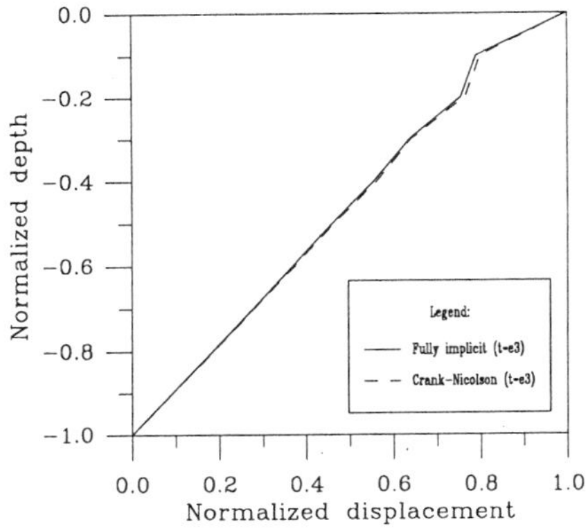

Figure 11. Spatial displacement (t=1000 s).

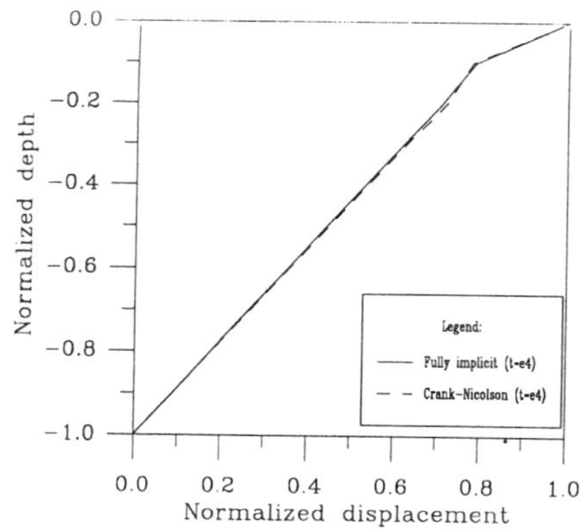

Figure 12. Spatial displacement (t=10000 s).

8. CONCLUDING REMARKS

The system of equations for the analysis of a consolidation scenario may be ill-conditioned as a result of significant difference in magnitude between solid displacement and fluid pressure. This situation may become aggravated for the multiple coupling where the media are naturally fractured. The conditioning of the system of equations may be investigated through evaluation of the condition number. The condition of the system of equations may be viewed as poor for large condition numbers even though the computational residual is small.

Alternative treatments of an ill-conditioned system for a poroelastic system may be summarized as follows:

(a) If numerical instability is due to round-off errors, pivoting is useful to improve the condition of the system of equations. For a coupled formulation, however, pivoting may not be cost effective. The problem might be due to the inability of the pivoting to preserve the symmetric nature of the displacement-pressure system.

(b) Similar to the pivoting technique, normalization of the system of equations appears to be useful to reduce the numerical oscillation. For a coupled formulation, it is essential to maintain compatibility in magnitudes for all coefficients in the system of equations. This compatibility may be achieved by manipulating the units to make similar magnitudes in the displacement or pressure components.

(c) For consolidation problems, it is found that the numerical oscillation often occurs immediately after imposition of the external load, or at the location close to the loading surface. This type of oscillation may be minimized either by increasing mesh density globally, or increasing the time step.

(d) For a coupled formulation, a stable time discretization scheme should be adopted to minimize the adverse time influence. Using a higher order scheme such as the Crank-Nicolson's method may improve the accuracy of the calculation, but may result in more violent oscillation, especially at larger time. For the smaller time calculation, the methods of time discretization appear to exert less impact on the result.

ACKNOWLEDGMENT

Support of the NSF/State/Industry under the joint S/IUCRC program and under the contract EEC-9209619, is gratefully acknowledged.

REFERENCES

Bai, M., A. Bouhroum and D. Elsworth 1994. Some aspects of solving advection dominated flows, Kohle-Erdgas-Petrochemie vereinigt mit Brennst-off-Chemie, *J. Hydrocarbon Tech.*, 47: S11-17.

Brady, B.H.G. 1979. A direct formulation of the boundary element method of stress analysis for complete plane strain, *Int. J. Rock Mech. Min. Sci. and Geomech. Abstr.*, 15: 21-28.

Elsworth, D. and M. Bai 1992. Coupled flow-deformation response of dual porosity media, *J. Geotech. Eng., ASCE.*, 118: 107-124.

Ghaboussi, J. and E.L. Wilson 1973. Flow of compressible fluid in porous elastic media, *Int. J. Num. Methods in Eng.*, 5: 419-442.

Huyakorn, P.S. and G.F. Pinder 1983. *Computational Methods in Subsurface Flow*, Academic Press, New York.

Rice, J.R. 1981. *Matrix Computations and Mathematical Software*, McGraw-Hill Book Company, 248 p.

Sandhu, R.S., S.C. Lee and H.I. The 1985. Special finite elements for analysis of soil consolidation, *Int. J. Numer. and Analy. Meth. in Geomech.*, 9: 125-147.

Sudicky, E.A. 1989. The Laplace transform Galerkin technique: A time continuous finite element theory and application to mass transport in groundwater, *Water Resour. Res.*, 25: 1833-1846.

5 Numerical algorithms: Formulation and performance

Numerical Models in Geomechanics, Pietruszczak & Pande (eds) © 1997 Balkema, Rotterdam, ISBN 90 5410 886 X

A comparison of linear and nonlinear programming formulations for lower bound limit analysis

A.V.Lyamin & S.W.Sloan
Department of Civil, Surveying and Environmental Engineering, University of Newcastle, N.S.W., Australia

ABSTRACT: Linear programming is often used to perform limit analysis of large scale problems in continuum mechanics. In this approach, a finite element discretisation is employed in conjunction with a linearised yield surface to generate the optimisation problem. Unfortunately, the linearisation process replaces a single nonlinear inequality by a series of linear inequalities and may result in a large number of yield constraints. Moreover, it is often difficult and cumbersome to perform a linearisation of acceptable accuracy for 3D yield functions. Both of these shortcomings can be removed by adopting nonlinear programming schemes such as the iterative feasible point algorithm. This paper compares the performance of a new two stage quasi-Newton feasible point method with the performance of a steepest edge active set method. The latter is one of the most efficient linear programming algorithms developed for solving plasticity problems. The comparison shows that the nonlinear programming strategy is vastly superior in terms of accuracy and speed. It exhibits fast convergence to the optimum solution and, unlike the linear programming scheme, the number of iterations is largely independent of the problem size.

1 INTRODUCTION

The lower bound theorem of classical plasticity states that the limit load calculated from a statically admissible stress field, which satisfies equilibrium, the stress boundary conditions and the yield condition, is a lower bound on the actual collapse load. This theorem assumes a rigid plastic soil model and provides a method for computing a safe estimate of the true collapse load.

Deriving statically admissible stress fields by hand is difficult, except for simple problems, and it is usually necessary to employ numerical methods for practical cases involving complicated geometries, inhomogeneous soils and complex loading. The most commonly used numerical implementation of the lower bound theorem is based on a finite element discretisation of the rigid plastic soil mass. This results in a finite-dimensional optimisation problem with large, sparse constraint matrices. By adopting linear finite elements and a polyhedral approximation of the yield surface, the optimisation problem is one which can be solved using classical linear programming techniques. This type of approach has been widely used and is described in detail, for example, in Bottero *et al* (1980) and Sloan (1988a). The approximation of a nonlinear yield surface by a polyhedron with a reasonable number of sides can be done only for simple types of deformation (such as those that arise in 2D and axisymmetric loading) and for simple yield functions. Moreover, this linearisation process may generate very large numbers of constraints, thereby increasing the cost of solution. Nonlinear programming algorithms, such as those developed by Zouain *et al.* (1993), avoid the need to linearise the yield constraints and can thus be used for 3D stability analysis.

This paper describes the steps involved in a two-stage quasi-Newton algorithm for the solution of lower bound optimisation problems. The performance of this nonlinear programming strategy is then compared against that of an active set linear programming formulation.

2 DISCRETE FORMULATION OF LOWER BOUND THEOREM

Any discrete formulation of the lower bound theorem leads to a constrained optimisation problem which can be stated in the following form

$$\text{maximise } Q(\mathbf{x}) \tag{1}$$

subject to

$$a_i(\mathbf{x}) = 0, \quad i \in I = \{1,\ldots,m\}$$
$$f_j(\mathbf{x}) \leq 0, \quad j \in J = \{1,\ldots,p\}$$
$$\mathbf{x} \in \mathrm{R}^n$$

where $\mathbf{x}$ is an n-dimensional vector of stress and unit weight variables and Q is the collapse load. The equalities defined by the functions a_i come from equilibrium, inter-element, boundary and loading conditions, while the inequalities defined by the functions f_j arise because of yield and unit weight constraints.

3 LINEAR PROGRAMMING APPROACH

After replacing the nonlinear yield surface constraints by a fixed number of linear constraints, and adopting a linear finite element model, the general optimisation problem (1) becomes a linear programming problem of the form (Bottero *et al.*, 1980)

$$\text{maximise } \mathbf{c}^{\mathrm{T}}\mathbf{x} \tag{2}$$

subject to

$$\mathbf{A}\mathbf{x} = \mathbf{b}$$
$$\mathbf{A}_i\mathbf{x} \leq \mathbf{b}_i$$
$$\mathbf{x} \in \mathrm{R}^n$$

where $\mathbf{c}$ is a vector of objective function coefficients of length n, $\mathbf{A}$ is an $m \times n$ matrix of equality constraint coefficients, $\mathbf{A}_i$ is an $r \times n$ matrix of inequality constraint coefficients ($r = p_y \times NSID + p_Y$, where p_y and p_Y are the number of yield and unit weight constraints, respectively, and *NSID* is the number of sides in the linearised yield surface), $\mathbf{b}$ and $\mathbf{b}_i$ are vectors of length m and r, respectively, and $\mathbf{x}$ is a vector of length n which is to be determined.

The constraint matrices $\mathbf{A}$ and $\mathbf{A}_i$ are extremely sparse and usually have many more rows than columns. One of the most robust linear programming schemes for solving the problem (2) is the steepest edge active set algorithm (Sloan, 1988b). A detailed description of this procedure will not be repeated here. Briefly, each iteration of the algorithm consists of three distinct steps:

1. The determination of the search direction and test for optimality.
2. The determination of the maximum feasible step size and new active constraint.
3. Update of the solution and active constraint matrix.

At each iteration the objective function decreases (or in the case of zero maximum feasible step size remains the same) and the algorithm moves along an edge of the feasible region from one vertex to another. Iteration is halted once it is no longer possible to decrease the objective function further by moving to an adjacent vertex.

4 KUHN-TUCKER OPTIMALITY CONDITIONS

If the functions Q and f_j are convex and the functions a_i are affine (which is certainly the case when linear finite elements are employed) then problem (1) is equivalent to the system of Kuhn-Tucker optimality conditions:

$$\left.\begin{aligned} -\nabla Q(\mathbf{x}) + \sum_{i\in I}\mu_i\nabla a_i(\mathbf{x}) + \sum_{j\in J}\lambda_j\nabla f_j(\mathbf{x}) &= \mathbf{0} \\ a_i(\mathbf{x}) = 0, &\quad i \in I \\ \lambda_j f_j(\mathbf{x}) = 0, &\quad j \in J \\ f_j(\mathbf{x}) \leq 0, &\quad j \in J \\ \lambda_j \geq 0, &\quad j \in J \\ \mathbf{x} \in \mathrm{R}^n & \end{aligned}\right\} \tag{3}$$

Introducing the Lagrange function $L(\mathbf{x}, \boldsymbol{\mu}, \boldsymbol{\lambda})$ defined by

$$L(\mathbf{x}, \boldsymbol{\mu}, \boldsymbol{\lambda}) = -Q(\mathbf{x}) + \sum_{i\in I}\mu_i a_i(\mathbf{x}) + \sum_{j\in J}\lambda_j f_j(\mathbf{x})$$

we can cast the first equation of (3) in the form

$$\nabla_{\mathbf{x}} L(\mathbf{x}, \boldsymbol{\mu}, \boldsymbol{\lambda}) = \mathbf{0}$$

Zouain *et al.* (1993) have proposed a general feasible point algorithm for solving limit analysis problems arising from a mixed formulation. Although the optimisation problem considered by these authors is different from problem (1), their algorithm can be modified to solve the optimisation problem considered here. In brief, the algorithm of Zouain *et al.* (1993) uses a quasi-Newton iteration formula for the set of all equalities in the optimality conditions. At each stage, this is followed by a small "deflection" in order to preserve feasibility with respect to the inequality conditions. Our algorithm uses a different deflection strategy and some additional modifications to account for the nature of the optimisation problem that is generated by the lower bound formulation. The new procedure can deal with arbitrary loading and also permits optimisation with respect to the weight of the material. A detailed description of the two stage quasi-Newton strategy for solving the system (3) is given in the next section.

5 TWO STAGE QUASI-NEWTON ALGORITHM.

In Newton's method, the nonlinear equations in the vicinity of the current point "k" are linearised and the resulting system of linear equations is solved to obtain a new point "k+1". The process is repeated until the governing system of nonlinear equations is satisfied. Application of this procedure to the equalities of the system (3) leads to

$$\nabla^2_{\mathbf{x}} L(\mathbf{x}^k, \boldsymbol{\mu}^k, \boldsymbol{\lambda}^k)(\mathbf{x}^{k+1} - \mathbf{x}^k) + \sum_{i \in I} \mu_i^{k+1} \nabla a_i(\mathbf{x}^k) + \sum_{j \in J} \lambda_j^{k+1} \nabla f_j(\mathbf{x}^k) = \nabla Q(\mathbf{x}^k) \quad (4)$$

$$\nabla a_i(\mathbf{x}^k)(\mathbf{x}^{k+1} - \mathbf{x}^k) = - a_i(\mathbf{x}^k), \ i \in I \quad (5)$$

$$\lambda_j^k \nabla f_j(\mathbf{x}^k)(\mathbf{x}^{k+1} - \mathbf{x}^k) + \lambda_j^{k+1} f_j(\mathbf{x}^k) = 0, j \in J \quad (6)$$

where the superscript k denotes the iteration number. Assuming that at point $\mathbf{x}^k$ the equalities $a_i(\mathbf{x}^k) = 0$ hold for $i \in I$ and λ_j^k is positive for any $j \in J$ (as will be forced by the updating rule), it is possible to rewrite (4),(5) and (6) in the more compact form

$$\mathbf{Hd} + \mathbf{A}^{\mathrm{T}}\boldsymbol{\mu} + \mathbf{G}^{\mathrm{T}}\boldsymbol{\lambda} = \mathbf{c} \quad (7)$$

$$\mathbf{Ad} = \mathbf{0} \quad (8)$$

$$\mathbf{Gd} + \mathbf{F}\boldsymbol{\lambda} = \mathbf{0} \quad (9)$$

where

$$\begin{aligned} \mathbf{H} &= \nabla^2_{\mathbf{x}} L(\mathbf{x}^k, \boldsymbol{\mu}^k, \boldsymbol{\lambda}^k) \quad (10) \\ \mathbf{A}^{\mathrm{T}} &= \left[\nabla a_1(\mathbf{x}^k) ... \nabla a_m(\mathbf{x}^k)\right] \\ \mathbf{G}^{\mathrm{T}} &= \left[\nabla f_1(\mathbf{x}^k) ... \nabla f_p(\mathbf{x}^k)\right] \\ \mathbf{F} &= \mathrm{diag}\left(f_j(\mathbf{x}^k)/\lambda_j^k\right) \\ \mathbf{c} &= \nabla Q(\mathbf{x}^k) \end{aligned}$$

and

$$\mathbf{d} = \mathbf{x}^{k+1} - \mathbf{x}^k, \quad \boldsymbol{\mu} = \boldsymbol{\mu}^{k+1}, \quad \boldsymbol{\lambda} = \boldsymbol{\lambda}^{k+1}$$

If we denote

$$\mathbf{T} = \begin{bmatrix} \mathbf{H} & \mathbf{A}^{\mathrm{T}} & \mathbf{G}^{\mathrm{T}} \\ \mathbf{A} & \mathbf{0} & \mathbf{0} \\ \mathbf{G} & \mathbf{0} & \mathbf{F} \end{bmatrix}, \quad \mathbf{y} = \begin{Bmatrix} \mathbf{d} \\ \boldsymbol{\mu} \\ \boldsymbol{\lambda} \end{Bmatrix}, \quad \mathbf{v} = \begin{Bmatrix} \mathbf{c} \\ \mathbf{0} \\ \mathbf{0} \end{Bmatrix}$$

then the system defined by equations (7)–(9) may be rewritten in the even more compact form

$$\mathbf{Ty} = \mathbf{v} \quad (11)$$

Now we have a highly sparse, symmetric system of $n + m + p$ linear equations which can be solved for the unknowns $\mathbf{y}$. In the case of linear finite elements the functions Q and a_i are linear, so that the only nonzero contributions to the matrix $\mathbf{H}$ are the Hessians of the yield and unit weight constraints according to

$$\mathbf{H} = \sum_{j \in J} \mathbf{H}_j = \sum_{j \in J} \lambda_j^k \nabla^2 f_j(\mathbf{x}^k) \quad (12)$$

For most applications, the unit weight constraints are also linear and typically consist of one nonnegativity constraint for each unit weight variable. This implies that the nonzero terms in (12) are usually due only to the nonlinear yield constraints on the stresses. So, if each plastic constraint depends on an uncoupled set of stress variables, then $\mathbf{H}$ is block diagonal and $\mathbf{G}$ also consists of disjoint blocks. The size of each block in $\mathbf{H}$ is equal to the number of stress variables in the corresponding yield constraint (with a maximum of six in the 3D case) and $\mathbf{H}^{-1}$ can be computed very quickly in a node by node manner.

For some yield functions, the Hessian of equation (12) is not a positive definite matrix. In these cases, we can apply a small pertubation $\varepsilon\mathbf{I}$ to restore positive definiteness according to

$$\mathbf{H} = \sum_{j \in J} \lambda_j^k \left[\nabla^2 f_j(\mathbf{x}^k) + \varepsilon_j \mathbf{I}_j \right]$$

5.1 *Stage 1: Increment estimate*

In the first stage of the algorithm, estimates for $\mathbf{d}$, $\boldsymbol{\mu}$ and $\boldsymbol{\lambda}$ are found by exploiting the above mentioned features of the matrices $\mathbf{H}$ and $\mathbf{G}$. From (7) we have

$$\mathbf{d}_0 = \mathbf{H}^{-1}\left(\mathbf{c} - \mathbf{A}^{\mathrm{T}}\boldsymbol{\mu}_0 - \mathbf{G}^{\mathrm{T}}\boldsymbol{\lambda}_0\right) \quad (13)$$

where the subscript 0 is introduced to indicate that the solution is from stage one of the algorithm.

Substituting (13) into (9) gives

$$\boldsymbol{\lambda}_0 = \mathbf{W}^{-1}\mathbf{Q}^{\mathrm{T}}\left(\mathbf{c} - \mathbf{A}^{\mathrm{T}}\boldsymbol{\mu}_0\right) \quad (14)$$

where the diagonal matrix $\mathbf{W}$ and the matrix $\mathbf{Q}$ are computed from

$$\mathbf{W} = \mathbf{G}\mathbf{H}^{-1}\mathbf{G}^{\mathrm{T}} - \mathbf{F}$$

$$\mathbf{Q} = \mathbf{H}^{-1}\mathbf{G}^{\mathrm{T}}$$

Next we can eliminate $\boldsymbol{\lambda}_0$ from (13) using (14), which gives

$$\mathbf{d}_0 = \mathbf{D}\left(\mathbf{c} - \mathbf{A}^{\mathrm{T}}\boldsymbol{\mu}_0\right) \quad (15)$$

with

$$\mathbf{D} = \mathbf{H}^{-1} - \mathbf{Q}\mathbf{W}^{-1}\mathbf{Q}^{\mathrm{T}}$$

Multiplying both sides of (15) by $\mathbf{A}$ and taking into account (8) we obtain

$$\mathbf{K}\boldsymbol{\mu}_0 = \mathbf{ADc} \quad (16)$$

where

$$\mathbf{K} = \mathbf{ADA}^{\mathrm{T}} \quad (17)$$

is $m \times m$ symmetric positive semi-definite matrix whose density is roughly double that of $\mathbf{A}$ (Zouain *et al.*, 1993). To start the computational sequence, $\boldsymbol{\mu}_0$ is first found from (16). The quantities $\boldsymbol{\lambda}_0$ and $\mathbf{d}_0$ are then computed, respectively, using (14) and (15).

This computational strategy is approximately ten times faster than solving (11) by direct factorisation of the matrix **T**. It should be noted, however, that the condition number of the matrix **K** is equal to the square of the condition number of the matrix **T** as the optimum solution is approached. Fortunately, experience suggests that double precision arithmetic is sufficient to get an accurate solution using (16) before the matrix **K** becomes extremely ill-conditioned or numerically singular.

5.2 *Stage 2: Computation of a deflected feasible direction*

It is clear from (9) that the vector $\mathbf{d}_0$ is tangential to the active constraints (where $f_j(\mathbf{x}^k) = 0$), so we must deflect it to make the search direction feasible. This can be done by setting the right side of every row in (9) to some negative value which is known as the deflection factor. Zouain *et al.* (1993) suggested that the same deflection factor, θ, should be used for each yield constraint. It was found, however, that this strategy is not effective when the curvatures of the active constraints differ greatly at the current iterate $\mathbf{x}^k$. We describe here another technique which has a simple geometrical interpretation and works well for all types of yield criteria.

Consider the component-wise form of equation (9). Letting $\mathbf{d}_j$ denote the part of the vector **d** which corresponds to the set of variables of the function f_j, and using the notation $\bar{\lambda}_j$ instead of λ_j^k, we can rewrite (9) in the form

$$\nabla f_j^{\mathrm{T}} \mathbf{d}_j + \left(f_j / \bar{\lambda}_j\right) \lambda_j = 0 \tag{18}$$

Using the geometrical definition of the scalar product, this may also be expressed as

$$\left\|\nabla f_j\right\| \left\|\mathbf{d}_j\right\| \cos \varphi_j + \left(f_j / \bar{\lambda}_j\right) \lambda_j = 0$$

where φ_j is the angle between the normal to the j-th constraint and the vector $\mathbf{d}_j$. If we now introduce the deflection factor as the product $\left\|\nabla f_j\right\| \left\|\mathbf{d}_j\right\| \theta_{dj}$, then θ_{dj} is equal to $\cos \varphi_j$ for all active constraints. If an equal value of θ_d is used for these cases, then this implies that the angle between the vector $\mathbf{d}_j$ and the tangential plane to the yield surface at the point $\mathbf{x}_j^k$, which we term the deflection angle ϑ, is identical for all active constraints. For many yield surfaces used in limit analysis, the curvature at a given point $\mathbf{x}^k$ is strongly dependent on the direction $\mathbf{d}_j$. Therefore, it is better to make θ proportional not only to the length of the vector $\mathbf{d}_j$, but also to the curvature $\varkappa_j$ of the function f_j in the direction $\mathbf{d}_j$. The latter may be estimated as the norm of the derivative of ∇f_j in the direction $\mathbf{d}_j$ multiplied by R_j, where R_j is the distance from the point $\mathbf{x}_j^k$ to the axis of the yield surface in the octahedral plane. Combining all of the above factors leads us to rewrite (18) as

$$\nabla f_j^{\mathrm{T}} \mathbf{d}_j + \left(f_j / \bar{\lambda}_j\right) \lambda_j = -\theta_d \left\|\nabla f_j\right\| \left\|\mathbf{d}_{0j}\right\| \varkappa_j \tag{19}$$

where

$$\begin{aligned}
\theta_d &\in (0, 1) \\
\varkappa_j &= \max\left(\varkappa_j^0 / \varkappa_{\max}, \varkappa_{\min}\right) \\
\varkappa_j^0 &= \frac{\left\| \nabla^2 f_j(\mathbf{x}^k)\, \mathbf{d}_{0j} \right\| R_j}{\left\|\nabla f_j\right\| \left\| \mathbf{d}_{0j} \right\|} \\
\varkappa_{\max} &= \max_{j \in J} \varkappa_j^0 \\
\varkappa_{\min} &\in (0, 1)
\end{aligned}$$

Note that θ_d is equal to the sine of the maximum deflection angle $\vartheta_{\max}$ and can be related analytically to the increment in the objective function $\mathbf{c}^{\mathrm{T}} \mathbf{d}_0$ (as will be shown later).

Replacing (9) by the matrix form of (19) furnishes the system of equations for computing the new feasible direction **d** as

$$\begin{aligned}
\mathbf{H d} + \mathbf{A}^{\mathrm{T}} \boldsymbol{\mu} + \mathbf{G}^{\mathrm{T}} \boldsymbol{\lambda} &= \mathbf{c} \\
\mathbf{A d} \qquad\qquad &= \mathbf{0} \\
\mathbf{G d} \qquad + \mathbf{F} \boldsymbol{\lambda} &= -\theta_d \mathbf{u}
\end{aligned}$$

where **u** is a p-dimensional vector with components

$$u_j = \left\|\nabla f_j\right\| \left\|\mathbf{d}_{0j}\right\| \varkappa_j$$

Applying the same computational sequence as was implemented for the solution of the system (7)–(9), we finally have

$$\mathbf{d} = \mathbf{H}^{-1}\left(\mathbf{c} - \mathbf{A}^{\mathrm{T}} \boldsymbol{\mu} - \mathbf{G}^{\mathrm{T}} \boldsymbol{\lambda}\right) \tag{20}$$

$$\boldsymbol{\lambda} = \mathbf{W}^{-1} \mathbf{Q}^{\mathrm{T}}\left(\mathbf{c} - \mathbf{A}^{\mathrm{T}} \boldsymbol{\mu}\right) + \theta_d \mathbf{W}^{-1} \mathbf{u} \tag{21}$$

$$\mathbf{K} \boldsymbol{\mu} = \mathbf{A D c} - \theta_d \mathbf{A Q W}^{-1} \mathbf{u} \tag{22}$$

To derive an expression for θ_d, we substitute (21) into (20) to give

$$\mathbf{d} = \mathbf{D}\left(\mathbf{c} - \mathbf{A}^{\mathrm{T}} \boldsymbol{\mu}\right) - \theta_d \mathbf{Q W}^{-1} \mathbf{u} \tag{23}$$

Multiplying both sides of (23) by $\left(\mathbf{c}^{\mathrm{T}} - \boldsymbol{\mu}_0^{\mathrm{T}} \mathbf{A}\right)$ and using (8), (14) and (15) we obtain

$$\theta_d = \frac{\mathbf{c}^{\mathrm{T}} \mathbf{d}_0 - \mathbf{c}^{\mathrm{T}} \mathbf{d}}{\boldsymbol{\lambda}_0^{\mathrm{T}} \mathbf{u}}$$

If we impose the condition that the *Stage 1* and *Stage 2* increments of the objective function are related by a fixed parameter $\beta \in (0, 1)$ such that $\mathbf{c}^{\mathrm{T}} \mathbf{d} = \beta \mathbf{c}^{\mathrm{T}} \mathbf{d}_0$, then θ_d can be computed as

$$\theta_d = (1 - \beta) \frac{\mathbf{c}^{\mathrm{T}} \mathbf{d}_0}{\boldsymbol{\lambda}_0^{\mathrm{T}} \mathbf{u}}$$

5.3 *Initialisation*

It has been assumed so far that the current solution $\mathbf{x}^k$ is a feasible point. Therefore, to start the iterations, we need a procedure which will give us an initial feasible point $\mathbf{x}^0$. First of all we have to solve

$$\begin{bmatrix} \mathbf{I} & \mathbf{A}^{\mathrm{T}} \\ \mathbf{A} & \mathbf{0} \end{bmatrix} \begin{Bmatrix} \mathbf{x}^0 \\ \boldsymbol{\mu} \end{Bmatrix} = \begin{Bmatrix} \mathbf{0} \\ \mathbf{b} \end{Bmatrix}$$

to obtain $\mathbf{x}^0$ satisfying all linear equality constraints. In this equation, $\mathbf{b}$ is merely a vector of right hand sides which is comprised of any constant terms in the equality constraints. We then substitute $\mathbf{x}^0$ into the inequalities and compute the scalar

$$\alpha = \max_{j \in J} f_j(\mathbf{x}^0)$$

If $\alpha \le 0$, the initial search direction $\mathbf{d}_0$ can be found by solving

$$\begin{bmatrix} \mathbf{I} & \mathbf{A}^{\mathrm{T}} \\ \mathbf{A} & \mathbf{0} \end{bmatrix} \begin{Bmatrix} \mathbf{d}^0 \\ \boldsymbol{\mu} \end{Bmatrix} = \begin{Bmatrix} \mathbf{c} \\ \mathbf{0} \end{Bmatrix}$$

We then set $\bar{\lambda}_j = 1$ for all $j \in J$ and start the iteration process. If $\alpha > 0$ it is necessary to solve the phase one problem

$$\text{minimise } \alpha_{p+1} \qquad (24)$$

subject to

$$\begin{aligned} a_i(\mathbf{x}_a) &= 0, & i \in I \\ \alpha_{j+1} - \alpha_j &= 0, & j \in J \\ f_j(\mathbf{x}_a) - \alpha_j &\le 0, & j \in J \\ -\alpha_{p+1} &\le 0 & \\ \mathbf{x}_a &\in R^{n+p} & \end{aligned}$$

where $\mathbf{x}_a^{\mathrm{T}} = \{\mathbf{x}^{\mathrm{T}}, \alpha_1, \ldots, \alpha_{p+1}\}$. Because of the similarity of problem (24) to problem (1) it can be solved using the same two stage algorithm.

5.4 *Convergence criterion, line search and updating*

A comprehensive description of the line search and updating steps of the algorithm can be found in Zouain *et al.* (1993) and will not be repeated here. A few minor modifications to these components of the procedure have been made and are incorporated in the step-wise outline of the algorithm which follows.

5.5 *Algorithm 1(Two stage quasi–Newton algorithm)*

Step 1.0 – (*Initialisation*)

1.0.0 Set $\beta = 0.7$ and $\delta_f^0 = 0.01$.

1.0.1 Read $\varepsilon_c, \varepsilon_\lambda, \varepsilon_f, \delta_\lambda^0$ and $\varkappa_{\min}$

1.0.2 Set $\mathbf{x} = \mathbf{x}^0$ and $\mathbf{d} = \mathbf{d}^0$

1.0.3 Find $s = \min_{j \in J} s_j \mid f_j(\mathbf{x}+s_j\mathbf{d}) = \delta_f^0 f_j(\mathbf{x})$

1.0.4 Set $\mathbf{x} = \mathbf{x}+s\mathbf{d}$

1.0.5 For all $j \in J$ set $\bar{\lambda}_j = 1$

Step 1.1 – (*Increment estimate*)

1.1.0 For all $j \in J$

compute $\nabla f_j(\mathbf{x})$ and $\mathbf{Z}_j = \nabla^2 f_j(\mathbf{x}^k)$

compute $\mathbf{H}_j^{-1} = \left[\bar{\lambda}_j\left(\varepsilon_j \mathbf{I}_j + \mathbf{Z}_j\right)\right]^{-1}$

compute $\mathbf{Z}_j = R_j \mathbf{Z}_j$

compute $f_j(\mathbf{x})/\bar{\lambda}_j$ and mount in $\mathbf{F}$

mount $\mathbf{H}_j^{-1}$ in $\mathbf{H}^{-1}$ and $\nabla f_j(\mathbf{x})$ in $\mathbf{G}$

1.1.1 Compute matrices

$\mathbf{Q} = \mathbf{H}^{-1}\mathbf{G}^{\mathrm{T}}$

$\mathbf{W} = \mathbf{G}\mathbf{Q} - \mathbf{F}$

$\mathbf{D} = \mathbf{H}^{-1} - \mathbf{Q}\mathbf{W}^{-1}\mathbf{Q}^{\mathrm{T}}$

$\mathbf{K} = \mathbf{A}\mathbf{D}\mathbf{A}^{\mathrm{T}}$

1.1.2 Decompose $\mathbf{K}$

1.1.3 If $\mathbf{K}$ becomes singular go to Step 1.2.3

1.1.4 Solve $\mathbf{K}\boldsymbol{\mu}_0 = \mathbf{A}\mathbf{D}\mathbf{c}$

1.1.5 Compute $\boldsymbol{\lambda}_0 = \mathbf{W}^{-1}\mathbf{Q}^{\mathrm{T}}(\mathbf{c} - \mathbf{A}^{\mathrm{T}}\boldsymbol{\mu}_0)$

1.1.6 Compute $\mathbf{d}_0 = \mathbf{D}(\mathbf{c} - \mathbf{A}^{\mathrm{T}}\boldsymbol{\mu}_0)$

Step 1.2 – (*Convergence check*)

1.2.0 If $\|\mathbf{c} - \mathbf{A}^{\mathrm{T}}\boldsymbol{\mu}_0 - \mathbf{G}^{\mathrm{T}}\boldsymbol{\lambda}_0\| > \varepsilon_c\|\mathbf{c}\|$ go to Step 1.3

1.2.1 Compute $\lambda_{\max} = \max_{j \in J} \lambda_{0j} \big| \lambda_{0j} > 0$

1.2.2 For all $j \in J$

if $\lambda_{0j} < -\varepsilon_\lambda \lambda_{\max}$ go to Step 1.3

if $\lambda_{0j} > \varepsilon_\lambda \lambda_{\max}$ and $f_j(\mathbf{x}) < -\varepsilon_f$ go to Step 1.3

1.2.3 Exit with optimal solution $\mathbf{x}^*$

Step 1.3 – (*Deflection*)

1.3.0 For all $j \in J$

compute $\varkappa_j^0 = \dfrac{\|\mathbf{Z}_j \mathbf{d}_{0j}\|}{\|\nabla f_j\| \, \|\mathbf{d}_{0j}\|}$

1.3.1 Compute $\varkappa_{\max} = \max_{j \in J} \varkappa_j^0$

1.3.2 For all $j \in J$

compute $\varkappa_j = \max\left[\varkappa_j^0/\varkappa_{\max}, \varkappa_{\min}\right]$

compute $u_j = \|\nabla f_j\| \, \|\mathbf{d}_{0j}\| \varkappa_j$

1.3.3 Compute $\theta_d = \min\left[(1-\beta)\dfrac{\mathbf{c}^{\mathrm{T}}\mathbf{d}_0}{|\boldsymbol{\lambda}_0^{\mathrm{T}}\mathbf{u}|}, 1\right]$

1.3.4 Solve $\mathbf{K\mu} = \mathbf{ADc} - \theta_d \mathbf{AQW}^{-1}\mathbf{u}$

1.3.5 Set $\mathbf{d} = \mathbf{D}(\mathbf{c} - \mathbf{A}^T\mathbf{\mu}) - \theta_d \mathbf{QW}^{-1}\mathbf{u}$

Step 1.4 – (*Line search*)

1.4.0 Set $\delta_f = \min\left[\delta_f^0, \dfrac{|\mathbf{c}^T\mathbf{d}_0|}{|\mathbf{c}^T\mathbf{x}|}\right]$

1.4.1 Find $s = \min_{j \in J} s_j \mid f_j(\mathbf{x}+s_j\mathbf{d}) = \delta_f f_j(\mathbf{x})$

Step 1.5 – (*Updating*)

1.5.0 Set $\mathbf{x} = \mathbf{x}+s\mathbf{d}$

1.5.1 Set $\delta_\lambda = \min\left[\delta_\lambda^0, \dfrac{\|\mathbf{d}\|}{\|\mathbf{x}\|}\right]$

1.5.2 For all $j \in J$

set $\bar{\lambda}_j = \max\left[\lambda_j^0, \delta_\lambda \lambda_{\max}\right]$

1.5.3 Go to Step 1.1

5.6 *Algorithm 2 (Two phase algorithm for lower bound limit analysis)*

Step 2.0 – (*Managing equalities*)

2.0.0 Factorise $\mathbf{AA}^T$

2.0.1 If detected, delete redundant rows from matrix $\mathbf{A}$ and go to Step 2.0.0

2.0.2 Solve $\mathbf{AA}^T\mathbf{\mu} = -\mathbf{b}$

2.0.3 Compute $\mathbf{x}^0 = -\mathbf{A}^T\mathbf{\mu}$

2.0.4 Solve $\mathbf{AA}^T\mathbf{\mu} = -\mathbf{Ac}$

2.0.5 Compute $\mathbf{d}^0 = \mathbf{c} - \mathbf{A\mu}$

Step 2.1 – (*Managing inequalities*)

2.1.0 Compute $\alpha^0 = \max_{j \in J} f_j(\mathbf{x}^0)$

2.1.1 If $\alpha^0 \leq 0$ go to Step 2.6

Step 2.2 – (*Form phase one problem*)

2.2.0 Expand $\mathbf{x}$ to $\mathbf{x}_\alpha^T = \left\{\mathbf{x}^T, \alpha_1, \ldots, \alpha_{p+1}\right\}$

2.2.1 Expand $\mathbf{x}^0$ to $\mathbf{x}_\alpha^{0T} = \left\{\mathbf{x}^{0T}, \alpha_1^0, \ldots, \alpha_{p+1}^0\right\}$

2.2.2 Add equalities $\alpha_{j+1} - \alpha_j = 0, \quad j \in J$

2.2.3 Modify inequalities according to

$f_j(\mathbf{x}_\alpha) - \alpha_j \leq 0, \quad j \in J$

2.2.4 Add inequality $-\alpha_{p+1} \leq 0$

2.2.5 Set $d_{\alpha i}^0 = 0$ for $i = 1, \ldots, n$ and

$d_{\alpha i}^0 = -1$ for $i = n+1, \ n+p+1$

2.2.6 Form objective function vector as

$c_{\alpha i} = 0$ for $i = 1, n+p$

$c_{\alpha n+p+1} = 1$

Step 2.3 – (*Solve phase one problem*)

2.3.0 Solve phase one problem using Algorithm 1 with modified constraints and $\mathbf{x}_\alpha^0, \mathbf{d}_\alpha^0, \mathbf{c}_\alpha$ as initial data

2.3.1 Exit with optimal solution $\mathbf{x}_\alpha^*$

Step 2.4 – (*Check for feasible solution*)

2.4.0 If $x_{\alpha\, n+p+1}^* \neq 0$ print message that initial problem has no feasible solution and stop

Step 2.5 – (*Extract initial data for phase two*)

2.5.0 Restore all constraints to originals

2.5.1 Set $x_i^0 = x_{\alpha i}^*$ for $i = 1, \ldots, n$

Step 2.6 – (*Solve phase two problem*)

2.6.0 Solve phase two problem using Algorithm 1 with original constraints and $\mathbf{x}^0, \mathbf{d}^0, \mathbf{c}$ as initial data

2.6.1 Exit with optimal solution $\mathbf{x}^*$

6 NUMERICAL RESULTS

To verify the performance of the new lower bound algorithm, we consider the collapse of a smooth rigid strip footing resting on a cohesive-frictional Mohr-Coulomb soil layer. The exact collapse pressure for this problem has been derived by Prandtl and is equal to $p/c = 46.14$ for a soil with a friction angle of $\phi = 35°$. Three different meshes are adopted for the lower bound finite element study. The stress boundary conditions and material properties used in the analyses, together with one of the meshes, are shown in Figure 1. All runs use the three-noded linear stress triangle and special extension elements to extend the stress field in the infinite half space (Sloan, 1988a). For the nonlinear programming formulation, the Mohr-Coulomb yield criterion is expressed in the form

$$f = \sqrt{(\sigma_x - \sigma_y)^2 + 4\tau_{xy}^2} + (\sigma_x + \sigma_y)\sin\phi - 2c\cos\phi = 0$$

To solve the sparse symmetric system of equations (17) efficiently, we use the MA27 multifrontal code developed at Harwell. This code is a direct solution method and employs threshold pivoting to maintain stability and sparsity in the factorisation.

The results presented in Table 1 compare the performance of the nonlinear two-stage algorithm with that of the linear programming formulation described by Sloan (1988a, 1988b). The new algorithm demonstrates fast convergence to the correct limit load and, most importantly, the number of iterations is essentially independent of the problem size. For the coarsest mesh, the nonlinear formulation is nearly four times

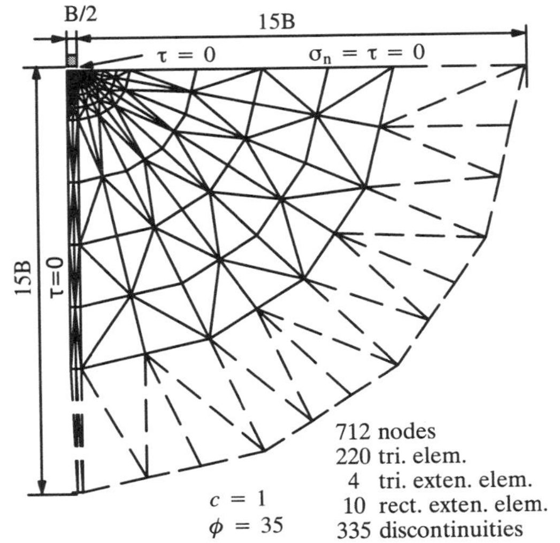

Figure 1. Lower bound mesh

faster than the linear programming formulation and gives a lower bound limit load which is two percent better. For the largest problem, the speed advantage of the nonlinear procedure is dramatic, with a 55 fold reduction in the CPU time. In this case, the lower bound limit load is one percent below the exact limit load and the analysis uses just 13 seconds of CPU time. Because the number of iterations with the new algorithm is essentially constant for all cases, the CPU time grows only linearly with the problem size. This implies that the CPU time saving, relative to that of the linear programming formulation, will become larger as the problem size increases.

The tests in Table 1 were all run with the convergence tolerances set to $\varepsilon_c = \varepsilon_\lambda = \varepsilon_f = \varepsilon = 0.01$. The influence of changing these values, for the mesh illustrated in Figure 1, is shown in Table 2. Clearly a uniform tolerance value of 0.01 is sufficiently accurate for most practical calculations with the nonlinear algorithm. For the linear programming formulation, at least 24 sides need to be used in the yield surface linearisation. As expected, the linear and nonlinear programming methods give the same collapse load when the former is used with an accurate linearisation.

7 CONCLUSIONS

A new algorithm for performing lower bound limit analysis has been presented. Numerical results show that the new approach is vastly superior to the commonly used linear programming formulation, especially for large scale problems.

Table 1 Results for smooth strip footing on cohesive -frictional soil obtained with linear and nonlinear code. (* All times are for a SUN Ultra Model 1 170 operating under UNIX with the optimising FORTRAN 77 compiler.)

Mesh	Quantity	Linear. Prog. NSID=24	Nonlinear Prog. ε=0.01	LP/ NLP
228 nod 74 elm 101 dis	Limit load	36.70	37.64	0.98
	No. of iter	288	21	13.7
	CPU (sec)*	4.17	1.10	3.79
	Time/iter	0.015	0.052	0.29
	Error	10^{-7}	10^{-5}	0.01
712 nod 234 elm 335 dis	Limit load	42.52	43.79	0.97
	No. of iter	1845	33	56
	CPU (sec)*	90.55	6.02	15.0
	Time/iter	0.049	0.18	0.27
	Error	10^{-6}	10^{-6}	1.0
1452 nod 478 elm 697 dis	Limit load	43.99	45.53	0.97
	No. of iter	5552	29	191
	CPU (sec)*	695.5	12.68	54.9
	Time/iter	0.13	0.38	0.34
	Error	10^{-5}	10^{-4}	0.1

Table 2 Results for smooth strip footing problem with different convergence tolerances $\varepsilon = \varepsilon_c = \varepsilon_\lambda = \varepsilon_f$ (* matrix $\mathbf{K}$ is numerically singular)

Linear Prog.				Nonlinear Prog.			
CPU	Iter	NSID	Load	Load	ε	Iter	CPU
111	2172	36	43.20	43.24	0.1	14	2.82
160	2565	54	43.55	43.55	0.05	19	3.66
672	5505	168	43.79	43.79	0.01	33	6.02
1183	7507	240	43.80	43.80	0.005	39	7.05
2355	10866	360	43.81	43.81	0.001	51*	9.06

REFERENCES

Bottero, A., Negre, R., Pastor, J. and Turgeman, S. (1980). 'Finite element method and limit analysis theory for soil mechanics problems', *Comp. Meth. Appl. Mech. Eng.*, 22, 131–149.

Sloan, S.W. (1988a). 'Lower bound limit analysis using finite elements and linear programming', *Int. J. Numer. Anal. Methods Geomech.*, 12, 61–77.

Sloan, S.W. (1988b). 'A steepest edge active set algorithm for solving sparse linear programming problems', *Int. J. Numer. Methods Eng.*, 26, 2671–2685.

Zouain, N., Herskovits, J., Borges, L.A. and Feijóo, R.A. (1993), 'An Iterative Algorithm for Limit Analysis with Nonlinear Yield Functions', *Int. J. Solids and Structures,* Vol. 30, 10, 1397–1417.

Numerical Models in Geomechanics, Pietruszczak & Pande (eds) © 1997 Balkema, Rotterdam, ISBN 90 5410 886 X

A rate-type mixed f.e.m. for large transformations of non-viscous materials

P. Royis & H. Royis
Département Génie Civil et Bâtiment, Ecole Nationale des Travaux Publics de l'Etat, Vaulx-en-Velin, France

ABSTRACT: The paper deals with finite element modelling of quasistatic large transformations (deformations and rotations) of non-viscous continua. A rate-type mixed variational formulation of the time-discretized problem, involving both velocity and objective stress rate, is first developed. It avoids the inversion of the constitutive equations considered. The time integration of the various fields, which is totally separated from the finite element computations, is then tackled. Finally, some numerical results are presented.

1 INTRODUCTION

Let Ω be a materially simple body with non-viscous (i.e. rate-independent) behaviour, the motion of which is studied over the time interval $[0,T]$. We denote by Ω_t the configuration of Ω relating to the time t. We assume that Ω_t is an open, bounded and simply-connected region of the physical space IR^3, with Lipschitz-continuous boundary Γ_t, and we denote by $\mathbf{n}_t$ the outer unit normal to Γ_t. The successive configurations of Ω are observed with respect to the same fixed orthonormal frame, and the time $t=0$ is chosen as the reference value. The positions of the material particles of Ω are given by the vectors $X \in \Omega_0$ at time $t=0$, and by the vectors $x \in \Omega_t$ at the current time t. The transformation $\Im_t : X \rightarrow x$ relating to that time is then a one-to-one mapping from Ω_0 onto Ω_t. We denote by $\mathbf{F}(t,X)$ the gradient tensor of $\Im_t$ at point X, and by $J(t,X)$ the determinant of $\mathbf{F}(t,X)$.

Let $\mathbf{b}(t,.):\Omega_t \rightarrow \mathrm{IR}^3$ be the vector field at time t of the body forces acting per mass unit in Ω_t. We denote by Γ_{t1} the part of Γ_t on which we have, at time t, the essential boundary conditions $\mathbf{u}|\Gamma_{t1} = \mathbf{u}^1$, where $\mathbf{u}^1(t,.):\Gamma_{t1} \rightarrow \mathrm{IR}^3$ are the values of the displacement $\mathbf{u}(t,.)$ given on Γ_{t1}, and by Γ_{t2} the part of Γ_t on which the values of the stress vector are prescribed at the same time. We assume that Γ_{t1} and Γ_{t2} constitute, at every time t, a partition of Γ_t such that Γ_{t1} has at least three points, and we denote by $\mathbf{g}(t,.):\Gamma_{t2} \rightarrow \mathrm{IR}^3$ the values of the stress vector given on Γ_{t2}. Eventually, we consider only quasistatic problems, for which the acceleration may be ignored.

Let then $\mathbf{v}(t,.):\Omega_t \rightarrow \mathrm{IR}^3$ be the velocity field of the material body Ω at time t. We denote by $\varepsilon(\mathbf{v})$ the strain rate tensor, symmetric part of the velocity gradient $\mathbf{grad}_x(\mathbf{v})$, and by $\mathbf{W}$ the spin tensor, that is to say the skew-symmetric part of $\mathbf{grad}_x(\mathbf{v})$. We assume that the behaviour of the continuum Ω is non-viscous (i.e. rate-independent) and governed by constitutive relations taking the following form

$$\varepsilon(\mathbf{v}) = \mathbf{H}(\dot{\sigma}^{obj}, \aleph) \tag{1}$$

where $\dot{\sigma}^{obj}$ is an objective rate of the Cauchy stress tensor σ, and $\mathbf{H}$ a tensorial function depending on the set $\aleph$ of the memory parameters at the material point considered. $\mathbf{H}$ is positively homogeneous of degree one with respect to $\dot{\sigma}^{obj}$ since the behaviour of the medium is rate-independent. The constitutive law (1) is hypoelastic if $\mathbf{H}$ is linear with respect to $\dot{\sigma}^{obj}$, and anelastic if not. In that last case we obtain the class of classical elastoplastic constitutive relations expressing the strain rate tensor $\varepsilon(\mathbf{v})$ as a function of an objective derivative of the Cauchy stresses, but also the set of rate-independent laws based upon the same principle of expression of $\varepsilon(\mathbf{v})$ and called 'incremental laws involving interpolations', which in particular differ from the previous ones on account of the absence of an elastic component. In other words the resulting constitutive equations are irreversible even for very small stress levels. The use of such laws for describing the non-linear behaviour of geomaterials such as soils is well established and many rheological models have been issued from this formalism (Darve 1978, Royis 1989,...).

The aim of the following section is to build a sequence of time-discretized mixed variational problems involving the fields of both velocity and objective stress rate, which avoids the inversion of the constitutive equations (1) while dissociating the time integration of the various fields from the rate-type finite element computations.

2 VARIATIONAL FORMULATION OF THE TIME-DISCRETIZED PROBLEM

Let $N \in \mathbb{N}^*$ and let $t_0, t_1, \ldots, t_N$ be an increasing sequence of time values, such that $t_0 = 0$ and $t_N = \mathrm{T}$. We assume that the choice of these values is made in such a way that the partition of Γ_0 defined by $\Im_t^{-1}(\Gamma_{t1})$ and $\Im_t^{-1}(\Gamma_{t2})$ remains time-independent over each of the intervals $[t_{n-1}, t_n[$, $n \in \{1, \ldots, N\}$.

Since our double aim is to avoid the inversion of the constitutive equations (1) and dissociate the time integration of the various fields from the rate-type finite element computations, we take as unknowns of the problem the two fields appearing in these equations, that is to say the velocity $\mathbf{v}$ and the objective derivative $\dot{\sigma}^{obj}$ of the Cauchy stresses (Royis 1994). We are then interested, for $n \in \{0, \ldots, N-1\}$, in finding the fields $\mathbf{v}$ and $\dot{\sigma}^{obj}$ relating to the time value t_n. After solving this problem and independently of it, the time integration of the various fields is carried out, so as to allow the resolution of the problem relating to time t_{n+1}. In all the following we shall put $\mathbf{v}_n(x) = \mathbf{v}(t_n, x)$, $\forall n \in \{0, \ldots, N\}$ and $\forall x \in \Omega_{t_n}$, as well as analogous notations for σ, $\dot{\sigma}^{obj}$, $\aleph$, $\mathbf{b}$, $\mathbf{g}$ and for the other fields which will be introduced subsequently (as regards the Lagrangian fields, that definition becomes $\forall X \in \Omega_0$). Finally we denote by $L^2(\Omega_{t_n})$ the space of real square integrable functions defined on Ω_{t_n}, and by $H^1(\Omega_{t_n})$ the Sobolev space of real square integrable functions defined on Ω_{t_n} with square integrable generalized derivatives of order one.

Let then $n \in \{0, \ldots, N-1\}$, $V_n = [L^2(\Omega_{t_n})]^9_{sym}$, $M_n = [H^1(\Omega_{t_n})]^3$, and let M_{0n} denote the closed subspace of M_n constituted by the fields $\mathbf{v}$ such that $\mathbf{v}|\Gamma_{t_n 1} = 0$. We obtain the first part of the mixed variational formulation of the time-discretized problem relating to time t_n by doing the inner product of $\mathbf{s} \in V_n$ by the constitutive equations (1) written at that time, and then by integrating the resulting expression on Ω_{t_n}. We get

$$\int_{\Omega_{t_n}} \mathbf{H}(\dot{\sigma}_n^{obj}, \aleph_n) : \mathbf{s} \, d\Omega_{t_n} - \int_{\Omega_{t_n}} \mathbf{s} : \varepsilon(\mathbf{v}_n) \, d\Omega_{t_n} = 0 \tag{2}$$

As to the second part of that formulation, it is obtained by considering an objective time derivative of the equations coming from the balance principle of linear momentum. The inner product of this derivative by $\mathbf{w} \in M_{0n}$ is first performed, before integrating it on Ω_{t_n}. After integration by parts and use of the Gauss integral identity we obtain, independently of the objective derivative considered (Royis 1996), the following relation

$$\begin{cases} \displaystyle\int_{\Omega_{t_n}} \left[\dot{\sigma}_n^{TR} : \varepsilon(\mathbf{w}) + \left(\mathbf{grad}_x(\mathbf{v}_n).\sigma_n\right) : \mathbf{grad}_x(\mathbf{w})\right] d\Omega_{t_n} \\ \displaystyle\qquad = \int_{\Omega_{t_n}} \rho_n \dot{\mathbf{b}}_n . \mathbf{w} \, d\Omega_{t_n} + \int_{\Gamma_{t_n 2}} \mathbf{w}.\mathrm{D}_t\left(\mathbf{g}_n \, d\Gamma_{t_n}\right) \end{cases} \tag{3}$$

where ρ denotes the mass density, where the operator D_t as well as the point above a given field represent the material derivation, and where $\dot{\sigma}^{TR}$ is the Truesdell objective derivative of σ defined by

$$\dot{\sigma}^{TR} = \dot{\sigma} - \mathbf{grad}_x(\mathbf{v}).\sigma - \sigma.^t\mathbf{grad}_x(\mathbf{v}) + \sigma \, \mathrm{div}_x(\mathbf{v}) \tag{4}$$

Let now $\mathbf{v}_n^1$ be the rate of the displacement given on $\Gamma_{t_n 1}$, $M_{0n}^{\perp}$ be the orthogonal set of M_{0n} in M_n, and let $\mathbf{v}_{1n}$ be the unique element of $M_{0n}^{\perp}$ such that $\mathbf{v}_{1n}|\Gamma_{t_n 1} = \mathbf{v}_n^1$. Then the mixed variational problem (P_n) relating to time t_n takes the following abstract form (Royis 1994)

$$(P_n) \begin{cases} \text{Find } (\dot{\sigma}_n^{obj}, \mathbf{v}_{0n}) \in V_n \times M_{0n} \text{ such that :} \\ a_n(\dot{\sigma}_n^{obj}, \mathbf{s}) - b_n(\mathbf{s}, \mathbf{v}_{0n}) = h_n(\mathbf{s}) \quad \forall \mathbf{s} \in V_n \\ b_n(\dot{\sigma}_n^{obj}, \mathbf{w}) + c_n(\mathbf{v}_{0n}, \mathbf{w}) = l_n(\mathbf{w}) \quad \forall \mathbf{w} \in M_{0n} \end{cases} \tag{5}$$

where $a_n(\dot{\sigma}^{obj},\mathbf{s})$ is defined by the first integral of equality (2), $b_n(\mathbf{s},\mathbf{v}_n)$ by the second integral of the same equality, and $h_n(\mathbf{s})$ by $h_n(\mathbf{s})=b_n(\mathbf{s},\mathbf{v}_{1n})$. As to the expressions of the bilinear form c_n and of the linear form l_n, they are dependent on the choice of the objective derivative $\dot{\sigma}^{obj}$ as well as on the kind of boundary conditions prescribed on $\Gamma_{t_n 2}$. We shall consider the two following classical cases: $\dot{\mathbf{g}}_n$ given on $\Gamma_{t_n 2}$, and $\mathbf{g}_n=-p_n\mathbf{n}_{t_n}$ with $\dot{p}_n$ given on $\Gamma_{t_n 2}$. $c_n(\mathbf{v}_n,\mathbf{w})$ is then obtained by adding the appropriate boundary integral to a volume integral depending on the choice of $\dot{\sigma}^{obj}$. If $\dot{\mathbf{g}}_n$ is known on $\Gamma_{t_n 2}$ the boundary contribution to $c_n(\mathbf{v}_n,\mathbf{w})$ is given by the following expression

$$-\int_{\Gamma_{t_n 2}}[\mathrm{div}_x(\mathbf{v}_n)-\mathbf{n}_{t_n}.\varepsilon(\mathbf{v}_n).\mathbf{n}_{t_n}]\mathbf{g}_n.\mathbf{w}\,\mathrm{d}\Gamma_{t_n} \qquad (6)$$

and when $\dot{p}_n$ is prescribed, that contribution becomes

$$\int_{\Gamma_{t_n 2}}p_n[\mathrm{div}_x(\mathbf{v}_n)\mathbf{n}_{t_n}-{}^t\mathbf{grad}_x(\mathbf{v}_n).\mathbf{n}_{t_n}].\mathbf{w}\,\mathrm{d}\Gamma_{t_n} \qquad (7)$$

The volume contribution to $c_n(\mathbf{v}_n,\mathbf{w})$ is given by relation (8) if $\dot{\sigma}^{obj}=\dot{\sigma}^{TR}$,

$$\int_{\Omega_{t_n}}[\mathbf{grad}_x(\mathbf{v}_n).\sigma_n]:\mathbf{grad}_x(\mathbf{w})\,\mathrm{d}\Omega_{t_n} \qquad (8)$$

and by the following one when the classical Jaumann derivative $\dot{\sigma}^J=\dot{\sigma}+\sigma.\mathbf{W}-\mathbf{W}.\sigma$ is used.

$$\begin{cases}\displaystyle\int_{\Omega_{t_n}}[\mathbf{grad}_x(\mathbf{v}_n).\sigma_n+\sigma_n\mathrm{div}_x(\mathbf{v}_n)]:\mathbf{grad}_x(\mathbf{w})\,\mathrm{d}\Omega_{t_n}\\ \displaystyle-\int_{\Omega_{t_n}}[\varepsilon(\mathbf{v}_n).\sigma_n+\sigma_n.\varepsilon(\mathbf{v}_n)]:\varepsilon(\mathbf{w})\,\mathrm{d}\Omega_{t_n}\end{cases} \qquad (9)$$

Analogous contributions corresponding to other objective derivatives of the Cauchy stresses can be found in the reference (Royis 1996). For instance, the contribution generated by the use of the Cotter-Rivlin derivative $\dot{\sigma}^{CR}=\dot{\sigma}+\sigma.\mathbf{grad}_x(v)+{}^t\mathbf{grad}_x(v).\sigma$ is

$$\begin{cases}\displaystyle\int_{\Omega_{t_n}}[\mathbf{grad}_x(\mathbf{v}_n).\sigma_n+\sigma_n\mathrm{div}_x(\mathbf{v}_n)]:\mathbf{grad}_x(\mathbf{w})\,\mathrm{d}\Omega_{t_n}\\ \displaystyle-2\int_{\Omega_{t_n}}[\varepsilon(\mathbf{v}_n).\sigma_n+\sigma_n.\varepsilon(\mathbf{v}_n)]:\varepsilon(\mathbf{w})\,\mathrm{d}\Omega_{t_n}\end{cases} \qquad (10)$$

As to $l_n(\mathbf{w})$, it is obtained by subtracting $c_n(\mathbf{v}_{1n},\mathbf{w})$ from the second integral of equality (3), before adding the boundary contribution (11) if $\dot{\mathbf{g}}_n$ is given on $\Gamma_{t_n 2}$,

$$\int_{\Gamma_{t_n 2}}\dot{\mathbf{g}}_n.\mathbf{w}\,\mathrm{d}\Gamma_{t_n} \qquad (11)$$

and the following one when $\dot{p}_n$ is prescribed.

$$-\int_{\Gamma_{t_n 2}}\dot{p}_n\mathbf{w}.\mathbf{n}_{t_n}\,\mathrm{d}\Gamma_{t_n} \qquad (12)$$

The existence and uniqueness of the solution of the mixed variational problem (P_n) has been established (Royis 1994) for a given class of incremental constitutive laws involving interpolations and for the Jaumann derivative $\dot{\sigma}^J$. That result can be easily generalized to other constitutive equations taking the form (1) and other objective derivatives of the Cauchy stresses. Problem (P_n) can then be solved after building finite element spaces $V_{nh}\subset V_n$ and $M_{0nh}\subset M_{0n}$. Some considerations relating to the resolution of the resulting systems of non-linear algebraic equations are detailed in the references (Laouafa 1996) and (Royis 1996).

3 TIME INTEGRATION

The time integration of the various fields between t_n and t_{n+1}, which is required by the numerical resolution of problem (P_{n+1}), is based upon two assumptions. The first is a classical one and consists in assuming that the velocity $\mathbf{v}(t,X)$ of each particle X remains constant over the time interval $[t_n,t_{n+1}[$. Thus, after putting $\tau=t_{n+1}-t_n$, we have $x_{n+1}=x_n+\tau\,\mathbf{v}_n$ and $\mathbf{F}_{n+1}=\mathbf{F}_n+\tau\,\mathbf{grad}_X(\mathbf{v}_n)$, and the computation of the strains relating to t_{n+1} follows immediately. The second hypothesis is about the Piola-Kirchhoff stress tensor $\Pi=J\mathbf{F}^{-1}.\sigma.{}^t\mathbf{F}^{-1}$, the variations of which as a function of time are assumed to be linear over the time interval $[t_n,t_{n+1}]$. This hypothesis respects the principle of objectivity, and therefore that of incremental objectivity (Charlier 1987). If the Truesdell objective derivative $\dot{\sigma}^{TR}$ is used it leads to the following expression of σ_{n+1}

$$\sigma_{n+1}=J^{-1}_{n+1,n}\mathbf{F}_{n+1,n}.\left[\sigma_n+\tau\dot{\sigma}^{TR}_n\right].{}^t\mathbf{F}_{n+1,n} \qquad (13)$$

where $\mathbf{F}_{n+1,n} = \mathbf{F}_{n+1}.\mathbf{F}_n^{-1}$ and $J_{n+1,n} = J_{n+1} J_n^{-1}$.

If the Jaumann derivative $\dot{\sigma}^J$ is chosen that relation becomes

$$\begin{cases} \sigma_{n+1} = J_{n+1,n}^{-1} \mathbf{F}_{n+1,n}.\left[\sigma_n + \tau\dot{\sigma}_n^J - \tau\varepsilon(\mathbf{v}_n).\sigma_n \right. \\ \qquad \left. - \tau\sigma_n.\varepsilon(\mathbf{v}_n) + \tau\sigma_n \mathrm{div}_x(\mathbf{v}_n)\right]. {}^t\mathbf{F}_{n+1,n} \end{cases} \tag{14}$$

Eventually, the expressions of σ_{n+1} relating to other choices of objective derivatives of the Cauchy stresses are given in the reference (Royis 1996). For instance, the following relation is obtained when the Cotter-Rivlin derivative $\dot{\sigma}^{CR}$ is used

$$\begin{cases} \sigma_{n+1} = J_{n+1,n}^{-1} \mathbf{F}_{n+1,n}.\left[\sigma_n + \tau\dot{\sigma}_n^{CR} - 2\tau\varepsilon(\mathbf{v}_n).\sigma_n \right. \\ \qquad \left. - 2\tau\sigma_n.\varepsilon(\mathbf{v}_n) + \tau\sigma_n \mathrm{div}_x(\mathbf{v}_n)\right]. {}^t\mathbf{F}_{n+1,n} \end{cases} \tag{15}$$

4 NUMERICAL RESULTS

To begin with, let us consider the homogeneous plane distortion of an elastoplastic cube with unit edge. The kinematic, which is completely imposed, is defined by $x_1 = X_1 + \alpha t X_2$, $x_2 = X_2$ and $x_3 = X_3$, with $\alpha = 7.5\ 10^{-4}\,\mathrm{s}^{-1}$ and $t \in [0,\mathrm{T}]$, where $\mathrm{T} = 10^4\,\mathrm{s}$.

For a given objective derivative $\dot{\sigma}^{obj}$ of the Cauchy stresses, the constitutive law takes the form (1) and is elastoplastic with isotropic linear hardening. The values of the mechanical parameters are as follows: Young's modulus $E = 1000$ MPa, Poisson's ratio $\nu = 0.3$, yield stress for uniaxial tension test $\sigma_e = 1$ MPa and hardening modulus $H = 3$ MPa. For the purposes of the following we also need to define the elastic shear modulus μ_e and the tangent shear modulus μ_t given by

$$\mu_e = \frac{E}{2(1+\nu)} \qquad \mu_t = \frac{EH}{3E + 2(1+\nu)H} \tag{16}$$

On the other hand we consider the Jaumann derivative $\dot{\sigma}^J$, the Truesdell derivative $\dot{\sigma}^{TR}$ together with the Cotter-Rivlin derivative $\dot{\sigma}^{CR}$ defined in the previous section.

For the Jaumann derivative the solution of the problem considered is elastic as long as $\alpha t \le a_0$ and becomes elastoplastic if $a_0 < \alpha t \le a_1$, with the following expressions of a_0 and a_1

$$\begin{cases} a_0 = 2 \arcsin \dfrac{\sigma_e}{2\sqrt{3}\mu_e} \\ a_1 = \arctan\left[\cot a_0 - \dfrac{\mu_e}{(\mu_e - \mu_t)\sin a_0}\right]^{-1} + \Pi \end{cases} \tag{17}$$

For $\alpha t > a_1$, the system of differential equation with initial conditions in $\dot{\sigma}$ becomes ill-posed. I $\alpha t \in [0, a_0]$ the nonzero components of the Cauchy stresses are given by

$$\begin{cases} \sigma_{12}(t) = \mu_e \sin \alpha t \\ \sigma_{11}(t) = -\sigma_{22}(t) = \mu_e (1 - \cos \alpha t) \end{cases} \tag{18}$$

and if $\alpha t \in [a_0, a_1]$ the following relations hold

$$\begin{cases} \sigma_{12}(t) = \mu_e \sin \alpha t - (\mu_e - \mu_t)\sin(\alpha t - a_0) \\ \sigma_{11}(t) = -\sigma_{22}(t) = \mu_e (1 - \cos \alpha t) \\ \qquad\qquad - (\mu_e - \mu_t)(1 - \cos(\alpha t - a_0)) \end{cases} \tag{19}$$

For the Truesdell and Cotter-Rivlin derivatives, the solution remains elastic if $\alpha t \le a_0$ and becomes elastoplastic if $\alpha t > a_0$, with

$$a_0 = \sqrt{\frac{3}{2}} \left(\sqrt{1 + \left(2\sigma_e / 3\mu_e\right)^2} - 1 \right)^{\frac{1}{2}} \tag{20}$$

Regarding the Truesdell derivative the nonzero components of σ are then given by the relations (21) if $\alpha t \in [0, a_0]$ and by the (22) ones if $\alpha t \ge a_0$,

$$\begin{cases} \sigma_{12}(t) = \mu_e \alpha t \\ \sigma_{11}(t) = \mu_e (\alpha t)^2 \end{cases} \tag{21}$$

$$\begin{cases} \sigma_{12}(t) = \mu_e a_0 + \mu_t (\alpha t - a_0) \\ \sigma_{11}(t) = \mu_e a_0^2 + 2\mu_e a_0 (\alpha t - a_0) + \mu_t (\alpha t - a_0)^2 \end{cases} \tag{22}$$

whereas the nonzero components of the solution corresponding to the Cotter-Rivlin derivative are obtained from these same relations by substituting $-\sigma_{22}$ for σ_{11}.

For each of the three objective derivatives considered, the numerical computations have been carried out by cutting the time interval $[0,\mathrm{T}]$ into 10^4 increments of 1s and the unit square into two equal triangles, with linear shape functions for the velocity, and constant ones for the stress rate. The diagrams in figure 1 giving the nonzero components of the stresses as a function of the distortion $\gamma_{12} = \alpha t$ show a good agreement between numerical results and the corresponding analytical values.

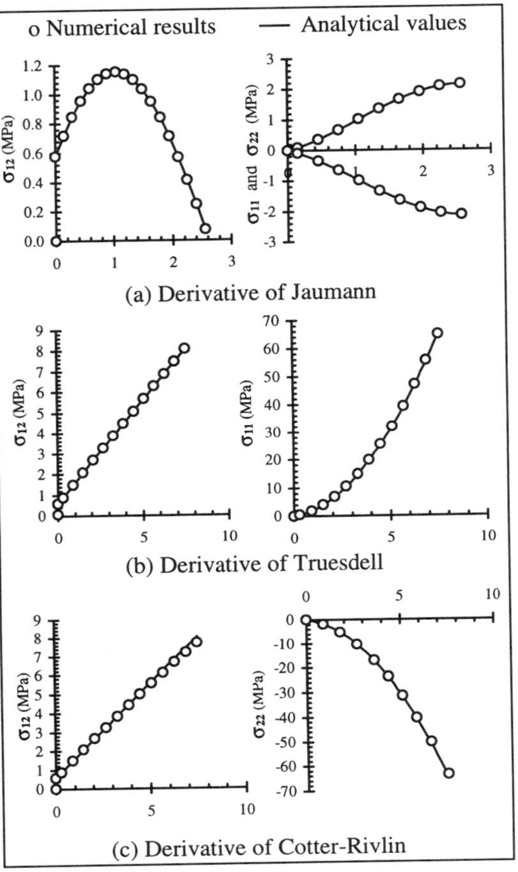

Figure 1. Nonzero stress components versus distortion

We are now interested in a computational structure test which had been proposed within the scope of the GRECO (Groupe de Recherches Coordonnées) 'Large strains and damaging'. This test consists in the plane-strain three-point bending of a cantilever beam with height h and span $L = 3h$ (Braudel 1986). As previously the constitutive law is elastoplastic with isotropic linear hardening, and the values of the mechanical parameters are: $E = 20000\,\text{daN.mm}^{-2}$, $\nu = 0.3$, $\sigma_e = 40\,\text{daN.mm}^{-2}$ and $H = 100\,\text{daN.mm}^{-2}$. The beam is discretized into 48 quadrilaterals divided up into two equal triangles, with quadratic shape functions for the velocity and linear ones for the stress rate. The total vertical displacement at the end of the beam, which is worth h, is imposed with the help of 1000 increments of 0.1 s. Besides the three objective derivatives $\dot{\sigma}^J$, $\dot{\sigma}^{TR}$ and $\dot{\sigma}^{CR}$ defined in the previous section, we consider the Jaumann-Cescotto derivative $\dot{\sigma}^{JC}$, the Oldroyd derivative $\dot{\sigma}^{O}$ as well as the Cotter-Rivlin-Royis derivative $\dot{\sigma}^{CRR}$, which are defined from the three previous ones by the following relations

$$\begin{cases} \dot{\sigma}^{JC} & = & \dot{\sigma}^{J} + \sigma \operatorname{div}_x(\mathbf{v}) \\ \dot{\sigma}^{O} & = & \dot{\sigma}^{TR} - \sigma \operatorname{div}_x(\mathbf{v}) \\ \dot{\sigma}^{CRR} & = & \dot{\sigma}^{CR} + \sigma \operatorname{div}_x(\mathbf{v}) \end{cases} \tag{23}$$

Figure 2 shows the shape of the beam at the end of the loading.

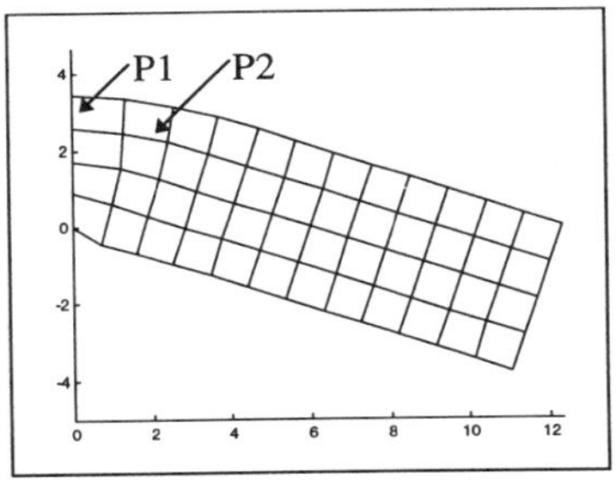

Figure 2. Final shape of the beam

Figures 3 and 4 illustrate the variations of the horizontal normal stress σ_{11} and of the shear stress σ_{12} at point P1 shown on figure 2, for each of the six objective derivatives considered.

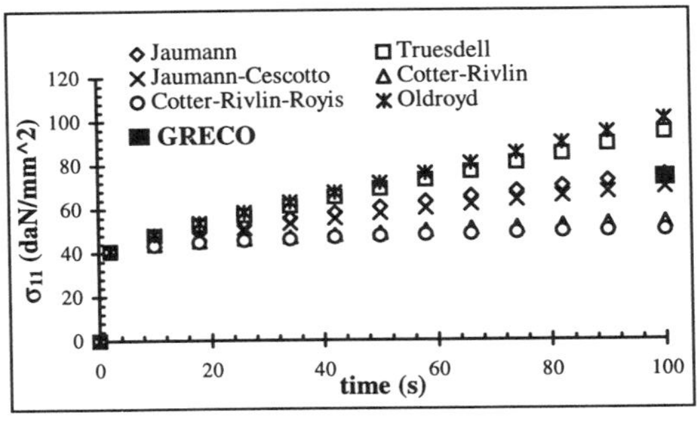

Figure 3. Variations of the horizontal normal stress at point P1

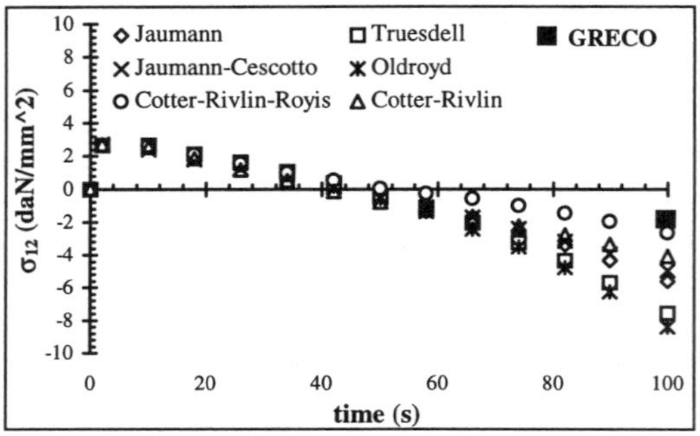

Figure 4. Variations of the shear stress at point P1

The black-filled square drawn on these two figures gives the values of those stresses obtained at the end of the loading by the various laboratories of the GRECO, with the help of the Jaumann derivative.

As to figures 5 and 6, they illustrate the variations of the same stresses at point P2 shown on figure 2.

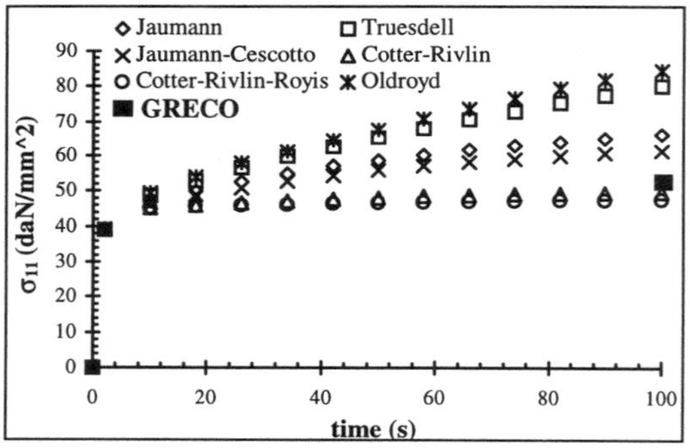

Figure 5. Variations of the horizontal normal stress at point P2

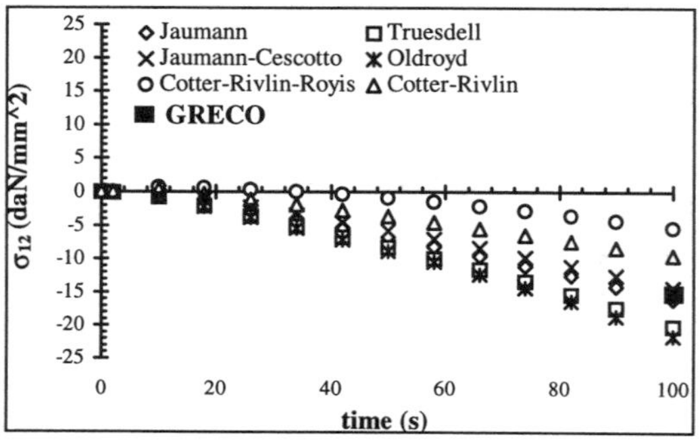

Figure 6. Variations of the shear stress at point P2

Our numerical results are in keeping with those obtained by the various laboratories of the GRECO for the horizontal normal stress at point P1 and for the shear stress at point P2, whereas slight discrepancies can be observed as concerns the shear stress at point P1 as well as the horizontal normal stress at point P2.

5 CONCLUSION

In this paper we presented a mixed finite element formulation involving the fields of both velocity and objective stress rate, for a class of non-viscous continua subjected to large transformations. This rate-type mixed approach allows to avoid the inversion of the constitutive equations considered while dissociating the time integration of the various fields from the rate-type finite element computations. The non-linearities of the resulting algebraic problems are of a pure rheological nature.

REFERENCES

Braudel, H.J. 1986. Modélisation des grandes transformations élastoplastiques d'un solide isotrope par la méthode des éléments finis: Application à la forge à froid des métaux. *Thèse de Doctorat ès Sciences*. Université Claude Bernard de Lyon.

Charlier, R. 1987. Approche unifiée de quelques problèmes non linéaires de mécanique des milieux continus par la méthode des éléments finis. *Thèse de Doctorat en Sciences Appliquées*. Université de Liège.

Darve, F. 1978. Une formulation incrémentale des lois rhéologiques: Application aux sols. *Thèse de Doctorat ès Sciences Physiques*. Université Scientifique et Médicale de Grenoble.

Laouafa, F. 1996. Analyse et développement de méthodes d'éléments finis: Unification et application aux lois incrémentales de type interpolation. *Thèse de Doctorat en Génie Civil*. Institut National des Sciences Appliquées de Lyon.

Royis, P. 1989. Interpolations and one-to-one properties of incremental constitutive laws: A family of incrementally non-linear constitutive laws. *European Journal of Mechanics. A/Solids*, Vol 8, 5 :385-411. Paris: Gauthier-Villars.

Royis, P., H. Royis & F. Laouafa 1994. Implementation of a two fields f.e.m. for incremental laws in large deformations. *Proceedings of the third European Conference on Numerical Methods in Geotechnical Engineering, Manchester, 7-9 September 1994* : 115-120. Rotterdam: Balkema.

Royis, H. 1996. Contribution à l'analyse numérique des équations de l'élastoplasticité incrémentale en transformations finies: Une formulation mixte en vitesses. *Thèse de Doctorat en Génie Civil*. Institut National des Sciences Appliquées de Lyon.

Numerical Models in Geomechanics, Pietruszczak & Pande (eds) © 1997 Balkema, Rotterdam, ISBN 90 5410 886 X

A robust formulation for FE-analysis of elasto-plastic media

A.Truty
Cracow University of Technology, Poland (Presently: LSC-DGC Swiss Federal Institute of Technology, Lausanne, Switzerland)

T.Zimmermann
LSG-DGC Swiss Federal Institute of Technology, Lausanne, Switzerland

ABSTRACT: The problem of element performance in finite element computations of elasto-plastic media exhibiting incompressible or dilatant plastic flow has been investigated recently in many publications. Most of them are concerned with mixed formulations derived from Hu-Washizu variational principle and extended strain approximation being the base of the so called Enhanced Assumed Strain Method (EAS) (Simo & al. 1990, Groen 1994, de Borst & al.1995). Unfortunately this method does not (yet) apply to many well known finite elements (like linear triangles in 2D or linear thetraedrons in 3D) and moreover requires specific design of the enhanced part of the strain field for each element separately. The goal of this paper is to propose and extend another philosophy. It is based on a mixed continuous displacement-pressure formulation using discontinuous plastic multipliers approximations, and complementing the Galerkin scheme by least-square terms enhancing its stability.

1. Introduction

The extension of the mixed u-p formulation to elastoplastic analysis has already been proposed in several publications (Sussman & al. 1987, Pinsky 1987, Pastor & al. 1995,1997). In all of them only J_2 type plasticity has been considered. The simplicity of J_2 plasticity models allows uncoupling of the volumetric and deviatoric behavior as well as preserving a constant mean stress during plastic return. Moreover, any mean stress is plastically admissible thus the plastic consistency condition can still be satisfied at any integration point in an exact sense. More general plastic models may however require weak enforcement of the consistency condition. The problem of weak enforcement of the plastic consistency condition in the context of mixed stress-displacements formulations has been studied already by (Nyssen & al.1984, Pinsky 1987, Simo & al.1989). In this paper an approach for a wide class of rate independent multisurface and generally nonassociated elasto-plastic models is proposed, extending the standard mixed u-p formulation to mixed displacement-pressure-plastic multpliers (u-p-γ) formulation, following the methodolgy proposed earlier by (Pinsky 1987) for a single surface J_2 type hardening plasticity model. The need for the introduction of the third field is induced by the fact that mean pressure field may no longer be plastically admissible (Nyssen & al.1984).

A second aspect developed in this paper is the finite element technology necessary for this type of formulations. In most papers concerned with u-p formulations a continuous displacement/ discontinuous pressure approach is used since it is known that equal order interpolation for both fields leads to violation of the Brezzi-Babuska condition (Franca 1987), which must be satisfied to obtain stable methods. This condition excludes a lot of possible approximations (Zienkiewicz & al.1989). To overcome the problem a family of stabilized methods have been proposed (Franca 1987, Franca & al.1987). These methods have already been used with great success for solving incompressible Navier-Stokes flow. In the field of solid mechanics the first attempt in the same direction has been done by (Pastor et al.1995,1997), but it is restricted to J_2 type plasticity and linear triangular elements, as the authors are using the stabilization technique proposed by (Brezzi & Pitkaranta 1984).

In stabilized methods a least-square term is added to the standard Galerkin formulation. The following form of least square term is proposed in (Franca & al. 1992):

$$R^{GLS}(\bar{\mathbf{u}}^h, \bar{p}^h, \bar{\gamma}^h, \mathbf{u}^h, p^h, \gamma^h) = -\sum_{e=1}^{n_{el}} \int_{\Omega^e} \left(\mathbf{L}^T\sigma(\bar{\mathbf{u}}^h, \bar{p}^h, \bar{\gamma}^h)\right)^T \tau\left(\mathbf{L}^T\sigma(\mathbf{u}^h, p^h, \gamma^h) + \mathbf{f}\right) d\Omega \tag{1.1}$$

where the differential operator $\mathbf{L}$ takes the form:

$$\mathbf{L}^T = \begin{bmatrix} \frac{\partial}{\partial x} & 0 & \frac{\partial}{\partial y} & 0 & \frac{\partial}{\partial z} & 0 \\ 0 & \frac{\partial}{\partial y} & \frac{\partial}{\partial x} & 0 & 0 & \frac{\partial}{\partial z} \\ 0 & 0 & 0 & \frac{\partial}{\partial z} & \frac{\partial}{\partial x} & \frac{\partial}{\partial y} \end{bmatrix} \tag{1.2}$$

The corresponding stress σ is represented by the vector:

$$\left(\sigma_x . \sigma_y, \tau_{xy}, \sigma_z, \tau_{xz}, \tau_{yz}\right) \tag{1.3}$$

$\mathbf{u}^h, p^h, \gamma^h$ are the approximations of the solution, and $\mathbf{f}$ is the body force.

The operator acting on the test functions $\bar{\mathbf{u}}^h, \bar{p}^h, \bar{\gamma}^h$ can be reduced by taking into consideration only selected components of the stress. In (1.1) the stabilization factor τ appears, which depends on element size and stiffness parameters. For more generality this factor should be replaced by a matrix (Shakib 1988) in order to include effects of element distorsion and stiffness anisotropy.

2. Summary of the elastoplastic constitutive theory

The proposed formulation is designed for a class of rate independent multisurface elastoplastic models. The constitutive equations for the deviatoric and volumetric parts of the stress tensor are as follows:

$$\begin{aligned} \mathbf{s} &= \mathbf{D}^e \ \mathbf{dev}\ (\varepsilon - \varepsilon^p) = \bar{\mathbf{D}}^e(\varepsilon - \varepsilon^p) \\ p &= K\ \mathrm{tr}(\varepsilon - \varepsilon^p) \end{aligned} \tag{2.1}$$

where ε is the total strain, ε^p is the plastic strain, deviatoric stress is denoted by $\mathbf{s}$ and the mean stress (positive in tension) by p. The deviatoric projection $\bar{\mathbf{D}}^e$ of the elastic stiffness matrix $\mathbf{D}^e$ and the elastic bulk modulus K are defined as follows:

$$\begin{aligned} \bar{\mathbf{D}}^e &\doteq \mathbf{D}^e\left(\mathbf{I} - \frac{1}{3}\mathbf{1}\,\mathbf{1}^T\right) \\ K &= \frac{1}{9}\mathbf{1}^T\mathbf{D}^e\mathbf{1} \end{aligned} \tag{2.2}$$

where symbol $\mathbf{1}$ is the vector representation of the Kronecker delta.

The flow rule and the hardening law take the following form:

$$\dot{\varepsilon}^p = \sum_{\alpha=1}^{M} {}^\alpha\dot{\gamma}\ {}^\alpha\mathbf{r}(\sigma, \mathbf{q}) \quad \forall_{\alpha \in Jact} \tag{2.3}$$

$$\dot{\mathbf{q}} = \sum_{\alpha=1}^{M} {}^\alpha\dot{\gamma}\ {}^\alpha\mathbf{h}(\sigma, \mathbf{q}) \quad \forall_{\alpha \in Jact} \tag{2.4}$$

where the plastic surface index is denoted by α, the flow direction by $\mathbf{r}$, the hardening parameters by $\mathbf{q}$, hardening functions by $\mathbf{h}$, and the summation is performed only for active surfaces ($\alpha \in$ Jact).
The loading/unloading conditions:

$${}^\alpha f \le 0\ ,\quad {}^\alpha\dot{\gamma} \ge 0\ ,\quad {}^\alpha\dot{\gamma}\ {}^\alpha f \le 0 \tag{2.5}$$

where f designates yield surfaces.

3. A mixed variational formulation

The weak form of the momentum equation, constitutive equation for the mean stress (2.1) and plastic consistency condition (2.5) can be expressed in the following form:

$$R_1(\bar{\mathbf{u}}, \mathbf{u}, p, \gamma) = \int_\Omega \varepsilon(\bar{\mathbf{u}})^T\left[\bar{\mathbf{D}}^e\left(\varepsilon(\mathbf{u}) - \varepsilon^p(\mathbf{u}, p, \gamma)\right) + \mathbf{1}\,p\right] d\Omega - \int_\Omega \bar{\mathbf{u}}^T\mathbf{f}\, d\Omega - \int_\Gamma \bar{\mathbf{u}}^T\bar{\mathbf{t}}\, d\Gamma = 0 \tag{3.1}$$

where $\bar{\mathbf{t}}$ is the imposed boundary traction,

$$R_2(\bar{p},\mathbf{u},p,\gamma)=\int_\Omega \bar{p}\left[\mathrm{tr}\ (\varepsilon(\mathbf{u})-\varepsilon^p(\mathbf{u},p,\gamma))-p/K\right]d\Omega=0 \tag{3.2}$$

$$^\alpha R_3(^\alpha\bar{\gamma},\mathbf{u},p,\gamma)=-\int_\Omega {}^\alpha\bar{\gamma}\left\langle {}^\alpha f(\sigma(\mathbf{u},p,\gamma),\mathbf{q}(\mathbf{u},p,\gamma)\right\rangle d\Omega=0 \tag{3.3}$$

where $\langle . \rangle$ denotes Macauley bracket.

The minus sign in (3.3) is introduced to get formally the same signs for appropriate terms of the linearized equations (3.1) and (3.3).

The above equations can be written in combined form:

$$R(\bar{\mathbf{u}},\bar{p},\bar{\gamma},\mathbf{u},p,\gamma)=R_1(\bar{\mathbf{u}},\mathbf{u},p,\gamma)+R_2(\bar{p},\mathbf{u},p,\gamma)+\sum_{\alpha=1}^{M}{}^\alpha R(\bar{\gamma},\mathbf{u},p,\gamma)=0 \tag{3.4}$$

Since the pastic flow rule (2.3) and the hardening law (2.4) are formulated in rate form they need to be integrated first and then substituted into the variational equations. A fully implicit integration scheme has been adopted and it results in the following formuli for plastic strains and hardening parameters at time step n+1:

$$\varepsilon^p_{n+1}=\varepsilon^p_n+\sum_{\alpha=1}^{M}{}^\alpha\gamma_{n+1}\,{}^\alpha\mathbf{r}(\sigma_{n+1},\mathbf{q}_{n+1})\quad \underset{\alpha\in Jact}{\forall} \tag{3.5}$$

$$\mathbf{q}_{n+1}=\mathbf{q}_n+\sum_{\alpha=1}^{M}{}^\alpha\gamma_{n+1}\,{}^\alpha\mathbf{h}(\sigma_{n+1},\mathbf{q}_{n+1})\quad \underset{\alpha\in Jact}{\forall} \tag{3.6}$$

The corresponding stress state defined at step n+1 can be expressed as follows:

$$\sigma_{n+1}=\overline{\mathbf{D}}^e(\varepsilon_{n+1}-\varepsilon^p_{n+1})+\mathbf{1}p_{n+1} \tag{3.7}$$

The system of variational equations (3.1-3.3) combined in (3.4) is nonlinear in arguments $\mathbf{u}$,p,γ. It can be solved with the aid of Newton's method but first it has to be linearized. Consistent linearization is an essential step to get an efficient iterative scheme. Also, inserting the integrated formuli for state variables $\sigma_{n+1},\varepsilon^p_{n+1},\mathbf{q}_{n+1}$ into variational equations prior to linearization should preserve the quadratic convergence rate of the resulting Newton scheme.

A comprehensive explanation of the linearization procedure for the above formulation will not be presented here due to the limited scope of the article. Some details for a single surface J_2 type plasticity model with hardening can be found in (Pinsky 1987).

4. Finite element approximation

The proposed variational formulation consists of three independent fields eg. displacements, mean pressure and plastic multipliers. For the sake of numerical efficiency a continuous approximation is assumed only for displacements and mean pressure while for plastic multipliers a discontinuous approximation is the only one which is appropriate. The discontinuous approximation allows static condensation of plastic multipliers degrees of freedom at the element level.

The displacement and pressure field is interpolated within each finite element domain by standard shape functions via formuli:

$$u^h=N^u_i u^h_i \qquad \bar{u}^h=N^u_i \bar{u}^h_i \tag{4.1}$$

$$p^h=N^p_i p^h_i \qquad \bar{p}^h=N^p_i \bar{p}^h_i \tag{4.2}$$

An upper index ^{u,p} for shape functions is introduced for the sake of generality. This way we include use of classes of elements in which different shape functions interpolate both fields (for ex. Q9/Q4 nine nodes for displacements and four nodes for pressures).

The problem of approximation of the field of plastic multipliers is more complex. A comprehensive assessment of several proposals is given by (Pinsky 1987). Considering his conclusions the following approximations have been assumed:

$$^\alpha\gamma={}^\alpha N^\gamma_i\,{}^\alpha\gamma_i \qquad {}^\alpha\bar{\gamma}={}^\alpha N^\gamma_i\,{}^\alpha\bar{\gamma}_i \tag{4.3}$$

where summation is performed over integration points and the shape functions $^\alpha N^\gamma_i$ at any point are defined as follows (after Pinsky 1987):

$$^\alpha N^\gamma_i(\xi_j)=\begin{cases}0 & \text{if } i\neq j\\ 0 & \text{if } i=j\ \wedge\ \alpha\notin Jact\\ 1 & \text{if } i=j\ \wedge\ \alpha\in Jact\end{cases} \tag{4.4}$$

5. Stabilization technique

The stabilization term added to the scheme expressed by equation (3.4) is meant to enhance its stability. Expression (1.1) is the most general form of least-square term which can be applied. The general aim of adding this term to (3.4) is to prevent the effect of volumetric locking, reduce the effect of stress oscillations and finally to render possible the use equal order continuous u-p interpolation (including linear triangles) and moreover to enable mixing of different types of elements to fit the desired geometry of the domain.

In order to define the form of stabilization term currently considered let us split the stress state into deviatoric and volumetric parts, and then rewrite the (1.1) in more detailed form as follows:

$$R^{GLS}(\bar{\mathbf{u}}^h, \bar{p}^h, \bar{\gamma}^h, \mathbf{u}^h, p^h, \gamma^h) = -\sum_{e=1}^{n_{el}} \int_{\Omega^e} \left(\mathbf{L}^T \left(\bar{\mathbf{D}}^e \left(\varepsilon(\bar{\mathbf{u}}^h) - \varepsilon^p(\bar{\mathbf{u}}^h, \bar{p}^h, \bar{\gamma}^h) \right) + \mathbf{1} p^h \right) \right)^T \tau \left(\mathbf{L}^T \sigma(\mathbf{u}^h, p^h, \gamma^h) + \mathbf{f} \right) \tag{5.1}$$

Omitting one term from the weighting part (in our case the term related to plastic strains) the following final stabilization term can be obtained:

$$R^{*GLS}(\bar{\mathbf{u}}^h, \bar{p}^h, \bar{\gamma}^h, \mathbf{u}^h, p^h, \gamma^h) = -\sum_{e=1}^{n_{el}} \int_{\Omega^e} \left(\pm \bar{\mathbf{D}}^e \mathbf{L}^T \varepsilon(\bar{\mathbf{u}}^h) + \mathbf{L}^T \mathbf{1} \bar{p}^h \right)^T \tau \left(\mathbf{L}^T \sigma(\mathbf{u}^h, p^h, \gamma^h) + \mathbf{f} \right) d\Omega \tag{5.2}$$

The simplest version is obtained by taking as a weighting operator only the gradient of the pressure and only this formulation will be used in all following applications . It should be noted that for some linear elements the first weighting term disappears automatically (eg. for linear triangles). This term added to (3.4) after introduction of the finite element approximation has to be linearized and finally brings a contribution to both the left- and right-hand side of the resulting linearized system of equations.

For Navier-Stokes problems the stabilization factor τ is defined as follows (Franca 1987):

$$\tau = \frac{\delta_1 h^2}{2G} \tag{5.3}$$

where δ_1, called optimal stabilization factor, should be a constant for a given element type and can be obtained by a convergence analysis; G is the shear modulus. However, it has been pointed out already by (Shakib 1988) that τ should be represented by a matrix in order to account for element distorsion. In our approach we followed the design of τ proposed by (Shakib 1988).

A last aspect of the proposed stabilization term is the problem of its global sign. The global sign is selected negative. The argument is that increasing the value of δ_1 monotonically from zero up to large values we never change signs of diagonal terms corresponding to pressure degrees of freedom in the generalized stiffness matrix. The opposite is however adopted in (Pastor & al. 1995,1997).

6. Numerical results

The first test is the thick cylinder test, which has an analytical solution (see Hill 1950). The test has been performed using a mesh which consists of both T3 and Q4 stabilized elements (fig.6.1). The internal and external radii of the cylinder are 100 and 200 mm. Young's modulus E=21000 dN/mm^2 and Poisson's ratio ν=0.49999. Huber-Mises' criterion is used with yield stress $\sigma_Y = 24\,\text{dN}/\text{mm}^2$. Isolines of the displacements norm, corresponding to aproximately 95% of the ultimate pressure are shown in fig.6.2.

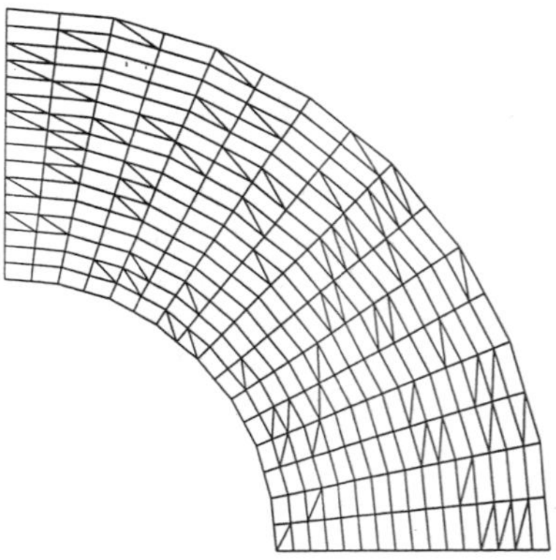

Fig.6.1

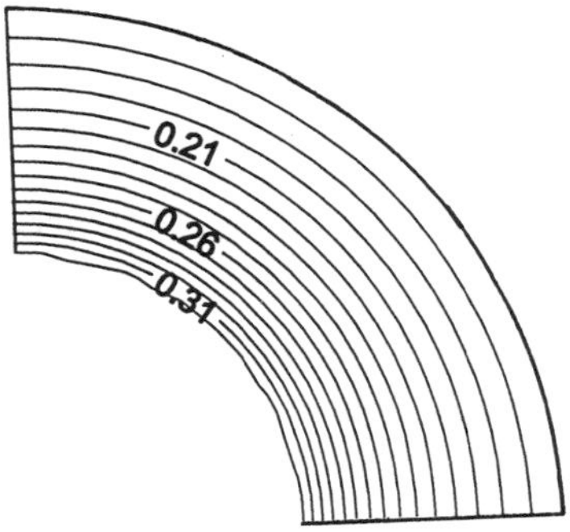

Fig.6.2

Observe that no disturbance of the displacements field due to mixing of different element types appears.

The next two tests are concerned with the safety analysis of a 4m high cut modelled with a Drucker-Prager plastic model with two different plastic flow rules, i.e deviatoric and dilatant. The material properties are as follows: E=10000 kPa, ν=0.3, c=16 kPa, $\phi=30^o$,γ=20 kN/m3 and $\psi=0^o/30^o$. In both cases a plane strain failure size adjustment to the Coulomb-Mohr criterion is used. The failure mode for deviatoric flow at safety factor SF=1.30 is shown in fig.6.3. The same corresponding to the associated flow, also at SF=1.30, which matches the simplified analytical solution, is shown in fig.6.4.

The last test is a plane strain ultimate load analysis of a footing. The width of the footing is 2.0 m. The soil is modelled by a Drucker-Prager plastic model with the same adjustment and nonassociated/ associated flow rules. Material properties are as

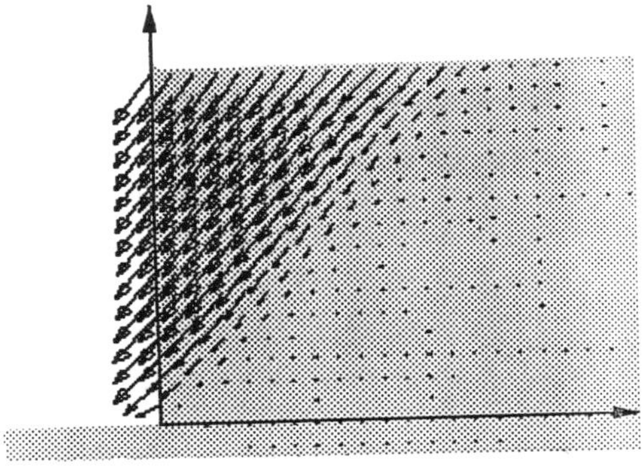

Fig.6.3

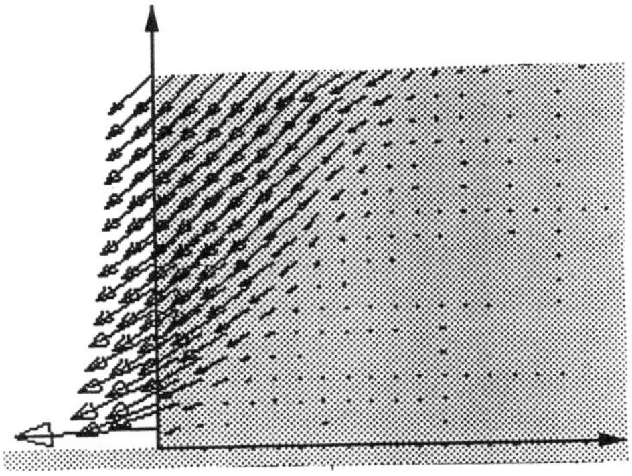

Fig.6.4

follows: E=10000 kPa, ν=0.3, c=1 kPa, $\phi=20^o$, $\psi=0^o/20^o$. The domain is discretized using only T3 (non cross-diagonal) elements .

Fig 6.5 shows the failure mode at the ultimate load value q=18.0 kN/m2 for the case of deviatoric flow. The failure mode at the ultimate load value q=18.4 kN/m2 for the case of associated flow is shown in fig. 6.6. The analytical solution from (Matar & al. 1979) for this case is q=15.6 kN/m2.

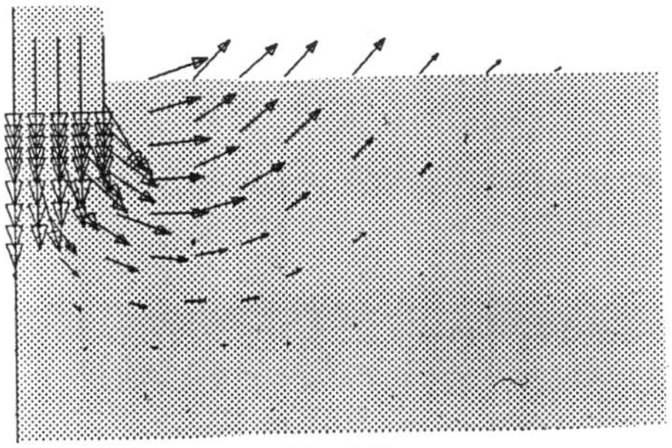

Fig.6.5

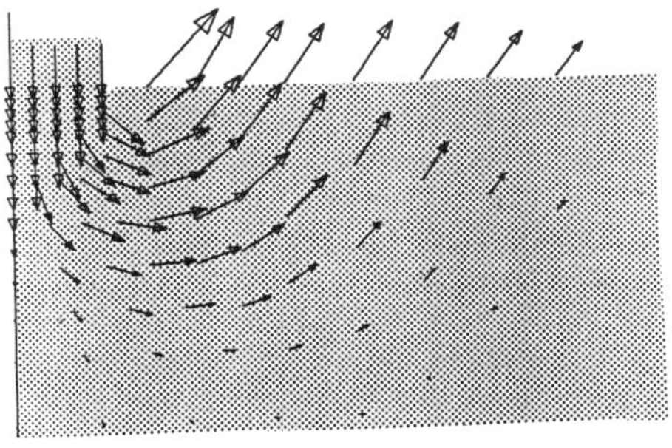

Fig.6.6

7. Conclusions

The proposed stabilized scheme seems to be a promising direction for development of efficient methods for analysis of elasto-plastic media exhibiting incompressible or dilatant plastic flow. The final form of the stabilization term and the influence of other weighting factors, as in the most general definition given by (1.1), has to be further investigated. The method is general for any type of finite elements. It has however to be adapted to a wide class of finite elements in order to define the optimal values of stabilization factors with matrix form. The most promising aspect is that this formulation allows to use the simplest linear elements. Moreover it is possible to mix different element types. A detailed derivation of the proposed mixed formulation will be given in further publications.

Acknowledgments

The authors acknowledge the financial support of the CTI (Commission for Technology and Innovation, Switzerland) for the first author.

The authors also would like to thank T.J.R. Hughes (Stanford University) for advice offered at the early stage of the project and T.Tezduyar (University of Minnesota) for useful discussions.

8. Literature

Brezzi F., Pitkaranta J. 1984. *On the stabilization of finite element approximations of the Stokes problem.* Efficient Solutions of Elliptic Problems. Notes on Numerical Fluid Mechanics. Vol.10, pp.11-19, Vieweg, Wiesbaden.

de Borst R., Groen A.E. 1995. *Some observations on element performance in isochoric and dilatant plastic flow.* International Journal for Numerical Methods in Engineering. Vol. 38, pp.2887-2906.

Franca L.P.1987. *New mixed finite element methods.* Ph.D. Thesis, Stanford University, USA.

Franca L.P., Frey S.L., Hughes T.R.J. 1992. *Stabilized finite element methods: I. Application to the advective-diffusive model.* Computer Methods in Applied Mechanics and Engineering. Vol.95, pp.253-276.

Groen A.E. 1994. *Improvement of low order elements using assumed strain concepts.* TU-Delft report Nr 25.2.94.203.

Hill R. 1950. *The mathematical theory of plasticity.* Oxford University Press.

Matar M., Salençon J. 1979. Capacité portante des semelles filantes. Revue Française de Géotechnique, No. 9.

Nyssen C., Beckers P. 1984. *A unified approach for displacement, equilibrium and hybrid finite element models in elastoplasticity.* Computer Methods in Applied Mechanics and Engineering. Vol. 44, pp.131-151.

Pastor M., Quecedo M., Zienkiewicz O.C. 1997. *A mixed displacement-pressure formulation for numerical analysis of plastic failure.* Computers and Structures vol.62, No 1, pp.13-23.

Pastor M., Quecedo M., Zienkiewicz O.C. 1995. *Localization problems under volume contraints: A mixed formulation approach.* Proceedings of the fifth International Conference on Numerical Models in Geomechanics-NUMOG V, Pande & Pietruszczak (eds) pp.403-409.

Pinsky P.M.1987. *A finite element formulation for elastoplasticity based on a three-field variational equation.* Computer Methods in Applied Mechanics and Engineering.Vol.61, pp.41-60.

Shakib F. 1988. *A Finite Element Analysis of the Compressible Euler and Navier-Stokes Equations.* Ph.D. Thesis, Stanford University, USA.

Simo J.C., Kennedy J.G., Taylor R.L. 1989. *Complementary mixed finite element formulations for elastoplasticity.* Computer Methods in Applied Mechanics and Engineering. Vol. 74, pp. 177-206.

Simo J.C., Rifai M.S. 1990. *A class of mixed assumed strain methods and method of incompatible modes.* International Journal for Numerical Methods in Engineering. Vol. 29, pp.1595-1638.

Sussman T., Bathe K.J.1987. *A Finite Element Formulation for Nonlinear Incompressible Elastic and Inelastic Analysis.* Computers and Structures. Vol.26, No 1/2, pp.357-409.

Zienkiewicz O.C., Taylor R.L. 1989. *The Finite Element Method.* Vol. I. McGraw Hill, New York.

Numerical Models in Geomechanics, Pietruszczak & Pande (eds) © 1997 Balkema, Rotterdam, ISBN 90 5410 886 X

Evaluation of a stress point algorithm for a hypoplastic constitutive model for granular materials

C.Tamagnini
Università degli Studi di Roma La Sapienza, Italy

G.Viggiani
Laboratoire 3S, Grenoble, France & Università della Basilicata, Italy

R.Chambon & J.Desrues
Laboratoire 3S, Grenoble, France

ABSTRACT: A class of implicit one-step stress point algorithms is used to integrate the incrementally non-linear constitutive model CLoE, recently developed for granular soils. The paper discusses the performance of the numerical procedures on some particular strain paths, focusing on the effects of strain increment size and direction, as well as initial stress level and anisotropy.

1. INTRODUCTION

In recent years, a new class of constitutive models for granular materials has been developed in order to capture some aspects of observed behavior of granular soils, such as irreversibility and incremental non-linearity. These models are generally referred to as *hypoplastic* (Kolymbas 1991), or incrementally non-linear models.

The practical application of these models to the solution of engineering problems requires their implementation in numerical methods like the finite element method. This entails the definition of a solution strategy to solve the discrete non-linear equilibrium equations, a fundamental element of which is the development of an appropriate stress point algorithm to integrate the rate equations over a finite strain increment. There are of course many other important computational ingredients in the overall procedure, such as the linearization of the relevant algorithm to evaluate the appropriate tangent operator in implicit methods. However, the accuracy of the solution obtained does essentially depend on the integration scheme.

A particular hypoplastic model, the CLoE model (Chambon 1989, Chambon *et al.* 1994) is considered in this work. Its basic structure is given by the following equation:

$$\dot{\sigma} = \boldsymbol{A}\dot{\varepsilon} + \boldsymbol{b}\|\dot{\varepsilon}\| \qquad (1)$$

In eq. (1), the (objective) stress rate $\dot{\sigma}$ is evaluated as a continuously non-linear function of the strain rate $\dot{\varepsilon}$ (note that here and from now on, all stresses are effective stresses, unless otherwise stated). The two constitutive tensors $\boldsymbol{A}$ and $\boldsymbol{b}$ (of rank four and two, respectively) are given homogeneous functions of degree one of the stress tensor, to be obtained through interpolation from the material responses determined on suitable (and experimentally accessible) basic paths.

Differently from the most recent hypoplastic models developed in Karlsruhe (Gudehus 1996, Niemunis & Herle 1997), the (effective) stress state represents the only state variable in CLoE. This implies material isotropy and precludes the possibility to model cyclic behavior. An essential feature of CLoE is that the constitutive response in stress space is bounded by an explicitly assumed limit surface, which implies further requirements for the values of $\boldsymbol{A}$ and $\boldsymbol{b}$ (*consistency condition*).

In the following, a class of implicit one-step stress point algorithms is used for the integration of CLoE. It is worth noting that the algorithms presented herein are applicable without modifications to other hypoplastic models as well (e.g., Kolymbas *et al.* 1995, Gudehus 1996). Two integration schemes are tested, discussing their performance on some particular strain paths (axisymmetric compression and plane-strain simple shear). The paper focuses on the effect of such variables as strain increment size and direction, and initial stress state.

2. STRESS POINT ALGORITHM

The problem to be solved numerically consists in the integration of the CLoE rate equation (1) for any assigned finite strain increment $\Delta\varepsilon_{n+1}$ in the time interval $[t_n; t_{n+1}]$, given the stress state σ_n of the material at time t_n. Following an original proposal by Charlier (1987), in this work the following class of implicit one-step algorithms is considered:

$$\begin{aligned}
&\sigma_{n+1} = \sigma_n + \Delta\sigma_{n+1} \\
&\Delta\sigma_{n+1} \cong \boldsymbol{A}_{n+\alpha}\Delta\varepsilon_{n+1} + \boldsymbol{b}_{n+\alpha}\left\|\Delta\varepsilon_{n+1}\right\| \\
&\boldsymbol{A}_{n+\alpha} := \boldsymbol{A}(\sigma_{n+\alpha}) \qquad \boldsymbol{b}_{n+\alpha} := \boldsymbol{b}(\sigma_{n+\alpha}) \\
&\sigma_{n+\alpha} := (1-\alpha)\sigma_n + \alpha\sigma_{n+1} \qquad 0 < \alpha \le 1
\end{aligned} \tag{2}$$

which is generally referred to as Generalized Trapezoidal Method (GTM). In the following two particular choices are considered, $\alpha = 1/2$ (Crank-Nicolson, CN) and $\alpha = 1$ (backward Euler, BE). For any given α, the evaluation of the stress increment $\Delta\sigma_{n+1}$ requires an iterative procedure to determine the generalized midpoint stress $\sigma_{n+\alpha}$.

In order to improve the algorithm accuracy, a substepping scheme is adopted in which the given strain increment is subdivided in n_s substeps. For each substep, the strain increment is computed as:

$$\Delta\Delta\varepsilon_{n+1} = \frac{1}{n_s}\Delta\varepsilon_{n+1} \tag{3}$$

A full description of the stress-point algorithm is given in Chambon *et al.* (1995), and summarized in Fig. 1. First, the stress is initialized at the beginning of first substep (k = 1). Then, the generalized midpoint stress $\sigma_{k+\alpha}^{(j)}$ and the stress increment $\Delta\sigma_k^{(j)}$ are computed for substep k at the iteration j. The iteration process is repeated until a relative norm of the residual $\boldsymbol{r}^{(j)}$ is less then a prescribed tolerance *TOL*, or j exceeds the prescribed maximum number of iterations *JMAX*.

1.	$k = 1$
2.	$\sigma_k\big\|_{k=1} = \sigma_n \quad \Delta\sigma_0 = \boldsymbol{0}$
3.	$j = 1$
4.	$\sigma_{k+\alpha}^{(j)} = \begin{cases} \sigma_k + \alpha\Delta\sigma_{k-1} & \text{if } j = 1 \\ \sigma_k + \alpha\Delta\sigma_k^{(j-1)} & \text{otherwise} \end{cases}$
5.	$\Delta\sigma_k^{(j)} = \boldsymbol{A}_{k+\alpha}^{(j)}\Delta\Delta\varepsilon_{n+1} + \boldsymbol{b}_{k+\alpha}^{(j)}\left\|\Delta\Delta\varepsilon_{n+1}\right\|$
6.	$\boldsymbol{r}^{(j)} := \begin{cases} \Delta\sigma_k^{(j)} - \Delta\sigma_{k-1} & \text{if } j = 1 \\ \Delta\sigma_k^{(j)} - \Delta\sigma_k^{(j-1)} & \text{otherwise} \end{cases}$
7.	$\left(\left\|\boldsymbol{r}^{(j)}\right\| / \left\|\Delta\sigma_k^{(j)}\right\| < TOL\right).OR.\left(j > JMAX\right)$? yes $\Rightarrow \begin{cases} \Delta\sigma_k = \Delta\sigma_k^{(j)} \\ \text{GO TO 8} \end{cases}$ no $\Rightarrow \begin{cases} j = j+1 \\ \text{GO TO 4} \end{cases}$
8.	$\sigma_{k+1} = \sigma_k + \Delta\sigma_k$
9.	$k = n_s$? yes $\Rightarrow \begin{cases} \sigma_{n+1} = \sigma_{k+1} \\ \text{EXIT} \end{cases}$ no $\Rightarrow \begin{cases} k = k+1 \\ \text{GO TO 4} \end{cases}$

Figure 1. One-step implicit stress point algorithm.

3. PERFORMANCE OF THE ALGORITHM

An extensive program of numerical tests has been performed in order to assess the performance of this class of algorithms for different strain paths and initial conditions, investigating the effect of substep size, strain increment direction, and initial stress state. While only a selection of the analyses performed is described in the following, a complete account of the results is the subject of a forthcoming paper.

Hereafter, the stress state will be described in terms of either its components σ_{ij} or the following invariant quantities:

$$p := \frac{1}{3}\operatorname{tr}(\sigma) \quad q := \sqrt{\frac{3}{2}\boldsymbol{s}\cdot\boldsymbol{s}} \quad \boldsymbol{s} = \sigma - p\boldsymbol{1} \tag{4}$$

As far as the material parameters are concerned, a typical set for dense Hostun sand is adopted in all the analyses.

To assess the effect of the assumed tolerance *TOL* in the local iteration, several preliminary tests have been performed, including oedometric (1-d) and constant volume axisymmetric compression. While the number of local iterations required to achieve convergence increased with decreasing tolerance, the assumed value for *TOL* had almost no effect on the computed value of σ_{n+1} within the investigated range ($TOL = 10^{-4} \div 10^{-1}$). Strictly, this result applies for the substep magnitudes considered (on the order of 10^{-4}), while larger differences are likely to appear for larger substeps. However, for the given material parameters and initial state, the investigated range of strain increments can be considered representative of the average strain increments occurring over a load step in the FE solution of typical BVP's. All the results described in the following have been obtained assuming $TOL = 10^{-4}$.

3.1 Effect of substep size in isotropic compression

To illustrate the effect of the substep size, the case of isotropic compression is first considered, due to the strong nonlinear character of the response of the

model along this particular loading path. Since isotropic compression from an isotropic initial state is one of the paths for which a closed form solution is available (Chambon *et al.* 1994), the accuracy of numerical solutions can be directly assessed.
Starting from an initial state (p, q) = (100.0, 0.0) kPa, a total volumetric strain increment $\Delta\varepsilon_v = 0.009$ is applied in 10 steps with $n_s = 1$ and $n_s = 100$. When a sufficiently large number of substeps is used, both BE and CN algorithms give a quite accurate solution, as shown by the compression curves in the plane p:ε_v, plotted in Fig. 2.
As the number of substeps decreases, the Crank-Nicolson scheme performs much better, due to its higher order of accuracy. In particular, for $n_s = 1$ only the solution obtained for $\alpha = 1/2$ remains sufficiently close to the exact solution in the strain increment range considered. This is a combined consequence of the linear dependence of the stiffness on the stress level and the large variations in mean effective stress experienced in isotropic compression paths. The BE algorithm results in a stiffer response than CN since it evaluates the stiffness of the material at the final state of stress, rather than at mid-point stress. Similar results have been obtained in 1-d compression tests, while smaller differences between the two schemes have been observed for other strain paths (e.g., constant volume axisymmetric compression).

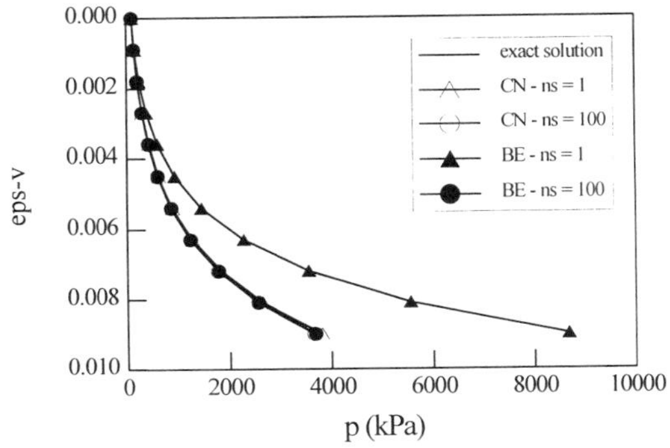

Figure 2. Isotropic compression test results.

3.2 *Stress response envelopes*

Due to the incrementally non linear character of the constitutive equation, it is of interest to evaluate the performance of the algorithms as the loading direction $\Delta\varepsilon/\|\Delta\varepsilon\|$ is varied.
A series of tests has been performed in which the strain increment magnitude $\|\Delta\varepsilon\|$ is held constant while its direction θ is continuously varied between 0 and 2π in the strain increment space.
The following two cases have been considered:

- axisymmetric loading ($\Delta\varepsilon_{11} = \Delta\varepsilon_{33}$):

$$\sqrt{2}\Delta\varepsilon_{11} = \|\Delta\varepsilon\|\cos\theta \quad \Delta\varepsilon_{22} = \|\Delta\varepsilon\|\sin\theta \qquad (5)$$

- simple shear ($\Delta\varepsilon_{11} = \Delta\varepsilon_{33} = 0$):

$$\sqrt{2}\Delta\varepsilon_{12} = \|\Delta\varepsilon\|\cos\theta \quad \Delta\varepsilon_{22} = \|\Delta\varepsilon\|\sin\theta \qquad (6)$$

Test results can be conveniently represented by means of response envelopes in the stress increment space (SRE, Gudehus 1979). It is worth noting that since the applied strain increments are finite, the obtained response envelopes are *algorithmic* rather than constitutive (i.e. obtained for infinitesimal strain increments).
Figure 3 shows the SRE's obtained in axisymmetric loading, starting from two different stress states:

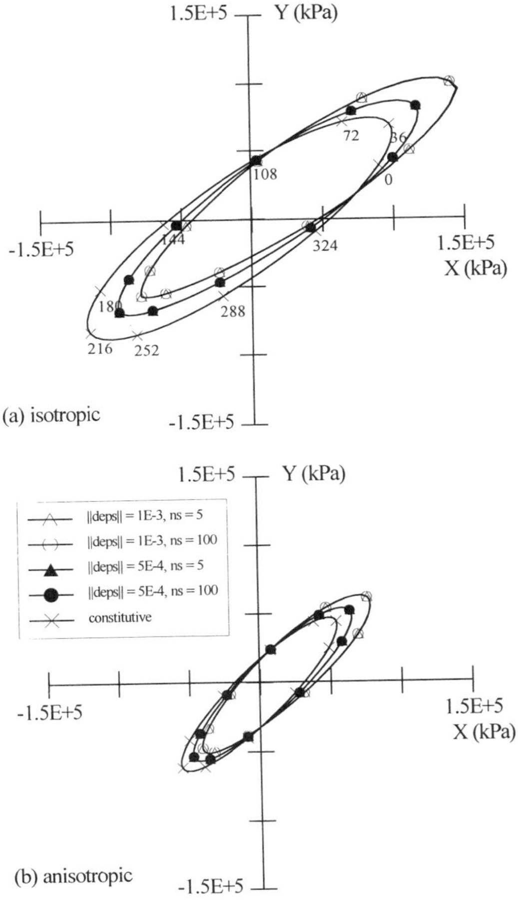

Figure 3. Stress response envelopes for axisymmetric compression.

isotropic with p = 100.0 kPa (Fig. 3a) or anisotropic (K_o) with p = 100.0 kPa, q = 81.27 kPa (Fig. 3b). The Crank-Nicolson integration scheme has been adopted, applying two different strain increments, with $\|\Delta\varepsilon\| = 5 \cdot 10^{-4}$ and $1 \cdot 10^{-3}$, in 5 and 100 substeps, respectively.

To better compare the results obtained with different strain increment magnitudes, the envelopes are plotted in terms of the two following quantities:

$$X = \sqrt{2}\,\Delta\sigma_{11}/\|\Delta\varepsilon\| \qquad Y = \Delta\sigma_{22}/\|\Delta\varepsilon\| \tag{7}$$

The constitutive SRE's, as obtained for infinitesimal strain increments ($\|\Delta\varepsilon\| \to 0$), are also plotted in the figure for comparison. Labels on data points in Fig. 3a give the direction of the strain increment vector, expressed by the angle θ (in degrees). From a physical point of view, the particular choice of the response space is such that the radius connecting the origin with each point on the SRE can be considered as a directional secant stiffness.

All envelopes in Fig. 3a have a pseudo-elliptical shape, with the major axis aligned with the space diagonal, indicating a stiffer material response in isotropic loading (θ ≅ 35° and 215°) than under any other direction.

The envelopes computed for 5 and 100 substeps are almost coincident, indicating that, in the range considered for the strain increment magnitude, the solution is sufficiently accurate for all the directions in the incremental loading space, even when obtained with a relatively small number of substeps.

The algorithmic nature of the SRE's is apparent from their progressive shift along the space diagonal as $\|\Delta\varepsilon\|$ is increased. The computed response along a given (finite) strain increment is affected by the variation of material stiffness with the stress state along the integration path. The increase in mean effective stress associated to loading directions contained in the upper right quadrant causes an increase of directional stiffness with increasing strain increment size. The opposite is true for loading directions associated with a decrease in p. Note that this effect is essentially independent on the number of substeps, as indicated by the practical coincidence of the SRE's computed with n_s = 5 and 100.

For an initially anisotropic stress state (Fig. 3b) similar considerations can be made. The envelopes are still pseudo-elliptical and tend to shift along the space diagonal with increasing $\|\Delta\varepsilon\|$. However, the SRE's are significantly smaller compared to the previous case (Fig. 3a), as a result of the detrimental effect of the deviatoric component of the initial stress on the stiffness of the material. Also, the major axes of the ellipses are not parallel to the space diagonal but form a positive angle with it in counterclockwise direction. Finally, the envelopes appear shifted along their minor axis with respect to the origin of the stress increment space. In fact, a large difference in response is observed between purely deviatoric strain paths in loading (θ ≅ 125°) and unloading (θ ≅ 305°) as an effect of the incremental non-linearity of the model.

As for the effect of the number of substeps, similar conclusions as in the case of initial isotropic stress state can be drawn.

Figure 4 shows the SRE's obtained in simple shear, starting from the same two stress states as for the axisymmetric loading. Again, the values of $5 \cdot 10^{-4}$ and $1 \cdot 10^{-3}$ have been considered for the strain increment size and the Crank-Nicolson integration scheme has been adopted, using 5 and 100 substeps. In this case, each of the strain increment vectors re-

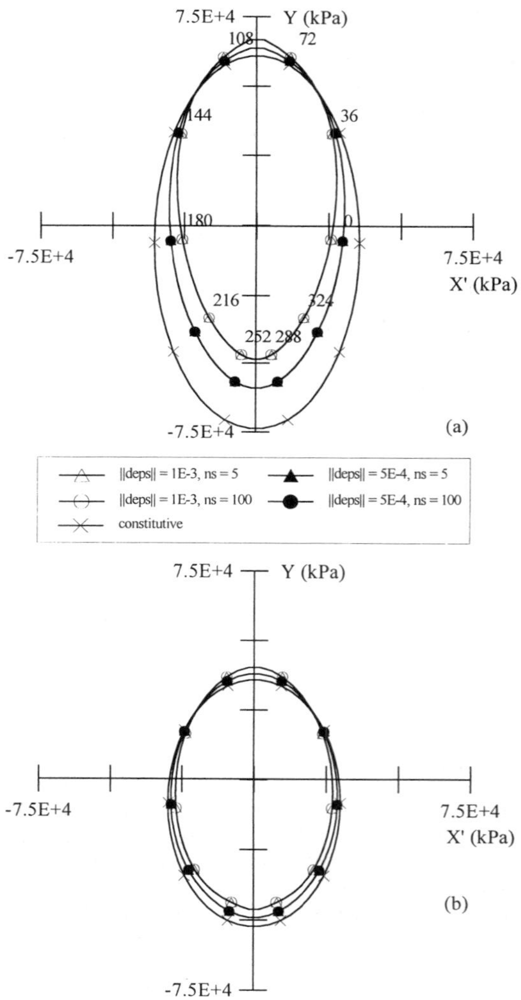

Figure 4. Stress response envelopes for simple shear.

sults in a variation of all the non-zero components of the stress tensor. Therefore, the SRE's are surfaces in a four dimensional space, and several 2-d representations of them can be obtained by projection on different planes. Here, the results are represented in terms of the two following quantities:

$$X' = \sqrt{2}\Delta\sigma_{12}/\|\Delta\varepsilon\| \quad Y = \Delta\sigma_{22}/\|\Delta\varepsilon\| \tag{8}$$

The envelopes in Figs. 4a and 4b refer to initial isotropic and anisotropic states, respectively. In both cases, constitutive SRE's are also plotted for comparison.

As in axisymmetric compression, the envelopes computed for 5 and 100 substeps are almost coincident, that is, the solution is sufficiently accurate even when obtained with a relatively small number of substeps. The response envelopes have an elliptical shape with the major axis corresponding to the Y axis. This indicates that stiffness is higher in 1-d compression than under shear. The symmetry of the envelopes about the vertical axis indicates the equivalence between the responses to positive and negative shear strain increments. This is a consequence of the stress state being the only state variable in CLoE. As observed for the axisymmetric tests, the SRE's referring to an initial isotropic state are larger than those referring to initial anisotropic conditions. Again, an upward translation and a distortion of the envelopes are apparent as the strain increment magnitude $\|\Delta\varepsilon\|$ increases. This effect, due to the variation in mean effective stress in 1-d loading, is more pronounced when starting from an isotropic stress state.

3.3 *Isoerror maps*

The above results provide a complete picture of the performance of the Crank-Nicolson algorithm as the strain increment direction is varied. In this section the investigation is extended to a full range of substep sizes, considering both CN and BE schemes.

A number of axisymmetric loading paths have been applied in a single substep, with magnitude $\delta = \|\Delta\Delta\varepsilon\|$ in the range $[1.0\cdot10^{-4}, 8.5\cdot10^{-4}]$ and direction θ in the range $[0, 2\pi]$. At each point (δ, θ), the accuracy of the numerical solutions is assessed by means of the following relative measure of the integration error:

$$E = \|\sigma - \hat{\sigma}\|/\|\hat{\sigma}\| \tag{9}$$

where σ is the integrated stress, and $\hat{\sigma}$ represents the corresponding "exact" solution, computed numerically with $n_s = 1000$.

In Figure 5 various isoerror maps, representing the contour lines of E in the δ:θ plane, are reported.

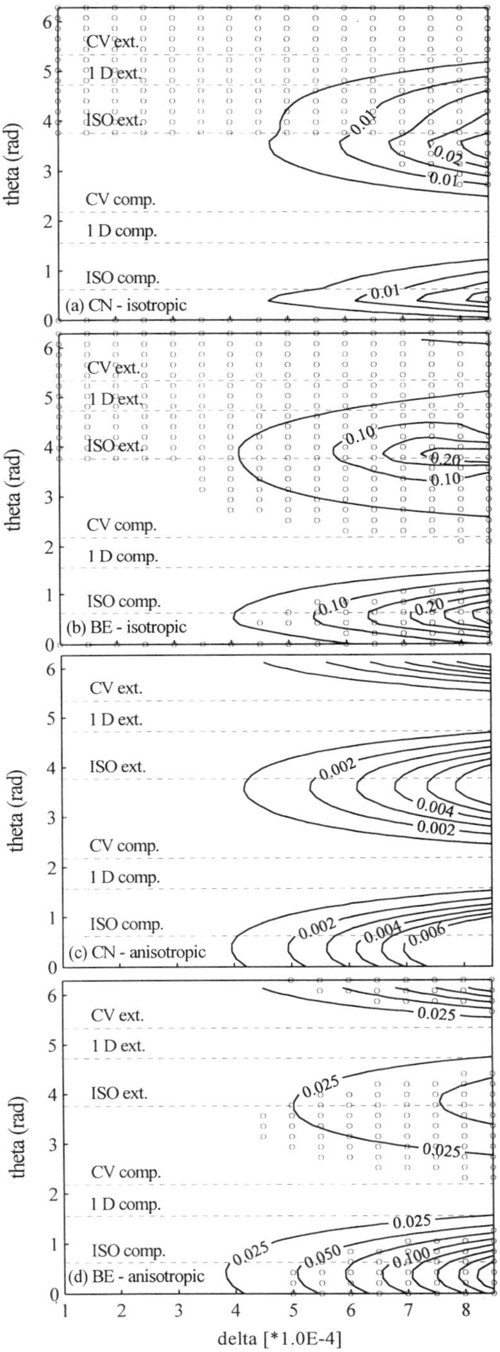

Figure 5. Isoerror maps.

Results have been obtained with $\alpha = 1/2$ (Figs. 5a and 5c) and $\alpha = 1$ (Figs. 5b and 5d). The initial stress state considered is either isotropic with p = 100 kPa (Figs. 5a and 5b) or anisotropic with p = 100.0 kPa, q = 81.27 kPa (Figs. 5c and 5d). Six particular values of θ are noted on the figures, corresponding to one-dimensional, isotropic, and constant volume (CV) compression and extension paths.

The results indicate that Crank-Nicolson scheme is quite accurate in all the investigated range, with a maximum error of approximately 2.5% in a small portion of the δ:θ plane in Fig. 5a. The integration error given by the backward Euler scheme is approximately one order of magnitude larger, with a maximum value of about 30%. As expected, the accuracy of the solution decreases with δ. However, results also show a major effect of the strain increment direction. The largest errors are obtained along isotropic loading paths. For a given d, the error decreases with decreasing volumetric component of the strain increment. Indeed, the solution is always more accurate along constant volume paths. These observations can be explained with the larger variations in mean effective stress associated to the former loading directions. Comparison of Figs. 5a and 5c, and Figs. 5b and 5d, shows that, for any given loading path, the accuracy of the solution is generally higher when starting from an anisotropic state.

The relative efficiency of the two algorithms can be assessed by the number of cases where the iterative procedure failed to converge in the prescribed maximum number of iterations. These cases are indicated in the figures with an open circle on the relevant (δ, θ) point. The results show that, while the lack of convergence does not affect very much the accuracy of the solution in all cases considered, Crank-Nicolson scheme is always more efficient than backward Euler. Both algorithms appear more efficient when starting from an anisotropic state, and in compression rather than extension paths.

4. CONCLUSIONS

The above results provide a complete picture of the performance of the GTM algorithm as the substep size and loading direction are varied. In fact, the results reported in sections 3.1 and 3.2 can be regarded as special cases of the more complete investigation discussed in section 3.3. The isotropic compression tests in Fig. 2 investigate a θ = const. section of the relevant isoerror maps (Figs. 5a and 5b), while the stress response envelopes in Fig. 3 consider two δ = const. sections of the δ:θ plane.

Results indicate that, when the size of the substep is sufficiently small (say, less than $2.0 \cdot 10^{-4}$ for the material parameters and initial conditions examined) both schemes are sufficiently accurate for all the directions in the incremental loading space. However, when larger values of δ are considered, only Crank-Nicolson gives satisfactory results. Solutions obtained with the backward Euler scheme can be affected by large errors for strain increments with a large isotropic component. This is a combined consequence of the linear dependence of the stiffness on the stress level and the large variations in mean effective stress experienced in isotropic and 1-d compression paths. The results also indicate a higher efficiency of the Crank-Nicolson over the backward Euler scheme.

The effect of the initial stress state on the numerical solution has been investigated by considering two different conditions, isotropic and anisotropic. Overall, the results suggest that both accuracy and efficiency are increased in the latter case.

5. REFERENCES

Chambon R. (1989) - Une classe de lois de comportement incrémentalement nonlinéaires pour les sols non visqueux, résolution de quelques problèmes de cohérence. *C. R. Acad. Sci. Paris*, **308**(II):1571-1576.

Chambon R., Desrues J., Hammad W., Charlier R. (1994) - CLoE, a new rate-type constitutive model for geomaterials. Theoretical basis and implementation. *Int. Journ. Num. Anal. Methods in Geomechanics*, **18**:253-278.

Chambon R., Desrues J., Hammad W., Charlier R. (1995) - CLoE. Consistance et localization explicite: une loi incrémentale non linéaire. *Internal report*, Laboratoire 3S, UJF Grenoble.

Charlier R. (1987) - Approche unifiée de quelques problèmes non linéaires de mécanique des milieux continus par la méthode des éléments finis. *Thèse de doctorat*, Université de Liège, 1987.

Gudehus G. (1979) - A comparison of some constitutive laws for soils under radially symmetric loading and unloading. *Proc. 3rd Int. Conf. Num. Meth. in Geomech., Aachen, W. Wittke ed.*, Balkema, **4**:1309-1324.

Gudehus G. (1996) - A comprehensive constitutive equation for granular materials. *Soils and Foundations*. **36**:1-12.

Kolymbas D. (1991) - An outline of hypoplasticity. *Arch. Appl. Mech.*, **61**:143-151.

Kolymbas D., Herle I., v. Wolffersdorff P.-A. (1995) - Hypoplastic constitutive equation with backstress. *Int. Journ. Num. Anal. Methods in Geomechanics*, **19**:415-446.

Niemunis A., Herle I. (1997) - Hypoplastic model for cohesionless soils with elastic strain range. *Mech. Cohesive-Frictional Materials*, in print.

Numerical Models in Geomechanics, Pietruszczak & Pande (eds) © *1997 Balkema, Rotterdam, ISBN 90 5410 886 X*

New method of material modeling using neural networks

J. Ghaboussi & D. E. Sidarta
Department of Civil Engineering, University of Illinois at Urbana-Champaign, Ill., USA

ABSTRACT: In earlier papers Ghaboussi and his co–workers (Ghaboussi, Garrett and Wu, 1990, 91; Wu, 1991; Ghaboussi, 1992a, 92b; Wu and Ghaboussi, 1992, 93; Ghaboussi, Lade and Sidarta 1994) have proposed a neural network based methodology for modeling of complex constitutive behavior of materials, including geomaterials. In this paper, we introduce nested adaptive neural networks, a new type of neural network developed by Ghaboussi and his co–workers, and apply this neural network in modeling of the constitutive behavior of geomaterials. Nested adaptive neural networks take advantage of the nested structure of the material test data, and reflects it in the architecture of the neural network. This new neural network is applied in modeling of the drained and undrained behavior of sand in triaxial tests. Experimental data from a series of triaxial tests (Lee, 1965; Lee and Seed, 1967a, 1967b) were used in this study.

1 INTRODUCTION

A new method of constitutive modeling using neural networks was originally proposed by Ghaboussi, Garrett and Wu (1990, 91). This methodology was then applied to constitutive modeling of concrete (Wu, 1991; Ghaboussi, 1992a, 92b; Wu and Ghaboussi, 1992, 93), sand (Ghaboussi, Lade and Sidarta, 1994) and composite materials (Zhang, 1996 and Ghaboussi, Pecknold, Zhang and HajAli, 1997). Other researchers have also applied neural networks in modeling of constitutive behavior of materials (Penumadu, et. al., 1994; Ellis, et. al., 1995).

Unlike the conventional modeling methods, which use mathematical expressions to approximate the experimentally observed behavior of materials, neural networks offer a fundamentally different approach in modeling of constitutive behavior of materials; they use the learning capabilities of neural networks. Neural networks are massively parallel computational models for knowledge representation and information processing. They have unique learning capabilities which can be used in learning complex nonlinear relationships. The learning capabilities allow neural networks to be directly trained with the results of experiments. They learn the constitutive material behavior present in the experimental data and store that knowledge in their connection weights. If the training data contains the relevant information on material behavior, then the trained neural network can generalize from its training data to novel cases. Such a trained neural network can be used similar to any other material model in finite element analysis of boundary value problems.

In this paper we present a new neural network to take advantage of the nested structure of the material data. The new nested adaptive neural networks were originally developed to deal with the path dependency of the material behavior. However, it is also useful in dealing with other types of nested structure present in the material behavior data, and indeed, it can take advantage of the nested structure in many other types of data other than material data.

This new neural network modeling methodology is applied to a series of drained and undrained triaxial tests on Sacramento River sand (Lee, 1965). The nested neural network approach is used in modeling the test results. The same test results were also used in a companion paper in this proceedings (Sidarta and Ghaboussi, 1997).

2 REVIEW OF NEURAL NETWORKS IN MATERIAL MODELING

Multi–layered feed–forward neural networks are most suitable for development of neural network material models. In these neural networks, the artificial neurons (also referred to as processing units or nodes) are arranged in layers, with all the neurons in each layer having connections to all the neurons in the next layer. The operation of the neural network consists of a forward pass through the neural network. Signals are received at the input layer, pass through the hidden layers and reach the output layer, producing the output of the neural network. In neural network material modeling, similar to other engineering applications of neural networks, three important issues must be addressed. First, the composition of the input and output layers must be determined. This is called the representation problem. As will be seen in the next section, this is the most important step, because the composition of the input and output layer determines how the material modeling problem is represented in the neural network. Second, it must be determined how to train the neural network material model; there are important issues related to the composition of the test data which is used for training. Finally, the neural network architecture, primarily the configuration of the hidden layers, must be addressed; a method for adaptive determination of the network architecture during the training will be described.

A new notation is introduced to facilitate the discussion of neural networks in material modeling and their incorporation in computational mechanics. The notation is general enough that it can also be used in other engineering applications of neural networks. The general form of the notation is:

$$\boldsymbol{\sigma} = \boldsymbol{\sigma}\,\mathbf{NN}\,(\{\text{input parameters}\} : \{\text{NN architecture}\}) \qquad (1)$$

The symbol **NN** is introduced to denote the output of a multi–layer feed forward neural network, and the notation indicates that the stress vector $\boldsymbol{\sigma}$ is the output of the neural network. For example, the following equation describes a specific neural network material model for a two dimensional state of stress.

$$\boldsymbol{\sigma} = \boldsymbol{\sigma}\,\mathbf{NN}\,(\boldsymbol{\epsilon} : 3 \mid 2\text{–}6 \mid 2\text{–}6 \mid 3\,) \qquad (2)$$

The first argument field indicates that the strain vector is the input. The second argument field describes the neural network architecture, i.e. the number of processing units in each layer and the adaptive training history. Eq. (2) indicates that the neural network has four layers: an input layer with three processing units; two hidden layers; and an output layer with three processing units. The three processing units in the input layer represent two normal strains and one shear strain; the three processing units in the output layer represent the corresponding stresses. Each of the two hidden layers has two processing units at the start of the training process, which grows six units at the completion of training (denoted by 2–6). An adaptive method which allows the automatic addition of processing units in the hidden layers during the training process has been used; this method is described in Joghataie, Ghaboussi, and Wu (1995).

The knowledge acquired by a neural network during its training is stored in its connection weights, which are the numerical values assigned to the connections between processing units. The initial connection weights are selected randomly. A *learning rule* is used to iteratively modify the connection weights during the training process. A number of methods are available for modification of the connection weights (Hertz et. al., 1991). The training method used in this study is the modified version of the Delta Bar method (Jacobs, 1988) and its variant proposed by Ochiai and Usui (1993). In this study a sigmoidal activation function which has the following form was used.

$$f(x) = \frac{2}{1 + e^{-\gamma x}} - 1 \qquad (3)$$

It will become clear later when introducing the nested neural network that an activation function is required to map a zero net input into a zero activation value, and this sigmoidal activation function satisfies this requirement.

Feed–forward neural networks consist of a number of layers of artificial neurons or processing units. Two hidden layers are sufficient in most applications and, if the problem can be modeled with one hidden layer, that is generally preferred. The number of processing units in the input and the output layers is dependent upon how the problem is formulated for neural network representation. The number of processing units in the hidden layers determines the capacity of a neural network, which in turn is related to the complexity of the underlying knowledge base in the training data. However, the degree of complexity of the problem cannot easily be quantified, and its relation to the size of the neural network is not very well understood at the present. Trial and error is one method of architecture determination in current use. Adaptive determination of network architecture has played an important role in previous neural network material modeling studies (Ghaboussi, Garrett and Wu, 1990, 1991; Wu, 1991; Wu and Ghaboussi, 1992, 1993, 1995; Ghaboussi, Lade and Sidarta, 1994). The method for adaptive evolution of network architecture was first developed by Wu (1991) and has been further developed and refined (Joghataie, 1994; Joghataie, Ghaboussi and Wu, 1995).

3 NESTED STRUCTURE OF MATERIAL DATA

The new neural network based modeling approach proposed in this paper is based on a fundamental observation on the nature of material behavior data. Of course, this fundamental observation is general and it is applicable to data in many other fields. In this paper we are concentrating on the material behavior data, the type of data which can be generated from material tests and is used in training of neural network material models. The fundamental observation is that most data have an inherent structure, and that this structure can be exploited in developing special neural networks. One type of inherent internal structure in data is the nested structure.

In the case material data there are several types of nested structure. One type of nested structure in the material data concerns the dimensionality. Consider a general constitutive relation expressed by the following equation.

$$\dot{\sigma}_j = f_j(\sigma_j, \varepsilon_j, \dot{\varepsilon}_j) \; ; \; j = 1, 3, 4, 6 \tag{4}$$

In this equation j denotes the number of terms in the stress and strain vector. The values of j refer to the number stress and strain terms in 1–D, 2–D, axisymmetric and 3–D problems.

The constitutive functions $\mathbf{f}_j$ can be considered to be members of function spaces $\mathbf{F}_j$, which are nested.

$$f_j \in \boldsymbol{F}_j \; ; \; \boldsymbol{F}_1 \subset \boldsymbol{F}_3 \subset \boldsymbol{F}_4 \subset \boldsymbol{F}_6 \tag{5}$$

This equation states that one dimensional constitutive behavior is a subset of the constitutive behavior in two–dimensional plane strain problems, which in turn is a subset of the constitutive behavior in axisymmetric problems, which in turn is a subset of the constitutive behavior in three–dimensional state of stress. Eq. 5 clearly shows the nested structure of the material data of various dimensionalities.

In absence of shear stresses, as in standard triaxial tests and true triaxial tests, we also have a nested structure. The material behavior in standard triaxial tests is two–dimensional, represented by the stress vector $\boldsymbol{\sigma} = \{\sigma_1, \sigma_2\} = \{\sigma_{11}, \sigma_{33} = \sigma_{22}\}$. The material behavior in true triaxial tests is three–dimensional, represented by the stress vector $\boldsymbol{\sigma} = \{\sigma_1, \sigma_2, \sigma_3\} = \{\sigma_{11}, \sigma_{22}, \sigma_{33}\}$. The constitutive function $\bar{\mathbf{f}}$ has been used to denote these states. The following equations show the nested structure of these constitutive models.

$$\dot{\sigma}_j = \bar{f}_j(\sigma_j, \varepsilon_j, \dot{\varepsilon}_j) \; ; \; j = 1, 2, 3 \tag{6}$$

$$\bar{f}_j \in \overline{\boldsymbol{F}}_j \; ; \; \boldsymbol{F}_1 \subset \overline{\boldsymbol{F}}_2 \subset \overline{\boldsymbol{F}}_3 \subset \boldsymbol{F}_6 \tag{7}$$

A neural network which is trained with a higher dimensional material test data should also be able to represent the lower dimensional material data. The manner in which this nested structure of the data can be represented in the structure of the neural network will be described in the next section.

Another form of nested structure in the material data arises from the inherent path dependency of the material behavior. The stress–strain relations shown in Eqs. 4 and 6 express the stress rate in terms of strain rate, using the current state of stresses and strains. Clearly, such relations can only be approximate. In the past, Ghaboussi and his co–workers (Ghaboussi, Garrett and Wu, 1990, 91; Wu, 1991; Wu and Ghaboussi, 1992, 93; Ghaboussi, Lade and Sidarta, 1994) have included states of stresses and strains at several points along the stress path in the immediate past history of loading. These have been referred to as the history points. In previous studies it was found that the uniaxial cyclic behavior of materials can be represented by neural networks if three history points were included. In general, it can be stated that as more history points are included the material behavior can be represented more accurately. Moreover, the material behavior represented with a given number of history points is always a subset of the material behavior represented with a larger number of history points. These relationships can be expressed through the following equations.

$$\dot{\sigma}_j = f_{jk}(\sigma_{j,0}, \varepsilon_{j,0}, \sigma_{j,-1}, \varepsilon_{j,-1}, \ldots, \sigma_{j,-k}, \varepsilon_{j,-k}, \dot{\varepsilon}_j)$$
$$j = 1, 3, 4, 6 \; ; \; k = 1, \ldots . \tag{8}$$

$$\dot{\sigma}_j = \bar{f}_{jk}(\sigma_{j,0}, \varepsilon_{j,0}, \sigma_{j,-1}, \varepsilon_{j,-1}, \ldots, \sigma_{j,-k}, \varepsilon_{j,-k}, \dot{\varepsilon}_j)$$
$$j = 1, 2, 3 \; ; \; k = 1, \ldots . \tag{9}$$

$$f_{j,k} \in \boldsymbol{F}_{j,k} \; ; \; \bar{f}_{j,k} \in \overline{\boldsymbol{F}}_{j,k} \tag{10}$$

$$\boldsymbol{F}_{j,k} \subset \boldsymbol{F}_{j,k+1} \; ; \; \overline{\boldsymbol{F}}_{j,k} \subset \overline{\boldsymbol{F}}_{j,k+1} \tag{11}$$

A third form of nested structure in the material data is related to the drained or undrained condition. The underlying material behavior in the undrained condition is the same as the material behavior in drained condition, with the additional condition of no drainage superimposed on it. Therefore, it can be stated that the drained behavior of sand is a subset of the undrained material behavior. This type of nested data structure is superimposed on the nested structure resulting from the history dependence in the following equations.

$$\{\dot{\sigma}_j, \dot{u}\} = f^u_{jk}(\sigma_{j,0}, \varepsilon_{j,0}, u_{j,0}, \sigma_{j,-1}, \varepsilon_{j,-1}, u_{j,-1}, \ldots, \sigma_{j,-k}, \varepsilon_{j,-k}, u_{j,-k}, \dot{\varepsilon}_j)$$
$$j = 1, 3, 4, 6 \; ; \; k = 1, \ldots . \tag{12}$$

$$\{\dot{\sigma}_j, \dot{u}\} = \bar{f}^u_{jk}(\sigma_{j,0}, \varepsilon_{j,0}, u_{j,0}, \sigma_{j,-1}, \varepsilon_{j,-1}, u_{j,-1}, \ldots, \sigma_{j,-k}, \varepsilon_{j,-k}, u_{j,-k}, \dot{\varepsilon}_j)$$
$$j = 1, 2, 3 \; ; \; k = 1, \ldots . \tag{13}$$

$$f^u_{j,k} \in \boldsymbol{F}^u_{j,k} \; ; \; \bar{f}^u_{j,k} \in \overline{\boldsymbol{F}}^u_{j,k} \tag{14}$$

$$\boldsymbol{F}_{j,k} \subset \boldsymbol{F}^u_{j,k} \; ; \; \overline{\boldsymbol{F}}_{j,k} \subset \overline{\boldsymbol{F}}^u_{j,k} \tag{15}$$

In this section we have described the three types of nested structure inherent in the data on the constitutive behavior of materials. This nested structure does not play an important role in the conventional methods of mathematical modeling of constitutive behavior and, as such, it has not attracted much attention. However, as will be seen later in this paper and in the companion paper in this proceedings (Sidarta and Ghaboussi, 1997), this nested structure can play an important role in development of comprehensive three dimensional material models using neural networks. In this paper we have proposed new neural networks which reflect the nested structure of material data.

4 NESTED ADAPTIVE NEURAL NETWORKS

Taking advantage of the nested structure of the material data described in the previous section allows a stepwise method of building and training a neural network to represent the complex behavior of materials. Nested adaptive neural networks, introduced in the paper, takes advantage of the nested structure of the data.

A nested neural network consists of several modules. The starting point of building a nested adaptive neural network is to develop and train a *base module* to represent the material behavior in the lowest function space in the data structure. The base module is a standard multi–layer feedforward neural network. However, it may be trained adaptively so that the number of nodes in the hidden layers are determined automatically during the training. The base module is successively augmented by attaching *added modules* to form higher level NANNs. A first level NANN is composed of the base module and a first level added module; the second level module is composed of a first level module and a second level added module, and so on. The process is theoretically open ended and more and more modules can be added. The added modules themselves are also standard multi–layer feedforward neural networks. In attaching a new added module to a lower level NANN only one way connections are used; all the nodes in each layer of the new added module are connected to all the nodes in the next layer of the lower level NANN. There are no connections from the lower level NANN to the new added module. The reason for this pattern of connections will be explained later.

We have chosen to illustrate the nested adaptive neural networks on the simplest one–dimensional material model in which the added modules represent history points. The base module is given by the following equation.

$$\Delta\sigma_j = \Delta\sigma_j \mathbf{NN}_1 \; (\Delta\epsilon_j, \; \sigma_j, \; \varepsilon_j \; : \;) \tag{16}$$

Higher level NANNs are obtained by successively adding history point module. The general form of the NANN with n modules is as follows.

$$\Delta\sigma_j = \Delta\sigma_j \mathbf{NN}_{n+1} \; (\; [\; \Delta\epsilon_j, \; \sigma_j, \; \varepsilon_j \;], \; [\; \sigma_{j-1}, \; \varepsilon_{j-1} \;], \ldots, \; [\; \sigma_{j-n}, \; \varepsilon_{j-n} \;] : \; | \; | \; | \;) \tag{17}$$

The reason for the one way connections used in NANNs becomes clear if we consider that a new added module should not have any effect on the function space represented by the lower level module. In the case the history point modules, more recent past should not influence the effect of the more distant past on the current material behavior. A consequence of this observation is that the lower level NANNs should be retrievable from the higher level NANNs. This can be formalized in the following equation.

$$\mathbf{NN}_n \; (\; [\; \Delta\epsilon_j, \; \sigma_j, \; \varepsilon_j \;], \; [\; \sigma_{j-1}, \; \varepsilon_{j-1} \;], \ldots, \; [\; \sigma_{j-(n-1)}, \; \varepsilon_{j-(n-1)} \;] : \; | \; | \; | \;) = \mathbf{NN}_{n+1} \; (\; [\; \Delta\epsilon_j, \; \sigma_j, \; \varepsilon_j \;], \; [\; \sigma_{j-1}, \; \varepsilon_{j-1} \;], \ldots, \; [\; \sigma_{j-(n-1)}, \; \varepsilon_{j-(n-1)} \;], \; [\; 0, \; 0 \;] : \; | \; | \; | \;) \tag{18}$$

A full two way connection between the added modules and the lower level modules would violate this condition. A series of NANNs represent a sequence of nested function spaces, which can be expressed by the following equation.

$$\mathbf{NN}_n \subset \mathbf{NN}_{n+1} \tag{19}$$

Another necessary condition for satisfying Eqs. 18 and 19 is related to the way the NANNs are trained. Each level NANN is trained up to a satisfactory level of accuracy. After a new added module is attached all the connections of the lower level NANN are frozen and only the weights of the new connections are trained.

This type of nested adaptive neural network has been shown to model considerably well uniaxial plain concrete behavior under cyclic loading and biaxial plain concrete behavior under monotonic loading and unloading (Zhang, 1996).

5 MODELING OF DRAINED AND UNDRAINED BEHAVIOR OF SAND

Nested adaptive neural networks were applied in modeling of the drained (Table 1) and undrained (Table 2) behavior of sands using the results of the triaxial compression tests on Sacramento River sand (Lee, 1965). These tests were divided into two groups. The first group of tests were used in training the neural network, while the second group, identified with star signs, were used to test the performance of the trained neural networks in generalizing to novel cases.

The objective is to use the test data to train a neural network to model the material behavior in both drained and undrained conditions for a range of initial void ratios and initial confining pressures. A base module was developed to represent the drained and undrained behavior, and then the history point modules were added to create the next two NANNs. These three nested adaptive neural networks are give in the following equations

$$\Delta \mathbf{s}_j = \Delta \mathbf{s}_j \mathbf{NN}_{S,1} \; (\; \Delta\boldsymbol{\epsilon}_j, \; \mathbf{s}_j, \; \boldsymbol{\epsilon}_j, \; e_0 : \; 8 \; | \; 2\text{–}4 \; | \; 2\text{–}4 \; | \; 3 \;) \tag{20}$$

$$\Delta \mathbf{s}_j = \Delta \mathbf{s}_j \mathbf{NN}_{S,2} \; (\; [\; \Delta\boldsymbol{\epsilon}_j, \; \mathbf{s}_j, \; \boldsymbol{\epsilon}_j, \; e_0 \;], \; [\; \mathbf{s}_{j-1}, \; \boldsymbol{\epsilon}_{j-1}, \; e_0 \;] : \; 8, \; 6 \; | \; 2\text{–}4, \; 2\text{–}9 \; | \; 2\text{–}4, \; 2\text{–}9 \; | \; 3 \;) \tag{21}$$

$$\Delta \mathbf{s}_j = \Delta \mathbf{s}_j \mathbf{NN}_{S,3} \; (\; [\; \Delta\boldsymbol{\epsilon}_j, \; \mathbf{s}_j, \; \boldsymbol{\epsilon}_j, \; e_0 \;], \; [\; \mathbf{s}_{j-1}, \; \boldsymbol{\epsilon}_{j-1}, \; e_0 \;], \; [\; \mathbf{s}_{j-2}, \; \boldsymbol{\epsilon}_{j-2}, \; e_0 \;] : \; 8, \; 6, \; 6 \; | \; 2\text{–}4, \; 2\text{–}9, \; 2\text{–}14 \; | \; 2\text{–}4, \; 2\text{–}9, \; 2\text{–}14 \; | \; 3 \;) \tag{22}$$

In these equations $\mathbf{s} = \{\boldsymbol{\sigma}, u\}$, $\boldsymbol{\sigma} = \{p', q\}$, u = pore pressure, $\boldsymbol{\varepsilon} = \{\varepsilon_v, \varepsilon_d\}$, e_0 = void ratio, $p' = (\sigma'_1 + 2\sigma'_3)/3$

Table 1 The list of the drained triaxial compression tests on Sacramento River Sand[1]

Initial void ratio e_o	Consolidation pressure p_o (kg/cm^2)
0.642	1.00
0.635	3.00
0.620	6.00
0.613	10.50
0.602	20.00 *
0.620	29.90
0.614	40.10 *
0.722	1.00 *
0.710	3.40
0.710	5.87 *
0.710	8.44
0.710	10.52
0.710	20.00
0.710	29.90
0.710	40.10
0.780	1.00
0.780	3.00
0.780	6.00
0.780	12.65 *
0.780	20.00
0.780	29.90 *
0.780	40.10
0.870	0.94
0.870	2.00
0.870	3.00 *
0.870	4.50
0.870	12.65
0.870	20.00
0.870	29.90

1. Lee, K. L. (1965)
* Experimental data from these tests were treated as novel cases and were used to test the trained NANNs.

Table 2 The list of the undrained triaxial compression tests on Sacramento River Sand[1]

Initial void ratio e_o	Consolidation pressure p_o (kg/cm^2)
0.610	20.20
0.610	29.90 *
0.610	40.10
0.710	3.00
0.710	8.44
0.710	9.00 *
0.710	14.40
0.710	20.00
0.710	29.90
0.710	40.10 *
0.780	7.00
0.780	8.44 *
0.780	15.10
0.780	20.00 *
0.780	29.90
0.780	40.10
0.870	3.00
0.870	3.00 *
0.870	5.00
0.870	6.00 *
0.870	10.90
0.870	12.65 *
0.870	20.00
0.870	29.90
0.870	40.10

1. Lee, K. L. (1965)
* Experimental data from these tests were treated as novel cases and were used to test the trained NANNs.

= mean effective stress, $q = \sigma_1 - \sigma_3$ = deviatoric stress, $\varepsilon_v = \varepsilon_1 + 2\varepsilon_3$ = volumetric strain, $\varepsilon_d = \varepsilon_1 - \varepsilon_3$ = deviatoric strain, σ_1 = axial stress, $\sigma_3 = \sigma_2$ = lateral stress, ε_1 = axial strain, $\varepsilon_3 = \varepsilon_2$ = lateral strain. As was described in previous section, after adding a new history point module the previous connection weights were frozen and only the new connection weights were trained. It is interesting to note that the hidden layers of the second history point added module grew from 2 to 14 nodes during the training. This is an indication that the effect of history points on the material behavior becomes increasingly more difficult for the neural network to learn. The sequence of three NANNs are shown in Figure 1.

Some selected results from the trained NANNs are shown in Figure 2. As can be seen, although the base module is able to model the material behavior reasonably well, addition of one history point module has significantly improved the material model. For the drained behavior the improvement is more pronounced for low consolidation pressures since the behavior for high confining pressure is represented reasonably well in the base module. A possible explanation is that the behavior of sand at low confining pressures exhibits more nonlinearity and is more path dependent than at high pressures. The addition of the second history point module improves the modeling of the undrained behavior of sand.

The results for all the training and testing cases are summarized in Tables 3. The values given in these tables are the error indices, defined by the following equation.

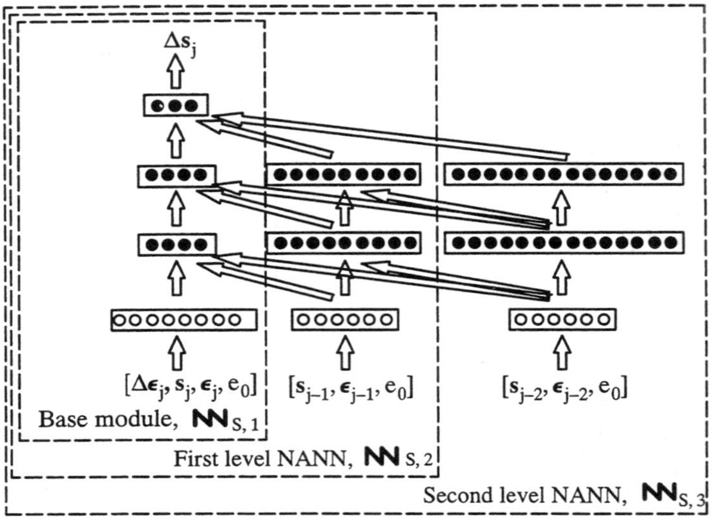

Figure 1. The sequence of nested adaptive neural networks developed for modeling of drained and undrained behavior of sand.

Table 3 The error indices for the NANN material models

	Drained	Undrained	Combined
	Training Data		
$NN_{S,1}$	0.0763	0.0643	0.0710
$NN_{S,2}$	0.0449	0.0430	0.0441
$NN_{S,3}$	0.0416	0.0375	0.0398
	Testing Data		
$NN_{S,1}$	0.0714	0.0720	0.0717
$NN_{S,2}$	0.0518	0.0484	0.0499
$NN_{S,3}$	0.0499	0.0512	0.0506

$$\text{Error Index} = \frac{1}{N} \sum_{j=1}^{j=N} \frac{\| s_j - s_j^m \|}{\| s_j \|} \tag{23}$$

where s_j and s_j^m are neural network and measured values, respectively, for the j–th load step, and N is the number of load steps. For the drained tests the pore pressure is set to zero. This error index is a measure of the accuracy of NANNs in modeling the material behavior.

The error indices are presented separately for the drained and undrained tests. The third column shows the combined error index for all the drained and undrained tests. It can be observed that all error indices are small. This indicates that even the base module is able to learn the material behavior reasonably well.

All the cases show that the addition of one history point module significantly reduces the error index. The addition of the second history point module does not appear to have a significant influence.

6 CONCLUDING REMARKS

A new type of neural network architecture has been introduced. The new nested adaptive neural networks are developed to take advantage of the nested structure of the material data. Several types of nested structures in material data has been identified. The two types of nested structure in material data which were utilized in this study were the nested modularity of the history points resulting from the path dependence of the material behavior and the drained and undrained behavior of sand. The new nested adaptive neural networks can be successively built by adding new modules, which are connected to the lower level NANN through one way connections, to create the next level NANN. This process can be continued as needed. The one way connections assure that at each stage all the lower level NANNs can be retrieved. This methodology has been applied to some experimental results from drained and undrained triaxial tests on Sacramento river sand. The results show that this type of neural network is capable of learning the drained and undrained behavior of sand for a range of initial void ratios and confining pressures.

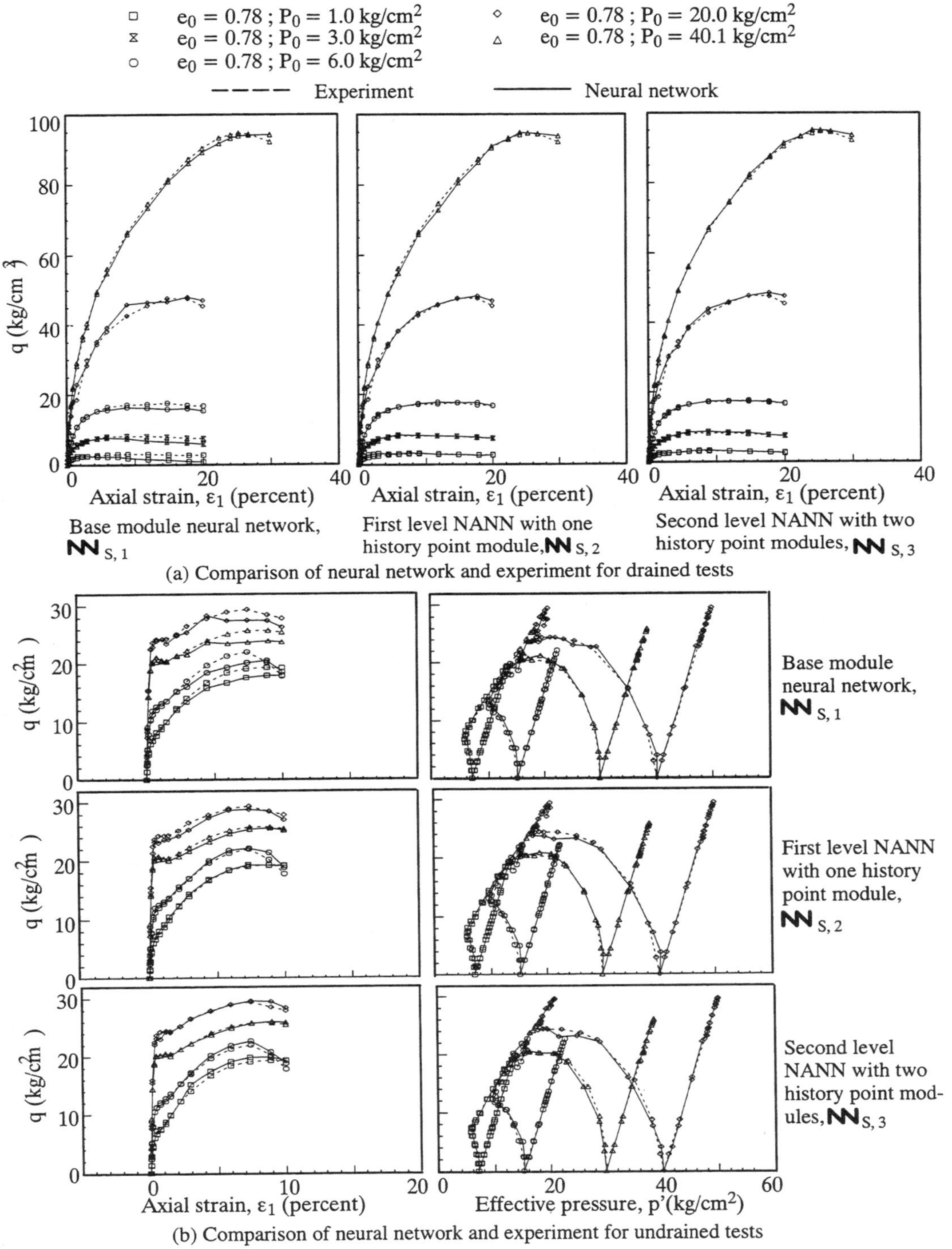

Figure 2. Performance of the trained nested adaptive neural network on (a) drained triaxial tests, and (b) undrained triaxial tests.

7 ACKNOWLEDGEMENTS

The research reported in this paper is part of a research project with analytical and experimental components. Professor Poul Lade of The Johns Hopkins University is developing special new experimental techniques for neural network applications. Both parts of this research project are funded by National Science Foundation under Grant number CMS–95–03462. This support is gratefully acknowledged.

8 REFERENCES

Ellis, G. W., Yao, C., Zhao, R., and Penumadu, D. (1995). "Stress–Strain Modeling of Sands Using Artificial Neural Networks," *Journal of Geotechnical Engineering*, ASCE, 121(5), 429–435.

Ghaboussi, J. (1992a). "Potential Applications of Neuro–Biological Computational Models in Geotechnical Engineering," *Proceedings, Fourth International Symposium on Numerical Models in Geomechanics*, Swansea, U. K.

Ghaboussi, J. (1992b). "Neuro–Biological Computational Models with Learning Capabilities and Their Applications in Geomechanical Modeling," *Proceedings, Workshop on Recent Accomplishments and Future Trends in Geomechanics in the 21st Century*, Norman, Oklahoma.

Ghaboussi, J., Garret, J. H., Jr., and Wu, X. (1990). "Material Modeling with Neural Networks," *Proceedings of the International Conference on Numerical Methods in Engineering : Theory and Applications,* Swansea, U. K., 701–717.

Ghaboussi, J., Garret, J. H., Jr., and Wu, X. (1991). "Knowledge–Based Modeling of Material Behavior with Neural Networks," *Journal of Engineering Mechanics Division*, ASCE, 117(1), 132–153.

Ghaboussi, J., Lade, P. V., and Sidarta, D. E. (1994). "Neural Networks Based Modelling in Geomechanics," *Proceedings of the 8th International Conference on Computer Methods and Advances in Geomechanics*, Morgantown, West Virginia.

Ghaboussi, J., Pecknold, D. A., Zhang, M., and HajAli, R. M. (1997). "Autoprogressive Training of Neural Network Constitutive Models," submitted for publication.

Ghaboussi, J. Zhang, M., Wu, X. and Pecknold, D. A., (1997) "Nested Adaptive Neural Networks: A New Architecture", submitted to the International Conference on Artificial Neural Networks in Engineering, St. Louis, Missouri.

Hertz, J., Krogh, A., and Palmer, R. G. (1991). *Introduction to the Theory of Neural Computations*, Addison–Wesley, Redwood City, California.

Jacobs, R. A. (1988). "Increased Rates of Convergence through Learning Rate Adaptation," *Neural Networks*, 1, 295–307.

Joghataie, A., Ghaboussi, J., and Wu, X. (1995). "Learning and Architecture Determination through Automatic Node Generation," *International Conference on Artificial Neural Networks in Engineering*, ANNIE '95, St. Louis, Missouri.

Lee, K. L. (1965). "Triaxial Compressive Strength of Saturated Sands under Seismic Loading Conditions," Ph.D. thesis, University of California, Berkeley, California.

Lee, K. L. and Seed, H. B. (1967a). "Drained Strength Characteristics of Sands," *Journal of Soil Mech. and Found. Eng. Div.*, ASCE, 93(SM6), 117–141.

Lee, K. L. and Seed, H. B. (1967b). "Undrained Strength Characteristics of Cohesionless Soils," *Journal of Soil Mech. and Found. Eng. Div.*, ASCE, 93(SM6), 333–360.

Penumadu, D., Jin–Nan, L., Chameau, J.–L., and Sandarajah, A. (1994). "Anisotropic Rate Dependent Behavior of Clays Using Neural Networks," *Proc. XIII ICSMFE*, New Delhi, India, vol. 4, 1445–1448.

Ochiai, K. and Usui, S. (1993). "Improved Kick Out Learning Algorithm with Delta–Bar–Delta–Bar Rule," *Proc. of IEEE ICNN '93*, IEEE, San Francisco, California, 28 March – 1 April 1993, 269–274.

Sidarta, D. E. and Ghaboussi, J. (1997) " NN Constitutive Modeling Using Non–Uniform Material Tests", *Proceedings of the Sixth International Symposium on Numerical Models in Geomechanics*, Montreal, Quebec, Canada.

Sidarta, D. E. and Ghaboussi, J. (1997) " Constitutive Modeling of Geomaterials from Non–Uniform Material Tests", Submitted for publication in Computers and Geotechnics.

Wu, X. (1991). "Neural Network–Based Material Modeling," Ph.D. thesis, University of Illinois at Urbana–Champaign, Urbana, Illinois.

Wu, X. and Ghaboussi, J. (1992). "Neural Network–Based Modelling of Composite Materials with Emphasis on Reinforced Concrete," *Proceedings of 8th ASCE Conference on Computing in Civil Engineering*, Dallas, Texas.

Wu, X. and Ghaboussi, J. (1993). "Modelling Unloading Mechanism and Cyclic Behavior of Concrete with Adaptive Neural Networks," *Proceedings, Second Asian–Pacific Conference on Computational Mechanics*, Sydney, Australia.

Wu, X. and Ghaboussi, J. (1995). "Neural Network Based material Modeling", Report No. SRS–599, Department of Civil Engineering, Univ. of Illinois, Urbana, Illinois,

Zhang, M. M. (1996). "Determination of Neural Network Material Models from Structural Tests," Ph.D. thesis, University of Illinois at Urbana–Champaign, Urbana, Illinois.

Numerical Models in Geomechanics, Pietruszczak & Pande (eds) © 1997 Balkema, Rotterdam, ISBN 90 5410 886 X

NN constitutive modeling using non-uniform material tests

D.E. Sidarta & J. Ghaboussi
Department of Civil Engineering, University of Illinois at Urbana-Champaign, Ill., USA

ABSTRACT: As the name implies, a non–uniform material test has a non uniform distribution of deformation within the specimen. An example is a triaxial test with end friction. In a conventional material test, ideally, all the points within the specimen follow the same stress path, whereas, the specimen in a non–uniform material test is subjected to widely varying stress paths in different regions. Consequently, a non–uniform material test contains far more information than a uniform (conventional) material test. To date, to the authors' knowledge, there is no method available to develop a constitutive material model from a non–uniform material test. In this paper, a method to extract the material constitutive behavior from a non–uniform material test is presented. This method employs the use of a neural network material model in a finite element analysis. To illustrate this method, it was applied to a series of triaxial compression tests with end friction on sand. Each test was simulated by a finite element model and the material behavior was modeled by a neural network. The proposed method was used to train a neural network material model during finite element analysis. The trained material model was subsequently used to simulate a similar triaxial test but without end friction.

1 INTRODUCTION

In a companion paper in this proceedings (Ghaboussi and Sidarta, 1997a and 1997b) a new neural network type was proposed for modeling of the constitutive behavior of materials. The proposed nested adaptive neural networks (NANN) takes advantage of the nested structure of the material data and results in a sequence of nested neural networks. This work and the previous research by Ghaboussi and his co–workers on constitutive modeling using neural networks (Ghaboussi, Garret, and Wu, 1990 and 1991; Wu, 1991; Ghaboussi, 1992a, 92b; Wu and Ghaboussi, 1992, 93; Ghaboussi, Lade, and Sidarta, 1994; Zhang, 1996; Ghaboussi, Pecknold, Zhang, and HajAli, 1997; Ghaboussi, Zhang, Wu, and Pecknold, 1997) was based on using the results of conventional material tests which impose a uniform state of stresses within the sample. This is indeed the case in all constitutive modeling, whether mathematical formulation or neural networks are used. In conventional material tests the specimen represents a material point, and boundary conditions are imposed in a way that all the material within the sample is subjected to the same stress path. In order to produce a comprehensive data set for training neural network material models, a large number of tests with different stress paths are needed.

A non–uniform material test is defined as causing a non–uniform distribution of stresses within the sample. Early triaxial tests in which the end frictions were not eliminated (Lee, 1965; Lee and Seed, 1967) are examples of non–uniform material tests. Special non–uniform triaxial tests are being performed and will be used in future studies (Lade, Ghaboussi, Inei, and Yamamuro, 1994). In these tests the sample is not only subjected to end friction, but it is also subjected to torsion. An attraction of the non–uniform tests is that a single test contains information on material behavior under a wide variety of stress paths. If that information could be extracted, then the results of a single test may be sufficient to train a neural network. A method for training of neural network material models from the results of non–uniform material tests has been developed (Ghaboussi, Pecknold, Zhang, and HajAli, 1997). In this methodology, which is called the autoprogressive training method, the neural network material model is trained within a finite element model of the experiment. Autoprogressive method is used in this study to train neural network material models for behavior of sand using the results of triaxial tests with end friction (Lee, 1965). These tests were treated as non–uniform material tests. Finite element models of these tests were used in the autoprogressive training. The trained neural networks were first used in finite element analysis of actual tests with end friction. Next they were used in finite element analysis of hypothetical test with no end friction.

2 AUTOPROGRESSIVE METHOD OF TRAINING NEURAL NETWORK MATERIAL MODELS

The training of a neural network material model associated with this algorithm is called autoprogressive training, which was coined to describe a process in which the neural networks is itself an integral part of the iterative algorithm that is used to create stress–strain training cases from the global response data; thus, in a sense, the neural networks 'bootstraps' itself (Ghaboussi, Pecknold, Zhang, and HajAli 1997). The quality of the training cases are gradually improved during the training process. This process may be considered as evolution of the training cases. Initially these training cases do not accurately represent the material constitutive behavior, then as the training progresses they are getting closer to the true material constitutive behavior.

Autoprogressive method requires measured boundary displacements and boundary forces. The autoprogressive training gradually imposes the measured boundary conditions on the finite element model of the sample in which the constitutive model at the integration points is represented by a neural network material model. The boundary conditions are divided into two complementary sets, and each set includes either the displacement or the force at a boundary degree of freedom. During an incremental nonlinear finite element analysis, the iterations at each load step consist of two finite element analyses. In the first FE analysis one set of boundary conditions is applied, and in the second FE analysis the complementary set of boundary conditions is enforced. The two FE analyses produce new training cases which are used to further train the neural network material model before repeating the two analyses. This process is repeated iteratively until the neural network material model learns the material behavior and the finite element analysis is able to match both sets of measured boundary conditions with sufficient accuracy.

2.1 Autoprogressive Algorithm

Before starting the autoprogressive method, a neural network material model is pre–trained to learn the linearly elastic constitutive behavior of the material. Pre–training is needed to avoid having to start the autoprogressive method with a randomly initialized neural network.

The autoprogressive algorithm uses a finite element model of the sample. The pre–trained neural network is used to represent the constitutive model at all the integration points in all the elements. In the following, we have assumed that the first sets of the boundary conditions to be applied in the first FE analysis includes the boundary forces, and the second set of boundary conditions to be enforced in the second FE analysis includes the differences between the measured boundary displacements and those computed from the first FE analysis.

At the nth load increment, the measured load increment vector $\Delta \boldsymbol{P}_n$ is applied and the displacement increment $\Delta \boldsymbol{U}_n$ is computed.

$$\boldsymbol{P}_n = \boldsymbol{P}_{n-1} + \Delta \boldsymbol{P}_n \tag{1}$$

The solution for this load increment is obtained by using modified Newton–Raphson iterations. At the jth iteration the following equations are used in the first FE analysis with the applied boundary forces.

$$\boldsymbol{K}_t \, \delta \boldsymbol{U}_n^j = \boldsymbol{P}_n - \boldsymbol{I}_n^{j-1} \tag{2}$$

In this equation $\boldsymbol{K}_t$ is the tangent stiffness matrix, and the internal resisting force vector $\boldsymbol{I}_{n-1}$ is computed from the stress vector $\boldsymbol{\sigma}_{n-1}$ at the end of the previous iteration which is determined using the neural network material model

$$\boldsymbol{I}_n^{j-1} = \Sigma \int \boldsymbol{B}^T \left[\sigma_n^{j-1} \mathbf{NN}\left(\epsilon_n^{j-1}, \; \ldots \; : \; \ldots \right)\right] dv \tag{3}$$

In Eq. (3) $\boldsymbol{B}$ is the matrix which relates the strains to the element nodal displacements, and the integral is over each element. The new notation **NN** for neural networks was introduced in Ghaboussi and Sidarta (1997a and b) The displacement increment δU_n^j is used to update the strains, stresses and the internal resisting force vector. The Newton–Raphson iterations are continued until a convergence criterion, expressed in terms of the norm of δU_n^j or the norm of the residual force vector is satisfied.

In solving the system of equations in Eq.(2), the tangent stiffness matrix $\boldsymbol{K}_t$ need not be formed, if conjugate gradient method is used. Otherwise, the tangent stiffness matrix can be formed using the tangent constitutive matrix, $\mathrm{C_{ij}}$, generated by the neural network material model. The columns of the tangent constitutive matrix are determined by using small strain probes $\delta\varepsilon$.

$$C_{ij} = [\, \sigma_{ni} \, \mathbf{NN}\left(\epsilon_n + \delta\epsilon \; e_j, \; \epsilon_{n-1}, \; \sigma_{n-1}, \; \ldots \; : \ldots \right) - \sigma_{ni} \, \mathbf{NN}\left(\epsilon_n, \; \epsilon_{n-1}, \; \sigma_{n-1}, \; \ldots \; : \ldots \right)] / \, \delta\epsilon \tag{4}$$

In this equation $\boldsymbol{e}_j$ are the columns of the identity matrix.

At this point the boundary displacement error, $\delta\overline{\boldsymbol{U}}_n$, at the second set of boundary conditions is determined as the difference between the measured boundary displacements $\overline{\mathcal{U}}_n$ and the boundary displacement $\overline{\boldsymbol{U}}_n$

computed from the first FE analysis.

$$\delta \overline{U}_n = \overline{\mathcal{U}}_n - \overline{U}_n \tag{5}$$

As long as $\delta\overline{U}_n$ is larger than a convergence criterion, it is applied as imposed boundary conditions in a second FE analysis. This displacement controlled analysis is carried out in the same manner as the load controlled forward analysis. The results of the second analysis are considered as corrections to the first analysis. The incremental strain correction $\delta\varepsilon_n$ computed from the second analysis, along with the results of the first analysis can then be used to train the neural network material model. The input vector $\varepsilon_n+\delta\varepsilon_n$ and the output vector σ_n are used in training the neural network. The dual FE analysis in the nth load increment, and the training of the neural network material model, is repeated until the displacement error $\delta\overline{U}_n$ is reduced to below a specified convergence criterion. Then, we proceed to the next load increment and repeat the whole process. This is continued until all the load increments are applied, constituting a *load pass*. Often, one load pass is not sufficient to adequately train a neural network material model and several load passes may be needed.

3 AUTOPROGRESSIVE ALGORITHM APPLIED TO A TRIAXIAL TEST

A triaxial compression test with end friction is a non–uniform material test. Different points in the specimen undergo different stress and strain paths. Three of the drained triaxial tests on Sacramento River sand (Lee 1965) were selected for this study. All these tests were performed with end friction. The relative densities ranged from dense to medium dense to loose, with initial void ratios of 0.613, 0.71, and 0.87 and corresponding initial consolidation pressures of 10.5, 10.52, and 12.65 kg/cm^2 (149, 149, and 180 psi). It was further assumed that the friction at the ends of the sample effectively prevented radial displacements.

In addition to initial void ratio and confining pressure, the measurements included axial load, axial strain, and volumetric strain. Thus, the measured boundary forces, which included the axial forces and confining pressures, were directly from the test data. Using the volumetric strains, determined from the total volume change, and the axial strains, the radial displacements of the outer surface of the sample were determined by assuming a parabolic distribution. These measured force and displacement boundary conditions were used in the autoprogressive method.

These tests were performed in two stages. First the sample was consolidated under all around pressure, then it was sheared under axial loads.

The sample was modeled by a finite element mesh with 2D–axisymmetric elements. Taking advantage of the symmetry only a quarter of the sample was modeled as shown in Fig. 1. The height and diameter of the specimen are 8.64 cm (3.4 in) and 3.56 cm (1.4 in), respectively. In order to model the effect of end platens and their friction with sample, all the radial displacements at the top of the finite element mesh were constrained, and the vertical displacements were constrained to a single degree of freedom.

The constitutive model at all the integration points in all the finite elements is represented by a neural network material model, expressed by the following equation.

$$\sigma_j = \sigma_j \ \mathsf{NN} \left(\epsilon_j : \ldots \right) \tag{6}$$

When the material behavior is strongly path dependent, then history point modules may be required to form a nested adaptive neural network. The following equation shows a neural network material model which has historical states as part of its input vector.

$$\sigma_j = \sigma_j \ \mathsf{NN} \left([\epsilon_j\ ],\ [\epsilon_{j-1},\ \sigma_{j-1}],\ \ldots : \ldots \right) \tag{7}$$

As an illustration, the process of the autoprogressive training of one of the tests is presented in detail. The sand specimen had an initial void ratio of 0.71 and a consolidation pressure of 10.52 kg/cm^2 (149 psi). First, a neural network material model was pre–trained with a linear elastic, isotropic behavior. The result of the adaptive pre–training of the material model is given by the following equation.

$$\sigma = \sigma \ \mathsf{NN} \ (\ \epsilon,\ e_o\ :\ 5 \mid 2\text{–}7 \mid 2\text{–}7 \mid 4\ ) \tag{8}$$

Of the 5 input nodes, 4 nodes represented the strain components, $\boldsymbol{\varepsilon} = \{\varepsilon_{rr}, \varepsilon_{zz}, \varepsilon_{\theta\theta}, \varepsilon_{r\theta}\}$, and the fifth node represented the initial void ratio, and the 4 output nodes represented the stress components. During the pre–training process, the number of hidden nodes grew from 2 to 7.

Then three *load passes* of autoprogressive training were performed, each *load pass* covered 10 load increments. During this training, no additional hidden layer nodes were needed. Subsequently, one history point module was added, and another *load pass* was performed. In the fourth pass, the autoprogressive training was able to cover one additional load increment. The final architecture of the neural networks material model is given by the following equation.

$$\sigma_j = \sigma_j \ \mathsf{NN} \ (\ [\epsilon_j,\ e_o],\ [\epsilon_{j-1},\ \sigma_{j-1}]\ : \\ 5,\ 8 \mid 2\text{–}7,\ 2 \mid 2\text{–}7,\ 2 \mid 4\ ) \tag{9}$$

The first history module had 8 input nodes, and the hidden layers started the autoprogressive training with two nodes, and no additional nodes were needed.

Fig. 2 shows the global response of the specimen which was obtained by performing a forward analysis using the trained neural network material model after completing each of the four passes of the autoprogressive training. This figure shows a comparison of the experimental and the computed results for deviatoric stress versus axial strain and volumetric strain versus axial strain. Both plots show that the result of elastic pre–training is only close to the measured data at the beginning of the loading. As the autoprogressive training continues, the computed boundary displacements try to match the enforced measured displacements, and in the process the neural networks material model evolves to represent the material behavior more accurately.

The radial displacements of the outer surface of the specimen at several stages in the forward analysis are shown in Fig. 3. The neural networks material mode after the fourth pass was used in this analysis. In gen eral, the computed boundary displacements were rela tively close to the measured boundary displacement (assumed parabolic distribution). During the simula tion of the consolidation portion of the test the end fric tion constrains the inward radial displacement at th top and bottom of the sample, thus, producing conve deformed shape. The opposite occurs during shearin under increased axial load, when the end frictio constrains the outward radial displacement at the to and bottom of the sample, which results in bulging o the sample at the mid–height.

Similar autoprogressive training was used for th two other tests; one on loose sand with an initial voi ratio of 0.87 and a consolidation pressure of 12.65 kg cm^2 (180psi), and the other on dense sand with an ini tial void ratio of 0.61 and a consolidation pressure c 10.5 kg/cm^2 (149psi). The trained neural network ma

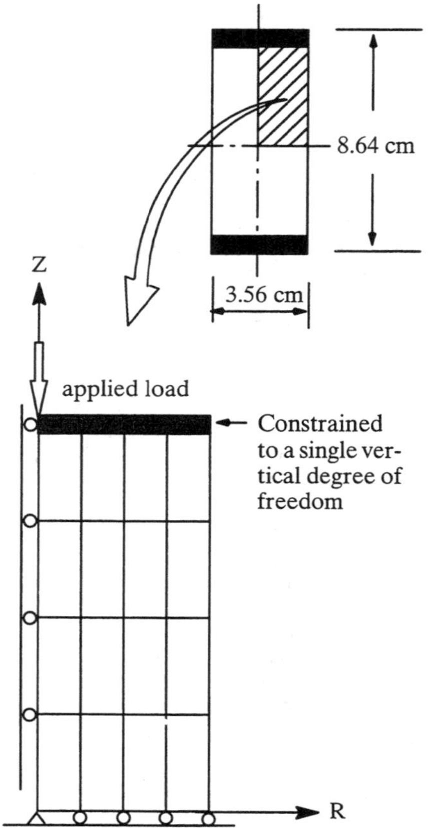

Figure 1. Finite element model of one–fourth of the sample used in the autoprogressive training of the neural network material model.

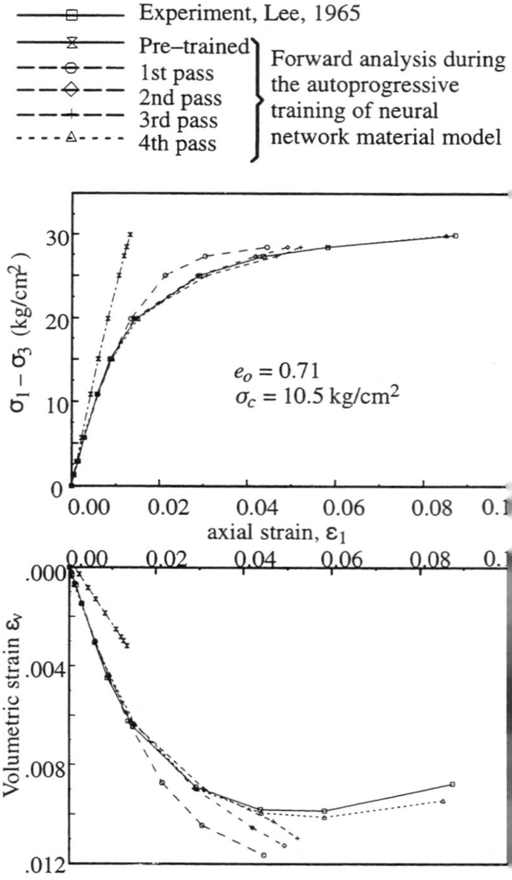

Figure 2. Forward analysis of the triaxial test wit end friction using neural network material mode after various autoprogressive training passes.

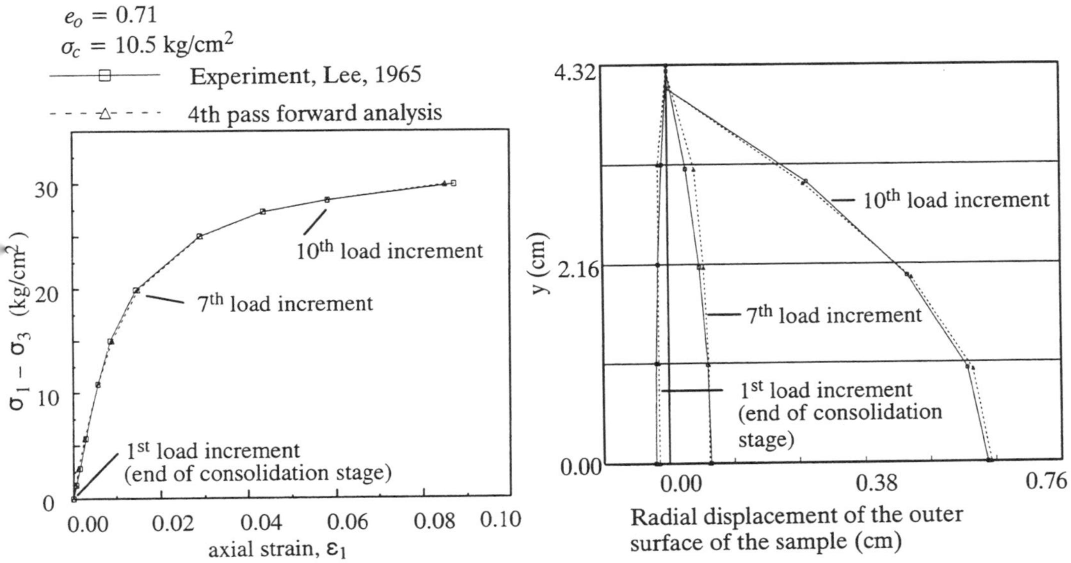

Figure 3. Radial displacement of the outer surface of the sample during the triaxial test with end friction

Table 1. Log of the autoprogressive training of the neural network material models.

		loose sand [1]	med. dense sand [2]	dense sand [3]
Pre–training	hidden nodes	2–9	2–7	2–12
Base module	load passes	2	3	2
	hidden nodes	9	7	12
	load inc.	9 * 15**	10	6
First level NANN (first history module)	load passes	—	1	4
	hidden nodes	—	2	2
	load inc.	—	11	8
Second level NANN (second history module)	load passes	—	—	3
	hidden nodes	—	—	2–6
	load inc.	—	—	9

1. $e_0 = 0.87, \sigma_c = 12.65$ kg/cm^2
2. $e_0 = 0.71, \sigma_c = 10.52$ kg/cm^2
3. $e_0 = 0.61, \sigma_c = 10.5$ kg/cm^2

* First pass
** Second pass

terial models are given in the following equations,

$$\sigma \;=\; \sigma\, \mathbf{NN}\,(\, \epsilon,\; e_o \;:\; 5 \mid 2\text{–}9 \mid 2\text{–}9 \mid 4\,) \qquad (10)$$

$$\sigma_j \;=\; \sigma_j\, \mathbf{NN}\,([\epsilon_j,\; e_o],\; [\epsilon_{j-1}, \sigma_{j-1}],\; [\epsilon_{j-2}, \sigma_{j-2}] \;:\; 5,\; 8,\; 8 \mid 2\text{–}12,\; 2,\; 2\text{–}6 \mid 2\text{–}12\;, 2,\; 2\text{–}6 \mid 4) \qquad (11)$$

and the history of the autoprogressive training for all the three cases are given in Table 1. The architecture of the three trained neural network material models are shown in Fig. 4. The results of the forward finite element analyses with the trained neural network material models for the three tests are shown in Fig. 5 and compared with the experimental results. It can be seen that the neural network material models have learned the material behavior well, and are able to match the global response of the experiments.

These three tests had been performed under similar confining pressures. However, their relative densities varied from loose to medium dense to dense. The dilatancy of the sand increases with density. In denser sands the transition from contractive to dilatant behavior occur earlier, thus leading to increasing path dependency. This is observed in term of the difficulty of the learning of the material behavior. The log of the autoprogressive training passes and history modules for the NANN material models for the three tests are given in Table 1. For the loose sand, with the initial void ratio of 0.87, two load passes with the base module was sufficient. The medium dense sand, with the initial void ratio of 0.71, required one history module and a total of four load passes. The dense sand, with the initial void ratio of 0.61, required two history modules and

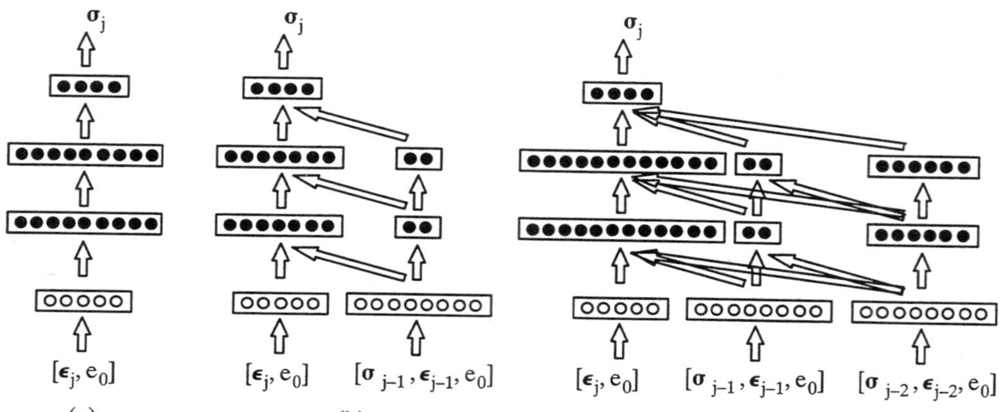

$$\sigma = \{ \sigma_{rr}, \sigma_{zz}, \sigma_{\theta\theta}, \sigma_{rz} \}$$
$$\epsilon = \{ \epsilon_{rr}, \epsilon_{zz}, \epsilon_{\theta\theta}, \epsilon_{rz} \}$$

Figure 4. The autoprogressively trained nested adaptive neural network material models for the three tests: (a) loose sand, e_0=0.87; (b) medium dense sand, e_0=0.71; (c) dense sand, e_0=0.61.

a total of 9 load passes. This clearly indicates that the material behavior becomes increasingly more complex and path dependent with the increasing density.

4 SIMULATION OF TRIAXIAL COMPRESSION TESTS WITH NO END FRICTION

A trained neural network material model can be used like any other material model in finite element analysis of any boundary value problem. We have chosen to use the autoprogressively trained neural network material models in finite element analysis of a problem which is of some interest in material testing. The experiments that have been used in the autoprogressive training had been performed with end friction. It is interesting to determine the effect of the end friction on the observed behavior of the material. This is accomplished by using the trained neural network material model in a forward analysis of hypothetical test of the same sample but without end friction.

The results of simulating the triaxial tests without end friction are presented in Fig. 6. The major effects of the end friction can only be observed at higher stresses. It can be seen that the axial displacements are much larger when the end friction is eliminated, and the volumetric strains are lower. These differences are no doubt due to the effect of the dilatancy under the shear stresses at the ends of the sample caused by the end friction. However, the overall effect of the end friction seems to be rather small.

5 CONCLUDING REMARKS

The material in a non–uniform test experiences a wide variety of stress paths, and, thus, non–uniform tests inherently contain far more information on the constitutive behavior of material than the conventional uniform material tests. The autoprogressive method is able to extract the constitutive information from the results of non–uniform material tests through a series of dual finite element analyses of the experiment during which the measured boundary conditions are imposed on a finite element model of the sample.

In this paper we have applied the autoprogressive method in training of neural network material models from the results of non–uniform material tests. Nested adaptive neural networks (Ghaboussi and Sidarta 1997) were used in the autoprogressive training. Subsequently the trained neural network material models were used in forward analyses of the experiment with and without the end friction. The results of the forward analysis with end friction matched those of the actual experiment well. The results of the forward analysis of the hypothetical tests with no end friction showed some revealing differences with the actual experimental results.

6 ACKNOWLEDGEMENTS

The research reported in this paper is part of a research project with analytical and experimental components. Professor Poul Lade of The Johns Hopkins University is developing special new experimental techniques for neural network applications. Both parts of this research project are funded by National Science Foundation under Grant number CMS–95–03462. This support is gratefully acknowledged.

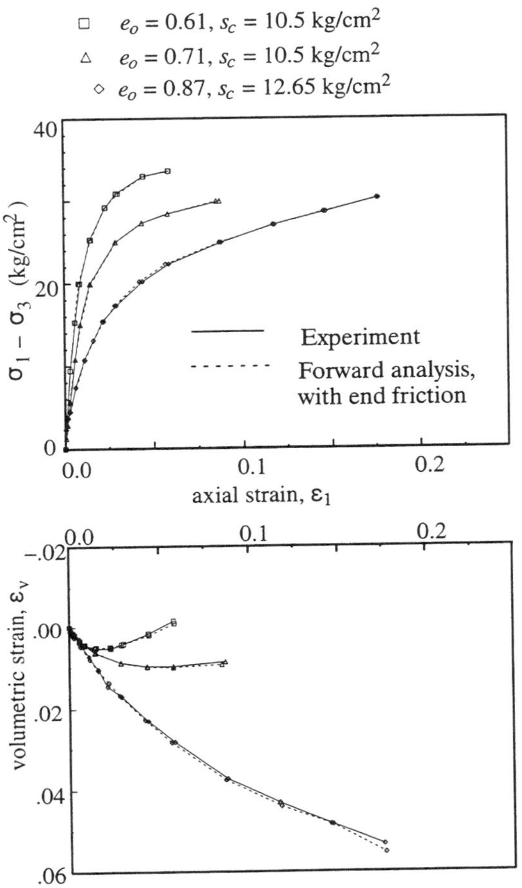

Figure 5. Forward analyses of triaxial tests with end friction, using the neural network material model trained with autoprogressive method.

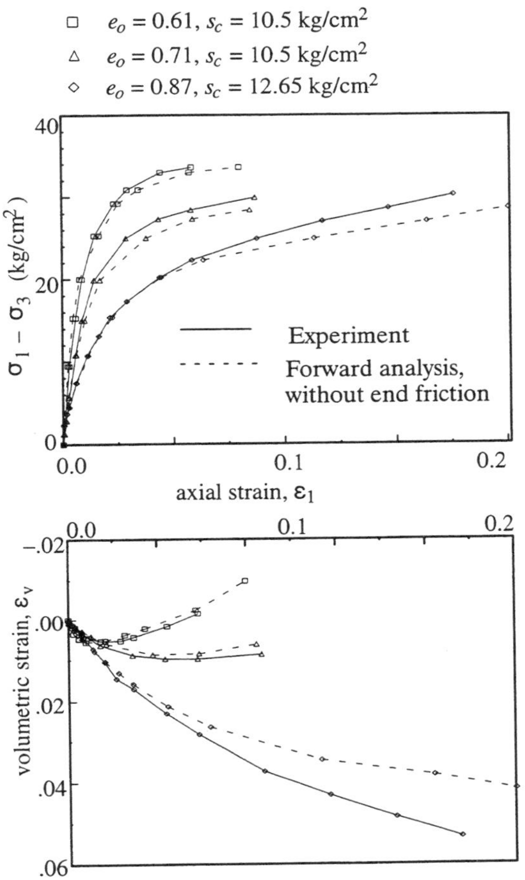

Figure 6. Forward analyses of triaxial tests without end friction, using the neural network material model trained with autoprogressive method.

7 REFERENCES

Ghaboussi, J. (1992a). "Potential Applications of Neuro–Biological Computational Models in Geotechnical Engineering," *Proceedings, Fourth International Symposium on Numerical Models in Geomechanics*, Swansea, U. K.

Ghaboussi, J. (1992b). "Neuro–Biological Computational Models with Learning Capabilities and Their Applications in Geomechanical Modeling," *Proceedings, Workshop on Recent Accomplishments and Future Trends in Geomechanics in the 21st Century*, Norman, Oklahoma.

Ghaboussi, J., Garret, J. H., Jr., and Wu, X. (1990). "Material Modeling with Neural Networks," *Proceedings of the International Conference on Numerical Methods in Engineering : Theory and Applications,* Swansea, U. K., 701–717.

Ghaboussi, J., Garret, J. H., Jr., and Wu, X. (1991). "Knowledge–Based Modeling of Material Behavior with Neural Networks," *Journal of Engineering Mechanics Division*, ASCE, 117(1), 132–153.

Ghaboussi, J., Lade, P. V., and Sidarta, D. E. (1994). "Neural Networks Based Modelling in Geomechanics," *Proceedings of the 8th International Conference on Computer Methods and Advances in Geomechanics*, Morgantown, West Virginia.

Ghaboussi, J., Pecknold, D. A., Zhang, M., and HajAli, R. M. (1997). "Autoprogressive Training of Neural Network Constitutive Models," submitted for publication.

Ghaboussi, J. and Sidarta, D. E. (1997a). "New Nested Adaptive Neural Networks(NANN) for Modeling of Constitutive Behavior of Materials", submitted for publication in *Computers and Geotechnics*.

Ghaboussi, J. and Sidarta, D. E. (1997b). "New Method of Material Modeling Using Neural Networks", *Proceedings, NUMOG VI*, Montreal.

Ghaboussi, J. Zhang, M., Wu, X. and Pecknold, D. A., (1997) "Nested Adaptive Neural Networks: A New Architecture", submitted to the International Conference on Artificial Neural Networks in Engineering, St. Louis, Missouri.

Lade, P. V., Ghaboussi, J., Inei, S. and Yamamuro, J. A., "Experimental Determination of Constitutive Behavior of Soils", *Proceedings, International Conference on Numerical Methods and Advances in Geomechanics*, May 1994.

Lee, K. L. (1965). "Triaxial Compressive Strength of Saturated Sands under Seismic Loading Conditions," Ph.D. thesis, University of California, Berkeley, California.

Lee, K. L. and Seed, H. B. (1967). "Drained Strength Characteristics of Sands," *Journal of Soil Mech. and Found. Eng. Div.*, ASCE, 93(SM6), 117–141.

Wu, X. (1991). "Neural Network–Based Material Modeling," Ph.D. thesis, University of Illinois at Urbana–Champaign, Urbana, Illinois.

Wu, X. and Ghaboussi, J. (1992). "Neural Network–Based Modelling of Composite Materials with Emphasis on Reinforced Concrete," *Proceedings of 8th ASCE Conference on Computing in Civil Engineering*, Dallas, Texas.

Wu, X. and Ghaboussi, J. (1993). "Modelling Unloading Mechanism and Cyclic Behavior of Concrete with Adaptive Neural Networks," *Proceedings, Second Asian–Pacific Conference on Computational Mechanics*, Sydney, Australia.

Zhang, M. M. (1996). "Determination of Neural Network Material Models from Structural Tests," Ph.D. thesis, University of Illinois at Urbana–Champaign, Urbana, Illinois.

Numerical Models in Geomechanics, Pietruszczak & Pande (eds) © 1997 Balkema, Rotterdam, ISBN 90 5410 886 X

Implementation of adaptive mesh refinement in MONICA

M.A. Hicks
University of Manchester, UK

ABSTRACT: MONICA is a finite element based numerical algorithm, encompassing the Monot double-hardening constitutive model. An adaptive mesh refinement (AMR) version of this algorithm has been developed. This is based on a smoothed stress-strain formulation and a scheme for regenerating unstructured mesh configurations at regular intervals during nonlinear computations. Details are given of the general MONICA formulation and of modifications made to include mesh adaptivity. The performance of the revised algorithm is demonstrated for the problem of localisation in drained biaxial compression.

1 INTRODUCTION

MONICA, short for 'MONot Incremental Computer Algorithm', enables the soil model Monot to be used in the analysis of practical boundary value problems. Detailed accounts have been given of its development and validation (e.g. Hicks, 1995a). Other publications (e.g. Hicks & Smith, 1986, 1988) have described various applications and, in so doing, have demonstrated the wide range of material behaviours modelled by Monot—i.e. responses ranging from liquefiable to strongly dilative.

Hicks & Mar (1994a) discussed the benefits of combining MONICA with AMR—e.g. accurate simulation of soil stress-strain response and adequate mesh discretisation in regions of high strain gradient, enabling realistic computations of shear banding. Example AMR analyses were reported in this and several subsequent publications (e.g. Hicks & Mar, 1994b; Hicks, 1995c).

This paper describes various aspects of the AMR implementation. A brief description of Monot is followed by details of the basic MONICA formulation. The adaptive mesh version of MONICA is summarised by reviewing important new equations and by describing how these equations fit in with the original formulation. Algorithm performance is demonstrated for a simple boundary value problem and brief details are then given of the most recent algorithm developments.

2 CONSTITUTIVE MODEL MONOT

Monot (Molenkamp, 1981) is illustrated in Figure 1, in which a section has been taken through the model in the s'-t plane. (s' and t refer to isotropic and deviatoric stress invariants, respectively.) The figure shows that the model is characterised by a failure surface and two isotropically hardening yield surfaces: a non-associated deviatoric surface for modelling plastic strains due to shearing; and an associated compressive surface for modelling plastic strains due to increasing confining pressure. There is also a nonlinear elastic model component.

Figure 1 shows that for a stress path starting from point P, there are four possible types of stress-strain response, depending on how many, if any, of the yield surfaces are activated during an increment of stress.

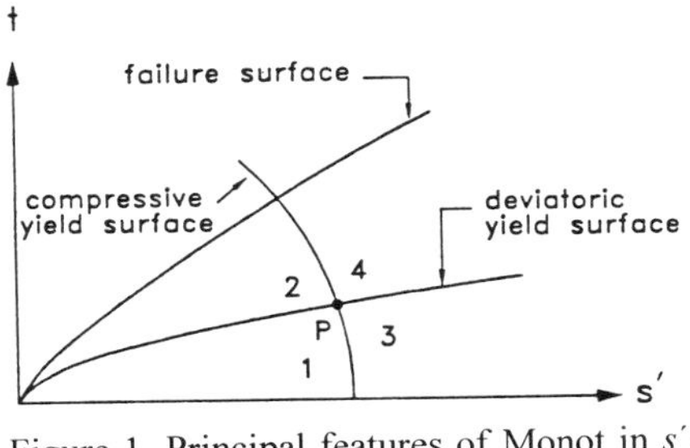

Figure 1. Principal features of Monot in s'-t plane

3 THE BASIC MONICA FORMULATION

MONICA contains specialised techniques for coping with complicated constitutive models—especially those of the double-hardening variety. For example, the methods used for predicting the correct stress-strain response and for updating yield surface positions are particularly relevant. Details of these and other features are given by Hicks (1992, 1995a, 1995b). The present summary is confined to more general aspects of the formulation.

The basic equation is

$$[KG]\{dr\}_n^m = \{df\}_n^{m-1} \tag{3.1}$$

in which $\{df\}$ and $\{dr\}$ are the incremental load and displacement fields, respectively, and $[KG]$ the global coefficient matrix. Subscript and superscript denote increment and equilibrium iteration numbers, respectively. Convergence is achieved when the iterative change in $\{dr\}$ is sufficiently small.

3.1 *Uncoupled equations*

Perfectly drained and undrained responses are modelled using an uncoupled approach, in which $[KG]$ equals the elastoplastic stiffness $[KEC]$. This matrix is formed only from the elastic and plastic compressive components of Monot, and is therefore symmetric. $[KEC]$ is a function of stress state and stress region estimate and is normally reformed at the start of each increment. However, for many analyses $[KEC]$ may be updated less frequently, with little or no loss of accuracy, and for the example in this paper it has only been reformed following each mesh-update. Perfectly undrained conditions are modelled by including a large bulk modulus to represent the pore fluid.

Equation (3.1) may therefore be rewritten as

$$[KEC]\{dr\}_n^m = \{df\}_n^{m-1} \tag{3.2}$$

in which

$$\{df\}_n^{m-1} = \{df_{ext}\}_n + \{df_{err}\}_n + \{df_{body}\}_n^{m-1} \tag{3.3}$$

$\{df_{ext}\}$ is the vector of incremental external loads, which remains constant throughout the iteration loop, while $\{df_{body}\}$ contains the incremental out-of-balance forces which 'drive' the iteration process. For a given iteration, these are computed by integrating the incremental excess 'plastic' stresses $\{d\sigma^p\}$ over the mesh domain Ω. Hence,

$$\{df_{body}\}_n^{m-1} = \int_\Omega [B]^T \{d\sigma^p\}_n^{m-1} d\Omega \tag{3.4}$$

in which $[B]$ contains shape function derivatives and

$$\{d\sigma^p\}_n^{m-1} = \{d\sigma^{ec}\}_n^{m-1} - \{d\sigma\}_n^{m-1} \tag{3.5}$$

$\{d\sigma^{ec}\}$ is the stress increment computed using the 'modified' stress-strain relationship which was used in forming $[KEC]$, and $\{d\sigma\}$ is the stress increment based on the complete constitutive model.

In equation (3.3), $\{df_{err}\}$ are the unconverged incremental out-of-balance forces from the end of the previous increment: i.e. the difference in $\{df_{body}\}$ for the last 2 iterations in that increment. Obviously, the magnitude of $\{df_{err}\}$ depends on the convergence tolerance specified for the analysis. The inclusion of this vector means that relatively large convergence tolerances can be used, leading to rapid convergence (typically, in 2 iterations). The error introduced into the analysis is small, because the unconverged forces are not ignored—they are instead applied in the next increment as an extra set of external loads. The rate of convergence is further enhanced by the first iteration of an increment using the contents of $\{df_{body}\}$ from the end of the previous increment as the initial estimate for the out-of-balance forces.

3.2 *Coupled equations*

For the coupled formulation, equations (3.2) and (3.3) are replaced by

$$\begin{bmatrix} [KEC] & [C] \\ [C]^T & -\theta\Delta t[KP] \end{bmatrix} \begin{Bmatrix} \{dr\} \\ \{du\} \end{Bmatrix}_n^m = \{df\}_n^{m-1} \tag{3.6}$$

$$\{df\}_n^{m-1} = \{df_{ext}\}_n + \{df_{fluid}\}_n + \{df_{err}\}_n + \{df_{body}\}_n^{m-1} \tag{3.7}$$

in which $\{du\}$ are the incremental excess pore pressures, $[C]$ the fluid connectivity, $[KP]$ the fluid 'stiffness', Δt the time step size, and θ the time stepping scheme: in MONICA a fully implicit method is adopted—hence $\theta=1$. The additional fluid component in equation (3.7) is given by

$$\{df_{fluid}\}_n = \Delta t[KP]\{u_0\} \tag{3.8}$$

in which $\{u_0\}$ are the accumulated excess pore pressures at the start of the increment.

4 ADAPTIVE MESH VERSION OF MONICA

The AMR algorithm allows for mesh updating at user-defined intervals during an analysis. For each mesh-update increment: (a) a new mesh is created; and (b) information is transferred from old to new meshes so that the analysis can continue. The algorithm follows a sequence of smoothing, error estimation and mesh generation (Mar, 1993).

4.1 *Smoothing of internal variables*

Internal variables are smoothed using a form of the global 'stress projection' smoothing method of Zienkiewicz & Zhu (1987). Hence,

$$\int_{\Omega}[N^*]^T[N^*]\mathrm{d}\Omega\{\bar{q}^*\}=\int_{\Omega}[N^*]^T\{q\}\mathrm{d}\Omega \qquad (4.1)$$

in which the unsmoothed integration point quantities $\{q\}$ are smoothed over the mesh domain Ω using the smoothing shape functions $[N^*]$, to give the smoothed nodal quantities $\{\bar{q}^*\}$. In the present formulation, $\{q\}$ may contain a range of variables: e.g.

$$\{q\}=\begin{Bmatrix}\{\mathrm{d}\sigma\}\\ \{\mathrm{d}\sigma^{ec}\}\\ \{\mathrm{d}\varepsilon\}\\ \{\mathrm{d}u\}\\ \{\kappa\}\end{Bmatrix} \qquad (4.2)$$

in which $\{\mathrm{d}\sigma\}$ and $\{\mathrm{d}\sigma^{ec}\}$ are the incremental stresses from equation (3.5), $\{\mathrm{d}\varepsilon\}$ the incremental strains, $\{\mathrm{d}u\}$ the incremental excess pore pressures, and $\{\kappa\}$ the hardening parameters defining the locations of the two yield surfaces.

$\{\mathrm{d}\sigma\}$ is smoothed on an iterative basis, with equilibrium between smoothed and unsmoothed incremental stresses being ensured through an extension of the method proposed by Cantin et al. (1978). Specifically, the incremental plastic stresses are here computed using a modified version of equation (3.5):

$$\{\mathrm{d}\sigma^p\}_n^{m-1}=\{\mathrm{d}\sigma^{ec}\}_n^{m-1}-\{\mathrm{d}\sigma^*\}_n^{m-1} \qquad (4.3)$$

Hence, the equilibrium iterations are driven by the difference between the unsmoothed 'modified' stress increment, as computed using the stress-strain relationship used in the stiffness matrix, and the smoothed 'complete' stress increment $\{\mathrm{d}\sigma^*\}$, as computed using the complete version of Monot.

$\{\mathrm{d}\varepsilon\}$ and $\{\mathrm{d}u\}$ are smoothed on an incremental basis and, along with $\{\mathrm{d}\sigma^*\}$, are used to update the vectors containing the accumulated smoothed quantities once convergence has been achieved. Obviously, $\{\mathrm{d}u\}$ does not need to be smoothed for the coupled formulation, because in that case the spatial variation of excess pore pressure is already continuous.

$\{\mathrm{d}\sigma^{ec}\}$ and $\{\kappa\}$ are only smoothed for mesh-update increments, to enable these quantities to be mapped to the new mesh following mesh generation. An estimate for $\{\mathrm{d}\sigma^{ec}\}$, as given by its smoothed counterpart, is needed at the new integration points, so that $\{\mathrm{d}f_{err}\}$ and $\{\mathrm{d}f_{body}\}$ can be reformed for inclusion in either equation (3.3) or (3.7) at the start of the next increment. The yield surface locations need to be mapped to ensure that the correct stress-strain responses continue to be computed.

4.2 *Error estimation*

Error estimates are represented by the L_2 norm of the error in the incremental shear strain invariant (Mar, 1993). Hence,

$$\|e_{\mathrm{d}\gamma}\|_{L_2}=\left(\int_{\Omega_e}e_{\mathrm{d}\gamma}e_{\mathrm{d}\gamma}\mathrm{d}\Omega_e\right)^{1/2} \qquad (4.4)$$

in which the element error norm $\|e_{\mathrm{d}\gamma}\|_{L2}$ is computed for the element domain Ω_e, using the error in the incremental shear strain invariant:

$$e_{\mathrm{d}\gamma}=\mathrm{d}\gamma^*-\mathrm{d}\gamma \qquad (4.5)$$

in which $\mathrm{d}\gamma$ is the unsmoothed incremental shear strain invariant, given by

$$\mathrm{d}\gamma=\left(\frac{1}{3}\left((\mathrm{d}\varepsilon_1-\mathrm{d}\varepsilon_2)^2+(\mathrm{d}\varepsilon_2-\mathrm{d}\varepsilon_3)^2+(\mathrm{d}\varepsilon_3-\mathrm{d}\varepsilon_1)^2\right)\right)^{1/2} \qquad (4.6)$$

and $\mathrm{d}\gamma^*$ the equivalent invariant based on smoothed incremental strains.

4.3 *Mesh generation*

Unstructured meshes of triangular elements are generated using an advancing front mesh generation scheme similar to that of Peraire et al. (1987). The element errors for the previous (i.e. background) mesh are used, in conjunction with a user-specified 'aiming error', to define a distribution of new nodal spacings $\{\delta\}$. The aim is to use $\{\delta\}$ to produce a mesh in which: (a) element errors are as near equal as possible—this leads to the optimal mesh configuration, as demonstrated by Mar & Hicks (1996); and (b) the prescribed aiming error is satisfied.

The new mesh is produced using an iterative scheme, in which the mesh-update increment is repeated a number of times. Each iteration uses a different mesh configuration, which has been generated by averaging the $\{\delta\}$ distributions from all previous iterations of that increment. Mesh 'convergence' is quickly achieved—typically, in 3 or 4 iterations. Note that when no iterations are used (i.e. the next mesh is based only on one $\{\delta\}$), the tendency is for mesh over-refinement—so the extra time needed for performing the iterations is more than offset by the saving in time through using a cruder (more efficient) mesh in all increments up to the next mesh-update.

At the end of each mesh-update increment, the internal variables (e.g. stresses, strains, hardening parameters, etc.) are mapped across to the new mesh via nodal interpolation.

5 EXAMPLE OF ALGORITHM PERFORMANCE

Figures 2 and 3 summarise results for a drained biaxial compression test on a loose, non-dilative, sand. A plane strain rectangular specimen, 1.0m×2.0m, is subjected to an initial effective confining pressure of 1000kPa and then axially compressed between perfectly rough and rigid end-platens. The peak friction angle of the sand in triaxial compression is 30.5°, giving an equivalent Mohr-Coulomb friction angle of 37° for plane strain conditions. (In the π-plane, the Monot failure surface lies outside that for Mohr-Coulomb.)

Figure 2 shows mobilised earth pressure coefficient K_P versus axial strain ε_a, in which: $K_P=F/F_0$; F is the backfigured load at any instant during loading; and F_0 is the value for F at the start of the test (due to the initial stress state). The analysis uses 160 equal-sized axial strain increments, with mesh updating taking place every tenth increment and in increment two, as indicated in the figure. Hence, there are 17 mesh-updates in total. Figure 2 shows that the maximum value obtained for K_P is 3.8, which may be compared with the theoretical value of K_P=4.0 for ϕ=37°.

Figure 3 illustrates the developing failure mechanism, by showing various mesh-updates and contour plots. The starting mesh uses 128 equal-sized triangular elements (3-noded), as shown in Figure 3(a)—note that double-symmetry has been assumed, so that only one quarter of the problem needs to be analysed. Figure 3(b) shows that a fairly uniform mesh, comprising 552 elements, is produced at the first mesh-update. Subsequent mesh-updates produce similar configurations, until the peak load is reached in increment one hundred. At this stage, stress states in the central region of the specimen are on, or near to, the failure surface and localised strains start to appear, leading to the change in mesh appearance shown in Figure 3(c). Figures 3(d) to 3(f) show the next three mesh-updates, while Figures 3(g) to 3(i) demonstrate stress and strain fields in the specimen at the time of the 15th mesh-update (Figure 3(j)).

In detail, Figure 3(g) shows contours of mobilised shear stress ratio t_f—a dimensionless quantity, varying linearly from zero on the hydrostatic axis to unity on the failure surface. These contours demonstrate that stiff elastic unloading zones have developed since the onset of localisation. Indeed, the only parts of the domain still at failure are the shear bands themselves. Figures 3(h) and 3(i) show contours of accumulated and incremental shear strain invariants, $\gamma*$ and $d\gamma*$ respectively, while Figure 3(j) shows the new mesh. Note that this revised mesh configuration is based on estimates of error in the quantity shown in Figure 3(i)—hence, the similarity in appearance between these two figures. Figures 3(g) to 3(j) show that the shear bands are inclined at approximately 54° to the horizontal, thereby agreeing with the value given by $45°+(\phi+\psi)/4$, in which ψ, the dilation angle, is zero at peak stress states for this material.

6 MORE RECENT DEVELOPMENTS

Goldsmith (1997) has described improvements made to the mesh generation stage of the AMR process. In particular, various forms of data structure have been implemented, along similar lines to those proposed by Löhner (1988)—e.g. the quality of the mesh has been enhanced through using a heap list to store the advancing front. This enables the smallest elements in the front to be constructed first (see Figure 4), thereby eliminating the possibility of large elements being constructed too close to strain localisations. This usually causes a small increase in the number of elements used—i.e. compared with

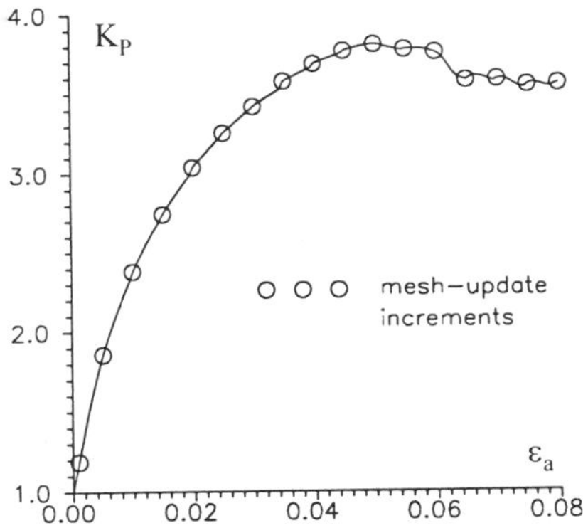

Figure 2. Mobilised earth pressure coefficient versus axial strain for drained biaxial compression

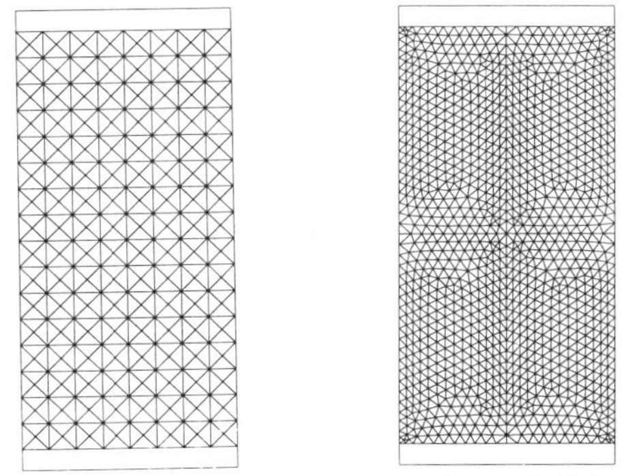

(a) ε_a=0.000
starting mesh
(128 elements)

(b) ε_a=0.001
1st mesh-update
(552 elements)

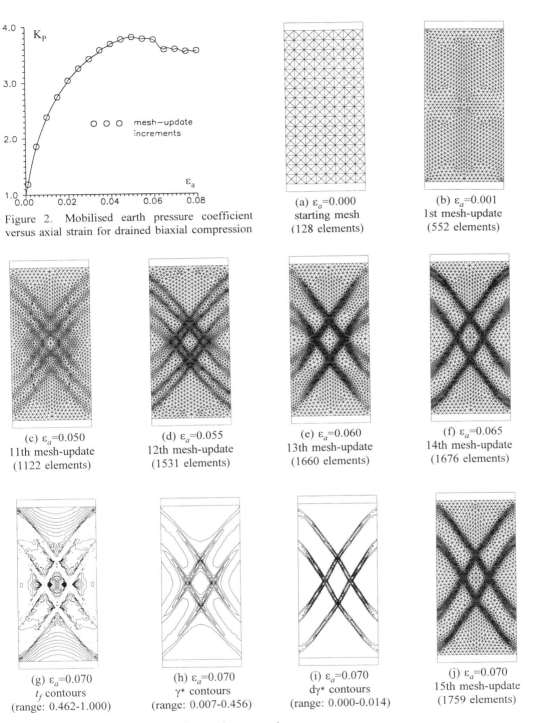

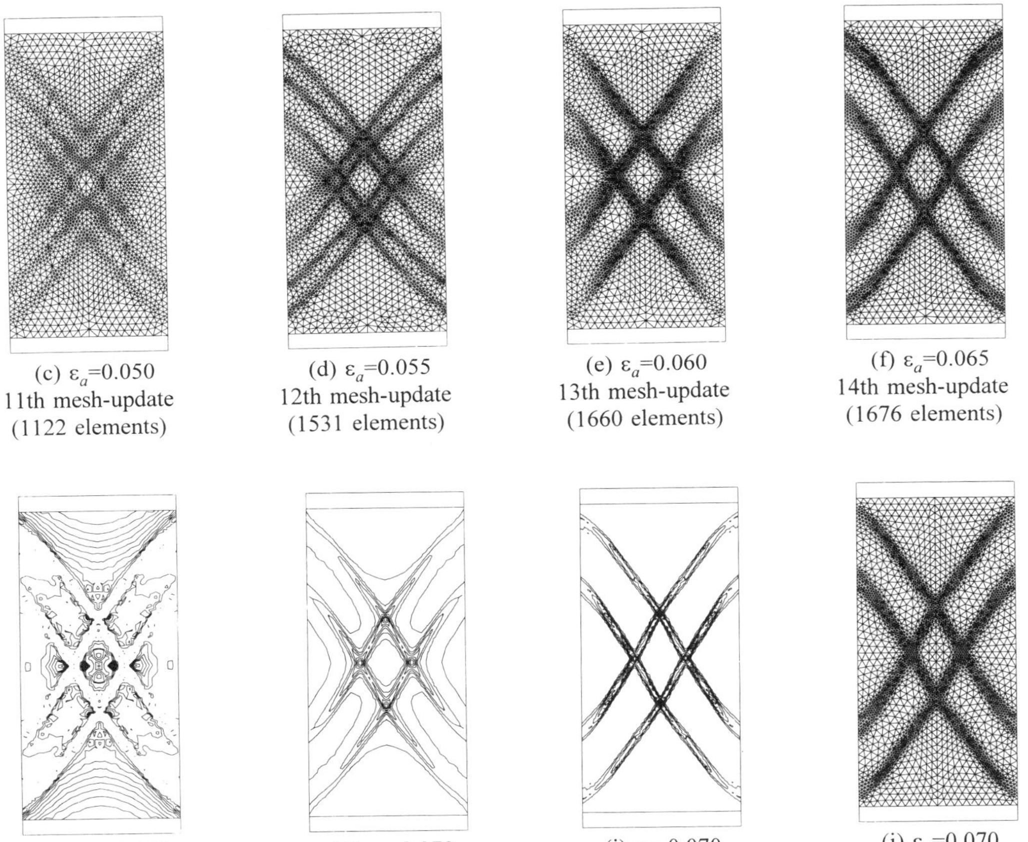

(c) ε_a=0.050
11th mesh-update
(1122 elements)

(d) ε_a=0.055
12th mesh-update
(1531 elements)

(e) ε_a=0.060
13th mesh-update
(1660 elements)

(f) ε_a=0.065
14th mesh-update
(1676 elements)

(g) ε_a=0.070
t_f contours
(range: 0.462-1.000)

(h) ε_a=0.070
γ^* contours
(range: 0.007-0.456)

(i) ε_a=0.070
$d\gamma^*$ contours
(range: 0.000-0.014)

(j) ε_a=0.070
15th mesh-update
(1759 elements)

Figure 3. Updated mesh configurations and contour plots

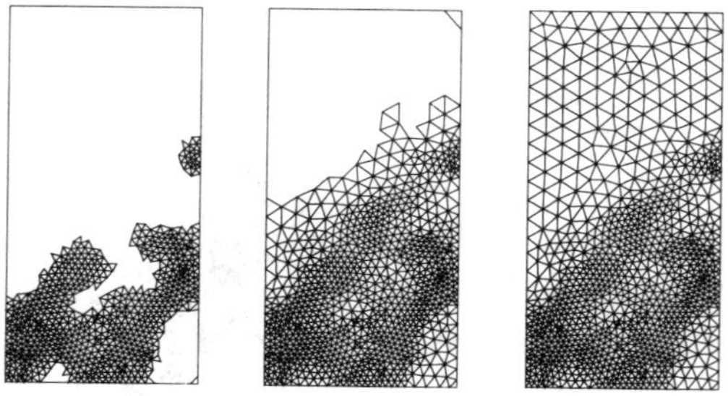

Figure 4. Example of advancing front mesh generation using a heap list

the standard inward spiralling approach, which was used in the example above.

Algorithms have also been developed for increasing the speed of searching the previous mesh. Searching is necessary for: (a) interpolating nodal spacings during construction of the new mesh; and (b) mapping internal variables from old to new meshes at the end of each mesh-update increment. Various forms of quadtree searching have been implemented, including a form of tree incorporating the latest programming features available in Fortran 90.

7 CONCLUDING COMMENTS

The AMR algorithm enables repeated mesh updating during nonlinear computations. Its efficient use of elements means that large problems may be solved on a PC. For the numerical example presented, realistic computations have been obtained for peak load, shear band orientation and failure mechanism.

REFERENCES

Cantin, G., Loubignac, G. & Touzot, G. 1978. An interactive algorithm to build continuous stress and displacement solutions. Int. J. Num. Meth. Eng. 12: 1493-1506.

Goldsmith, M.J. 1997. Data structures and other techniques for improving the performance of advancing front mesh generators. M.Sc. thesis: University of Manchester.

Hicks, M.A. 1992. Numerical implementation of a double-hardening soil model for use in boundary value problems. Proc. 7th Int. Conf. Comp. Meth. Advances Geomech. 3: 1705-1710. Cairns, Queensland. Rotterdam: Balkema.

Hicks, M.A. 1995a. MONICA—A computer algorithm for solving boundary value problems using the double-hardening constitutive law Monot: I. Algorithm development. Int. J. Num. Anal. Meth. Geomech. 19: 1-27.

Hicks, M.A. 1995b. MONICA—A computer algorithm for solving boundary value problems using the double-hardening constitutive law Monot: II. Algorithm validation. Int. J. Num. Anal. Meth. Geomech. 19: 29-57.

Hicks, M.A. 1995c. Computation of localisation in undrained soil using adaptive mesh refinement. Proc. 5th Int. Symp. Num. Models Geomech: 203-208. Davos, Switzerland. Rotterdam: Balkema.

Hicks, M.A. & Mar, A. 1994a. Mesh adaptivity for geomaterials using a double-hardening constitutive law. Proc. 8th Int. Conf. Comp. Meth. Advances Geomech. 1: 587-592. Morgantown, West Virginia. Rotterdam: Balkema.

Hicks, M.A. & Mar, A. 1994b. A combined constitutive model—adaptive mesh refinement formulation for the analysis of shear bands in soils. Proc. 3rd European Conf. Num. Meth. Geotech. Eng: 59-66. Manchester, UK. Rotterdam: Balkema.

Hicks, M.A. & Smith, I.M. 1986. Influence of rate of porepressure generation on the stress-strain behaviour of soils. Int. J. Num. Meth. Eng. 22: 597-621.

Hicks, M.A. & Smith, I.M. 1988. Class A prediction of Arctic caisson performance. Géotechnique 38: 589-612.

Löhner, R. 1988. Some useful data structures for the generation of unstructured grids. Comm. Appl. Num. Meth. 4: 123-135.

Mar, A. 1993. Adaptive mesh refinement for nonlinear problems in geomechanics using an advanced constitutive law. Ph.D. thesis: University of Manchester.

Mar, A. & Hicks, M.A. 1996. A benchmark computational study of finite element error estimation. Int. J. Num. Meth. Eng. 39: 3969-3983.

Molenkamp, F. 1981. Elasto-plastic double hardening model Monot. LGM Report CO-218595: Delft Geotechnics.

Peraire, J., Vahdati, M., Morgan, K. & Zienkiewicz, O.C. 1987. Adaptive remeshing for compressible flow computations. J. Comp. Phys. 72: 449-466.

Zienkiewicz, O.C. & Zhu, J.Z. 1987. A simple error estimator and adaptive procedure for practical engineering analysis. Int. J. Num. Meth. Eng. 24: 337-357.

Numerical Models in Geomechanics, Pietruszczak & Pande (eds) © *1997 Balkema, Rotterdam, ISBN 90 5410 886 X*

Comparison of convergence of 8-node and 15-node elements

F. Molenkamp
Department of Civil and Structural Engineering, University of Manchester, Institute of Science and Technology, UK

S. Kay
Fugro Engineers BV, Leidschendam, Netherlands

ABSTRACT: It is shown that in the analysis of plane problems with elasto-plastic materials models the rate of convergence of equilibrium iterations using the global elastic stiffness matrix can differ significantly between 8-node and 15-node finite elements. The convergence limit as experienced with 15-node elements seems to be related to the magnitude of the allowed tolerance of the stress with respect to the yield surfaces implemented in the numerical code of the constitutive models. Apparently 15-node elements require significantly smaller tolerances than 8-node elements to satisfy the same global measure of accuracy.

1. INTRODUCTION

The aim of the paper is to demonstrate the different convergence rates of initial stiffness iterations for the calculation of equilibrium of plane problems with elasto-plastic material models using 15-node finite elements as compared to 8-node elements. The benchmark problem of a smooth passive wall studied by Hicks (1995), who used 8-node finite elements with 4-point reduced integration and the double hardening elasto-plastic Monot model (Molenkamp, 1980) are used herein for drained deformation. In this paper the convergence rate of a finite element program using this 8-node element and the 15-node element with 12-point integration (Sloan, Randolph, 1982) are compared. Two constitutive models are used, namely the classical purely elastic, purely plastic Mohr-Coulomb model and the Monot model (Molenkamp, 1980).

First the equivalence of the parameters used for both models are demonstrated by simulating a plane-strain compression test. The passive wall problem is then analysed for the above element types and constitutive models and several levels of the convergence criterion. Radically different convergence rates are observed for both material models. The cause of the slow convergence for the 15-node finite element is discussed in detail.

2. PLANE-STRAIN COMPRESSION WITH MOHR-COULOMB AND MONOT MODELS

The material parameters used for the double hardening elasto-plastic Monot model (Molenkamp, 1980) are identical to those used by Hicks (1995). It represents the properties of a loose sand. To illustrate the effect of these parameters a drained plane strain compression test is simulated at a lateral effective stress of 5 kN/m^2, starting at the isotropic stress state. This applied lateral effective stress equals the average vertical effective stress in front of the passive wall considered later. The test is loaded to 5 % strain and then unloaded. The stress-strain response according to the Monot model is shown in figure 1 together with that of the equivalent Mohr-Coulomb model. In figure 2 for both constitutive models the volumetric strain is shown as a function of the enforced compressive strain.

For the approximately equivalent purely elastic, purely plastic Mohr-Coulomb model the following parameters are used, namely the isotropic linear elastic shear modulus G = 685 kN/m^2 and bulk modulus K = 1660 kN/m^2, the friction angle ϕ = 36 degrees, the cohesion c = 0 kN/m^2 and the dilatancy angle ψ = 0 degrees.

From the unloading parts in figure 1 it can be

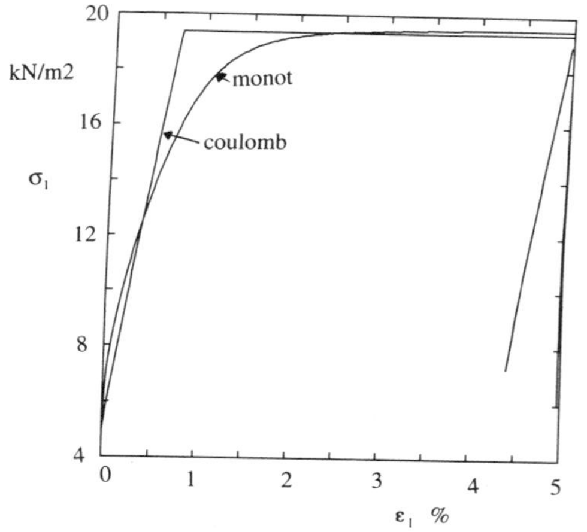

Figure 1: Major principal stress versus enforced compressive strain.

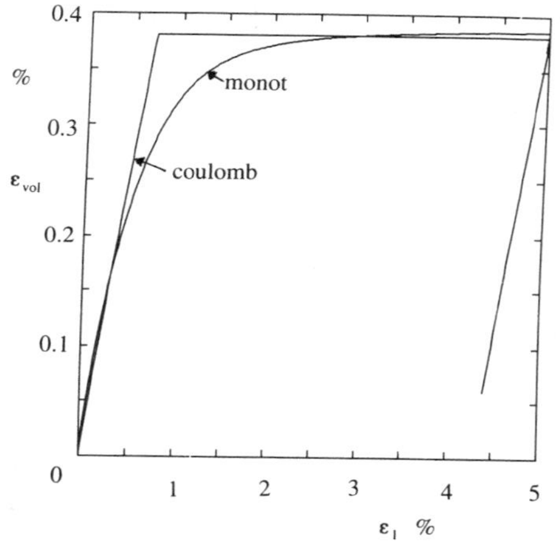

Figure 2: Volumetric strain versus enforced compressive strain.

observed, that the Monot model is much stiffer in unloading than the Mohr-Coulomb model, for which the loading and unloading stiffnesses are identical. The numerical data indicate a ratio of about 30 between the stiffnesses. This number is useful when later comparing the convergence rates.

3. DRAINED PASSIVE WALL PROBLEMS USING 8-NODE AND 15-NODE ELEMENTS.

The smooth passive wall problem (Hicks, 1995) involves a block of soil with unit width, a height of 2 metres and a length of 5 metres. The applied rough

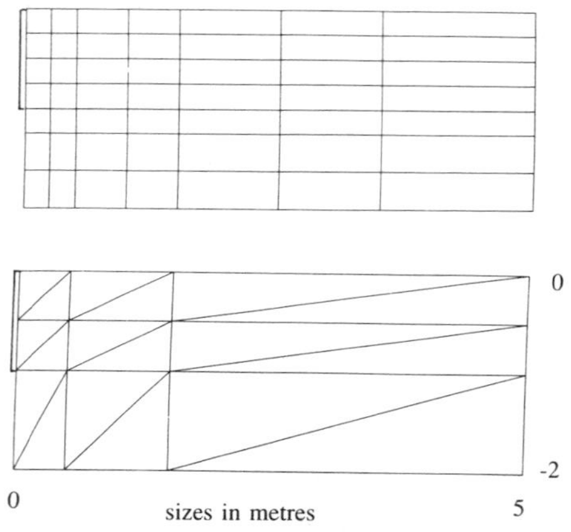

Figure 3: Finite element meshes of 8-node and 15-node elements.

8-node and 15-node finite element meshes are compared in figure 3. The smooth wall with a height of one metre is pushed horizontally 50 mm using displacement control with an enforced displacement of 0.2 mm per step. On the lower part of this left boundary the initial traction is maintained constant. The bottom and right boundaries are fixed.

The initial effective stress state is due to the submerged volumetric weight of 10 kN/m2 and a coefficient of lateral pressure at rest of $K_0 = 1$. In the Monot model the soil is assumed to be normally consolidated with the initial isotropic stress state located on both yield surfaces.

For convergence the following measure r for the relative error of the equilibrium is used, namely

$$r = \sqrt{\frac{\Sigma_i \Sigma_j U_{ij}^2}{\Sigma_p \Sigma_q H_{pq}^2}} \qquad \textbf{(1)}$$

in which U_{ij} is the component of the unbalance load vector on node j of the mesh in direction i and H_{pq} is the component of the nodal load vector on node q of the wall in direction p. Here only the horizontal load on the wall is considered thus p=1. The symbol Σ indicates the sum over all relevant quantities.

The unbalance load U is defined as follows

$$U_{ij} = \int \gamma_i N_j \, d\Omega + F_{ij} - \int B_{ijkl} \sigma_{kl} \, d\Omega \qquad \textbf{(2)}$$

in which γ_i is the component of the volumetric

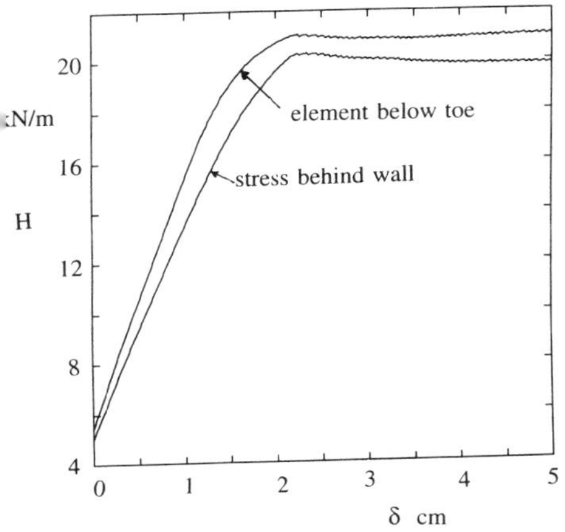

Figure 4: Horizontal force versus horizontal displacement of wall for Mohr-Coulomb model and 8-node finite elements with 4-point integration.

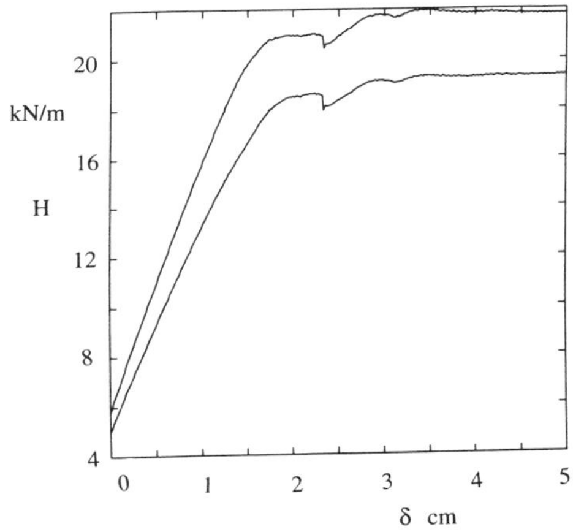

Figure 5: Horizontal force versus horizontal displacement of wall for Mohr-Coulomb model and 15-node finite element mesh.

weight in the i-th direction of the Cartesian basis, N_j is the shape function of the j-th node, F_{ij} is the external load on node j in the i-th Cartesian direction and B_{ijkl} is the strain-nodal displacement matrix with k and l indicating the components of the stress matrix σ_{kl}.

The calculations were done for several magnitudes of the allowable relative error. Typical horizontal load - horizontal displacement curves for the Mohr-Coulomb model for r = 0.02 and the 8-node element with 4-point integration and the 15-node element are shown in figures 4 and 5 respectively. Two curves are shown; the lower curve is due to the stress in front of the wall and the upper curve due to the nodal loads of the wall also involving the element below the toe of the wall. The curves for the 8-node element mesh in figure 4 are rather smooth. However, significant irregularities occur in figure 5 for the 15-node element mesh. A maximum of 32 iterations are needed for the 15-node element mesh at about 23 mm displacement as shown in figure 6 whereas for the 8-node element at maximum 3 iterations suffice.

In the calculations with the Monot model the elastic unloading material stiffness was used to compose the global stiffness matrix. A similar difference in convergence behaviour was observed. The load-displacement curve for the 8-node mesh in figure 7 is again rather smooth and the upper curve is equal to that published by Hicks (1995). The curves for the 15-node mesh in figure 9 show more significant irregularities at the state of failure. These

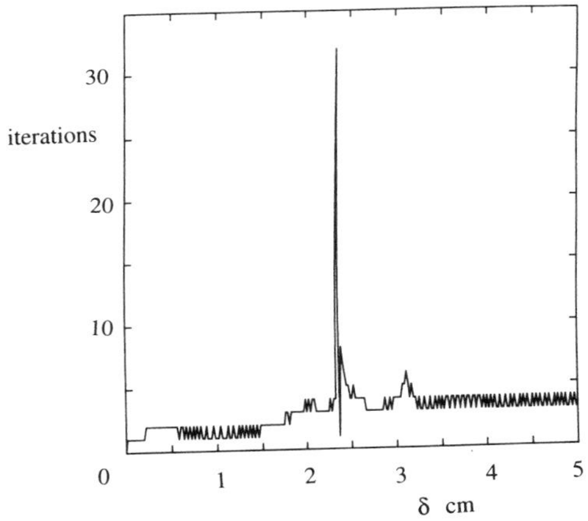

Figure 6: Number of iterations versus horizontal displacement of wall for Mohr-Coulomb model and 15-node finite element mesh.

irregularities are again supported by the differences in the convergence rates as shown in figures 8 and 10 respectively. For the 15-node element mesh the maximum number of iterations is about 280, while for the 8-node elements at most 70 iterations are needed.

Similar results were obtained for the other smaller allowable relative errors. The results in terms of the horizontal load on the wall at 0.05 m displacement have been collected in table 1. Table 2 contains the corresponding total number of iterations required.

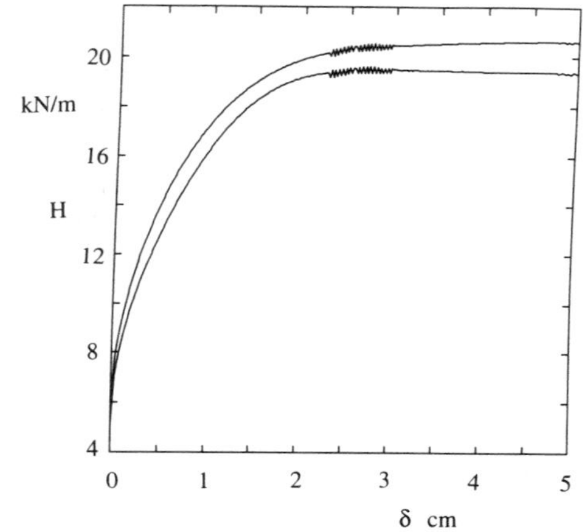

Figure 7: Horizontal force versus horizontal displacement of wall for Monot model and 8-node finite elements with 4-point integration.

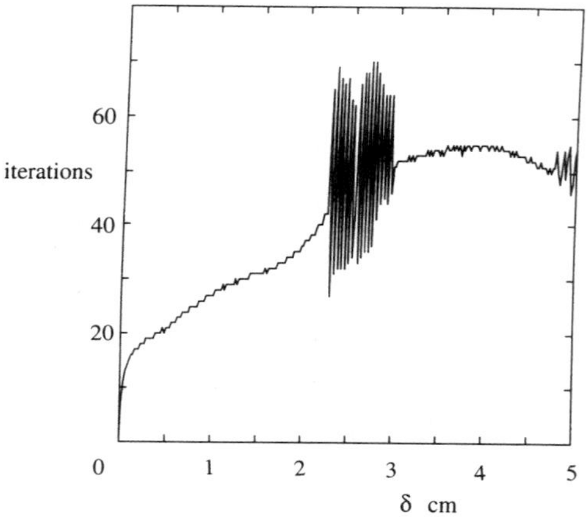

Figure 8: Number of iterations versus horizontal displacement of wall for Monot model and 8-node finite elements with 4-point integration.

From table 1 it is seen that the wall load due to the stress behind the wall decreases with the allowable relative error. For a given relative error the load is larger for the 8-node mesh than for the 15-node mesh. Table 2 shows that the number of iterations increases with decreasing allowable error. For both constitutive models and the same relative error the number of iterations is smallest for the 8-node element and largest for the 15-node element.
As indicated in table 1 convergence could not be reached in some cases. With reducing allowable relative error this occurred first at r = 0.01 for the

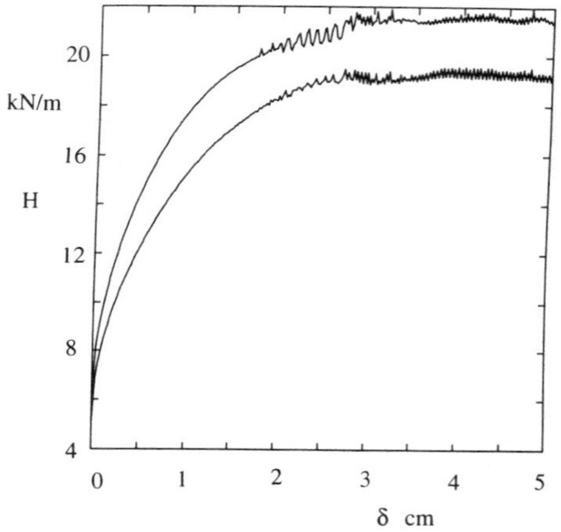

Figure 9: Horizontal force versus horizontal displacement of wall for Monot model and 15-node finite element mesh.

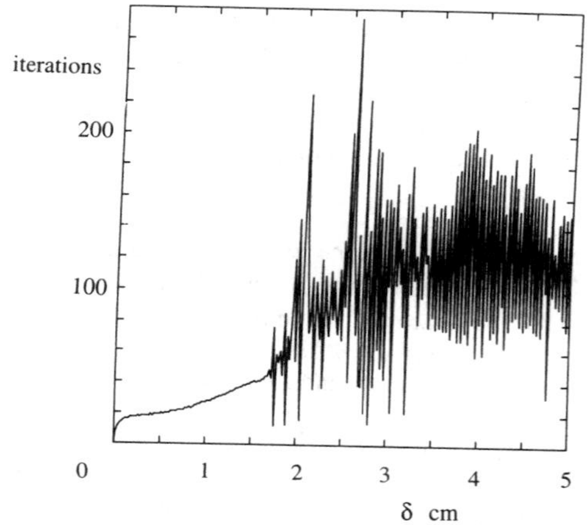

Figure 10: Number of iterations versus horizontal displacement of wall for Monot model and 15-node finite element mesh.

Monot model using the 15-node element. In fact no convergence was achieved within 10000 iterations within the 129-th displacement increment to reach a horizontal wall displacement of 0.0258 m.

4. CONVERGENCE WITH 15-NODE ELEMENT.

The problem with convergence with the 15-node element and the Monot model at the allowable relative error r = 0.01 at displacement step 129 is illustrated in figure 11 in terms of the relative error

Table 1. Horizontal load on wall in [kN/m] due to stress behind wall at wall displacement of 5 cm.

Allowable error	Horizontal load [kN/m]			
	Mohr-Coulomb model		Monot model	
	8-node element 4-point integration	15-node element 12-point integration	8-node element 4-point integration	15-node element 12-point integration
0.04	20.62	19.79	19.80	19.78
0.02	19.79	19.35	19.31	19.31
0.01	19.43	18.84	19.10	?
0.005	19.26	18.61	19.01	?
0.0025	19.17	18.55	18.98	?

Table 2. Total number of initial stiffness iterations at wall displacement of 5 cm

Allowable error	Total number of iterations			
	Mohr-Coulomb model		Monot model	
	8-node element 4-point integration	15-node element 12-point integration	8-node element 4-point integration	15-node element 12-point integration
0.04	261	332	5043	7890
0.02	503	748	10308	20361
0.01	984	1612	19924	?
0.005	1693	3305	36819	
0.0025	2922	6519	53731	

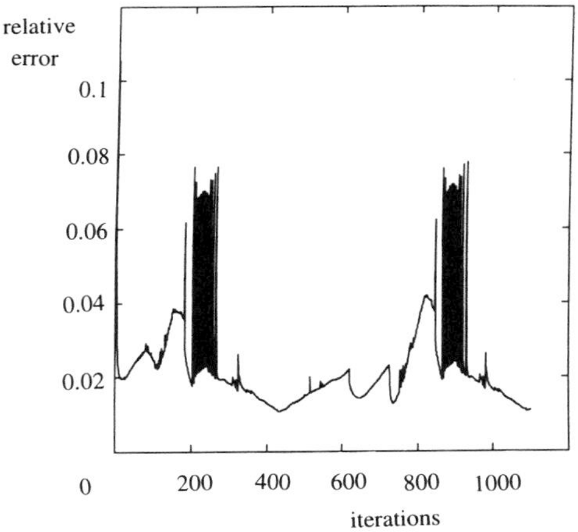

Figure 11: Relative error versus 1100 iterations in displacement step 129 for Monot model and 15-node finite element mesh.

versus the iteration step number up to 1100 iterations. Characteristic minimum and maximum relative errors occur at iteration steps 434, 820 and 1092 iterations. In fact this error has a cyclic nature with about 16 cycles per 10000 iterations.

Detailed comparison of the stress states occurring at these characteristic iterations showed that in the shear zone starting from the toe of the wall and propagating inclined upward behind the wall these stress states vary between two extreme states. One extreme stress state occurs at subsequent maximum relative error and the other one at the minimum errors. The stress difference between these extreme stress states is only significant in this shear zone. It is distributed in the form of parallel zones of increased and reduced principal pressure. These zones are parallel to the shear zone direction and alternate in the normal direction to the shear zone at subsequent neighbouring integration points.

5. CONCLUSIONS.

The convergence of finite element calculations with elasto-plastic material models can be problematic with plane 15-node elements, while 8-node elements produce satisfactory results. From the results given in this paper this convergence limit seems to be related to the applied allowable tolerance of the stress from the yield surface as applied in the numerical implementation of the elasto-plastic model.

6. REFERENCES.

Hicks, M. A. (1995)
MONICA - A computer algorithm for solving boundary value problems using the double-hardening constitutive law MONOT: II. Algorithm validation. Int. J. Numer. Anal. Methods in Geomechanics, Vol. 19, 29-57.

Molenkamp, F. (1980)
Elasto-plastic double hardeningmodel MONOT. Report of Laboratorium voor Grondmechanica, Delft. Second revision. November 1983.

Sloan, S. W., Randolph, F. M. (1982)
Numerical prediction of collapse loads using finite element methods, Int. J. Numer. Anal. Methods in Geomechanics, Vol. 6, 47-76

Numerical Models in Geomechanics, Pietruszczak & Pande (eds) © *1997 Balkema, Rotterdam, ISBN 90 5410 886 X*

Finite elements and parallel computation in geomechanics

I.M.Smith
School of Engineering, University of Manchester, UK

M.A.Pettipher
Computing Centre, University of Manchester, UK

INTRODUCTION

This paper describes how the improved syntax and array handling facilities of Fortran 90 (Fortran 95) lead to neat coding which in turn highlights the scope for vectorisation and parallelisation of finite element algorithms. The stage has now been reached where typical finite element programs for geomechanics problems can make very good use of a range of current parallel architectures. The intention has been to develop a strategy which is not specific to any single problem type or any specific parallel computer.

BRIEF RESUME OF PAST WORK

Parallel finite element computations began to be reported about 10 years ago (eg Farhat and Wilson, 1987). Until relatively recently, the diversity of target machines and unavailability of general purpose software meant that isolated communities of users worked in their own specific environments.

Despite the basic problems of parallelisation being identified - communication versus computation, gather and scatter, scalability, load-balancing and so on, relatively little of general applicability emerged. For example Topping and Khan (1996) describe a body of work targeted at transputer architectures while a special volume of "Computer Methods in Applied Mechanics and Engineering" - v119, November 1994 - reports on a Symposium on Parallel Finite Element Computations targeted almost exclusively at the CM-5. Neither of these platforms is particularly viable today.

Papadrakakis (1997) describes recent work in the area.

Perhaps the "Grand Challenge" strategy has also not helped the production of general purpose parallel algorithms for finite element work. While a lot may have been done in CFD for example, much finite element expertise continues to reside in the general area of solid and geomechanics which has not been designated as a Grand Challenge area.

Interesting work on solid mechanics has been featured in the EUROPORT project, in which industry-standard codes have been ported to parallel architectures regardless of efficiency of implementation.

MOTIVATION

In other areas of solid mechanics, particularly crash dynamics, very large problems are routinely solved by finite elements. Typically, several million degrees of freedom may be involved. Run times of hundreds of hours on a workstation (which is what industry can afford) are too long to influence the design cycle effectively. Hence there is a motivation to reduce the run time to, say, 10 hours by running 20 workstations in parallel with an efficiency of 50%.

In geomechanics, there are plenty of comparable problems to be solved. At present these are routinely approximated in two dimensions but some are inherently three dimensional. For example a piled raft fully modelled in 3-d as shown in Figure 1 or nailed soil wall, curved in plan as shown in Figure 2 require a full 3-d finite element representation. In both cases roughly 1 million degrees of freedom are involved (Smith and Wang, 1994). We seek a portable strategy for solving such problems in parallel.

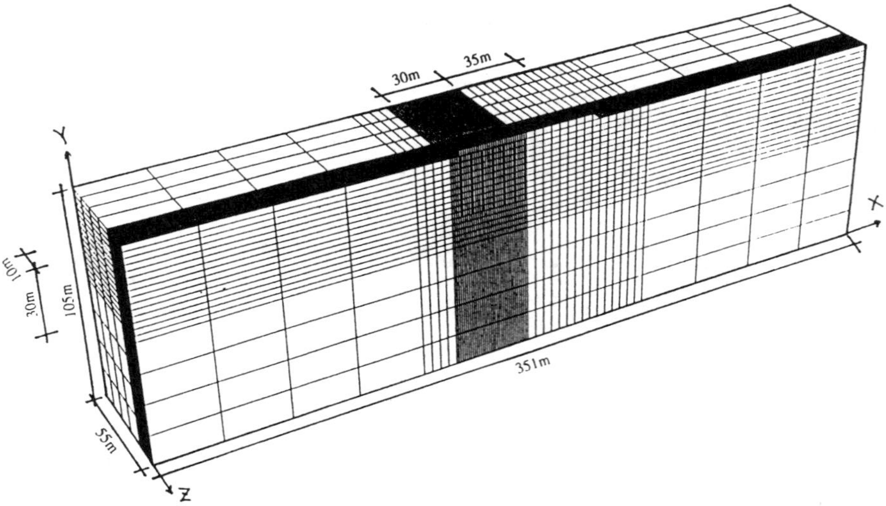

Figure 1 Piled raft

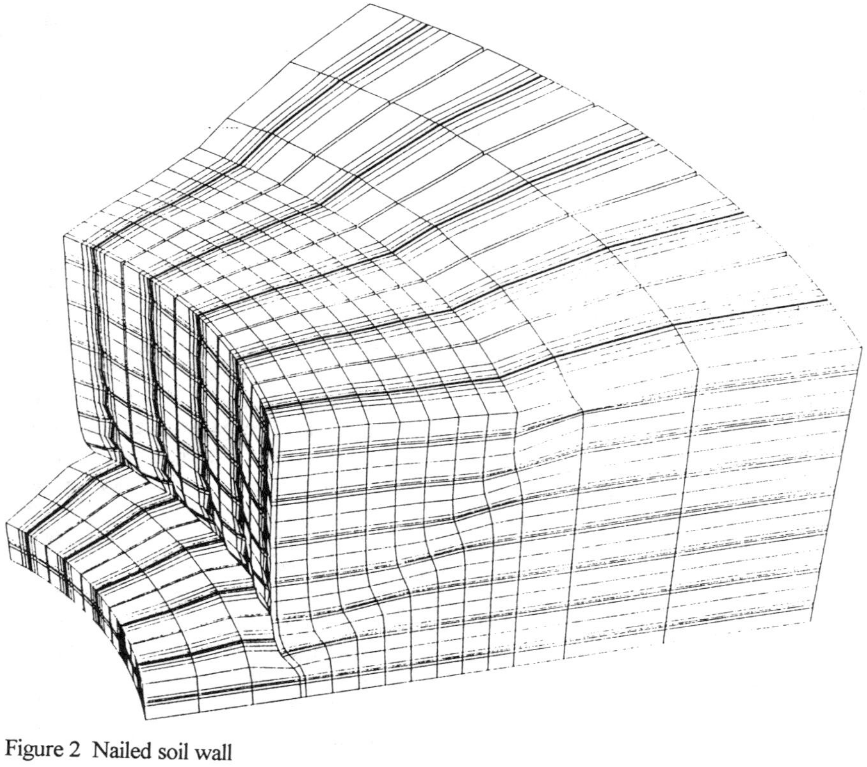

Figure 2 Nailed soil wall

PRESENT STRATEGY FOR GENERAL PURPOSE FINITE ELEMENT ANALYSIS

Much finite element analysis methodology can be seen to be concentrated in three main areas (Smith and Griffiths, 1997). Broadly speaking, these are

(i) Static equilibrium : solution of linear simultaneous equations
(ii) Dynamic equilibrium : solution of linear eigenvalue equations
(iii) Evolutionary equilibrium : solution of first or second order ordinary differential equations in time.

As a strategy for area (i) we have chosen (somewhat arbitrarily) preconditioned conjugate gradients. This arose as the most natural finite element strategy and avoids "domain decomposition" or "tearing". The (Fortran 90) algorithm has been given many times (Smith, 1992) and takes the form:

```
PROGRAM PRECON
USE LIBRARY
   allocatable array declarations
        read data
   allocate arrays

  build and store the element stiffness
        matrices (in blocks) and the
          preconditioner
!----- PRECONDITIONED C.G. ITERATIONS -----
   ITERATIONS : DO
      ITERS = ITERS+1
   gather from the right hand side vector
K-matrix *vector or K-matrix *matrix multiply
   scatter back to right hand side vector
!------------- PCG EQUATION SOLUTION ---------
  various dot products, saxpy and
     convergence testing
     END DO ITERATIONS
!----------------- STRESS RECOVERY -------------
END PROGRAM PRECON
```

Area (ii) we attack by Lanczos-type algorithms which have much in common with pcg. At their core is the gather/matrix multiply/scatter sequence seen above. Lack of space precludes a full description here but see Smith and Griffiths (1997).

Area (iii) may be regarded "implicitly" as a sequence of linear simultaneous equation solutions at every chosen time interval and hence pcg could be used again but a more interesting comparative algorithm is the "explicit" one taking the form (which naturally embraces nonlinearity):

```
          PROGRAM EXPLICIT
          USE LIBRARY
     allocatable array declarations
          read data
      allocate arrays
!----------- EXPLICIT INTEGRATION LOOP ------
          TIME_STEPS : DO
     TIME = TIME + TIME_INCREMENT
          ELEMENTS : DO
     gather from the right hand side vector
      build the element B matrix
        B-matrix *vector multiply
  scatter back to right hand side vector
          END DO ELEMENTS
update accelerations, velocities, displacements
          END DO TIME_STEPS
          END PROGRAM EXPLICIT
```

It should be clear that the same programming strategy carries over from area (i) through area (ii) to area (iii).

THE PARALLELISATION STRATEGY (HPF)

Since all programs are in Fortran 90, and utilise its features such as whole array operations, array intrinsics and so on, the natural parallelisation route is via HPF, which may be viewed as an extension to Fortran 90.

Before illustrating this, it is worth mentioning the attractions of Fortran 90 for the gather/matrix multiply/scatter operations seen as the essential core of the above algorithm (Smith, 1993, 1994).

If INDICES is an integer vector containing the index components of a large real vector GLOBAL, gathering and scattering can simply be accomplished via GLOBAL (INDICES) in Fortran 90. The matrix by vector or matrix by matrix multiplication coming between the gather and scatter is done by an intrinsic procedure MATMUL in Fortran 90. If, as can happen, the manufacturer's implementation of MATMUL is disappointing, the BLAS equivalent can be substituted with minimum disruption. The HPF extension to program PRECON for linear equation solution by preconditioned conjugate gradients is shown below:

```
PROGRAM PRECON_HPF
USE LIBRARY
        allocatable array declarations
!HPF$ directives : PROCESSORS, ALIGN
        WITH, DISTRIBUTE etc
            allocate arrays

        build and store the element stiffness
                matrices (in blocks) and the
                  preconditioner
                (involves FORALL loops)
!------ PRECONDITIONED C.G. ITERATIONS ------!
HPF$ INDEPENDENT
   ITERATIONS : DO
        ITERS = ITERS+1
        gather from the right hand side vector
                (involves FORALL)
K-matrix *vector or K-matrix *matrix multiply
        scatter back to right hand side vector
                (involves FORALL)
!----------- PCG EQUATION SOLUTION -----------
                various dot products, saxpy and
                   convergence testing
               (involves FORALL)
                 END DO ITERATIONS
!--------------- STRESS RECOVERY ----------------
END PROGRAM PRECON_HPF
```

This algorithm has been run successfully on a cluster of Silicon Graphics workstations and on a T3D. It is not efficient on either, but the very fact that it works, having involved minimal alterations to the original Fortran 90, is believed to be highly significant for the future.

THE PARALLELISATION STRATEGY (MESSAGE PASSING)

Pending efficient vendor implementations of HPF, the main parallelisation strategy which has been explored is via message passing. This has included "portable" solutions such as PVM and MPI and also machine-specific software such as Cray's SHMEM routines. Initially, only PVM was available via a small Cray EL98 emulator, but subsequently most effort has been expended on MPI. This has been run successfully on the following machines:

- Cray T3D
- IBM SP2
- Fujitsu VPP

and so is believed to be essentially portable. The underlying coding is that of PRECON and PRECON_HPF described previously but the MPI calls are intrusive and it remains to be seen how well these can be "hidden" to provide workable solutions for general purpose users.

FINITE ELEMENT "GRANULARITY"

Given the iterative program strategies for areas (i),(ii) and (iii) described above there is no need to assemble elements at all. In the limit (eg van Rietbergen et al, 1996) all elements are the same (in their case identifiable with screen voxels) but more generally (Smith and Wang, 1995) one can anticipate groups made up of parent elements which consist either of identical elements or of parents whose children differ only by a scalar (for example stiffness) or vector from one another. When there are groups of elements, the matrix*vector multiply inherent in all algorithms becomes matrix*matrix and in general this produces faster programs (probably due to cache structure). If single element matrix*vector or matrix*matrix is too small to achieve effective vectorisation (for example on a VPP) elements are merely assembled into blocks of whatever size is optimal for that machine. The factor dominating algorithm speed is then not Fortran 90 MATMUL but whatever sparse matrix*vector multiplication is chosen. In any event, the "hot spots" will be a few Fortran statements which can be optimised or replaced by BLAS equivalents.

RESULTS

The Tables given below refer to computations on a Cray T3D using up to 64 processors. The problem is from area (i) and results are presented for an "ideal" linear calculation and for a nonlinear calculation both with approximately 100,000 equations solved by pcg.

The first table of results is for an MPI version with some machine-specific SHMEM contributions.

The results are encouraging. In Table 1 sustained speeds of 3.2 Gflop/s have been obtained on 64 processors. Note that the peak single node performance was only 40% of machine peak. Despite increasing remote/local accesses, the overall performance scaled well.

Table 1 : Linear Problem

Processors		1	4	16	32	64
Total	cpu	30.9	8.78	3.02	2.83	2.93
Broadcast and allocate	cpu	0.10	0.14	0.22	0.33	0.53
Element Computation	cpu	2.29	1.25	0.88	1.49	1.86
Iterations Section	cpu	28.3	7.33	1.90	1.02	0.53
	effic	100	96	93	86	81
	Mflop/s	62	238	914	1702	3234
Remote/Local Accesses		0	0.11	0.63	2.14	14.99
Machine Peak	Mflop/s	150	600	2400	4800	9600

The second table of results is for a pure MPI version.

Table 2 : Nonlinear Problem

Processors		16	32	64
Total	cpu	228.5	116.4	66.2
Broadcast and allocate	cpu	0.26		0.57
Element integration	cpu	201.8	101.3	50.1
PCG iterations	cpu	13.5	7.3	3.8
	Mflop/s	5.5	960	1833
Peak	Mflop/s	2400	4800	9600

In Table 2, element integration computation was more significant and scaled perfectly as one would hope.

Peak speed achieved was about 1.8 Gflop/s on 64 processors. In general, scaling was reasonable and provides plenty of incentive for further developments.

REFERENCES

Farhat, C. and Wilson, E. (1987) A New Finite Element Concurrent Computer Architecture, IJNME, 24, 9, pp 1771-1792.

Papadrakakis, M. (1997) Parallel Solution Methods in Computational Mechanics. Wiley Series in Solving Large-Scale Problems in Mechanics, John Wiley.

Smith, I.M. (1992) Programming in Fortran 90, John Wiley.

Smith, I.M. (1993) The Impact of Fortran 90 on Computational Mechanics Codes, Proc ACME Symposium, Swansea.

Smith, I.M. (1994) Experience with Three Fortran 90, Compilers, Proc ACME Symposium, Manchester, pp 31-37.

Smith, I.M. and Wang, A. (1995) Finite Elements, Fortran 90 and Parallelisation, Proc ACME Symposium, Glasgow, pp 29-32.

Smith, I.M. and Griffiths, D.V. (1997) Programming the Finite Element Method, 3rd edition, John Wiley.

Topping, B.V. and Khan, A. (1996) Parallel Finite Element Computations, Saxe Coburg Press.

van Rietbergen, B., Wemans, H., Huiskes, R. and Polman, B.J.W. (1996) Computational Strategies for Iterative Solutions of Large FEM Applications Employing Voxel Data, IJNME, 39, 16, pp 2743 -2788.

Numerical Models in Geomechanics, Pietruszczak & Pande (eds) © *1997 Balkema, Rotterdam, ISBN 90 5410 886 X*

Computer-aided seismic analysis of soils

H. Modaressi, E. Foerster & A. Mellal
BRGM, Direction de la Recherche, France

ABSTRACT : It is now commonly admitted that when encountering soft soils, the seismic ground motion is largely affected. To study such aspects a computer-aided design tool called *CYBERQUAKE*, in which the behaviour of soils is introduced by a cyclic elastoplastic model, has been recently developed. It is a user-friendly program written in an *object-oriented* language with integrated graphics and in which several tools necessary for earthquake engineers are included. In this paper some specific features of this software and two illustrative examples are presented.

1 INTRODUCTION

Anti-seismic design of structures is of paramount interest in earthquake prone areas. For this purpose, numerous accelerometer networks have been installed all around the world. Moreover, it is now commonly admitted that on soft soils, the seismic motion on the ground is largely affected by local conditions. Non linear behaviour of these materials with respect to distortion induced by seismic loading makes methods such as empirical Green's functions inappropriate for driving strong ground motion from recorded weak ground motion.

That is the reason why since the last two or three decades it has been recommended that strong motions should be recorded on outcropping bedrock and that the seismic motion on soil deposits should be evaluated taking into account the geometry and mechanical behaviour of subsurface.

For complex subsurface structures, i.e. sediment-filled valleys, sophisticated numerical techniques should be used. In the absence of lateral heterogeneities, a one-dimensional multilayered system may be considered. The most commonly-used technique in this case is the one based on viscoelastic multilayered soil with equivalent-linear model for sediments (Idriss and Seed 1968,1970). This equivalent-linear model may not be considered as a constitutive model able to represent the soil's behaviour and several drawbacks have been already listed in the literature.

In fact, this technique based mainly on the hysteretic behaviour of soils is unable to predict irrecoverable displacement and strain generation and consequently the liquefaction. Moreover, as this technique represents a total stress analysis, it is also unable to take into account the degradation of the stiffness resulting from the pore-pressure build-up in the soil. So it might overestimate the intensity of the shaking.

The pore-pressure develops in saturated undrained sands during seismic loading. It is caused by the fact that the presence of nearly incompressible water prevents plastic volumetric strain to occur. Hence, the load is transferred to the water resulting in an increase in pore-water pressure. Therefore the behaviour of the material is highly coupled with the pore-water pressure and an effective stress approach with an appropriate constitutive model which takes accurately into account the monotonous and cyclic behaviour of soils is needed.

Several general-purpose softwares, mainly based on the Finite Element method are now available for modelling the seismic behaviour of dry or saturated soils, e.g. GEFDYN, DYNAFLOW, SAWNDYNE. The difficulty in the use of such programs arises from the lack of data on some major factors as the initial stress state, the seismic input motion and the laboratory and *in situ* investigations for calibration of constitutive models. Their use is also limited to well-trained engineers familiar to numerical modelling techniques and able to handle nonlinear multiphase dynamic analyses.

In addition, the cost of parametric studies is high, while different hypotheses with respect to geometry, the hydro-mechanical behaviour, input motion, initial stress state, hydraulic boundary conditions, impedance contrast between the soil profile and bedrock, and finally the assumption of continuous *vs.* discontinuous medium have to be analysed.

Henceforth, simpler tools are necessary for current applications and some well-known softwares as *SHAKE* (Schnabel *et al.* 1972) based on equivalent-linear assumption are available for such purposes. Simple geometry assumption, when applicable, is the only simplification which does not introduce any supplementary uncertainty in the computed results.

We have recently developed an interactive software in which one-dimensional geometry and either two or three-dimensional kinematics have been assumed. Nevertheless, all other features of the complex behaviour of soils under seismic loading have been preserved. In addition, all aspects related to the numerical modelling technique are handled automatically and do not require any specific knowledge from the user. This tool is written with an *object-oriented language (Visual C++)* and is run on micro-computer platforms.

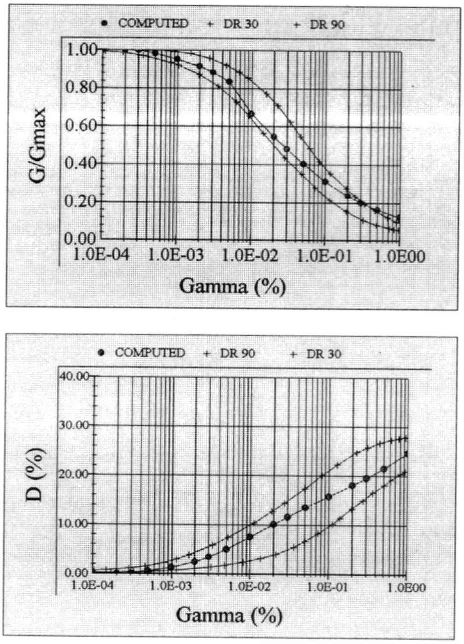

Fig. 1. Simulated variation of the (secant) shear modulus and damping ratio with distortion for a medium sand

2 THEORETICAL BACKGROUND

The transient analysis is based on the resolution of dynamic equation of motion and mass conservation completed by the theory of elastoplasticity for the constitutive law. In this section, we will briefly present some specific aspects of the constitutive model and the governing equations.

2.1 *Constitutive Modelling*

An appropriate constitutive model is the keystone of a reliable tool for evaluating the ground seismic response. For soft soils subjected to strong motion, the capacity of the model to simulate the response in a wide deviatoric strain range (from 10^{-6} to 10^{-2}) is essential. In the Soil Dynamics context, the variation of the secant shear modulus and the damping ratio with respect to this quantity usually called G-D/γ curves is considered as being of utmost importance. Pande and Pietruszczak (1986) compared several constitutive models and reported that almost all of them, except those based directly on this property, were unsuccessful in reproducing this feature of the soil behaviour. The elastoplastic cyclic model incorporated in the software is successful in reproducing coherent results as may be shown in figure 1. This model is based on the original model developed by Aubry *et al.* (1982).

2.2 *One/Two-phase Dynamic Modelling*

In the presence of a water table in the ground to be studied, it is assumed that the pore pressure is always nil above the water table (drained or one-phase condition), whereas the evolution is supposed to be either totally undrained or partly drained (two-phase) below the water table. The totally undrained assumption is correct in the case of very short term loading, such as earthquakes for soils with small to moderate permeability, and avoids solving the pore pressure diffusion equation. The partly drained condition is more realistic in long-term analysis as it simulates the pore-pressure dissipation after the earthquake. It is also more appropriate for soils with high permeability or for thin layers in immediate contact with highly draining layers.

3 NUMERICAL IMPLEMENTATION

3.1 *Transient Analysis*

The governing equations are discretized by finite elements and numerically integrated with respect to time by an explicit predictor-corrector Newmark scheme with an automatic choice of the time step. This latter is computed with respect to stability and precision requirements for the integration scheme.

3.2 *Equivalent Linear Analysis*

The equivalent linear analysis may be also carried out using this software. The computations are performed in the frequency domain.

3.3 *Control Point and Deconvolution*

Even though it is recommended to record seismic motion on outcropping bedrock, results are sometimes only available on soft soil deposits. Thus, it is necessary to deconvoluate the motion to an underlying bedrock and then perform computations directly from bedrock to the ground surface. The deconvolution is performed using the same approach as in the equivalent-linear analysis.

4 USER INTERFACE

The user interface plays an important role in a software aimed at a large professional environment. A screen hardcopy is presented in figure 2, showing that several soil profiles may be studied during the same session for either transient or equivalent-linear analyses. Only meaningful physical properties and the thickness of each layer are to be introduced. It is possible to identify the constitutive model parameters from experimental results using an interactive *Driver* integrated into the software. Several complementary tools necessary for soil-dynamic and seismological analyses are also included in the software. Fast Fourier direct and inverse transformations with low/high-pass filters, smoothing procedures using sliding windows, arithmetic operations on different functions, base line correction algorithms, response spectrum calculation are available. Graphics and on-line Help are also included. A Run-time display as presented in figure 3 permits a continuous control of computation progress, and one can interrupt/restart or break a computation at any time.

5 ILLUSTRATIVE EXAMPLES

Two applications using this software are presented here. In the first one a liquefiable soil is studied. In

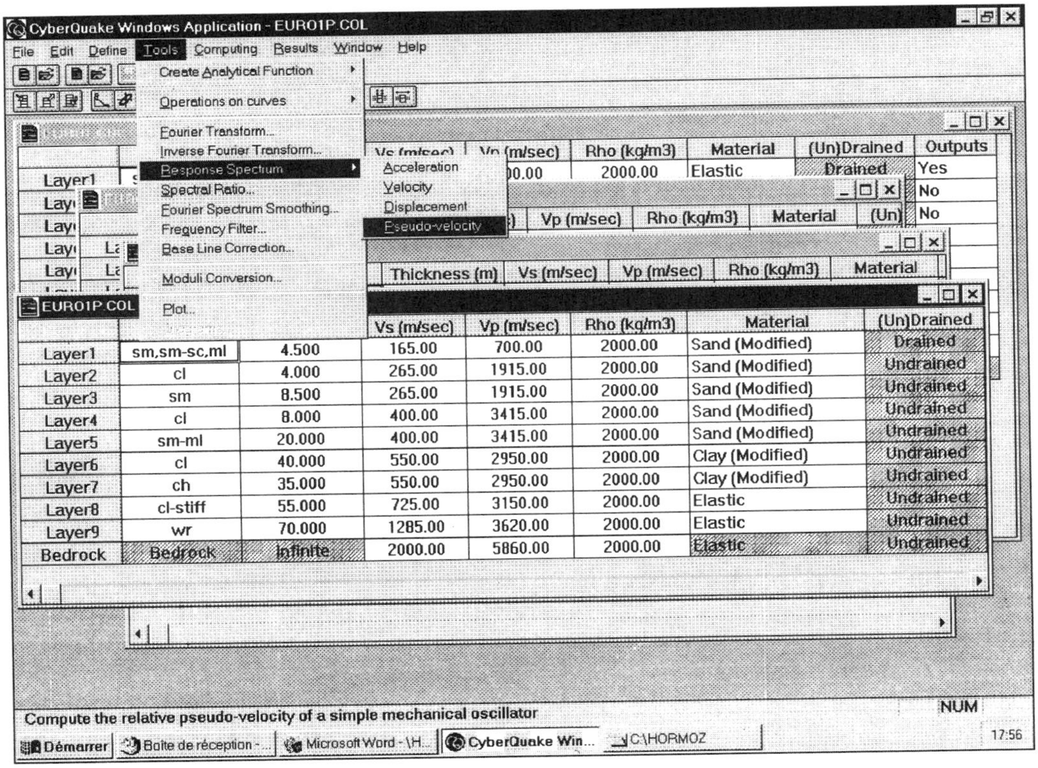

Fig. 2. Interactive analysis in Windows environment

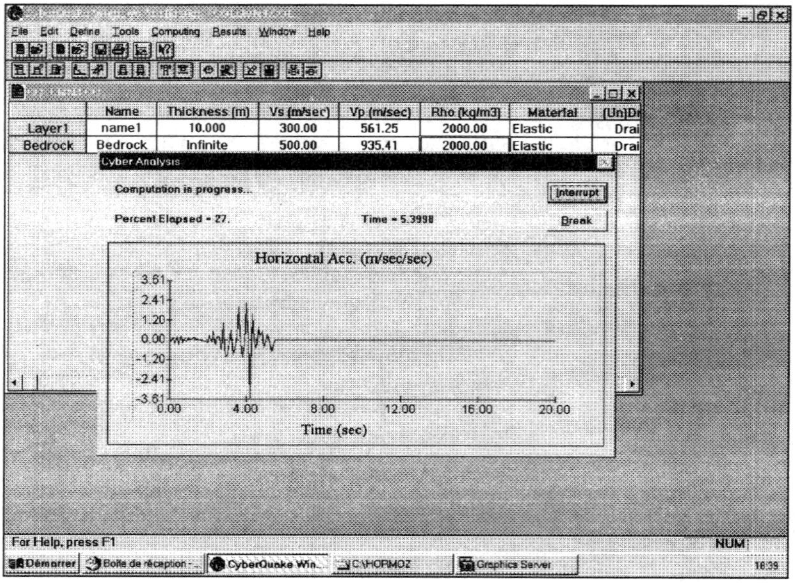

Fig. 3. Run-time display for continuous control of computations

Table 1. Soil profile for the first example

	height m	Vs m/s	Vp m/s	ρ kg/m³	Material	
Layer 1	4.0	150.	280.	2000	Sand	Drained
Layer 2	16.0	150.	280.	2000	Sand	Undrained
Bedrock	Infinite	500.	935.	2000	Elastic	Undrained

the second example the case of coupling between different components of the seismic motion due to plasticity is studied.

5.1 - *Liquefaction*

The soil profile characteristics for this example are given in Table 1.

The results presented in figures 4 through 8 shed light on some important features of the nonlinear behaviour of saturated soft soils subjected to seismic loading. The horizontal acceleration vanishes after the liquefaction of the soil at -12m. Moreover, the maximum acceleration is not obtained at the ground surface. This is conforming to observations during Kobe earthquake (see Mohammadioun 1997). The pore-pressure build-up and the effective stress are presented in figure 9 for the point located at -12m where the liquefaction has occurred.

The liquefaction induces large irreversible horizontal displacements as may be seen in figure 10. This large value is explained by the absence of any resistance assumed for the soil after the liquefaction in this computation.

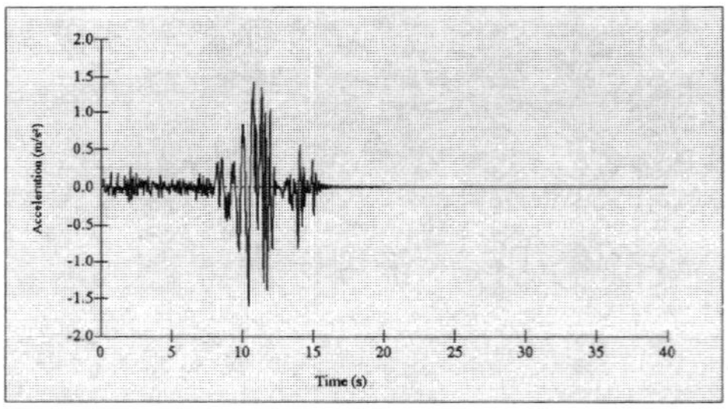

Fig. 4. Computed horizontal acceleration at the top

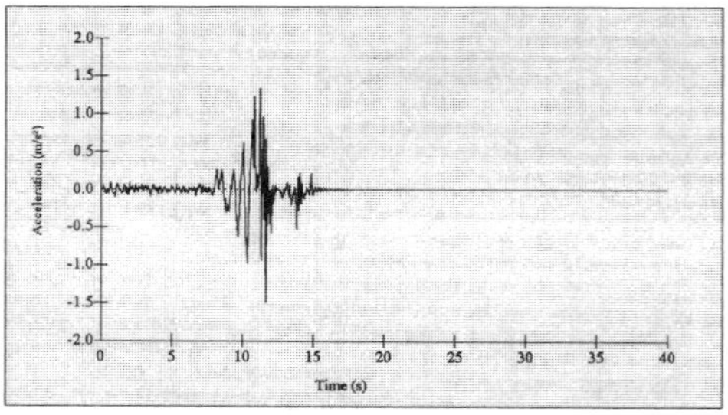

Fig. 5. Computed horizontal acceleration at -4m

5.2 - *Plasticity-induced coupling effects*

The non-linearity associated to the elastoplasticity highlights the flaw of computations assuming the

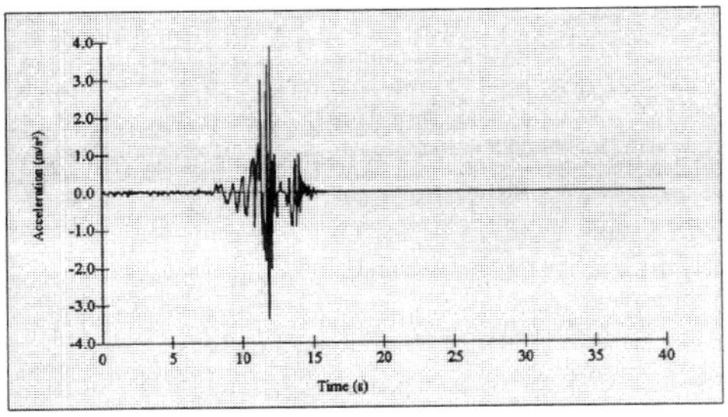

Fig. 6. Computed horizontal acceleration at -8m

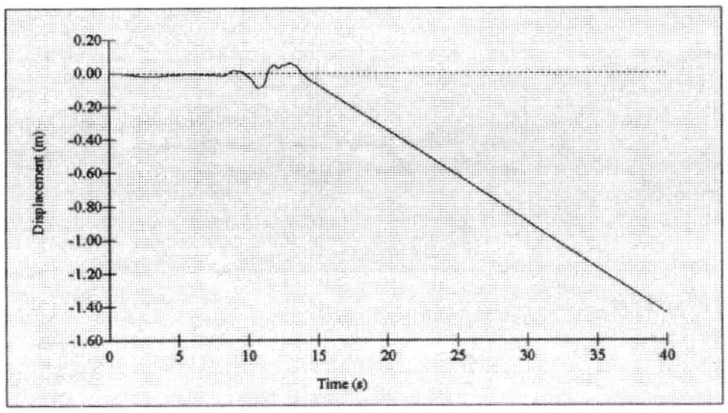

Fig. 10. Computed horizontal displacements at -12m

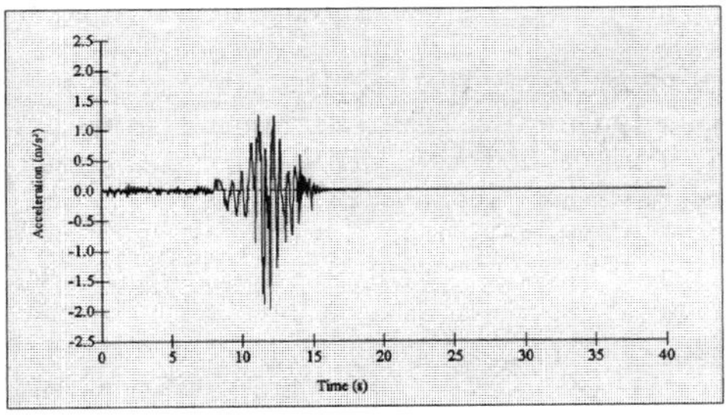

Fig. 7. Computed horizontal acceleration at -12m

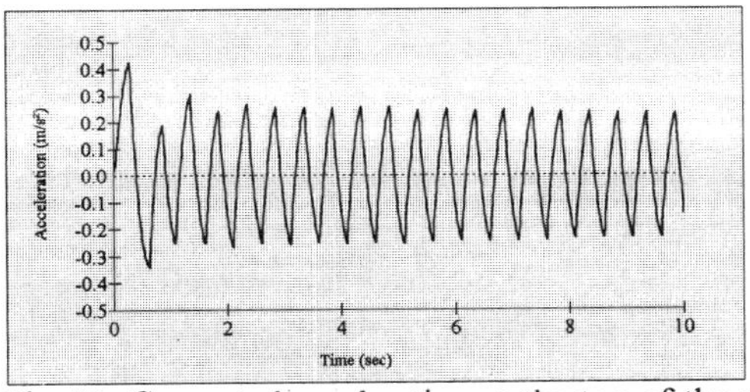

Fig. 11. Computed acceleration at the top of the soil column loaded only in the x direction

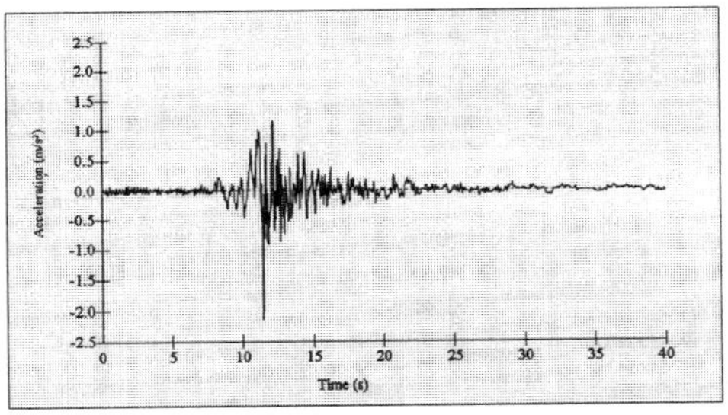

Fig. 8. Computed horizontal acceleration at -16m

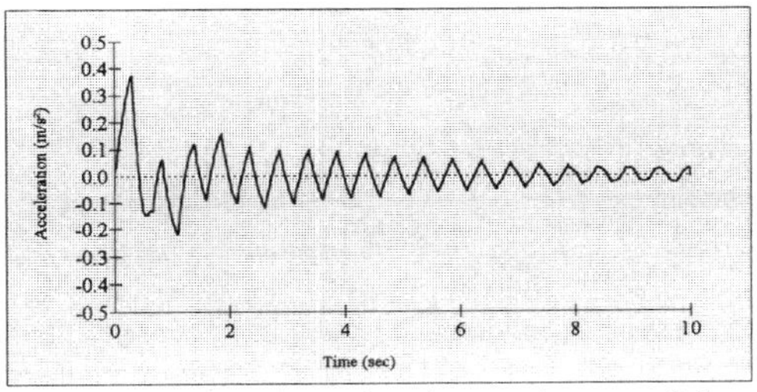

Fig. 12. Computed acceleration at the top of the soil column loaded in the x and y directions simultaneously

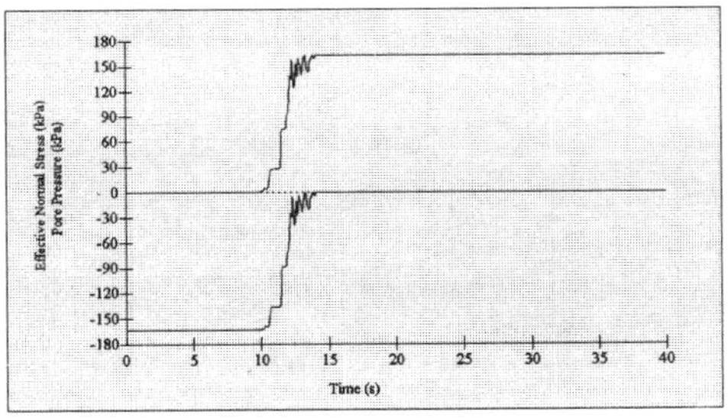

Fig. 9. Computed pore-pressure build-up (top) and effective stress (bottom) at -12m

body-wave decomposition hypothesis generally admitted in seismic analysis. Two calculations have been performed. In the first one, a soil profile is subjected to a harmonic signal applied only in the horizontal x-direction. In the second one, the same signal is applied in two orthogonal horizontal directions (x and y) and results are always presented in the x-direction. The computed acceleration at the ground surface shows a significant difference (see figures 11 and 12).

The time history of pore-pressures (Fig. 13 and 14) explains such a difference. In fact, in the case of simultaneous loading, the pore-pressure build-up results in the liquefaction of the soil layer. Consequently, the horizontal acceleration decreases while the horizontal displacement increases.

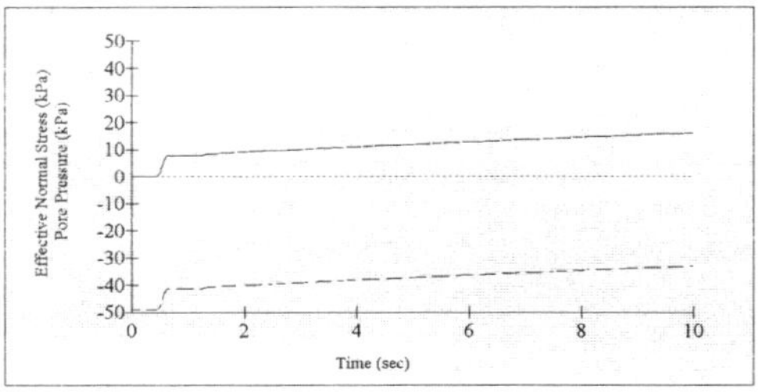

Fig. 13. Pore-water pressure build-up (top) and the effective normal stress (bottom) for the loading in the x direction only

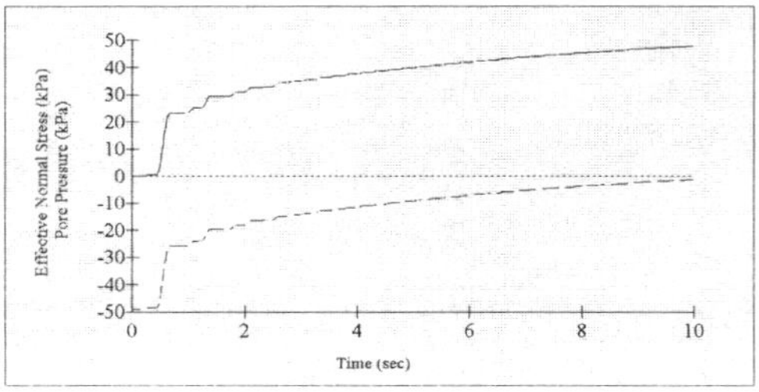

Fig. 14. Pore-water pressure build-up (top) and the effective normal stress (bottom) for the loading in the x and y directions simultaneously

6 CONCLUSION

The use of a more realistic analysis including the elastoplastic behaviour of soft soils does not enjoy the deserved popularity in engineering design. The difficulties associated to such computations have been possibly one of the main obstacles. The software presented in this paper aims at partly remedying this problem.

All numerical aspects for such analysis are treated internally and the attention of engineers may be only focused on the choice of physical properties. The main hypothesis concerning the geometry simplifies computations and allows for fast parametric studies. This hypothesis is encountered in many situations where the lateral heterogeneities are almost negligible. The effectiveness of this assumption may be examined once the multidimensional resonances assessed prior to one-dimensional computations using performant numerical codes based on linear behaviour assumption.

The software is user-friendly and is run on micro-computers. It is then considered as an individual package for use in office or *in situ* by engineers involved in Soil Dynamics studies.

REFERENCES

Aubry, D., Hujeux, J.C., Lassoudière, F. & Meimon, Y. 1982, A double memory model with multiple mechanisms for cyclic soil behaviour, *Int. Symp. Num. Mod. Geomech.*, Zurich.

Idriss, I.M. & Seed, H.B. 1968, Seismic response of horizontal soil layers, *J. Soil Mech. Found. Div.* ASCE, 94:1003-1031.

Mohammadioun, B. 1997, Non-linear response of soils to horizontal and vertical bedrock earthquake motion, *J. Earthquake Eng.*, vol. 1

Pande, G.N. & Pietruszczak, S. 1986, A critical look at constitutive models for soils, Geomechanical Modelling in Engineering Practice, (Ed. R. Dungar and J. Astuder), 369-395.

Schnabel P.B., Lysmer J. & Seed H.B., 1972, SHAKE: A computer program for earthquake response analysis of horizontally layered sites, *Earthquake Engineering Research Center*, Report No. UCB/EERC 72-12, Univ. of California, Berkeley, CA.

Seed, H.B. & Idriss, I.M. 1970, Soil moduli and damping factors for dynamic response analysis of horizontally layered sites, *Earthquake Engineering Research Center*, Report No. UCB/EERC 70-10, Univ. of California, Berkeley, CA.

Numerical Models in Geomechanics, Pietruszczak & Pande (eds) © 1997 Balkema, Rotterdam, ISBN 90 5410 886 X

Finite element method and limit analysis for geotechnical design

A. Antao, J. P. Magnan, E. Leca, P. Mestat & P. Humbert
Laboratoire Central des Ponts et Chaussées, Paris, France

ABSTRACT : This paper presents an application of finite element analyses which seems to be promising for the practice of geotechnical engineering : the regularized method of kinematic limit analysis (subroutine LIMI of the finite element code, CESAR-LCPC, developed by the Laboratoire Central des Ponts et Chaussées). This method was applied to two- and three-dimensional problems of shallow foundations, embankments on soft soils, unstable slopes, slope nailing and tunnels. The results were compared to existing analytical solutions and observed failures. Several examples of calculations are presented and discussed in the paper.

1 INTRODUCTION

The finite element method is widely used in two areas of geotechnical engineering : seepage through soils and dam design, on the one hand, and tunnel design, on the other hand. Research on earth structures, retaining structures and shallow and deep foundations makes frequent use of this type of calculation methods as well, but finite element analyses are not carried out systematically in every day practice, for these types of structure.

Design of such geotechnical structures relies on relatively simple, experience-based methods, the parameters of which cannot be used straightforwardly in a complete analysis within the framework of continuum mechanics. Nevertheless, the design of complex structures (e.g. interaction of various types of structures, built independently on soil exhibiting time dependent deformations, etc.) has led engineers to investigate the conditions for practical use of the finite element method, and as a result has produced experience for many types of problems.

This paper presents an application of finite element analysis which seems to be promising for the practice of geotechnical engineering : the regularized method of kinematic limit analysis. This approach has been applied to many geotechnical problems of stability in realistic situations (foundations, slopes, embankments, dams, underground openings, retaining structures).

2. LIMIT ANALYSIS

Limit analysis allows to obtain a bracketed estimate of the loads acting on a structure at failure, by providing an upper bound, on the basis of the kinematic analysis of failure mechanisms, and a lower bound, which is derived from a static analysis of the structure (Salençon 1983). The smaller the difference between the upper and lower bounds, the better the estimate of limit loads. Accurate solutions can even be determined for simple geotechnical problems.

For more complex problems, failure loads can only be estimated by means of this approach ; in such case, the upper bound theorem, which involves the implementation of a kinematically admissible failure mechanism, appears to be more straightforward than the lower bound theorem, which requires the determination of a statically admissible stress field.

The kinematic approach has been extensively used in France by Frémond & Salençon (1973), Salençon (1983), de Buhan (1986) and others, and has provided a valuable means for obtaining estimates of collapse loads for a variety of geotechnical structures. One shortcoming of this approach, however, is that it provides estimates of the loads at failure, which are on the unsafe side (upper bound solutions). Therefore, one essential issue to be addressed relates to positioning the upper bound solution with respect to the actual value of collapse load and optimizing the upper bound solutions derived from a series of kinematically admissible velocity fields.

Many efforts have been made to develop techniques that would allow an automatic determination of failure mechanisms. One approach consists in using a viscoplastic-regularization scheme, by means of the finite element method.

3 FINITE ELEMENT METHOD OF KINEMATIC LIMIT ANALYSIS

The finite element kinematic limit analysis approach is based on the Norton-Hoff model (Norton 1929) and on a variational form, which was originally proposed by Friaâ (1978) and Guennouni (1982), and more recently improved by Jiang (1992) and Antao (1997). The latter two authors have implemented a numerical scheme, within the subroutine LIMI of CESAR-LCPC, a general purpose finite element computer code which has been developed since the early 1980s at the Laboratoire Central des Ponts et Chaussées (Humbert 1989 ; Magnan & Mestat 1992).

This method is based on the theoretically proved result that a series of viscoplastic solutions can be found that converge towards the rigid-plastic solution of the kinematic problem under study.

The type of visco-plastic model used is derived from the Norton-Hoff viscoplastic law.

The procedure used for determining the kinematic solution is described in Figure 1 : a first viscoplastic calculation is made, using a prescribed value of the viscosity parameter p of the model. The displacement field within the soil is obtained through an iterative process, until the calculated stresses comply with the plasticity criterion. Then, the internal work rate associated with the viscoplastic strain rate is compared with the external work rate of the loads under the ground displacement field. The kinematic solution is considered stable (respectively unstable) if the internal work is larger (respectively smaller) than the work of external forces.

In practice, the comparison is made by means of two functions J_p and G_p :

$$J_p = \textit{Internal work} - \textit{External work} \quad (1)$$

$$G_p = \left(\frac{-pJ_p}{(p-1)\int_{\Omega} d\Omega} \right)^{\frac{p-1}{p}} \quad (2)$$

where Ω is the volume of the whole soil model.

The convergence criterion is usually based on the latter function values :

- $G_p < 1$ for stable systems ;
- $G_p > 1$ for unstable systems.

The calculation is repeated for decreasing values of viscosity parameter p, until the condition $G_p \cong 1$ is reached.

A regular homogeneous finite element mesh can be used, taking into account the actual soil layering (several soil layers with different mechanical properties), as well as boundary conditions of the displacement type, for fixed boundary nodes.

This approach requires no a priori assumption on the failure mechanism to be analyzed, as the kinematics of failure is derived from the calculation process.

Soil masses and structures can be modelled using a regular mesh. Input parameters include the soils unit weights and shear strengths : a variety of failure criteria can be used (Tresca, von Mises and Mohr-Coulomb are the most currently used).

On the other hand, the kinematically admissible condition (incompressible condition for Tresca and von Mises yield criteria) or the flow rule is directly introduced in the minimization problem by means a linear relationship between principal strain rates.

It has been demonstrated (Nagtegall et al. 1974 ; Jiang, 1994, among others) that finite elements cannot totally be arbitrary for incompressible or dilatant problems, if unrealistic failure mechanisms are to be avoided.

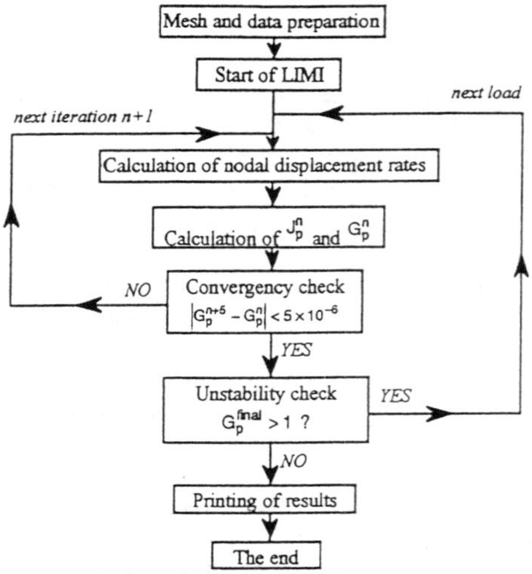

Figure 1. Iterative scheme for the resolution of kinematic limit analysis problems using LIMI.

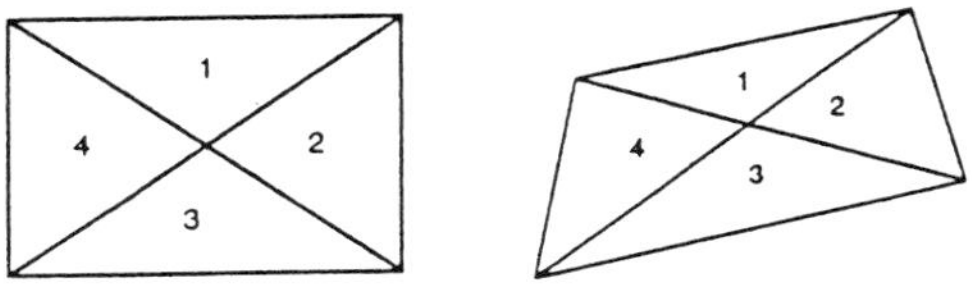

Figure 2. Two-dimensional analysis (linear triangular elements).

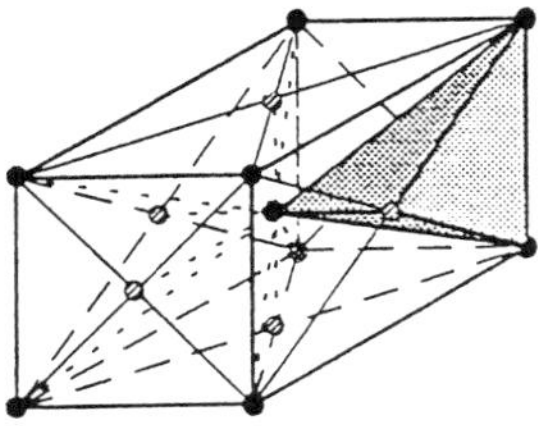

Figure 3. Three-dimensional analysis (linear tetrahedral elements).

4. EXAMPLES

The subroutine LIMI of CESAR-LCPC was first applied to two- and three-dimensional problems related to shallow foundations (Jiang 1992), embankments on soft soils (Jiang & Magnan, to be published), and has been more recently used to study unstable slopes and slope nailing (Sassi 1996 ; Sassi & Magnan 1996).

4.1 Embankment stability

This approach has been in particular applied to the analysis of a trial embankment at Sallèdes LPC experimental site (in the Central Massif), where a granular fill was built on an unstable slope consisting of clay.

In practice, for problems of complex geometry and boundary conditions, analytical and numerical experiences show that some types of elements are more efficient : one can choose four constant triangular elements which form a quadrilateral and its diagonals (Figure 2) or twenty four constant tetrahedral elements which form a prism and its diagonals (Figure 3).

The results of a two-dimensional short term analysis of the embankment are presented in Figure 4a. The undrained shear strengh of the clay is equal to s_u = 20 kPa, except in the lower layer where s_u = 19 kPa. The granular fill can be characterized with a Mohr-Coulomb criterion (c' = 0 kPa and φ' = 35°).

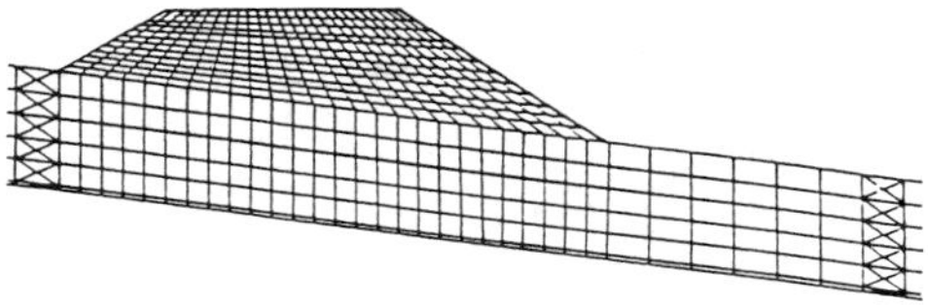

a. Finite element mesh

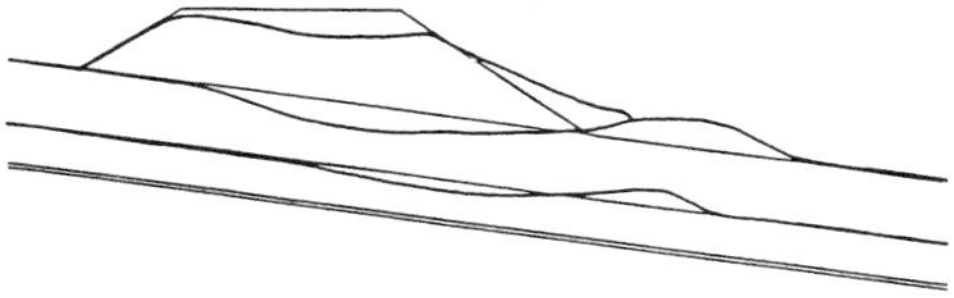

b. Shape of the failure mechanism

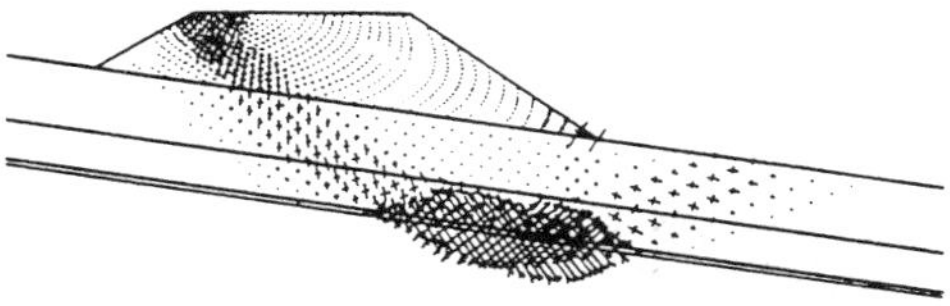

c. Principal strain rates

Figure 4. Two-dimensional analysis of Sallèdes embankment A (Sassi 1996).

The computed failure mechanism (Figure 4b) mainly consists in the displacement of a a rigid block along a deep slip surface, and is very similar to the observed one (Sassi 1996). The extent of failure zone can be more accurately evidenced in Figure 4c, where the principal strain rates in the embankment and in the slope have been plotted.

4.2 Underground opening stability

In this paper, more emphasis is put on works completed over the past five years in the field of tunnel and underground opening stability. Figures 5 to 7 show some of the two-dimensional results obtained for circular tunnels constructed at different depths in homogeneous soil strata (for symmetry reasons, only half of the geometry was considered in the analyses). The Tresca criterion was used in the first two cases and the Mohr-Coulomb criterion in the last case.

The mesh used in the analysis, the distribution of principal strain velocities and velocity fields obtained at failure are displayed for all three cases in Figures 5 to 7.

Figures 5 and 6 relate to a tunnel excavated in a

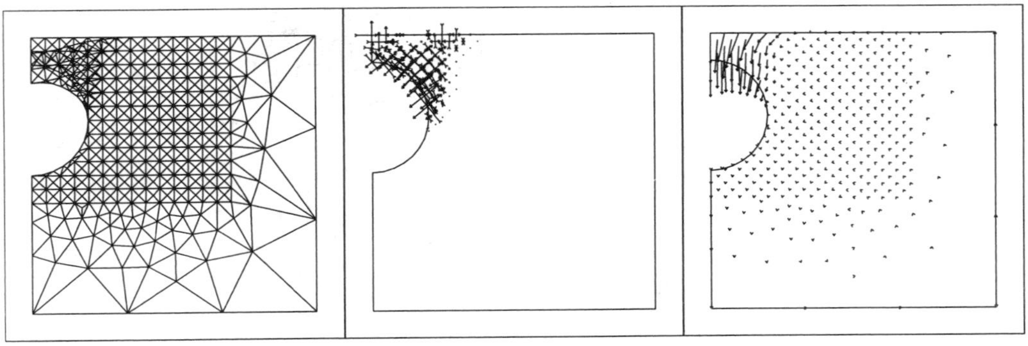

a. Deformed mesh b. Strain velocities c. Displacement velocities

Figure 5. Kinematic calculation of stability in Tresca's material (C/D=0.25)

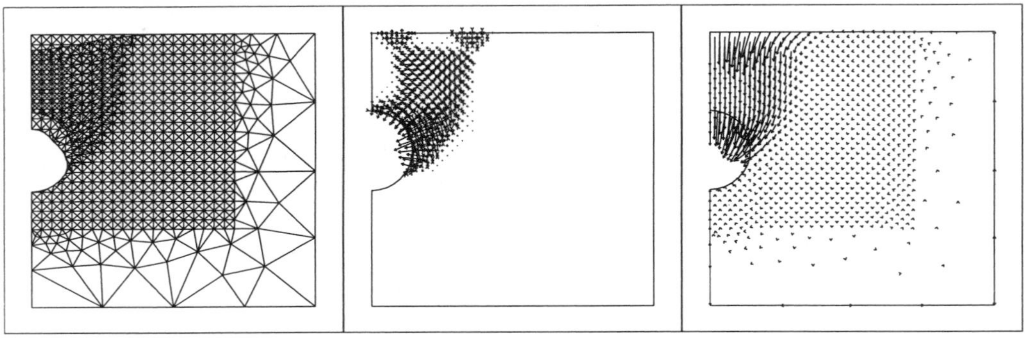

a. Deformed mesh b. Strain velocities c. Displacement velocities

Figure 6. Kinematic calculation of stability in Tresca's material (C/D=1)

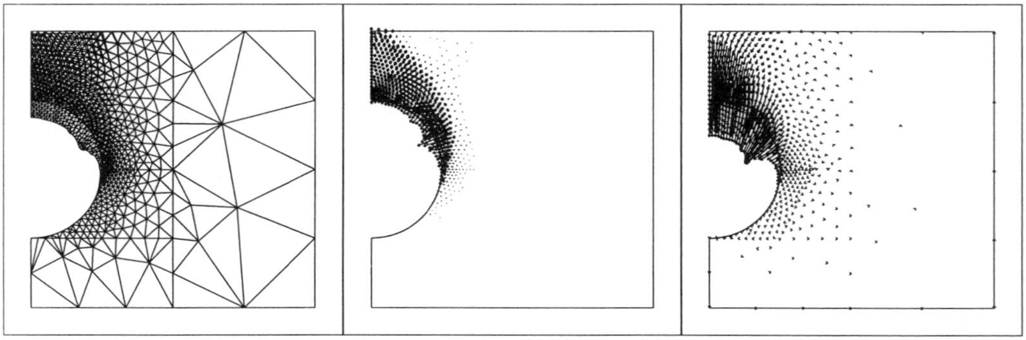

a. Deformed mesh b. Strain velocities c. Displacement velocities

Figure 7. Kinematic calculation of stability in Mohr-Coulomb's material (C/D=0.5)

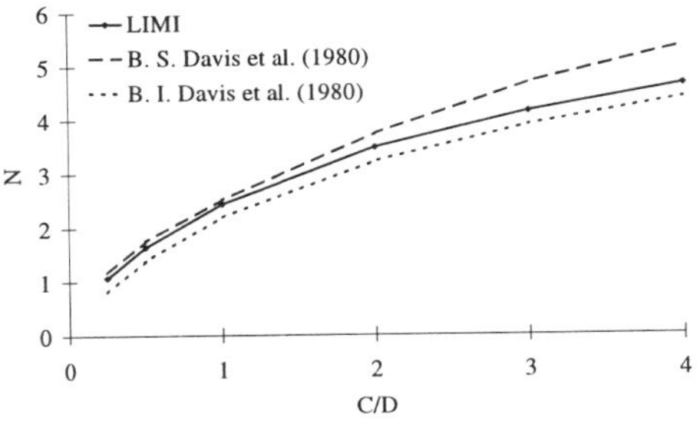

Figure 8. N vs. C/D values at failure for a plane strain tunnel in a purely cohesive material.

purely cohesive soil. In this case, tunnel stability can be characterized by means of the overload factor introduced by Broms & Bennermark (1967) and Peck (1969) :

$$N = \frac{\sigma_S + \gamma . H - \sigma_T}{s_u} \tag{3}$$

where σ_S represents a uniform load applied at the ground surface and σ_T the internal pressure used to support the tunnel walls. The surcharge load σ_S was solely considered in the cases displayed in Figures 5 and 6 (i.e. the computations were run for an unsupported tunnel in a weightless ground), with two different values of the ratio between the depth of cover C and tunnel diameter D : C/D = 0.25 (Figure 5) and C/D = 1 (Figure 6).

The distributions of principal strain velocities allow to identify the geometry of the failure mechanisms obtained with LIMI : it can be noticed that failure is restricted to an area located over and to the side of the tunnel, and consists in the movement of a rigid block over the tunnel crown (which is separated from the stable ground by high shear strain lines). Values of σ_S at failure were found to be in close agreement with the results of an earlier study using limit analysis principles, completed by Davis et al. (1980).

A more comprehensive summary of results obtained for the plane strain tunnel in a purely cohesive material is provided in Figure 8, where values of the overload factor at failure are plotted with respect to the depth of cover to diameter ratio for C/D values ranging between 0.25 and 4. It is apparent that upper bound solutions obtained with LIMI lay between the best upper and lower estimates found by Davis et al. (1980) for the same problem, therefore providing an improvement to existing published upper bound solutions (Antao et al., 1995).

The results displayed in Figure 7 were obtained for a cohesionless material of friction angle φ' = 40° and a depth of cover to diameter ratio of C/D = 1. Two loading parameters (σ_T and γ) were considered. The geometry obtained at failure with LIMI was consistent to results published by Atkinson & Potts (1977) on the basis of limit analysis and centrifuge testing studies.

The results for cohesionless soils can be analyzed using the ratio $\frac{\sigma_T}{\gamma . D}$. The upper bound solution obtained with LIMI ($\frac{\sigma_T}{\gamma . D}$ = 0.142) was found to lay between the upper ($\frac{\sigma_T}{\gamma . D}$ = 0.104) and lower ($\frac{\sigma_T}{\gamma . D}$ = 0.228) bounds provided by Atkinson & Potts (1977).

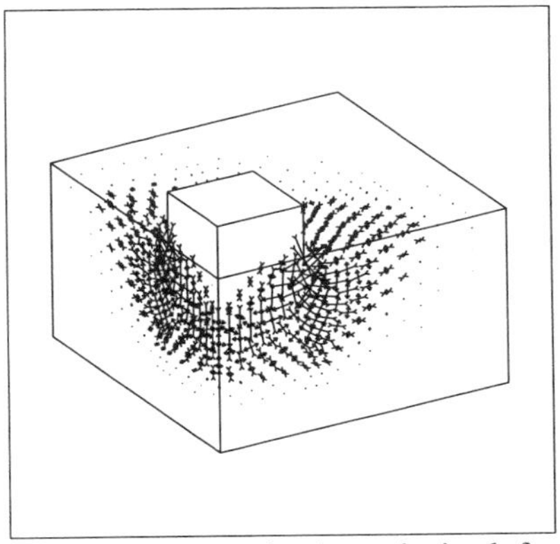

Figure 9. Failure mechanism obtained for the bearing capacity analysis of a shallow foundation.

Three-dimensional computations can also be completed using the same approach, with both a Tresca or Mohr-Coulomb yield criterion. Figure 9 shows the failure mode obtained for the bearing capacity of a shallow foundation laying on a uniform purely cohesive ground. Both the geometry and computed load at failure are consistent with earlier results published for the punching failure case.

Further work is underway to extend our approach to the analysis of tunnel face stability, with due consideration to three-dimensional effects.

5. CONCLUSIONS

The regularized approach used in the subroutine LIMI of the finite element computer code CESAR-LCPC for the determination of upper bound solutions of failure loads acting on geotechnical structures has been applied to the solution of complex geotechnical stability problems over the last seven years at LCPC. The results presented in this paper illustrate some of the advantages of this method : no a priori knowledge of the failure mechanism is needed and the results compare well with those obtained from other studies based on limit analysis principles, as well as failures observed in situ or through centrifuge model tests.

It is therefore our intention to investigate more extensively the use of this technique for a variety of geotechnical structures, including slope stability problems, retaining structures, tunnels and underground openings and other categories of complex earth structures. In the future, this method could be used as a supplement to conventional design methods in everyday practice, since it can be easily performed, using a finite element computer code.

ACKNOWLEDGEMENTS

This research has been partly supported by the Portugese "Junta Nacional de Investigação Cientifica e Technológica (J.N.I.C.T.)".

REFERENCES

Antao, A. 1997. Analyse de la stabilité des ouvrages souterrains par une méthode cinématique régularisée. Thèse de doctorat ENPC, Paris (to be published).

Antao, A., E. Leca & J.P. Magnan 1995. Finite element determination of upper bound solutions for the stability of shallow tunnels, based on a regularization principle. *Xème Conférence Europe-Danube de Mécanique des Sols et des Travaux de Fondations, Mamaia, Constanta (Roumanie)* (3):479-486.

Atkinson, J.H. & D.M. Potts 1977. Stability of a shallow circular tunnel in cohesionless soil, *Géotechnique*, 27(2):203-215.

Broms, B.B. & H. Bennermark 1967. Stability of clay at vertical openings. *J. Soil Mech. Found. Div. ASCE*, 93(SM1):71-94.

de Buhan P. 1986. *Approche fondamentale du calcul à la rupture des ouvrages en sols renforcés.* Thèse de Doctorat-ès-sciences, Université P. et M. Curie, Paris VI.

Davis, E.H., M.J. Gunn, R.J. Mair & H.N. Seneviratne 1980. The Stability of Shallow Tunnels and Underground Openings in Cohesive Material, *Géotechnique*, 30(4):397-419.

Frémond M. & J. Salençon 1973. Limit analysis by finite element method. *Symp. on the Role of Plasticity in Soil Mech.*, Cambridge:297-309.

Friaâ, A. 1978. Le matériau de Norton-Hoff généralisé et ses applications en analyse limite. *C. R. Acad. Sci. Paris* 286(A):953-956.

Guennouni T. 1982. Matériau de Norton-Hoff pour divers critères de plasticité de la mécanique des sols, Thèse de Docteur-Ingénieur, ENPC, Paris.

Humbert, P. 1989. CESAR-LCPC, un code général de calcul par éléments finis. *Bull. liaison Labo. P. et Ch.*, 160:112-116.

Jiang, G. L. 1992. *Application de l'analyse limite à l'étude de la stabilité des massifs de sol*, Thèse de Doctorat, ENPC, Paris, France.

Jiang, G. L. 1994. Regularized Method in Limit Analysis. *J. Eng. Mech.* 120:1179-1197.

Jiang G.L. & J.P. Magnan Stability Analysis of Embankments : Comparison of Limit Analysis with Methods of slices (*accepted for publication in Geotechnique*).

Magnan, J.P. & Ph. Mestat 1992. Utilisation des éléments finis dans les projets de géotechnique, *Annales de l'ITBTP*. Série Sols et Fondations, 6:80-107.

Nagtegall J.C., F.M. Parks & J.R. Rice 1974. On numerically accurate finite element solutions in the fully plastic range. *Comp. Methods in Appl. Mech. and Engng.*, 15(2):153-177.

Norton F.H. 1929. *The creep of steel at high temperature*. McGraw Hill, New York, N.Y.

Peck R.B. 1969. Deep excavations and tunneling in soft ground, *7th Int. Conf. on Soil Mechanics and Foundation Engineering*, Mexico, State-of-the-Art (1):225-290.

Salençon, J. 1983. *Calcul à la rupture et analyse limite*. Presses de l'ENPC, Paris, France, 366 p.

Sassi, K. 1996. *Contributions à l'étude des mécanismes de déformations des pentes instables*. Thèse de Doctorat, INSA, Lyon, France.

Sassi, K. & J.P. Magnan 1996. Finite element kinematic limit analysis of Sallèdes trial embankment A. *Lausanne Symposium*.

Numerical Models in Geomechanics, Pietruszczak & Pande (eds) © *1997 Balkema, Rotterdam, ISBN 90 5410 886 X*

Densification by the successive crushing of grains

D. Robertson & M. D. Bolton
Cambridge University, UK

ABSTRACT: Discrete element methods have been widely used to investigate the behaviour of assemblies of granular materials under applied stresses and deformations. Usually the particles are represented by elliptical or ellipsoidal elements which may deform at contact points, but which are not permitted to fracture. McDowell, Bolton and Robertson[3] demonstrated some of the principles which could be used to explore the statistical fracture mechanics of a granular aggregate. An iterative scheme is used in which particles, represented by rigid, close-packed, triangular elements are assigned finite probabilities of fracturing. The probabilities depend on the value of increasing macroscopic stress, particle size and co-ordination number. No attempt is made to assess local equilibrium or to include contact deformations. This earlier work is now extended to include the explicit presence of voids. As stress increases, particles begin to fracture and a kinematic rule permits broken fragments to fall, so that the voids ratio reduces. The forms of the derived plots of voids ratio against the logarithm of the macroscopic stress are discussed.

INTRODUCTION

McDowell, Bolton and Robertson[3] (1996) introduces a two-dimensional numerical model, using an extension of Weibull statistics, in which a granular medium under stress evolves as a function of the behaviour of individual grains. An initial sample of uniformly sized grains is represented by an array of identical right-angled triangles, each of which might fracture to create two right-angled triangles in its place. Starting from an initial level of macroscopic stress a pass is made of the triangles, fracturing them on the basis of the proposed probability equation. When a sufficiently small proportion of the triangles fracture during a single pass, the level of stress is increased and the process repeated. For each increment of stress the corresponding reduction of voids is deduced from the amount of breakage occurring, using a work equation. A simulated compression curve is generated, showing the voids ratio versus the logarithm of the macroscopic stress. In this case, however, the voids are not explicit in the geometry of the aggregate.

Earlier work has referred to the fractal nature of a debris of crushed fragments (Sammis, King and Beigel[5] (1987); Palmer and Sanderson[4] (1991); Turcotte[6] (1996)). Bolton and McDowell[1] (1996) advance the hypothesis that both sands and clays adopt such a fractal geometry on their "normal consolidation line" and propose mechanisms of "clastic yielding" followed by "clastic hardening". Clastic yielding is the onset of fracture of the weakest particles of the aggregate. Clastic hardening occurs because the fracture strength of the broken fragments exceeds that of their progenitors. Self-similarity can emerge in the hardening, crushing aggregate, and particle size distributions can display some fractal properties.

The new simulation first creates a "loose" assembly of triangles with voids, and the migration of fragments. The resulting compression curve is therefore solely a function of statistics and kinematics.

INITIALISATION - CREATING A SAMPLE

The purpose of the initialisation process is to generate an initially stable assembly of equally sized, triangular elements, containing voids, so that the reduction of these voids within a region can be observed when a fracturing process is introduced. In reference to the sample, the terms "porosity" and "void ratio" of a region are taken to mean the proportion of area not covered by any triangle, and the ratio of the areas of voids and triangles. As a first step, a solid array of 200 triangles containing no voids is created (e.g. Fig. 1)

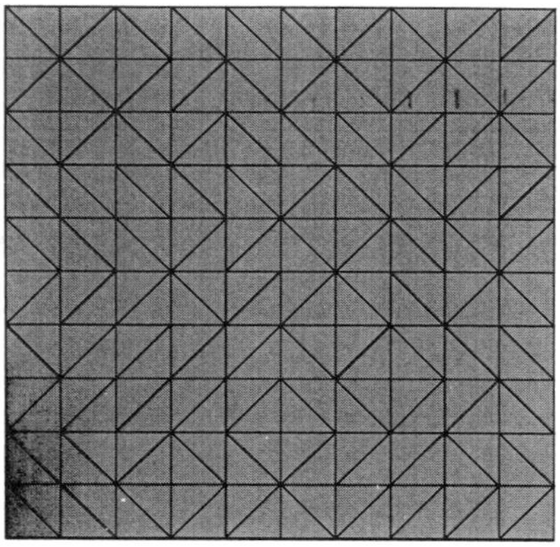

Fig. 1 - Solid array of triangles.

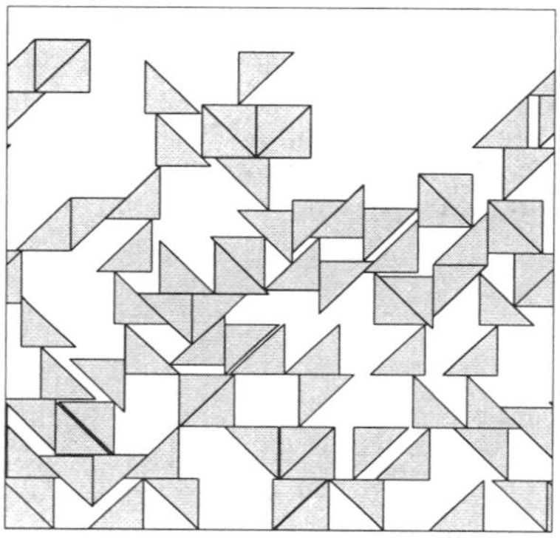

Fig. 2 - Generation of "initial" sample.

Particle removal and filling

Particles are next removed, one at a time, opening up voids within the sample. Each particle in the current sample has an equal probability of being removed. Following each removal, the remaining particles are permitted to fall and slide according to simple kinematic rules, so that some of the space created migrates towards the top of the sample.

Whenever a sufficient amount of space becomes available at the top of the sample, a fresh block of 20 triangles is introduced and allowed to fall, to replenish those which have been removed.

The aim is to repeat this process a sufficient number of times until a region has been formed within the sample in which the density of the triangles might acceptably be described as "loose" (e.g. Fig. 2).

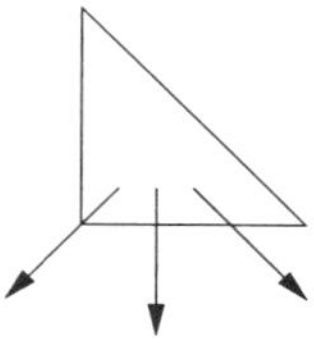

Fig. 3 - Direction in which particles may fall.

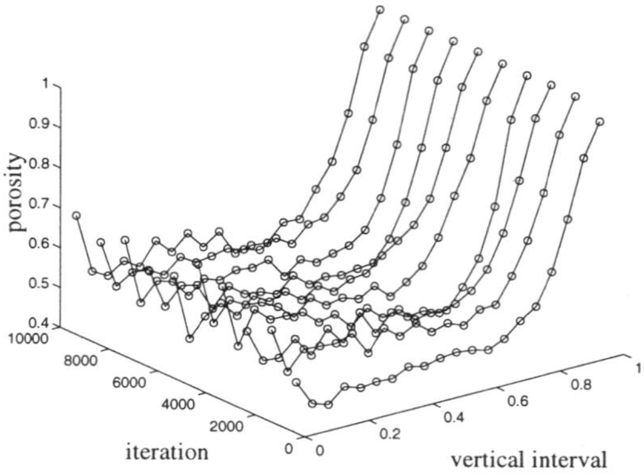

Fig. 4 - Average porosity.

Kinematics

Triangles are not allowed to rotate and are permitted to fall in only three directions: vertically, or downwards at an angle of 45 degrees (Fig. 3). Triangles may fall whenever their movement in a particular direction is not blocked by another particle.

A horizontal "wrap-around" is implemented so that particles which fall out or protrude from one side of the sample are allowed to fall in or intrude into the other side. This wrap-around is taken into account when the co-ordination numbers of particles is being calculated. No corresponding vertical "wrap-around" is defined. Particles which come into contact with the base of the sample are not permitted to fall further.

A triangle which is able to move, moves according to the following rules. First, the particle is allowed to fall as far vertically as it can until it comes into contact with another triangle or reaches the base of the sample. Assuming it has not reached the base of the sample, it is then allowed to fall in one of the diagonal directions, selected at random, until it again comes into contact with another triangle or reaches the base of the sample. If the triangle is unable to move any distance in the diagonal direction selected, then it is allowed to fall as far as it can in the other diagonal direction. Having now fallen diagonally, the triangle is again permitted to fall as far vertically as it can and the process repeated until the triangle is no longer able to move, being constrained in the three possible direcions by other triangles or the base of the sample.

For the purposes of determining how far a triangle may move, or with which other triangle it is in contact, the sample space is considered to be divided into a grid. Each triangle maintains a list of the boxes of the grid which it overlaps and each box maintains a list of the triangles by which it is overlapped. The grid is chosen so that the width of its squares are slightly larger than the size of the initial triangles so that each triangle need maintain at most 4 boxes in its list. In calculating the co-ordination number of a triangle, only the triangles belonging to these boxes need be examined. Triangles which share more than one box are counted only once.

A triangle may be prevented from moving in one of the directions by more than one triangle. However, to simplify storage, each triangle stores the identities of one constraining triangle for each of the three directions. These are updated whenever a triangle falls, or attempts to fall.

If a triangle falls, or is fractured, or is removed, a pass is made through those triangles with which it previously shared a box. Those which had recorded the triangle as a constraint are appended to a list of potentially unstable triangles. Each triangle on this list is then considered in turn and tested to determine whether it can now fall. If it is still prevented from moving, its constraints are updated and it is removed from the list, otherwise it is allowed to fall. This may or may not result more triangles being appended to the list. When all triangles have been removed from the list, every triangle in the sample is again constrained in each direction. When a triangle splits, the resulting fragments are automatically appended to the list of potentially unstable triangles.

The boxes are also used when determining the maximum distance a triangle may fall in a given direction. The distance is initially set to the that required for the triangle to come into contact with the base of the sample. Starting form those boxes to which a triangle currently belongs, those boxes which might contain an intervening triangle are then considered in turn. The triangle is tested against all of the triangles belonging to the current box. This is continued until the maximum distance found is zero, or less than the distances to the boxes still to be considered.

Statistics/Sampling

Initially, as triangles are removed from the sample the overall density of triangles decreases.

The voids ratio calculated during the fracturing process is based on a sub-sample within the assembly of triangles created during the initialisation process. The proportion of area occupied with triangles is considered within 20 equally sized horizontal layers through the sample. Fig. 4 shows the porosity of the 20 layers averaged out over periods of 1000 iterations within an initialisation period of 10,000 iterations. From this graph, and the trend and scatter from each successive iteration a sub-sample was chosen extending from 10% to 60% of the total height of the sample, from the base.

An initialisation period of 1,000 iterations was found to be sufficient to ensure that the assembly of triangles was sufficiently random. In general, the porosity in the sub-sample rises rapidly to a reasonably stable level long before this.

Once a "loose" sample has been generated in this way, the fracturing process is applied.

STATISTICS OF FRACTURE OF A GRAIN

The model for breakage is based on Weibull[7] statistics of failure. Weibull postulated that the probability of the survival of a specimen, of a given size, subjected to a tensile stress σ could be written in the form:

$$P_s(\sigma) = exp\left\{-\left(\frac{\sigma}{\sigma_0}\right)^m\right\} \tag{1}$$

where m, the Weibull modulus, is a measure of the variability in the strength of the material, increasing with decreasing variability, and σ_0 is the value of stress at which approximately 37% of the samples survive (Fig. 5).

Weibull also recognises that the survival of a block of material under tension requires that all its constituent parts remain intact. The mean fracture strength decreases as the size of the specimen increases. For a block of material of volume V, under an applied tensile stress σ, the survival probability of the block is given by:

$$P_s(V) = exp\left\{-\frac{V}{V_0}\left(\frac{\sigma}{\sigma_0}\right)^m\right\} \tag{2}$$

McDowell, Bolton and Robertson[3] used a modified survival equation which introduces an extra term to account for the variable co-ordination number of the particles, and so that the maximum tensile stress induced within a particle reduces as the number of contacts increases. The survival probability of a 2D grain is then given by

$$P_s(d) = exp\left\{-\left(\frac{d}{d_0}\right)^2\frac{(\bar{\sigma}/\sigma_0)^m}{(C-2)^a}\right\} \tag{3}$$

where d is the size of the particle, d_0 is the size of the original particles and the factor a can be used to vary the degree to which the co-ordination number C affects the probability of fracture.

The same equation is used in this simulation. In the original work no triangles were allowed to move and the co-ordination number was taken as the number of triangles sharing an edge. The minimum possible co-ordination number was therefore 3. Here the particles may move and for ease of calculation the co-ordination number is taken to be the number of triangles sharing any contact: edge-to-edge, edge-to-corner or corner-to-corner. Two triangles are considered to be in contact if the distance between them is below a very small tolerance value, based on the size of the smaller of the two triangles. Since it is now possible for a triangle to have only one or two neighbours, the value of C is arbitrarily set to a value of 10 in the probability equation if the particle has less than 3 neighbours, to give these particles a very small chance of fracturing.

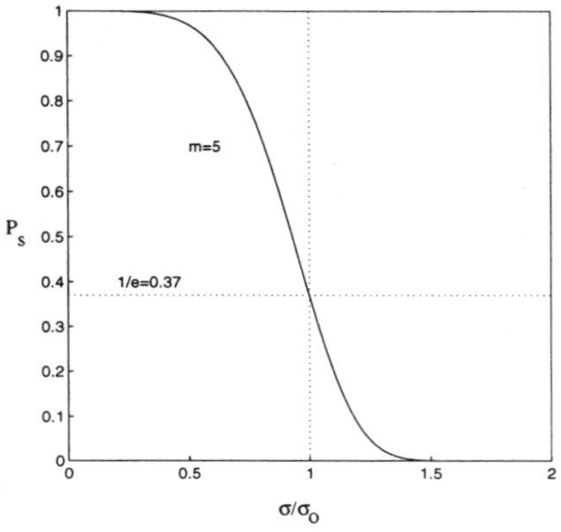

Fig. 5 - The Weibull function.

COMPRESSION

McDowell, Bolton and Robertson[3] did not explicitly represent voids in the model. Fig. 6 shows the fractal geometry of the broken fragments after a significant increase in stress. It was argued that such a figure could be regarded as creating a logical map of co-ordination numbers and particle sizes. An implicit voids ratio e was then inferred by applying a work equation. The work done per unit volume by the macroscopic stress $\bar{\sigma}$ was written $\bar{\sigma}de/(1+e)$ and equated to the work absorbed in fracture and frictional rearrangement.

The following relationship was derived to calculate the change in voids ratio associated with each increment in stress:

$$de \propto \frac{\Gamma}{(1-\mu)\bar{\sigma}} \sum_{r=0}^{r=s} B_r d_r \tag{4}$$

where the orders of the particle sizes are listed as r=0 for the largest particle (size d_0) and $r = s$ for the smallest (size d_s), and B_r is the number of particles of size d_r which are splitting. Γ is the critical strain energy release rate, a measure of the "material toughness", and μ is a function solely of the internal angle of friction.

In calculating the value of de in the simulation, equation 4 was simplified to:

$$de = \frac{K}{\bar{\sigma}} \sum_{r=0}^{r=s} B_r d_r \tag{5}$$

where K is taken to be a constant.

Fig. 7 shows a simulated compression curve generated using this approach. In the current simulation the voids ratio is calculated directly from the area of triangles within the sub-sample rather than from the work equation.

A stress level is initially calculated such that a single particle fractures. After each set of fractures the earlier kinematic rule, used to generate the initial sample, is invoked and the crushing and subsequent falling of particles is allowed to continue until less than 2% of particles fracture in a single pass of the triangles. The stress is then increased by 1% and the process continued.

Fig. 8 shows the sample after 1193 iterations and Fig. 9 shows the corresponding compression curve. The form of this curve differs from the curve generated using the work equation. During the first stages of fracturing the sample has undergone significant compression, corresponding to the fall in the first part of the compression curve. However, larger triangles with relatively large survival probabilities then begin to form a stable structure through which the smaller particles are unable to fall. Although fracturing continues to occur, the tendency is for the smaller particles to continue crushing and the curve begins to level out. The grading curve of the sub-sample, is shown in Fig. 10. Fig. 11 shows the porosity within 20 vertical slices through the entire sample space. This show the values within the chosen sub-sample to be reasonably consistent.

The statistical rules adopted in the new kinematic simulation lead to something of a paradox. Heavily fractured debris tends to continue crushing even though it appears to lie on the "floor" of a "cave" formed between larger blocks. The answer lies in relaxing the condition that all particles should carry a fair share of the macroscopic stress. The best way of achieving a lifelike soil structure, and a reasonable variation in local stress, is under investigation. The theoretical ideal would be to capture the detailed equilibrium and kinematics of every grain. This

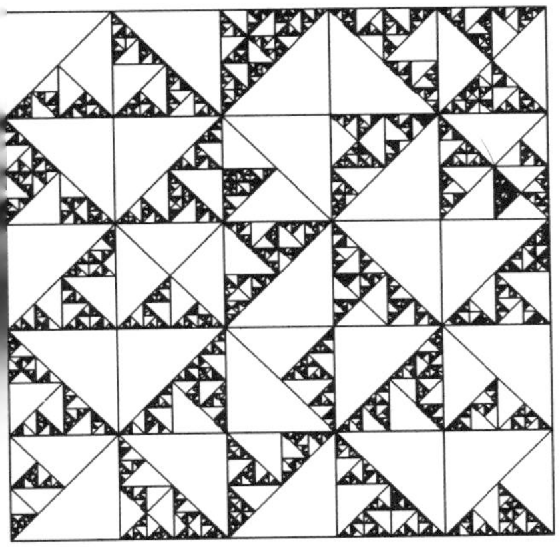

Fig. 6 - Fractal geometry without explicit voids.

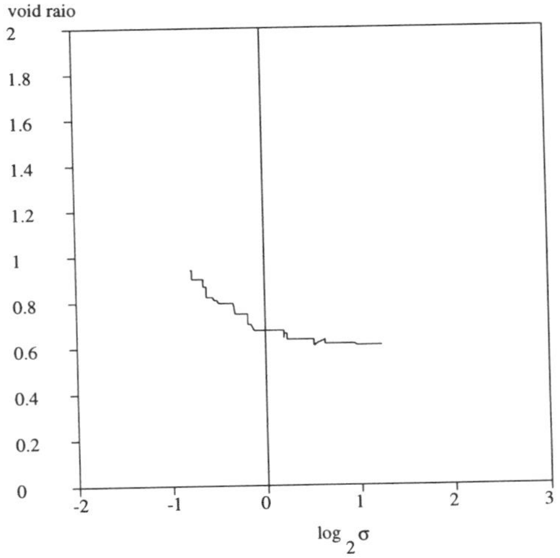

Fig. 9 - Compression curve calculated from area.

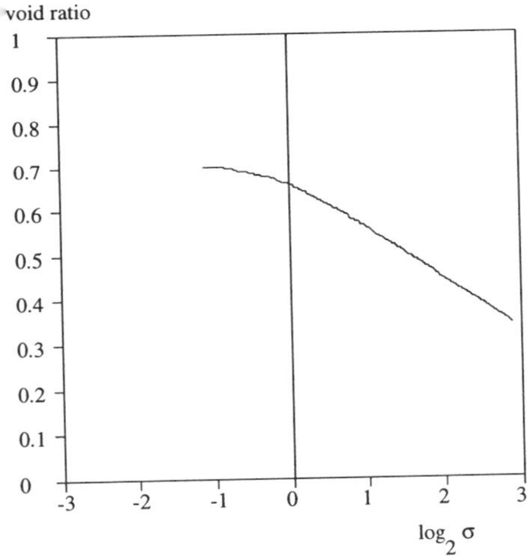

Fig. 7 - Compression curve using the work equation.

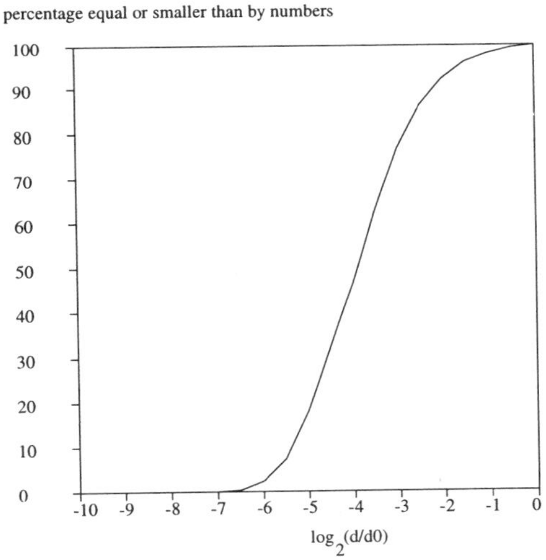

Fig. 10 - Distribution of particle sizes.

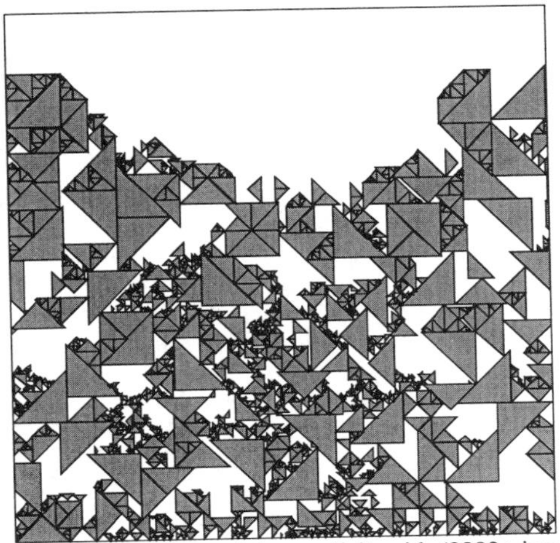

Fig. 8 - Sample containing explicit voids (2083 triangles).

seems not to be a realistic computational proposition if, as proposed, the successive creation of many generations of fragments is an essential pre-requisite to the modelling of "plastic" compression in granular media. The statistical approach, adopted here, appears to have some promise.

CONCLUSIONS

1) Techniques are available to obtain simulated compression curves using statistics of fracture of rigid clasts.
2) Voids can be created by particle removal, and a

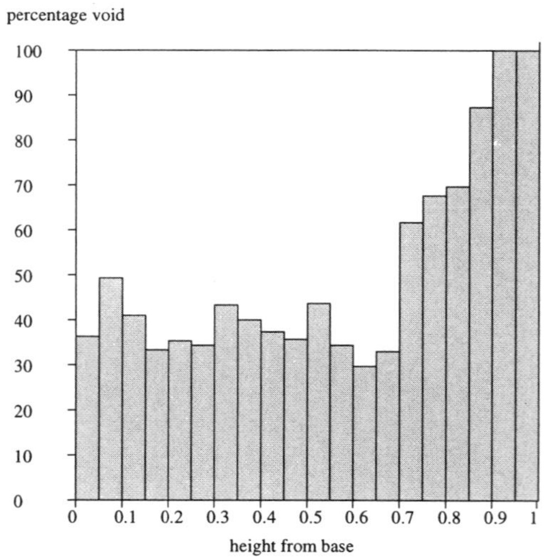

Fig. 11 - Porosity within vertical slices.

statistically reliable "loose" soil can be generated.
3) Stress analysis and Weibull statistics lead to successive fracture, rearrangement, wide gradings, efficient packing, and the irrecoverable reduction of voids.
4) Work is in hand relating the kinematics of broken fragments to earlier observations of fractals, and the use of a work equation to derive changes of voids ratio.

REFERENCES

1) M. D. Bolton and G. R. McDowell, 'Clastic Mechanics', *IUTAM Symposium on Mechanics of Granular and Porous Materials* (1996)(in press)
2) P. A. Cundall and O. D. L. Strack, 'A discrete numerical model for granular assemblies', *Geotechnique*, **29**, 47-65 (1979)
3) G. R. McDowell, M. D. Bolton and D. Robertson, 'The fractal crushing of granular materials', *Int. J. Mech. Phys. Solids*. **44**, 2079-2102 (1996)
4) A. C. Palmer and T. J. O. Sanderson, 'Fractal crushing of ice and brittle solids', *Proc. R. Soc. Lond. A* **433**, 469-477 (1991)
5) C. Sammis, G. King and R. Beigel, 'The kinematics of gouge deformation', *Pure Appl. Geophys.* **125** 777-812 (1987)
6) D. L. Turcotte, 'Fractals and fragmentation', *Journal of Geophysical Research* **91**, 1921-1926 (1986)
7) W. Weibull, 'A statistical distribution of wide applicability', *J. Appl. Mech.* **18**, 293-297 (1951)

Numerical Models in Geomechanics, Pietruszczak & Pande (eds) © 1997 Balkema, Rotterdam, ISBN 90 5410 886 X

How an iterative solver can affect the choice of f.e. method

D.J. Naylor
University of Wales, Swansea, UK

ABSTRACT: Due to the need to use an iterative solver for finer meshes in 3D analyses, the question arises as to what effect this may have on the choice of the finite element method. One direct (a skyline solver) and two iterative solution techniques (relaxation and conjugate gradient) are compared using tangential and constant stiffness finite element formulations. The comparisons are based on the times required to analyse the two quite different 3D boundary value problems of the loading of a footing and the construction of an embankment dam. These are analysed for different mesh refinements using both linear elastic and elasto-plastic idealisations. It is shown that the constant stiffness method is superior to the tangential irrespective of the type of solver. The relative merits of the three solvers depend on whether the analysis is linear or nonlinear. Conclusions are drawn about this and on how the iterative solvers compare.

1. INTRODUCTION

It is well established that iterative solution techniques become preferable to direct solvers when large sets of sparse banded equations are involved (Crisfield, 1986, Griffiths and Smith, 1991). This is especially the case with large 3D finite element analyses.

In such analyses the number of multiplications required to solve the equations varies with nearly the cube of the number of nodes (Press *et al.*, 1992). If an in-core solver is used the storage requirements also increase very rapidly as the mesh is refined. This rules out the use of a direct solver (at least on a PC) for finer meshes. Iterative in-core solvers require less storage since only the non-zero coefficients need be stored. This contrasts with a skyline solver (as is used here) where storage locations for all the coefficients under the skyline need to be reserved for use during the elimination stage. The difference is very significant and, as will be seen, is an order of magnitude for some of the meshes studied.

The solution procedures used for nonlinear finite element problems divide into "tangential" methods in which the stiffness is calculated at least once in each load step and "constant stiffness" (or "load transfer") methods in which there is an initial formulation and assembly of usually elastic stiffnesses followed by iterations within each load step. (The intermediate category of the Modified Newton Raphson method in which new stiffnesses are calculated at the start of each load step and then constant stiffness iterations carried out is not considered here.)

Consequently if an iterative equation solver is used in conjunction with a constant stiffness method there will be iterations at two levels: the constant stiffness iterations and within each of these a set of equation solving iterations.

Two important questions arise from this:

(1) How are the relative merits of the tangential and constant stiffness methodologies affected by the choice of equation solver? ie if, say, a constant stiffness method is superior to a tangential method using a direct solver does this still apply if an iterative solver is used?

(2) How do direct and iterative techniques compare?

The answer to question (2) is already known to the extent that as the mesh gets finer so an iterative solver becomes increasingly favourable. (1) however is an open question. It is complicated by the fact that there is considerable variation in the detail of the two finite element method classifications given above and also there are a variety of iterative equation solving procedures. Attention here is restricted to one of each of the former and to two iterative equation solvers.

The aim of this study is to answer (1) as far as possible within the scope of the work carried out, and to throw further light on (2) by quantifying the difference between the two techniques.

A conjugate gradient and a relaxation method are used as the iterative equation solvers. These and the two nonlinear methodologies are described briefly below.
The study is based on two quite different problems: A stiff footing on soft clay analysed using an elasto-plastic Tresca idealisation as would be suitable for an undrained analysis in terms of total stress, and the construction of an embankment dam idealised by a critical state model.
The hoped for outcome is that the investigation will give a clear pointer to the most suitable combinations of methods for 3D static finite element analyses of such problems.

2. EQUATION SOLVERS

2.1 Direct solver

The skyline direct solver used was written in house but is thought to be fairly conventional, on the lines of that described in Chapter 5 of Crisfield (1986). The stiffness coefficients under the skyline are stored in a long vector in core. The solver is restricted to symmetrical matrices thereby nearly halving the storage which would otherwise be required.

2.2 Iterative solvers

The conjugate gradient iterative equation solver is based on the formulation given by Jennings (1977) (which is also given in the texts by Griffiths and Smith (1991) and Smith (1995)).
The relaxation solver was derived by the Author. It has two versions. The first is a basic relaxation solver and is used in the dam analyses. Details are given by Naylor (1996b). It is a simple technique which can be shown (Naylor, 1996a) to be almost identical to the classic Gauss-Seidel method. The only difference is that the *change* in the unknown displacement vector per iteration is updated in each iteration rather than the *complete* vector.
The second version is a modified form of the basic method to allow block relaxation of designated rigid zones. Instead of relaxing each degree of freedom within the rigid zone one at a time (to restore equilibrium by the removal of conceptual clamps) the complete zone is relaxed as a rigid body. In 3D this involves up to six relaxations, one for each of the six rigid body movement modes. Rigid body relaxations are alternated with basic relaxations to allow deformation to take place within the rigid zones as these will not be completely rigid. Details of the method are given by Naylor (1996c). It is used in the footing analyses.
For both the conjugate gradient and the basic (but not the rigid zone) relaxation solvers the equations are first normalised so that the stiffness matrix diagonals (a_{ii}) become unity. This is done as follows. Let b_i and x_i denote the nodal force and displacement components for degree of freedom i (i =1,n) respectively. The assembled stiffness equations are then:

$$\sum_{j=1}^{n} a_{ij}x_j = b_i \qquad (i=1,n) \qquad (1)$$

The normalisation involves replacing a_{ij} by $\bar{\bar{a}}_{ij} = a_{ij} / \sqrt{a_{ii}a_{jj}}$, x_j by $\bar{x}_j = x_j\sqrt{a_{jj}}$ and b_i by $\bar{b}_i = b_i / \sqrt{a_{ii}}$ so that the equations become:

$$\sum_{j=1}^{n} \bar{\bar{a}}_{ij}\bar{x}_j = \bar{b}_i \qquad (i = 1,n) \qquad (2)$$

Not only is $\bar{\bar{a}}_{ii} = 1$ but symmetry is also retained.
The x_i, b_i transformation is carried out at the start of the iterative solver routine and then reversed before returning to the main program.
It has the advantage that it makes the solver more robust for problems in which there are large stiffness variations across the mesh. In the case of the conjugate gradient method it also converts the Jennings-Smith formulation into a *pre-conditioned* formulation, albeit one of the simplest forms of pre-conditioning.

3. FINITE ELEMENT METHODS

3.1 Tangential method

The formulation for this study uses what is strictly an *incremental* stiffness. The stiffness is based on an estimate of the stress at the middle of a forthcoming load step. This estimate is achieved by extrapolating displacements from the previous load step. There is just one solution per load step. Accumulation of residual error is reduced by the long established practice of adding the force residuals from a completed step to the new right hand sides.
Consideration was given to a Newton Raphson approach. This idea was however rejected on account of the difficulty in evaluating a consistent tangential matrix for complex elasto-plastic material laws such as the critical state model used for the dam analyses. (The yield surface for this model has a non-circular cross-section in the deviatoric plane. This makes the expressions for the various stress derivatives relatively complex algebraically.)

3.2 Constant stiffness method

This involves a single evaluation of the elastic stiffness matrix at the start of the analysis. It is essentially the old fashioned "initial stress" or "load transfer" (Naylor *et al.*, 1981) method but with the important addition that, just as in the tangential method, displacements are extrapolated. In this case the extrapolated displacements provide a good estimate of the "equivalent load", ie the load to be added to the right hand side to make the equivalent elastic body have the correct deformation for the forthcoming load step. An over-relaxation factor of 1.5 was used for the footing analyses. The dam analyses, however, converged better without it. The number of constant stiffness iterations was limited to four per load step as it was found that by then most of the convergence had taken place.

3.3 Step size selection

A novel feature used in both the tangent method and the constant stiffness method is a procedure for choosing near optimal load increments. As these are not known before hand a preliminary analysis is run using fairly large steps - ten say. A record is kept of the r.m.s. force residual divided by the r.m.s. of the total applied loads at the end of each step. A "residual ratio" is then defined as this normalised residual divided by a dimensionless tolerance (tolr). A routine is called at the end of the analysis which calculates a recommended new step size on the basis that the number of subivisions per step is proportional to the square root of the residual ratio. Thus if the ratio turned out to be four the step size would (ideally) be halved for a repeat analysis. Similarly if the residual ratio was less than one the step size would be increased.

In the case of the constant stiffness method if convergence was achieved in less than four iterations provision was made to increase the step size, otherwise a new step subdivision was based on the residual ratio exactly as for the tangential method.

The procedure worked well and it was often sufficient to make just one trial analysis.

4. ANALYSES

4.1 General

The combinations of f.e. method, equation solving technique, mesh refinement and problem studied are shown in Table 1.

Ten noded linear-strain tetrahedral elements were used throughout. Four sampling points within each element were used for the numerical integration.

The numbers of elements, nodes and free degrees of freedom are given in Table 2.

The same tolerances were used for all the analyses involving iterations. These were 0.001 for tolr (see 3.3 above) and 0.0002 for the corresponding dimensionless parameter "tolq" for the relaxation and conjugate gradient iterative equation solvers.

Table 1. The analyses
"D" stands for Direct, "G" for Conjugate Gradient and "R" for Relaxation solver.

		Footing			Dam	
		Coarse	Med.	Fine	Coarse	Med.
Lin. el.	D	√	√	-	√	-
	G	√	√	√	√	√
	R	√	√	√	√	√
Nonlin. constant stiffness	D	√	√	-	√	-
	G	√	√	√	√	√
	R	√	√	√	√	√
Tang. s.	D	√	√	-	√	-

Table 2 Mesh data
n_f = free degrees of freedom.

	Footing			Dam	
	Coarse	Med.	Fine	Coarse	Med.
Elements	90	540	2025	128	1041
Nodes	227	1075	3566	302	1914
n_f	422	2406	8855	429	3951

4.2 Footing analyses

The footing comprised a hypothetical 4m x 3m x 2m deep concrete footing located centrally in a 16m x 12m x 6m deep block of soil. Figure 1 shows a perspective of the fine mesh with the soil cut away to show two sides of the footing. The medium and coarse meshes are the same except that the broken lines showing the tetrahedral element edges differ from mesh to mesh.

The stiffness parameters for all the analyses were: Young's modulus (E) = 25 MPa, Poisson's ratio (ν) = 0.25, and for the Tresca analyses cohesion (c_u) = 60 kPa ($\phi_u = 0$).

The footing was loaded by a distributed pressure which varied linearly over its surface so that the loading was symmetrical about AB (Figure 1), ie contours of equal pressure were normal to AB viewed in plan. The intensity was varied from zero at A to 900 kPa at B. This gave an eccentric resultant vertical load of 5400 kN. The average pressure was 450 kPa, or $7.5c_u$. This value was chosen so that the footing would be close to failure at the end of loading.

4.3 Dam analyses

The dam is an idealisation of the Canales dam in southern Spain near Granada. It is a central clay core rockfill dam with a maximum height of 156m. It is sited in a steep sided canyon. Figure 2 shows a perspective of the mesh. The shaded area is the near

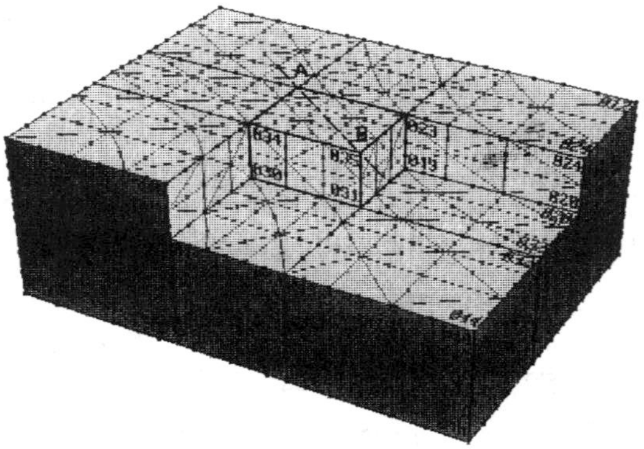

Figure 1 Footing fine mesh

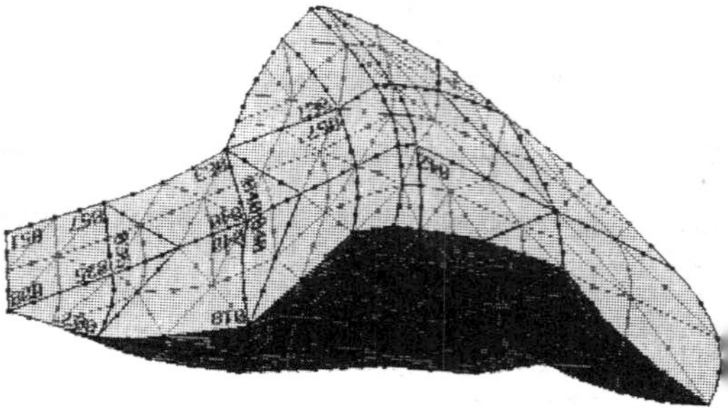

Figure 2. Dam medium mesh

vertical left abutment contact. The reservoir (not shown) is to the right.
The construction of the dam was modelled in three stages of approximately equal height. The self weight was applied to each layer as a distributed body force on to initially unstressed material.
Different values of Young's modulus were used in the elastic analyses according to both position and the stage in the analysis. Details are given in Naylor (1995) which describes an analysis of this dam using the same coarse mesh as used here.
A version of the critical state model similar to that described in Chapter 7 of Naylor *et al.* (1981) was used for the nonlinear analyses. It differs from conventional c.s. models in that the yield surface is a hexagon in the deviatoric plane as in the Mohr Coulomb model. Other differences are that a small cohesion c_{cs} has been introduced to ensure that a zero stress point is inside the yield surface, and the yield surface above and to the left of the critical state line is a "flattened" ellipse (Naylor, *et al.*, 1997).The analyses were in terms of effective stress. The core was treated as being partly saturated and loaded "undrained". This was achieved by assigning a pore fluid equivalent bulk modulus K_f (Naylor, *et al*, 1981) and a small initial suction to the soil. (See Chap. 12 of Maranha das Neves, 1990)
Parameters typical for the materials were selected. No attempt has been made to match them to the measured performance. (This is planned for later.) Details are not relevant to the present study and are not given.

5. RESULTS

5.1 Footing

Figure 3 shows the relationships between the average vertical settlement (δ) and the rotation (Ω) of the footing about an axis normal to AB (Figure 1.) against load. The differences between the results obtained using the direct (D), conjugate gradient (G) and relaxation (R) solvers were too small to show up at the scale of the plot. They generally agreed within about 0.1% and nowhere exceeded 1%.
Figure 4. shows the computed yielded regions for the three meshes on a vertical section through the diagonal AB of the footing. The shaded region is based on the 0.99 over-stress ratio (osr) contour where osr = $\sigma_d/(2c_u)$ and $\sigma_d = \sigma_1 - \sigma_3$.

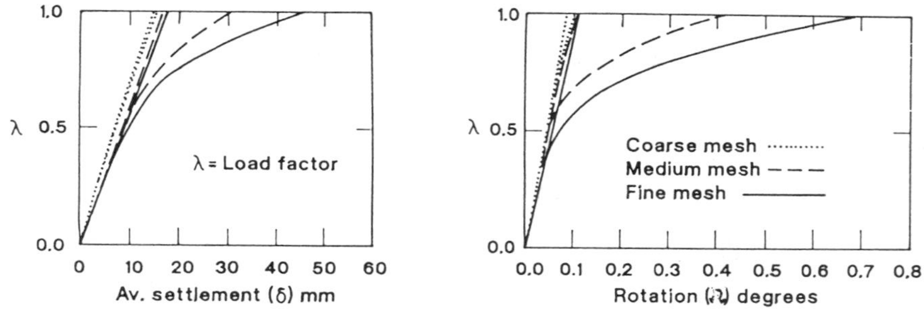

The straight lines are the linear elastic, the curved the Tresca analysis.

Figure 3. Load on footing v. settlement and rotation.

(a) Coarse mesh   (b) Medium mesh   (c) Fine mesh

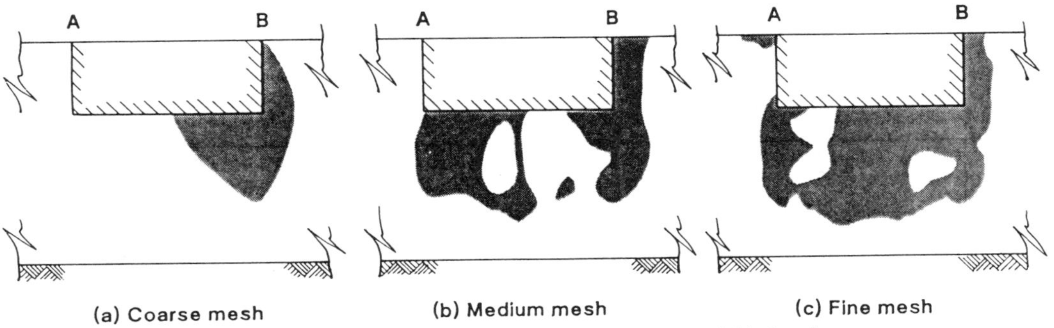

Figure 4. Diagonal section through footing showing yielded regions.

(a) Linear elastic   (b) Nonlinear

	Lin.el.	Const. stiff.	Tangential
Legend Direct	●	▲	■
Conj. G.	○	△	
Relax'n	×	+	

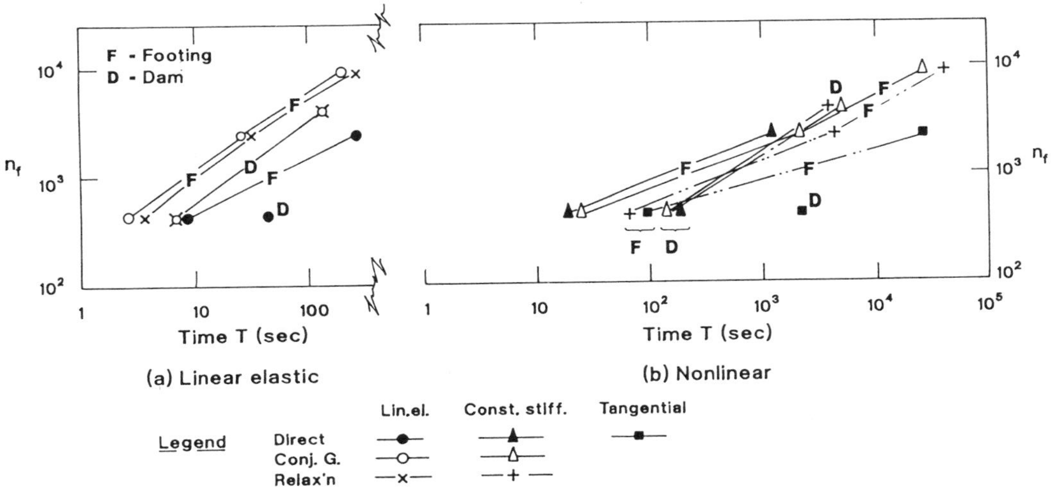

Figure 5. Times for analyses.

Table 3. Times for footing analyses (seconds)

		Coarse mesh			Medium mesh			Fine mesh		
	nincs	t_p + t_s = T		nincs	t_p + t_s = T		nincs	t_p + t_s = T		
Lin. el. D	1	1.8 + 7.0 = 8.8		1	51 + 223 = 274		-	-		
G	1	1.2 + 1.5 = 2.7		1	14 + 13 = 27		1	122 + 86 = 208		
R	1	1.2 + 2.6 = 3.8		1	14 + 19 = 33		1	126 + 153 = 279		
Const. D	15	8.7[(i)] + 10.4 = 19.1		70[(ii)]	271[(i)]+ 997 = 1268		-	-		
stiff- G	9	1.2 + 24.3 = 25.5		90[(iii)]	14 + 2176 = 2190		150[(vi)]	124 + 28060 = 28184		
ness R	9	1.2 + 67.0 = 68.2		70[(iv)]	14 + 4466 = 4480		125[(vii)]	124 + 42053 = 42177		
Tang. D	15	1.8 + 97.5 = 99.3		120[(v)]	52 + 26784 = 26836		-	-		

Notes:

[(i)] t_p also includes gauss elimination which in the other cases is included in t_s.

[(ii) - (vi)] Tabled nincs and times estimated to make the average residual norm (resn) approx. = tolr. The actual analyses gave the following:

[(ii)] nincs = 60, resn/tolr = 1.1; [(iii)] nincs = 75, resn/tolr = 1.2; [(iv)] nincs = 60, resn/tolr = 1.1; [(v)] nincs = 90, resn/tolr = 1.7; [(vi)] nincs = 90, resn/tolr = 3.0; [(vii)] nincs = 90, resn/tolr = 2.2.

Table 4. Times for dam analyses (seconds)

	Coarse mesh		Medium mesh	
	nincs	t_p + t_s = T	nincs	t_p + t_s = T
Lin. el. D	1	12.8 + 34.6 = 47.4	-	-
Lin. el. G	1	5.6 + 1.4 = 7.0	1	117 + 24 = 141
Lin. el. R	1	5.6 + 1.6 = 7.3	1	120 + 24 = 144
Const. stiffness D	60	47(i) + 153 = 199	-	-
Const. stiffness G	60	6 + 142 = 148	200(iii)	118 + 5068 = 5186
Const. stiffness R	60	6 + 136 = 142	180(iv)	118 + 3764 = 3882
Tang. D	210(ii)	13 + 2320 = 2333	-	-

Notes:

(i) t_p also includes gauss elimination which in the other cases is included in t_s.

(ii) - (iv) Tabled nincs and times estimated to make the average residual norm (resn) approx. = tolr. The actual analyses gave the following:

(ii) nincs = 150, resn/tolr = 1.4; (iii) nincs = 120, resn/tolr = 1.8; (iv) nincs = 120, resn/tolr = 1.6

The total times (T) for the analyses are plotted against the number of free degrees of freedom (n_f) in Figure 5. These times are listed in Table 3 which also gives the number of increments (nincs) used and the breakdown between the "preliminary" times t_p needed to carry out once-and-for-all operations at the start of the analyses and the subsequent "solution" times t_s. In order to make the comparisons as meaningful as possible the times for nonlinear analyses are based on an estimate of the nincs values which would give a force residual equal to the tolerance (tolr). The actual nincs and average residual norms/tolr are given in the table footnotes.

5.2 Dam

The times for the dam analyses are given in Table 4 and plotted in Figure 5. The times for the three stages have been added together. As with the footing analyses (and as indicated in the footnotes to Tables 3 and 4) some of the analyses did not achieve the target residual and an extrapolated time has been used.

6. DISCUSSION

The first objective of the study was to find out if the use of an iterative solver would change the balance of preference between a tangential and a constant stiffness finite element formulation for the nonlinear analyses. A secondary purpose was to assess the relative merits of the three solution schemes: direct, conjugate gradient and relaxation. To assist in these comparisons linear as well as nonlinear analyses of the two geometries have been carried out.

The first main finding is that the constant stiffness method used is superior to the tangential irrespective of the type of solver. This is shown clearly in Figure 5b where an order of magnitude improvement in times is indicated. (Compare the three pairs of solid squares and triangles.) No attempt was made to couple an iterative solver to the tangential method. This was because it was suspected that relatively slow convergence would result in the solution iterations due to the effect of the flow rule constraint on the stiffness matrix. This would certainly affect the relaxation method.

It could be that had a *consistent* tangent stiffness matrix been used this relatively poor performance would not have occurred. As mentioned above this was not a practical possibility due to the complexity of the formulation, at least for the critical state model.

The constant stiffness formulation had two features which made it work so well. One was the relatively short time required to compute the new right hand sides in each constant stiffness iteration. This was two orders of magnitude faster than the time required for the once-off gauss elimination and other preliminaries. The second was the procedure used to extrapolate the displacement increments at the end of a completed load step to provide an estimate for the next one so that relatively few iterations were required to complete the step.

The relative merits of the three solution procedures were different between the elastic and nonlinear analyses. In the former the iterative were in all cases more efficient than the direct. (See Figure 5a) This was expected for the finer meshes but it was slightly surprising that this also applied to the coarse mesh analyses.

The expected economy in core storage over direct solvers has been confirmed. This is quantified in Figure 6 where the storage is related to the number of nodes in the mesh (n). The increasing relative advan-

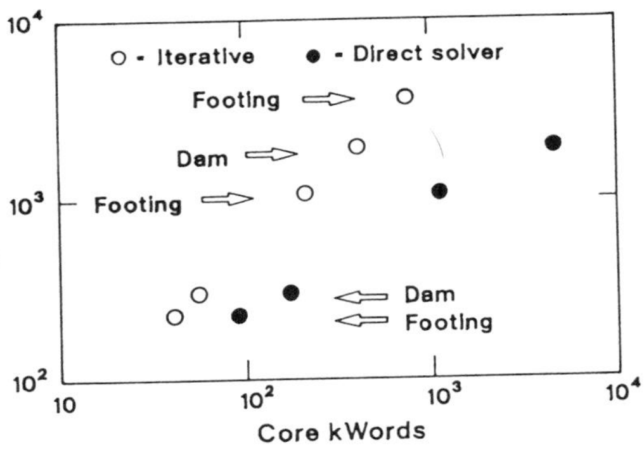

Figure 6. core storage v. No. of nodes

tage of the iterative over the direct solvers as the mesh is refined is clearly shown both in time (Figure 5) and in core storage (Figure 6).

There is not much to choose between the conjugate gradient and the relaxation solver. The former performed better in the footing analyses, more so in the nonlinear where it was up to twice as fast. This was because the more complicated form of "rigid zone relaxation" (Naylor, 1996c) had to be used here because of the high relative stiffness of the footing. For the dam there was no significant difference between the two iterative solvers in the linear analyses, whereas for the nonlinear (c.s.m.) analyses the relaxation was slightly faster. Here the basic relaxation technique, which as mentioned above is in effect the Gauss-Seidel method, was used. It is encouraging that such a simple well established procedure, which can readily be interpreted physically, still has a role.

7. CONCLUSIONS

1. The superiority of iterative solvers over a skyline in-core direct solver for linear 3D analyses of the finer meshes studied has been confirmed. They have also been shown to be superior for the coarse meshes.
2. The constant stiffness method was found to be superior to the tangential for the nonlinear analyses irrespective of the type of solver. This is attributed to the very short relative time needed to perform the iterative re-solutions in the former and also to the displacement predictor technique used. The relatively poor performance of the tangential method may in part be due to the impracticality of using a *consistent* tangent formulation.
3. In the constant stiffness nonlinear analyses the direct solver was faster than both the iterative solvers. This contrasts with the linear analyses where the iterative were superior. The difference was not however very great and the iterative solvers in conjunction with the constant stiffness method have a clear role for fine meshes where storage requirements rule out the use of a direct solver.
4. There is little to choose between the conjugate gradient and relaxation methods tested. Overall the former behaved a little better but the relaxation method produced converged solutions more quickly in the nonlinear dam analyses.
5. Since these studies have been based on the two quite different problems of a footing and an embankment dam, and have used linear elastic, Tresca and Critical state constitutive laws the findings are believed to have quite wide ranging validity in the context of 3D geotechnical f.e. analyses.

REFERENCES

Crisfield, M.A. 1986. *Finite Elements and Solution Procedures for Structural Analysis*, Pineridge Press, Swansea.

Griffiths, D.V. and Smith, I.M. 1991. *Numerical Methods for Engineers*, Blackwell.

Jennings, A. 1977. *Matrix Computation for Engineers and Scientists*, Wiley.

Maranha das Neves (editor) 1990. *Advances in Rockfill Structures*, Kluwer Academic Publishers.

Naylor, D.J. 1995. Canales Dam - a 3D collapse settlement study, Int. report No. CR/902/95, Univ. of Wales, Swansea.

Naylor, D.J. 1996a. The equivalence of old fashioned relaxation and Gauss-Seidel, Int. report No. CR/910/96, Univ. of Wales, Swansea.

Naylor, D.J. 1996b Iteration and finite elements - some experiences, Part I: Basic relaxation and the conjugate gradient method, Int. report No. CR/933/96, Univ. of Wales, Swansea.

Naylor, D.J. 1996c Iteration and finite elements - some experiences, Part II: Rigid zone relaxation, Int. report No. CR/934/96, Univ. of Wales, Swansea.

Naylor, D.J., Maranha, J.R., Maranha das Neves, E. and Veiga Pinto, A.A. 1997 A back analysis of Beliche Dam, *Geotechnique* (expected to appear in March, 1997 issue).

Naylor, D.J., Pande, G.N., Simpson, B. and Tabb, R. 1981. *Finite Elements in Geotechnical Engineering*, Pineridge Press, Swansea.

Press, W.H., Teukolsky, S.A., Vetterling, W.T., Flannery, B.P. 1992. *Numerical Recipes*, Cambridge University Press.

Smith, I.M. 1995. *Programming in Fortran 90*, Wiley.

6 Application of numerical techniques to practical problems

6.1 Tunnels and underground structures

Numerical Models in Geomechanics, Pietruszczak & Pande (eds) © 1997 Balkema, Rotterdam, ISBN 90 5410 886 X

Application of a new shotcrete model in a 3-D-FE-analysis of a tunnel excavation

H.Walter
IGT, Geotechnik und Tunnelbau, Consulting Engineers, Salzburg, Austria

ABSTRACT: A new shotcrete model is presented in this paper. The main component of the constitutive law is the Meschke-Mang model, based on the concept of multisurface-viscoplasticity. The main material properties are time dependent. The different material properties of shotcrete at different states of stress are considered. One parameter describes creep at high stress levels. In order to achieve a greater flexibility in the adaptation to experimental creep and relaxation data, additional creep terms have been added to the original model.
The constitutive law is applied to the 3D-FE-analysis of a tunnel excavation. An excavation sequence with crown, bench and invert has been modelled. From experimental data parameters for the shotcrete models have been obtained. Some results, containing deformations, stress, and plastic strains for both rock and shotcrete are shown for characteristic steps and compared to the results of a simple linear elastic material law for shotcrete.

1 INTRODUCTION

A research project focusing on the numerical simulation of the excavation of tunnels and caverns was carried out by the consulting firms Geoconsult, Glitzner, IGT, Technodat and Verbundplan Consulting Engineers in Salzburg, Austria (Walter et al., 1996). Experimental data were supplied by the Mining University in Leoben, Austria. The Finite element program MARC served as the analysis tool. This general purpose package was adapted and extended in order to be suited for the demands for static analyses of tunnels. Some of the results of the the research work will be presented here.

One major issue in the analysis of tunnels excavated according to the principles of NATM (New Austrian Tunnelling Method) is the design of the shotcrete lining. Until recently, the shotcrete lining has been modelled as having linear elastic properties. Usually, two values for the Young's modulus of the shotcrete have been used; one for young shotcrete, and a second one for hardened shotcrete. In many cases the analysis yielded far too conservative values for the amount of reinforcement required. According to engineering judgement and experience, usually much less reinforcement was put into the shotcrete lining than was required by the analysis with linear elastic material properties.

In order to provide a more rational basis for the design of the shotcrete lining, a more sophisticated model was required. The shotcrete model of Meschke and Mang (Meschke, 1996) considers all of the important properties of shotcrete and is implemented in a very efficient manner. The main shortcoming of the model is the existence of only one parameter for the design of creep and relaxation. Using a model developed earlier for fitting experimental data (Schubert, 1988), the Meschke-Mang-model has been extended and now shows a more realistic long-time behaviour.

In the following the properties of the shotcrete model will be presented in more detail. The application of the model for the 3-D discretization of a tunnel excavation (which was performed using NATM with a crown-bench-invert excavation sequence) shows its applicability in engineering practice.

2 MATERIAL MODEL FOR SHOTCRETE

The main constituent of the material law for shotcrete is the Meschke-Mang model. It models accurately the short-time behaviour as well as the time dependent properties such as the modulus of elasticity or the compressive stress. However, the flexibility of the model for the description of creep and relaxation had to be increased in order to match experimental data. This was accomplished by adding additional creep terms to the constitutive law.

2.1 *Meschke-Mang-model*

Detailed descriptions of the Meschke-Mang-shotcrete model can be found in the literature (Meschke, 1996; Meschke, Kropik, Mang, 1996). The main features of the model are:

A strain-hardening Drucker-Prager loading surface with a time dependent hardening parameter to account for the compressive regime. Cracking of maturing shotcrete is accounted for in the framework of the smeared crack concept by means of three Rankine failure surfaces, perpendicular to the axes of principal stresses. Two independent hardening and softening mechanisms control the constitutive behaviour of shotcrete subjected to compressive and tensile stresses, respectively.

The increase of elastic stiffness during hydration of shotcrete as well as the time-dependent increase of compressive strength, tensile strength, and yield surface are all considered..

The extension of the inviscid elastoplastic model for aging shotcrete to viscoplasticity is based upon the model by Duvant and Lions.

A numerically efficient algorithmic formulation of multisurface viscoplasticity in principal axes results in a robust implementation for engineering applications.

2.2 *Refined model*

The Meschke-Mang-model describes creep effects with one single parameter, the viscosity. At stress levels within the current yield surface, no creep or relaxation occurs. Therefore, the adaptability to experimental creep or relaxation data is limited.

In the course of the research project mentioned above, experiments with shotcrete specimens were conducted at the Mining University Leoben, Austria. In the tests, shotcrete prisms were loaded at an age of only 6 hours. The specimens were subjected to a large variety of load paths, some with load control and some with displacement control.

In order to match the test results, it was necessary to extend the Meschke-Mang-model. A simple engineering approach was chosen:

A model already used for the description of the long-term shotcrete behaviour (Schubert, 1988) is used to define additional creep terms: It is based on the rate of flow-method (England, Illston, 1965).

For each of the principal stresses creep strain increments for the time interval (t_i, t_{i+1}) are calculated as

$$\Delta\varepsilon_{CR,u} = \sigma_{t_{i+1}} \cdot \Delta C(t) \cdot e^{k\cdot\sigma_{t_{i+1}}} + \Delta\varepsilon_d \tag{1a}$$

with

$$\Delta\varepsilon_d = \left(\sigma_{t_{i+1}} \cdot C_{d\infty} - \varepsilon_{d,t_i}\right) \cdot \left(1 - e^{\frac{-\Delta C(t)}{Q}}\right), \tag{1b}$$

$$\Delta C(t) = C(t_{i+1}) - C(t_i) \tag{1c}$$

and

$$C(t) = A \cdot t^{\frac{1}{3}}. \tag{1d}$$

In equations (1) denote

$\Delta\varepsilon_{CR,u}$	the increment of creep strains in the time interval from $t_1$to t_{i+1}
$\sigma_{t_{i+1}}$	the principal stress at time t_{i+1}
and	
$\Delta\varepsilon_d$	the increment of 'delayed elastic strain', ε_d, i.e. of the reversible creep component, in the interval from t_i to t_{i+1}

The other variables are model parameters:

A	describes the amount of (irreversible) creep
$C_{d\infty}$	is an ultimate value for the reversible creep compliance
Q	describes how fast the delayed elastic strains develop
k	is used to introduce a nonlinearity for high stress levels.

With the help of a Poisson's ratio for creep, ν_{CR}, the creep strains in the principal directions are related to each other:

$$\varepsilon_{1,CR} = \varepsilon_{1,CR,u} - \nu_{CR}\left(\varepsilon_{2,CR,u} + \varepsilon_{3,CR,u}\right) \tag{2a}$$

$$\varepsilon_{2,CR} = \varepsilon_{2,CR,u} - \nu_{CR}\left(\varepsilon_{1,CR,u} + \varepsilon_{3,CR,u}\right) \tag{2b}$$

$$\varepsilon_{3,CR} = \varepsilon_{3,CR,u} - \nu_{CR}\left(\varepsilon_{1,CR,u} + \varepsilon_{2,CR,u}\right) \tag{2c}$$

In equations (2) the first index of the strain values denotes the index of the principal direction.

For the back transformation of the principal values of the creep components into global directions, the same transformation matrix as for the principal stresses is used.

The three-dimensional extension of the creep law is somewhat arbitrary. Due to the lack of experimental data for creep under multiaxial stress states this simple generalisation seems to be justified. As a first guess, the value of the creep Poisson's ratio has been set equal to the Poisson's ratio for shotcrete.

As was previously mentioned, the extension of the Meschke-Mang-model is thought of as a simple engineering approach which allows a better fitting of experimental data. The additional terms are treated as if they were temperature or shrinkage strains and not stress dependent. The terms are evaluated at the beginning of each calculation step with the converged stresses of the preceding increment. Within the iteration scheme of an increment the additional terms are kept constant.

If the current time increment has large stress changes, the additional terms are inaccurate because the stresses at the beginning of the time increment are used for the calculation of the additional terms. A proper choice of the sequence of load and time increments helps to keep the inaccuracies small. Considering the scatter of the shotcrete properties which are inevitable in tunnelling, the inaccuracies of the model are of very little significance. More important is the fact that the additional terms have been introduced in a way which does not disturb the quadratic convergence properties of the implementation of the Meschke-Mang-model.

The refined model contains shrinkage terms according to the relation

$$\varepsilon_{sh} = \varepsilon_{sh,\infty} \cdot \frac{t}{B+t} \quad (3)$$

with ε_{sh} being the volumetric shrinkage strains at time t, $\varepsilon_{sh,\infty}$ the ultimate value of the shrinkage strains and B a parameter describing the development of the shrinkage strains with time.

2.3 *Parameters based on experimental data*

The experimental data were obtained from prismatic specimens with the size of 0,15 m x 0,15 m x 0,30 m. The specimens were made from dry mix shotcrete at a tunnel construction site close to the university. After preparation, the specimens were loaded, usually in the direction normal to the spraying direction. Metal sheets and lubrication were used to minimise disturbances of the uniaxial stress state. A large variety of load and displacement histories was investigated. For each load path, five specimens were loaded, while a sixth remained unloaded and its shrinkage deformations were measured.

The experimental results used for the determination of the model parameters are shown as solid lines in figure 1. The shrinkage strains have already been subtracted from the total strains. Two characteristic curves were selected in order to show the degree of accuracy that a single set of parameters can fit the data. Some of the short-time parameters are based on additional test results based on uniaxial compression tests.

The dashed lines in figure 1 show the creep and relaxation curves using the refined model. The match is not perfect. However, the scatter of the experimental data is much larger than the deviations of experimental and analytical curves.

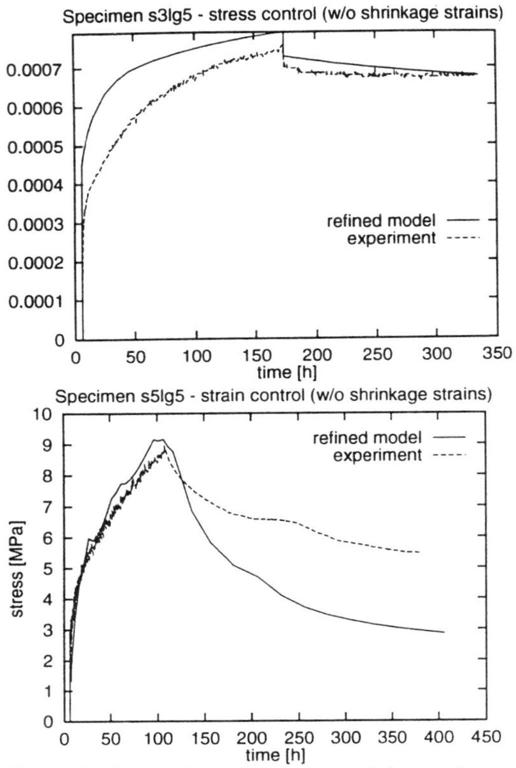

Figure 1: Comparison of experimental data and calculated values

3 3-D-TUNNEL ANALYSIS

3.1 *Data of the model*

3.1.1 *Geometry and FE-discretization*

An Austrian road tunnel with an excavation area of about 78.4 m² has been chosen for the analysis. A cross section has been selected where the overburden amounts to about 200 m. The primary lining has a thickness of 0,15 m and the secondary lining is 0,3 m thick.

Neglecting asymmetries due to the geological conditions one half of the tunnel and the surrounding rock has been investigated. Only a part of the overburden (27 m above the crown) have been discretizised; the remaining cover is simulated as an external load.

The finite element mesh spans 21 m in the horizontal direction normal to the longitudinal tunnel axis, 72 m in the longitudinal direction, and 61 m in the vertical direction. Symmetry conditions at both end surfaces simulate the virtually infinite extension in the longitudinal direction.

The shotcrete lining shell is modelled with the help of layered shell elements. A total of 5 layers was used. The model consists of 60 'slabs' in the longitudinal direction, each comprising 198 solid elements and 13 shell elements.

3.1.2 *Material properties*

The rock was a quartzitic mica slate with banded soft mylonitic faults. The Mohr-Coulomb constitutive law has been applied using the following parameters:

modulus of electricity E = 500 MPa

Poisson's ratio	$\nu = 0.25$
specific weight	$\gamma = 22$ kN/m³
angle of friction	$\varphi = 25°$
cohesion	c = 1 MPa

For the shotcrete the parameters derived from the experimental data in section 2.3 have been used.

For comparison, analyses with linear elastic material properties have been performed as well.

3.1.3 *Excavation sequence and discretization in time*

The cross section was excavated in three parts: crown, bench and invert. It would have been too time consuming to model the excavation of each round separately. Instead, fictitious rounds with a thickness of 3 m were simulated. Each round consisted of 3 'slabs' of the FE-model). The distance between the excavation of crown and bench, and bench and invert, was defined to be 6 m. (In reality the distances are usually larger; small distances result in more pronounced interdependencies of the part excavations.)

3.2 *Results and their interpretation*

In order to show the influence of the types of material model three analyses have been compared:

- one with linear-elastic shotcrete and elastic rock
- one with linear-elastic shotcrete and Mohr-Coulomb model for rock
- one with viscoplastic shotcrete model and Mohr-Coulomb model for rock

In Figure 2 the normal stresses in the middle layer of the shotcrete shell are shown for a node at the roof of the tunnel. The node is situated in the cross section in the middle between the two end surfaces. There is a large difference in stress depending on whether the rock behaves in a linearly elastic or plastic manner. (The difference of the displacements is much smaller.) The viscoplastic concrete model yields a pronounced reduction of the normal stresses. The creep and relaxation effects are rather severe due to the experimental data which was used. Therefore the drop of the normal stresses is probably higher than it would be in reality. Along with the reduction of the normal stresses, the displacements increase by about 30% if the viscoplastic model is used.

Figure 3 shows the normal stress in the middle layer of the shotcrete shell for a characteristic excavation step. The highest compression stresses occur in the crown and at the foot of the bench. It can easily be seen that the stresses decrease with increasing distance from the heading. (The stresses at the end (i.e. the symmetry plane) are not realistic because they are disturbed by the boundary conditions.) Similar plots for the time increments after the end of the excavation show that the normal stresses are reduced to a low percentage with time. This confirms that the chosen model parameters overestimate the stress relaxation. On the other hand, no creep has been modelled for the rock. Depending on the geological conditions some soil pressure could build up with time and cause additional stress in the shell.

By examining the vertical stresses in the rock, stress concentrations in front of the heading are clearly visible. The building of a load carrying arch causes stress concentrations about half a diameter sideways of the tunnel. Figure 4 shows the deformed configuration of the rock for one excavation step together with the zones having plastic strains. It is clear that the stress reduction in the shotcrete shell is accompanied by an increase of the plastic strains as

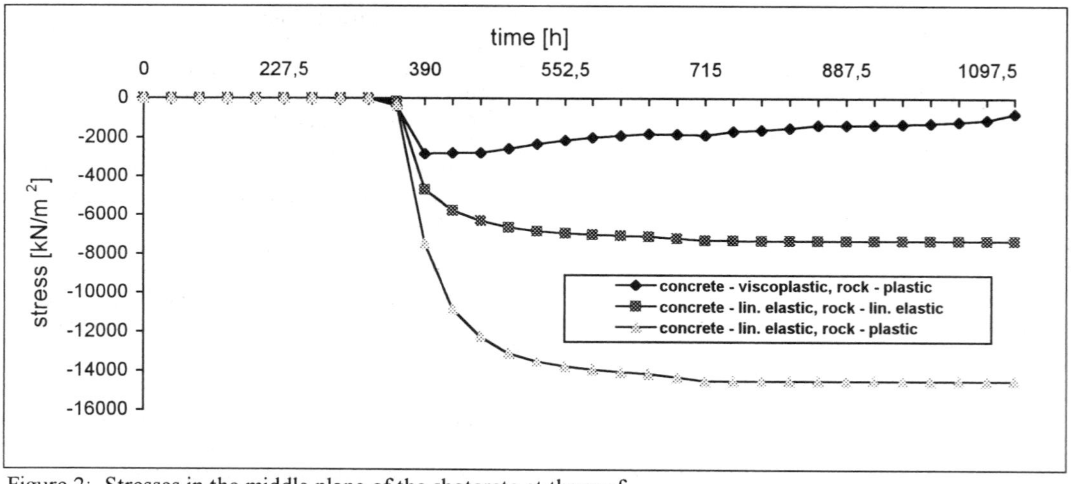

Figure 2: Stresses in the middle plane of the shotcrete at the roof

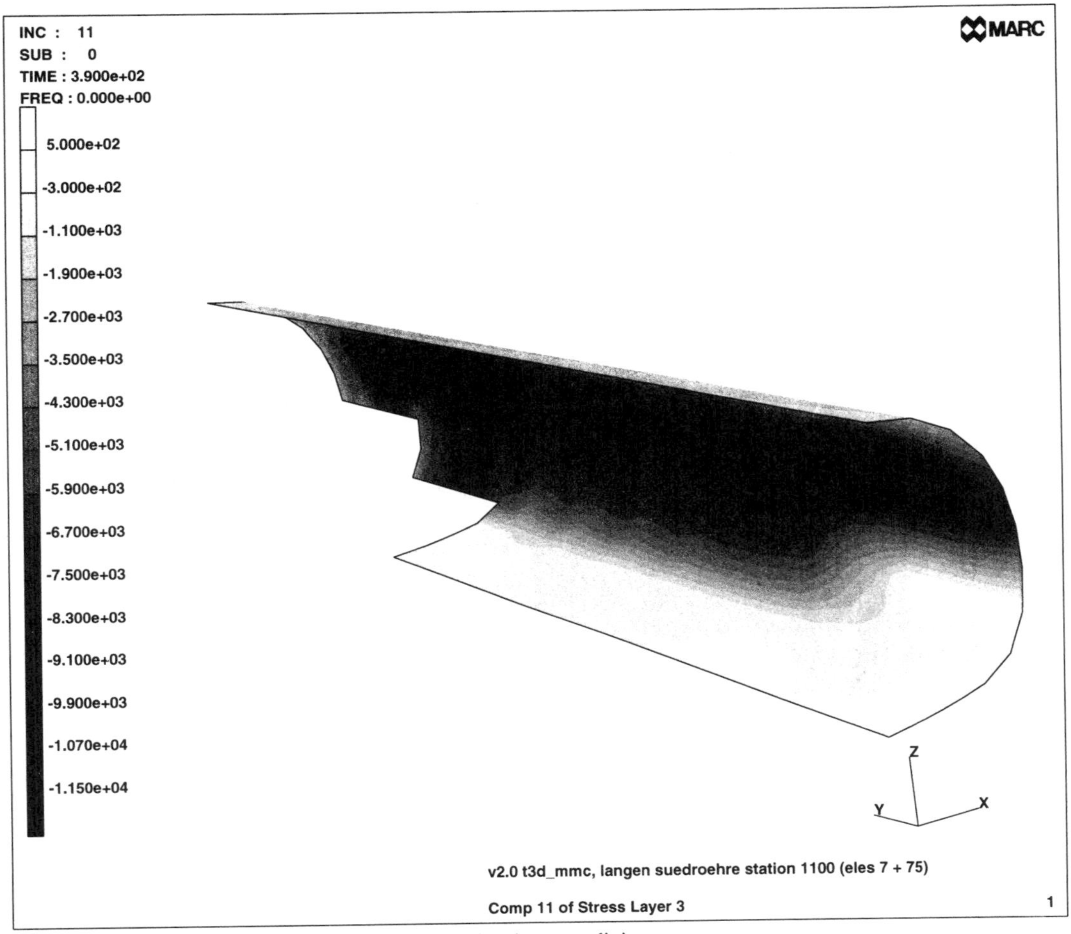

Figure 3: Normal stresses in the middle layer of the shotcrete lining

the distance from the heading is increased.

4 CONCLUSIONS

It has been shown that the refined shotcrete model is suited for fitting experimental shotcrete data. The numerical implementation is stable, even if used with unrealistic high creep rates. The implementation is also numerically efficient: a single time step per round yielded reasonable results. The low strength of shotcrete subjected to tension and the reduction of stress peaks is accounted for by the Meschke-Mang-model, as well as the development of material parameters with time. In addition, the refined model is capable of describing creep and relaxation at moderate stress levels.

The application of the refined shotcrete model together with an elastoplastic rock model for a 3-D-tunnel excavation yielded reasonable results with a moderate effort in preparation and interpretation. The total turn-around time at a minicomputer was about 2 days.

ACKNOWLEDGEMENTS

The entire research team in Salzburg and Leoben has contributed to the results presented herein. Much of the implementation of the model has been accomplished by Michaela Kofler and Helmut Schweiger. Markus Brandtner performed most of the 3-D-analyses. Günther Meschke of the Institute of Strength of Materials of the Technical University Vienna provided a 2-D and a 1-D-version of his shotcrete model. The author expresses his appreciation and thanks to all of them.

Financial support of the project by the Austrian

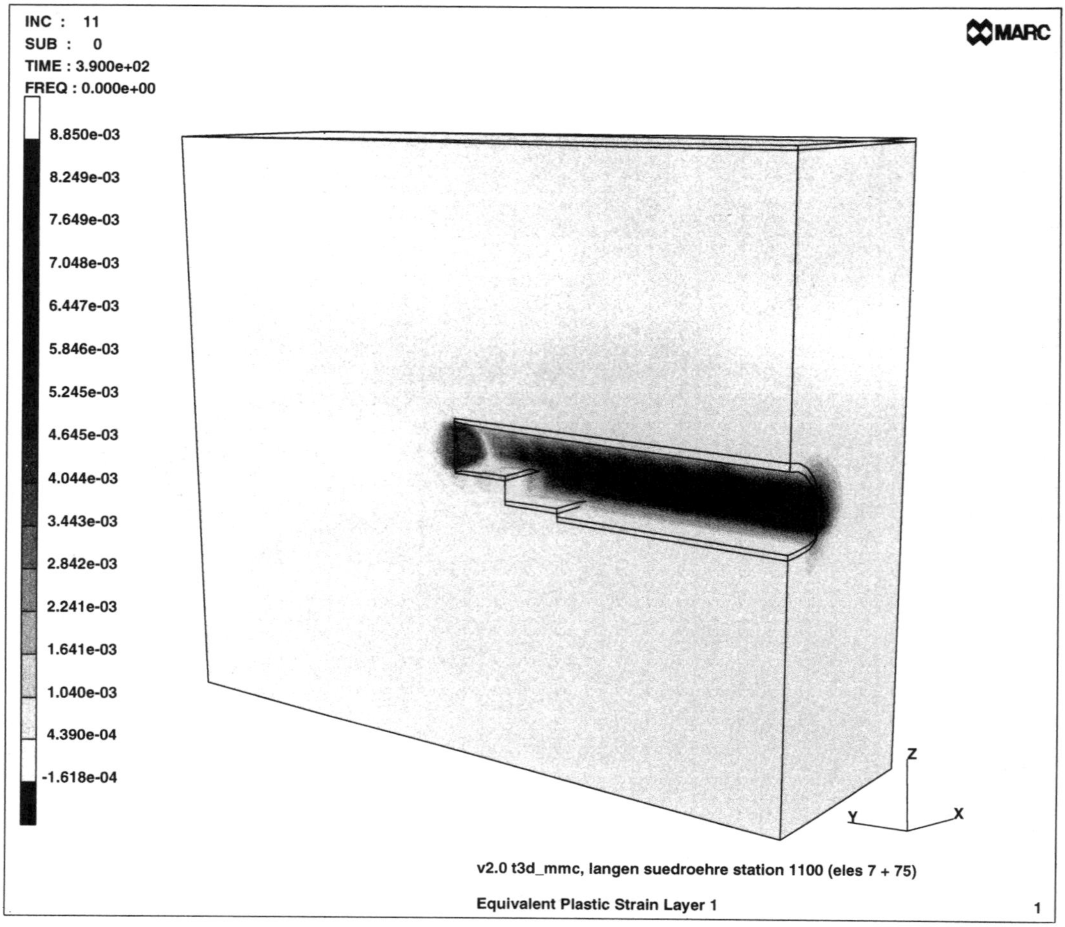

Figure 4: Equivalent plastic strains in the rock

'Forschungsförderungsfonds der gewerblichen Wirtschaft' and the country of Salzburg is gratefully acknowledged.

REFERENCES

England, G.L., Illston, J.M. 1965. Methods of Computing Stress in Concrete from a History of Measured Strain. *Civil Engineering and Public Works Review*, 60, pp. 513-517, 692-694, 846-847.

Meschke, G. 1996, Consideration of Ageing of Shotcrete in the Context of a 3-D Viscoplastic Material Model. *Int. Journal for Numerical Methods in Engineering* 39, 3123-3143.

Meschke, G. & Kropik, C. & Mang, H.A. 1996. Numerical Analyses of Tunnel Linings by Means of a Viscoplastic Material Model for Shotcrete. *Int. Journal for Numerical Methods in Engineering* 39, 3145-3162.

Schubert, P. 1988. Beitrag zum rheologischen Verhalten von Spritzbeton. *Felsbau* 6, 150-153.

Walter, H. (ed.) 1996. Praxisorientiertes Rechenmodell für Tunnel und Kavernen. *Final report to the Forschungsförderungsfonds der gewerblichen Wirtschaft*, Vienna, Austria.

Numerical Models in Geomechanics, Pietruszczak & Pande (eds) © 1997 Balkema, Rotterdam, ISBN 90 5410 886 X

Recent developments and application of the boundary element method in geomechanics

G. Beer
University of Technology Graz, Austria

O. Sigl & J. Brandl
Geoconsult, Salzburg, Austria

ABSTRACT: Recent applications of the boundary element method (BEM) and coupled boundary element/finite element method (BE/FE) in geomechanics are shown. The examples of application include the analysis of caverns of the CERN particle accelerator project and the simulation of the processes which occur during the detonation of explosives in a bore hole. It is concluded that substantial savings can be made over conventional analysis methods in the time required to discretize the problems and to compute the results.

1 INTRODUCTION

Whereas the Finite Element Method (FEM) has received wide acceptance among geotechnical engineers and has been extensively used to model a wide range of problems from tunnelling to foundations the application of the Boundary Element Method (BEM) and the coupled FEM/BEM has been the exception rather than the rule. However, since the semi-infinite or infinite domain is automatically included in the solution and only a surface mesh is needed these methods offer substantial savings in the effort spent in generating a mesh and in computing results.
Reasons for the reluctance to use these methods are the fact that there are few commercial programs available which offer BEM and coupled analysis capabilities and that there is a common misconception that the BEM can only be used for elastic, homogeneous and isotropic domains.
It is the purpose of this paper to dispel some of these misconceptions by showing recent practical applications which were carried out using BEM or coupled FEM/BEM discretizations and which used substantially fewer resources than conventional analyses.

2 THE BOUNDARY ELEMENT METHOD

As the BEM is still not widely known and accepted it may be appropriate to start off with a basic discussion of the method and to point out certain advantages and shortcomings of the method.

In contrast to the Finite Element Method the Boundary Element method works with „shape functions" which are fundamental solutions of the governing differential equations. Therefore - unlike the FEM - the displacements and stresses computed by the method satisfy equilibrium and compatibility conditions exactly and approximations only occur on the boundary of the domain. (In the FEM equilibrium is only satisfied in terms of nodal point forces but may be violated locally). In the BEM the boundary conditions have to be satisfied only and therefore only boundaries have to be discretized, i.e. divided into boundary elements. This essentially means that the discretization effort is reduced by an order of magnitude: line elements are used fro 2-D problems and surface elements for 3-D problems. Because there is no meshing inside the domain the method is able to model infinite and semi-infinite domains without having to revert to mesh truncation (in the case that boundary surfaces extend to infinity infinite boundary elements can be used).
Using the reciprocal theorem by Betti (Beer & Watson 1992) a system of equations can be obtained which relates the traction values acting on the nodes of the boundary elements, **t**, to the displacements at these nodes, **u**:

$$\mathbf{A}\mathbf{u} = \mathbf{B}\mathbf{t} \tag{1}$$

A and **B** are fully populated and unsymmetrical coefficient matrices. The fundamental solutions used are the tractions and displacements due to a unit concentrated static force or a unit concentrated impulse (in the case of dynamics).
The BEM as presented here has several drawbacks which limits its application to geomechanical problems:

1. A fundamental solution of the governing differential equation must exist. Such solutions are usually only available for elastic, homogeneous materials. Anisotropic materials can be dealt with but fundamental solutions are more complicated than for isotropic materials.
2. Although body forces may be included in the Betti theorem these have not been considered in Eq (1) because volume integrals would occur. However, some kinds of body forces, such as gravitational and centrifugal forces can be dealt with in a relatively straightforward way by transferring the associated volume integrals into boundary integrals (Brebbia 1984).
3. Nonlinear material behavior can also be considered with the BEM but a volume integration has to be performed. The additional volume discretization needed may mean that the method looses some of its attraction.

2.1 *Multi-region BEM*

To be able to model inhomogeneous materials the domain can be subdivided into subregions each of which can be assigned different material properties. For each region a system of equations can be set up which relates the tractions and the displacements at the boundary nodes of the region. At the nodes on the interfaces between regions both the tractions and the displacements are unknown.

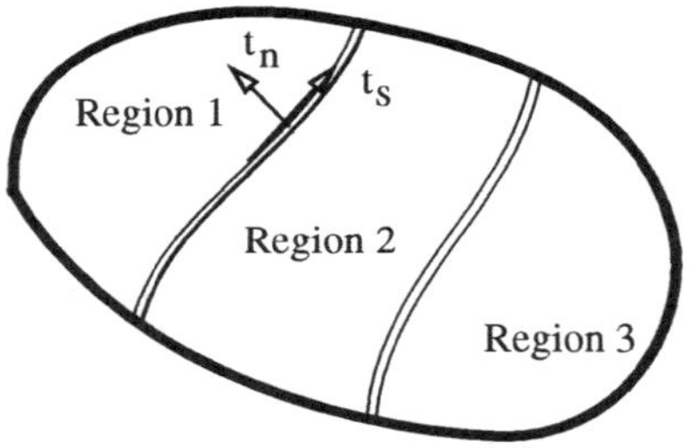

Fig.1. Multi-region Problem and interface tractions

Using compatibility and equilibrium conditions at the interfaces the following system of equations can be obtained (Beer 1993).

$$\hat{\mathbf{K}}\mathbf{u}^{i} = \mathbf{t}^{i} - \mathbf{t}^{i0} \quad (2)$$

In the above equation $\mathbf{t}^{i}$ is a vector containing applied tractions at all interface nodes, $\mathbf{t}^{i0}$ is a vector containing tractions at the interface nodes for the case where $\mathbf{u}^{i}$ (the interface displacements) are zero. $\hat{\mathbf{K}}$ is a „stiffness matrix" whose coefficients are **traction** values for unit interface displacements.

After solving for the unknown interface displacements $\mathbf{u}^{i}$, the displacements of the non-interface nodes (free nodes) can be computed by backsubstitution of these displacements into the BE equations for the region.

The method just explained can now handle piecewise inhomogeneous regions and as we will see in the next section is ideally suited for the modelling of faults and joints.

2.2 *Faults and joints*

Because there is a relationship between **tractions** and displacements (in contrast to the FEM where there is a relationship between Nodal Forces and displacements) the method is ideally suited for modelling of nonlinear material behavior such as slip and separation on joint or fault planes. Once Equations (2) have been solved for interface displacements $\mathbf{u}^{i}$ then the interface nodal traction vector can be computed for example for the interface of region I by:

$$\mathbf{t}^{iI} = \hat{\mathbf{K}}\mathbf{u}^{iI} + \mathbf{t}^{i0I} \quad (3)$$

The traction vector is usually defined in the cartesian coordinate frame but can be easily converted to components of tractions normal t_n and tangential t_{s1}, t_{s2} to the interface. The Mohr-Coulomb yield condition can be written as:

$$F_s = \tau + t_n \tan\phi - c \quad (4)$$

Where $\tau = \sqrt{t_{s1}^2 + t_{s2}^2}$, ϕ is the angle of friction and c is the cohesion. Alternatively it would be possible to compute the stress tensor or stress concentration factors at the boundary nodes and use these as criteria for crack opening or slip. As outlined in some detail elsewhere (Beer, 1993) efficient algorithms which control slip/dilation at the joint plane can be implemented by simply disconnecting/connecting appropriate degrees of freedom at the interface.

These procedures have been developed some time ago for statics and are found to work well even for extreme cases where the FEM has difficulty (Day & Potts 1994).

They have also been applied recently to dynamic problems involving very short pulses such as they occur in blasting and were found to work well under certain circumstances (Tabatabai-Stocker & Beer 1997). However further work is needed with regards to improving their efficiency with respect to the computation time and the disk storage requirement.

The multi-region BEM with joint capability offers a significant step forward in making the method suitable for problems in rock mechanics as now nonlinear ehavior - at least the one concentrated in a fault or oint plane can be modelled without internal meshing. Iowever the method as explained is still not able to leal with elasto- (visco)plastic behavior in the rock nass, sequential excavation and construction and the pplication of nonuniform body forces. Here we can esort to the coupling of the BEM and the FEM thus ısing all the advantages of the FEM while retaining he advantages of the BEM.

COUPLING WITH FINITE ELEMENTS

The coupling of the BEM and FEM is treated the ame way as multi-region problems except that the odal tractions have to be converted to nodal point orces. Using the finite element shape functions and he virtual work principle (Beer & Watson 1992) the Nodal Point forces $\mathbf{F}^e$ of a Finite Element may be xpressed in terms of nodal point tractions $\mathbf{t}^e$ by:

$$\mathbf{F}^e = \mathbf{M}^e \mathbf{t}^e \tag{5}$$

vhere $\mathbf{M}^e$ is a matrix of Integrals of shape function roducts. Using (5) the actual stiffness matrix (i.e the one relating nodal point **forces** to nodal displacenents) of the BE region is obtained as:

$$\mathbf{K}_{BE} = \mathbf{M}\hat{\mathbf{K}} \tag{6}$$

vhere $\mathbf{M}$ is a matrix assembled from element contriutions $\mathbf{M}^e$ (see Beer & Watson 1992).

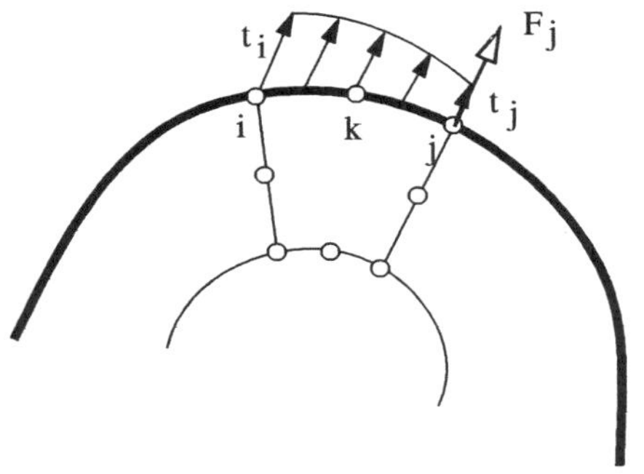

Fig.2. Relationship between nodal point forces and nodal tractions.

The coupled analysis involves two steps:

1. Computation of the Stiffnessmatrix of the nodes of the BE region which connect to Finite Elements.
2. Assembly with Finite Elements in the usual way
3. Solution for the displacements at the Finite Element nodes. Creep and visco-plasticity can be handled with a relatively small system of equations only involving the nodes of the finite elements and the nodes on the interfaces with boundary elements.
4. If required the displacements and stresses may be computed inside the region using the boundary traction and displacement values.

Since $\hat{\mathbf{K}}$ is not symmetric it follows that the stiffness matrix $\mathbf{K}_{BE}$ is also not symmetric. As shown in (Beer & Watson 1992) a symmetric $\mathbf{K}_{BE}$ can be obtained if the principle of minimum potential energy is used instead of virtual work and thus all the advantages of symmetric FE solvers can be retained. Symmetric stiffness matrices were found to work well especially for boundary elements with linear variation of displacement and traction (where there is not a marked unsymmetry in the virtual work approach anyway).

4 EXAMPLES OF APPLICATION

Several examples of application are shown here in order to demonstrate that the numerical solution procedures outlined above lead to efficient and accurate solutions for practical problems in geomechanics. The first is an example of the multi-region BEM with faults. The second of a coupled FEM/BEM analysis and the third of a dynamic multi-region analysis.

4.1 Tunnel approaching fault

The first example is taken from tunneling and relates to the detection of faults ahead of the tunnel face using *in situ* measurements. In tunneling very often there is limited geological exploration and not all geological features can be mapped. When a tunnel approaches an unmapped fault zone then this may be potentially dangerous. Using modern measurement techniques it is possible to perform very accurate absolute or relative displacement measurements in a very short time. If anomalies in the measurements can be correlated to the presence of fault zones then this would be a definite advantage.

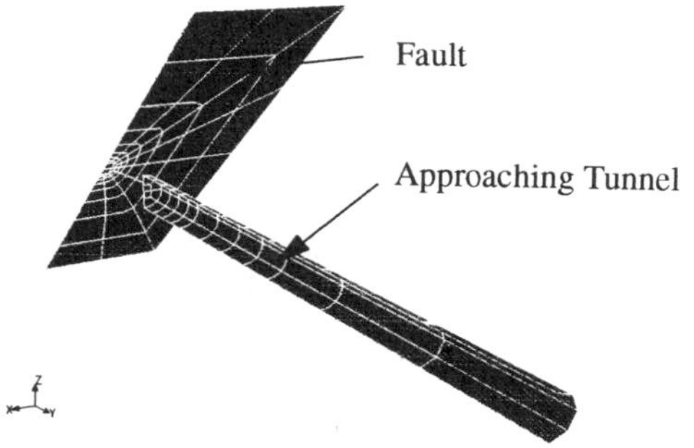

Fig.3. Boundary Element mesh of tunnel and fault

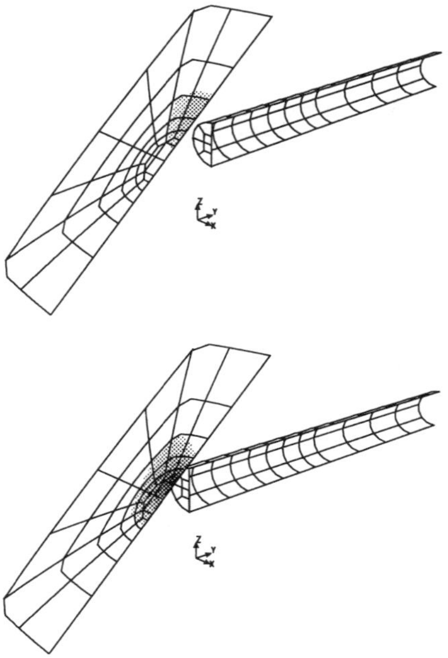

Fig.4. Contours of slip at the fault plane for different locations of tunnel face (darker = higher slip)

The numerical model used to simulate the advance of a tunnel to a fault zone is shown in Figure 3. A plane of symmetry was assumed and therefore only half of the tunnel and fault surface had to be discretized. Infinite boundary elements are used to simulate a tunnel and fault of infinite extent. The main feature of the model is that nonlinear behavior is concentrated in the fault plane whereas the rest of the rock mass behaves linearly elastic. This was thought to be realistic enough a model for the purposes of this study. The advantage of the multi-region BEM is that the time required to build the model is considerably smaller than with the FEM and that computer resources required are small (a run took typically 10 minutes on a Pentium PC).
The computed values of slip on the fault plane for different positions of the tunnel face are shown in Fig.4. For more details see (Schubert & Budil 1995).

4.2 *CERN LHC Project*

The CERN organization in Geneva is planning to extend at several locations (points) the existing 27 km long Large Electron Positron Collider (LEP) to install a new particle collider machine with considerably higher collision energy, the Large Hadron Collider (LHC).
In the joint venture which was awarded the design contract for Point 5 of the LHC, Geoconsult is responsible for the design of the underground structures. At Point 5, a scheme of two large caverns (UXC55 and USC55) with spans of up to 26 m and more than 30 m height, two large new shafts and a number of auxiliary tunnels have to be constructed in a geological sequence of marls and sandstones (Molasse) underneath a major aquifer. The large caverns are separated by a 7 m thick concrete pillar. Especially, the design of the secondary lining structure implied a particular technical challenge due to the geometrical complexity of the structure and the sort of load cases to be considered. In the past, the design calculations for secondary lining structures were carried out by choosing appropriate equivalent two-dimensional FEM and/or beam element models. Three-dimensional analyses, if any at all, were limited to intersections with tunnels or shafts. In these cases, FEM shell analyses were carried out with the distinct simulation of rock mass interaction by springs with a given stiffness. However, in the case of the LHC underground structures at Point 5, a comprehensive three-dimensional approach employing coupled BEM/FEM is chosen due to the complex interactions between rock mass, concrete lining, concrete pillar, shaft and cavern end walls, all considering lining weight, swelling and rock mass creep load cases. Although, the rock mass at Point 5 is layered and therefore anisotropic, the drawback #1 identified in Chapter 2 above can be relativized for the secondary lining design. Since the rock support design is carried out based on other numerical models using distinct element approach, the behavior of the rock mass is not anymore of immediate interest during the design of the secondary lining. The main goal for the numerical lining design calculations is to properly simulate the rock mass/concrete lining interaction. Therefore, if the rock mass can be represented in the calculation model such that proper stiffness response to lining deformations is ensured, disregarding the anisotropic rock behavior can be regarded as an acceptable simplification.

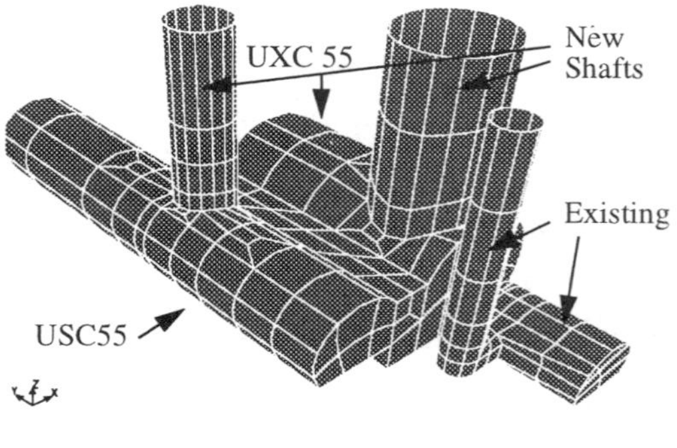

Fig.5. Layout of underground excavations

the best of our knowledge, this is the first applica-n of the coupled BEM/FEM to a problem of this e. The coupled analysis is carried out to study the ects of swelling and rock mass creep on the sec-dary lining structure and to derive relevant sec-nal forces for concrete design purposes. The XC55 and the USC55 caverns are modelled sepa-tely using symmetry conditions to simulate the luence of the caverns on each other. The mesh con-sts of Boundary Elements which model the infinite ck mass and Finite Elements which model the sec-dary concrete lining. The thickness of the second-y lining is 800 mm for the UXC55 and 500 mm for e USC55 cavern. The Finite Elements of the con-ete lining are connected to the rock Boundary Ele-ents by joint elements (Beer & Watson 1982) which ve a very low shear stiffness assigned. These inter-ces are essential for the proper rock mass lining teraction simulation. In addition, the model com-ises also the concrete pillar and a zone to apply elling and rock mass creep loads. Both zones are odelled by Finite Elements. See Figure 6 showing e mesh used for the analysis of the USC55 cavern. ne swelling and the rock mass creep is assumed to confined to a zone of weak marl which is approxi-ately 7 m thick and located 4 m above the crowns of e caverns. Swelling is simulated by a temperature uivalent approach, rock mass creep by using a real eep model dependent on the actual stresses. For the ck mass creep simulation it was therefore necessary introduce to the rock mass *in situ* stress field condi-ons. For the evaluation of necessary concrete rein-rcement, the Finite Elements representing the ncrete lining were degenerated to shell elements hich allow computation and display of sectional rces in a local co-ordinate system. Figure 7 shows a articular result of the analysis, namely the computed ternal sectional forces in the concrete shell for a articular load case. Note that if one were to model is problem with Finite Elements only, the whole ck mass which is represented in the coupled BEM/ EM case by the boundary element region would ave to be filled up with solid elements up to a dis-nce sufficiently far away from the excavation so at the influence of artificially applied boundary con-itions on the results is negligible. As a matter of fact, ıch an FEM analysis was carried out by another esigner for another point of the LHC Project (Point) and recently published (Sloan 1996). Although the ctual number of finite elements used was not speci-ed in that paper a mesh of approximately 20 000 ele-ents was indicated. A CRAY supercomputer omputer was required to run the model. In contrast, ne coupled BEM/FEM mesh used for the Point 5 nderground design and presented above has less an 2000 Boundary and Finite Elements and took pproximately 2 hours to run on a Pentium Pro PC.

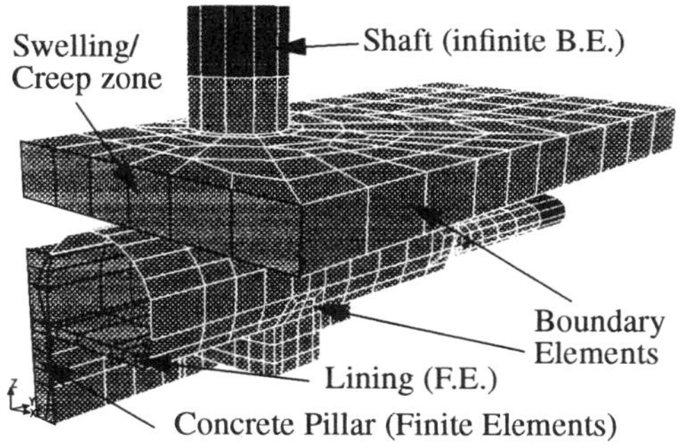

Fig.6. Coupled FEM/BEM mesh for cavern USC55

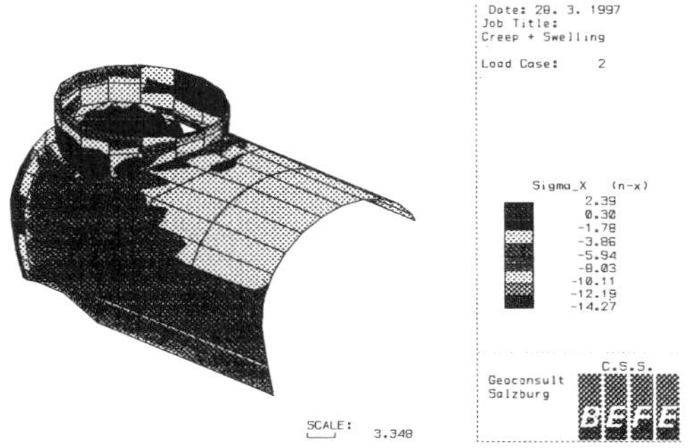

Fig.7. Distribution of normal force in concrete shell

4.3 Borehole Blasting

To be able to optimize the blasting process in order to achieve maximum of rock breakage with a minimum of blasting energy it is important to better understand the physical processes which occur during blasting. The BEM is ideally suited for modelling these processes because no internal meshing is required and the infinite domain is implicitly considered in the analysis without having to resort to mesh truncation. Mesh truncation is a more critical problem in dynamics because of the need to introduce nonreflecting boundaries. For boundary loading which is applied in an extremely short time as in blasting the results will be very sensitive to the fineness of the discretization. If finite elements are used then the volume discretization would have to be very fine. The analysis presented here is an example is the modelling of crack propagation due to the detonation of a charge in a circular borehole. The explosive charge is simulated as time varying pressure on the borehole walls. To simplify the analysis of this complex problem in order to gain experience with the method a two-dimensional (plane strain) analysis was performed. Furthermore it was assumed that cracks can only propagate on predefined interfaces at 45 degrees as shown in Figure 8.

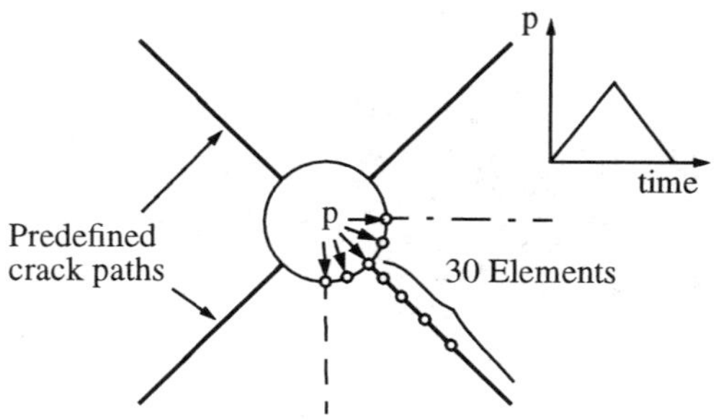

Fig.8. Borehole blasting: problem statement and boundary element mesh used.

Because of the axisymmetry of the problem only one quarter of the domain needs to be discretized. A quarter of the borehole wall was discretized into 4 boundary elements with quadratic variation of traction and displacements, whereas the predefined crack surface was divided into 30 elements. The loading was applied as a triangular pulse (Fig 5) in $24 \cdot 10^{-7}$ sec, the maximum pressure was assumed to be 690 N/mm^2. Further details are given in (Tabatabai-Stocker 1997).

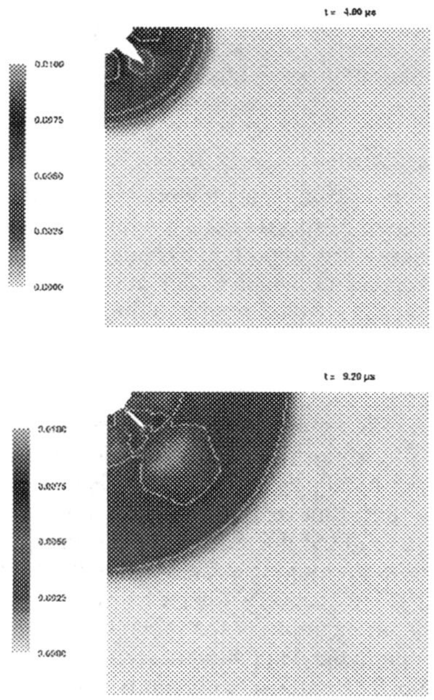

Fig.9. Contours of radial displacement plotted on displaced shape at 4.8 and 9.2 microseconds after detonation of explosive charge (lower right quarter shown)

One result of the analysis is shown in Fig. 9. It show the contours of radial displacement plotted on th deformed shape at 4.8 and 9.2 microseconds after th detonation of the explosive charge. The opening/clos ing of the crack and the advance of the pressure wav can be clearly seen.

5 SUMMARY AND CONCLUSIONS

Examples of application of the Boundary Elemen and the coupled Boundary Element/Finite Elemen methods in geomechanics were presented. All of th examples involved nonlinear material behavior con fined to certain areas of the soil/rock mass thus dis pelling the common assumption that the BEM is onl applicable to elastic problems. Although isotropi rock mass behavior has been assumed in one exampl this is not a limitation of the method but rather a cur rent limitation of the software. Fundamental solution for orthotropic domains can be implemented withou too much difficulty.

The examples presented here show that the applica tion of boundary element and coupled finite elemen boundary element discretizations to problems in geo mechanics can result in a substantial reduction in th effort required to generate a mesh and the compute solution time. This is particularly true for three dimensional problems. The fact that the discretizatio dimension is reduced by one order of magnitude (lin elements for 2-D problems and surface elements fo 3-D problems) not only considerably reduces th mesh generation effort but also increases the accurac of the results because the solutions inside a regio satisfy equilibrium exactly at every point (in the FEM only the equilibrium of the equivalent nodal poin forces is satisfied).

For problems in transient dynamics the fact that n internal discretization is needed is of particular bene fit because the results of the high frequency analyse are very much dependent on the finite element siz used. Furthermore in geomechanics we are dealin with infinite or semi-infinite domains which can b modelled in the FEM only by truncating the mesh an by applying some artificial boundary conditions. Fo dynamic problems special non-reflecting boundar conditions have to be applied. In the BEM the fac that a domain is finite, infinite or semi-infinite is of n consequence as this implicitly considered in the solu tion.

The static analyses were performed with the com puter program BEFE (C.S.S. 1997) the first commer cially available computer program to completel integrate the finite and the boundary element method The dynamic 2-D analysis has been performed usin the program MUR (Tabatabai-stocker 1997) espe cially developed for this purpose.

It is planned to extend this program to three dimen

sions and to integrate its capabilities into the BEFE program.

6 ACKNOWLEDGEMENTS

The research on borehole blasting was supported by the Austrian National Science fund (Project P10326-GEO). Thanks are due to B.Tabatabai-Stocker for supplying the results of borehole blasting on Figure 9.

REFERENCES

Beer,G. & J.O.Watson 1992. *Introduction to Finite and Boundary Element Methods for Engineers,* Chichester: Wiley.

Brebbia C.A., J.C.F. Telles & L.C. Wrobel 1984. *Boundary Element Techniques,* New York: Springer.

Beer, G. 1993. An efficient numerical method for modelling initiation and propagation of cracks along material interfaces. *Int. j. numer. methods eng.* 36: 3579-94.

C.S.S 1997. BEFE Version 6.1w: Users and Reference Manual, Computer Software and Services, Carnerigasse 10/45, GRAZ, Austria

Day, R.A. & D.M.Potts 1994. Zero thickness interface elements - numerical stability and application. *Int. j. numer. anal. methods geomech.* 18: 689-708.

Schubert W. & A.Budil 1995. The importance of longitudinal deformation in tunnel excavation. *Proc. 8th Int. Congr. on Rock Mechanics - Vol. III.* Rotterdam: Balkema.

Sloan, A. , D. Moy & D.Kidger 1996. 3-D modelling for underground excavation at point 1, CERN. In G. Barla (ed.) EUROCK 96: 957-963. Rotterdam: Balkema.

Tabatabai-Stocker B. & G.Beer 1997. A bounday element method for modelling cracks along material interfaces in transient dynamics. submitted to: *Int. j. numer. anal. methods geomech.*

Numerical Models in Geomechanics, Pietruszczak & Pande (eds) © 1997 Balkema, Rotterdam, ISBN 90 5410 886 X

Prediction of soil restraint to a buried pipeline using interface elements

P.C.F.Ng
Department of Civil Engineering, Queen's University of Belfast, UK (Formerly: University of Sheffield, UK)
I.C.Pyrah
Department of Civil & Transportation Engineering, Napier University, UK (Formerly: University of Sheffield, UK)
W.F.Anderson
Department of Civil & Structural Engineering, University of Sheffield, UK

ABSTRACT: Numerical analyses have been carried out to simulate an *in situ* pipe loading test in an attempt to better understand pipeline response to lateral ground loading. A two stage analysis technique for the modelling of laterally loaded pipelines has been adopted. The first stage uses a 2-D plane strain FE analysis to predict the restraining effect of the soil as a function of pipe displacement. These predictions are then used in stage 2 which models the behaviour of the pipeline, using an elastic beam on elastic foundation program. From previous analyses, it was found that the assumed tensile behaviour of the soil at the back of the pipe plays a significant part in the determination of the *P-y* curve. In this paper, the stage 1 analysis is extended by using interface elements to model slip and separation at the soil/pipe interface and at the interface between the backfill and the natural ground.

1 INTRODUCTION

Numerical models have been used by many researchers in an attempt to better understand pipeline response to lateral ground loading for use in design and risk analysis. Although the three-dimensional finite element method is the most direct and comprehensive way to analyse the behaviour of laterally loaded pipelines, it is too expensive to be used as a day-to-day analytical tool. On the other hand, the elastic beam on elastic foundation approach is more economic and is a widely used method, despite it being developed over a century ago. The approach is relatively simple when compared with the 3-D FE method, but still captures the essential features of the problem.

British Gas has developed a two stage analysis technique for the modelling of laterally loaded pipelines. The first stage uses a finite element program to predict the restraining effect of the soil as a function of pipe displacement. This is accomplished by performing a 2-D plane strain FE analysis. These predictions are then used in stage 2 which models the behaviour of pipelines, employing an elastic beam on elastic foundation program. Booth (1991) and Ng, Pyrah & Anderson (1994) have used the above methodology to numerically simulate a full scale lateral pipe loading test, performed by British Gas in 1988. It was found that, in stage 1, the assumed tensile behaviour of the soil at the back of the pipe plays a significant part in the determination of the *P-y* curve (pressure-displacement relationship).

Booth (1991) tackled this problem by using a highly simplified method which considered only the reaction at the front of the pipe and ignored the reaction at the back. This assumption implies that the behaviour of the soil at the front (in compression) is independent of the soil at the back (in tension) of the pipe, which is not true when compared to the results of Ng et al. (1994). The straining of soil and tensile stress generated at the back of the pipe in a soil model that allows tension could influence the behaviour of the soil at the front. Also when the pipe is moving obliquely upward as shown by Ng et al. (1994), the distribution of reaction force is not symmetrical about the vertical centre line of the pipe and this leads to error when using the method propounded by Booth (1991). The approach used by Ng et al. (1994) is to use a "no tension" soil model, such that virtually no tensile stress can occur. However, this approach still ignores the possible slip and separation at the soil/pipe interface which may influence the predicted *P-y* curve.

The purpose of this paper is to investigate the suitability of using interface elements in stage 1 of the analysis to model slip and separation at the soil/pipe interface and at the interface between the backfill and the natural ground. The predicted results from the numerical model will be compared with the results of the field test.

2 RIGID CAVITY MODEL

This is the same model used by Ng et al. (1994). The geometry of the FE mesh is based on the *in situ* measurements as shown in Fig. 1. The pipe is not included in the FE mesh but is modelled as a rigid cavity. The 2-D plane strain FE mesh, as shown in Fig. 2, consists of 184 8-noded quadrilateral elements. The lower boundary of the mesh was fixed at the surface of the stiff clay stratum (Material zone 3 in Fig. 1).

The analyses were performed using a modified version of the finite element program CRISP90 developed in British Gas. Both the backfill and the natural ground were modelled using an Elasto-plastic soil model developed in British Gas and its behaviour was described by Ng (1994). A prescribed horizontal pipe displacement of 0.2m was applied in the analyses.

In order to give a true representation of the pipe movement, possible vertical movement (upward or downward) needs to be assessed. The method that Booth (1991) and Ng et al. (1994) used to determine the possible vertical movement is to carry out a number of analyses each having a displacement of 25mm in the *X*-direction (horizontal) but different displacements in the *Y*-direction (vertical). By plotting the total *Y*-reaction of the pipe against *Y*-displacement for each analysis, the actual vertical pipe movement for zero vertical reaction could be estimated. The modified version of CRISP90 has a new "auto-float" option, developed by Cambridge University, so that the vertical movement can be determined in one analysis. By assuming an initial vertical movement of the pipe, the program calculates the total vertical reaction of the pipe at the end of each increment. Adjustment to the vertical movement is then made within the program in the next increment according to the sign of the reaction force. The vertical reaction will slowly converge to zero with an automatically calculated vertical movement of the pipe. Thus the rigid pipe will "float" up or down during the analysis to maintain a zero vertical reaction for a prescribed horizontal movement.

Two analyses were performed, one with normal full tension model (Analysis 1) and the other with no tension model (Analysis 2). The displacement vectors shown in Figs 3 and 4 clearly demonstrate the effect of the tension at the back of the pipe. It can be seen that a clear flow pattern has developed for the full tension model (Fig. 3). In the no tension model (Fig. 4), the soil movement of the backfill behind the pipe is reduced and the overall mechanism changed. It should also be noted that the soil movement in the natural ground behind the pipe

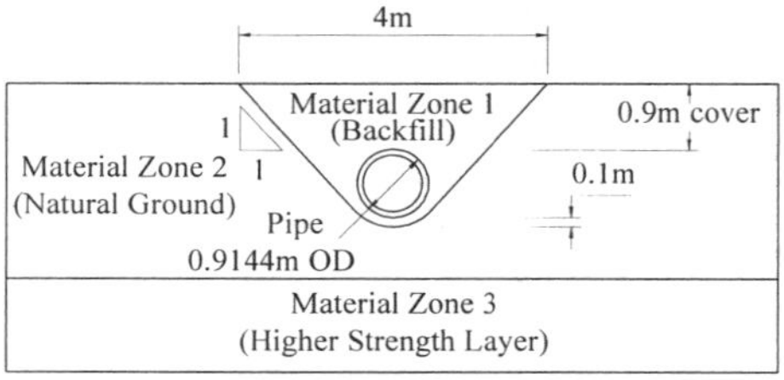

Fig. 1. Cross-section of the trench.

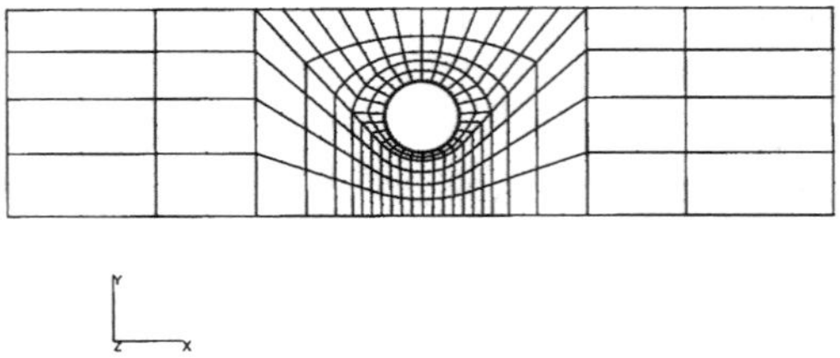

Fig. 2. 2-D finite element mesh.

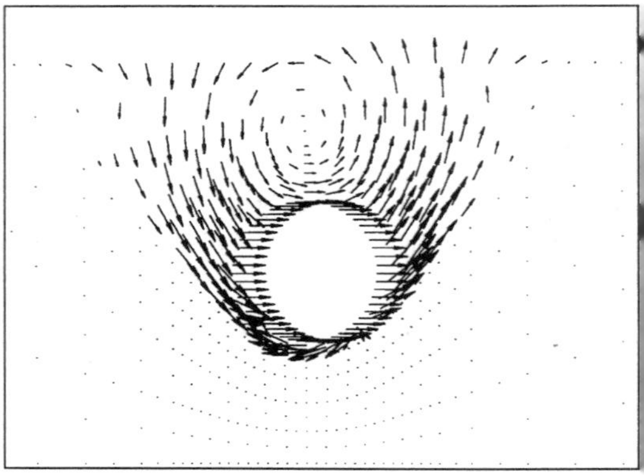

Fig. 3 Displacement vector plot for rigid cavity model with full-tension (Analysis 1).

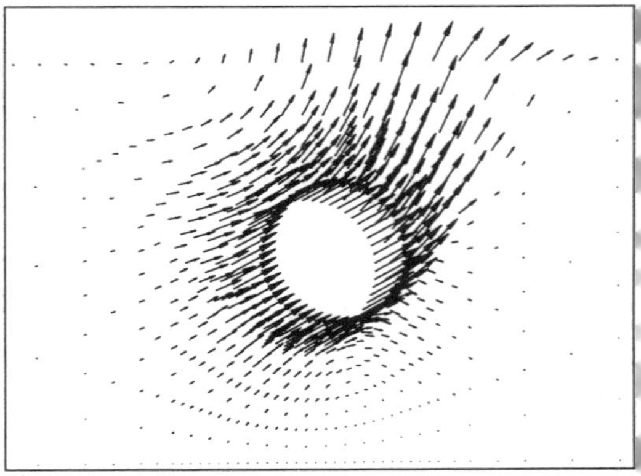

Fig. 4. Displacement vector plot for rigid cavity model with no tension (Analysis 2).

has increased with the no tension model because the soil in this region is subject to tensile stresses and the no tension criterion reduces its strength and stiffness.

3 MODEL WITH INTERFACE ELEMENTS AROUND THE PIPE

In order to model the separation at the soil/pipe interface due to the tensile stress developed at the back of the pipe, a second FE model with interface elements was used. The geometry of this model is similar to the previous rigid cavity model but the pipe ring has been included in the mesh. A layer of modified CRISP90 interface elements (Ng, Pyrah & Anderson 1997) was placed between the pipe ring and the soil. The interface elements can model the separation at the soil/pipe interface due to the tensile stress and any possible slip between the soil/pipe interface due to high shear strain. Isotropic *in situ* stresses were used in the analysis to ensure the interface elements are initially in compression and to avoid separation occurring at the onset of the analysis.

The displacement vectors are shown in Figs 5 and 6 for the analyses using the Elasto-plastic model with full tension (Analysis 3) and no tension (Analysis 4) respectively. In both analyses, the displacement patterns are very similar. The soil in front of the pipe displaces laterally and upward, following the sloping side of the trench. Small movement of the soil occurs behind the pipe for the no tension model because of the reduced strength and stiffness but virtually no movement for the full tension model. Separation at the interface elements behind the pipe indicates that a gap has formed. Slip also occurs between the soil and the pipe at the top and bottom of the pipe.

Since the interface elements behind the pipe have separated, very little tensile stress can be transferred to the soil. This is evident in the distribution of mean total stress for the full tension analysis; only small amounts of tensile stress have developed (Ng 1994).

4 MODEL WITH INTERFACE ELEMENTS AROUND THE PIPE AND ALONG THE TRENCH SIDES

4.1 *Displacement controlled analysis*

The previous two sets of analyses demonstrated the importance of the tensile stress developed in the soil at the back of the pipe, and the roughness of the soil/pipe interface. In this model, a layer of the modified interface elements is placed along the V-shaped trench between the backfill and the natural

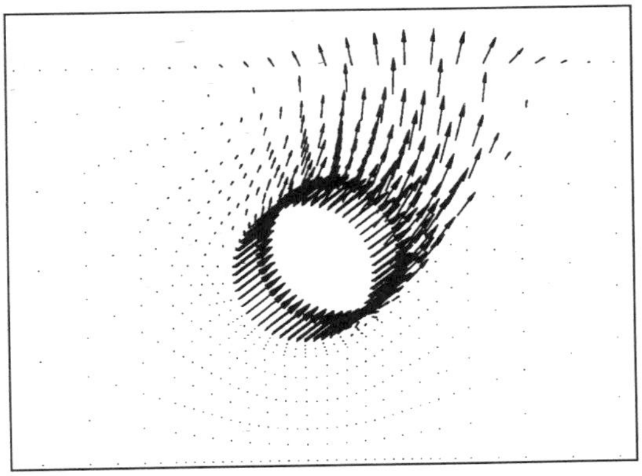

Fig. 5. Displacement vector plot for model with interface elements around the pipe and full tension (Analysis 3).

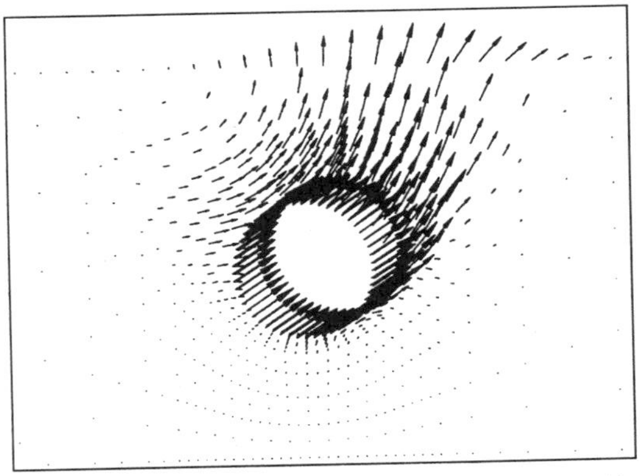

Fig. 6. Displacement vector plot for model with interface elements around the pipe and no tension (Analysis 4).

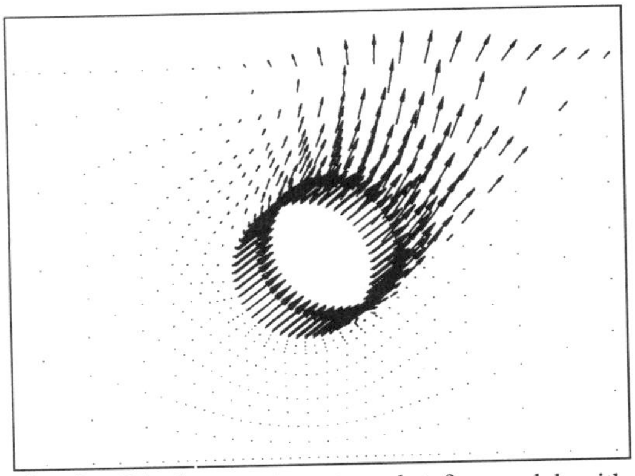

Fig. 7. Displacement vector plot for model with interface elements around the pipe and along the trench sides, full tension (Analysis 5).

ground to model any possible slip due to high shear strain developed between the two materials.

There is not much difference in the displacement patterns (Figs 7 and 8) between the two analyses and the results are similar to the corresponding analyses presented in Section 3. A gap has formed behind the pipe and slip has occurred at the soil/pipe interface. In addition, with this model slip has also occurred between the backfill and the natural ground in front of the pipe.

4.2 *Load controlled analysis*

All of the above analyses were displacement controlled and it was assumed the pipe moves as a rigid circular shape. However, this assumption is not strictly correct since ovalization of the pipe cross-section has been measured during the pipe loading test (Ng 1994). This change in shape may influence the overall behaviour, including the predicted *P-y* curve since the increased diameter of the pipe leads to a larger projected area of the pipe. For this reason, a load controlled analysis is performed using the same Elasto-plastic model (with full tension) as before to examine the effect of the flexibility of the pipe cross-section.

A uniformly distributed pressure is applied horizontally at the back of the pipe over the entire diameter. The uniform pressure is converted into nodal forces and their magnitudes are proportional to the projected vertical length between two nodes.

The overall displacement pattern for the load controlled analysis (Fig. 9) is essentially the same as for the displacement controlled analysis (Fig. 7). However, the displacement vectors clearly show that, as well as moving obliquely upward, the pipe has also rotated as the vectors at the back of the pipe are in different directions as the vectors at the front of the pipe. The back of the pipe has separated from the soil, and slip occurred along the trench and at the bottom part of the soil/pipe interface. Squashing of the pipe cross-section has also occurred but the magnitude is too small to be noticed in Fig. 9. The vertical diameter of the pipe ring has increased by 10mm (≈ 1.1%) which agrees well with the measured value of 1% in the field test.

The predicted *P-y* curve is plotted in Fig. 10 together with the *P-y* curves predicted by all the displacement controlled analyses. All curves exhibit a strain hardening behaviour and show no clear peak strength. The *P-y* curves for the models with interface elements (Sections 3 and 4) are very similar (Analyses 3 to 7), while the rigid cavity model (Section 2) predicted much stiffer *P-y* curves, especially with full tension (Analysis 1). The maximum pressure of the load controlled analysis (Analysis 7) is approximately 9% greater than the equivalent displacement controlled analysis (Analysis 5).

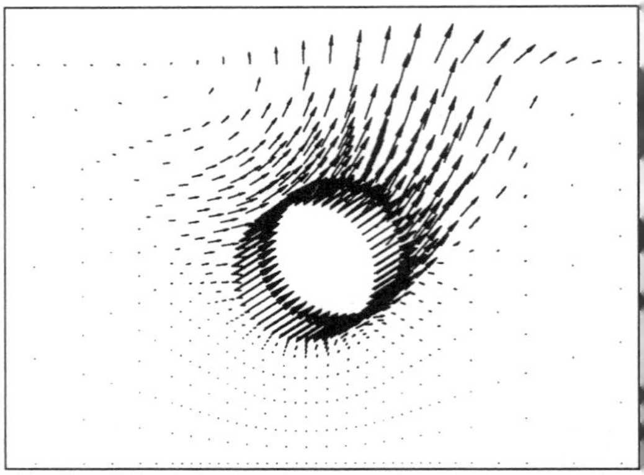

Fig. 8. Displacement vector plot for model with interface elements around the pipe and along the trench sides, no tension (Analysis 6).

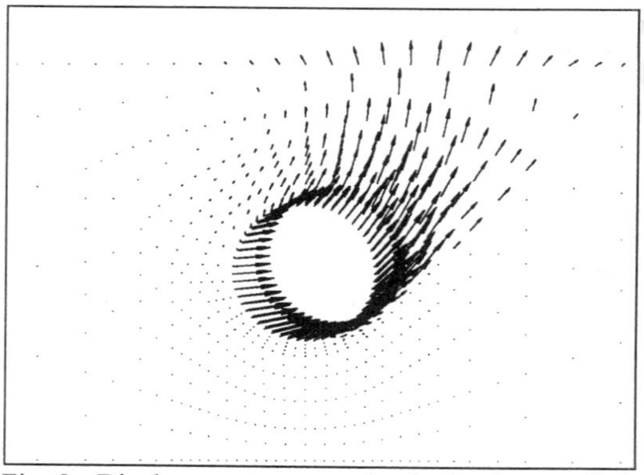

Fig. 9. Displacement vector plot for load controlled analysis of the model with interface elements around the pipe and along the trench sides, full tension (Analysis 7).

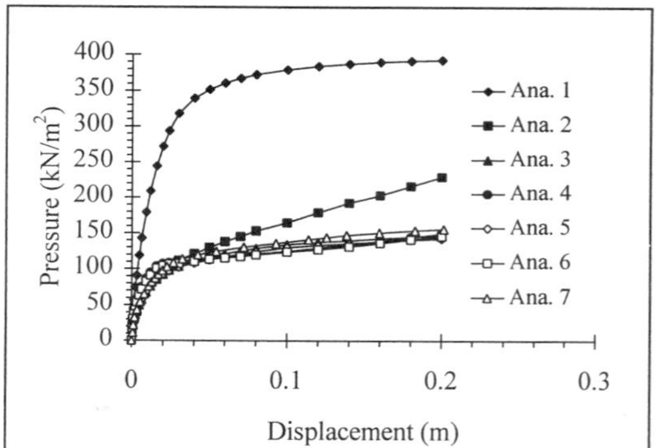

Fig. 10. *P-y* curves predicted by different analyses.

5 ELASTIC BEAM ANALYSIS

After obtaining the *P-y* curves, the second stage of the analysis is to simulate the pipe loading test in an elastic beam on elastic foundation program WOMOD. The original program has been modified (Ng, Pyrah & Anderson 1995) to include the effects of the plastic behaviour of the pipe material, ovality of the pipe cross-section and the shear deformation of the pipe. The analyses performed in this paper were using the modified version of the program. The *P-y* curves predicted in Fig. 10 were used to represent the non-linear behaviour of the soil springs in the program. The results are shown in Figs 11 and 12. The displacement profiles (Fig. 11) and the bending strain profiles (Fig. 12) of the pipe all show that the predictions by the models with interface elements have the best agreement with the field test results. The rigid cavity model, on the other hand, gave poor agreement due to the much stiffer *P-y* curves, especially Analysis 1 which does not use the no tension procedure or interface elements to prevent the build up of tensile stress at the back of the pipe.

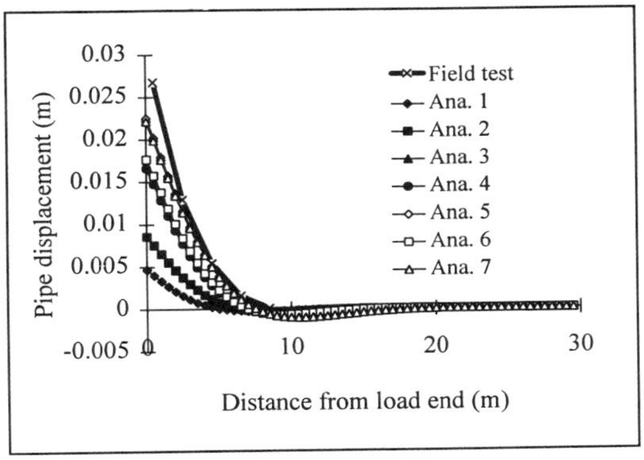

Fig. 11. Predicted distributions of lateral pipe displacement at load = 289 kN.

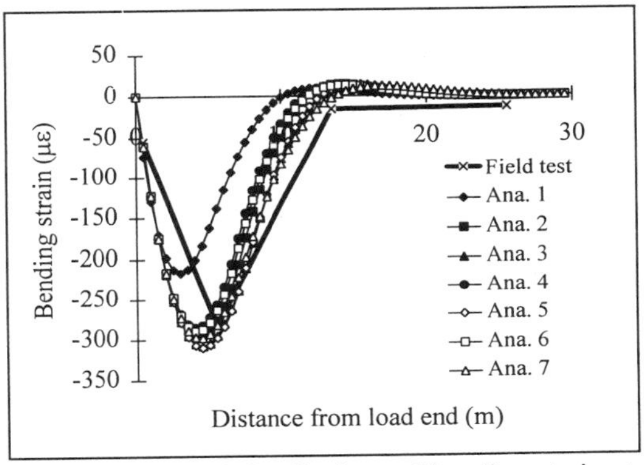

Fig. 12. Predicted distributions of bending strain at load = 289 kN.

6 DISCUSSION

From the photographs taken during the field test, separation occurred between the soil and the back of the pipe. The no tension procedure, which is a simplification of the real behaviour, cannot model this separation nor any slip between the soil and the pipe surface which may occur. A better option is to use interface elements between the pipe and the backfill and the backfill and natural ground to allow for any possible slip. When the interface elements are used, the results for "full tension" and "no tension" are very similar because the separation of the interface elements has prevented any tensile stress being transferred to the soil. However, the no tension procedure reduces the strength and stiffness of the soil in the tensile region (behind the pipe) and the soil displacement at this region may be over estimated in the analyses with no tension.

Additional analyses have been carried out (see Ng 1994) to examine the effect of roughness and limiting adhesion of the soil/pipe interface in the 2-D FE analysis for predicting the *P-y* relationship. The results, not shown in here due to space limitation, show that the roughness of the pipe could significantly affect the predicted *P-y* relationship (up to 20% difference). Thus an accurate description of the properties of the soil/pipe interface is important in this kind of analysis.

A change in the shape of the pipe was recorded during the field loading test; this will influence the stresses around it. The main effect in the 2-D FE analysis is the increase in the projected area used in the determination of the average soil pressure. Therefore the average pressure results calculated assuming a rigid pipe (all displacement controlled analyses) are over-estimated. On the other hand, the increased projected area attracted more soil reaction, and the total soil reaction is increased. The net effect is dependent on the soil properties and the stiffness of the pipe cross-section. The load controlled analysis reported in Section 4.2 allows the shape of the cross-section of the pipe to change and is therefore more realistic. The analysis predicted a maximum pressure 9% greater than with the displacement controlled analysis. The authors feel that the load controlled analysis is more realistic in terms of the displacement pattern than the displacement controlled analyses and makes a slightly better prediction in the elastic beam on elastic foundation analysis.

However, the load controlled analysis is more difficult to carry out because the horizontal pressure needs to be converted into nodal forces beforehand. Moreover, all the available post-processing programs for obtaining the P-y relationship were developed based on displacement controlled analysis. Extra work was needed to obtain the P-y relationship from load controlled analysis at this moment. Modifications to the existing post-processing programs are needed if load controlled analysis is to be used regularly.

7 CONCLUSIONS

i. The two stage method of analysis used in this research has been shown to be valid and gives good results for predicting the behaviour of buried pipes subjected to lateral loading.
ii. Soil restraint to a buried pipeline can be predicted by suitable modelling with the finite element method.
iii. The modified interface element can be used to model the separation and slip between the soil and the pipe surface. It is theoretically more sophisticated than the no tension procedure in tackling the problem of tensile stress behind the pipe.
iv. The change in the shape of the pipe cross-section during the FE analysis can affect the prediction of the pressure-displacement relationship of the pipe. A load controlled analysis can be used instead of a displacement controlled analysis to take into account this factor but existing post-processing programs will have to be modified to obtain a pressure-displacement relationship from load controlled analysis.

ACKNOWLEDGEMENT

The Authors wish to thank British Gas for providing the computer hardware and software to perform the analyses. The raw data from the pipeline load test and guidance on its processing was provided by Mr. Eric Middleton. The first Author was financially supported by a British Gas Research Scholarship and by the Overseas Research Students Awards Scheme whilst undertaking research towards a PhD degree at the University of Sheffield.

REFERENCES

Booth, J. (1991). *Numerical Prediction of Soil Restraint to Pipe Movement*, BEng project report, Dept. of Mechanical Eng., Newcastle-upon-Tyne Polytechnic.

Ng, C. F. (1994). *Behaviour of Buried Pipelines Subjected to External Loading*, PhD Thesis, University of Sheffield.

Ng, C. F., Pyrah, I. C. & Anderson, W. F. (1997). Assessment of three interface elements and modification of the interface element in CRISP90. Submitted to *Computers and Geotechnics*.

Ng, C. F., Pyrah, I. C. & Anderson, W. F. (1995). Modelling of Laterally Loaded Pipelines Using Elastic Beam on Elastic Foundation Approach. In *Developments in Computational Techniques for Structural Engineering*, Proceedings of the Sixth International Conference on Civil and Structural Engineering Computing, Cambridge, UK, 28–30 August 1995, pp. 71–76.

Ng, C. F., Pyrah, I. C. & Anderson, W. F. (1994). Lateral soil restraint of a buried pipeline. *Proceedings of the Third European Conference on Numerical Methods in Geotechnical Engineering*, 7–9 September 1994, Manchester, UK, ed. Smith, I. M., pp. 215–220.

Numerical Models in Geomechanics, Pietruszczak & Pande (eds) © 1997 Balkema, Rotterdam, ISBN 90 5410 886 X

Subsidence prediction above gas storage cavity fields

J.E.Quintanilha de Menezes
Faculdade de Engenharia, DEC, Geotecnia, Oporto, Portugal

D.Nguyen-Minh
L.M.S., Ecole Polytechnique, Palaiseau, France

ABSTRACT: This study deals with a comparative analysis on subsidence over two leached salt cavern storage sites in France: Tersanne and Etrez. The two fields are somehow similar, with fourteen cavities each located at the same depth, but with a different spatial distribution. This study confirms and enhances previous conclusions of studies by the authors on Tersanne field, and applies the same analysis method to the Etrez field. In a first step, one verifies that the numerical predictions agree with the subsidence measurements done by Gas de France on the two sites. This is obtained when considering that mechanical parameters of rock salt are given from GDF calibration on cavities convergence measurements over ten years, while unknown elastic parameters for non saline strata are chosen arbitrarily in a reasonable range. In the second step, the numerical model is used to predict up to 100 years of surface subsidence.

1 INTRODUCTION

For more than ten years, Gaz de France has performed surface subsidence observations on the Tersanne and Etrez gas storage fields, located in France, for control, surveillance as well as environmental purposes, by monitoring of elevation measurements by a network of survey benchmarks (Durup, 1990). This paper is concerned with an instructive parallel subsidence analysis of the two sites of Tersanne and Etrez. These sites have the same number of cavities but with a different distribution pattern (Figure 1).

The site of Tersanne has been object of several former reports and publications (Durup, 1990, 1991; Nguyen Minh et al., 1993; Menezes, 1996). It has been shown that rock salt played a major role in the subsidence phenomena, while the cover strata behavior, which characteristics are not well known, was of secondary significance. With Etrez site, where the salt behavior is very different of the one from Tersanne, one is going to have the opportunity of confirming such a result.

For some cases this three-dimensional problem under consideration can be approximated by a superposition of single cavities in infinite medium calculations. The need of an accurate calculation of the subsidence above a single cavity in infinite medium, using a two-stage finite element procedure is avoided herein, thanks to a newly developed mixed finite and boundary element code (Menezes, 1996). The characterisation of a very large layer on the surface does not present a serious problem. Indeed, in the case of Etrez, the former two-stage calculation would be heavy to carry out due to multiple salt layers.

2 STORAGE CAVITY FIELDS

Tersanne and Etrez cavern fields present some remarkable differences concerning space distribution and operation dates (Figures 1 and 2). The first storage cavities in Tersanne were leached with a spacing of about *400 m*, and this distance has been increased for cavities, operated more recently, after years 80. On the other hand, at the Etrez site, the distribution of cavities is different in space and time, probably following the experience acquired in Tersanne (the first cavity of Etrez was put in service 10 years after the first cavity of Tersanne), but also, maybe, due to a less intricate topography. Etrez cavities are thus displayed on a more regular hexagonal centred pattern than in Tersanne, with a larger spacing (*600 m*). The former cavities are surrounded by more recent ones. Besides the

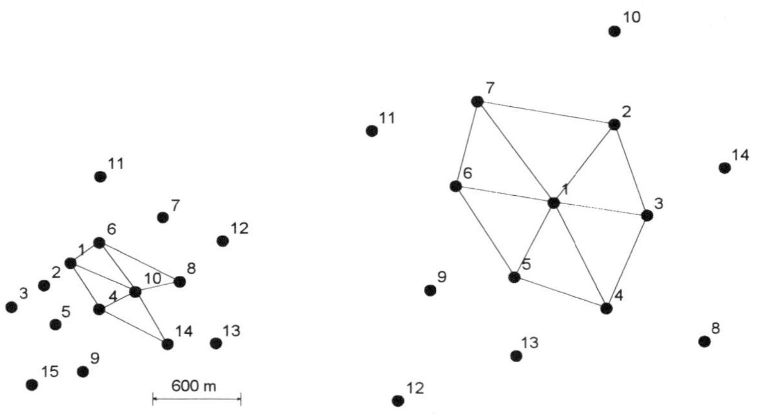

Figure 1. Surface distribution of storage cavities from Tersanne (left) and Etrez (right).

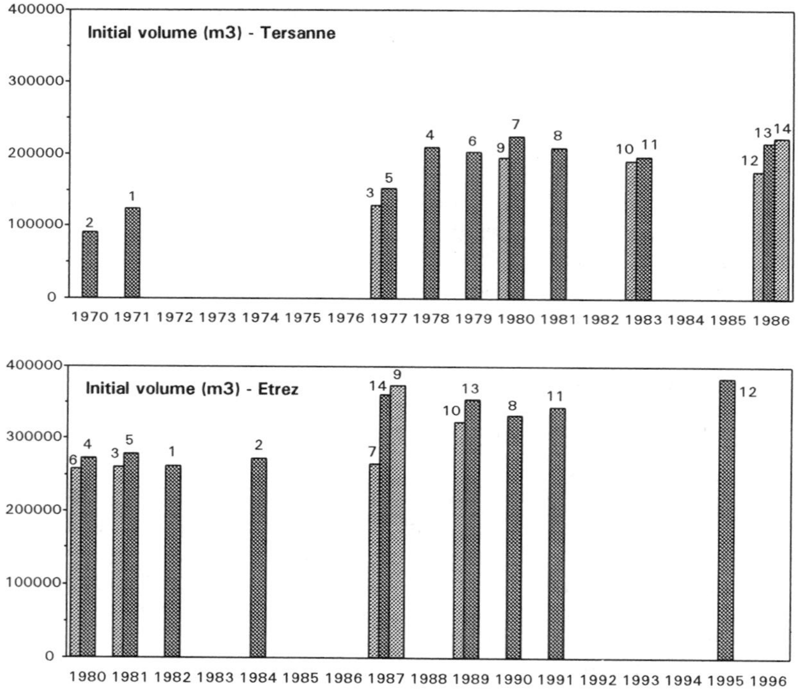

Figure 2. Time distribution of first year operation and initial volume of cavities at Tersanne and Etrez.

different Etrez salt mechanical characteristics, this strategy of cavity distribution at Etrez certainly contributes to the lower measured subsidence. Figure 2 shows histograms of the initial volumes of Tersanne and Etrez caverns at the first year in operation.

At Tersanne maximal subsidence measures after the same period of time are about ten times those from Etrez. Maximal subsidence rate, in the centre of Tersanne cavity field, is about 10 mm/year.

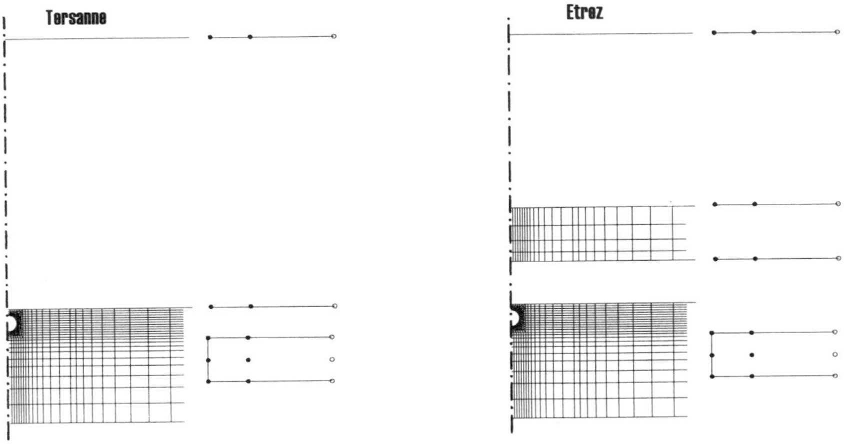

Figure 3. Mixed discretizations of Tersanne and Etrez isolated cavity.

At Etrez subsidence measured values are at least fifty per cent lower.

3 DESCRIPTION OF NUMERICAL MODEL

The subsidence analysis of the 14 cavity field is a three-dimensional problem. However, simplified analysis using a superposition procedure with calculations of single cavities in infinite medium can be used. Theoretically, the superposition should not be valid according to the non linear nature of the problem. But validity of this method holds for the specific conditions of the problem, involving linear elastic cover strata and largely spaced cavities (typically *400-600 m*).

The geological differences on the two sites are basically reflected in the rock mass modelisation for a single cavity model (Figure 3). For Tersanne, rock salt layer is *650 m* thick, and lays over a rigid rock bed and under a cover strata *1350 m* thick. The basic computation involves a single spherical cavity, situated *1500 m* deep, with a volume of *268000* m^3 (radius *40 m*), located inside the salt layer and where cover strata is located *110 m* above the roof of the spherical cavity.

For Etrez, the basic computation considered a different strata pattern with four layers: two of rock salt, with thicknesses *270 m* and *570 m*, the last being deeper and containing the cavity, and two other insoluble strata, one *210 m* thick, separating salt layers, and the other being the cover strata with *870 m* thick.

The numerical model used for the basic computation uses the integral equation and boundary element method, associated with the finite element method. Besides, one introduces infinite elements in order to simulate infinite remote boundaries (Pande *et al.*, 1990). The surface discretization reaches a maximal length of *12000 m* in order to obtaining a precise solution of field displacements on the surface. As cover elastic strata are discretized with boundary elements such distance does not constitute an excessive weight for computer memory. The discretization extension reduces to *6000 m* at the level of salt strata. In this way one obtains a mixed mesh which develops following a greatly bigger distance in reaching the soil surface.

The mixed discretizations used for the computation of single isolated cavities at Tersanne and Etrez are shown in Figure 3. At Tersanne one considers two layers: the rock salt strata with elastoviscoplastic behavior and the cover layer with elastic behavior. Between the two domains one has the possibility of placing joint elements. At Etrez there are four layers, two of salt intercalated by two elastic strata.

Calculations were extended to a period of 100 years. For such duration, convergence in cavity may become too big to still admit valid the hypothesis of small transformations. Consequently, numerical analysis can be redone using the software ability to count on the structure geometry variations. However, this procedure increases the computation time by a factor of two to three.

The real internal pressure history, after the cavity

leaching phase, was not simulated in numerical calculations. One admits that all the cavities have the same internal pressure constant all the time. Time intervals between the different years of operation of the different cavities are followed during the calculations (Figure 2).

The viscoplastic laws chosen to represent the salt rheology are given by a same formulation. They are the Norton-Hoff law, that admits a stationary creep, and the Lemaitre-Menzel-Schreiner law. The viscoplastic deformation rates depend only on the second deviatoric invariant, therefore the material has an incompressible viscoplastic behavior. The viscoplastic behavior law can be expressed in a general form as:

$$\dot{\varepsilon}_{ij}^{vp} = -10^{-6}\frac{3}{2}\frac{\sigma_{ij}^{'}}{\sqrt{3J_2^{'}}}\frac{d}{dt}(\xi^{\alpha})\,;\quad \dot{\xi} = \left(\frac{\sqrt{3J_2^{'}}}{K}\right)^{\beta/\alpha} \tag{1}$$

The internal pressure was determined in such a way that the volume loss of one spherical cavity situated at *1500 m* depth would be the same as one real cavity with a real pressure history at Tersanne. According to GDF, this pressure would be *16.5 MPa* with the L-M-S model. For the case of Etrez, one considered the same internal pressure, although the salt presents a different rheology from the one of Tersanne.

The rheological parameters used in the numerical calculations, identified according to observation periods of ten years average, are given in Table 1.

Table 1. Rheological parameters.

	Law	α	β	*K*(MPa)	Temp.
Tersanne	LMS	0.5	3.6	0.85	70°C
	NH	1.0	3.0	2.60	
Etrez	LMS	0.36	2.98	1.1	50°C
	NH	1.0	3.0	6.34	
	LMS	0.44	3.93	1.95	55°C

One admits a zero cohesion for the salt. Several relaxation tests of rock salt samples detected a cohesion accordingly weaker as test rate was lower (Fauveau *et al.*, 1986).

In order to take into account the influence of temperature increase with the depth and therefore the reduction of viscosity, one introduces in the expression (1) of behavior law a temperature dependent factor:

$$a_o\, e^{-\frac{U}{RT}} \tag{2}$$

with the following parameters (for temperatures in Table 1):

$a_0 = 1$
$U = 55$ *kJ/mole* - activation energy
$R = 8.325$ *J/K/mole* - perfect gas universal constant
T - temperature in Kelvin degree

The cover strata has a compressible elastic behavior. Rock salt has an initial compressible elastic behavior with incompressible viscous deformations (Table 2).

Table 2. Elastic parameters.

	ν	*E* (MPa)
Rock salt	0.25	25000
Cover	0.4	6000

4 SUBSIDENCE SIMULATION RESULTS

After performing a single calculation for each cavity of the fields represented in Figure 1, considering their real location and initial volume, a superposition procedure is performed.

In Figure 4 measured subsidence values during the periods of 1985-1990, 1985-1995 and 1990-1995, are compared with the subsidence curves calculated for 5, 10 years, and the difference between these two. The subsidence curves were obtained after performing the superposition of the displacements fields for the 14 cavities arranged as indicated on the Figure 1. Notice the different displacement scale in Figure 4 for Tersanne and Etrez.

One finds a good correspondence between measured and computed subsidence values at Tersanne, for the periods of 5 and 10 years. It is necessary to notice once more that one supposes that all 14 cavities have the same interior pressure constant in time. However, for Etrez such correspondence is not so good, which means that rheological parameters should be reviewed. Nevertheless the smallness of measured and computed subsidence guaranties that there will not be any structural problems at the surface in the future.

Numerical calculations with variable geometry are of slow performance, especially with CPU time. We get a diminution of convergence values more evident on the surface.

Surface movements predicted for 50 and 100 years are displayed on Figure 5. Such results are important to determine which areas around the storage field

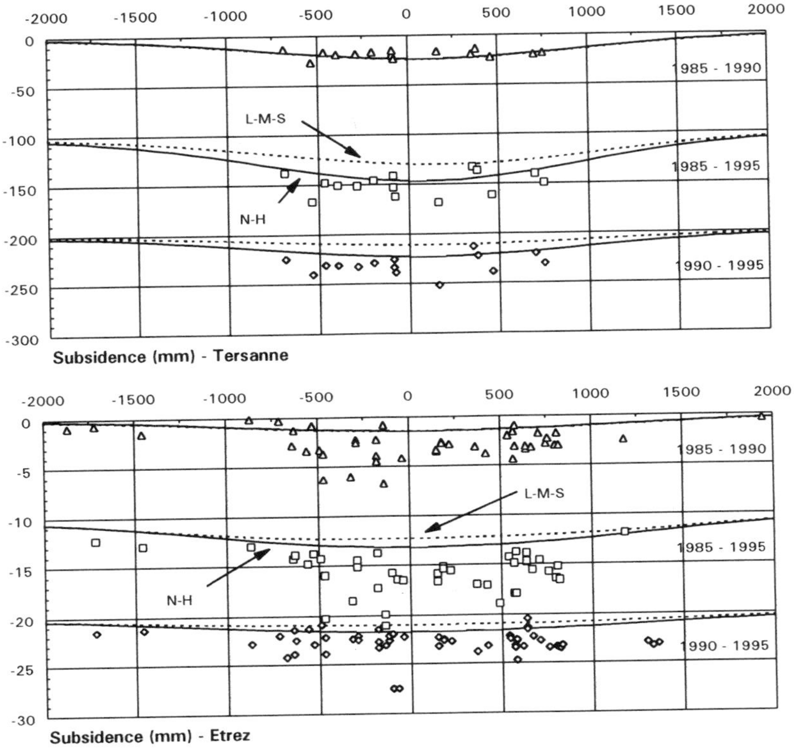

Figure 4. Comparison between measured subsidence and calculated values.

correspond to compression or traction situations. One could say that the subsidence situation in Etrez after a certain period of time (50 years) will be the situation in Tersanne after a period of time reduced by a factor of 10 (5 years).

5 CONCLUSIONS

The survey of Tersanne and Etrez sites confirms the idea that rock salt behavior controls subsidence phenomena. Indeed, one knows that Etrez salt is much less viscous than the one in Tersanne, and this agrees with subsidence at Etrez being much inferior to those of Tersanne. The agreement of the numerical model with the observed subsidence, although elastic cover parameters remains arbitrary, definitely permits the confirmation of such result.

The prediction for 100 years by the numerical model induced us to consider two alternative behavior models, the L-M-S and the N-H law, this one with stationary creep, where parameters were determined after the observation of the structure evolution until present time, to give a reasonable superior limit.

Some correlation could be found between subsidence at Tersanne and Etrez as more measurements and predictions by numerical models are available, for a given behavior law. Thus it appears, with the antecedent creation of Tersanne site and the quite larger convergence rates, that the Tersanne case constitutes a "projection" in the future of Etrez site. This is an interesting prediction element for Etrez, knowing that the present registered subsidence in Tersanne have not caused any surface problem, and may be an equivalent situation at Etrez in a time period 10 times greater.

The superposition procedure was used in a global way, but remains reserved to specific conditions of spacing between cavities and assumed elasticity of cover layers.

The numerical model used can be taken as a specific tool of subsidence analysis, especially with the improvements brought with the utilisation of

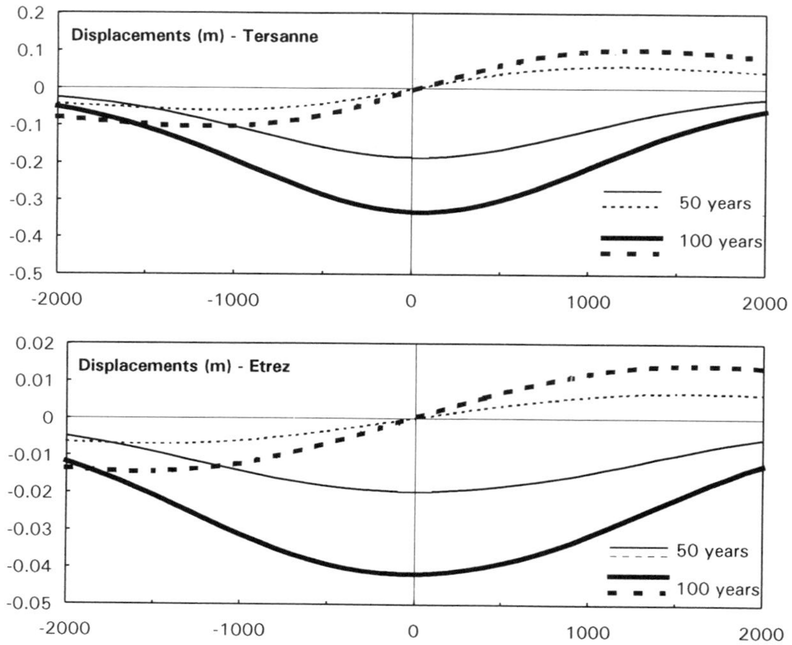

Figure 5. Subsidence and horizontal displacements predictions above Tersanne and Etrez fields.

boundary and infinite elements for the semi-infinite elastic layers modelisation, where it is essential a good precision necessary to the surface subsidence evaluation. This model avoids the two step subsidence determination for an isolated cavity, as such procedure would have been of difficult application to the case of Etrez, where there are several rock salt layers.

The utilisation of boundary elements becomes particularly useful in subsidence computations. This because one can discretize a very large surface extension, essential to obtain a good precision in the displacements, without having too much costs concerning the time and necessary memory to calculations.

ACKNOWLEDGEMENTS

The authors are sincerely grateful to Mr. J.G.Durup and C.Rollin from Gaz de France, for providing essential field data.

REFERENCES

Durup, J.G., 1990. Surface subsidence measurements on Tersanne cavern field. Solution Mining Research Institute, Fall Meeting 1990, Paris, Oct. 14-19.

Durup, J.G., 1991. Relationship between subsidence and cavern convergence at Tersanne (France). Solution Mining Research Institute, Spring Meeting 1991, Atlanta, April 28-30.

Fauveau, M. & P. Le Bitoux, 1986. Progrès récents de la connaissance du comportement mécanique des cavités salines. Assoc.Tecnh. Ind. du Gaz en France, Congrès 1986.

Menezes, J.E. 1996. *Méthodes analytiques et numériques pour l'étude des cavités salines profondes. Application au stockage du gaz naturel.* PhD Thesis, Ecole Polytechnique, Palaiseau, Paris, France.

Nguyen Minh, S. Braham & J.G. Durup. 1993. Surface subsidence over deep solution mined storage cavern field. 3rd Int. Conference on Case Histories in Geotechnical Engineering, St Louis, Miss., USA..

Pande, G.N., G. Beer & J.R. Williams, 1990. *Numerical Methods in Rock Mechanics.* John Wiley & Sons Ltd, Chichester, England.

Numerical Models in Geomechanics, Pietruszczak & Pande (eds) © 1997 Balkema, Rotterdam, ISBN 90 5410 886 X

Nonlinear three-dimensional analysis of closely spaced twin tunnels

I. Shahrour & H. Mroueh
Laboratoire de Mécanique de Lille (URA CNRS), Ecole Centrale de Lille, Villeneuve d'Ascq, France

ABSTRACT : This paper includes an analysis of the interaction between closely spaced tunnels. It is achieved using a three dimensional elastoplastic finite element calculation. After a description of numerical developments used in this analysis, we present a study of a reference case (closeness ratio = 1D) which shows the influence of the construction of a tunnel close to an existing one on (i) vertical and lateral displacements induced in the soil mass and on (ii) axial force and bending moment generated in the lining. Influence of closeness is then analysed for a twin tunnels with a closeness ratio equal to 0.5D.

1. INTRODUCTION

The extension of transport infrastructure in urban areas needs the construction of shallow tunnels, in particular closely spaced tunnels. Since the interaction between the second tunnel and the existing one may be important for both existing structures and tunnels, it is of major interest to perform an analysis of this problem. This analysis requires the solution of a three dimensional non linear problem. Non linearity results mainly from geomaterials behaviour and construction procedure, while the three dimensional aspect is due to the tunnel stress release at the tunnel face. Soliman, Duddeck and Ahrens (1993) dealt with this problem using simplified configurations (elastic behaviour for soil and linings, deep tunnels, etc..). Their results showed that single-tunnel solutions can be used in the calculation of double-tubes tunnels.

In this paper we propose a three-dimensional non-linear finite element analysis of this problem. Firstly, we present numerical developments used in this analysis which are based on the use of iterative method for the solution of large scale non symmetrical systems arising from the finite element discretization of elastoplastic problems with non associative constitutive relation. Then we analyse the interaction between closely spaced shallow twin tunnels.

2. NUMERICAL PROCEDURES

Numerical modelling of the construction of tunnels is realised using the finite element program *PECPLAS* (Shahrour, 1992, Gorbanbeighi 1995, Mroueh, 1997) which permits the solution of three-dimensional problems encountered in geomechanics.

Tunnel driving is modelled with respect to the construction procedure. At each step, soil elements corresponding to the excavated area are removed and concrete elements corresponding to the lining are activated (figure 1). In the case of cohesionless soils, a pressure is applied on the tunnel face to ensure stability. Its value is equal to the geostatic horizontal stress at the centreline of the tunnel.

The solution of the elastoplastic problem corresponding to tunnel construction is performed using the Newton Raphson method. At each step, the stiffness matrix is calculated using the tangent elasto-plastic matrix. An iterative method is used for the solution of the large sparse linear system. Since the use of non-associative elastoplastic constitutive relations leads to non symmetric linear systems and the presence of hard (concrete) and soft (soil) materials gives rise to very ill-conditioned matrices, we have checked the performance of well-known preconditioned iterative methods, particularly the gradient-like family methods, based on the non-symmetric Lanczos procedure (Axelsson 1994, Bathe et al. 1993, Papadrakakis 1993, Pommerel 1992, Tong 1992). Numerical tests showed a very good performance of the *QMR-CGSTAB* method (Chan et. Al. 1994) which is based on the *bi-CGSTAB* method (Van der Vorst 1992) and the Quasi-Minimisation Residual (QMR) algorithm in combination with appropriate preconditioner (*SSOR* preconditioning method, Pommerell 1992) and stopping criterion. This method was used in the following numerical simulations.

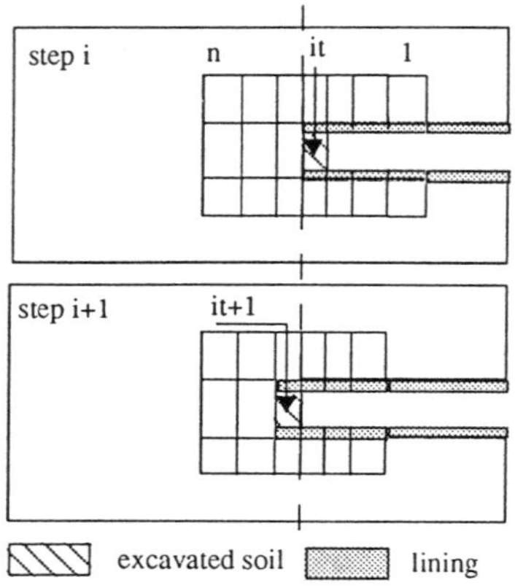

Figure 1. Construction procedure of lined tunnels

3. APPLICATION TO A CLOSELY SPACED TWIN TUNNELS

3.1 Presentation of the Reference Example

Figure 2 shows the characteristics of the reference example. It consists of a shallow twin tunnels with a depth ratio (H/D) equals to 1.75. The closeness ratio (a/D) was assumed to be equal to 1, which is generally considered significant for the interaction between twin tunnels. The soil material was assumed to be cohesionless. Its behaviour was described using an elastic perfectly-plastic constitutive relation based on the Mohr Coulomb criterion with a non-associative flow rule. The behaviour of the tunnel lining (concrete) was assumed to be linear. The outside diameter is equal to the excavated circular hole. The stiffness ratio between the lining and the soil is assumed to be high (stiff linings). The mechanical properties of the soil and the lining materials are given in table 1.

3.2 Finite Element Modelling

The finite element mesh used in the analysis is illustrated in figure 3. It consists of 3752 twenty-nodes isoparametric hexahedral elements which give rise to 16897 nodes representing 50691 degrees of freedom. In order to avoid disturbing boundary effects, extensions of the soil massif are defined as : 12D in lateral axis, 9D in longitudinal axis, and 6D in vertical axis. Tunnels axis depth is equal to 1.75D.

The number of the non-zero terms under the skyline of the stiffness matrix is *8,461,258 64*-bit words, corresponding to 2.9% of stored terms under the skyline. This shows the attractive property of the sparse storage method used, which permits a substantial economy in storage requirements. Computation was performed on an ALPHA DIGITAL 5/266 MHz workstation, with 128 MB of RAM memory. The average CPU time used for the calculation of one step was about 1.6875h.

The finite element analysis is performed in two stages:
(i) The first stage includes the simulation of the excavation of the first tunnel using the construction procedure illustrated in figure 1. Numerical simulation was performed up to a tunnel length of 9D. Calculation of this stage is performed in 11 steps.
(ii) The second stage consists in the simulation of the construction of the second tunnel with the primary stresses obtained at the end of the first stage. Numerical simulation was performed up to a tunnel length of 5D using 9 steps.

Table 1. Properties of constitutive materials

Material	E (MPa)	ν	c' (MPa)	φ' (°)	γ (MN/m^3)
Soil	30	0.3	0.01	30	0.02
Lining	20000	0.25			0.025

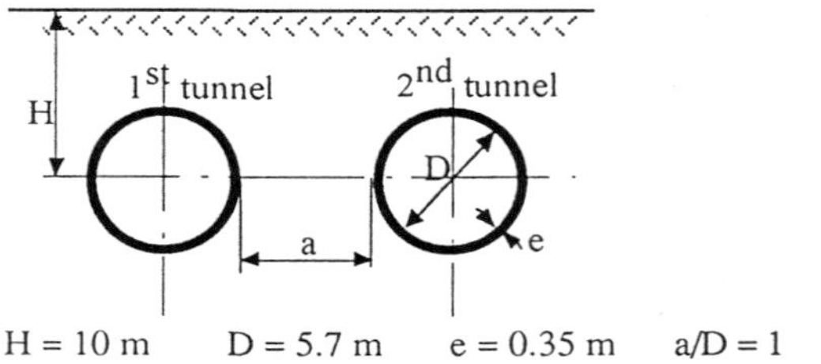

Figure 2. Illustration of the reference example

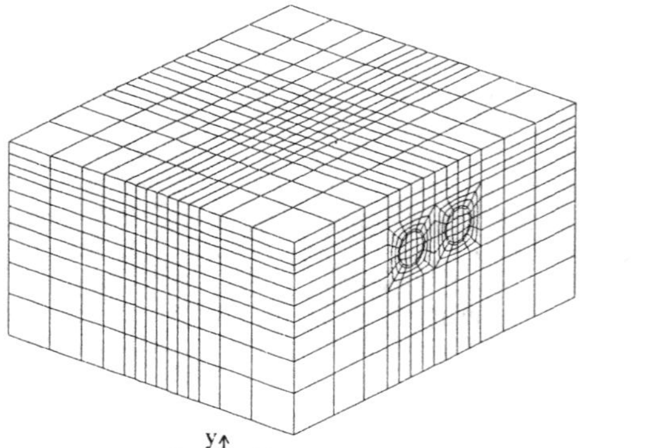

Figure 3. Three-dimensional finite element mesh for the twin tunnels analysis (a / D = 1) : 3752 20-nodes isoparametric hexahedral elements ; 16897 nodes

3.3 *Construction of the First Tunnel*

Results corresponding to the first tunnel excavation are presented in figure 4. Figure 4a shows the evolution of the soil settlement at the surface above the centreline. Stabilisation is observed after an excavation length of about 4D with a corresponding settlement u_∞ = 0.095%D. It can be noted that the major part of the settlement is induced in the vicinity of the tunnel face. Settlement at the open face of the tunnel (u_0) (represented by an arrow in figure 4a) is approximately constant, and is equal to 37% of the stabilised settlement (u_∞). Figures 4b and 4c show the settlement pattern at the soil surface : the three-dimensional character is clearly observed, with a concentration of the settlement above the centreline and the tunnel face. Figure 4d shows the evolution of the vertical displacement along the crown axis during excavation. The stabilised vertical displacement v_∞ is observed after an excavation of about 4D. It can be observed that the vertical displacement is mainly obtained at and back to the tunnel face. This aspect is different from that observed at the soil surface where significant displacement was observed ahead of the tunnel face. The settlement at the tunnel face (v_0) is equal to 48% of (v_∞), while at 1D ahead the tunnel face we observe only 6% of total settlement.

3.4 *Construction of the Second Tunnel*

3.4.1 Vertical displacement

The settlement pattern at the soil surface is shown in figure 5a. We can observe that settlement induced by the construction of the second tunnel is concentrated between the centrelines of the two tunnels. Settlement in the transverse section (Z = 0) is illustrated in figure

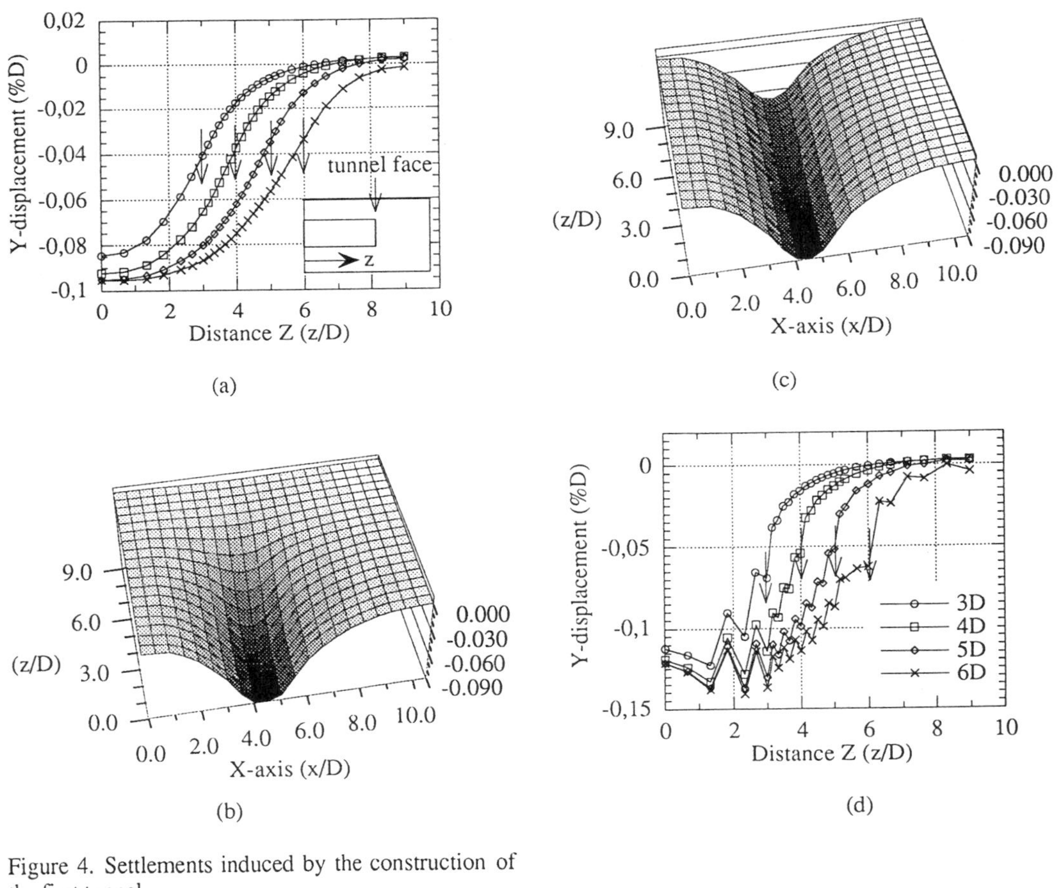

Figure 4. Settlements induced by the construction of the first tunnel
(a) At the surface above the centreline of the 1st tunnel
(b) 3D settlement pattern (tunnel face at Z = 4D)
(c) 3D settlement pattern (tunnel face at Z = 9D)
(d) Vertical displacement at the crown tunnel

5b. Comparison of settlement profile resulting from the first stage to that obtained at the end of the second stage indicates an important increase of settlement above the first tunnel. In order to evaluate this increase, we give in figure 5c the evolution of the vertical displacement at the surface (u_∞) and at the crown (v_∞) obtained in section (Z = 0). We can observe that settlement at the surface is higher than that obtained at the tunnel crown. After an excavation of about 5D, the increase of vertical displacement at the tunnel crown (resp. soil surface) is about 29% (resp. 40%). Figure 5d shows vertical displacement induced by the construction of the second tunnel along the axis of the first one (Z = 0 indicates the location of the second tunnel face). We can observe that these displacements are important beyond the face of the second tunnel. In front of this face, we observe a rapid decrease of the induced displacement.

3.4.2 Lateral displacement

Lateral displacement induced by the construction of the twin tunnels is shown in figure 6. During the first stage (figure 6a), ground responds slightly to excavation when the tunnel face is 3D behind the explored section. Noticeable lateral displacements begin to occur when the tunnel face approaches the explored section. The most important effect is located close to the shoulder, where we observe a ground driving back. At the surface, soil moves toward the centreline of the tunnel ; this is due to the soil loose above the tunnel excavation. Once the tunnel passes the explored section, we observe a rapid development of lateral displacement. Figure 6b shows lateral displacement induced by the construction of the second tunnel in the explored section. It can be

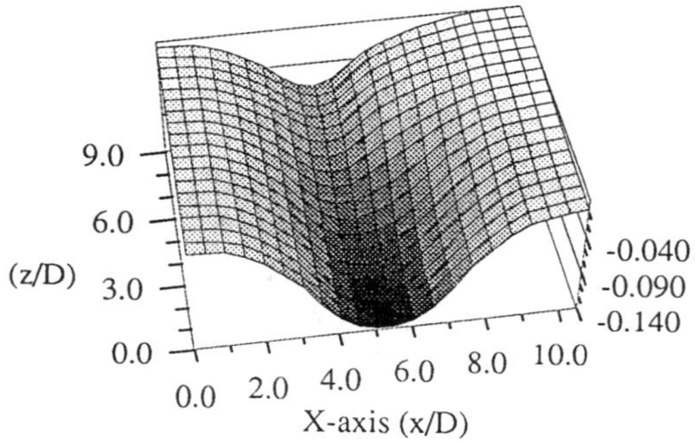

(a)

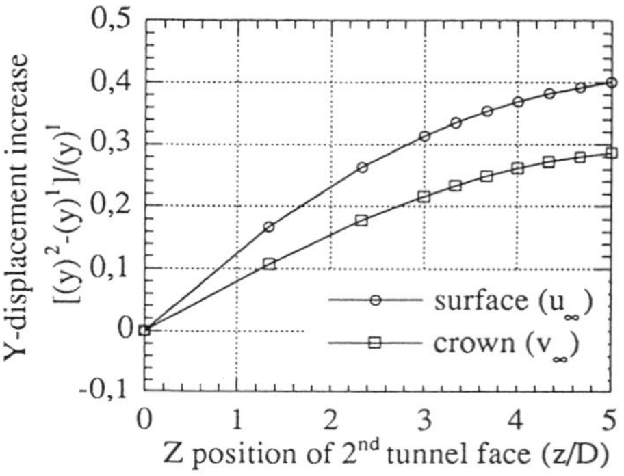

(c)

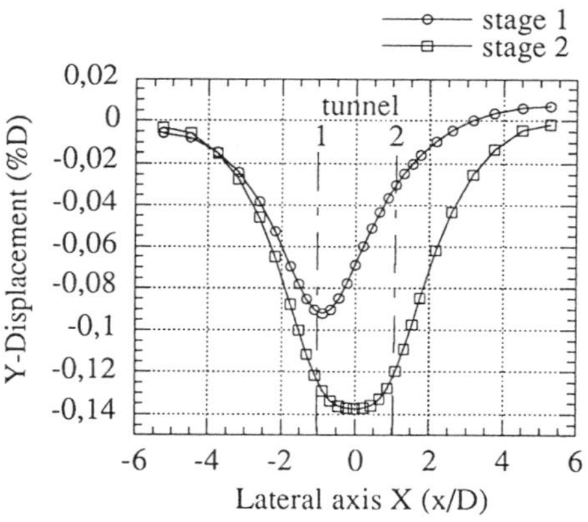

(b)

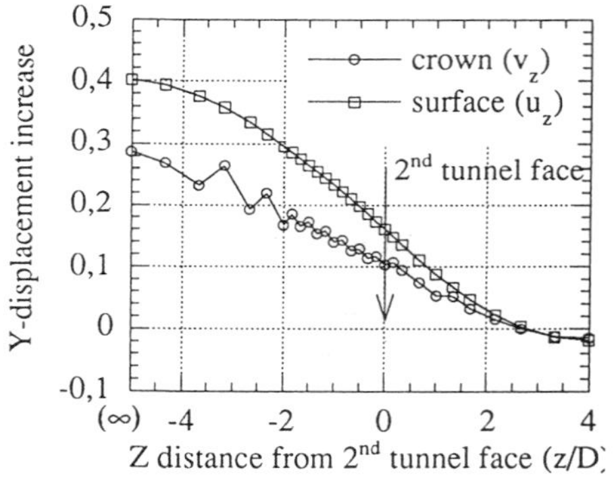

(d)

Figure 5. Settlement induced in the first tunnel by the construction of the second tunnel

(a) 3D settlement pattern (tunnel face of the second tunnel at Z = 5D)

(b) Settlement profile at the section (Z = 0)

(c) Displacement induced in the section (Z = 0)

(d) Displacement induced along the centreline

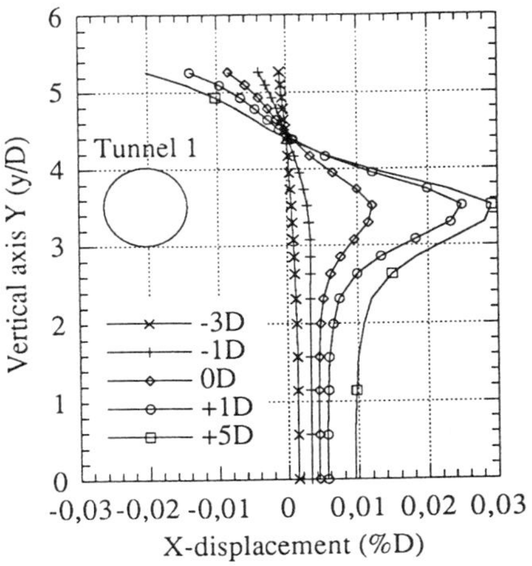

(a)

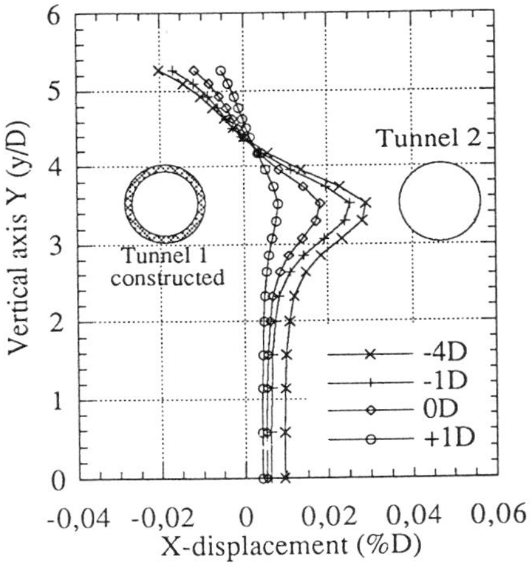

(b)

Figure 6. Lateral displacement induced by the construction of the twin tunnels
(a) of the first tunnel
(b) of the second tunnel one

observed that this displacement is oriented towards the fist tunnel axis (opposite to that obtained in the first stage). It occurs mainly around the tunnel opening.

3.4.3 Tunnel support

Figure 7a shows radial displacement induced at the first tunnel lining in section (Z = 0) (crown corresponds to θ = +90°). We observe an increase of this displacement of about 75% (resp. 25%) at the

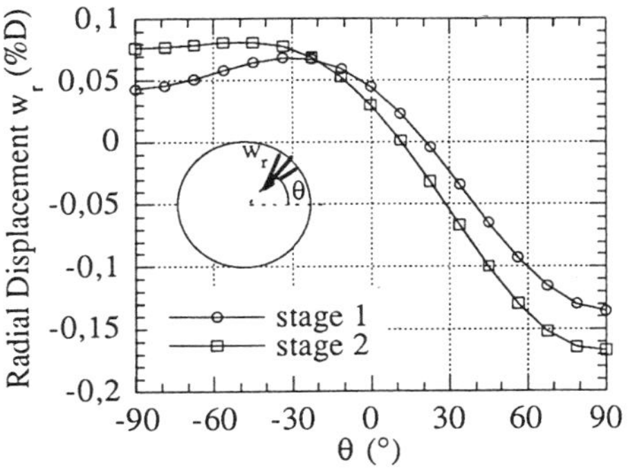

(a)

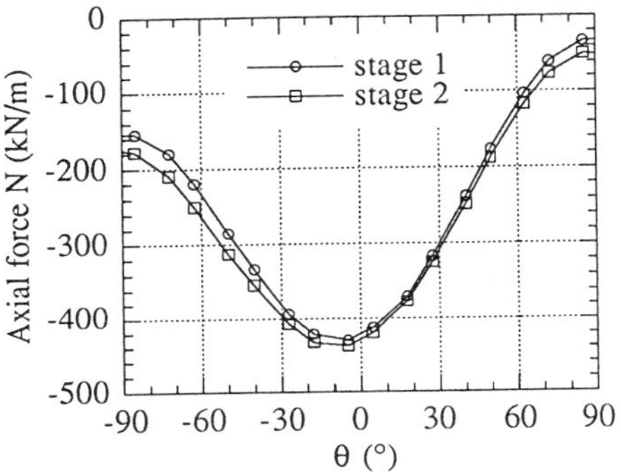

(b)

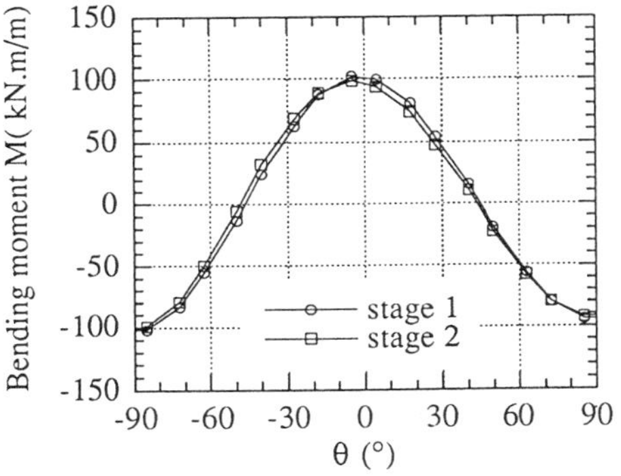

(c)

Figure 7. Influence of the construction of the second tunnel on the first tunnel lining of the section (Z = 0)
(a) Radial displacement
(b) Axial forces (N)
(c) Bending moment (M)

tunnel invert (resp. crown.). Figures 7b and 7c show respectively the distribution of the axial force (N) and bending moment (M) induced in the lining of the first tunnel (section Z = 0). It can be observed that the

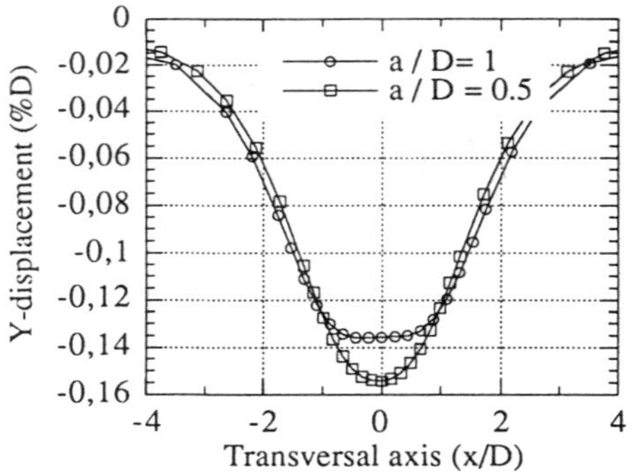

(a)

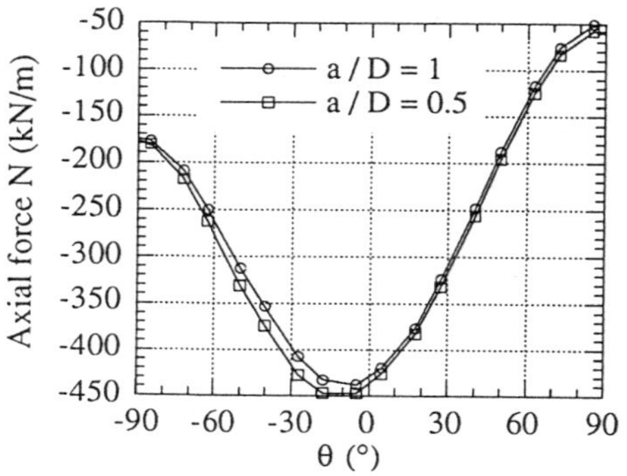

(b)

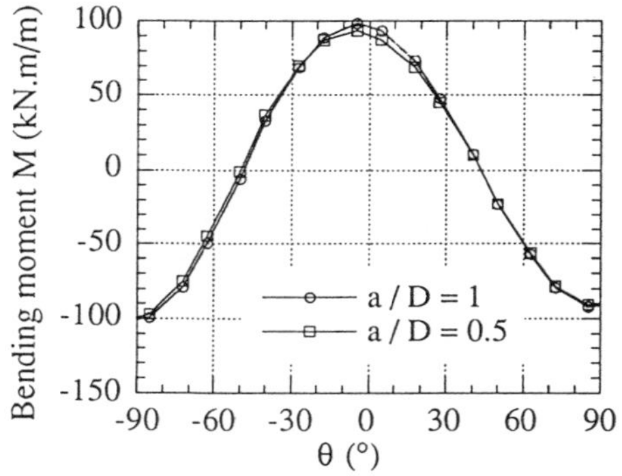

(c)

Figure 8. Influence of the closeness ratio at the end of the construction of the twin tunnels
(a) On soil settlement profile
(b) On axial forces (N)
(c) On bending moment (M)

influence of the construction of the second tunnel is very moderate on both N and M (values induced by the second stage are less than 10% of the first one).

3.5 Influence of the Distance Between Tunnels

In order to study the influence of closeness ratio in the interaction of a twin tunnels, we analyzed a second case with a/D = 0.5. This case can be encountered in sites with severe conditions. Figures 8a -c show a comparison of results obtained with this configuration to that obtained with the reference case (a/ D = 1). It can be observed that the decrease of the closeness ratio from 1 to 0.5 induces only a moderate increase of the vertical displacement above the tunnels axis (14%).

3.6 Comparison with Empirical Approach

It is of some interest to evaluate settlement profiles in terms of normal probability (Gaussian) distribution. Based on Peck (1969) formulas, for unlined tunnels, some particular parameters can be defined. The pattern of displacement near the ground surface may be characterised by the three parameters *Vs*, *i* and *L*. *Vs* is the volume of the transverse settlement trough per unit distance of tunnel advance, obtained by the equation $Vs = (2\pi)^{1/2}\, 2\, i\, u_{max}$; *i* is the distance from the centreline of the tunnel to the point of inflexion, where settlement $u_i = 0.606\, .u_{max}$; *L* is the total width of the profile, approached by $L = 5\, i$.

Using the above relationships, *i*, *L* and *Vs* for the two study cases (a = D) and (a = 0.5D) can be determined. Values presented in table 2 confirm that the surface settlement profile is more concentrated in the second case : *i* is reduced of about 15%, while the total volume moved in the surface is slightly reduced (-5%).

4. CONCLUSION

In this paper, we presented a three-dimensional finite element analysis of a closely spaced twin tunnels constructed in cohesionless soil. Analysis of numerical results shows that the construction of a tunnel close to an existing one (closeness ratio a/D = 1) induces an important soil settlement which in our example approaches 40% of that obtained by the construction of the first tunnel. It shows also that this construction induces a moderate increase of the bending moment and axial force in the lining of the first tunnel.

Table 2 : Settlement induced by the construction of twin tunnels : Parameters for the normal distribution

case	u_{max} (%D)	i/D	L/D	Vs/D
a = D	-0.141	1.73	8.65	1.22
a = 0.5D	-0.155	1.51	7.57	1.17

REFERENCES

AXELSSON O., 1994, "Iterative solution methods", *Cambridge Univ. Press.*

BATHE K. J. et al., 1993, "Some recent advances for practical finite element analysis", *Computers & Structures*, Vol. 47, N° 4/5, pp. 511-521.

CHAN T. F. et al., 1994, "A Quasi-Minimal Residual variant of Bi-CGSTAB algorithm for non symmetric systems", *SIAM J.. Stat. Comput.*, Vol. 15, N°2, pp. 338-347.

GORBANBEIGHI S., 1995, "Développement et validation d'un programme tridimensionnel pour le calcul des ouvrages souterrains : Application aux tunnels peu profonds", Thèse de Doctorat, USTL.

MROUEH H., 1997, "Modélisation tridimensionelle de l'interaction creusement de tunnels et ouvrages souterrains en site urbain", Thèse de Doctorat, USTL.

PAPADRAKAKIS M., 1993, "Solving large-scale problems in mechanics", John Wiley & sons Ltd, England.

PECK R. B., 1969, "Deep excavations and tunnelling in soft ground" Proceedings, 7th Int. Conf. on Soil Mech. and Found. Eng., Mexico City, pp. 225-290.

POMMEREL C., 1992, "Solution of large unsymmetric systems of linear equations, Vol. 17, *Series in Micro-electronics*", Hartung-Gorre, Konstanz.

SHAHROUR I., 1992, "PECPLAS : A finite element program for the resolution of geotechnical problems", Actes du Colloque ENPC 92, *Géotechnique et Informatique*, Paris.

SOLIMAN E., DUDDECK H. and AHRENS H., 1993, "Two- and three-dimensional analysis of closely spaced double-tube tunnels", Tunnelling and Underground Space Technology, Vol. 8, No. 1, pp. 13-18.

TONG C. H., 1992, "A comparative study of preconditioned Lanczos methods for non symmetric systems", Technical Report SAND91-8240B, Sandia National Laboratories.

VAN DER VORST H. A., 1992, "Bi-CGSTAB : A fast and smoothing converging variant of bi-CG for solution of non symmetric linear systems", SIAM J.. Stat. Comput., Vol. 13, No 2, pp. 631-644.

Numerical Models in Geomechanics, Pietruszczak & Pande (eds) © *1997 Balkema, Rotterdam, ISBN 90 5410 886 X*

Numerical and analytical modelling of a new gallery in the Boom Clay formation

V. Labiouse
Belgian Nuclear Research Centre (SCK·CEN), Mol, Belgium

ABSTRACT: Among the extension works of the Underground Research Laboratory of SCK•CEN at Mol, the construction of a 80 m length gallery is planned. The paper presents preliminary calculations of this opening and focuses on the comparison between finite difference results (FLAC code) and predictions got from "convergence-confinement" analyses. Within the framework of this analytical tool, a special attention is paid to the decompression of the medium ahead of the excavation face and to the appraisal of the "initial" convergence u_{ri0} undergone by the gallery walls before the installation of the lining. An improvement is proposed that leads to a better agreement between analytical and numerical results.

1 INTRODUCTION

1.1 *Framework*

As part of the Belgian R&D programme on the disposal of radioactive waste, the Belgian Nuclear Research Centre (SCK•CEN) has initiated studies to demonstrate the technical feasibility and the long-term safety of geological disposal of reprocessed High-Level radioactive Wastes (HLW). As an acceptable repository site is the Tertiary Boom Clay formation below the Mol/Dessel nuclear site, SCK•CEN has built an underground research facility (called HADES) at 223 m depth where in-situ experiments can be performed in representative disposal conditions (Bonne & al, 1992).

In the summer of 1997, extension works of this underground laboratory will start. They first consist in drilling a second access shaft down to 230 depth, and then in excavating a 80 m length gallery at level 223 m to connect this shaft with the existing infrastructure.

Although this connecting gallery is not fully representative of the future HLW disposal galleries, it is worthwhile to study the hydro-mechanical disturbances that will develop in its surroundings. Indeed, it is foreseen to conduct the works in order to reduce the damaged zone to a minimum, with the aim of demonstrating the feasibility of galleries that have a small effect upon the integrity of the Boom Clay formation.

The preliminary calculations that are presented in this paper were performed to work out an efficient instrumentation programme around the connecting gallery and to appraise the importance of the hydro-mechanical disturbances arising during its construction

1.2 *Analysis methods*

To study the mechanical and hydraulic disturbances that develop in the surroundings of galleries driven in the Boom Clay formation, our analyses are performed in the framework of the porous media theory (Coussy, 1991). This means that the response of the clayey medium is explained by distinguishing the parts taken by the soil matrix and by the pore water contained in the void spaces (as in the effective stress approach), but making as well allowances for the deformabilities of both constituents.

The excavation of a tunnel is a coupled three-dimensional problem which involves two different structures: the soil medium and the lining. This can be studied in several ways. However, owing to the *in situ* conditions of the Boom Clay formation (continuous medium and nearly isotropic stress field) the analyses are mainly based on two methods: numerical calculations with the FLAC finite difference code, and analytical solutions by means of the "convergence-confinement" method (Labiouse & Bernier 1997).

1.3 *Modelling assumptions*

For the purpose of this study, the calculations will merely focus on the gallery construction and will not evaluate the hydro-mechanical changes arising from the pore water pressure dissipation after excavation. Owing to the very low permeability of the Boom Clay formation and to a fast tunnel advancing rate, the response of the medium to the excavation has been idealized as undrained (no flow of water).

Since the supporting system and the working sequence of the connecting gallery haven't yet been decided, some usual techniques have been assumed for the calculations (Fig. 1). They are hereafter described:

1. the lining consists of cast iron segments, 1.75 m internal radius, 0.16 m thick and 0.5 m wide. Their general arrangement includes jacking points allowing an expansion of the ring against the excavation walls. As a consequence, a perfect contact of the outer face of the lining with the excavation walls is assumed.

2. the entire cross section of the gallery is excavated in one single step by semi-mechanical means (backhoe of roadheader type of machinery). The rounds are 0.5 m long and, after each round, a 0.5 m wide ring is installed. The unlined distance between the front and the segments varies within 0.25 and 0.75 metre.

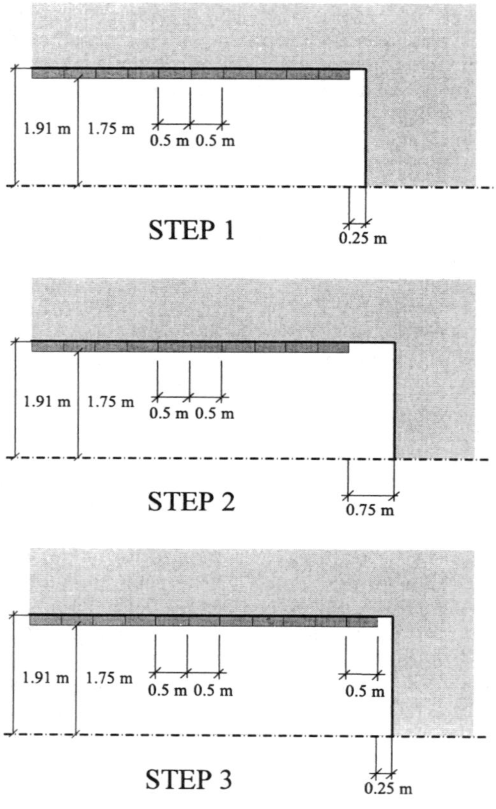

Fig. 1: Sequential tunnelling and lining process.

For this preliminary modelling, the behaviour of the soil matrix has been idealized as linear-elastic perfectly-plastic with a Mohr-Coulomb strength criterion. The chosen set of drained soil properties and pore water data is summarized in table 1.

Table 1. Characteristics taken for the calculations.

Young's modulus	E'	$3\ 10^5$	kPa
Poisson's coefficient	ν'	0.125	-
Cohesion	c'	300	kPa
Friction angle	φ'	18	°
Dilation angle	ψ	0	°
Permeability coefficient	k	$4\ 10^{-12}$	m/s
Porosity	n	0.39	-
Water bulk modulus	K_w	$2\ 10^6$	kPa

With regard to the in situ conditions of the Boom Clay formation at -223 m depth, realistic assumptions are made to simplify the problem: homogeneity and isotropy of the soil mass, as well as isotropic *in situ* stress field : total stress p_0 of 4500 kPa (overburden) and pore pressure p_{p0} of 2250 kPa (water column).

2 FINITE DIFFERENCE CALCULATIONS

2.1 *Modelling sequence*

The numerical calculations are carried out with the FLAC code (1995). Two-dimensional axisymmetric simulations were preferred to one-dimensional plane-strain axisymmetric ones. Indeed, although the latter are much easier and faster to perform (the number of elements is greatly reduced), they are too simplistic to account correctly for the three-dimensional response of the medium near the excavation face as well as for the support laying distance behind the front.

The geometry at the end of the run and boundary conditions are schematically represented in Fig. 2. The axis of the tunnel is an axis of radial symmetry, and the external boundary is selected at 20 m from it (i.e. ± five tunnel diameter). To model accurately the infinite extent of the medium, a special tool has been used: instead of fixed displacement or fixed pressure boundary conditions, the variation of both values are linked by a simple linear relationship. This is achieved introducing springs as external boundary and determining their stiffness from the boundary radius and the elastic characteristics of the medium (Labiouse, 1993).

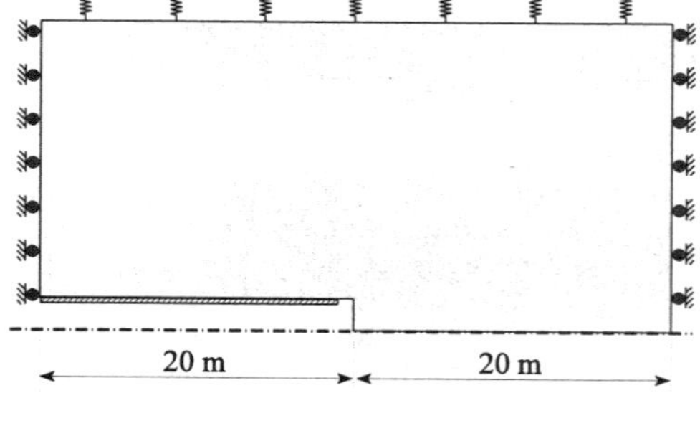

Fig. 2: Geometry and boundary conditions

With regard to the lateral boundaries, they are located 20 m on both sides from the final position of the tunnel face, and are constrained by rollers in the axial direction. As it has been noted in the results, they are largely far enough to avoid any influence on the soil-lining equilibrium state.

The modelling steps account for the sequential tunnelling process illustrated in Fig. 1 : for each round of 50 cm, two rows of 25 cm width are successively excavated (removal of elements of the grid) and then a 50 cm wide lining segment is installed (adding of elements). The equilibrium is achieved after each step.

2.2 *Numerical results*

Figure 3 is a filled contour plot of the displacement magnitudes in the vicinity of the connecting gallery face. It shows that a significant part of the convergence occurs before the lining installation and even ahead of the excavation face. Then, owing to the expansion of the rigid lining against the excavation walls, the displacements stabilize very quickly behind the support laying distance.

The mechanical disturbances that develop in the surroundings of the future connecting gallery are illustrated in Fig. 4: the plastic zone extends 4 m ahead of the front and has a thickness of 2.5 m beyond the excavation wall (→ 9 m diameter damaged zone). Let us however point out a slight elastic recompression of the medium following the lining set-up: the points that were characterised by a plastic behaviour before the support installation (i.e. the yielded region represented by the darker zone) find an incremental elastic behaviour again (grey zone). Indeed, owing to the very high stiffness of the cast-iron segments and to the absence of an overexcavation gap, there is an important build-up in confining pressure, which leads to a reduction of the difference between the major and minor principal stresses in the medium (resp. the circumferential and radial stresses).

The filled contour plot of Fig. 5 represents the drop of pore water pressure arising from the mechanical deformations undergone by the clay mass during tunnelling. One notices that this hydraulic disturbance only occurs in the damaged zone around the gallery (Fig. 4) and that suctions down to -1.5 MPa develop close to its face. Since this value exceeds the air-entry value for Boom Clay (assessed to 1 MPa by Horseman & *al*, 1996), air will move into the gallery front causing localized desaturation.

Analysing in Fig. 3 to 5 the shape of the mechanical and hydraulic disturbances that develop in the medium ahead of the gallery front, one clearly notices that the state of stress, pore pressure and displacement can be reasonably approximated by a spherical symmetry response. This fact had already been established by Egger (1978) from the orientation of displacement vectors around the heading of a tunnel mock-up.

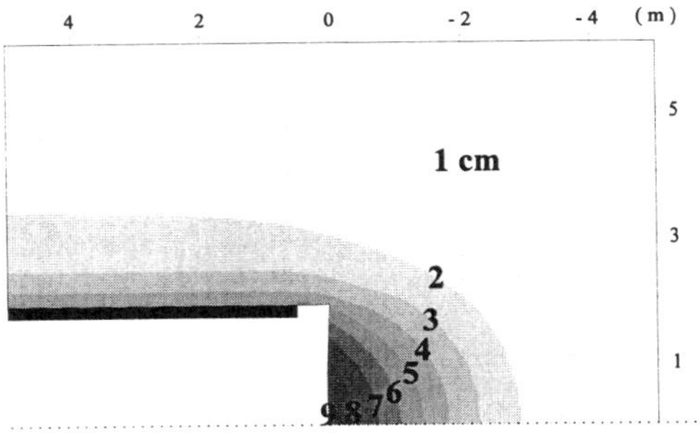

Fig. 3: Displacement magnitudes

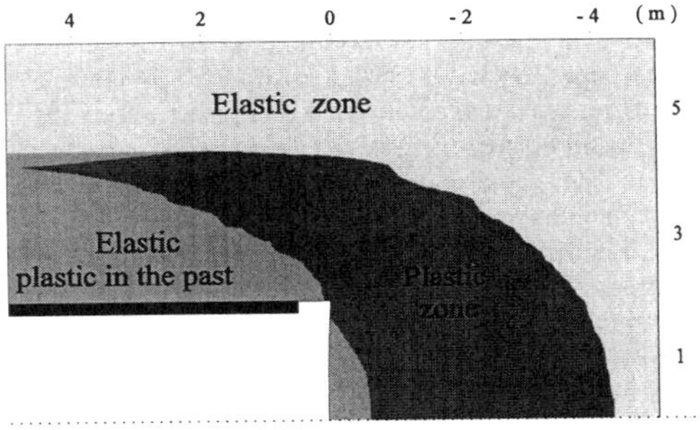

Fig. 4: Disturbed zone extent

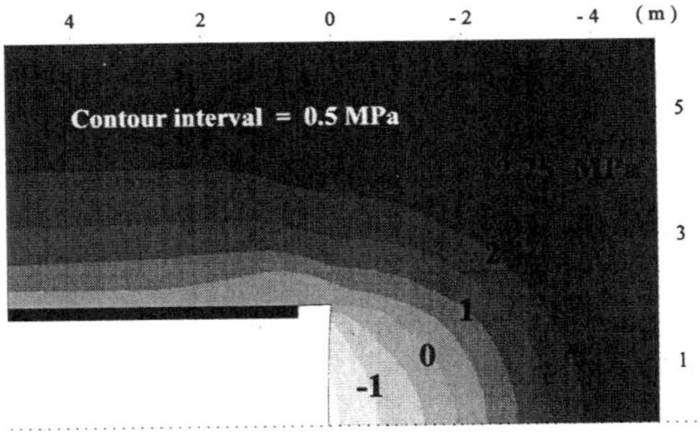

Fig. 5: Pore water pressures

3 ANALYTICAL SOLUTIONS

3.1 *Convergence - confinement method*

Within the framework of the analytical solutions, the reference method is the well known "convergence-confinement" method (or characteristic lines method) which replaces the three-dimensional problem by the plane strain and axisymmetric study of a tunnel section. The principle of this convergence-confinement method (C_V - C_F) is usually best illustrated in a (u_{ri}; p_i)

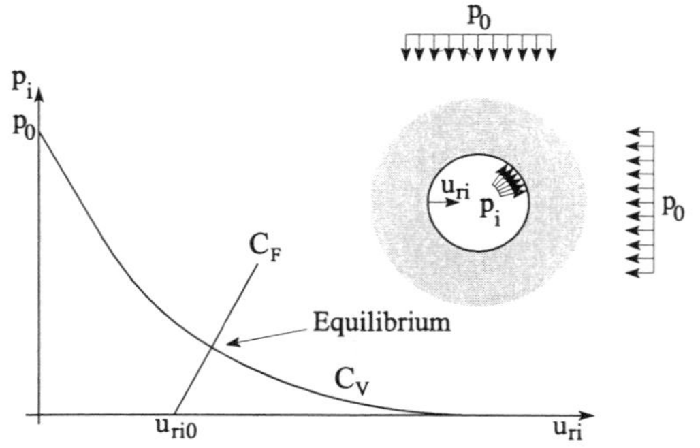

Fig. 6: Convergence - confinement method

diagram, with u_{ri} the radial wall displacement and p_i the applied pressure (Fig. 6). The "convergence" curve C_V is representative of the ground response and the "confinement" curve C_F of the support reaction. The soil-lining equilibrium is reached at their intersection point. The slope of the support characteristic line as well as its origin u_{ri0} have an influence on the equilibrium state. The former depends on the support stiffness and the latter on the unsupported span and the overexcavation gap. Indeed, u_{ri0} represents the radial displacement already undergone by the gallery wall before its contact with the liner.

The solutions published in the literature are written in terms of total stresses and undrained soil characteristics (Corbetta 1990; Panet 1995). For our part, to account for the pore water characteristics and the more intrinsic drained soil properties, we have developed analytical solutions for the short term response of a poro-elasto-plastic medium due to the excavation of circular and spherical cavities (Labiouse & Bernier 1996, 1997). The equations emphasize that the plastic zone extent as well as the stresses and pore pressures around those openings depend not only on the strength parameters of the soil (as predicted by the common total stress approach), but as well on the deformability characteristics of both soil matrix and pore water.

3.2 *Initial convergence u_{ri0}*

The correct assessment of the hydro-mechanical disturbances induced by the excavation of underground structures strongly depends on a good appraisal of the decompression undergone by the medium before the lining installation, and more particularly ahead of the excavation face. To account for that three-dimensional aspect in the "convergence-confinement" method, stress must be laid on a good estimate of the radial displacement u_{ri0} already undergone by the excavation walls when the support is installed. Indeed, as already mentioned, this value represents the origin of the lining characteristic line and has a major influence on the equilibrium state (Fig. 6).

Till 1990, the C_V - C_F method was known to remain imprecise for the determination of u_{ri0} for tunnels driven in non-elastic media. Since then, strides have been made and new methods that use principles similar to those of the C_V - C_F method have been proposed to get a better estimation of this initial convergence :

1. the New Implicit Method (NIM) developed by Bernaud (1991) from 2D-axisymmetric finite element calculations.
2. the Self Similarity Principle method (SSP) and an extension of it (ESSP) proposed by Corbetta (1990) and Nguyen Minh (1992) as two bounding solutions.

Another frequently-used method is based on the establishment that the state of stress and displacement ahead of the excavation face of a circular gallery can be reasonably approximated by a spherical symmetry response (Egger 1974). In this tool, hereafter-called Spherical Decompression of the Face method (SDF), the initial closure u_{ri0} undergone by the gallery walls before the set-up of the lining is assessed by the inner radial displacement of an unlined spherical cavity.

3.3 *Improvement for short support laying distances*

Comparing the predictions of this "Spherical Decompression of the Face" method with the results of the finite element calculations (cf. subheading 2.2), they were found to overestimate the decompression of the medium and conversely to underestimate the soil-lining equilibrium pressure.

On the other hand, in such instances where a rigid lining is laid very close to the excavation face (within ¼ of the gallery radius), an improvement can be made. It consists in fixing the symmetry centre of the spherical cap half a radius behind the tunnel heading (Fig. 7) and in assessing the initial convergence u_{ri0} by the displacement that takes place at 1.5 radius distance of this centre. By this way, the estimates of stresses, pore water pressures, displacements and plastic zone extent better match the FLAC results, not only in the region ahead of the excavation face, but also in sections located far behind the face.

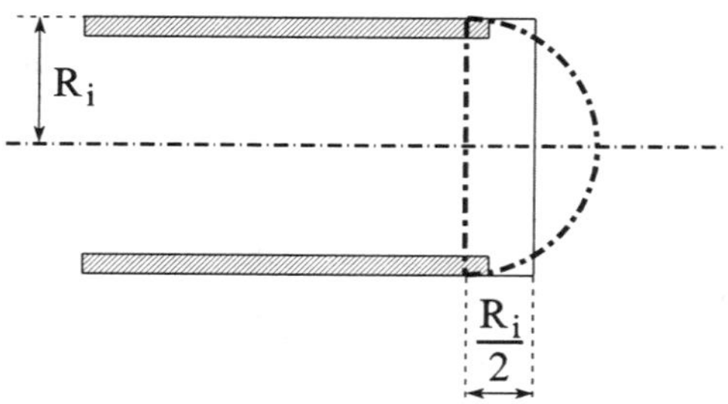

Fig. 7: Improvement by fixing the symmetry centre of the spherical cap half a radius behind the face

3.4 *Numerical results*

Figure 8 shows the distributions of total and effective stresses (axial σ_a, tangential σ_θ), pore water pressure and displacement along the symmetry axis ahead of the face. One notices the good agreement between the results got from the FLAC code (dashed lines) and the predictions with the "improved" method (solid lines).

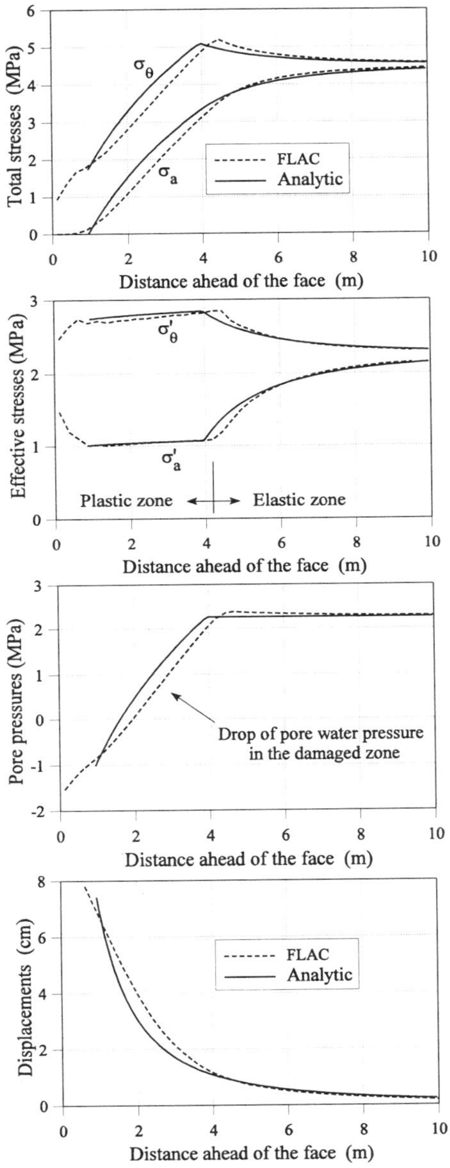

Fig. 8: State ahead of the excavation face

Figure 9 represents the distribution of total and effective stresses (radial σ_r, tangential σ_t and longitudinal σ_l), pore water pressure p_p and displacement u_r in the clay mass in a section located 10 m behind the front. The dashed lines are related to the FLAC calculations and the solid lines to the predictions got from the closed-form solutions and the improved appraisal of initial convergence u_{ri0} proposed in subheading 3.3.

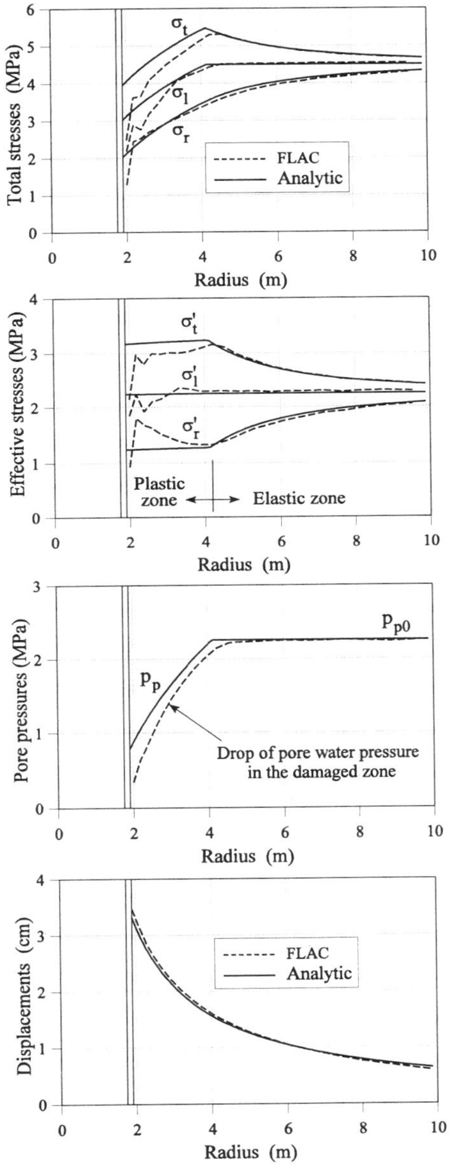

Fig. 9: Equilibrium state 10 m behind the front

The figure clearly illustrates the disturbed zone that develops in the surroundings of the gallery and the significant drop of pore water pressure in this zone. It shows as well the good agreement of both methods, except for the stresses in the damaged zone. Analysing this discrepancy, one notices that the difference between the tangential (major) and radial (minor) stresses is smaller in FLAC than in the analytical solution. This confirms that the damaged zone which was previously characterised by a plastic behaviour finds an incremental elastic behaviour again owing to the important build-up in lining pressure.

4 FURTHER MODELLING WORKS

It is essential to bear in mind some assumptions that have been made for the above-presented calculations:

1. perfect contact of the outer face of the lining with the excavation walls (due to the expansion of the ring).
2. use of a simple constitutive model : linear-elastic perfectly-plastic with a Mohr-Coulomb criterion.

Because these assumptions have a major influence on the magnitude of both mechanical and hydraulic disturbances, some complementary modelling works are now being performed in the framework of a European Commission contract, called CLIPEX.

5 CONCLUSIONS

The paper presents preliminary calculations of a 80 m length gallery to be built in the Boom Clay formation. It focuses on the comparison of two modelling methods : numerical calculations with the FLAC finite difference code, and analytical solutions by means of the "convergence-confinement" method.

In the framework of the latter method, the correct assessment of the hydro-mechanical disturbances arising during construction strongly depends on a good appraisal of the decompression undergone by the soil before the lining installation, and more particularly ahead of the excavation face. One tool is based on the establishment that the state of stress and displacement ahead of the front of a circular gallery can be reasonably approximated by a spherical symmetry response.

In the case of short support laying distances (within ¼ of the gallery radius), this method is improved for a better appraisal of the decompression as well as of the initial convergence u_{ri0} undergone by the gallery walls before the installation of the lining. It leads to a better agreement between analytical and numerical results.

Acknowledgements

Most of the results laid in this paper are taken from a contractual work for the E.I.G. PRACLAY managed by ONDRAF/NIRAS and SCK•CEN (Bernier & *al*, 1997). Their support is greatly acknowledged.

REFERENCES

Bernaud D. (1991): *Tunnels profonds dans les milieux viscoplastiques. Approche expérimentale et numérique.* Ph. D. Thesis, Ecole Nationale des Ponts et Chaussées, Paris, France.

Bernier F., Labiouse V., Verstricht J. 1997. *Praclay project : Geotechnical modelling and instrumentation programme of the connecting gallery.* Report R-3150 to NIRAS/ONDRAF, SCK•CEN, Mol.

Bonne A., Beckers H., Beaufays R., Buyens M., Coursier J., De Bruyn D., Fonteyne A., Genicot J., Lamy D., Meynendonckx P., Monsecour M., Neerdael B., Noynaert L., Voet M., Volckaert G. 1992. *The HADES demonstration and pilot project on radioactive waste disposal in a clay formation.* Final report to EC, EUR 13851.

Corbetta F. (1990): *Nouvelles méthodes d'étude des tunnels profonds. Calculs analytiques et numériques.* Ph. D. Thesis, Ecole Nationale Supérieure des Mines de Paris, France.

Coussy O. 1991. *Mécanique des milieux poreux.* Editions Technip, Paris, France.

Egger P. (1974): *Gebirgsdruck im Tunnelbau und Stützwirkung der Orksbrust bei Überschreiten der Gebirgsfestigkeit.* Proceedings of the 3rd Congress of the International Society of Rock Mechanics, Vol. 1, pp. 1007-1011.

Egger P. (1978): *Déformations au front de taille et détermination de la cohésion du massif rocheux.* Journée d'études Paris, 26 oct. 1978, Tunnels et Ouvrages Souterrains, n°32, pp. 93.

FLAC 1995. Manuals of the FLAC code. Itasca Consulting Group, Minneapolis, USA.

Horseman S.T., Higgo J.J.W., Alexander J., Harrington J.F. (1996): *Water, gas and solute movement through argillaceous media.* Report CC-96/1 for Nuclear Energy Agency.

Labiouse V. (1993): *Etudes analytique et numériques du boulonnage à ancrage ponctuel comme soutènement de tunnels profonds creusés dans la roche.* Ph. D. Thesis, Université Catholique de Louvain, Louvain-la-Neuve, Belgium.

Labiouse V., Bernier F. 1996. *Influence of pore water compressibility on the short and long term stability of nuclear waste disposal galleries in clay.* Geomechanics'96, Roznov, Czech Republic, 6 pp.

Labiouse V., Bernier F. 1997. *Hydro-mechanical disturbances around excavations.* Feasibility and Acceptability of Nuclear Waste Disposal in the Boom Clay Formation, SCK•CEN, Mol, pp. 15-26.

Nguyen-Minh D., Corbetta F. (1992): *New methods for rock-support analysis of tunnels in elastoplastic media.* Proceedings of the International Symposium on Rock Support in Mining and Underground Construction, Sudbury, Canada, pp. 83-90.

Panet M. (1995): *Le calcul des tunnels par la méthode convergence-confinement.* Presses de l'Ecole Nationale des Ponts et Chaussées, Paris, France.

Numerical Models in Geomechanics, Pietruszczak & Pande (eds) © 1997 Balkema, Rotterdam, ISBN 90 5410 886 X

The assessment of the stability of a very old tunnel by discrete finite element method (DFEM)

N.Tokashiki
Department of Civil Engineering, University of Ryukyu, Japan

Ö.Aydan
Department of Marine Civil Engineering, Tokai University, Japan

Y.Shimizu
Department of Civil Engineering, Meijyo University, Japan

T.Kawamoto
Department of Civil Engineering, Aichi Institute of Technology, Japan

ABSTRACT : This paper presents the assessment of the stability of a very old tunnel by discrete finite element analysis and the experiments on the model tunnels in the laboratory. The research program includes site investigations, numerical analysis and labolatory model tests. Several scenarios were considered and each scenario was studied through numerical analysis by the discrete finite element method (DFEM) and laboratory model tests. On the basis of this research program, a procedure how to rehabilitate the existing tunnel was proposed. The authors describe the results of this research program with an emphasis on numerical analyses by DFEM.

1 INTRODUCTION

Assessing the stability of existing tunnels presents a challenging problem to rock engineers. The recent tunnel portal collapse in Hokkaido Island of Japan, which killed 20 peoples in a bus and a car passing through at the time of collapse, resulted in a great public concern on this problem. As a result, the authorities initiated a program to assess the stability of existing tunnels. The authors were asked to assess the stability of a very old unsupported tunnel, which was constructed about 70 years ago in carstic sandstone, in Okinawa Island.

To investigate the stability of the tunnel, we considered the research program for site investigations, numerical analysis and model tests in laboratory. From site investigations, it was found that well defined discontinuities existed around the tunnel. Therefore, several scenarios were considered in numerical analyses and model tests. These scenarios include the spatial orientation and position of discontinuties in the rock mass, and the strength and the displacement characteristics of the discontinuty.

The discrete finite element method (DFEM) (Aydan et al, 1996) was used to analyse the stability of the tunnel in discontinuous rock mass. This method is essentialy based on the finite element method together with the contact element. It consists of mechanical models to present the deformable rocks and contacts models that specify the interaction among them.

The laboratory model tests were carried out to check the validity of the stability analysis of the tunnel by DFEM. The base-friction test technique for the model tests was employed using the base-friction apparatus (Kawamoto et. al., 1983).

The site investigation program involves regional geological investigations, and boring for structural geology and physical and mechanical properties of intact rocks and joints.

2 SITE INVESTIGATIONS

The investigated tunnel is located about 60 km North of Naha in Okinawa Island (Japan). The tunnel is 15 m long with an overburden of aproximately 10 m. Rock around the tunnel is mainly carstic sandstone and 7 layers having different grain size were distinguished as shown in Fig. 1. Bedding planes dips at an angle of 4 - 6 from horizontal toward North-West. Besides 4 distinct discontinuities were recognised. Among them, discontinuities numbered 2 and 3 were considered to be critical for the stability of the tunnel.

Three boreholes of 6 m long were drilled for geological investigations and for obtaining cores for physical and mechanical characteristics of rock. Typical physical and mechanical properties of samples obtained from boreholes are given in Table1. As it is noted from this table rock around the tunnel can be classified as a weak rock.

3 MODEL TESTS

Model tests were carried out using a base friction apparatus as shown in Fig. 2. The geometrical scale of models was 1/50. It is possible to reduce the mechani-

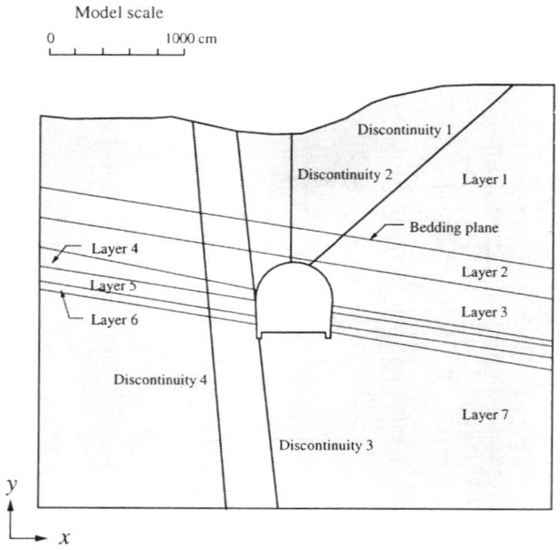

Fig 1. Geological cross section.

Table 1. Physical and mechanical properties of rock sample.

Property		Boreholes		
		Borehole 1	Borehole 2	Borehole 3
Young's modulus	(kgf/cm^2)	3,100	3,400	10,000
Poisson's ratio		0.259	0.310	0.277
Density	(gf/cm^3)	1.83	1.81	1.90
Porosity	(%)	30.2	30.0	32.3
Elastic wave velocity	(km/s)			
V_p		1.59	1.43	1.83
V_s		0.64	0.57	0.75
Uniaxial strength	(kgf/cm^2)			
σ_c		8.02	7.77	16.83
σ_t		1.24	0.86	1.31
Friction angle	(Degree)	40.1	40.1	40.8
Cohesion	(kgf/cm^2)	1.40	1.10	0.83

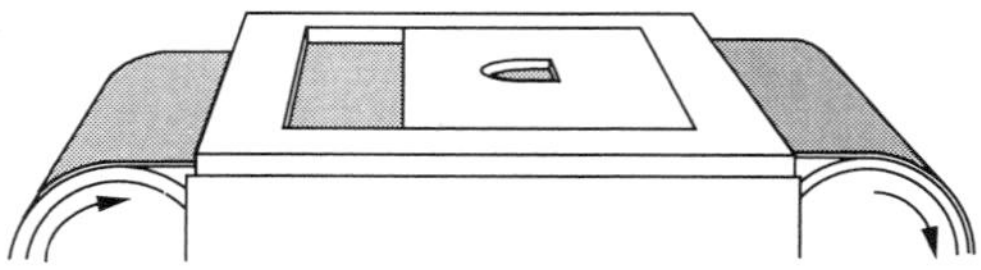

Fig 2. An illustration of the base friction apparatus.

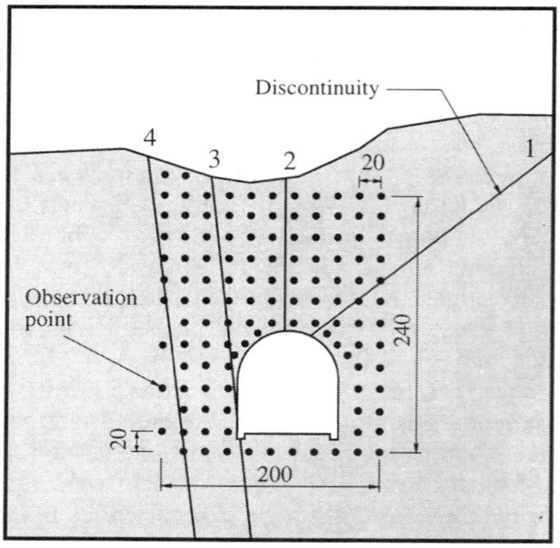

scale : 1/50 (mm)

Fig 3. Configuration of model.

cal properties of model material in a similar manner to the geometrical scale through the variation of consolidation pressure applied on the model material. However, such a strict scaling was not followed in model tests since the discontinuities were considered to be the most critical element in the stability of the tunnel. Discontinuities numbered 1-4 were explicitly were modelled in tests (Fig. 3). For obtaining deformation and straining around the tunnel, observation points were set. Displacements of points were obtained by using digitised images of the model at different time steps.

Three different scenarios were considered:

CASE 1: Friction angle of all discontinuities was same and it had a value of 35° (no support).

CASE 2: Friction angle of discontinuities 2-3 was reduced to 4-6 by placing double teflon sheets (no support).

CASE 3: Friction angle of discontinuities 2-3 was reduced to 4-6 by placing double teflon sheets (with support).

No failure was observed in CASE 1. However, the tunnel was stable when the friction angle of discontinuities 2-3 was reduced. The block bounded by these discontinuities sliddown into the model tunnel as shown in Photo. 1. Figure 4 shows displacement vectors of selected points.

4 DISCRETE FINITE ELEMENT METHOD

The discrete finite element method is based on the principles of finite element method (Aydan et al. 1996). It can handle deformable blocks and contacts that specify the interaction among them. Small displacement theory is applied to intact blocks while blocks can take finite displacement. Blocks are polygons with an arbitrary number of sides which are in contact with neighboring blocks, and are idealized as a single or multiple finite elements. Block contacts are represented by a contact element.

Starting from the moment of momentum and employing the conventional finite element discretisation procedure yields the following set of equations for a typical finite element.

$$\boldsymbol{M\ddot{U}} + \boldsymbol{C\dot{U}} + \boldsymbol{KU} = \boldsymbol{F} \qquad (1)$$

Where

$$M = \int_{\Omega_e} \rho N^T N d\Omega \ ,$$

$$C = \int_{\Omega_e} B^T D_v B d\Omega \ ,$$

$$K = \int_{\Omega_e} B^T D_e B d\Omega \ ,$$

$$F = \int_{\Omega_e} N^T b d\Omega + \int_{\Gamma_{te}} \overline{N}^T t d\Gamma \ .$$

Blocks are modelled by solid elements and contact elements are employed for modelling block contacts or discontinuities.

There are a number of procedures to treat the problem of time discretization. If the central difference method is used Eq.(1) takes the following form

$$\overline{\boldsymbol{K}}\boldsymbol{U}_{n+1} = \overline{\boldsymbol{F}}_{n+1} \qquad (2)$$

Where

$$\overline{K} = \frac{1}{\Delta t^2} M + \frac{1}{2\Delta t} C \ ,$$

$$\overline{F}_{n+1} = \left(\frac{2}{\Delta t^2} M - K \right) U_n - \left(\frac{1}{\Delta t^2} M - \frac{1}{2\Delta t} C \right) U_{n-1} + F_n \ .$$

When Eq.(1) is solved, oscillations may occur depending upon the rate-dependent and rate independent parameters of constitutive models. When the inertia time is considered, contacts and solid blocks are assumed to behave elasto-visco-plasticaly and visco-elasticaly, respectively. On the other hand, if the inertia term is omitted, then the behaviours of contacts and blocks are assumed to be elasto-plastic and elastic, respectively.

In this study, a pseudo-dynamic version of this method is employed. Intact blocks and block contacts can behave elasto-plastically. The method of analysis is a pseudo time stepping incremental procedure. To model the elasto-plastic response of materials and contacts in numerical analysis, the secant method or initial stiffnes method can be selected together with the use of *Updated Lagrangian Scheme*.

5 NUMERICAL ANALYSES AND COMPARISONS

The scenarios analysed by the discrete finite element method were the same as those for model tests. Typical FE model of unsupported tunnel is shown in Fig. 5. Mechanical and physical properties of layers and discontinuities used in analyses are given in Table 2.

Fig. 6 shows the computed deformed configurations of the tunnel and its surrounding for CASE 1. As noted from the figure, the tunnel was stable for CASE 1. The friction angle of discontinuities 2-3 was reduced to 0. Fig. 7 shows the computed deformed configurations of the tunnel for CASE 2. As seen from the figure, the block bounded by discontinuities 2-3 slide down into the tunnel. This results are consistent with

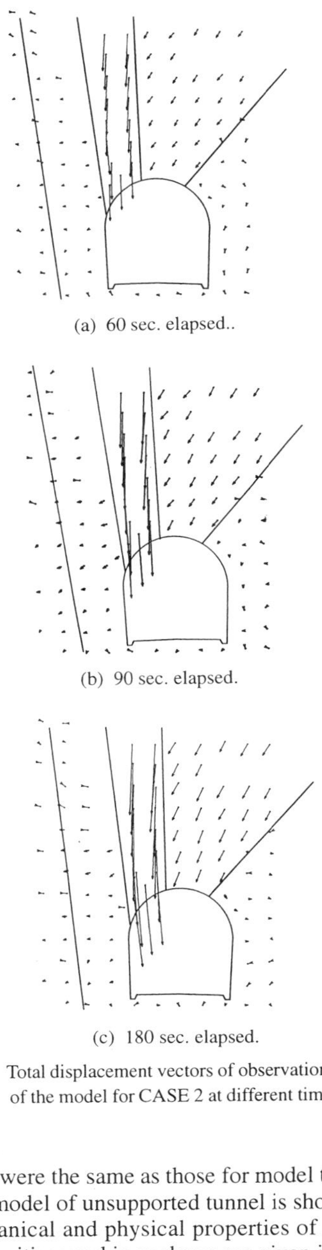

(a) 60 sec. elapsed..

(b) 90 sec. elapsed.

(c) 180 sec. elapsed.

Fig. 4 Total displacement vectors of observation points of the model for CASE 2 at different time steps.

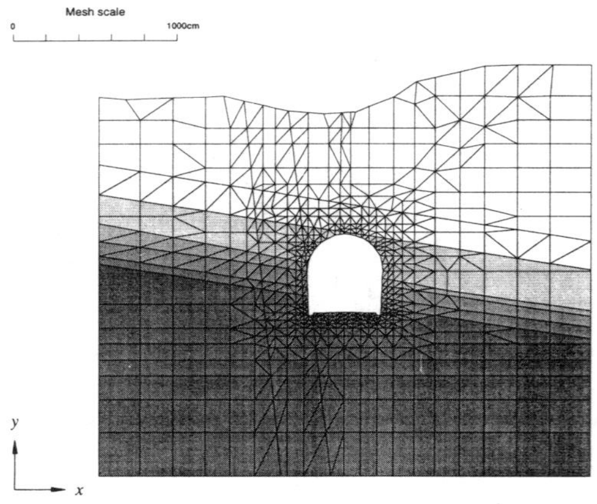

Fig. 5 The FE model for CASE 1 and CASE 2.

Table 2. Physical and mechanical properties of layers and discontinuity.

Material	Property		
	E (kgf/cm^2)	ν	γ (gf/cm^3)
Layers			
Layer 1	5,000	0.282	1.85
Layer 2	10,000	0.282	1.85
Layer 3	3,500	0.282	1.85
Layer 4	2,500	0.282	1.85
Layer 5	4,500	0.282	1.85
Layer 6	3,500	0.282	1.85
Layer 7	1,500	0.282	1.85
Discontinuity	$E = 1{,}500$ (kgf/cm^2), $G = 585$ (kgf/cm^2), $\nu = 0.282$, $t = 0.6$ (cm), $\sigma_t = 0.1$ (kgf/cm^2),		
CASE 1	$c = 0.0$ (kgf/cm^2), $\phi = 35.0^\circ$ (unsupported)		
CASE 2	$c = 0.0$ (kgf/cm^2), $\phi = 0.0^\circ$ (unsupported)		
CASE 3	$c = 0.0$ (kgf/cm^2), $\phi = 0.0^\circ$ (supported)		
CASE 4	$c = 0.0$ (kgf/cm^2), $\phi = 35.0^\circ$ (supported)		

model test results. Although such a drastic reduction of the friction angle of discontinuities is unrealistic, the computed results show how the tunnel may fail if there should be an instability problem.

To see the effect of concrete lining as a rehabilitation measure on the stability of tunnel, analyses ware carried out and computed deformed configurations of the tunnel and stress state in the concrete lining and its close vicinity are shown in Fig. 8. The placement of the concrete lining stabilize the tunnel even for CASE 3. On the basis of numerical analyses and model tests, a rehabilitation of the tunnel including placement of fiber-glass rockbolts, 100 mm shotcrete and 300 mm concrete lining has been decided and this work will be shortly implemented.

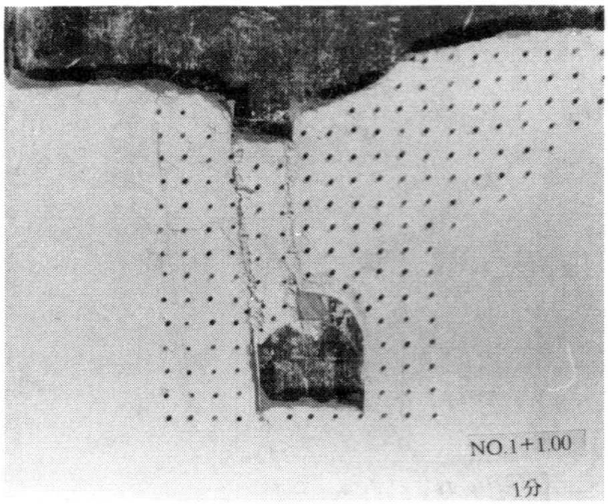

(a) 60 sec. elapsed.

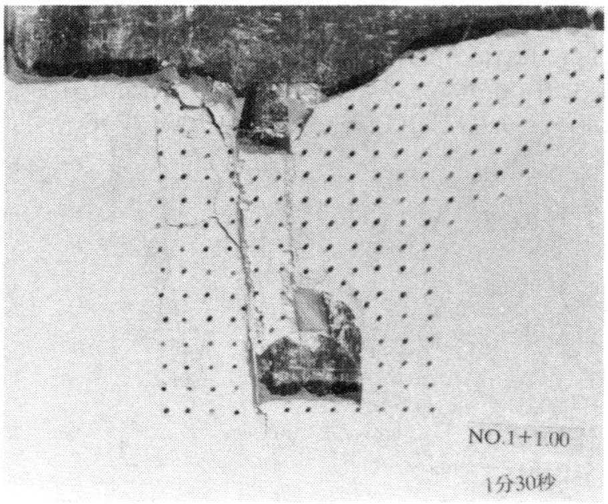

(b) 90 sec. elapsed.

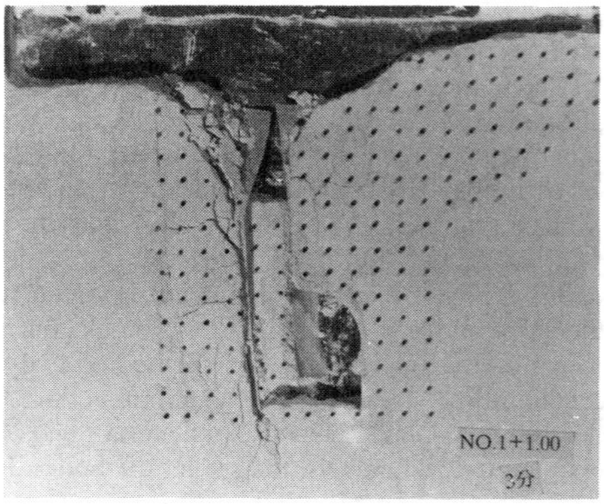

(c) 180 sec. elapsed.

Photo. 1 The result of the model for CASE 2 at different time steps.

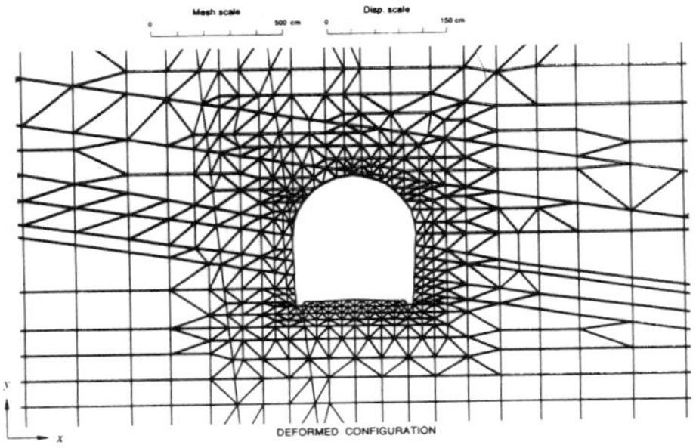

(a) Deformed configuration at iteration 15.

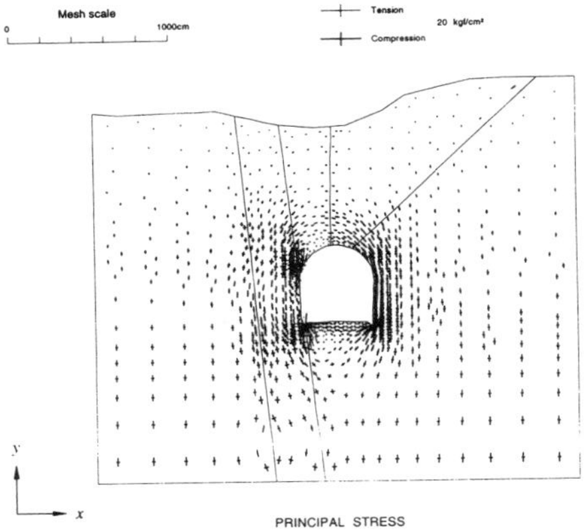

(b) Principal stress at iteration 15.

Fig.6 Computed deformed configuration and principal stress distribution for CASE 1.

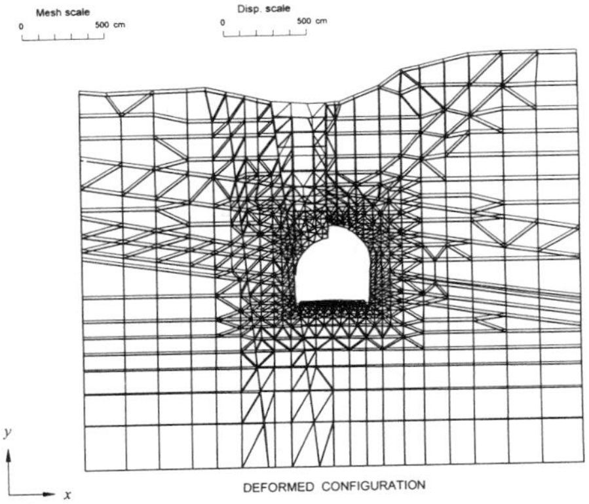

(a) Deformed configuration at iteration 10.

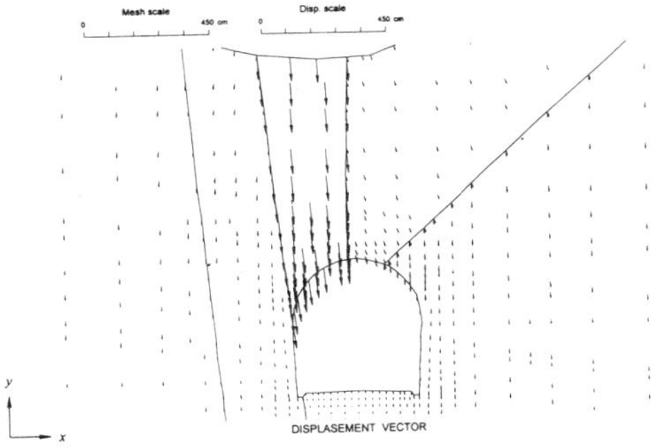

(b) Total displacement vectors at iteration 10.

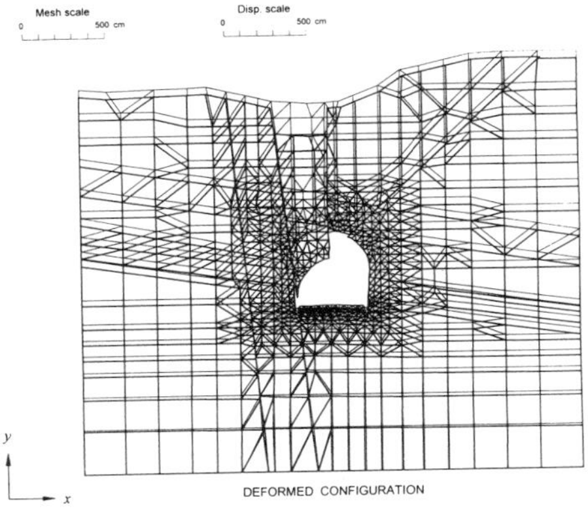

(c) Deformed configuration at iteration 20.

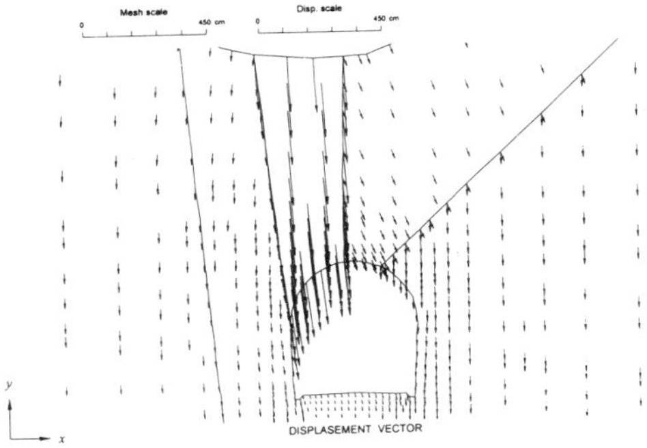

(d) Total displacement vectors at iteration 20.

Fig.7 Computed deformed configurations and total displacement vectors for CASE 2.

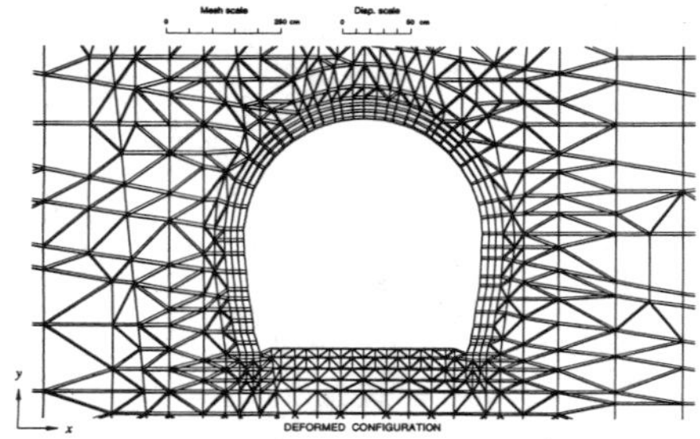

(a) Deformed configuration for CASE 3.

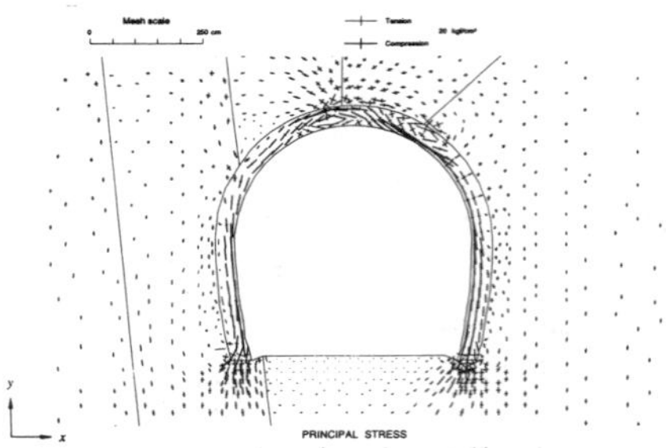

(b) Principal stress for CASE 3.

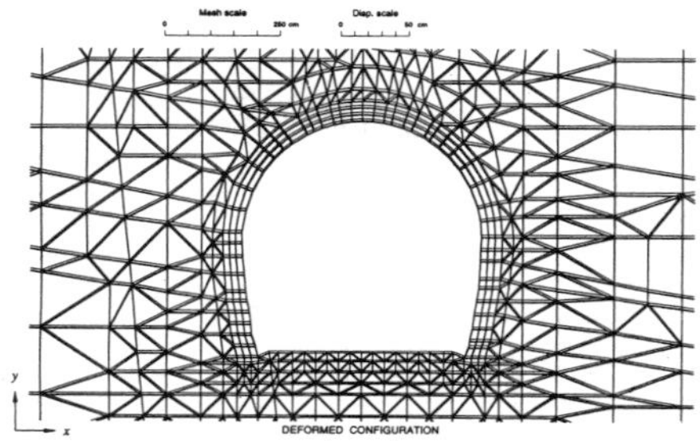

(c) Deformed configuration for CASE 4.

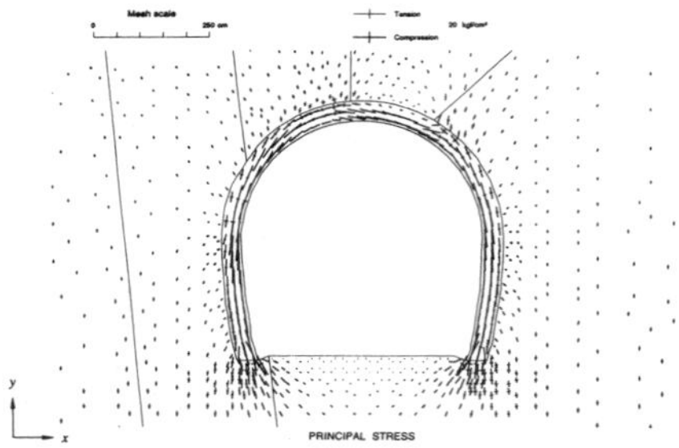

(d) Principal stress for CASE 4.

Fig. 8 Computed deformed configurations and principal stress distributions of the supported tunnel.

6 CONCLUSIONS

In this paper, the stability of a very old unsupported tunnel was investigated by model tests and discrete finite element method. Three different scenarios were considered and the stability of the tunnel was checked. The computed results and model test results showed that the tunnel should be stable unless there is no sudden reduction of the friction angle of discontinuities 2-3. Even if such a reduction occurs, the concrete lining to be constructed as a rehabilitation measure will prevent the instability of the tunnel.

REFERENCES

Aydan, Ö., Mamaghani, I.H.P, & Kawamoto, T. 1996. Application of discrete finite element method (DFEM) to rock engineering structures, *NARMS'96*, 2039-2046.

Kawamoto, T., Obara, Y. & Ichikawa, Y. 1983. A base-friction apparatus and mechanical properties of model material (in Japanese), *J. of Min. and Metall, Inst. of Japan 99*, 1-6.

Numerical Models in Geomechanics, Pietruszczak & Pande (eds) © 1997 Balkema, Rotterdam, ISBN 90 5410 886 X

Dynamic analysis of the failure behavior of a tunnel face

M. Hisatake
Kinki University, Osaka, Japan

T. Murakami
Chizaki Kogyo, Ltd, Tokyo, Japan

T. Tamano
Osaka Sangyo University, Japan

ABSTRACT: To clarify and estimate the failure behavior of a tunnel face in sandy ground with little cohesion, 3-dimensional model experiments with a high-speed video camera and a 2-dimensional numerical analysis were conducted. The movement of the failure ground in the experiments is compared with that in the analysis. The results observed support the numerical analysis.

1 INTRODUCTION

Recently mountain tunneling methods such as NATM are being applied to urban areas with low strength. When a ground consists of sand with little cohesion, maintaining the stability of the tunnel face becomes a very important task. The objective of this research is to estimate the stability and failure phenomena of a shallow tunnel face in a sandy ground, through interplay between physical and numerical model analyses. Three dimensional static and dynamic model experiments are conducted for the physical analysis. A high-speed video camera, with a maximum speed of 40,500 pictures per second, is employed to analyze the dynamic movement of the ground around the tunnel face. The numerical analysis is carried out with Contact Element Method (CEM)[1)-3)] to the conditions of the dynamic model experiments with dry sand. Values for the parameters involved in this analysis are reasonably determined through identification of the mechanical behavior of the experimental material. The dynamic ground movement analyzed by the CEM is compared with that analyzed by the dynamic experiments.

2. DYNAMIC MODEL EXPERIMENTS [4),5)]

2.1 Experimental device and method

Figure 1 shows the experimental device used. A circular tunnel is excavated in dry sand with a horizontal ground surface. Due to the symmetrical conditions of the vertical plane through the tunnel axis, the tunnel is half the size of the actual tunnel geometry. The front part of the device is composed of an acrylic plate on which vertical and horizontal black lines are drawn, and through the plate ground movement on the symmetrical plane can be observed. After a semicircular retaining plate (outer diameter (D) = 14 cm, inner diameter = 13 cm) is set in the steel lining of the half cylinder, sand is dropped from a height of 30 cm and the initial ground is produced. In order to catch the ground movement easily, the sand is painted black and scattered 3 cm apart in a vertical direction only at the contact surface with the acrylic plate. Figure 2 is a grain size accumulation curve of the sand used. The method used for the dynamic experiments is as follows:

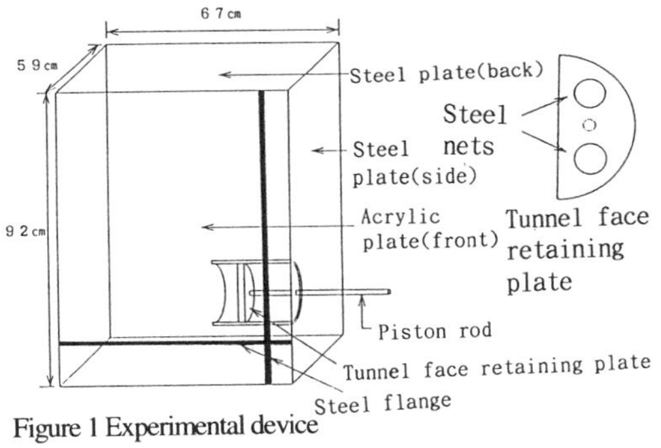

Figure 1 Experimental device

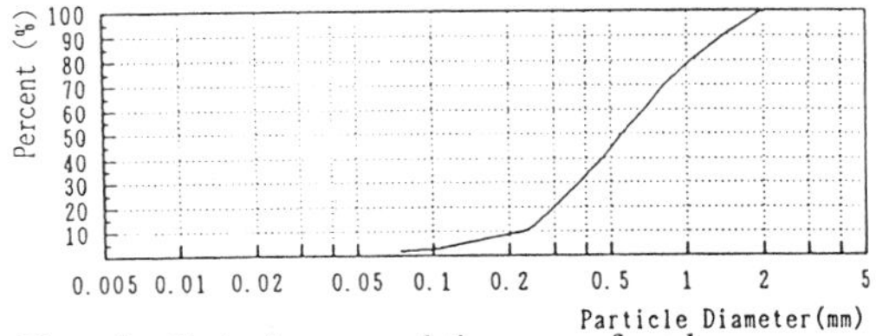

Figure 2 Grain size accumulation curve of sand

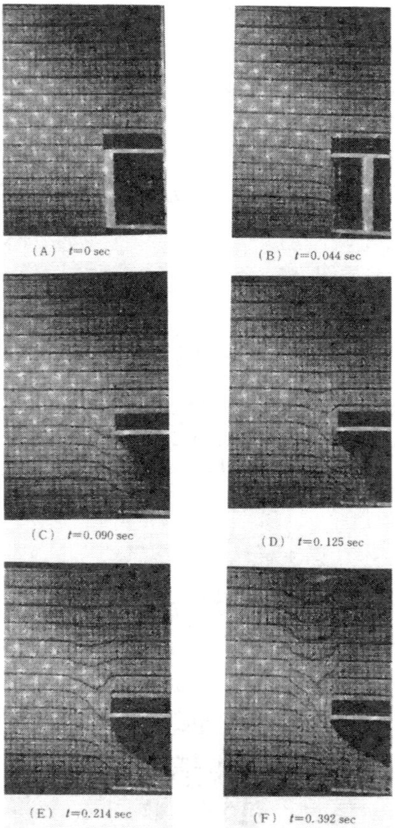

Figure 3 Results of the dynamic failure experiments (H/D=2)

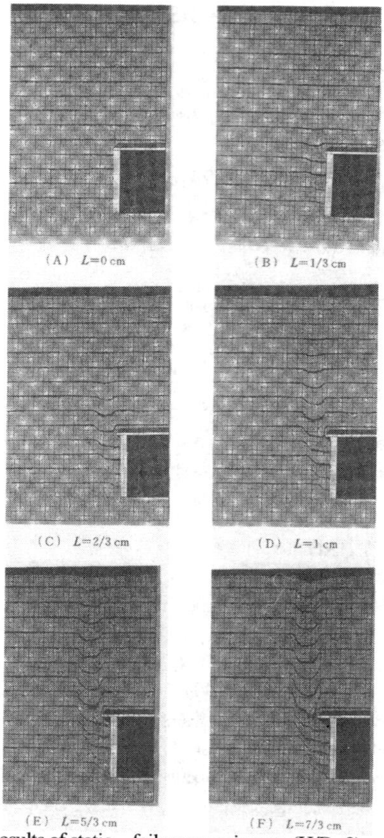

Figure 4 Results of static failure experiments (H/D=2)

After the initial ground is produced, the retaining plate is moved in the direction of the tunnel entrance at a speed of 147 cm/sec, and the dynamic movement of the ground is taken by a high-speed video camera. In order to avoid the occurrence of a vacuum between the tunnel face and the retaining plate, two circular holes (diameter = 3 cm) are perforated in the retaining plate, which are covered by steel nets. In the static experiments, the retaining plate is slowly moved with a velocity of less than 0.017 mm/sec.

2.2 Results

Figures 3 and 4 show dynamic and static failure phenomena, respectively, where H is the overburden and L is the distance of the retaining plate from its initial position. By comparing the results, it is recognized in the static experiments that the sand in the region surrounded by both the retaining plate and a slip line ahead of the retaining plate move horizontally. In the dynamic experiments, on the other hand, the face remains in a vertical state for a moment directly after the retaining plate is removed, and then the sand near the face moves down the incline because of horizontal stress release and gravity. Such movement is not observed in the static experiments.

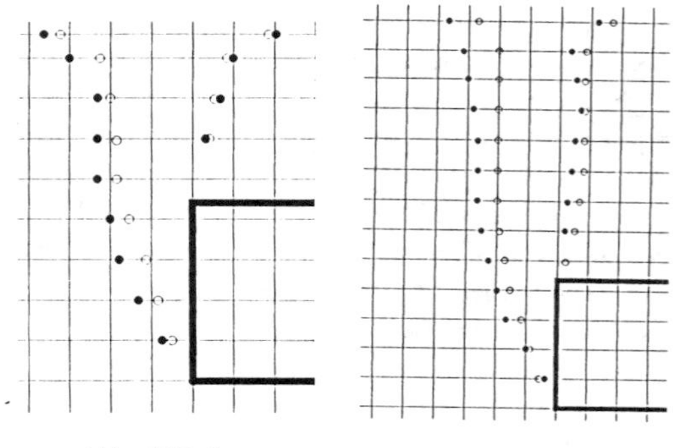

Figure 5 Static(○) and dynamic(●) slip lines measured

Figure 5 illustrates the difference in the position of the slip

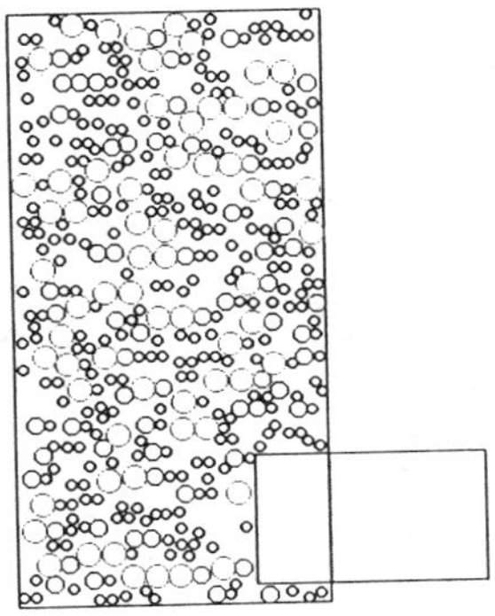

Figure 6 Elements randomly produced

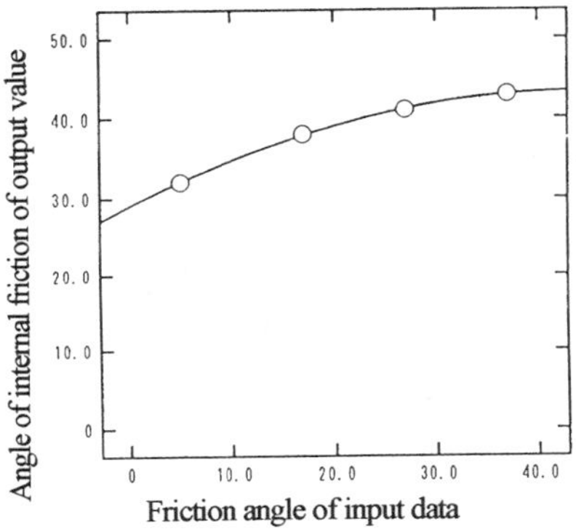

Figure 7 Relationship between friction angle of input data and angle of internal friction of output value (unit:degree)

lines between the static and the dynamic experiments when the retaining plates in both experiments move the same distance from the initial position. In the dynamic experiments, the slip line observed in the front of the retaining plate is situated farther away than that in the static experiments, and the difference in slip lines increases with the growth of the overburden.

3.NUMERICAL ANALYSIS OF DYNAMIC EXPERIMENTS

3.1 Production of the model ground

Three kinds of circular elements (418 elements) are produced randomly in the analytical region, as shown in Figure 6. The weight ratio and the radius ratio of the CEM elements correspond to three points (partial diameter = 0.38 mm, 0.56 mm, 0.80 mm) in the grain size accumulation curve

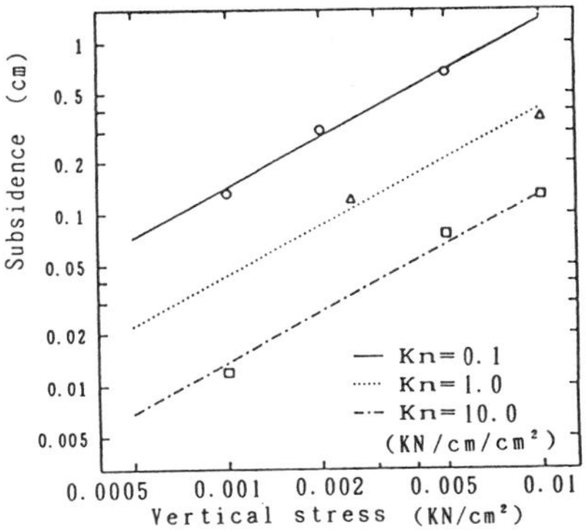

Figure 8 Relationship between subsidence and vertical stress

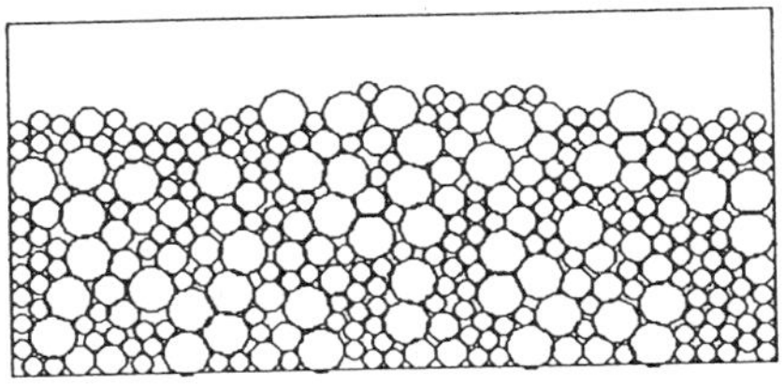

Figure 9 Analytical model

in Figure 2. By letting the elements fall, the initial ground is produced with an overburden 1.0D.

3.2 Determination of CEM parameters involved

The elements used in this CEM analysis are circular in two dimensions. In reality, however, the model ground is composed by grains with three dimensional geometry. Thus, it is impossible to directly apply the value of the angle of internal friction measured experimentally to the CEM analysis. In short, it is necessary in the CEM analysis to use the friction angle which corresponds to the angle of internal friction measured experimentally. Figure 7 is the relationship between the friction angle of the input data and the angle of internal friction of the output value, in the two dimensional compression analysis with CEM elements. By applying the angle of internal friction (38 degrees) measured experimentally to Figure 7, the input value of the friction angle is determined as 20 degrees.

The constant value of spring kn in a normal direction is determined as follows:

Figure 8 shows the relationship between subsidence and vertical stress when vertical stress is applied to the analytical model (Figure 9).

The incline of the three lines is almost equal, and these lines

Table 1 Values of Q and A to respective kn

kn(KN/cm/cm²)	Q	A
0.1	120	0.98
1.0	42	0.98
10.0	11	0.98

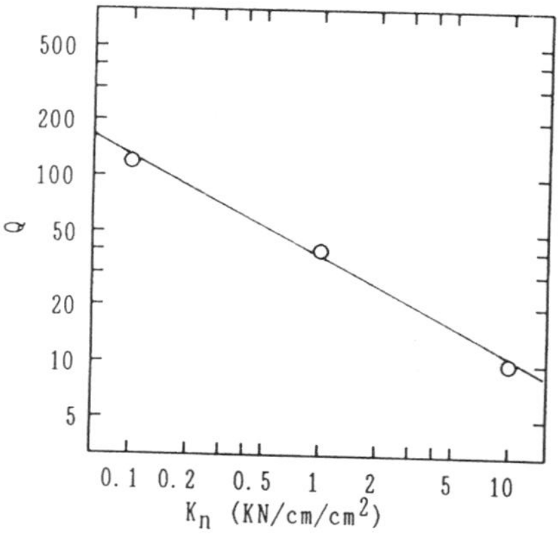

Figure 10 Relationship between Q and kn

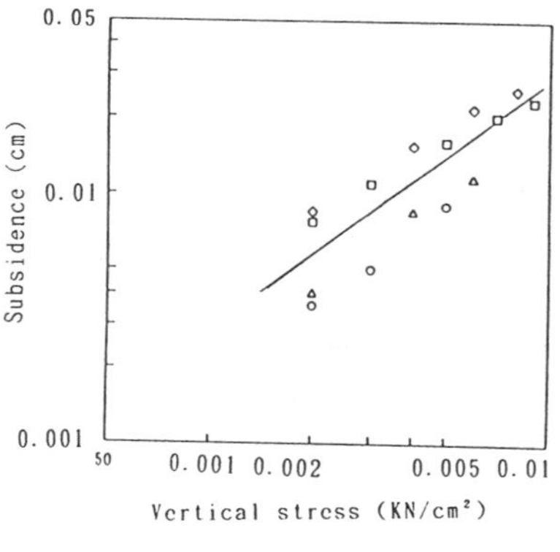

Figure 11 Experimental relationship between subsidence and vertical stress

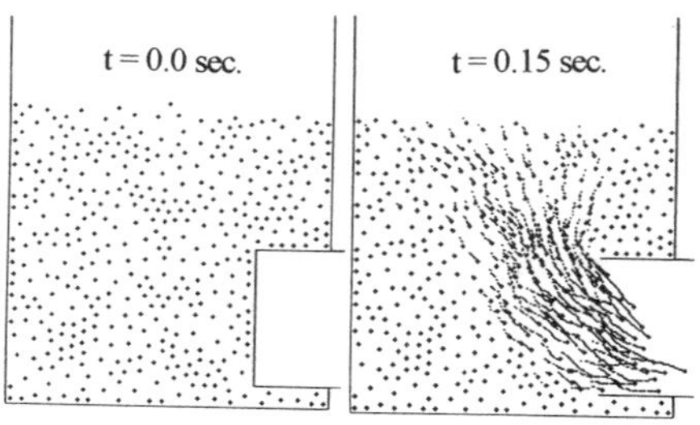

Figure 13 The loci of the element centers

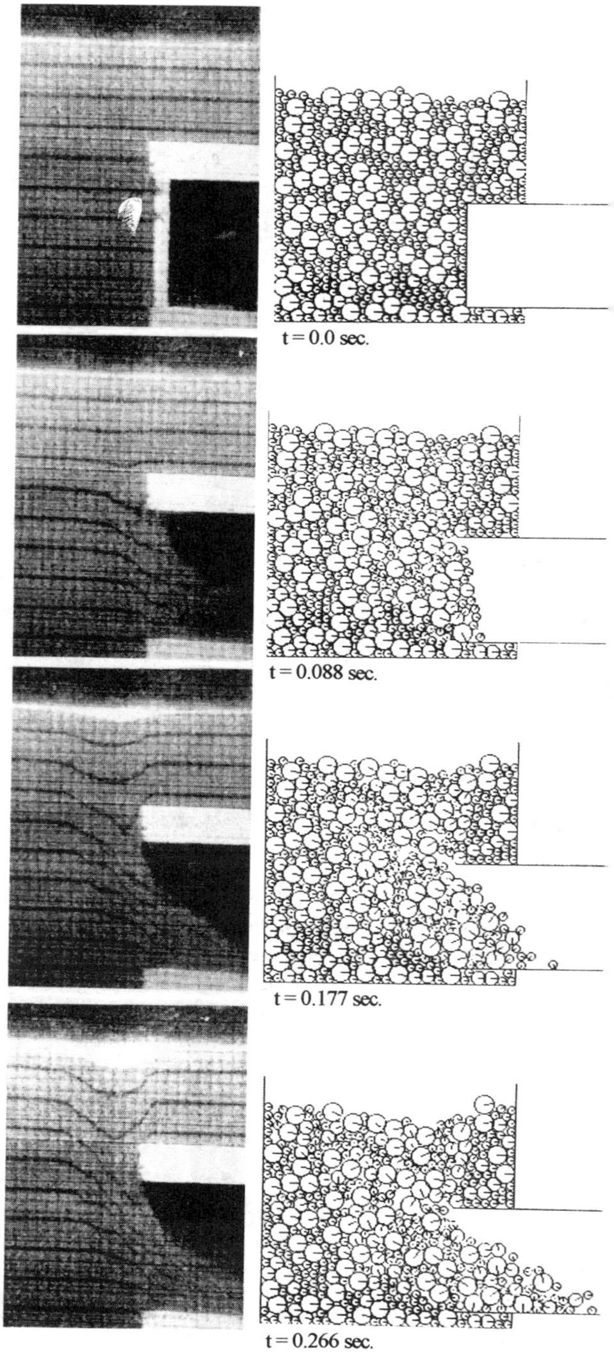

Figure 12 Comparison of experimental and analytical failure movements

can be expressed by the following equation,

$$V = QS^A \qquad (1)$$

where V(cm) is the subsidence, S(KN/cm^2) is the vertical stress, and Q and A are constants. The values for Q and A can be estimated to the respective kn values as shown in Table 1. From the results in Table 1, the relationship between Q and kn can be expressed in Figure 10, and this equation is

$$Q = 37kn^{0.51} \qquad (2)$$

The substitution of equation (2) into equation (1) leads to the equation

$$V = 37kn^{0.51}S^{0.98} \qquad (3)$$

On the other hand, the experimental relationship between subsidence and vertical stress is shown in Figure 11, and the straight line in Figure 11 is

$$V = 2.7S^{0.98} \qquad (4)$$

The analytical subsidence should coincide with the experimental one. Thus from equations (3) and (4), the kn value for the sand used in the experiment here is determined as

$$kn = 170 \ (KN/cm/cm^2) \qquad (5)$$

The constant value of spring ks in a tangential direction is determined as kn/4, because this value does not influence the element movements much. Values for the damping constant are also fixed as 0.5 by considering the results on slope stability analysis previously conducted [2].

3.3 Comparison of model test results with analytical ones

Figure 12 illustrates the experimental and analytical results at the time measured after the retaining plate is removed. The loci of the center of the CEM elements are shown in Figure 13. The experimental results at t = 0.088 sec are slightly affected by air passing through the two holes in the retaining plate. As the side wall of the tunnel lining cannot be taken into account in a two dimensional analysis, the failure region in the CEM analysis is wider than that in the experiments. Although there is some difference between the two, it is recognized that the movement of the sand in the experiment almost corresponds to that in the CEM analysis.

4. CONCLUSION

Static and dynamic mode experiments for a tunnel in a sandy ground are conducted with a three dimensional test device. The static movement of the ground is taken by photographs, and a dynamic one is caught by a high speed video camera. It is recognized that there is a big difference in results between the two experiments. In particular, the failure region in the dynamic experiment is wider than that in the static one, and this result should be considered in the construction of tunnels.

A CEM analysis is carried out for the dynamic experiments, and comparatively good correspondence can be found between the two, even though the failure region in the CEM analysis is wider than that in the experiment. This is because the side wall of a tunnel lining cannot be taken into account in a two dimensional analysis.

REFERRENCES

1)Hisatake, M. & Murakami, T.:Unified analysis of continuous and discontinuous behavior of the ground by CEM, Proc. Int. Symp. on Assessment and Prevention of Failure Phenomenon in Rock Eng., pp.915-920, 1993.

2)Hisatake, M. & Murakami,T.:An Analytical Method of Continuous and Discontinuous Behavior of the Ground, Proc. Japan Soci. of Civil Eng., No.523, pp.175-180, 1995(in Japanese).

3)Murakami, T., Hisatake, M. & Sakurai, S.: Modeling of the Ground by CEM and Determination of its Input Parameters, Proc. Japan Soci. of Civil Eng., No.529, pp.11-18, 1995(in Japanese).

4)Hisatake, M. & Murakami,T. & Eto, T.:Stability and failure mechanisms of a tunnel face with a shallow depth, Proc. 8th Int. Cong. on Rock Mech.., Vol.2, pp.581-591, 1995.

5)Hisatake, M.:Tunnel Face Behavior at Sandy Shallow Ground, Proc. Japan Soci. of Civil Eng., No.517, pp.105-115, 1995(in Japanese).

Numerical Models in Geomechanics, Pietruszczak & Pande (eds) © *1997 Balkema, Rotterdam, ISBN 90 5410 886 X*

Interface elements to model subgrade reactions for tunnel design

K.J. Bakker
Delft University of Technology, CUR/COB, Centre for Underground Construction & Rijkswaterstaat, Netherlands
H.J. Lengkeek
Delft University of Technology & Witteveen + Bos, Consultants, Netherlands
P.G. Bonnier
Plaxis B.V., Rhoon, Netherlands

ABSTRACT: A subgrade reaction model (SRM), ~~for the analysis of tunnel deformation behaviour~~ and stress analysis, was developed, applying techniques based on Finite Element modelling. The influence of tunnel depth (ratio) and direction of loading with respect to the vertical axis was examined, applying a slightly modified version of the superposition method, such as published by Sagaseta. Simple relations for the subgrade reaction modulus, as a function of depth ratio and inclination to the vertical axis where derived and compared to 2D FEM results. Finally, the bending moments calculated with FEM , with SRM, standard and modified where compared for a topology representative for a Dutch soft soil situation.

1 INTRODUCTION

For the design of the lining for bored tunnels, Subgrade Reaction Models (SRM) are often used. In the near future however, it is to be expected, that practice will "upgrade" to 2D Finite Elements (FEM). Hence there is a need to understand the relation between FEM results and SRM results. A comparison between SRM and FEM models was described by Ahrens et/al [1982]. Ahrens however maintains the choice proposed by Duddeck to model a stiffness free wedge on the upper 'quarter' of the tunnel. Such a choice can be understood for the analysis of the ultimate limit state design. For the analyis of deformations, for the evaluation of the serviceability limit state, such a concept is in our opinion not feasible.

Therefore a renewed version of a 'ring' SRM was made operational using the Element Library of "Plaxis" (1995). In this paper, a short description will be given how with Curved Mindlin Beam Elements, and Interface elements, a concept was derived feasible for the modelling of a double ring interactive model. A decisive parameter in the SRM is the subgrade reaction modulus and its distribution. In order to examine the influence of the depth (ratio h/R) of the centre of the tunnel, the direction with respect to the vertical, an analysis was made using a slightly modified version of the concept as published by Sagaseta (1987)

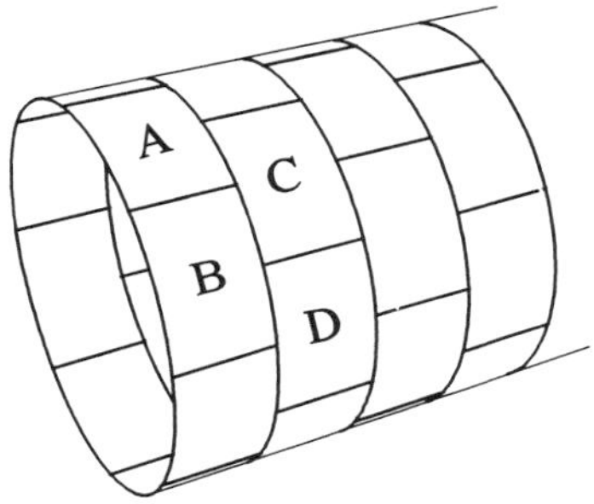

Figure 1 Segmented tunnel.

2 THEORY

2.1 *Subgrade reaction model with Mindlin beam- and interface-elements*

The usual modelling of the Duddeck solution consists of straight 2-noded bending elements for the tunnel lining in combination with springs to model the subgrade reactions. In case of segmented tunnels, a usual approach is to use two rings with a slightly different diameter and additional springs to couple the two rings. Here we use a different approach. Higher order curved beam elements with 3 or 5 nodes per element are used to model the lining(s) and 6- or 10-noded interface elements are

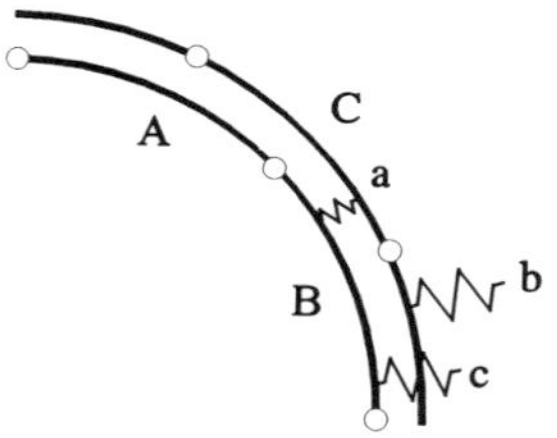

Figure 2 Finite element model of a segmented tunnel; the rings have the same coordinates.

used to model the subgrade reaction and the interaction between the linings. In Fig 1, a normal configuration of segments is shown for a bored tunnel. The tunnel consists of rings of segments in a brick configuration.

For the 2D-simulations we model 2 rings of elements. Each ring consists of a number of segments. Between these segments (i.e. "A" and "B") we use hinges, implying that no bending moments can be transferred. Between two consecutive rings there is interaction in the sense that the rings cannot move independently. To model this, interface, elements are used to limit the difference in the displacements (both radial and tangential). These interface elements are placed for instance between segments "A" and "C".

The finite element model looks like, see Fig 2. Where we again have the segments "A", "B" and "C" from the previous figure. As mentioned, there is interaction between for instance segments A and C, in the graph this is indicated by interface element "a". The interaction between the segments and the surrounding soil is modelled using inter face elements "b" and "c". Each segment consists

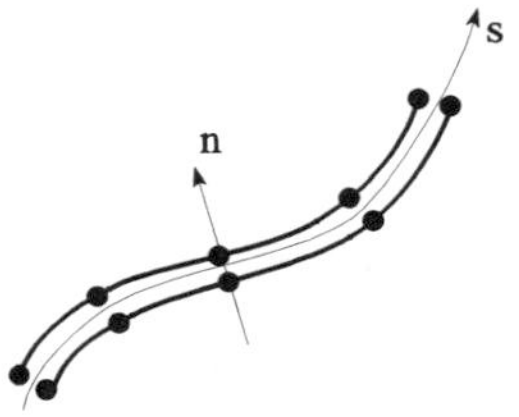

Figure 4 Ten-noded interface element

Figure 3 Five-noded beam element

of an even number of beam elements to ensure compatibility.

For the segments we used curved beam elements, see Bathe, (1982/1996), capable of describing normal, bending and shear deformations see Fig 3. For the interaction between segments "a" and the interaction with the surrounding soil we used zero-thickness-elements or interface elements, see Fig 3.

For the strains in the interface elements, see Fig 4, a virtual thickness, $l_{virtual}$, is used.
in combination with the, difference in displacements of opposite sides of the element, Δu.

$$\dot{\varepsilon}_n = \frac{\Delta \dot{u}_n}{l_{virtual}} \quad ; \quad \dot{\varepsilon}_s = \frac{\Delta \dot{u}_s}{l_{virtual}} \qquad \textbf{(1)}$$

For the calculation of the stresses we use equation (2).

$$\begin{Bmatrix} \dot{\sigma}_n \\ \dot{\sigma}_s \end{Bmatrix} = \begin{bmatrix} k_n & 0 \\ 0 & k_s \end{bmatrix} \begin{Bmatrix} \dot{\varepsilon}_n \\ \dot{\varepsilon}_s \end{Bmatrix} = \frac{1}{l_{virtual}} \begin{bmatrix} k_n & 0 \\ 0 & k_s \end{bmatrix} \begin{Bmatrix} \Delta \dot{u}_n \\ \Delta \dot{u}_s \end{Bmatrix} \qquad \textbf{(2)}$$

For the outer interface elements, we directly want to enter the soil stiffness. This can be reached by setting the virtual thickness to unity and fixing the displacements of the outer interface nodes. The soil stiffness can now directly be entered in k_n and k_s., where the suffix $_n$ stands for 'normal' and the suffix $_s$ stands for shear.

It should be noted that when assembling the global stiffness matrix for 2D-problems, we normally assume unit thickness in the third direction. In this case, for a unit thickness we have only half of the stiffness of the segments and half of the stiffness of interface elements "b" and "c" contributing to the total stiffness. Also when calculating equivalent nodal forces we should only use a weighting factor of one half. The interface element "a" should be treated specially because the stiffness of these elements should account for bending and load transfer in the direction perpendicular to the tunnel.

One way of calculating the values of the stiffness of interface "a" is to consider different mechanisms. For the normal stiffness one can consider the rigid rotation of two segments (i.e. "A" and "C"). Assuming that the length of a segment is L_s, see Fig 5., and assuming 3 or 5 notches with a stiffness K we can calculate the bending moment for a given rotation.

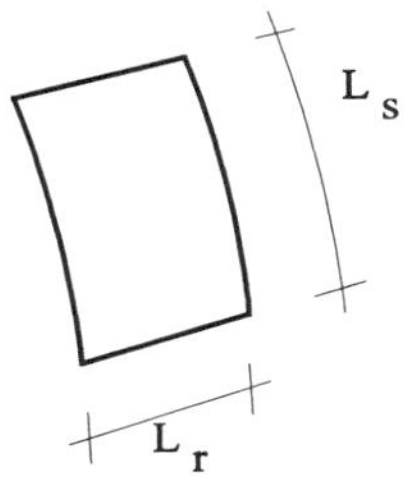

Fig. 5 Segment dimensions

$$M = \frac{1}{8} K L_s^2 \alpha \tag{3}$$

for an interface element with continuous stiffness k_n, it is found that

$$m = \frac{1}{48} k_n L_s^3 \alpha \tag{4}$$

which works over half the ring so

$$M = \frac{1}{48} k_n L_s^3 \alpha \frac{L_r}{2} =$$
$$= \frac{1}{96} k_n L_s^3 L_r \alpha = \frac{1}{8} K L_s^2 \alpha \tag{5}$$
$$k_n = \frac{12 K}{L_s L_r}$$

The shear stiffness can easily be calculated from sliding of two segments, averaging the stiffness of the notches and divide by half the depth of a segment.

$$k_s = \frac{2}{L_r} \frac{n K}{L_s} \tag{6}$$

where n is the number of notches on a segment side.

2.2 *Derivation of subgrade reaction modulus*

The value of 'spring' constants, such as needed for

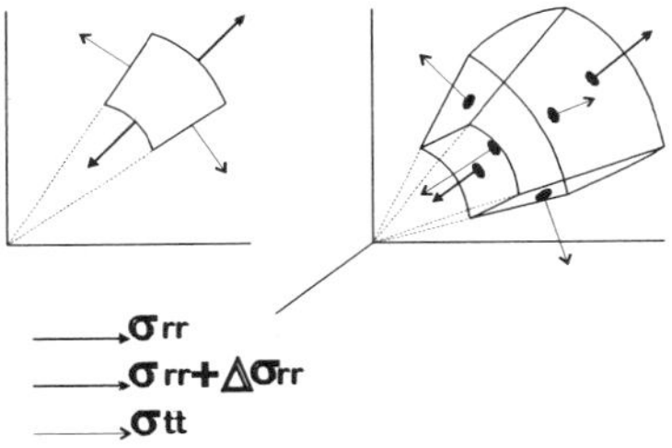

Figure 6 Stress divergence

the subgrade reaction models, can be determined either by analytical or numerical analysis. Analytical studies for tunnels such as published by Sagaseta (1987), and Ahrens et. al (1982).

A typical condition for the validation of these is volume loss due to the excavation. In most of the studies done thus far displacement-displacement or force-force study is done see Sagaseta (1987). The subgrade reaction of a cavity is determined for a finite linear elastic homogeneous medium. The basis of which is an assumption of stress divergence in space and integration of this in the radial directions, see Fig 6.

The divergence in space can be two dimensional modelled as a cylinder or three dimensional, applying a spherical model. Here in this paper only two dimensional divergence of stresses is considered. The radial Subgrade Reaction Modulus is calculated then acc. to:Where E^* is an Elasticity variable

$$k_r = \frac{\Delta \sigma}{\Delta u} = \frac{p}{\int_R^\infty (g[f(p)]) \cdot \partial r} = \frac{E^*}{L} \tag{7}$$

and L is a length.

Practically the deformation u, the integral over the strains is derived by assuming the stresses, as a function of the surface loading. The strains related with this are found assuming a material law, and the deformation is calculated, integrating the strains.

$$\bar{\sigma} = f(p) \qquad \varepsilon_r = g(\bar{\sigma}) \tag{8}$$

Where $\bar{\sigma}$ stands for the stresses, ε_r stands for the radial strain, and p for the loading at the inner surface of the cavity.

For a cavity expansion according to elasticity theory, the analytical solution, Timoshenko (1934), yields:

$$\sigma_r = -p \frac{R^2}{r^2} \qquad \sigma_t = p \frac{R^2}{r^2} \tag{9}$$

The relation between stress and strain used is according to:

$$\begin{Bmatrix} \epsilon_r \\ \epsilon_t \\ \epsilon_z \end{Bmatrix} = \frac{1}{E} \begin{bmatrix} 1 & -\nu & -\nu \\ -\nu & 1 & -\nu \\ -\nu & -\nu & 1 \end{bmatrix} \begin{Bmatrix} \sigma_r \\ \sigma_t \\ \sigma_z \end{Bmatrix} \qquad \textbf{(10)}$$

considering that $\epsilon_t = 0$, the subgrade reaction modulus k_r can than be calculated to yield:

$$k_r = \frac{E}{(1+\nu)R} = \frac{2G}{R} \qquad \textbf{(11)}$$

As can be seen from equation (11), for the length parameter the radius (R) of the tunnel is found, whereas, for the elasticity parameter of the ground 2G has to be taken. The aforementioned relation is only valid for the situation of a first order fourier loading; a contraction or an expansion. For the situation of a 2nd order fourier loading or deformation such as more characteristic for the ovalisation such as mostly calculated for tunnels, Ahrens describes the following relation:

$$k_r^{(2)} = \frac{3(1-\nu)}{(3-4\nu)} \cdot \frac{2G}{R} \qquad \textbf{(12)}$$

Sometimes 2G as in equation (11) is replaced by the Young's modulus E or by E_{oed}, the Elasticity parameter as derived by an Oedometer test. For practical purposes this is considered to be acceptable as the accuracy with which this parameter can be derived in practice, does not permit to be too strict.

In the model adopted here, the subgrade reaction modulus is determined based on a constant radial force condition according to the analytical solution. In a symmetrical plane strain situation a tunnel contraction (displacement condition) or radial force around the tunnel (force condition) have the same results.

Next to the radial SRM a tangential SRM has to be considered. In this paper this will not be given further attention, as here we want to focus in on the influence of depth and inclination on the SRM

The influence of the depth to radius ratio (d=H/R) can be determined using a similar derivation as applied by Sagaseta (1987), see Fig 7. The influence of the surface has been taken into account using a virtual tunnel. The original tunnel is contracted and the virtual tunnel is expanded because of a stress free surface. The shear forces are ignored because they are of no interest for the vertical equilibrium. This assumption is considered to be acceptable for urban areas where the surface

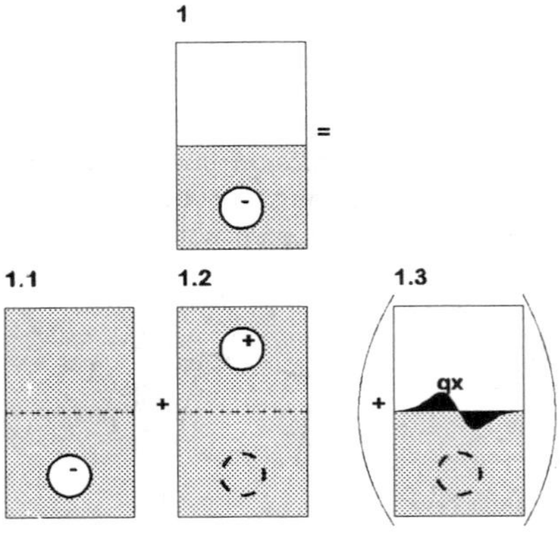

Figure 7 Tunnel in half space; 'superposition'

can be constrained in horizontal direction.

The modified subgrade reaction modulus can be rewritten as a function of the depth to radius ratio 'd' and its direction to the vertical axis θ see Fig 8, using a model-factor $C_r(d,\theta)$.

The modelfactor c_r is determined by summing up the influence (stress and displacement) of the original and virtual tunnel on the original tunnel.

By multiplying the normal subgrade reaction modulus with the modelfactor c_r the new Subgrade Reaction modulus is found.

$$c_r(d,\theta) = \frac{\left(1 - \dfrac{(\cos^2\theta - \sin^2\theta)}{(2d - \cos\theta)^2}\right)}{\left(1 + \dfrac{\cos\theta}{(2d - \cos\theta)}\right)} \qquad \textbf{(13)}$$

$$k_r(d,\theta) = c_r \cdot \frac{2G}{R} \qquad \textbf{(14)}$$

Depending on the depth to radius ratio the c_r is calculated at the top (θ=0°), at half the tunnel height (θ=90°) and at the tunnel bottom, (θ=180°).

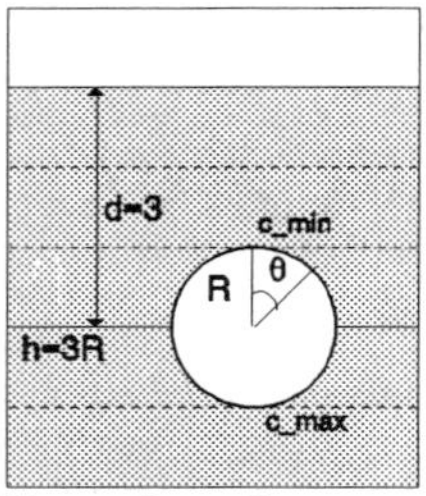

Figure 8 Geometry of the tunnel

Tabel I Subgrade reaction modulus, as a function of depth ratio

ratio	Modelfactor		
d=h/R	top $\theta = 0°$	side $\theta = 90°$	bottom $\theta = 180°$
2	0.667	1.063	1.200
3	0.800	1.028	1.143
4	0.857	1.016	1.111
5	0.889	1.010	1.091
6	0.909	1.007	1.077

2.2.1 *Comparison with Finite Element Analysis (PLAXIS)*

Equation (13) is subsequently compared to the soil stiffness as found using a EEM analysis with plaxis. For a tunnel geometry in a 2d elastic medium, a small contraction is imposed. The imposed contraction is modelled
as a volume loss of 1%.

The radial subgrade reaction modulus is taken 2G/R = 2500 kN/m^3. In table I the analytic results of the determination of the subgrade reaction moduli are presented. The results of the analytical and finite element analysis are quite similar as can be seen in table II. In case of undrained behaviour of the soil the SMR as calculated according to the Duddeck model; E_{oed} is taken instead of 2G which yields for the SRM:

$$K_r = \frac{2G(1-\nu)}{R(1-2\nu)} \quad \textbf{(15)}$$

which is not realistic, for high values of ν
For the numerical comparison the following input data has been chosen: Shear modules: G = 5000; Poisson's ratio: ν = 0, 0.25, 0.45; Radius of the tunnel: R = 4; Relative depth : d=3;

In the analysis with PLAXIS it was among other things found that deformation of the tunnel in a 'free contraction' is different to that for a 'fixed' contraction. As described by Verruijt (1997), a tunnel in an elastic half plane will produce a vertical translation which influences the subgrade reaction. Whereas a fixed contraction will not produce vertical equilibrium, a FEM analysis implicitly does. This fact cannot be neglected. To make the results comparable, the tunnel in the FEM analysis was fixed vertically.

For practical purposes it is further assumed that this vertical correction does not gives a significant difference in the bending moments. This aspect is a topic for further analysis. The results are assembled in Table II.

Table II Verification results

d=3	k_r [kN/m^3]	k_r FEM	k_r Dud-deck	k_r Analytical
ν=0	top	1935	0	2000
	middle	2480	2500	2570
	bottom	2500	2500	2857
ν=0.25	top	1945	0	2000
	middle	2495	3750	2570
	bottom	2545	3750	2857
ν=0.45	top	1915	0	2000
	middle	2580	27500	2570
	bottom	2710	27500	2857

In the next section the
derived relation for radial subgrade reaction modulus, k_r according to equation (15) is implemented in the model as described in section 2.1

3 VERIFICATION

The influence of the depth to radius ratio and inclination angle has been studied by implementing it into a computational model. One simple case has been chosen to compare the maximum moments according to FEM analysis, with results for a SRM, first conventional, and secondly with the modified relation K_r is calculated according to equation (13) and (14). The input data was: Shear mod.: G = 2500 kN/m^2; Poisson's ratio: ν = 0.35; Tunnel radius: R = 5 m; Lining thickness: t = 0.4 m; Relative depth: d=3; Phreatic surface is equal to the soil surface; Volumetric weight of the ground equals 15 kN/m^3; Soil friction angle 25° Material model: Mohr-Coulomb; Because of the water pressure and the elastic relaxation, the tunnel has a vertical heave of several centimetres.
To compensate for this effect, and considering that for ovalisation a stiffer reaction might be expected, see equation (12), the elasticity of the soil is taken equal to the Young's modulus "E" (6750) instead of "2G" (5000). The moments presented in table III are derived after 'excavation and dewatering' of the tunnel, and a contraction of 0.5% to model the volume loss.

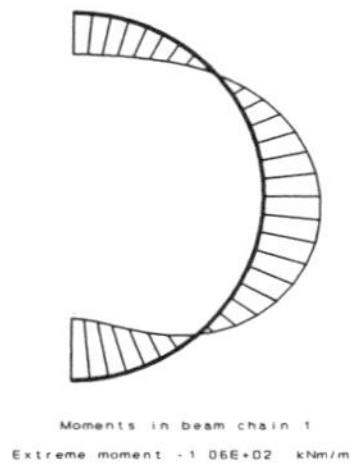

Fig 9 Verification results PLAXIS

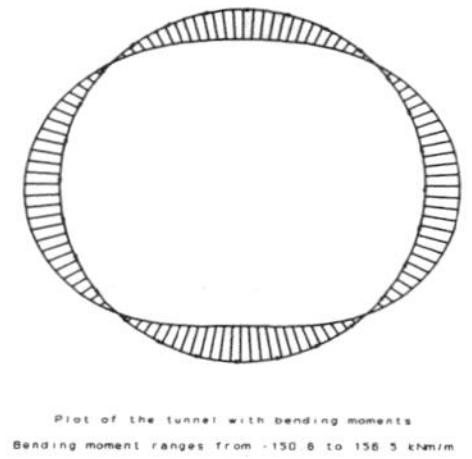

Fig 10 Verification results standard SRM

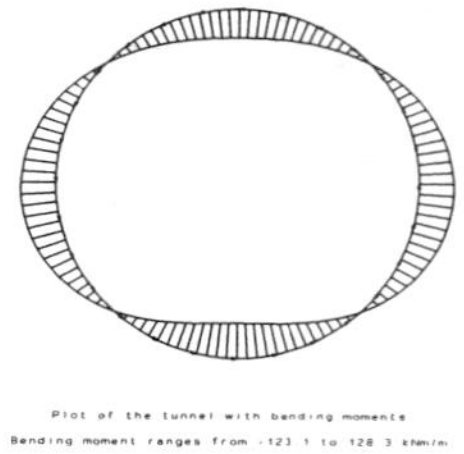

Fig 11 Verification results mod. SRM

Table III Verification results

method	FEM	SRM	SRM modified
moment [kNm]	106	157	128

The EEM results present the lowest value for the max. bending moment. The modification of k_r results in a max. bending moment which is better in agreement with EEM.

4 CONCLUDING REMARKS

One of the aspects observed as a side issue, is that the soil reaction related vertical equilibrium of the tunnel, as indicated in section 2.2, is significantly less stiff in comparison the integration of k_r and k_t resp. A factor of 4-10 less stiff than would be expected is found. This observation, although of less importance for the cross sectional stress distrubution, cannot be disregarded if the beam action of a tunnel becomes important. As discussed by Bakker (1997), it is considered that for soft soil conditions the beam action may not be disregarded.

As in Ahrens (1982) described, Duddeck state's that the SRM has to be disregarded in the upper angular segment (90°) of the tunnel. For the analysis of the Ultimate Limit State of a tunnel this can be considered a conservative assumption and thus a practical solution. With respect to the Service-Ability Limit State this assumption is less practical. For this situation further research on SRM's might be considered to deliver practical tools. On the other hand the limits to what might be derived with an improvement of SRM are limited. A rigorous coupling of beam action and ring action has to be doubted on the fact that basically there is too much difference in the mechanisms underlying the effective moduli. Therefore more detailed analysis with double ring models in the context of FEM analysis is considered to offer a better prospect.

REFERENCES

Ahrens, H., Lindner, E, and Lux, K.H (1982)
Zur dimensionierung von Tunnelausbauten nach den "Emphelungen zur Berechnug von Tunneln im Lockergestein (1980) Die Bautechnik 8 und 9 1982

Bakker, K.J., Berg, P van den, & Rots, J. (1997)
Monitoring soft soil tunnelling in the Netherlands; an inventory of design aspects. Proc XIV ICSMFE, Hamburg, sept '97 (to be published)

Sagaseta, C. (1987)
Analysis of undrained soil deformation due to ground loss, Geotechnique, 37, 301-320,

Timoshenko, S.P. and Goodier, J.N. (1934)
Theory of Elasticity
McGraw-Hill, 1934, 1970

Verruijt, A and Booker, J.R. (1996)
Surface settlements due to deformation of a tunnel in an elastic half plane
Geotechnique 46, No 4 753-756

Verruijt, A (1997)
A complex variable solution for a deforming circular tunnel in an elastic half plane
Int. J. Num. & Analy. Meth. in Geomechanics vol 21, 77-89

Vermeer, P.A. ed., 1995,
PLAXIS; Finite Element Code for Soil and Rock Analysis, Version 6 Balkema, Rotterdam

Numerical Models in Geomechanics, Pietruszczak & Pande (eds) © *1997 Balkema, Rotterdam, ISBN 90 5410 886 X*

Stress measurement in elastoplastic ductile geomaterials

M.Quiertant & J.F.Shao
Mechanics Laboratory of Lille, EUDIL, University of Lille, Villeneuve d'Ascq, France

C.Trentesaux
ANDRA, Châtenay Malabry, France

ABSTRACT : The overcoring technique is an experimental method used for the determination of *in situ* stresses. In this technique, the strains caused by a local stress relaxation are measured. In a classic approach, the recorded strains are then back-analysed using the elasticity theory to calculate natural stresses. However, experimental investigations conducted on geomaterials show an important inelastic behavior. So, the parameters of an extented Drucker Prager constitutive model were calibrated with data obtained on a marle, and a numerical process, including this elastoplastic behavior, was developed to simulate the experimental procedure of the overcoring. The elastoplastic solutions obtained were compared with those given by the theory of elasticity. Large differences were noticed. In order to take into account non linear constitutive behavior and surabondant experimental data, a probalistic approach, using the inverse problem theory, is then proposed to obtain a better interpretation of the stress measurement in non-linear media.

1 INTRODUCTION

In the evaluation of the stability of underground construction, the knowledge of the *in situ* stress is a primordial data. For this reason, *in situ* stress mesuring techniques are numerous (hydraulic fracturing, borehole slotting, overcoring, ...). A state-of-the-art in stress measuring techniques has been established by Leeman (Leeman 1971) and by Corthésy (Corthésy and all 1993).

In this paper, we give our interrest to the overcoring technique, performed with a C.S.I.R.O. cell (Commonwealth Scientific and Industrial Research Organisation) and more particularly to consequences of the usual assumption that rock exibits isotropic elastic material behavior during overcoring. Indeed, this hyphothesis simplifies the back-analysis of experimental results, but it is well known that most rocks have inelastic and/or anisotropic behaviors. The evaluation of the influence of anisotropy has been widely investiged by many authors (see for exemple Amadei 1983, Amadei and Goodman 1982, Rahn 1984 or Borsetto and all 1984) but the works concerning effects of the non-linear behavior of rock mass in the experimental results of stress measuring, as proposed by Corthésy (Corthésy and all 1993) are less numerous. The purpose of this paper is to present some numerical results showing the importance of the plastic deformation in the values given by strain gauges and to propose a numerical method that allows all kind of constitutive behaviors to be taken into account.

2 DESCRIPTION OF THE OVERCORING TECHNIQUE

The first step of the overcoring technique consists in drilling a preliminary hole at the chosen depth of measurement (Fig. 1.a), then a second hole, with a lower radius, is made in the prolongation of the same axis (Fig. 1.b). The strain cell, containing twelve strain gauges (C.S.I.R.O. cell), is installed in this second hole (Fig. 1.c) and is sticked on the borehole. The overcoring (Fig. 1.d) causes a stress relief and the borehole deformations induced are recorded.

This technique gives an over-determined system of twelve linear equations for six unknown values (all six components of stress tensor) usually resolved by use of the least squares method.

3 DESCRIPTION OF THE NUMERICAL PROCESS OF THE OVERCORING SIMULATION

The numerical method used in this work to simulate the overcoring is based on a finite element code (CESAR L.C.P.C.) modified to calculate the ortho-radial strain in the location of the experimental gauges, for an adapted mechanical behavior. The first step of a finite element modeling is to choose appropriate mesh and boundary condition. The purpose of this section is to present this modelling.

The rock mass is initially in equilibrium under the pre-existing stresses. Then the excavation (Fig. 2.a) and the overcoring (Fig. 2.b) induce two succesive stress reliefs on the two borehole surfaces (named a and b in Fig. 2).

We limit our study to the two dimentionnal analysis with plane strain condition being assumed in

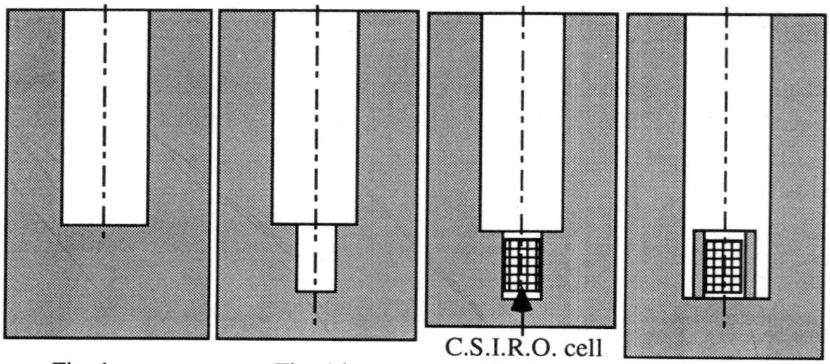

Figure 1. Different steps of an overcoring test.

the axial direction. In the case of an isotropic behavior, the symmetry condition of the problem allowed us to consider only one quater of the complet area (Fig. 2.c.). It should be emphasised that the presence of high strain gradients around the excavation necessitate an important refinement of the mesh.

The simulation of the excavation leads to make
the border of the excavated area free of normal stress. In a bidimentionnal finite element modeling, forces normal to the surface of excavation are calculated to restore the initial equilibrium. The gradual decreasing of these forces simulates the drilling process. Such a procedure is used for the first excavation (surface a Fig. 2) and for the overcoring (surface b Fig. 2) after removing elastic coeficients of the overcored part (clear part of the mesh Fig. 2.c).

The plane approach adopted in this work only allowed the calculation of the strains recorded perpendicular to the borehole axis (i.e. in the XY-plane). For this reason only four ortho-radial strain measurement gauges are taking into account. These gauges are usually called B, E, C and F and are assumed to be respectively oriented to 0°, 30°, 60° and 90° from the major principal stress of the XY-plane σ_H ($|\sigma_H| > |\sigma_h|$).

N.B. : In all this work, the principal stress orientation is assumed to be known , and we suppose that the vertical stress (σ_V) is a principal component of the stress tensor, equal for example to the overburden pressure.

In the hypothesis of an elastic behavior, analytical solutions were proposed by Kirsch (Kirsch 1898) for the problem of a circular excavation made in a medium under a known initial state of stress. Comparisons of the values of strain calculated by our code in the location of the differents gauges with the analytical solutions showed good agreements.

The authors are aware of the fact that a lot of experimental errors are produced during the stress measurement, for example it is well known that some thermal effects can influence the results (thermal expansion of the rock due to heating of the borer, variation in resistivity of the gauges caused by some thermal changes in the electrical system, difference of temperature of the drilling water, of the cell and of the rock mass ...). But we assume, for the sake of simplicity, and as a first approach that all these effects are negligible. For the same reason, no hydromechanical effects are taking into account.

An other origin of experimental errors comes from the effects of small scale heterogeneities and from heterogeneous geological conditions. The reason of the first type of problem is that only a relatively small volume of rock is involved (approx. 1 m^3) in the overcoring. The best results are provided by an overcoring made with a big diameter (252 mm), as proposed by ANDRA (See Bigarre 1993). For the second origin of errors, Wiles and Kaiser (Wiles and Kaiser 1990) have demonstrated that a minimum of ten different points of measurement are required to obtain a representative "averaged" value of the stress state (and more than ten for a jointed rock mass). For this reason, the probabilistic approach proposed in section 6, is based on the hypothesis of an over number of experimental results.

4 PRESENTATION OF THE ELASTO-PLASTIC MODEL

In order to accurately describe the mecanical behavior of rock during the overcoring test, a constitutive model proposed by Khan (Khan et al 1992 & 1993), was implanted in the numerical code. We briefly describe in this section the main feature of this model.

This model is based on the hypothesis of infinitesimal deformation, and the total deformation (ε) is assumed to consist of two parts, an elastic(ε^e) and a plastic one (ε^P) :

$$\varepsilon = \varepsilon^e + \varepsilon^P$$

The elastic deformation is assumed to be isotropic linear.

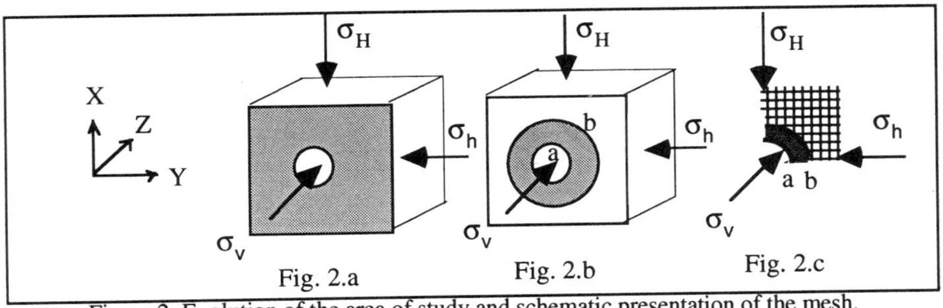

Figure 2. Evolution of the area of study and schematic presentation of the mesh.

The initial yield surface is represented by the function :

$$F(\sigma,\beta) = \sqrt{2J_2} - \beta\left(b_0 + b_1 I_1 + b_2 I_1^2 \right)$$

Where I_1 is the first invariant of the stress tensor,and J_2 second invariant of the stress deviator.

The growth of this surface is described by the hardening function :

$$\beta = \beta_{max} - \left(\beta_{max} - 1 \right) e^{-M\zeta}$$

$$\text{with } \zeta = \sum \left(d\varepsilon^{P} : d\varepsilon^{P} \right)^{1/2}$$

N.B. : $1 \le \beta \le \beta_{max}$

We make the hypothesis of an associated flow rule :

$$\overset{\circ}{\varepsilon}{}^{P} = \overset{\circ}{\lambda} \frac{\partial F}{\partial \sigma}$$

The parameters of this model were calibrated from data obtained in conventional triaxial tests performed on a marl in the Mechanics Laboratory of Lille. The values of the material constants are given below :

E=13 500 MPa, υ=0.18,
b_0=1.87, b_1=0.395, b_2=-0.0012 MPa^{-1},
M=730, $\beta_{max} = 2$

The model was used to simulate the behavior of marl under triaxial compression. An exemple of this simulation is given in Fig. 3.

N.B. : The sign convention of rock mechanics is used in this section (the compression stress is positive). However, in the following of this paper the convention of continuous media is chosen.

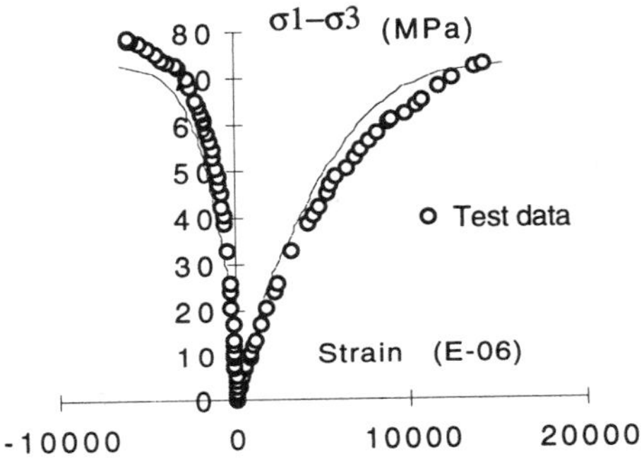

Figure 3. Comparison between predicted and observed stress-strain curves (confining pressure=-40 MPa).

5 INFLUENCE OF MATERIAL BEHAVIOR ON THE ORTHO-RADIAL STRAIN

A series of numerical simulations of the overcoring in the marl was performed for a wide range of stress states. In this simulations, ortho-radial strains were calculated at Gauss points corresponding to the positions of strain gauges on the cell. For each simulation we consider the mechanical behavior to be elastic in a first calculation and then elastoplastic for a second calculation. The elastoplastic solutions were compared with the elastic one and large differences were noticed.

A significant example is provided in Fig. 4. a, where λ represents the deconfining rate (λ is growing from 0 to 1 while stress removing is in progress during the first excavation, and from 1 to 2 during overcoring). Opposite algebraical signs of the evolution of strain (during overcoring) are obtained with the two different material behaviors. It should be emphasized that such results are obtained for gauges positioned where plastic strain occurs during the first excavation (See Fig. 4. a), but appreciable changes can also be noticed at gauges that are not affected by plasticity during the excavation (for the case of an anisotropic state of stress).

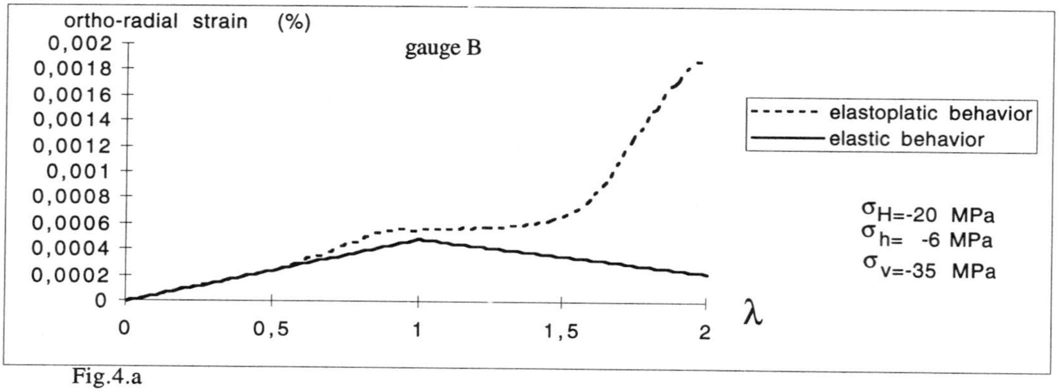

Fig.4.a

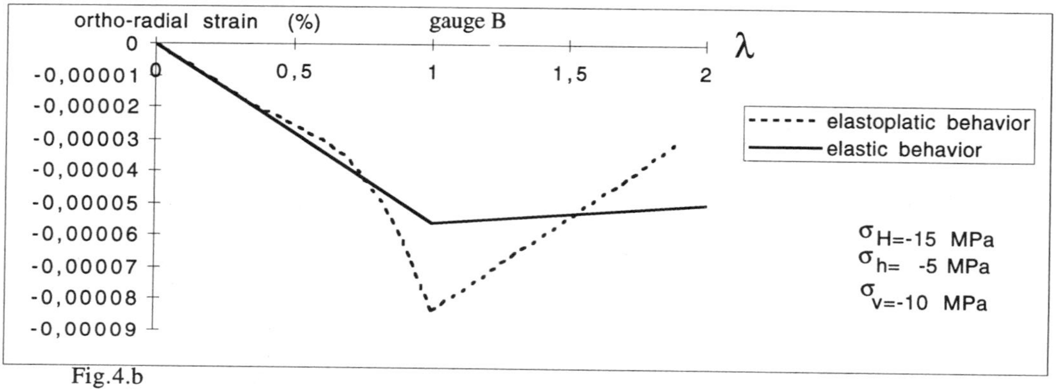

Fig.4.b

Fig. 4- Evolution of the strain during overcoring test .
($0 \geq \lambda \geq 1$ during the first excavation, $1 \geq \lambda \geq 2$ during overcoring)

In fact, the plastic behavior modifies all the stress distribution around the excavation and then it is the global response of the rock mass during the overcoring that is influenced by plascity (see Giraud et al 1996). Therefore, contrary to the classic approach neglecting the effects of plasticity during the first excavation, it is necessary to take these into account during the overcoring phase, even if these responses are linear. As it can be seen in Figure 4.b, the rock behavior can be completely elastic during stress relief, but can be strongly affected by the stress field produce by plasticity around the first hole.

Many authors propose different formulations expressing the relation between the regional stress field (here called *in situ* stress) and the local stresses caused by the excavation. For example in the case of an anisotropic behaviour, analytical solutions were published by Barly & Wane (Barly & Wane 1970), but for complex non-linear behavior no close form solutions exist and only numerical analyse can provide realistic evaluation of the stress field after excavation.

Since the stress field due to an excavation can not be determined by the linear elasticity, it is easy to understand that no accurate description of the stress relaxation can be provided in this hypothesis. Consequently, it is clear that the strain associated to the stress relief can be back-analysed only using an appropriate material behavior.

6 APPLICATION OF INVERSE PROBLEM THEORY

The problem consisting in modelling the overcoring test, knowing the initial *in situ* stresses, is considered here as a direct problem in the sense that, with an suitable numerical process, it is possible to calculate the strain caused by the stress remove.

Let $\vec{P}$ represents a set of parameters that completely characterize the direct problem. This parameters can be representative of the initial geometry of the first hole, of the mechanical behavior and of the initial state of stress. Resolving the direct problem is to find $\vec{R}$, a set of results (here the ortho-radial strains calculated at the location of the strain gauges).

The direct problem is described by the relation :

$$\vec{R} = F(\vec{P}) \qquad (1)$$

where F is an operator associating $\vec{P}$ to $\vec{R}$.

In fact the strain back-analysis procedure involves to sole :

$$\vec{P} = F^{-1}(\vec{R}) \qquad (2)$$

where only the initial state of stress is unknown in $\vec{P}$ (The geometry and plastic parameters are supposed to be known).

The solution of (2) consists in finding the set of parameters $\vec{P}$ such that :

$$F(\vec{P}) = \vec{R}_{num} = \vec{R}_{exp}.$$

where $\vec{R}_{num}$ are numerical solutions of (1), and $\vec{R}_{exp}$ is the vector of the strain values recorded in an overcoring test.

Due to the different possible errors, the solution is such that :

$$\vec{R}_{num} \approx \vec{R}_{exp}$$

or : $\| \vec{R}_{num} - \vec{R}_{exp} \| < e$

where e is the chosen precision.

Due to the experimental scatters, we consider $R_{exp}{}^{i}$, as the average value of strains recorded by gauge i in the different overcoring tests (in the hypothesis that several tests are performed), with which we associate the dispersion S_R^i. The recorded strains are then estimated to be random variables assumed to below to the Gaussian probability density (Tarantola 1987) which is denoted by :

$$\Pi 1 = A\exp[-\frac{1}{2}[{}^{t}(\vec{R}_{num} - \vec{R}_{exp}) \left[C_R\right]^{-1} (\vec{R}_{num} - \vec{R}_{exp})]]$$

A is a constant and $[C_R]$ is the covariance matrix calculated as :

$$[C_R]^{-1} = \begin{bmatrix} C_R^1 & 0 & . & 0 \\ 0 & C_R^2 & . & 0 \\ . & . & . & 0 \\ 0 & 0 & 0 & C_R^n \end{bmatrix}$$

with $C_R^i = \left(\frac{1}{S_R^i}\right)^2$ and where n is the number of different gauges.

The problem is then to find $\vec{P}$ maximising $\Pi 1$. The resolution of such a problem can provide different solutions. To be sure that the solution has got a physical meaning, we defined a new probability density $\Pi 2$ which ensures that $\vec{P}$ belongs to an a priori data space. For example, for the overcoring problem, we can assume that the stresses are strictly negative (compression), and not too "further" from a first estimation of the solution (called $\vec{P}_o$) obtained from a first rough analysis of the data of the problem. $\Pi 2$ is given by :

$$\Pi 2 = B \exp[-\frac{1}{2}[{}^{t}(\vec{P} - \vec{P}_o) \left[C_p\right]^{-1} (\vec{P} - \vec{P}_o)]]$$

where B is constant and $[C_p]$ is the covariance matrix of parameters calculated with the admissible dispertion of parameters (s_P^i).

Resolution of the inverse problem involve to find $\vec{P}$ maximising $\Pi 1$ in a physically admissible space (maximising $\Pi 2$). So, we define the joint (global) density function :

$$D(\vec{P}) = \Pi 1 \, \Pi 2 = A \, B \exp(-S)$$

With:`

$$S = \frac{1}{2}[{}^{t}(\vec{R}_{num} - \vec{R}_{exp}) \left[C_r\right]^{-1} (\vec{R}_{num} - \vec{R}_{exp}) + {}^{t}(\vec{P} - \vec{P}_o) \left[C_p\right]^{-1} (\vec{P} - \vec{P}_o)]$$

The minimisation of S gives the most possible *in situ* stresses from back-analysis of strain values recorded in overcoring test. The Gauss-Newton method is choosen to solve this minimisation problem. There is no warranty that the maximum likehood point is unique, or that a given point wich is locally minimum, is the absolute minimum. To be sure that the solution is not a local solution, we have to begin the iterative process from different points ($\vec{P}_o$).

7 PRESENTATION OF THE CODE MIDAS

In order to adapte the inverse problem method to overcoring problem, a numerical code called M.I.D.A.S. was developed.To valid this code, the direct problem was first solved for a choosen state of stress, and the values of strain calculated for the different gauges were recorded. Then, this values were injected and back-analysed by MIDAS. Two types of validation were made, a first series with elastic behavior in the direct and inverse calculation (see Table 1), and a second with the elasto-plastic behavior (see Table 2.). Good agreement between real solutions and those obtained by MIDAS was found.

For the case where the direct problem was solved with an elasto-plastic behaviour, we tried an back analysis in the hypothesis of an elastic behavior, like it is usually done in the conventional technique (with the used of the least squares method). This comparison allows to show good performance of our approach that provide an average percentage of errors of 28% when the classical back-analysis gives 49% of errors (see Table 2).

8 CONCLUSION :

In the design of underground structures, *in situ* stresses are an important parameter generally difficult to be accurately evaluated. The back analysis usually using in the hypothesis of elasticity is not applicable when the rock mass exibits a non-linear behavior.

We have proposed here a simple approach allowing an accurate stress calculation taking into account an appropriate behavior of the rock, the over number of test results and the scatters of experimental investigations. This approach is general and can be conducted with different non-linear behaviors of a wide class of rocks.

Table 1- Validation of MIDAS for the elastic behavior

Initial solution $\vec{P}_o$	Numerical Solutions	Exact Solutions
σ_H=-30 MPa	σ_H=-39,99 MPa	σ_H=-40 MPa
σ_h=-30 MPa	σ_h=-11,99 MPa	σ_h=-12 MPa

Table 2- Comparison of the back-analysis made by MIDAS and by the classic method

Exact Solutions	Solutions of MIDAS	Conventional back-analysis
Inversion with $\vec{P}_o$: (σ_H=-8 MPa, σ_h=-8 MPa)		
σ_H=-40 MPa	σ_H=-45.7 MPa	σ_H=-48.5 MPa
σ_h=-12 MPa	σ_h=-16.8 MPa	σ_h=-21.2 MPa
Inversion with $\vec{P}_o$: (σ_H=-20 MPa, σ_h=-8 MPa)		
σ_H=-40 MPa	σ_H= -46.3 MPa	σ_H=-48.5 MPa
σ_h=-12 MPa	σ_h= -17 MPa	σ_h=-21.2 MPa

REFERENCES

Amadei, B. 1983. Rock anisotropiy and the theory of stress measurements. Lecture Notes in Engineering Series. Springer, New York.

Amadei, B., and Goodman, R. E. 1982. The influence of anisotropy on stress measurements by overcorring techniques. Proc. 23rd US Symp. Rock Mech., Berkeley, pp.157-167 .

Barly, G. and Wane M. T. 1970. Stress relief method in anisotropic rocks by means of gauges applied to the end of a borehole. Int. J. Rock Mech. Min. Sci. 7, 171-182.

Bigarre, P. 1993. Mesures de contraintes dans les marnes par surcarottage en gros diamètre, Rapport ANDRA 621 RP INE 93 002.

Borsetto, M., Martinetti, S. and Ribacchi, R.1984. Interpretation of in situ stress measurements in anisotropic rocks with the doorstopper method., Rock Mech. Rock Engng 17, 167-182.

Corthésy R., Gill D. E. and Leite M. H. 1993. An integrate approach to rock stress measurement in non linear elastic anisotropic rocks. Int. J. Rock Mech. Min. Sci. & Geomech. Abstr. 30, 395-411.

Corthésy R., Gill D. E. and Ouellet J. 1993. Méthodes de mesures de contraintes dans les massifs de roches dures, P. 101. Centre de recherches minérales, Québec.

Giraud, A., Shao, J.F., Homand, F. 1996. Interprétation des mesures de contraintes par surcarottage en milieu élastoplastique, Sensibilité aux chemins de chargement, à la plasticité et aux contraintes anisotrope , Rapport A.N.D.R.A. (identification: B RP O ENG 96-010/A).

Khan, A.S., Xiang, Y. and Huang, S., 1992. Behavior of berea sandstone under confining pressure part I : : Yield and Failure Surfaces, and non-linear Elastique Response, Int. J. Plast., Vol 7, pp. 607.

Khan, A.S., Xiang, Y. and Huang, S., 1992. Behavior of berea sandstone under confining pressure part II : Elastic-plastic Response, Int. J. Plast., Vol 8, pp. 209.

Kirsch, G., 1898. Die theorie der elastizitat und die bedurfnisse der fertigkeitslhre , Veit. Ver. Deut. Ing., Vol 28, 797.

Leeman E. R. 1971. The measurement of stress in rock : A review of recent developments (and bibliography). Proc. Int. Symp. on the Determination of Stresses in Rock Masses, Lisbon, pp. 200-229.

Rahn W. 1984. Stress concentration factors for the interpretation of doorstopper stress measurements in anisotopic rocks. Int. J. Rock Mech. Min. Sci. & Geomech. Abstr. 21, pp. 313-326.

Tarantola, A., 1987. Inverse problem theory. Methodes for Data Fitting and Model Parameter Estimation, Elsevier.

Wiles, T. D. and Kaiser P. K. 1990. A new approach for statistical treatment of stress tensors. Special conference on Stress, Ottawa, Ontario.

6.2 Piles and foundations

Numerical Models in Geomechanics, Pietruszczak & Pande (eds) © 1997 Balkema, Rotterdam, ISBN 90 5410 886 X

Reliability analysis of rigid piles subjected to lateral loads

W. Puła
Institute of Geotechnics and Hydrotechnics, The Technical University of Wrocław, Poland

ABSTRACT: Computations of rigid piles by Brinch Hansen method demonstrates high sensitivity of ultimate lateral loading to precise determination of the rotation centre of the pile under consideration. The position of the centre is affected by some random factors, for example random variability of soil properties and loading applied. To investigate effects of such a variability a reliability approach was suggested. Final results show a large effect of the friction angle random fluctuations on the variation of the ultimate lateral loading. The necessity of the use of high values of total safety factors when the Brinch Hansen approach is applied was demonstrated.

1. INTRODUCTION

Due to modernisation of main railway lines in Poland there is a need to change the existing overhead electrical transmission lines together with their supports and foundations of the supports. One of possible ways to construct a new foundation of such support is the direct connection of the support with a single pre-cast concrete pile embedded in soil. This kind of foundations was successfully applied in some important railway lines in Europe. A special equipment allows to install these piles in a fast way, which is an important advantage (idle time reduction) in comparison with traditional block (massive) foundations. In the design of such pile foundations lateral forces and moments are of the vital importance. The piles used are usually short, then in certain soil conditions have to be treated as rigid piles and the ultimate soil resistance must be considered. For rigid piles one of the most precise methods to evaluate ultimate lateral resistance is the procedure recommended by Brinch Hansen (1961). In this procedure, the centre of rotation of a rigid pile has to be found. In can be demonstrated that the value of the ultimate horizontal loading H_u, which can be applied in the head of the pile is extremely sensitive to a position of the rotation centre z_r (Figure 1). Then the accuracy of determining the rotation centre is very important for a precise computation of an allowable horizontal force H_a. On the other hand the position of the centre is affected by some random factors. Among them the friction angle of the subsoil as well as the uncertainty of geotechnical recognition (the number of borehols along the railway track can be not sufficient enough) are of the prime importance. Then a vital problem is to investigate for such a value of the total safety factor which can guarantee a small probability that the applied load H_a exceeds the ultimate loading H_u. This problem is a typical one within the framework of the structural reliability theory. However, due to the nature of the solution of the equilibrium equations, which could not be written in a closed analytical form, the existing structural reliability procedures could not be applied straightforward. The solution of the reliability problem will be demonstrated and discussed in the section 4 of this paper.

2. BASIC ASSUMPTIONS

The following general assumptions are imposed (see Figure 1):

- A short rigid pile is considered with an unrestrained head. Any elastic deformations within the pile material are neglected.
- In the state of failure the pile is assumed to rotate as a rigid body about a rotation centre at the depth z_r.
- The pile is subjected to a horizontal force H_a and the ultimate soil pressure at any depth z below the soil surface is p_u.

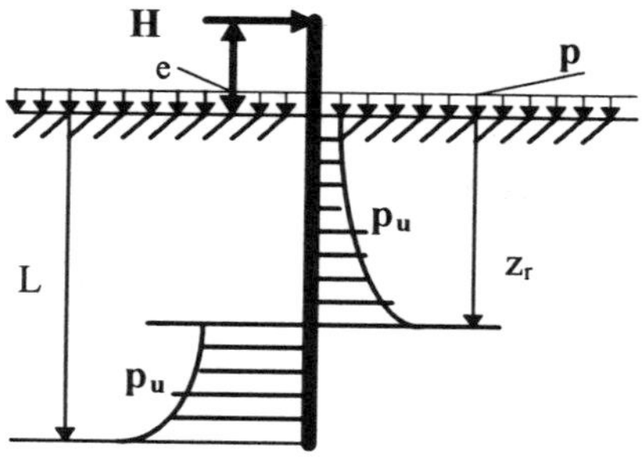

Figure 1: Laterally-loaded rigid pile

• The limiting values H_u and M_u , to cause failure - that is, to mobilise the ultimate soil resistance along the pile, may be obtained by considering equilibrium of horizontal forces and moments, and solving the following resulting simultaneous equations for the unknown depth of rotation z_r , and the ultimate horizontal load H_u :

$$H_u = \int_0^{z_r} p_u(z)Ddz \quad - \quad \int_{z_r}^{L} p_u(z)Ddz \ , \tag{1}$$

$$M_u = H_u e = - \int_0^{z_r} p_u(z)Dz\,dz + \int_{z_r}^{L} p_u(z)Dz\,dz \tag{2}$$

where L is the embedding of the pile in soil, D is its diameter or width and e is the eccentricity of loading.

According to the Brinch Hansen approach (Brinch Hansen 1961) the resultant (passive minus active) ultimate lateral soil resistance (per unit area), on an arbitrary depth z can be expressed by the following equation:

$$p(z) \ = \ q(z)K_q(z) + cK_c(z) , \tag{3}$$

where $q(z)$ is the effective vertical overburden pressure at the depth z

$$q(z) = p + \gamma z_d + \gamma' z_s \quad ; \quad z = z_d + z_s \tag{4}$$

γ, γ' are unit weights above ground water table and below, respectively and c is the cohesion of the soil. The coefficients $K_q(z)$, $K_c(z)$ are the pressure coefficients which depend on the friction angle φ of the soil, diameter of the pile D and the depth z. They can be obtained by means of non-linear interpolation formulae, where solutions for the ground level and solutions at great depth, based essentially on limit state and earth-pressure theory, are utilised. The dependencies of coefficients K_q and K_c on friction angle φ are very complex. However Brinch Hansen has found them in a closed analytical form (Brinch Hansen, 1961).

As a measure of safety a „total" safety factor is considered, which is defined as the ratio of the ultimate lateral force H_u and the applied lateral force H_a

$$F = \frac{H_u}{H_a} \ . \tag{5}$$

3. SOME RELIABILITY CONCEPTS

Reliability problems are usually described by the so called limit state function $g(\mathbf{v})$. The argument $\mathbf{v}$ of the function g is a random vector $\mathbf{V} = (V_1, V_2, \dots, V_n)$ consisting of basic random variables defining loads, material properties, geometrical quantities, *etc*. The function $g(\mathbf{v})$ is formulated in such a way that $g(\mathbf{v}) > 0$ defines the safe state of a structure, whereas the case of $g(\mathbf{v}) < 0$ stands for failure. The hypersurface $g(\mathbf{v}) = 0$ is called the limit state surface. For settlement analysis, for example the function g can be written as follows:

$$g(\mathbf{v}) \ = \ u_{\max} \ - \ U(V_1, V_2, \dots, V_n) \ , \tag{6}$$

where $u_{\max}$ stands for a maximal allowable settlement (selected arbitrarily), U is the settlement of a given point affected by a set of random parameters $V_1, \dots V_n$ like subsoil model parameters, loads, geometrical properties, *etc*. In this case the failure ($g(\mathbf{v}) < 0$) means exceeding of the given threshold $u_{\max}$ by random settlement at a selected point (Brząkała and Puła 1996).

As a measure of risk the probability of failure is used

$$p_F \ = \ \int_{\{g(\mathbf{v})<0\}} f_{\mathbf{V}}(\mathbf{v})d\mathbf{v} \, . \tag{7}$$

Here $f_{\mathbf{v}}$ denotes a mutidimensional joint probability density function (p.d.f.) of the random vector $\mathbf{V}$.

Among the methods of an assessment of the probability (7) the FORM (first-order reliability method) and the SORM (second-order reliability method) are most commonly in use (Rackwitz and Fissler 1978; Hohenbichler et al 1987).

To evaluate the probability p_F, it is convenient to transform the variables **V** into the standard normal space:

$$\mathbf{V} = \mathbf{Y}(\mathbf{V}). \tag{8}$$

Such a transformation is known as the probability transformation or Rosenblatt transformation (Rosenblatt 1952). The mapping of the limit state surface onto the standard normal space can be described by

$$g(\mathbf{v}(\mathbf{y})) \equiv G(\mathbf{y}) = 0. \tag{9}$$

The probability of failure then equals

$$p_F = \int_{\{G(\mathbf{y})<0\}} \phi(\mathbf{y})d\mathbf{y}\ , \tag{10}$$

where ϕ is the standard normal p.d.f. of **Y**. Note that the effective analytical evaluation of the integral (10) is very toilsome or even impossible, unless G is a linear function (multidimensional hyperplane).

Two important properties of the multidimensional standard normal distribution are the reasons for the use of probability transformation of the type (8). The primary one is the rotational symmetry of the standard normal p.d.f. about the origin axes. Furthermore, the standard normal density ϕ exponentially decreases with the square of the distance from the origin. Hence, most of the contribution to p_F in equation (10) comes from the neighbourhood of the point on the limit state surface which is nearest to the origin.

In the FORM approximation (see Figure 2), the limit state surface in the standard normal space is replaced with the tangent hyperplane $\nabla G(\mathbf{y} - \mathbf{y}^*) = 0$ at the point $\mathbf{y}^*$ with the minimum distance from the origin (the so called design point) and the probability of failure is approximated as

$$p_F \approx \int_{\{\nabla G(\mathbf{y}-\mathbf{y}^*)<0\}} \phi(\mathbf{y})d\mathbf{y} = \Phi(-\beta)\ , \tag{11}$$

where β, called the reliability index, is the minimum distance from the origin and Φ denotes the one-dimensional standard normal cumulative probability function. In the SORM approximation, the limit state surface is fitted with a quadratic surface in the vicinity of the design point $\mathbf{y}^*$ and the right-hand side of equation (11) is multiplied by a certain correction factor (Breitung 1984). The most important problem in the

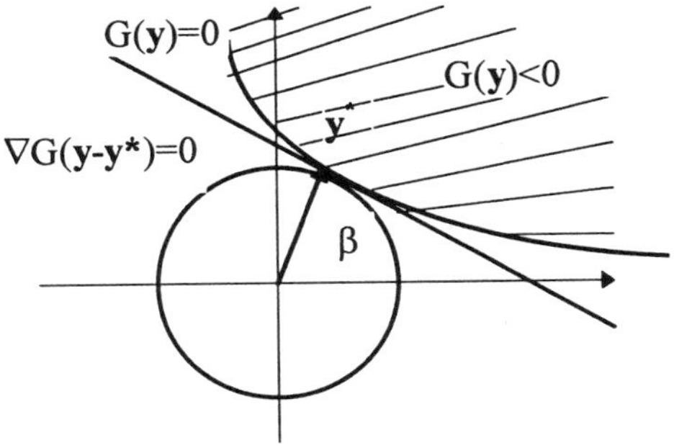

Figure 2: Schematic presentation of the FORM approximation

FORM and the SORM lies in finding the minimum-distance point $\mathbf{y}^*$, i.e. the design point. The problem can be formulated in terms of a constrained optimisation as follows.

$$\textit{mininimize}\ \|\mathbf{y}\|,\quad \textit{subject}\ G(\mathbf{y}) = 0, \tag{12}$$

where $\|\,.\,\|$ denotes the Euclidean norm. An important feature of the FORM and SORM methods is the necessity of the explicit formulation of the limit state function $g(\mathbf{v})$ in a closed analytical form, which can be easily seen from equations (7) - (11). Furthermore complete probability information about random vector **V** in the form of probability density function (or cumulative probability function) must be known. Otherwise some simpler reliability measures and methods have to be applied (Ditlevsen 1988).

4. RELIABILITY ASSESSMENT IN THE CASE OF RIGID PILE

Applying the Brinch Hansen method in ultimate lateral resistance estimation it can be easily observed a high sensitivity of the ultimate loading value H_u to a position of pile's rotation centre z_r. Even small fluctuations in the position of the centre z_r can produce large variation in the value of the ultimate lateral force H_u (due to strongly non-linear dependencies of rotation centre on friction angle). An example of such a phenomenon is presented in the Table 1.

This imposes that the coordinate z_r must be determined with high precision by solving the equation (2). Of course it can be done with the help of a computer. On the other hand it is well known that random fluctuations of soil properties in natural deposits are usually significant and very important in the context of engineering computations (Biernatow-

Table 1: Changes of the ultimate lateral loading H_u due to friction angle variability

Angle of internal friction φ [rad] ; [°]	Position of the rotation centre z_r [m]	Ultimate lateral loading H_u [kN]
0.46 ; 26.35	2.106	13.30
0.48 ; 27.52	2.110	14.61
0.50 ; 28.63	2.113	16.07
0.52 ; 29.78	2.117	17.67
0.54 ; 30.93	2.121	19.45
0.56 ; 32.10	2.125	21.43
0.58 ; 33.22	2.129	23.62

ski and Puła, 1988).

If some soil properties are a subject of random variability then a natural question arises how reliable the total safety factor (eqn. (5)) is. Appropriate reliability problem can be formulated as follows:

Find the probability p_F that the applied loading H_a exceeds the ultimate lateral loading H_u :

$$p_F = P\{H_a > H_u\} = P\left\{\frac{H_u}{H_a} < 1\right\} = P\{F < 1\} \quad (13)$$

(both values H_u and H_a are positive).

The probability p_F in (13) can be understood as a probability of failure as considered in the previous section (eqn. (10)). As it has been already mentioned, to evaluate probability (13) by means of the methods described in section 3, the explicit dependence of the value H_u on parameters assumed as random (for example, friction angle, cohesion, unit weight) must be known. The dependencies of H_u on soil properties are given by eqns. (1) - (4) and by formulae for K_q and K_c (Brinch Hansen 1961). It can be proved that a closed formula for these dependencies can not be obtained. To overcome some difficulties the *MATEMATICA* system can be applied which is able to perform some symbolic mathematical transformations. With the help of this system the equation for the rotation centre z_r can be transformed (under dependencies suggested by Brinch Hansen) to the following non- integral form:

$$a_3 z_r^3 + a_2 z_r^2 + a_1 z_r + a_0 + \ln(b + h z_r) = 0, \quad (14)$$

where z_r is the unknown coordinate of the rotation centre and $a_0,\ldots,a_3,b,h$ are functions of parameters of the problem (both random and deterministic) but independent of z_r . To get a closed formula for z_r and consequently for H_u, the logarithmic term in eqn. (14) was expanded into a power series usually up to the term of degree three. The resulting formula is very complex and long, however all transformations are performed automatically by *MATEMATICA* in symbolic way. The resulting equation can be transferred to a code executing reliability computations according to FORM and SORM procedures. This way the probability (13) could be evaluated. Finally we are able to search for such a value of the factor F (eqn. (5)), which guarantees a small probability that the applied load H_a exceeds the ultimate loading H_u .

5. NUMERICAL EXAMPLE

Consider a pile subjected to lateral loading as in Figure 3. Some parameters of the problem under consideration are collected in the Table 2. The computation were carried out in the following way:

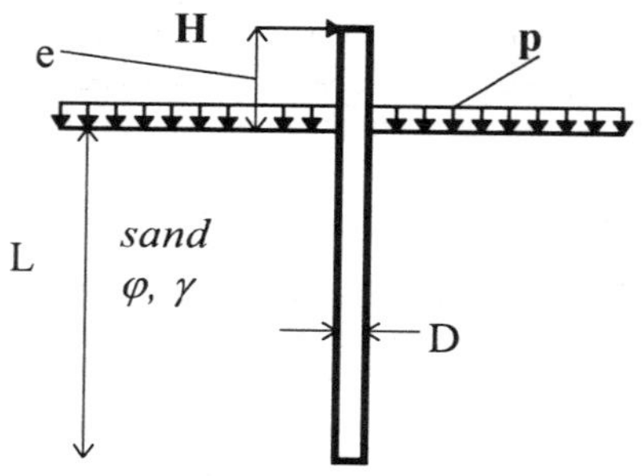

Figure 3: Rigid pile analysed in the example

1. Select a value of the safety factor F.
2. For fixed value of F and keeping values of all random variables as constant and equal to their expected values, compute the value of the allowable lateral loading H_a corresponding to selected safety factor F.
3. Treating the allowable loading H_a as a random variable of mean value computed in 2 and considering other random parameters (according to Table 2) compute the probability of failure p_F in the form of (13) and the corresponding reliability index β.
4. Repeat computations for anther value of F.

Table 2: Parameters of the problem

Parameter	Probability distribution	Expected value	C.O.V. (%) *
Friction angle φ	Lognormal	33.6 °	10; 15
Applied load H_a	Normal	8 - 25 kN	10; 15; 20
Surface overburden p	Normal	8.8 kNm^{-2}	5
Eccentricity e	Normal	8.64 m	5
Unit weight γ	Constant	20.15 kNm^{-3}	
Pile's length L	Constant	2.9 m	
Pile's diameter D	Constant	0.36 m	

*C.O.V. = coefficient of variation =
= (standard deviation / expected value) 100%

The reliability computations were carried out with the help of *STRUREL* system. Some results are summarised in Figure 4.

In the references (Ditlevsen and Madsen 1996) some safety classification and requirements concerning values of reliability index β are available. But it is commonly believed that for a safe structure its reliability index β has to be at least 3. It is easy to observe (Figure 4) that in the case of the friction angle variation coefficient of $v_\phi = 15\%$ the total safety factor F must be assumed at least 3 to guarantee the reliability index β on the level 3. For the value of $v_\phi = 10\%$, total safety index of a range 2.1-2.4 can be satisfactory. However it is worth mentioning that values less than 15% of v_ϕ are rather rarely observed in laboratory testing. The variability of friction angle in natural soil deposits is usually grater than 15% (≈ 25% for cohesive soils). This shows the necessity of the use of high values of total safety factors F when the Brinch Hansen method is applied in engineering computations.

In order to evaluate effects of variability of other quantities some parameters studies were carried out. The friction angle φ was kept as random variable as shown in the Table 2. Example of such a parameter study is presented in the Figure 5. It can be observed non-linear and rapid changes of the reliability index β as a function of applied lateral force H (Fig. 5A.) as well as a function of eccentricity e (Fig. 5 C.).

6. CONCLUDING COMMENTS

Random fluctuations of soil properties can cause significant changes in the value of ultimate lateral loading determined according to the Brinch Hansen method. Then reliability analysis can be an important tool indicating a save way in selecting parameters for design. The difficulties in applying commonly used structural reliability methods, due to complex nature of solution of equilibrium equations, can be overcome by applying the symbolic computations by

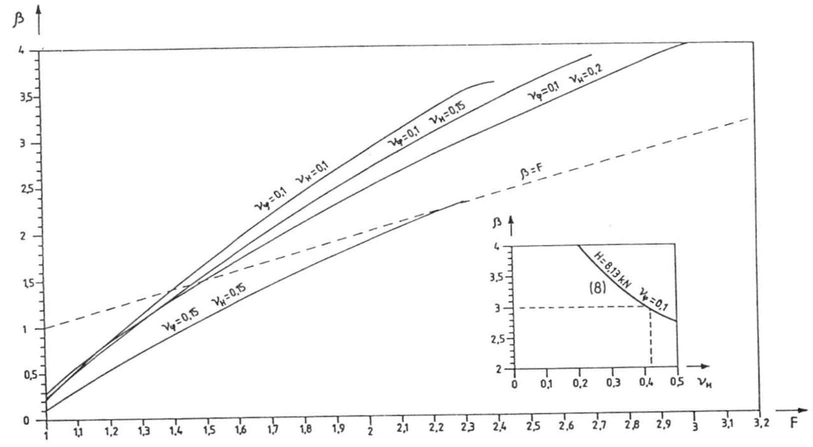

Figure 4: Reliability index β versus safety factor F (results of the numerical example) (Luźna 1996)

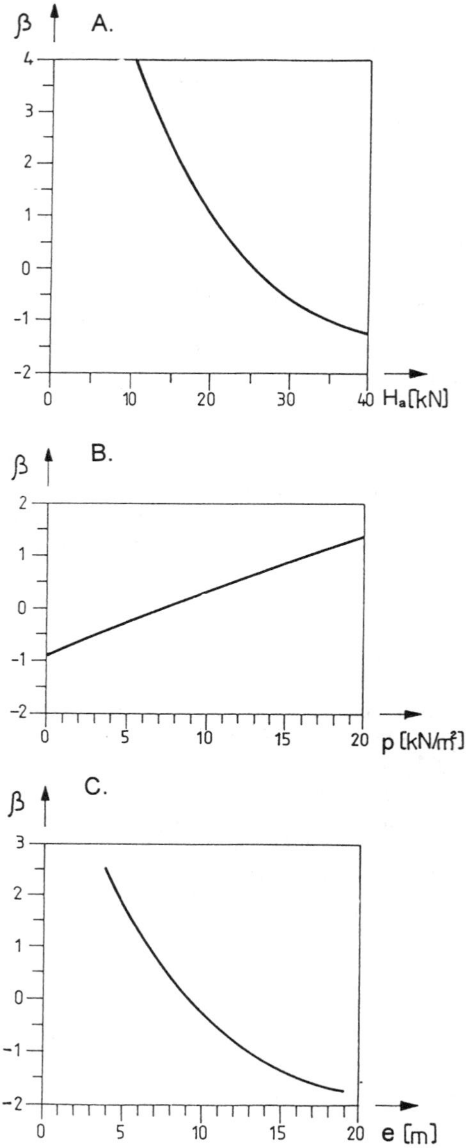

Figure 5: Results of the parameters studies:
A. Variability of lateral force;
B. Variability of surface overburden;
C. Variability of eccentricity.

MATEMATICA system in conjunction with structural reliability code (*STRUREL*, for example).

The computations show a large effect of the friction angle random fluctuations on the variations of the ultimate lateral force. Hence rather low values of the reliability index β are obtained.

The results show the necessity of the use of high values of total safety factors F if the Brinch Hansen method is applied.

REFERENCES

Biernatowski, K. and W. Puła 1988. Probabilistic analysis of the stability of massive bridge abutments using simulation method. *Structural Safety* 5, 1-15.

Breitung, K. 1984. Asymptotic approximation for multinormal integrals. *J.Engng. Mech., ASCE* 110, 357-366.

Brinch Hansen J. 1961. The ultimate resistance of rigid piles against transversal forces. *The Danish Geotechnical Institute, Bulletin No.12*, Copenhagen.

Brząkała, W. and W. Puła 1996. A probabilistic analysis of foundation settlements. *Computers and Geotechnics,* 18, 291-309.

Ditlevsen O. and H.O. Madsen 1996. *Structural Reliability Methods*. J.Wiley & Sons.

Hohenbichler, M. Gollwitzer S. , Kruse, W. and R. Rackwitz 1987. New light on first- and second- order reliability methods. *Structural Safety* 4, 267-284.

Luźna, D. 1996. *Reliability of pile foundation* (in Polish). Msc. Thesis, Institute of Geotechnics and Hydrotechnics, The Technical University of Wrocław.

MATEMATICA 1993. A system for doing mathematics by computer. Ver.2.2.1., Wolfram Research Inc.

Rackwitz R. and B. Fiessler 1978. Structural reliability under combined random loads sequences. *Computers and Structures*. No. 9, 489-494.

Rosenblatt M. 1952. Remarks on a Multivariate Transformation. *Ann. of Mathematical Statistics,* vol.23, 470-472.

STRUREL 1992. Theoretical Manual. RCP, Munich.

Numerical Models in Geomechanics, Pietruszczak & Pande (eds) © 1997 Balkema, Rotterdam, ISBN 90 5410 886 X

Sensitivity analysis of deep foundations in heterogeneous media

B.B. Budkowska
Department of Civil and Environmental Engineering, University of Windsor, Ont., Canada

ABSTRACT: The paper presents the investigations of deep foundations, penetrating homogeneous/heterogeneous media, from sensitivity theory perspective. In theoretical formulation the design variables are assumed to be of continuous type. This fact allows one to conduct the analysis in the framework of variational calculus. The virtual work theorem is used with respect to variations of generalized displacement and internal forces caused by the variations of bending stiffness and the modulus of subgrade reaction. The dummy load method is applied to the adjoint structure concept. The final form of the derived sensitivity equations is modified on account of heterogeneity of the soil medium of layered type. The mathematical formulation is illustrated by some numerical examples.

1. INTRODUCTION

The sensitivity analysis of structural systems with respect to variations of their parameters is one of primary importance in the design/redesign process. The main concepts of sensitivity analysis of structural responses to the changes of the design variables are well documented in several publications (Arora & Haug, 1979; Dems & Mroz, 1985; Haug, Choi & Komkov, 1986).

In general, the sensitivity of structural response to the changes of the design parameters is measured by the partial derivatives of some response functions with respect to system parameters. These partial derivatives can be computed by some methods which are effective for a certain type of the design variables. The analytical, semi-analytical or finite difference methods are well suited for the design variables of scalar type. All these methods enable one to establish a measure to control the changes of the structural response due to the changes in the design parameters in the vicinity of their initial values.

When the design variables are the spatial functions, the convenient and effective way of sensitivity analysis of structural systems can be performed in the scope of variational calculus. The application of sensitivity theory to pile structures under different load conditions for continuous distribution of the design variables was presented in (Budkowska & Szymczak, 1993).

The paper presents the sensitivity analysis of pile structure subject to bending embedded in layered heterogeneous soil. The design variables are taken as the bending stiffness of the pile structure and the modulus of subgrade reaction. Due to heterogeneity of the soil medium surrounding the pile, the modulus of subgrade reaction has continuous distribution within each soil layer varying in discontinuous fashion when entering from one layer to another. The determination of the sensitivity of arbitrary component of kinematic or static field employs the concept of virtual energy applied with respect to appropriate variations imposed on the original pile structure.

During analysis, the constitutive equations have been extended for the purpose of sensitivity theory such that, besides the dependence on the state variables of the system, explicit dependence on the design variables was also introduced. It was obtained through the formulation of the variational form of constitutive relationships.

The derived sensitivity equations for the continuous distribution of the design variables are also extended to discontinuous types of the design variables which are used to describe the behaviour of heterogeneous soil of layered type. The applicability of the obtained equations has been tested in comparative analysis of some numerical examples of practical importance.

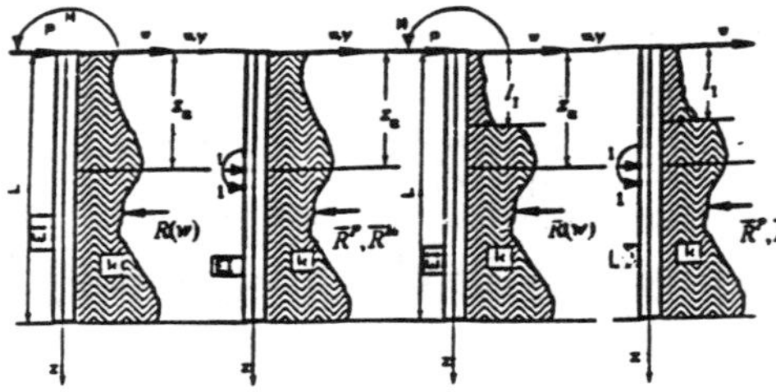

Figure 1. Geometry, applied load, soil support, dummy loads and the appropriate adjoint structure.

2. SENSITIVITY OF HORIZONTAL DISPLACEMENT AND ANGLE OF BENDING OF THE PILE

The pile subject to bending, having length l and bending stiffness EI is embedded in a soil. It is modelled as one dimensional beam element. The interaction between pile structure and the surrounding soil medium is described by the elastic foundation of Winkler type having modulus of subgrade reaction k (Fig. 1).
The horizontal displacement w_α and the angle of bending φ_α at the cross-section $z = z_\alpha$ define the components of deformation field. The behaviour of the pile structure is described by means of the following equations:

$$M(z) = -EIw'' \tag{1}$$

$$R(z) = kw \tag{2}$$

where $M(z)$ = the bending moment, $R(z)$ = the soil reaction, w, w'' = the generalized displacements.

The changes of generalized displacements and internal forces are defined by means of variations. The design variables components are: bending stiffness of the pile EI and the modulus of subgrade reaction k.

The variations of horizontal displacement δw_α and the angle of bending $\delta\varphi_\alpha$ at the cross-section $z = z_\alpha$ due to variations $\delta(EI)$ and δk are sought. The dummy load method combined with adjoint structure concept is employed. At the same time, the appropriate variations of generalized displacements are imposed on the primary structure which is subjected to constant bending loads.

The virtual work theorem applied with respect to kinematic variations imposed on the primary structure together with dummy loads applied to the adjoint structure gives (Washizu, 1974):

$$\bar{1}\delta w_\alpha = \int_0^l (\overline{M}^P \delta w'' - \overline{R}^P \delta w) dz \tag{3}$$

where $\overline{M}^P$ = the bending moment of the adjoint structure, $\overline{R}^P$ = the soil reaction at the adjoint structure, $\bar{1}$ = horizontal dummy load applied to the adjoint structure at the cross-section $z = z_\alpha$.

The variations δw and $\delta w''$ required by Eq. (3) can be determined from Eqs. (1) and (2). Bearing in mind the fact that in Eqs. (1) and (2), the internal forces depend on the deformation field components as well as the design variables, then the following relationships are satisfied:

$$\delta w'' = -\frac{\delta M}{EI} + \frac{M}{(EI)^2}\delta(EI) \tag{4}$$

$$\delta w = \frac{\delta R}{k} - \frac{R}{k^2}\delta k \tag{5}$$

Substitution of Eqs. (4) and (5) into Eq. (3), and employing the fact that the primary structure is subjected to constant type of load, we arrive at the following equation:

$$\delta w_\alpha = \int_0^l \left(\frac{M\overline{M}^P}{(EI)^2}\delta(EI) + w\overline{w}^P \delta k \right) dz = \int_0^l (w''\overline{w}^{P''}\delta(EI) + w\overline{w}^P \delta k)\, dz \tag{6}$$

where $\overline{w}^P$, $\overline{w}^{P''}$ = the generalized displacements of the adjoint structure subject to dummy lateral load $\bar{1}$ at the cross-section $z = z_\alpha$.

The expression under the second integral of Eq. (6) is obtained by substituting the relationship (1) (which is valid for the primary as well as for adjoint structure) into Eq. (6). It represents the sensitivity equation of w_α due to the changes of bending stiffness of the pile and the modulus of subgrade reaction.

Similarly, the first variation of the angle of bending $\delta\varphi_\alpha$ at the cross-section $z = z_\alpha$ can be determined. In this case, the virtual work theorem requires the adjoint structure at the same cross-section is subjected to dummy bending moment $\overline{M} = \bar{1}$. At the same time, some kinematic variations are imposed on the primary structure which is permanently subject to unchangeable load. Thus, the virtual work theorem applied to kinematic variations of primary structure combined with suitable dummy load of the adjoint structure gives:

$$\bar{1}\delta\varphi_\alpha = \int_0^l (\overline{M}^M \delta w'' - \overline{R}^M \delta w) dz \tag{7}$$

where $\overline{M}^M$, $\overline{R}^M$ = the bending moment and the soil reaction of the adjoint structure subject to dummy bending moment $\overline{M} = \overline{1}$ applied at the cross-section $z = z_\alpha$.

Substituting Eqs. (4) and (5) with modifications for constant load conditions into Eq. (7) and then employing constitutive relationships (1) and (2) for the primary and adjoint structure gives:

$$\overline{1}\,\delta\varphi_\alpha = \int_0^l (w''\overline{w}^{M''}\delta(EI) + w\overline{w}^M\delta k)\,dz \qquad (8)$$

where $\overline{w}^M$, $\overline{w}^{M''}$ = the generalized displacement of the adjoint structure subject to dummy bending moment $\overline{M} = \overline{1}$ at the cross-section $z = z_\alpha$.

It represents sensitivity equation for $\delta\varphi_\alpha$ due to the changes of *(EI)* and *k*.

3. SENSITIVITY OF BENDING MOMENT AND SHEAR FORCE OF BENT PILE

Similar concept that has been used in determination of sensitivity of generalized displacement components due to variations of the design variables is employed in sensitivity investigations of the bending moment M_α and the shear force T_α at the cross-section $z = z_\alpha$. The guidance of virtual work theorem indicates that the variation of bending moment δM_α at the cross-section $z = z_\alpha$ requires the application of dummy rotational distortion $\Delta\overline{\varphi} = \overline{1}$ imposed on the adjoint structure at the cross-section $z = z_\alpha$. At the same time, the original pile structure is subjected to additional variations of bending moment δM distributed along the pile axis such that at the cross-section $z = z_\alpha$, δM is equal to δM_α. Thus:

$$\overline{1}\,\delta M_\alpha = \int_0^l (\delta M\overline{w}^{\varphi''} - \delta R\overline{w}^{\varphi})\,dz \qquad (9)$$

where $\overline{w}^{\varphi}$, $\overline{w}^{\varphi''}$ = the generalized displacement of the adjoint structure subject to dummy bending rotation $\Delta\overline{\varphi} = \overline{1}$ at the cross-section $z = z_\alpha$.

For the purpose of sensitivity analysis, Eqs. (1) and (2) can be modified to contain the variations of all variables. This means that the following relationships are satisfied:

$$\delta M = -EI\,\delta w'' - w''\delta(EI) \qquad (10)$$

$$\delta R = k\,\delta w + w\,\delta k \qquad (11)$$

Substitution of Eqs. (10) and (11) into Eq. (9) gives:

$$\overline{1}\,\delta M_\alpha = \int_0^l ((-EI\overline{w}^{\varphi''}\delta w'' - k\overline{w}^{\varphi}\delta w) - (w''\overline{w}^{\varphi''}\delta(EI) + w\overline{w}^{\varphi}\delta k))\,dz \qquad (12)$$

$$= \int_0^l ((\overline{M}\delta w'' - \overline{R}\delta w) - (w''\overline{w}^{\varphi''}\delta(EI) + w\overline{w}^{\varphi}\delta k))\,dz$$

The expression within the first bracket under integral of Eq. (12) must vanish since it represents the work of internal forces of the adjoint structure on the variations of the generalized displacements of the primary structure. However, the adjoint structure is subjected to no external loads; therefore, the expression in the first bracket of Eq. (12) is equal to zero. Finally, the sensitivity of the bending moment at $z = z_\alpha$ due to variations of bending stiffness and the modulus of subgrade reaction can be determined from the following equation:

$$\overline{1}\,\delta M_\alpha = -\int_0^l (w''\overline{w}^{\varphi''}\delta(EI) + w\overline{w}^{\varphi}\delta k)\,dz \qquad (13)$$

The sensitivity investigations of the shear force T_α at the cross-section $z = z_\alpha$ due to the changes of the design variables can be obtained with aid of dummy unit shear distortion $\Delta\overline{w} = \overline{1}$ imposed on the adjoint structure at the cross-section $z = z_\alpha$. This dummy shear distortion induces the generalized lateral displacements $\overline{w}^{\Delta}$, $\overline{w}^{\Delta''}$ in the adjoint structure. At the same time, the variations of shear force δT are imposed on the primary structure in such a way, that at the cross-section $z = z_\alpha$, it is equal to δT_α. The virtual work theorem for variation of shear force δT_α allows us to write the following relationship:

$$\overline{1}\,\delta T_\alpha = \int_0^l (\delta M\,\overline{w}^{\Delta''} - \delta R\,\overline{w}^{\Delta})\,dz \qquad (14)$$

Substitution of Eqs. (10) and (11) into Eq. (14) results:

$$\overline{1}\,\delta T_\alpha = \int_0^l ((-EI\overline{w}^{\Delta''}\delta w'' - k\overline{w}^{\Delta}\delta w) - (w''\overline{w}^{\Delta''}\delta(EI) + w\overline{w}^{\Delta}\delta k)) = \int_0^l ((\overline{M}^{\Delta}\delta w'' - \overline{R}^{\Delta}\,\delta w) - (w''\overline{w}^{\Delta''}\,\delta(EI) + w\overline{w}^{\Delta}\,\delta k))\,dz \qquad (15)$$

where $\overline{M}^{\Delta}$, $\overline{R}^{\Delta}$ = the internal forces of the adjoint structure which is subjected to unit dummy shear distortion $\Delta\overline{w} = \overline{1}$ at the cross-section $z = z_\alpha$.

Taking into account the fact that the expression within the first bracket under integral of Eq. (15) will vanish, since it represents the work of internal forces of the adjoint structure (which is free of external load) on the variations of generalized displacements of the original structure, the final form of sensitivity equation for shear force due to $\delta(EI)$ and δk is as follows:

$$\delta T_\alpha = -\int_0^l (w''\overline{w}^{\Delta''}\delta(EI) + w\overline{w}^{\Delta}\,\delta k)dz \tag{16}$$

It is worth noting that in the resulting sensitivity Eqs. (6), (8), (13), and (16), the bending stiffness (EI) is represented as multiplication of two components, i.e., Young's modulus E and the moment of inertia I, each of them can be considered to be an independent design variable. The consequence of this fact is the possibility of investigation of δw_α, $\delta\varphi_\alpha$, δM_α and δT_α due to the changes of E and I separately. For this case, the variation of $\delta(EI)$ is divided into two terms according to the following formula:

$$\delta(EI) = I\delta E + E\delta I \tag{17}$$

Since $\delta(EI)$ in Eqs. (6), (8), (13), and (16), is connected with $(w''\,\overline{w}'')$, this means that assumption on independence of the design variables E and I according to Eq. (17), leads to the following modifications in Eqs. (6), (8), (13), and (16):

$$w''\overline{w}''\delta(EI) = Ew''\overline{w}''\delta I + Iw''\overline{w}''\delta E \tag{18}$$

This equation clearly indicates the difference in computation of sensitivity of w_α, φ_α, M_α and T_α due to the changes of $\delta(EI)$, δE and δI.

4. SENSITIVITY ANALYSIS OF THE PILE STRUCTURE EMBEDDED IN HETEROGENEOUS SOIL

The sensitivity equations derived for w_α, φ_α, M_α and T_α are valid for the pile structure penetrating homogeneous soil with arbitrary but continuous variability of the modulus of subgrade reaction. For heterogeneous soil of layered type, when the modulus of subgrade reaction is continuous within each layer (i), however it has discontinuities on the boundaries between layers $(i-1)$, (i) and $(i+1)$, the sensitivity equations for w_α, φ_α, M_α and T_α due to δk should be modified in the following manner:

$$\delta w_\alpha = \int_0^l w''\overline{w}^{P''}\delta(EI)dz + \sum_{i=1}^{n}\int_{l_{i-1}}^{l_i} w\overline{w}^{P}\delta k_i\,dz \tag{19}$$

$$\delta\varphi_\alpha = \int_0^l w''\overline{w}^{M''}\delta(EI)dz + \sum_{i=1}^{n}\int_{l_{i-1}}^{l_i} w\overline{w}^{M}\delta k_i\,dz \tag{20}$$

$$\delta M_\alpha = -\int_0^l w''\overline{w}^{\varphi''}\delta(EI)dz + \sum_{i=1}^{n}\int_{l_{i-1}}^{l_i} w\overline{w}^{\varphi}\delta k_i\,dz \tag{21}$$

$$\delta T_\alpha = -\int_0^l w''\overline{w}^{\Delta''}\delta(EI)dz + \sum_{i=1}^{n}\int_{l_{i-1}}^{l_i} w\overline{w}^{\Delta}\delta k\,dz \tag{22}$$

where $l_0 = 0$.

5. NUMERICAL EXAMPLES

The theoretical formulation presented in previous Sections has been used in analysis of numerical examples of practical importance. The investigated cases are shown in Fig. 2.

The variations that have been derived indicate that sensitivity analysis is performed in the vicinity of the initial solution with respect to variations of the design variables. This means that the same amount of changes of the design variables affects the quantity under consideration differently, depending on the solutions of the problem represented by generalized displacement of primary and adjoint structure.

As far as the sensitivity analysis of piles is concerned, the initial values of the generalized lateral displacements of the primary structure for the same load conditions are affected by the soil support geometry, type of soil, as well as the bending stiffness of the pile. In particular, in order to examine the former factors, the different soil support conditions shown in Fig. 2 are involved to sensitivity analysis. The first two cases shown in Fig. 2 and denoted as A and B are dealing with homogeneous clayey soil, while

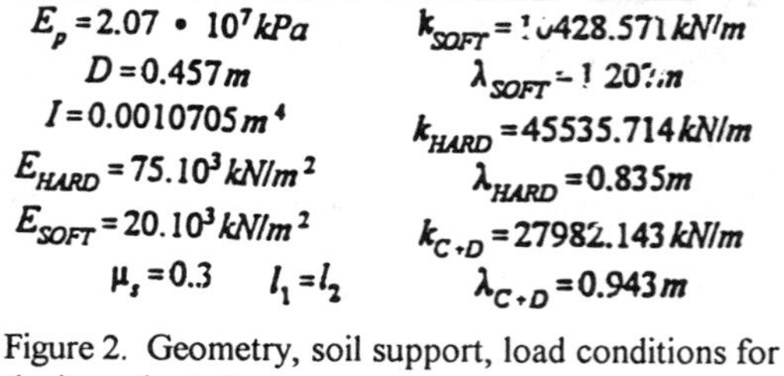

$E_p = 2.07 \cdot 10^7\,kPa$ $\quad$ $k_{SOFT} = 10428.571\,kN/m$

$D = 0.457\,m$ $\quad$ $\lambda_{SOFT} = 1.20?\,m$

$I = 0.0010705\,m^4$ $\quad$ $k_{HARD} = 45535.714\,kN/m$

$E_{HARD} = 75.10^3\,kN/m^2$ $\quad$ $\lambda_{HARD} = 0.835\,m$

$E_{SOFT} = 20.10^3\,kN/m^2$ $\quad$ $k_{C+D} = 27982.143\,kN/m$

$\mu_s = 0.3$ $\quad$ $l_1 = l_2$ $\quad$ $\lambda_{C+D} = 0.943\,m$

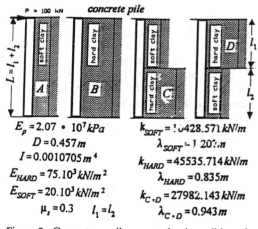

Figure 2. Geometry, soil support, load conditions for the investigated cases.

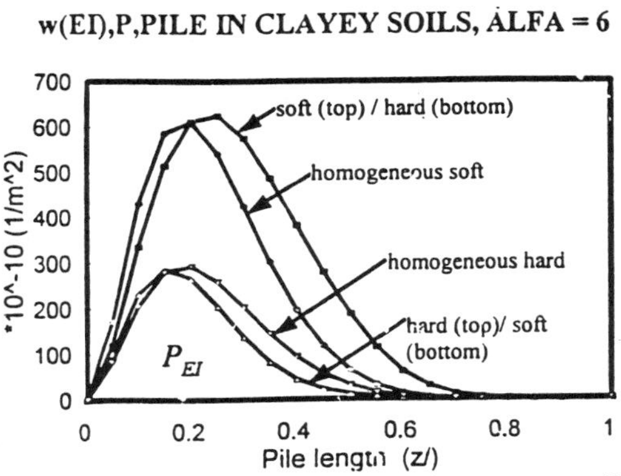

Figure 3. Distribution of sensitivity operators P_{EI} affecting the change of maximum displacement due to the changes of (EI) for homogeneous and heterogeneous soil support.

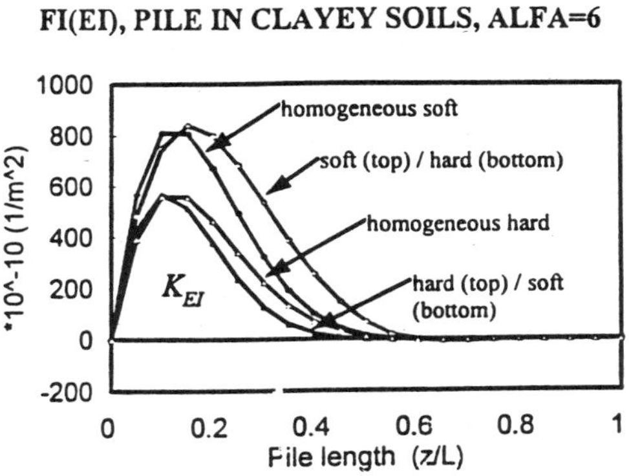

Figure 5. Distribution of sensitivity operators K_{EI} affecting the changes of maximum angle of bending due to the changes of bending stiffness (EI) for homogeneous and heterogeneous soil support.

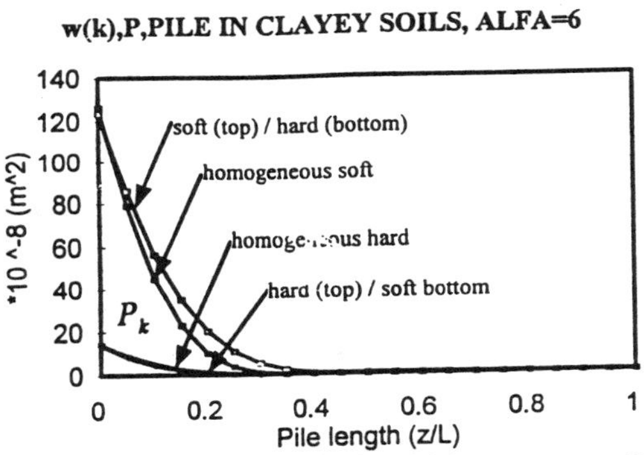

Figure 4. Distribution of sensitivity operators P_k affecting the change of maximum displacement due to the changes of modulus of subgrade reaction k for homogeneous and heterogeneous soil support.

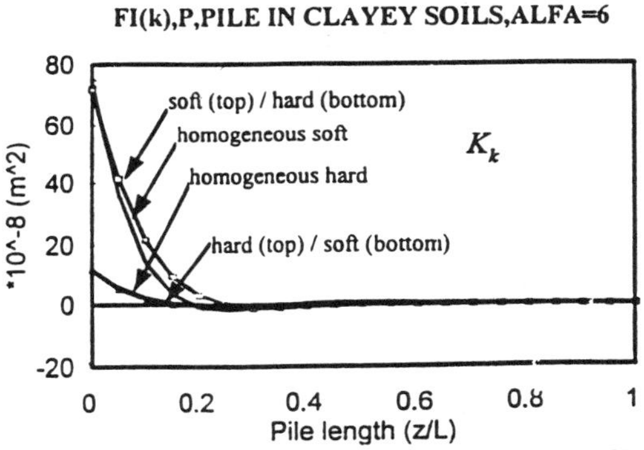

Figure 6. Distribution of sensitivity operators K_k affecting the changes of maximum angle of bending due to the changes of modulus of subgrade reaction k for homogeneous and heterogeneous soil support.

cases C and D are connected with heterogeneous soil of layered type. The modulus of subgrade reaction was taken (after Vesic, 1961) as:

$$k = 0.65 \sqrt[12]{(E_s D^4)/(E_p I_p)}\; E_s/(1-\mu_s^2) \qquad (23)$$

where E_s, μ_s = Young's modulus and the Poisson's ratio for the soil, D = the diameter of the pile, $E_p I_p$ = the bending stiffness of the pile.

The constant value of modulus of subgrade reaction is used in definition of the characteristic length λ of the pile, which is then used in determination of the length of the pile (Matlock & Reese, 1960; Das, 1990). Thus:

$$\lambda = \sqrt[4]{(EI)/k} \qquad (24)$$

When the length of pile is equal to or bigger than five multiples of the characteristic length, then the pile is called long and it behaves as flexible during bending. In the investigation the length of the pile $l = 6\lambda$.

The majority of piles (Poulos & Davis, 1980) are designed for maximum deformations. Taking this fact into account, the attention is focused on the sensitivity of maximum values of horizontal displacement w_0 and the angle of bending φ_0, both located at the top cross-section.

The Eqs. (19) and (20), modified for the purpose of numerical analysis, are taken as follows:

$$\delta w_0 = \int_0^l w'' \overline{w}^{P''} \delta(EI)\,dz + \sum_{i=1}^{2} \int_{l_{i-1}}^{l_i} w \overline{w}^{P} \delta k_i \, dz = \int_0^l P_{EI}\, \delta(EI)\,dz + \sum_{i=1}^{2} \int_{l_{i-1}}^{l_i} P_k \delta k_i \, dz \qquad (25)$$

$$\delta\varphi_0 = \int_0^l w''\overline{w}^{M''}\delta(EI)\,dz + \sum_{i=1}^{2}\int_{l_{i-1}}^{l_i} w\overline{w}^{M}\delta k_i\,dz = \int_0^l K_{EI}\;\delta(EI)\,dz + \sum_{i=1}^{2}\int_{l_{i-1}}^{l_i} K_k\delta k_i\,dz \qquad (26)$$

where $l_0 = 0$, P_{EI}, P_k = sensitivity operators affecting the change of w_0 due to the changes of (EI) and k, K_{EI}, K_k = sensitivity operators affecting the change of φ_0 due to the changes of (EI) and k.

The determined sensitivity operators are presented in Figs. 3, 4, 5, and 6 for the pile structure subject to horizontal force.

6. CONCLUSIONS

The numerical results of sensitivity analysis presented are valid with the following assumptions:

A) The formula for the characteristic length of the piles embedded in homogeneous clay depends on the constant value of the modulus of subgrade reaction k. For the heterogeneous clayey soil of layered type, the characteristic length was determined with the aid of mean weighted value of k for hard and soft clay.

B) The sensitivity analysis of kinematic and static field components of long piles due to variations of the design variables vector components is formulated for the constant value of load conditions.

Bearing in mind these facts, the numerical results of sensitivity analysis performed for long pile embedded in homogeneous and heterogeneous clayey soil of layered type allow us to draw the following conclusions:

1. The sensitivity of maximum values of kinematic field components due to bending stiffness of the pile, as well as modulus of subgrade reaction, depends basically on the magnitude of the modulus of subgrade reaction of the clay located in the top part of the pile.
2. The modulus of subgrade reaction of the clay soil located in the lower part of the pile has small effect on the sensitivity of maximum values of kinematic field components.
3. The above cited conclusions allow us to classify the investigated examples into two categories. The first category comprises the long piles penetrating homogeneous soft clay and heterogeneous clayey soil with soft clay layer located in the top part of the piles. The second category contains the long piles embedded in homogeneous hard clay and heterogeneous clayey soil having hard clay layer localized in the top part of the pile.
4. The sensitivity of maximum values of lateral displacement and the angle of bending due to the changes of the bending stiffness of the pile and the modulus of subgrade reaction for the first category is much more pronounced than for the second category.
5. The appropriate distributions of sensitivity operators affecting the changes of maximum values of w_0 and φ_0 due to the changes of (EI) and k, form two distinct families of curves associated with each category described above.

7. ACKNOWLEDGEMENT

The financial support of the Natural Sciences and Engineering Research Council of Canada under Grant #OGP 01102G2 is gratefully acknowledged.

REFERENCES

Arora, S.J. & Haug, E.J. 1979, Methods of design sensitivity analysis in structural optimization. *AIAA J.* 17, 9: 970-974.

Budkowska, B.B. & Szymczak, C. 1993. Sensitivity analysis of axially loaded piles. *Archives of Civil Engineering, Polish Academy of Sciences.* XXXIX, 1: 93-105.

Das, B.M. 1990. *Principles of Foundation Engineerring.* Boston: PWS - Kent Publishing Co.

Dems, K. & Mroz, Z. 1985. Variational approach to first and second order sensitivity analysis of elastic structures. *Int. J. Solids and Structures.* 21: 637-661.

Frank, P.M. 1978. *Introduction to system sensitivity theory.* New York: Academic Press.

Haug, E.J., Choi, K., & Komkov, V. 1986. *Design sensitivity analysis of structural systems.* Orlando: Academic Press.

Matlock, K. & Reese, L.C. 1960. Generalized solution for laterally loaded piles. *J. Soil Mech. Found. Div. ASCE.* 86: SM5, 63-91.

Poulos, K.G. & Davis, E.K. 1980. *Pile foundation analysis and design.* New York: John Wiley and Sons.

Vesic, A.S. 1961. Bending of beams resting on isotropic elastic solids. *J. Eng. Mech. Div. ASCE.* 87: EM2, 35-53.

Washizu, K. 1974. *Variational methods in elasticity and plasticity.* Oxford: Pergamon Press.

Numerical Models in Geomechanics, Pietruszczak & Pande (eds) © *1997 Balkema, Rotterdam, ISBN 90 5410 886 X*

A numerical model for large deformation problems in driving of open-ended piles

D.S. Liyanapathirana, A.J. Deeks & M.F. Randolph
Department of Civil Engineering, The University of Western Australia, Nedlands, W.A., Australia

ABSTRACT: An 'Eulerian-like' finite element model is presented to analyse driving of open-ended piles subjected to multiple hammer blows, including large soil deformation. For each hammer blow, results are obtained using a small strain finite element model and, at the end of each blow, material flow is taken into account with reference to a fixed mesh. Residual stresses calculated at the Gauss integration points of the deformed mesh are mapped on to the fixed mesh and used as initial stresses for the next hammer blow. Numerical integrations for stiffness and mass are carried out for the volume of material remaining inside the fixed mesh. Results obtained with and without allowing material to flow through the fixed mesh are compared for several hammer blows. Soil flow around the pile is presented for unplugged, partially-plugged and plugged cases.

1 INTRODUCTION

Driving of an open-ended pile creates a soil column inside the pile. In the past, small strain finite element formulations have been used (Simons; 1985 and Smith *et al.*; 1986) to simulate the driving process when the pile is subjected to multiple hammer blows. However, when the pile is subjected to multiple hammer blows, large deformations occur. Therefore, for an accurate study of the dynamics of the problem, a large strain finite element formulation is essential.

There are three main approaches to deal with finite element problems associated with large deformations. They are the total Lagrangian, updated Lagrangian and Eulerian formulations. Recently, another method known as arbitrary Lagrangian-Eulerian (ALE) has been developed (Haber, 1984).

In Lagrangian methods, the finite element mesh coincides with the same set of material points throughout the computation. In the total Lagrangian formulation all static and kinematic variables are referred to the initial configuration, whereas in the updated Lagrangian formulation the current configuration is considered (Bathe *et al.*, 1974). In the arbitrary Lagrangian-Eulerian and Eulerian formulations, element distortions can be avoided by uncoupling of nodal point displacements and material displacements (Van den Berg, 1991). In the ALE formulation, the extent to which material flows through the fixed finite element mesh (Eulerian), or the mesh moves with the material (Lagrangian) may be varied arbitrarily.

In the Eulerian method, the movement of the material is considered with respect to a fixed mesh. In solid mechanics, less attention has been paid to Eulerian formulations compared with Lagrangian formulations. Van den Berg (1991) applied the Eulerian method to analyse a cone penetration problem.

Although there are several methods available to deal with large deformation problems, each method has its own advantages and disadvantages depending on the nature of the application. For example, in driving of open-ended piles, relative motion between the external soil and the pile is large but, between the soil plug and the pile, it depends on whether the soil plug is unplugged, partially-plugged or plugged. If a Lagrangian method is adopted, it can lead to highly deformed interface elements. Therefore in this paper an Eulerian-like finite element model is presented.

In this method a fixed finite element mesh is used throughout the analysis. A small strain finite element model is used to obtain results within each hammer blow. At the end of each hammer blow, flow of material properties and stresses through the fixed finite element mesh is taken into account to avoid element distortions due to large deformations.

The organisation of the paper is as follows. In Section 2 the numerical model is briefly described. Then in Section 3 the numerical procedure is given in point form. Numerical results are presented in

Section 4 and in Section 5 conclusions are drawn.

2 NUMERICAL MODEL

For the following analysis, an axisymmetric finite element model as shown in Figure 1 is used. The soil field surrounding the pile shaft is truncated by the use of boundary equations developed by Deeks and Randolph (1994) for axisymmetric shear and dilation waves. Soil at the bottom is truncated by applying a standard viscous boundary. Soil is modelled using eight-node rectangular elements while the pile is modelled using three-node tube elements.

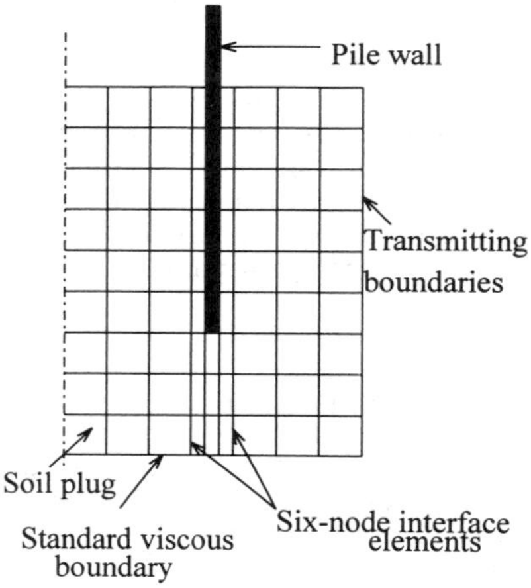

Figrue 1. Finite element mesh.

The interface between pile and soil is modelled using a thin-layer interface element (Desai *et al.*, 1984). Mass of the interface element is neglected to avoid high natural frequencies present in elements with high stiffness and low mass. Numerical integration for the interface elements is performed at two Gauss integration points on the centre line of the element. The width of an element is taken equivalent to one tenth of the adjacent soil elements and the same soil properties are used as for eight-node rectangular elements.

The force exerted on top of the pile by the falling hammer is modelled using the ram-cushion-anvil model (Deeks and Randolph; 1993).

Time integration is performed using the constant-average-acceleration method. The time step used in the numerical model is kept less than 0.5dz/c, where dz is the node spacing and c is the wave velocity. Since the pile has the highest wave velocity, the time step is calculated by taking dz and c of the pile.

Nodal spacing is selected by considering shear wave velocity in the horizontal direction and dilation wave velocity in the vertical direction. The nodal spacing is determined by ensuring that the rising portion of the applied load is spread over six nodes.

The soil has been modelled by an elastic, perfectly plastic, Von Mises model. Near the pile shaft, the normal and radial stresses are small compared with the shear stress. Hence the shear strength of the bond between pile shaft and soil is equivalent to Von Mises yield stress/√3. The interface elements adjacent to the pile shaft are allowed to slip when the shear stress reaches the shear strength of the bond between the pile and soil.

Since the pile is considered as thin-walled, thickness is not included in the finite element mesh. To facilitate the penetration of the pile, six-node interface elements adjacent to the pile wall are extended down to the bottom of the mesh. Three-node tube elements are also extended down to the bottom of the mesh but, below the pile tip level, zero density and modulus is assigned. Properties of the pile are assigned to these elements depending on the amount of pile penetration. This numerical model accurately simulates the penetration of a thin-walled pile into the soil.

3 NUMERICAL PROCEDURE

In the proposed numerical method, the movement of material with respect to the fixed finite element mesh is considered at the end of each hammer blow. The solution procedure comprises the following steps:

1) From the small strain finite element model, time integration is performed for a single hammer blow.
2) Residual stresses are calculated at the Gauss integration points of the deformed finite element mesh by equating velocity and acceleration to zero. It is assumed that the behaviour is elastic at the end of time integration.
3) Smoothed nodal stresses are calculated for the deformed mesh using the method given by Zienkiewicz and Zhu (1992). Here a patch of four elements are used and a polynomial is assumed for the stress distribution within the patch. The polynomial terms are identical to those occurring in the shape functions and are given by

$$P = [1, r, z, rz, r^2, z^2, r^2 z, z^2 r] \quad (1)$$

where r and z are the normalised coordinates for the element patch. At every node, the average nodal stress is calculated from all elements that are connected to it. This gives a continuous stress field and within each element, the stress distribution can be written as

$$\sigma = \sum_{i=1}^{8} N_i \sigma_i \tag{2}$$

where, N_i are the shape functions and σ_i are the nodal stresses.

4) In the previous step, nodal stresses are calculated for the deformed mesh. They should be transferred to the fixed mesh by taking material flow into account. As shown in Figure 2, if 'A' is a nodal point of the fixed mesh inside an element of the deformed mesh, the co-ordinates of A can be written as

$$x_A = \sum_{i=1}^{8} N_i x_i \tag{3}$$

$$y_A = \sum_{i=1}^{8} N_i y_i \tag{4}$$

where, N_i are the shape functions at point A with respect to an element of the deformed mesh, and x_i, y_i are the co-ordinates of nodes of the enclosing element in the deformed mesh.

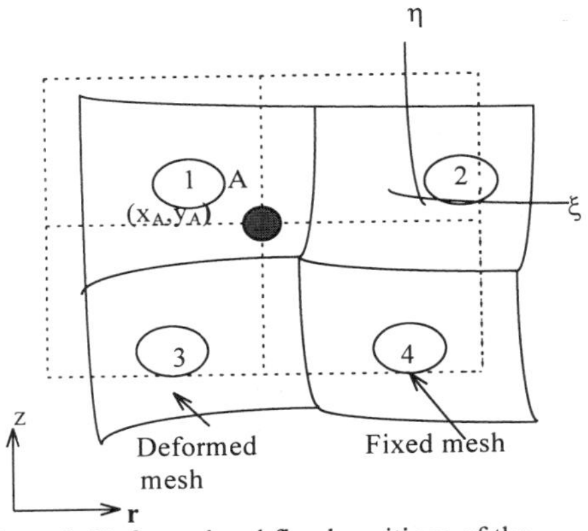

Figure 2. Deformed and fixed positions of the mesh after a single hammer blow

Each N is a function of normalised co-ordinates ξ and η. Since (x_A, y_A) and (x_i, y_i) are known, (ξ, η) corresponding to (x_A, y_A) can be found from equation (3) and (4) by solving the system of equations iteratively using a Newton-Raphson approach. If (ξ, η) lies in between -1.0 and 1.0, (x_A, y_A) lies within the element considered. The stress at point A is then given by

$$\sigma_A = \sum_{i=1}^{8} N_i \sigma_i \tag{5}$$

where σ_i are the nodal stresses for the element in the deformed mesh.

5) After calculating nodal stresses for the fixed mesh, stresses at the integration points can be obtained from

$$\sigma_{\text{int}} = \sum_{i=1}^{8} \sigma_i N_i, \tag{6}$$

where, N_i are the shape functions at the Gauss integration points and σ_i are the nodal stresses of the fixed mesh.
These stresses are used as initial stresses for the next hammer blow.

6) After the first hammer blow, material and the nodes of the fixed mesh are decoupled. New positions of the material particles, initially at nodes, can be computed by adding nodal displacements to the co-ordinates of the fixed mesh as:

$$x_1 = x_0 + \Delta u_1 \tag{7}$$
$$y_1 = y_0 + \Delta v_1 \tag{8}$$

where, (x_1,y_1) is the position, and Δu_1, Δv_1 are the displacements of (x_0,y_0) after the first hammer blow.
At the end of the second hammer blow, it is necessary to find out the element of the fixed mesh, within which (x_1,y_1) lies. Then using shape functions and displacements of the nodes, displacement of (x_1,y_1) can be calculated. Along the free surface, vertical movements of the points corresponding to x coordinates of the fixed mesh are required. Therefore, polynomials are fitted to the new locations of five nodes of each of the two adjoining elements along the free surface and new nodal positions are calculated for the x coordinates of the fixed mesh.

7) Stiffness and mass matrices are recalculated by considering the new position of the soil free surface, pile head and pile tip. The space outside the material volume but lying within the fixed mesh is filled with a soft material. For this soft material, the density and Young's modulus are selected to avoid ill-conditioning of the global matrices. In other words, this soft material does not affect the strength of soil and steel remaining within the fixed mesh. When calculating stiffness and mass, depending on the position of soil and steel, a factor is calculated for each element. For an eight-node rectangular element, if it lies inside the soil, the factor is unity and if it lies outside the soil, the factor is taken as zero. For the elements along the boundary, average density, Young's modulus

and yield stress are calculated depending on this factor. A fully smooth transition may be achieved by this method.

8) Depending on the pile penetration, the position of the applied load is decided. After the first blow, load is applied on the node just beneath the free surface. Also, at the level of the pile tip, restraining condition of the pile node and adjacent soil nodes are changed depending on the amount of pile penetration. Always the pile nodes and soil nodes adjacent to the pile, above the pile tip level, are restrained against horizontal movement.

4 NUMERICAL RESULTS

Numerical results obtained with and without material flow are compared for a partially-plugged pile. The parameters are given in Table 1.

Table 1. Values of parameters for the example

Properties of Hammer	
Ram mass (kg)	264.0
Anvil mass (kg)	132.0
Cushion stiffness (GN/m)	1.65
Initial velocity of ram (m/s)	6.0
Properties of Pile	
Impedance (MNs/m)	0.264
Diameter (m)	0.24
Length (m)	2.0
Soil plug height (m)	1.5
Wall thickness (m)	0.0082
Young's modulus (GN/m^2)	200.0
Density (kg/m^3)	8000.0
Poisson's ratio	0.29
Properties of Soil	
Shaft friction (kN/m^2)	200.0
Poisson's ratio	0.45
Shear modulus (MN/m^2)	45.0
Density (kg/m^3)	2000.0

The soil field surrounding the pile shaft is truncated at a distance equivalent to three times the diameter of the pile. At the bottom, the soil field is truncated at a distance equivalent to 6.25 times the diameter of the pile. This avoids yielding of the Gauss integration points adjacent to the wave transmitting boundaries.

Yielded zones for the two cases mentioned above are shown in Figure 3 during the tenth hammer blow. Around the pile shaft, only Gauss integration points adjacent to the pile shaft are yielded due to attenuation of the stress waves travelling in the radially outward direction. Inside the soil plug, all Gauss integration points are yielded due to magnification of the stress waves travelling in the radially inward direction. Around the pile tip, the yielded zone extends in the outward direction, and a clear difference can be seen in the results obtained for the two cases. When the material flow is not taken into account, a wider zone of yielded points occurs around the pile shaft close to the pile tip. This is because in the analysis with material flow, stresses stream through the finite element mesh with the material, but in the other case, stresses are accumulated at the initial Gauss integration points throughout the analysis.

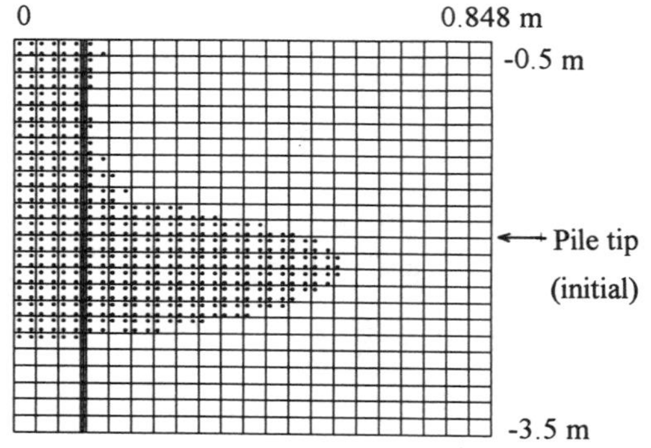

Figure 3 (a). Yielded zone without material flow.

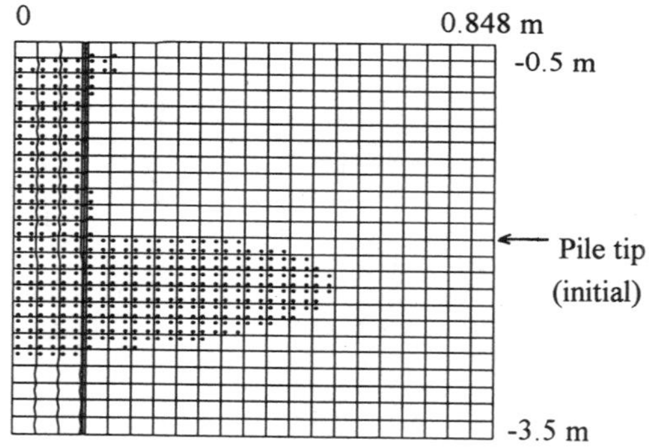

Figure 3(b) Yielded zone with material flow.

Figure 4 shows the variation of Incremental Filling Ratio (IFR) (Brucy *et al.*; 1991) along the soil plug. At the end of the first hammer blow, the IFR at the pile tip is higher than the IFR at the top of the soil plug due to compression of the soil plug. During subsequent hammer blows, the IFR tends to increase towards the top of the soil plug and remains nearly the same along depth. At the steady state, the IFR at the pile tip should be less than that observed at the pile tip after the first hammer blow due to build up of residual stresses. This may be observed only when the material flow is included.

Figure 5 shows the residual vertical stress distribution around the pile at the end of the tenth hammer blow. Smoothed nodal stresses are plotted in both figures. When stresses are streamed through the mesh, along the free surface at the top of the soil plug, zero vertical stresses are observed as in reality. In addition, stress distribution with respect to the position of the pile after several hammer blows can be obtained.

Figure 6 shows the variation of IFR at the pile tip with the number of hammer blows. The difference between the first and the second hammer blows is due to the assumption that the soil is initially unstressed. However, a significant difference can be seen in the results obtained for the two analyses because the new model with material flow is closer to reality than the case without material flow.

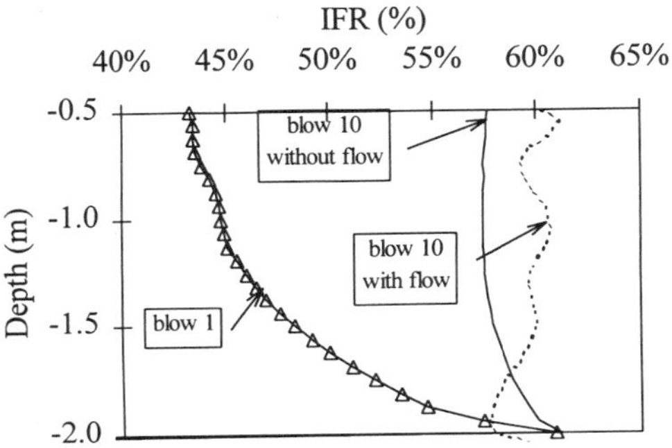

Figure 4. IFR obtained with and without material flow after hammer blows 1 and 10

4.1 *Soil flow around the pile tip*

From the new finite element model, soil flow around the pile tip was studied. Similar soil flow has been observed when driving plugged and partially-plugged piles. Therefore, in Figure 7, results are presented only for unplugged and partially-plugged piles. To obtain the unplugged condition, a pile with 720 mm diameter was selected (Liyanapathirana *et al.*; 1996). Hammer properties and dimensions of the pile are chosen by keeping dimensionless terms identical to the partially-plugged case given in Section 4, except the dimensionless pile radius (Liyanapathirana *et al.*; 1996).

In Figure 7, displacements are 15 times larger than the mesh size and the flow for ten hammer blows is shown around the pile tip. According to Figure 7 (a), when the pile is plugged or partially-plugged, soil is displaced in the radially outward direction, and inside the soil plug, displacements are uniform across the radius.

When the pile is unplugged, soil displaced radially outward is dragged in the downward direction. Inside the soil plug, highest displacements are observed at the centre of the soil plug with lower displacements towards the pile wall. Closer to the pile wall, displacements are increased but less than the displacements at the centre of the soil plug.

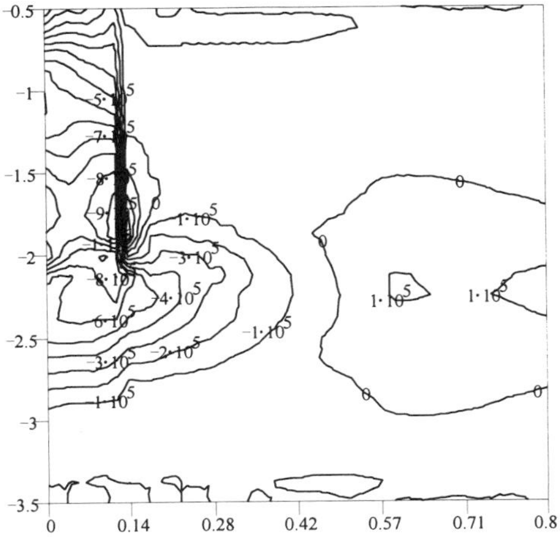

A

Figure 5 (a) Residual vertical stress distribution around the pile after the 10th hammer blow without material flow.

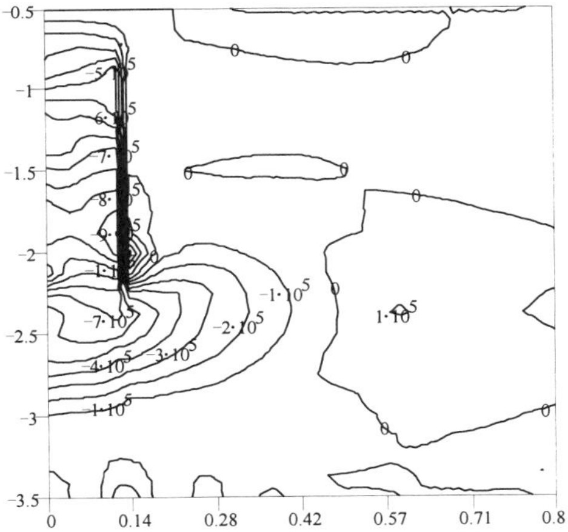

B

Figure 5 (b) Residual vertical stress distribution around the pile after the 10th hammer blow without material flow.

5 CONCLUSIONS

The numerical technique described in this paper provides an accurate and direct method of solution for the pile driving problem involving large deformations. The main advantage of this method is that it can be incorporated in a standard "small-strain" finite element analysis computer program.

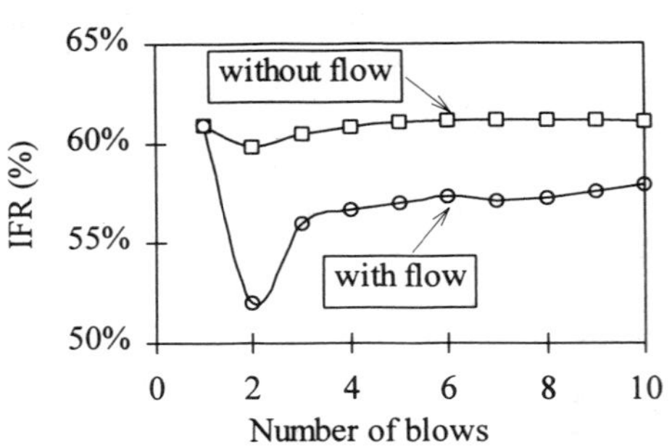

Figure 6. Variation of IFR with number of hammer blows at the pile tip

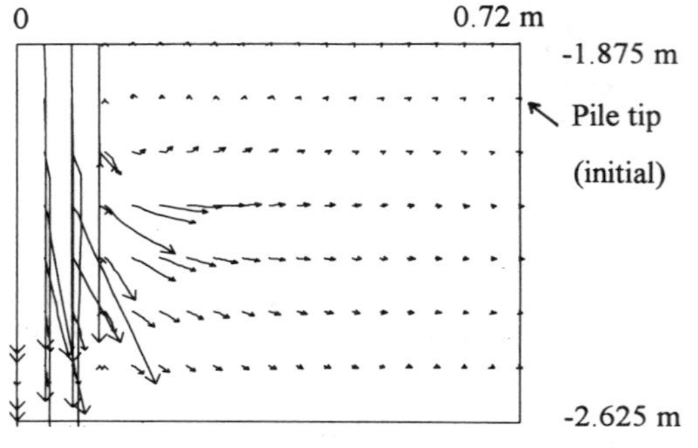

Figure 7 (a) Soil flow around a partially-plugged pile

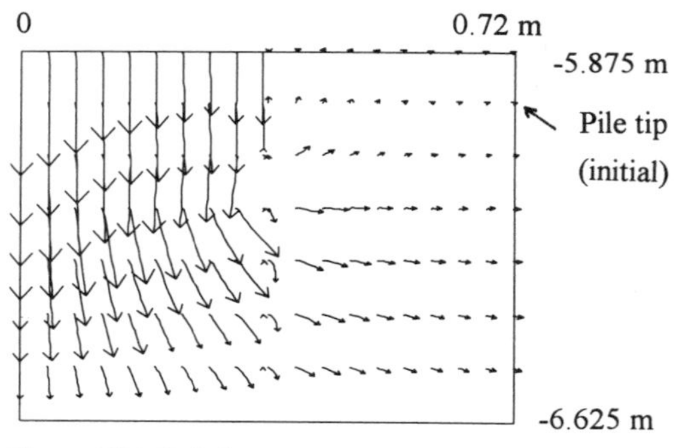

Figure 7(b) Soil flow around an unplugged pile

There is a significant improvement in the displacements and the stresses obtained when the material flow through the fixed mesh is considered. This method can be used to analyse the driving of a pile from the soil surface which cannot be carried out with a standard "small-strain" finite element analysis.

A clear difference was observed in the soil flow around the pile, when the pile is unplugged or partially-plugged. In the unplugged case, the soil displaced radially outward is dragged in a downward direction, while in the partially-plugged case soil is displaced only in the radial direction. It can be concluded that in the unplugged case, at the point of failure, there is an interaction between the soil around the shaft and the soil below the shaft.

At the end of first hammer blow, IFR at the pile tip is higher than that observed at the top of the soil plug. After several hammer blows, IFR remains nearly the same along depth due to compression of the soil plug. At the steady state, IFR at the tip is less than that observed at the pile tip due to built up of residual stresses. This may be observed from the numerical model presented.

REFERENCES

Bathe, K.J., H.Ozdemir and E.L.Wilson 1974. *Static and dynamic geometric and material nonlinear analysis*, University of California, Division of structural Engineering and structural mechanics, Berkeley, No. 74-4.

Brucy, F., J.Meunier and J.F.Nauroy 1991 'Comparison of static and dynamic tests of piles in sand.', *Int. Conf. on Deep Foundations,* Paris March 1991.

Deeks, A.J. and M.F.Randolph 1993. 'Analytical modelling of hammer impact for pile driving.' *Int. J. Numer Anal. Methods Geomech*, **17**, 279-302.

Deeks, A.J. and M.F.Randolph 1994. 'Axisymmetric time-domain transmitting boundaries.' *J. Eng. Mech.* ASCE, **120**, 25-41.

Desai, C.S., M.M.Zaman, J.G.Lightner and H.J.Siriwardena 1984. 'Thin-layer element for interfaces and joints.', *Int. J. Num. Methods in Geomechanics*, **8**, 19-43.

Haber, R.B. 1984. 'A mixed Eulerian-Lagrangian displacement model for large deformation analysis in solid mechanics', *Comp. Methods Appl. Mech. Eng.*, **43**, 277-292.

Liyanapathirana, D.S., A.J.Deeks and M.F.Randolph 'Numerical Analysis of soil plug behaviour inside open-ended piles during driving', submitted to *Int. J. Num. and analytical Methods in Geomechanics* also *University of Western Australia Research Report No. G 1224.*

Simons, H.A. 1985. *A Thoeretical Study of Pile Driving*, PhD thesis, University of Cambridge.

Smith, I.M., P.To and S.M.Wilson 1986. 'Plugging of Pipe Piles' *Third Int. Conf. on Numerical Methods in Offshore Piling,* Nantes.

Van den Berg, P. 1991. 'Numerical Model for Cone Penetration', *Proc. Int. Conf. on Computer Methods and Advances in Geomechanics*, **3**, 1777-1782.

Zienkiewicz, O.C. and J.Z.Zhu 1992. 'The superconvergent patch recovery and a posteriori error estimates Part I: The Recovery Technique', *Int. J. Num. Meth. in Eng.,* **33**, 1331-1364.

Numerical Models in Geomechanics, Pietruszczak & Pande (eds) © 1997 Balkema, Rotterdam, ISBN 90 5410 886 X

he behavior of pile foundation using three-dimensional elasto-plastic FEM ıalysis

Gose
ΊΙ Engineering Co. Ltd, Tokyo, Japan
.Ugai
epartment of Civil Engineering, Gunma University, Japan
.Ochiai
epartment of Civil Engineering, Kyushu University, Fukuoka, Japan

ABSTRACT: A field test was performed on a driven group of steel piles to determine the horizontal load fraction carried by each pile. Three-dimensional elasto-plastic FEM analysis was conducted to evaluate the results of this field test. The test was used to ascertain the validity of the 3-D FEM. Good agreement was obtained. The FEM analysis was further used to study the effects of soil properties and the 3-D geometrical configuration of the pile group. The results from the study indicate that it is possible to use the FEM analysis for predicting the bending moment distribution and load distribution ratio from extremely small to large horizontal load quite accurately. The analytical results also indicate the load distribution ratio is considerably larger at the front row of piles as compared with the back piles at large horizontal loads. The phenomenon was also confirmed by the field test.

. INTRODUCTION

ⱥ pile foundation, when subjected to a horizontal load t its footing, will resist the impact, carrying the ıduced load with its piles. This resisting force is a ombined force of the stiffness of the pile itself and the trength of the soil around each pile. The stiffness of iles within the same footing is normally consistent, but ıow much of the horizontal load each pile is to carry 'aries depending largely on the resisting properties of he soil that surrounds each pile. For example, it has een shown that the soil exhibits different resisting roperties between piles on the periphery and piles near he center of the footing, or between "front" piles and back" piles ("front" and "back" are used here to mean he relative locations of pile groups when they are 'iewed in the direction of load application). Therefore, t is also reasonable to propose that the load listribution ratio will also vary among these pile groups [),2)].

3ased on the above understanding and on the results of ı field test [2)] we conducted a three-dimensional elasto-ılastic FEM analysis in order to introduce a elationship between the load-carrying force of each ɔile and the applied horizontal load. The purpose of the ınalysis was to demonstrate the possibility of predicting the changes of each pile's load-carrying force for different soil types and stiffness by a numerical approach, by introducing field test conditions that involve a variety of soil conditions into the analysis.

2. NUMERICAL APPROACHES

Because the purpose of this analysis was to demonstrate the difference between the load distribution ratios at pile heads and the bending moment distributions of pile shafts, the soil conditions were set so that the numerical results agreed with the test results of a single pile. Using these soil constants, the pile foundation was simulated. Key items to be studied were the load exerted on each pile head and its displacement, and the pile shaft bending moment distribution. We chose this method for determining the soil conditions for our analysis for the following reason: the initial soil constants as determined by boring and laboratory testing represented the physical properties of the soil before piling, and the soil surrounding the piles could be changed by subsequent pile driving and compacting processes. Furthermore, the spacing between piles was 2.5D (D=pile diameter) which was relatively small. It appeared quite reasonable that under such limited pile spacing condition, the soil properties could change more readily but in a more complicated manner than in a single pile condition.

Therefore, in this paper we will present a numerical approach that can evaluate the tendency of load distribution ratio changes by taking into account the soil elasto-plastic conditions, this approach will be adopted rather than focusing on the evaluation of the absolute values of each pile's bending moment and shear force distribution at pile heads obtained at the moment when the surrounding soil becomes plastic.

3. GENERAL CONSIDERATION

When the soil is assumed to be elastic, then the shear force distribution ratios at pile heads should theoretically be symmetric. In actual situations, these ratios are not symmetric; when viewed in the direction of load application, the ratios for piles in the foremost row are highest, and decrease towards the piles in the rearmost row. Soil element failure and its development have been shown to play a significant role in this process [3]. Therefore, it is necessary to consider soil plasticity in the analysis.

In this paper the relationship of soil stress and its strain is assumed to be that of an elasto-plastic body, while real soils soften or harder under strain. This assumption (i.e. that the soil elements are elasto-plastic) has an advantage in that elasticity and plasticity can be identified as separate properties on a stress-strain curve, allowing the effects of either property on the soil or foundation deformations to be assessed independently of the other [4].

When defining materials such as soil, the yield requirement for elasto-plastic models corresponds to that of failure. The Mohr-Coulomb's equation was used to determine yield conditions as it is suitable for soils. To determine the plastic potential, the Drucker-Prager's equation was used. To account for the dilatancy effect associated with soil deformation, the non-associated flow rule was applied by replacing the internal soil friction angle ϕ used in the Mohr-Coulomb's equation with dilatancy angle ψ. The plastic potential (flow rule) was determined on a combined MC-DP model [5] utilizing the Drucker-Prager's equation disregarding any singular points.

4. NUMERICAL MODEL

Figure 1 shows the pile embedment (14.5 m) related to a field-investigated ground cross-sectional view. The pile arrangement is given in Fig. 2. The pile foundation

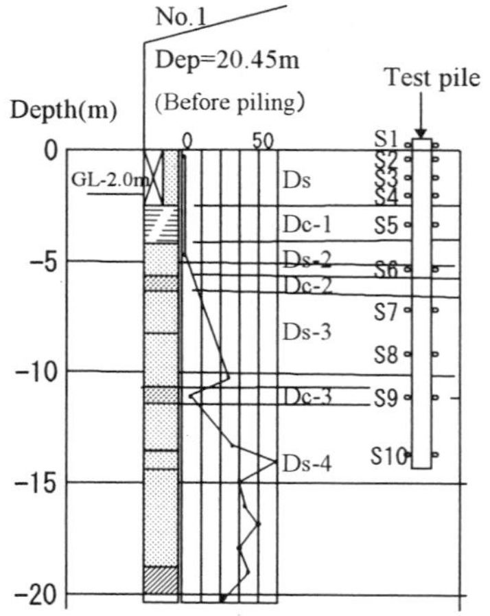

Fig.1 Ground cross-sectional view

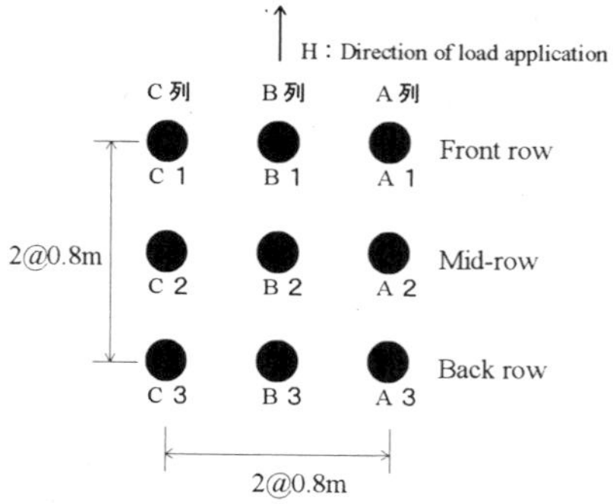

Fig.2 Pile arrangement

consisted of nine 318.5-mm-diameter steel piles, eac 6.9-mm thick. The symbols S_1 through S_{10} shown i Fig. 1 indicate the location and number of each strai gauge used. The initial soil constants (c', ϕ', and γ_t for a single pile configuration were determined fror laboratory testing, The Young's modulus E_{50} wa determined from the N value (28 • N) of the borin column diagram, and Poisson's ratio ν was assumed Values thus obtained are shown in Fig. 3, wherei symbols A_1 through A_5 denote analytical zones. Sinc no laboratory data was available for layer Dc-1, th value of layer Ds-2 was used instead as the value o layer Dc-1. In order to adjust the soil conditions so tha the numerical predictions correspond with th experimental measurements, it would be most to selec a relatively shallow ground and to adjust its constants because the horizontal loads normally exert the greates influence on flexural piles in the range 4 to 6 times th

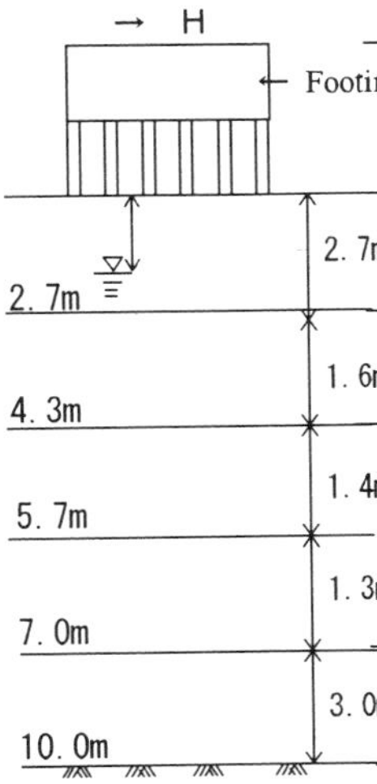

Thickness	Layer	C' (tf/m²)	φ'	ψ	γ or(γ')	E_{50} (tf/m²)	ν
2.7m	A1	0.17	34.4°	0°	1.44 (0.85) (Ave.1.29)	140	0.3
1.6m	A2	0	34.4°	0°	(0.88)	1450	0.3
1.4m	A3	2.8	34.4°	0°	(0.88)	1650	0.3
1.3m	A4	7.6	0°	0°	(0.75)	265	0.49
3.0m	A5	0.86	40.5°	10°	(0.90)	2000	0.3

Fig.3 Soil conditions for analysis(initial values)

pile diameter D. In this study, the initial soil conditions were set so that the measured load-displacement relationship of a single pile configuration agreed with that predicted by analysis.

Pile specifications, pile head conditions and loading point are shown in Fig. 4. As shown by the figure, the pile foundation is 30 cm above the ground surface and the load-carrying point is 130 cm above the ground surface. The vertical load of 200 tf represents the load from the upper structure. Strains in the pile shaft were measured with strain gauges, that were protected from any damage by the channel method. The pile shaft thickness was t=2.16 cm, to ensure that its flexural rigidity was uniform. The Yield stress of the pile shaft was assumed as 24000 tf/m².

Figure 5 shows a mesh diagram of the pile foundation analysis model. The model is symmetric on the plane, and the pile heads were of rigid material and coupled firmly together as shown in Fig. 6. Therefore, the concrete member of the footing was regarded as a rigid body. The pile shaft was regarded as a hollowed solid element rather than a beam element so as to account for the dimensional effects of the pile shaft on the surrounding soil. The sides, except the symmetric plane, and the bottom were fixed. One of the elements used in the model was a hexahedron with 20 nodes, composed of isoparametric elements.
To introduce discontinuity in the model, a thin element having a thickness of 1.5 mm and a wall friction angle of 25° was added around the piles. These values were selected based on labolatory testing on sand and aluminum pipes.

In general, when piles deform laterally, the front piles form a passive area while the rear piles create an active area. To preclude the possible generation of excessive tensile stress in the active area, a no-tension analysis [6] was performed. An additional technique was used to gradually reduce the modulus of elasticity for soil elements around piles under tensile stress. The initial static ground pressure coefficient Ko was set as 0.5.

5. RESULTS FOR A SINGLE PILE

The soil for a single pile configuration was analyzed by a parametric approach using the boring column diagram and dynamic laboratory test results as

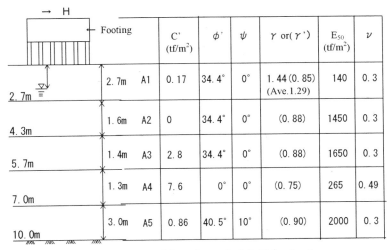

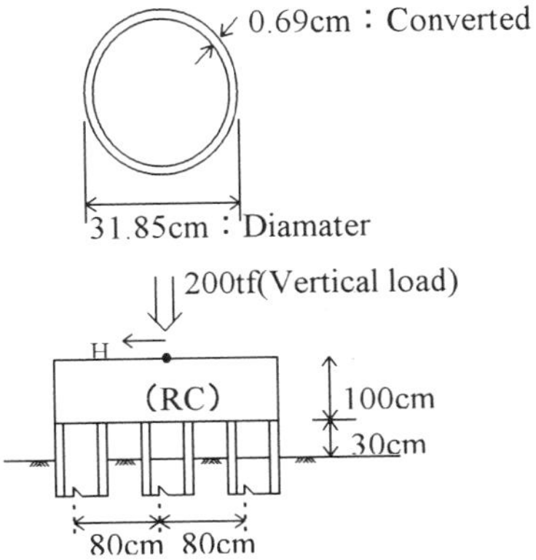

Fig.4 Loading point and pile head (pile foundation)

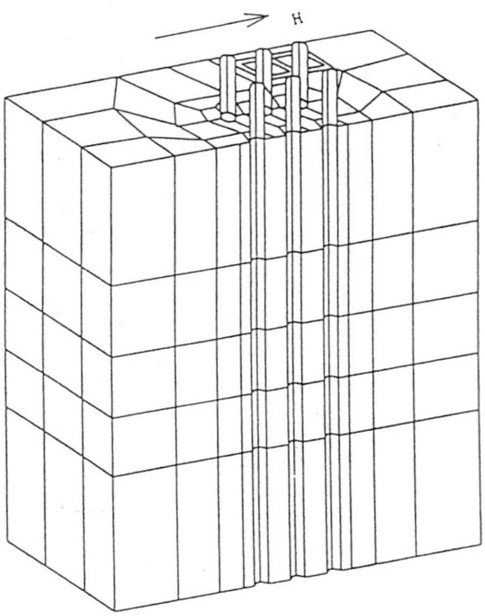

Fig.5 Finite element model for a pile foundation (symmetry plane:free;sides and bottom:fixed)

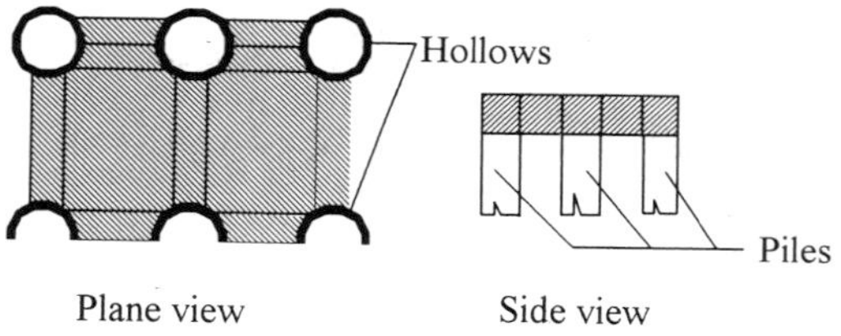

Fig.6 Footing

reference data. Numerical conditions were determined from the relationship between horizontal load and displacement at the head of the single pile, and by approximating the distribution of pile shaft bending strains across the depth. After examination, the soil conditions were adjusted as follows:

(1) Young's modulus E for the sand layer should be 1.8 times the initial value.

(2) Tan ϕ' for the sand layer should be 1.2 times the initial value, and dilatancy angle should be $\psi=\phi$'-30°.

(3) For the surface layer, C and ϕ should be adjusted as C=3.0 tf/m^2 and ϕ=20°, respectively, and Young's modulus E should be changed to 900 tf/m^2.

(4) Thin elements (between pile and soil) should be ϕ'=25° ψ=0°.

(5) Layers A1 and A2 alone should be regarded as no-tension layers (see Fig. 3).

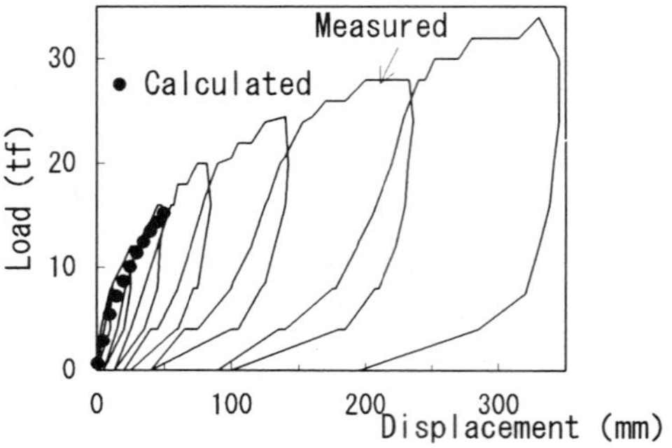

Fig.7 Load vs.horizontal displacement

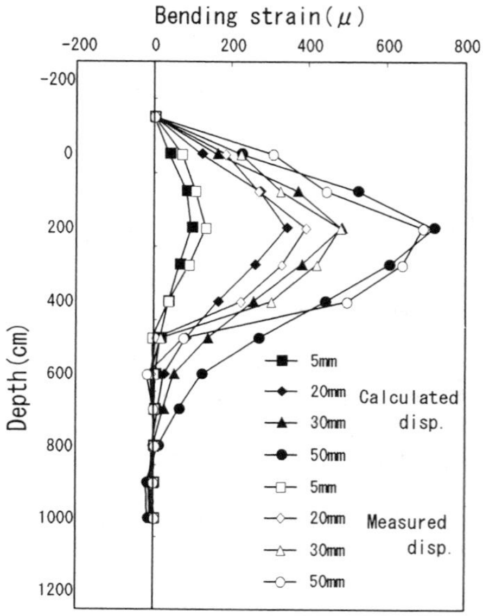

Fig.8 Comparison of measured and calculated bending strains

Items (1) and (2) above take into account the effect of embedment. Item (3) accounts for the effect of non-saturation.

Figure 7 shows the relationship between load on the pile head and horizontal displacement. While the predicted deformation is slightly larger than that measured, the predicted load-displacement relationship appears to be consistent with the measured relationship.

Figure 8 compares the predicted and measured pile head bending strain distributions across the depth. It is clear that the predicted data coincide with the measured ones.

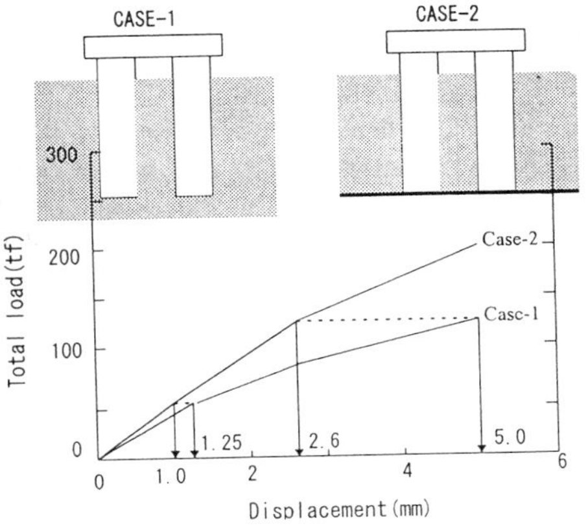

Fig.9 Load-displacement relationship for piles with different Pile tip fixing conditions

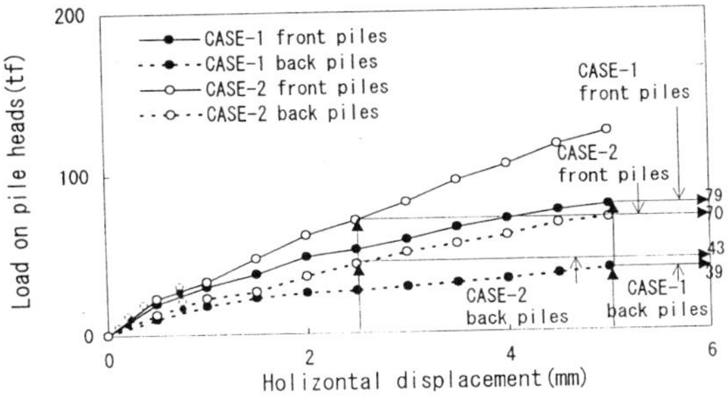

Fig.10 Comparison of loading on pile heads

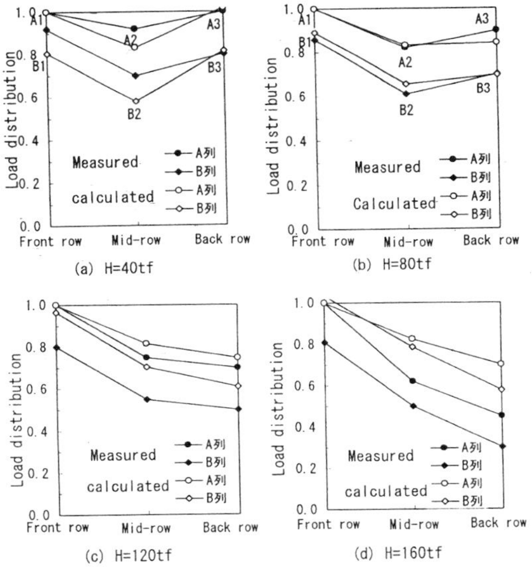

Fig.12 Comparison of calculated and measured load distributions

6. NUMERICAL RESULTS FOR A PILE FOUNDATION

This section compares the predicted values with the measured values in terms of the bending moment distributions of the pile shaft and its load distribution

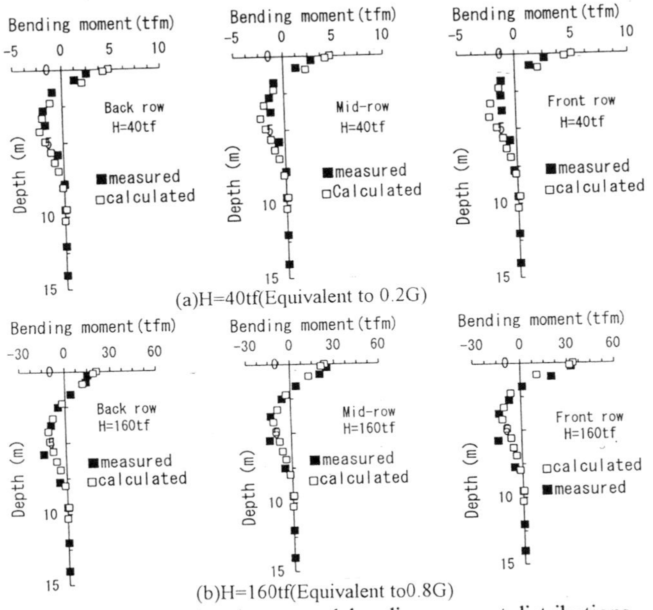

Fig.11 Compdrison of calculated and measured bending moment distributions

ratio. It should be noted that pile bottoms were fixed by hinges unlike the actual fixing condition. Figure 9 compares the load on the pile head and displacement relationships for two 2-pile foundations of different pile bottom conditions, hinged (Case 2) and free (Case 1). Under the same load, Case 1 exhibits displacement about twice as large as Case 2. In Fig. 10, the load on each pile head and its displacement are plotted utilizing the relationship taken from Fig. 9. A displacement of 5 mm in Case 1 corresponds to a displacement of 2.6 mm in Case 2 under the same load. In this condition, the load induced on each pile head was obtained from Fig. 10 as follows:

Case 1: 79 tf on front pile, 39 tf on back pile
Case 2: 70 tf on front pile, 43 tf on back pile

From these values it can be seen that the pile tip condition is not very reactive to the pile head's load distribution ratio. In reality, pile tips are elastically fixed and their reactions are not known. In the following paragraphs, a comparison with measured values will be made in terms of the relationship with loads acting on the pile heads fixed by hinges.

Pile shaft bending moment distributions of measured and calculated values are shown in Fig. 11 (a) and (b). The terms "front row", "mid row" and "back row" represent the average of piles in the front row, mid row and back row as seen in the direction of load application. Loads were 40 tf for (a) and 160 tf for (b). It is clear that the predicted absolute values of bending moments and their distribution tendency are quite similar to those obtained with the measured values.

Figure 12 plots the load distribution ratios at individual pile heads against the load (= 1.0) on pile A1. Both measured and calculated results show that the load distribution ratios for front and back piles are the same when the load is not severe, and seem to support the load distribution patterns obtainable from elasticity theory. However, as the load is increased the load distribution ratios for the front piles increase more than for the back piles. This indicates that this tendency can be simulated comparatively well by the numerical approach presented in this paper.

7. CONCLUSION

A horizontal load application test was simulated by a three-dimensional elasto-plastic FEM analysis regarding the load distributions at pile heads, using full-size pile foundations. The results obtained indicate that bending moment distributions and load distribution can be simulated numerically with comparatively good accuracy, whether the load is small or large. The data obtained indicates that the numerical approach presented in this paper could be used to assess the phenomena that the load distribution under minor loading differs greatly from that under severe loading, and that back piles carry less load than front piles when viewed in the direction of load application.

REFERENCES

1) Ogasawara, M., Hanko, M., Gose, S. and Kawaguchi ,M. : A study of proof stress of pile foundations, Structural Engineering Letters, Vol. 37A, pp.1467-1477, Mar. 1991
2) Saito, A., Hanko, M., Gose, S. and Yi ,F. : A Study of pile foundation behavior at large-scale horizontal deformation, Structural Engineering Letters, Vol. 39A, pp.1395-1407, Mar. 1993
3) Ugai, K., Saito, A., Gose, S. and Wakai ,A. : Horizontal loading tests and analysis of a single pile, Proc. of 7th Computed Dynamics Symposium, pp.223-228, Nov. 1993
4) Tanaka, C., Ugai, K., Kawamura, M., Sakajo, S. and Otsu ,H. : Analysis of the ground by the three-dimensional elasto-plastic FEM, Chapters 3 and 4, Maruzen, Mar. 1996
5) Tanaka, C.: Deformation and stability analysis by FEM; Principles of Soil Mechanics, The JGS, first edition, pp.109-154, 1992
6) Zienkiewicz, O. C., Valliappan, S. and King. I. P. Stress analysis of rock as a 'no tension' material,G'eotechnique,18,pp.56-66,1968.

Numerical Models in Geomechanics, Pietruszczak & Pande (eds) © 1997 Balkema, Rotterdam, ISBN 90 5410 886 X

Seismic evaluation of pile foundations

M. Kimura
Kyoto University, Japan

F. Zhang
Chuo Fukken Consultants Co. Ltd, Japan

In the dynamic analysis of structures with pile foundation or direct foundation, it is very important to consider the nonlinear interaction between soil and the foundation. In this paper, a three dimensional elasto-plastic finite element analysis is conducted in order to properly evaluate such an interaction. Based on the analysis, the equivalent springs required for the dynamic analysis are evaluated. A nonlinear structural dynamic analysis is then conducted to verify the safety of a bridge and its pile foundation.

1 INTRODUCTION

It is known that a dynamic analysis is very useful to seismic design. In geomechanical engineering, the seismic design is not only related to the structures but also to the foundations and the ground on which the structures are built. It is desirable to consider a full system in the dynamic analysis which consists of a structure, a foundation and a ground. It is not easy, however, to deal with a full system in the analysis when the nonlinearity of both the structure and the ground must be considered. A normal way to simplify the problem is to evaluate the interaction between the ground and the foundation by a static analysis and then equip it with equivalent springs so that only the springs and the structure are involved in the dynamic analysis. In such a case, it becomes very important to properly evaluate the nonlinear interaction between the ground and the foundation.

In this paper, a three-dimensional elasto-plastic finite element analysis (GPILE-3D) is first conducted to evaluate the nonlinear interaction between a pile foundation and the ground. The mechanical behavior of the ground in the analysis is supposed to be elasto-perfect plastic with the associated flow rule of Drucker-Prager. The nonlinearity between the bending moment and the curvature of the piles is simulated by a trilinear relation, taking into consideration the influence on the yielding moment M_{yd} and ultimate moment M_{ud} by the axial force of the pile. Based on the results of the analysis, the structure-foundation-ground system is simplified to a frame-spring system for which the nonlinear dynamic analysis can be easily conducted in a time domain with the direct integration method.

Secondly, a nonlinear dynamic analysis is conducted on a toll bridge with pile foundation using the simplified frame-spring model. From the dynamic analysis, it is possible to evaluate the maximum forces acting on the equivalent springs.

Thirdly, the three-dimensional elasto-plastic finite element analysis (GPILE-3D) is again conducted on the pile foundation using the maximum forces, which are acting on the equivalent springs and evaluated by the dynamic analysis as the external forces, to verify if the pile foundation is safe.

2 EVALUATION OF EQUIVALENT SPRINGS WITH A 3-DIMENSIONAL FEM (GPILE-3D)

2.1 SIMPLIFIED FRAME-SPRING MODEL

The structure-foundation-ground system is, in this paper, simplified to a frame-spring system, as shown in Fig.1. The stiffness matrix of the three-Dimensional equivalent springs is given in the following equation as,

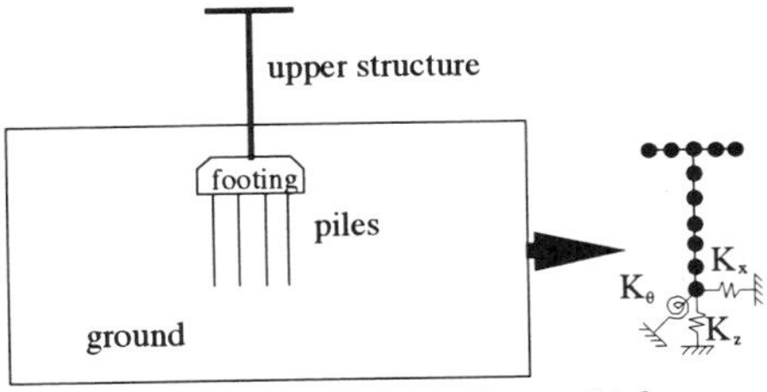

(a) structure-foundation-ground system (b) frame-springs system

Fig.1 Simplification of the model in a dynamic analysis

$$\begin{Bmatrix} F_x \\ F_y \\ F_z \\ M_x \\ M_y \end{Bmatrix} = \begin{bmatrix} k_x & 0 & 0 & 0 & k_{x\theta y} \\ 0 & k_y & 0 & k_{y\theta x} & 0 \\ 0 & 0 & k_z & 0 & 0 \\ 0 & k_{y\theta x} & 0 & k_{\theta x} & 0 \\ k_{x\theta y} & 0 & 0 & 0 & k_{\theta y} \end{bmatrix} \begin{Bmatrix} \delta_x \\ \delta_y \\ \delta_z \\ \theta_x \\ \theta_y \end{Bmatrix} \quad (1)$$

where kx, ky, kz, $k\theta x$, $k\theta y$, $ky\theta x$, and $kx\theta y$ are equivalent springs in the simplified frame-spring model. These springs usually show nonlinear behavior and should be evaluated before the dynamic analysis is carried out. In the following sections, we will introduce a numerical way with static finite element analysis to evaluate the springs.

2.2 NONLINEAR BEHAVIOR OF THE GROUND AND THE PILES

Kimura et al. (1991) developed a three-dimensional finite element analysis program, GPILE-3D, in which the stress-strain relation of the ground was supposed to be elasto-plastic and the associated flow rule of Drucker-Prager's yielding criteria was adopted. Later, various improvements were added to the program (Adachi et al., 1994).

It is known that in order to estimate the behavior of group-pile foundations to a very large lateral deformation, the nonlinearity of both the piles and the ground should be properly considered. The stress-strain relation of the ground is supposed to be elasto-perfect plastic. Drucker-Prager's yielding criteria is expressed as follows:

$$f_y = (J_2)^{1/2} - 3\alpha\sigma_m - \kappa_s = 0 \quad (2)$$

where J_2 is the second invariant of the deviatoric stress tensor and σ_m is the mean stress. α and κ_s are material constants which can be expressed by the average $\overline{N}$ value, ϕ, the frictional angle, and by c, the cohesion of the ground, as

$$\alpha = \frac{\sin\phi}{(9+3\sin^2\phi)^{1/2}} \quad (3)$$

$$\kappa_s = \frac{3c\cos\phi}{(9+3\sin^2\phi)^{1/2}} \quad (4)$$

$$c = 10/16\,\overline{N} \quad (\text{tf/cm}^2) \text{ for clay} \quad (5)$$

$$\phi = 0.3\,\overline{N} + 27 \quad (\text{degree}) \text{ for sand} \quad (6)$$

The nonlinearity of the pile is also considered. For reinforced concrete piles, the relation between the bending moment and the curvature of the piles is simulated by a trilinear relation, taking into consideration the influence on the yielding moment M_{yd} and ultimate moment M_{ud} by the axial force of the pile.

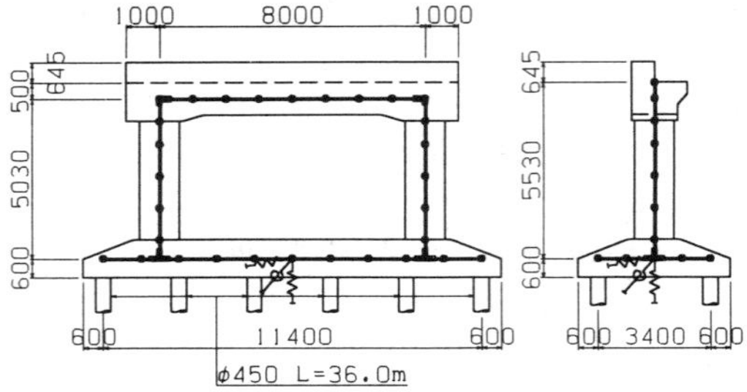

Fig.2 Structure of the bridge and its pile foundation

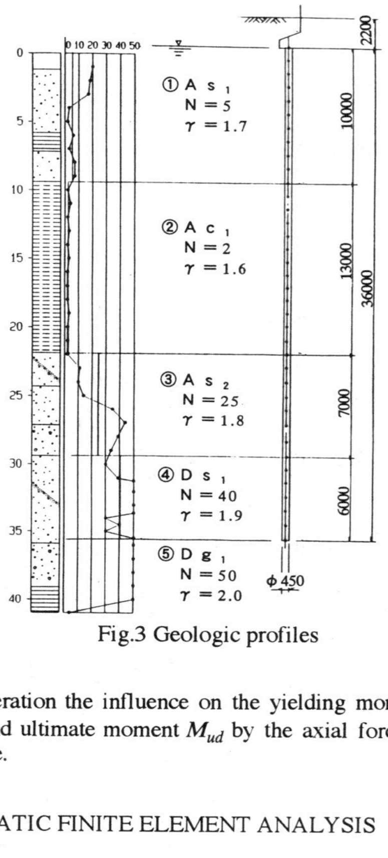

Fig.3 Geologic profiles

2.3 STATIC FINITE ELEMENT ANALYSIS

A static finite element analysis was conducted on a toll bridge, taking into consideration the nonlinearity of both the ground and the pile as discussed in section 2.2. The bridge was supported by a pile foundation with 2×6 pre-cast reinforced concrete piles as shown in Fig.2. The geologic profile of the ground is shown in Fig.3.

Fig.4 shows the three-dimensional finite element mesh of the group-pile foundation used in the analysis, with slide boundaries on the side surfaces and fixed boundaries at the bottom. By taking into consideration the symmetric geometric and loading

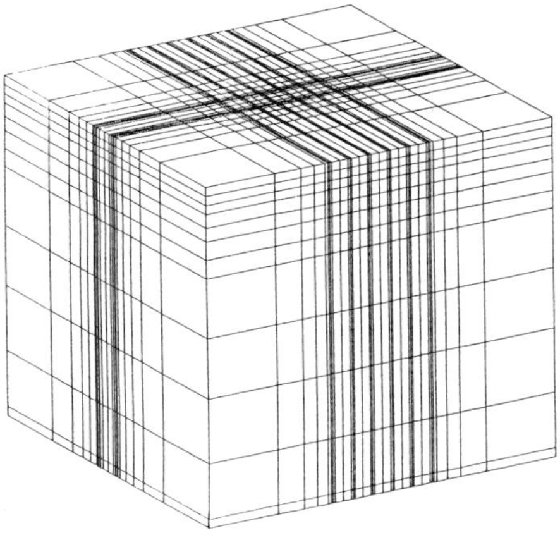

Fig.4 Three-dimensional finite element mesh

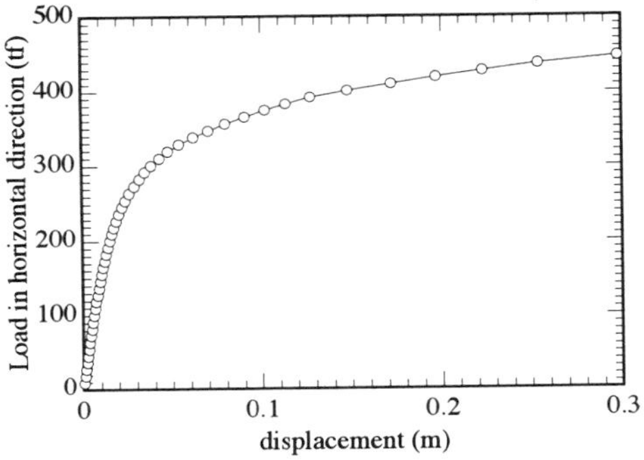

Fig.5 Relation between the Load and the displacement of pile foundation

conditions, only half of the domain shown in Fig.4 is used in the analysis. The mesh has 11466 (the half domain is 6006) nodes and 9880 (the half domain is 4940) 8-node iso-parametric solid elements. In order to evaluate the values of the equivalent springs, the calculation should be done at least five times. In the calculation, the load can be applied by either prescribed loads or prescribed displacements. In the present study, prescribed displacements are used in the calculation so that both the horizontal spring, k_x, and the associated spring, $K_{x\theta y}$, can be obtained in one calculation.

Fig.5 shows the horizontal load-displacement relation of the group-pile foundation under the condition that the rotating angle of the footing is fixed and only the prescribed displacement along the x-direction is applied. From the figure, it is clear that the interaction between group-pile foundation and the ground shows a nonlinear behavior. Table 1 shows the stiffness of the equivalent springs obtained from the above-mentioned analysis.

Table 1 The values of the equivalent springs

K_z (tf/m)	$K_{\theta x}$ (tf*m/rad)	$K_{\theta y}$ (tf*m/rad)	$K_{x\theta y}$ (tf/rad)	$K_{y\theta x}$ (tf/rad)
1.45×10^5	2.17×10^6	4.27×10^5	1.46×10^5	1.75×10^5
$K_x = K_y$ (tri-linear)				
δ(cm)	0.0	1.76	14.06	40.50
P(tf)	0.0	190.0	300.0	400.0

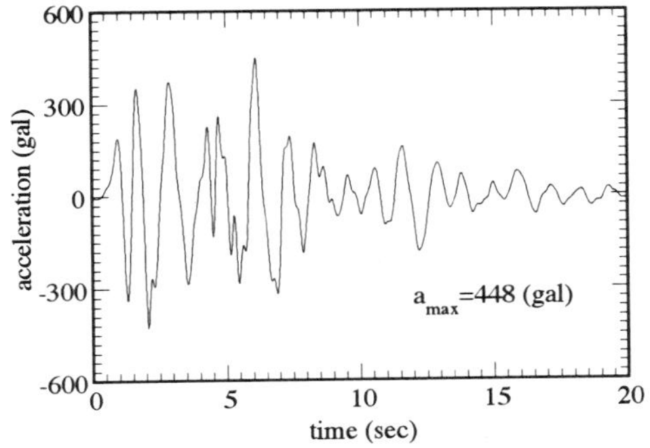

Fig.6 Input earthquake wave

3 SEISMIC EVALUATION OF THE PILE FOUNDATION

Based on the results of the equivalent springs obtained in previous section, a dynamic analysis with a simplified frame-spring model is conducted considering the nonlinearity of both the upper structure and the equivalent springs. The nonlinearity of the upper structure is simulated with a trilinear model based on the strength and deformation condition of reinforced concrete. The direct integration method is adopted for the dynamic analysis. The input earthquake wave (with 20.0 seconds of main vibration) with a maximum acceleration of 448 gal in horizontal direction, as shown in Fig.6, is used in the dynamic analysis.

It is found from the analysis that the response of the maximum forces acting simultaneously on the equivalent springs are 267.0 tf in horizontal direction and 1179.0 tfm in rotating direction. If the nonlinearity of the equivalent springs is not considered in the dynamic analysis of the simplified frame-spring model, the maximum forces are 399.8 tf in horizontal direction and 1670.0 tfm in rotating direction. It is obvious from the difference that the nonlinearity of the interaction between the pile foundation and the ground cannot be ignored or disregarded in a major earthquake.

Using the maximum forces, which are acting on the equivalent springs and evaluated by the dynamic analysis considering the nonlinearity of the springs,

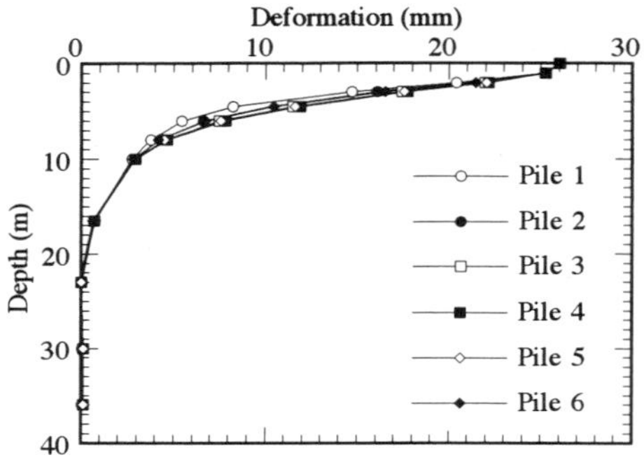

Fig.7 Deformation of the piles

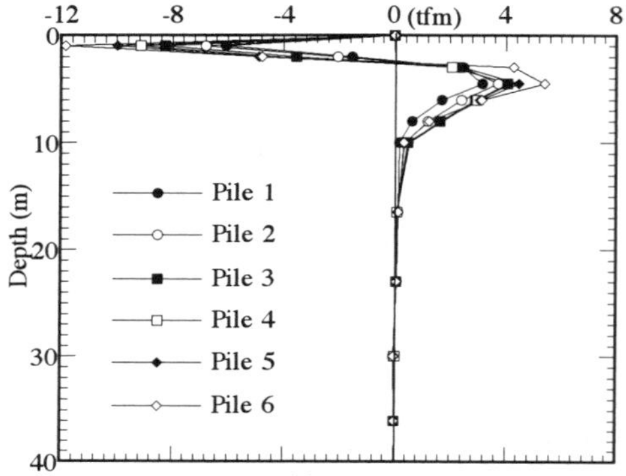

(a) moment

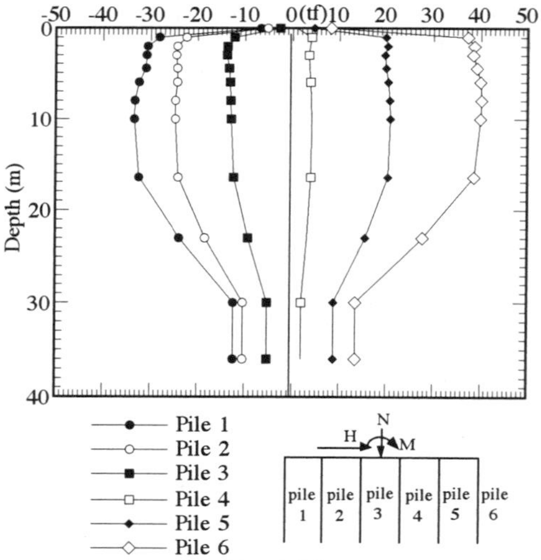

(b) axial force

Fig.8 Distribution of the sectional forces in piles

as the external forces, the three-dimensional elasto-plastic finite element analysis (GPILE-3D) is again conducted on the pile foundation to calculate the

Table 2 Verification of the safety of the piles

Pile No	Sectional forces			Strength			Md/Myd	Vd/Vyd	Md/Mud
	Axial force (tf)	Moment Md(tfm)	shear force Vd(tf)	Myd (tfm)	Vyd (tf)	Mud (tfm)			
①	-28.18	6.07	4.53	2.30	21.02	4.07	2.64	0.22	1.49
②	-22.33	6.79	4.73	3.23	21.87	5.20	2.10	0.22	1.31
③	-12.28	8.27	4.72	4.87	23.00	7.06	1.70	0.21	1.17
④	4.07	9.14	4.79	7.40	24.07	9.79	1.24	0.20	0.93
⑤	19.91	9.98	5.09	9.69	24.55	12.34	1.03	0.21	0.81
⑥	37.25	11.85	7.08	12.04	24.91	15.01	0.98	0.28	0.79

deformation and the sectional forces of the piles. Fig.7 shows the deformation pattern of the piles. Fig.8 shows the distribution of the sectional forces in the piles. It is found that though the deformations of the piles show no much difference, the bending moments in each pile are quite different. Because of the influence of the axial force on the M-ϕ relation of the pile, the bending strength of a compressive pile is larger than the one of a extensive pile. As a result, the front piles(compressive piles) bear more load than back piles(extensive piles).

Table 2 shows the verification of the safety of the piles. It is found that the bending moments in the extensive piles exceed the ultimate strength. Considering the cyclic loading condition of a earthquake, it is clear that the pile foundation is lack of seismic capacity for the input earthquake.

4 CONCLUSION

In this paper, a new way to evaluate the nonlinear interaction between a pile foundation and the ground is proposed to simplify the model in a dynamic analysis when both the nonlinearity of the upper structure and the ground are to be considered. Based on this new method, it is possible to evaluate the influence of the nonlinear interaction between the foundation and the ground on responding values of a structure in earthquakes. It is also found by a case study that the nonlinear interaction between the foundation and the ground plays an very important rule in the seismic evaluation of pile foundation in a major earthquake.

REFERENCES

Kimura. M., Yashima, A. and Shibata, T.
Three-dimensional finite element analysis of laterally loaded piles, *Proc. of 7th Int. Conf. Computer Method and Advance in Geomechanics*, Cairns, A.A. Balkema, Vol. 1, pp. 145-150, 1991.

Adachi, T., Kimura, M., Zhang, F.
Analyses on ultimate behavior of lateral loading cast-in-place concrete piles by three-dimensional elasto-plastic FEM, *Proc. of 8th Int. Conf. Computer Method and Advance in Geomechanics*, Morgantown, A.A. Balkema, Vol. 3, pp. 2279-2284, 1994

Numerical Models in Geomechanics, Pietruszczak & Pande (eds) © 1997 Balkema, Rotterdam, ISBN 90 5410 886 X

3-D model for the analysis of rectangular machine foundations on a soil layer

M.Z.Aşık
Department of Engineering Sciences, Middle East Technical University, Ankara, Turkey

ABSTRACT: In order to analyze the rectangular machine foundations subjected to a vertical harmonic force, a simple mathematical model based on variational principle is developed. A surface footing is resting on a layer of soil deposit with a non-compliant rock or rock-like material at the base. Governing equations of the problem are obtained through the minimization of energy using Hamilton's principle. Equations are nondimensionalized and written in terms of nondimensional parameters which are very useful in engineering practice. Some of the graphs which are very useful and necessary for the design are generated for the first time to be used by the engineers.

1 INTRODUCTION

Machine foundations constructed or faced with in engineering practice are generally in rectangular shape that requires a three dimensional analysis. .Footings resting on a layered medium have a widespread application rather than the footings resting on an elastic half-space and on which most of the analytical efforts were concentrated.

In the mathematical formulation of the problem, the effect of the dynamic interaction through the footing and the soil, the dissipation of energy due to geometric damping and the continuity of the medium should be included in order to represent true behavior of the physical problem.

For the well known simple models, parameter k for one-parameter model -which is the classical Winkler Model- or parameters k and s (where s represents shear effects in the system) for two-parameter model known as the Pasternak (1954) Model were being determined experimentally. In order to improve the Winkler Model, Barkan and Ilyichev (1977) added viscous dampers to the system parallel to the elastic springs regarding energy dissipation, and dashpot constant was determined through the dynamic tests . Vlasov and Leont'ev (1966) derived formulae which relate Pasternak parameters to the subgrade displacement profile by recoursing to a virtual work principle. Vallabhan and Das (1988a, 1988b, 1991) in their studies for static case, developed a new three-parameter model by introducing a new parameter "γ" and named their model the "Modified Vlasov Model". For dynamic analysis, the simulation of energy dissipation in the system is still a problem.

The Lamb's solution of the three dimensional wave propagation problem, -known as "Dynamic Boussinesq Problem"- formed the basis for the study of oscillation of footings resting on a surface of half-space (Reissner, 1936; Sung, 1953; Quinlan, 1953; Richart et al., 1970). A state of uniform stress assumption by Reissner (1936) under the footing directed the attention on radiation damping (or geometric damping) that had not been known until that time.

Arnold (Arnold et al., 1955), Bycroft (1956) and Warburton (1957) have done some studies on the problem by assuming a static stress distribution under the footing. But Bycroft was the first who presented the solution for vibration of a footing on a layered medium.

In the development of the models, Lysmer (1965), Hsieh (1962), Hall (1967) and Richart and Withman (1967) have made important contributions to the vibration analysis of the footings. Later, Luco (1974), Gazetas (1975), Gazetas and Roesset (1976, 1979). developed analytical solutions for different types of footings

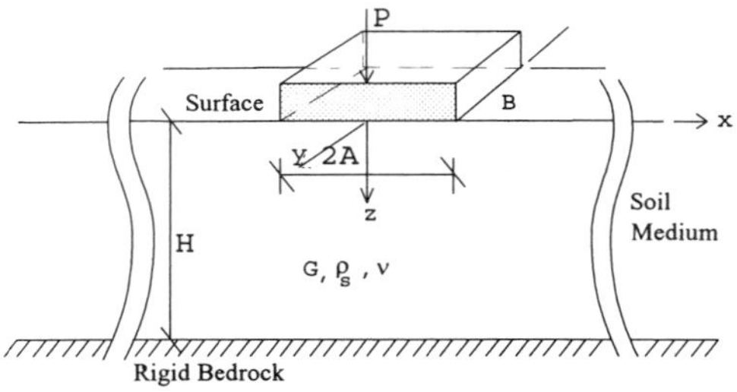

Fig. 1. Cross-section of the Soil-Footing System

on the surface of a layered medium with rigid rock as the last layer or a layered medium lying on a half space. But it was the very rare to find an analytical solution for the rectangular footings.

In the present study, a new model which is able to simulate the energy dissipation in the system by geometry is introduced to solve the problem of vibration of rectangular footings resting on a soil layer.

2 MATHEMATICAL MODEL

Fig. 1. shows the physical problem that has been studied in this paper. Harmonic load P is assumed to be applied at the center of the footing.

Hamilton's principle states that

$$\delta \int_{t_1}^{t_2} (T - V)dt = 0, \tag{1}$$

where, T is the kinetic energy of the footing and soil, and V is the potential energy of the footing and soil. By assuming that the footing and soil experience only vertical vibration (i.e., $u(x,y,z,t) = 0$ in the soil), defining vertical displacement in soil at any point $\overline{W}(x,y,z,t) = W(x,y,t)\phi(z)$, where $\phi(z) = 1$ at $z = 0$ and $\phi(z) = 0$ at $z = H$ and taking the variations in W and ϕ, the following field equations and boundary conditions are obtained through Hamilton's principle:

1. For $-A \le x \le A$ and $-B \le y \le B$ (soil surface under the footing)

$(\rho_f h + \overline{m})\ddot{W} + kW = \overline{q}$ with boundary conditions (BC's) $2t\dfrac{\partial W}{\partial x}\delta W\Big|_{-B}^{B} = 0;$ (2)

2. For $x \le -A$ & $x \ge A$ and $y \le -B$ & $y \ge B$ (soil surface outside the footing)

$\overline{m}\ddot{W} - 2t\nabla^2 W + kW = 0$ with BC's

$$2t\frac{\partial W}{\partial x}\delta W\Big|_{-\infty}^{-A} = 0 \quad \text{and} \quad 2t\frac{\partial W}{\partial x}\delta W\Big|_{A}^{\infty} = 0;$$

$$2t\frac{\partial W}{\partial y}\delta W\Big|_{-\infty}^{-B} = 0 \text{ and } 2t\frac{\partial W}{\partial y}\delta W\Big|_{B}^{\infty} = 0 \tag{3}$$

3. For $0 \le z \le H$ (inside the soil) and $\delta\phi \ne 0$

$$\frac{d^2\phi}{dz^2} - \left(\frac{\gamma}{H}\right)^2\phi = 0 \text{ with BC's } m\frac{d\phi}{dz}\delta\phi\Big|_{0}^{H} = 0; \tag{4}$$

In the above equations,

$$\overline{m} = \int_0^H \rho_s \phi^2 dz = \rho_s H c_t,$$

$$k = \frac{2(1-\nu)}{(1-2\nu)}\int_0^H G\left(\frac{d\phi}{dz}\right)^2 dz = G\frac{2(1-\nu)}{(1-2\nu)}\frac{c_k}{H},$$

$$2t = \int_0^H G\phi^2 dz = GHc_t,$$

in which $c_k = H\int_0^H \left(\dfrac{d\phi}{dz}\right)^2 dz$, $c_t = \dfrac{1}{H}\left(\int_0^H \phi^2 dz\right)$;

and,

$$m = \frac{2(1-\nu)}{(1-2\nu)}\int_{-\infty}^{\infty}\int_{-\infty}^{\infty} GW^2\,dxdy,$$

$$n = \int_{-\infty}^{\infty}\int_{-\infty}^{\infty} G(\bar{\nabla}W \bullet \bar{\nabla}W)dxdy$$

and

$$\left(\frac{\gamma}{H}\right)^2 = \frac{n - \int_{-\infty}^{\infty}\int_{-\infty}^{\infty}\rho_s\left(\frac{\partial W}{\partial t}\right)^2 dxdy}{m}.$$

Solution of the differential Eq. 4 is

$$\phi(z) = \frac{1}{1-e^{2\gamma}}\left(e^{\frac{\gamma}{H}z} - e^{2\gamma}e^{-\frac{\gamma}{H}z}\right). \tag{5}$$

The differential Eq. 3 is solved for the medium outside the footing area. By considering the equilibrium of forces (including soil reactions etc.) acted upon footing and by recoursing to the wave propagation in soils (Lysmer and Kuhlemeyer, 1969; Gazetas, 1987), the following nondimensional force-displacement relationship for a massless rigid footing without material damping in the soil is obtained as follows:

$$\frac{W_0 GA}{P_0} = \frac{1/4}{\left\{\begin{matrix}\dfrac{2(1-\nu)}{(1-2\nu)}\dfrac{c_k}{(H/A)} - (B/A)(H/A)c_t a_0^2 - \\ (H/A)c_t + i(H/A)c_t(1+(B/A)+4/\alpha)a_0\end{matrix}\right\}} \tag{6}$$

where $a_0 = \dfrac{\Omega A}{V_s}$ is the nondimensional frequency,

$$\left(\frac{\alpha}{B}\right)^2 = \frac{k - \overline{m}\Omega^2}{2t}$$

Using the correspondence principle of viscoelasticity (Gazetas (1983), Dobry and Gazetas (1985)) and defining β as the material damping ratio, the impedance of the soil can be written as

$$K = K_s(k + ica_0)(1 + 2i\beta) \quad (7)$$

where K_s is the static stiffness coefficient with $a_0 = 0$. The displacement functions are $F = [K]^{-1}$ and $f_1 = \mathrm{Re}[F]$, $f_2 = \mathrm{Im}[F]$.

The dynamic equilibrium equation of the system in compact form can be written as $M\ddot{W} + R(t) = P(t)$, where $P = P_0 e^{i\Omega t}$ is the harmonic excitation; R(t) is the soil reaction; and the third term is the inertia force of the footing. For a harmonic excitation, the reaction R(t) from the soil, will also be harmonic (i.e., $R(t) = R_0 e^{i\Omega t}$), where Ω is the operational frequency of the machine and P_0 is the amplitude of the force. In general, P_0 can be assumed to be a constant or equal to $m_e e \Omega^2$ which is created by the vibratory machine, where m_e is the unbalanced mass and e is the radius of eccentricity. M is the total mass of the foundation and the machine(s) on this foundation. The uniform harmonic displacement under the footing can now be written in terms of the soil reaction R_0 and the displacement functions f_1 and f_2 is $W = \frac{R_0}{GA}(f_1 + if_2)e^{i\Omega t}$. Then, the following nondimensional amplitude of vibration is obtained:

$$\tilde{W}_0 = \frac{W_0 GA}{P_0} = \left(\frac{f_1^2 + f_2^2}{(1 - b_0 a_0^2 f_1)^2 + (b_0 a_0^2 f_2)^2} \right)^{1/2} \quad (8)$$

where b_0 is a dimensionless mass ratio, $b_0 = \frac{M}{\rho_s A^3}$.

3 NUMERICAL RESULTS

At first, the results obtained for circular footing modeled (Asik (1996)) like rectangular footing in this study are compared with the results from Warburton's model (Warburton, 1957) in Fig. 2 for the sake of verification. There are differences between the two solutions due to different assumptions made in formulating these two models. As it is known, Warburton assumed a stress distribution under the footing as for a static case, whereas the stress distribution under the footing depends on the frequency as in the new model. For H/R=1, the difference is around 5% and

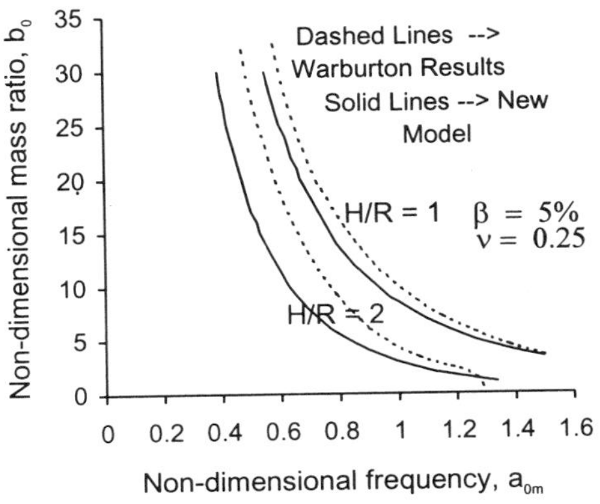

Fig. 2. Comparison of Resonant Frequencies

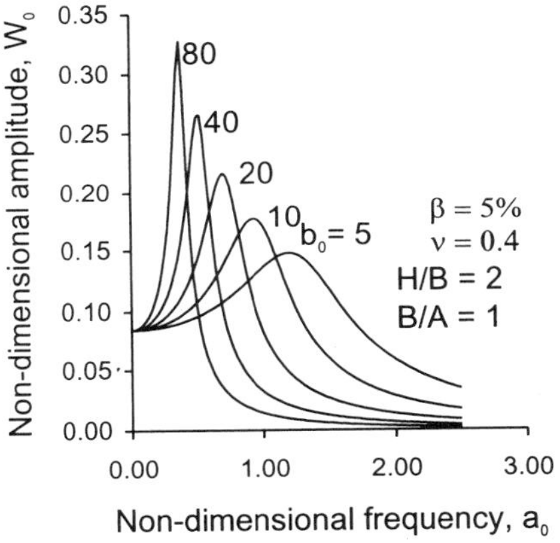

Fig. 3 Comparison of Results for Different Mass Ratios

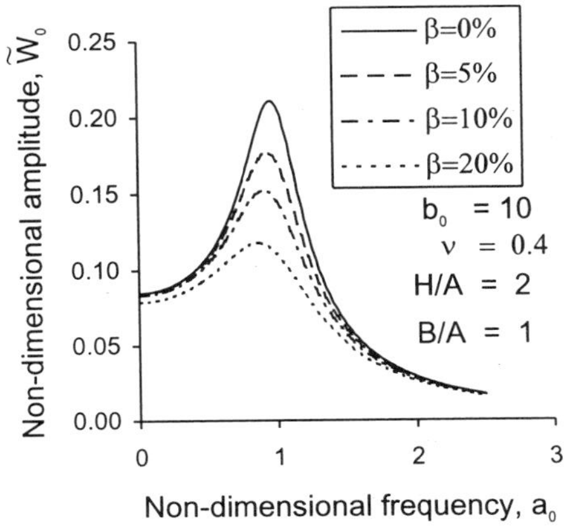

Fig. 4 Effect of Damping Ratio on the Response of Footing

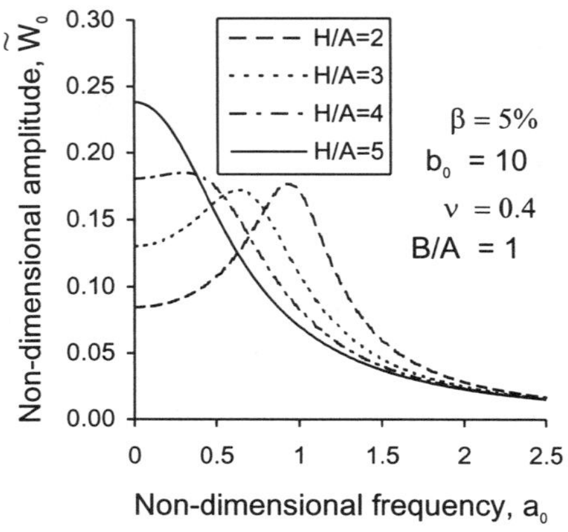

Fig. 5 Effect of H/A Ratio on the Response of Footing

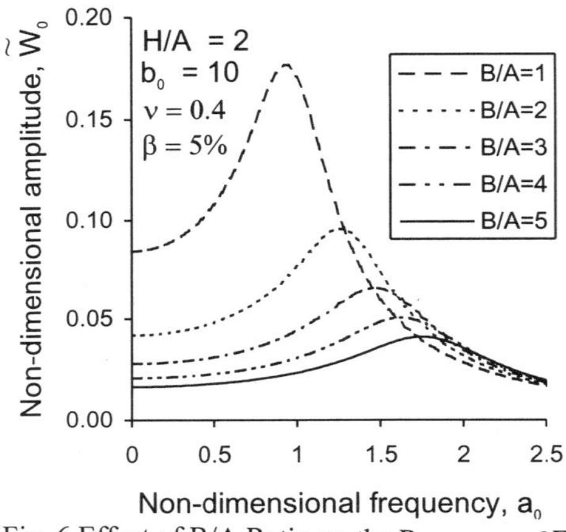

Fig. 6 Effect of B/A Ratio on the Response of Footing

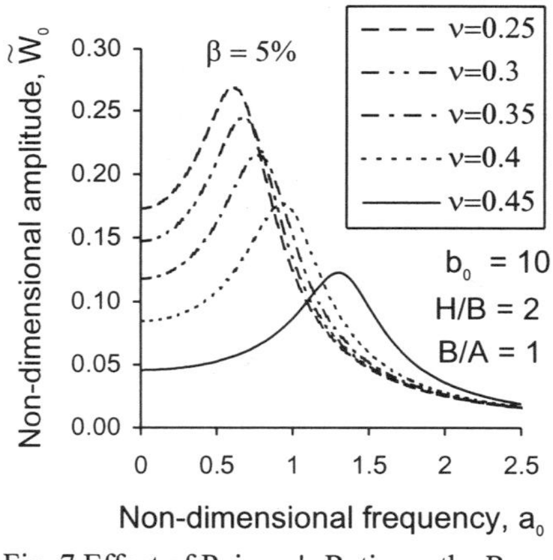

Fig. 7 Effect of Poisson's Ratio on the Response of Footing

for H/R=2, it is around 10%. Also, Warburton used a relaxed boundary condition, which is the assumption of zero secondary (horizontal) stress between the footing and the soil, whereas, in the present case, there is a secondary stress under the footing.

Fig. 3 shows the nondimensional amplitude of vibration $\tilde{W}_0$ versus the nondimensional frequency a_0 for different mass ratios b_0 with fixed parameters $\beta = 5\%$, $\nu = 0.4$ and $H / A = 2$.

Fig. 4 is a classical plot in vibration problems. It shows the effect of the material damping ratio β. As it is known, the amplitude of vibration decreases when the material damping ratio increases. The resonant frequency of the system is not affected very much by the damping ratio.

The geometry of the system has an important effect on the resonant frequency of the system. Fig. 5 shows the effect of the ratio of the soil depth to the half-length of the footing, H/A on the behavior of the system. This ratio has an especially strong effect on the resonant frequency. As the H/A ratio increases, the resonant frequency of the system decreases. The amplitude of the vibration is not affected very much by the H/A ratio in a shallow layer. For high H/A ratios (deep soils), such as H/A equal to 5, the nondimensional amplitude versus the nondimensional frequency curves show the behavior of half space response curves as expected. Fig. 6 is the depiction of the effect of B/A ratio. As it is observed, stiffer solution is obtained and resonant frequency increases when B/A ratio increases.

Fig. 7 shows the effect of Poisson's ratio of the soil on the behavior of the system. It is observed from the curves that Poisson's ratio affects both the amplitude and the resonant frequency of vibration. The amplitude of vibration decreases as Poisson's ratio increases, but the resonant frequency increases as Poisson's ratio increases.

4 CONCLUSIONS

A new mathematical model for the response of rectangular footing resting on a soil layer is developed by using Hamilton's principle combined with a variational approach. This method is semi-analytical and uses an improved form of the classical Vlasov Model for elastic foundations to solve steady-state dynamic problems. Using variational principles, the governing differential equations are developed for the surface disturbance (displacement) $W(x,y,t)$ and vertical decay function, ϕ, which are coupled through parameters k, $2t$, m, and n that describe the characteristics of the soil medium. Two additional parameters α and γ are introduced which are dependent on the properties of the overall system and the operating frequency, Ω. An iterative procedure is applied to find these α and γ parameters.

In the model, the horizontal displacement u is assumed to be zero in the entire soil medium that this assumption leads to a stiffer system. It is also known that, the variational formulation with an assumed displacement function makes the system stiffer, too. If a system is stiffer, it gives higher resonant frequencies and smaller amplitudes than the actual system.

Geometric damping which is important in open systems experiencing the propagation of waves, is represented by edge and corner shear forces. Edge shear forces are calculated from the elastic wave theory by making them inversely proportional to the particle velocity. Therefore, edge shear forces and part of the reaction forces at the corners are considered as the viscous energy absorbing mechanisms.

5 ACKNOWLEDGMENTS

The author of this paper acknowledges the support from the AFP (Research Fund for Projects in Middle East Technical University), Grant AFP-96-03-10-01 for research on vertical vibration analysis of footings on a soil layer.

REFERENCES

Arnold, R. N., Bycroft, G. N. and Warburton, G. B., (1955), "Forced Vibrations of a Body on an Infinite Elastic Solid," J. Appl. Mech. ASME, 22,391.

Asik, M. Z., C. V. G. Vallabhan and Y. C. Das (1996), "Vertical vibration analysis of rigid circular footings on a soil layer with a rigid base", April 1996, SECTAM,Volume XVIII, Alabama, 'Developments in Theoretical and Applied Mechanics'.

Barkan, D. D. and Ilyichev, V. A., 1977, "Dynamics of Bases and Foundations," Proc. 9th ICSMFE, Tokyo,2,630.

Bycroft, G. N., 1956, "Forced Vibrations of a Rigid Circular Plate on a Semi-Infinite Elastic Space and an Elastic Stratum," Phil. Trans. Royal Soc. Lond., A248, 327.

Dobry, R. and Gazetas, G., 1985, "Dynamic Stiffness and Damping of Foundations by Simple Methods," Vibration Problems in Gėotechnical Engineering, ed. G. Gazetas and E. T. Selig, ASCE, 75-107.

Gazetas, G., 1987, "Simple Physical Methods for Foundation Impedances," Dynamic Behavior of Foundations and Buried Structures: Devp. Soil. Mech. Found. Engng.-3, Ed. P. K. Banerjee and R. Butterfield, Elsevier Applied Science, London and New York, 45-93.

Gazetas, G., 1983, "Analysis of Machine Foundation Vibrations: State of the Art," Journal of Soil Dynamics and Earthquake Engineering, 2(1),2-42.

Gazetas, G. and Roesset, J. M., 1979, "Vertical Vibrations of Machine Foundations," J. Gėotech. Engng. Div., ASCE, 105,GT12,1435.

Gazetas, G. and Roesset, J. M., 1976,"Forced Vibrations of Strip Footings on Layered Soils," Meth. Struct. Anal., ASCE, 1,115.

Gazetas, G., 1975,"Dynamic Stiffness Functions of Strip and Rectangular Footings on Layered Soil," S. M. Thesis,, MIT.

Hall, J. R., Jr., 1967, "Coupled Rocking and Sliding Oscillations of Rigid Circular Footings," Proc. International Symposium on Wave Propagation and Dynamic Properties of Earth Materials, Albuquerque, N.M., Aug.

Hsieh, T. K., 1962, "Foundation Vibrations," Proc. Inst. Civil Engrs., 22,211.

Luco, J. E., 1974, "Impedance Functions for a Rigid Foundation on a Layered Medium," Nucl. Engng. Des., 31,204.

Lysmer, J. and Kuhlemeyer, R. L., 1969, "Finite Dynamic Model for Infinite Media," J. Engng. Mech. Div., ASCE, 95, EM4, 859-877.

Lysmer, J., 1965, "Vertical Motions of Rigid Footings," Ph. D. Thesis, University of Michigan, Ann Arbor.

Pasternak, P. L., 1954, "On a New Method of Analysis of an Elastic Foundation by Means of Two Foundation Constants(in Russian)," Gosudarstvennogo Izatestvo Literaturi po Stroitelstvu i Architekutre, Moskow.

Quinlan, P. M., 1953, "The Elastic Theory of Soil Dynamics, Symp. on Dyn. Test. of Soils, ASTM-STP, 156, 3-34.

Reissner, E., 1936, "Stationare Axialsymmetrische, Durch eine Schut-Telnde Masse Erregte Schwingungen eines Homogenen Elastischen Halbraumes," Ing. Arch.,7,381.

Richart, F. E. and Whitman, R.V., 1967, "Comparison of Footing Vibration Tests with Theory," J. Soil Mech. Fdn. Engng. Div., Proc. ASCE, 93, SM6,143-168.

Richart, F. E., Woods, R. D. and Hall, J. R., 1970, Vibrations of Soils and Foundations, Prentice-Hall, 1970.

Sung, T. Y.,1953, "Vibrations in Semi-Infinite Solids due to Periodic Surface Loadings," Symp. on Dyn. Test. of Soils, ASTM-STP, 156, 35-64.

Vallabhan, C. V. G. and Das, Y. C., 1991, "Modified Vlasov Model for Beams on Elastic Foundations," J. Geotech. Engng., ASCE, 117(6), 956-966.

Vallabhan, C. V. G. and Das, Y. C., 1988a, "An Improved Model for Beams on Elastic Foundations," Proc. of the ASME Pressure Vessel Piping Conf.

Vallabhan, C. V. G. and Das, Y. C., 1988b, "A Parametric Study of Beams on Elastic Foundations," J. Engng. Mech.,ASCE, 114(2), 2072-2082.

Vlasov, V. Z. and Leont'ev, N. N., 1966, Beams, Plates and Shells on Elastic Foundations, Israel Program for Scientific Translations, Jerusalem, Israel.

Warburton, G. B., 1957, "Forced Vibration of a Body on an Elastic Stratum," J. Appl. Mech., Trans. ASME, 24, 55-58.

Numerical Models in Geomechanics, Pietruszczak & Pande (eds) © 1997 Balkema, Rotterdam, ISBN 90 5410 886 X

A study of seismic behavior of a caisson quay wall

H. Liu
Geotechnical Engineering Research Institute, Hohai University, Nanjing, People's Republic of China (Presently: Port and Harbour Research Institute, Ministry of Transport, Yokosuka, Japan)

S. Iai, K. Ichii & T. Morita
Port and Harbour Research Institute, Ministry of Transport, Yokosuka, Japan

ABSTRACT In order to identify the seismic behavior of the caisson quay wall, the effective stress analysis of eight types of caissons is performed in the paper. The constitutive model used in this study is a multiple shear mechanism type in strain space and can take into account the effect of rotation of principal stress axis directions. The earthquake motion recorded at a depth of 32 m at the Port island during the 1995 Great Hanshin earthquake is used as input bedrock accelerations. The results of analysis show that, by employing reasonable type of caisson, the residual displacement can be improved from 30 to 50 percent. Finally, several types of caissons having improved response during earthquake are suggested.

1 INTRODUCTION

Caisson quay walls are widely used in port area of many places in the world. During the 1995 Great Hanshin earthquake, many caisson type quay walls were damaged in Kobe Port area of Japan. There are maximum 5m, average 3 m seaward displacements, 1 to 2m settlements induced for the caissons measured after earthquake. The typical cross section and deformed shape of a caisson wall is shown in Fig.1.

The numerical simulation study made by Iai et al(1995) revealed that, for a standard square caisson, the magnitude of deformation is significantly affected by liquefaction induced in the reclaimed soil behind

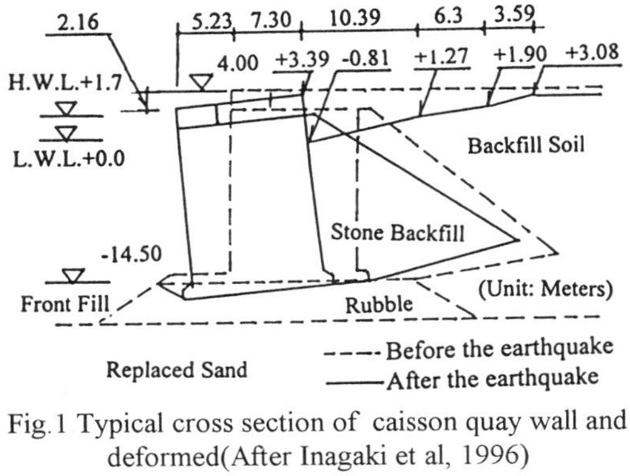

Fig.1 Typical cross section of caisson quay wall and deformed(After Inagaki et al, 1996)

the caisson and the increasing pore water pressure in the replaced sand under the caisson. However, some other factors like caisson shape, unit weight distribution of caisson which are presented as the probable countermeasures, have not been taken into account in the past research. In order to study the deformation mechanism of caissons under the effect of these factors, finite element analysis of eight types of caisson is conducted by using an effective stress analysis, and the results obtained are discussed.

2 EFFECTIVE STRESS ANALYSIS

The effective stress numerical procedure is based on a set of dynamic equations for two phase soil(Zienkiewicz et al. 1982). The model used in the analysis is a multiple shear mechanism type which takes into account the effect of rotation of principal stress axis directions(Iai et al 1992). A static analysis is performed first to simulate the stress distributions which take the effect of gravity into account. The same constitutive model is used as in the static analysis. With these initial conditions, the earthquake response analysis is performed on the caisson with the undrained conditions (Zienkiewicz et al. 1982). The numerical time integration is made by using of Wilson-θ method(θ=1.4), the time interval is 0.01 seconds. Rayleigh damping (α=0 and β=0.003) is used to ensure stability of numerical

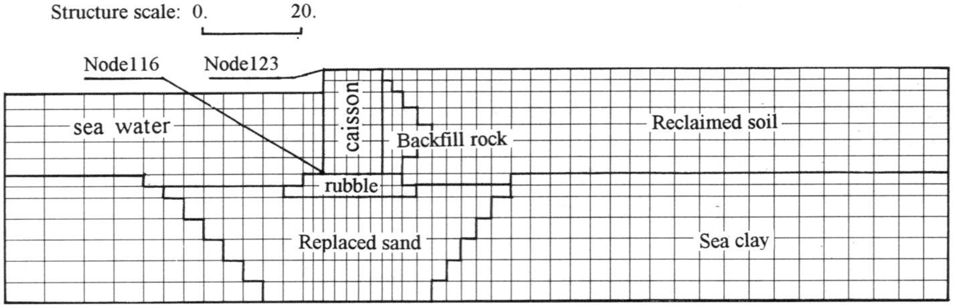

Fig.2 Finite element mesh

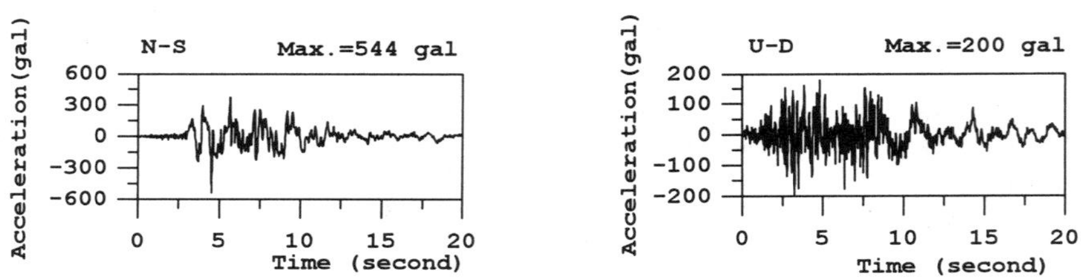

Fig.3 Earthquake motion at 1995 Great Hanshin earthquake recorded at GL-32 meters on Port Island

solution process, and the damping matrix is updated at every time step in accordance with the current tangent stiffness matrix.

In order to study the dynamic failure mechanism of caisson wall, finite element analyses of eight types of caisson are performed. The finite element mesh shown in Fig.2 is used for type one named as standard cross section in here under plane strain condition. There are 705 nodal points, 1204 elements. Four types of elements are used in the analysis, in which, linear element for caisson, nonlinear element for sand and clay, liquid element for sea water and joint element for soil-structure interaction. The sea water is simulated as incompressible fluid and formulated as an added mass matrix based on the equilibrium and continuity of fluid at the solid-fluid interface(Zienkiewicz, 1977). Also shown in Fig.2 are nodes 123 at the top of caisson and 116 at the bottom of caisson which will be discussed later in the paper.

At the side boundaries of cross section, free field response motions are given through the transmitting boundaries to approximate in-coming and out-going waves to and from the finite element region. At the bottom boundary, the record of January 17,1995 Great Hanshin earthquake at the depth 32 meters on Port Island is used as the bedrock motion, the maximum horizontal acceleration is 0.54g, while the maximum vertical acceleration is 0.2g. The recorded accelerations in time domain are shown in Fig.3.

The other seven types of caisson shown in Fig.4 change merely the shape or the distribution of unit weight of caisson but keep all the other soil properties similar to standard cross section. In which, inclination bottoms of caisson are designated as types two and three inclining to seaside and landside respectively, three angles of inclination are set for each type. For type four, five and six shown in Fig.4, the shape of caisson are changed but the area of caisson keep the same to the standard cross section as type one. For types seven and eight, the shape are keeps the same to standard cross section but the distribution of unit weight of caisson are changed.

There are ten parameters in the present model, two of which K_a and G_{ma} characterize elastic properties of soil; another three ϕ_f, ϕ_p and H_m specify plastic shear behavior, and the rest p_1, p_2, w_1, s_1 and c_1 characterize dilatancy of soil. In this analysis, the model parameters are calibrated by referring to in-situ velocity measurements and the cyclic triaxial test results of in-situ frozen samples(Kobe, 1995). All the calculated parameters are shown in Table 1.

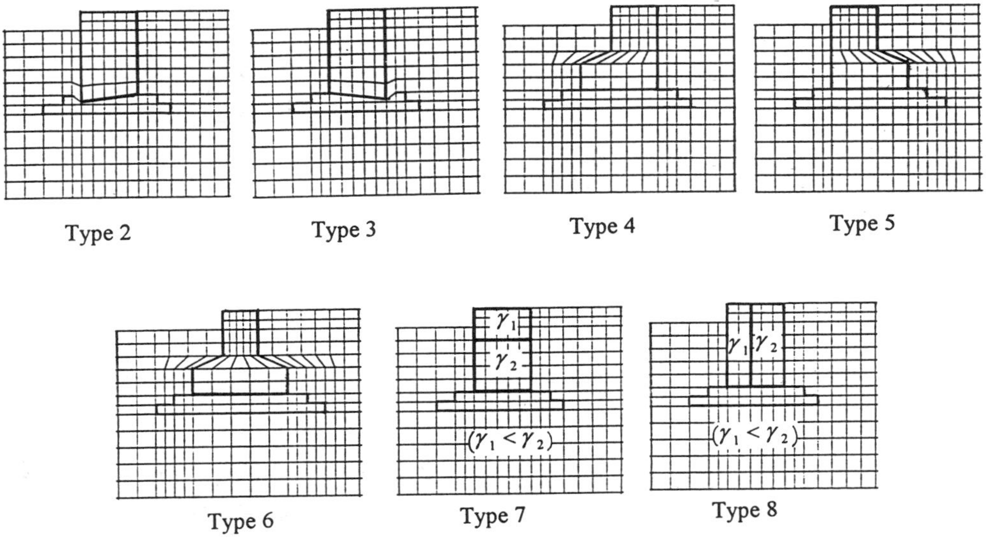

Fig.4 Types of caisson from two to eight

Table 1 Calculated parameters

Layer	γ (tf/m^3)	G_0 (tf/m^2)	$-\sigma_m$ (tf/m^2)	ϕ'_f (deg)	ϕ'_p (deg)	μ	Dilatancy parameters S_1 w_1 P_1 P_2 c_1
Sea clay	1.70	7497.0	14.30	30.0		0.33	
Replaced sand	1.80	5832.0	10.60	37.0	28	0.33	0.005, 5.50, 0.6, 0.6, 2.30
Reclaimed soil	1.80	7938.0	6.300	36.0	28	0.33	0.005, 6.00, 0.5, 0.8, 2.43
Backfill rock	2.00	18000	9.800	40.0		0.33	
Founda. rubble	2.00	18000	9.800	40.0		0.33	

3 ANALYSIS RESULTS

The results of caisson quay wall predictions are shown in Fig.5 through Fig.8 and table 2. Shown in Fig.5 are the time histories of displacement of Node 123 at the top of caisson for type one, it appears that the displacements are gradually increased for about 15 seconds and then kept it at the residual values until the end of earthquake. The whole deformation distribution of type one is shown in Fig.6. The computed horizontal residual displacements is 3.55m while the settlement is 1.67m on the top of caisson. The deformation induced mainly around the caisson including in the foundation. It is speculated that the failure mechanism is similar to the one of circular slip analysis assumed for evaluating the bearing capacity of foundation. The order of magnitude of these results is consistent with the observed deformation after earthquake(Inagaki et al. 1996).

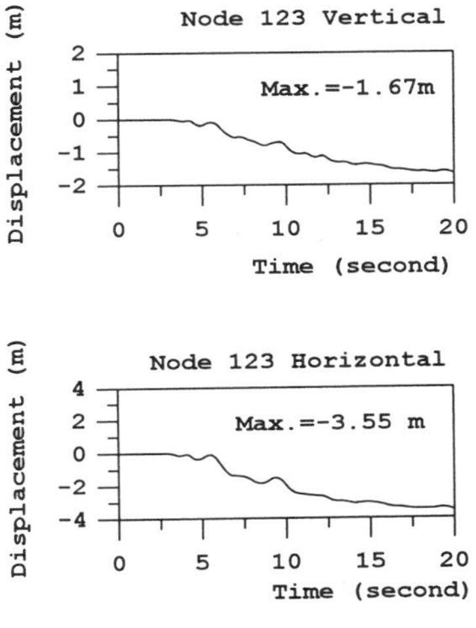

Fig.5 Time history of residual displacement

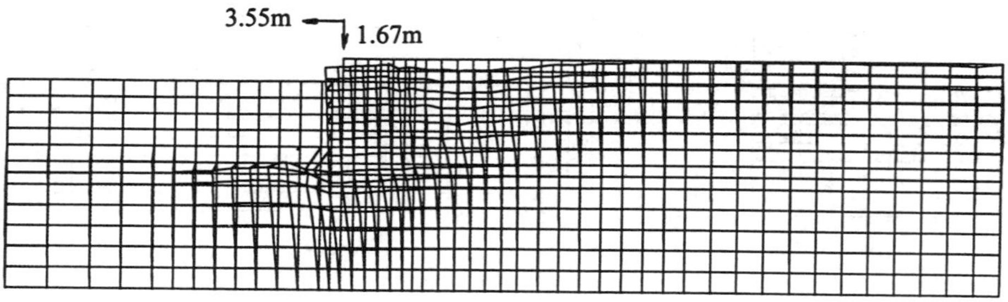

Fig. 6 Computed residual displacement of type one

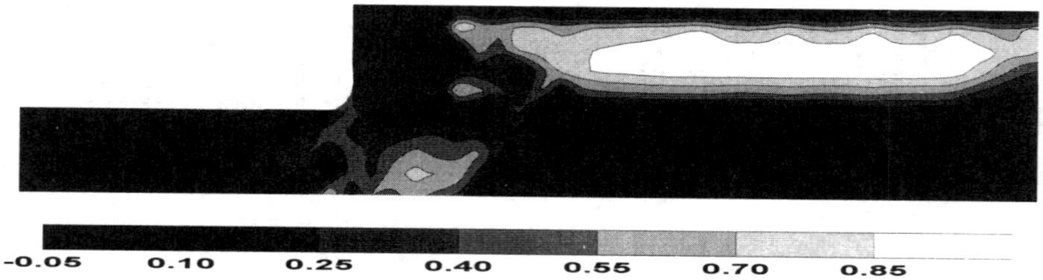

Fig.7 Computed excess pore water pressures of type one

The excess pore water pressure ratio is computed and specified in terms of $(1-\sigma'_m/\sigma'_{m0})$, where σ'_m, σ'_{m0} are current and initial mean effective stresses respectively. The computed excess pore water pressures contourof type one is shown in Fig.7. It is seen that high excess pore water pressure is induced in the reclaimed soil behind the caisson and in the replaced sand under the caisson.

In order to reveal the influence affected by excess pore water pressure, the reclaimed soil behind the caisson and the replaced sand are idealized as non-liquefiable soil which has the same properties as those used in the aforementioned analysis but lacks the characteristics of dilatancy. The computed horizontal displacement is 1.63m and the settlement is 0.64m at the top of caisson at the end of earthquake. It is seen that, the excess pore water pressure increasing in the reclaimed soil and replaced sand is to increase the deformation of caisson about 2 times as large as that purely caused by the seismic inertia force, it is significantly affected the performance of residual deformation of caisson. Therefore, these two phenomenas (i.e. liquefiable and non-liquefiable) will be discussed for the rest of the analyses.

Shown in Fig.8 is the relation between residual deformation and case numbers. The results are also listed in Table 2. It is seen that, for type two(which set the inclining bottom of caisson to seaside), when the inclining angle changes from 0.0 to 7.13 degrees, there is only a little but not a big difference in deformation, reducing about 1.8 to 3.7%. For type three which set the inclining bottom of caisson to landside, the residual deformation is reduced from 4.0 to 7.6% when the inclining angle changes from 0.0 to 7.13 degrees. Even improved for type two, but it is not yet a good countermeasure.

For type four to six(which change the standard cross section to other style but keep the same area of caisson), the horizontal residual deformation is improved from 34 to 37% while the vertical deformation is improved from 48 to 51% for type four and six, but only about 5% for type five. It is speculated in here that, comparing to standard cross section of type one, setting the center of weight of type four deviate to landside or enlarging symmetrically the width of caisson bottom can restrain the residual deformation.

The unit weight distribution of caisson is studied on types seven and eight. For type seven, both the weight and center point of weight of caisson are decreased, the residual deformation only reduced 1.4%. However, the displacement can be reduced to 20% for horizontal and 36% for settlement of type

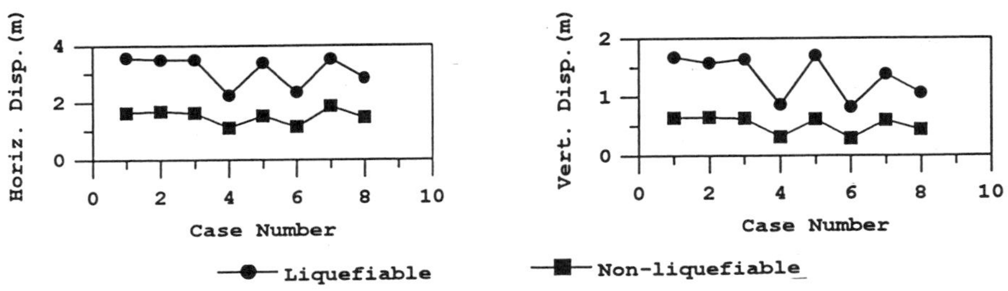

Fig. 7 The relations between residual deformation and case numbers

Table 2 The computed displacement of caisson(unit: meter)

	Caisson type	Bottom Inc. angle (deg.)	Top of caisson		Bottom of caisson	
			Hori. Disp.	Settle.	Hori. Disp	Settle.
Liquefiable	One	0.00	3.548	1.672	2.714	1.672
	Two	2.38	3.483	1.640	2.640	1.640
		4.76	3.482	1.570	2.537	1.570
		7.13	3.416	1.557	2.528	1.557
	Three	2.38	3.404	1.610	2.688	1.610
		4.76	3.484	1.637	2.760	1.637
		7.13	3.275	1.477	2.617	1.477
	Four	0.00	2.236	0.860	2.570	0.739
	Five	0.00	3.369	1.697	2.576	1.697
	Six	0.00	2.337	0.817	2.335	0.818
	Seven	0.00	3.498	1.381	2.540	1.381
	eight	0.00	2.832	1.062	2.814	1.062
Non-liquefiabel	One	0.00	1.628	0.643	1.090	0.643
	Two	2.38	1.657	0.644	1.054	0.644
		4.76	1.684	0.648	1.050	0.648
		7.13	1.684	0.657	1.012	0.658
	Three	2.38	1.612	0.632	1.097	0.632
		4.76	1.619	0.628	1.100	0.628
		7.13	1.625	0.622	1.097	0.622
	Four	0.00	1.104	0.313	1.242	0.263
	Five	0.00	1.520	0.610	1.006	0.610
	Six	0.00	1.130	0.287	1.139	0.284
	Seven	0.00	1.848	0.596	1.083	0.596
	Eight	0.00	1.452	0.434	1.230	0.434

eight. From the view of construction, it is possible to change the distribution of unit weight, for example, using the quenched blast furnace slag or light cement treated soil in the part area of caisson.

If the reclaimed soil behind the caisson and the replaced sand are taken into account for the unliquefy soil, it is seen from Fig.7 and Table 2 that, the computed residual displacement can be improved about 30 to 32% for horizontal and 51 to 55% for vertical for type four and six, and 11 to 32% for type eight. Therefore, by employing the reasonable type of caisson like type four, five and eight, the residual deformation of caisson quay wall may be restrained.

4 CONCLUSIONS

The numerical simulation for eight types of caisson quay wall is performed in the paper, the lessons may be summarized as follow:

1. For the standard cross section as type one, the computed horizontal displacement and settlement at the top of caisson are 3.55m and 1.67m respectively. High excess pore water pressure induced in the reclaimed soil behind the caisson and in the replaced sand under the caisson, and it is significantly affected the performance of residual deformation of caisson. The results of analysis are in good agreement with the observed deformation after earthquake.

2. For type two and three which set the inclining bottom of caisson, there is only a little difference in deformation induced under different inclining angles. So setting the sloping bottom of caisson may be not considerated as an effective countermeasure.

3. For type four and six which change the standard cross section to new styles but keep the same area of caisson, the residual deformation will be more improved by 30 to 50%, but only a little difference in deformation induced for type five. Therefore, setting the center of weight deviate to landside as type four and enlarging symmetrically the width of caisson bottom as type six may be suggested to improve the earthquake response.

4. The unit weight of caisson is also studied. If we change the distribution of unit weight as in type eight(for example, using the quenched blast furnace slag or light cement treated soil in the part area of caisson), the residual deformation can also be reduced from 20 to 36%, so it can also be used as a countermeasure.

5. For the reasonable type of caisson just suggested by numerical simulation, further studies like shaking table test for these new types of caissons is highly recommended.

ACKNOWLEDGMENT

The authors gratefully acknowledge the supports of the STA Fellowship of Japan and the National Science Foundation of China.

REFERENCE

Iai, S., Matsunaga, Y and Kameoka, T. 1992, Strain space plasticity model for cyclic mobility, *Soils and Foundations,* 32(2):1-15.

Iai, S., Ichii, K., and Morita, T. 1995, Effective stress analysis on a caisson type quay wall, *Technical Note of the Port and Harbour Research Institute*, 813:253-279(In Japanese).

Inagaki, H., Iai, S., Sugano, T., Yamazaki, H., and Inatomi, T. 1996, Performance of caisson type quay walls at Kobe port, *Soils and Foundations, Special Issue,* 119-136.

Kobe Development Beauty, Report of geotechnical behavior investigation of reclaimed soil for Port and Rokko Islands, 1995.

Ministry of Transport ed. 1993, *Handbook on Liquefaction Remediation of reclaimed land*, 243-245(in Japanese).

Zienkiewicz, O.C. 1977, *The Finite Element Method, 3rd edition*, McGraw-Hill Book Co.

Zienkiewicz, O.C. and Bettess, P. 1982, Soils and other saturated media under transient, dynamic conditions, *Soil Mechanics-Transient and Cyclic Loads*, Pande & Zienkiewicz eds., John Wiley and Sons, 1-16.

Numerical Models in Geomechanics, Pietruszczak & Pande (eds) © 1997 Balkema, Rotterdam, ISBN 90 5410 886 X

Numerical simulation of soil-foundation interaction behavior during subsoil liquefaction

Y.Taguchi & A.Tateishi
Taisei Corporation, Tokyo, Japan

F.Oka & A.Yashima
Gifu University, Japan

ABSTRACT: The soil-foundation interaction behavior during liquefaction was investigated by using effective stress based two and three-dimensional finite element method. The behavior of a foundation structure and its surrounding ground near the quaywall during a strong seismic motion was discussed in detail. The numerical result by two-dimensional analysis and that by three-dimensional analysis were compared in terms of earth pressure as well as pore water pressure acting on the foundation. The ground displacements in both analyses were also compared. A three-dimensional analysis was found to be able to evaluate the influence of the multi-directional seismic loading as well as the large lateral flow of liquefied soil around the foundation.

1 INTRODUCTION

In designing the foundation structure in a liquefiable ground, it is very important to predict earth pressure as well as pore water pressure acting on the foundation during an earthquake. A two-dimensional finite element analysis has been mostly carried out to get these informations. In the real situation, however, it is very difficult to model soil-foundation interaction behavior in two dimensional plane strain condition. The informations obtained from the two-dimensional analysis are thought to be not quantative but qualitative. In the present study, attempts were made to investigate the soil-foundation interaction behavior during liquefaction by using two and/or three-dimensional finite element method. A foundation structure (bridge pier with caisson foundation) and its surrounding ground near the quaywall was modeled. An analytical behavior during a strong seismic motion was discussed. The numerical result by two-dimensional (2D) analysis and that by three-dimensional (3D) analysis were compared in terms of earth pressure as well as pore water pressure acting on the foundation. The ground displacements in both analyses were also compared.

2 NUMERICAL METHOD

The constitutive model in the present study was based on the concept of the non-linear kinematic hardening rule, which had originally been used in the field of metal plasticity (Chaboche and Rousselier; 1983). Oka et al. (1992) derived a cyclic elasto-plastic constitutive model for sand through use of the concept of non-linear kinematic hardening. Tateishi et al. (1995) incorporated a new stress dilatancy relationship and cumulative strain dependent characteristic of the shear modulus to modify the original model. Such a constitutive model was formulated under the three-dimensional stress conditions in the present study. The validity of the constitutive model had been verified by the experimental evidence from the hollow cylindrical torsion tests under various stress conditions (Tateishi et al.; 1995).

This constitutive model was then incorporated into a coupled finite element-finite difference (FEM-FDM) numerical method (Oka et al.; 1994) for the liquefaction analysis of a fluid-saturated ground. Using a **u-p** (displacement of the solid phase and pore water pressure of the liquid phase) formulation, the numerical method was developed. The finite element method was used for the spatial discretization of the equilibrium equation, while the finite difference method was used for the spatial discretization of the conservation law of mass. Newmark's β-method was used for the time discretization of both equations. The applicability of the proposed numerical method had been already verified by the previous studies (e.g., Taguchi et al.;1995, Taguchi et al.;1996).

3 NUMERICAL MODEL AND CONDITION

In the present study, the foundation structures and its surrounding ground near the quaywall was modeled. The numerical model and material properties are shown in Fig. 1 and Table 1, respectively. A caisson foundation, pier, quaywall and back-fill were modeled as elastic materials. On

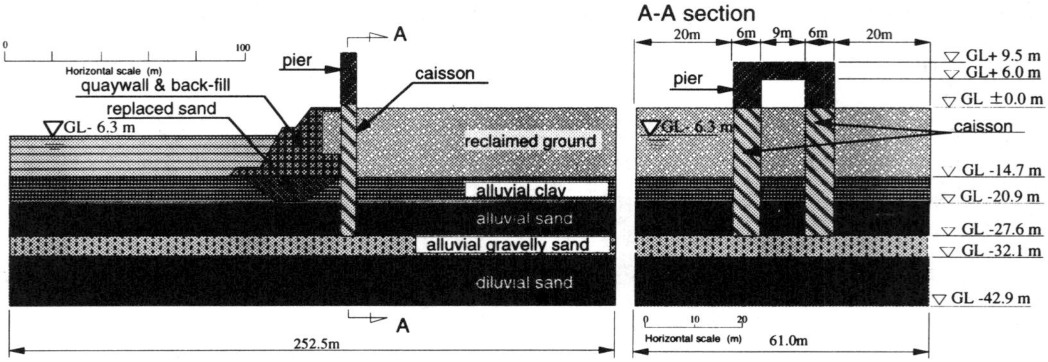

Fig. 1 Numerical model

Table 1 Material properties

Depth (GL m)	Soil type <No.>	Unit weight tf/m³ (SPT)	Void ratio e	shear wave Velocity Vs m/sec	Poisson ratio ν	Coefficient of permeability m/sec	G_0/σ'_{m0}	Phase transformation angle degree (stress ratio M_m*)	Failure angle degree (stress ratio M_f*)	liquefaction strength N_c=10	liquefaction strength N_c=30	Plastic modulus parameters upper G_{max}^P mid. G_{min}^P lower Cf*	Dilatancy parameters upper: D_0* lower: n	Reference strain upper: γ^*_{DAr} lower: γ^*_{DAr}
0.0 – -6.3	<6> sandy gravel (reclaimed)	1.8 (13)	0.88	150	0.25	—	2000	28.0 (0.91)	29.0 (0.94)	—	—	6000 600 2000	0.0 —	∞ ∞
-6.3 – -14.7	<5> sandy gravel (reclaimed)	1.8 (13)	0.88	150	0.25	3.0×10^{-6}	800	28.0 (0.91)	29.0 (0.94)	0.20	0.16	2400 240 2000	0.5 3.0	0.25% 0.10%
-14.7 – -20.9	<4> alluvial clay	1.5 (4)	1.20	135	0.30	1.0×10^{-8}	250	28.0 (0.91)	44.5 (1.49)	—	—	3000 300 2000	0.0 —	∞ ∞
-20.9 – -27.6	<3> alluvial sand	1.8 (26)	0.75	215	0.25	1.0×10^{-5}	600	28.0 (0.91)	34.7 (1.15)	0.36	0.28	4200 420 2000	0.5 5.0	0.15% 0.10%
-27.6 – -32.1	<2> diluvial sand with gravel	1.9 (35)	0.70	315	0.25	1.0×10^{-5}	1100	30.0 (0.98)	37.9 (1.26)	0.60	0.45	7000 700 2000	0.1 5.0	0.05% 0.05%
-32.1 – -42.9	<1> diluvial sand	1.9 (33)	0.70	315	0.25	1.0×10^{-5}	800	30.0 (0.98)	37.2 (1.24)	0.60	0.45	6000 600 2000	0.1 5.0	0.05% 0.05%
	<5> replaced sand	1.8 (13)	0.88	150	0.25	3.0×10^{-6}	800	28.0 (0.91)	29.0 (0.94)	0.20	0.16	2400 240 2000	0.5 3.0	0.25% 0.10%

Structural material <No.>	Unit weight tf/m³	Elastic modulus tf/m²	Poisson's ratio ν
<7> Caisson foundation	2.6	7.5×10^5	0.25
<8> backfill soil (gravelly sand)	2.0	2.3×10^5	0.25
<9> Revetment	2.6	7.5×10^5	0.25

Note 1. Initial stress in every soil element was calculated by the previous initial stress analysis.
2. The values of over consolidation ratio in all soil materials were assumed as 1.2.

the other hand, soil deposits including reclaimed ground and replaced sand were modeled as elasto-plastic materials mentioned above. Material parameters shown in Table 1 were determined from the previous studies based on the ground information at coastal area in Hanshin district. Note that the material parameter of replaced sand was assumed to be the same with that of reclaimed ground.

Table 2 shows the list of the analytical cases carried out in the present study. Four kind of cases were conducted and compared with each other. Case-2D was the two-dimensional analytical case, i.e., the plane strain model on the section crossing the caisson structure to the transverse direction against the quaywall line. Case-3D-1, Case-3D-2 and Case-3D-2' were three-dimensional analytical cases. Case-3D-1 and Case-3D-2 had the same analytical model except for the input acceleration property. The input acceleration in Case-3D-1 had uni-directional horizontal component in the transverse direction against the quaywall line, which was able to compare with Case-2D. It is noted, however, that analytical model in Case-3D-1 did not perfectly correspond to that in Case-2D, because finite element mesh in Case-3D-1 was coarser than that in Case-2D.. The input acceleration in Case-3D-2 had multi-directional

Table 2 Analytical cases

case name	model dimension	input horizontal acceleration	number of finite elements (nodes)
Case-2D	two (plane strain)	uni-directional	724 (777)
Case-3D-1	three	uni-directional	1183 (1608)
Case-3D-2	three	multi-directional	1183 (1608)
Case-3D-2'	three	multi-directional	2074 (2662)

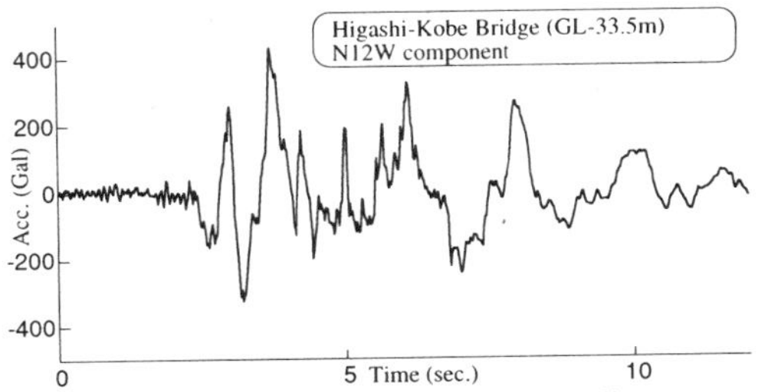

(a) transverse direction to the quaywall line

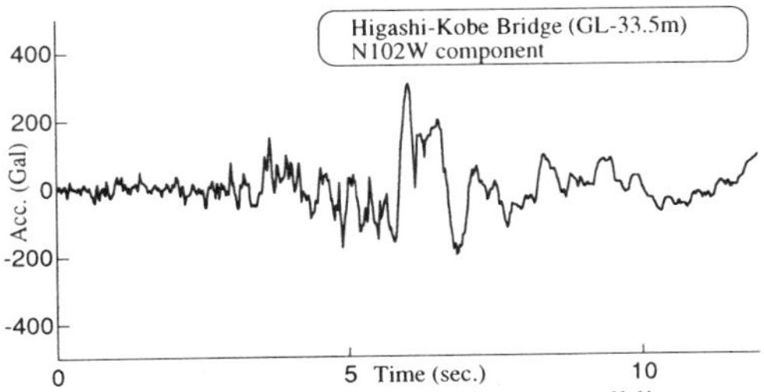

(b) longitudinal direction to the quaywall line

Fig. 2 Input horizontal acceleration

horizontal components in the longitudinal direction as well as transverse direction against the quaywall line. Comparing the analytical result in Case-3D-2 with that in Case-3D-1, the influence of multi-directional input motion were investigated. And the finite element mesh in Case-3D-2' was much finer than that in Case-3D-2. These two cases were compared to investigate the effect of the mesh size and interaction behavior between the foundation and its surrounding soils.

The time histories of input acceleration used in the present study is shown in Fig. 2. These were observation records obtained at GL-33.5m below the Higashi-Kobe Bridge during the 1995 Hyogoken-Nambu Earthquake. Input acceleration in the transverse direction against the quaywall line shown in Fig. 2(a) was the N12W component, and input acceleration in the longitudinal direction to the quaywall line shown in Fig. 2(b) was the N102W component. Twelve seconds including the strong motions were used. The vertical acceleration was not considered in the present study.

Boundary conditions are as follows. The node displacements at bottom surface were fixed, and the lateral boundary on the transverse section against the quaywall line was a equi-displacement boundary condition. The lateral boundary on the longitudinal section to the quaywall line, on the other hand, was dealt with that the free field seismic response behavior could be reproduced. The drained boundary condition was assumed on the seabed surface (GL-6.3m) at the front side of the quaywall and on the ground water level (GL -6.3m) at the back side of the quaywall. Other boundaries were assumed undrained boundary condition. 5% of Rayleigh damping, and β=0.3025,

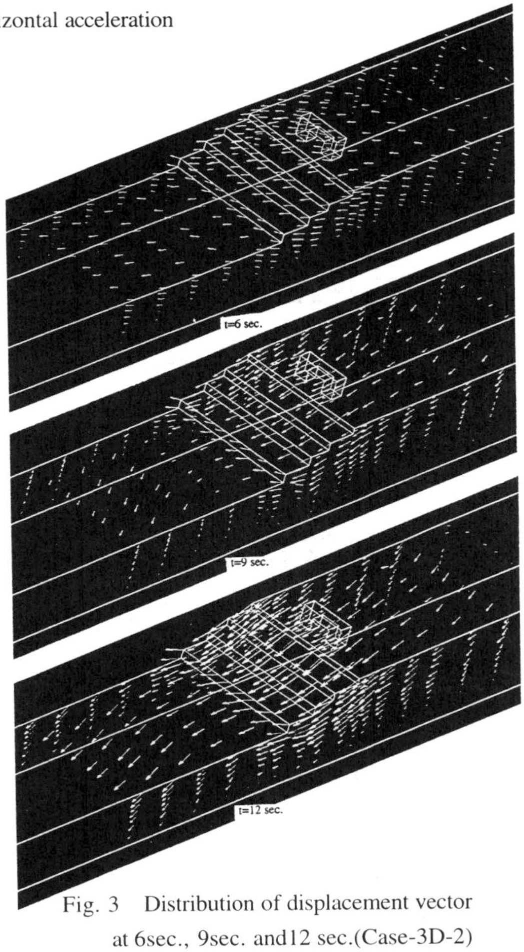

Fig. 3 Distribution of displacement vector at 6sec., 9sec. and12 sec.(Case-3D-2)

γ=0.6 in Newmark's β-method parameters were used for the numerical stability.

4 ANALYTICAL RESULTS AND DISCUSSIONS

Analytical results of four cases shown in Table 2 are indicated and discussed in this section. Fig. 3 shows the displacement distribution in Case-3D-2 at 6 sec., 9 sec. and 12 sec.. Fig. 4 summerizes the comparison of analytical results in Case-2D, Case-3D-1 and Case-3D-2, which are time histories of horizontal as well as vertical displacement at the top of quaywall, and time histories of excess pore water pressure ratio at the alluvial sand layer just below the quaywall (GL-24m). It is found from Fig. 3 that the large displacement vectors form a circular slip surface around the quaywall including the caisson foundation. Although Fig. 3 shows the results in Case-3D-2 (three-dimensional multi-directional analytical case), the displacement mode looks like almostly two-dimensional behavior. From the above reason, a significant difference could not be found in each analytical case as shown in Fig. 4.

As shown in Fig. 4(e) and (f), the excess pore water pressure was sharply increasing between 3 sec. and 4 sec. due to the strong input motion in the transverse direction (Fig. 2(a)). The horizontal displacement as well as vertical displacement (settlement) were also significantly increasing at the same time. As for the horizontal displacement, it could be seen that 50 cm ~ 80 cm of residual displacement toward the sea side was obtained in the transverse direction against the quaywall line. On the other hand, little residual displacement was occurred in the longitudinal direction as shown in Fig. 4(b). As for the vertical displacement, it can be seen from Fig. 4(c) and (d) that 10 cm ~ 30 cm of residual settlement was obtained.

Comparing the result in Case-2D with that in Case-3D-1, qualitative trend almostly coincided with each other. However, the displacement as well as the build-up of pore water pressure in Case-2D was generally larger than that in Case-3D-1 as shown in Fig. 4(a), (c) and (e). It is considered that the difference in detail of the modeling method between Case-2D and Case-3D-1 would be occurred such difference in both analytical results.

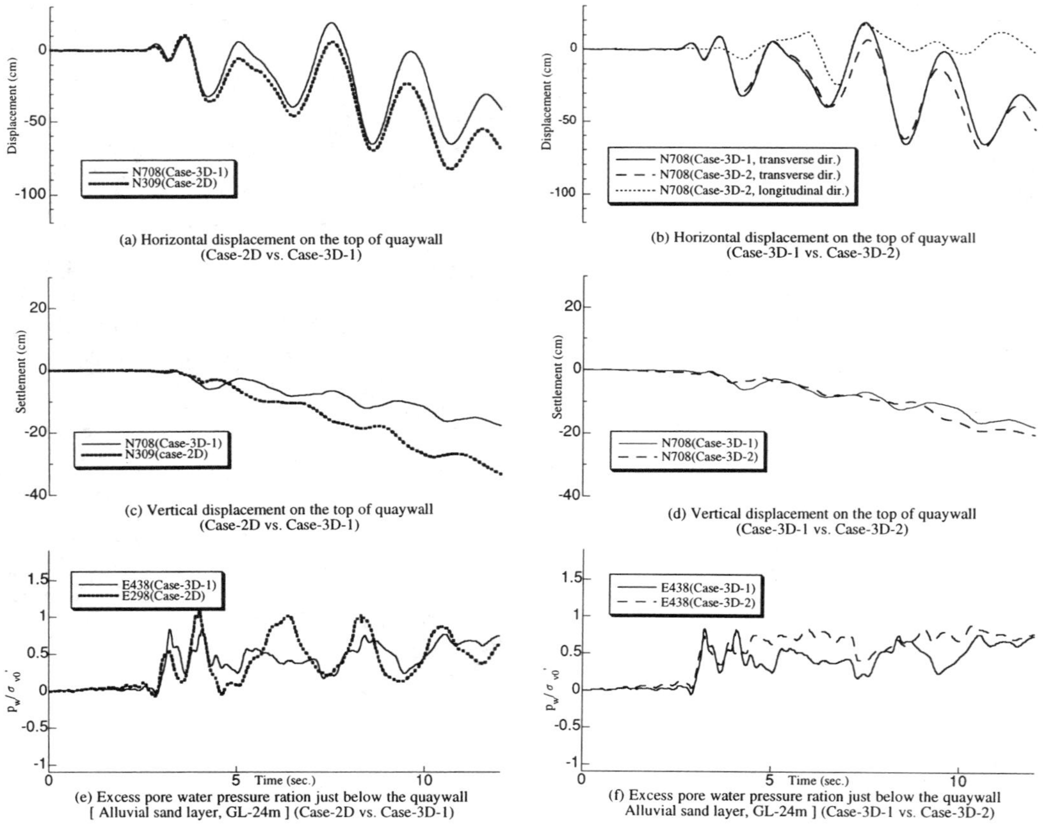

(a) Horizontal displacement on the top of quaywall (Case-2D vs. Case-3D-1)

(b) Horizontal displacement on the top of quaywall (Case-3D-1 vs. Case-3D-2)

(c) Vertical displacement on the top of quaywall (Case-2D vs. Case-3D-1)

(d) Vertical displacement on the top of quaywall (Case-3D-1 vs. Case-3D-2)

(e) Excess pore water pressure ration just below the quaywall [Alluvial sand layer, GL-24m] (Case-2D vs. Case-3D-1)

(f) Excess pore water pressure ration just below the quaywall Alluvial sand layer, GL-24m] (Case-3D-1 vs. Case-3D-2)

Fig. 4 Analytical results of Case-2D, Case-3D-1 and Case-3D-2 (time histories at any points)

Comparing the result in Case-3D-1 with that in Case-3D-2 as shown in Fig. 4(b), (d) and (f), significant difference could not be found from these figures. It is noted that the residual horizontal displacement , the residual settlement and the excess pore water pressure ratio in Case-3D-2 was slightly larger than that in Case-3D-1, which is the influence of the multi-directional loading. In the present analytical cases, the influence of multi-directional loading was not so remarkable. Two reasons can be considered as the above findings. One is concerned with the input acceleration. It can be seen from Fig. 2 that the input motion in the longitudinal direction was much smaller than that in the transverse direction. The input motion component in the longitudinal direction must have been originally minor to the pore water pressure build-up and deformation. The other reason is related to the failure mode. As mentioned before, the circular slip surface was generated in the present analysis as shown in Fig. 3. Such failure mode occurs without three-dimensional but with two-dimensional behavior.

Fig. 5 shows the horizontal as well as the vertical surface displacement of the ground behind the quaywall. It can be seen from this figure that the ground close to the quaywall was influenced by the deformation of the quaywall especially in the range of 50 m. This analytical result coincides with the observed data during the 1995 Hyogoken-Nambu Earthquake investigated by Ishihara et al.(1996).

Fig. 6(a) shows the vertical distribution of lateral pressure (earth pressure and water pressure) acting to the foundation. Fig. 6(b) shows the vertical distribution of K, which is the ratio of the horizontal earth pressure (including excess pore water pressure) against the vertical earth pressure (including excess pore water pressure). In these figures, the vertical distribution at every two seconds are indicated. A short dotted line in Fig. 6(a) indicates the static lateral pressure under K=1.0, which corresponds to the perfect liquefaction. As shown in Fig. 6(a), however, exceeded values beyond the short dotted line in lateral pressure were obtained during strong motion, e.g., at 4 sec., 6 sec. and 8 sec. The long dotted line indicates

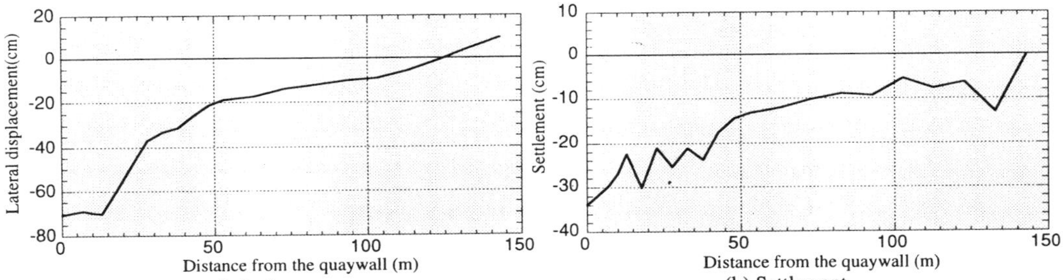

(a) Lateral displacement

(b) Settlement

Fig.5 Ground surface displacement on the back side ground of the quaywall

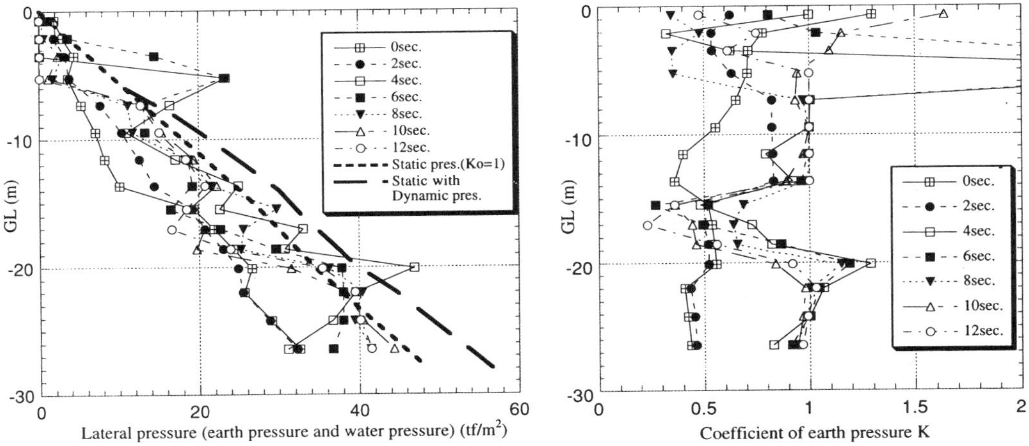

(a) Lateral pressure distribution

(b) K-distribution

Fig. 6 Lateral pressure distribution acting to the foundation at any time section

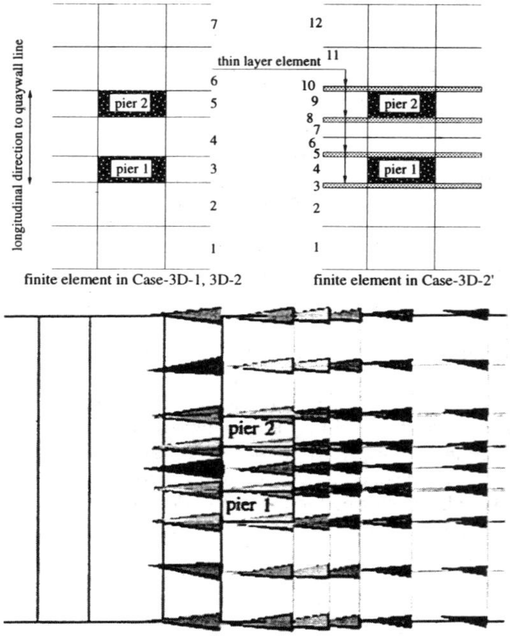

Fig. 7 Relative displacement on the ground surface (GL-0m) between foundation and soil (Case-3D-2', t = 12 sec.)

the lateral pressure considering the dynamic water pressure with the static pressure. As the dynamic water pressure, following equation was assumed.

$$p_d = \frac{7}{8} k \gamma_{sat} \sqrt{H \cdot Z_u} \tag{1}$$

where, k is the design seismic intensity, γ_{sat} is the saturated unit weight, H is the height of the foundation below the ground water level and Z_u is the distance from the ground water level. In the present study, it was assumed that k was 0.24 and H was 21.3 m. Most of the lateral pressure distribution in any time were below the value of the long dotted line.

From the designing view point, it is appropriate to take into account the dynamic increment with the static lateral pressure for design load acting to the foundation.

All these discussions were made for Case2D, Case-3D-1 and Case-3D-2. The number of divided finite elements in Case-3D-1 and Case-3D-2 is not neccesarily enough to investigate the detail interaction between the foundation and soils. Just one element exists between two foundation elements (pier 1 and pier 2) as shown in Fig. 7. On the other hand, four elements including thin layer elements were arranged between two foundations in Case-3D-2' to investigate the difference of displacement between foundations and soils. Fig. 7 shows the plane view of ground surface displacement at 12 sec. It can be seen from this figure that relatively larger displacement of soils than that of foundation was obtained. It is noted that the groval behavior in Case-3D-2' was almostly same with that in Case-3D-2.

5 CONCLUDING REMARKS

The soil-foundation interaction behavior during liquefaction was investigated by using effective stress based two and three-dimensional finite element method. The behavior of a foundation structure and its surrounding ground near the quaywall during a strong seismic motion was discussed in detail. The numerical result by two-dimensional analysis and that by three-dimensional analysis were almostly same in the present study.

From the designing view point, it was found to be appropriate to consider the dynamic increment with the static lateral pressure as the design load acting to the foundation.

The effect of the multi-directional loading, the difference between two-dimensional model and three-dimensional model, and interaction behavior between foundation and its surrounding soils should be throughly investigated as the future work.

ACKNOWLEDGMENTS

The authors wishes to express their sincere thanks to Hanshin Express Way Public Corporation for giving them the precious informations.

REFERENCES

Chaboche, J.L. and Rousselier, G. 1983. On the pladtic and viscoplastic constitutive eqations. J. of Pressure Vessel Technology. ASME. 105:153-164.

Ishihara, K., Yasuda, S. and Nagase, H. 1996. Soil characteristics and ground damage. Special Issue of Soils and Foundations. JGS.:109-118.

Oka, F., Yashima, A., Kato, M. and Sekiguchi, K. 1992. A constitutive model for sand based on the non-linear kinematic hardening rule and its application. Proc. of 10th WCEE. :2529-2534.

Oka, F., Yashima, A., Shibata, T., Kato, M. and Uzuoka, R. 1994. FEM-FDM coupled liquefaction analysisof a porous soil using an elasto-plastic model. Applied Scientific Research. 52: 209-245.

Taguchi, Y., Tateishi, A., Oka, F. and Yashima, A. 1995. A cyclic elasto-plastic model for sand based on the generalized flow rule and its application. Proc. of NuMoG V.: 47-52.

Taguchi, Y., Tateishi, A., Oka, F. and Yashima, A. 1996. Three-dimensional liquefaction analysis method and array record simulation in Great Hanshin Earthquake. Proc. of 11th WCEE.: No.1042 Pergamon.

Tateishi, A., Taguchi, Y., Oka, F. and Yashima, A. 1995. A elasto-plastic model for and its application under various stress conditions. Proc. of IS-TOKYO '95.: 399-404.

6.3 Slopes, embankments and dams

Numerical Models in Geomechanics, Pietruszczak & Pande (eds) © 1997 Balkema, Rotterdam, ISBN 90 5410 886 X

Use of joint elements in the behaviour analysis of arch dams

P. Divoux
Electricité de France / National Hydro Engineering Center, Savoie Technolac, Le Bourget du Lac & Laboratoire 3S, Université Joseph Fourier, Grenoble, France

E. Bourdarot
Electricité de France / National Hydro Engineering Center, Savoie Technolac, Le Bourget du Lac, France

M. Boulon
Laboratoire 3S, Université Joseph Fourier, Grenoble, France

ABSTRACT : This paper presents the results of a series of calculations performed on the Puylaurent arch dams. Joint elements are used to model the non-linearities in the behaviour of the dam-foundation interface and the joints between the construction blocks. The construction of the dam, the block joint grouting and the filling of the reservoir are modeled.

1. INTRODUCTION

The objective of this article is to demonstrate the advantage of using joint elements to model the behaviour of arch dams. At present, joint elements are commonly used in finite element analysis (Boulon [1995]). We focused on the behaviour of the dam itself without taking into account any possible discontinuities in the foundation. Prat&al (1994), Gens&al (1995), and Alonso, Carol&al (1996) propose a similar approach to model such discontinuities in the rocky foundation of the Canelles dam.

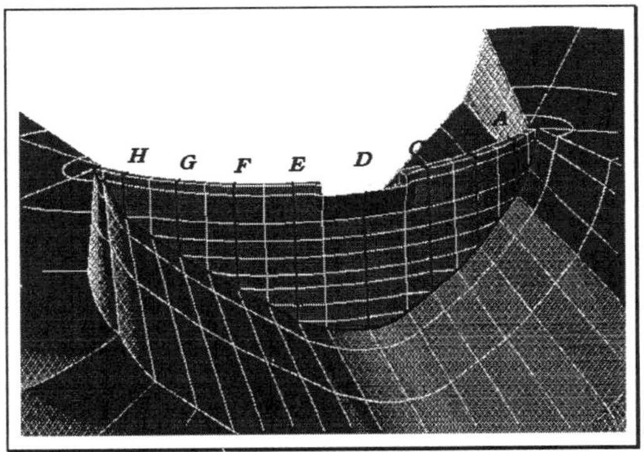

Figure 1 -Mesh of the Puylaurent dam

Due to the type of construction and the characteristics of the concrete (thermal shrinkage, low tensile strength), an arch dam cannot really be considered a structure with elastic linear behaviour. This is consistently demonstrated by finite element calculations. For example, on a monolithic structure, the calculation of the structure's own weight produces major tensile stresses on the banks, which have no real existence.

The behaviour of arch dams is greatly influenced by the non-linear behaviour at the discontinuities, i.e. the block joints, the concrete lift joints and the concrete-rock contact. In order to take these non-linearities into account, we chose to use 3D hydromechanical interface elements. This is a different approach that is complementary to the one using volume models such as poro-damage models (Bourdarot (1994a and 1994b), Bary (1996)).

The study described in this article was performed on the Puylaurent dam. In addition to obtaining results close to those obtained in situ, we tried to determine the influence of phenomena such as thermal shrinkage, joint grouting and the propagation of uplift at the concrete-rock contact on the dam's behaviour.

2. PRESENTATION OF THE STRUCTURE AND THE MODEL

The Puylaurent dam, used as an example in this study, is an arch dam 73 m high, and 220 m long at the crest, creating a 70 000 m^3 reservoir. It was built under a partnership between EDF and the *départements* of Ardèche and Lozère, and impounded in 1996. Figure 1 shows its 3D mesh.

It is composed of parabolic volume elements (hexahedrons with 20 nodes, pentahedrons with 15

nodes and tetrahedrons with 10 nodes). Parabolic interface elements (triangular with 12 nodes and quadrangular with 16 nodes) allow modelling of the contacts between each of the eight blocks as well as of the dam-foundation contact. The behaviour of the contacts between blocks ***A*** to ***H*** is exclusively mechanical, as the "water-stop" joints installed during the dam's construction are designed to block uplift propagation. However, we will measure the influence of uplift at the dam-foundation contact on the dam's behaviour, using coupled hydromechanical interface elements. A drainage curtain is located at the downstream part of the dam-foundation contact. Uplift is considered nil beyond this drainage curtain.
The rocky foundation matrix and the dam's concrete are materials that are assumed to be linear elastic. The structure's non-linearities are entirely concentrated at the block joints and the concrete-rock contact, which obey an elastoplastic constitutive law of the Mohr Coulomb type.
Calculations were performed with the GEFDYN code.

3. INTERFACE BEHAVIOUR

A simple elastoplastic constitutive law was chosen to model the joints' mechanical behaviour, in order to focus on the most important phenomena.

3.1. Mechanical behaviour

- Opening and closing criterion:

A closed joint opens when the normal stress σ_n disappears, and it can no longer withstand the tensile stress.
An open joint closes when the normal relative displacement δ_n is less than δ_{nf} , where δ_{nf} is an historic variable representing the normal displacement below which the joint closes.

- Behaviour of the open joint:

The normal and tangential stresses are nil when the joint is open.

$$\sigma_n = \sigma_s = \sigma_r = 0 \qquad (1)$$

- Behaviour of the closed joint:

Under compression, the joint's constitutive law is elastoplastic with a hyperbolic plasticity criterion ***f*** (2) (cf. Hohberg (1995), Prat&al (1994)) and a flow law ***g*** of the Mohr-Coulomb type (3).

$$f = \sqrt{\sigma_s^2 + \sigma_r^2} - \sqrt{\sigma_n^2 . \tan^2 \varphi - 2 . \sigma_n . \frac{c}{\tan \varphi}} \qquad (2)$$

$$g = \sqrt{\sigma_s^2 + \sigma_r^2} - \sigma_n . \tan \psi \qquad (3)$$

The joint's behaviour can be completely defined from five criteria:
- k_n normal elastic rigidity,
- k_s tangential elastic rigidity,
- φ a friction angle,
- c cohesion,
- ψ a dilatancy angle,

and an historic variable:
- δ_{nf} the closing criterion.

- Joint grouting:

A grouting law was developed to model block joint grouting, with the grouting stress σ_{nc} as the only parameter. This law is complementary to the elastoplastic law described above.
The normal and tangential stresses are calculated using the mechanical elastoplastic law. If the normal stress σ_n is greater than σ_{nc}, then the grouting law imposes the following:

$$\begin{cases} \sigma_n = \sigma_{nc} \\ \delta_{nf} = \delta_n - \sigma_{nc}/k_n \end{cases} \qquad (4)$$

In the opposite case, the stress response remains unchanged and equal to that obtained with the elastoplastic law.

3.2. Flow laws

A very simple flow law was used to model the uplift field below the structure, and is expressed mathematically by equation (4). k_j represents the joint's constant permeability.

$$q_s . \delta_n = k_j . \frac{\partial p}{\partial s} \text{ and } q_r . \delta_n = k_j . \frac{\partial p}{\partial r} \qquad (4)$$

Although quite simplified, this law allows the mechanical effects of an uplift field at the concrete-rock contact under the structure to be taken into account.
The cubic law was developed based on the mathematical solution of the laminar flow of a viscous fluid between two parallel plates (Witherspoon (1980)).

4. EVALUATION METHOD AND CONVERGENCE CRITERIA

4.1. The modified Newton-Raphson method

The modified Newton-Raphson method was used to perform all the calculations whose results are presented in this article.
Considering the shear and the openings calculated, equilibrium was reached for certain stages after a very large number of iterations. The number for the

model's degree of freedom is 7787 for the coupled model and 7647 for the mechanical model. Modelling from the construction of the dam up to the end of impounding takes approximately 24 hours on a SUN SPARC 1000 work station.
To accelerate the convergence of the calculations, we tested the modified Newton-Raphson and the Newton-Raphson algorithms. We also worked on the choice of the auxiliary rigidities. The dam calculations performed with the Newton-Raphson algorithm systematically diverged. If, at a given integration point, the joint closes and the normal auxiliary rigidity is nil, the solids interpenetrate each other. The residual strength calculated at the following iteration is then quite significant and leads to divergence of the calculation. By observing the residual strength calculated at each of the integration points, one can observe that divergence in calculation is often caused by a poor choice of auxiliary rigidity at a single integration point. Tests performed on stages with a lesser load reached the same logical conclusion.
For this reason, the modified Newton-Raphson algorithm would seem very robust, although not very effective.
In general, very few calculation stages are necessary to perform quasi-static calculations, and this algorithm is sufficient to model the behaviour of joints.

4.2. Convergence criteria

For the mechanical calculations, convergence is reached if the two criteria for strength and displacement given in Table 2 are complied with. For coupled calculations, both the mechanical criteria and the flow and pressure criteria must be complied with.

Tableau 2 - Convergence criteria for the different phases of calculation.

	Relative residual strength	Relative residual displacement
Construction	0.1%	0.1%
Thermal shrinkage	0.1%	0.1%
Grouting	1%	0.1%
Impounding	1%	0.1%

	Relative residual flow	Relative residual pressure
Impounding	1%	0.1%

Tableau 3 - Physical properties of materials

Foundation rock	
Density	$\rho_f = 2400$ kg/m^3
Young module:	$E_f = 18$ GPa
Poisson coefficient:	$\nu_f = 0.2$
Concrete	
Density:	$\rho_c = 2400$ kg/m^3
Young module:	$E_r = 36$ GPa
Poisson coefficient:	$\nu_r = 0.2$
Thermal expansion coefficient:	10^{-5} °C^{-1}
Block joints	
Normal rigidity:	$k_n = 10$ Gpa/m
Tangential rigidity:	$k_s = 0$ Gpa/m before grouting
	$k_s = 5$ Gpa/m after grouting
Friction angle:	$\varphi = 45$ °
Cohesion :	$c = 0$ Pa
Dilatancy angle:	$\psi = 0$ °
Concrete-rock contact	
Normal rigidity:	$k_n = 10$ Gpa/m
Tangential rigidity:	$k_s = 5$ Gpa/m
Friction angle:	$\varphi = 45$ °
Cohesion :	$c = 0$ Pa
Dilatancy angle:	$\psi = 0$ °

5. MATERIAL PROPERTIES

The physical properties of the materials are given in Table 3.
As far as the flow in the dam-foundation contact joint is concerned, permeability k_j is assumed to be constant over the entire joint surface. Therefore, the dam's uplift profile and subsequent mechanical behaviour are not influenced by the values of k_j.

6. SOLICITATIONS

In order to take into account different cases of load, several calculations were done. We tried to determine the influence on the dam of factors such as thermal shrinkage, joint grouting and the uplift field at the dam-foundation contact.

6.1. Dam construction

The dam is built in a single stage during which the structure's own weight is applied. The tangential rigidity of the joints is assumed to be nil. In this way, the blocks behave individually, without interpenetrating each other. This method for modelling the dam's construction may appear too simplified, but it will be subsequently justified. We will see that thermal shrinkage of the concrete leads to separation of each of the blocks from the others.

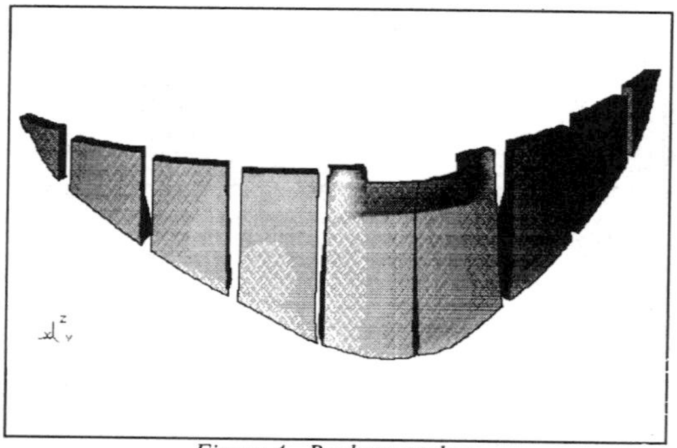

Figure 4 - Puylaurent dam
Thermal shrinkage - Deformed shape x 1000

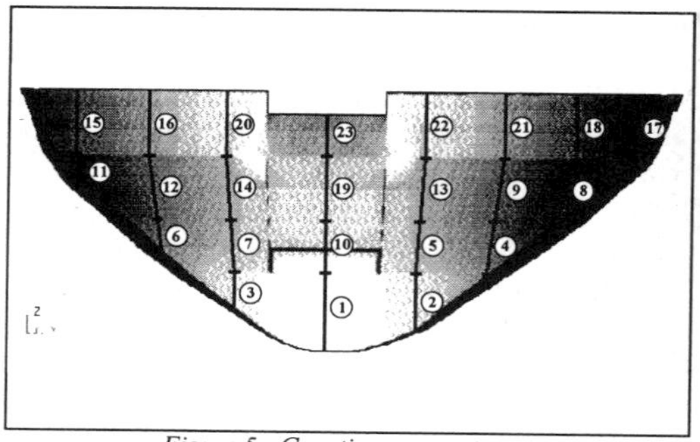

Figure 5 - Grouting compartments

6.2. Thermal shrinkage

Shrinkage after construction, due to heat development in the concrete, is modelled by lowering the concrete temperature to 15°C.

Figure 4 shows the displacements caused by thermal shrinkage. The maximum opening of the joints reaches 4.8 mm at the top of joints ***C*** and ***E***. The central joint ***D*** remains slightly closed at the top, and normal stresses are less than 0.12 MPa. The maximum normal stress calculated at the upstream toe is 3.8 MPa. At the downstream toe, the weight of the spillway is represented by a compression stress of 1 MPa.

Thus, the calculations and observations made in situ show that near the end of construction and after the thermal shrinkage of the concrete, the block joints are open. It is the grouting which provides block linkage and the transfer of forces via the arch effect.

6.3. Block joint grouting

Near the end of construction, joint grouting via the injection of cement slurry allows the structure to become monolithic. The grouting law presented above allows modelling of the pressurising and the filling of joints. The simplified grouting scenario represented in the order of joint injection (figure 5) takes into account 23 compartments (or grouting stages) out of 27. The grouting pressure chosen for each compartment is given in Table 6.

Table 6 - Grouting pressure per compartment.

Compartment:	Grouting pressure:
1-2-3-4-5-6-7-10	1 MPa
8-9-11-12-13-14-19	0.5 MPa
15-16-17-18-20-21-22-23	0.2 MPa

It is indispensable to take into account a grouting scenario closely resembling reality in order to represent bending and initial contact of the blocks during joint grouting.

Near the end of grouting, the maximum displacement measured at the crest in the upstream-downstream direction is 1.5 cm.

Figure 7 shows the results obtained at the joint of central block ***D***. Grouting of a compartment causes opening of the upper portion of the compartment below it. Due to bending of the block, and considering the grouting scenario, the normal stress in compartment ① is in some places greater than the grouting stress.

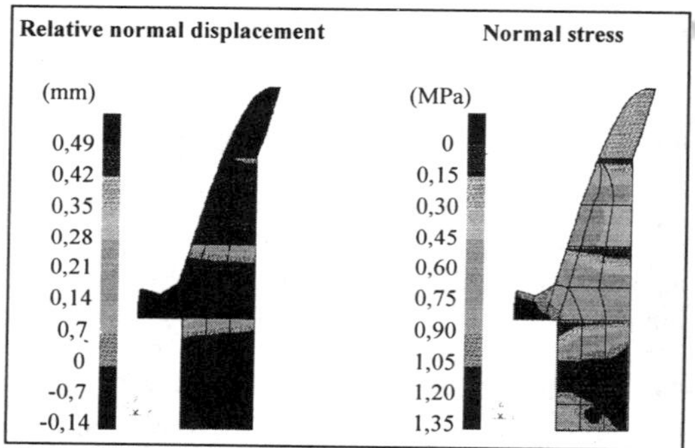

Figure 7 - Relative normal displacement and normal stress in central block D after joint grouting.

6.4. Impounding

The impounding of the dam was modelled in eight stages. The reservoir level is heightened 8 meters at a time for each of the first seven stages. In the last stage it reaches 938.8 , i.e. the normal reservoir level.

Three different impounding calculations were performed once the block joint grouting had been performed, with a grouting pressure equal to zero.

The study consisted in evaluating the influence of

uplift at the concrete-rock contact, as well as the influence of the concrete's residual thermal shrinkage on the structure's behaviour during impounding.

	Uplift at the concrete-rock contact	Residual thermal shrinkage
Calculation n° 1	No	No
Calculation n° 2	Yes	No
Calculation n° 3	Yes	Yes

Uplift profile:

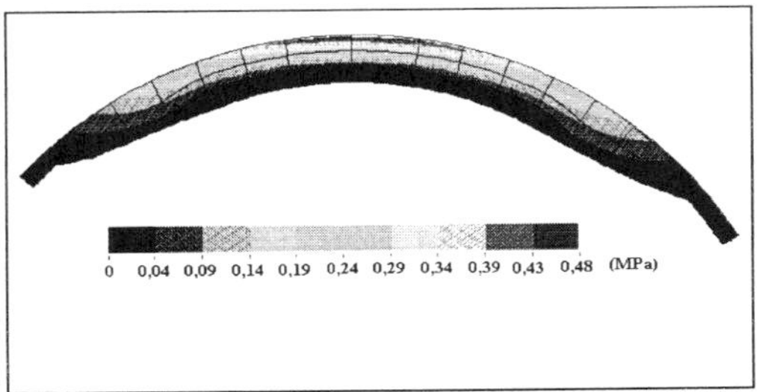

Figure 8 - Uplift in the joint at the dam-foundation contact calculated for the reservoir at a level of 922 mngf - Aerial view

Calculations were performed with a flow law that assumes constant permeability in the concrete-rock contact joint. Figure 8 shows the uplift profile for the reservoir at a level of 922 m. Pressure is nil beyond the drainage profile.

Residual thermal shrinkage:

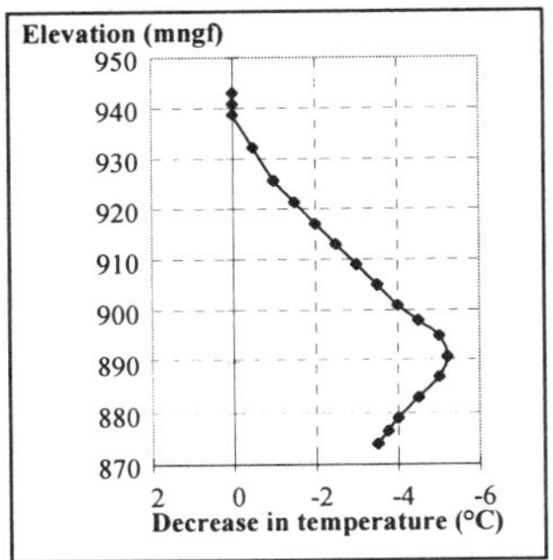

Figure 9 - Residual thermal shrinkage

A residual thermal shrinkage is simulated. The curve in Figure 9 gives the variation in temperature of the concrete between the beginning and the end of impounding, depending on elevation.

Seasonal thermal variations were not taken into account.

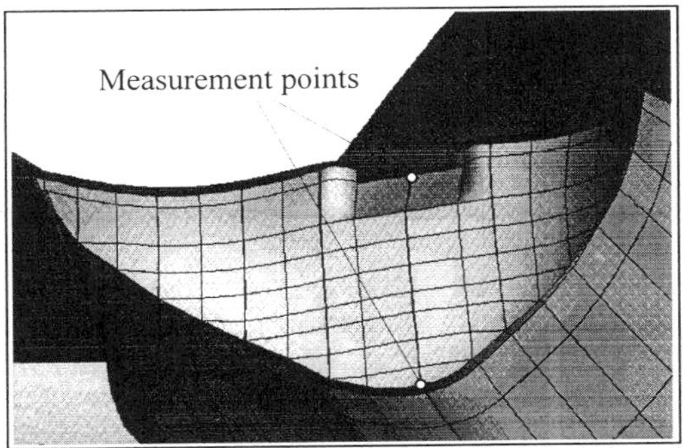

Figure 10 - Results of calculation n° 1 - Deformed shape (x 1000) of the Puylaurent dam near the end of impounding - reservoir level: 938,8 mngf

The following results mainly concern the study of the behaviour of the dam-foundation contact joint and the study of the displacements at the crest. The curves are provided for the two points of measurement shown in Figure 10.

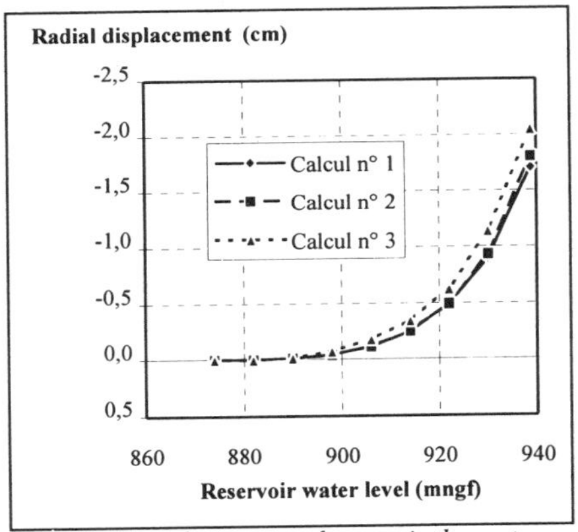

Figure 11 - Displacement at the crest in the upstream-downstream direction Y

The influence of uplift (calculation n°2) and the residual thermal shrinkage (calculation n°3) on the displacements along the crest is slight.

The openings calculated for the dam-foundation contact joint at the toe of the structure (figure 12) are quite different in calculations 1, 2 and 3.

At the toe of the structure, the arch is 13 m thick. When the reservoir is full, the length of cracking is estimated at 6.5 m in calculation n°1 and 8.2 m in calculation n°2. This length is about 9 m in calculation n°3, i.e. close to the drainage profile.

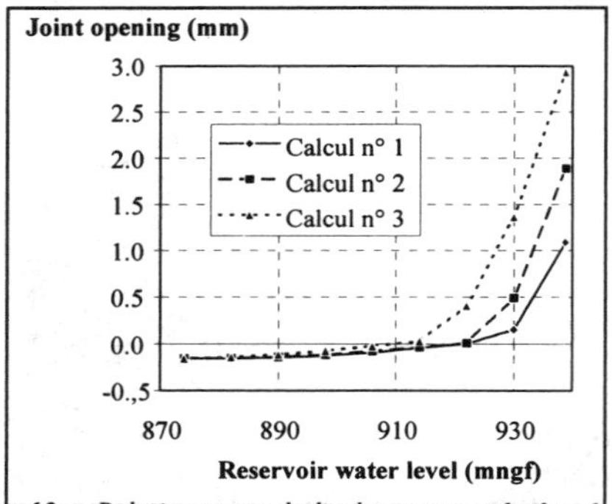

Figure 12 - Relative normal displacement calculated at the upstream toe of the Puylaurent dam during impounding.

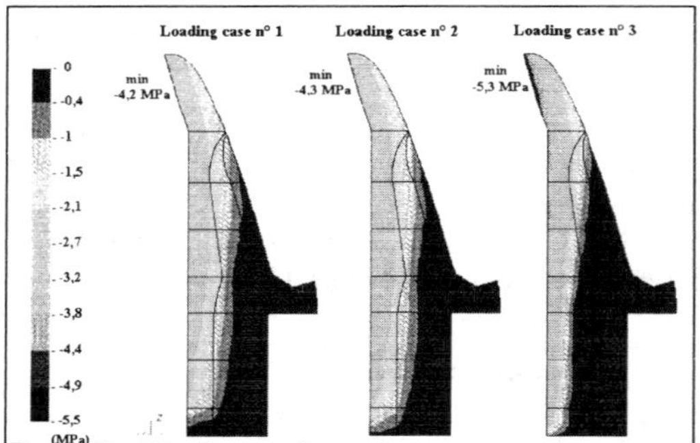

Figure 13 - Distribution of normal stress in the joint of central block D near the end of impounding - Reservoir level: 938,8 mngf

These results were interpolated from the results at the integration points of the joint elements.

The last stage of the impounding greatly contributes to the opening of the joint and to the increase in the displacements at the crest.

Figure 13 demonstrates that the concrete's residual thermal shrinkage modifies the distribution of forces due to hydrostatic thrust. The normal stress is more significant at the upper arches.

Near the end of impounding, the considerable lengths of opening calculated at the toe of the structure and the low normal stress of the block joints downstream can be partly linked to the pessimistic hypothesis chosen whereby the impounding calculations are performed after having grouted the joints with grouting pressure σ_{nc} equal to zero.

7. CONCLUSION

The advantage to the modelling method for arch dams presented in this article is that it takes into account the main non-linearities of behaviour created by block-type construction. The structure's response can be calculated for several different construction, grouting and impounding scenarios. Different calculations allowed estimation of the influence of thermal shrinkage and uplift at the concrete-rock contact on the structure's behaviour.

This study could be pursued, taking into account seasonal thermal factors and improving the flow law in the dam-foundation contact joint.

REFERENCES

ALONSO E., CAROL I., DELAHAYE C., GENS A., PRAT P., 1996, *"Evaluation of safety factors in discontinuous rock."*, Int. J. Rock Mech. Min. Sci. & Geomech. Abstr., Vol. 33, N° 5, pp 513-537

BARY B., 1996, *"Etude du couplage hydrau.-mécanique dans l béton endommagé."*, Thèse de doctorat du Lab. de Mécanique et Technologie, Cachan, France

BOULON M., GARNICA P., VERMEER P. A., 1995, *"Soil-structure interaction: FEM computations."*, Studies in applied mechanics 42, Mechanics of geomaterial interfaces, ELSEVIER, pp. 147-172

BOURDAROT E., BARY B., 1994a, *"Effects of temperature and pore pressure in the non-linear analysis of arch dams."*, Proc. Int. Worksh. on Dam Fracture and Damage, Edt Balkema, pp 187-195

BOURDAROT E., 1994b, *"Analyse du comportement thermomecanique du barrage du Gage 2."*, COGECH 1994

CASTAING P., HABAUZIT J.P., 1996, *"Barrage de Puylaurent - Etudes et travaux de clavage."*,COGECH 1996

GENS A., CAROL I., ALONSO E., 1995, *"Rock joints : FEM implementation and applications."*, Studies in applied mechanics 42, Mechanics of geomaterial interfaces, ELSEVIER, pp. 395-420

HOHBERG J. M., 1995, *"Concrete joints."*, Studies in applied mechanics 42, Mechanics of geomaterial interfaces, ELSEVIER, pp. 421-448

PRAT C., DELAHAYE C., GENS A., ALONSO E.E., 1985, *"Safety evaluation of an arch dam founded on fractured rock based on a 3D non linear analysis with joint elements."*, Proc. Int. Worksh. on Dam Fracture and Damage, Edt Balkema, pp 211-221

P. WITHERSPOON, J. WANG, K. IWAI, J. E. GALE, 1980, *"Validity of cubic law for luid flow in a deformable rock fracture."*, Water Resources Research, Vol. 16, N° 6, pp. 1016-1024

Numerical Models in Geomechanics, Pietruszczak & Pande (eds) © 1997 Balkema, Rotterdam, ISBN 90 5410 886 X

Earthquake induced joint opening in arch dams

G. Zenz & E. Aigner
Verbundplan GmbH, c.o. Tauernplan Consulting GmbH, Salzburg, Austria

ABSTRACT: The behaviour of an arch dam under earthquake loading is investigated. Due to the nature of earthquake acceleration under linear system assumptions high tensile stresses occur and are transmitted across the block joints. Within this analysis the influence of block joint opening and sliding during earthquake excitation is studied. During a fraction of seconds the joints can open and therefore a change in the bearing behaviour takes place. The forces are transmitted by bending moments and introduce higher stresses on dam's upstream and downstream faces. The calculated results of linear and nonlinear analyses are discussed.

INTRODUCTION

The calculation of the earthquake response of an arch dam due to seismic excitation is important to judge upon dam's safety.
Different linear and nonlinear methods are used to analyze such problems dependent on excitation and structural behaviour. In this investigation two methods of analyzing the response are used:

- a frequency response analysis and
- a nonlinear calculation in the time domain.

Within direct time integration analysis it is possible to investigate a fully nonlinear system behaviour. This time consuming calculation procedure is necessary when considering the opening of block joints in arch dams due to the occurrence of tensile stresses.

CALCULATION MODEL

The whole finite element model and the dam with the block joints are shown in figure 1 and 2. The normal cross section of the 30 m high arch dam with the water loading definition is shown in figure 3.
The subsoil is modeled with 1682 elements and the arch dam with 1021. For the calculation the subsoil is considered massless and the boundaries are excited dynamically. The elastic material properties and the viscous damping ratio ζ for the dam concrete are:
$E_c = 34.8$GPa, $\nu_c = 0.2$, $\zeta = 5\%$, $\gamma_c = 24{,}0$kN/m^3 and

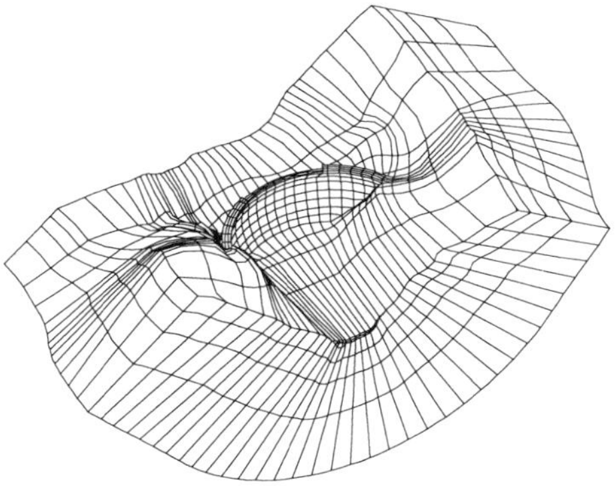

Figure 1. Overview of considered finite element mesh

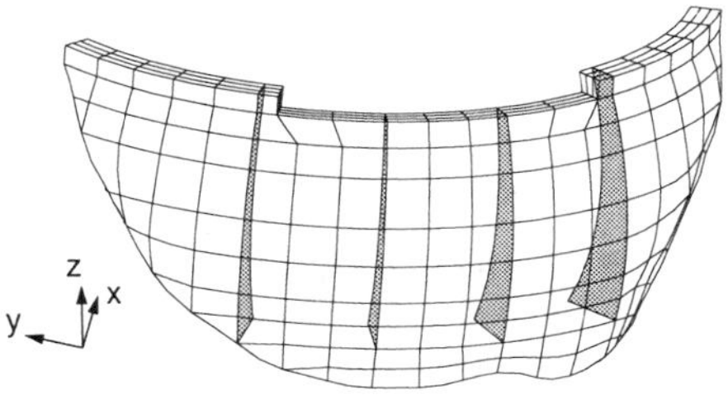

Figure 2. Arch dam elements - 4 block joints introduced

for the subsoil $E_c = 2.0$GPa, $\nu_c = 0.1$.
The analysis is started with closed joints prior applying any load. Due to dead weight and water loading the joints are allowed to slide or open.

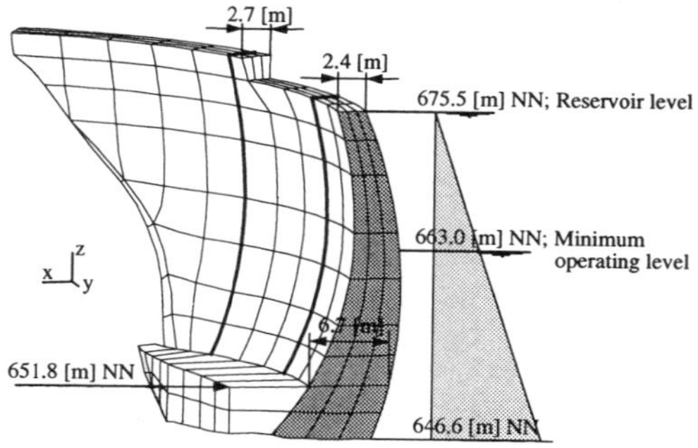

Figure 3. Cross section, geometry and load definition

The friction factor for the contact surface behaviour is set to $\mu = 1.0$.
An implicit time integration scheme is employed to solve for the nonlinear dynamic system behaviour, and a comparison is made with the frequency response spectrum method. To account for the dynamic interaction of the dam structure and the reservoir an added mass approach is considered being sufficient for the current dam height.
The program used for this investigation is Abaqus/Standard Version 5.4-1.

APPLIED LOADING HISTORY

Linear static loading is applied prior the nonlinear earthquake analysis starts. For this case, the dead weight is applied onto a system of free standing cantilevers. The dead weight bearing is not at all effected by the arch behaviour. After activating the dead weight and the hydrostatic water pressure, prestressing the arch, the nonlinear earthquake analysis is carried out.
The input time history with its maximum acceleration of about 0.22g and the corresponding response spectra in case of linear system behaviour for 2,5 and 10% damping are shown in figure 4 and 5.

NUMERICAL PROCEDURE

Throughout the analysis quadratic, isoparametric elements are used. These are 27 noded brick elements in the vicinity of the 9 noded interface elements and 20 noded brick elements throughout the dam body. At the abutment 15 noded wedge elements are used. The system mass matrix is based on a consistent formulation for the continuum elements.
The water added masses are applied with a lumped mass matrix approach at the water surface boundary nodes according to H.M.Westergaard.

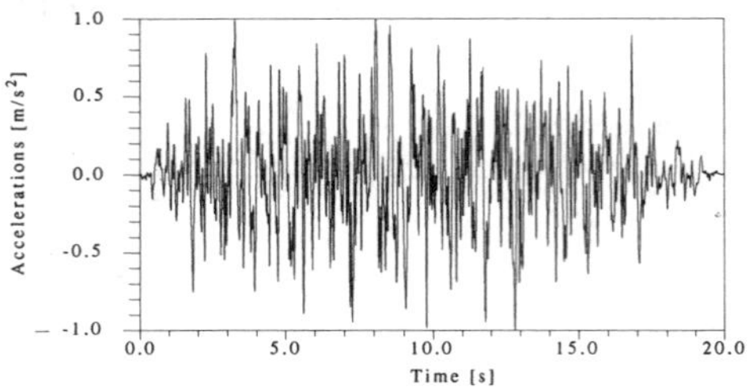

Figure 4. Time history applied in X-Direction

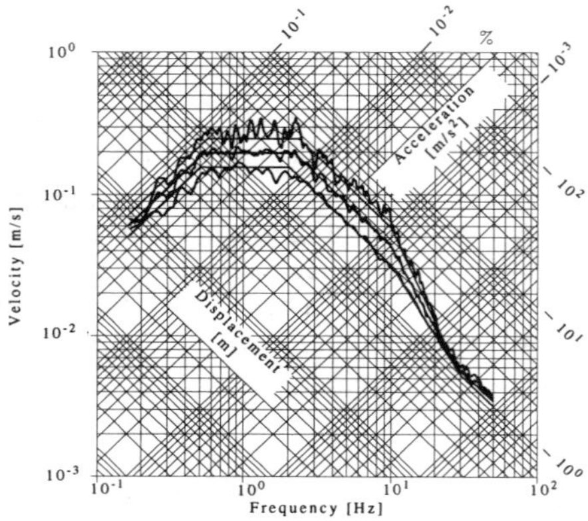

Figure 5. Response spectra for 2, 5 and 10 % damping

The contact procedure required has to account for opening, closing and for frictional behaviour in the defined interface. For frictional sliding:

- isotropic and anisotropic Coulomb friction
- user defined friction model

are available in the program context. For this calculation we use isotropic Coulomb friction achieved with the use of the penalty method. During an increment at which contact is established (after a step under “open-condition”), the frictional effect is neglected. During execution of this analysis the penalty parameter is defined with an elastic slip of $\gamma_{elastic} = 0.5$[mm]. During initial state of contact (no sliding should occur) with some pressure stress "p" acting, the allowable shear stress transmitted within the contact surface is calculated with

$$\tau_{max} < \mu * p$$

for sticking state. If the current shear stress τ equals τ_{max}, the sticking state changes to sliding. The procedure may model the behaviour rigorous and “accurate” by introducing Lagrange multiplier, but this might cause convergence problems during iteration. A weaker, but more time effective formulation is the introduction of penalty parameters and therefore used herein.

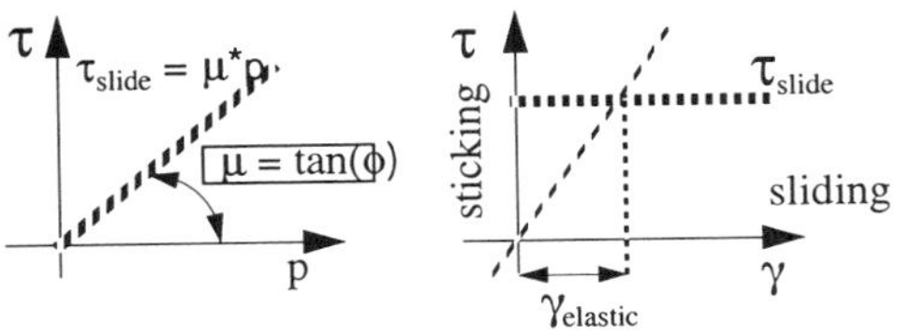

Figure 6: Condition between friction, pressure and shear stress

This gives some additional sliding during sticking state. This sliding can be controlled with some allowable $\gamma_{elastic}$ value and gives the program the value to choose the penalty parameter (see figure 6).
The penalty parameter for this mode is found iteratively from an allowable sliding during sticking from:

$$K_p < \mu^* p / \gamma$$

Sudden changes of the penalty value may lead to convergence problems; therefore some amount of elastic sliding improves convergence performance. The sensitiveness of the solution, depending on the elastic slip has to be verified. The smaller the sliding value the larger the computation time.
The contact algorithm is implemented via Lagrange multiplier method for opening and closing condition. Two different methods to account for contact are available, these are

softened contact and

hard contact.

The used softened contact condition allows the numerical procedure to calculate an interpenetration of two surfaces, which however is chosen to be very small. The value c_0 of 0.1[mm] allows a transmission of pressure stresses, although rigorouse interpreted the surfaces are not in contact at all. These procedure dependent numerical parameter are affecting the physical contact behaviour but we see their influence on the results as "physically allowable"; we can tolerate this kind of performance. During softened contact the change of velocity of two surface nodes are determined by their amount of interpenetration. This gives no sudden changes in velocity, no momentum balance is required being calculated.
Throughout this benchmark we use softened contact with definitions for friction parameter μ=1.0, with a pressure "clearance-relationship" for assumed closed condition as it is chosen for c_0=0.1[mm] and an amount of applied contact pressure for zero closing of p_0 = 0.1[MPa] (see figure 7).

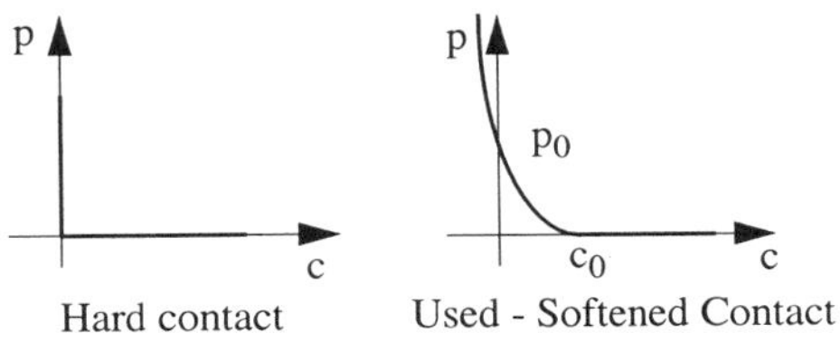

Figure 7. Pressure "clearance-relationship"

STRUCTURAL DYNAMIC BEHAVIOUR

For identifying the main dynamic system response an eigenvalue extraction analysis is carried out. The system with the impounded reservoir and the added mass approach for structure reservoir interaction gives eigenfrequencies from the first pair of about 9[Hz], the third of about 13[Hz] and for the fourth of about 16[Hz].
Within the context and the output provided by the program Abaqus (scaling of calculated eigenvalues and participation factors) it is possible to calculate for the linear dynamic system response in one direction k for one mode (α) the physical unknown R_α^k by the summation over:

- participation factor $\Gamma_{\omega j}$ related to eigenvalue $\alpha(\omega)$ in direction j
- and the related spectral value $S_j(\omega,\zeta)$ for the frequency ω and the assumed damping value ζ
- scaling of spectral value with factor c_j
- and premultiplied with the eigenvalue ϕ_α^k.

The appropriate physical variable reads:

$$R_\alpha^k = \phi_\alpha^k \cdot \left(\sum_{j=1}^{3} c_j S_j(\omega, \zeta) \Gamma_{\omega j} \right)$$

Different summation procedures are more or less appropriate to estimate the maximum system response. For well distinguishable eigenvalues the SRSS (square root of the sum of the squares) summation method is appropriate and therefore used for comparison with our results calculated with the direct time integration method.
Carrying out the frequency response analysis and considering the eigenmodes one gets for the frequency response analysis under SSRS summation assumption the maximum acceleration in x direction for point 3 of $a_{3,freq} = 16.4[ms^{-2}]$.
From the frequency response analysis it can be seen, that the first 6 eigenmodes are appropriate and able to represent the structural response. The highest frequency considered is 20[Hz] for the linear frequency response analysis. The influence of the last modes are very small compared to the

significant first modes. From this point it seems to be sufficient and appropriate to assume 30[Hz] as the maximum frequency of major interest for calculating the total structural response. The higher frequencies are not significant in the overall system response but however they are naturally included throughout the analysis but with decreasing accuracy as frequency content increases. This has to be recognized in relation to the chosen time step size.

TIME INTEGRATION ISSUES

Within the context of the finite element program different procedures are available to solve for dynamic excitation. These are for

- purely linear systems - eigenmode based analysis
- mildly nonlinear systems - modal projection method
- nonlinear system - implicit time integration procedure.

Due to the nonlinear nature of the calculation the implicit time integration procedure is used. The implicit time integration uses the time integration operator suggested by Hilber, Hughes and Taylor [6]. This is a modification of Newmark β method and gives for α=0 the trapezoidal rule:

$\beta = 0.25(1-a)^2$ and $y = 0.5-\alpha$ with $-\frac{1}{3} \le \alpha \le 0$

with a parameter α to adjust for controllable numerical damping. This damping is slowly growing at low frequencies but increases with a more rapid growth at higher frequencies. Within a nonlinear dynamic analysis due to changing of time step size throughout the calculation an artificial ringing is introduced into the model. With the small amount of introduced default damping $\alpha = -0.005$ high frequency oscillation is damped out.

The time step size of the integration procedure (Bathe) has to be chosen dependent on the frequency range of interest and the error of the solution which can be tolerated. The procedure itself is unconditional stable. The Newmark method introduces only a period elongation but no amplitude decay after increasing the time step size. For a time step size of

$$\Delta t = \frac{T}{10} = \frac{1}{10f}$$

with the highest frequency of interest f in [Hz] the period (T is the time period of one cycle in [sec]) elongation after one cycle is of about 3%. Increasing the time step size to an amount of 20% of the period T of interest would increase the error to an amount of 10% in period elongation. Assuming a highest frequency of interest to be 30Hz this will give a time step size of $\Delta t = \frac{T}{10} = 0,003$ [sec].

Higher frequencies will be integrated with lower accuracy but however they are not at all of major interest in the overall solution. Abaqus uses the calculation of a so called "half-step-residual" to

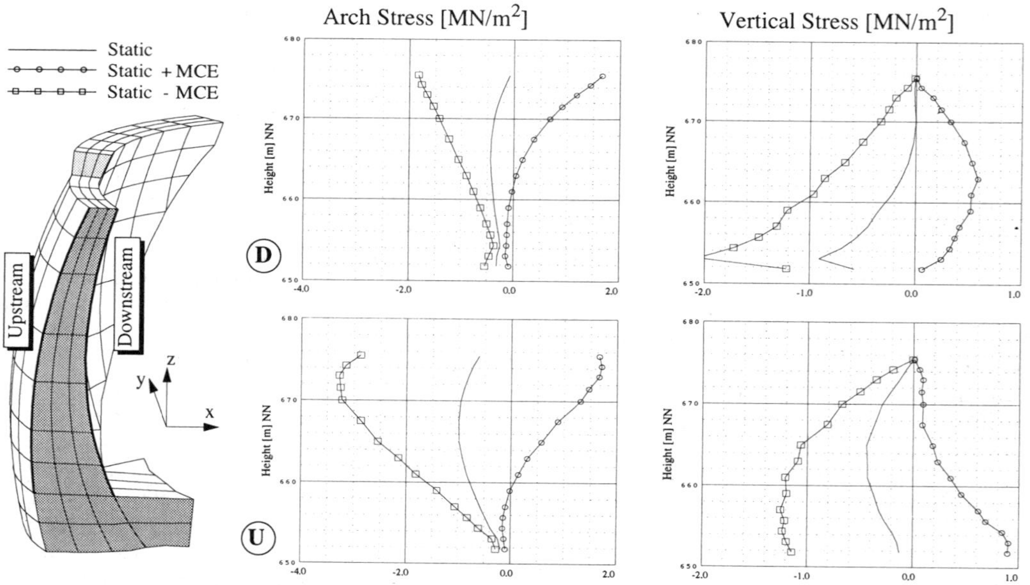

Figure 8: Vertical and Arch Stresses for Upstream and Downstream - Response Spectrum Analyses

account for rapid changes in acceleration. Though the value is purely empirical this approach works quite well for problems such as impulsively loaded systems. The time step size is increased during the solution after the high frequency response is dissipated. In our example we start therefore with an initial time step of 0,02[sec], which is adjusted by the program accordingly to account for the high frequency content.

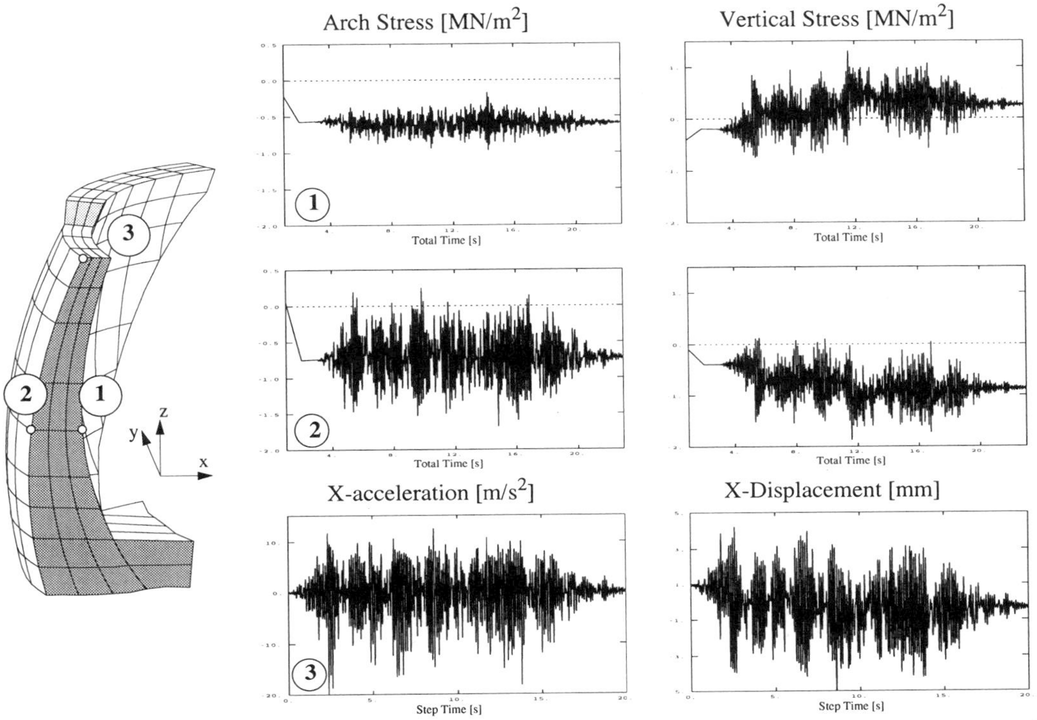

Figure 9: Vertical and Arch Stresses for Point 1 and 2 ; Horizontal Accelerations and Relative Displacements for Point 3

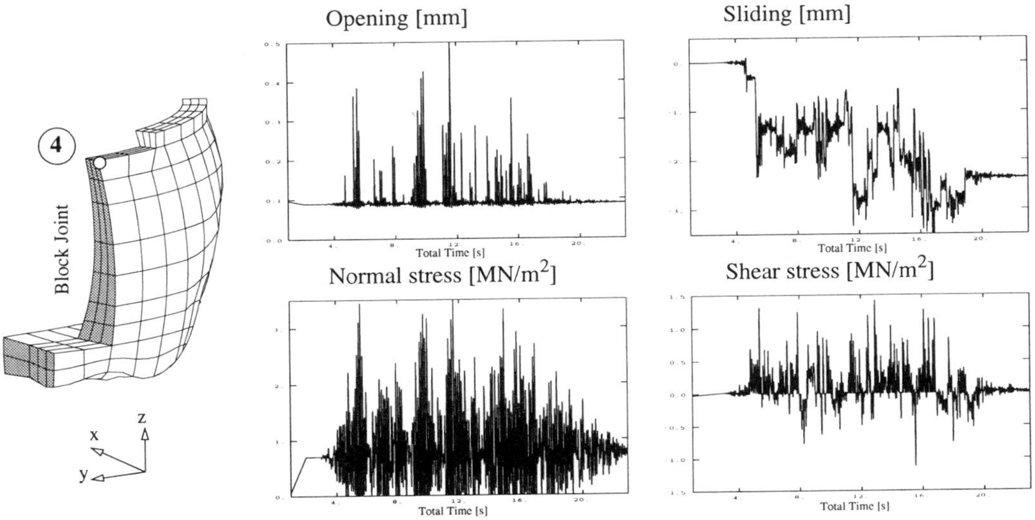

Figure 10: Opening, Sliding, Normal Pressure and Shear Stress for Point 4

DISCUSSION OF RESULTS

Some results of the frequency response analysis are displayed in table 1. The results for the non-linear calculation are shown in figure 9 and 10. The selected points show the significant differences between both methods. Because of the possibility of the block joints to move, it causes a change in the bearing behaviour. For the middle block joint the vertical stresses are different at the beginning and at the end of the calculation. This happens, because the block moves of about 2.3 [mm] into upstream direction. This causes an increase of the tensile stresses in the vertical direction on the downstream side of the dam.

One additional aspect is the required calculation time:

frequency response analysis takes 68 [min]
the non-linear calculation 7 762 [min].

In table 1 the main results of the different numerical calculations are shown. The structural response with joint elements doesn't exhibit tensile stresses at the interface. The application of dead weight leads to an opening of the block joints.

Due to the locally contact-impact behaviour within the block joint spurious high frequencies are monitored within the resultant acceleration time histories. Therefore a frequency reduction is carried out with postprocessing by a digital filter (cut off frequency 30[Hz]).

Table 1: Results [MPa] - Compression [-]

Results	Response Spectra		Direct Time Integration	
	max	min	max	min
Stress σ_y - point 1	-0.08	-0.76	-0.17	-0.97
Stress σ_y - point 2	0.15	-1.83	0.25	-1.68
Stress σ_z - point 1	0.52	-0.96	1.31	-0.75
Stress σ_z - point 2	0.36	-1.23	0.12	-1.87
Acceleration - point 3	16.4	-16.4	12.5	-20.2
Displacement - point 3	5.7	-3.7	4.2	-5.4
Interface results - point 4				
Opening [mm]			0.54	0.00
Sliding [mm]			0.11	-3.65
Normal stress [MN/m^2]			3.57	0.00
Shear stress [MN/m^2]			1.42	-1.13

The dynamic structural behaviour is reported within time history plots. From evaluated results we recognize that the maximum tensile stresses in vertical direction are smaller, if the system itself is acting as an arch dam structure, allowing to transmit certain tensile stresses across the block joints. For this case the maximum tensile stresses are at level 661.0[m] of about 0.5[MPa]. For the nonlinear case with temporary open block joints the maximum tensile stresses are of about 1.3[MPa].

It can be seen from the minimal and maximal stress values reported in table 1 (this is shown for point 1 and 2), that for a short duration these stresses are acceptable. However, not only maximum values but also their time of occurrence, their interaction (tensile and/or pressure) and their distribution in the dam body itself are substantial for deciding on the overall safety for a specific loading assumption.

ACKNOWLEDGMENT

For using Wiederschwing Arch Dam's geometry and model to carry out this numerical investigation the support of KELAG - Electricity Supplier of Carinthia - is gratefully acknowledged.

REFERENCES

Nackler K., Neuschitzer F. [1994]: Adaptation of Appurtenant Works to Increase Safety Requirements at a 40-Year Old Concrete Arch Dam, Wiederschwing Arch Dam, ICOLD, Q68, R66, pp 1153 - 1165, Durban.

Abaqus [1995]: Theory and User's Manual; Lecture Notes on "Contact in Abaqus Standard", Hibbit, Karlsson & Sorenson, Inc., Version 5.4-1.

Hohberg, J.H. [1992]: A Joint Element for the Non-linear Dynamic Analysis of Arch Dams. Ph.D.-Thesis, ETH-Zürich No. 9651.

Lei, X.Y., Swoboda, G. & Zenz, G. [1992]: New contact elements for the simulation of faults and cracks in arch dams. In Proc. of Int. Conf. on Arch Dams. Hohai University, Nanjing, China, Oct. 17 - 20, pp 245-250.

Stäuble,H. & Widmann,R. [1992]: Design Considerations for High Arch Dams. In Proc. of Int. Conf. on Arch Dams. Hohai University, Nanjing, China, Oct. 17 - 20.

Zenz,G. Aigner,E. Obernhuber,P. [1996]: Dynamic arch dam analysis considering nonlinear block joint behaviour; IVth ICOLD Benchmark Workshop on Numerical Analysis of Dams.

ICOLD [1994]: Third Benchmark Workshop on 'Numerical Analysis of Dams', "ad-hoc" Committee on Computational Aspects of Dam Analysis and Design; Paris, (1994).

Westergaard H.M. [1933]: Water Pressures on Dams During Earthquakes, Transaction, ASCE, Vol. 98.

Valliappan S.& Y.C. Wang [1994]: Advances in computational mechanics applied to wave propagation problems, Computer Methods and Advances in Geomechanics, Siriwardane & Zaman.

Hilber H.M., T.J.R. Hughes, and R.L. Taylor [1978]: Collocation, Dissipation and 'Overshoot' for Time Integration Schemes in Structural Dynamics, Earthquake Engineering and Structural Dynamics, Vol. 6,pp. 99-117.

Celep Z., Z.P. Bazant [1983]: Spurious reflection of elastic waves due to gradually changing finite element size, Int. Journal for Numerical Methods in Engineering, Vol. 19, 631-646.

Bathe K.J. [1982]: Finite Element Procedures in Engineering Analysis, Prentice-Hall.

Numerical Models in Geomechanics, Pietruszczak & Pande (eds) © 1997 Balkema, Rotterdam, ISBN 90 5410 886 X

On the influence of the constitutive model on the numerical analysis of deep excavations

M.G. Freiseder & H.F. Schweiger
Institute for Soil Mechanics and Foundation Engineering, Technical University Graz, Austria

ABSTRACT: A major problem in the analysis of deep excavations employing classical limit equilibrium methods is the prediction of deformations and displacements in the ground. This is one of the reasons why numerical methods are increasingly employed. In the proposed paper a finite element analysis is carried out for a deep excavation in soft soil which has been chosen such that it can be regarded as a simplified calculation of a real construction site. A parametric study was performed to investigate the influence of interface element properties on the interaction between soil layers and diaphragm wall and on the deformation behaviour. The effect of the constitutive law on the deformation behaviour is shown by comparing results obtained from a Mohr-Coulomb model, an Advanced Mohr-Coulomb model and a Soft Soil model. The assumptions to be made in these methods are critically assessed.

1 INTRODUCTION

A number of researchers have adressed the problems of analysing deep excavations and numerical studies have been presented ranging from fundamental earth pressure studies (e.g. Potts & Fourie 1986) to more practical oriented analysis (e.g. Powrie & Li 1991, Whittle et al. 1993). One important question of numerical analysis of deep excavationis besides the constitutive modelling is the simulation of the interaction wall/surrounding soil. In order to investigate the effects of interface element properties on the deformation behaviour a comprehensive study was performed. In addition results obtained from applying different constitutive laws are presented.

2 DESCRIPTION OF PROBLEM ANALYSED

The excavation which has been chosen for the analysis is shown in Figure 1. The construction steps, prop levels and a simplified soil profile are also indicated. It follows from this Figure that the excavation is 30 m wide and 12 m deep. Figure 2 shows the mesh which consists of 200 15-noded triangle elements. The finite element code PLAXIS (Vermeer & Brinkgreve 1995) is used. The wall is assumed to be in place i.e. the construction of the wall has not been modelled. The diaphragm wall was considered as linear elastic material and the relevant parameters are given in Table 1. The diaphragm wall is supported by two props indicated in Figure 2. These props are simulated by using fixed-end anchor elements which act like fixed-ending springs.

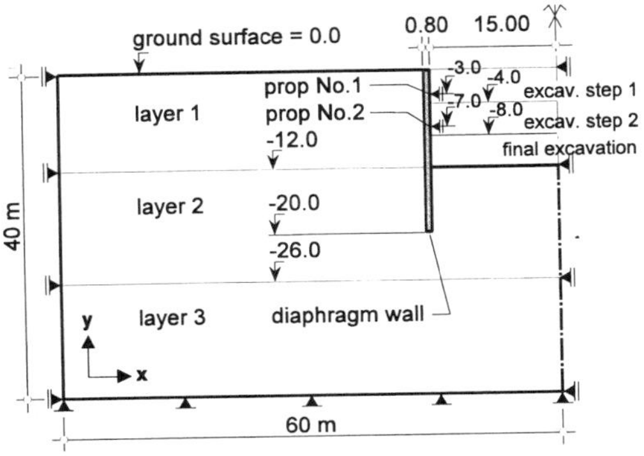

Figure 1. Cross section of excavation

Table 1. Stiffness parameters for diaphragm wall

diaphragm wall	E (MPa)	ν (-)
concrete	21 000	0.15

The strength and stiffness parameters for the soil layers used in a drained, elastic perfectly plastic analysis using a Mohr-Coulomb failure criterion, which serves as reference calculation, are given in Table 2.

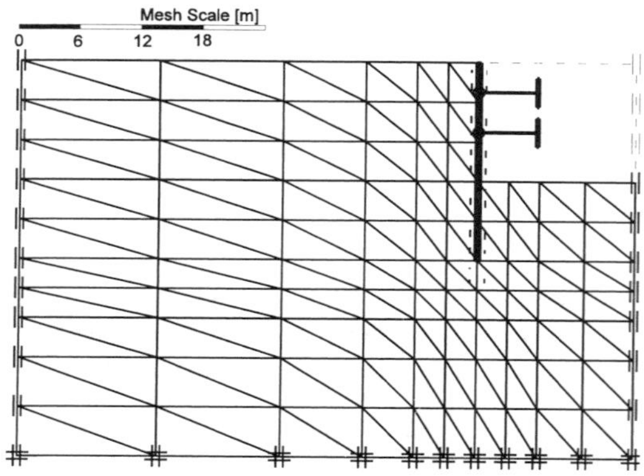

Figure 2. Finite element mesh

Table 2. Strength and stiffness parameters

soil layer	friction angle ϕ (°)	cohesion (kPa)	E (MPa)	ν (-)
layer 1	35.0	2.0	20	0.40
layer 2	26.0	10.0	12	0.40
layer 3	26.0	10.0	80	0.30

The following computational steps have been performed:

- step 1: initial conditions (layer 1: $k_o = 0.5$, layer 2+3: $k_o = 0.65$) and self weight of diaphragm wall
- step 2: excavation to a depth of 4.0m
- step 3: excavation to a depth of 8.0m, prop in -3.0m in place
- step 4: excavation to a depth of 12.0m, props in -3.0 and -7.0m in place.

3 INFLUENCE OF INTERFACE ELEMENTS ON THE DEFORMATION BEHAVIOUR

A number of different joint or interface elements suitable for finite element calculations have been developed (e.g. Goodman et. al. 1968, Pande & Xiong 1982, Gens et al. 1988). In the code used for this study the coordinates each pair of nodes are identical which means that the element has physically a zero thickness. However, each interface element has assigned to it a 'virtual thickness', i.e. an imaginary dimension used to obtain the material properties of the interface. The default value of the 'virtual thickness' is one tenth of the average length of interface elements along the interface chain. An elastic-plastic model is used to describe the behaviuor of the interface elements. A Coulomb relation is used to distinguish between elastic behaviour, where small displacements can occur within the interface, and plastic interface behaviour. The roughness of the structure is modelled by choosing a suitable value for the strength reduction factor. This strength reduction factor R is used to specify the interface element properties as indicated below:

$$c_{interface} = R \cdot c_{soil} \tag{1}$$

$$\tan \varphi_{interface} = R \cdot \tan \varphi_{soil} \tag{2}$$

$$G_{interface} = R^2 \cdot G_{soil} \tag{3}$$

$$K_{interface} = \frac{(1 - \nu_{interface})}{(1 - 2 \cdot \nu_{interface})} \cdot R^2 \cdot G_{soil} \tag{4}$$

$$E_{interface} = 2 \cdot (1 + \nu_{interface}) \cdot R^2 \cdot G_{soil} \tag{5}$$

$$\textit{elastic gap displacement} = \frac{\sigma \cdot t_{interface}}{K_{interface}} \tag{6}$$

$$\textit{elastic slip displacement} = \frac{\tau \cdot t_{interface}}{G_{interface}} \tag{7}$$

where:

$c_{interface}$ = *the adhesion at the interface*

$\varphi_{interface}$ = *the interface friction angle*

$G_{interface}$ = *the shear modulus of the interface*

$K_{interface}$ = *the bulk modulus of the interface*

$E_{interface}$ = *the Young's modulus of the interface*

$\nu_{interface}$ = *the Poisson's ratio of the interface*

$t_{interface}$ = *the 'virtual thickness' of the interface*

Obviously, if R is set to unity, the interface properties are the same as those in the surrounding soil. The value of $\nu_{interface}$ is fixed at 0.45 and the value of the dilation angle $\psi_{interface}$ is set to 0.

A parametric study was carried out to investigate the influence of interface properties on the deformation behaviour. The strength reduction factor R was varied from 0.45 to 1.0.

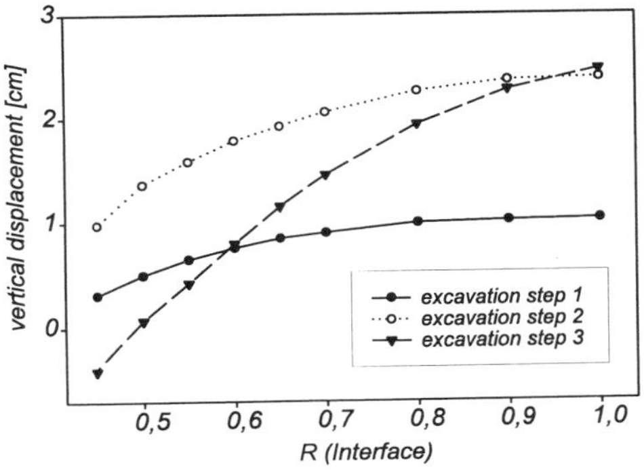

Figure 3. Vertical displacement - top of diaphragm wall

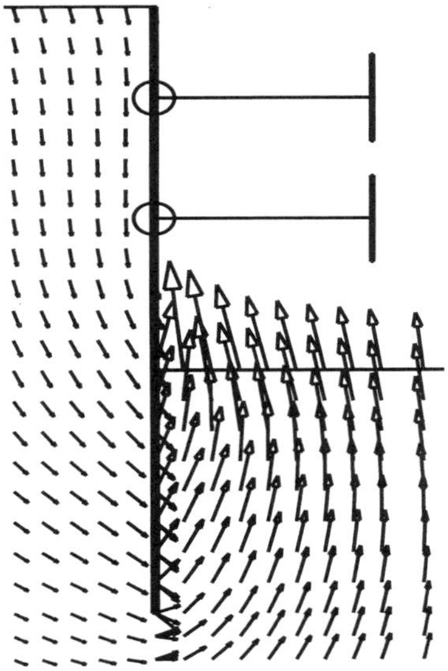

Figure 4. Incremental displacement field

Employing simple constitutive laws finite element calculations usually exhibit somewhat unrealistic vertical displacements of the wall which is not observed in practice. Therefor this aspect is addressed in this paper in particular. The effect of the strength reduction factor R on the vertical displacement of the wall is indicated in Figure 3. The first two excavation steps lead to a heave of the wall. As expected the higher the value of the factor R is, the more heave is calculated because in this case a fully bonded interface is modelled. However, in the last excavation step (both props built-in) the wall settles, especially in the case of low wall friction corresponding to the movement of the sourrunding soil as indicated in Figure 4, which compares qualitatively well which

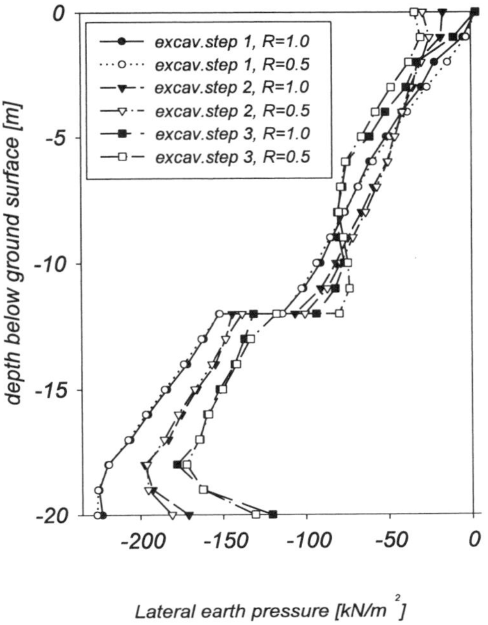

Figure 5. Lateral earth pressure

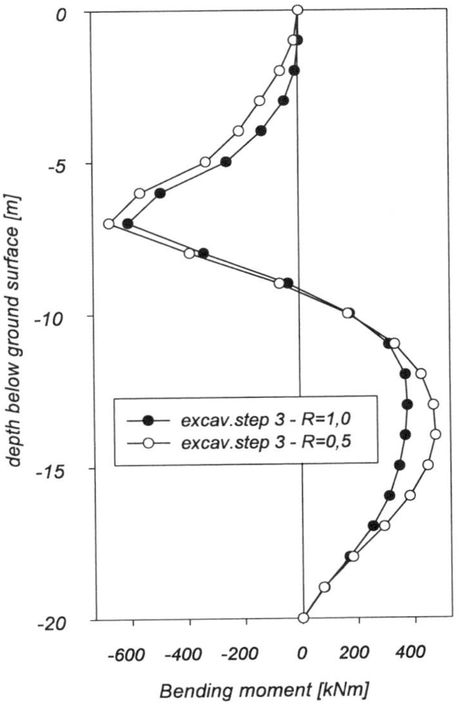

Figure 6. Bending moments for excavation step 3

observations in the field. The influence of the factor R on the horizontal displacements of the toe of the wall is not as pronounced which has also been observed by Mangels (1996). The horizontal movements of the top of the wall increase with lower factor of *R*. As expected lateral earth pressure is only slightly influenced by the stiffness reduction factor *R* (Figure 5).

The effect on the bending moments is pointed out in Figure 6. If wall friction is reduced bending moments increase, the maximum difference being less than 30% for the final excavation step for *R* ranging from 0.5 to 1.0.

4 INFLUENCE OF DIFFERENT CONSTITUTIVE MODELS

This section of the paper deals with the influence of different constitutive models on the deformation behaviour of deep excavations. In order to investigate the effects of different models which are suitable for describing the stress-strain behaviour of soft soils deformations obtained from a simple Mohr-Coulomb model are compared with an Advanced Mohr-Coulomb model and a so called Soft Soil model.

4.1 Advanced Mohr-Coulomb model

Modelling errors inherent in the usual Mohr-Coulomb model are associated with two main simplifications. The crudest simplification concerns the use of a single, constant stiffness modulus. An additional shortcoming is usually the simulation of the dilatancy. In order to model the influence of the stress level on the soil stiffness, a simple power law for the shear modulus is introduced in a so called Advanced Mohr-Coulomb model:

$$G = G^{ref} \cdot \left(\frac{p^{*}}{p^{ref}} \right)^{m} \qquad (8)$$

$$p^{*} = -\frac{1}{3} \cdot (\sigma'_{xx} + \sigma'_{yy} + \sigma'_{zz}) + c \cdot \cot\varphi \qquad (9)$$

where:

G^{ref} = *Reference shear modulus (corresponding to refernce pressure* p^{ref})

m = *constant*

The power m determines the stress dependency. For m = 0, there is no stress dependency at all and for m = 1 the obtained stiffness is proportinal to the stress level. Typical values for sandy soils are about 0.5, whereas for clays a more appropiate value would be 1. In addition dilatancy cut-off is included (Vermeer 1984). After extensive shearing, dilating materials arrive in a state of critical density where no further volume change is observed. In the used Advanced Mohr-Coulomb model dilatancy cut-off is controlled by the parameters n_{init} and n_{max}, where n_{init} is the initial porosity and n_{max} is the maximum porosity. The porosity is related to the volumetric strain, ε_v by the relationship:

$$\varepsilon_v^{e} - \varepsilon_v^{e0} = \ln\left(\frac{1 - n^{0}}{1 - n} \right) \qquad (10)$$

where: ε_v is positive for dilatant behaviour

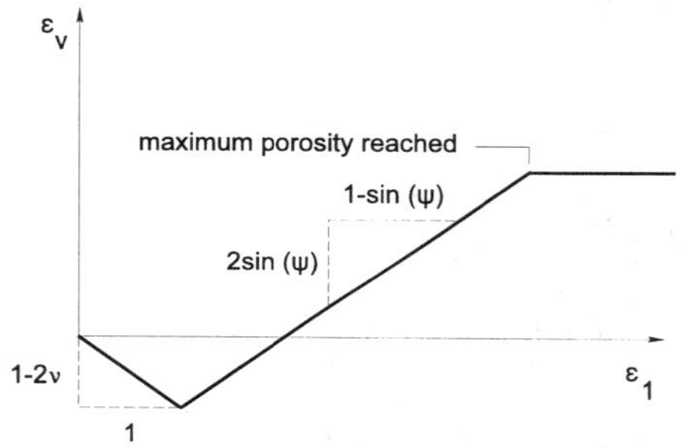

Figure 7. Idealisation of dilatancy cut off

As soon as the volume change results in a state of maximum porosity, the dilatancy angle is automatically set back to zero (see Figure 7).

It should be noted, that it is known from preceeding analysis, that the dilatancy has only a slight influence on the deformation behaviour for the problem investigated here. The parameters used in the Advanced Mohr-Coulomb model are given in table 3.

Table 3. Parameters for Advanced Mohr-Coulomb model

soil layer	G^{ref} (MPa)	p^{ref} (kPa)	m (-)	n_{init} (-)	n_{max} (-)
layer 1	7.69	40	0.3	0.35	0.60
layer 2	4.28	250	0.9	0.40	0.70
layer 3	28.57	420	1.0	0.45	0.75

The dilatancy angle was assumed to be equal to the friction angle, i.e. an associated flow rule was applied.

4.2 Soft Soil model

The Soft Soil model is a Cam Clay model, which resembles the Modified Cam-Clay model (Burland 1965, Burland 1967). Some of the characteristics of the Soft Soil model are:

- Stress dependent stiffness
- Distinction between primary loading and unloading/ reloading
- Memory of preconsolidation stress
- Failure behaviour according to the Mohr-Coulomb criterion.

The parameters κ* and λ* can be obtained from one-dimensional compression tests. A relationship exists with the parameters for one-dimensional compression and unloading. This relation to the Compression index C_c and the Swelling index C_s is given below:

$$\lambda^* = \frac{C_c}{2.3\ (1 + e)} \tag{11}$$

$$\kappa^* \approx 1.3\ \frac{1 - \nu_{ur}}{1 + \nu_{ur}}\ \frac{C_s}{(1 + e)} \tag{12}$$

where:

λ^* = *Modified swelling index*

κ^* = *Modified compression index*

ν_{ur} = *Poisson's ratio for unloading/reloading*

The parameters used in the Soft Soil model for layer 2 and layer 3 are given in table 4.

Table 4. Parameters for Soft Soil model

soil layer	λ* (-)	κ* (-)	ν_{ur} (-)
layer 2	0.090	0.0214	0.15 / 0.25
layer 3	0.024	0.0056	0.15 / 0.25

For layer 1 in the Soft Soil example a Mohr-Coulomb failure criterion is used with the same parameters as preliminary defined because this layer represents a gravel type material. For layers 1 and 2 the Over-Consolidation Ratio OCR and the strength reduction factor *R* was assumed to be 1.0.

The vertical displacements of the diaphragm wall for the Advanced Mohr-Coulomb model are smaller than the displacements for the Mohr-Coulomb model, as it can be seen in Figure 8. The vertical

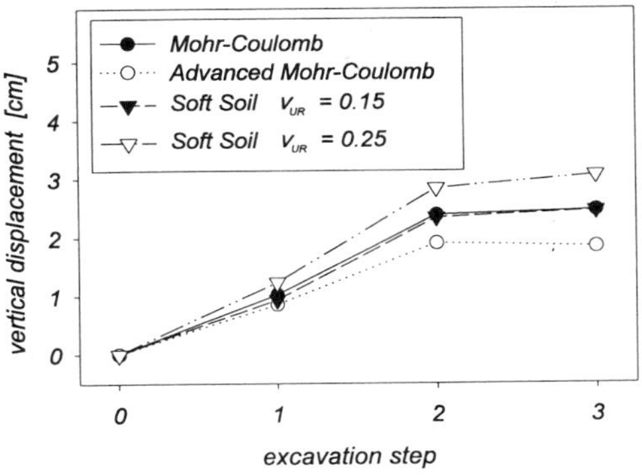

Figure 8. Vertical displacement of diaphragm wall

displacements of the wall of the Soft Soil model are significantly influencd by the Poisson's ratio for unloading / reloading ν_{ur}.

The heave inside the excavation is compared in Figure 9. Layer one is modelled in all of the four considered examples as a Mohr-Coulomb material.

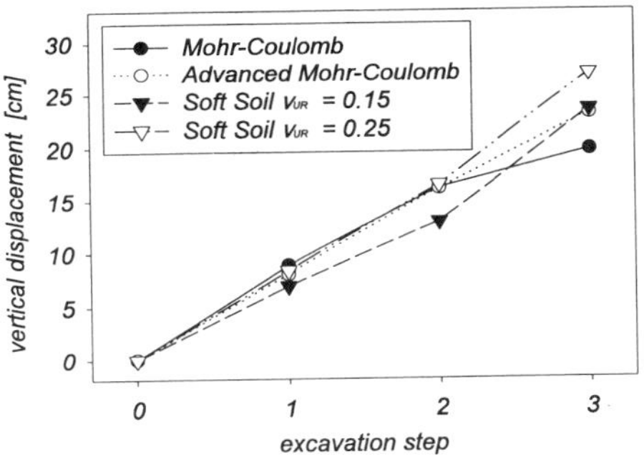

Figure 9. Heave inside the excavation

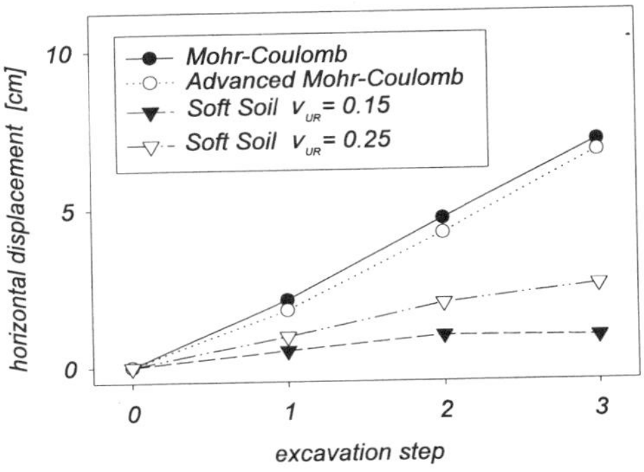

Figure 10. Horizontal displacements of the toe of diaphragm wall

Due to that circumstance the heaves show the biggest differences for excavation step three, wherein layer one is completely removed from the inside of the excavation.

Horizontal displacements of the toe of the diaphragm wall resulting from the Soft Soil model are significantly smaller then the displacements obtained from the Mohr-Coulomb model (see Figure 10) and in situ observations under similar conditions suggest that these lower values are closer to the actual behaviour in the field.

5 CONCLUSION

A numerical study of a deep excavation in soft soil has been presented. The application of interface elements allows the simulation of the interaction between the wall and the surrounding soil in a more appropriate way. By considering vertical displacements of the diaphragm wall the results are more realistic as in analysis without interface elements. These elements have only a slight influence on the earth pressure and the bending moments of the wall. To improve on the shortcomings of a Mohr-Coulomb model the use of the Advanced Mohr-Coulomb model allows to introduce a stress dependent stiffnes behaviour. A further advantage is the improved simulation of the dilatancy.

The Soft Soil model resembles the Modified Cam-Clay model. The displacements of this model are strongly influenced by the assumed Poisson's ratio for unloading and reloading. For the choosen parameters the horizontal deformations of the toe of the daiphragm wall resulting from the Soft Soil model are significantly smaller then those obtained from a Mohr-Coulomb model and more likely to represent the behaviour in situ.

Nevertheless it should be noted that the use of more sophisticated models is only justified when high quality input data available.

REFERENCES

Burland, J.B. 1965. The Yielding and Dilation of Clay. *Geotechnique,*.Vol. 15: 382-405.

Burland, J.B. 1967. Deformation of Soft Clay. Dissertation. Cambridge University

Gens, A., I. Carol & E.E. Alonso 1988. An interface element formulation for the analysis of soil-reinforcement interaction. *Computers and Geotechnics 7:* 133-151.

Goodman, R.E., R.L.Taylor & T.L. Brekke 1968. A model for the mechanics of jointed rock *J. Soil Mech. and Found. Div. A.S.C.E. 94:* 637-659.

Mangels, J. 1996. Anwendung der Methode der finiten Element auf ausgewählte Beispiele in der Geotechnik. *Diploma Thesis.*Bergische Universität Gesamthochschule Wuppertal

Pande, G.N. & W. Xiong 1982. An improved multi-laminate model of jointed rock masses. Int. Symp. on Numerical Models in Geomechanics: 218-226. Zürich

Potts, D.M. & A.B. Fourie 1986. A numerical study of the effects of wall deformation on earth pressures. *Int. J. Num. Anal. Meth. Geom.* 10: 382-405.

Powrie, W. & E.S.F. Li 1991. Finite element analyses of an in situ wall propped at formation level. *Geotechnique* 41: 499-514.

Vermeer, P.A. & R.B.J. Brinkgreve 1995. PLAXIS Finite Element Code for Soil and Rock Analyses. Rotterdam: Balkema.

Whittle, A.J., Y.M.A. Hashash & R.V. Whitman 1993. Analysis of deep excavation in Boston. *J. of Geotech. Engineering* 119: 69-90.

Numerical Models in Geomechanics, Pietruszczak & Pande (eds) © 1997 Balkema, Rotterdam, ISBN 90 5410 886 X

Finite element slope stability analysis – Why are engineers still drawing circles?

P.A. Lane
Department of Civil and Structural Engineering, UMIST, Manchester, UK

D.V. Griffiths
Division of Engineering, Colorado School of Mines, Colo., USA

ABSTRACT: Limit equilibrium methods of slope stability analysis can be wholly superseded by finite element analysis for any potential slope failure mechanism. Reluctance to use the finite element method for slope stability analysis in practice has been partly due to concerns that it is complex and computationally time consuming. A finite element (FE) program to model slope failure mechanisms has been developed which allows rapid analysis of a wide range of slope stability problems including the influence of water and submerged loading conditions. Mesh density has been balanced with accuracy and economy as essential criteria for the successful engineering application of FE programs. The program has been validated against traditional slip circle analyses and documented case histories.

1 INTRODUCTION

Slope stability analyses are usually conducted for one of the following reasons: assessment of an existing natural slope; design of a proposed embankment or cut; or back-analysis of a failed or failing slope. Any method must therefore be capable of analysis using in-situ conditions; forecast of potential behaviour using design conditions and back-analysis of previous conditions. Any such analysis should be as insensitive as possible to 'a priori' assumptions or constraints. Traditional method of slope stability analysis, based on the slip circle approach, are governed by the imposed circle (or modified arc) and the detail of the analysis method. Finite element analyses are less dependent on detail of method, requiring only general classification of drained/undrained behaviour; cohesive or predominantly granular material and possibly large-strain considerations in extreme cases. The pre-condition of an assumed failure mechanism is not required.

1.1 *The influence of water*

Given the crucial influence of water on the stablity of a slope and the possible variation of this effect over time it is also vital that any method can accommodate the full range of effects and in a realistic manner. Efforts have been made to incorporate such conditions into traditional slip circle methods but all are constrained by the imposition of the mechanism and the importance of its assumed location in the calculation. Comparison with the work of Bishop and Morgernstern (1960) and Lambe and Silva (1995) illustrates the importance of the correct modelling of pore pressure variation beyond that of a global r_u value.

1.2 *Submerged and drawdown conditions*

The ability of the FE method to allow an almost infinite range of properties to be accommodated dependent only on the mesh density and machine capability is especially important in allowing a fine variation in pore pressures to be included and the modelling of submerged and *rapid drawdown* conditions allows utilisation of the program for the most extreme and usually critical cases. The charts of Morgernstern (1963) are themselves based on traditional slip circle analysis and therefore suffer the same constraints. The FE program allows the full range of conditions to be tested without constraint and the most critical case pinpointed.

2 THE FE PROGRAM: FEEMB1LG

The program is an expanded version of 'FE-EMB1' developed by Griffiths (1996) for 2-dimensional slope stability analysis by finite elements using 8-node quadrilateral elements of elastic-perfectly plastic soil with a Mohr-Coulomb failure criterion. The primary development has been the inclusion of free-surface and/or external reservoir loading. The soil's self-weight is modelled by a gravity 'turn-on' procedure (Smith and Griffiths, 1988) with nodal loads added in the first increment.

2.1 *Factor of Safety and 'Failure'*

The Factor of Safety (FoS) for the slope is defined by division of the original shear strength parameters where:

$$c'_f = c'/FoS \tag{1}$$

$$\phi'_f = \arctan(\frac{\tan\phi'}{FoS}) \tag{2}$$

Failure of the slope can be defined in different ways (Abramson et al, 1996). The program uses the failure of the visco-plastic algorithm to converge within an iteration limit (usually 250), with a nodal displacement criterion on successive iterations. This is considered to be a physically real criterion. The FoS 'at failure' lies between the FoS at which the iteration limit is reached and the immediately previous value. By comparison, the FoS generated by traditional methods represents the ratio between the driving and restoring forces.

Piezometric surfaces, pore water pressure conditions and submerged effects can be included by definition within the data file. As well as the numerical results, displaced mesh and displacement vector plots are produced to assist in the determination of the failure mechanism. Figure 1. shows such a plot.

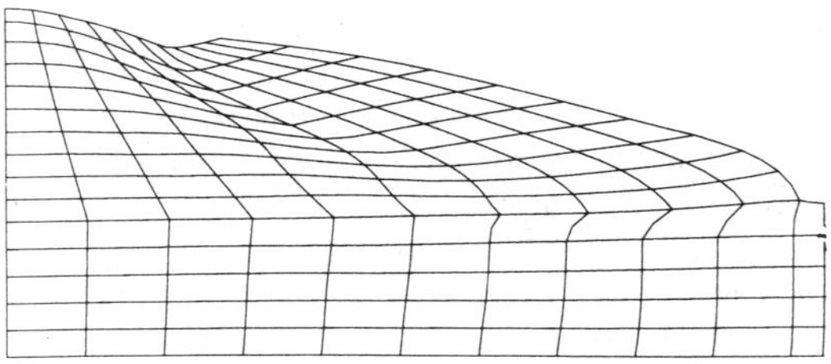

Figure 1. Displaced mesh plot from FEEMB1LG

3 BENCHMARKING WITH BISHOP'S SLOPES

3.1 *Bishop and Morgernstern's slopes*

The classic papers of Bishop and Morgernstern were taken as benchmarks for the finite element approach. The example calculations for the slope in Figure 15 of their joint 1960 paper where reproduced giving FoS of 1.65 (for $r_u = 0.5$) and 3.0 for $r_u = 0$, compared to their Stability Coefficients results of 1.65 and 3.06. Similarly the results for Figure 2 of their paper were recalculated giving excellent agreement as shown in Figure 2 here.

3.2 *'The Bishop slope'*

Lambe and Silva (1995) re-analysised 'the Bishop slope' of his 1955 paper for a critique of the Ordinary Method of Slices (OMS). Analysis by FEEMB1LG gave results that lay between the FoS values reported by them for OMS and their suggested correction method.

Table 1. Comparison of results for 'the Bishop slope'

r_u	Lambe and Silva	FEEMB1LG
0	2.6 - 2.5	2.5
0.6	0.7 - 1.0	0.83

But they commented '..engineers can average $\bar{B}$ to produce a constant r_u value. Our experience has never shown a section with a constant r_u.... A constant r_u simplifies analysis but does not make good sense and could provide misleading results.'

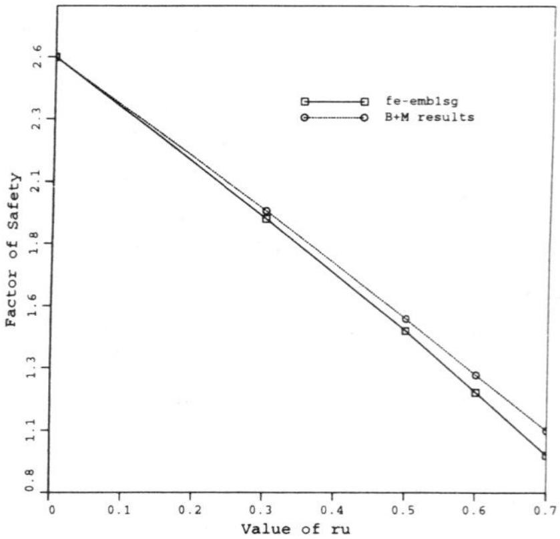

Figure 2. Comparison of Linear Relationship of r_u and FoS with Bishop and Morgernstern result.

3.3 *The Lodalen slope*

Bishop and Morgernstern (1960) attempted to account for this by averaging r_u values over the predicted slip circle. They used the 'Lodalen slope' (Sevaldson, 1956) as an example of the suggested method. Using their averaged value of r_u of 0.28 in FEEMB1LG gave a FoS of 1.07 compared to their result of 1.08. Utilising the original paper of Sevaldson to insert the observed water table levels and the measured r_u range of 0 to 0.49 in the program gave a FoS of 1.0, i.e. the actual slope failure reported at Lodalen. For maximum flexibility FEEMB1LG allows individual element specification of $\bar{B}$ and independent specification of water table levels.

4 SLOPE STABILITY UNDER SUBMERGED CONDITION

The effect of submergence on the slope is included in FEEMB1LG by the addition of nodal loads on the slope face. These are calculated automatically for a given water level but allows for the possibility of other loading sources through the use of a K_0 value. The imposition of water to the slope face is handled separately from the piezometric water levels. This allows total flexibility in the specification of conditions and, especially, the modelling of *rapid drawdown*. Taking the Example 6.14 of Smith and Griffiths (1988), the effect of varying the submerged water level was analysed and the results shown in Figure 3 for two sets of shear strength parameters.

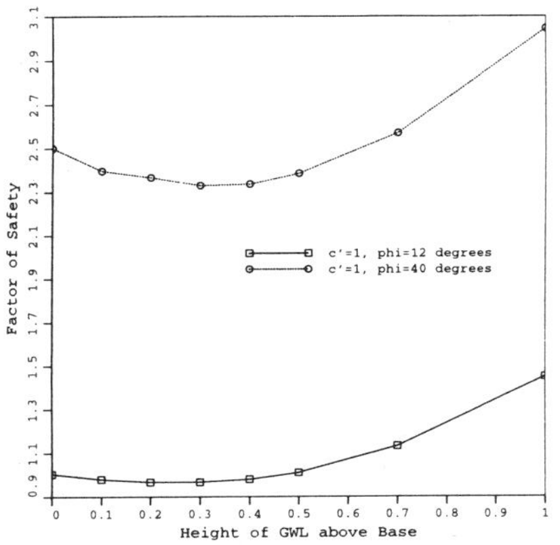

Figure 3. Submerged slope stability for variable water level.

The $\phi' = 40^o$ result shows a minimum FoS of about 2.3 with 0.3m of water above the toe. The case with $\phi' = 12^o$ also gives a minimum at 0.3m where the slope becomes unstable whilst it is stable either with no water or fully submerged.

These examples were run in a traditional slope stability package using slip circle searching techniques. Agreement was excellent for this simple case as the mechanism is predominately circular. However, the traditional method requires manual intervention in the searching strategy. The finite element approach generates the critical circle automatically and requires only a crude mesh ($5x5$ elements) as shown in Figure 4.

5 RAPID DRAWNDOWN

The most critical condition for most submerged slopes is the *rapid drawdown* case. The internal pore water pressures from the submerged condition cannot dissipate at the same speed as the external water level is reduced in a fine-grained material. Morgernstern's 1963 paper presented stability charts based on parametric studies using slip circle analysis automated on then available computers. He assumed $\bar{B}$ was unity and that no dissipation occurred during drawdown.

In FEEMB1LG the piezometric surface is specified as per the original water level but the face loads are based on the specified water surface level which in this case is below that of the piezometric values. Morgernstern's charts are non-dimensional for various values of:

$$\frac{c'}{\gamma H} \tag{3}$$

and interpolation can be used for other values. Figures 5 and 6 illustrate the comparison of results between Morgernstern and FEEMB1LG for a range of cases. Excellent agreement was obtained although the finite element results are slightly lower especially over the higher drawdown ratios.

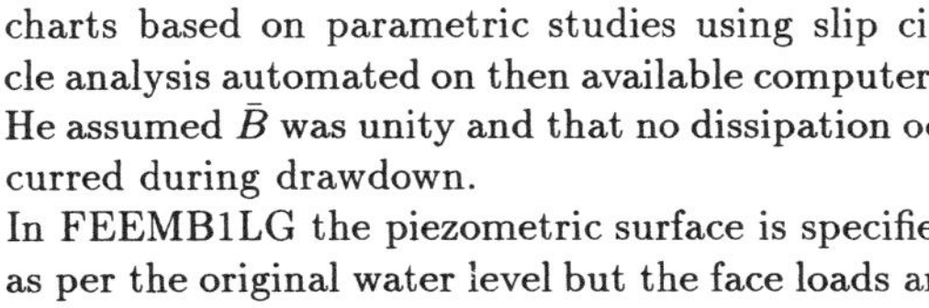

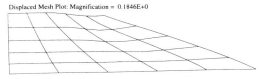

Figure 4. Displaced mesh for Example 6.14

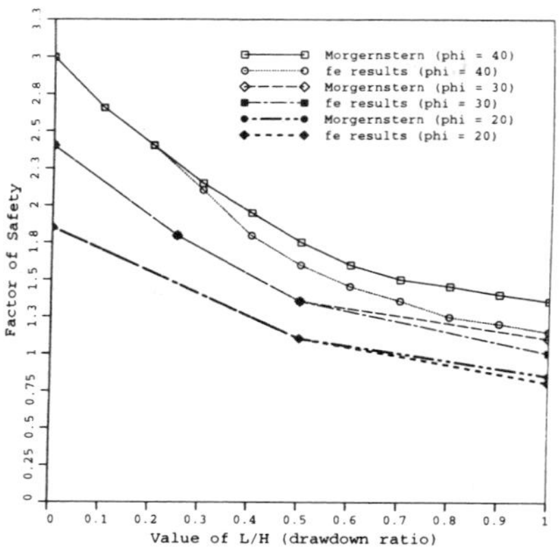

Figure 5. Comparison with Morgernstern results for a 2 : 1 slope and $c'/\gamma H = 0.05$

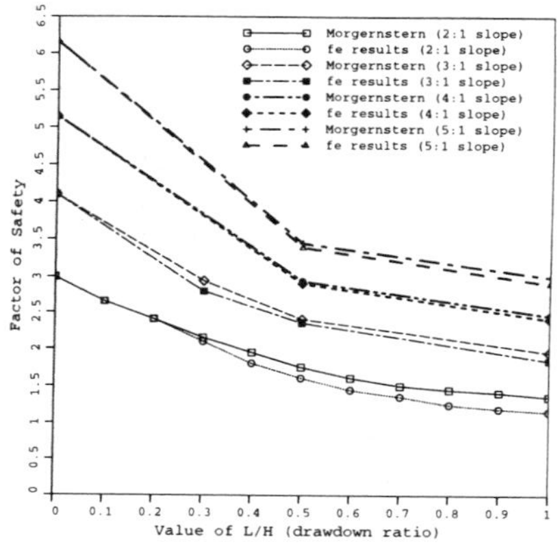

Figure 6. Comparison with Morgernstern results for $\phi' = 40^o$

The drawdown ratio (L/H) is the ratio between the slope height (H) and the depth below the crest to which the water level falls (L). All slopes are assumed to be initially fully submerged ($L = 0$).

6 CASE STUDY

An existing dam (WC) was being investigated for stability. The particular concern is for partially submerged conditions. A traditional slip circle analysis had been conducted and the Bishop and Morgernstern Stability Coefficient calculation was also performed. The dam had two layers of material - the embankment itself with $\phi' = 40^o$ and the stronger foundation layer with $\phi' = 55^o$. For various conditions for the dam the FoS results were . The worst case condition was found to be that of emptying from a partially submerged condition. This corresponds well with the earlier results of the case of lowest FoS at partial submergence and the *rapid drawdown* effect. The traditional slip circle approach produced consistently higher estimates of FoS although the trend between conditions was the same. Further investigations are continuing as the difference between these results is important in deciding on the acceptability of its current condition.

Table 2. Results comparison for WC Dam

Condition	Bishop etc	Slip Circle	FEEMB1LG
Dry slope	1.78		1.85
Partially submerged		2.36	1.7
Fully submerged		2.56	1.95
Empty with internal pwp		2.11	1.65

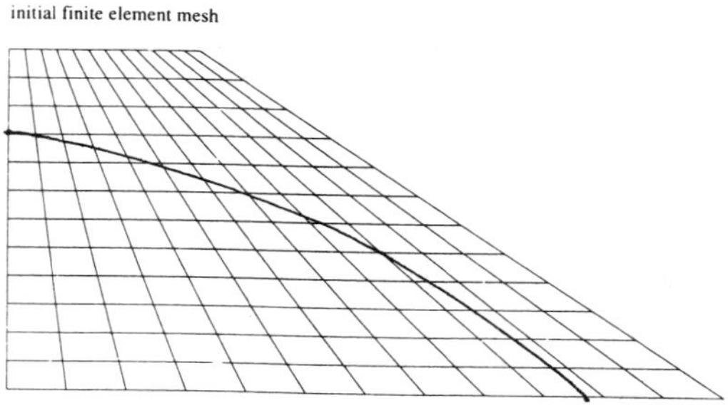

Figure 7. Initial 12×12 mesh for Lambe and Whitman Example 24.3

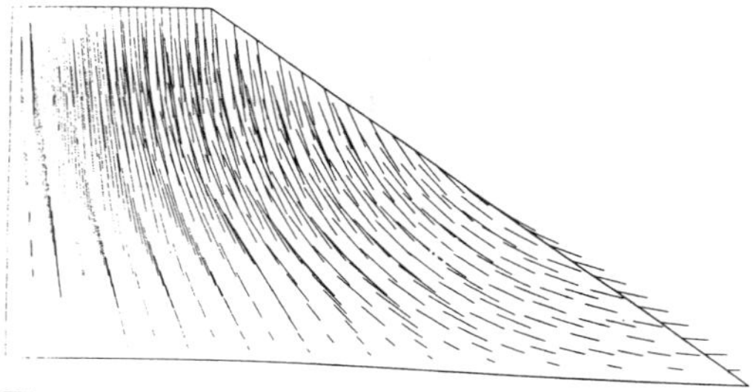

Figure 8. Displacements for Example 24.3

7 MESH DENSITY

A study of mesh density was carried out to determine its' impact on the results. Lambe and Whitman's (1979) Example 24.3 was considered with mesh densities of 5 × 5, 10 × 10 and 12 × 12. The FoS for 250 iterations was taken and the result was found to vary by less than 2% between the meshes (from 1.23 to 1.21). Figures 7 and 8 show the initial mesh and phreatic surface for the Example 24.3 and the displacement plot for a 12 × 12 mesh.

The greatest difference between the meshes is the clearer illustration of the denser meshes in terms of identifying the failure mechanism. The numerical results are only marginally affected. The finite element result is in good agreement with the slip circle result, although slightly lower.

8 CONCLUSIONS

A finite element program has been shown to give consistent results over a wide range of the most critical slope stability problems. The finite element method has fewer constraints and initial assumptions than traditional slope stability analysis methods and automatically identifies the critical failure mechanism without the need for manual intervention. With commonly available computer power the finite element method is readily accessible to practicing engineers. Traditional slip circle methods suffer their own inherent problems which make them susceptible to misuse. The finite element method, with its greater potential, should should now be the basis of engineering analysis and design.

REFERENCES

Abramson, L.W., Lee, T.S., Sharma, S., Boyce, G.M., 1996. *Slope stability and stabilization mthods.* New York: Wiley.

Bishop, A.W., 1955. The use of the slip circle in the stability analysis of slopes. *Geotechnique*, Vol. 5, No. 1, pp. 7-17.

Bishop, A.W., Morgernstern, N.R., 1960. Stability Coefficients for earth slopes. *Geotechnique*, Vol. 10, No. 1, pp 129-150.

Griffiths, D.V., 1996. FE-EMB2 and FE-EMB1: Slope stability software by finite elements. *Geomechanics Research Center Report*, GRC-96-37, Colorado School of Mines.

Lambe, T.W., Silva, F., 1995. The Bishop slope revisited. *Geotechnical News*, Vol. 13, No. 4., North Dakota: BiTech.

Lambe, T.W., Whitman, R.V., 1979. *Soil Mechanics*, New York: Wiley.

Morgernstern, N.R., 1963. Stability charts for earth slopes during rapid drawdown. *Geotechnique*, Vol. 13, No. 1, pp. 121-131.

Sevaldson, R.A., 1956. The slide in Lodalen, October 6th 1954. *Geotechnique*, Vol. 6, No. 4, pp. 167-182.

Smith, I.M., Griffiths, D.V., 1988. *Programming the Finite Element Method*, 2. Ed. Chichester: Wiley.

Acknowledgement

Dr. Lane was a Visting Assistant Professor at the Colorado School of Mines for the period of this research. Support is gratefully acknowledged to CSM and to the Peter Allen Scholarship Fund of UMIST for financial support.

Numerical Models in Geomechanics, Pietruszczak & Pande (eds) © 1997 Balkema, Rotterdam, ISBN 90 5410 886 X

Stability of crushed mudstone embankments with the occurrence of 'slaking'

M. Nakano, D. Constantinescu & A. Asaoka
Department of Civil Engineering, Nagoya University, Chikusa-ku, Aichi, Japan

ABSTRACT: In this paper the authors study the long term stability of an embankment made of the Tertiary mudstones which easily give rise to slaking. In order to explain the mechanism of slaking behavior, the one-dimensional compression tests and the triaxial compression tests for the crushed mudstones are carried out. The slaking behavior of the crushed mudstones, being saturated with water, can be idealized as the softening behavior of the heavily overconsolidated soils due to swelling above the critical state line in $q : p'$ space during shearing, which reaches the normally consolidated state. The reason for the failure of mudstones specimens is the formation of "double structure" in the mudstone. The mudstone pebbles by themselves swell with water, i.e. slaking, due to large shear stresses at contact surfaces between pebble boundaries even under constant loading.

1. INTRODUCTION

Large quantities of Tertiary mudstones are resulting as excess material in some of the major construction sites in Japan. They are currently disposed as a waste material due to its susceptibility to slaking. From the view point of recent environmental concerns, this soil is expected to be used as a construction material. However, due to their problematic slaking behavior, the mudstones are not yet widely used.

The main objective of this study is therefore to explain the mechanism of slaking behavior of the Tertiary mudstones based on the Critical State Soil Mechanics and to suggest a concept for the long term stability of embankments made of crushed mudstone. To interpret the mechanism of slaking behavior in terms of elasto-plastic mechanics, saturated mudstone is idealized as the heavily overconsolidated clay. Through a numerical simulation on heavily overconsolidated clay behavior, we bring forth a definition for the mechanism of slaking. To interpret the behavior of an embankment made of mudstone, firstly is essential to understand the behavior of the assembly of crushed mudstones. In this study we call the assembly of crushed mudstones as "double structure". To understand the assembly which has "double structure", we carry out one-dimensional compression tests and triaxial compression tests.

2. SIMULATION OF SLAKING, SOFTENING BEHAVIOUR OF HEAVILY OVER-CONSOLIDATED CLAYS

The behavior of heavily overconsolidated clay can be well described by the subloading surface Cam-clay model which is obtained by applying the subloading surface concept to the original Cam-clay model. A numerical simulation based on the subloading surface Cam-clay model on the conventional drained triaxial compression test of the overconsolidated KAWASAKI clay was firstly carried out . For details of the simulation , refer Asaoka, Nakano and Noda (1997).

Figs.1(a) and (b) show the results of the numerical simulation on the element-wise behavior, in which the locations of the soil elements are also indicated. Both of the two designated elements show the hardening behavior above the critical state line during early stages of shearing. Subsequently, softening with swelling occurs above the critical state line. These plots(Figs.1(a),(b)) further show that the elements have already reached the normally consolidated state. Especially, the element shown in Fig.1(b) swells with water to a greater extent than the element shown in Fig.1(a) and make a "comeback" to normally consolidated state, of which pre-consolidation pressure is about one fourth of its initial pre-consolidation pressure of 1764kPa.

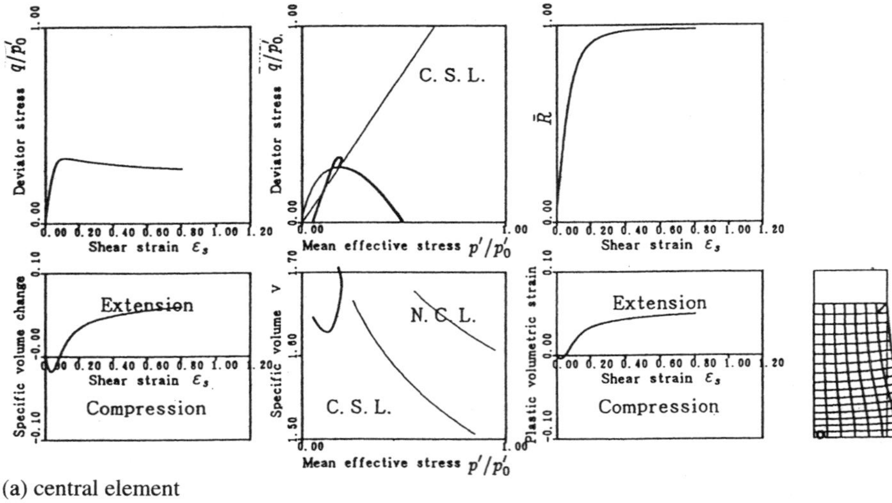

(a) central element

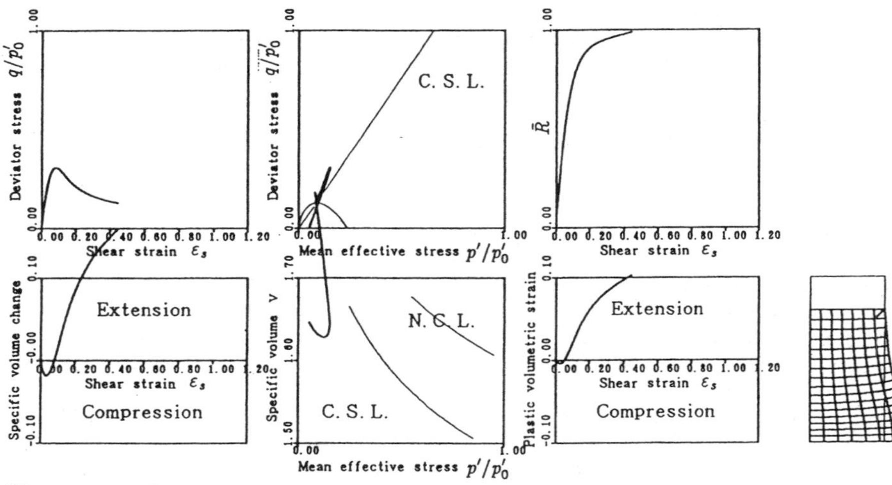

(b) outermost element

Fig.1 Numerical simulation on element-wise behavior

Fig. 2 shows two loading processes through which the overconsolidated clay can be brought to normally consolidated clay in $e\sim \ln p$ space. One is to apply a large stress to the overconsolidated clay and the other is to shear the clay with supplying water. The latter process shows the slaking phenomenon as seen in the element wise behavior in Fig.1. The slaking behavior of mudstones can therefore be idealized as the process of reaching normally consolidated states from heavily overconsolidated states not by applying a large compressive stresses, but by swelling during shearing.

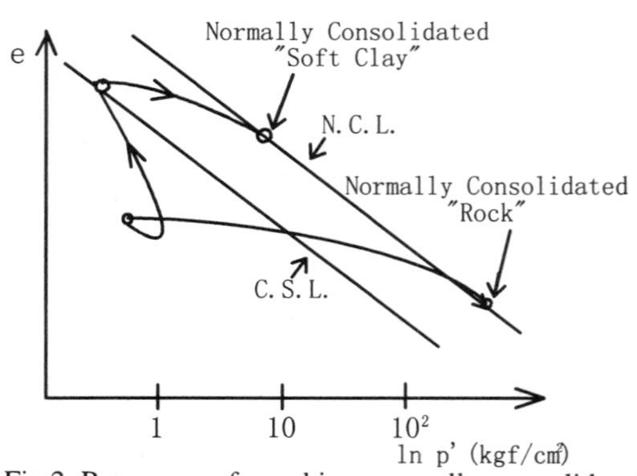

Fig.2 Processes of reaching normally consolidated states from heavily overconsolidated states

3. DEFINITION OF DOUBLE STRUCTURE

In order to make safe designs of embankments made of mudstones, we should firstly understand the characteristic behavior of the crushed mudstone.

Fig.3 shows three states in which the mudstones can exist. First one is the stiff solid state which is generally found underlying the alluvial deposits. This state can be considered to be a heavily overconsolidated state. Second one is the crushed state of mudstones, which is usually appearing after excavation. It is packed loosely or densely and thus is the assembly of solid mudstone pebbles, which consist of pebbles and void spaces.

The pebble itself, which is stiff and solid, consists of soil grains and voids like saturated heavily overconsolidated clays. Void spaces between pebbles are fully saturated with water. In this study, therefore, the structure of this assembly of the mudstone pebbles is called a "double structure". It is most important for the appropriate design of an embankment made of mudstones, to grasp the behavior of the "double structure". This is because the "double structure" causes the slaking rather easily. In the following sections, experimental results are described in this regard.

The third state is the remolded normally consolidated state. If the slaking occur perfectly in all mudstone pebbles, the mudstone would reach this state.

4. ONE DIMENSIONAL COMPRESSION PROPERTIES OF CRUSHED MUDSTONES

4.1. *Experimental procedure*

One-dimensional compression tests of the crushed TOKONAME mudstones were carried out to examine the compression characteristics of "double structure". The saturated sample was obtained from a depth of 50m below the sea level. The index properties of mudstones are shown in Table 1.

The one-dimensional compression specimens consisted of mudstone pebbles of whose average diameters were between 19.1mm~37.5mm or 4.76mm~9.50mm. The test specimen had a diameter of 150mm and a height of 100mm. They were made in the CBR mold by compaction. As shown in Table 2, by two methods of compaction, loose (Samples-A3, B3, C3, D3, H5, I5, J5, M6 and N7) and dense (Sample-E3) specimens were obtained. They were then subjected to the constant vertical stress in the CBR mold under full saturation with water for about five days. The magnitudes of vertical stress applied are also listed in Table 2.

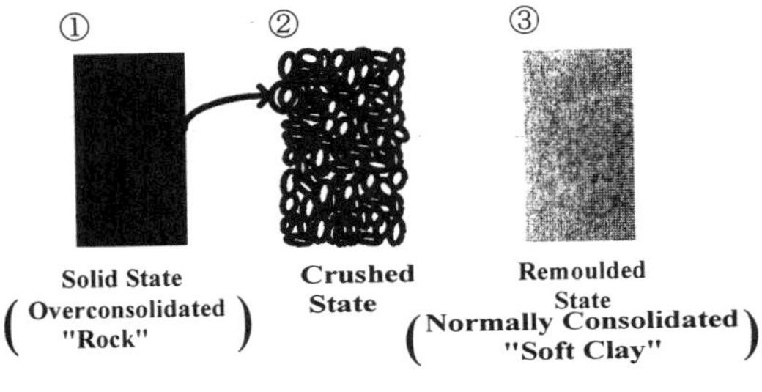

Fig.3 Three states of mudstones

4.2. *Experimental results*

Fig.4 shows the results of the test on the mudstone specimens plotted as the relationship between void ratio e and vertical compressive stress σ_v for all mudstone specimens. Fig.4 also shows the normal consolidation line (N.C.L.) which is obtained from

Table 1 Index proprieties of TOKONAME mudstone

Natural Water Content(%)	Specific Gravity	Water Limit(%)	Plastic Limit(%)	Dry Density (kgf/cm3)
24.3478	2.688	43.63	30.33	1.5976

Table 2 Initial and test conditions of the mudstone sample

Test	Loading (kgf/cm2)	Initial Void Ratio	Initial Grain Size (mm)	Final Settlement (cm)
A3	3.0	1.349	19.1-37.5	0.2504
B3	3.0	1.411	19.1-37.5	0.6260
C3	3.0	1.540	19.1-37.5	0.5884
D3	3.0	1.562	4.76-9.50	0.6017
E3	3.0	0.729	19.1-.5	0.0030
H5	5.0	1.451	19.1-37.5	0.4052
I5	5.0	1.310	19.1-.5	0.8338
J5	5.0	1.827	4.76-9.50	0.9410
M6	6.5	1.419	4.76-9.50	0.6628
N7	7.0	1.330	19.1-37.5	0.4864

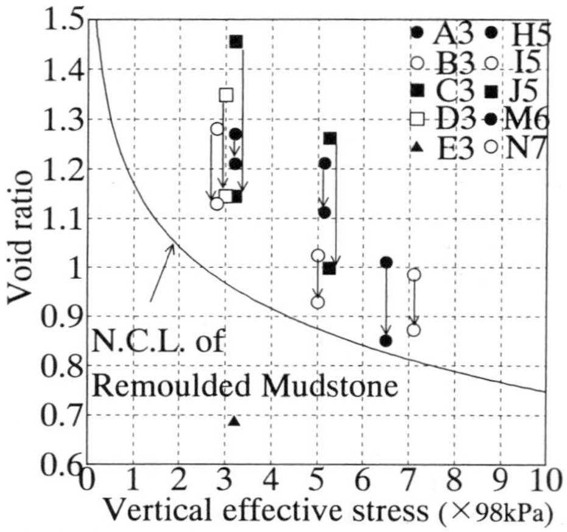

Fig.4 Change of void ratio during testing

the one-dimensional compression test on fully remolded mudstone specimen. The states of loosely packed mudstone specimens are initially located above the N.C.L. in $e\sim\sigma_v$ space because they can have large void space between pebbles owing to the "double structure". They move down toward the N.C.L. during testing. The result shows that due to slaking the mudstones change to remolded normally consolidated state. Since the crushed mudstone form of the "double structure", the excess concentrated shear stresses occur at the contact surfaces between pebbles' boundaries and they make pebbles soften with water from void spaces even under one-dimensional loading.

The state of dense mudstone specimen E3, however, is initially located below the N.C.L. and it does not change during testing. Even though the crushed mudstone form the "double structure", it has little saturated void space between pebbles and the excess concentrated shear stresses hardly occur at the contact surfaces between pebbles' boundaries. Consequently the slaking hardly occurs for dense mudstone specimen. Fig.4 further shows that the N.C.L. is a demarcation line of the states of mudstone specimens for compression to be possible or not. Through this results it is possible to determine the degree of packing of the crushed mudstones in the field to reduce its compression.

5. SHEAR PROPERTIES OF CRUSHED MUDSTONES DUE TO SLAKING

In order to examine the shear characteristics of crushed mudstone specimens which form a "double structure", both the strain controlled and the load controlled drained triaxial compression tests were carried out.

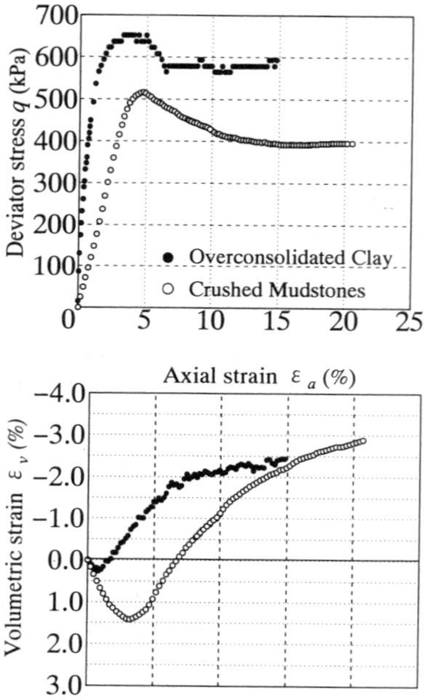

Fig.5 Drained behavior of mudstone and overconsolidated clay

5.1. *Strain controlled drained triaxial compression test*

The triaxial specimens were packed densely with saturated mudstone pebbles of which the average diameters were between 2.0 and 4.75mm. The test specimen had a diameter of 50mm and a height of 100mm and was filled with water.

The mudstone specimens were isotropically compressed to 98kPa, in which the cell pressure was 294kPa, and the back pressure was 196kPa. They were then subjected to the strain controlled drained compression test with the strain rate of 7.8×10^{-3}%/min, in which the cell pressure was held constant. Fig.5 shows the relationships between $q\sim\varepsilon_a$ and $\varepsilon_v\sim\varepsilon_a$ for both the crushed mudstone specimen and the overconsolidated KAWASAKI clay specimen.

The behavior of the crushed mudstone specimen is different from the one of the overconsolidated clay specimen. The crushed mudstone hardens only with drainage and softens only with swelling, while the overconsolidated clay still hardens even with swelling. This is because the mudstone specimen forms the "double structure". Since the specimen is packed densely, it has a dilatancy characteristics which seems to be the same as the overconsolidated clay or dense sand. The mudstone pebbles in it, however, soften due to slaking and then transform to remolded mudstone state during shearing and as a result it does not show hardening with swelling.

5.2. *Strain controlled undrained triaxial compression test*

The undrained triaxial compression tests were performed on loose saturated samples. Two types of mudstones specimens were used: one with grain size between 1.7~2.0mm(LSU1 and LSU2) and another between 0.25~2.0mm(LSU3).

The specimens were isotropically compressed to 98kPa and 294kPa respectively, and the back pressure was 196kPa.

Fig.6(a) shows the results of the test plotted as the relationship between the deviator stress q and the mean effective stress p'. Fig.6(a) also shows the critical state line (C.S.L.) which is obtained from the remolded mudstone specimens. It should be noticed that the samples LSU1 and LSU2 show softening below the C.S.L. These peaks are easily noticed on the q~ ε_a graph in Fig.6(b). The mudstones with higher initial effective mean stress exhibit softening to a larger extent. This is because the specimens form the "double structure" of the mudstones and the slaking occur for pebbles. On the contrary, the specimen which has a large range of grain size (LSU3) presents only hardening below and above C.S.L. It has a behavior similar to that of remolded clay. This is because the slaking hardly occurs for specimen LSU3. Since the mudstone specimen, which has a wide range of pebbles' grain size, has more contact surface between pebbles' boundaries, the excess concentrated shear stress hardly occur there.

5.3. *Load controlled drained triaxial compression test*

The mudstone specimens were isotropically compressed to 98kPa, in which the cell pressure was 294kPa, and the back pressure was 196kPa.

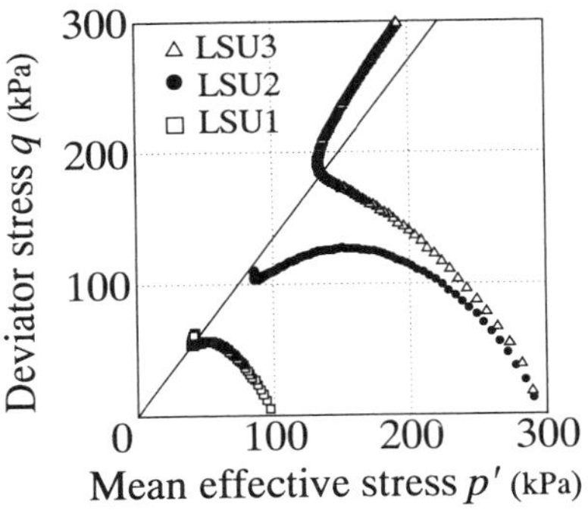

(a) deviator versus mean effective stress

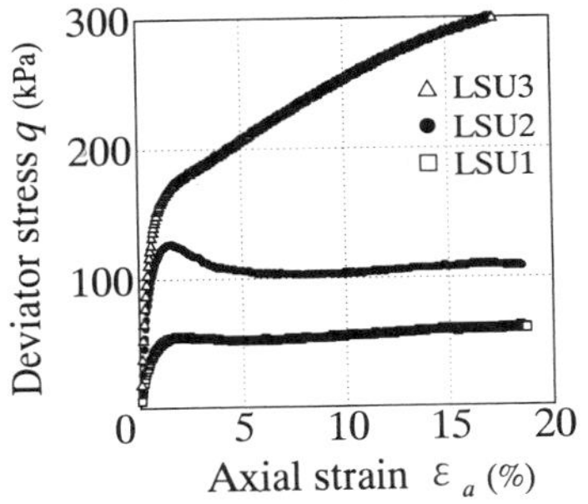

(b) deviator stress versus axial strain

Fig.6 Undrained behavior of loose mudstone sample

They were then subjected to the load controlled drained compression test with the loading rate of 4.9kPa/min, in which the cell pressure was held constant. The test was carried out with two constant load magnitudes, one was 441kPa between "peak strength" and "residual strength" of the strain controlled drained test for the crushed mudstones, and the other was 294kPa between "residual strength" for the crushed mudstones and the drained strength of the strain controlled drained test for remolded mudstone. Fig.7 shows the relationships between q~ ε_a and ε_v~ ε_a. Under the constant load of 441kPa, the specimen failed progressively with swelling, while under the constant load of 294kPa, neither the axial strain nor the volumetric strain changed and the specimen did not reach failure. The reason for the failure under the constant load of 441kPa is that the large shear stress concentration at contact surfaces between pebbles' boundaries cause slaking. This raises concerns for the long term stability of the embankments and suggests that the use of "peak strength" or the "residual strength" given by the triaxial test on crushed mudstones will not guarantee

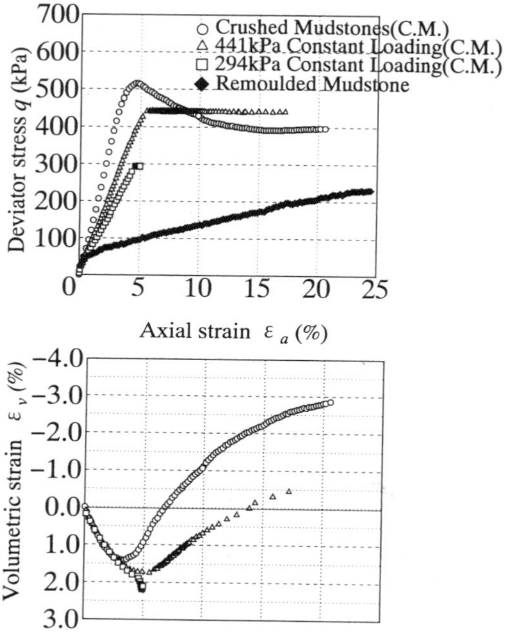

Fig.7 Progressive failure of mudstone

the safety of embankment. The soil parameters obtained from the remolded mudstone state should be used as the design parameter for the long term stability of an embankment made of mudstones.

6. CONCLUSIONS

One-dimensional compression tests and triaxial compression tests on the crushed TOKONAME mudstones were carried out to investigate the slaking phenomenon. The crushed mudstones form a "double structure", hence slaking of pebbles give rise to large settlement one-dimensionally and progressive failure under constant loading. The following conclusions are drawn through this study:

1) The slaking behavior of mudstones, being saturated with water, can be idealized as the softening due to swelling of the heavily overconsolidated clays
2) The states of loose mudstone specimens, which lie above the N.C.L. determined from the remolded mudstone, continue to move down towards the N.C.L. To the contrary, the states of dense mudstone specimens show no change.
3) To avoid occurrence of large settlements it is necessary to compact the mudstones so that their void ratio is below the N.C.L
4) The soil parameters obtained from remolded normally consolidated mudstone should be used as the design parameter for long term stability of an embankment made of mudstones.

REFERENCES

Asaoka, A., Nakano, M. and Noda, T. (1995). Annealable behavior of saturated clay, an experiment and simulation. *Soils and Foundations* 35:49-20.

Asaoka, A., Nakano, M. and Noda, T. (1997). Soil-water coupled behavior of heavily overconsolidated clay near/at critical state. *Soils and Foundations* 37:113-39.

Hashiguchi, K. and Ueno, M. (1977). Elasto-plastic constitutive laws of granular materials. *Proc. Spec. Session 9 of 9th Int. Conf. SMFE.*73-82.

Hashiguchi, K. (1989). Subloading surface model in unconventional plasticity,*Int. J. of Solids and Structures,* Vol. 25,pp.917-945.

Schofield, A.N. and Wroth, C.P.(1968): Critical State Soil Mechanics, McGRAW-HILL.

Numerical Models in Geomechanics, Pietruszczak & Pande (eds) © 1997 Balkema, Rotterdam, ISBN 90 5410 886 X

Strain localization in embankment foundations

T. Noda, G. S. K. Fernando & A. Asaoka
Department of Civil Engineering, Nagoya University, Chikusa-ku, Aichi, Japan

ABSTRACT: This study describes the embankment foundation behavior under plane strain condition during load application with particular emphasis on the strain localization that leads to failure of the embankment. The embankment is founded over the lightly overconsolidated clay foundation of OCR 2. The soil-water coupled analysis using the finite element method, which is based on the finite deformation theory has been used with the subloading surface Cam-clay model for the soil. The numerical results are described within the Critical State Soil Mechanics framework.

The area below the embankment slope reaches the normally consolidated state first during load application. With gradual reaching of the normally consolidated state during loading, the shear deformation tends to concentrate to a relatively confined zone in this area leading to physical failure of the structure with no further load application. The soil elements in the localized region show non-dilatant characteristics with stress states remaining on the critical state.

INTRODUCTION

Localization of the deformation is a naturally occurring phenomenon in a broad class of materials. In the geotechnical engineering perspective, the occurrence of localized deformations, which may finally lead to failure, is commonly observed in the triaxial tests. Such obseıvations are sometimes used to give conceptual descriptions of the behavior of prototype structures, e.g. slip failure of embankment foundations.

Localization of the deformation of a soil mass can be understood as an onset of instability with the continuing deformations under external loads. This is attributed to both the material non-linearity and the geometric non-linearity. Having realized these facts, any numerical approach to such localization problems should address both the non-linearities appropriately.

Based on the "Critical State Soil Mechanics" (Schofield and Wroth, 1968) framework, the elasto-plastic behavior of saturated clay can be well explained by taking the plastic volume change as a hardening/softening parameter. This leads to a rational linking of strength with consolidation. The shear behavior of saturated clay mass is therefore closely related to the volume change behavior of the soil: In fact not only to a particular soil element concerned but to the surrounding soil elements too. This is known as the "non-local" nature of the saturated soil. In this respect, the deformation problems in which the saturated soils are involved are to be solved as soil-water coupled boundary value problems.

In this research our main purpose is to illustrate the occurrence of strain localization as a naturally occurring phenomenon through the soil-water coupled finite deformation analysis. In the analysis we have only considered the soil-water coupled finite deformation regime without introducing any artificial conditions; e.g. imperfections (Simo et al, 1992). An embankment founded over a lightly overconsolidated clay foundation is considered as an example problem. The governing equations used are based on the finite deformation theory. For the overconsolidated soils, the subloading surface (Hashiguchi and Ueno, 1977) Cam-clay model (Asaoka et al. 1997a) is employed.

NUMERICAL PROCEDURE

In this section only the vital points are highlighted. Comprehensive descriptions can be found in references. 1, 2, 3 and 4. Since the initiation and progression of localized shear zones occur during large deformation, the geometry change during deformation should be appropriately accounted in the computation. Thus, all the governing equations used in this research were based on the finite deformation theory (Asaoka et al, 1994, 1997b). The effect of body forces is appropriately considered in deriving the governing equations. Linear rate type constitutive equation, relating Green and Naghdi (1965) objective stress rate to the stretching was used. In order to cater for the changes in traction force due to subsiding embankment below the ground water table, the traction force is divided into two components: the force due to embankment and the force due to static water pressure over the boundary. With the use of first Piola Kirchhoff stress, the stresses induced by the embankment are transformed into undeformed original configuration to avoid any numerical difficulty due to changing traction boundary geometry. The coupling equation to be solved together with equilibrium equation is treated in terms of total head consisting pressure head and elevation head. The finite element technique with the updated Lagrangian scheme is used in the solution procedure. The convergence during iterations within an incremental step is checked with respect to the effective stresses.

For the non-linear material response of overconsolidated clays, the subloading surface Cam-clay model in which the subloading yield surface concept (Hashiguchi and Ueno 1977) is applied to original Cam-clay model (Roscoe et al, 1963) is used. In this model it is assumed that a subloading surface exists within the normal yield surface of the Cam-clay. These origin convex two surfaces, that expand or contract are similar in shape with the origin in p'~q space as the center of similarity. The subloading surface always passes through the current stress point not only in loading but also in unloading. When the soil is at normally consolidated state, these two surfaces become-identical and the soil behaves as if it was the Cam-clay. The complete description of the model can be found in Asaoka et al, 1997a.

Table 1 MATERIAL PARAMETERS

Compression index	0.02
Swelling index	0.04
Critical state parameter	1.53
Permeability constant	$1x10^{-7}$cm/s
Poisson's ratio	0.30
Voids ratio on NCL at unit pressure	3.0
Subloading model constant (ν_2)	10.0
Earth pressure at rest K_0	0.7
Saturated density	18.5kN/m^3

ANALYTICAL CONDITIONS

A homogeneous clay foundation of overconsolidation ratio (OCR) 2 was considered in the analysis. The plane strain analytical (half) domain with necessary boundary conditions was as shown in Fig. 1. In order to visualize the localization process, the domain was descritized into smaller elements totaling 370. The static ground water table was assumed to be at the ground surface and no variation was considered in the computation. The embankment load was applied gradually at a rate of 1kN/m^2/day until the foundation reaches failure. The assumed material parameters are listed in Table 1.

DEFORMATION BEHAVIOR OF THE FOUNDATION

Deformed mesh

The computation could be proceeded only up to 190

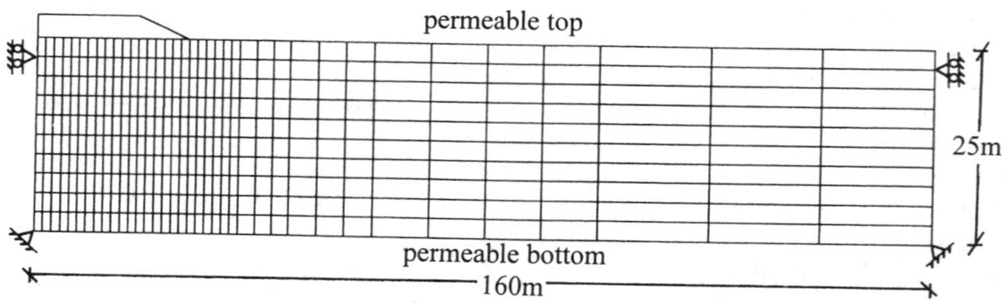

Fig. 1 Domain for the analysis and boundary conditions

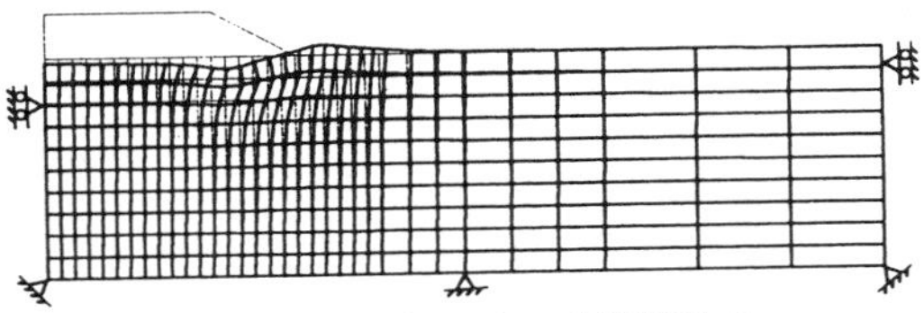

Fig. 2 Deformed mesh at 190kN/m²

kN/m² at the selected loading rate of 1 kN/m²/day due to the non-convergence of stress components during iterations for the next incremental step. Fig. 2 illustrates the deformed mesh at 190kN/m². By examining the deformation pattern of the area in the vicinity of the embankment slope it is clear that such a deformation pattern should correspond to the failure condition of the embankment. However, when the load~settlement curve (at the center of the embankment or at point A under the slope) is plotted, as shown in Fig. 3, a limit load with continuing settlement cannot be found.

Shear strains in the foundation

The soil elements under the slope have undergone significant distortion implying the occurrence of strain localization. Thus, shear strain contours were plotted at 100, 150, 180 and 190 kN/m² load levels during load application to examine the localization process. Figs. 4a, b, c and d illustrate the development of shear strain at the load levels mentioned before. The shear strain here refers to Eulerian strain and the numerals in the figures

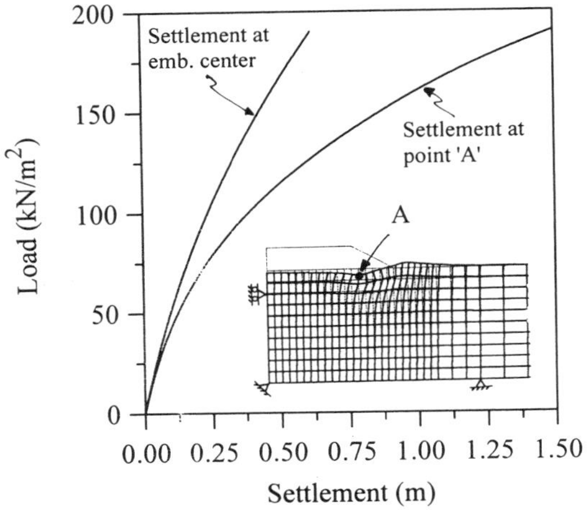

Fig. 3 Load - Settlement behavior

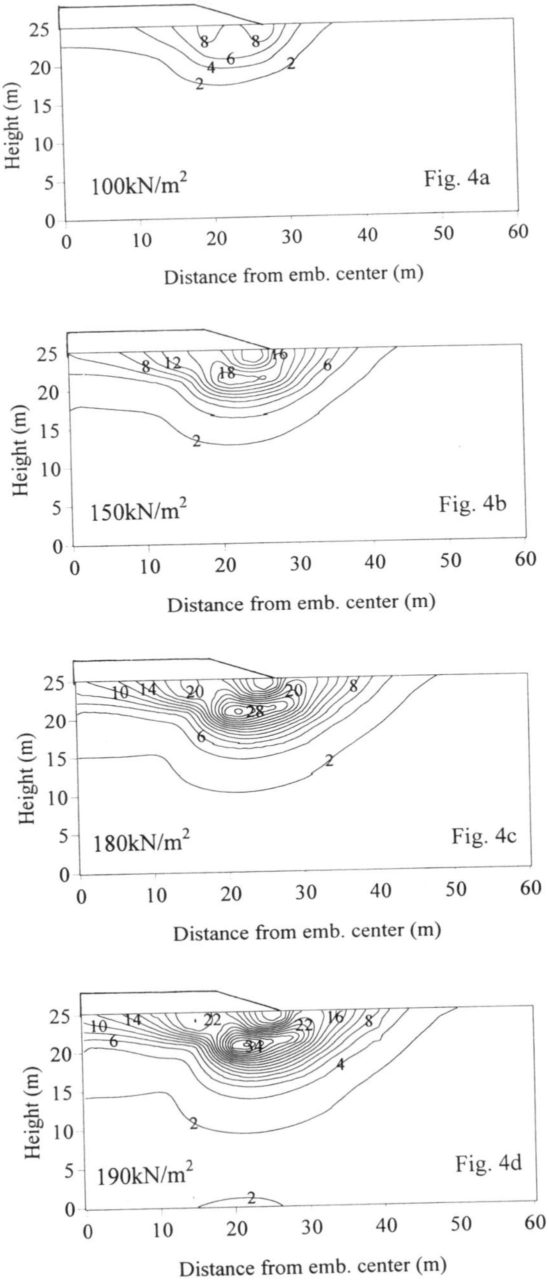

Fig. 4 Shear strain localization during loading

represent 100 times the actual magnitude. Shear deformation starts to concentrate in the area below the embankment slope around 100kN/m² load level

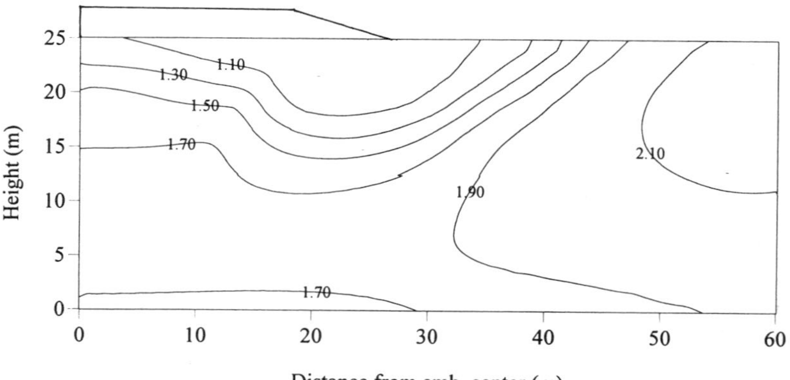

Fig. 5 Overconsolidation ratio distribution at 190kN/m²

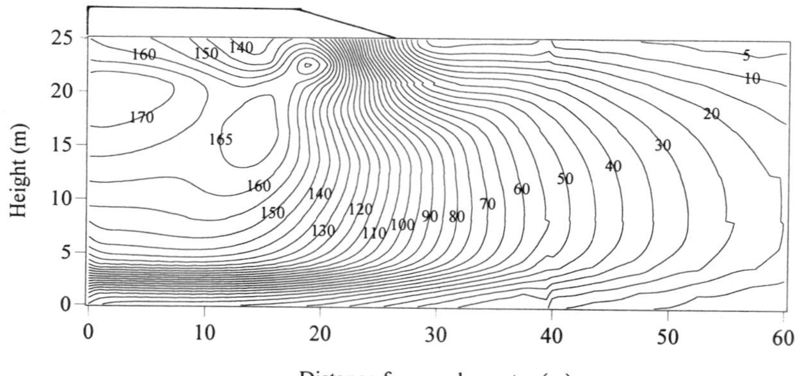

Fig. 6 Excess pore water pressure distribution at 190kN/m²

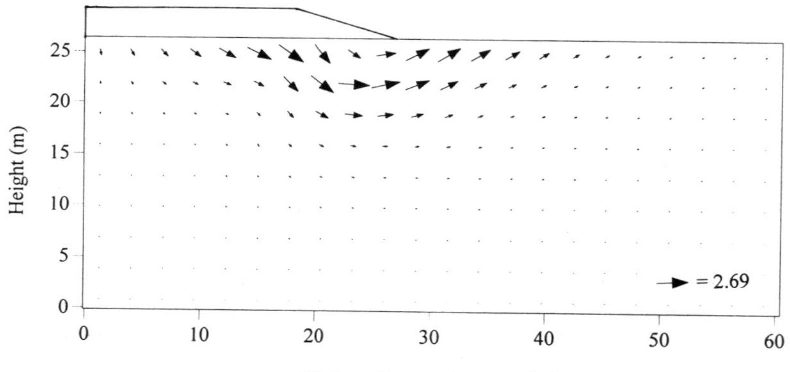

Fig. 7 Velocity field at the last load step (190kN/m²)

and then gradually proceeded to a significant concentration at 190kN/m^2.

Overconsolidation ratio distribution

Fig. 5 shows the overconsolidation ratio distribution at 190 kN/m^2. Through this figure it is clear that the area under the embankment slope has reached the normally consolidated state at the last incremental load step (at 190kN/m^2). Further, the highly sheared area falls within this normally consolidated region. This evidences that in this embankment case the strain localization directly corresponds to the development of the normally consolidated state.

Excess pore pressure distribution

The excess pore water pressure distribution at 190kN/m^2 is shown in Fig. 6 (the magnitudes shown are in kN/m^2). A very high pore pressure gradient can be seen under the embankment slope. The existence of such a high pore pressure gradient in this area obviously makes the foundation unstable leading to a circular slip like failure.

Velocity vectors

The velocity vector field at the last incremental load step (190kN/m^2), plotted in Fig. 7, confirms the occurrence of a slip like failure of the foundation. In this plot the velocity components are exaggerated by 100 times the actual magnitude. It is to be noted here that in this computation the interaction between the embankment and the foundation has not been considered. In fact there can be an influence by the stiffness of the embankment on the deformation behavior.

Element-wise Soil Behavior

The soil behavior of a typical element in the highly sheared zone is illustrated in Fig. 8. As illustrated in Fig. 8a, when the applied load is nearing 190kN/m^2 shear strains continue to develop under unchanging deviator stress. Such a stress-strain behavior illustrates that the soil element has already reached the limiting shear resistance. For the chosen permeability value of 1x10^{-7}cm/s and loading rate of 1kN/m^2/day, this element deforms under perfectly

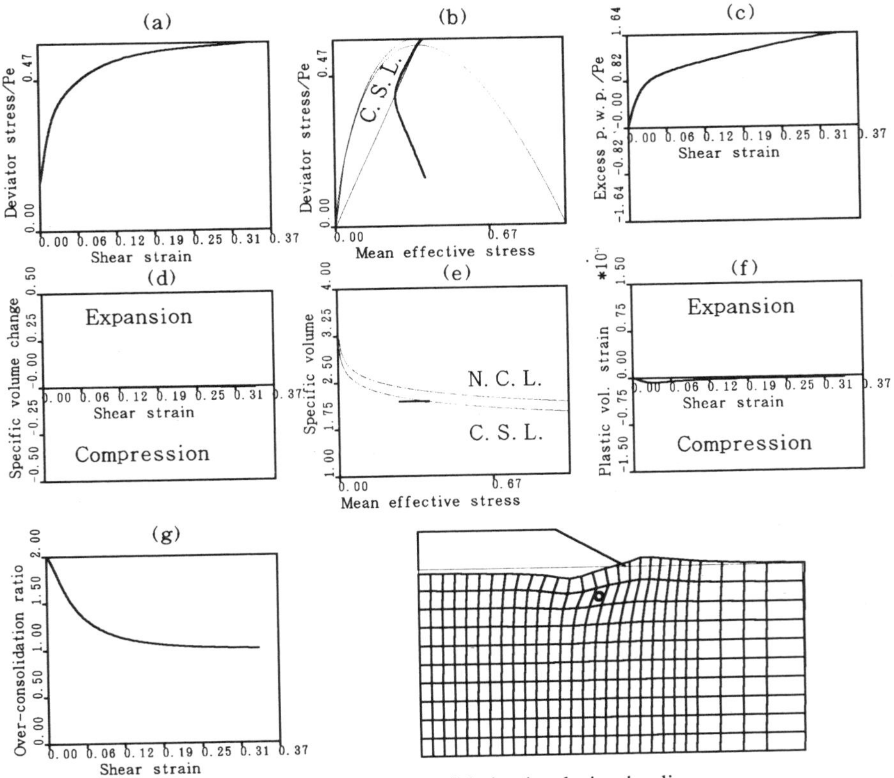

Fig. 8 Element-wise soil behavior during loading

undrained condition (Fig. 8d). The effective stress path, that starts from the initial K_0 condition moves above the Critical State Line (CSL) and then moves upwards near the CSL (Fig. 8b). When the total embankment load is nearing 190kN/m^2, the effective stress path of this element retains on the critical state line exhibiting perfectly plastic behavior. As described by the subloading surface Cam-clay model (Asaoka et al, 1997) there should be a plastic volume expansion when the effective stress path moves above the CSL. This fact is confirmed through the Fig. 8f. Even though there exists some plastic volumetric compression during early stages of loading, the counter balancing elastic volumetric strain component makes the specific volume unchanged. It is to be noted here that similar behavior is observed for the elements in the highly sheared zone.

CONCLUSIONS

The deformation behavior of lightly overconsolidated clay foundation under embankment loading has been investigated using the soil-water coupled finite deformation analysis. Through this study the following conclusions are drawn:

The area under the embankment slope reaches the normally consolidated state first during load application. This area undergoes strong strain localization which can lead to physical failure of the structure. The soil elements in the localized region show non-dilatant characteristics with stress states remaining on the critical state

REFERENCES

1. Asaoka, A., Nakano, M. and Noda, T. (1994), Soil-water coupled behavior of saturated clay near/at critical state, *Soils and Foundations* Vol.34, No.1, pp. 91-105.
2. Asaoka, A., Noda, T. and Fernando, G.S.K. (1995), Effects of changes in geometry on deformation behavior under embankment loading, *NUMOG V, Davos, Switzerland,* pp. 545-550.
3. Asaoka, A., Nakano, M. and Noda, T. (1997a), Soil-water coupled behavior of heavily overconsolidated clay near/at critical state, *Soils and Foundations,* Vol.37, No.1, pp.13-28
4. Asaoka, A., Noda, T. and Fernando, G.S.K. (1997b), Effects of changes in geometry on the linear elastic consolidation deformation, *Soils and Foundations* Vol.37, No.1, pp. 29-39.
5. Carter, J.P., Booker, J.R. and Small, J.C. (1979), The analysis of finite elasto-plastic consolidation, *Int. J. Numer. Analy. Methods in Geomech.*, Vol.3, pp. 107-129.1.
6. Green, A.E. and Nagdhi, P.M. (1965), A General theory of elasto-plastic continuum, *Arch. Rat. Mech. Analy.*, Vol. 18, pp. 251-281.
7. Hashiguchi, K. and Ueno, M. (1977), Elasto-plastic constitutive laws of granular materials, Constitutive equations of soils, (*Proc. Spec. Session 9th Int. Conf. SMFE, Murayama, S. and Schofield, A.D. eds.*), Tokyo, JSSMFE, pp. 73-82.
8. Hashiguchi, K. (1989), Subloading surface model in unconventional plasticity, *Int. J. of Solids and structures*, Vol.25, pp.917-945.
9. Roscoe, K.H., Schofield, A.N. and Thurairajah, A. (1963), Yielding of clays in states wetter than critical, *Geotechnique*, Vol. 13, No. 1, pp. 211-240.
10. Schofield, A.N. and Wroth, C.P. (1968), *Critical State Soil Mechanics*, McGraw Hill, England
11. Simo, J.C. and Meschke, G. (1992), New alogorithms for multiplicative plasticity that preserve the classical return mappings, Application to soil mechanics, *Computational plasticity, Fundamentals and Application*, Owen, D.R.J. (ed.), Pineridge Press, Part 1, pp.765-790.

Numerical Models in Geomechanics, Pietruszczak & Pande (eds) © 1997 Balkema, Rotterdam, ISBN 90 5410 886 X

One dimensional consolidation behaviour of soft ground under low embankment

H.Shirako
Construction Project Consultants Co., Ltd, Japan

M.Sugiyama & M.Akaishi
Department of Civil Engineering, University of Tokai, Japan

T.Takeda
Onoda Chemico Co., Ltd, Japan

ABSTRACT: This paper is concerned with finite difference analysis of one-dimensional consolidation taking account of the effect of long term creep settlement. From the laboratory observations, it is shown that the actual rate of consolidation differs from the predicted consolidation rate based on Terzaghi's theory. This difference is attributed to the secondary consolidation effect influenced by the load increment ratio. It is shown that the proposed analysis is appropriate for the practical prediction of consolidation settlement time curve including secondary consolidation and the Terzaghi's theory of the primary consolidation is not applicable for the field consolidation problem with the small load increment ratio.

1 INTRODUCTION

For the one-dimensional consolidation analysis of soft grounds, a large load increment ratio is used in a standard consolidation test. According to Ramiah et al.(1959), it was found that e-logp curves depend on the load increment ratio. Leonards et al.(1961) reported that the compression tests of different load increment ratios would give different consolidation curves respectively.

Barden(1948) and Newland et al.(1960) also pointed out that the load increment ratio would have an influence on the occurrence of secondary consolidation, and Crawford(1964) even stated that the standard consolidation test had the load increment ratio set at "1", which would be excessively large as compared with those usually employed at the construction sites, and that it therefore was a testing method in order to comply with the Terzaghi's theory.

Recently, there have been many instances of constructing low embankments on thickly deposited soft grounds. When laboratory consolidation tests are carried out using the field load increment ratios, the obtainable consolidation curves deviate entirely from the previous studies based on the Terzaghi's theory. Furthermore the necessary soil constants for analysis are not always obtainable.

In this paper, in the context of low embankment works on soft grounds, the consolidation tests using the design load and standard consolidation tests are carried out to compare the respective consolidation behaviour of each, and the effects of the load increment ratio on the behaviour of one-dimensional consolidation are discussed. In addition, the effects of time dependent strain (secondary consolidation) are included in the analysis of one-dimensional consolidation.

2 SAMPLES AND TEST PROCEDURES

Soils used in the experiments were collected from the alluvial ground in the suburbs of Tokyo. Physical properties of the samples are shown in Table 1. The soil profiles for the undisturbed cohesive soil sample *A* are shown in Figure 1. Except for some organic matters admixed in GL-6m - GL-9m, the soil is composed of a generally homogeneous silt layer up to GL-24m. Sample *A* was set on consolidation test apparatus, and 5 types of one-dimensional consolidation tests were carried out (Test A-E).

Test A: Standard consolidation test with the load increment ratio $dp/p_0 = 1$.

Test B: After pre-consolidation, due to an effective overburden pressure of the respective soil samples, a design embankment load dp=10kPa was applied, and the consolidation settlement was measured for 2 days. In this Test B, the load increment ratio is varied from 0.1 to 0.08 according to the effective overburden pressure.

Test C: For sample *A-4* nearly at the central part of the silt layer under an effective overburden pressure of 140kPa, five load increments (dp=10, 20, 40, 80, 160kPa) were applied, and the effect of the load increment ratio was examined by measuring the changes with time of the consolidation settlements for 2 days.

Test D: Sample *B* was remolded at a water content

Table 1. Physical properties of samples.

Sample	Gs	Wn (%)	WL (%)	Wp (%)	Grading(%) Clay	Silt	Sand
A-1	2.585	109.5	115.2	45.6	62.3	12.6	25.1
A-2	2.589	84.5	105.0	49.2	70.2	26.0	3.8
A-3	2.612	92.4	98.6	43.3	68.3	29.2	2.5
A-4	2.603	96.6	92.1	56.1	50.4	33.8	15.8
A-5	2.658	60.3	59.3	36.2	15.1	30.3	54.6
B	2.710	110.0	79.9	36.3	38.9	43.6	17.1

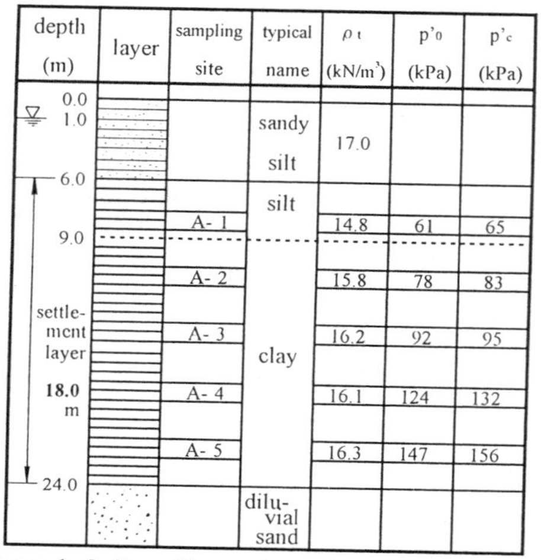

depth (m)
layer
sampling site
typical name
ρt (kN/m³)
p'0 (kPa)
p'c (kPa)
0.0
1.0
6.0
9.0
settle-ment layer
18.0 m
24.0
sandy silt
17.0
silt
A- 1
14.8
61
65
A- 2
15.8
78
83
A- 3
16.2
92
95
clay
A- 4
16.1
124
132
A- 5
16.3
147
156
dilu-vial sand

Figure 1. Soil profiles and conditions for sample A.

over the liquid limit and consolidated at a pre-consolidating load of p_0=20kpa for 2 days, then test specimens of 6cm diameter and 2cm height were prepared. Consolidating again at p_0 for 2 days, one-dimensional consolidation tests having consolidation load increments of dp=10, 20, 40, 60, 80kPa applied stepwise at an interval of 2 days to a predetermined consolidation load were conducted, and the changes with time of the consolidation settlement were measured.

Test E: In Test D, dp was changed with p_0 fixed, but here p_0 was changed to 20-140kPa while dp was fixed at 20kPa, assuming a ground of different depth might affect the result. The test conditions, consolidation time, etc. were the same with Test D.

3 RESULTS AND DISCUSSIONS

Figure 2 shows the coefficient of volume compressibility m_v and the coefficient of consolidation c_v as related to the mean consolidation pressure $\overline{p}$ from the result of Test A. The ranges of these coefficients at the design load are m_v=4~9kPa^{-1} and

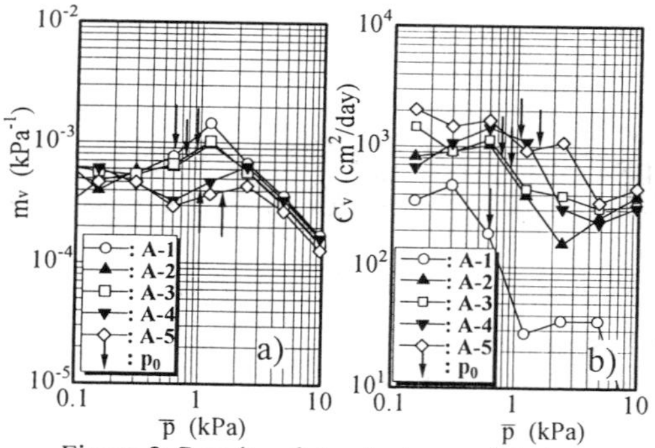

Figure 2. Results of standard consolidation test (Test A) : a) p versus m_v , b) p versus c_v.

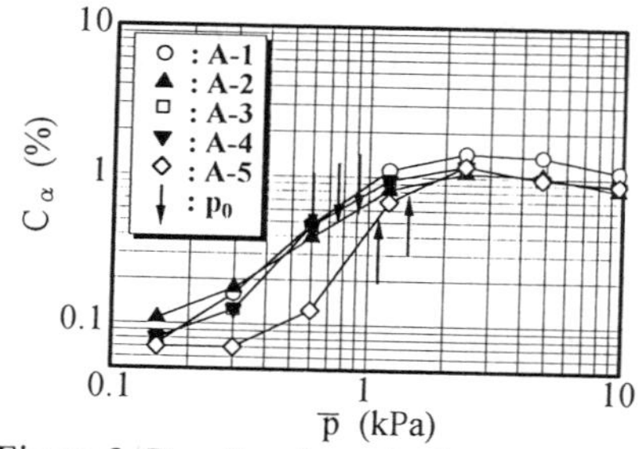

Figure 3. Results of standard consolidation test (Test A) : $\overline{p}$ versus c_α.

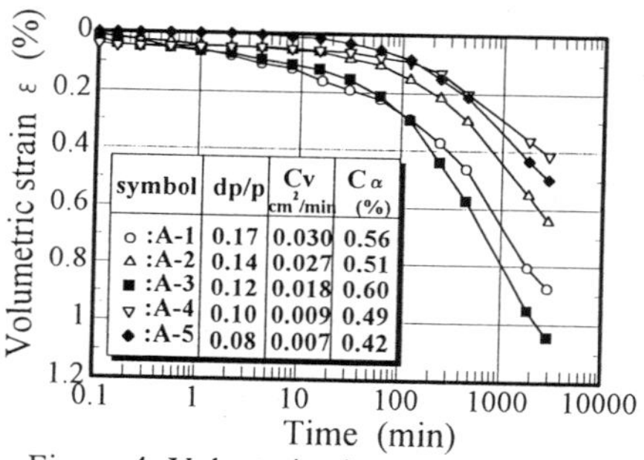

Figure 4. Vol. strain-time curves (Test B)

c_v=10^2~10^{3}cm^2/day. The coefficient of secondary consolidation c_α is shown as a function of $\overline{p}$ in Figure 3. The value of c_α in the normal consolidation region is around 1.0% regardless of the magnitude of $\overline{p}$.

Figure 4 indicates the volumetric strain ε-time curves of samples A-1~5 at each depth. In each of the samples, the consolidation is occurring in proportion to the logarithm of the time after 200 minutes. The consolidation curves show a form as if a secondary consolidation occurs after the consolidation time of 2 days because it is in the course of the primary consolidation, but the consolidation at 200 minutes and after is considered to be a secondary consolidation from the results of the standard consolidation test in Test A. It is considered that, as the consolidation load increment *dp* is small, the primary consolidation becomes smaller than the secondary consolidation to present the strain-time curve shown in Figure 4. Sample *A-3* presented a large settlement, and as a cause, the effects of physical properties or disturbances are considered. The value of c_v obtained by somewhat forcibly applying a curve ruler method to the initial part of the consolidation is noted in Figure 4, but a very small c_v value is obtained as compared with the value in Figure 2.

Using the results of Tests A and B in Table 2 and the soil conditions in Figure 1, a conventional one-dimensional settlement calculation was executed, with the result shown in Figure 5. In this figure, although great differences in both consolidation settlement and settlement speed occurred, it is considered that the one-dimensional consolidation by embankment on soft ground of a small load increment ratio causes these phenomena. Therefore, it will be necessary to clarify the effect of the load increment ratio on the form of the consolidation curve.

Figure 6 shows consolidation curves of the undisturbed sample *A* being consolidated at a consolidation load p_0=140kPa which is greater than the consolidation yield stress then having different sizes of consolidation load increments dp(=10, 20, 40, 80, 160kPa) (Test C). As the value of *dp* is varying, the consolidation also varies as a matter of course, but the coefficient of secondary consolidation is in the range of 1.1-1.4%, and the difference due to the magnitude of *dp* is not recognized. However, the forms of the consolidation curves vary greatly on account of *dp* as seen in Figure 6. The diagram also shows the values of c_v similarly which were obtained by the curve ruler as in Figure 4.

Table 2. Results of Test A and Test B.

Sample	Test A		Test B	
	mv(kPa^{-1})	Cv(cm^2/day)	mv(kPa^{-1})	Cv(cm^2/day)
A-1	8.98	125	8.21	43
A-2	8.06	620	5.61	39
A-3	9.39	575	9.75	26
A-4	4.90	1100	3.70	13
A-5	4.08	990	4.40	10

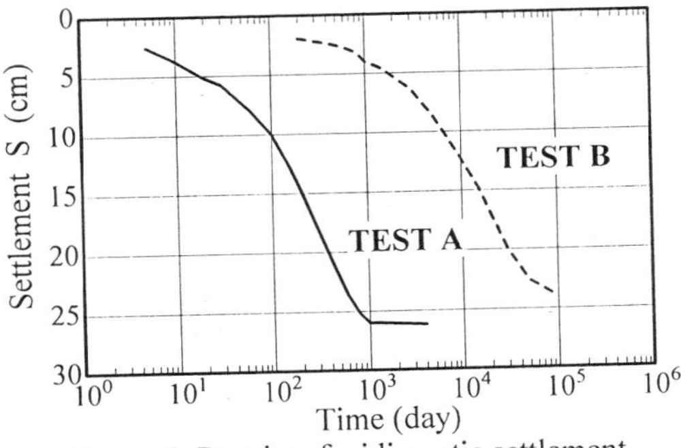

Figure 5. Results of a idiomatic settlement caluculation in Test A and Test B.

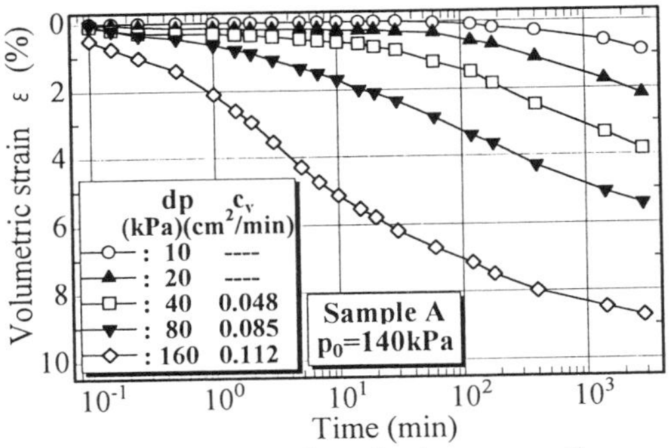

Figure 6. Vol. strain-time curves for undisterbed sample A (Test C)

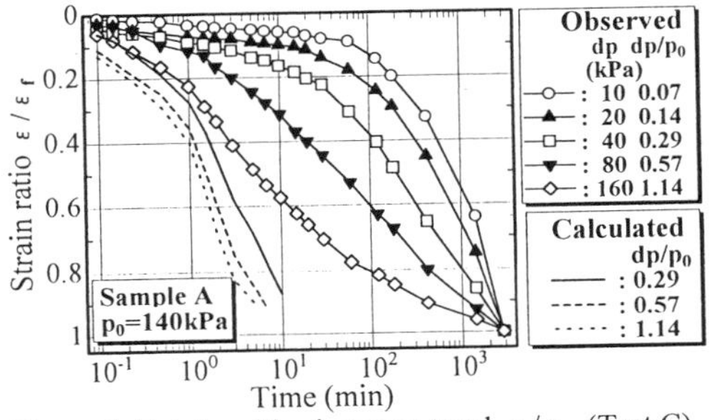

Figure 7. Relationships between t and $\varepsilon / \varepsilon_f$ (Test C)

Consolidation curves of a load increment ratio *dp/po* at 1 or less include the part where the consolidation is proportional to the logarithm of time at the later stage of consolidation, and such part is considered to be secondary consolidation to obtain c_α, but, from the form of the consolidation curves, it is very difficult to distinguish from the primary consolidation. Dividing the volumetric strain ε each time by the volumetric strain ε_f after 2 days to obtain the strain ratio $\varepsilon / \varepsilon_f$, the change with time of the strain ratio is shown in Figure 7. Using the coefficient of consolidation c_v due to the Terzaghi theory as shown in Figure 6, the strain-time relation was calculated and was also shown in Figure 7. From a comparison of the results of calculations and experiments with each other, the smaller the value of *dp/po*, the greater the proportion of the secondary consolidation in the whole settlement, and so it is obvious that the difference between both will increase.

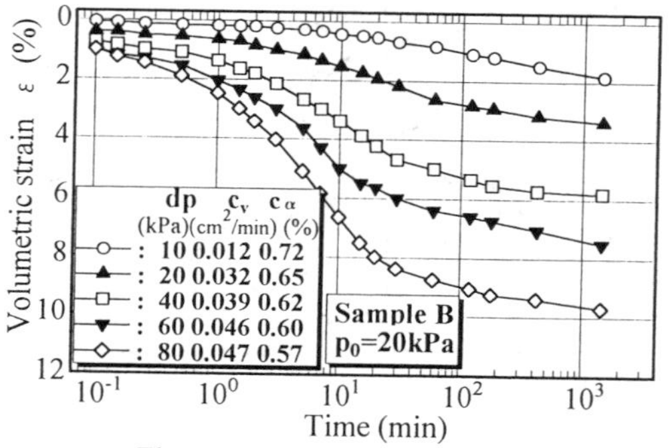

Figure 8. Vol. strain-time curves for remoulded sample B (Test D)

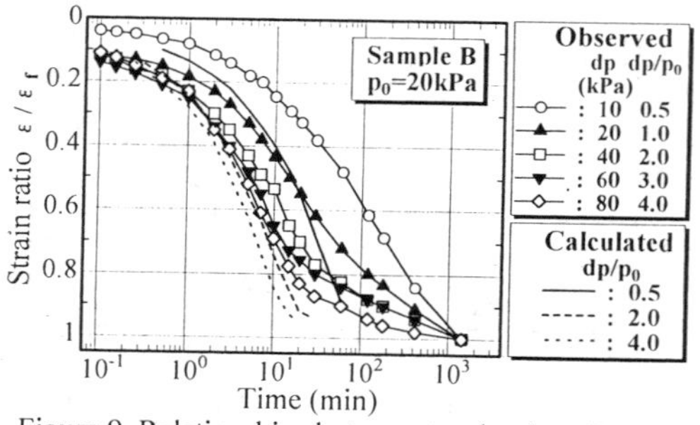

Figure 9. Relationships between t and $\varepsilon / \varepsilon_f$ (Test D)

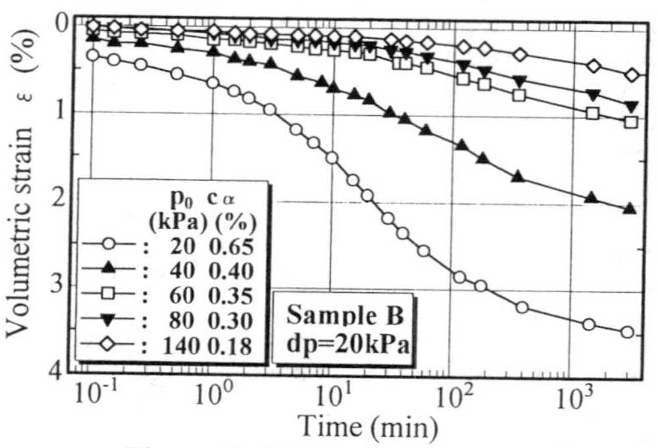

Figure 10. Vol. strain-time curves for remoulded sample B (Test E)

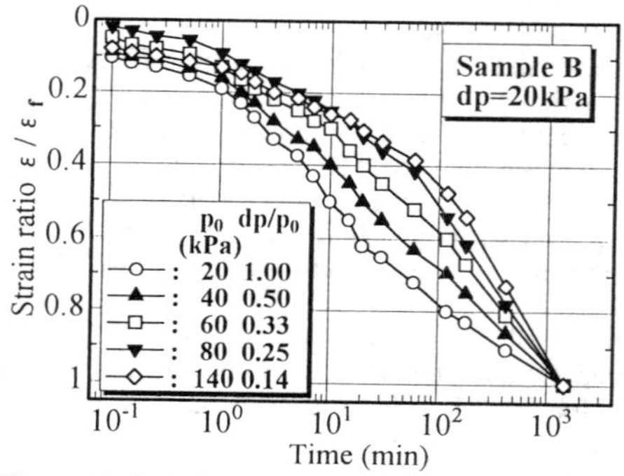

Figure 11. Relationships between t and $\varepsilon / \varepsilon_f$ (Test E)

Figure 8 and 9 show the consolidation and the strain ratio-time curves of the remolded sample *B* consolidated by consolidation load : p_0=20kPa, load increments : dp(=10, 20, 40, 60, 80kPa) (Test D). As seen through the values of c_v and c_α shown in Figure 8, the c_v tends to increase while the c_α slightly decrease with increasing dp/p_0. Strain ratio-time curves calculated in use of three values of c_v(=0.047, 0.039, 0.012cm^2/min) in Figure 8 are shown in Figure 9. It will be found that when the value of dp/p_0 is 1 or more, the result of the calculation is considerably similar, unlike Figure 7, to the result of experiment.

Figures 10 and 11 are the consolidation curves and the strain ratio-time curves when the pre-consolidation load p_0(=20~140kPa) is changed to be equal to the consolidation load increment dp (=20Pa) (Test E). With decreasing dp/p_0, changes in the strain ratio-time curves are observed as in Figure 6 and 7.

From the foregoing experiment results, it is obviously recognised that the primary consolidation is proportional to the consolidation load increment dp but that the coefficient of secondary consolidation is scarcely affected by the magnitude of dp or dp/p_0. When the load increment ratio dp/p_0 is 1 or more, the rate of the primary consolidation increases, but when dp/p_0 becomes smaller than 1, the rate of the primary consolidation decreases, and so the shape of the strain-time curves changes as shown in Figure 12.

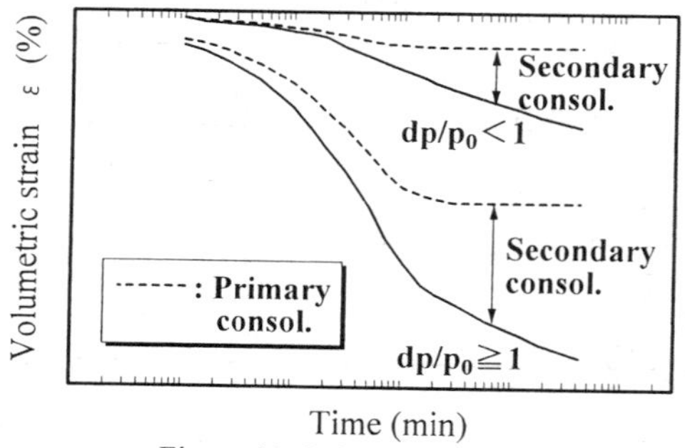

Figure 12. Schematic diagram for vol. strain-time curves.

4 ONE-DIMENSIONAL CONSOLIDATION ANALYSIS INCLUDING SECONDARY COMPRESSION

4.1 *Stress-strain-time relations and governing equation*

In one-dimensional consolidation, if the volumetric

strain ε is expressed as a function of the effective stress σ' and time t, the volumetric strain increment is equal to a sum of instant and creep strain components.

$$d\varepsilon = \left(\frac{\partial \varepsilon}{\partial \sigma'}\right)_t \cdot d\sigma' + \left(\frac{\partial \varepsilon}{\partial t}\right)_{\sigma'} \cdot dt \qquad (1)$$

The first term represents the instant strain followed by increasing the effective stress, while the second term represents the creep strain associated with time. These strains are expressed as follows :

$$\left(\frac{\partial \varepsilon}{\partial \sigma'}\right)_t = m_i \qquad (2)$$

$$\left(\frac{\partial \varepsilon}{\partial t}\right)_{\sigma'} = \frac{0.434 \cdot a \cdot \sigma'}{t} \qquad (3)$$

where,

$$a = c_\alpha / \sigma' \qquad (4)$$

m_i is the coefficient of volume compressibility. a is constant related to secondary consolidation.

From the principle of continuity of mass and Darcy's Law, the equations are expressed as follows :

$$\frac{\partial \varepsilon}{\partial t} = -\frac{k}{\gamma_w} \cdot \frac{\partial^2 u}{\partial z^2} \qquad (5)$$

$$d\sigma' = d\sigma - du \qquad (6)$$

where, k is permeability, γ_w is the unit weight of water, u is the excess pore water pressure, σ is the total stress and z is the vertical distance, respectively. Then substituting Eq.(1) and Eq.(2) into Eq.(5), the governing equation used in this study is given as

$$\left(\frac{\partial \varepsilon}{\partial \sigma'}\right)_t \frac{\partial u}{\partial t} = \frac{k}{\gamma_w} \cdot \frac{\partial^2 u}{\partial z^2} + \left(\frac{\partial \varepsilon}{\partial t}\right)_{\sigma'} + \left(\frac{\partial \varepsilon}{\partial \sigma'}\right)_{\sigma'} \frac{\partial \sigma}{\partial t} \qquad (7)$$

The finite difference approximation for the governing equation (8) is represented by substituting Eq.(2), (3) and (4) into Eq.(7) as follows:

$$u_{z,t+dt} = u_{z,t} + \alpha \cdot (u_{z+dz,t} - 2 \cdot u_{z,t} + u_{z-dz,t})$$

$$+ \frac{0.434 \cdot a \cdot \sigma'_{z,t} \cdot dt}{m_i \cdot t_z} + \sigma_{z,t+dt} - \sigma_{z,t} \qquad (8)$$

where, $\alpha = dt \cdot c_v / dz^2 = 0.5$, $c_v = k / \gamma_w / m_i$.

4.2 *Finite difference analysis*

The effects of secondary consolidation were examined by executing the calculation of the one-dimensional consolidation with some applied load increments. The soil constants used are shown in Table 3. The constant a was changed so that secondary consolidation would keep constant regardless of dp, but the other constants were kept unchanged.

Table 3. Parameters of samples.

Length of drainage H (cm)	c_v (cm^2/min)	m_v (kPa^{-1})	Parameter a		
			dp=100	dp=50	dp=10
1.0	0.1	0.001	0.02	0.04	0.1

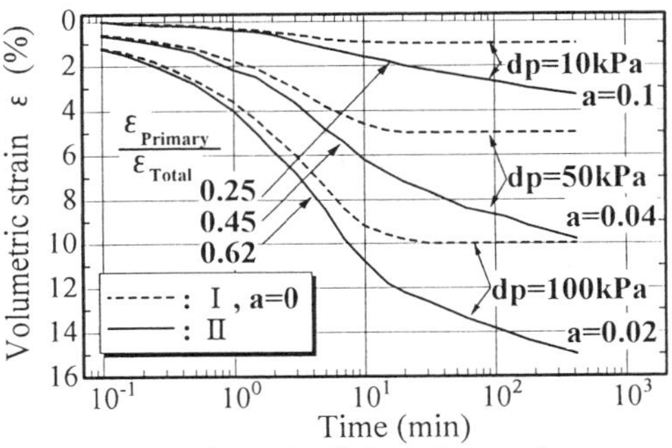

Figure 13. Culculation results.

The curve I in Figure 13 represents the case where the constant a is equal to zero. The consolidating phenomena are calculated as the Terzaghi's one-dimensional consolidation theory. In the case of $a \neq 0$, the process of one-dimensional consolidation including secondary consolidation is calculated. Namely the difference between the curves I and II is influenced by secondary consolidation. Obviously, as dp becomes smaller, the rate of the primary consolidation decreases, and the shape of the strain-time curves changes greatly.

5 CONCLUSIONS

Assuming small load increments for a low embankment on soft grounds, the one-dimensional consolidation behaviour was experimentally examined. The results are summarized as below.

1. Coefficient of secondary consolidation c_α is not influenced by the effects of the consolidation load increment and load increment ratio.
2. When the load increment ratio decreases below 1, the primary consolidation decreases, and the shape of consolidation curves change greatly.
3. From the one-dimensional consolidation analysis including the secondary consolidation, the secondary consolidation affects the shape of consolidation curves.

Where the load increment ratio is small as in low embankments, consolidation analysis including the secondary consolidation is required.

REFERENCES

Leonards, G.A. & B.K. Ramiah 1959. Time effects in the consolidation of clay. *Am. Soc. Test. Mater., STP. No.*254-116.

Leonards, G.A. & P. Girault 1961. A study of the one-dimensional consolidation test. *Proc. 5th int. Conf. soil. mech.:* 1-213.

Barden, R. A. 1948. Consolidation of fine grained soils by drain wells. *Trans. Am. Proc. Soc. Civ.Engs.:* 113-718.

Newland, P.L. & B.H. Allely 1960. A study of the consolidation characteristics of a clay. *Geotechnique:* 10-62.

Crawford, C.B. 1964. Interpretation of the consolidation test. *Proc. Am. Soc. Civ. Engs. 90SM: 5-87.*

6.4 Other applications

Numerical Models in Geomechanics, Pietruszczak & Pande (eds) © *1997 Balkema, Rotterdam, ISBN 90 5410 886 X*

Discrete element modelling of a strip anchor embedded in a brittle geomaterial

K. Sepehr & A. P. S. Selvadurai
Department of Civil Engineering and Applied Mechanics, McGill University, Montreal, Que., Canada

ABSTRACT: The present paper examines the problem of a rigid strip anchor which is embedded in a geomaterial susceptible to brittle fragmentation. The analysis is achieved through a discrete element approach which incorporates viscoplastic effects in the fragmentable geomaterial and relatively simple criteria for fragment development The analysis also accounts for size dependency of the strength of the brittle fragments. The interaction between individual fragments is modelled by nonlinear stiffness constraints which include Coulomb frictional behaviour. The numerical procedure is used to examine both the extent to which fragmentation zones occur and to establish the load-displacement behaviour of the rigid strip anchor.

INTRODUCTION

The conventional approaches for the modelling of the working load range behaviour of anchors embedded in geomaterials, are based on the classical theory of elasticity [see, for example, Selvadurai (1976),(1979),(1980),(1984),(1993); Selvadurai and Nicholas (1979); Rowe and Booker (1979); Selvadurai and Au; Selvadurai et al. (1990)]. Although elastostatic studies in this area have largely focused on isolated disc anchors with perfect interface bonding, the subject matter has been extended to include the influence of debonding (Hunter and Gamblen (1975); Keer (1975)) and crack development (Selvadurai (1989)) at the boundaries of the circular anchor. The paper by Selvadurai (1989), in particular, considered the problem of the axial load behaviour of a rigid circular anchor plate embedded in bonded contact with the surfaces of a penny-shaped crack. The work was subsequently extended to include interaction of a circular disc anchor and a single surface of a circular crack (Selvadurai, 1994a, 1994b). The modelling of the performance of anchors in the ultimate load range has to take into consideration the processes such as yielding of the geomaterial (Rowe and Davis (1982). With brittle geomaterials, the modelling has to accommodate the possible transition from an initially continuum region to a discontinuum. This transition process can be most conveniently handled through the application of discrete element techniques. The provision for accommodating both the fragment initiation from an initial continuum state, and the continued evolution of fragments are regarded as essential ingredients for the successful modelling of processes which are observed within the brittle geomaterial in the vicinity of the anchor-geomaterial interaction zone. In this paper, the basic concepts and methodologies advocated in discrete element modelling procedure are extended to examine the anchor-geomaterial interaction problem. The first modification relates to the introduction of the concept of combined viscoplastic failure and brittle fragmentation where viscoplastic flow can occur in regions where the state of stress is "predominantly compressive" and brittle fragmentation can occur in regions where the state of stress is "predominantly tensile". Experimental observations tend to support such processes. The second aspect of the study introduces the concept of a size-dependent strength of evolving fragments. Such size dependency is regarded as a characteristic feature in the fragment development in most brittle geomaterials (e.g. Jahns (1966); Beniawski (1968); Pratt et al. (1972)). In two dimensional plane strain modelling, the initial elastic behaviour of the brittle geomaterial is assumed to be isotropic and linear. The fragmentation of brittle geomaterial in zones of high stress gradients is modelled by considering a critical tensile stress approach based on elementary Mohr-Coulomb and limited tension failure criteria. The modified discrete element model is used to establish the time history of the load carrying capacity of the embedded anchor which is subjected to a uniform velocity (Figure 1). Cases of full `interface bounding` (Figure 2.a) between the embedded anchor and surrounding geomaterial and `interface debonding` at the base of the anchor (Figure 2.b) have been studied. The corresponding development of fragmentation within the brittle geomaterial due to interaction is also documented.

CONSTITUTIVE MODELLING

Brittle Geomaterial

The constitutive modelling of the brittle geomaterial takes into consideration elastic and viscoplastic effects. The elastic behaviour is characterized by isotropic linear elasticity and viscoplasticity is characterized via a model of the type proposed by Perzyna (1966) which contains a yield function (F) and a flow function $\Phi(F)$ and a flow parameter (γ). In the present study the yield function is characterized by Mohr-Coulomb behaviour with cohesion (c) and internal friction (ϕ). The model can also accommodate a limiting tensile strength defined by a tension cut off (σ_T).

Initiation of fragmentation

The process of fragment development during compressive shear failure is prescribed by considering a Mohr-Coulomb criterion. This criterion can be written in terms of the principal stresses in the form

$$\sigma_1 \geq \sigma_c + \sigma_3 \tan^2(45° + \phi/2) \qquad (1)$$

where σ_c is the unconfined strength in compression, ϕ is the angle of internal friction and σ_1 and σ_3 are, respectively, the maximum and minimum principal stresses. The unconfined compressive strength σ_c is related to the shear strength parameters c and ϕ associated with the Mohr-Coulomb failure criterion according to

$$\sigma_c = 2c\{(\tan^2 \phi + 1)^{1/2} + \tan\phi\} \qquad (2)$$

Considering the compression failure mode, there are two possible conjugate orientations of fragmentation inclined at equal angles to the direction of the stress on either side of it. In two dimensions, these are defined by

$$\alpha = \tan^{-1} \{(\sigma_1 - \sigma_{xx})/\sigma_{xy}\} \pm (90° - \phi)/2 \qquad (3)$$

where α is the angle between the fragmentation plane and the positive global axis, σ_{xx} , is the x-component of the stress tensor and σ_{xy} is the xy shear component of the stress tensor. In the computational scheme the sign of the second term on the right hand side of (3) is determined by using a `random number generator`. This allows for the development of a relatively random fragmentation pattern.

The tensile fragmentation criterion assumes that the material will fragment when the minimum principal stress σ_3 reaches the tensile strength of the material, σ_T. In two dimensions, the criterion can be written as:

$$\sigma_3 \geq \sigma_T \qquad (4)$$

The orientation of the fragmentation plane is given by

$$\alpha = \tan^{-1} \{(\sigma_1 - \sigma_{xx})/\sigma_{xy}\} \qquad (5)$$

where, in two dimensions, α is the angle between the fragmentation plane and the positive global x-axis. In this case, the α-direction is aligned with the maximum principal stress direction.

Viscoplastic Flow vs. Fragmentation

In the current study, fragment development and failure initiation for viscoplastic flow are described by the Mohr-Coulomb criterion specified in terms of c and ϕ. The following procedure is adopted in the computational schemes:

The intact geomaterial will experience brittle fragmentation only for situations where any single principal stress component or both are in the tensile mode and viscoplastic flow will occur for only stress states involving purely compressive stresses for both principal stresses. In situations where viscoplastic flow occurs first, there is provision for subsequent fragmentation in tension; this subsequent fragmentation will be governed by the prescribed post peak strength characteristics of the geomaterial, which can include softening.

Size dependency of strength

Geomaterials have an inherent internal structure consisting of crystals and grains in a fabric arrangement that can include random distributions of defects such as cracks and fissures. When the size of a tested specimen is small in comparison to a length scale of the microstructure, such that only relatively few defects are present, failure is forced to involve new crack growth; whereas a material loaded through a larger volume will encounter pre-existing weak planes at critical locations. Geomaterial strength can therefore be size-dependent. In particular, materials such as coal, granitic rocks and shale are found to exhibit size dependency in the strength; the ratio of laboratory to field strengths sometimes attaining values of 10 or more. A few definitive studies have been made to examine the influence of size effect on the strength over a broad range of specimen sizes. Bieniawski (1968) reported tests on prismatic in-situ coal specimens up to 1.6m x 1.6m x 1.0m as shown in Figure 3. Jahns (1966) reported results of similar tests conducted on cubical specimens of calcareous iron ore (Figure 3). Available data from these investigations are insufficient to reach a conclusive recommendation valid for all types of geomaterials, however, it does appear that there is generally a limiting size such that larger specimens suffer no further decrease in strength. Figure 3 also shows test results obtained by Pratt et al. (1972) for fissured quartz diorite, where results confirm the size dependency of strength. The brittle geomaterial investigated in this paper is assumed to exhibit the general trends of size dependency indicated in the above tests. In this study it is assumed that the tensile and compressive strengths vary according to the following criteria:

$$\sigma_T = (0.3125/A) \text{ MPa if } 0 \le A \le 0.625 \text{ m}^2;$$
$$\sigma_T = 0.5 \text{ MPa if } A > 0.625 \text{ m}^2$$

where A is the area of the two dimensional fragment in m^2; d the equivalent edge length (i.e. $d = \sqrt{4A/\pi}$ and the compressive strength $\sigma_C = (0.9375/A)$ MPa if $0 \le A \le 0.625$ m^2; and $\sigma_C = 1.5$ MPa if $A > 0.625$ m^2.

Fragment Interaction Responses

The interaction between fragments is characterized by responses which relate the normal (n) and shear (s) forces at the interface to their respective differential displacements at the contacting surfaces (see, Selvadurai and Boulon, 1995), e.g.,

$$dF_i = k_j(du_i^{(1)} - du_i^{(2)}) \qquad (6)$$

where $i = n,s$, dF_i are the incremental changes in the contact force per unit length between the contacting planes, du_i^1 and du_i^2 are displacement increments at the contact plane between regions (1) and (2) and k_j are stiffness coefficients defined in the normal and shear directions of the average plane at the point of contact. The stiffnesses themselves could be a function of the differential displacements $(du_i^1 - du_i^2)$. In this study, the generated inter-fragment shear behaviour is characterized by a Coulomb friction (μ_f) and an adhesion (c_f).

For this response,

$$k_s = k_s^* \; ; \; |d\tau_s| < c_f + \mu_f\sigma_n \; ; \text{ and } k_s = 0 \; ; \; |d\tau_s| = c_f + \mu_f\sigma_n \qquad (7.a)$$

where τ_s is the shear stress, σ_n is the normal stress at the inter-fragment contact plane. For the interaction response in the normal direction, it is reasonable to assume elastic behaviour provided the contact force between fragments is compressive. The interactive stiffness will vanish when the fragments separate, i.e.

$$k_n = k_n^* \; ; \; d\sigma_n \le 0$$
and
$$k_n = k_s = 0 \; ; \; d\sigma_n > 0 \qquad (7.b)$$

Constitutive Parameters

The constitutive modelling options include (i) elastic behaviour in the intact state, (ii) provision for

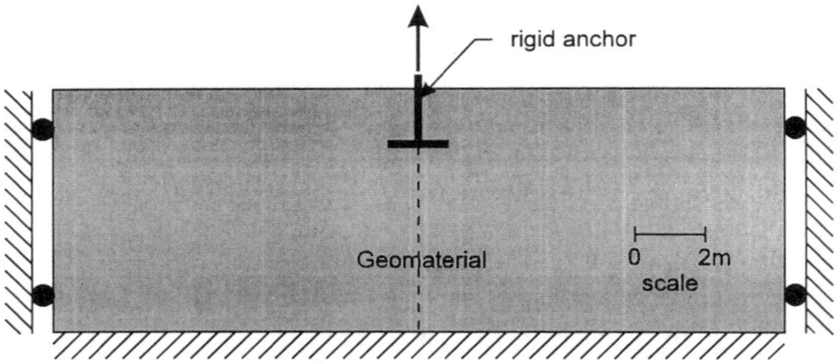

Figure 1. Geometry of the strip anchor problem

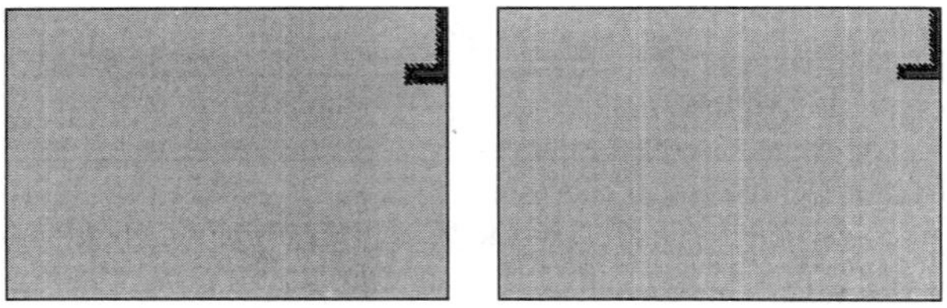

Figure 2.a Figure 2.b

Figure 2: Boundary conditions between the anchor and geomaterial.

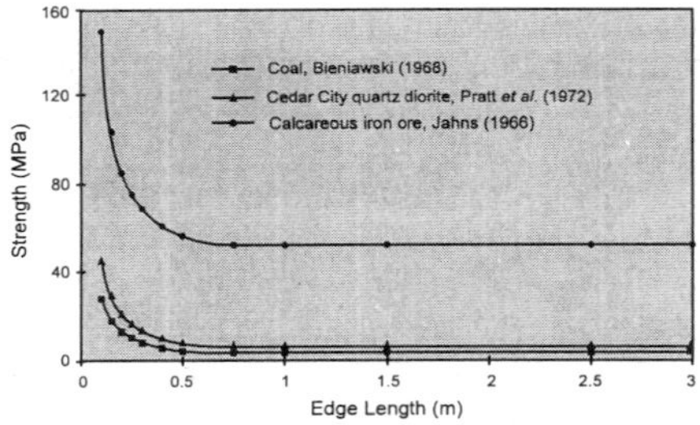

Figure 3: Dependence of the tensile strength of geomaterials on the specimen size.

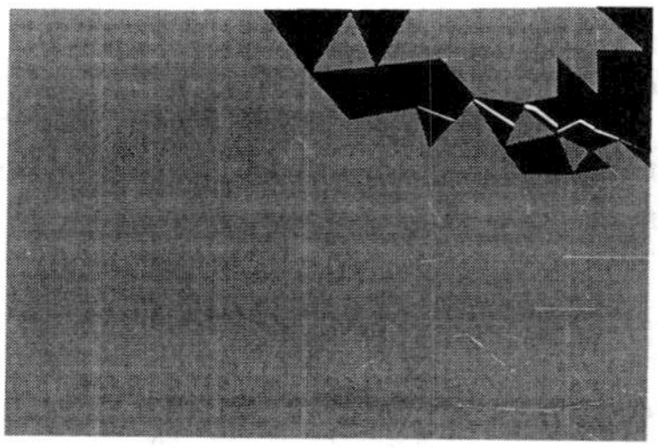

Figure 5: Extent of fragmentation at the end of the computational process

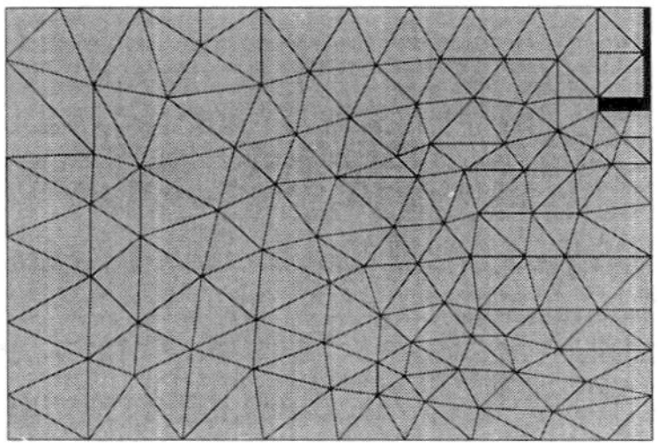

Figure 4:The discrete element mesh.

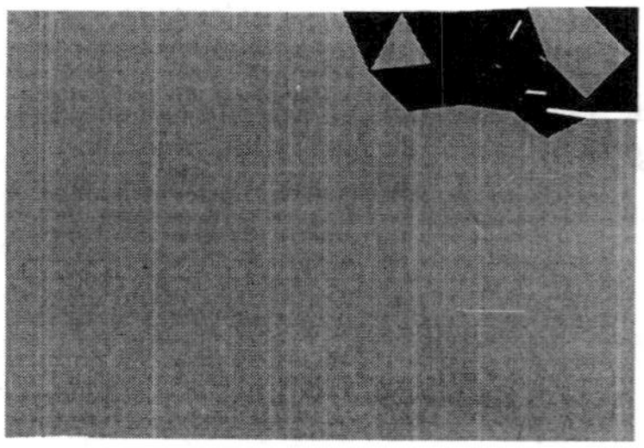

Figure 7: Extent of fragmentation at the end of the computational process

fragment development in the tension mode and viscoplastic flow in the compression mode, and (iii) inter-fragment interaction responses which are characterized by a Coulomb frictional response.

For the purpose of numerical modelling, the material parameters are selected to recfect the behaviour of a typical geomaterial such as ice or brittle rocks.

Intact geomaterial:

Young's Modulus=3500 MPa, Poisson's ratio=0.35
Cohesion = 1.5 MPa, Angle of friction = 30°
Tensile Strength = 0.5 MPa

Fragmented geomaterial:
Young's Modulus=3500 MPa, Poisson's ratio=0.35
Cohesion=1.5 MPa, Angle of friction=30°
Tensile strength=0.3125/A* MPa

Inter-Fragment Interaction:
Normal stiffness coefficient=1000 MPa,
Shear stiffness coefficient=1000 MPa
Cohesion=0.0, Angle of friction=30°

Computational procedure for viscoplasticity:

The complete details of the computational procedures used in the modelling are documented by Selvadurai and Sepehr (1996). The accuracy of the computational scheme is governed by the time increments associated with the treatment of viscoplastic effects and dynamic phenomena. With viscoplasticity, the time integration scheme is unconditionally stable if $\Omega \geq 1/2$; i.e. the procedure is numerically stable but does not ensure accuracy of the solution. (i.e. even for values of $\Omega \geq 1/2$, limits must be imposed on the selection of the time step to achieve a valid solution). For viscoplasticity problems which are based on an associative flow rule ($Q = F$); a linear flow function $\Phi(F)$ ($= F$) and where F corresponds to the Mohr Coulomb yield criterion; the recommended limit for the time increment is

$$\Delta t \leq [4(1+\nu)(1-2\nu)c\cos\phi]/[\gamma(1-2\nu+\sin^2\phi)E] \qquad (8)$$

With regard to interpretation of the dynamical equations, the stability condition for the time increment Δt, which employs an explicit-explicit partition (applicable to linear systems), takes the form

$$\Delta t \leq 2\{(1+D^2)^{1/2} - D\}/\omega_{max} \qquad (9)$$

where ω_{max} is the maximum frequency of the combined system involving both rigid body motion and deformability of the system; and D is the fraction of critical damping at ω_{max}. Other stability criteria have been proposed in the literature on computational methods for transient dynamic analysis; (Owen and Hinton, 1980), these include

$$\Delta t \leq 2/\omega_{max} \quad \text{and}$$
$$\Delta t \leq \eta L[\rho(1+\nu)(1-2\nu)/E(1-\nu)]^{1/2} \qquad (10)$$

where η is a coefficient which depends on the element type used and L is the smallest length between any two nodes. This constraint is found to be suitable for application to non-linear systems.

Rigid Anchor - Brittle Geomaterial Interaction

In this section, we briefly discuss the application of the discrete element approach, which incorporates fragmentation and viscoplasticity phenomena which are stress-dependent and size dependent to examine the dynamic interaction of a rigid strip anchor embedded in a brittle geomaterial bonded to a rigid stratum. The rigid anchor is subjected to a constant prescribed velocity of 0.1 m/s.

Figure 4 illustrates the initial discrete element mesh discretization used in the computation. Two types of interface conditions are assumed for the rigid anchor-brittle geomaterial interface. In the first analysis, the rigid anchor is assumed to experience complete adhesion at the interface. In the second analysis, the steel anchor is assumed to be debonded at the base of the anchor section. Figure 5 illustrates the extent of fragmentation that takes place after 1.0 second, for the case where the rigid anchor is fully bonded to the geomaterial. As can be observed extensive fragmentation of the brittle geomaterial has occurred in the vicinity of the embedded strip anchor both at the side and beneath the base of the rigid anchor. Figure 6 illustrates the load-displacement behaviour of the embedded anchor. A peak total pull-out force of 2.1 MN per unit length is predicted for this case. As the displacements increase the total pull-out force per unit length reduces, assuming a nearly steady value of 1.5

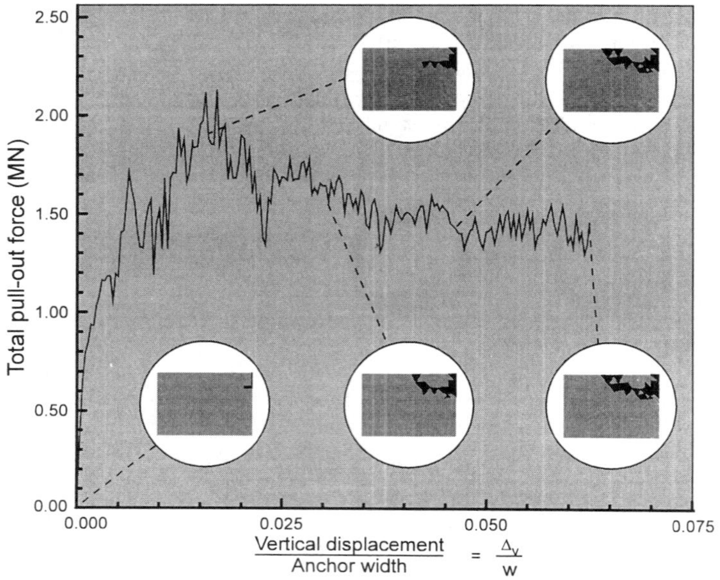

Figure 6: Time history of the total pullout force per unit length.

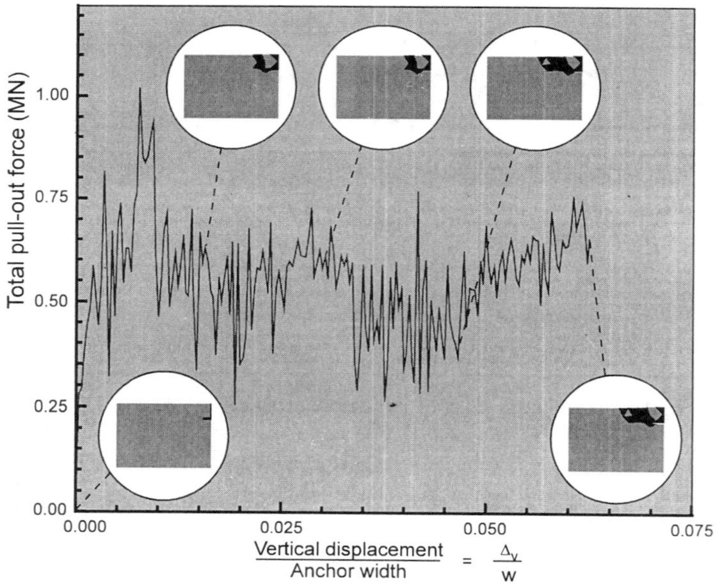

Figure 8: Time history of the total pullout force per unit length.

MN per unit length. Figure 7 illustrates the extent of fragmentation that takes place after 1.0 second, for the case where the base of the anchor is unbonded to the geomaterial. A different pattern of fragmention is observed. Complete separation of the anchor and geomaterial occurs at the lower interface between the two materials. Figure 8 illustrates the load-displacement behaviour of the embedded anchor. As can be observed a peak total pull-out force per unit length of 1.0 MN is generated which, as expected, is lower than for the case where there is full adhesion at the anchor-geomaterial interface.

CONCLUDING REMARKS

The transition from an initially continuum region to a fragmented medium can be most conveniently handled through the application of a discrete element technique. Such a methodology can be used to examine the anchor-geomaterial interaction in the presence of brittle fragmentation and viscoplastic flow. The results of the numerical simulations are intended for demonstration purposes only. The accurate application of the modelling requires the identification of the elastic and strength parameters applicable to both intact and fragmented geomaterial region. It is demonstrated that the load-displacement behaviour of the anchor can be determined using the discrete element model. The modelling adopted in this paper considers the possible influences of fragmentation of the brittle geomaterial during the anchor-geomaterial interaction. The results of the analysis is achieved through the extension of an existing discrete element techniques to include viscoplastic effects in the medium and a relatively simple fragmentation criteria. The time history of the uplift loads for a constant rate of movement are indicative of the responses that could be obtained in brittle geomaterial which undergo fragmentation and separation during uplift movement of the anchor loads.

ACKNOWLEDGEMENT

The acquisition of the DECICE Discrete Element Code was made possible by a Research Grant awarded to McGill University, by National Energy Board, Canada. The authors are grateful to Dr. Ibrahim Konuk of the National Energy Board for his continued interest in supporting research initiatives in Ice Mechanics.

REFERENCES

Bieniawski, Z.T., (1968), The effect of specimen size on compressive strength of coal. Int. J. Rock Mech. Mining Science, Vol 5, pp. 325-335.

Hunter, S.C. and Gamblen, D. (1974). The theory of a disc ground anchor buried in an elastic soil either with or without adhesion. J. Mech. Phys. Solids Vol. 22, pp. 371-399.

Hobbs, D.W., (1967), Rock Tensile Failure and its Relationship to a Number of Alternative Measures of Rock Strength, Int. J. Rock Mech. Min. Sci., Vol. 4, pp. 115-127.

Jahns, H., (1966), Measuring the strength of rock in-situ at increasing scale. Proc. 1st Cong. ISRM(Lisbon), Vol. 1, pp. 477-482.

Keer, L.M. (1975), Mixed boundary value problems for a penny-shaped cut. J. Elasticity, Vol. 5, pp. 89-98.

Owen, D.R.J. and Hinton, E., 1980, Finite Elements in Plasticity: Theory and Practice, Pineridge Press Ltd., Swansea, UK.

Perzyna, P., (1966), Fundamental Problems in Viscoplasticity. In: Recent Advances in Applied Mechanics, Academic Press, New York.

Pratt, H.R. Black, A.D. Brown, W.D. and Brace, W.F., (1972), The effect of specimen size on the mechanical properties of unjointed diorite. Int. J. Rock Mech. Mining Sci., Vol. 9, pp. 513-530.

Rowe, R.K. and Booker, J.R. (1979). A method of analysis of horizontally embedded anchors in an elastic soil. Int. J. Numer. Anal. Meth. Geomech. Vol. 3, pp. 187-203.

Rowe, R.K. and Davis, E.H. (1982). The behaviour of anchor plates in clay. Geotechnique, Vol. 32, pp. 9-23.

Selvadurai, A.P.S. (1976). The load-deflexion characteristics of a deep rigid anchor in an elastic medium. Geotechnique, Vol. 26, pp. 603-612.

Selvadurai, A.P.S. (1979), An energy estimate of the flexural behaviour of a circular foundation embedded in an isotropic elastic medium. Int. J. Numer. Anal. Meth. Geomech. Vol. 3, pp. 285-292.

Selvadurai, A.P.S. (1980). The eccentric loading of a rigid circular foundation embedded in an isotropic elastic medium, Int. J. Numer. Anal. Meth. Geomech. Vol. 4, pp.121-129

Selvadurai, A.P.S. (1984). Elastostatic bounds for the stiffness of an elliptical disc inclusion embedded at a transversely isotropic bi-material interface. J. Appl. Math. Phys. Vol. 35, pp. 13-23.

Selvadurai, A.P.S. (1989). The influence of boundary fracture on the elastic stiffness of a deeply embedded anchor plate. Int. J. Numer. Anal. Meth. Geomech. Vol. 13, pp. 159-170.

Selvadurai, A.P.S. (1993). The axial loading of a rigid circular anchor plate embedded in an elastic halfspace. Int. J. Numer. Anal. Meth. Geomech. Vol. 17, pp. 343-353.

Selvadurai, A.P.S. (1994). A contact problem for a smooth rigid disc inclusion in a penny-shaped crack. J. Appl. Math. Phys. Vol. 45, pp. 166-173

Selvadurai, A.P.S. and Au, M.C. (1986). Generalized displacements of a rigid elliptical anchor embedded at a bi-material geological interface. Int. J. Numer. Anal. Meth. Geomech. Vol. 10, pp. 633-652.

Selvadurai, A.P.S. and Au, M.C. and Singh, B.M. (1990). Asymmetric loading of an externally cracked elastic solid by an in-plane penny-shaped inclusion. Theort. Appl. Fracture Mech. Vol. 14, pp. 253-266.

Selvadurai, A.P.S. and Nicholas, T.J. (1979), A theoretical assessment of the screw plate test. Proc. 3rd Int. Conf. on Num. Methods in Geomech., Vol. 3, pp. 1245-1252. Aechen.

Selvadurai , A.P.S. (1994a), On the problem of a detached anchor plate embedded in a crack, Int. J. Solids Structures, Vol 31, No. 9, pp. 1279-1290.

Selvadurai, A.P.S. and Sepehr, K., (1996), The role of discrete element modelling in the analysis of ice-structure interaction: Two-dimensional plane strain analysis, Department of Civil Engineering and Applied Mechanics, McGill University.

Selvadurai, A.P.S. (1994b), Analytical methods for flat anchor problems in geomechanics, Proc. 8 th Int. Conf. Comp. Meth. Adv. Geomech. Morgantown W. Va. (H.J. Siriwardane and M.M. Zaman Eds.). pp. 305-321

Numerical Models in Geomechanics, Pietruszczak & Pande (eds) © *1997 Balkema, Rotterdam, ISBN 90 5410 886 X*

Discrete element simulation of interface behaviour

E. Evgin & T. Fu
Department of Civil Engineering, University of Ottawa, Ont., Canada

ABSTRACT: The mechanical behaviour of an interface between sand and a steel plate is investigated. The comparisons between the observed behaviour of the interface and the discrete element simulations are presented. The results of the numerical simulations and laboratory experiments compare well when units of two bonded particles, called peanuts, are used in the calculations. Particle translations and rotations within the shear box are presented in order to identify the thickness of the interface zone.

1 INTRODUCTION

Field observations and theoretical studies indicate that the characteristics of interfaces influence the overall behaviour of soil-structure systems. The interface behaviour is taken into account in the numerical analysis of many soil-structure interaction problems. Zero-thickness interface elements or thin-layer finite elements with an arbitrarily assumed thickness are used to represent interfaces. In most cases, only some of the factors influencing the interface behaviour are considered in the modelling. In depth experimental investigations are necessary to evaluate the validity of the existing interface models and also to develop new mathematical models for interfaces.

A comprehensive study by Uesugi et al. (1989) showed that sand particles roll on rough interfaces and displacement localization takes place in a thin layer of soil next to the contact surface. This layer is referred to as the interface. An X-ray technique was used by Tejchman and Wu (1995) to measure particle movements. Boulon and Hassan (1993) used an image processing method to measure particle translations and rotations. These in-depth studies showed that it would be necessary to specify the rotations as well translations to describe the kinematics of particles in the interface zone.

In the numerical modelling of the interface behaviour, various Cosserat continuum models were used for two dimensional problems to account for the rotations and translations of particles in the interface layer (Vardoulakis and Unterreiner, 1995; Tejchman and Wu, 1995). As compared with the finite element method, the discrete element method is, in principle, a more realistic tool for simulating the behaviour of granular materials. The discrete element method has been used by Cundall and Strack (1979) and many others, yet, there are still various problems existing in this method.

In the present study, an attempt was made to explore the advantages and disadvantages of using the discrete element method to simulate the behaviour of an interface between a granular soil and a steel plate. Circular particles, as well as units of two circular particles bonded together, i.e., peanuts, were used in various calculations. Both the strength-deformation behaviour of the interface and particle movements during shearing were evaluated.

2 INTERFACE TEST

An interface test was conducted using a multi-axial cyclic interface testing apparatus (Evgin and Fakharian, 1996). The soil container was a 25 mm thick, hollow aluminium box with an inside area of 100 mm × 100 mm. The box was placed on a steel plate with a roughness of 25 μm.

A two dimensional monotonic test with a constant normal stress of 100 kPa was conducted on an air-dry Silica sand which had D_{50} of 0.6 mm, γ_{max} of 16.05 kN/m^3 and γ_{min} of 12.88 kN/m^3. The sample had a measured relative density of 81% after the application of constant normal stress. The results of this test are presented later with the numerical results.

3 DISCRETE ELEMENT SIMULATIONS

3.1 *Discrete Element Method*

In this study, a two-dimensional discrete element code, PFC^{2D}, was used for simulations. The constitutive model available in the code for the behaviour of contact points consists of three parts related to: (1) stiffness, (2) slip, and (3) bonding. The stiffness part of the model is an elastic relation between the contact force and relative displacement both in the normal and tangential directions. The slip part enforces a limit on the ratio between the shear force and the normal force. The two particles in contact may slip relative to one another if this ratio tends to exceed the coefficient of friction. The third part of the model provides bonds between particles, if desired, and also sets limits for the total tensile normal force and for the shear force that a contact can carry.

Previous experience confirms that circular particles have fewer number of contact points with neighbouring particles and they rotate with ease. On the other hand, elliptical particles have greater number of contacts and a larger resistance against rotation. Consequently, the strength of elliptical particle assemblies is higher (Rothenburg and Bathurst, 1992).

In some of the calculations of this study, bonds were generated at the contact points of circular particles in order to investigate the effect of the particle shape on the results. A strong bond between two particles forced them to stay together without slip during shearing of the sample. As a result, two bonded particles behaved like one particle with the shape of number 8 which was referred to as "peanut" in this study.

The following model parameters are required as input in the analysis: density, interparticle friction coefficient $\tan\phi_p$, normal and shear interparticle contact bond strengths c_n and c_s, normal and shear interparticle stiffnesses k_{np} and k_{sp}, wall-particle normal and shear stiffnesses k_{nw} and k_{sw}, wall-particle friction coefficient $\tan\phi_w$, and a timestep.

3.2 *Simulations*

The interface test was simulated by a model with dimensions of 15 mm × 40 mm as shown in Fig.1. Two stationary vertical walls represented the sides of the shear box. The top horizontal boundary which was mobile in the vertical direction represented the

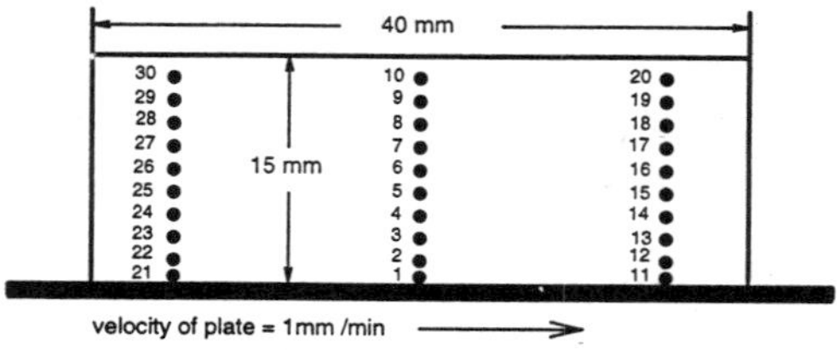

Figure 1. Geometry of interface model.

platen used for the application of normal load. The position of this boundary was adjusted by a numerical servo-control technique in order to maintain the normal stress constant during shearing. The bottom horizontal boundary was free to move only in the horizontal direction and it represented the steel plate. A velocity was specified, rather than specifying the force, to move the bottom boundary in the horizontal direction.

The number of particles varied between 650 and 1300 in different runs. Calculations involved samples with same size particles and samples with different size of particles.

At the beginning, all particles within the area of the shear box were generated with radii smaller than the actual values. Then, the radii were increased in stages in order to achieve a desired porosity in the sample. The procedure of sample preparation influenced the the initial fabric of the sample and consequently affected its strength and deformation behaviour. After this stage, gravity was applied. Once a stable state (i.e., unbalanced forces became small) was achieved under the gravity, the desired normal stress was applied at the top boundary of the sample. The shearing started subsequently by moving the bottom boundary horizontally at a certain velocity.

In this study, the model parameters were determined from previous numerical studies and physical experiments. According to the suggestion made by Cundall and Strack (1979), the ratio of k_s/k_n was set to 2/3. In the calculations, both k_{np} and k_{nw} were equal to 1.5E7 N/m^2, and both k_{sp} and k_{sw} were equal to 1.0E7 N/m^2. The coefficient of friction, $\tan\phi_w$, was assigned the following values. For the side walls, it was 0.017 and for the top boundary, it was 1.0. For the bottom boundary, the value of $\tan\phi_w$ was chosen 0.75 because that was a measured value in the experiment. Very high values were assigned to the normal and shear bonding strengths, i.e., c_n and c_s =1.E60 N and therefore the bonds could not be broken during shearing. The density of particles was equal to 2650 kg/ m^3. The horizontal velocity of the bottom boundary was related to the tangential displacement rate used in the laboratory test, i.e., 1 mm/min.

The calculated cases with their specific characteristics are listed in Table 1. The computed results are discussed in the sections to follow.

Table 1. Description of all calculated cases

Case No.	Particle radius (mm)	Particle number	$\tan\phi_p$	Bonding
D01	0.5	650	0.5	none
D02	0.5	675	0.5	none
D03	0.5	650	1.0	none
D04	0.5	675	1.0	none
D05	0.5	665	1.0	none
D06	0.5	670	1.0	none
D07	0.5	665	1.0	after applying gravity
D08	0.2→0.5	1225	1.0	none
D09	0.2→0.5	1225	1.0	after applying gravity
D10	0.2→0.5	1225 5 layers	1.0	after applying gravity
D11	0.2→0.5	1225 5 layers	0.5	after applying gravity
D12	0.2→0.5	1225 5 layers	1.0	before applying gravity
D13	0.2→0.5	1250 5 layers	1.0	after applying gravity
D14	0.2→0.5	1300 5 layers	1.0	after applying gravity

4 INTERFACE STRENGTH AND DEFORMATION CHARACTERISTICS

All calculated cases were divided into two groups.

4.1 *Group-I: Particles with Same Radius*

In Group-I, all particles had 0.5 mm radius. The computed results for this group are shown in Fig.2. In this group, Case D05 gave the best simulation for the shear stress development. However, the calculated deformations almost in all cases of this group were different than the measured results represented by the heavy solid line in Fig.2. The following general observations were made:

- When the interparticle coefficient of friction was assigned a value of 0.5, the failure took place on a curved surface inside the sample rather than at the contact surface. The average shear strength of the interface was much lower than the measured value (Case D01 and Case D02). It was noted that the measured values of $\tan\phi_p$ reported in the literature were usually around 0.5 or below that value (Misra, 1995).
- It was necessary to increase $\tan\phi_p$ to 1.0 in order to bring the calculated value of the average shear strength of the interface up to a comparable value with the measured strength (Case D03 through Case D07).
- The fabric of the sample had a strong influence on the calculated strength and deformation behaviour.
- The calculated vertical displacements were not satisfactory.

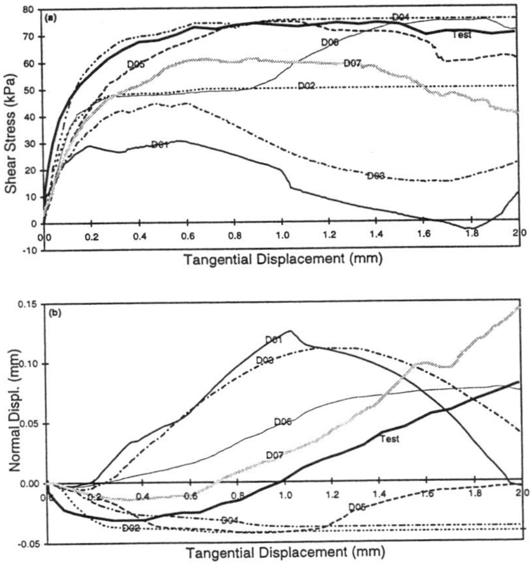

Figure 2. Simulations using uniform size particles (r = 0.5 mm).

4.2 *Group-II: Particles with Various Radii*

In the second group, the samples had particles with various radii uniformly distributed within a range between 0.2 mm to 0.5 mm. In addition, peanut shape particle units were used in the analysis except in Case D08. The bonding was formed either before or after the application of gravity.

The computed results are shown in Fig.3. In this group, Case D10 gave the best simulation in both shear stress development and deformation behaviour (i.e., there was an initial contraction and subsequently continuing dilation as the shearing progressed).

The following observations were made in this group of calculations:

- The use of particles with various radii improved

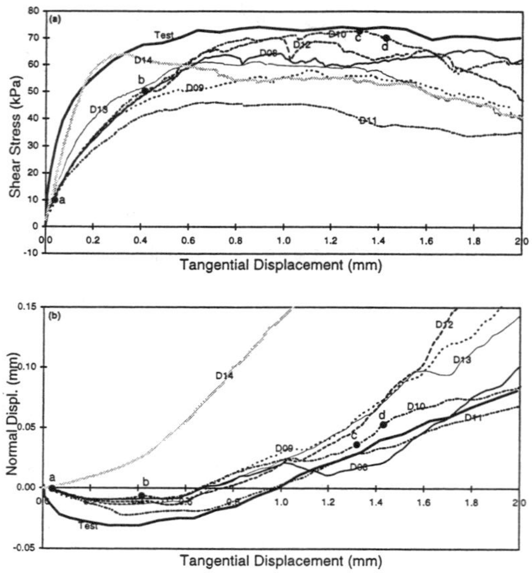

Figure 3. Simulations using variable size particles (r = 0.2 mm - 0.5 mm).

the deformation simulation, although the peak shear stress remained low (Case D08).

- Peanuts improved the simulations (Case D10).
- When the particles were generated in layers, the interface shear strength increased.

All these cases showed the effect of fabric on sample behaviour. However, when the coefficient of friction between particles was reduced again to 0.5, the interface shear strength became very low in spite of the improved fabric by using variable size peanut shape particles.

An increase in the number of particles made the sample stiffer as in Case D13 and Case D14, but the failure took place before the expected peak shear strength could be developed. In these two cases, the sample dilated excessively. These cases indicated that increasing particle number (using the same size of particles) was not always a good approach to make the sample stiffer.

5 PARTICLE MOVEMENTS

Considering both the shear stress - tangential displacement and normal displacement - tangential displacement behaviour, Case D10 gave the best simulation. The initial particle arrangement in an enlarged portion of the sample for Case D10 is shown in Fig. 4. In the following, particle

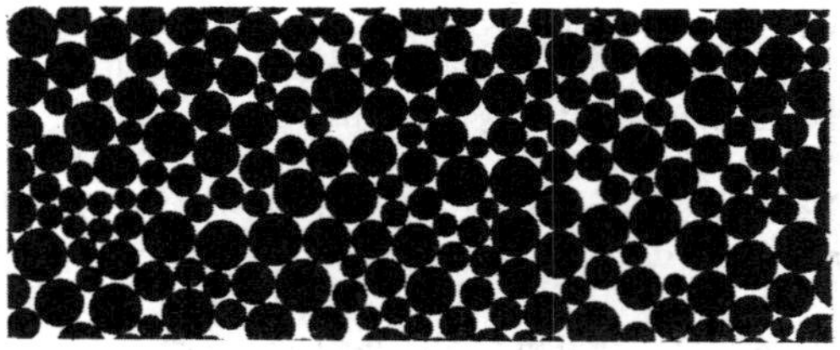

Figure 4. Arrangement of peanuts in an enlarged portion of sample.

movements in Case D10 are discussed in order to provide information about what happens inside the sample during shearing.

The movements of particles nearest to the points labelled from 1 to 10 (see Fig.1) were traced. Horizontal, vertical, and angular displacements of particles in Case D10 are shown in Figs. 5 through 7, respectively. Fig. 8 shows the particle positions traced for the specific moments indicated in Fig. 3. These plots showed an obvious pattern for particle movements. The particles closer to the contact surface had larger horizontal displacements. The horizontal displacements of the top four particles were small. The difference between the heavy solid line (indicating the horizontal displacement of the plate) and the medium solid line (indicating the horizontal displacement of particle 1) represented the relative displacement, i.e., sliding between the steel plate and the particles at the bottom of the sample. The relative displacement started to increase at about 0.2 mm horizontal displacement of the plate and gradually increased during shearing. These horizontal displacements remained almost unchanged after the peak shear stress was reached at about 1.3 mm.

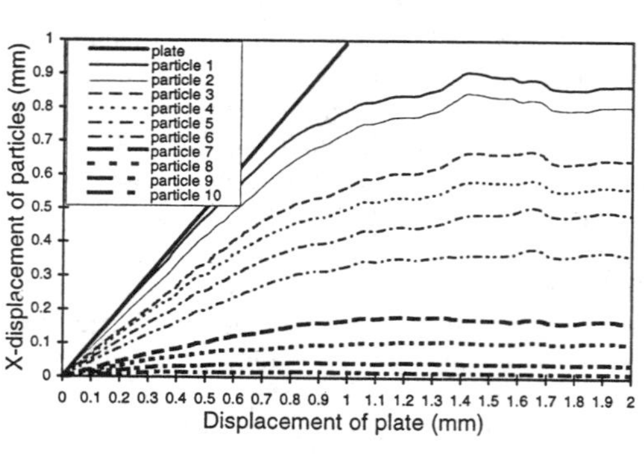

Figure 5. Horizontal displacements of particles.

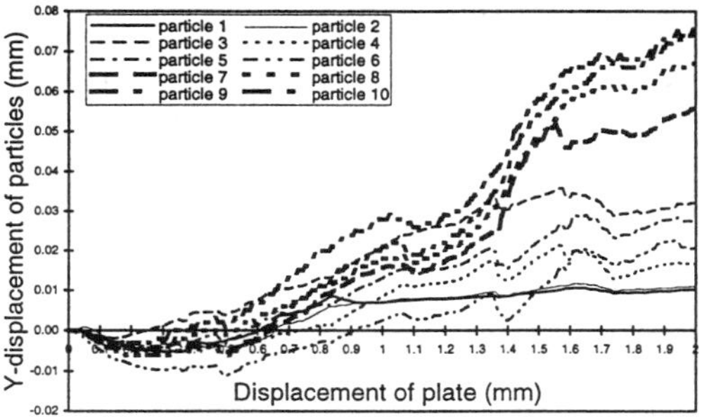

Figure 6. Vertical displacements of particles.

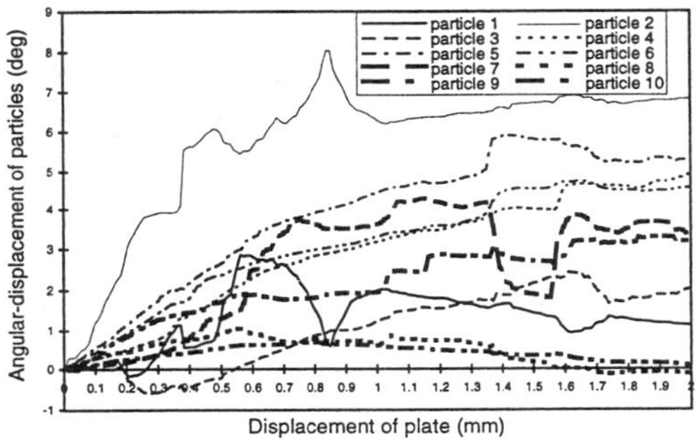

Figure 7. Angular displacements of particles.

Figures 6 and 8 show the vertical displacements of the particles. The top four particles moved the same amount vertically after the peak shear stress was reached. This observation suggested that the dilation of the sample resulted from the accumulation of the relative upward displacements of the particles within an approximately 8 mm to 10 mm thick layer next to the contact surface. This height coincided with the height where the horizontal displacements of particles were significant (Fig. 5).

Fig. 7 shows that the particles closer to the contact surface had larger angular displacements during shearing. Generally, all particles had positive angular displacements (representing counter-clockwise rotation) within the range of 1 degree (for particles away from the contact surface) to 7 degree (for particles near the contact surface).

The pattern of particle movements described above was not be the same for the particles near the side walls of the soil container due to the boundary effects. This problem was examined by tracing the positions of those particles near the side walls. It was clear that the side walls of direct shear type soil container had a definite influence on the movement patterns of particles.

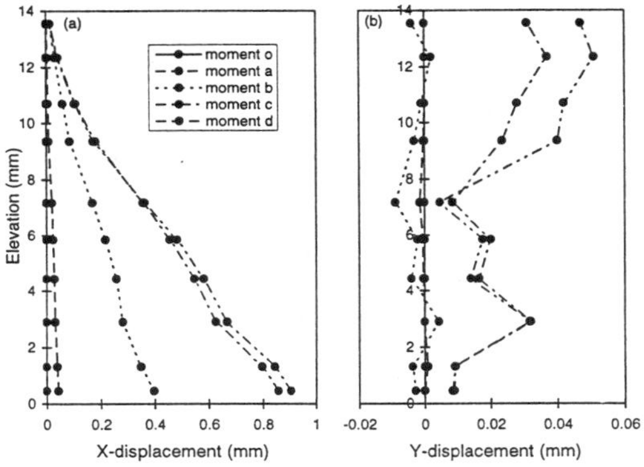

Figure 8. Horizontal and vertical displacements of particles at various moments.

An overall picture of the particle kinematics in the shear box is presented by velocity plots at different times during the test as shown in Fig. 9. The velocities of particles shown in this figure were normalized with respect to the largest velocity of particles. At moment "a", the particles moving with a relatively large horizontal velocity were concentrated in an approximately 8 mm thick layer next to the contact surface. This layer was a little thiner than the one obtained from the plots of displacements (Fig. 5). The particles close to the right wall climbed up, and the particles close to the left wall moved downwards. In addition, some whirlpool type movements were observed. At moment "b", the velocities of particles were reduced due to sliding at the contact surface. However, the movements of the particles were still following a pattern in the central portion of the sample. At moment "c", upward movements started to developed. Only a small number of particles near the contact surface were still moving horizontally. At moment "d", almost all particles moved upward. Generally, non-uniformities of particle movements depended on the loading stage and they could be found even in the middle part of the sample.

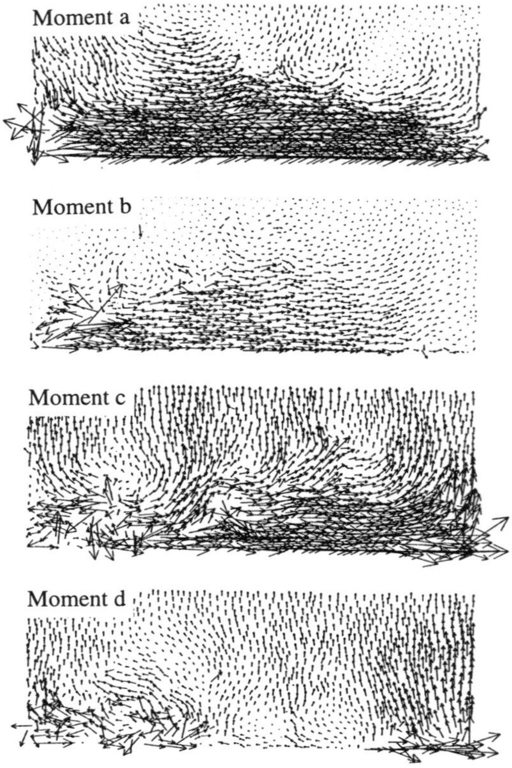

Figure 9. Velocities of particles at various moments.

6 CONCLUSIONS

Discrete element simulations using the peanut shape particle units produced satisfactory results for both strength and deformation behaviour of an interface. The horizontal displacements were localized within a layer of 8-10 mm above the contact surface. The thickness of the interface was about 7 times the average dimension of bonded particle units. The dilation of the sample resulted from the relative upward displacements of the particles in the interface layer. The amount of particle rotation varied between 1 and 7 degrees. Particles near the contact surface rotated most. The side walls of the soil containers had a significant influence on the movement patterns of particles near the walls.

ACKNOWLEDGEMENT

Financial support was provided by the Natural Sciences and Engineering Research Council of Canada and National Energy Board of Canada.

REFERENCES

Boulon, M. and Hassan, H. 1993. Development of a visualisation of the movements of the grains within a soil-structure interface. *CEC Science Program* No. 659.

Cundall, P.A. and Strack, O.D.L. 1979. A discrete numerical model for granular assemblies. *Geotechnique*, Vol. 29, No. 1, pp. 47-65.

Evgin, E. and Fakharian, K. 1996. Effect of stress paths on the behaviour of sand-steel interfaces. *Can. Geotech. J.*, Vol. 33, No. 6, pp. 853-865.

Fakharian, K. and Evgin, E. 1996. An automated apparatus for three-dimensional monotonic and cyclic testing of interfaces. *Geotechnical Testing J.*, ASTM, Vol. 19, No. 1, pp. 22-31.

Misra, A. 1995. Interfaces in particulate materials. *Mechanics of Geomaterial Interfaces*, Selvadurai, A.P.S. and Boulon, M. (eds.), Elsevier Science B.V., Amsterdam, pp. 513-536..

PFC^{2D} - Particle Flow Code in 2 Dimensions, Itasca Consulting Group, Inc., 1994.

Rothenburg, L. and Bathurst, R.J. 1992. Micromechanical features of granular assemblies with planar elliptical particles. *Geotechnique*, Vol. 42, No. 1, pp. 79-95.

Tejchman, J. and Wu, W. 1995. Experimental and numerical study of sand-steel interfaces. *Int. J. for Num. and Anal. Meth. in Geomechanics*, Vol. 19, pp. 513-536.

Uesugi, M., Kishida, H., and Tsubakihara, Y. 1989. Friction between sand and steel under repeated loading. *Soils and Foundations*, Vol. 29, No. 3, pp. 127-137.

Vardoulakis, I. and Unterreiner, P. 1995. Interfacial localisation in simple shear tests on a granular medium modelled as a Cosserat continuum. *Mechanics of Geomaterial Interfaces*, Elsevier Science Publishers, Amsterdam, pp. 487-512.

Numerical Models in Geomechanics, Pietruszczak & Pande (eds) © 1997 Balkema, Rotterdam, ISBN 90 5410 886 X

Load transfer mechanisms in ground anchorages

R.I.Woods & K.Barkhordari
University of Surrey, Guildford, UK

ABSTRACT: The ultimate capacity of a ground anchorage is governed by the manner in which bond stresses are mobilised along the fixed anchor length. The different elastic stiffnesses of the tendon, grout and ground mean that significant extensions may occur at the proximal end of the fixed anchor before any load reaches the distal end. This gives rise to a process of progressive debonding along the tendon-grout and/or grout-ground interfaces as the applied load is increased. Nonlinear finite elements have been used successfully to model this phenomenon using joint elements with a strain-softening plasticity constitutive law to represent the interfaces. A simple mathematical model for load transfer and debonding has also been devised and implemented in a spreadsheet. This appears to be capable of replicating observed behaviour particularly well, and has the potential to become a useful design tool.

1 INTRODUCTION

Ground anchorages typically comprise a steel tendon inserted into a pre-drilled hole, grouted into the surrounding soil or rock along part of its length (the "fixed length") and debonded along the remainder (the "free length"), Figure 1. Current design methods (e.g. BS 8081: 1989) estimate the ultimate capacity of low-pressure grouted anchorages from equations which are essentially of the form:

$$T = \lambda \tau_s L \qquad (1)$$

where T = ultimate capacity; L = fixed anchor length; τ_s is the bond stress; and λ lumps together many other factors - including fixed anchor diameter, in-situ stress, installation method, and soil grading.

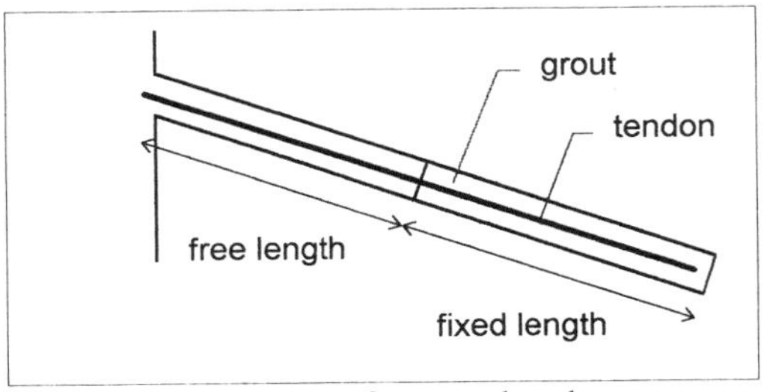

Figure 1. Components of a ground anchorage

Equation (1) suggests that ultimate capacity is linearly dependent on fixed anchor length, which in turn implies a uniform average bond stress at the grout-ground interface. However, the evidence from instrumented anchors is that bond stress distributions are highly nonuniform. This nonuniformity becomes more pronounced with increasing fixed length, with a considerable reduction in anchor efficiency. The solution is either to develop anchor systems which use the fixed length more effectively (such as the single bore multiple anchor or SBMA - Barley 1995) or to attempt to predict the true bond stresses in long anchors with greater accuracy. This paper describes numerical studies directed towards an improved understanding of load transfer mechanisms, and a more reliable estimation of ultimate capacity.

2 OBSERVED BOND STRESS DISTRIBUTIONS

It was the field work of Ostermayer and Scheele (1977) on instrumented anchors in sand which first showed a pronounced concentration of bond stress at the proximal end of the fixed anchor, under working conditions, Figure 2. A significant portion of the fixed length towards the distal end is effectively unloaded at first, picking up load only as applied anchor head force approaches ultimate capacity- by

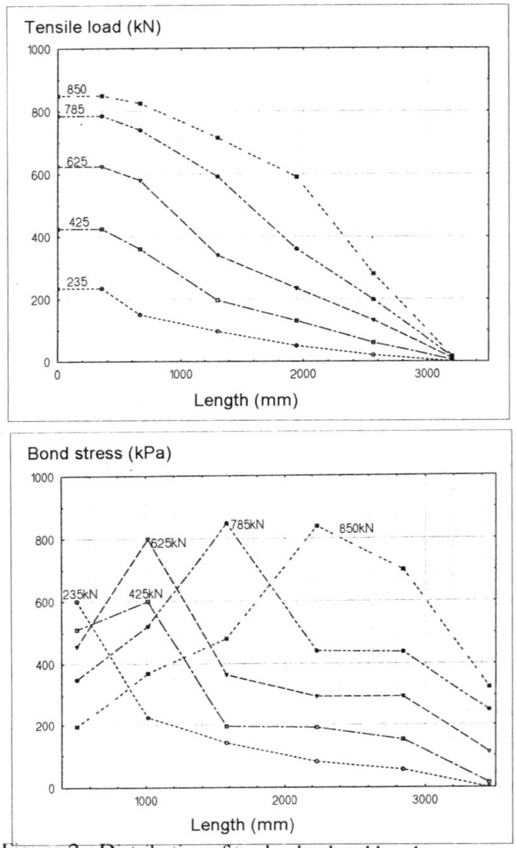

Figure 2. Distribution of tendon load and bond stress along fixed length (after Ostermayer and Scheele, 1977)

which time bond stresses at the proximal end have dropped well below their peak values. These data imply that the average mobilised bond stress τ_{ave} reduces with increasing fixed anchor length. More recent work (e.g. Weerasinghe, 1993) has confirmed the phenomenon of progressive debonding, and its existence is now generally accepted.

3 FINITE ELEMENT MODEL

The axial tendon load and bond stress distributions in ground anchorages have been investigated using the VISAGE finite element system (VIPS, 1996). A two-phase model has been studied, representing grout and surrounding ground, and the interface between them (Barkhordari, 1995). A typical finite element mesh is shown in Figure 3.

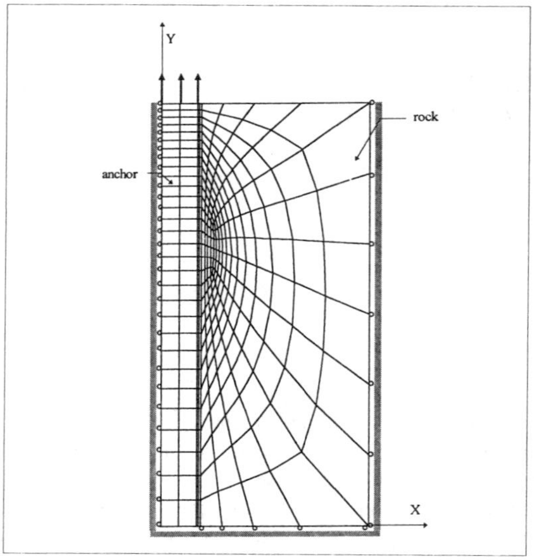

Figure 3. Typical finite element mesh

3.1 *Interface behaviour*

This is a vital part of the finite element model, as the behaviour of the interface will determine how well the model can capture progressive debonding. Earlier analyses conducted with an FE package employing standard interface elements (Goodman *et al.*, 1968), was not particularly successful. It was decided to use VISAGE because it offered a more sophisticated representation of interface behaviour, via elements originally developed for rock joints.

The joint/interface behaves elastically until peak bond stress is mobilised, based on a Mohr-Coulomb criteria. Plastic yield and flow then take place, accompanied by strain softening - in which joint cohesion is reduced to a residual value over a specified shear strain interval. The constitutive model for these joint elements requires conventional normal and shear stiffness, angles of friction and dilation, peak and residual cohesion, and residual strain. Other attributes relevant only to actual rock joints (e.g. asperity height) were suppressed.

Experimental data from full-scale model anchor tests (Weerasinghe 1993; Adams 1995) are currently being used to calibrate the joint element parameters. To overcome problems with "aliasing" (i.e. where a particular computed result may be obtained by several combinations of input parameters) use has been made of an optimisation routine based on a genetic algorithm (Goldberg, 1989). The algorithm uses a process of "natural selection" and shows much promise in such back-analysis applications.

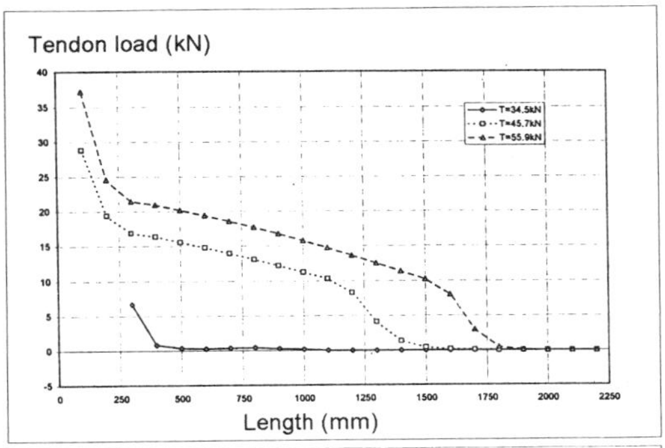

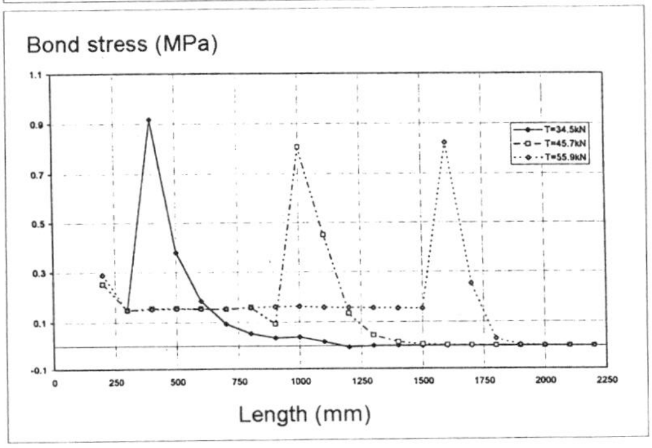

Figure 4. Typical tendon load and bond stress distributions computed by VISAGE

3.2 *Typical results*

Figure 4 shows typical tendon load and bond stress distributions along the fixed anchor at the grout-ground interface, at selected levels of applied load. Debonding can be seen quite clearly at the highest load level, with residual bond stress operating over 2/3 of the fixed length - and yet with no axial load having reached the distal end of the tendon.

4 SPREADSHEET MODEL

The finite element model is capable of predicting the bond stress distribution at a particular load level, and the load-displacement curve of the tendon at the proximal end of the fixed anchor. However, it is far from being a practicable design tool and there is still a need for something to fill the gap between sophisticated analysis and crude empiricism. To address this, the authors have devised a simple mathematical model for progressive debonding and load transfer, which can easily be implemented in a spreadsheet and used to predict the load-extension behaviour and ultimate capacity of a ground anchor.

4.1 *Assumptions*

The spreadsheet model assumes that:

- no debonding occurs at the tendon-grout interface
- as load is applied to the proximal end of the tendon, it will strain elastically and the surrounding grout will deform with it in a pseudo-compatible fashion, with discoid cracking of the grout column
- slip will occur at the grout-ground interface when a "threshold" axial strain in the tendon/grout column is exceeded; prior to this, bond stress increases with strain up to an ultimate value τ_{ult}; thereafter it softens to a residual value τ_{res} with continued straining.

4.2 *Computational steps*

The fixed anchor is divided into a number of short segments of length Δx, within which bond stress, anchor load, axial strain etc. are constant. Load and bond stress distributions are obtained as follows:

a) Zero the distribution of bond stress $\tau(x)$, axial strain $\varepsilon(x)$, and anchor load $P(x)$ at all points along the grout-ground interface.
b) Apply a small initial load Po to the proximal end of the tendon.
c) Using the current distribution of bond stress $\tau(x)$ calculate the axial load $P(x)$ in the tendon-grout column from $P(x) = Po - \Sigma\tau.\pi.d.\Delta x \quad \geq 0$.
d) From the axial load distribution $P(x)$, estimate the axial strains $\varepsilon(x)$ along the tendon-grout column from $\varepsilon(x) = P(x) / EA$.
e) If necessary, adjust the bond stress distribution $\tau(x)$, based on the current axial strains $\varepsilon(x)$.
f) Repeat steps (c) - (e) until a converged solution is obtained (15-20 iterations normally adequate).
g) Estimate the extension of each segment using the average axial strain for that segment, and sum to obtain the total extension of the tendon at the proximal end i.e. $\delta o = \Sigma\, \varepsilon.\Delta x$
h) Increment the applied load Po and repeat steps (c) - (g) until the ultimate capacity of the anchor is reached - when the area under the bond stress distribution curve ($= \Sigma\tau.\Delta x$) is a maximum.

The steps have been laid out sequentially to illustrate the method, but the spreadsheet actually deals with an implicit loop of expressions; the "cells" holding expressions for $P(x)$ depend on $\tau(x)$, which depend on $\varepsilon(x)$, which in turn depend on $P(x)$. The iterations implied in step (f) are performed using the recalculation function repeatedly.

4.3 *Bond stress : strain relationship*

Fairly complex relationships between τ and ε can be handled by the spreadsheet. The 4-part composite relationship shown in Figure 6 provides a reasonably comprehensive and flexible definition.

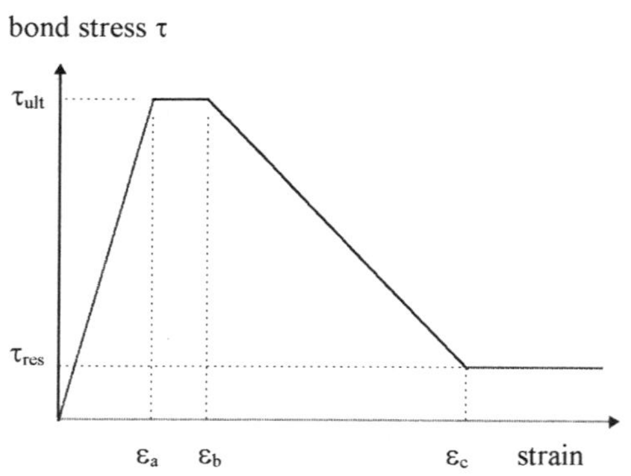

$$\tau = \tau_{ult}\,(\varepsilon/\varepsilon_a) \qquad \varepsilon \le \varepsilon_a$$

$$\tau = \tau_{ult} \qquad \varepsilon_a < \varepsilon \le \varepsilon_b$$

$$\tau = \tau_{ult} - (\tau_{ult}-\tau_{res})(\varepsilon-\varepsilon_b)/(\varepsilon_c-\varepsilon_b) \qquad \varepsilon_b < \varepsilon \le \varepsilon_c$$

$$\tau = \tau_{res} \qquad \varepsilon > \varepsilon_c$$

Figure 6. Bond stress : axial strain relationship

4.4 *Analysis of a trial anchor*

To illustrate the spreadsheet, the performance of a trial anchor in London Clay at Hampton has been modelled, using the parameters in Table 1.

Table 1. Parameters for trial anchor

Property		Value
Fixed length	L	10 m
Hole diameter	D	105 mm
Tendon area	A	1105 mm^2
Tendon Young's modulus	E	200 GPa
Ultimate bond stress	τ_{ult}	200 kPa
Residual bond stress	τ_{res}	50 kPa
Strain to reach ultimate bond	ε_a	0.050 %
Strain to begin softening	ε_b	0.075 %
Strain to reach residual bond	ε_c	0.150 %

Figure 7(a) shows the development of axial load in the tendon as the applied load is increased, up to an ultimate value of 340 kN. Figure 7(b) gives the bond stress profiles at the same load levels, showing progressive debonding very clearly. Finally, the predicted load-extension curve for the proximal end is shown in Figure 7(c), evidencing an initial linear response up to 150 kN and the onset of debonding thereafter. This compares well with the observed ultimate capacity of 370 kN at a fixed length displacement of about 15 mm.

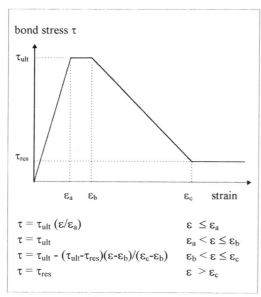

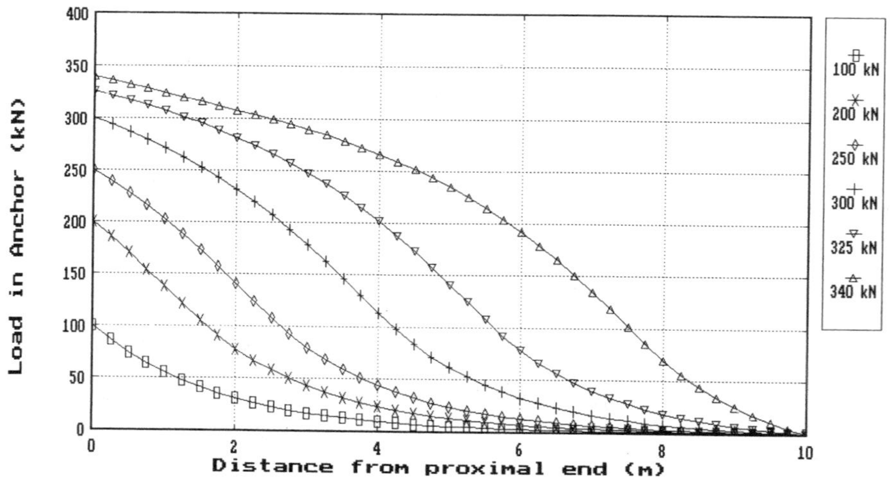

(a)

Figure 7 Predicted response of test anchor using spreadsheet model

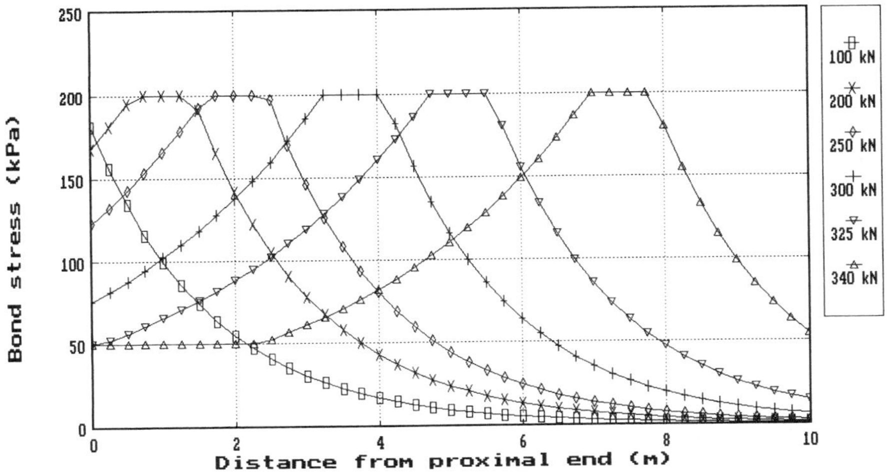

(b)

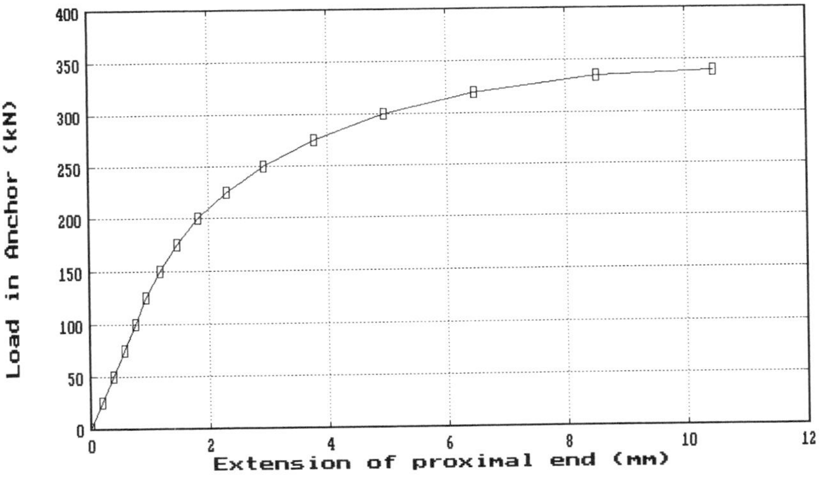

(c)

Figure 7 Predicted response of test anchor using spreadsheet model (contd.)

4.5 *Sensitivity study*

The bond stress values were based on undrained strength data, whereas the strain levels were best estimates. To investigate the sensitivity of the spreadsheet model to these strain parameters, ε_c was varied between 0.076 to 0.3%, giving a range for ε_c - ε_b (the range over which softening occurs) from 0.01 to 0.225%. Figure 8 shows ultimate capacity T plotted against (ε_c - ε_b) for two different values of residual bond τ_{res}, 20 and 50 kPa. Clearly the strain range over which softening takes place has a major effect on T, causing a variation in excess of 200 kN over the range considered. The ultimate capacity observed in the field corresponds to a value of ε_c = 0.175%, and the spreadsheet could easily be used in back-analysis mode to determine this - and thus improve forward predictions. Residual bond τ_{res} would seem to have rather less effect on T, causing slightly larger differences at lower (ε_c - ε_b) values.

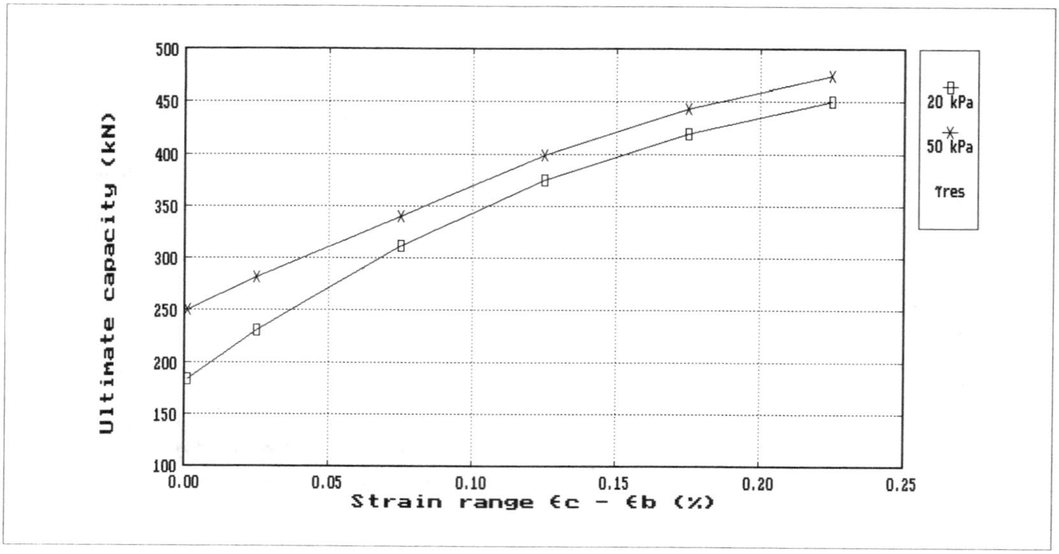

Figure 8 Sensitivity of predicted ultimate capacity to selected input parameters

5 CONCLUSIONS

Bond stress distributions along fixed anchor lengths can be highly nonuniform. This is due to progressive debonding which, although widely acknowledged to occur, has not yet been incorporated into routine design methods. It is principally long anchors which are affected, prompting the development of multiple anchor systems. However, there is still a need for improvement in design methods for long anchors.

Finite element methods can be used successfully to model progressive debonding and load transfer. Special attention must be paid to the behaviour of the interface, and whilst sophisticated joint elements and constitutive laws offer high levels of control, obtaining material parameters is not straightforward.

Additional insight into the mechanisms of load transfer and debonding are offered through a simple analytical model which can be implemented in a spreadsheet. The input parameters are of a modest number, and most of them can be fixed with some confidence. The strain range over which softening is allowed to occur has a big influence on results, and further materials testing and calibration would seem necessary. The spreadsheet model clearly has potential as a simple yet powerful design tool.

REFERENCES

Adams, D. 1995. Private communication.

Barkhordari, K. 1995. Design methods for ground anchors. *Internal Research Report* R/M/CVL/95/8, University of Surrey.

Barley, A.D. 1995. Theory and practice of the Single Bore Multiple Anchor system. In Widmann (ed.), *Anchors in Theory and Practice:* 293-301. Rotterdam: Balkema.

BS 8081: 1989. *British Standard Code of Practice for Ground Anchorages*. London: BSI.

Goodman, R.E., R.L. Taylor & T.L. Brekke 1968. A model for the mechanics of jointed rock. *J. Soil Mech. Found. Engg Div., ASCE,* 94(SM3), 637-659.

Goldberg, D.E. 1989. *Genetic Algorithms in Search, Optimization and Machine Learning*. Reading, MA: Addison-Wesley.

VIPS 1996. *The VISAGE System - User Manual*. Kingston-on-Thames: Vector International Processing Systems Ltd.

Ostermayer, H. & F. Scheele. 1977. Research on ground anchors in non-cohesive soils. *Revue Française de Géotechnique*, 3, 92-97.

Weerasinghe, R.B. 1993. *The behaviour of anchorages in weak mudstone*. PhD Thesis, University of Bradford.

Numerical Models in Geomechanics, Pietruszczak & Pande (eds) © 1997 Balkema, Rotterdam, ISBN 90 5410 886 X

Numerical simulation of rock block impacts on rigid sheds

F.Calvetti, R.Genchi, L.Nesta & R.Nova
Milan University of Technology (Politecnico), Italy

ABSTRACT: Road and railway protection against rock-falls is one of the main problems involved with transportation engineering in Alpine areas. A safe design of rock sheds has to be based on the correct evaluation of dynamic impact loads. In this paper, a 2-D DEM code (UDEC) is used to simulate the impact of rock blocks on a rigid plate covered by an absorbing cushion of gravel-like material. Relevant impact quantities, i.e. block kinematic variables and plate reaction, are compared to the corresponding values measured at LMR (EPFL-Lausanne) during a series of similar laboratory tests.
First, numerical parameters are calibrated to reproduce results of laboratory impacts, and the influence of mechanical parameters on impact kinematics and plate reaction is evaluated. Next, simulations under different design conditions (block mass, block speed and absorbing cushion thickness) are analysed, and a relation between impact energy and plate reaction is proposed for design purposes.

1 INTRODUCTION

The protection of roads and railways against rock-falls involves the design of adequate sheltering structures (Fig. 1). In general, such structures consist of a reinforced concrete roof-plate (1) supported by a bearing structure (2) (namely a retaining wall on the mountain side, and concrete columns on the other). The roof-plate can be covered by an absorbing cushion (3), to fulfil the following tasks: to reduce rock-fall impact loads on the roof-plate and on the bearing structure; to protect the roof-plate from damage due to direct impact; to dissipate the kinetic energy of impacting blocks, avoiding rebounds. Such requirements are accomplished by using a gravel-like material layer, which is highly deformable and dissipative. Rock fragments deriving from the construction of the shelter are often used because of low-cost.

The structural design of the shelter is based on the evaluation of dynamic loads exerted during impacts, which are far larger than permanent. To this purpose, it is necessary to recall the fact that the stress distribution within the whole shelter depends on stiffness, mass and plastic behaviour of each component. From a structural standpoint, the input load derives from the inertia forces acting on the block when it slows down as impact takes place, since inertial block forces are in equilibrium with the block-to-layer interaction forces.

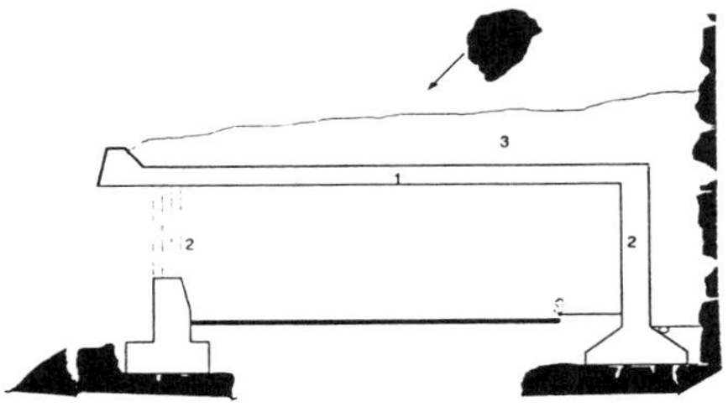

Fig. 1: Schematic section of a sheltering structure

Indeed, the absorbing cushion plays a double fold influence on plate loads, as it will be shown in the paper: first of all, the presence of the layer determines the input load, as it damps out the block deceleration; moreover, the behaviour of the layer determines the load transfer to the plate.

The assessment of the design loads can be performed with different techniques, namely experimental or numerical. In the latter case, the use of the Distinct Element method seems to be particularly adequate, for the following reasons: impacts are dynamic phenomena, involving relevant inertia forces, large displacements and rearrangement in the absorbing cushion; the absorbing cushion is formed by a discrete frictional material, which implies that energy dissipation is mainly linked to grain relative displacements.

2 DISTINCT ELEMENT MODEL

2.1 Reference laboratory tests

The employed numerical model is designed to reproduce a series of tests performed at LMR-EPFL of Lausanne (Labiouse et al. (1994)), that will be used as a reference for the model calibration. A numerical analysis of these impacts was presented by Labiouse et al. (1995), who used a one-dimensional spring and dashpot model to reproduce block kinematics.

The device employed at LMR will be briefly described first. A reinforced concrete plate (3.4x3.4x.2 m) rests on four bilateral supports, and is covered by an absorbing cushion with a thickness varying from .35 to 1 m. Three soil types are used (gravel or rock fragments), with d_{10} ranging from .06 to 6 mm, and internal friction angle ranging from 41° to 47°. The system is loaded by concrete blocks impacts (mass: 100, 500, 1000 kg; falling height: .25÷10 m). The available measurements, sampled at 1.2 kHz, regard block and plate kinematics (acceleration of the blocks, and plate displacements), and loads on the plate (normal stress on the upper face of the plate, and support reaction). The last two measurements differ, because of plate inertia.

A detailed description of the experiments is presented in the given references. Here, it is interesting to report the fact that laboratory impacts are only slightly influenced by the material forming the absorbing layer. This is of main importance for the numerical simulations, since it means that the key-point is the discrete frictional nature of the cushion, despite the different grain-size curves and friction angles.

2.2 The numerical model

The numerical model employed to reproduce laboratory tests represents a particular application of a commercial 2-D DEM code (UDEC), which is commonly used to model blocky-rock systems (Cundall and Hart (1985)). The model (Fig. 2), consists of a rigid and fixed plate (a), on which a discontinuous material layer (b) rests. This layer is composed of gravel-shaped elements (up to about 6000 for the thicker layer), obtained from an intact block, which is intensely fractured. The system is loaded by an impact block (c) with two fixed blocks on its sides (d), which prevent rotation of the impacting block. The blocks have the same weight and impact surface of the corresponding laboratory ones. At the beginning of the numerical simulation, the impact block is just in contact with the upper surface of the absorbing cushion, and it is given a vertical speed computed on the basis of the simulated free-fall height, assuming zero air resistance: therefore, the initial potential energy is equal to the kinetic energy at impact, that will be called impact energy. The numerical model provides data about block kinematics, which are directly compared to the corresponding laboratory measurements, and plate reaction, which is compared to the sum of support reactions.

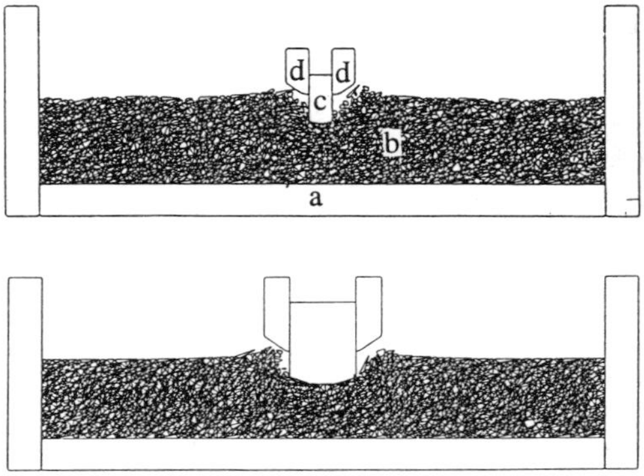

Fig. 2:The DEM model at the end of two impacts block mass: a) 100 kg; b) 500 kg

Before discussing the results, it is necessary to point-out some of the model features, that may influence them in a relevant manner.

First of all the use of a 2-D code has several consequences: on the impact kinematics first, and therefore on the inertia forces which raise inside the granular layer during impacts; moreover, the three-dimensional behaviour of the cushion and, in particular, the load transfer mechanism could be substantially different from the model ones.

Next, all elements are rigid, and the behaviour of the system is ruled by contact properties only (linear elastic compliance and Coulomb's friction). In particular, the use of a rigid and fixed plate does not allow to incorporate the effects of plate bending and translational vibration, and inhibits direct evaluation of stresses inside the plate, which is of main importance for design. Finally, since no numerical damping is used, energy dissipation is entirely due to inter-particle and block-particle sliding.

In fact, the numerical model is very simple if compared with the laboratory set-up; therefore, the possibility of properly reproduce experiments is due to the ability of DEM to catch the basic aspects of the phenomenon.

3 PARAMETER CALIBRATION

As previously stated, the only numerical parameters available for the model calibration are related to inter-particle properties, i.e. to normal and tangential contact stiffness (jkn = jks) and to contact friction (jf); in particular, all contact properties are given fixed values, except for contacts within the absorbing layer. The parameters are chosen to make the model fit the overall behaviour of the experimental set-up, focusing attention on block kinematics and plate reaction. A preliminar study of the individual influence of parameters was conducted. This step, that was useful to gain a qualitative sensibility about the model behaviour, was necessary because a direct evaluation of DEM parameters is not possible when micro-mechanical measurements of the individual soil particles properties are not available. As it will be shown later on, a check on the validity of the chosen set of parameters was subsequently made by comparing the experimental and numerically determined friction angles of the cushion material.

3.1 Parametric analysis

In the following, we will show the importance of the absorbing cushion and the influence of contact parameters on the overall behaviour of the numerical model, trying to establish some links between micro-properties and the corresponding global ones (elastic stiffness and friction angle of the equivalent material). To this purpose, a reference impact (block mass 100 kg; layer thickness .5 m; falling height 10 m) is analysed under the combination of contact parameters summarised in the following table.

Tab. 1: contact parameters variation range

stiffness jkn=jks (Pa/m)	1e8÷1e11
friction coefficient jf	0.2÷1.0 (11°÷45°)

note: 1eN = 1×10^N

First of all, it is necessary to point out that the plate reaction and the block deceleration are considerably reduced by the presence of the cushion: this evidence is clearly shown in Fig.3a,b, where the results obtained in the reference impact using different values of contact parameters are compared to the corresponding values for an impact on a layer of vanishing thickness, referred to as direct impact.

The data presented in Fig.3a,b show the influence of contact stiffness and friction: the peak value of the plate reaction increases with both contact stiffness and contact friction (Fig.3a); the maximum block deceleration (Fig.3b) depends on contact stiffness alone, and is only slightly affected by friction coefficient.

For a given input load (controlled by the block deceleration, which is determined by contact stiffness alone) the friction dependency of the plate reaction is linked to a different mechanism of load transfer within the layer. Indeed, the cushion acts as a signal

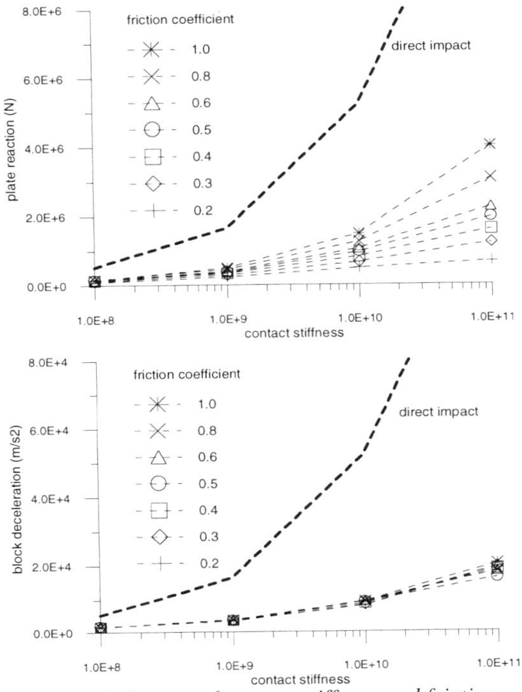

Fig.3: Influence of contact stiffness and friction a) plate reaction; b) block deceleration

processor, which first determines, and then amplifies or reduces the input load.

The load transfer to the plate can be described introducing a parameter (internal damping efficiency, *i.d.e.*) defined in the following way:

i.d.e.=input load/plate reaction (1)

The actual value of *i.d.e.* (Tab.2) depends on the relative influence of inter-particle sliding (energy dissipation) and particle inertia, which may increase the plate reaction, once the cushion elements are accelerated by impact. Of course, an *i.d.e.* value less than one (Tab.2, shaded area), that was measured in laboratory experiments too, does not mean that the load on plate would be lower for a direct block-plate impact, because the primary effect of the cushion is that of reducing the input load, as previously shown.

Tab. 2: internal damping efficiency

		jf				
		0.2	0.4	0.6	0.8	1.0
	1e8	1.75	1.18	1.09	1.04	1.05
jkn	1e9	1.41	1.04	0.91	0.80	0.74
jks	1e10	1.67	1.07	0.91	0.74	0.64
	1e11	2.86	1.16	0.80	0.58	0.50

The increase of *i.d.e.* under decreasing inter-particle friction values is linked to the progressive activation of local energy dissipating sliding mechanisms, which sets a threshold for tangential force transmission within the layer: as a counterpart, the decreased value of friction forces is compensated by the increase of block penetration into the soil.

In Fig.4 we show the delay between impact and force transmission to the plate, which is linked to the speed of the compression wave in the layer. To this purpose, a comparison with the effect of Young modulus of an elastic layer is instructive: in fact, for an elastic layer, the delay is inversely proportional to the square root of the elastic modulus. Determining the proportionality constant in such a way that the average delay corresponding to a contact stiffness of 1e8 is fitted (point A in Fig.4), the thick line traced in Fig.4 is obtained. Such a result allows to conclude that the contact stiffness plays the role of an elastic stiffness of the material, as could be expected.

On the whole, the obtained results show that loose layers are more efficient than dense ones; in fact, Fig.3a shows that plate reaction decreases with decreasing friction angle and contact stiffness (i.e. material stiffness and friction angle of the layer, for the mentioned reasons).

3.2 Reproduction of experimental results

First, contact properties are chosen to reproduce six laboratory impacts (block mass 100 kg); then, the retained choice of parameters is verified reproducing four impacts with different block masses.

The peak value of the loads on plate is properly reproduced (Fig.5): the difference between the numerical and the experimental results is quite acceptable (less than 20%), with the exception of two impacts (number 2 and 7), for which the numerical model overestimates by 30% and 40% the experimental data, respectively.

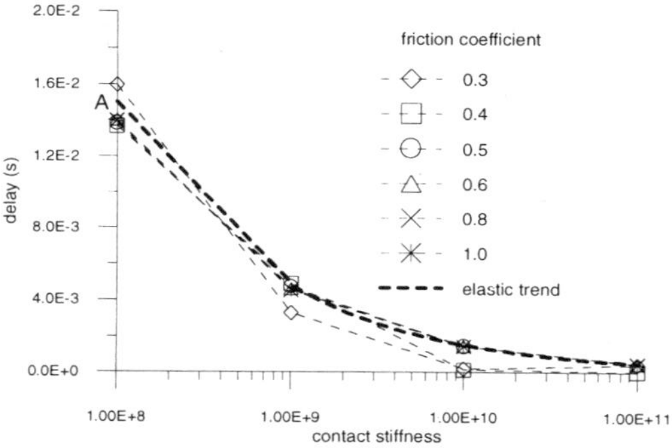

Fig.4: Delay between impact and force transmission: influence of contact parameters

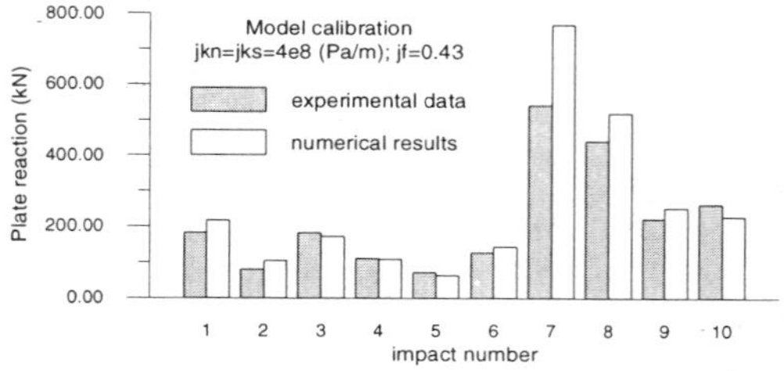

Fig.5: Experimental and numerical results. Calibration impacts (1-6); verification (7-10)

In fact, the dynamic nature of loading requires that both peak values and time history of the measured quantities should be matched as close as possible: in the following, we detail the results of a calibration impact (impact number 5). In fact, block kinematics is very well reproduced (Fig.6a), which is of main importance, since it determines the input load of the system, as previously discussed. The peak value of the plate reaction and the duration of the impulsive load are reproduced, as well (Fig.6b), even though a remarkable difference exists between the numerical and experimental curves, the latter showing a rebound not predicted by the model. This discrepancy is believed to be entirely due to the unrealistic assumption that the plate was rigid and fixed.

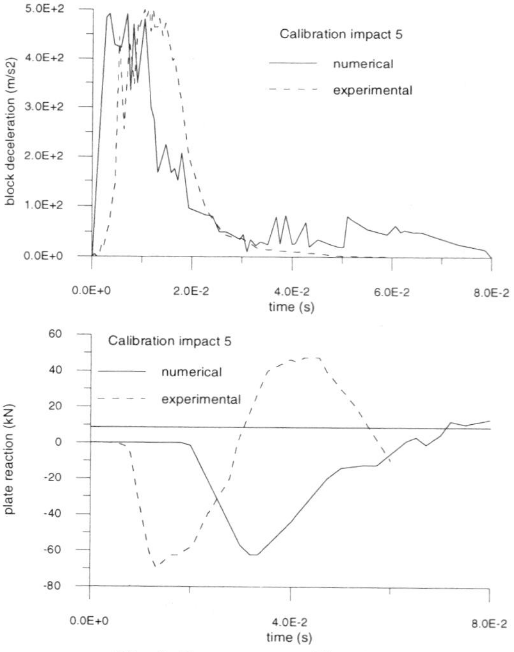

Fig.6: Parameter calibration
a) block deceleration; b) loads on plate

After the calibration tests, the internal friction angle of the numerical layer was measured by performing a biaxial test on a layer sample with the retained values of parameters. The obtained value was 53°, not too far from the laboratory range (41°÷47°); anyway, this difference is quite acceptable, if we consider the fact that laboratory impacts results are only slightly influenced by the different soils employed to form the cushion.

4 DESIGN LOAD ANALYSIS

In this section the numerical model will be used to assess the load exerted on the plate during impacts characterised by different block energy and layer thickness. Numerical soil parameters are given values determined in the previous section. The ranges of the impact characteristics examined are shown in the following table.

Tab. 3: impact parameters

layer thickness	.25÷1 m
block mass	100÷1000 kg
falling height	.25÷10 m

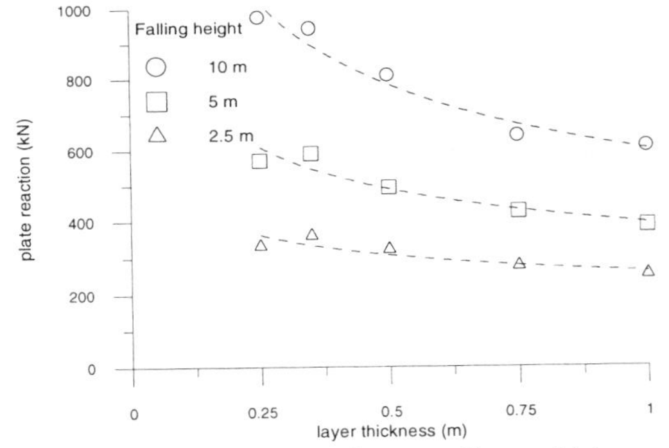

Fig. 7a: Plate reaction: influence of layer thickness (block mass 500 kg)

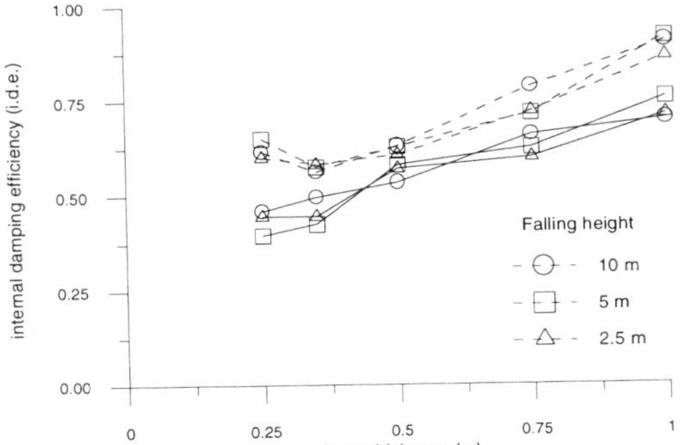

Fig. 7b: Damping efficiency vs. layer thickness (block mass: 500 kg, dashed; 1000 kg solid line)

4.1 Layer thickness influence

The influence of the layer thickness on plate reaction is considered first, gathering results of impacts with different energies. As expected, an increase of layer thickness reduces the load transferred to the plate (Fig.7a), even though this effect becomes less important for thicker cushions (see §4.3 for a quantitative evaluation).

This result is related to an increase of the absorbing properties of the layer, represented by the parameter *i.d.e.* (Fig.7b), since it is found that an increase of thickness does not affect in a significant way the maximum block deceleration. In fact, it is interesting to note that the internal damping efficiency does not depend on impact velocity (i.e. falling height), and it is slightly higher for the lighter block (Fig.7b); this tendency is confirmed if the 100 kg block is considered.

This result is probably related to the different size of the blocks, which implies a different ratio between impact surface and layer thickness (see Fig.2). An increase of the impact surface makes the loading conditions closer to the oedometric ones and, as a consequence, the mobilised friction angle decreases; from a micro-mechanical viewpoint, this fact means that the tangential contact forces are lower, which reduces contact sliding and energy dissipation.

4.2 Influence of block mass and falling height

Results of impacts with varying block mass and falling height are easily interpreted plotting the maximum value of the plate reaction versus the impact energy (Fig.8).

In fact, the individual influence of impact speed and block mass is not easily found, which means that the only relevant parameter is impact energy. This evidence has a practical importance, since it validates the usual design procedure, which is based on the assessment of a representative event, characterised by an estimated impact energy.

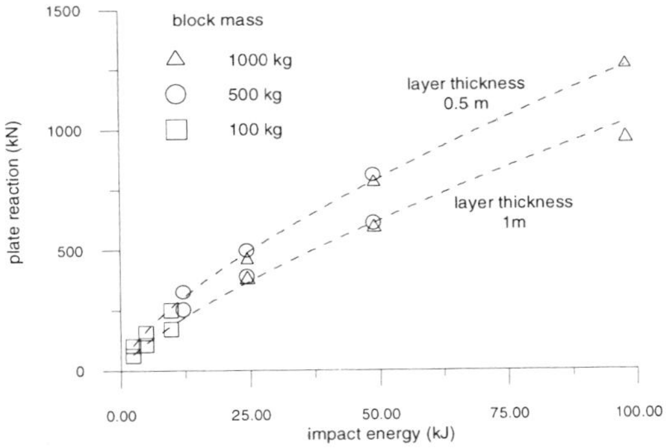

Fig. 8: Plate reaction: influence of impact energy

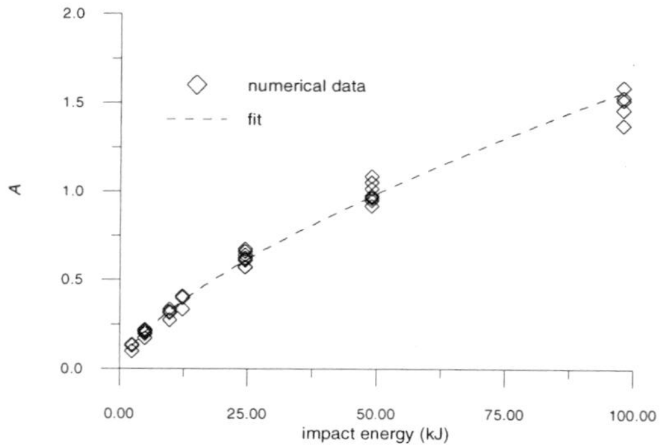

Fig. 9: Impact parameter A vs. impact energy

4.3 Synthesis of results

The data plotted in Fig.8 show, in an alternative way, the effect of a layer thickness variation.
Starting from these data, a quantitative evaluation of the thickness influence is made introducing the impact parameter A, defined in the following way:

$$A = F/F_0 \cdot \sqrt[3]{s/s_0} \qquad (2)$$

where F is the plate reaction (kN) and s is the layer thickness (m); the terms with index 0 correspond to a reference impact ($F_0 = 800$ kN, $s_0 = 0.5$ m).
In fact, when parameter A, instead of plate reaction, is plotted versus the impact energy (Fig.9), the data obtained for different values of thickness collapse, with acceptable scattering, to a common locus, that can be fitted by the following equation:

$$A = \left(\frac{E}{E_0}\right)^{0.68} \qquad (3)$$

in which E (kJ) is the impact energy and $E_0 = 48$ kJ its reference value.
Of course, the obtained relation in part depends on the model employed, and its validity is limited to the investigated design parameter variation range. Therefore, a certain care is required before a direct extension to the design of a real rock-shed can be made; anyway, the relationship represented by *Eq.* 3, which has the advantage of putting in evidence, in a clear way, the influence of the different design parameters, could guide the first design of a sheltering structure, or, from another point of view, the execution of a series of real scale tests.

5. FINAL REMARKS

The research presented in this paper is intended as a first step towards a real-scale modelling of a sheltering structure. An extremely simplified DEM model of the shelter roof-plate is found to be able to reproduce laboratory impact tests in a way that can be considered satisfactory. The numerical model, was used to assess a simple relation between impact energy, cushion thickness and plate reaction, for design purposes.
These encouraging results, obtained despite the several differences between the numerical and experimental devices, are probably due to the ability of the Distinct Element Method in providing the appropriate mechanical behaviour of the absorbing layer.
The numerical model is currently under development, and some aspects of the laboratory set-up will be incorporated: in particular, the most relevant difference between the laboratory device and the numerical model has been removed, substituting the rigid and fixed plate with an elastic plate set on two elastic supports.

ACKNOWLEDGEMENTS

This research is part of the activities of the network ALERT-Geomaterials supported by EU in HCM programme.
The authors acknowledge Sara Montani (LMR-EPFL) for the information delivered about the experimental set-up and the useful discussion of results.
Financial contribution from MURST is also gratefully acknowledged.

REFERENCES

Cundall P.A., Hart R.D. (1985). *Development of generalised 2-D and 3-D Distinct Element Programs for modeling jointed rock.* ITASCA Report to U.S. Army Engineering Waterways Experiment Station. Misc. Paper SL-85-1. U.S. Army Corps of Engeneers.

Labiouse V., Descoeudres F., Montani S., Schmidhalter C.A. (1994). *Etude expérimentale de la chute de blocs rocheux sur une dalle en béton armé recouverte par des matériaux amortissants.* Revue Française de Géotechnique 69, 4, 41-62.

Labiouse V., Descoeudres F., Montani S (1995). *Numerical analysis of rock blocks impacting a soil cushion.* Proceedings NUMOG V, G.N. Pande and S. Pietruszczak eds., 645-650.

Numerical Models in Geomechanics, Pietruszczak & Pande (eds) © *1997 Balkema, Rotterdam, ISBN 90 5410 886 X*

Numerical analysis of rock blocks impacting a rock shed covered by a soil layer

S. Montani & F. Descoeudres
Laboratoire de Mécanique des Roches, Ecole Polytechnique Fédérale de Lausanne, Switzerland
K. Bucher
Ernst Basler & Partners Ltd, Zollikon, Switzerland

ABSTRACT: Rock sheds are usually covered by a layer used as a shock absorbing cushion. To acquire a reasonable estimation of the impulsive design load and thus to obtain a design guideline, an experimental study has been first carried out. Weights simulating falling rock blocks were dropped from various heights on a reinforced concrete slab covered by different filling materials. To extend the range of the parameters a numerical analysis of the phenomenon has been conducted using the code AUTODYN.

1 INTRODUCTION

In mountainous areas, highways are frequently laid out following steep slopes. Since they are exposed to avalanches and falling rocks, they are usually protected at the most hazardous places by rock sheds. These structures are generally characterised by a reinforced concrete roof slab, which is covered by a soil layer used as a shock absorbing cushion (Fig. 1).

To have a better knowledge of the damping abilities of the covering cushion, and thus to acquire a reasonable estimation of the impulsive load due to a rock fall, an experimental study has been carried out at the rock mechanics laboratory of the Swiss Federal Institute of Technology Lausanne [1].

As at the laboratory the parameters used are limited (e.g. falling height) a numerical study to extend the range of parameters has been conducted by Ernst Basler & Partners using the code AUTODYN [2] completed by a structure dynamic analysis supplemented by structural dynamic analyses using the finite element method. The purpose of this study was to produce simple design guidelines for the design of rock sheds.

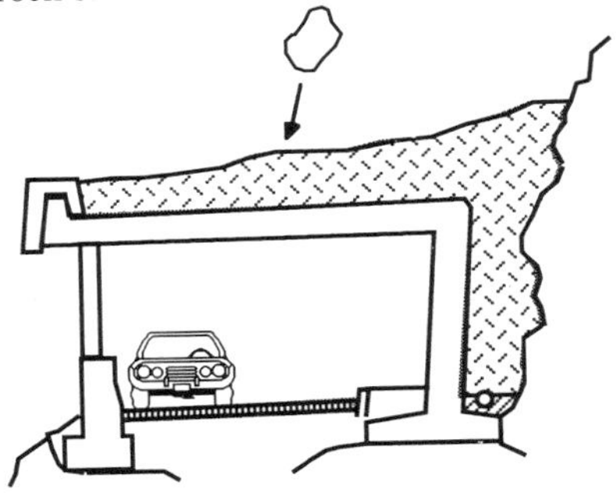

Fig. 1: Rock shed

2 DESCRIPTION OF THE PROBLEM

The impulsive forces F on the structure and the penetration depth d are mainly governed by three kinds of factors (Fig. 2):

- the weight, shape and falling height of the rock blocks;
- the slope, thickness and material properties of the covering cushion;
- the dynamic structural characteristics of the rock shed (mass, stiffness, strength,...).

The range of the parameters used for the experimental and the numerical analysis is summarised in Table 1.

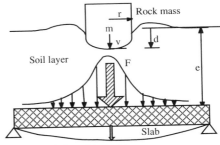

Fig. 2: Description of the problem

Table 1: Range of parameters

Parameter	Experiments	Numerical analysis
Mass of block m [t]	0.1, 0.5, 1	1, 5, 10, 20
Radius of block r [m]	0.21, 0.36, 0.45	0.451, 0.771, 0.972, 1.225
Impact velocity v [m/s]	4.4 ÷ 14	10, 20, 30, 50
Soil layer thickness e [m]	0.35, 0.5, 1.0	0.5 ÷ 4.0
Value of M_E [kPa]	2000÷20'000	2000÷50'000

3 EXPERIMENTAL STUDY

3.1 *General layout*

The tests have been carried out in a 5 m diameter and 8 m deep shaft (Fig. 3). At its bottom, a reinforced concrete slab rested on four supports and was then covered by a soil layer. The experiments were performed by impacting this set-up, varying the parameters according to Table 1.

The shape of falling weights was cylindrical with a spherical bottom, and they were made of steel shells filled with concrete.

Three kinds of soil materials have been used. First, the tests were performed with a concrete aggregate 3/32. Afterwards, they were carried on with filling materials that can be economically laid on real structures, for example materials from alluvial cones or scrap rocks from tunnelling excavations.

Some geotechnical properties of the three tested absorbing cushions are shown in Table 2.

3.2 *Measuring device*

The scientific study of such an intricate dynamic problem requires systematic and well instrumented tests. For this purpose, several transducers were installed (Fig. 4):

- an accelerometer (A) on the falling blocks,
- five earth pressure gages (P1 to P5) fixed in the concrete slab,

Table 2: Geotechnical properties of the uncompacted soils

Soil nature	Density (t/m^3)	Angle of friction	Cohesion	M_E [kN/m^2]
Gravel	1,650	41	0	2500
Alluvial cone	1,890	45	0	1000
Scrap rocks	1,790	47	0	600

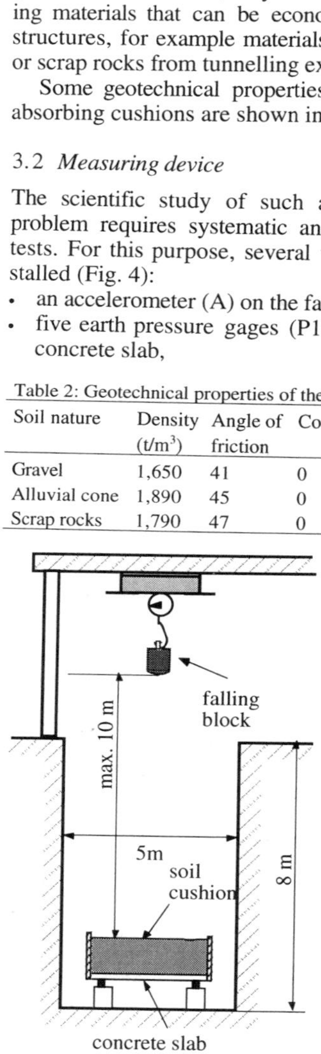

Fig. 3: Elevation of the testing shaft

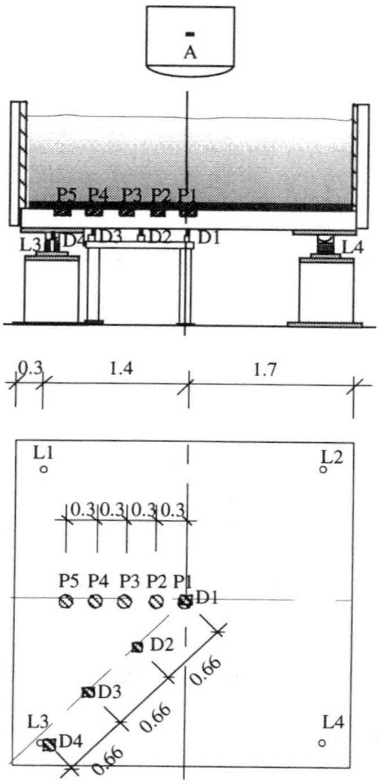

Fig. 4: Layout of measuring devices in plan and elevation

- four displacement transducers (D1 to D4) located underneath the slab,
- four load cells (L1 to L4) on the supports.

Before and after every testing series, plate bearing tests were carried out on the filling surface to determine the stiffness characteristics of the covering material. These were expressed by the standardised moduli of subgrade reaction (used for the design of road pavements).

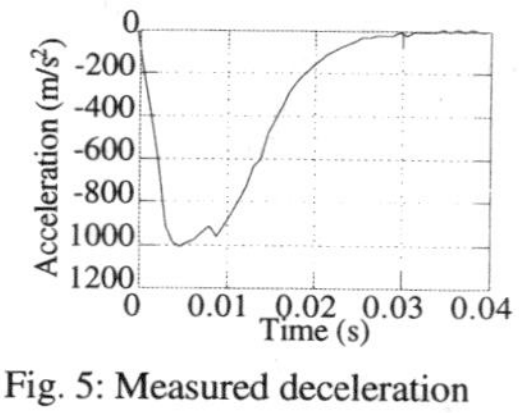

Fig. 5: Measured deceleration of falling block

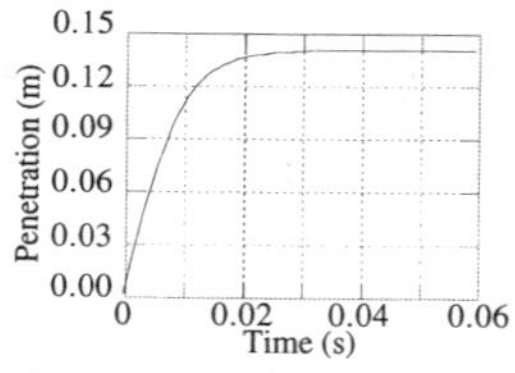

Fig. 6: Penetration of the falling block

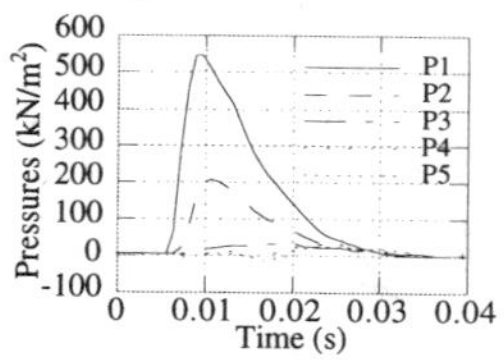

Fig. 7: Soil pressure on the slab

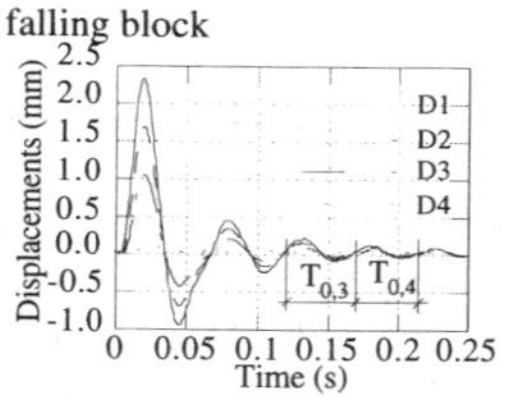

Fig. 8: Vertical slab displacements

3.3 *Typical results*

From the accelerometer located on the falling block, measures were taken of the deceleration of the weight during the impact (Fig. 5). Then a first integration enabled to determine the decrease of velocity with time, and a second one the block penetration into the soil layer (Fig. 6).

Fig. 7 represents the evolution of the earth pressures (measured by gauges P1 to P5) acting upon the surface of the structure roof.

Assuming a perfectly centred impact (above P1 pressure meter) and an axisymmetrical distribution of earth pressure, an integration easily enables to evaluate the resultant interaction force acting on the slab.

The slab deflection induced by the falling weight is illustrated in Fig. 8. The displacements measured by the transducers D1 to D4 are in phase, showing therefore that the slab is vibrating at its first natural frequency.

3.4 *Maximum impulsive force*

Several formulae based on simple theories or experimental facts have been proposed to evaluate the impulsive force due to rock falls. One of them, derived from Hertz's elastic contact theory, assumes that a rigid sphere impacts a half-infinite elastic medium. This last formulation has been slightly modified to take into account the block geometry (not spherical) as well as the available geotechnical data (moduli of subgrade reaction):

$$F_{max} = 1.765\ M_E^{2/5}\ R^{1/5}\ (mv^2/2)^{3/5} \qquad (1)$$

where

F_{max}	(kN)	Maximal impulsive force
M_E	(kPa/m^2)	Modulus of subgrade reaction obtained from a standardised plate bearing test on the soil layer
R	(m)	Radius of the block part in contact with the soil layer
m	(kN)	Mass of the falling rock
v	(m)	Impact velocity

4 NUMERICAL ANALYSIS

4.1 *Numerical Method*

The phenomenon of a rock impacting a soil layer is highly non-linear and dynamic with respect to deformations and to material behaviour. To simulate this kind of impact problem the commercially available computer code AUTODYN was used. This program is a so called hydrocode which is using a Lagrangian and an Eulerian discretisation together with a finite difference scheme. A wide range of constitutive models is available to simulate complex material behaviour.

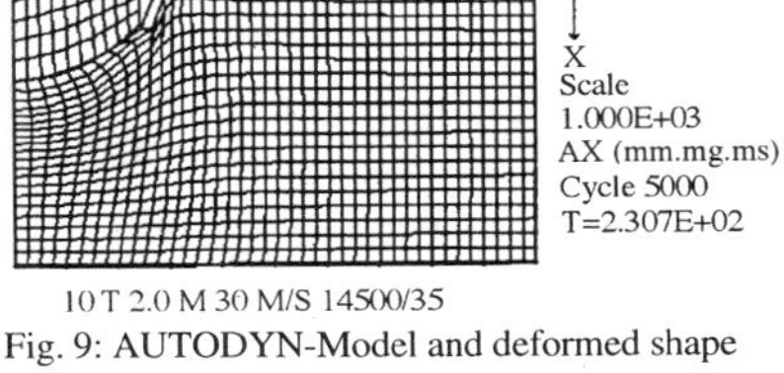

Fig. 9: AUTODYN-Model and deformed shape

4.2 *Models*

Two different models were used. The first one is an axisymmetric approximation of the experimental set-up (Fig. 3) to verify the code by comparing the results with those of the tests.

The second model corresponds to actual cases of rock sheds for a double lane road (Fig. 1). The model is axisymmetric about the vertical axis. It consists of a rock mass with the same geometrical proportions as in the tests, a soil layer with a depth varying among 0.5 to 4.0 m and a lateral extension of 4 m. The radius of 4 m corresponds to about half the span of a roof slab of a typical gallery. A typical model and its deformed shape is shown in Fig. 9. The analysis was done for a soft, a medium and a stiff soil as described later. Both the rock and the soil layer are modelled as Lagrangian grids. The slab is represented by thin walled shell elements having a thickness of 800 mm. This thick slab is considered as rather stiff and was used for the whole series of runs. The influence of the slab stiffness' on the interaction forces between the soil and the slab was also investigated. The contact zone between the rock and the soil and the soil and the slab was modelled by so called gap-elements which take care of the contact or loss of it and the friction. The size of the mesh necessary to get „accurate" results was found by trial and error. It varied between 50 mm in the impact zone to a maximum of 100 mm in remote zones. Along the vertical lateral boundary the soil was fixed in the lateral direction and the shell in the vertical one.

4.3 *Material models*

The most difficult part in the numerical analysis is the choice of an appropriate material model and its properties. The rock was modelled as a linear elastic material with infinite strength. The mass corresponds to a rock material of about 2.6 t/m^3. For the slab linear elastic properties corresponding to concrete were used.

The soil was represented by the so called porous material model together with the Mohr-Coulomb strength model. The porous model is based on a compaction curve which describes the relationship between the density ρ and the pressure p. This relationship can be deduced from Oedometer or triaxial tests. Typical diagrams to describe the soil are shown in Fig. 10. Three sets of soil parameters were used besides the ones needed for the verification runs.

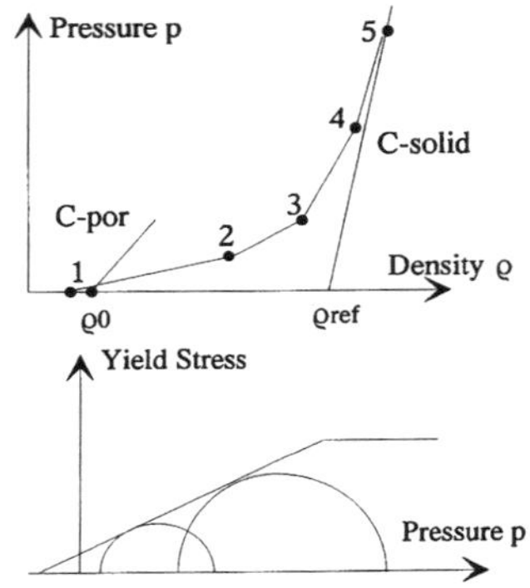

Fig. 10: Soil model, compaction curve, Mohr-Coulomb relationship

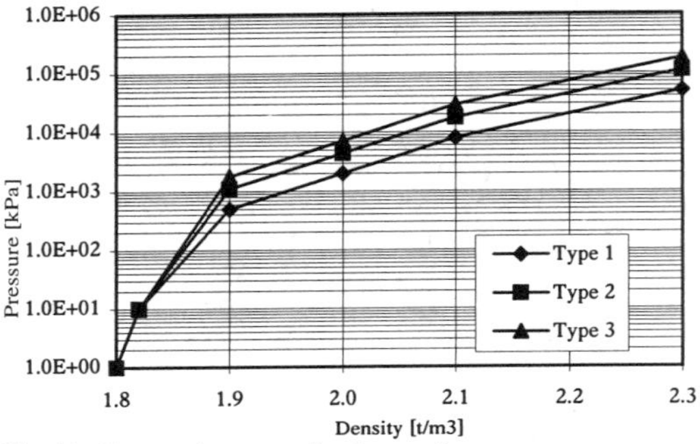

Fig. 11: Compaction curve for three soil types

These three sets are assumed to be most likely encountered in practical cases. The reference parameters to determine the compaction curve were the seismic velocity c_p and the static constrained modulus M_E ($\approx M_E$ Modulus of subgrade reaction) defined by the first and second point of the compaction curve in Fig. 10. The remaining part of the curve had to be chosen somewhat arbitrarily. Fig. 11 shows the whole set of curves used in the analysis in a log-log scale. The log-log scales facilitates the definition of a consistent set of compaction curves. In Table 3 the full set of soil data is shown including the shear modulus G which is chosen to correspond with the static constrained modulus by assuming a Poisson's Ratio of 0.275. For the Mohr-Coulomb strength model a angle of friction of 35, 38 and 42° and a very small amount of tensile strength and cohesion were used (Fig. 10).

Table 3: Soil parameters for three soil types

	Type 1	Type 2	Type 3
Bulk Modulus [kPa]	9082	20002	32742
Poisson's Ratio	0.275	0.275	0.275
M_E-Modulus [kPa]	15492	34121	55854
G-Modulus [kPa]	4808	10589	17334
c_p [m/s]	80	120	150
c_s [m/s]	1000	1000	1000
Ref-Density [t/m³]	2.2	2.2	2.2

4.4 *Procedure of calculation*

The calculation was carried out in two steps. In the first step the gravity was applied using a dynamic relaxation scheme. The necessary damping value to achieve rapid convergence to equilibrium was determined by trial and error. In the second step the rock mass was provided with mass and with an impact velocity and the calculation was then continued. The calculations were carried out for the range of parameters given in Table 1 and various combinations thereof. This led to about 300 computer runs each taking about half an hour on a PC-Pentium.

4.5 *Results*

As relevant results the motion of the rock mass, the interaction stresses at equal radial intervals in the cells just above the slab and the motion of the slab were recorded. Other variable such as the energy distribution, the state of material and others are also determined. Some typical results for a 10 t block, a 2 m thick layer, an impact velocity of 30 m/s and soft soil parameters are shown in the Fig. 12 to Fig. 17. Of these results, the maximum penetration depth and the

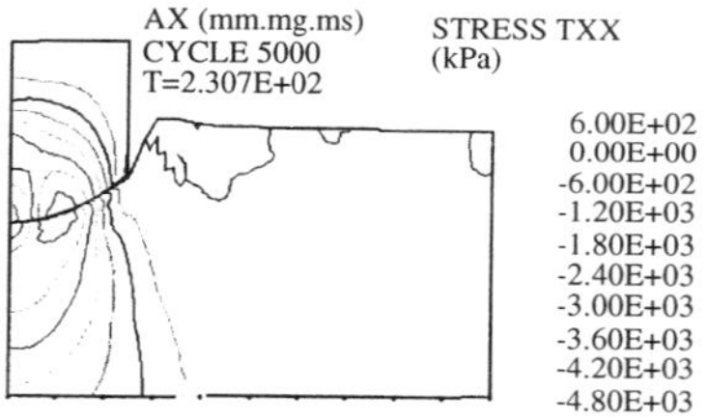

Fig. 12: Pressures contours

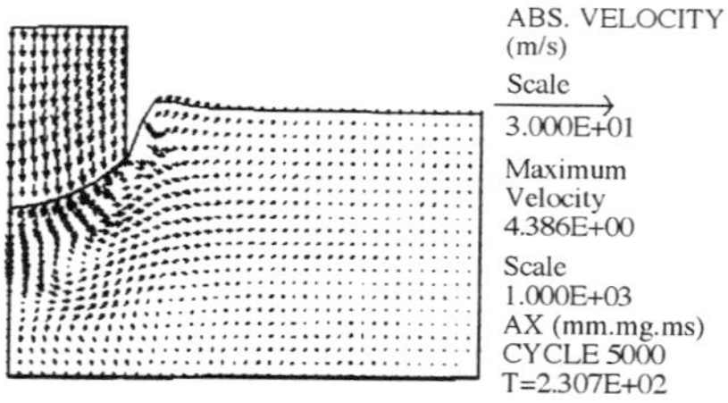

Fig. 13: Velocity vectors

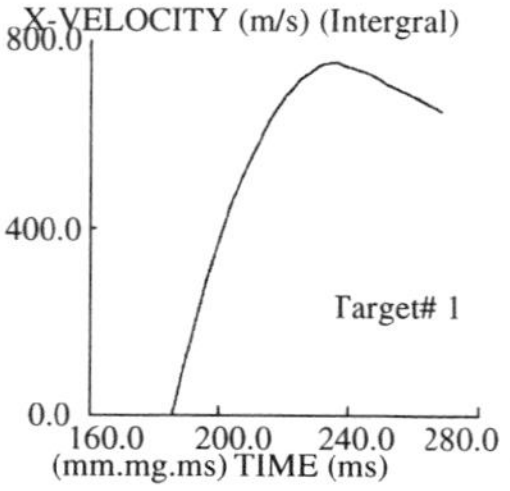

Fig. 14: Penetration depth

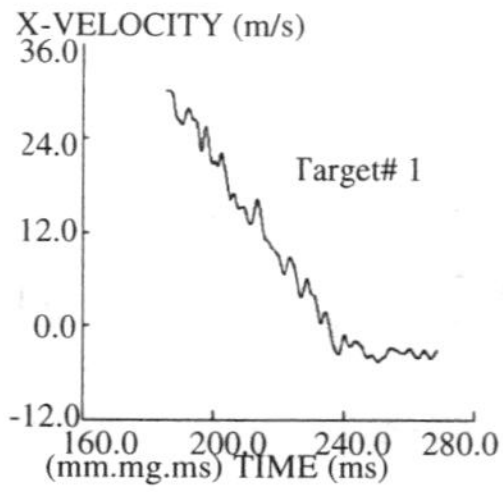

Fig. 15: Impact velocity

stress distribution on top of the slab were evaluated. The latter were integrated over the loaded area of the slab to obtain the total interaction force F as a function of time and its maximum.

4.6 *Comparisons with test results*

The computer code was verified using a selected number of test results as a basis for comparison. The maximum deviation in the interaction force was found to be less than 30 % and less than 20 % for the penetration depth.

4.7 *Influence of slab flexibility on the interaction forces*

In the verification phase the effect of the slab flexibility on the interaction forces between the soil and the slab was investigated. For this purpose the slab stiffness was varied from 400 to 1200 mm. It was found that over the range considered here variations of the maximum interaction forces up to 30 % are possible. The stiffer the slab the larger is the maximum force. As a consequence it was decided to neglect the influence of the slab flexibility for the further evaluation of the results. This means that the impact force is considered as decoupled from the structural behaviour. With the rather stiff slab of 800 m [sic] thickness the interaction forces are rather conservative.

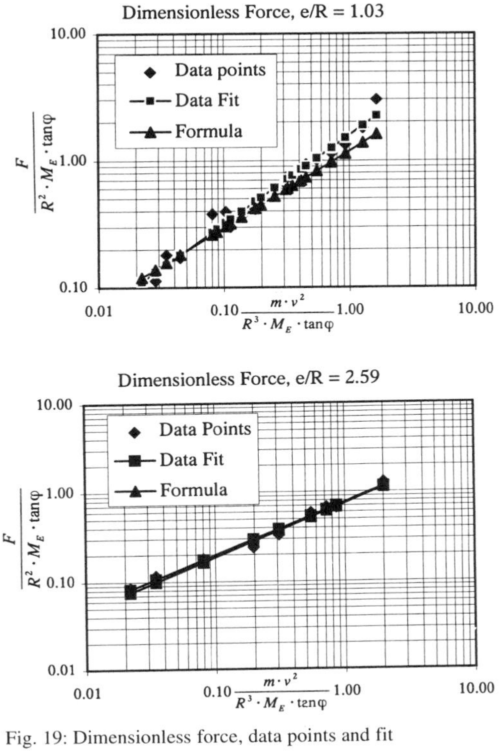

Fig. 19: Dimensionless force, data points and fit

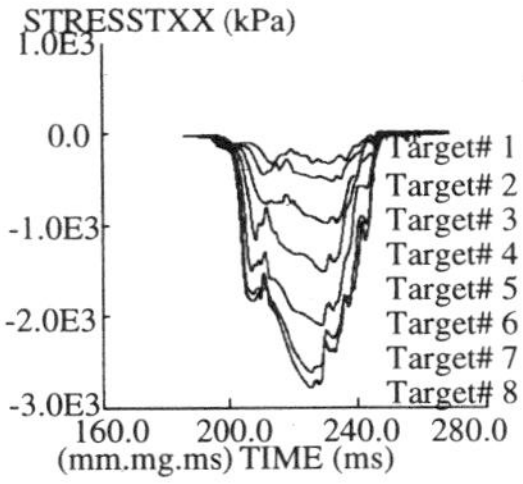

Fig. 16: Pressures ont the slab at radial intervals of 300 mm

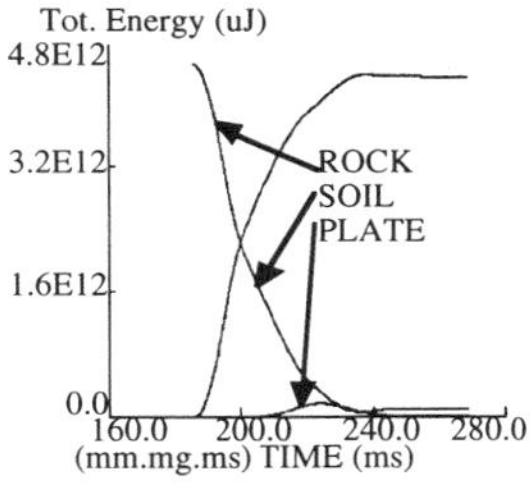

Fig. 17: Energy in the rock, soil and slab

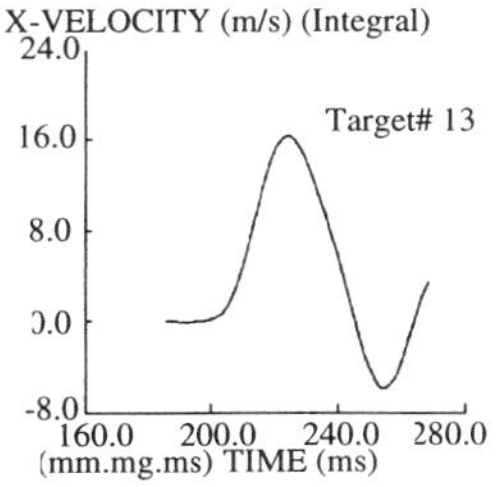

Fig. 18: Vertical displacement of the slab

5 DIMENSIONAL ANALYSIS AND DATA REDUCTION

To evaluate the large amount of results produced by the nearly 300 computer runs and to develop simple relationships to describe the impact forces and the

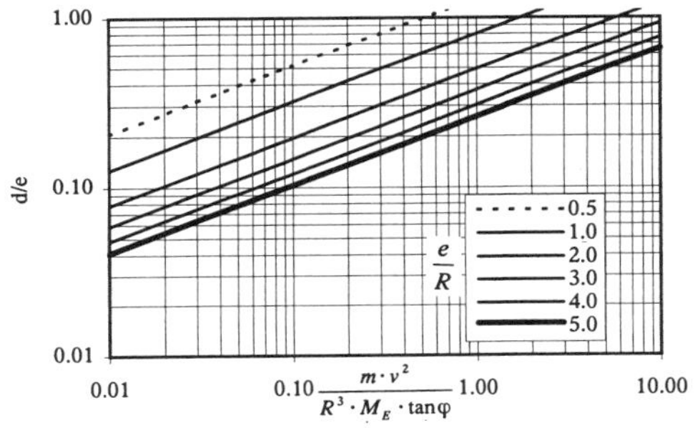

Fig. 20: Design chart for the dimensionless penetration depth d

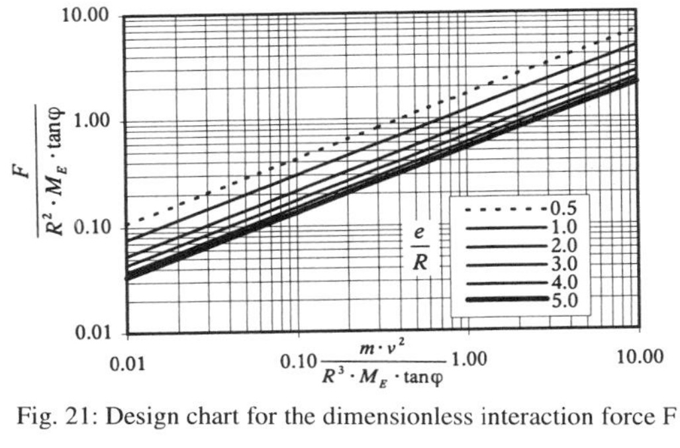

Fig. 21: Design chart for the dimensionless interaction force F

penetration depth the dimensional analysis and statistical methods are very helpful. The dimensional analysis is described extensively in various text books. In essence the theory says that for a set of n variables used in the analysis or the tests and m dimensions, there is a set of k = n - m dimensionless parameters that can describe the problem. Neglecting the effect of the slab stiffness the following relevant independent variables were used: Rock mass m, radius of projected area of the rock R, impact velocity v, layer thickness e, static constrained modulus M_E and the angle of friction ϕ. The dependent variables are the interaction force F and the penetration depth d. With the angle of friction ϕ i.e. $\tan\phi$ considered as a dimensionless quantity we have a set of six variables for the force and for the penetration depth, respectively. Having three dimensional unit [t], [m] and [s], the following three dimensionless parameters can be found:

The dependent parameters are $\pi_1 = \dfrac{F}{R^2 M_E \tan\phi}$ and $\dfrac{d}{e}$, respectively.

The independent parameters are:

$\pi_2 = \dfrac{e}{R}$ and $\pi_3 = \dfrac{mv^2}{R^3 M_E \tan\phi}$

Other combinations of the variables can also be chosen.

Having reduced all the data to these three parameters and plotted in log-log scale, a linear trend can be found. A linear regression analysis of these data leads to very simple and consistent diagrams and formulae. Two sets of data for a given π_2 and the linear fit is shown in Fig. 19. The final diagrams are shown in Fig. 20 and Fig. 21. They can also be expressed by the following formulae:

$$\frac{d}{e} = 0.8 \cdot \left(\frac{e}{R}\right)^{-0.7} \cdot \left(\frac{mv^2}{R^3 M_E \tan\phi}\right)^{0.4} \tag{2}$$

$$\frac{F}{R^2 M_E \tan\phi} = 1.25 \cdot \left(\frac{e}{R}\right)^{-0.3} \cdot \left(\frac{mv^2}{R^3 M_E \tan\phi}\right)^{0.6} \tag{3}$$

$$F = \frac{mv^2}{d} \tag{4}$$

The formulae look similar to the ones that Hertz developed, the difference being that Hertz' theory for the contact problem assumes a linear elastic material and a radius defined as that of a sphere. In the formulae above the material behaviour is non linear and the penetration depth can be large so that the relevant radius is taken to be the projected area of the rock mass. Consequently the formulae apply for any kind of prismatic rock mass with deep penetration. However the influence of different shapes of the contact zone was not investigated. Using data from weapon tests the penetration depth for a flat contact zone has to be reduced by a factor of about 1.2 or increased for a longish (sharp) nose by about 2. The interaction force can then be determined according to equation (4).

For the dynamic analyses of structures the impact force can be approximated by a half-sine function. The impact duration was not explicitly evaluated in this study. It can be estimated by the formula:

$$t_d \approx 2.0 \cdot \frac{d}{v} \pm 30\% \tag{5}$$

6 STRUCTURAL RESPONSE

For the decoupled system, as is assumed here the structural response can be considered separately. For the design guidelines static analysis procedures were envisaged in contrast to fully dynamic and nonlinear analyses. Therefore, so called dynamic load factors (DLF) were determined. The DLF is defined as the ratio between the maximal dynamic to the static response for a defined load case and for selected moments or forces in the structure. In this study linear elastic dynamic analyses for two representative rock shed galleries were carried out by the finite element method. Various impact locations and impact times (20 to 80 ms) were analysed and the DLF determined. Similarly this can be done for nonlinear structures. The values are not given here as they depend on the design philosophy of the national codes.

7 CONCLUSIONS

Based on both tests and numerical analyses, some simple design charts and formulae for the determination of the penetration depth and the interaction forces with a structure due to a rock impacting a soil layer were developed. The accuracy achieved is considered as sufficient for design purposes in view of the fact that the soil properties are usually not known exactly and that they can change over the years.

8 ACKNOWLEDGEMENTS

This work was initiated and funded by the Swiss Federal Department for Road Construction and the Swiss Federal Railway (SBB).

9 REFERENCES

[1] Montani S., Descoeudres F. Etude expérimentale de la chute de blocs impactant une dalle en béton armé recouverte par des matériaux amortissants, Mandat 98/92 *Publ. OFR N° 524* (Ecole Polytechnique Fédérale, 1996).

[2] AUTODYN, Century Dynamics, Ltd, Horsham, West Sussex, RH13, 5BA, England

Numerical Models in Geomechanics, Pietruszczak & Pande (eds) © 1997 Balkema, Rotterdam, ISBN 90 5410 886 X

Determination of constitutive parameters from pressuremeter tests in lightly cemented soils

N.C.Consoli, F.Schnaid, A.Thomé & F.M.Mántaras
Federal University of Rio Grande do Sul, Porto Alegre, Brazil

ABSTRACT: Cohesive-frictional materials pose a complex geotechnical problem due to the difficulties in predicting soil parameters representative of *in situ* conditions and the lack of a comprehensive constitutive model to describe soil behavior. The present study makes use of a numerical technique adopted to bypass these limitations. A simple non-linear elastic constitutive model was implemented in a finite element code which was used to produce a curve fitting analysis of field pressuremeter tests. Constitutive soil parameters assessed from this approach produced realistic predictions of load-settlement response of field plate loading tests. Comparatively, parameters obtained from triaxial compression tests with internal measurements of strains failed to produce the same accuracy given by pressuremeter data. A suggestion is made to interpret pressuremeter test results numerically to assess soil data on a regular basis. Comparison of prediction and performance of plate loading test results in lightly cemented homogeneous residual soils demonstrate the potential application of the proposed approach to foundation design problems.

1 INTRODUCTION

Soil investigation based on triaxial tests carried out on undisturbed samples may not give realistic information on stiffness of lightly cemented soils as results often reflect the effect of confining stresses in the breakage of the bonds of the soil [*e.g.* Bressani & Vaughan (1989); Leroueil & Vaughan (1990)]. Lack of constitutive models that effectively describe observed cemented soil behaviour makes it difficult to implement any solution from routine laboratory tests. An alternative approach is to make a more direct measurement of soil properties *in situ*. A methodology is here suggested based on the curve fitting analysis of pressuremeter tests. This procedure enables a simple constitutive model to be used once the assessment of the stress - strain - strength characteristics in lightly cemented homogeneous soil deposits is obtained throughout the calibration against field pressuremeter data. The routine uses the finite element formulation for the interpretation of pressuremeter test results. The proposed approach allows checking of all constitutive parameters needed for geotechnical design. The applicability of the formulation is here demonstrated by the simulation of plate loading tests carried out at the geotechnical experimental field of Federal University of Rio Grande do Sul (UFRGS), located in southern Brazil.

2 EXPERIMENTAL PROGRAMME

A comprehensive experimental testing programme was carried out at the testing site of UFRGS, on a homogeneous lightly cemented residual soil strata. The programme comprises the accomplishment of laboratory tests, as well as *in situ* tests such as penetrometer tests (CPT, SPT) to determine soil profile, Ménard pressuremeter at several depths and plate loading tests.

The site investigation revealed features that are common to almost every site which has been subjected to severe weathered conditions. The degree of weathering varies with depth in distinctive layers. Two subsoil layers were clearly identified by visual inspection, a sand-silty red clay of about 3.5 m of depth overlaying a stratum of red silty clay with grey and yellow intrusions. The soil profile at test site is presented in Figure 1, in which measurements of cone tip resistance q_c and SPT blow count number N are plotted against depth. The upper surface layer, in which all tests were carried out, is easily identified due to homogeneity presented in all results of *in situ* penetration testing. The representative SPT index within the first three meters is 6 blows/0.3 m. The average electrical cone tip resistance within this layer is 1.2 MPa. The water table in the site is at

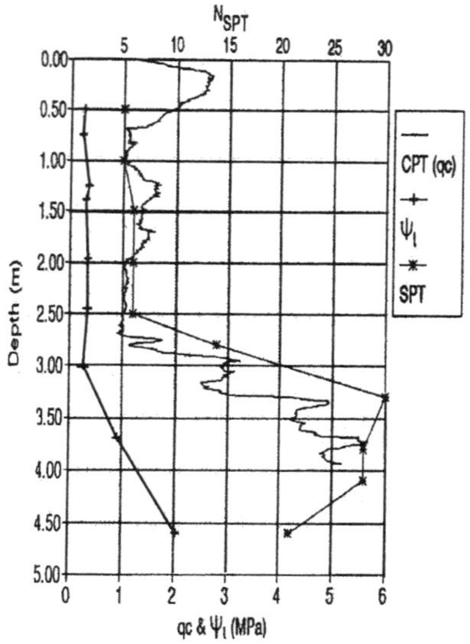

Figure 1. Soil profile at test site

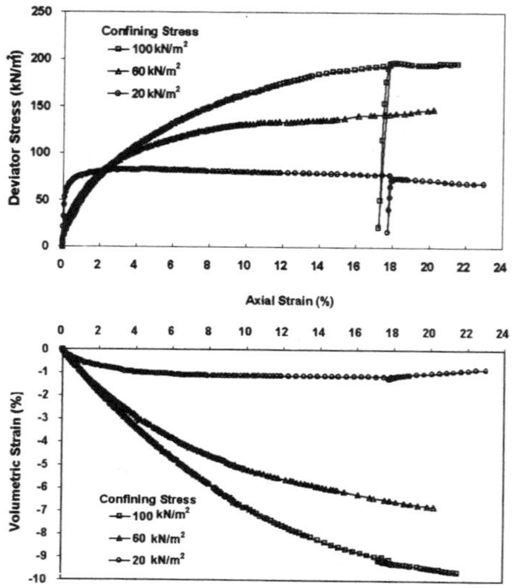

Figure 2. Drained triaxial test results

approximately 4.0 m of depth. A summary of laboratory and *in situ* test results is described below; detailed information is given by Schnaid et al (1995) and Mantaras et al (1995).

2.1 Laboratory test results

The upper homogeneous layer in which all loading tests were carried out was investigated by index tests. The soil is classified as low plasticity clay (CL) according to Unified Soil Classification System. Grain size data indicate 6% of medium sand, 38% of fine sand, 32% of silt and 24% of clay. The average bulk unit weight (γ_t) of this upper layer was 17.9 kN/m^3, the moisture content (ω) was typically 24.5%, the degree of saturation (S) ranged around 78% and the voids ratio (e) varied between 0.80 and 0.86. Atterberg limits of the material were: Liquid Limit (ω_L) of 42.8% and Plastic Limit (ω_P) of 22.6% which yields a Plasticity Index (PI) of 20.2%. Through the technique of pressure plate apparatus [Fredlund and Rahardjo (1993)], the relation suction versus water content was obtained and for the field interval of water content the suction reached values of less than 10 kPa. The low measured suction values indicate that soaking would have little effect on settlements of the residual soil at the test site.

Isotropically consolidated drained saturated triaxial tests were performed to evaluate the stress-strain-strength behaviour of the lightly cemented soil. The tests were carried out with internal measurement of deformations using Hall Effect sensors [Clayton and Khatrush (1986) and Clayton et al. (1989)]. The applied consolidation pressures were 20 kPa, 60 kPa and 100 kPa. Results of the triaxial tests are plotted in Figure 2, which shows the measured stress-axial strain-volumetric strain curves. Reduction of soil stiffness due to the effect of breakage of cemented bonds caused by increasing of confining stresses is directly noticed [Leroueil and Vaughan (1990)]. An angle of friction (ϕ') of 26^0 and a cohesion intercept (c') of 17.0 kPa were determined from peak stress triaxial test analysis.

2.2 In situ test results

2.2.1 Ménard pressuremeter tests

Pressuremeter tests were carried out through the cemented soil strata at the testing site employing a Ménard-type probe. A description of equipment and procedures is found in detail elsewhere [e.g. Schnaid, Sills & Consoli (1996)]. Typical Ménard pressuremeter pressure-expansion curves have S-shape.

Figure 3 shows four pressure-injected volume curves for tests carried out from 1.26 m to 2.45 m depth. There is no significant variation of pressuremeter limit pressure (ψ_l) with depth [see Figure 1] which gives further support to the evidence provided by penetrometer data that the top layer is

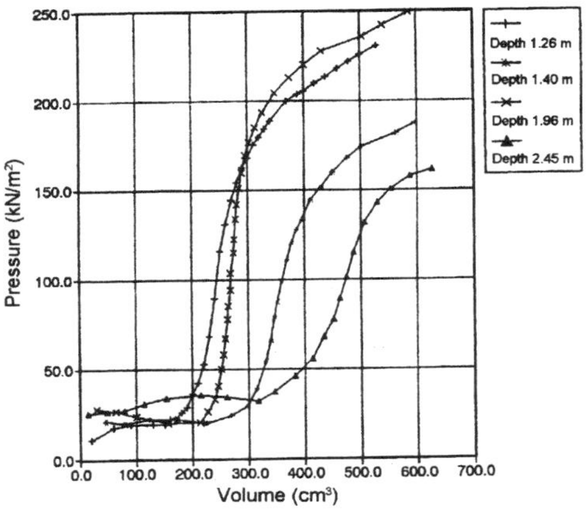

Figure 3. Pressuremeter test results

reasonably homogeneous. Values of the Ménard pressuremeter modules are also practically constant with depth up to 3.0 m, with a representative average value of 3.2 MPa. This evidence indicates that the homogeneity of the upper soil layer is not restricted to strength but can be extended to the deformation viewpoint. Through pressuremeter tests analysis, K_0 values where obtained for the whole upper layer length, conveying to a consistent value of 2.1.

2.2.2 Plate loading tests

A load testing programme was carried out at the test site to study the behaviour of shallow foundations on cemented soils. Rigid circular steel plates of 0.3 m, 0.45 m and 0.6 m diameter were tested at a depth of 1.20 m below ground level. Tests were in accordance with American Society for Testing and Materials standard ASTM D1194-72 (1987). Detailed information can be found in Schnaid et al. (1995).

3 CONSTITUTIVE PARAMETERS

An interpretation method has been developed to incorporate non-linear soil behaviour to the analysis of pressuremeter test results. Duncan & Chang's (1970) model was implemented on the finite element program CRISP90 [Britto & Gunn (1991)]; seven constants are considered in this model: the moduli number for loading and unloading-reloading, respectively K *and* K_{ur}, the exponent determining the rate of variation of initial modulus with confining stress n, the Poisson ratio ν, the failure ratio R_f and the Mohr-Coulomb strength parameters c' and ϕ'.

The method is evaluated on a two step analysis. First a comparison among parameters assessed from pressuremeter data and triaxial data is made. Secondly, an attempt is made to use both pressuremeter and triaxial parameters to predict load - settlement behaviour of footings.

A number of assumptions is intrinsically made on the proposed analysis: (a) the soil stress-strain behaviour can be represented by a hyperbolic function; (b) the finite element approach is able to simulate the excavation of the borehole followed by a cylindrical expansion; (c) boundary effects due to the finite length of the pressuremeter can be evaluated; (d) the strain field remains essentially in axisymetrical conditions and (e) strains are considered to be small.

The following steps are prescribed for the interpretation methodology: (a) the numerical technique is used to simulate the loading part of the pressuremeter tests using the parameters measured in triaxial tests; (b) an optimisation is made by a curve fitting test procedure to assess soil parameters representative of field conditions and (c) recognising the limitations of laboratory tests to give reliable information of soil stiffness due to structure breakdown, only the values of the deformability parameters K and n at the structured condition are optimised. The influence of the other five parameters has been investigated in detail by Mantaras et al (1995), and support the assumptions made in this paper.

The finite element mesh and the boundary conditions used for the numerical analysis of pressuremeter tests are presented in Figure 4. The simulation was made in two steps: (a) excavation of the borehole and (b) application of pressure increments in the pressuremeter cells. The *in situ* total vertical stress state was obtained by adopting a total unit weight of 17.9 kN/m^3. The *in situ* horizontal stresses were estimated from K_0=2.1.

Figure 5 shows the measured and simulated pressure-expansion response on a typical pressuremeter test result, in which the effectiveness of the solution to back-analyse a pressuremeter test is clearly demonstrated. For the homogeneous lightly cemented soil strata the optimisation approach yielded parameters presented in Table 1. For comparison the constitutive parameters assessed directly from triaxial tests are also listed in this table [Mantaras (1995)]. There are two constitutive parameters based on triaxial results which are questionable: n which is negative due to structure breakdown meaning that the increase in confining stresses reduces stiffness and K which has a non-representative value also in consequence of bonds collapse. The negative value of n takes the numerical analysis to inconsistent results; for example, during the simulation of the excavation of a footing trench the soil located right beside it becomes stiffer with decreasing confining pressure.

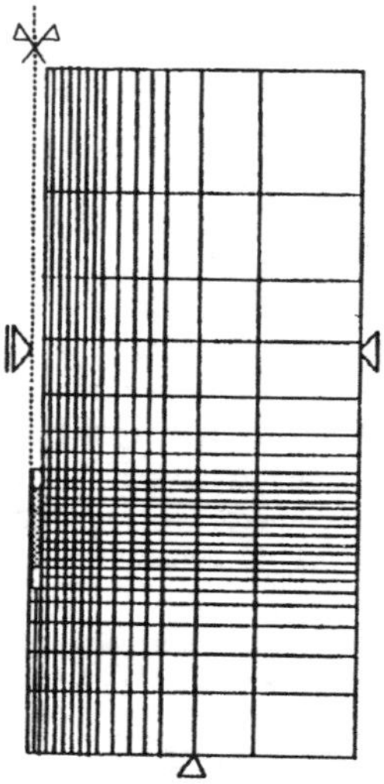

Figure 4. Finite element mesh of pressuremeter tests

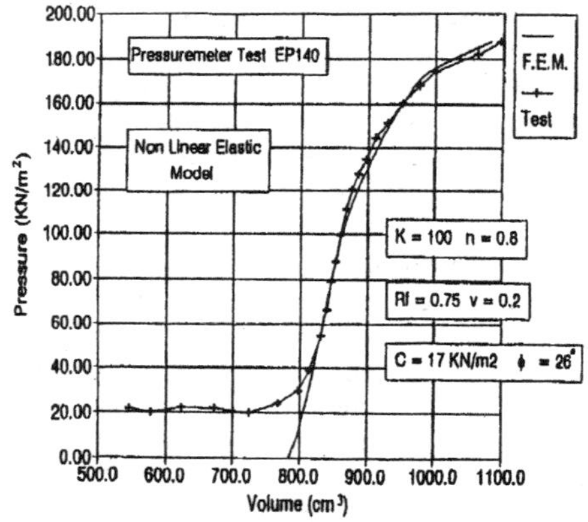

Figure 5. Comparison between measured and simulated pressuremeter response

Table 1. Constitutive parameters calibrated from drained compression triaxial tests and pressuremeter tests

Calibration Test	K	n	K_{ur}	ν	c' *(kPa)*	ϕ'	R_f
Triaxial	42.5	-1.56	825	0.2	17	26^0	0.75
Pressuremeter	100	0.8	825	0.2	17	26^0	0.75

4 APPLICATION TO DESIGN

Several plate loading tests and real scale concrete footings have been carried out at the UFRGS test site [e.g. Schnaid et al (1995)]. Two typical results are presented in Figure 6, which are used in this paper to highlight some aspects of footing design. The Finite Element Method described for pressuremeter interpretation is also used for predicting the load-settlement behaviour of the plate loading tests shown in Figure 6. The finite element mesh of plate loading tests is shown in Figure 7.

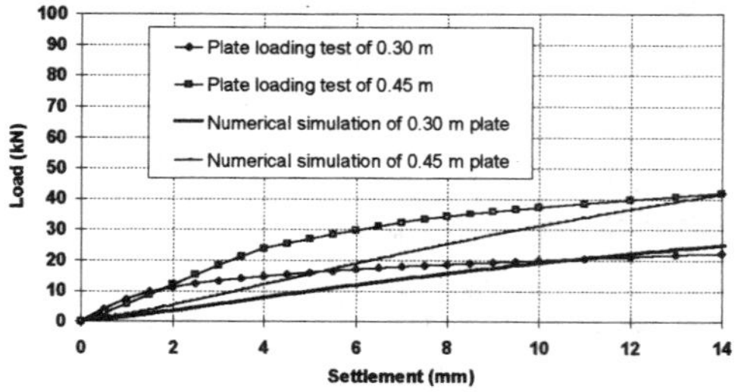

Figure 6. Comparison between measured and simulated load-settlement plate loading test results

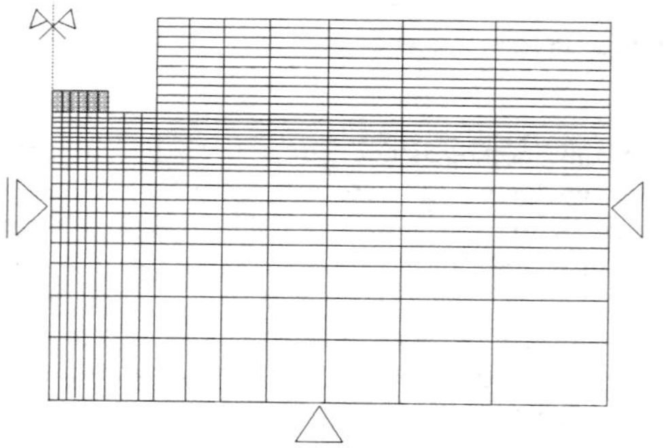

Figure 7. Finite element mesh of plate loading tests

The simulation was made considering the soil as a non-linear elastic material with constitutive parameters assessed from the back-analysis of pressuremeter tests. Comparison between measured load-displacement data and numerical simulations is also shown in Figure 6. Initial settlements were grossly overestimated by the numerical simulation. It is suggested that the different stress path followed by pressuremeter tests and vertically loaded plates partially explains the observed behaviour, as already noticed on field tests carried out in sands [Leonards and Frost (1986)]. Failure conditions, on the other hand, are well represented by the numerical simulation. That is not surprising since failure parameters from triaxial tests fit directly pressuremeter results as shown by the simulation.

A back-analysis of the same plate load tests indicates

that a K=300 would provide a reasonable curve fitting at small shear strain level. The magnitude of the stiffness in this case is compatible to the stiffness measured in the triaxial test before the occurrence of the effects of structure breakdown. However, any improvement in the prediction would imply a more sophisticated approach with the lost of simplicity.

5 CONCLUSIONS

A method for calculating the load-settlement behaviour of shallow foundations bearing on lightly cemented soils has been proposed. The soil is described as non-linear elastic and constitutive parameters are obtained from the back-analysis of pressuremeter data. Numerical simulation underestimate the initial settlement of plate loading tests and failure conditions were well matched. Despite of the fact that this approach does not give an excellent assessment of initial settlements of vertically loaded footings, it is believed that this numerical approach combined with the adopted calibration procedures convey to the most adequate design method that has been proposed to date for non-textbook cohesive-frictional soils, mainly if it is considered that a simple constitutive model has been adopted in the numerical analysis.

ACKNOWLEDGEMENTS

The authors are grateful to the financial support of Brazilian research agency CNPq and technical assistance from CEEE, the power company of the province of Rio Grande do Sul, Brazil.

REFERENCES

ASTM D1194-72 1987. *Standard test method for bearing capacity of soil for static load and spread footings.* American Society for Testing and Materials, Philadelphia, USA.

Bressani, L. A and P.R. Vaughan 1989. Damage to soil during triaxial testing. *Proceedings of the XII International Conference on Soil Mechanics and Foundation Engineering.* A. A. Balkema, Rotterdam, The Netherlands, 1, 17-20.

Britto, A. and M. Gunn 1991. *CRISP90 - User and Programmer's Guide.* Cambridge University; UK.

Clayton, C. R. I. and S.A. Khatrush 1986. A new device for measuring local axial strain on triaxial specimens.. *Geotechnique,* 25, 657-670.

Clayton, C. R. I.; S.A. Khatrush; A.V.D. Bica, and A. Siddique 1989. The use of Hall effect semiconductor in geotechnical instrumentation. *Geotechnical Testing Journal,* 12, 69-76.

Duncan, J.M. and C.Y. Chang 1970. Non-linear analysis of stress - strain in soils. *Journal of the Soil Mechanics and Foundation Engineering Division,* 96 (SM5), 1625-1653.

Fredlund, D.G. and H. Rahardjo 1993. *Soil mechanics for unsaturated soils.* John Wiley & Sons, New York, NY.

Gens, A. and R. Nova 1993. Conceptual bases for a constitutive model for bonded soils and weak rocks. *Geotechnical Engineering of Hard Soils-Soft Rocks,* A. A. Balkema, Rotterdam, The Netherlands, 485-494.

Leonards, G.A. and J.D. Frost 1986. Settlement of shallow foundations on granular soils, *Journal of Geotechnical Engineering, ASCE,* 114 (7), 791-809.

Leroueil, S. and P.R. Vaughan 1990. The general and congruent effects of structure in natural soils and weak rocks. *Geotechnique,* 40 (3), 467-488.

Mantaras, F.M. 1995. *Numerical analysis of pressuremeter tests applied to shallow foundations of unsaturated soils (in Portuguese),* M.Sc. Dissertation, Federal University of Rio Grande do Sul, Brazil.

Mántaras, F. M.; N.C. Consoli and F. Schnaid 1996. Assessment of the properties of unsaturated soils by the pressuremeter test, *Proceedings of X Panamerican Conference of Soil Mechanics and Foundation Engineering,* Mexico, 277-288.

Schnaid, F.; N.C. Consoli; R.O. Cudmani and J. Milititsky 1995. Load-settlement response of shallow foundations in structured unsaturated soils. *Proceedings of the First International Conference on Unsaturated Soils,* A. A. Balkema, Rotterdam, The Netherlands, 999-1004.

Schnaid, F.; G.C. Sills and N.C. Consoli 1996. Pressuremeter tests in unsaturated soils, *Proceedings of Advances In Situ Investigation Practice,* Thomas Telford, London, 586-597.

Numerical Models in Geomechanics, Pietruszczak & Pande (eds) © *1997 Balkema, Rotterdam, ISBN 90 5410 886 X*

Stress and strain fields around a penetrating cone

D. Sheng, K. Axelsson & O. Magnusson
Department of Civil and Mining Engineering, Luleå University of Technology, Sweden

ABSTRACT: In this paper, finite element analyses for cone penetration tests (CPT) is carried out. The soil mass will be treated as an elasto-plastic body and the generalised Cam clay model will be used to simulate the soil behaviour. Interaction between soil-cone and soil-shaft will be accounted for by introducing contact elements along the interface and finite sliding and friction will be allowed along the interface. To study the pore pressure development coupled hydro-mechanical analyses will be performed.

The numerical analyses will be utilized to study the deformation and failure pattern of soils surrounding the cone and the shaft, to study the influences on test results of soil parameters, the lateral stress factor K_o and the penetration manner and speed, and to back-estimate soil properties from CPT results.

1. INTRODUCTION

Soil properties as stiffness, strength parameters and bearing capacity can be estimated by studying the response of the soil to penetrating rigid objects. A typical such method is the cone penetration test (CPT). For correct interpretation of results from such tests as well as for improvement of testing methods, knowledge of penetration effects on the surrounding soil is essential.

Penetration of a rigid object into a soil may cause complex deformation and strain since

- the disturbance is a complex combination of shearing, compression and cutting,
- the soil often behaves as a complex elasto-visco-plastic multi-phase material, and
- excess pore pressure may develop and dissipate with time.

Moreover, the deformation of the soil is strongly influenced by the manner and the speed of penetration and the geometry of the penetrating object. Due to these complexities, the mechanism of cone penetration into soil is poorly understood and, consequently, reliable methods for analysing the deformation-stress pattern of the process are very limited. The existing interpretation methods for CPT can broadly be divided into two classes, namely bearing capacity theories and cavity expansion theories. Very often the results of plane strain bearing capacity calculations are modified empirically for application to the axi-symmetric problems. Some numerical solutions to axi-symmetric problems using the plasticity theory have also been presented (e.g. Cox et al., 1961, Houlsby & Wroth 1983). Adaptation of the bearing capacity theory to deep penetration problems is much more difficult and often involves adoption of boundary conditions which are not appropriate for the cone penetration (Koumoto & Kaku 1982).

Cavity expansion theories have been suggested as being appropriate for the deep penetration problem (Vesic 1972, Torstensson 1977, Collins & Yu 1996). The limit solution for the cylindrical cavity pressure is widely assumed to be applicable to the estimation of the radial stress on the shaft, and the limit pressure for spherical cavity expansion is relevant to the end bearing pressure of the cone. However, it has been pointed out (e.g. Baligh 1986) that one of the inconsistencies with this approach is that it does not model correctly the strain paths followed by soil elements.

The recently developed "strain path method", pioneered by Baligh (1985), is particularly applicable to problems of deeply buried structures where the pattern of deformation can be described by steady state flow compatible with the boundary conditions. For undrained (incompressible) flow this may be conveniently expressed in terms of a stream function, from which the velocities are determined by differentiation. Certain simple problems can be solved using particularly simple stream functions. The strain rates can be determined from the velocities at every point. The strain history of any material point as it moves along a streamline is then available. Given stress boundary conditions in the upstream direction it is now possible to integrate a constitutive (stress-strain) law along each streamline to give the stresses throughout the soil. The resulting stresses will not be in equilibrium unless, by chance, the first guess of the streamlines happened to be correct. It is at this stage that the method requires further development, and various possibilities exist for the next step, Baligh (1985), Houlsby and Withers (1988) and Collins & Yu 1996.

The problem of cone penetration has also in recent years been analysed by the finite element methods (DeBorst & Vermeer 1984, Kiousis et al. 1988, Gupta 1992, Yu 1993). In these analyses the cone has effectively been introduced into a pre-bored hole with the surrounding soil still in its in-situ stress state. An incremental plastic collapse calculation is then carried out and the limit value identified as the indentation pressure. This interpretation is not entirely correct however. During the real penetration of the cone, very high lateral and vertical stresses develop adjacent to the shaft of the penetration cone. These higher soil stresses will influence the conditions around the cone tip, resulting in higher cone penetration pressures than are predicted by the analysis of the cone in a pre-bored hole. In the analysis of the cone penetration, a careful distinction is necessary between a plastic collapse solution and a steady state penetration pressure.

The current analysis methods all suffer certain limits (Sandven 1990). The traditional methods i.e. bearing capacity theories and cavity theories can only approximately yield the limit pressures of cone penetration with no correct account of the pattern of displacement development, whereas the strain path method does not result in equilibrium stresses and the current finite element methods only simulate the indentation effect of a cone in a pre-bored hole, but not the real penetration process. A real penetration process involves cutting and pushing of the soil near the cone tip and frictional sliding of soil particles along the shaft. Excess pore pressure may develop in the surrounding soil. Therefore a complete analysis of the CPT process should take into consideration the governing equations like equilibrium and continuity equations, a well-defined constitutive model for the soil behaviour and a realistic description of the interaction between the soil and the penetrometer.

2. THE SOIL

A realistic analysis of the penetration problem should include a realistic soil constitutive model accounting for the most important features of soil behaviour. During the last years the Cam clay model has been accepted to be a simple and representative model for cohesive soils. The model is described by a yield function dependent on stress invariants, an associated flow assumption to define the plastic strain rate, a critical state where unrestricted, purely deviatoric, plastic flow of the soil skeleton occurs under constant effective stress, and a strain hardening theory that determines the size of the yield surface according to the plastic volumetric strain. Detailed description of the model can be found elsewhere e.g. Muir Wood (1990).

Since a soil is a multi-phase medium, deformation of the soil is often accompanied by development of excess pressure of pore liquid. The pore pressure change is governed by the continuity equation which is coupled to the equilibrium equation through the rate of pore volume change and the effective stress theory. This is also the theoretical base of soil consolidation. The continuity and equilibrium equations have been given in references e.g. Small et al. (1976) and Zienkiewicz & Taylor (1989).

3. NUMERICAL SIMULATION

3.1 Soil-Cone Interaction

As mentioned above, penetration of a cone into a soil involves cutting of the soil at the cone tip and pushing of the soil due to the cone injection as well as frictional sliding at the soil-cone and soil-shaft interface. The problem would be readily represented by an axi-symmetric model (Fig. 1), if it was not due to the cutting. In the case of a sharp cone ($\leq 60^{\circ}$), the soil particles in the vicinity of the very tip of the cone are reorientated due to the very high stresses in the area, so that the very tip of the cone is more likely hanging in a pore space than residing on a solid particle. The soil is "cut" through separating soil particles at the cone tip. If the soil has no tensile strength, which is the case for most of soils, the resistance to the penetrating cone is only due to the surface friction T and the vertical component of the normal force N, Fig. 1. Therefore the real penetration problem can be closely simulated by the axi-symmetric model.

To effectively simulate the soil-cone interaction, we treat the cone and the shaft as a rigid body and the soil as a deformable body whose constitutive relationships are described by the Cam clay model, see above. Points on the surface of the rigid body may penetrate into the deformable surface. Sliding as well as separation of finite amplitude may arise at the contact interface. Special contact elements are generated at both the deformable surface and the rigid surface. At each integration point these elements construct a measure of overclosure (penetration) and measures of relative shear sliding. These kinematic measures are then used to introduce surface interaction theories. If A is a point on the deformable surface, with current coordinates $\boldsymbol{x}_A$, Fig. 2, the closest distance h from A to the rigid surface is determined according to

$$\boldsymbol{n}h = \boldsymbol{x}_A - \boldsymbol{x}_C - \boldsymbol{r} \quad (1)$$

where $\boldsymbol{n}$ is the unit normal of the rigid surface at point

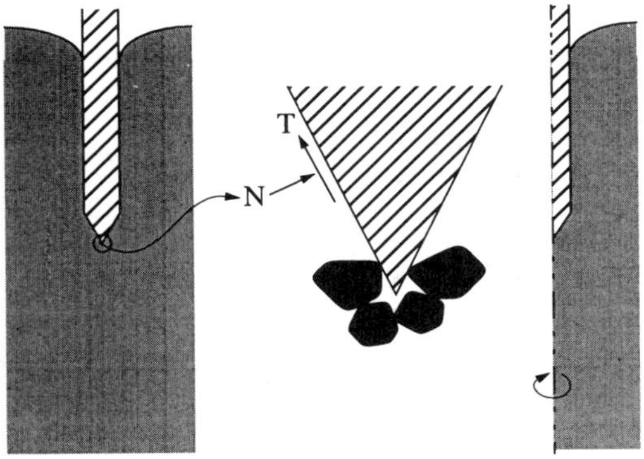

Fig. 1 Representation of cone penetration and soil-cone interaction.

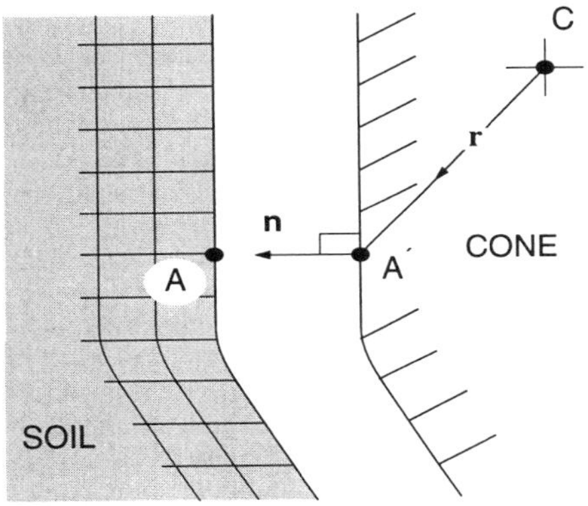

Fig. 2 Soil-cone contact simulation.

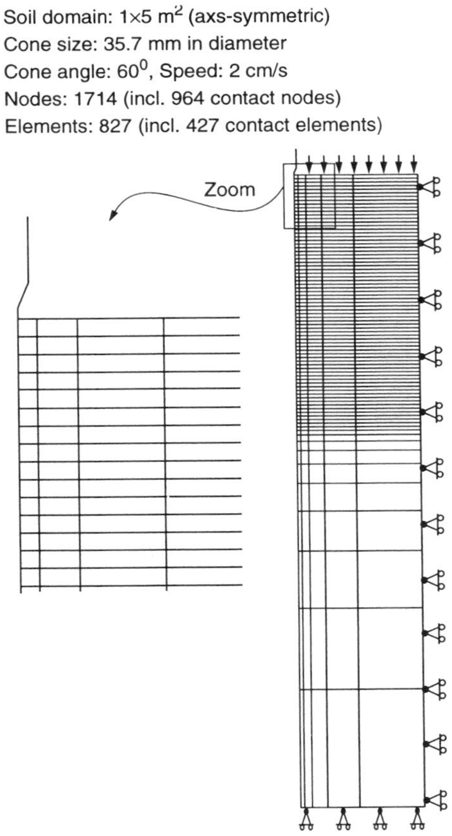

Fig. 3 The finite element model.

A' which is the closest point to A, $\boldsymbol{x}_C$ the coordinates of the reference node C of the rigid body, and $\boldsymbol{r}$ the vector from C to A'. If $h>0$, there is no contact between the surface at A, and no further surface interaction calculations need be done at this point. If $h \leq 0$, the surfaces are in contact and any positive pressure can be transmitted between A and A'. To implement the contact constraint the first and second variations of h are needed and they are given e.g in ABAQUS Manual (1994).

3.2 Finite element model

An axi-symmetric soil domain, 1 m in radius and 5 m in depth, is discretized into 400 8-node biquadratic elements with pore pressure as the third degree freedom at corner nodes, Fig. 3. At the potential contact interface, additional 427 contact elements are generated. The mesh is designed so that the elements near the contact surface are in a similar size as the cone tip, see the zoomed mesh in Fig. 3. At the bottom boundary, no vertical displacement is allowed. The top boundary is drained and subjected to a distributed load of 10 kPa, which corresponds to an in-situ depth of about half a meter. The geostatic stress state is used as the initial condition.

The soil properties are as follows, the dry unit weight $\gamma_d=14$ kN/m^3, the initial void ratio $e=1.0$, the slope of the critical state line $M=1.2$ in a p'-q diagram (corresponding to a friction angle $\phi'=30^o$), the slope of the normal compression line in a lnp'-e diagram $\lambda=0.3$, the slope of the unloading and reloading line in a lnp'-e diagram $\kappa=0.03$, the Poisson ratio $\mu=0.3$, the preconsolidation pressure $p_o=200$ kPa, and the permeability $k=10^{-9}$ m/s. The diameter of the cone shaft is 35.7 mm and the cone angle is 60°. Two values of the friction angle between the cone shaft and the soil are tested. One is 0 and the other 10°. The speed of cone penetration is 2 cm/second.

4. RESULTS AND DISCUSSION

4.1 Displacement, stresses and strains

A deformed finite element mesh at an early time is

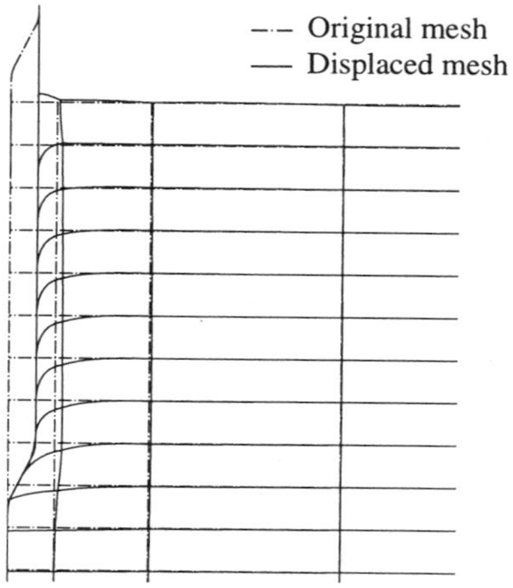

Fig. 4 Deformed finite element mesh, with interface friction θ=10°.

shown in Fig. 4. The general deformation pattern of the soil around the cone shaft is characterized by horizontal compression and vertical tension, see also Fig. 7. Indentation effect is not pronounced but can be noticed at the tip of the cone. Around the shaft a layer of soil of a thickness of about the shaft radius undergoes plastic deformation (results not shown in the paper).

The stress situation in the soil is strongly influenced by the cone penetration. Around the cone tip and the cone shaft the effective radial (horizontal) stress increases and the maximum increase occurs in the vicinity of the upper end of the cone tip, Figs. 5a, 6a. The stress increment increases as the penetration increases and as the interface friction increases (result not shown here). Just beneath the cone tip the stress may undergo a decrease and tensile (positive) horizontal stress may appear, Figs. 6a.

The maximum increase in the effective vertical stress occurs beneath the cone tip. In Figs. 5b, 6b, we can clearly see stress bulbs developed beneath the cone tip. The stress bulbs expand as the penetration increases. The interface friction also causes a higher increase in vertical stress (not shown here).

The maximum shear stress occurs in the vicinity of the upper end of the cone tip, regardless of the interface friction, Figs. 5c, 6c. Close to the maximum shear stress, an upper active (negative) shear zone and an lower passive (positive) shear zone develop. These two shear zones expand as the penetration increases and as the interface friction increases.

4.2 Excess pore pressure

Development of excess pore pressure is one of most important factors needed to be considered in cone penetration tests. In Fig. 8, we note that both positive and negative excess pore pressure develop. Around the cone tip a zone of high positive excess pore pressure is created. In a small zone above the upper end of the cone tip, where the overconsolidated soil experiences extensive shearing, negative excess pore pressure develops. Comparing Fig 8b and Fig. 8c, we note that the interface friction causes a higher negative excess pore pressure (-380 kPa when θ=10° compared to -193 kPa when θ=0) and a lower positive pore pressure (558 kPa when θ=10° compared to 596 kPa when θ=0). The maximum values of positive and negative excess pore pressure primarily depends on soil permeability, penetration speed and deformation of the soil.

The occurrence of both positive and negative excess pore pressure is of significance for determining the filter's position in a penetrometer. In most penetrometers the filter is placed just above the upper end of the cone tip, where, according to this analysis, the excess pore pressure varies dramatically from positive to negative. To measure the maximum positive excess pore pressure, where is of more interest e.g. for estimation of the cone resistance, the filter should be placed at the middle height of the cone tip.

4.3 Total resistance

The total resistance, which is the sum of the sleeve resistance (friction) and the cone resistance, is equivalent to the vertical force acting on the rigid cone in unit area of the cross section of the cone shaft. Plotting this total resistance against the penetration depth gives Fig. 9. It can be noted that the total resistance q_t for the case without interface friction increases somewhat as penetration increases. In the case with interface friction, the sleeve friction also increases as the penetration increases, due to the increase of the contact pressure (horizontal stress). Therefore the increase in q_t with increasing depth is more pronounced in this case. The cyclic changes of the total resistance have not yet been fully understood by the authors.

4.4 Sensitivity analysis

Influences of various parameters on the total resistance can be studied through a sensitivity analysis. In such an analysis, a parameter to be studied is changed in a range of interest while all the other parameters are kept constant. The basic values of parameters used in the sensitivity analyses are kept the same as in section 3.2, except the overburden pressure at the top boundary is changed from 10 to 100 kPa, which corresponds to a depth of about 5 meter. A finer element mesh of a smaller soil volume, 0.5 m in radius and 1 m in depth, is used in order to save CPU time. The maximum penetration depth is 20 cm. The total resistance q_t is determined from the vertical pressure acting on the rigid cone when it penetrates 20 cm into the soil.

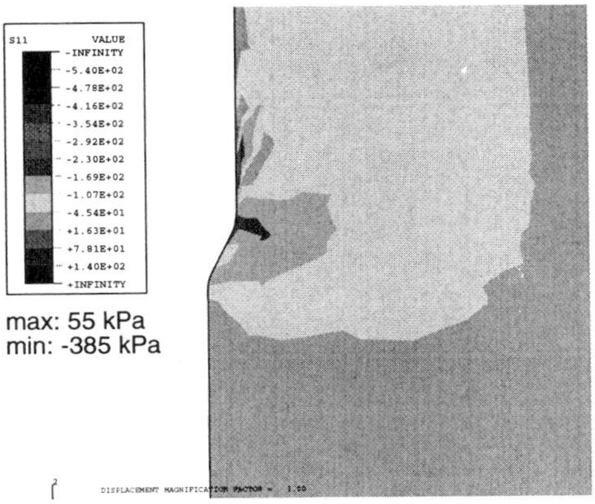

a) σ_{11}': Effective radial stress, kPa

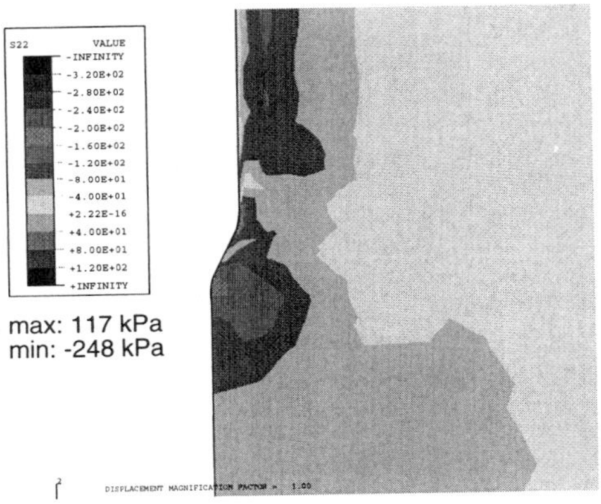

b) σ_{22}': Effective vertical stress, kPa

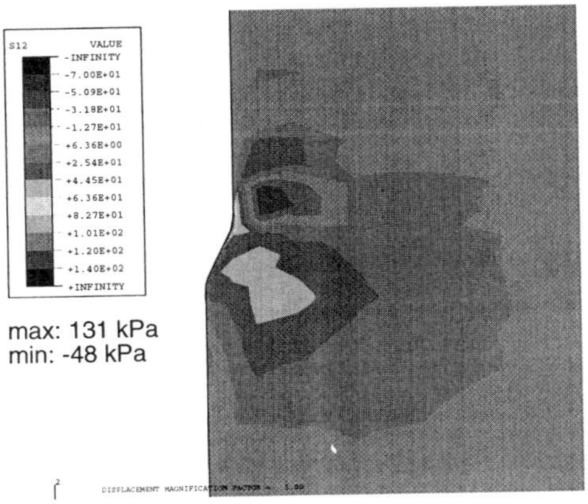

c) σ_{12}: Shear stress, kPa

Fig. 5 Effective stresses near the cone tip at time 100 seconds, penetration depth: 2 m, interface friction angle $\theta=10^{o}$.

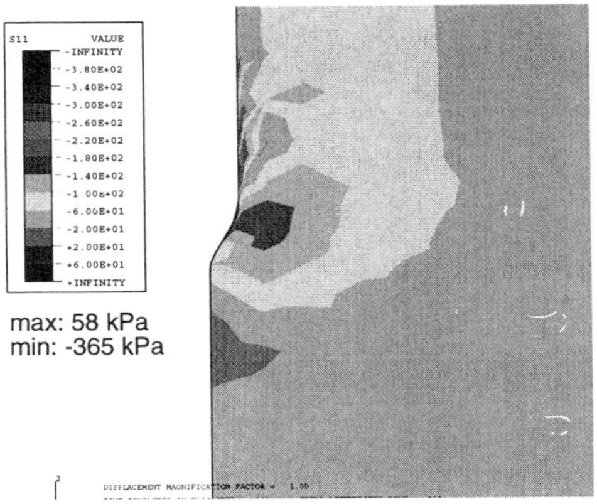

a) σ_{11}': Effective radial stress, kPa

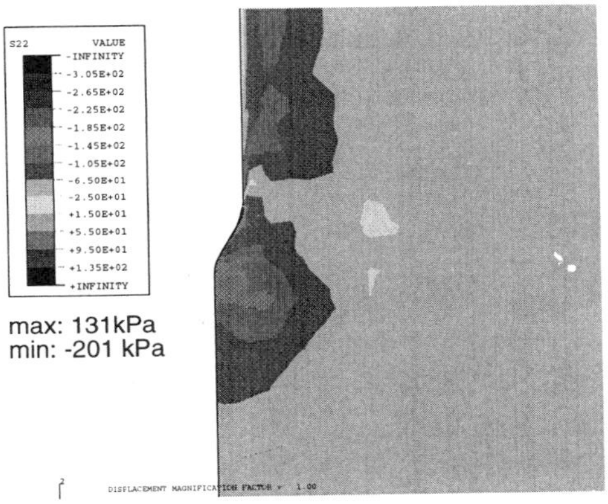

b) σ_{22}': Effective vertical stress, kPa

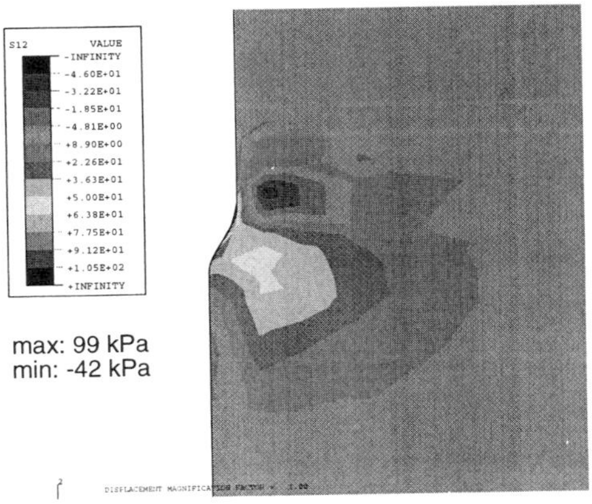

c) σ_{12}: Shear stress, kPa

Fig. 6 Effective stresses near the cone tip at time 100 seconds, penetration depth: 2 m, interface friction angle $\theta=0$

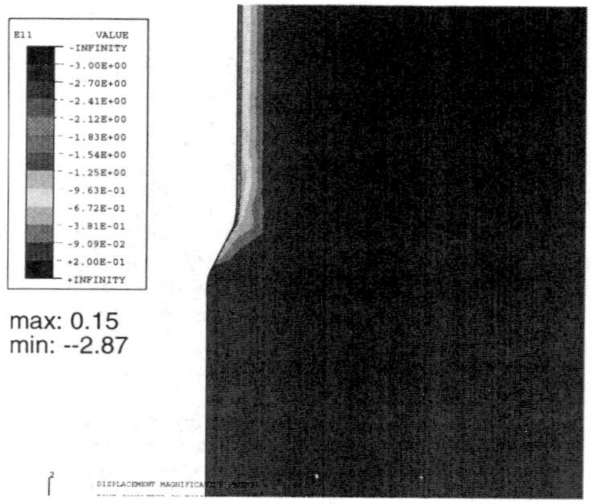

a) ε_{11}: Radial strain

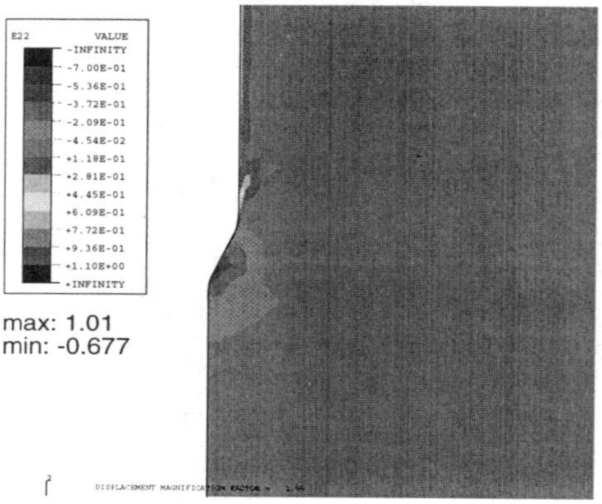

b) ε_{22}: Vertical strain

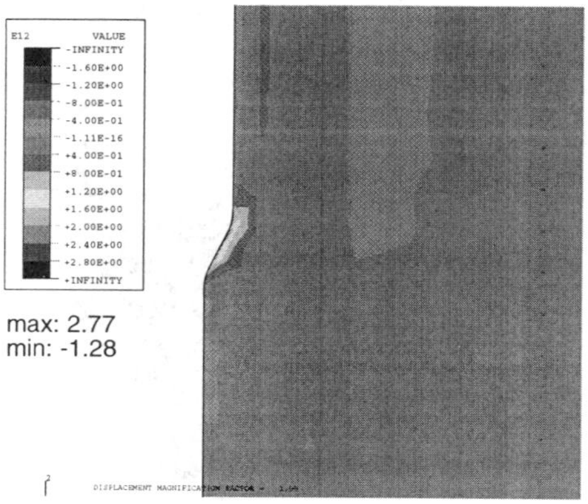

c) ε_{12}: Shear strain

Fig. 7 Strains near the cone tip at time 100 seconds, penetration depth: 2 m, interface friction angle $\theta=10^{o}$.

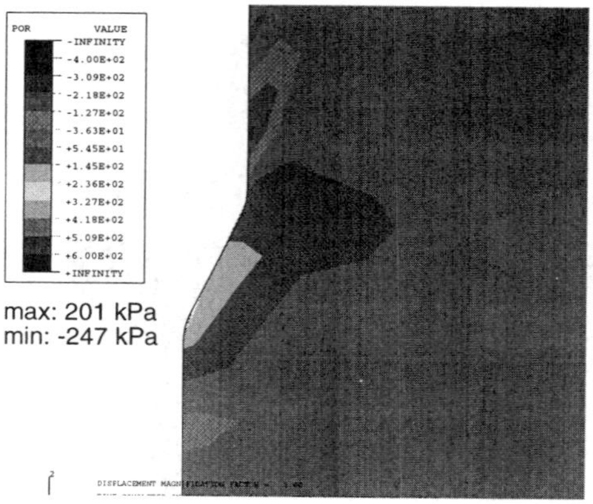

a) Time 54 seconds (1.08 m), $\theta=10^{o}$

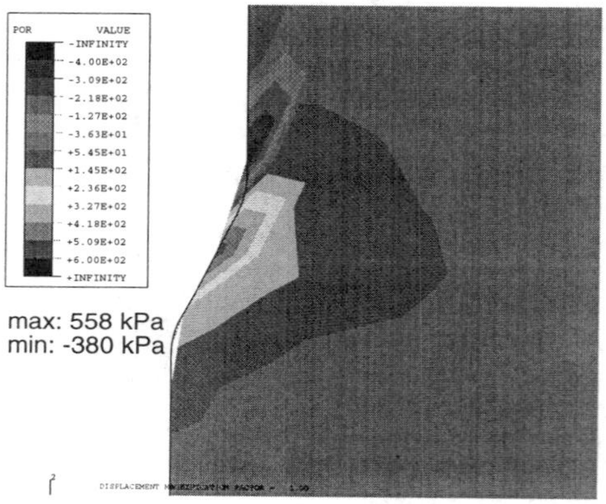

b) Time 100 seconds (2.0 m), $\theta=10^{o}$

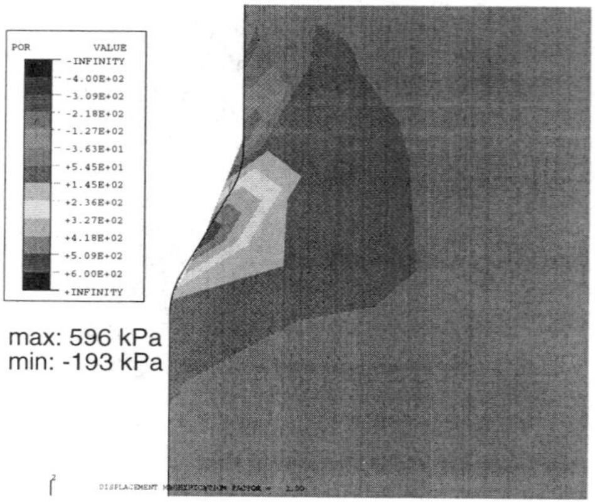

c) Time 100 seconds (2.0 m), $\theta=0^{o}$

Fig. 8 Excess pore pressure (POR, in kPa) near the cone tip, with or without interface friction.

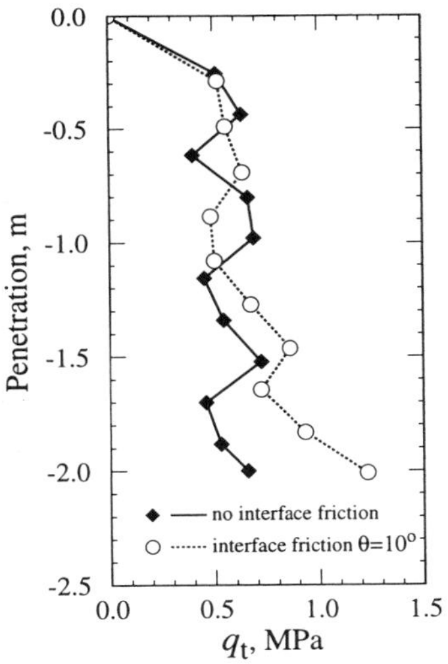

Fig. 9 Total resistance versus penetration depth.

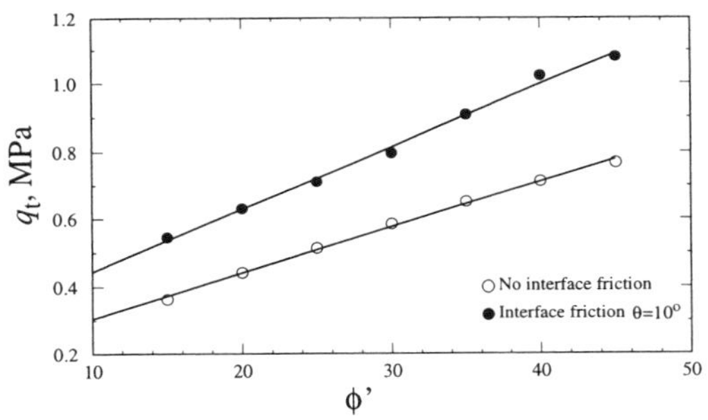

Fig. 10 Relationship between the total resistance and the soil friction angle ϕ'.

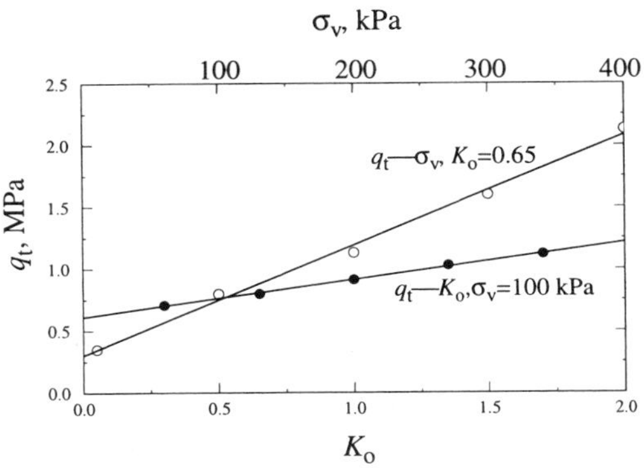

Fig. 11 Total resistance versus the stress coefficient K_o and the vertical stress σ_{22} respectively.

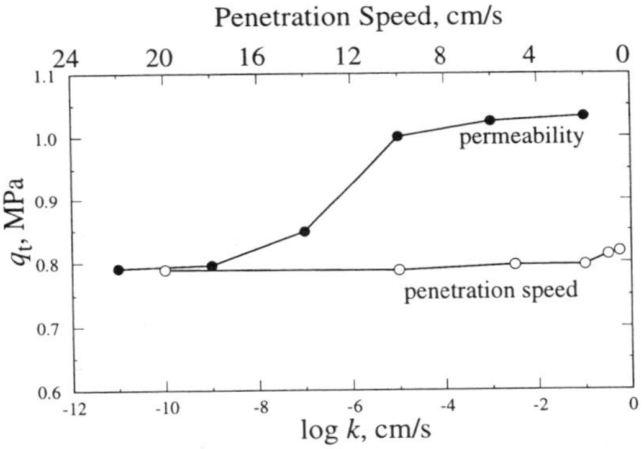

Fig. 12 Total resistance versus the permeability and penetration speed.

The parameters studied include the internal friction angle ϕ' of the soil, the stress coefficient K_o, the total vertical stress, the soil permeability and the penetration speed. According to Figs. 10, 11, the total resistance q_t increases linearly with increasing ϕ', increasing K_o and increasing vertical stress σ_{22}. The influences of the soil permeability and the penetration speed are however not pronounced, Fig. 12. Influences of other soil parameters like shear modulus and compressibility may be of importance but have not yet been studied.

5. CONCLUSIONS

Results from the coupled finite element analysis carried out in this paper have shown that

- around the cone and the shaft a layer of soil of thickness equivalent to the radius of the shaft undergoes plastic deformation
- the maximum increases in the effective radial stress and the shear stress occur in the vicinity of the upper end of the cone tip, and the maximum increase in the effective vertical stress occurs beneath the cone tip.
- the excess pore pressure can be positive and negative, depending on the position along the cone. Around the cone tip a zone of high positive excess pore pressure develops and in a small zone above the upper end of the cone tip negative excess pore pressure occurs.
- the total resistance increases linearly with increasing friction angle ϕ' of the soil, increasing stress coefficient K_o and increasing vertical stress.

The studies presented in this paper are preliminary. In future, more comprehensive studies for different soil types, stratified soil profiles, different cone geometry, and even different soil constitutive models should be carried out. Such numerical studies should also be compared with in situ CPT results. In addition, systematic sensitivity analyses on various soil parameters, which can lead to the possibility of back-estimation of such parameters, should be carried out.

6. REFERENCES

Allersma, H.G.B. (1982). Photoelastic investigation of the stress distribution during penetration, *Proc. 2nd Europ. Symp. on Penetration Testing*, Balkema, Rotterdam, 2:411-418.

Baligh, M.M. (1985). Strain path method. *J. Geotech. Engng*, ASCE, 111:1109-1135.

Baligh, M.M. (1986). Undrained deep penetration, I. shear stresses.*Geotechnique*, 36, 471-485.

Bergdahl, U. & U. Eriksson (1983). *Bestämning av jordegenskaper med sondering - en litteraturstudie*, SGI Rapport 22, Linköping.

Collins, I.F. & H.S. Yu (1996). Undrained cavity expansions in critical state soils, *Int. J. Num. Anal. Methods in Geomech.*, 20:489-516.

Cox, A.D., G. Eason & H.G Hopkings (1961). Axially symmetric plastic deformation in soils. *Trans Royal Soc.*, London, Ser. A, 254-210.

De Borst, R. & P.A. Vermeer (1982). Finite element analysis of static penetration tests, *Proc. 2nd Europ. Symp. on Penetration Testing*, Balkema, Rotterdam, 2:457-462.

De Borst, R. & P.A. Vermeer(1984). Possibilities and limitations of finite element for limit analysis, *Geotechnique*, 34(2):199-210.

Gupta, R.C. (1992). Finite element analysis for deep cone penetration, *J. Geotech. Engng*, ASCE, 177(10):11610-1630.

Houlsby, G.T. & N.J. Withers (1988). Analysis of the cone pressuremeter test in clay, *Geotechnique*, 38(4):575-587.

Houlsby, G.T. & C.P. Wroth (1983). Calculation of stresses on shallow penetrometers and footings, *Proc. IUTAM/IUGG Symp. on Seabed Mechanics*, Newcastle, 107-112.

Kiousis, P.D., G.Z. Voyiadjis & M.T. Tumay (1988). A large strain theory and its application in the analysis of static penetration mechanism. *Int. J. Num. Anal. Methods in Geomech.*, 12:45-60.

Koumoto, T. & K. Kaku (1982). Three-dimensional analysis of static cone penetration into clay, *Proc. 2nd Europ. Symp. on Penetration Testing*, Balkema, Rotterdam, 2:457-462.

Larsson, R. (1993). *CPT-sondering, Utrust-ning - utförande - utvärdering*, SGI Information 15, Linköping.

Magnusson, O., Y. Yu & K. Axelsson (1994). *Sättningsuppfölning av 5-våningshus grundlagda med platta på sulfidjord*, Research Report TULEA 1994:18, Luleå University of Technology.

Meigh, A.C. (1987). *Cone penetration testing*, Butterworths, London.

Muir Wood, D. (1990). *Soil behaviour and critical state soil mechanics*, Cambridge.

Sandven, R. (1990). *Strength and deformation properties of fine-grained soils obtained from piezocone tests*. Doctoral Thesis, Norwegian Technical Institute, Tronheim.

Small, J.C., J.R. Booker & E.H. Davis (1976). Elasto-plastic consolidation of soil, *Int. J. Solids and Struct.*, 12:431-448

Torstensson, B.A. (1977). The pore pressure probe. *Proc. NGM*, Norwegian Geotechnical Society, Oslo. Paper 34.

Vesic, A.S. (1972). Expansion of cavities in infinite soil mass *J. Soil Mech. Found. Eng. Div.*, ASCE, 98:265-290.

Yu, Y. (1993). *Three-dimensional finite element analysis of pile-group foundations,* Research Report TULEA 1993:42, Luleå University of Technology.

Zienkiewicz, O.C. & R.L. Taylor (1991). *The finite element method*, Vol. 2, McGraw-Hill.

Numerical Models in Geomechanics, Pietruszczak & Pande (eds) © 1997 Balkema, Rotterdam, ISBN 90 5410 886 X

Finite difference simulation of cylindrical cavity expansion

I.C.Pyrah
Napier University, Edinburgh, UK

W.F.Anderson
University of Sheffield, UK

R.Hashim
University of Malaya, Malaysia

ABSTRACT: The paper describes a finite difference technique for modelling cylindrical cavity expansion. Modified Cam-clay is used to model the stress-strain behaviour of the soil and the Singh-Mitchell deviatoric creep model is used to model the time-dependent behaviour of the soil. The finite difference technique has been used to simulate different types of pressuremeter test and the method was found to be efficient both in terms of development time and computing resources. The analyses can model either true idealised undrained behaviour or fully-coupled (Biot) consolidation for stress-controlled and strain-controlled loading. The results have been particularly useful as an independent check on solutions obtained using another method.

1 INTRODUCTION

In its simplest form, the pressuremeter test can be modelled as the expansion of an infinitely long cylindrical cavity in an infinite medium. The material is assumed to be elastic or elastic-perfectly plastic and numerous analytical and semi-analytical solutions are available. Solutions are usually based on small deformations although large strain solutions are available and recent investigations have provided semi-analytical solutions for cavities expanding from zero initial radius in materials exhibiting more realistic behaviour such as critical state soil models. Such solutions, however, neglect time effects and assume that the soil is a single phase material. For problems involving the prediction of soil behaviour both during and after expansion, e.g. in a holding test, the soil must be modelled as a two-phase material and use of a suitable numerical technique is necessary. Numerical techniques allow realistic effective stress soil models to be used and time-dependent behaviour such as consolidation and creep to be examined (Pyrah et al 1985).

The finite element method is the most commonly used numerical technique and previous work by the authors has included the simulation of pressuremeter tests using this method. The research has also involved laboratory-scale hollow-cylinder tests and full-scale pressuremeter tests carried out in a calibration chamber (Pyrah & Anderson 1990). As the tests were performed in a normally consolidated clay, the modified Cam-clay model was used to represent the soil behaviour in the analyses. One of the main objectives was to examine the effect of rate of testing on the test results and it was important that consolidation and creep of the soil skeleton were included in the analysis. A conventional (Biot) two phase representation of the soil enabled consolidation to be modelled whilst undrained creep was modelled using the approach suggested by Singh & Mitchell (1968).

The FE simulations of the "undrained" hollow-cylinder tests included an examination of the effects of consolidation, an examination of the effects of creep and an examination of the combined effects of consolidation and creep for tests carried out at different rates of strain. All simulated pressuremeter curves were interpreted as though they were undrained tests and the stress-strain curves for the soil derived using the method proposed by Palmer (1972). The irregular nature of the stress-strain curves derived from the simulated tests including both consolidation and creep were in marked contrast to the smooth curves obtained when consolidation and creep were considered separately (Pyrah et al 1985). The discrepancies between the test data and the results of the simulations, together with the irregularities in the derived stress-strain curves, caused some concern because although similar patterns of behaviour were obtained the agreement was not entirely convincing. These discrepancies could have been due to the soil models not being truly representative of the real soil behaviour, numerical approximations introduced in the load/time descretization or errors in the implementation of the models in the FE code. Because of these uncertainties it was decided to develop an alternative method to provide solutions which would act as an independent check on the FE solutions. The technique adopted, and described below, is a form of finite differences (FD) and software was developed to solve the problem previously modelled using finite elements.

2 CYLINDRICAL CAVITY EXPANSION

The problem is idealised as the plane strain expansion of a cylindrical cavity in a fully saturated soil and entails the solution of two partial differential equations representing radial equilibrium and solid-fluid continuity. The basic assumptions involved in the analysis are:

i. the principle of effective stress is applicable throughout the soil body at all times,
ii. the soil is saturated with an incompressible fluid which flows through the soil according to Darcy's law,
iii. radial, plane strain conditions exist, and
iv. the stress-strain relationship that governs the soil behaviour may be written as a relationship between the rates of change (or increments) of effective stress and strain, i.e.

$$d\sigma' = D d\varepsilon \tag{1}$$

where D is a soil properties matrix which depends on the stress history and current stress state of the soil.

2.1 *Governing Equations*

In the idealised pressuremeter test the internal pressure is raised from an initial pressure equivalent to the 'at rest' earth pressure (p_o) to p while the external pressure remains constant at p_o. Initially, in its 'at rest' state, the cylinder has an internal diameter of $2a_o$ and an external diameter of $2b_o$. Cylindrical polar co-ordinates (r, θ, z) are adopted and the radial displacement of any point is denoted by x. The stresses acting on a typical element of the soil, σ_r, σ_θ and σ_z are principal stresses; initially, $\sigma_r = \sigma_\theta = \sigma_z = p_o$, but as cavity expansion occurs, the radial stress increases to become the major principal stress and the circumferential stress decreases to become the minor principal stress while the vertical stress is the intermediate principal stress. Due to axial symmetry and plane-strain conditions the problem reduces to a one-dimensional problem in polar co-ordinates.

Employing the principle of effective stress, the radial equilibrium equation, neglecting any body force, is given by:

$$\frac{\partial \sigma_r'}{\partial r} + \frac{\partial u}{\partial r} + \frac{\sigma_r' - \sigma_\theta'}{r} = 0 \tag{2}$$

where

σ_r' = effective radial stress
σ_θ' = effective circumferential stress
u = pore pressure

For a two-phase material continuity must exist between the solid and fluid phases and, for a soil saturated with an incompressible fluid, the volumetric strain of the soil skeleton must be equal to the outflow of the fluid from the pores. Assuming Darcy's law to be valid, this condition is expressed mathematically as

$$\frac{k}{\gamma_w}\left\{\frac{\partial^2 u}{\partial r^2} - \frac{\partial u}{\partial r}\frac{1}{r}\right\} = -\frac{\partial \varepsilon_v}{\partial t} \tag{3}$$

where

k = coefficient of permeability
ε_v = volumetric strain (= $\varepsilon_r + \varepsilon_\theta + \varepsilon_z$)
γ_w = unit weight of the fluid

The stress-strain relationship is in incremental form and it is necessary to define the incremental strains in terms of incremental displacements. Considering a small change in the radial displacement, Δx, the in-plane strain increments are:

$$d\varepsilon_r = -\frac{\partial (\Delta x)}{\partial r} \tag{4}$$

$$d\varepsilon_\theta = -\frac{(\Delta x)}{r} \tag{5}$$

At the beginning of each load increment the displacements are known and hence so too are the total strains. After the application of the incremental load a new total strain is produced which can be split into two components, namely, the strain prior to application of the incremental load and the incremental strain caused by the change in load i.e.

$$\varepsilon_i = \varepsilon_o + d\varepsilon \qquad (6)$$

where:

ε_i = the total strain

ε_o = the strain prior to the application of the incremental load

$d\varepsilon$ = the incremental strain

The changes in the stresses are calculated directly using Equation (1) and the updated stresses and displacements at the end of an increment are found by adding the incremental changes to the values at the beginning of the increment.

3 SOIL BEHAVIOUR

3.1 *Effective Stress-Strain Behaviour*

The stress-strain relationship used in this work is that of modified Cam-clay model (Roscoe & Burland 1968). The model is a strain hardening, elastoplastic model based on critical state concepts and requires the specification of five parameters:

λ = the slope of the virgin consolidation line in e-ln p' space,

κ = the mean slope of the expansion-recompression line in e-ln p' space,

e_{cs} = the value of e at unit p' on the critical state line in e-ln p' space,

M = the value of the stress ratio q/p' at the critical state condition, and

G = the elastic shear modulus.

The symbol e represents the current void ratio of the soil and the quantities p' and q are given by:

$$p' = (\sigma_r' + \sigma_\theta' + \sigma_z')/3 \qquad (7)$$

$$q = \sqrt{(\sigma_r' - \sigma_\theta')^2 + (\sigma_\theta' - \sigma_z')^2 + (\sigma_z' - \sigma_r')^2} \qquad (8)$$

In addition, the description of the material behaviour requires the specification of the coefficient of permeability, k, the unit weight of the pore water, γ_w, and values of the in situ stresses.

Within the yield surface the deformations are determined by the elastic bulk modulus, K, and the shear modulus, G, where:

$$K = (1+e)p'/\kappa \qquad (9)$$

and G is held constant. The criterion for yielding is:

$$q^2 - M^2\left[p'(p_c' - p')\right] = 0 \qquad (10)$$

p_c' is defined by the intersection of the current ellipsoidal yield locus and the p' axis in principal effective stress space.

The consolidation analysis employed is of the Biot type and calculation of the pore pressure dissipation is achieved using an incremental, time marching process.

3.2 *Creep behaviour*

In general, creep strains may conveniently be considered as volumetric and deviatoric creep but as the duration of a pressuremeter test is relatively short the volumetric creep strain is likely to be small and is ignored in the analysis. For the Singh and Mitchell (1968) model the deviatoric creep strain rate, assuming the reference time is unity, is defined as:

$$\dot{\varepsilon}_c = A\exp(\alpha D)[1/t]^m \qquad (11)$$

where:

A, α and m = Singh-Mitchell creep parameters

D = the deviatoric stress level

t = the creep age

Equation (11), which relates the axial creep strain to the soil parameters and the deviator stress, is strictly only valid for triaxial loading conditions in which the axial and radial stresses are kept constant. A more general form, which allows the individual components of creep strain to be evaluated, has been proposed by Chang et. al. (1974). It is assumed that the volume change due to creep is zero and the principal shear strain rates are proportional to the corresponding shear stresses. For the cavity expansion problem, as ε_r, ε_θ and ε_z are principal strains, the rates of creep strain in the r, θ and z directions can be defined as:

$$\dot{\varepsilon}_{cr} = (3\dot{\varepsilon}_c/2q)(\sigma_r' - p') \qquad (12)$$

$$\dot{\varepsilon}_{c\theta} = (3\dot{\varepsilon}_c/2q)(\sigma_\theta' - p') \qquad (13)$$

$$\dot{\varepsilon}_{cz} = (3\dot{\varepsilon}_c/2q)(\sigma_z' - p') \qquad (14)$$

Assuming plane strain conditions for creep, $\dot{\varepsilon}_{cz} = 0$, and for no volume change, $\dot{\varepsilon}_{cr} = -\dot{\varepsilon}_{c\theta}$, thus

$$(\sigma_r' - p') = -(\sigma_\theta' - p') \quad (15)$$

and

$$p' = (\sigma_r' + \sigma_\theta')/2 \quad (16)$$

Substituting for p'

$$\dot{\varepsilon}_{cr} = (3\dot{\varepsilon}_c / 4q)(\sigma_r' - \sigma_\theta') \quad (17)$$

$$\dot{\varepsilon}_{c\theta} = (3\dot{\varepsilon}_c / 4q)(\sigma_\theta' - \sigma_r') \quad (18)$$

where $\dot{\varepsilon}_c$ is defined by Equation. (11).

The creep strains occurring in a particular time increment (Δt) are obtained by multiplying the above creep strain rates by (Δt). These creep strains are then treated in the program as initial strains as described by Zienkiewicz (1977).

4 FINITE DIFFERENCE EQUATIONS

The technique is based on dynamic relaxation (Otter et al 1966) using interlaced grids which are particularly suitable for problems involving two dependent variables. As shown in Fig. 1, the stresses and the displacements are defined at separate grid points with the strains defined at the same points as the stresses.

With this grid system the radial equilibrium condition is satisfied at the displacement points and the compatibility condition is satisfied at the stress points. There are two fictitious stress points which lie outside the soil region and these are used to establish the correct boundary conditions in the numerical modelling.

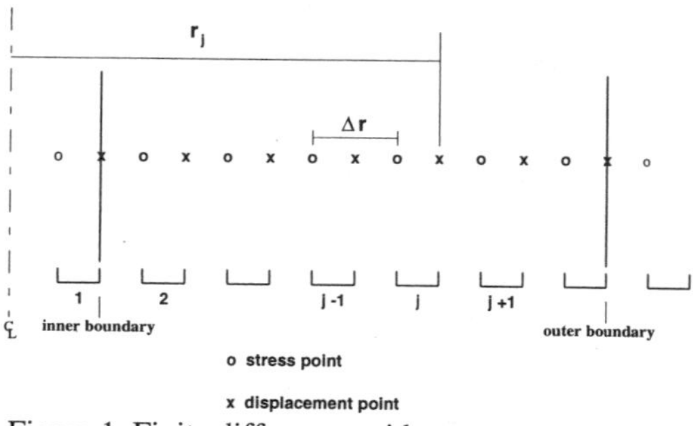

Figure 1 Finite difference grid

The governing equations outlined previously are solved using FD approximations. The radial equilibrium, Equation 2, expressed in FD form at point (j) is

$$\frac{\sigma_r'(j+1) - \sigma_r'(j) + u(j+1) - u(j)}{\Delta r} + \frac{\sigma_r'(j+1) + \sigma_r'(j) - \sigma_\theta'(j+1) - \sigma_\theta'(j)}{2r(j)} = 0 \quad (19)$$

The solid-fluid continuity equation (Equation 3) multiplied by Δt (to avoid dividing by zero when $\Delta t = 0$) and expressed in terms of finite differences for normal stress point (j) is:

$$\frac{k\Delta t}{2\gamma_w}\left[\frac{u(j+1) - 2u(j) + u(j-1) + u_o(j+1) - 2u_o(j) + u_o(j-1)}{(\Delta r)^2} - \frac{u(j+1) - u(j-1) + u_o(j+1) - u_o(j-1)}{2r(j)\Delta r}\right] + \Delta\varepsilon_v(j) = 0 \quad (20)$$

where u_o is the pore pressure at time t and u is the pore pressure at time t + Δt.

The pore pressures are assumed to vary linearly with time during each time increment. The overall problem consists of a series of such equations which have to be solved for all values of u(j) and x(j) so that the radial equilibrium and the continuity equations (Equations 19 & 20) are satisfied. The stresses are the summation of successive incremental stresses as defined by Equation 1. The total strains for normal stress point (j) are,

$$\varepsilon_r(j) = -\frac{x(j) - x(j-1)}{\Delta r} \quad (21)$$

$$\varepsilon_\theta(j) = -\frac{x(j) + x(j-1)}{2r(j)} \quad (22)$$

The strains used in the stress-strain equations are the total strains minus the strains prior to the application of the load increment Equation 6.

5 SOLUTION PROCEDURES

Pressuremeter tests may be performed in either a stress-controlled or a strain-controlled manner and a number of procedures were developed to model different types of cavity expansion. For the two forms of loading, where the cavity boundary conditions are defined in terms of applied stress or displacement, different solution procedures were adopted. In addition, the simulated tests can be modelled for idealised undrained conditions.

For stress-controlled expansion the FD equations are solved using a block version of the Gauss-Seidel iterative technique in which the equations are placed in a specific order and then solved in groups of two. Each pair consists of a radial equilibrium and a continuity condition allowing a particular radial displacement and adjacent pore pressure value to be calculated. Starting from the inside boundary and progressing outwards, a solution is obtained at each iteration by solving for the variables at points throughout the soil. At the end of each iteration there will be errors in the equilibrium and continuity equations and the iterations are continued until these errors are reduced to within acceptable limits. This technique is used whether the boundaries are impervious or free-draining.

For strain-controlled expansion, the general drained and idealised undrained cases use different solution procedures. In the idealised undrained case i.e. zero volume change at each strain/stress point, a direct method for solving the simultaneous equations is used. As the cavity displacement is known so too are the displacements at every grid point in order to satisfy the no volume change condition. No iterations are required and the procedure is very efficient.

For the more general analysis where some drainage may occur the no volume change condition is replaced by the soil-fluid continuity condition and the governing equations can no longer be solved by the above procedure and an iterative technique is required. The continuity condition is used to calculate the radial displacements for each nodal point and these displacements are then used to calculate the strains which in turn are used to calculate the effective stresses. The procedure starts at the inside boundary and moves progressively outwards. The reverse direction is used when calculating the pore pressures using the equilibrium equations; at the end of each iteration the equilibrium conditions are satisfied. The errors in the continuity equations are checked and the iterations continued until the errors are sufficiently small.

The FD technique has been implemented in computer programs written in FORTRAN and the programs used in the simulation of pressuremeter tests as described below; further details of the solution procedures may be found in Hashim (1989).

6 RESULTS OF SIMULATIONS

For the purpose of illustrating its applicability, and to check solutions obtained using another method, the technique has been used to model laboratory hollow cylinder tests. The cylindrical cavity, which has a diameter of 25 mm, is centrally located in a cylindrical soil specimen 150 mm in diameter. Expansion is assumed to take place in a saturated, homogeneous soil and the assumed soil parameters are given in Table 1. Both undrained and drained expansion have been modelled and the results compared with those obtained previously (Pyrah et al 1985) using an FE technique. The laboratory tests and the numerical simulations form part of a study into the effect of test procedure on a variety of pressuremeter test. The present work is examining the strain holding test (SHT) used to determine the consolidation characteristics of a soil.

6.1 *Results of simulations without creep*

The expansion curves predicted by both methods (FD and FE) for undrained stress-controlled expansion are shown in Figure 2 and the stress distributions around the expanding cavity (at a cavity pressure of 140 kN/m^2 corresponding to approximately 2% radial strain) are shown in Figure 3a. For these simulations creep has been suppressed by the use of appropriate soil parameters (i.e. A=0) and it can be seen that the solutions produced by the two methods are very similar. In fact, for most of the points in the figure it is impossible to differentiate between the two solutions as the FD data points are masked by the FE results.

The stress distributions for undrained strain-controlled expansion and partially drained strain-controlled expansion are shown in Figures 3b & 3c.

Table 1: Parameters for the numerical simulations

DATA TYPE	DETAILS
Modified Cam-clay soil properties	λ = 0.101 κ = 0.025 e_{cs} = 1.265 G = 6300 kN/m^2 M = 1.1
Fluid flow properties	k = 3.38×10^{-10} m/s γ_w = 9.807 kN/m^3
Singh-Mitchell creep parameters	A = 1.5×10^{-3} %/min m = 1.0 α = 5.7

These stresses are plotted for a radial strain of approximately 20%. Once again the agreement between the results using the two methods is very close.

The small discrepancies observed in the results can be explained by the differences in the modelling technique used in the two methods. In the FE method the nodal spacing is progressively increased with radius and the first node point is at the cavity boundary, whereas the FD method uses a regular grid with the first stress point within the soil. In addition, in the FD analyses true undrained behaviour is modelled whereas for the FE method the "undrained" expansion is achieved by using a low permeability and a small time increment in the analysis. It can be observed in Figure 3b that some dissipation occurs in the FE analysis despite the rapid rate of expansion.

6.2 *Results of simulations including creep*

The results of the simulations when creep is included in the analysis are shown in Figures 4 and 5. Figure 4 shows the expansion curves for both the FD and FE methods with and without creep while Figure 5 shows the pore pressure response for both cases. Whilst the expected trend is obtained for the expansion curves, i.e. a decrease in cavity pressure when creep is included, the FE method predicts a larger decrease than the FD method. For the pore pressure response, the FD method predicts a small decrease while the FE method predicts a large increase. The stress distribution curves for the FE analyses were examined and it was observed that as a result of creep the radial stress close to the cavity reduced significantly producing a much more uniform distribution across the cylinder. This was in

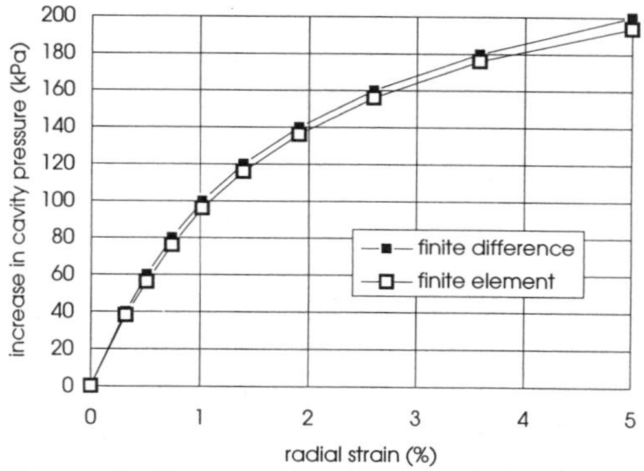

Figure 2 Expansion curves (undrained stress-controlled expansion, consolidation only)

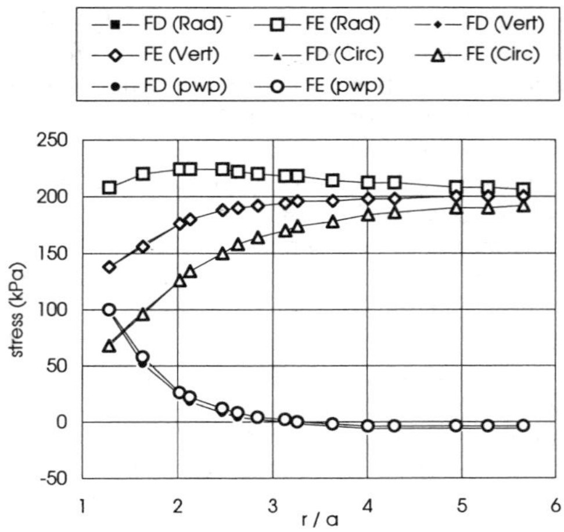

a) Undrained stress-controlled expansion

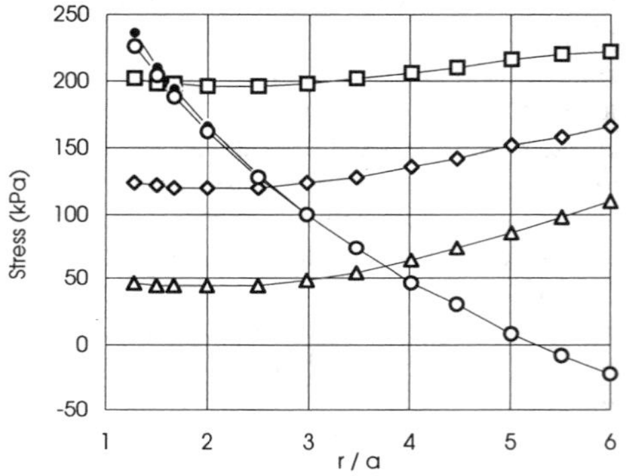

b) Undrained strain-controlled expansion

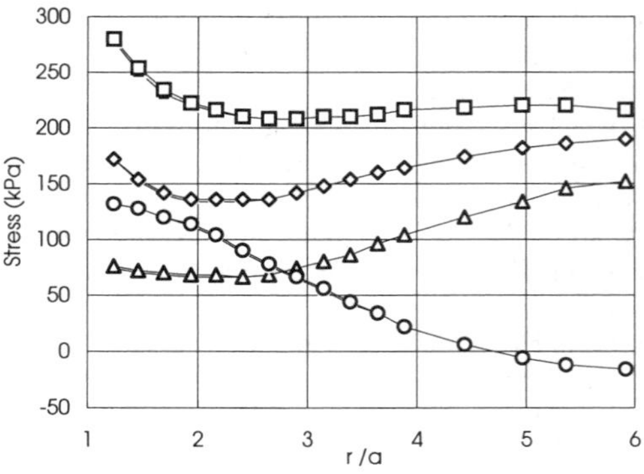

c) Partially drained expansion

Figure 3 Stress distributions around the expanding cavity (consolidation only)

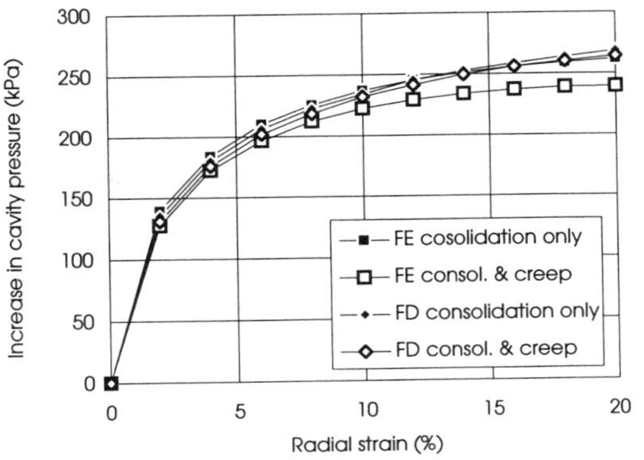

Figure 4 Effect of creep on expansion curve

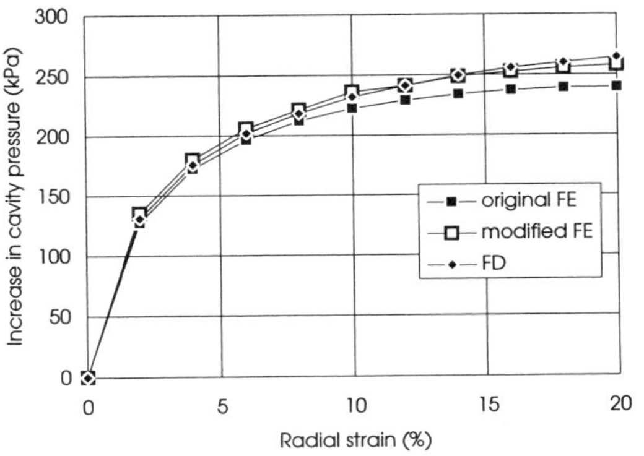

Figure 6 Predicted expansion curves using different methods (consolidation & creep)

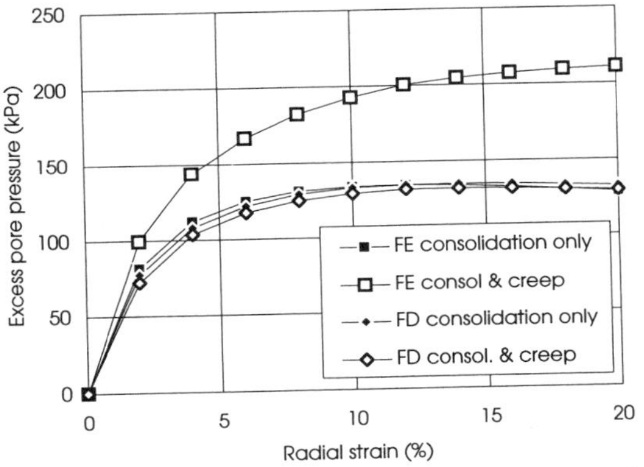

Figure 5 Effect of creep on development of porewater pressure

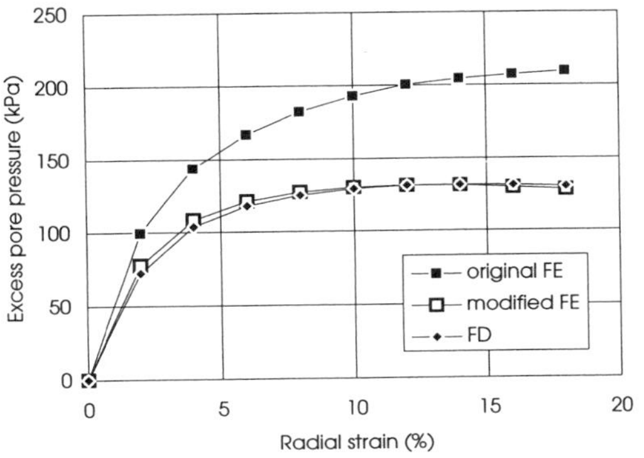

Figure 7 Predicted porewater pressure using different methods (consolidation & creep)

marked contrast to the FE solution without creep and, more importantly, significantly different from the FD solution including creep. The differences between the two methods were significant and unlikely to be simply due to a difference in the numerical techniques used as was the case for the analyses which considered consolidation alone.

This disagreement prompted further investigation and since the predictions using both methods agree for the analyses without creep it was concluded that the cause of the disagreement was due to the different way in which the creep model had been implemented in the two programs. The original work was checked and there appeared to be some errors in the original FE program due to confusion in the solution procedure adopted for dealing with creep. This is due to the apparent mixing of two methods for dealing with creep, namely the initial strain and the tangential (or modified) stiffness methods (Zienkiewicz 1977), and the easiest way to check this was to develop a new version of the FE code. To be consistent with earlier work the same FE program developed for consolidation only was used as the basis for the analysis and the new technique for incorporating creep was similar to that used in the FD program i.e. the initial strain method, and details of this work are given by Hashim (1989).

The results of the simulations using the modified FE program and the FD program are shown in Figures 6 and 7. Both the expansion curves and the pore pressure curves produced by the modified FE program and the FD method are in good agreement and the small differences are consistent with those obtained when consolidation only was considered. As a further check on the results another version of the FE program was developed in which creep was dealt with using the modified stiffness procedure.

The modified stiffness procedure is more expensive

than the simple initial strain method (where the stiffness remains constant) because the stiffness matrix has to be re-evaluated and solved for every time-step. This latter process, which has much similarity with the iterative procedures for dealing with plastic behaviour, is recommended by Zienkiewicz (1977) due to the improvement in accuracy. The results obtained for the pressuremeter holding test considered above were very similar to those produced using the initial strain method. The differences were less than 2% in the expansion curve and less than 3% in pore pressure predictions.

The results show that the original method adopted for the implementation of the Singh-Mitchell creep model in the FE analysis produces significantly different predictions than the method described in this paper. At this stage it is not possible to say with certainty which approach is more realistic although it is probable that previously published solutions using the original FE program overpredict the effect of creep. Nevertheless, solutions produced with the new implementation of the creep model indicate that creep of the soil skeleton does effect the values of the consolidation parameters derived from strain holding tests. Further work is required to validate this new approach.

6.3 *Further Discussion*

From the above results it can be seen that the predictions produced by both the FD and the FE methods are comparable when a similar method of implementation for creep is adopted. The work indicates the usefulness of an alternative method for obtaining solutions independent of those produced by another computer program or method. In addition to providing an independent check on FE solutions, finite differences can be an attractive alternative, especially for problems involving simple geometries. As the results presented in this paper illustrate, it is relatively easy to develop FD software even when realistic soil models, including multi-phase, non-linear, non-elastic behaviour, are used. In addition, the relative simplicity of the FD technique makes it suitable for assessing the effects of different soil models.

7 CONCLUSIONS

1. The finite difference technique has been used to check an existing FE program for simulating the expansion of a cylindrical cavity. The method models two-phase (Biot) consolidation and is based on modified Cam clay and the Singh-Mitchell model for deviatoric creep

2. The method, based on dynamic relaxation, is efficient both in terms of development and computing time. Its relative simplicity makes the method a useful alternative to other techniques particularly for verification purposes. For situations involving simple geometries the method may be used to assess the effects of different soil models.

3. The verification exercise has highlighted a discrepancy in the implementation of creep in the original program. Although previously published results may have overestimated the effect of creep, the current study confirms that consolidation characteristics derived from pressuremeter strain holding tests are affected by creep of the soil skeleton.

REFERENCES

Chang, C.Y., K. Nair & R.D. Singh 1974. Finite element methods for the non-linear and time-dependent analysis of geotechnical problems. *Proc. Conf. Analysis and Design in Geotechnical Engineering, Austin (Texas) 1 (1974)* 269-302.

Hashim, R. 1989. *Numerical study of time-dependent cavity expansion in a normally consolidated clay*. PhD Thesis: University of Sheffield, UK

Otter, J.R.H., A.C. Cassel, & R.E. Hobbs 1966, Dynamic relaxation, *Proc. ICE*. 35:633-656.

Pyrah, I.C., W.F. Anderson & F. Haji Ali 1985. The interpretation of pressuremeter tests: Time effects for fine-grained soils. *5th Int. Conf. Numerical Methods in Geomechanics*. 3:1629-1636.

Pyrah I.C. and W.F. Anderson 1990 Numerical assessment of SBP tests in a clay calibration chamber. *3rd Int. Symposium on Pressuremeters, Oxford, April, 1990*:179-188.

Palmer, A.C. 1972 Undrained plane strain expansion of a cylindrical cavity in clay. *Geotechnique*. 22:451-457.

Roscoe, K.H. and J.B. Burland 1968. On the generalised stress-strain behaviour of wet clays. In J. Heymann and F.A. Leckie (ed), *Engineering plasticity*: 535-609. Cambridge:Cambridge University Press

Singh, A. & J.K. Mitchell 1968. General stress-strain-time function of soils. *J. Soil Mech. Fdns., ASCE*. 94:21-46.

Zienkiewicz, O.C. 1977. The Finite Element Method in Engineering Science, London:McGraw Hill.

Numerical Models in Geomechanics, Pietruszczak & Pande (eds) © 1997 Balkema, Rotterdam, ISBN 90 5410 886 X

Elasto-plastic modeling of deformation of sands during earthquake-induced liquefaction

A. Anandarajah
Department of Civil Engineering, The Johns Hopkins University, Baltimore, Md., USA

ABSTRACT: An elasto-plastic model that is capable of describing the liquefaction behavior during a cyclic loading is described. The model is incorporated into a fully-coupled finite element procedure, which is used to analyse a centrifuge model of a horizontal deposit of saturated sand. The numerical results are compared with centrifuge observations.

1 INTRODUCTION

A problem that still poses a challenge to numerical modelers is modeling of the behavior of liquefiable soil deposits during earthquakes. Owing to the dominant effect of pore water pressure build-up on the behavior, the total stress based approaches have had limited success in realistically simulating the overall response in general and the deformation in particular. Presented in this paper is a fully-coupled, effective stress based finite element method of modeling earthquake-induced liquefaction behavior of geotechnical structures. The constitutive behavior of the soil is described using an elasto-plastic model that is capable of simulating the liquefaction behavior. The numerical results are compared with centrifuge data.

2 ELASTO-PLASTIC LIQUEFACTION MODEL

The constitutive model employed in the analysis is a simpler version of a generalized liquefaction model presented in Anandarajah (1993, 1994a, 1994b); a brief description is presented here. A surface $g = 0$, shown in Fig. 1 in a $p - q$ space, is used in separating *loading* from *stress reversal*, and in modeling the behavior during *loading*. A conical surface $T = 0$ is employed in defining the required directional properties for modeling the behavior during a *stress reversal*. Some ideas of the bounding surface theory (Dafalias and Herrmann, 1986) are utilized. In Fig. 1, point A represents a typical current state and point C represents its image, obtained by projecting radially from the origin. M_f is the slope of the critical state line.

An ellipse and distorted ellipse are used to represent $g = 0$ for $M_f \geq \frac{q^g}{p^g} \geq 0$ and $\frac{q^g}{p^g} \geq M_f$ respectively as as

$$\left[\frac{p_0 - Rp^g}{rp_0}\right]^{n_1} + \left[\frac{Rq^g}{M_f p_0}\right]^{n_2} = 1 \qquad (1)$$

with $r = 1$ for $\frac{q^g}{p^g} \geq M_f$ and $r = R-1$ for $\frac{q^g}{p^g} < M_f$, and $n_1 = n_2 = 2$ for $\frac{q^g}{p^g} < M_f$ and n_1 and n_2 are suitable constants for $\frac{q^g}{p^g} \geq M_f$. p^g and q^g are mean normal and deviatoric stresses derived from σ^g_{ij}. Note that the surfaces join smoothly at $\frac{q^g}{p^g} = M_f$ with the normal being parallel to $q-$axis (i.e., at Point G). Based on several simulations performed to date, the constants n_1 and n_2 associated with the distorted ellipse are assigned fixed values n_1=2.0 and $n_2 = 1.2$. The geometrical parameter R is shown in Fig. 1.

The generalized Hooke's law is used to relate the effective stress to the elastic strain, with the shear modulus defined as (Richart et al. 1970):

$$G = G_0 p_a \frac{(2.97 - e)^2}{1 + e} (\frac{p}{p_a})^{n_e} \qquad (2)$$

where G_0 and n_e are material parameters, e is the void ratio and p_a is the atmospheric pressure. The bulk modulus is computed by assuming a value for Poisson's ratio.

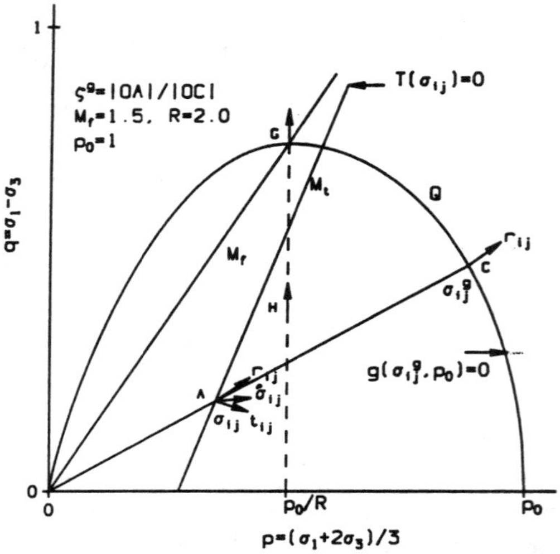

Fig. 1. Schematic of Bounding (g=0) and Conical (T=0) Surfaces

The rate of plastic strain is related to the rate of effective stress as

$$\dot{\epsilon}_{ij}^p = La_{ij} \quad (1)$$

where

$$L = \frac{1}{K_p}\dot{\sigma}_{k\ell}a_{k\ell} \quad (2)$$

and K_p is the plastic modulus. K_p and a_{ij} are to be chosen depending on whether a given stress probe is *loading* or *stress reversal*, which are defined as follows.

A stress probing is defined as *loading* when $L > 0$, *neutral loading* when $L = 0$ and *stress reversal* when $L < 0$. It is understood that when $K_p > 0$, a stress probing will either be loading or stress reversal depending on whether $\dot{\sigma}_{k\ell}a_{k\ell} > 0$ (i.e., the direction of $\dot{\sigma}_{k\ell}$ is directed outward from the surface that $a_{k\ell}$ is normal to) or $\dot{\sigma}_{k\ell}a_{k\ell} < 0$ (i.e., the direction of $\dot{\sigma}_{k\ell}$ is directed inward from the surface that $a_{k\ell}$ is normal to).

For loading, it is assumed that $a_{ij} = r_{ij}$, where r_{ij} is the unit normal to $g = 0$ as indicated in Fig. 1. Let K_b be the plastic modulus on the surface at any point such as point C. K_b is referred to as the bounding modulus (Dafalias and Herrmann 1986). The plastic modulus at a point within $g = 0$ (e.g., point A) is related to K_b as

$$K_p = K_b + H(\zeta^g, \sigma_{ij}) \quad (3)$$

The modulus H is referred to as the shape-hardening modulus (Dafalias and Herrmann, 1986), which is defined so that $H(\zeta^g = 1, \sigma_{ij}) = 0$. Having selected a hardening rule, K_b is determined by the consistency condition applied to the bounding surface $g = 0$. A suitable form of H is assumed.

The evolution of p_0 is assumed to depend on the volumetric plastic plastic strain as

$$\dot{p}_0 = \frac{1 + e_0}{\lambda^*}p_0\dot{\epsilon}_v^p \quad (4)$$

and $\lambda^* = \lambda - \kappa$, λ and κ are slopes of isotropic compression and rebound lines in the $e-\ell np$ space. e is the void ratio and e_0 is its initial value. The above equation gives rise to the usual volumetric (or density) hardening.

The consistency condition applied to $g = 0$ yields an expression for K_b as

$$K_b = -\frac{1 + e_0}{\lambda^*}\frac{\partial g}{\partial p_0}p_0 r_{kk} \quad (5)$$

where r_{kk} is the trace of r_{ij}.

The shape-hardening modulus H is assumed to be a function of p, ζ^g and $\eta_*^g = \frac{q}{M_f p}$ according to

$$H = h_0 K_{b0} H_1(p) H_2(\zeta^g) H_3(\eta_*^g) \quad (6)$$
$$H_1(p) = (\frac{p}{p_a})^{n_e} \quad (7)$$
$$H_2(\zeta^g) = < 1 - \zeta^g >^{m_1} \quad (8)$$
$$H_3(\eta_*^g) = 1 - (\eta_*^g)^{h_1} \text{ for } \eta_*^g \leq 1 \quad (9)$$
$$H_3(\eta_*^g) = 0 \text{ for } \eta_*^g > 1 \quad (10)$$

where $K_{b0} > 0$ is the bounding modulus (Eq. 7) evaluated at $q = 0$, and h_0, h_1, and m_1 are model parameters. n_e is the same parameter appearing in Eq. 2.

Each equation presented above has been discussed in detail in Anandarajah (1994a, 1994b). Also presented in these references are qualitative model behavior and extensive simulations of experimental triaxial behavior under monotonic loading conditions.

During a stress reversal, it is assumed that $a_{ij} = t_{ij}$, where

$$t_{ij} = -\frac{1}{\ell^t}\frac{\partial T}{\partial \sigma_{ij}} \quad (11)$$

and $\ell^t = (\frac{\partial T}{\partial \sigma_{ij}}\frac{\partial T}{\partial \sigma_{ij}})^{\frac{1}{2}}$. $T = 0$ is assumed to be a cone with its axis being the $p-$axis, and t_{ij} is taken to be the inward normal to $T = 0$ as shown in Fig. 1.

Table 1: Model Parameters for Nevada Sands of Dr=40% (N-Sand40)

Soil	λ^*	M_{fc}	G_0	n_e	m_1	h_0	h_1	h_r	α_r
N-Sand40	.014	1.354	90	1.0	0.2	10	8	1.0	.01

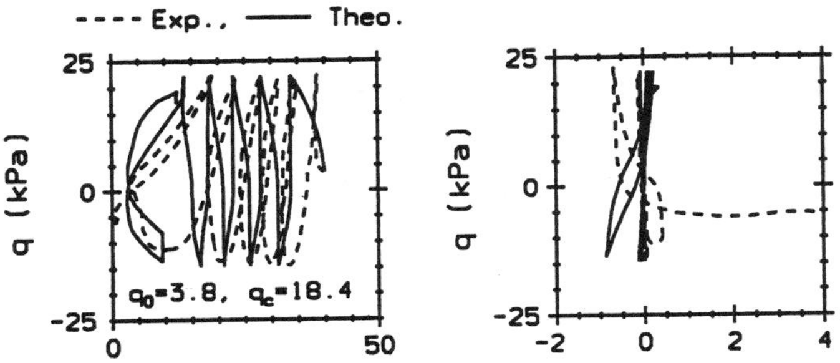

Fig. 2. Comparison of Model Behavior with Experimental Data of Cyclic Triaxial Behavior of Nevada Sand of Dr=40%at po=40 kPa

On the basis of an analysis, a suitable set of equations to use during a stress reversal for the plastic modulus K_p^*, bulk modulus K, and M_t are shown to be (Anandarajah, 1994b)

$$K_p^* = -h_r K_{b0} \quad (12)$$

$$K > K_{min} \quad (13)$$

$$K_{min} = \frac{K_{p0} M_{t0}}{(M_{t0} - M_f) t_{p0}^2} \quad (14)$$

$$\frac{1}{M_{t0}} = \frac{1}{1.111 M_f}(1 - e^{-\alpha_r \frac{\eta^{max}}{<1-\eta^{max}>}}) \quad (15)$$

$$M_t = \frac{M_{t0}}{\eta^f} \quad (16)$$

A summary of model parameters is presented in Table 1. Also, listed in this table are values of these parameters for a Nevada Sand that was used in the centrifuge model to be analysed. A sample model simulation of triaxial undrained cyclic test is presented in Fig. 2.

3 NUMERICAL PROCEDURE

The numerical procedure is based on the fully-coupled dynamic equations for two-phase porous media (Biot 1962a, 1962b). The finite element formulation employed is that of Zienkiewicz and Shiomi (1984). Specifically, the most general formulation, where the final matrix equations are in terms of solid and fluid displacements and pore pressure, is employed. The matrices involved in this form are symmetric. The constitutive model is based on the associated flow rule, and thus yields symmetric stiffness matrix. The entire analysis, therefore, involves symmetric global stiffness matrices. Two-dimensional finite element models are used to approximate the centrifuge models. Eight noded finite elements with solid and fluid displacement fields expressed in terms of eight nodal unknowns and pore pressure field expressed in terms of four unknowns at the corner nodes are employed. 3x3 and 2x2 Gauss integrations are employed for the respective matrices. One-dimensional (e.g., laminar box) condition is simulated by slaving appropriate degrees of freedom. A simple trapezoidal rule is employed in integrating the rate equations. Sub-increments are used to integrate the equations accurately. The time integration is performed using the Hilber-Hughes-Taylor method with a suitable value for numerical damping parameter. A computer program named HOPDYNE (1990), which, among many other features, has the above numerical aspects coded into it.

4 NUMERICAL PREDICTION OF CENTRIFUGE MODEL BEHAVIOR

The details of the centrifuge test may be found in Taboada and Dobry (1993). Fig. 3 presents a sketch of the model, along with the locations

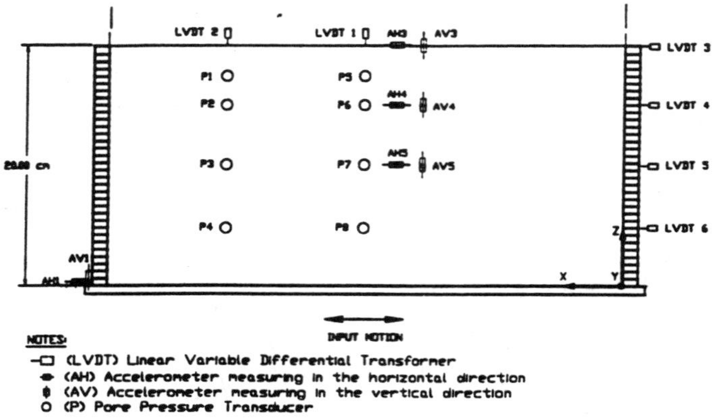

Fig. 3. Centrifuge Model

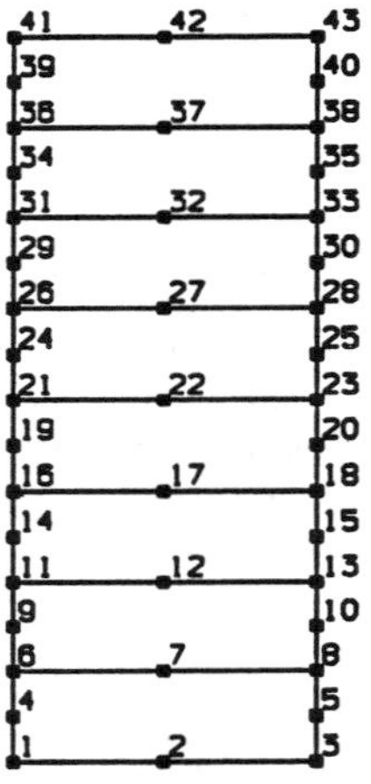

Fig. 4. Finite Element Mesh

of pore pressure transducers, accelerometers and LVDT. The test was conducted in a "laminar box" to simulate a horizontal deposit of prototype soil. The model was constructed using uniform Nevada sand at 45% relative density, and fully saturated. The prototype thickness of the soil layer was 10 meter. In addition to the soil parameters listed in Table 1, the parameters listed in Table 2 are needed for the analysis.

Assuming a one-dimensional condition, the finite element mesh shown in Fig. 4 was used to model this problem. The results of the "before the event" predictions are compared with the corresponding centrifuge observations in Figs. 5-7. The accelerations at the transducer locations AH1 (base), AH5 (mid-height) and AH3 (ground surface) are compared in Fig. 5. Note that the soil looses its ability to transmit the waves due to liquefaction during the shaking. The numerical predictions compare reasonably well experimental data. The long term pore pressure histories at the transducer locations P6, P7 and P8 (see Fig. 3) are compared with observed data in Fig. 6. The apparent good agreement indicates the ability of the constitutive model to capture the pore pressure response reasonably well. The rate of dissipation is predicted to be slower than the observation. The ground settlements are compared in Fig. 7. While the final settlements are reasonably close, the theory under predicts the settlement during the shaking. The observed higher rate of pore pressure dissipation and settlements are suspected to be due to increase in the permeability of soil during liquefaction.

5 CONCLUSIONS

Using an elasto-plastic constitutive model capable of simulating the liquefaction behavior of sands in a fully-coupled finite element analysis procedure, a computer code HOPDYNE was developed. The code was used to analyse a centrifuge model tested under a simulated earthquake loading. The numerical predictions of pore water pressure build-up, settlements and accelerations compare reasonably well with the observations.

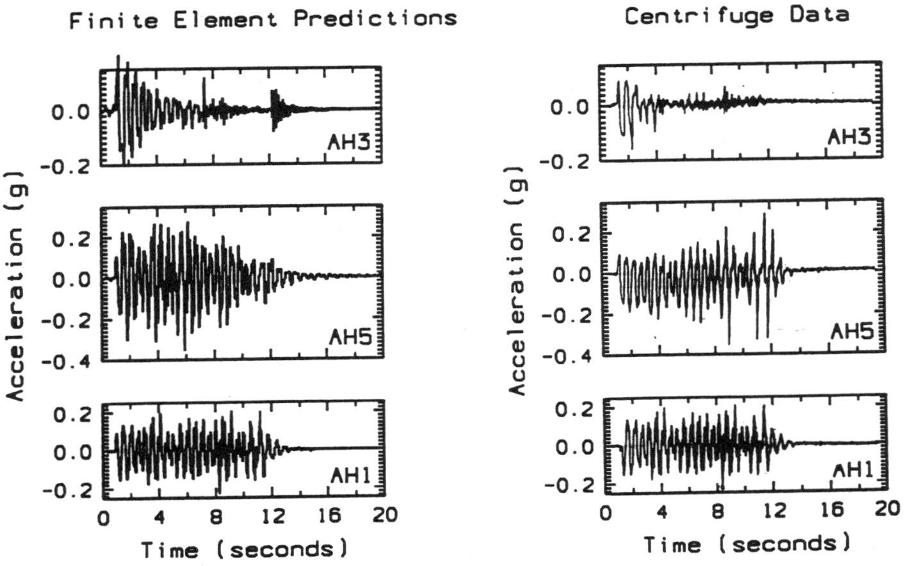

Fig. 5. Comparison of Horizontal Acceleration Histories

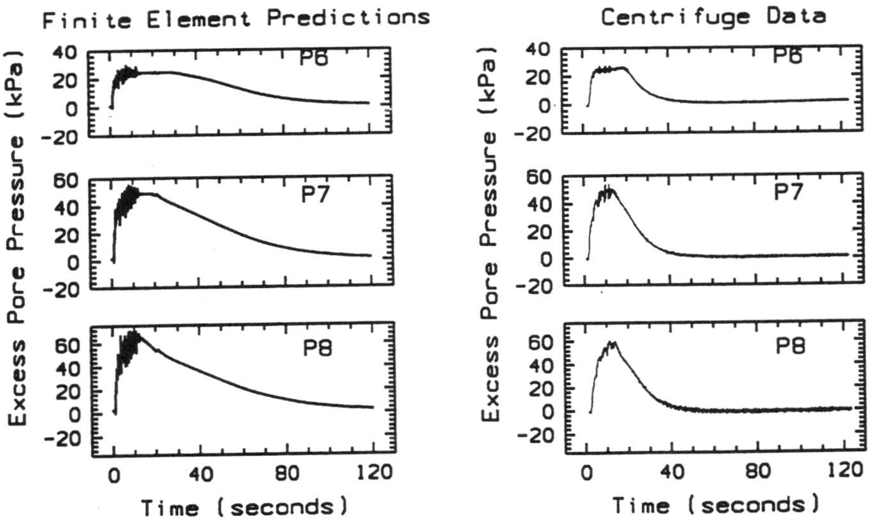

Fig. 6. Comparison of Long Term Pore Pressure Histories

Table 2: Parameters used in analysing Model 1 by HOPDYNE

k m/sec	n	ρ_s kN-sec^2/m^4	ρ_f kN-sec^2/m^4	K_0	γ' kN/m^3
3.3x10^{-3}	.420	2.7	1.0	.55	9.55

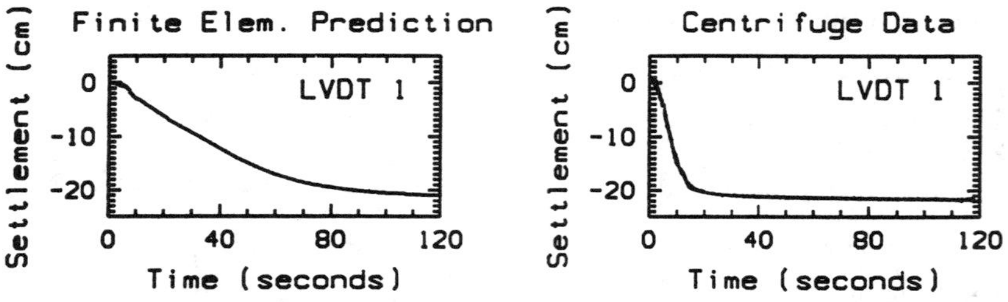

Fig. 7. Comparison of Ground Settlement Histories

References

[1] Anandarajah, A. 1990. HOPDYNE - A Finite Element Computer Program for the Analysis of Static, Dynamic and Earthquake Soil and Soil-Structure Systems. The Johns Hopkins University, Baltimore, Maryland.

[2] Anandarajah, A. 1993. VELACS Project: Elasto-Plastic Finite Element Prediction of the Liquefaction Behavior of Centrifuge Models Nos. 1, 3 and 4a. *Proc. Intl. Conf. on Verification of Numerical Procedures for the Analysis of Soil Liquefaction Problems*, Davis, California, Oct. 17-20, eds. K. Arulanandan and R. F. Scott, 1, 1075-1104.

[3] Anandarajah, A. 1994a. A Constitutive Model for Granular Materials Based on Associated Flow Rule. *Soils and Foundations*, Japanese Society of Soil Mechanics and Foundation Engineering, 34, 81-98.

[4] Anandarajah, A. 1994b. Procedures for Elasto-Plastic Liquefaction Modeling of Sands. *Journal of Engineering Mechanics Division*, ASCE, 120, 1563-1593.

[5] Biot, M. A. 1962a. Generalized Theory of Acoustic Propagation in Porous Media. *J. Acoustic Soc. of America*. 34, 1254-1264.

[6] Biot, M. A. 1962b. Mechanics of Deformation and Acoustic Propagation in Porous Media. *J. Appl. Phys.* 33, 1482-1498.

[7] Dafalias, Y. F. 1986. Bounding Surface Plasticity, Part 1: Mathematical Foundation and Hypoplasticity. *Journal of Engineering Mechanics Division, ASCE*. Vol. 112, No. 9, pp. 966-987.

[8] Richart, F. E., Hall, J. R., Jr., and R. D. Woods 1970. Vibrations of Soils and Foundations. *Prentice-Hall, Inc.* Inglewood Cliffs, New Jersey.

[9] Taboada, V. M. and Dobry, R. 1993. Experimental Results of Model 1 at RPI. *Proc. Intl. Conf. on Verification of Numerical Procedures for the Analysis of Soil Liquefaction Problems*, Davis, California, Oct. 17-20, eds. K. Arulanandan and R. F. Scott, 1, 3-18.

[10] Zienkiewicz, O. C. and T. Shiomi 1984. Dynamic Behavior of Saturated Porous Media: The Generalized Biot Formulation and Its Numerical Solution. *International Journal for Numerical and Analytical Methods in Geomechanics*. Vol. 8, pp. 71-96.

Numerical Models in Geomechanics, Pietruszczak & Pande (eds) © 1997 Balkema, Rotterdam, ISBN 90 5410 886 X

Liquefaction analysis by bounding surface model

Q.Li & K.Ugai
Department of Civil Engineering, Gunma University, Japan

S.Gose
CTI Engineering Co. Ltd, Japan

ABSTRACT: In this paper, the gUTS bounding surface model recently proposed by Crouch and Wolf is adopted in a FE program for the dynamic analysis of saturated soil. Here the saturated soil is treated as fluid-saturated porous media, as proposed by Zienkiewicz et al. A case study of a homogeneous horizontal sand layer is conducted, where the analysis focuses on the pore pressure generation and dissipation during dynamic excitation and the subsequent consolidation. It is expected that by utilizing the gUTS bounding surface model, the seismic behavior of the saturated soil could be satisfactorily simulated, and this would provide an effective approach in the liquefaction analysis of earth structure composed of saturated soil.

1 INTRODUCTION

The liquefaction analysis of saturated soils by the numerical approach still remains very important in the design of practical earth structures, such as embankments, dams and foundations. In Japan, after the Hanshin earthquake in January, 1995, it became an urgent need to provide civil engineers with such tools. Although it is well known that the mechanical behavior of saturated soils are governed by the interaction of their solid skeleton with the pore water, which can be handled by the generalized Biot's formulation of consolidation (Zienkiewicz et at., 1984), such an approach requires an efficient constitutive model to describe the behavior of solid skeleton under cyclic loading.

The gUTS bounding surface model proposed by Crouch and Wolf (1994), which is claimed to be capable of treating the sand, clay and silt under a single frame, regardless of their initial states and test paths, is adopted in our FE program. We propose that our work could provide a rational and practical approach to the liquefaction analysis of saturated soils.

2 THE BOUNDING SURFACE MODEL

The theory of bounding surface plasticity was first introduced by Dafalias, et al.(1980). It is especially effective to simulate the cyclic behavior of soil, concrete and other geomaterials. However, the early version of bounding surface model assumed only elastic unloading inside the bounding surface, which is not sufficient to simulate the behavior of soils, especially the sand under cyclic loading. The gUTS bounding surface model recently proposed by Crouch and Wolf (1994) removes such restrictions and gives a rational description of the continuous plastic response of soils along stress reversals.

Moreover, the model has the following features which are favorable for the numerical analysis of practical geotechnical problems:

(1) Combined use of the radial mapping and deviatoric mapping rule and the use of apparent normal consolidation line for sands; it provides a scheme to treat soils of different types and initial states under a single frame. Soft clays generally lie on the bounding surface or in the radial mapping region, whereas the behavior of over-consolidated clay and sands are described by the cone-like loading surface in the deviatoric mapping region. The movement of the projection center in the deviatoric mapping region allows a continuous plastic loading along stress reversals.

(2) Use of a non-associative flow rule where the plastic dilatancy surface is a 3-segment curve; The segment under the phase-transformation line can be adjusted to give a closer agreement with the observed plastic dilatancy rates.

To demonstrate the performance of the model, here we give two triaxial tests simulated by our program. The same tests have been modeled by Crouch and Wolf(1994) with the same parameters.

Fig.1 is a series of drained triaxial compression of Sacramento river sand. The evolution of volumetric

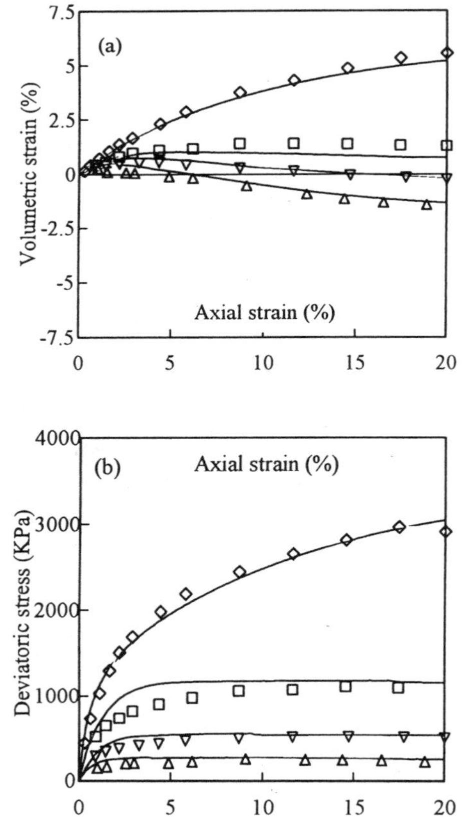

Fig.1 The drained triaxial compression of Sacramento river sand in loose state for four different cell pressures
(a) Volumetric strain versus axial strain;
(b) Deviatoric stress versus axial strain

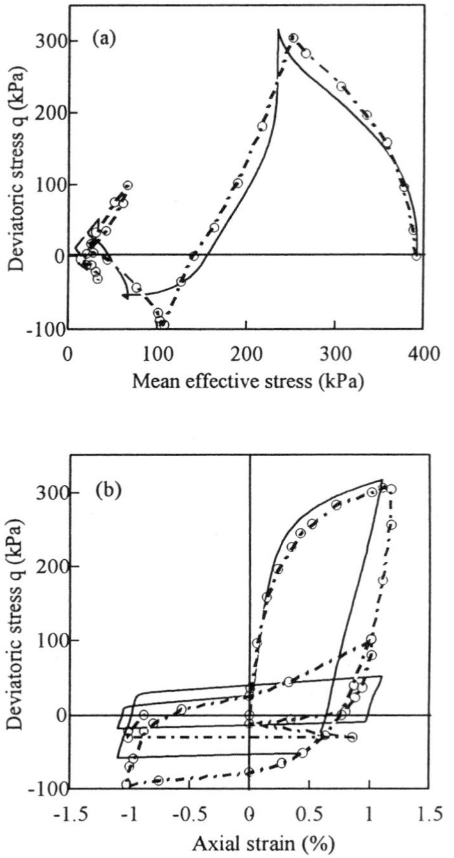

Fig.2 Cyclic undrained triaxial compression and extension of Davis silt (a) Effective stress path; (b) deviatoric stress versus axial strain

strain of the sand under different cell pressure is captured well by the model.

Fig.2 shows the strain controlled triaxial compression and extension of Davis silt. The mean effective stress decrease rapidly due to the accumulated volumetric plastic strain. The significant reduction of the peak deviatoric stress in the second cycle and the cyclic mobility near the stress origin is well simulated.

3 DYNAMIC ANALYSIS OF SATURATED SOIL

M. A. Biot introduced the theory to describe the consolidation of saturated soil. The soil was considered to be a fluid-saturated elastic porous medium. The theory was extended to describe the dynamic analysis of fluid-saturated porous media by Wilson et al.(1978) However, it was after the extensive study by Zienckiewicz et al.(1984), theat this method has been accepted by the subfield of soil dynamics to describe the dynamic behavior of saturated soil and of course the liquefaction.

The final FEM formulation in which soil displacement u and pore pressure p are taken as unknown variables, is:

$$\begin{aligned} &[M]\{\ddot{u}\} + [C]\{\dot{u}\} + [K]\{u\} - [Q]\{p\} = \{f_1\} \\ &[Q]^T\{\dot{u}\} + [S]\{\dot{p}\} + [H]\{p\} = \{f_2\} \end{aligned} \tag{1}$$

where [M] is the mass matrix of soil-water mixture; [C] and [K] are the damping matrix and tangent stiffness matrix of solid skeleton, respectively; [Q] is the coupling matrix; [S] and [H] are the compressibility and seepage matrices for pore water; $\{f_1\}$and $\{f_2\}$are the load vectors, corresponding to the mixture and pore water respectively. The details of these quantities can be found elsewhere (e.g., Zienkiewicz and Shiomi, 1984)

Table 1 The model parameters of sand

κ	I_1	ν	λ	e_0'
0.02	600kPa	0.35	0.14	1.94
ζ_{crc}	B_0	B_1	A_c	n_{gc}
0.252	0.95	0.97	0.0002	1.75
n_{ge}	A_{gc}	h_{1c}	h_{2c}	h_{1e}
1.75	0.2	0.06	0.06	0.03
h_{2e}	n_H	z_0	z_1	z_3
0.03	0.75	1.0	0.99	0

4 LIQUEFACTION ANALYSIS OF A ONE DIMENSIONAL SAND LAYER

As a case study, a one dimensional layer of saturated loose sand subjected to seismic loading is analyzed. The depth of the layer is 30m, the permeability coefficient of sand is $k=10^{-3}$m/s and the submerged unit weight of sand is 9.81kN/m^3. Parameters used in the gUTS bounding surface model are listed in table 1.

The El Centro record with maximum acceleration of 1.96m/s^2 is taken as the input, as shown in fig.3.

Under the seismic loading, the pore water pressure in the sand layer increases until the over-burden stress is reached, or in other words, the liquefaction occurs. This process starts near the surface and gradually spreads down to the bottom, as shown in fig.4.

The absolute acceleration response is shown in fig.5. The reduction of acceleration due to liquefaction in the upper part of the layer can be observed.

Fig.6 shows settlement of the layer. It can be noted that the settlement mainly occurs during the consolidation process.

The pore pressure generation and dissipation are shown in fig.7.

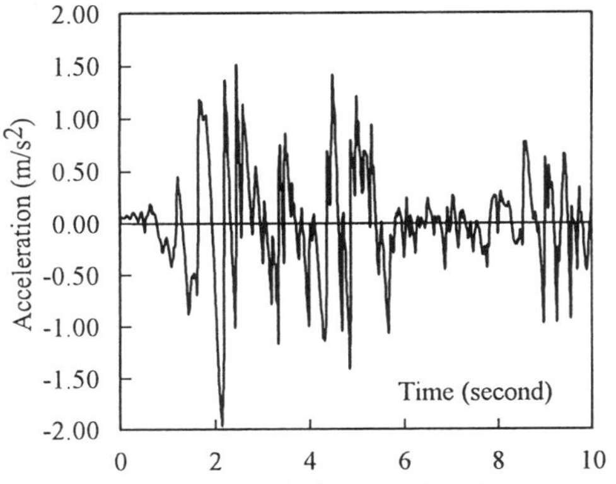

Fig.3 The input acceleration

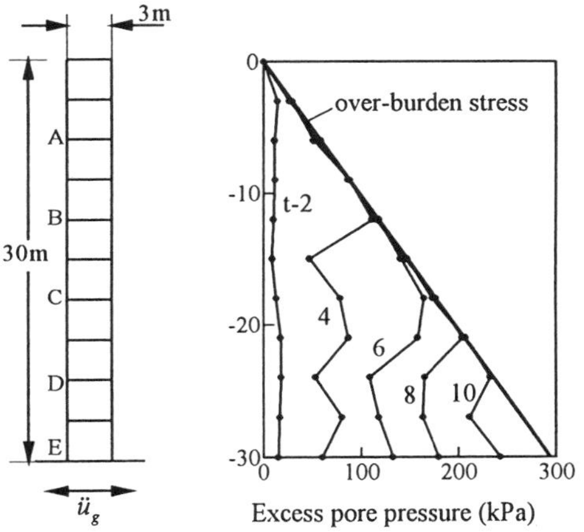

Fig.4 The pore pressure distribution

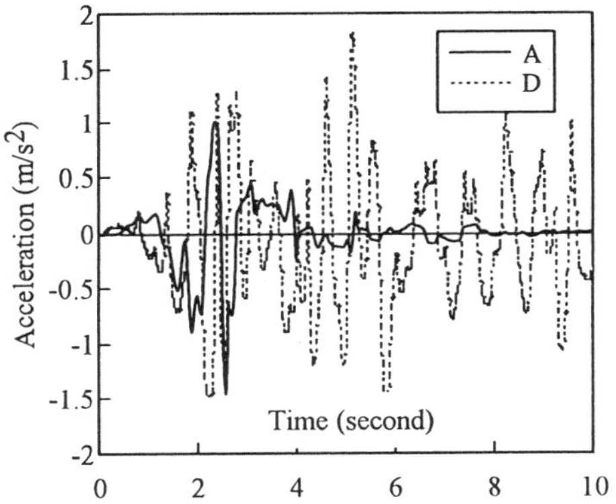

Fig.5 The response of absolute acceleration

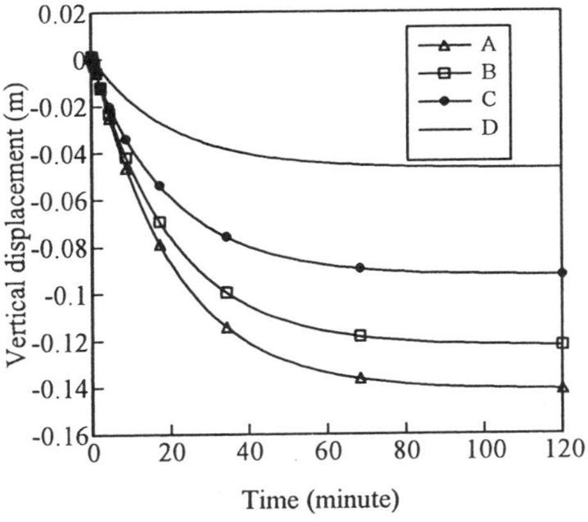

Fig.6 The response of vertical displacement

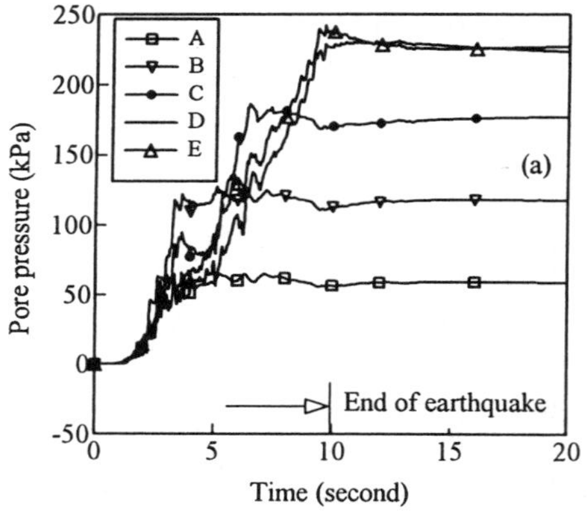

Fig.7a) The pore pressure generation

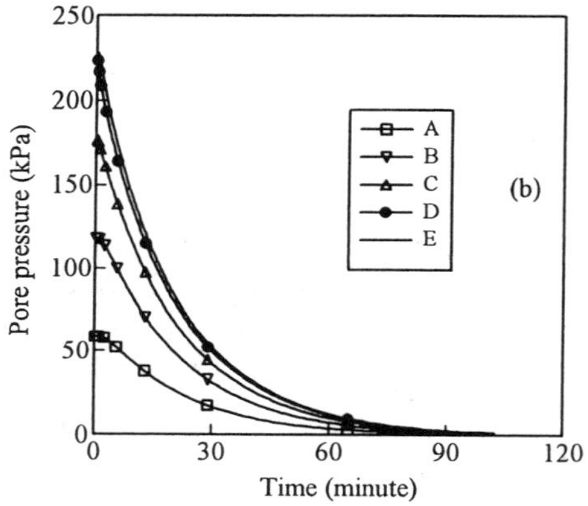

Fig.7b) The pore pressure dissipation during consolidation

Although this is a simple case, it can be expected that the dynamic behavior of saturated soil, including liquefaction, can be analyzed rationally by the incorporation of the bounding surface model into the theory of fluid-saturated porous media.

5 CONCLUSION

In this paper we incorporate the gUTS bounding surface model into a FEM program for the liquefaction analysis of saturated soil. The performance of the model is demonstrated by the simulation of two triaxial tests. A one dimensional saturated horizontal sand layer subjected to seismic loading is taken as the case study, showing that such an approach is effective in the engineering practice.

REFERENCES

1. Crouch, R. S. and Wolf, J. P.(1994): "Unified 3-D critical state bounding-surface plasticity model for soils incorporating continuous plastic loading under cyclic path. Part I and part II", pp 735-784, Vol.18, *Int. J. numer. anal. methods geomech.*, 1994.
2. Dafalias, Y. F. and Herrmann, L. R.(1980):"A bounding surface soil plasticity model" in G. N. Pande and O. C. Zienkiewicz (eds.), *Soils Under Cyclic and Transient Loading*, A. A. Balkema, 1980, pp335-345.
3. Sandhu, R. S. and Wilson, E. L.(1969):"Finite element analysis of flow in porous media", pp641-652, *Proc. ASCE*, Vol.95, 1965.
4. Zienkiewicz, O. C. and Shiomi, T. (1984): "Dynamic behavior of saturated porous media, the generalized Biot's formulation and its numerical solution", pp71-96, Vol.8, *Int. J. numer. anal. methods geomech.*, 1984.

Numerical Models in Geomechanics, Pietruszczak & Pande (eds) © 1997 Balkema, Rotterdam, ISBN 90 5410 886 X

Effects of input earthquake waves and hydraulic conditions on soil layer response

K. Suzuki
Saitama University, Urawa, Japan

J. H. Prevost
Princeton University, N.J., USA

J. Zheng
Kajima Technical Research Institute, Tokyo, Japan

ABSTRACT: Parametric studies are conducted on the response of a specific site subjected to earthquake shaking using nonlinear 2-phase FEM. Simulation of earthquake observation at the selected site is first performed to validate the numerical model, and then parametric studies are conducted by taking account of various types of input motions and hydraulic conditions. Emphasis is placed on the investigation of the effects of input earthquake motion and permeability condition on the responses of acceleration and excess pore water pressure in soil deposit. A new concept of input motion power contributed to the natural frequencies of soft deposit is proposed to be used as a quantitative index.

1 INTRODUCTION

Since the Niigata earthquake in 1964, various laboratory element tests have been successfully conducted for investigating liquefaction phenomenon. The major factors that affect soil liquefaction, such as the soil type, density, grain size and the confining pressure et al., have been examined in detail, and many important findings have been obtained. See, for example, Yoshimi (1980) and NRC (1985).

However, the liquefaction phenomena occurred at actual sites might be quite different from those observed in element tests. The reason is that the element tests could not reflect following effects (Adachi and Tatsuoka 1981):

(1) An actual site shows complicated frequency characteristics which remarkably depend on the natural frequencies of soil deposit and the change of natural frequency due to the reduction of the sand stiffness. Therefore, the response of soil deposit depends on the relation between its natural frequency and characteristics of input motion.

(2) Excess pore water pressure transfers from lower layer to upper one, and upper soil is easier to liquefy.

Therefore, to correctly predict the liquefaction phenomena at actual sites, it is necessary to investigate the effects of frequency characteristics of earthquake motion and the permeability condition. However, up to now, very few studies have been conducted on the effects of the input motion and the permeability.

One of the methods to study these effects is to perform numerical response simulation of saturated soil deposit at actual sites. Recently, a number of analysis procedure (e.g. Prevost 1981, Sunasaka et al. 1988) on 2-phase transient phenomena have been established. Though further verifications are still needed from a point of view of practical prediction, it has become possible to simulate the dynamic response of actual soil deposit during earthquakes, if detail site condition is available.

The purpose of this paper is to study the effects of input motion and permeability condition on the responses at soft alluvial soil deposit through 1-D nonlinear 2-phase analysis. The program code DYNAFLOW (Prevost 1981) is used for the calculation, and the constitutive model is the pressure sensitive multi-yield model (Prevost 1985). Owi site reported by Ishihara et al. (1981, 1987, 1989) is selected as the site to be studied, which has the recorded excess pore water pressures. The simulation of observed records is first conducted to verify the numerical model, and then parametric studies are carried out to investigate the relation of the site responses under various input motions and permeability conditions.

2 SIMULATION ANALYSIS AT OWI SITE

Before examining the effects of the input motion and the permeability at the selected site, the simulation analysis is executed to verify the validity of the material data and the other conditions for calculation.

Owi site is located in the west of the Tokyo Bay, and the observation at the site was conducted by the University of Tokyo and Tokyo Metropolitan Government. The soil profile is shown in Fig. 1. The

Table 1 Material parameters

Material division	Soil type	Constitutive model	Young's modulus (kPa)	Poisson's ratio	Porosity	Permeability (m/sec)	Failure stress ratio : comp : ext
1	Surface soil	Linear	41.39×10^4	0.33	0.5	1.0×10^{-4}	
2	Sandy silt	〃	14.23×10^4	〃	〃	〃	
3	Sandy gravel	〃	41.39×10^4	〃	〃	1.8×10^{-2}	
4	Fine sand with silt Silty fine sand	Pressure sensitive multi-yield	14.23×10^4	〃	0.448	1.0×10^{-3}	:1.2 :0.86
5	Silt with sand	〃	5.96×10^4	〃	0.5	〃	:1.2 :0.86
6	Silty clay Clay	〃	9.66×10^4	〃	〃	1.0×10^{-6}	
7	Silty fine sand	〃	〃	〃	0.417	1.0×10^{-4}	:1.2 :0.86
8	Silt with sand	〃	〃	〃	0.5	〃	:1.2 :0.86
9	Clay	〃	〃	〃	〃	1.0×10^{-6}	

soil down to the depth of 10.0 m is a reclaimed deposit. The whole soil deposit consists of sand and silt layers. The N-value is between 3 to 5 below the depth of 3.0 m. The location of seismograph and piezometers is also shown in Fig.1.

The data measured from The Mid-Chiba earthquake in 1980 (Ishihara et al. 1981) is accepted for the simulation analysis. The maximum horizontal acceleration at ground surface was 0.95 m/s^2 in the N-S direction, and excess pore water pressure is 7.35 kPa at the depth of 6.0 m, and 13.50 kPa at the depth of 14.0 m. These excess pore water pressures are below 20 % of the effective vertical stress.

The material parameters are shown in Table 1. As the sand investigated here is very loose one, the phase transformation stress ratio is set as the same as the failure stress ratio. The material division and the finite element model for 1-D analysis are also shown in Fig.1. The 1st and 2nd natural frequencies of the model are 1.68 Hz and 5.09 Hz, respectively. In dynamic response simulation, the time step Δt is set as 0.01 sec. The Newmark parameters in the integration scheme are chosen as γ=0.65, β =$(\gamma+1/2)^2/4$=0.33. This choice for γ introduces a slight numerical damping (γ=0.5 corresponds to no numerical damping) and the selected value for β maximizes high frequency numerical dissipation. No additional viscous physical damping is introduced. Since the acceleration at the depth of 20.0 m is not available, the input motion at the base is calculated

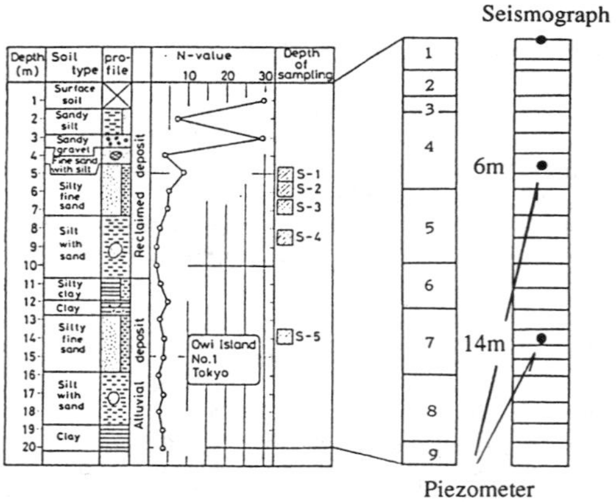

Fig.1 Soil profile and earthquake observation at Owi site (Ishihara et al. 1981)

using the observed acceleration at the surface of soil deposit by 1-D equivalent linear analysis. The maximum amplitude of the input motion is 0.30 m/s^2.

Fig.2 shows the comparison of calculated and observed responses. Computed responses agree well with recorded ones, especially for the generation and diffusion of the excess pore water pressure. Therefore, it could be concluded that the presented numerical model is able to represent the behavior of the actual site during earthquakes.

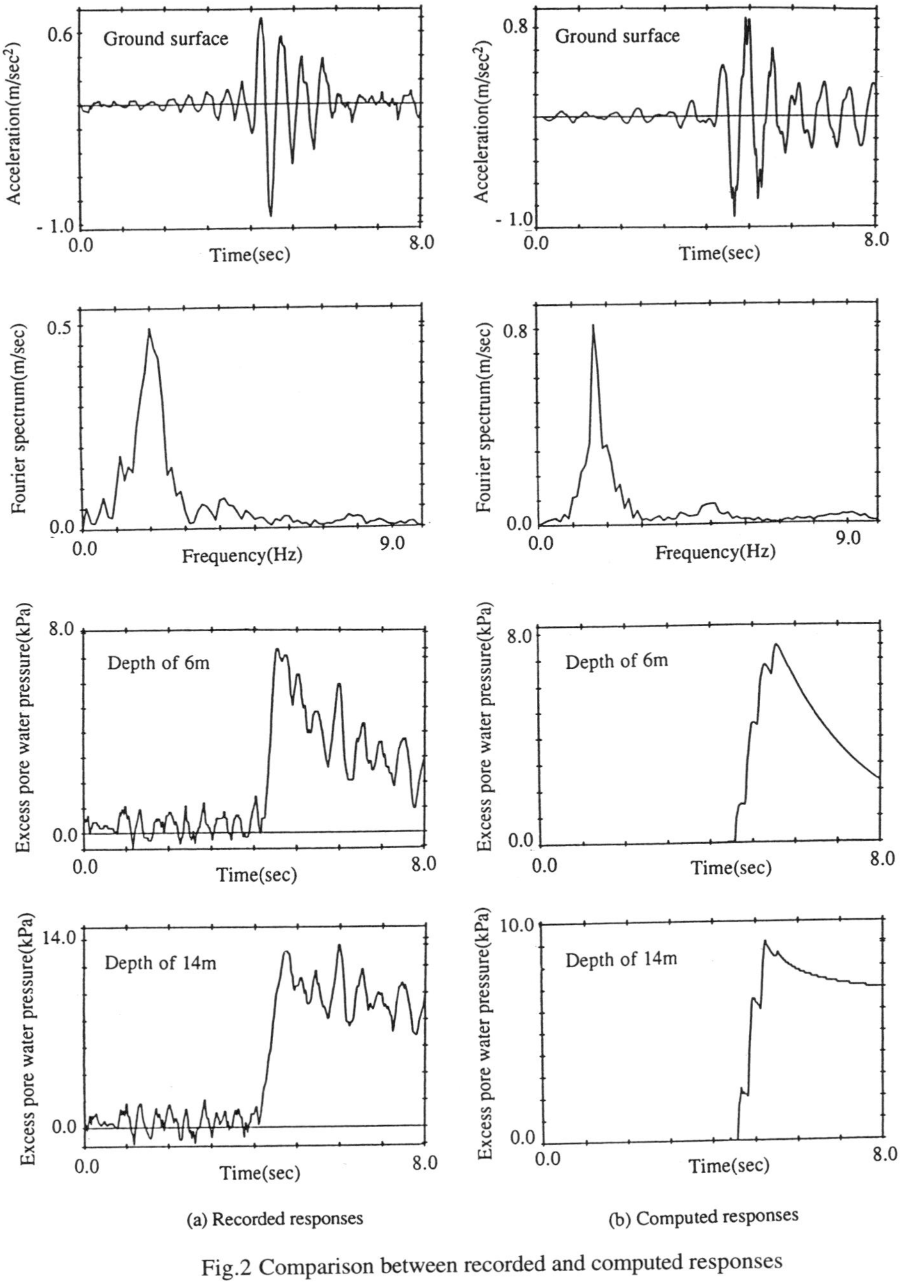

Fig.2 Comparison between recorded and computed responses

3 EFFECT OF INPUT EARTHQUAKE MOTION

Apparent relation exists between the frequency characteristics of the responses of ground and the frequency properties of input earthquake motion as shown by Iwasaki et al. (1977), Taga et al. (1987) and others. However, the researches in this field are still very few.

To study the effects of input motion, 17 cases, i.e., 15 cases using band-filtered white noise and 2 cases

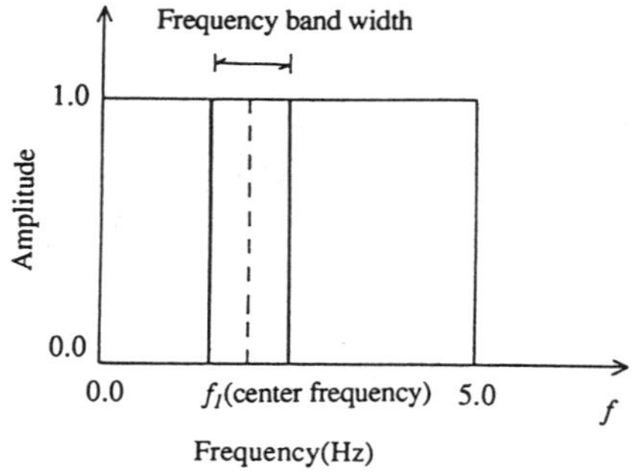

(a) Frequency band filter

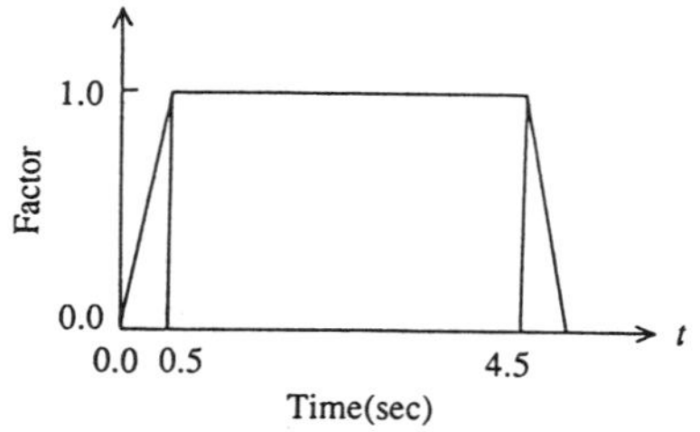

(b) Filter for smoothing input acceleration

Fig.3 Explanation of filters for generating input motion

using recorded earthquake motions as input at the base, are considered. The numerical model for calculation is the same as that used in the simulation of earthquake observation in the previous section.

The white noise is expressed by the following equation

$$a_i = \sum_{j=1}^{n=501} \sin\left(2\pi f_j t_i + \phi_j\right) \quad (1)$$

where a_i, t_i are the magnitude and the time at i-step, and f_j, ϕ_j are the j-th frequency and the j-th phase angle, respectively. The value of the j-th frequency is j-times of the frequency increment Δf=0.01 Hz. Only the phase angles are given as random numbers. The frequency range considered here is from 0.0 Hz to 5.0 Hz. The input motion for parametric study is determined by band-filtering the white noise described by Eq.(1). Fig.3(a) shows the example of frequency band-filter. The width and center frequency used for each case are arbitrarily chosen. The duration of input motions calculated by Eq.(1) is 5.0 seconds. In addition, to make the input acceleration wave smoothly, the beginning part (0.0-0.5 sec.) and the end part (4.5-5.0 sec.) of the acceleration are filtered as shown in Fig.3(b). The maximum amplitude of input acceleration is scaled to 0.3 m/s^2 in all cases.

As for the cases using recorded earthquake motion, the following two records are adopted as the input at

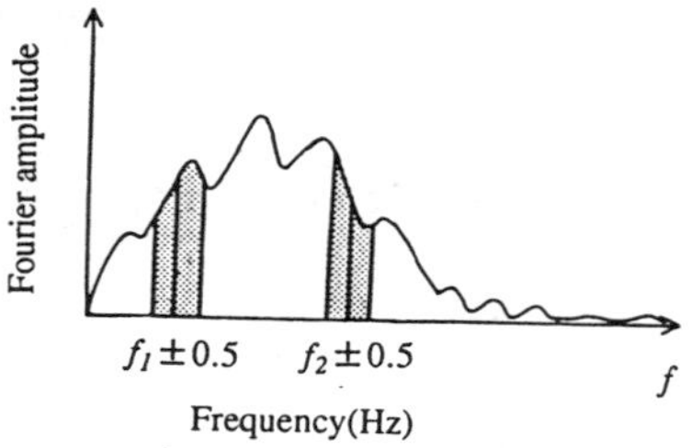

Fig.4 Filters for computing P_1 and P_2

the base: (1) acceleration recorded at Ohji site (Ohta et al. 1980), which is located in the north of Tokyo, during Miyagi pre. earthquake in 1978; and (2) acceleration recorded at Pacoima site (Prevost 1986), which is located in the south of California, during San Fernando earthquake in 1971. In the simulation, the maximum amplitude of the above two accelerations is also scaled to 0.30 m/s^2.

In order to investigate the relation between input motion and the responses, the input motion power (Katada et al. 1989) is introduced as following

$$P_0 = \sum_{i=1}^{n} a_i^2 \Delta t \quad (2)$$

where P_0 is the amount of power. And, the concept of the input motion power contributed to the 1st natural frequency of soil deposit P_1 (hereafter P_i indicates the input motion power contributed to the i-th natural frequency of soil deposit) is proposed in this paper. The value of P_1 is calculated as following: the input motion is firstly filtered using the band-filter with the center frequency of 1.68 Hz, which is the 1st natural frequency of soil deposit, and the width of band-filter is set as 1.0 Hz. Then P_1 is computed using the filtered input motion by Eq.(2). Fig.4 indicates the concept of the filter used for calculating P_1 and P_2.

Only the relation between the value of $P_1 + P_2$ and the computed responses are shown in Fig.5. The symbols in Fig.5 show the type of the input frequency. The excess pore water pressure ratio means the value of the ratio of the maximum excess pore water pressure to the initial effective vertical stress. Very clear correlation between the maximum responses and the value of $P_1 + P_2$ could be observed.

4 EFFECT OF PERMEABILITY

Generally speaking, it is very difficult to evaluate the exact permeability at a site. In order to examine the effect of the permeability, the parametric studies are conducted. Three types of permeability, namely $k_0 \times 10^1$, $k_0 \times 10^{-1}$ and $k_0 \times 10^{-2}$, are considered, here k_0 means the basic permeability for each material

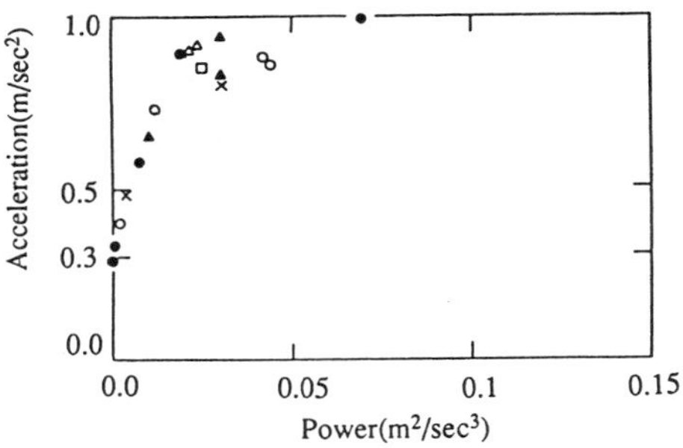

(a) Maximum acceleration at surface of soil deposit

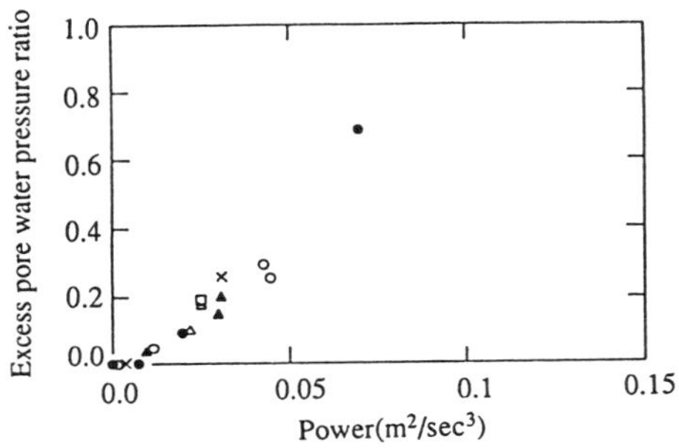

(b) Maximum excess pore water pressure at the depth of 6m

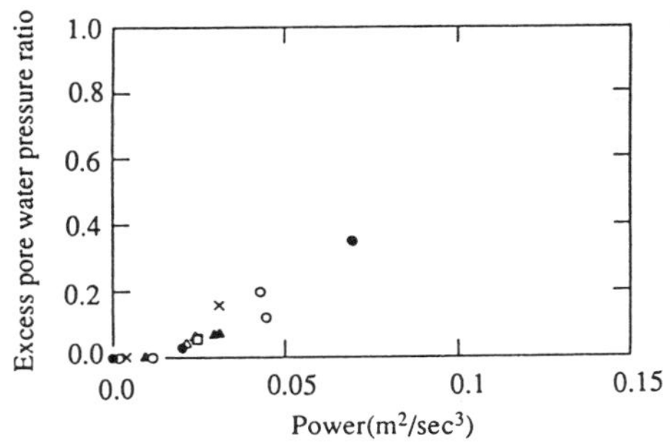

(c) Maximum excess pore water pressure at the depth of 14m

Fig.5 Relation between P_1+P_2 and computed responses

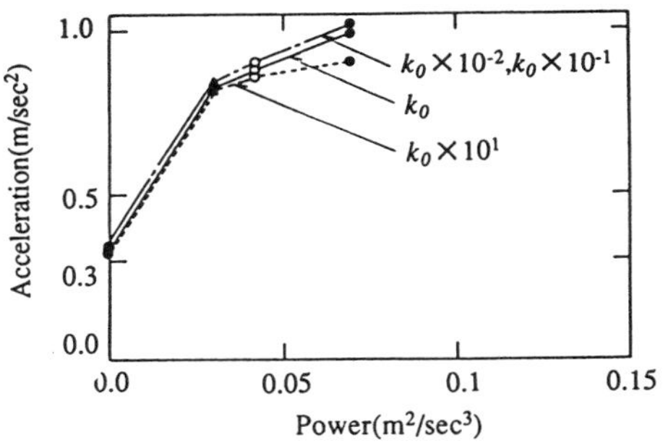

(a) Maximum acceleration at surface of soil deposit

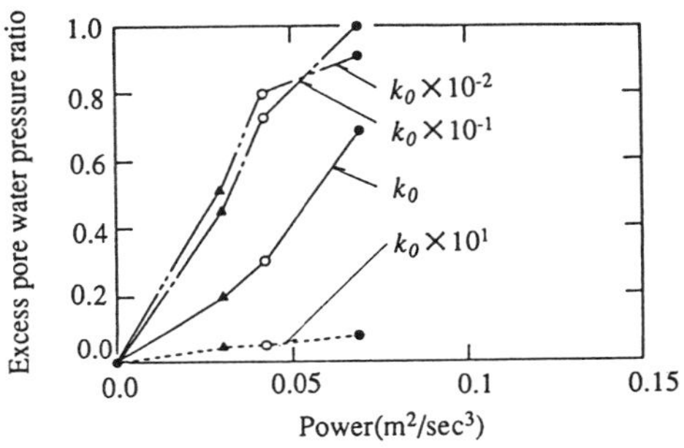

(b) Maximum excess pore water pressure at the depth of 6m

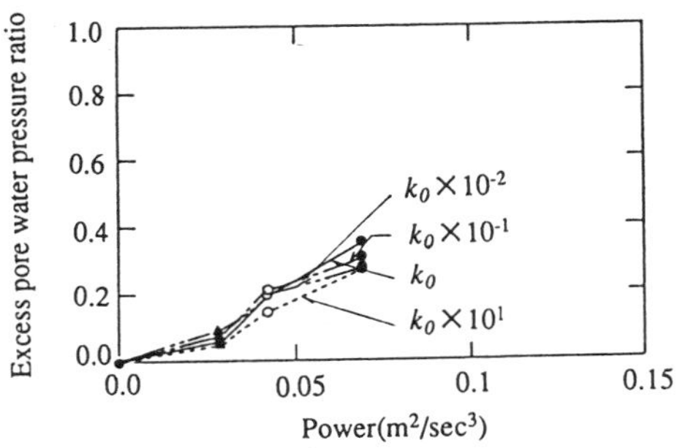

(c) Maximum excess pore water pressure at the depth of 14m

Fig.6 Relation between permeability and the computed responses

divisions as shown in Table 1.

Fig.6 indicates the relations among the responses of soil deposit, input motion power P_1+P_2 and permeability condition. The maximum acceleration increases as the permeability become smaller, but the difference is not remarkable. It should be noticed that the maximum acceleration depends on the condition before liquefaction, because the acceleration becomes nearly zero after liquefaction.

In the case of upper layer, the permeability has large effects on the maximum excess pore water pressure, and the effects are more significant when P_1+P_2 is large. On the other hand, as for the lower layer under the clay, the effects of permeability are relatively small, as the hydraulic condition of the lower layer is almost undrained because of the existing of upper impermeable clay.

As shown in Fig.6, the relation between the excess pore water pressure ratio and P_1+P_2 seems to have S-shaped curve, which is very similar to the relation between the measured excess pore water pressure ratio and the peak acceleration value observed in centrifuge tests (Arulanandan et al. 1983) under a sinusoidal excitation.

5 CONCLUSIONS

To investigate the effects of input motion and hydraulic condition on the responses of saturated soil deposit during earthquakes, parametric studies are conducted using 1-D nonlinear 2-phase FEM. The basic model of soil deposit is determined from an actual site, and the validation of the numerical model is verified through the simulation of earthquake observation. Parametric studies are performed by taking account of 17 types of input motions and 3 kinds of permeability conditions. The followings are the main results obtained from this study:

1) The effects of the input earthquake motion and the permeability could be evaluated by using a new concept of an input motion power contributed to the natural frequencies of soil deposit.

2) Both the maximum excess pore water pressure in soil deposit and the maximum acceleration at surface during earthquakes, depends on the input motion power contributed to the natural frequency of soil deposit.

3) From 2), it could be concluded that the behavior of soft deposit including nonlinear and liquefaction phenomenon strongly depends on the linear property of soil deposit before the reduction of sand stiffness due to the increasing of pore water pressure.

4) Small permeability leads to high pore water pressure and slightly large acceleration at surface. The effects of the permeability are more significant when the input motion energy contributed to the natural frequency of soil deposit is large. However, in the case of almost undrained condition like saturated sand under clay, the effects of permeability are small.

REFERENCES

Adachi, T. and F. Tatsuoka 1981. Soil mechanics (Ⅲ), *Gihoudou shuppan* (in Japanese).

Arulanandan, K., Anandarajah, A. and A. Abghari 1983. Centrifugal modeling of soil liquefaction susceptibility, *Proc. ASCE*, 109(GT3): 281-300.

Ishihara, K., Shimizu, K. and Y. Yamada 1981. Pore water pressures measured in sand deposits during an earthquake, *Soils and Foundations*, 1981, 21(4): 85-100.

Ishihara, K., Anazawa, Y. and J. Kuwano 1987. Pore water pressures and ground motions monitored during the 1985 Chiba-Ibaragi earthquake, *Soils and Foundations*, 27(3): 13-30.

Ishihara, K., Muroi ,T. and I. Towhata 1989. In-situ pore water pressures and ground motions during the 1987 Chiba-Toho-Oki earthquake, *Soils and Foundations*, 29(4): 75-90.

Iwasaki, T., Tatsuoka, F., Sakaba, Y. and H. Noma 1977. Model experiment regarding the effect of input frequency on liquefaction of sand, *Proc. 12th Conf. of JSSMFE*: 957-960 (in Japanese).

Katada, T., Orimoto, K. and T. Komuro 1989. Effects of input acceleration waves on absorbing process of hysteresis energy in soil, *Proc. JSCE*, 412(Ⅲ-12): 33-41 (in Japanese).

National Research Council 1985. Liquefaction of Soils During Earthquake, *National Academy Press*.

Ohta, T., Niwa, M. and N. Ujiyama 1980. Investigation of long-period component based on earthquake record observed at Ohji in Tokyo, *Annual Report of Kajima Institute of Construction Technology*, 28: 257-262 (in Japanese).

Prevost, J.H. 1981. DYNAFLOW: a nonlinear transient finite element analysis program, *Report 81-SM-1, Dept. of Civil Eng., Princeton University*.

Prevost, J.H. 1985. A simple plasticity theory for frictional cohesionless soils, *Soil Dyn. Earth. Eng.*, 4: 9-17.

Prevost, J.H. 1986. Effective stress analysis of seismic site response, *Int. J. Numer. Anal. Methods Geomech.*, 10: 653-665.

Sunasaka, Y., Yoshida, Y., Suzuki, K. and T. Matsumoto 1988. Comparison of the prediction methods for sand liquefaction, *Proc. 6th Int. Conf. on Num. Mech. in Geomech.*, Innsbruck: 1769-1773.

Taga, N., Kurimoto, O. and Togashi, Y. 1987. Liquefaction and seismic response of horizontal sand deposits coupled by the effect of progressive development of pore water pressure in drained condition, *Proc. of AIJ*, 374: 77-86 (in Japanese).

Yoshimi, Y. 1980. Liquefaction of sand ground, *Gihoudou shuppan* (in Japanese).

Numerical Models in Geomechanics, Pietruszczak & Pande (eds) © 1997 Balkema, Rotterdam, ISBN 90 5410 886 X

An analysis of lateral spreading of liquefied subsoil based on Bingham model

R.Uzuoka & M.Mihara
Hazama Corporation, Japan

A.Yashima
Department of Civil Engineering, Gifu University, Japan

T.Kawakami
Department of Marine Civil Engineering, Tokai University, Japan

ABSTRACT: This paper presents and verifies the numerical method to predict the lateral spreading of liquefied subsoil based on fluid dynamics. For this purpose, we use the fluid dynamic analysis code STREAM which incorporates the Bingham viscosity with the residual strength of liquefied subsoil. The numerical method is applied to the lateral spreading experiment with the shaking table and the inclined soil container. In this simulation, we determine the parameters for the Bingham model based on the results of several tests in which the coefficient of viscosity of liquefied subsoil was measured. As a result, the numerical method is found to qualitatively reproduce the time history of flow velocity of liquefied subsoil. There still, however, remains a room for further model improvement to reproduce the final configuration of ground surface.

1 INTRODUCTION

In past earthquakes, the lateral spreading of the liquefied ground has caused various damage to structures. In the 1995 Hyogoken-Nambu Earthquake, the structures located on the waterfront along the coastline and the river especially suffered serious damage by the lateral spreading (Hamada et al. 1995). In the future, the development of numerical method, which is able to predict the permanent displacement of subsoil and flow force subjected to underground structure, will become necessary to make a rational seismic design of the structures.

The liquefaction analysis based on the effective stress method has been used for simulation of the behavior during an earthquake. However, it is difficult in the current state to simulate the response during and after earthquake continuously by one analytical method. Therefore, the current lateral spreading analysis targets the behavior only after the earthquake. Various analytical methods have been proposed and they are classified into two methods which treat the liquefied ground as a solid or a fluid.

The estimation of "Brake" for the deformation of the liquefied subsoil is important to predict the observed phenomena. If the liquefied subsoil behaves like the Newtonian fluid, the ground level after flow will become horizontal. Moreover, the amount of the deformation becomes infinity if the liquefied subsoil is assumed a solid whose strength is 0. However, the ground level after the lateral spreading did not become completely horizontal and the deformation was limited in the observed phenomena.

Most of methods which assume the liquefied subsoil as a solid are based on the static deformation analysis using the finite element method. "Brake" is estimated in typical analytical methods by the following.

1. The infinitesimal deformation analytical method with the linear elastic model: The elastic modulus of the liquefied subsoil was approximated to 1/1000 of the initial one. According to this approximation, this method could consider the effect of the residual strength and the decrease in the initial shear stress due to the large deformation (Yasuda et al. 1992).

2. The large deformation analytical method with the nonlinear model: This method used the post-liquefaction stress-strain relation in consideration with the residual strength. The deformation stopped in the state where the redistributed initial shear stress balanced the residual strength (Finn et al. 1991).

On the other hand, the analytical method with the Newtonian fluid was proposed as the method which treats the liquefied subsoil as a fluid (Nakamura et al. 1991). However this method could not consider "Brake" for the deformation mentioned above.

In this paper, we propose the numerical method based on the fluid dynamics with the Bingham model by which "Brake" can be considered. We verify this numerical method based on the results of the lateral spreading experiment with the shaking table and the inclined soil container. In this analyses, the parameters of the model were determined based on several experiment results in which the coefficient of viscosity of the liquefied subsoil was measured.

2 ANALYTICAL METHOD

We used the general-purpose fluid analysis code STREAM (Software Cradle Co, Ltd 1990) . The characteristics of this code are as follows.

1. The governing equations consist of the mass conservation law, the momentum conservation law and the energy conservation law (when considering the change in temperature). The finite volume method which is one of the finite difference method is used for the spacial discretization.

2. This code can simulate the unsteady analysis with non-Newtonian fluid model. However, in each calculation step it solves Navier Stokes equations because the viscosity of non-Newtonian fluid model is treated to be constant.

3. The VOF method (Hilt & Nichols 1981) is used to express a free surface. This method obtains the liquid occupation ratio in the cell by using the state function which is 1 in case of the liquid fully occupied the cell and 0 in case of the gas fully occupied the cell.

We treated the liquefied subsoil as incompressible Bingham fluid. The Bingham viscosity is expressed as the equivalent Newtonian viscosity in each calculation step in STREAM.

In the pure shear condition, the shear stress-shear strain rate relation of the Bingham model can be shown by the following equation.

$$\tau = \eta \cdot \dot{\gamma} + \tau_f \tag{1}$$

where τ is the shear stress, η is the coefficient of viscosity after yield, $\dot{\gamma}$ is the shear strain rate and τ_f is the yield strength. This model behaves as a rigid body when the shear stress is smaller than the specified yield value. After the shear stress exceeds the yield value, it behaves as a Newtonian fluid. Up to now, the Bingham model has been used to analyze the phenomena to show the behavior like a visco-plastic flow (for example, rapid flow slides (Hunger 1995), fresh concrete flow (Tanigawa & Mori 1989) and so on).

On the other hand, the shear stress-shear strain rate relation of the Newtonian fluid can be shown by the following equation.

$$\tau = \eta' \dot{\gamma} \tag{2}$$

where η' is the equivalent coefficient of viscosity. Substituting equation (2) into (1), the equivalent coefficient of viscosity is obtained as follows:

$$\eta' = \eta + \frac{\tau_f}{\dot{\gamma}} \tag{3}$$

The relation of these equations (1) - (3) is shown in Figure 1. The equivalent coefficient of viscosity shows a secant coefficient of viscosity of the Bingham model

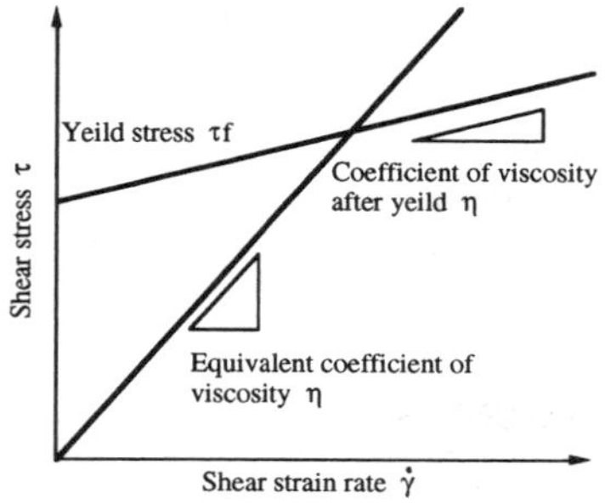

Figure 1. Bingham model

as shown in Figure 1. In the three-dimensional analysis with STREAM, $\dot{\gamma}$ is calculated as second invariant of the deviatoric strain rate tensor.

The yield value of the Bingham model is considered to be the residual strength of the liquefied ground in the flow phenomenon like the lateral spreading. Therefore, when the liquefied subsoil is modeled by the Bingham model, the following model characteristics can be taken into account.

1. The lateral spreading does not occur when the driving shear stress due to gravity is smaller than the residual strength. The liquefied ground behaves as a rigid body.

2. The lateral spreading is triggered when the driving shear stress due to gravity is greater than the residual strength. The liquefied ground behaves as a Newtonian fluid.

The method to determine the residual strength has been studied by the following procedures: the back analysis of the lateral spreading (Towhata 1995), the laboratory tests (Ishihara 1993) and so on. For example, Ishihara (1993) has proposed the following equation for the residual strength.

$$\frac{S_{us}}{\sigma_0'} = \frac{M}{2} \cdot \cos \phi_s \cdot \frac{p_s'}{p_c'} \tag{4}$$

where S_{us} is the residual strength, σ'_0 is the initial effective confining pressure, M is the stress ratio at the quasi-steady state, ϕ_s is the angle of phase transformation at the quasi-steady state, p'_s is the effective confining pressure at the quasi-steady state, p'_c ($=\sigma'_0$) is used instead of σ'_0 when referring to the point in the stress path.

According to equation (4), the residual strength depends on the effective confining pressure and varies at the different levels in the subsoil. Therefore, we assumed that the equivalent coefficient of viscosity is defined by the following equation in consideration of the dependency of the yield value τ_f on the confining pressure.

$$\eta' = \eta + \frac{R_r \cdot p}{\dot{\gamma}} \tag{5}$$

where R_r is the residual strength ratio and p is the hydraulic pressure (total confining pressure of the liquefied subsoil). The residual strength ratio shown by equation (4) is expressed as the ratio of the residual strength to the effective confining pressure. On the other hand, the residual strength ratio R_r of equation (5) is the ratio of the residual strength to the total confining pressure because the liquefied subsoil is assumed to be one phase material of the fluid.

3 LATERAL SPREADING EXPERIMENT

The lateral spreading experiments were carried out with the inclined soil container on the shaking table by Hamada et al. (1994). In this experiment, the shaking table was stopped after the model ground was liquefied, then the soil container was gradually inclined. The characteristic of this experiment was that the lateral spreading occurred without excitation.

The general view of the experiment device is shown in Figure 2. The foundation ground was prepared with Enshunada sand. The model ground was prepared by the boiling method to get the homogeneous loose condition. The measurements of the final displacement (by the marks in the ground), the subsoil displacement during the flow (by the rolling type displacement meters) and the excess pore water pressure during flow (by the pore pressure transducers) were made. The arrangement of the rolling type displacement meters and the pore pressure producers are shown in Figure 2.

Several cases of experiments were carried out by changing the conditions of the relative density of the sand, the inclined angle of the soil container and so on. In this paper we simulated the experiment case in which the relative density was 41% and the maximum inclination was 4.2% without non-liquefaction layer. The time history of the excess pore water pressure and the subsoil displacement in this case are shown in Figure 3 and Figure 4, respectively. The excess pore water pressure in the ground decreased a little when the soil container inclination started. The excess pore water pressure of PW2 (at the depth of 12cm) decreased suddenly when the flow nearly stopped (at about 10 seconds). On the other hand, the subsoil displacement was triggered at about 0.7 seconds after the soil container began to incline and it stopped about three seconds later.

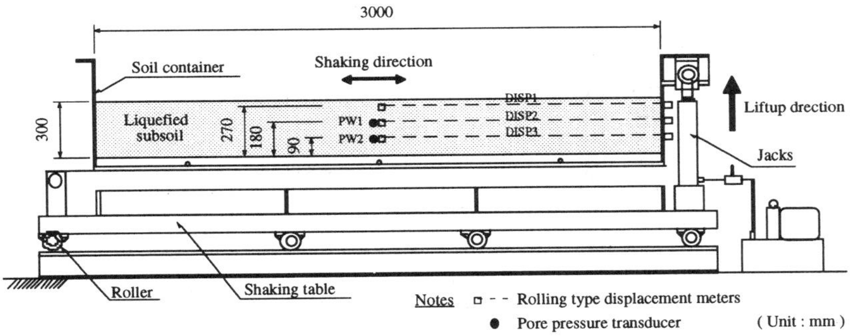

Figure 2. Shaking table and inclined soil container (Hamada et al. 1994)

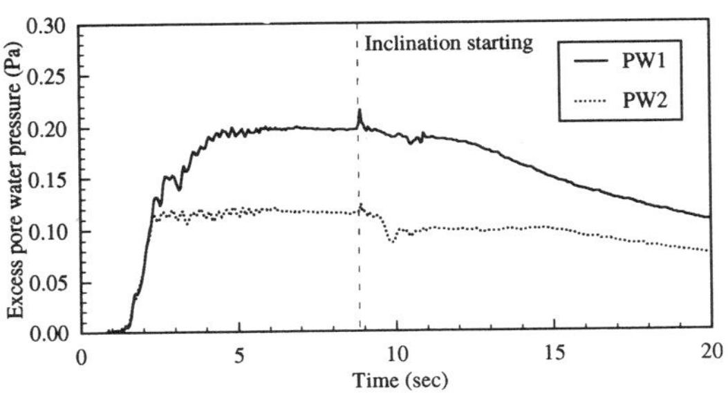

Figure 3. Time history of the excess pore water pressure (Hamada et al. 1994)

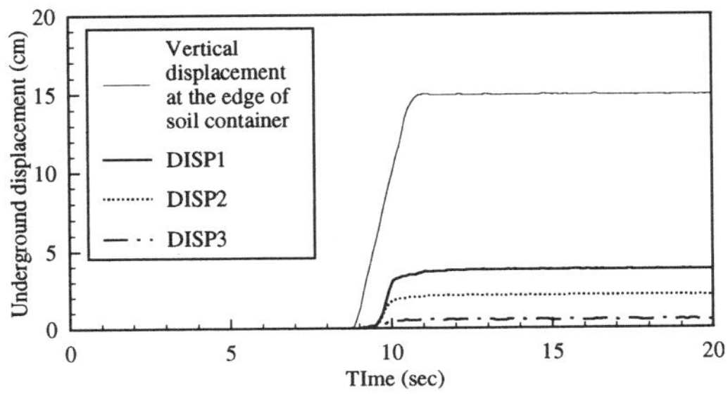

Figure 4. Time history of the subsoil displacement (Hamada et al. 1994)

4 ANALYTICAL CONDITION

In the beginning, we discuss the residual strength ratio R_r of the liquefied subsoil. It is assumed that the lateral spreading is triggered when the shear stress driven by the inclination of the soil container exceeds the residual strength. Based on this assumption, the residual strength ratio can be calculated from the angle of the soil container when the underground displacement was triggered by the following equation.

$$R_r = \tan \theta \quad (6)$$

where θ is the angle of the soil container inclination. Because the height at the edge of the soil container when the underground displacement was triggered was about 5.5cm, the residual strength ratio R_r became 0.018(=5.5/300).

Next, we discuss the equivalent coefficient of viscosity. The coefficients of viscosity measured by the lateral spreading experiment (Hamada et al. 1994), the load acting on sphere (Kawakami et al. 1994) and the viscometer (Kawakami et al. 1994) are shown with relation to the shear strain rate in Figure 5. The coefficients of viscosity for the analytical case 1-3 mentioned below are also shown in Figure 5.

The equivalent coefficient of viscosity was determined from these measurements shown in Figure 5. To understand the sensibility of the equivalent coefficient of viscosity to the result and the characteristic of the Bingham model, three cases of numerical simulation were carried out. The value of η and R_r in each cases are shown in Table 1. In the case 1, we used the residual strength ratio calculated from equation (6). In the case 2, we used the larger residual strength ratio than that of case 1. In case 3, we used the Newtonian viscosity. η was the lower limit value of the equivalent coefficient of viscosity, it was 1.0 in the case 1 and 2. The coefficient of viscosity in case 1 and 2 roughly reproduced the tendencies of various measurement results.

The relations between the shear strain rate and the equivalent coefficient of viscosity in each cases are shown in Figure 5. The relations between the shear strain rate and the shear stress in each cases are shown in Figure 6. The relations in case 1 and 2 were calculated in the central depth (at the depth of 15cm) of soil container.

Figure 7 shows the numerical model. The difference lattice was 2.0 cm pitch in the horizontal and vertical direction. In the experiment, the soil container was inclined gradually (up to 4.2%) after the excitation stopped. However, because it was difficult to simulate this condition of the experiment, the model was treated as the inclined ground (4.0%) in the inclined soil container in the simulation.

The boundary conditions of the soil container wall (right, left and bottom side) were nonslip boundaries

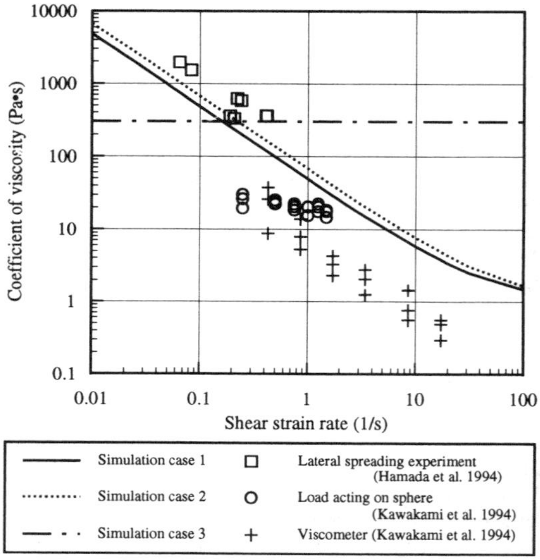

Figure 5. Measured and analytical coefficients of viscosity

Table 1. Analytical parameters in each cases

Case	η	Rr	Notes
1	1.0	0.018	Bingham fluid
2	1.0	0.025	Bingham fluid
3	300.0	0	Newtonian fluid

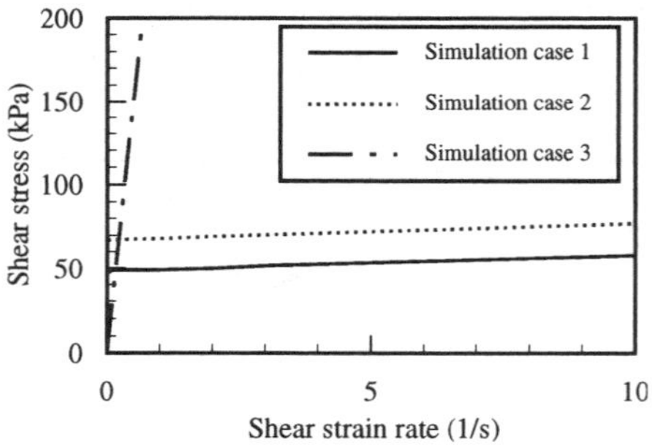

Figure 6. Relations between the shear strain rate and the shear stress in each cases

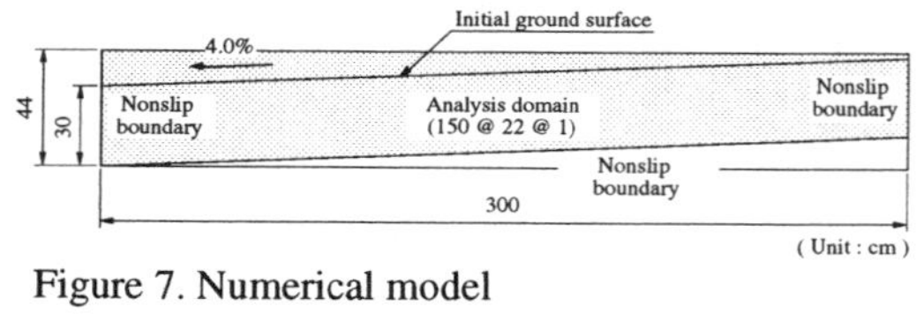

Figure 7. Numerical model

where the relative flow velocity to the wall is 0. The external force considered in the analysis was only gravity force, so we determined the body force from the saturated unit weight of the liquefied subsoil on the assumption that the ground was completely liquefied.

The simulations were carried out during 11 seconds after the soil container began to incline. The calculation time increment was determined to satisfy the stability condition automatically.

5 RESULTS OF SIMULATIONS

The time histories of the flow velocity at the central part of the container in the experiments and simulations are shown in Figure 8 for each case. The beginning time of the simulations was set to be the same time when the flow was triggered in the experiment. The following tendencies are obtained from Figure 8.

1. In the case 1, the simulation roughly reproduced the tendency of the experimental result that the flow was triggered and stopped rapidly. However, the simulated maximum flow velocity was considerably larger than the experimental one. Moreover, in terms of the distribution of the flow velocity in the depth direction, the simulation overestimated the flow velocity at the deeper measurement point.

2. In the case 2, the simulation also showed the tendency same as the case 1. The simulated maximum flow velocity was closer to the experimental one.

3. In the case 3, the simulated maximum flow velocity agreed with the experimental one. However, after 10.0 seconds when the velocity was maximum, the simulated flow velocity decreased slower than the experimental one and stopped at about 15 seconds.

In the case 1 and 2, one of the reasons why the simulated results overestimated the flow velocity was the difference between the analytical condition and the experimental condition. Although the soil container was inclined gradually in the experiment, the soil container was inclined with the final inclination angle from the beginning in the analysis. Therefore, as the initial shear stress driven in the analysis was larger than in the experiment, the faster flow was generated.

The final surface configuration of the experimental and the simulated results for each case are shown in Figure 9. The following tendencies are derived from Figure 9.

1. In the case 1, the final ground surface level became horizontal because the flow velocity was too large shown in Figure 8.

2. In the case 2, the inclined part remained a little in the final ground surface level. However, the surface configuration kept constant inclination angle. It was different from the experimental one with the inclination remained in the center part.

3. In the case 3, the final ground surface level became

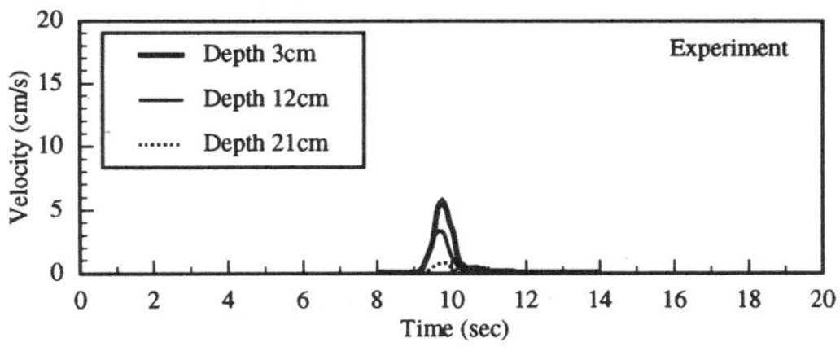

(a) Experiment (Hamada et al. 1994)

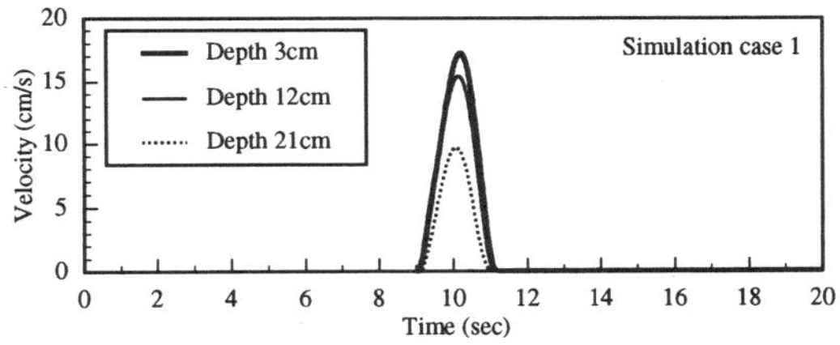

(b) Simulation (Case 1)

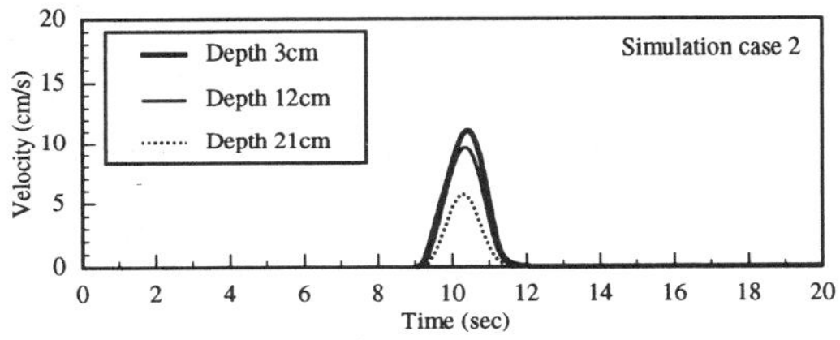

(c) Simulation (Case 2)

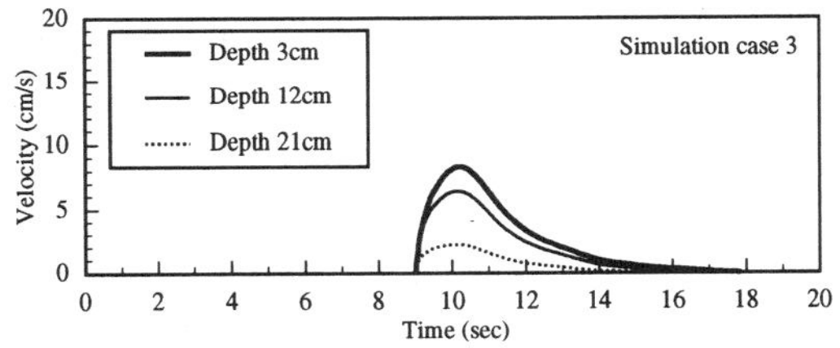

(d) Simulation (Case 3)

Figure 8. Time hisitories of the flow velocity

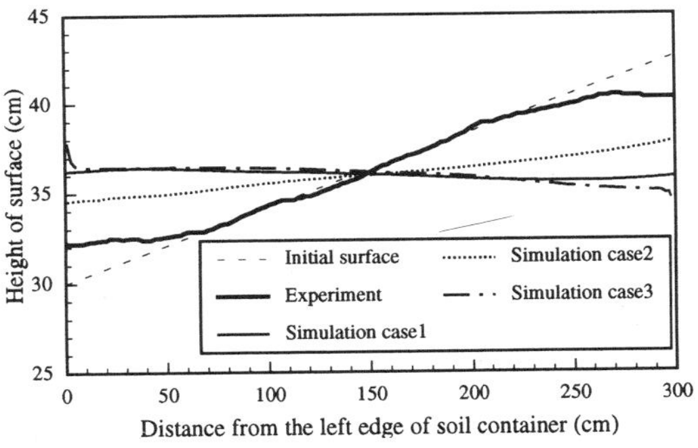

Figure 9. Final surface configurations

horizontal because the duration of the flow was too long shown in Figure 8.

Compared with the case 1 and 2, the simulated results were different in spite of a small difference in the residual strength ratio. The residual strength ratio is the most important parameter and it needs further consideration on the determination method of the residual strength in this analytical method.

6 CONCLUSIONS

This paper presented and verified the numerical method to predict the lateral spreading based on fluid dynamics. We used the fluid dynamic analysis code STREAM which incorporates the Bingham viscosity with the residual strength of liquefied subsoil. As a result of verification, the numerical method was found to qualitatively reproduce the time history of flow velocity of liquefied subsoil. However, because we could not simulate the experiment condition exactly, a quantitative estimation of the flow velocity was insufficient.

There is a room for further investigation about the experimental procedure which were not reproduced by the simulation. As Figure 4 indicates, the excess pore water pressure in the ground decreases gradually from the beginning of the lateral spreading, this decrease in the excess pore water pressure may produce the recovery of strength of the subsoil. However the recovery of the soil strength was not considered in this analytical method. There still remains a room for further model improvement.

ACKNOWLEDGMENTS

The authors would like to thank Professor Kyoji Sassa of Disaster Prevention Research Institute of Kyoto University for providing them with the materials about rapid landslides, Assistant Professor Hideo Sekiguchi of Kyoto University for valuable discussion about the lateral spreading mechanism and Professor Yuichiro Fujita of Gifu University for helpful suggestions about the fluid analysis method.

REFERENCES

Finn, W.D.L., Yogendrakumar, M., Ledbetter, R. and Yoshida, N. 1991. Analysis of liquefaction induced displacements, Proc. of 7th ICCMAG, pp.913-921.

Hamada, M., Sato, H. and Kawakami, T. 1994. A consideration of the mechanism for liquefaction-ralated large ground displacement, Proceedings from the Fifth U.S.-Japan Workshop on Earthquake Resistant Design of Lifeline Facilities and Countermeasures Against Soil Liquefaction, Technical Report NCEER-94-0026, pp.217-232.

Hamada, M., Isoyama R. and Wakamatsu, K. 1996. Liquefaction-induced ground displacement and its related damage to lifeline facilities, Soils and Foundations, SPECIAL ISSUE on Geotechnical Aspects of the January 17 1995 Hyogoken-Nambu Earthquake, pp.81-97.

Hilt, C.W. and Nichols, B.D. 1981. Volume of fluid method for the dynamics of free boundaries, Comp. Phys. J., No.39, pp.201-225.

Hunger, O. 1995. A model for the runout analysis of rapid flow slides, debris flows, and avalanches, Can. Geotech. J. , No.32, pp.610-623.

Ishihara, K. 1993. Liquefaction and flow failure during earthquakes, Geotechnique, Vol.43, No.3, pp.351-415.

Kawakami, T., Suemasa, N., Hamada, M., Sato, H. and Katada, T. 1994. Experimental study on mechanical properties of liquefied sand, Proceedings from the Fifth U.S.-Japan Workshop on Earthquake Resistant Design of Lifeline Facilities and Countermeasures Against Soil Liquefaction, Technical Report NCEER-94-0026, pp.285-299.

Nakamura, T., Sato, H. and Doi, M. 1993. Analysis of lateral flow due to liquefaction by BEM-FEM coupling model, Proc. of 10th Symposium on BEM's, pp.53-58 (in Japanese).

Software Cradle Co., Ltd 1990. Three dimensional heat-fluid analysis program STREAM.

Tanigawa, Y. and Mori, H. 1989. Analytical study on deformation of fresh concrete, J. Eng. Mec., Proc. of ASCE, Vol.115, No.3, pp.493-508.

Towhata, I. 1995. Liquefaction and associated phenomenon, First International Conference on Earthquake Geotechnical Engineering, Theme Lecture.

Yasuda, S., Nagase, H., Kiku, H. and Uchida, Y. 1992. The mechanism and a simplified procedure for the analysis of permanent ground displacement due to liquefaction, Soils and Foundations, Vol.32, No.1, pp.149-160.

Numerical Models in Geomechanics, Pietruszczak & Pande (eds) © 1997 Balkema, Rotterdam, ISBN 90 5410 886 X

A comparison between elastostatic and elastodynamic stochastic finite element analysis of pavement structures

M. Parvini & D. F. E. Stolle
McMaster University, Hamilton, Ont., Canada

ABSTRACT: A stochastic finite element model based on the perturbation technique is used to make comparison between the stochastic responses of a pavement system subjected to static and dynamic loads. It is shown that, for an approximated FWD load configuration and a two-layer pavement-subgrade system, the sensitivity of deflection variation to variation in the layer moduli does not change significantly when static analysis is used instead of a dynamic one, except for the case of shallow bedrock.

1 INTRODUCTION

Most of the recently developed, non-destructive, in-situ tests use a dynamic load in order to better simulate the effect of a moving wheel load on the pavement structure. However, a static approach to pavement deflection analysis and backcalculation of its system parameters is still common practice among the practitioners. There are many studies which compare the static and dynamic approaches for analyzing pavement structures or characterizing their properties in a backcalculation framework, see e.g. Mamlouk and Davies (1984), Ong et al. (1991), and Zaghloul et al. (1994). While these studies provide a good understanding of what happens to the expected values of the quantities under consideration if a dynamic analysis is used instead of a static one, they are not capable of commenting on the variation of the response or the uncertainty associated with backcalculated parameters.

The stochastic finite element method (SFEM) has been used to calculate the expected value of pavement deflection under a static load, along with its variation due to variation in layer moduli (Parvini and Stolle 1996). It was demonstrated through a forward calculation that the variation in deflection may be related to variation in layer moduli, thus allowing the engineer to infer on the uncertainty of backcalculated layer moduli from the possible variation in the pavement deflection measurements.

The objective of this paper is to compare an elastodynamic approach to pavement deflection analysis with an elastostatic one from the stochastic's point of view. For this purpose, a stochastic finite element method based on the second order perturbation technique is used. Emphasis is placed on the analyses of pavement systems subjected to loads typically used when simulating the FWD test.

2 STOCHASTIC FORMULATION

The results of a stochastic finite element analysis includes the expected value of a random quantity, as well as its higher order statistical moments. The stochastic equations leading to the statistical moments of a random function have been derived by Kleiber and Hien (1992). For the case of a structural system under static loads, these equations are relatively straightforward, while when dealing with dynamic loads, the level of complexity increases due to the interaction between the various modes of a response, or different time steps. However, as demonstrated below, if a periodic-load analysis approach to the dynamic problem is adopted, the formulation of the dynamic SFEM is similar to that of the static one, with the main exception that in the dynamic case computation is completed in the complex space due to the complex-valued nature of the dynamic stiffness matrix.

2.1 *General dynamic load*

The matrix form of the equation of motion for a multi degree of freedom (MDOF) system is given by

$$\mathbf{M}\,\ddot{\mathbf{u}}(\mathbf{t}) + \mathbf{C}\,\dot{\mathbf{u}}(\mathbf{t}) + \mathbf{K}\,\mathbf{u}(\mathbf{t}) = \mathbf{f}(\mathbf{t}) \qquad (1)$$

in which **u(t)** and **f(t)** are the displacement and force vectors, and **M**, **C**, and **K** are the mass, damping, and stiffness matrices of the system, respectively. A dot over **u** denotes differentiation with respect to time, t. For a system with a random property b, the mass, damping, stiffness and, at times, the applied force may be functions of the random variable. Therefore, the displacement, and consequently its derivatives with respect to time, are also functions of the random variable. Approximating the displacement by its truncated Taylor's expansion around the mean value b_0, and using the same procedure that is adopted for the static case, the expected value vector and covariance matrix of the displacement are obtained from

$$\mathbf{E}[\mathbf{u}(\mathbf{t})] = \mathbf{u}(\mathbf{t}) + \frac{1}{2}\mathbf{u}(\mathbf{t})^{,bb}\,\mathrm{var(b)} \qquad (2)$$

$$\mathbf{cov}(\mathbf{u}(\mathbf{t})) = (\mathbf{u}(\mathbf{t})^{,b})\mathrm{var(b)}(\mathbf{u}(\mathbf{t})^{,b})^{t} \qquad (3)$$

as described by Kleiber and Hien (1992). In these equations the superscript " ^{,b} " denotes differentiation with respect to b and all quantities are evaluated at b_0. The superscript "t" is used to denote the transpose of a vector, and var(b) implies the variance of random variable b.

To find the derivatives of displacement in eq.s 2 and 3, the equation of motion is differentiated two times with respect to random variable b. After rearranging the terms in the left-hand sides of the resulting equations and regrouping them, the following two equations are obtained

$$\mathbf{M}\ddot{\mathbf{u}}^{,b} + \mathbf{C}\dot{\mathbf{u}}^{,b} + \mathbf{K}\mathbf{u}^{,b} = \mathbf{F}^{,b} \qquad (4\text{-}1)$$

$$\mathbf{M}\ddot{\mathbf{u}}^{,bb} + \mathbf{C}\dot{\mathbf{u}}^{,bb} + \mathbf{K}\mathbf{u}^{,bb} = \mathbf{F}^{,bb} \qquad (4\text{-}2)$$

in which

$$\mathbf{F}^{,b} = \mathbf{f}^{,b} - (\mathbf{M}^{,b}\ddot{\mathbf{u}} + \mathbf{C}^{,b}\dot{\mathbf{u}} + \mathbf{K}^{,b}\mathbf{u}) \qquad (5\text{-}1)$$

$$\begin{aligned}\mathbf{F}^{,bb} = \mathbf{f}^{,bb} - (&\mathbf{M}^{,bb}\ddot{\mathbf{u}} + \mathbf{C}^{,bb}\dot{\mathbf{u}} + \mathbf{K}^{,bb}\mathbf{u} + \\ &2\mathbf{M}^{,b}\ddot{\mathbf{u}}^{,b} + 2\mathbf{C}^{,b}\dot{\mathbf{u}}^{,b} + 2\mathbf{K}^{,b}\mathbf{u}^{,b})\end{aligned} \qquad (5\text{-}2)$$

In order to evaluate the expected value and covariance of the displacement, eq.s 4-1 and 4-2 together with eq. 1 must be solved simultaneously to obtain **u**, $\mathbf{u}^{,b}$ and $\mathbf{u}^{,bb}$. This process can be simplified substantially for a periodic loading.

2.2 *Periodic load*

If **f(t)** is periodic, it can be expressed using a Fourier series expansion as

$$\mathbf{f}(\mathbf{t}) = \sum_{n=-\infty}^{+\infty} \tilde{\mathbf{f}}\,c_n e^{iw_n t} \qquad (6)$$

in which $\tilde{\mathbf{f}}$ is a frequency independent vector of load amplitudes, and c_n is the Fourier coefficient associated with the nth angular frequency, w_n. The displacement solution of such a system is defined by

$$\mathbf{u}(\mathbf{t}) = \sum_{n=-\infty}^{+\infty} \tilde{\mathbf{u}}_n c_n e^{iw_n t} \qquad (7)$$

in which $\tilde{\mathbf{u}}_n$ is the complex-valued vector of displacement amplitudes obtained from

$$\tilde{\mathbf{K}}_n\,\tilde{\mathbf{u}}_n = \tilde{\mathbf{f}} \qquad (8)$$

$\tilde{\mathbf{K}}_n$ is called the dynamic stiffness matrix and is defined by

$$\tilde{\mathbf{K}}_n = \mathbf{K} - \mathbf{M}\,w_n^2 + i\,\mathbf{C}\,w_n \qquad (9)$$

It may be noted that the matrix form given by eq. 8 is similar to that for the static case.

Using eq. 7, the expected value and covariance of **u(t)** are given by

$$\mathbf{E}[\mathbf{u}(\mathbf{t})] = \sum_{n=-\infty}^{+\infty} \mathbf{E}[\tilde{\mathbf{u}}_n] c_n e^{iw_n t} \qquad (10)$$

$$\mathbf{cov}(\mathbf{u}(\mathbf{t})) = \sum_{n=-\infty}^{+\infty}\sum_{m=-\infty}^{+\infty} c_n c_m e^{iw_n t} e^{iw_m t}\,\mathbf{cov}(\tilde{\mathbf{u}}_n, \tilde{\mathbf{u}}_m) \qquad (11)$$

For the case of an even periodic forcing function, the trigonometric form of eq. 6 can be written as

$$\mathbf{f(t)} = \sum_{n=0}^{\infty} \tilde{\mathbf{f}} a_n \cos(w_n t) \qquad (12)$$

in which $\frac{a_n}{2} = c_n = c_{-n}$ and $a_0 = c_0$.

Considering only the first N components of Fourier expansion and expressing the complex valued $\tilde{\mathbf{u}}_\mathbf{n}$ in terms of its real and imaginary form of

$$\tilde{\mathbf{u}}_\mathbf{n} = \mathbf{x}_\mathbf{n} + i\,\mathbf{y}_\mathbf{n} \qquad (13)$$

the expected value of **u(t)** is obtained from

$$\mathbf{E[u(t)]} = \sum_{n=0}^{N} \left\{ a_n [\mathbf{E[x_n]} \cos(w_n t) - \mathbf{E[y_n]} \sin(w_n t)] \right\} \qquad (14)$$

in which $\mathbf{E[x_n]}$ and $\mathbf{E[y_n]}$ are real-valued quantities which can be found directly be taking the expected value of eq. 13.

The covariance matrix of the displacement for such a system may be found from

$$\mathbf{cov(u(t))} = \sum_{j=0}^{N} \sum_{l=0}^{N} a_j a_l \{ \mathbf{cov(x_j, x_l)} \cos(w_j t) \cos(w_l t) + \mathbf{cov(y_j, y_l)} \sin(w_j t) \sin(w_l t) - \mathbf{cov(x_j, y_l)} \cos(w_j t) \sin(w_l t) - \mathbf{cov(y_j, x_l)} \sin(w_j t) \cos(w_l t) \} \qquad (15)$$

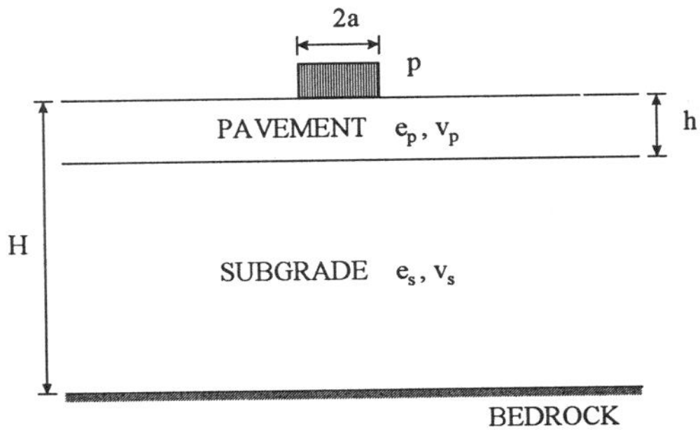

Fig. 1 Two-layer model of a pavement-subgrade system

in which a covariance matrix such as $\mathbf{cov(x_j , y_l)}$, which in this case accounts for the coupling between real and imaginary components, is defined by an equation similar to the equation used in the static case.

It should be noted that, the time-dependent variance of displacement at each degree of freedom is given by the associated diagonal member of **cov(u(t))**. To find the coefficient of variation of peak displacement at a point, the square root of variance at the time of maximum displacement is divided by the expected value of maximum displacement of that point.

3 NUMERICAL EXAMPLE

The results of the static and dynamic stochastic finite element analyses of a simplified pavement-subgrade model shown in Figure 1 are summarized in this section. The expected value of the pavement and subgrade elastic moduli are assumed to be e_p=4000 and e_s=100 MPa, respectively, with Poisson ratios of v_p=0.35 for the pavement and v_s=0.40 for the subgrade. The selected values are typical resilient moduli and Poisson ratios for hot mix asphalt pavement and fine-grained subgrade (Huang 1993). Since materials are assumed to be linear, the elastic and resilient moduli are identical. For both the pavement and subgrade layers, the density is considered to be 2.0 Mg/m^3 with a hysteretic damping ratio of 5 percent. Unless otherwise stated, the thickness of the pavement layer is selected as h=0.15 m with bedrock depth equal to H=7.35 m.

For the static case, a 40 kN static load was assumed to be uniformly applied to the surface of the pavement layer over an are a corresponding to a radius of a=0.15 m. In the case of dynamic loading, the FWD load history was modeled by a uniformly distributed, half-sine impact load with a duration of 0.03 second and a maximum amplitude of 40 kN. For such a load, most of the power is contained within the frequency range of 0 to 150 rad/s (Sebaaly et al. 1986). Given that the actual analysis was performed assuming a periodic load with a 0.1 second period, the load history was approximated using a Fourier series. The relatively long period allowed the pavement system to resume its stationary condition before the next period started. Ten harmonic components covering a frequency range of 0 to 500 rad/s were found to give a fairly good approximation to the loading function.

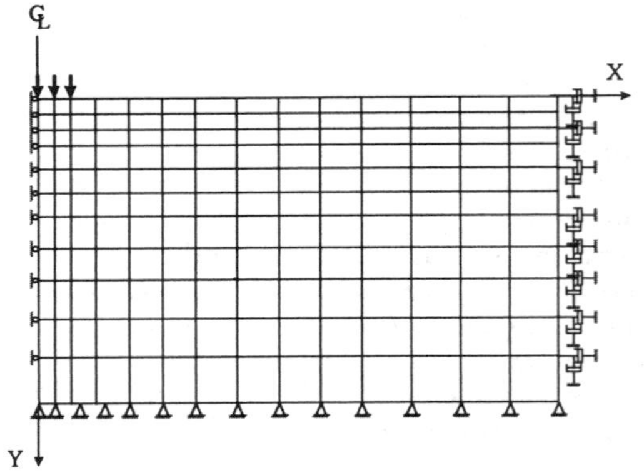

Fig. 2 Schematic of dynamic finite element model

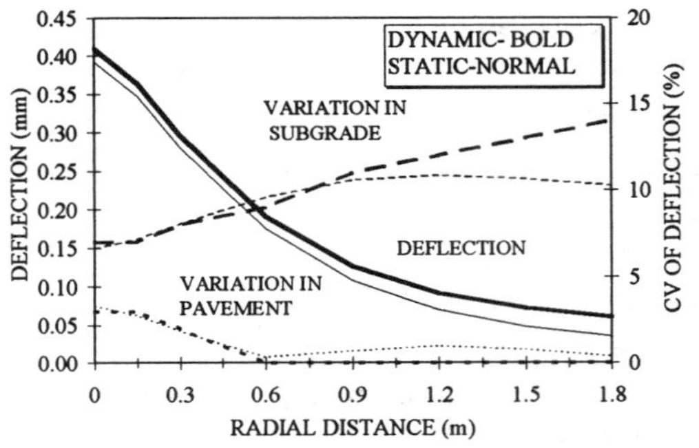

Fig. 3 Expected value and coefficient of variation of deflection in static and dynamic cases

The pavement-subgrade system was discretized using 4-noded, quadrilateral elements arranged in 30 rows and 40 columns. Therefore, the total number of elements in the mesh was 1200 with the total number of nodes equal to 1271. To meet the requirements of Lysmer and Kuhlemeyer (1969) the maximum element size in the region of interest was selected to be 0.15 m which was about 1/12 of the wavelength associated with the highest frequency component of the dynamic impact load. Solutions for dynamic problems, unlike for the static ones, are also extremely sensitive to the presence of artificial boundaries, even when such boundaries are located far from the region of interest. This is due to the reflection of energy back into the system at these boundaries. In order to simulate the effect of the outward radiation of energy at the artificial boundaries, transmitting boundaries in the form of dampers (Wolf 1988) were used for the dynamic case. The vertical transmitting boundaries were placed 11.85 m away from the point of load application, which corresponds to approximately 7 times the minimum wavelength associated with the propagating waves. Although less number of elements could have used to obtain accurate solutions for the static case, to be consistent with the dynamic analysis, the same mesh was used in both cases. A schematic of the finite element model used in the dynamic analysis is illustrated in Figure 2.

Figure 3 shows the deflection basin under the static load, along with the curve connecting peak deflections at different offsets under the dynamic load. For this pavement system configuration, the static deflections are slightly less than the dynamic ones. This figure also provides the coefficient of variation of deflection, CV(u), when there is a 10% variation in either the subgrade modulus, $CV(e_s)=10\%$, or the pavement modulus, $CV(e_p)=10\%$. For both cases, CV(u) at close offsets are almost the same for the static and dynamic cases. The fact that CV(u) corresponding to variations in subgrade modulus are much higher than those associated with variations in pavement modulus indicates that the surface deflections of a pavement-subgrade system are dominated by the properties of the subgrade. Also briefly summarized that, for the selected material characteristics and loading history, the sensitivity of deflection variation to variation in subgrade or pavement modulus are similar for dynamic and static analyses. This supports the belief that conclusions made regarding the uncertainty in backcalculated parameters via static analysis may be also applicable for dynamic cases involving FWD loading.

Figure 4 shows that, while in the static case the CV(u) due to $CV(e_s)$ does not change significantly as a result of a change in the bedrock depth, it does for the dynamic case. In the latter case, overall, CV(u) decreases as H decreases, however, there is not a consistent pattern at all the offsets. The reason for the sudden changes in the pattern of the curves in dynamic analysis may be attributed to the change in dynamic stiffness characteristics of the system as the depth to the bedrock changes. Figure 5 clearly shows that CV(u) due to variations in pavement modulus is insensitive to the bedrock depth for both static and dynamic loading. Once again, this may be attributed to the fact that the properties of the subgrade dominate the response of the pavement system.

Simulations were also carried out for pavement

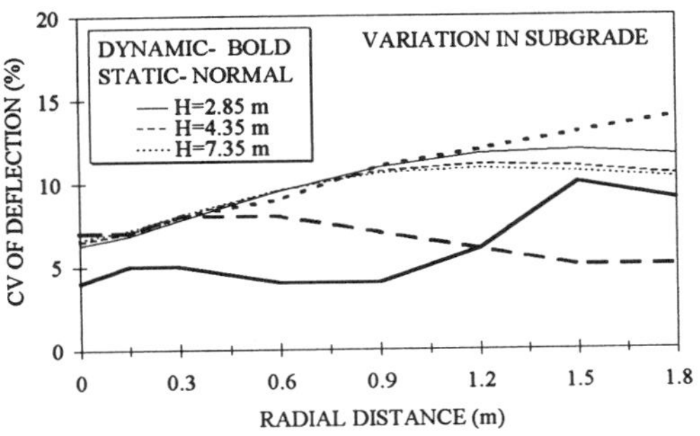

Fig. 4 Variation in deflection due to variation in subgrade modulus for different bedrock depths

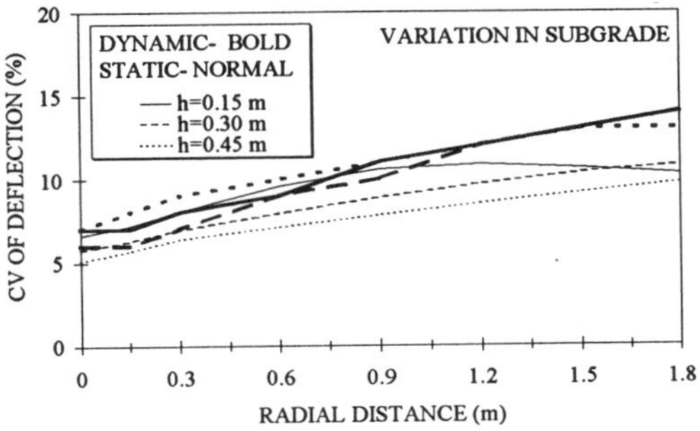

Fig. 6 Variation in deflection due to variation in subgrade modulus for different pavement thicknesses

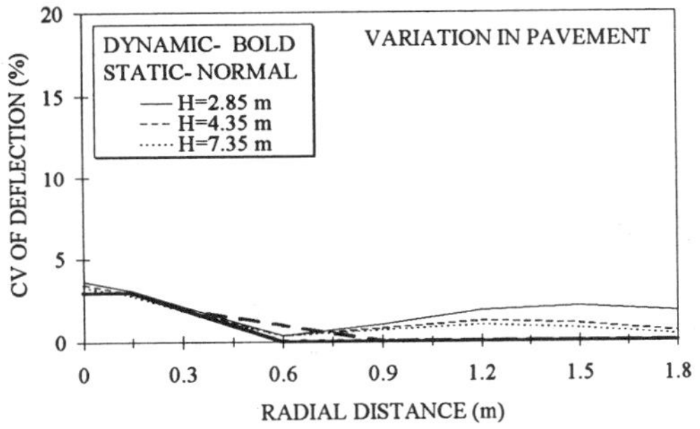

Fig. 5 Variation in deflection due to variation in pavement modulus for different bedrock depths

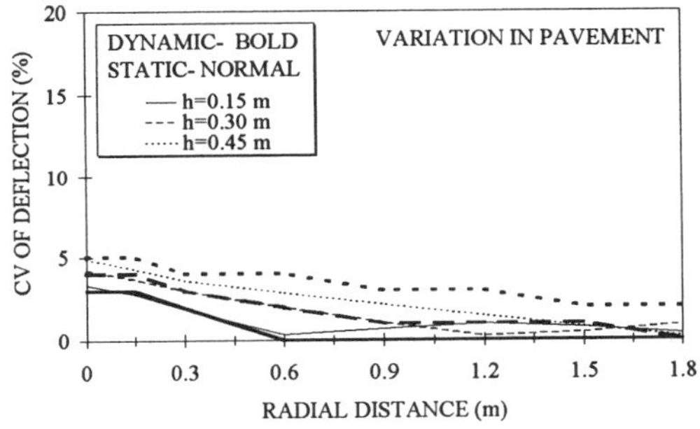

Fig. 7 Variation in deflection due to variation in pavement modulus for different pavement thicknesses

thicknesses of h=0.15, 0.30, and 0.45 m with results being summarized in Figures 6 and 7. Figure 6, which is for the case of subgrade variation, indicates that, for both static and dynamic loads, a change in the pavement thickness in the above range does not have an important effect on the sensitivity of the deflection variation to variation in the random modulus. This is also true for the case of variation in pavement modulus, as may be noticed in Figure 7.

In the foregoing discussion, either the pavement modulus was allowed to vary randomly or the subgrade modulus, but not both simultaneously. In order to obtain the CV(u) when both vary, without carrying out an additional analysis, two distinguished cases may be considered. In the first case, where there is a full correlation between the two moduli, CV(u) due to variation in both is simply the summation of CV(u) due to variation in each of them. In the second case, the moduli are not correlated and CV(u) is obtained by the square root of sum of the squares of variation in deflection due to variation in each layer modulus, the approach which is normally taken to find the simultaneous effect of independent random variables.

Figure 8 provides CV(u) based on the results of the previous simulations for the more realistic case of no correlation between the moduli. As illustrated, the accumulated CV(u) for both static and dynamic cases is approximately the same except for far points. Moreover, the value of CV(u) as a result of variation in both layer moduli is close to the value of CV(u) due to variation in only subgrade modulus, the observation which was expected from the insensitivity of deflection variation to variation in pavement modulus.

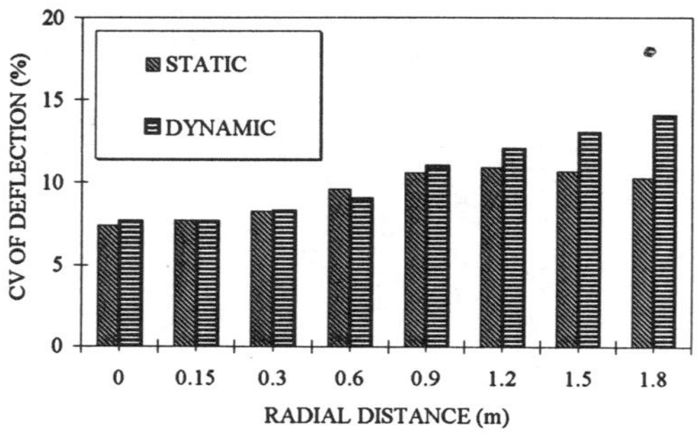

Fig. 8 Variation in deflection due to variation in both pavement layer and subgrade moduli

4 CONCLUDING REMARKS

Although neglecting the effect of inertia in the FWD test may have a significant influence on the expected value of the pavement deflection, and consequently, its backcalculated layer moduli (Sebaaly et al. 1986, Stolle and Sedran 1995), it has been demonstrated that, for typical pavement-subgrade properties, the effect of inertia on the coefficient of variation of deflection is not important for many cases. The largest difference between the results of the stochastic static and dynamic analyses was noticed when bedrock was relatively shallow and variation occurred in the subgrade modulus. This was attributed to a substantial change in dynamic stiffness characteristics of the system. The results from the other simulations suggested that, under the FWD dynamic load, the deflection sensitivity to variation in layer moduli was comparable to that of an equivalent static load. Similar observations have been made for other pavement-subgrade configurations and combination of material properties. Consequently it may be concluded that, the uncertainty level in backcalculated layer moduli, and therefore the accuracy of their estimation, may not change significantly due to the dynamic characteristics of the FWD test.

ACKNOWLEDGEMENTS

The authors wish to thank the Natural Sciences and Engineering Research Council of Canada and the Ministry of Culture and Higher Education of Iran for supporting this research.

REFERENCES

Huang, Y.H. 1993. *Pavement Analysis and Design*. Prentice Hall, Englewood Cliffs, N.J.

Kleiber, M., and Hien, T.D. 1992. *The Stochastic Finite Element Method: Basic Perturbation Technique and Computer Implementation*. John Wiley and Sons Inc., Chichester.

Lysmer, J., and Kuhlemeyer, R.L. 1969. Finite Element Model for Infinite Media. *Journal of the Engineering Mechanic Division*, ASCE, pp. 859-877.

Mamlouk, M.S., and Davies, T.G. 1984. Elasto-Dynamic Analysis of Pavement Deflections. *Journal of Transportation Engineering*, ASCE, 110(6), pp. 536-550.

Ong, C.L., Newcomb, D.E., and Siddharthan, R. 1991. Comparison of Dynamic and Static Backcalculation Moduli for Three-Layer Pavements. Transportation Research Board, *Record 1293*, TRB, National Research Council, Washington, D. C., pp. 86-92.

Parvini, M., and Stolle, D.F.E. 1996. Application of Stochastic Finite Element Method to Deflection Analysis of Pavement Structures. Transportation Research Board, *Record 1540*, TRB, National Research Council, Washington, D. C., pp. 64-70.

Sebaaly, B.E., Mamlouk, M.S., and Davies, T.G. 1986. Dynamic Analysis of Falling Weight Deflectometer Data. Transportation Research Board, *Record 1070*, TRB, National Research Council, Washington, D. C., pp. 63-68.

Stolle, D.F.E., and Sedran, G. 1995. Influence of Inertia on Falling Weight Deflectometer (FWD) Test Response. *Canadian Geotechnical Journal*, National Research Council of Canada, 32(6), pp. 1044-1048.

Wolf, J.P. 1988. *Soil-Structure-Interaction Analysis in Time Domain*. Prentice-Hall, Englewood Cliffs, N.J.

Zaghloul, S.M., White, T.D., Drnevich, V.P., and Coree, B. 1994. Dynamic Analysis of FWD Loading and Pavement Response Using a Three-Dimensional Dynamic Finite-Element Program. *Nondestructive Testing of Pavements and Backcalculation of Moduli* (Second Volume), ASTM STP 1198, H.L. Von Quintas, A.J. Bush III, and G.Y. Baladi, Eds., ASTM, Philadelphia, pp. 125-142.

Numerical Models in Geomechanics, Pietruszczak & Pande (eds) © 1997 Balkema, Rotterdam, ISBN 90 5410 886 X

Characterizing the durability of concrete pavement aggregate by neuronets

Y.M.Najjar & I.A.Basheer
Civil Engineering Department, Kansas State University, Manhattan, Kans., USA

ABSTRACT: The mechanisms of damage to concrete from repeated cycles of freezing and thawing are not well understood and continue to afford a considerable challenge in construction of many concrete projects. The durability of aggregate to frost action is commonly assessed experimentally by subjecting concrete beams containing the aggregate to repeated cycles of freezing and thawing. Due to high cost and long time (3-5 months) involved in completion of a single test, models that relate durability to easily measured physical properties of the aggregate are highly effective in pavement management studies. In this paper, backpropagation neural networks are developed from a huge database containing data pertinent to 750 different experimental investigations on concrete durability. The database was acquired from the Kansas Department of Transportation (KDOT). The networks are designed to enable determination of durability factor and percent expansion from five basic physical properties of the aggregate. The developed neural models were found to classify the aggregates with regard to their durability with a relatively high degree of accuracy. Issues related to development of the various neural networks and the items pertinent to examining prediction accuracy of the developed networks are also presented.

1 INTRODUCTION

Concrete pavements are the most widely used type of pavements and continue to constitute a high percentage of roadways worldwide. In 1990, an approximate total length of 130,000 miles (Huang 1993) of concrete pavements was estimated to cover a large area of the United States. The dependence on concrete pavements is attributed to their low maintenance costs and their capacity to withstand high volumes and excessive loads of traffic. In regions where concrete pavements is exposed to freezing and thawing, the pavement service life is more likely to be affected significantly. Therefore, durability of concrete masses and particularly the composition of the concrete matrix exhibit a primary concern in the design and construction of concrete pavements as well as other concrete structures.

One major component of concrete mixes for highway pavements is aggregate. In Kansas, D-cracking of concrete pavements containing limestone coarse aggregate is accounted for the noticeable mode of deterioration and maintain a continuously challenging problematic issue. Durability of limestone aggregate and related characteristics of the aggregate are presumed to be influential factors in the observed D-cracking of concrete pavements.

Commonly, the durability of concrete is measured by testing its resistance to rapid cycles of freezing and thawing (ASTM C 666-92). Aggregate suitability as a constituent in concrete mixes is checked by preparing concrete specimens (beams) of prespecified dimensions (approximate volume 0.60 ft^3) containing the aggregate under investigation. After curing for a certain period of time (up to 3 months), concrete beams are subjected to repeated cycles of freezing in air (or water) at 0 °F and thawing in water at +40 °F such that each cycle is completed in a period of time ranging from 2 to 5 hours. After a prespecified set of freezing and thawing cycles (not exceeding 36), the specimen is tested for fundamental transverse frequency and change in length. The specimen's relative dynamic modulus of elasticity after c cycles of freezing and thawing is calculated as the ratio of squared values of fundamental transverse frequency at c cycles and 0 cycles. The specimen is continued in the test until it has been subjected to 300 cycles or until its relative dynamic modulus of elasticity reaches 60% of its initial value, whichever occurs first. Alternatively, if the change in length is recorded then 0.10% expansion of the sample may be used as the criterion for termination of the test. A complete freezing and thawing test may *typically* last for up to 4.5 weeks, depending on the aggregate and integrity of concrete mix, with a preliminary 3 months of curing, therefore, a total time of 4 to 5 months is needed to complete one test.

In this paper, the durability of concrete is modelled by determining the durability factor based on experimentally determined physical characteristics of the aggregate used to prepare the beams. It is obvious that any prediction model will be beneficial to avoid the excessive testing time using freezing and thawing of concrete beams and the associated high costs. To the best of the authors' knowledge, little or no similar investigation and modeling effort is available in the literature. The experimental evaluation of the physical properties of aggregate are much less time consuming and involves considerably less amount of effort and cost as compared to freezing and thawing test. The physical properties of aggregate under consideration in this study includes modified freeze and thaw (soundness), absorption, specific gravity, and acid insoluble residue, requiring about 2 weeks, 12 hrs, 24 hrs, and 3 hrs, respectively, for completion for a typical limestone coarse aggregate. Neural network-based models are developed and compared for their accuracy in predicting the durability factors. The models were also manipulated to address several issues such as the influence of the various physical properties of the aggregate on its resistance to freezing and thawing cycles.

2 SPECIFICATIONS

For all types of concrete construction, according to KDOT specifications a coarse aggregate made of crushed limestone is designated as durable and resistant to freezing and thawing if the test results are as follows:

Durability Factor, DF=95 minimum
Expansion, EXP(%)=0.025 maximum
Modified freezing and thawing (Soundness), FT=0.85 minimum

3 NEURAL NETWORKS

In order to understand the development of the aggregate durability neural models, it is essential to briefly describe neural networks and discuss the methodology of neural computation. In this section, neural networks are discussed and various stages in development and design of neural models for predicting the durability factors and percent expansion of tested beams are presented.

Neural networks are simply defined as highly nodal structures copied after the nervous system of the mammalian cerebral, that when fed with appropriate data, can perform parallel computation in a fashion analogous to that of the brain. The nodal structure of an artificial neural network (NN) resembles a drastically oversimplified architecture of the biological structure of neurons in the nervous system. Although there are many variations of neural networks, the backpropagation neural network (BPNN) and its variants are currently the most widely used networks due to their apparent efficiency to model highly nonlinear problems and their flexibility to model a wide spectrum of applications in all research disciplines. BPNNs are multilayer structures constructed of three different types of layers. An input layer containing the input nodes receives information in form of signals from the external environment and process and forward them to a set of subsequent layers called hidden layers. The hidden layers receive the processed data and process and feed them forward towards an output layer where the solution of the problem is pursued. All nodes in one layer are connected (by links with appropriate and adjustable weight) to all nodes in adjacent layers but no side connections are permitted in such a networking scheme. One node receives signals from a number of nodes and integrate the signals as a weighted average. This weighted sum is simply the sum of all signals multiplied by their respective weights. The integrated signal at one receiving node is transformed to activation by using a transfer function such as the sigmoidal function. Similarly, the transformed signal is transmitted forward in a similar way to a following layer. The process is repeated until outputs are computed at the output side of the network. For one example (data set), the produced outputs are compared to actual (target) outputs and the error at the output side is used to calculate an error function. This function is used along with a learning rule (e.g. modified generalized delta rule) to propagate the error starting from the weights in the last layer and backward to other layers (hence the name backpropagation) in order to modify the weights. The procedure of forward activation of signals and the backpropagation of errors is repeatedly carried out until the error at the output side reduces to prespecified minimum. This process is also performed for all examples (data sets) used to train the network. For more details about the BPNN and other neural networks' algorithms and architectures, the readers are referred to the huge amount of literature on neural computing such as Simpson (1990) and Hassoun (1995).

4 MODEL DEVELOPMENT & ANALYSIS

Neural network development includes training a network on a number of data sets, and testing the accuracy of the developed network on sets never been used during the training process. Testing the network accuracy is performed using two sub-databases : testing database and validation database. The testing database contains all the test sets used to check the prediction capability of the network while the network is being trained on the training data. Alternatively, a validation database is used after selecting the most appropriate network in order to examine the network accuracy in predicting outputs for absolutely unseen cases. This procedure is actually similar to what happens after releasing the network (or any other predicting model) to the customer for use. For the aggregate durability database containing 750 data sets (examples), 548 sets were randomly selected to train the networks, 119 data sets were used to test the networks' generalization during the training process, and the remaining 83 sets were chosen to examine the validity of the developed networks. Although the selection of the sets in each sub-database was random, caution was practiced not to include in the testing or validation sets data sets irrelevant to the training domain. This was performed to prevent the developed networks from extrapolating beyond the domain on which they were trained.

In designing a BPNN, determination of the size of the hidden layer(s) constitutes a crucial step in network development. Unlike the input and output layers, the size of the hidden layer in terms of the number of hidden nodes it contains can not be known *a priori*. The backpropagation algorithm (Basheer and Najjar 1994; Najjar et al. 1996; Najjar and Basheer 1996, and Basheer and Najjar 1996) is encoded in a Fortran program with the ability to perform adaptive learning while building up the network architecture one hidden node at a time. Similarly, the number of epochs (iteration cycles) for one representative architecture (i.e. assuming a certain number of hidden nodes) is also determined during this adaptive learning process. The network adds one hidden node to the hidden layer or terminate iteration within certain architecture when the average prediction error obtained using the testing sets exceeds a prespecified minimum. Hence, in the designed adaptive learning backpropagation model, both the optimal architecture and optimal number of iteration cycles are determined in parallel. Once these are identified, the developed network is further examined for prediction of the validations sets.

In this study, three different types of BPNNs were developed in attempt to obtain the network(s) that generalize better. The three networks were varied in the size and type of their output layer. However, all networks attempted were fixed in the size of their input layer which included 5 input parameters; namely, FT, ABS, GSSD, GDRY, and ACIDR. For network 1, the output layer is designed to include both outputs; viz. DF and EXP. Conversely, networks 2 and 3 were designed to include one single output; namely DF and EXP, respectively.

Two schemes of examining the accuracy of prediction of the networks were adopted. In the first scheme, accuracy based on error incorporated in prediction of validation and testing sets is calculated. In the other scheme, which may be more useful for design purposes, agreement between the predicted output (EXP or DF) and the corresponding experimental values is examined based on whether the pair of data (i.e. predicted and experimental) meet the specifications of whether the aggregate passes or fails to qualify as a durable aggregate. To explain this further, consider the case where the *NN-predicted* DF and the *experimental* DF constitute the pair of data (98, 92). In this arbitrary example, the neural network model states that the aggregate qualifies as good aggregate according to the adopted specifications whereas the experimental results disqualifies this aggregate. Hence, the model is assigned "failure" for accurately predicting this particular case. On the other hand, if a pair of data for DF is represented as (85, 81) then the model is assigned "successful" to predict this case since both prediction and experimental values disqualify this aggregate as durable for construction. Similarly, a pair of data (98, 95) denotes a "successful" model prediction due to the model's capability to qualify the aggregate as good as the experimental investigation does.

It is clear that the first scheme of examining the accuracy of prediction based on the error in the output does not imply any provision for specifications. Hence, it may practically be assumed that the second scheme of testing the accuracy of model prediction can be quite advantageous in the light of specifications regarding the issue of whether to use or not to use a certain aggregate for constructing concrete structural elements. In the following sections, the three BPNNs are presented and comparisons regarding accuracy of prediction are also addressed.

4.1 DF-EXP Neural Network

In this network, both DF and EXP are included in the output layer. The input layer contains the 5 input parameters mentioned earlier. Using the designed adaptive learning BPNN, the performance of the current network was found to be optimum at one hidden node and 400 iteration cycles (epochs) with a momentum coefficient of 0.95 and a learning rate of 0.45. Hence, the neural network is denoted by 5-1-2 to refer to the number of nodes in each layer. The error at the outputs is determined as the Mean of the Absolute value of the Relative Error (MARE) between predicted and actual values for all data sets used in testing the accuracy. Table 1 summarizes the errors achieved using the 5-1-2 NN for predicting the data sets in training data, testing data, validation data, and all data combined. Using all data combined, DF was predicted with an average error of 5.78%, while EXP was predicted with a much less accuracy at a MARE of 298%. It is apparent that high values of MARE observed in EXP can be attributed to *Actual* values being close to 0.

Using the second scheme of examining the accuracy of prediction (i.e. in light of specifications). It was found that DF-EXP model was more accurate in predicting DF at high values and EXP at low values. On the average, the model was able to reflect (classify correctly) the experimental data at a rate=38% for DF and at 74.5% for EXP. Because DF and EXP have to meet specifications in order for a certain aggregate to qualify as durable material for construction, both DF and EXP were simultaneously considered to compute the *overall* accuracy of the DF-EXP neural model. The model was found to be able to accurately classify aggregates in only 24.8% of the instances.

4.2 DF Neural Network

In this network, the durability factor, DF, is used as the single output. Similar procedure of developing the DF-EXP-network was adopted to develop this particular model. Upon training on networks with variable architecture, the network with three hidden nodes was found to produce the optimal solution. Hence, this neural network is denoted as 5-3-1. The relative errors computed for the various databases are summarized in Table 1. As can be seen from Table 1, less errors are observed (compare to DF-EXP-network) when the DF was the single output of the model. This might be due to the potential deviation of the weight vector in the error hyperspace from the point of global minimum when DF is accompanied with another output (i.e. EXP) which might have contained large errors.

Using the other scheme of evaluating the performance of the network, the present network was found to provide two features better than the DF-EXP network. Firstly, the DF network was able to produce a wider resolution for DF at lower values. That is, the DF-EXP-network was only able to map the data in the region between 90 to 97 whereas the DF-network produced wider mapping that encloses data ranging from 82 to 100. Secondly, errors associated with qualification or disqualification (classification capability) of the aggregates are reduced using the DF-network as opposed to the errors of the DF-EXP-network. Using all data combined, the overall success rate of the DF-network model in performing accurate classification of the aggregates is estimated as 63.1% (compare to 38% using the DF-EXP-network). Therefore, the DF-network will be adopted herein as the final network for predicting the DF of aggregates.

4.3 EXP Neural Network

This network is designed particularly for prediction of percent expansion, EXP, from the 5 input parameters. Using the adaptive learning backpropagation network, the optimum network was found at 3 hidden nodes. Table 1 presents the errors of prediction of the various sub-databases. Using the second scheme of assessing the prediction accuracy, the *overall* success rate was calculated as 75.8%, slightly better than that of the DF-EXP-network. Therefore, the developed single output DF-network was adopted to predict EXP values.

Table 1: Error associated with prediction of DF and EXP

Network	DATA	No. Sets	MARE(%) (DF)	MARE(%) (EXP)
DF/EXP-Net	Training	548	5.75	81.60
(5-1-2)	Testing	119	5.35	111.10
	Validation	83	6.31	299.00
	All Data	750	5.78	298.00
DF-Net	Training	548	5.00	
(5-3-1)	Testing	119	4.58	
	Validation	83	6.42	
	All Data	750	5.09	
EXP-Net	Training	548		83.20
(5-3-1)	Testing	119		114.10
	Validation	83		34.10
	All Data	750		55.00

MARE(%) : Mean of Absolute Value of Relative Error (%)

5 RELIABILITY OF PREDICTION

Based on the results and analysis of the three networks developed, the DF-network and EXP-network will be used to predict, respectively, the DF and EXP of concrete specimens in order to assess the durability of a given aggregate based on some known physical properties of the aggregate. The scheme of evaluating the accuracy of prediction based on specification (i.e. accuracy of classification) can be used to associate the computed values of DF and EXP with reliability factors. In other words, this scheme is manipulated to furnish some information regarding the reliability (probability) that the aggregate under consideration will meet the durability specification(s) based on neural network-predicted values of DF and/or EXP.

Since both DF and EXP are commonly used as indicators of durable aggregate, reliability of prediction based on these two parameters combined is more reflective than the independent reliability of prediction of DF and EXP. The reliability of prediction via the durability neural model (composed of DF-NN and EXP-NN) was determined by observing the simultaneous accurate classification of both DF and EXP. Hence, reliability is dependent on the quality of prediction of both DF and EXP. A thorough inspection of the predicted data eventually yielded the prediction reliability matrix shown in Table 2. To illustrate use of Table 2, consider the case where a specific aggregate is estimated using the neural models to yield DF=96.0 and EXP=0.0150. Table 2 implies that the given aggregate is believed to be durable (or is recommended for construction) with 87% confidence. It is obvious from Table 2 that the reliability (probability) of aggregate being durable decreases with decreasing DF and increasing EXP.

6 CONCLUSIONS

The research presented in this paper has provided a simple means for evaluating the durability of aggregate in accord to its resistance to freezing and thawing from the knowledge of five physical properties of the given aggregate. The neural models developed are easy to use and can be easily retrained when new relevant

Table 2: Reliability of an aggregate to meet durability specifications

		% EXP				
		<0.0130	0.0130-0.0175	0.0175-0.0300	0.0300-0.0375	>0.0375
DF	>98.0	92%	-----	-----	-----	-----
	94.5-98.0	90%	87%	66%	-----	-----
	92.0-94.5	-----	76%	60%	42%	-----
	86.0-92.0	-----	-----	50%	33%	12%
	<86.0	-----	-----	-----	-----	10%

----- : No data observed in this class

experimental data become available. In most cases, the relatively accurate neural models developed herein can conservatively replace the experimental procedure (the only available tool of checking frost resistance of aggregate) which tend to be both costly and time-consuming; factors that hinder prompt assessment in pavement management studies. The developed models were encoded into simple computer program that can be run on personal computers.

ACKNOWLEDGEMENT

The authors would like to acknowledge the Kansas Department of Transportation for their financial support to conduct this research project.

REFERENCES

Basheer, Imad A. and Najjar, Yacoub M. (1994)," Designing and Analyzing Fixed-Bed Adsorption Systems with Artificial Neural Networks," J. Env. Sys., Vol. 23, No. 3, pp. 291-312.

Basheer, Imad A. and Najjar, Yacoub M. (1996)," Predicting Dynamics Response of Adsorption Columns with Neural Nets," J. Comp. Civ. Engrg., Vol. 10, No. 1, pp. 31-39.

Hassoun, M. (1995), Fundamentals of Artificial Neural Networks, MIT Press, Cambridge, MA.

Huang, Y.H. (1993), Pavement Analysis and Design, Perntice Hall, Englewood Cliffs, NJ.

Janssen, D.J. and Snyder, M.B. (1994), Resistance of Concrete to Freezing and Thawing, Strategic Highway Research Program, National Research Council, SHRP-C-391,Washington, DC.

Najjar, Yacoub M., Basheer, Imad A., and Naouss, Wissam A. (1996)," On the Identification of Compaction characteristics by Neuronets," Computer and Geotechnics, Vol. 18, No. 3, pp. 167-187.

Najjar, Yacoub M. and Basheer, Imad A. (1996),"Utilizing Computational Neural Networks for Evaluating the Permeability of Compacted Clay Liners," J. of Geotech. and Geol. Engrg.,Vol. 14, No.3, pp. 193-212.

Simpson, P. K. (1990), Artificial Neural Systems: Foundations, Paradigm, Applications, and Implementations, Pergamon Press, New York.

Numerical Models in Geomechanics, Pietruszczak & Pande (eds) © 1997 Balkema, Rotterdam, ISBN 90 5410 886 X

Track support model and its application to dynamic problems

G.P. Raymond
Department of Civil Engineering, Queen's University, Kingston, Ont., Canada

Z.Cai
Bombardier Inc., Transportation System Division, Kingston, Ont., Canada

ABSTRACT: The gradual introduction of heavier axle loads and faster speeds in railway train operations is resulting in track support overloading. Herein an analytical dynamic wheel/rail and track interaction model is used to study these effects on the wheel/rail forces, rail seat loads and ballast/subgrade pressures. The axle loads considered are typical of 50, 70, 100 and 125 ton freight cars in use in North America. Also included is a projection of the effects of 150 ton freight cars should these become acceptable. The traversing railway truck used in this study has a front wheel with a rounded flat (50 mm length x 0.4 mm depth) and a rear wheel of perfect shape. The track is composed of RE 136 rail (having a mass of 68.7 kg/m) supported by CN 55A concrete ties at 610 mm centre to centre and insulated with EVA tie pads. The solutions are easily extended to other rail/wheel size and shape defects.

1 INTRODUCTION

A major function of the track structure is to transmit and attenuate the wheel loads to the track bed. As train loads become heavier and speeds faster the high frequency and high magnitude dynamic wheel/rail forces and track responses associated with wheel, rail and track irregularities, become of paramount importance. These increased loadings have resulted in an increase in the deterioration of the track support including the ballast layer and subgrade leading to an increase in track research. Studies included research by Battelle Columbus Laboratories (Meacham and Ahlbeck, 1969, Ahlbeck et al., 1975, Ahlbeck, 1980, Ahlbeck and Hadden, 1985, Ahlbeck and Harrison, 1988), British Rail Research Division (Jenkins, et al., 1974, Newton and Clark, 1979, Clark et al., 1982, Tunna, 1988), Track Laboratory of Japan National Railways (Kuroda, 1973, Hirano, 1972, Sato,1977), and Cambridge University Engineering Department in collaboration with British Rail (Grassie et al., 1982a, 1982b, Grassie, 1984). These studies have been mainly limited to existing speeds and loadings. The effects of increased axle loads and faster speeds are rarely a previous subject of focus.

Reported is a continuation of earlier research (Cai

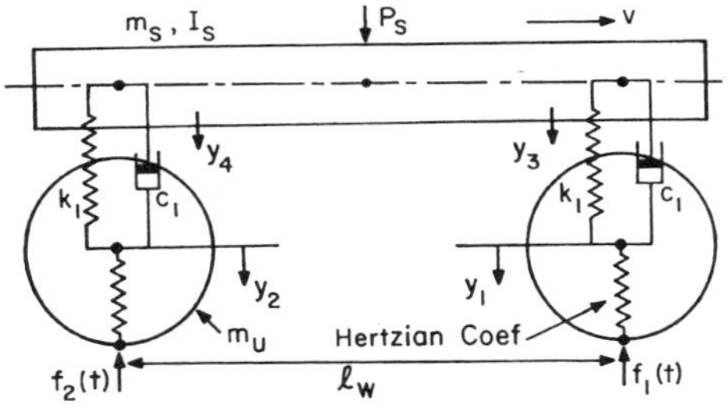

Figure 1. Wheelset model.

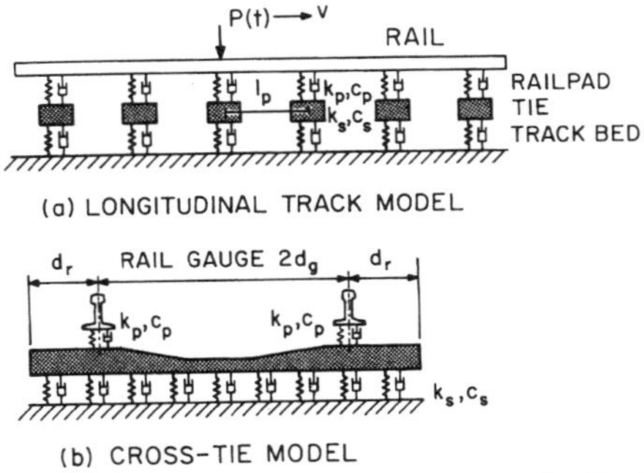

Figure 2. The track model. *top:* longitudinal; *bottom:* cross-tie model.

and Raymond, 1992a, 1992b, 1994) on-rail vehicle and track dynamics. Considered herein is the effect of wheel/rail impact forces and track responses due to wheel and rail irregularities in relation to a variety of axle loads and speeds, both related to existing and possible future usage.

2 WHEELSET AND TRACK MODELLING

The wheelset is modelled as four-degrees of freedom lumped mass as shown in Figure 1. The equations of motion of the wheelset system are as follows:

$$[M]\ \{\ddot{Y}\} + [C]\ \{\dot{Y}\} + [k]\{Y\} = \{f\} \qquad (1)$$

where

$$[M] = \begin{bmatrix} m_u\ , 0\ , & 0\ , 0 \\ 0\ , m_u\ , & 0\ , 0 \\ 0\ , 0\ , & \frac{m_u}{2}\ , \frac{m_u}{2} \\ 0\ , 0\ , & \frac{I_s}{l_w}\ , \frac{-I_s}{l_w} \end{bmatrix} \qquad (1a)$$

$$[C] = \begin{bmatrix} c_1\ , & 0\ , & -c_1\ , & 0 \\ 0\ , & c_1\ , & 0\ , & -c_1 \\ -c_1\ , & -c_1\ , & c_1\ , & c_1 \\ \frac{-c_1 l_w}{2}\ , & \frac{c_1 l_w}{2}\ , & \frac{c_1 l_w}{2}\ , & \frac{-c_s l_w}{2} \end{bmatrix} \qquad (1b)$$

m_u is the unsprung mass, m_s is the side frame mass, I_s is the side frame inertia, k_1 and c_1 are the suspension constants, f_1 and f_2 are the wheel/rail reaction forces. *[K]* takes the same form as *[C]* with c_1 replacing k_1. *[Y]* is the displacement vector and *{f}* the force vector.

The track is shown in Figure 2 and is formulated by representing the rail and the ties as elastic beams, the rail pads (for a concrete-tie track) as linear springs with viscous damping, and the stiffness and vibration absorbing effect of the underlying track bed as a continuous array of linear springs and viscous dashpots. The rail is assumed to have a finite length with the ends clamped, and is supported discretely on the tie beams at the rail seats. The effects resulting from the assumption of a finite length on the wheelset and track responses will be minimal at the mid-region away from the ends when the rail length is taken long enough.

The equations of the track are obtained by first solving the free vibration of the track and then by applying the method of modal analysis (Cai, 1992). The resulting set of equations are:

$$Q_n(t) + \sum_{N}^{k=1} 2\ \xi_{nk}\ \Omega_n\ Q_k(t) + \Omega_n^2\ Q_n(t) = f_n(t)$$
$$(n = 1, 2, \ldots\ldots\ldots, N) \qquad (2)$$

where $Q_n(t)$ is the modal time coefficient; N is the number of modes considered; Ω_n is the angular frequency of the track; ξ_{nk} is the coupled modal damping ratio; and $f_n(t)$ is the generalized modal force, expressed by

$$f_n(t) = \frac{1}{M_m}\left[W_n(vt)\ f_1(t) + W_n\ (vt - l_w)\ f_2(t)\right] \qquad (3)$$

where W_n is the *n-th* mode shape function of the rail, M_m is the corresponding generalized track mass, v is the train speed, and l_w is the axle spacing.

Equation 2 is a set of N coupled equations. The deflection of the rail is obtained using mode summation:

$$w(x,t) = \sum_{n=1}^{N} W_n(x)\ Q_n(t) \qquad (4)$$

3 HERTZIAN WHEEL/RAIL INTERACTION

The wheel/rail interaction is obtained by using the Hertzian contact theory commonly used in wheel/rail contact mechanics, and is expressed in the following form:

$$f(t) = G_H\ [\ y_w - w(x,t) - \delta(x)\ ]^{\alpha} \qquad (5)$$

where $f(t)$ is the wheel/rail contact force; y_w is the wheel displacement; $w(x,t)$ is the rail deflection at the wheel/rail contact point; $\delta(x)$ is the wheel or rail profile change; G_H is the Hertzian contact coefficient; and α is a constant (1.5 is used herein).

By coupling Equation [1] to [5], the wheel/rail interaction forces $f_1(t)$, $f_2(t)$ and the modal time coefficients $Q_n(t)$ may be solved. The track support

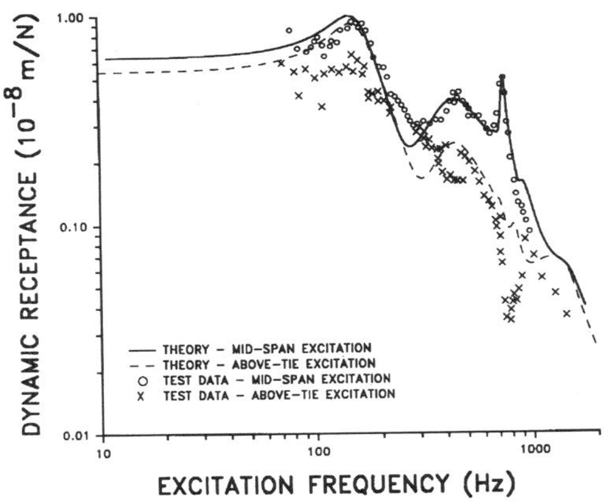

Figure 3. Comparison of dynamic receptance of model and field data.

responses are then obtained by using the principles of structural dynamics.

4 VALIDATION OF TRACK MODEL

Verification of the applicability of the track model to simulate field rail track vibration problems is illustrated by a comparison in Figure 3 of the predicted dynamic receptance characteristics of a typical British Rail track with field data obtained by Grassie (1984).

The theoretical results are based on the use of a non-uniform tie and may be seen to compare favourably with the field experimental data for both the mid-span and above-tie excitations. It was found that the track response is (a) dominated more by the rail span as a deep beam spanning between two ties when the excitation is at the mid-span, and (b) dominated more by the rail-pad-tie and ballast system when the excitation is above a tie. Figure 4 shows a comparison of the rail seat force prediction and the measured values of British Rail (Frederick el al. 1977) as the wheel approaches and travels across and past the rail seat location. The results typify the wheel/rail dynamic interaction at the rail seat.

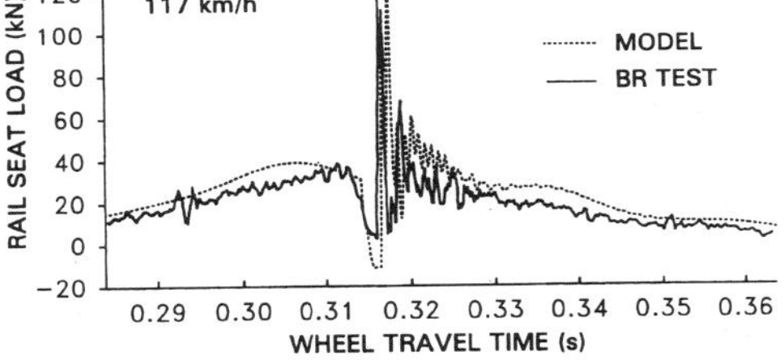

Figure 4. Comparison of theory and a British Rail field test.

5 WHEEL/RAIL IMPACT FORCES

Prior to the adoption of any increased speed or any increased axle loads it is useful to examine the effect of these factors to allow for preventive action to be undertaken or planned prior to their introduction. As an example of the type of examination the front wheel of the wheelset is given a rounded flat 50 mm long x 0.4 mm deep and the rear wheel as perfectly round. The flat is shown in Figure 5. The wheelset was run across a 40-tie track with concrete ties at various speeds up to 162 km/h. The track and wheel parameters assumed are given in Table 1 and 2 respectively.

Figure 6 shows the predicted peak wheel/rail impact loads resulting from the wheel flat for five freight cars of different capacity. The peak occurred from the wheel flat impacting directly above a tie. The 150 ton car response is a projection of what might be expected should 150 ton cars become acceptable. It is seen that the impact load depends highly on the speed. This is particularly so when the train speed exceeds 90 km/h.

At speeds up to 30 km/h the wheel/rail loads for all freight cars are primarily due to the net increase in the static wheel loads. At a higher speed, the dynamic effects of the wheelset and truck side-frame masses, coupling with the vertical vibration of the track become prominent. A first peak wheel/rail

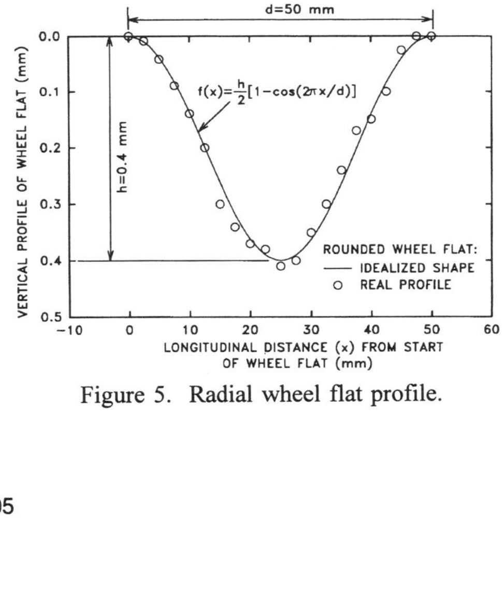

Figure 5. Radial wheel flat profile.

Table 1. Track Parameters.

Tie-(CN 55A) Parameters	
Elastic modulus E_t	50 GN/m^2
Poisson's ratio	0.30
Timoshenko shear coeft.	0.833
Tie spacing	0.61 m
Tie length	2.50 m
Tie width	0.25 m
Non-uniform section	
Mid-segment length	0.90 m
Mid-segment depth	0.14 m
End-segment length	0.80 m
End-segment depth	0.21 m
Rail gauge length	1.50 m
Tie end to rail seat	0.50 m
Pad Parameters	
Rail pad stiffness k_p	850 MN/m
Rail pad damping c_p	26000 N.s/m
Track-Bed Parameters	
Track bed stiffness k_s	50 MN/m/m
Track bed damping c_p	34000 N.s/m
Rail-(136 RE) Parameters	
Elastic modulus E_r	207 GN/m^2
Poisson's ratio	0.28
Timoshenko shear coeft.	0.34
Cross-sectional area	8610 mm^2
Second moment of area	3950 cm^4
Radius of gyration	67.7 mm
Bending rigidity EI_r	8.18 MN.m^2
Shear rigidity κAG_r	239.3 MN
Unit mass m_r	68.7 kg/m

Table 2. Wheel Parameters.

Car Name W (MN)	50 0.50	70 0.70	100 1.00	125 1.25	150 1.50
m_u (Mg)	1.02	1.10	1.42	1.59	1.79
m_s (Mg)	0.74	0.88	1.13	1.21	1.40
I_s (kg.m^2)	202	260	363	542	622
k_1 (MN/m)	1.50	1.79	2.14	2.80	3.10
D_w (mm)	762	762	762	965	965
l_w (m)	1.67	1.72	1.78	1.83	1.83
Hertz C (GN.m$^{-3/2}$)	81.9	81.9	81.9	86.8	86.8
W_w (kN)	94	122	146	178	203
Damping of primary suspension, c_1					9.9 kN.s/m

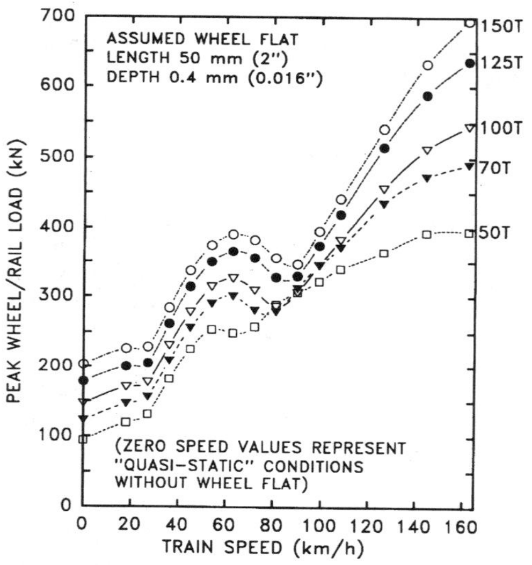

Figure 6. Predicted wheel/rail impact loads.

load is reached at about 60 km/h for all the cars. This is followed by a gradual drop in the peak load until a speed of 90 km/h is reached. Above 90 km/h the wheel/rail impact loads increase rapidly as the speed increases. As seen this is more pronounced for heavier capacity cars than for lower capacity ones. This is a warning to the rail industry should they continue to increase both speeds and axle loads.

6 RAIL SEAT LOAD

Figure 7 shows the corresponding peak dynamic rail seat loads directly below the rail seat in relation to the speed. The rail seat loads show similar trends to the wheel/rail load at speeds below 30 km/h. There are then sudden increases in the peak rail seat load for all the cars between 30 and 90 km/h. This is a direct result of the development of intense dynamic interactions between the wheel and the rail atop the tie, as is indicated in Figure 6 by the quick growth of the dynamic peak load within that speed range. The adverse effects are already apparent on North American concrete tie track in the form of excessive rail seat ware. Over 90 km/h, the increase in the peak rail seat load is moderate and begins to flatten and possibly decrease as the freight car axle load further increases. This level-off or drop of the rail seat load at higher speeds, despite the marked increase in the wheel/rail impact load shown in Figure 6, is believed to be primarily due to two reasons. The first is due to the shorter duration, and thus higher frequency of the wheel/rail impact forces

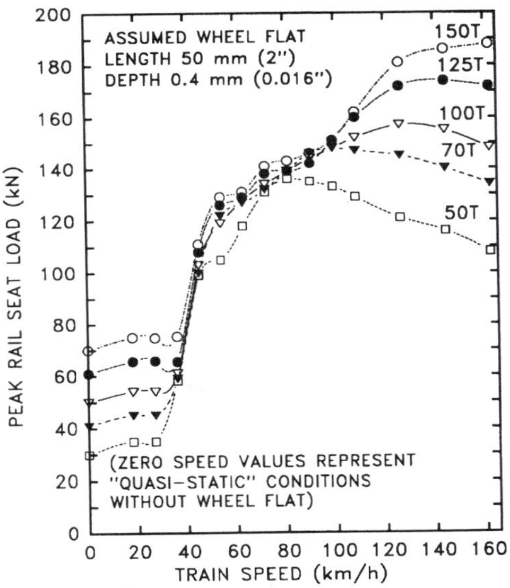

Figure 7. Peak rail seat load.

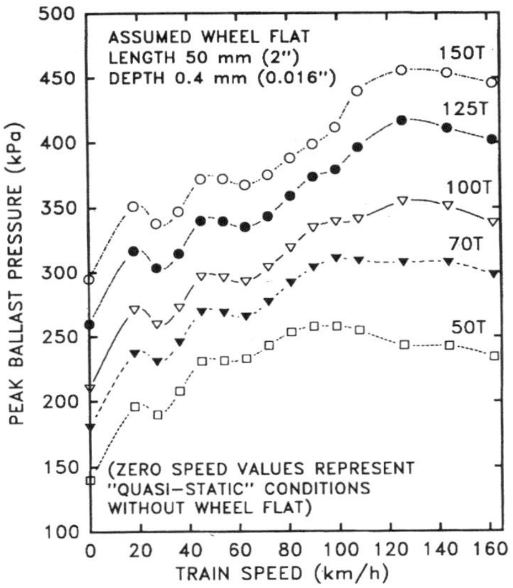

Figure 8. Peak ballast pressure.

at a faster speed. The second is due to the increased damping effect of the rail pad, which is more effective to attenuate high speed (or high frequency) vibrations.

7 BALLAST PRESSURES

Figure 8 shows the effect of increasing the freight car capacity on the increase in the ballast pressure. The effects are modest as compared with the effects of the wheel/rail impact forces and the rail seat loads. For example, the peak ballast pressure under the 125 ton car at 160 km/h is about 1.5 times its static value, whereas the peak rail seat load is 2.7 times its static value, and the peak wheel/rail load is 3.6 times its static value. The relatively parallel ballast pressure versus speed increase at low speeds would indicate that the initial increase in ballast pressure due to heavier capacity cars is to a larger extent attributed to the net increase in the static wheel load than to the increase in the wheel/rail load or in the rail seat load. The relatively moderate relationship between the peak ballast pressure and the speed occurs due to the vibration attenuating effects of the track structural components, namely the bending rigidity of the rail, the resilience and damping effects of the rail pad, the bending effect of the tie as an elastic beam, and the resilience and the damping effects of the ballast itself. At higher speed the adverse effects of high frequency vibrations are already apparent on North American track in the form of increased ballast powdering and breakage.

8 CONCLUSIONS

An analytical dynamic wheel/rail and track interaction model is presented and briefly shown to be able to reproduce measured field data. The model is then used to investigated the dynamic responses of wheel/rail impact loads, rail seat loads, and ballast pressures, produced by a wheel defect occurring on different capacity North American freight cars. Concluded from this investigation are:

For all freight cars the wheel/rail impact load, the rail seat load, and the ballast pressure is effected by train speed. The effect is largest for the wheel/rail impact load and least for the ballast pressure.

An increase in the car capacity increases the track support loading. At low speeds, the track loading increment is primarily due to the increase in the car weight. At high speeds, the dynamic interaction between the truck side frames, the wheelsets, and the track vertical vibration becomes dominant. This causes a track loading increment considerably greater than that due to the increase in the static car weight. This difference increases as speed increases and/or car capacity increases.

Increased car capacity and/or increased speed

results in a dynamic loading increment which is greatest for the wheel/rail impact increase, and least for the ballast pressure increase.

High frequency dynamic wheel/rail impact forces propagate into the ballast and subgrade. The ballast experiences high frequency pressure oscillations which are adverse to the ballast/subgrade integrity.

9 ACKNOWLEDGEMENT

Funding by the Natural Sciences and Engineering Council of Canada in the form of a grant awarded to G.P. Raymond is acknowledged.

10 REFERENCES

Ahlbeck, D.R. (1980) An Investigation of Impact Loads Due to Wheel Flats and Rail Joints, *ASME Paper No. 80-WA/RT-1*.

Ahlbeck, D.R. and Hadden, J.A. (1985) Measurement and Prediction of Impact Loads from Worn Railroad Wheel and Rail Surface Profiles, *J. Eng. for Ind.*, ASME, Vol. 107, May 1985, pp. 197-205.

Ahlbeck, D.R. and Harrison, H.D. (1988) The Effects of Wheel/Rail Impact Loading due to Wheel Tread Runout Profiles, *Proc. 9-th Int. Wheelset Congress*, Montreal, Paper 6-1.

Ahlbeck, D.R., Meacham, H.C. and Prause, R.H. (1975) The Development of Analytical Models for Railroad Track Dynamics, *Proc. Symp. on Railroad Track Mechanics*, Pergamon Press, pp. 239-260.

Cai, Z. and Raymond, G.P. (1992a) Theoretical Model for Dynamic Wheel/Rail and Track Interaction, *Proc. 10th. Int. Wheelset Cong.*, Sydney, Australia, pp. 127-131.

Cai, Z. and Raymond, G.P. (1992b) Dynamic Modelling of Parameters Controlling Railway Track Vibration, *Proc. 1st. Int. Conf. Motion and Vibration Control*, Yokohama, Japan, pp. 976-981.

Cai, Z. and Raymond, G.P. (1994) Modelling the Dynamic Response of Railway Track to Wheel/Rail Impact Loading, *J. Struc. Eng. and Mech.*, Vol. 2(1), pp. 95-112.

Clark, R.A., Dean, P.A., Elkins, J.A. and Newton S.G. (1982) An Investigation into the Dynamic Effects of Railway Vehicles Running on Corrugated Rails, *J. Mech. Eng. Sci.*, Vol. 24, pp. 65-76.

Frederick, C.O., Cannon, D.F. and Newton, S.G. (1977) Dynamic Rail Stresses due to Wheel Flat Impact, *AIT Symp. on Railway Dyn.*, 14p.

Grassie, S.L. (1982) Dynamic Modelling of Railway Track and Wheelsets, *Proc. of 2nd Int. Conf. on Recent Advances in Struc. Dyn.*, Petyt, M. and Wolfe, H.F. (eds.), ISVR, University of Southampton, 1984, pp. 681-698.

Grassie, S.L., Gregory, R.W. and Johnson, K.L. (1982a) The Dynamic Response of Railway Track to High Frequency Vertical Excitation, *J. Mech. Eng. Sci.*, Vol. 24, pp. 103-111.

Grassie, S.L., Gregory, R.W. and Johnson, K.L. (1982b) The Dynamic Loading of Rails at Corrugated Frequencies, *Proc. Contact Mech. and Wear of Rail/Wheel Systems*, Vancouver, pp. 209-225.

Jenkins, H.H., Stephenson, J.E., Clayton, G.A., Morland, J.W. and Lyon, D. (1974) The Effect of Track and Vehicle Parameters on Wheel/Rail Vertical Dynamic Forces, *Railway Eng. J.*, pp. 2-16.

Kuroda, S. (1973) Dynamic Variation of Wheel Load Attributed to Vertical Deformation of Rail End, *O. Rep.*, JNR., Vol. 14(3), pp. 143-144.

Hirano, M. (1972) Theoretical Analysis of Variation of Wheel Load, *O. Rep.*, JNR., Vol. 13(1), pp. 42-44.

Meacham, H.C. and Ahlbeck, D.R. (1969) A Computer Study of Dynamic Loads Caused by Vehicle-Track Interaction", *ASME Paper No. 69-RR-1*.

Newton, S.G. and Clark, R.A. (1979) An Investigation into the Dynamic Effects on the Track of Wheelflats on Railway Vehicles, *J. Mech. Eng. Sci.*, Vol. 21, pp. 287-297.

Sato, Y. (1977) Study on High Frequency Vibrations in Track Operated with High-Speed Trains, *O. Rep.*, JNR, Vol. 18(13), pp. 109-114.

Tunna, J.M. (1988) Wheel-Rail Forces due to Wheel Irregularities, *Proc. 9th Int. Wheelset Congress*, Montreal, Paper 6-2.

Numerical Models in Geomechanics, Pietruszczak & Pande (eds) © 1997 Balkema, Rotterdam, ISBN 90 5410 886 X

Seismic analysis of a bridge-ground system excited in the axial direction of the bridge

A.Wakai & K.Ugai
Gunma University, Kiryu, Japan

T.Matsuda
Japan Highway Public Corporation, Kobe, Japan

ABSTRACT: A rigid-frame bridge near Takamatsu in Japan that is due for construction in the near future, is analyzed against the earthquake motion which was observed at the time of the 1995 Great Kobe Earthquake in Japan. The analysis is done by the 3D dynamic elasto-plastic FEM, in which the dynamic interaction between the superstructures, piers, foundation piles, and the ground are properly considered. The calculated histories of acceleration and displacement at each point are discussed from the viewpoint of structural safety.

1 INTRODUCTION

The design of bridges in an earthquake zone such as Japan requires a reliable prediction method for evaluating the response of bridge-foundation-ground system during an earthquake. The 3D dynamic elasto-plastic FEM makes it possible to evaluate dynamic response of the system, in which the interaction between the ground and the structures are considered. In addition, the damage of structures can be predicted precisely, for example, by residual displacement or plastic strain caused by an earthquake.

There are a few reports related to the 3D dynamic FE analysis of a structure-ground system, which have considered the hysteretic characteristics of soils. The 3D effective stress analysis of a piled rigid structure, using a model extended from the modified Ramberg-Osgood model (Tatsuoka et al., 1978), has been done by Fukutake et al. (1995). It has discussed the influence of liquefaction on the total stability of the structures. Borja et al. (1994) has considered the 3D vibrations of footing, based on a simplified model from the bounding surface plasticity (Dafalias, et al., 1977). It assumed that the ground consists of incompressible soils with no elastic region.

In such analyses, the hysteretic characteristics on the stress-strain relations of each material are regarded as important. In this study, the elements of upper structures, piers and piles made of reinforced concrete are assumed to show trilinear hysteretic characteristics on which the recent Japanese design code is based. The soils in the ground are assumed to show newly proposed hysteretic characteristics that simulates the commonly observed $G-\gamma$ and $h-\gamma$ curves.

The purpose of this paper is to predict the plastic deformation of the system during an earthquake and to discuss it from the viewpoint of structural safety. The horizontal acceleration is applied to the base of the ground and the response of the system is analyzed by the 3D dynamic elasto-plastic FEM.

2 NUMERICAL MODEL

2.1 Descriptions of the problem

The 3D FE meshes for a bridge-foundation-ground system that is a half part of a continuous rigid frame bridge, consisting of superstructures, piers (P8, P9 and P10), footings, piles and the ground, are shown in Figure 1. Each pier is rigidly fixed at its top and connected to neighboring piers. The meshes consist of 20-node solid elements. The horizontal acceleration is applied to the base of the ground in the direction of the arrow, which is parallel to the axial direction of the bridge. In this paper an earthquake record obtained during the 1995 Great Kobe Earthquake will be utilized. The ground is assumed to be under undrained condition during excitation.

The front end (y=0m) in the meshes is assumed to be smooth because of the symmetry of the problem. The response of the back end (y=20.0m) in the meshes is assumed to be consistent with the result of the 1D free ground analysis, which is described in the following section.

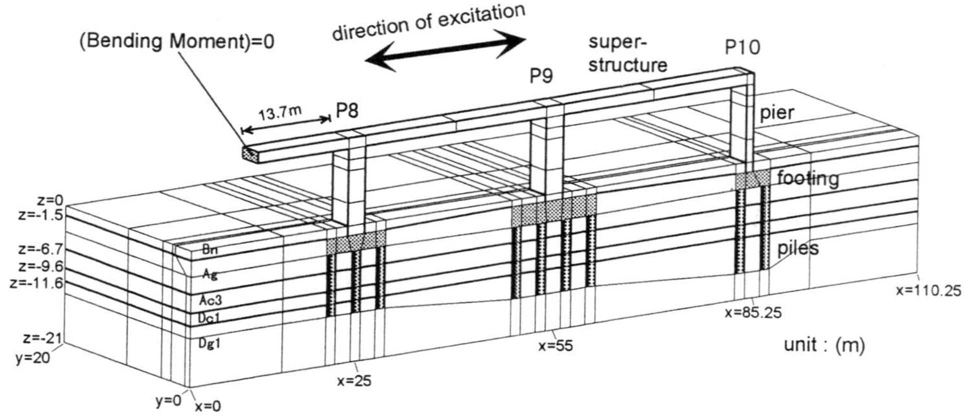

Figure 1. Half model of the system (FE meshes).

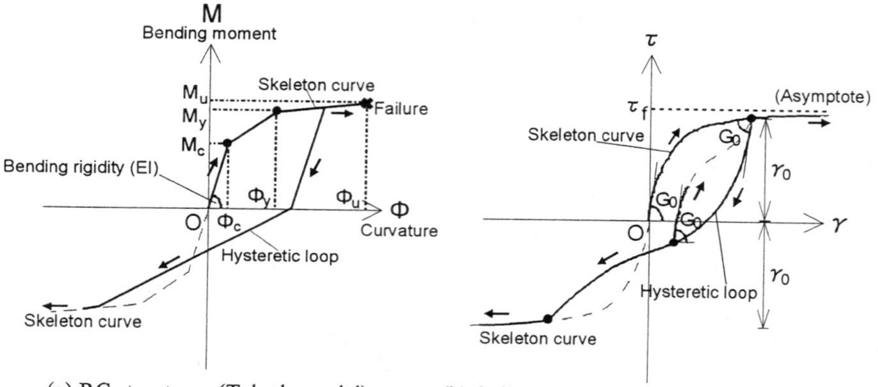

(a) RC structures (Takeda model) (b) Soils (Undrained cyclic loading model)

Figure 2. Examples of hysteretic curves of the constitutive models.

The bending moment at the leftmost end of the superstructure in the meshes is assumed to be zero.
The length of 13.7m corresponds to half of the distance from the neighboring pier. To simplify the meshes, a section of each pile is assumed to be square, so that its width is consistent with the diameter of each pile. The mass of superstructure is assumed to be uniformly distributed in the meshes.

2.2 Quasi-infinite boundaries

In the dynamic analyses, infinite boundaries are often used to eliminate reflected waves at the end of the analytical region. However, at a position far away from the structure, the motion of the ground can be approximated by the response of the free horizontal ground. In this study, the response of acceleration and displacement at each side of the analytical region (x=0, 110.25m or y=20.0m), except the symmetrical plane, was assumed to be given by the 1D shearing wave analysis of the free ground. The bottom of the analytical region is completely fixed.

2.3 Constitutive models

(a) RC Structures

The elements for the footings and the intersections of the superstructure and piers are assumed to be linearly elastic.

The superstructures, piers and piles made of reinforced concrete are assumed to show trilinear hysteretic characteristics on the bending moment - curvature relationships (Takeda et al., 1970). Figure 2(a) shows an example of hysteretic curves of the

Table 1. Material constants for the 3D analysis.

(a) RC structures (Takeda model) in pier P9

Materials (Name)	(Location)	M_c (MN·m)	M_y (MN·m)	M_u (MN·m)	Φ_c ($10^{-5}m^{-1}$)	Φ_y ($10^{-5}m^{-1}$)	Φ_u ($10^{-5}m^{-1}$)	Unit weight γ (kN/m^3)	Rayleigh damping α	β
Pier-A	**G.L.+10.3~12.0m**	42.6	97.8	113.6	10.3	98.1	3910.	4.01	0.114	0.00116
Pier-B	**G.L.+8.6~10.3m**	27.0	73.7	83.6	12.7	110.	3090.	4.01	0.114	0.00116
Pier-C	**G.L.+3.8~8.6m**	27.0	73.7	83.6	12.7	110.	3090.	2.50	0.114	0.00116
Pier-D	**G.L.+0.0~8.6m**	28.5	76.4	86.9	13.4	112.	2990.	2.50	0.114	0.00116
Piles	**G.L.-15.3~-4.3m**	1.56	2.83	3.91	24.3	199.	3300.	1.96	0.114	0.00116

(b) Elastic materials in pier P9

Materials	Young's modulus E (GPa)	Poisson's ratio ν	Unit weight γ (kN/m^3)	Rayleigh damping α	β
Intersection of pier and superstructure	8600.	0.152	59.9	0.286	0.002894
Footing	24.5	0.147	30.6	0.286	0.002894

(c) Soils (undrained cyclic loading model)

Materials	Depth (m)	Young's modulus E (MPa)	Poisson's ratio ν	G_0 (MPa)	τ_f (kPa)	$b\cdot\gamma_{G_0}$	n	Unit weight γ (kN/m^3)	Rayleigh damping α	β
Bn	**0.0 - 1.5**	95.9	0.438	33.4	19.6	3.0	1.7	1.70	0.114	0.001157
Ag	**1.5 - 6.7**	261.	0.493	87.3	88.3	10.	1.7	1.80	0.114	0.001157
Ac3	**6.7- 9.6**	138.	0.496	46.1	108.	5.0	1.7	1.60	0.114	0.001157
Dc1	**9.6 - 11.6**	336.	0.491	113.	118.	1.4	1.7	1.80	0.114	0.001157
Dg1	**11.6 - 21.0**	1520.	0.461	520.	343.	10.	1.8	2.00	0.114	0.001157

model. The inclination of the initial part of the skeleton curve corresponds to the flexural rigidity EI in the elastic region. Piers P8, P9 and P10 are supported by 9 (=3 × 3), 12 (=4 × 3) and 6 (=2 × 3) piles, respectively, as shown in Figure 1.

(b) Soils

The soils in the ground are assumed to show the newly proposed hysteretic characteristics that simulates the commonly observed $G-\gamma$ and $h-\gamma$ curves. It is assumed that liquefaction of the soil does not occur during excitation. G is the secant elastic modulus and h is the damping ratio. The concepts of these parameters are familiar.

Figure 2(b) shows hysteretic curves on the $\tau-\gamma$ relationships. τ is given by Eq. (1).

$$\tau = \sqrt{J_2}\sin\left(\frac{\pi}{3}+\Theta\right) \qquad (1)$$

where

$$\Theta = \frac{1}{3}\cos^{-1}\left(-\frac{3\sqrt{3}J_3}{2J_2^{3/2}}\right) \qquad (2)$$

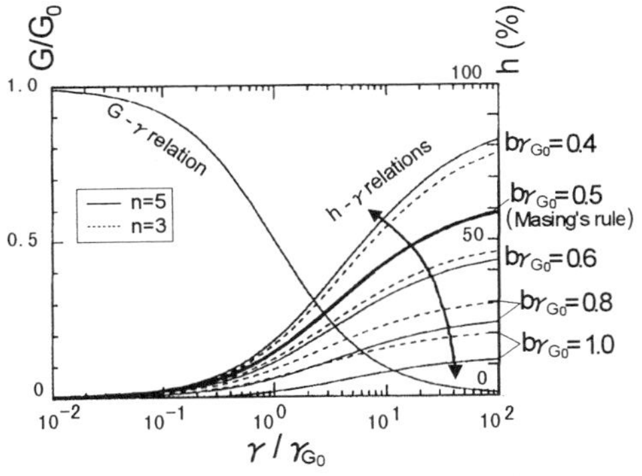

Figure 3. $G-\gamma$ and $h-\gamma$ relations of the model.

J_2 and J_3 are the local deviatoric stress invariants, which are defined in terms of local stress vector $\{\tilde{\sigma}\}=\{\sigma\}-\{\sigma_0\}$ where $\{\sigma\}$ and $\{\sigma_0\}$ defines the present and the initial value of each stress component, respectively.

γ is obtained from Eq. (1) after substituting the local strain vector for the local stress vector. The reversal of loading direction is judged by the sign of the γ increment, $d\gamma$.

The skeleton curves are given by the following hyperbolic equation.

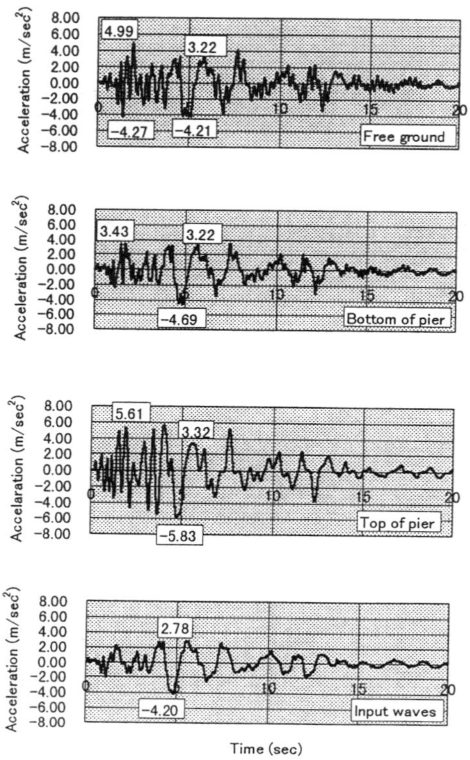

(a) Histories of horizontal acceleration

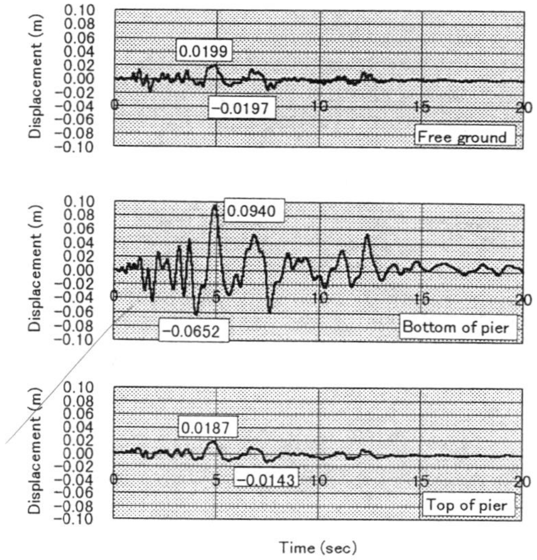

(b) Histories of horizontal displacement

Figure 4. Horizontal acceleration and displacement response at each point along pier P9 calculated by FEM.

$$\tau = \frac{G_0\gamma}{1+G_0\gamma/\tau_f} \tag{3}$$

As seen in Figure 2(b), G_0 is the inclination of the initial part of the skeleton curve and τ_f is the undrained shear strength. The hysteretic loops are defined as

$$\tau^* = \frac{a\gamma^{*n} + G_0\gamma^*}{1+b\gamma^*} \tag{4}$$

where b and n are material constants. The value of a is determined by b,n and the coordinates of the previous reversal point, so that the hysteretic loop is exactly closed.

The suffix '＊' attached to τ (or γ) indicates that the local deviatoric stress (or strain) invariants are defined in terms of local stress (or strain) vectors, $\{\sigma^*\} = \{\sigma\} - \{\sigma_R\}$ where $\{\sigma\}$ and $\{\sigma_R\}$ defines the present value and the value at the previous reversal point, respectively.

As shown in Figure 3, parameters b and n control the $h-\gamma$ relationship of the soil. γ_{G_0} in Figure 3 is given by Eq. (5).

$$\gamma_{G_0} = \frac{\tau_f}{G_0} \tag{5}$$

Even if the strain amplitude becomes larger, the value of h is usually a maximum of 20～30%. As seen in Figure 3, the value of h is overestimated in the cases where $b\cdot\gamma_{G_0}$ is relatively small. The case of $b\cdot\gamma_{G_0} = 0.5$, which corresponds to the case where the Masing's rule is adopted, is not reasonable in its $h-\gamma$ relation.

Table 1 shows the material constants for the 3D analysis. The Rayleigh damping is also adopted. The coefficient of static earth pressure K_0 is assumed to be 0.5 for all the layers in the ground. The $h-\gamma$ relationship of each soil is based on the experimental results of cyclic triaxial loading tests under undrained condition.

3 RESULTS OF THE 3D ANALYSIS

3.1 Responses at each point

Figure 4 shows time histories of horizontal acceleration and displacement responses calculated at each point of pier P9. The results concerning piers P8 and P10 are omitted here. The results of 'Free ground' indicate the response of the free ground surface which is given by the 1D analysis. It has been input at the end of the analytical region in the 3D analysis. The history noted as 'Input waves' is the horizontal acceleration input at the bottom of the ground.

The maximum acceleration at the free ground surface occurs around 2(sec) after the beginning of excitation. It shows a tendency different from the history of input acceleration. It seems that the natural vibration period of the ground is close to the prominent period of the waves in the initial part of the input acceleration. The maximum displacement at the free ground surface is about 2cm and it occurs around 5(sec) after the beginning of excitation. The time 5(sec) is consistent with the time of maximum input acceleration.

The results noted as 'Top of pier' indicate the response at the superstructure (G.L. +12.0m). Comparing the maximum accelerations at each point with the input waves, it is found that the response at both the superstructure and the bottom of pier are magnified. The maximum displacement at the superstructure is about 9cm, which is much larger than the one at the bottom of the pier. However, the residual displacement at the superstructure after the earthquake is found to be very small.

It has been shown that the 3D dynamic FEM makes it possible to evaluate the dynamic response of the system, in which the interaction between the ground and the structure is properly considered.

3.2 Hysteretic curves for RC structures

Figures 5 shows the hysteretic curves for bending moment - curvature relationships at each section of pier P9, given by the 3-D analysis. The results are output at the section near the top of the pile, and the bottom and the top of the pier. The ultimate curvatures Φ_u of the piles supporting pier P9 are 3300 ($10^{-5}m^{-1}$). Φ_u of the sections near the top and the bottom of pier P9 are 3910 and 2990 ($10^{-5}m^{-1}$), respectively.

It is found that both the pier and the piles are safe during the earthquake, because the maximum values of the curvatures have never reached the ultimate states. Large residual curvatures are not observed in the results. It is shown that the results obtained by the 3D dynamic elasto-plastic FEM are useful to evaluate the safety of the RC structures against a great earthquake.

4. CONCLUSIONS

The summary of this study is as follows:

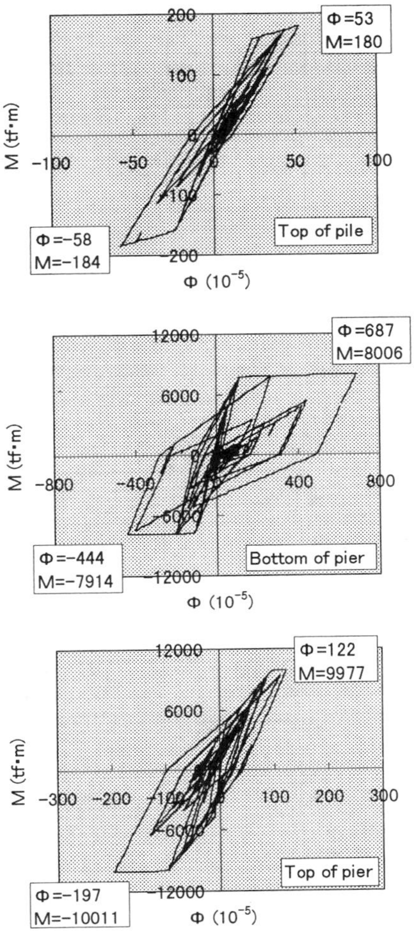

Figure 5. Histeretic curves for bending moment - curvature relationships at each section of pier P9, given by FEM.

(1) A continuous rigid-frame bridge in Japan is analyzed against a large earthquake motion, using the 3D dynamic elasto-plastic FEM in which the hysteretic characteristics of soils and RC structures are properly considered in the analysis.
(2) It is shown that the 3D dynamic elasto-plastic FEM makes it possible to evaluate the dynamic response of the system, which includes the interaction between the ground and the structures.
(3) The results obtained by the analysis are useful to evaluate the safety of RC structures against a great earthquake.

ACKNOWLEDGEMENTS

The authors wish to thank Mr. Mishina, Mr. Kawakami, Mr. Miura and Mr. Yoshizawa (Dainippon Consultant Co., Ltd., Japan) for their helpful advice.

REFERENCES

Borja, R.I. and Wu, W.H. (1994) : Vibration of foundations on incompressible soils with no elastic region, Proc. ASCE, Vol.120, No.GT9, pp.1570-1592.

Dafalias, Y.F. and Popov, E.P. (1977) : Cyclic loading for materials with a vanishing elastic region, Nuclear engineering and design 41, North-Holland, Amsterdam, The Netherlands, pp.293-302.

Fukutake, K., Ohtsuki, A. and Yoshimi, Y. (1995) : A new soil cement block system for protecting piles in liquefiable ground, Proc. of First International Conference on Earthquake Geotechnical Engineering, K. Ishihara ed., Tokyo, Vol.1, pp.605-610.

Takeda, T., Sozen, M.A. and Nielsen, N.N. (1970) : Reinforced concrete response to simulated earthquakes, Journal of Structural Division, ASCE, Vol.96, No.ST12, pp.2557-2573.

Tatsuoka, F. and Fukushima, S. (1978) : Stress-strain relation of sand for irregular cyclic excitation (in Japanese), Seiken-Kenkyu, Univ. of Tokyo, No.9, pp.356-359.

Numerical Models in Geomechanics, Pietruszczak & Pande (eds) © 1997 Balkema, Rotterdam, ISBN 90 5410 886 X

Numerical modeling of wellbore cavitation in coalbed methane reservoirs

H.H.Vaziri
Technical University of Nova Scotia, Halifax, N.S., Canada

I.D.Palmer
Amoco Exploration & Production Technology Group, Tulsa, Okla., USA

ABSTRACT: Using a consolidation-based finite element model, analyses are performed to assess the effectiveness of openhole cavity completion in a multi-seam coalbed methane deposit. Calibration of the model against the field measured wellbore cavitation profile provided a means of establishing the large-scale strength characteristics of the coalbed seams which were later validated using laboratory shear strength tests. Numerical modeling has shown that multi-seam cavity completions could potentially result in incomplete stimulation of those coal seams with higher native strengths due to the fact that seams with lower strength would cavitate earlier and act as "thief zones" during subsequent air injection/blowdown cycles. The sensitivity analyses performed indicate that the potential for cavitation and improvement in production increases with reduction in strength properties and the initial permeability. Stiffness properties do not influence cavitation or flowrate.

1 INTRODUCTION

This paper shows the application of a finite element consolidation model in analysis of a field project involving formation collapse around a wellbore. The wellbore under investigation was completed using the openhole cavity completion technique in a deep coalbed methane deposit composed of three distinct seams. The primary objective of the proposed analysis was to assess the effectiveness of openhole cavity completion for this project.

Openhole cavity completion technique uses subsurface pressure pumps to apply a number of pressure cycles (also referred to as surging cycles) to the pay-zone (this section of the wellbore is not cased) to induce cavity enlargement. Field data indicate that wellbores undergoing cavitation become much better producers (by several times) than those completed using gel-fracturing through cased wells (the more common technique of wellbore completion) as described in Palmer *et al* (1993).

In some cases, however, no appreciable improvement in production, relative to other modes of wellbore completion, have been noticed. Vaziri *et al* (1997a) have shown that shear strength characteristics of the formation play a key role in the performance of wells that employ the openhole cavity completion technique. Production enhancement is greatest in formations that are weak and dilatant. The weaker is the formation the greater is the potential for cavitation and the larger is the size of plastic zone around the enlarged cavity. The dilatancy within the plastic zone accounts for a significant portion of production boost. In strong formations, not only the potential for cavity enlargement reduces but also the associated plastic zone becomes smaller. Moreover, high strength properties result in permeability depression over an appreciable zone outside the plastic zone (this is due to elevation of the mean effective stress) which can yield a positive skin factor (i.e., a net reduction in permeability).

2 BACKGROUND OF THE FIELD PROBLEM

The proposed project, referred to as AMAX (Logan et al, 1995), is located outside the northwest corner of the fairway region in the San Juan Basin in Colorado. The site contains three different coal seams having the characteristics shown in Table 1. In this table c' is the cohesion intercept as defined in Fig. 1. The adopted bilinear failure envelope is shown to be most applicable to the coal in this region (Vaziri *et al*, 1997b).

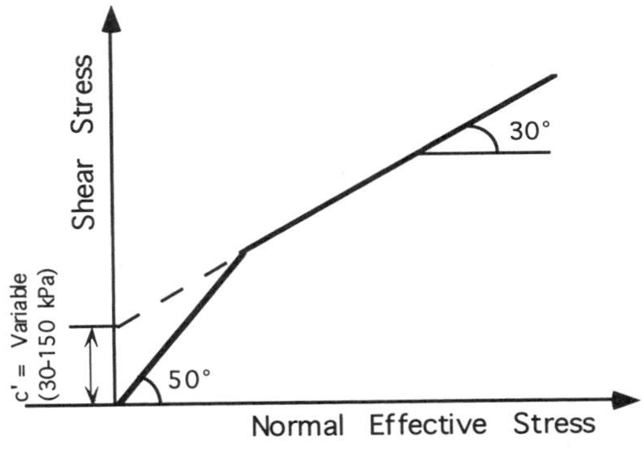

Figure 1. Adopted bilinear failure envelope for coal

Table 1. *In situ* conditions and properties

Coal seam unit	Thickness (m)	Depth to top (m)	Vert. eff. stress (kPa)	Horiz. eff. stress (kPa)	Pore pres. (kPa)	Permeability ($\times 10^{-15}$ m^2)	c' (kPa)
A	4.9	402	5126	3115	4030	25	35
B	5.5	409	5222	3170	4100	25	35
C	6.1	442	5643	2115	4430	12	100

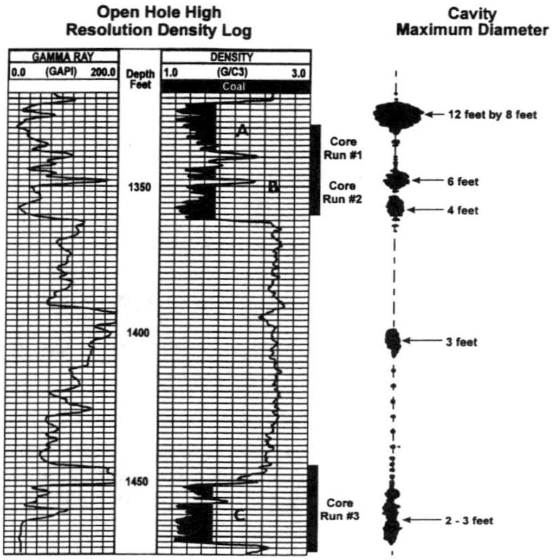

Figure 2. Post-stimulation cavity size

All three seams in this well were cavitated simultaneously (i.e., multi-seam cavity completion), resulting in an average cavity radius of 2.3 m, 1.5 m and 0.75 m in seams A, B and C as shown in Fig. 2.

3 THE NUMERICAL MODEL

The fully coupled flow and stress finite element code (*ENHANS-COAL*) that was used in this study is capable of time-dependent simulation of boundary conditions generally employed for openhole cavity completion. The principal features of the code relevant to this study include nonlinear elastic and elasto-plastic stress-strain relationships; linear and bilinear Mohr-Coulomb based failure criteria; various stress-porosity-permeability relationships; compressible fluid flow treatment (as required for gas pore fluid). The most significant, and unique, feature of the *ENHANS-COAL* for wellbore simulation studies is its ability to model the physical processes resulting in tensile failure and breakup of the wellface material and in maintaining stable analysis while accounting for the dramatic changes in formation properties such as fluid-like properties and several orders of magnitude increase in permeability (Vaziri, 1988). Key formulations, based on Biot's (1941) consolidation theory, and the methodology used to simulate wellbore cavitation, are as follows.

(a) Equilibrium equation:

$$\sigma_{ij,j} - F_i = 0 \tag{1}$$

where σ_{ij} is the total stress tensor and F_i is the body force vector.

(b) Effective stress strain relationship:

$$\sigma'_{ij} = \sigma_{ij} - p\delta_{ij} = D_{ijkl}\varepsilon_{kl} \tag{2}$$

where δ_{ij} is Kronecker's delta, ε_{kl} is the strain vector and D_{ijkl} is composed of tangent stiffness moduli whose variations are described by hyperbolic relationships described in Vaziri *et al* (1997a). For the proposed study of formation collapse, it was found that the adopted magnitude of Young's modulus and Poisson's ratio did not appreciably influence the results within a relatively wide range (E between 300 MPa and 1,500 MPa and ν between 0.25 and 0.4). The problem is primarily governed by the strength properties.

(c) Continuity condition for a compressible fluid:

$$-v_{i,i} + \dot{w}_{i,i} - \frac{\dot{p}}{B_e} = 0 \tag{3}$$

where v_i is the superficial velocity, w_i is the displacement vector and B_e is the equivalent bulk modulus of the compressible phases that constitute the formation:

$$\frac{1}{B_e} = \frac{n}{B_f} + \frac{1-n}{B_s} \tag{4}$$

where n is the porosity, B_f is the fluid bulk modulus (e.g., gas, water or a mixture of gas and water) and B_s is the bulk modulus of the solid grains.

$$v_i = -\frac{k_{ij}p_{,j}}{\mu} \tag{5}$$

where k_{ij} is the intrinsic permeability tensor and μ is the fluid viscosity. For this study, the fluid was assumed to be entirely composed of methane gas having a viscosity of 1.46×10^{-8} kPa.s. Based on uniaxial compression tests, for the San Juan coal, Seidle *et al* (1992) have proposed an empirical expression for relating permeability to the mean effective stress level:

$$\frac{k_i}{k_0} = e^{[-2.88\times10^{-3}(\sigma'_i - \sigma'_0)]} \quad (6)$$

where k_0 is the *in situ* permeability corresponding to stress level σ'_0 and k_i is the current permeability corresponding to σ'_i.

Using the above relationships, the equilibrium and continuity equations can be written as:

$$\left[0.5\left(w_{i,j} + w_{j,i}\right)D_{ijkl}\right]_{,j} + p_{,j} = F_i \quad (7)$$

$$\frac{(-k_{ij}p_{,j})_{,i}}{\mu} + \dot{w}_{i,i} - \frac{\dot{p}}{B_e} = 0 \quad (8)$$

The finite element forms of eqn (7) and (8) can be developed in a straight-forward manner.

Tensile failure, leading to cavitation, is presumed to occur when the minimum principal effective stress becomes equal to zero. Under these conditions, all the stresses at the gauss point that experiences tensile failure are set to zero and the out of balance stresses are redistributed to the other unfailed elements in an iterative manner. To account for other physical changes that are associated with an element in the tensile mode (which can be considered to have broken away from the intact coal and sloughed into the wellbore), stiffness moduli and permeability are changed to simulate the properties of an air-filled void (see Vaziri and Byrne, 1990 for more details). A schematic representation of different zones of failure around a wellbore is shown in Fig. 3.

Verification of the model with respect to matching the theoretical stress and pore pressure around a well in an elasto-plastic medium and the critical flow rate resulting in wellbore cavitation is shown in Vaziri (1995).

4 RESULTS

The applied well pressure employed in the field to stimulate the seams is shown in Fig. 4. Seven pressure surges are applied over a period of 8 hours. The blowdown is about -4000 kPa (resulting in a bottomhole pressure of 70 kPa at the top of Seam A) and the buildup pressure reaches about 3800 kPa.

4.1 *Calibration*

Using the aforementioned properties and relationships and the applied boundary condition in Fig. 4, the finite element analysis results in the cavitation response shown in Fig. 5 which is very close to that observed in the field. As it can be observed, the stimulation results in enlargement of the cavity size from an original wellbore radius of 0.15 m to about 2.3 m in Seam A, 1.5 m in B, and 0.75 m in C. Fig. 6 shows the corresponding size of the plastic zones.

Fig. 7 shows the pattern of failure within the three seams at the end of stimulation. Although seams A and B have the same strength and permeability characteristics, size of cavitation in seam A is larger than B. This is due to the higher pressure gradient

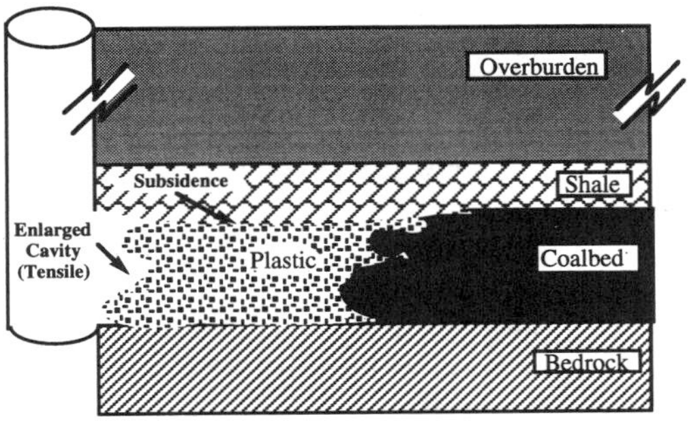

Figure 3. Schematic presentation of failure

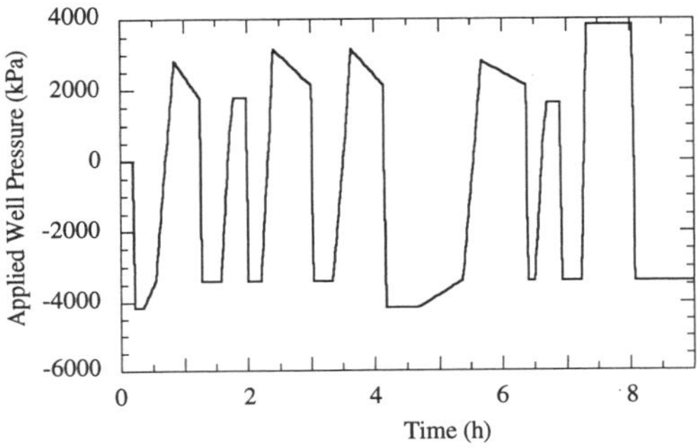

Figure 4. Applied wellbore pressure

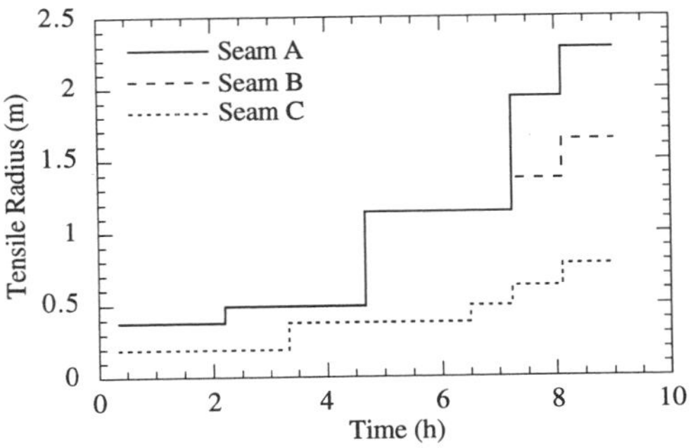

Figure 5. Computed cavitation response

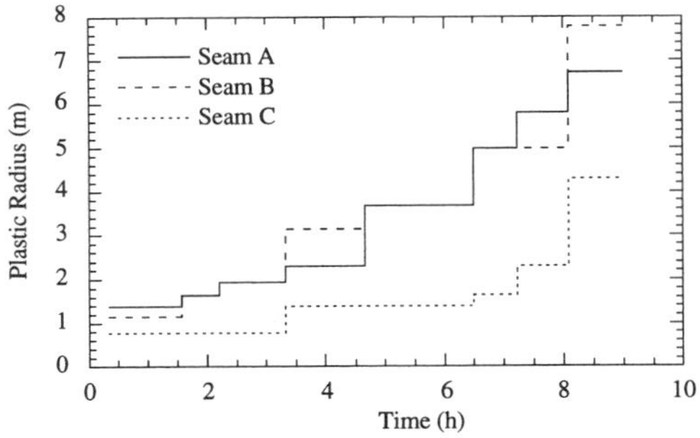

Figure 6. Computed size of plasticity

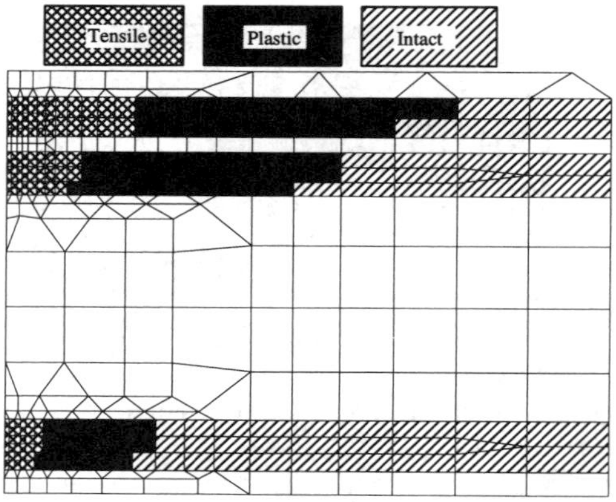

Figure 7. Failure pattern at the end of stimulation

that is generated in seam A (the initial pressure in seam A is 70 kPa less than B but both are subjected to the same well pressure as shown in Fig. 4).

It is to be pointed out that matching the observed size of seepage-induced cavitation requires capturing the correct value of both the strength and permeability characteristics. The magnitude of k affects failure as it controls the pressure gradient (more on this under 4.3 below). The k values used in this study were those obtained from field measurements. The strength properties were obtained by back-analysis. The applicability of the selected strength envelope was validated by performing laboratory shear strength experiments on representative intact coal samples as described in Khodaverdian *et al* (1996).

Fig. 8 shows the stress and pressure profiles at around the lower third section of Seam A at a stage corresponding to creation of the opening when the bottomhole pressure is about 200 kPa. Fig. 9 shows the conditions at the end of the pressure surging process. It can be seen that the enhanced permeability zone is increased from 1.7 m at the end of drilling to about 9.0 m by the end of surging.

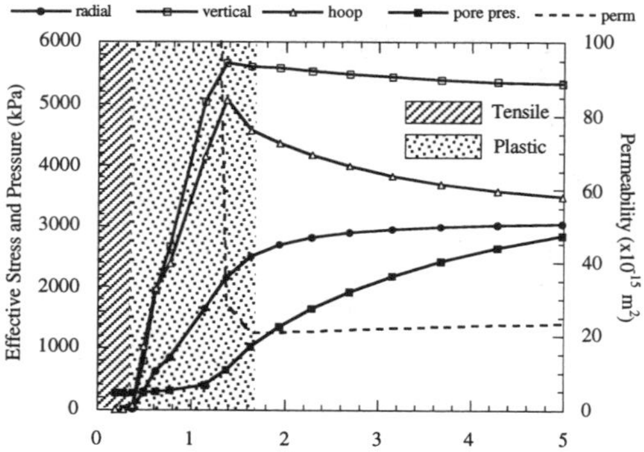

Figure 8. Conditions at the end of drilling in Seam A

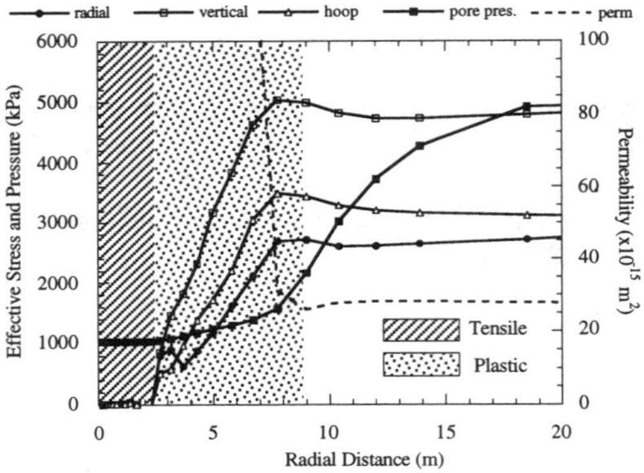

Figure 9. Conditions at the end of surging in Seam A

The difference in permeability profiles between the end of drilling and the end of surging for all seams is shown in Fig. 10.

The data in Fig. 10 suggest that the multi-seam cavitation procedure was generally less effective in stimulating Seam C, as compared with seams A and B. For example, in Seam C, an average permeability of 24 md (twice the *in situ* value) is estimated within a radius of 10 m. For seams A and B, however, an average permeability of 50 md (also twice the *in situ* value) is estimated within radial distances of 20 m and 30 m, respectively.

Fig. 11 illustrates the effectiveness of permeability improvement on production in Seam A. At the end of drilling and at a stage when the well pressure was brought down, the flowrate was about 1.06 m^3/d while at the end of surging under a similar well pressure, the flowrate increased by a factor of 6 to 6.5 m^3/d.

4.2 *Skin factor and flow efficiency analysis*

Steady state skin factors for the post-stimulation formation, were calculated using eqn (9).

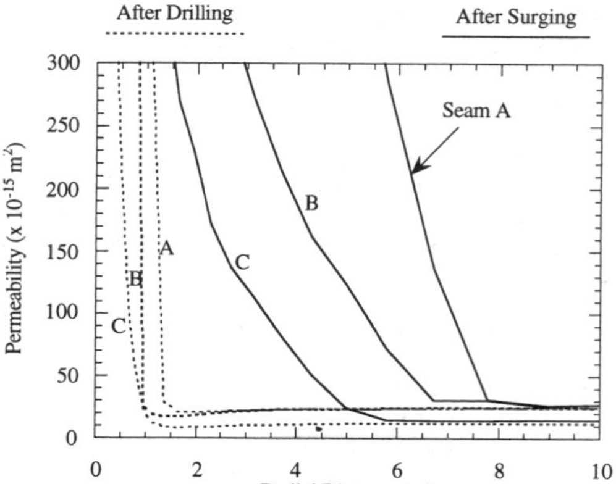

Figure 10. Stimulation-induced perm improvement

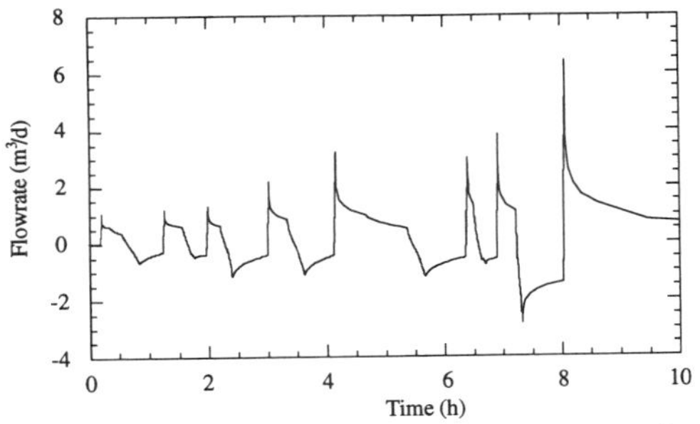

Figure 11. Flowrate through seam A during surging

$$S_t = \sum_{i=1}^{N}\left[\frac{k_u}{k_i} - 1\right]\ln\left[\frac{r_i}{r_{i-1}}\right] \quad (9)$$

where k_u is the unaltered formation permeability, the interval between r_{N-1} and r_N represents the outermost zone with altered permeability k_N, r_0 is the drilled diameter, and the interval between r_0 and r_1 is the cavitated zone wherein the perm is infinity ($k_1 = \infty$).

Application of eqn (9) to the post-stimulation data results in skin factors shown in Table 2. Also shown are the skin contributions arising from the enlarged cavities alone. The skin factors for the three seams suggest that multi-seam cavity completions could potentially result in incomplete stimulation of the higher strength coal seams.

Psuedo-steady-state flow efficiency (i.e., ratio of the reservoir fluid flow after stimulation to that of the unstimulated reservoir), for single-phase flow, is given by

$$R_q = \frac{\ln\left[\frac{0.473 r_e}{r_0}\right] + S_{ti}}{\ln\left[\frac{0.473 r_e}{r_0}\right] + S_t} \quad (10)$$

where r_e represents the drainage radius and S_{ti} represents the skin factor before stimulation.

Assuming a drainage radius of 650 m, a drilled radius of 0.15 m, and using skin factors of $S_{ti} = 0$ before stimulation and the preceding skins (-3.56, -3.81 and -2.73) after stimulation, the predicted flow efficiency for seams A, B and C are shown in Table 2. Eqn (10) can also be used to demonstrate the contribution of the permeability enhanced zone to the overall well productivity. Assuming $S_{ti} = -2.1$, -2.2 and -1.4 for seams A, B and C (skins due to the physical cavity) and $S_t = -3.56$, -3.81 and -2.73, the psuedo steady state flow efficiency can also be determined as shown in Table 2. The latter, in effect, indicates the level of production improvement due to increase in permeability.

Table 2. Effect of stimulation on the skin factor

Seam	Post Stimulation Skin	Enlarged Cavity Skin
A	-3.56	-2.1
B	-3.81	-2.2
C	-2.73	-1.4

Table 3 Predicted flow efficiency and improvement

Seam	Flow efficiency	Psuedo steady state efficiency	Production increase due to perm increase
A	1.88	1.36	36%
B	2.00	1.42	42%
C	1.56	1.27	27%

4.3 *Influence of initial permeability*

To quantify the influence of initial perm, sensitivity analysis comprised of three different k was performed on Seam B. Fig. 12 shows the radius of tensile and plastic zones at the end of surging. It can be seen that as the *in situ* permeability becomes larger, the radius of tensile and plastic zones reduces. Under transient conditions, permeability controls the pore pressure gradient. The smaller is the permeability the steeper will be the pore pressure gradient close to the wellbore and hence the greater will be the seepage forces. As the pressure gradient becomes steeper (or the seepage forces become larger), the effective stresses become smaller thus exacerbating failure. It can, therefore, be stated that as permeability becomes smaller the potential for cavitation increases. A corollary of this is that the improvement in production rate becomes greater as the *in situ* permeability decreases.

Fig. 13 illustrates the influence of initial k on the potential for production boost for Seam B. The issue that is of interest is the relative change in production. For the case of k = 250 x 10^{-15} m^2, there is almost no improvement in production due to surging; for the case of k = 2.5 x 10^{-15} m^2, however, the boost is quite appreciable. For the case of k = 2.5 x 10^{-15} m^2 the flowrate increases by over five times by the end of surging while for the case of k = 250 x 10^{-15} m^2 the increase is only about 25%. This result clearly shows the significance of cavitation (including the size of plastic zone) on production regardless of the parameter controlling failure (e.g., k or c').

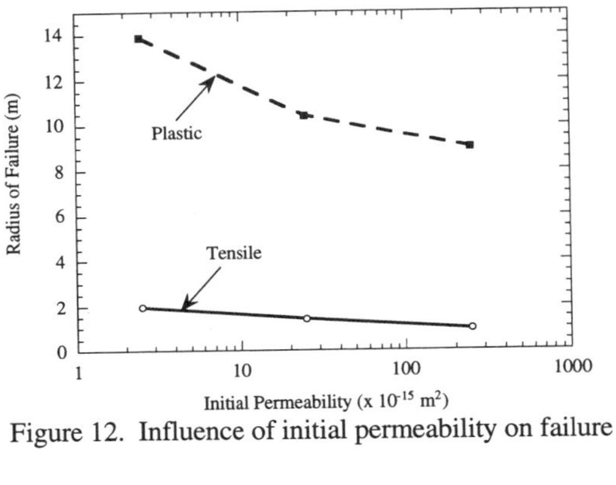

Figure 12. Influence of initial permeability on failure

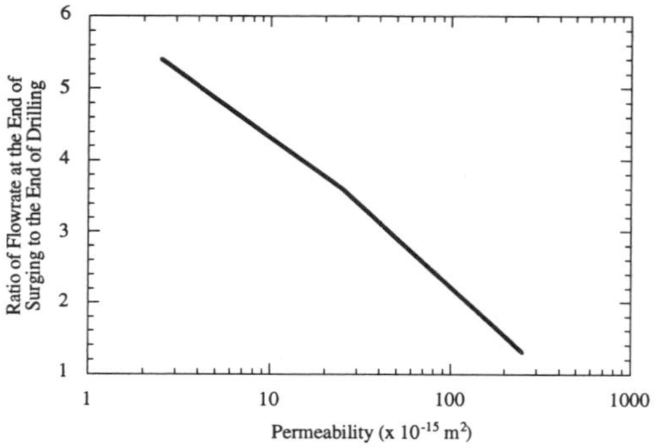

Figure 13. Increase in flowrate between end of drilling and surging as a function of initial k; seam B

5 SUMMARY AND CONCLUSIONS

The objective of this study was to determine the effectiveness of openhole cavity completion in the case of a multi-seam coalbed methane deposit. This was accomplished by calibrating a consolidation-based finite element numerical model against the measured enlarged cavity profile. In this calibration study, all the available field data were employed including the geometry, applied boundary conditions, initial stress and pressure conditions and permeability. The permeability-stress relationship used was based on laboratory tests performed on representative samples. The failure envelope used was established by history-matching the cavitation profile. We consider the adopted envelope to be representative as cavitation is a function of both permeability and strength characteristics (the latter values were based on field data). Independent shear strength laboratory tests that were performed later confirmed the validity of strength properties used.

Numerical modeling of the cavity completions for the AMAX site has shown that multi-seam cavity completions could potentially result in incomplete stimulation of those coal seams with higher native strengths. Predicted post-stimulation skins for the three seams at this site were determined to be -3.6, -3.8 and -2.7. These results were based on matching the predicted cavity diameters to those observed in the field. Although some stimulation is indicated, the results suggest that multi-seam cavitation should be avoided if the coal seams do not have similar properties which may be indicated by cleat spacing, rank and/or ash content. By inductive reasoning, these conclusions may also be extended to seams with significantly different permeabilities, *in situ* stress and depth.

The mechanisms potentially leading to incomplete stimulation during multi-seam cavity completions include: (1) the seams with lower strength coal would cavitate earlier, leading to larger cavity sizes and permeability enhanced zones than the higher strength coal seams, and (2) the better cavitated seams would act as "thief zones" during subsequent air injection/blowdown cycles, thus inhibiting significant stimulation in the other zones.

The sensitivity analyses performed indicate that the potential for cavitation and improvement in production increases with reduction in strength properties and the initial permeability. Stiffness properties do not appreciably influence cavitation or flowrate within a relatively wide range (i.e., within the scatter usually seen in stiffness measurements).

6 ACKNOWLEDGMENTS

The funding for this project was provided by the Gas Research Institute and NSERC. We appreciate the significant contributions of Dr. Khodaverdian and Dr. Wang in this study.

7 REFERENCES

Biot, M.A. (1941) General theory of three dimensional consolidation. Journal of Applied Physics, 12, 155-164.

Khodaverdian, M, McLennan, J, Palmer, I.D., Vaziri, H. and Wang, X. (1996) Cavity completions for enhanced coalbed methane recovery. Final report submitted to Gas Research Institute, GRI-95/0432, GRI Publication, Chicago, Illinoise, 215p.

Logan, T.L., Robinson, J.R., and Pratt, T.J. (1995) Cooperative research project AMAX Oil & Gas Inc., Southern Ute #5-7 Well San Juan Basin, Southern Colorado. Gas Reserach Institute Topical Report, GRI-93/0440, Chicago, IL.

Palmer, I.D., Mavor, M.J., Seidle, J.P, Spitler, J.L. and Volz, R.F. (1993). Openhole Cavity Completions in Coalbed Methane Wells in San Juan Basin. SPE, JPT, Nov issue, 1072 - 1080.

Seidle, J.P., Jeansonne, M.W. and Erickson, D.J. 1992. Application of matchstick geometry to stress dependent permeability in coals. SPE Rocky Mount. Meetg, Casper, WY, SPE 24361, 433 - 445.

Vaziri, H. (1988) Theoretical analysis of stress, pressure, and formation damage during production. The J. of Canad. Petrol. Tech, 27, No. 6, 111 - 117.

Vaziri, H. 1995. Analytical and Numerical Procedures for Analysis of Flow-Induced Cavitation in Porous Media. Int. J. of Computers & Structures, Vol 54, No. 2, 223-238.

Vaziri, H. and Byrne, P.M. (1990) Analysis of stress, flow and stability around wellbores. Geotechnique 40, No. 1, 63 - 77.

Vaziri, H., Wang, X. And Palmer, I. (1997a) Wellbore completion technique and geotechnical parameters influencing gas production. Canadian Geotechnical Journal, 34: Feb issue.

Vaziri, H., Wang, X., Palmer, I., Khodaverdian, M. and McLennan, J. (1997b) Back analysis of coalbed strength properties from field measurements of wellbore cavitation and methane production. In print, *Int. J. of Rock Mech. and Mining & Geomech. Abst* .

Numerical Models in Geomechanics, Pietruszczak & Pande (eds) © 1997 Balkema, Rotterdam, ISBN 90 5410 886 X

Modelling of the vacuum consolidation with prefabricated vertical drains

N. Puumalainen, A. Näätänen & P. Vepsäläinen
Helsinki University of Technology, Finland

ABSTRACT: Vacuum consolidation with vertical drains is a ground improvement method in which the atmospheric pressure replaces the preloading with surcharge. In 1996 the City of Helsinki and the Helsinki University of Technology established a test field in Arabianranta, Helsinki. The main target was to find out whether the vacuum method could be utilised for soil improvement in that area. During vacuum pumping the settlement behaviour was studied with observational and numerical methods. The aim of the analysis was to find a reliable method to estimate the behaviour of vertically drained ground in the course of vacuum pumping and after that. The parameters for the calculations were carefully predicted at the laboratory of Soil Mechanics and Foundation Engineering. The test site was well instrumented, including settlement plates and pore pressure transducers. The observations and the calculation results were compared.

1 INTRODUCTION

In many countries vacuum consolidation with prefabricated vertical drains has proved to be a feasible alternative to conventional soil improvement methods. In vacuum consolidation the surcharge load is replaced with atmospheric pressure. A vertically drained site is covered with an airtight but flexible membrane and a vacuum is created underneath it with effective pumps. In practise underpressure of 70 to 90 kPa, which equals a surcharge embankment of 4 to 5 meters of sand, has been achieved. Usually the actual pumping time varies between 3 and 6 months.

During vacuum consolidation the effective stress will increase due to dissipation of the pore water pressure as caused by the vacuum. Compared with the conventional surcharge load, the pore water pressure will not initially rise, thus preventing instability.

Following the international experience (Ye et al. 1991, Jacob et al. 1994, Technique Geosystems/ Cofra JV 1995) the vacuum settlement can be predicted by Terzaghi's, Taylor's and Biot's consolidation models. The underpressure is replaced with a corresponding surcharge load and the problem is solved as a traditional vertical drain and embankment consolidation process. For rough estimations it is acceptable to predict the settlement behavior of the vacuum field by using surcharge load instead of suction. However, many FEM-programs offer tools to simulate the development of pore pressure and water flow. These programs are very useful for more accurate and detailed modelling of the vacuum pumping process.

To predict the behavior of soil during the vacuum application a simple numerical model has been developed at Helsinki University of Technology. The reliability of the model has been tested by verifying the calculated results to the in-situ measurements obtained from two test fields, situated at the Torpparinmäki (Puumalainen & Vepsäläinen 1997) and Arabianranta regions in Helsinki. This article deals with the FEM-modelling of the Arabianranta test site.

2 TEST SITE

The City of Helsinki and Helsinki University of Technology have carried out the vacuum preloading project in the Arabianranta test field. Arabianranta is a former industrial area. Within a couple of years the area will be turned into a high quality housing and office area. The area has also been used as a dumping ground for the waste material produced by nearby factories. Above the soft clay and muddy soil the area comprises very inhomogeneous filling. The

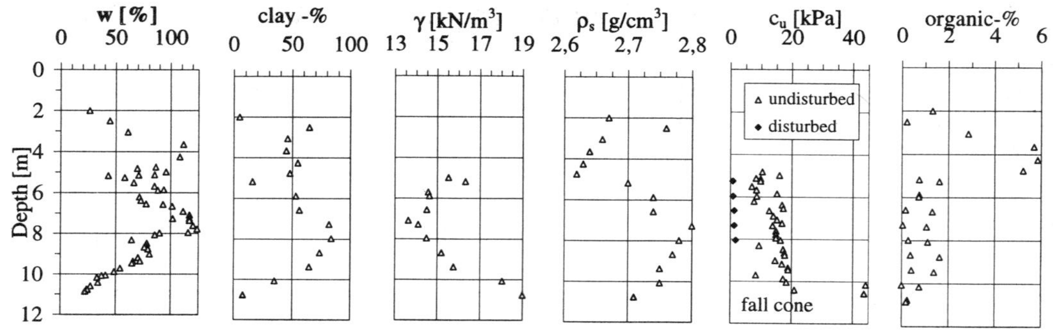

Figure 1. Arabianranta test site. Soil profile, classification characteristics.

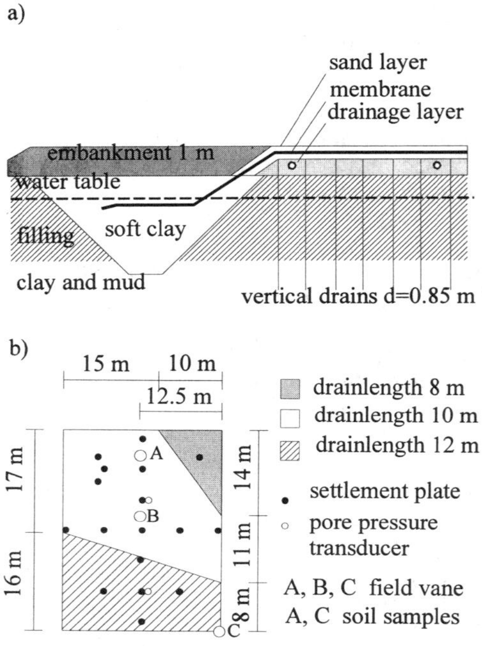

Figure 2. Arabianranta test site. a) The geometry and structure of the edge. b) The test field and instrumentation.

thickness of the filling varies between 2 and 8 meters. The filling has caused excess pore pressure (figure 3) and the consolidation process is still going on. The settlement rate is approximately 30-40 mm per year. The impermeable clay soil needs to be improved but the filling causes considerable problems to all commonly used techniques. Even though the site is not ideal for vacuum consolidation it was decided to test whether this method could be used in the Arabianranta area.

At the test site the thickness of the filling was approximately 3-3.5 m. The thickness of the mud and clay varied from 9 to 15 m. The water level lay one meter below the soil surface. It was estimated that during the next 20 years the present surcharge load would cause a surface settlement of 0.40 m.

Figure 1 presents the classification parameters of the test site. As the water permeability was poor, the vertical drain spacing was very dense, only 0.85 m. The drains were installed in a square grid. Three different drain lengths were used. The drains were anchored in the clay 0.5-1 m above the coarse silt and sand layers. Preholes were made through the filling to ease the installation. The site was isolated from the surrounding area with a peripheral trench that exceeded below the filling. The 3-4 m deep trench was filled with soft clay. It was estimated that the thick clay barrier would be sufficient to hinder suction via the inhomogeneous filling. The edges of the membrane were buried just below the ground water level. Figure 2 presents the structure and the instrumentation of the test field.

The vacuum loading started at the end of August, 1996. The designed consolidation time was three months. However, during the first month the vacuum load underneath the membrane did not reach the desired level. Therefore pumping was continued until January 1997. After 130 days the maximum settlement in the center of the drained area was 0.7 m, around the edges the deformations were smaller. The measured settlements at the center of the test field and the underpressure values are shown in figure 6.

3 CALCULATIONS

During the vacuum treatment the settlement behaviour was studied using analytical and numerical methods. Only the numerical methods are discussed in this article. The aim of the analysis was to find a reliable method to predict the behaviour of vertically drained ground in the course of pumping and thereafter.

3.1 Modelling and boundary conditions

Comparative calculations were done using the finite element programs CRISP (Britto & Gunn 1987, 1990, 1994) and Z-Soil V 3.1 (Z_Soil.PC 1995). Some of the programs' features relevant for the calculations discussed in this article are described here. The element type chosen for CRISP-calculations was the eight-noded linear strain quadrilateral with nine intergration points. The element mesh contained 80 elements. The element type in Z-Soil was the four-node isoparametric quadrilateral with bilinear interpolation function. The Z-Soil mesh contained 120 elements. Consolidation is calculated by using Biot's consolidation theory in both programs. In CRISP the bottom, right- and left-hand boundaries of the mesh were impermeable. In Z-Soil the pore water velocity was zero along the bottom and the right-hand boundaries.

Only single vertical drain was modelled, so the problem could be handled as axisymmetric. The unit cell was situated in the center of the vacuum field. The height of the element mesh corresponded to the drain length at that place. The equivalent radius of the unit cell, 0.48 m, is computed as a function of drain spacing by assuming that the volume of the cell equals the volume of soil served by each drain. Figure 3 presents the calculation model together with the geotechnical layers. In the figure the vertical drain is presented ruled. No special drainage element was available for the vertical drain.

Vacuum suction was simulated with pore pressure boundary conditions. In Z-Soil the suction was simulated by adding a linearly distributed pore pressure boundary condition to the symmetry axis of the vertical drain, along the left hand boundary of the element mesh. At the upper left corner of the mesh, the suction boundary condition corresponded to the measured underpressure. It was assumed, that in the vertical drain the suction at different depths would follow the theoretical pore pressure distribution (Cognon et al. 1994). The pressure decreased linearly with depth (see figure 4).

In CRISP it was possible to model the suction by adding negative pore pressure on top of the element mesh. Absolute excess pore pressure, that corresponded to the measured suction, was set between the upper two nodes at the right side of the mesh (see figure 4).

3.2 Soil parameters

In both calculations the isotropic hardening elasto-plastic material model was chosen for the clay layers lower filling. In CRISP Modified Cam Clay model and in Z-Soil extended Drucker-Prager yield criteria with cap-closure were used. The parameters were chosen so that the models gave analogous initial yield surfaces. The filling was modelled as linear elastic.

The soil parameters were determined from 27 oedometer and 4 triaxial tests. The input parameters for the filling were predicted from disturbed samples. Input parameters are shown in table 1. For all soil layers it was assumed that $k_h = 2*k_v$. During the consolidation process the permeability remained constant. The water permeability of the vertical drain both in horizontal and vertical directions was 100 m/day. The effects of well resistance and smear were neglected. The deformation properties of the drain were identical to soil properties at the same depth.

3.3 Settlements and pore pressure

Figure 5 graphically portrays the calculated pore pressure response after 1 day and after 130 days of vacuum application. The pressure responses mid-way between the drains (side II) were quite similar in CRISP and in Z-Soil. During pumping six pore pressure transducers measured the decrease of pore pressure at three different depths. At these depths the calculation results of both CRISP and Z-Soil corresponded well to the observed pore pressure values. The comparison of the computed and measured settlements with the corresponding loading histories is presented in figure 6. For both programs the correspondence between the actual and the calculated results was very good. The theoretical settlements were largest at layers four, seven and eight. No remarkable effect was found at layer six. At this depth the permeability was significantly smaller than elsewhere in the soil profile.

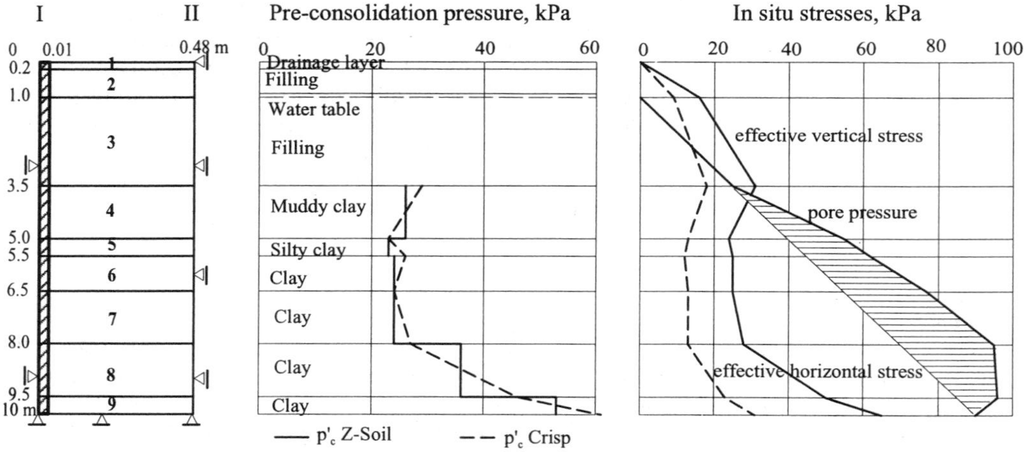

Figure 3. Arabianranta test site. Calculation model, geotechnical layers and in situ stresses.

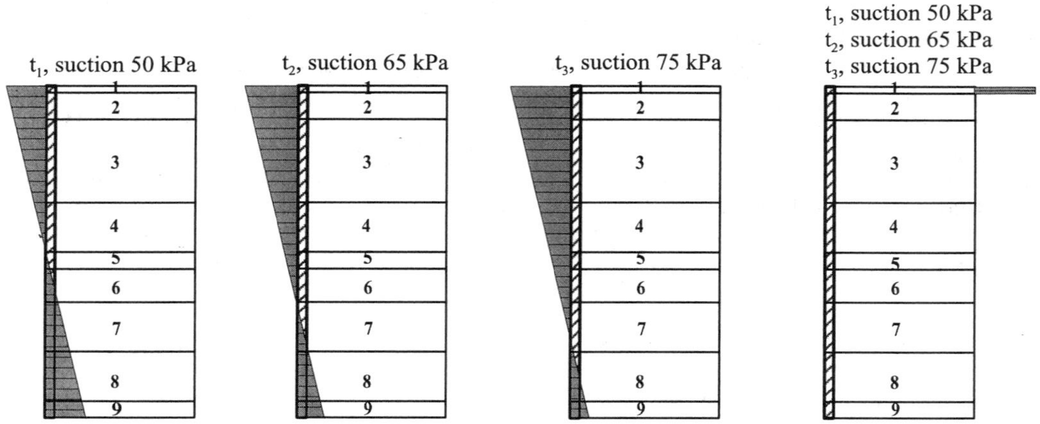

$t_1 = 0...40$ d, $t_2 = 40...55$ d, $t_3 = 55...130$ d

Figure 4. Arabianranta test site. Modelling of the suction in Z-Soil and CRISP94.

Table 1. Arabianranta test site. Calculation parameters.

Layer	γ kN/m^3	M	ϕ' °	ν'	E kPa	κ	λ	e_0	e_{cr}	k_y m/d
1	16,0			0,37	3000			1,00		100
2	16,0			0,37	1000			1,00		1,24E-03
3	16,0			0,37	1000			2,17		3,60E-04
4	15,5	1,10	27,7	0,35	4194*	0,06	0,70	2,69	4,12	1,46E-05
5	16,0	1,25	31,1	0,32	9207*	0,01	0,16	1,36	1,77	1,85E-05
6	14,5	1,10	27,7	0,35	4351*	0,03	0,40	2,06	3,17	2,05E-06
7	14,5	1,25	31,1	0,32	2621*	0,07	0,78	3,18	4,86	1,68E-05
8	15,5	1,25	31,1	0,32	2883*	0,06	0,37	2,20	3,23	1,74E-05
9	18,0	1,25	31,1	0,32	6545*	0,03	0,15	1,44	1,99	4,28E-05

* E corresponding to κ-value

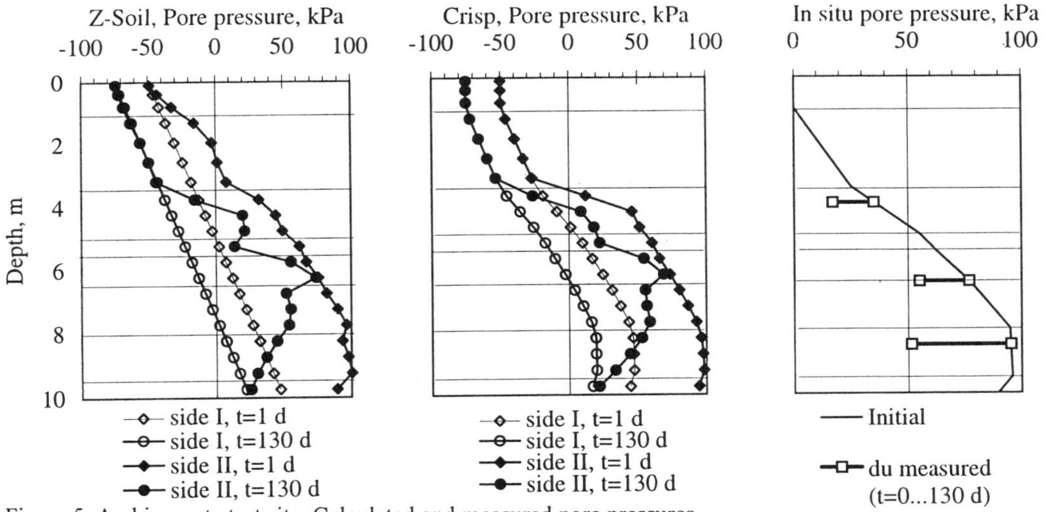

Figure 5. Arabianranta test site. Calculated and measured pore pressures.

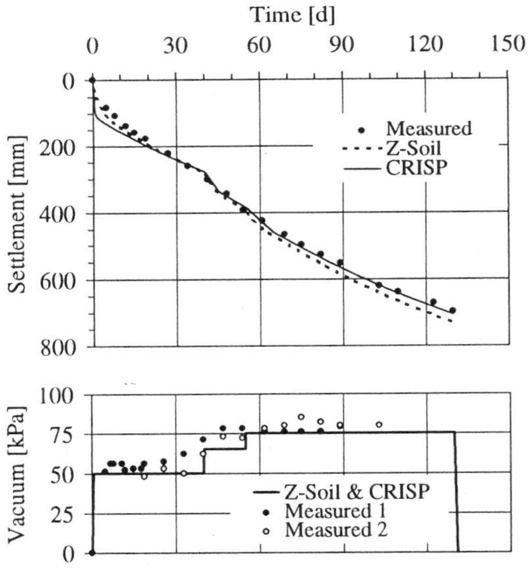

Figure 6. Arabianranta test site. Calculated and measured settlements and the variation of underpressure load with time.

3.4. Shear strength

Three field vane tests were done at the test area before vacuum pumping. Also numerous fall cone tests were made at point A. The initial field vane results at points A and B are presented in figure 7 together with the fall cone test values. By the time this article was written no field vane nor fall cone test results after vacuum loading were available.

The use of an isotropic hardening model like Modified Cam Clay enables the growth of shear strength due to vacuum consolidation to be estimated. The undrained shear strength can be calculated from the equation (1) (Wood 1990).

$$c_u = \frac{Mp'_c}{4} \qquad (1)$$

where c_u is undrained shear strength, M is the slope of critical state line and p'_c is the reference size of yield locus. Values of p'_c are taken from the CRISP calculations.

The calculated values of undrained shear strength at initial condition and after 130 days of consolidation are shown in figure 7. A quite remarkable growth of the undrained shear strength was reached at layers 4, 5 and 7-9. Layer 6 was the weakest and had the lowest permeability. At this depth the degree of consolidation was still low after 130 days, so no significant strengthening could be found.

4 CONCLUSIONS

By using finite element programs it is possible to take into account the layered soil structure and the variations of underpressure load with time. The vacuum consolidation at the Arabianranta test field was numerically simulated with programs CRISP

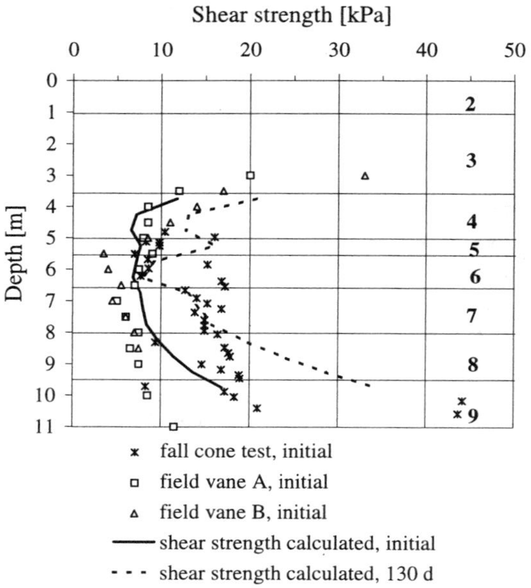

Figure 7. Arabianranta test site. Undrained shear strength profile, initial and calculated values.

and Z-Soil. Also, the already existing overpressure was included in the model. Excellent correspondence was found between the measured and calculated settlements. The development of pore pressure was successfully modelled.

Elasto-plastic hardening models can be used to estimate the growth of shear strength. However, special attention must be paid to the determination and selection of the parameters for the chosen model. At Arabianranta the MCC model was used and shear strength at the end of the test was estimated from calculation results. The accuracy of modelling cannot be valued until the final field vane and fall cone test results are available.

The calculation results show that in the test field the vacuum treatment of four months with an average suction of 70 kPa seems to be sufficient to remove the harmful settlements of the forecoming load in advance. The areal stability in Arabianranta is critical. The calculations suggest, that the development of shear strength in some layers might require longer pumping time.

When designing the vacuum application, the pumping time and the ratio between the final load and vacuum suction need to be considered separately for each field. Large settlements do not necessarily quarantee the desired growth of the shear strength in every layer. The calculation model presented here is very useful when detailed information about the soil behavior is needed.

REFERENCES

Britto, A.M. & Gunn, M.J. 1987. *Critical state soil mechanics via finite elements.* Chichester: Ellis Horwood Limited.

Britto, A.M. & Gunn, M.J. 1990. *CRISP90. User's and Programmer's Guide.* Volume 1. Cambridge University, Engineering Department (1/6/90).

Britto, A.M. 1994. *CRISP94. User's and Programmer's Guide.* Volume 3, PC-386/486 Version. Cambridge University, Engineering Department (1/6/94).

Cognon, J.M., Thevanayagam, S. 1994. Vacuum Consolidation Technology - Principles and Field Experience. *Vertical and Horizontal Deformations of Foundations and Embankments, Geotechnical Special Publication No. 40*: 1237-1248. New York: ASCE.

Jacob, A., Thevanayagam, S., Kavazanjian, E. (1994). Vacuum-assisted consolidation of a hydraulic landfill. *Vertical and Horizontal Deformations of Foundations and Embankments, Geotechnical Special Publication No. 40*: 1249-1261. New York: ASCE.

Puumalainen, N., Vepsäläinen, P. 1997. Vacuum preloading of a vertically drained ground at the Helsinki test field. *XIV ICSMFE, Hamburg 6-12th October 1997.* (to be published).

Technique Geosystems/ Cofra JV. 1995. *Design proposal for the execution of: Vacuum Consolidation Shah Alam Expressway Package A, Selangor, Malaysia.* Amsterdam.

Wood, D.M. (1990). *Soil behaviour and critical state soil mechanics.* Cambridge: Cambridge University Press.

Ye, B.R., Shang, S.Z., Ding, G.Q. 1991. Consolidation of Soft Soil by Means of Combined Vacuum Surcharge Preloading. *Proc. Geo-Coast '91, Yokohama*: 277-280. Yokohama.

Z_Soil™PC (1995). *Z-Soil V 3.1. User's guide.* Zace Services Ltd, Lausanne.

Numerical Models in Geomechanics, Pietruszczak & Pande (eds) © 1997 Balkema, Rotterdam, ISBN 90 5410 886 X

Seepage design charts for dams on heterogeneous media

Y.M. Najjar & W.A. Naouss
Department of Civil Engineering, Kansas State University, Manhattan, Kans., USA

ABSTRACT: The quantity of seepage under the foundation of most hydraulic structures is of great significance to dam engineers and designers since it plays an essential part in the dam design and safety. In this research study an efficient and accurate method for obtaining seepage estimates under hydraulic structures was developed. The finite element method was successfully implemented for providing seepage and exit gradient estimates under embedded dams and single sheet piles underlaid by heterogeneous media. The data obtained form the numerical model was compiled and plotted to form different sets of design charts. These charts could be used efficiently for obtaining seepage and exit gradient estimates under the aforementioned structures. The results obtained from the finite element model were compared and verified against existing relevant mathematical approximation and other techniques. Designers and practitioners in the field would benefit from these charts because of their simple yet accurate estimates.

1 INTRODUCTION

Seepage estimates play an essential role in designing water retaining structures such as rockfill dams, earth dams, concrete dams, or single embedded sheet piles. The size and shape of these hydraulic structures depend, to some extent, on the amount of seepage taking place under the hydraulic structure's foundation. Approximations of seepage could be obtained by graphical methods such as flownets and the method of fragments, in addition to more sophisticated numerical solutions. The method of fragments can be used for obtaining seepage estimates where the soil medium underlaying the hydraulic structure is of a homogeneous nature. In the case of flownets, the seepage estimate greatly depends on the accuracy of the drawn flownet. Precisely drawn flownets involve a tedious process of trial and error, and it is even more laborious to draw flownets for heterogeneous media. Numerical modeling has been successfully used to solve several complicated seepage problems. Selection of the modeling method (i.e. finite difference or finite element) is dependent on the complexity of the problem at hand. Moreover, the development of such numerical models requires good understanding of the modeling technique along with versatility in computer programming.

Design charts are mostly used to expedite the calculations or approximations of quantities that are usually unattainable by simple equations. In the area of dam design, design charts have been developed to size the dimensions of earth dams, estimate the seepage under flat bottom dams and single embedded sheet piles (Cedergren 1977, Polubarinova-Kochina 1962, Harr 1962 and Naouss and Najjar 1995). By making use of a mapping technique, Polubarinova-Kochina (1962) was successful in deriving a mathematical approximation for the seepage under flat bottom dams and single embedded sheet piles. Polubarinova-Kochina (1962) have developed design charts which furnish seepage estimates under flat bottom dams and exit gradient estimates for the case of embedded sheet piles. The design charts developed by Polubarinova-Kochina provide seepage and exit gradient estimates for the case where the two layers are of equal thickness. Although those charts account for variable dam width sizes, stipulations exist with regard to the thicknesses of the layers underlaying the dam or sheet pile. Seepage estimates could also be derived from the method of fragments which could be used where the soil medium consists of one homogeneous layer (Harr 1962 and Holtz and Kovacs 1981). A typical approach for obtaining

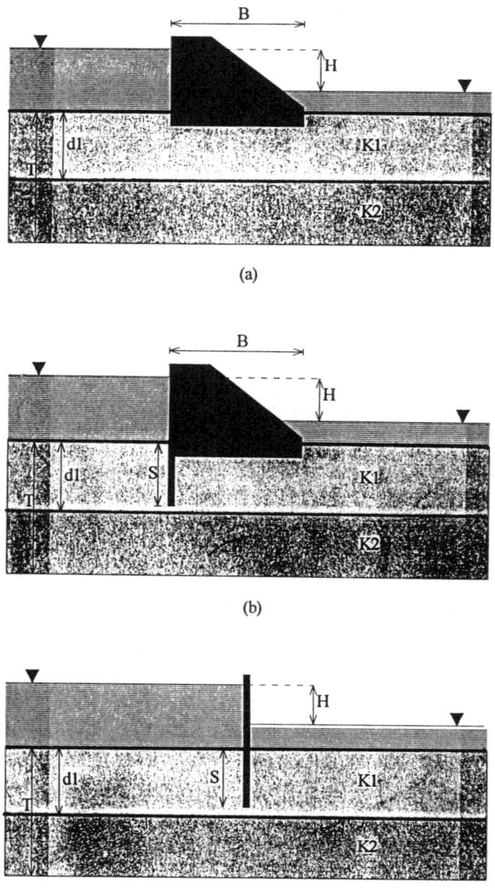

Figure 1 Schematics of the three hydraulic structures underlaid by heterogeneous media: a) embedded dam, b) embedded dam with sheet pile and c) single embedded sheet pile.

seepage estimates under hydraulic structures is to use the flownet technique. This method provides fairly reliable seepage and exit gradient estimates of seepage in homogeneous and layered media (Das 1983, Holtz and Kovacs 1981 and Har 1962). In order to attain accurate estimates of seepage from flownets, the net has to be drawn properly with reasonable number of flow channels. Numerical techniques could be put to use for solving the potential and stream functions which are needed in the construction of flownets (Naouss 1995). By making use of the finite element method, Christian (1980) outlined the major steps involved in solving

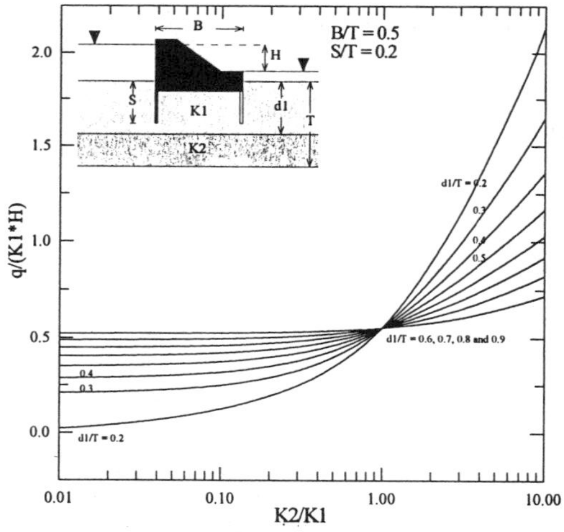

Figure 2 Seepage design chart for embedded dam with sheet piles where B/T = 0.5 and S/T = 0.2.

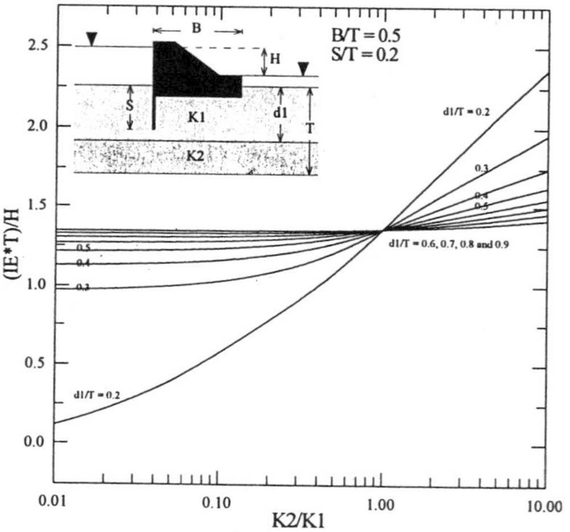

Figure 3 Exit gradient design chart for embedded dam with sheet piles where B/T = 0.5 and S/T = 0.2.

for the potential and stream functions that are needed in flownet construction.

In this study, a finite element model was developed to provide seepage and exit gradient estimates under different types of hydraulic structures that are built on layered media. The data obtained from the model was plotted to form different sets of seepage and exit gradient design charts. The design charts furnished in this study are considered as an accurate and efficient tool for estimating seepage and exit gradient as well as sizing of various hydraulic structures. These charts are aimed at providing quick and accurate seepage and exit gradient estimates compared to the lengthy process of other approximating techniques. The hydraulic structures considered in this study consists of embedded dams, embedded with a sheet pile positioned at the dam heel, and the case of single embedded sheet pile.

2 PROBLEM STATEMENT AND ASSUMPTIONS

Seepage and exit gradient estimates under hydraulic structures such as those depicted in Figure 1 were obtained by implementing the finite element method. Figure 1a shows the case of an embedded dam, Figure 1b shows sketch of an embedded dam with sheet pile installed under the heel side of the dam, and Figure 1c is a sketch of a single embedded sheet pile. In all three cases the soil medium is assumed to consist of two permeable regions of distinct hydraulic conductivities and is underlaid by an impervious layer. The thickness of each layer was varying from 0.1 to 0.9 of the total permeable thickness of the strata (T). Moreover, four dam width ratios (B/T = 0.5, 1.0, 1.5 and 2.0) were considered in this study, which are considered to be within the economical and practical range of design (Naouss 1995). Finally, the sheet pile was positioned at the heel side of the dam where it provides better reduction in seepage as compared to sheet piles placed somewhere in the middle of the dam (Harr 1962 and Naouss 1995).

As stated earlier, only four dam width to strata depth ratios (B/T) were considered (B/T = 0.5, 1.0, 1.5 and 2.0). It has been shown by Naouss (1995) that increasing (B/T) beyond 2.0 would not result in great reduction in the quantity of seepage.

The embedment depth of the dam was set at one tenth of the total permeable strata thickness. Several references in the literature have used an embedment depth (d') ranging from (1/8th) to (1/12th) of the total permeable strata thickness. The selection of an appropriate depth should be based upon the practical and economical aspects of the dam design. Therefore, an average (d'/T) ratio of (1/10th) was used throughout the entire analysis.

The finite element model (Reddy 1979 and Desai 1977) is designed to solve for seepage under dams having sheet piles at the heel side. The sheet pile depth was set to reach a maximum distance of one half of the total strata thickness. Although it is known that driving the sheet pile deep into the permeable stratum would greatly reduce the seepage quantity; doing so is not a common practice. In addition, the sheet pile depth was set to reach throughout the depth of the first layer without going through the second strata. This was based on the assumption that the bottom layer possesses low permeability compared to the top one, and hence seepage is retarded naturally by the second layer. After obtaining the numerical results, this assumption was proven to be valid and is discussed in greater detail in the discussion section of this paper.

Based on the design charts presented by Naouss and Najjar (1995), it was observed that as the permeability ratio [i.e. (K2/K1)] increases, the seepage and exit gradient values are increased. The increase in those quantities is more significant particularly as (K2/K1) ratio exceeds 10. Situations where (K2/K1) ratio are more than unity may not be suitable for dam design. Obviously, dams should not be build over soil strata where (K2/K1) ratio is beyond 10. Doing so is neither economical nor practical. As a result, all design charts developed herein were limited to a maximum value of 10 for (K2/K1) ratios.

The last decision (assumption) made before conducting the entire finite element analysis was to select the proper position, under the dam, for locating the sheet pile. Sheet piles are usually placed at the dam heel where they are anchored with the dam's foundation. They could also be placed at the dam toe or in combination with another one positioned closer to the upstream side. Preliminary investigations of this matter revealed that placing the same sheet pile at either the dam heel or the dam toe would produce the same amount of seepage reduction. The effect of sheet pile position was also analyzed to study its influence on the exit gradient. The analysis disclosed that sheet piles placed at the dam toe would result in greater reduction in exit gradient than sheet piles placed under the dam heel (Naouss 1995). Hence, it was decided to place the sheet pile at the dam heel. Even though, this position would give seepage and exit gradient estimates for sheet piles placed only at the dam heel. These estimates are regarded as conservative when the designer chooses to place the sheet pile under the dam toe.

3 VERIFICATION

Seepage and exit gradient results obtained from the finite element model were compared with Polubarinova-Kochina's (1962) solution for the case of single embedded sheet piles. Five different sheet

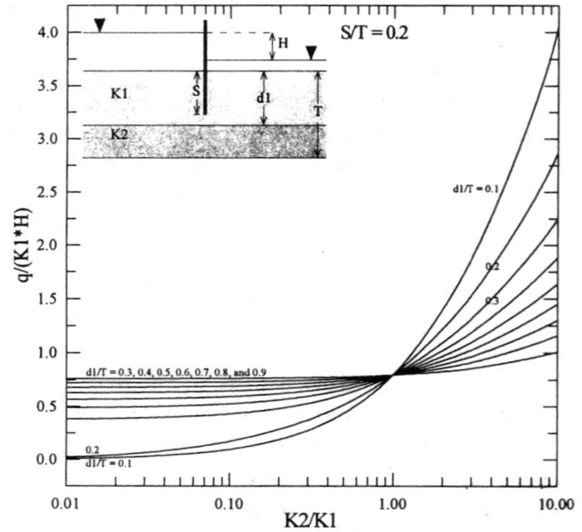

Figure 4 Seepage design chart for single embedded sheet piles where S/T = 0.2.

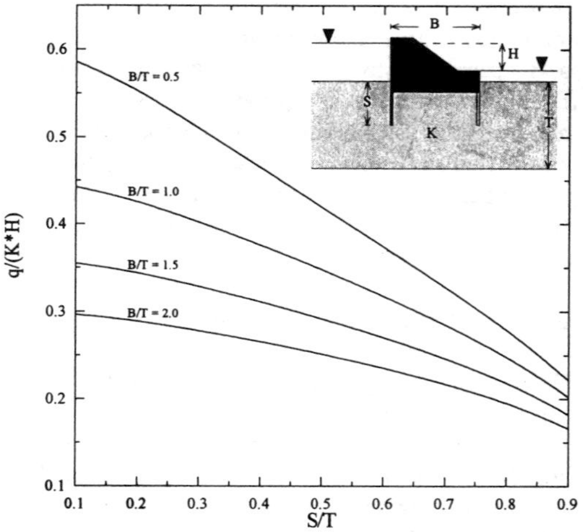

Figure 6 Seepage design chart correlating (q/K*H) estimates to (S/T) ratios for homogeneous stratum.

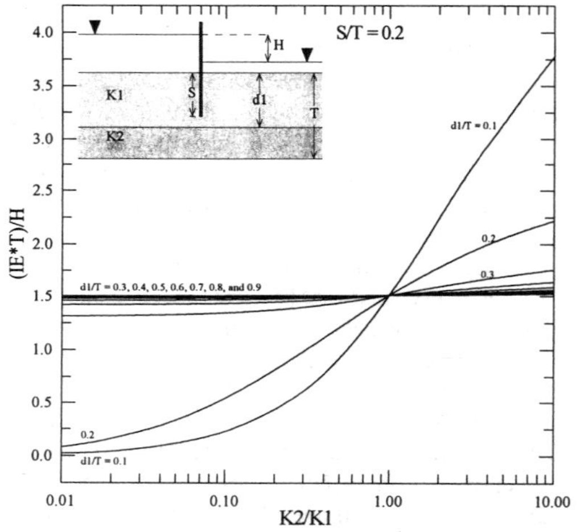

Figure 5 Exit gradient design chart for single embedded sheet piles where S/T = 0.2.

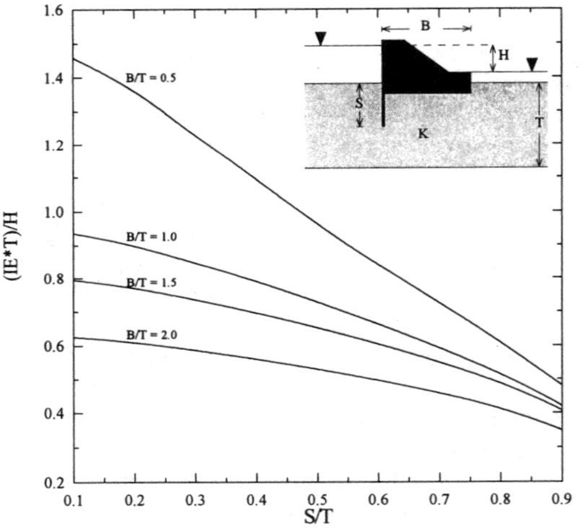

Figure 7 Exit gradient design chart correlating (IE*T/H) estimates to (S/T) ratios for homogeneous stratum.

pile depth ratios (S/T) were used for the comparison. It is noted from this comparison (Naouss 1995), that the finite element solution compares very well with Kochina's approximation. At high (K2/K1) ratios, (K2/K1) greater than 30, the model's estimates of seepage were below those obtained from Kochina's solution. This could be attributed to the sensitivity of the mapping technique at high permeability ratios that was adopted by Kochina. The exit gradient outputs were comparable to those of Kochina's for all (S/T) ratios even at high (K2/K1) ratios. In this case, all five (S/T) curves demonstrate an excellent agreement with the exit gradient solution derived by Polubarinova-Kochina (1962). In addition, the finite element (FE) model was used to generate flownets which in turn were compared to flownets found in literature. In these cases, the comparison revealed excellent agreement between flownets generated by the FE model and those found in the literature

(Cedergren 1977 and Holtz and Kovacs 1981).

Since the model's prediction of seepage under sheet piles were accurate, compared to Kochina's solution, it is reasonable to assume that the seepage and exit gradient estimates under embedded dams are also accurate. The only possible comparison for this case was accomplished through flownets estimates from several references in the literature. As mentioned earlier, very good agreement was found between both estimates. After establishing that the model's prediction of seepage and exit gradient is reliable, it was feasible to start generating data for the embedded dams (B/T= 0.5, 1.0, 1.5, and 2.0) and for the single embedded sheet piles. The thickness ratio (d1/T) varied from 0.1 to 0.9, and the sheet pile depth ratio (S/T) varied from 0.1 to 0.6 for the single embedded sheet pile case and from 0.1 to 0.5 for the embedded dam case. Seepage and exit gradient charts were generated to accommodate the aforementioned described situations. As a result, the number of charts generated in this case amounted to a total of 52 seepage and exit gradient design charts (Naouss 1995).

4 DISCUSSION OF DESIGN CHARTS

The finite element estimates of seepage and exit gradient under embedded dams and single embedded sheet piles were plotted to form different sets of design charts. One set provides the seepage quantity under embedded dams with sheet piles while the other gives exit gradient estimates under the same type of structure. Two other sets provide seepage and exit gradient estimates under single embedded sheet pile. It is not feasible to include all 52 charts in this paper, nonetheless few selected charts are presented and the complete set can be found in Naouss (1995). Figures 2 and 3 depict, respectively, selected examples of seepage and exit gradient charts for the case where B/T = 0.5 and S/T= 0.2. Also included in this paper are 2 seepage and exit gradient charts for the case of single sheet piles where S/T = 0.2 (Figures 4 and 5, respectively). Several conclusions could be drawn out of those charts regarding the effects of conductivity ratio, thickness of each permeable layer, and sheet piles depth on seepage and exit gradient values.

Inspecting the seepage design charts of embedded dams with sheet piles (only selected charts are given in this paper), it can be observed that the seepage quantity decreases with increasing dam width at the same sheet pile depth. At low (K2/K1) ratios, the seepage quantity does not show significant variation except at low (d1/T) ratios close to similar ratios of (S/T). This is illustrated in Figure 2 where at (d1/T = 0.2) the seepage quantity is almost zero. The lowest seepage values are obtained at (d1/T) ratios equal to (S/T). In other words, when the sheet pile is installed throughout the depth of the first layer, most significant reduction in seepage quantity is accomplished in that case. Driving the sheet pile beyond the first layer is not justifiable for reducing seepage except for cases where the conductivity ratio (K2/K1) is larger than unity. In this case. dams should not be recommended to be built over such strata. This behavior suggests the need for accurate geological investigations that can provide designers with valuable information on the conductivity ratio of the strata under consideration.

Similar conclusions are drawn from seepage design charts under single embedded sheet piles (Figures. 4 and 5). Small variations in seepage quantity are noticed at low (K2/K1) ratios. Similarly, driving the sheet pile through the depth of the first layer significantly reduces the seepage quantity compared to other situations where the sheet pile is partially embedded into the first layer. The second layer retards the flow if it possesses lower hydraulic conductivity than the top one. The assumption made earlier regarding the depth of the sheet pile is justified by this behavior and by the assumption that, in general, the bottom stratum would usually exhibit lower hydraulic conductivity than the top one.

The exit gradient behavior, as affected by the sheet pile depth, is similar to that of seepage quantity. A selected exit gradient chart for embedded dams with sheet piles is shown in Figure 3 which illustrates similar behavior. A noteworthy behavior is revealed from those charts in regard to the effect of thickness ratio on the value of exit gradient. At large (d1/T) ratios, small variations in exit gradient are noticed for most of the conductivity ratio range. This is specially true at low (B/T) ratios as demonstrated in Figure 3 where (B/T = 0.5). Similar observations are noticed in exit gradient charts under single embedded sheet piles especially for low (S/T) ratios. This behavior suggest that exit gradient values in these situations are not affected by the thickness or hydraulic conductivity of the lower layer. This is true since exit gradient estimates were obtained from an element at the top of the first layer.

Two additional charts were generated to depict the effect of sheet pile depth on seepage quantity under homogeneous conditions. Figures 6 and 7 demonstrate, respectively, both seepage and exit gradient behavior as it relates to pile depth and dam width. It is further, graphically, proven that increasing the dam width ratio (B/T) beyond 2.0 would not attain great reduction in seepage. One can check the validity of this hypothesis by examining Figure 6 at S/T ratio of 0.1 (absence of sheet pile). As the dam width ratio increases, the reduction in seepage quantity is diminishing which indicates increasing (B/T) beyond certain limits will not appreciably reduce seepage quantity.

5 CONCLUDING REMARKS

The finite element analysis of the seepage problem under consideration was successful in furnishing substantial data that was summarized in different sets of design charts. It is worthy to mention again that the seepage charts for embedded dams with sheet piles are useful for obtaining seepage estimates when the sheet pile is placed either under the dam toe or dam heel. Similarly, the exit gradient charts could be used by designers to provide estimates for the case where sheet piles are installed under the dam heel. Nevertheless, those charts could also be used to provide conservative exit gradient estimates if the sheet pile was to be positioned under the dam toe.

REFERENCES

Cedergren, H. R. 1977. *Seepage, drainage, and flow nets.* 2nd ed., John Wiley & Sons, New York, NY.

Christian, J. T. 1980. Flow nets from finite element data. *Int. J. for Numerical and Analytical Methods in Geomechanics*, Vol. 4, 191-196 (1980).

Das, B. M. 1983. *Advanced soil mechanics.* McGraw-Hill, New York, NY.

Desai, C. S. 1977. Flow through porous media. Chapter 14 in *Numerical Methods in Geotechnical Engineering*, McGraw Hill, New York, NY.

Harr, M. E. 1962. *Groundwater and seepage.* McGraw-Hill, New York, NY.

Holtz, R. D. and Kovacs, W. D. 1981. *An introduction to geotechnical engineering.* Prentice-Hall, Inc., Englewood Cliffs, NJ.

Naouss W. A. and Najjar, Y. M. 1995. Seepage designcharts for flat bottom dams resting on heterogeneous media. *Int. J. of Analytical and Numerical Methods in Geomechanics*, Vol. 19, pp. 637-651.

Naouss, W. A. 1995. Design charts for seepage estimates under hydraulic structures. *MS Thesis*, Department of Civil Engineering, Kansas State University, Manhattan, KS.

Polubarinova-Kochina, P. Ya. 1962. *Theory of ground water movement.* Princeton University Press, Princeton, NJ.

Reddy, J. N. 1993. *An introduction to the finite element method.* 2nd ed., McDraw Hill, New York, NY.

Numerical Models in Geomechanics, Pietruszczak & Pande (eds) © 1997 Balkema, Rotterdam, ISBN 90 5410 886 X

A model for deposition of sediments in reservoirs

R.F.Azevedo
UFV/MG, Viscosa, M.G. & PUC-Rio, Rio de Janeiro, R.J., Brazil

E.C.Silva & I.D.Azevedo
UFV/MG, Viscosa, M.G., Brazil

ABSTRACT: Modeling deposition of sediments in water reservoirs is important since the deposited material decreases the storage capacity of the reservoir. Also in problems related to the filling of tailings reservoirs, modeling the deposition process is fundamental for dimensioning the dam height. This paper describes a model for the deposition of sediments in water reservoirs considering their transport, sedimentation and consolidation. The transport-sedimentation theory is described. Subsequently, the consolidation theory used is briefly summarized. Finally, a parametric study is shown and some conclusions are made.

1. INTRODUCTION

Water resource engineers are concerned with the deposition of sediments in water reservoirs because it reduces the storage capacity of the reservoirs. On the other hand, in countries where mine exploration is intense (Brazil, for example), geotechnical engineers have to predict the storage capacity of tailings reservoirs. Both problems may be analyzed by models that conjugate theories of transport, sedimentation and consolidation of solids in water (liquid).

Sedimentation has been studied for many years. After the development of Stoke's law (Lamb, 1932) another important work on sedimentation was made by Kynch (1952) who formulated, as a transient process, the sedimentation of a uniform dispersion in a column. This formulation considers the solids continuity but ignores the continuity of the mixture. However, these studies only involve the sedimentation process. A literature review on transport and sedimentation shows empirical approaches, like the Empirical Area-Reduction Method (Borland and Miller, 1960), which do not consider flow characteristics, and more sophisticated developments, like the ones made by Paulet (1971) and Richards and Chang (1975), who considered some of the flow characteristics.

Soares et al. (1982b) developed a more general model in which the deposition is a function of the inflow and the outflow as well as the concentration of sediments in the inflow and in the outflow, the initial level in the reservoir and the grain-size distributions of transported sediment and reservoir bed. The inflow is treated as unsteady and non-uniform and the 3D geometry of the reservoir is also considered. However, this model does not consider the consolidation due to the self-weight of the deposited sediments. To do so, the well-known theory of consolidation formulated by Terzaghi (1927) is not applicable since it does not take into consideration the self-weight effects and is based on infinitesimal strains. A 1D finite strain consolidation theory that considers the self-weight of the material was formulated by Gibson et al. (1967).

Consoli (1991) developed a model similar to the one discussed in this paper. However, the transport and sedimentation model employed is less general than the one developed by Soares et al. (1982b). On the other hand, consolidation is made in 2D using finite elements and large deformations. This approach uses a constitutive relationship with parameters that are difficult to be obtained in the laboratory.

Considering what was mentioned above, the objective of this paper is to present a model for deposition of sediments in water reservoirs considering the transport-sedimentation model developed by Soares et al. (1982) and the 1D consolidation theory proposed by Gibson et al. (1967). In the following, the transport-sedimentation model is described, the consolidation theory used is summarized and results of a parametric study are shown. Finally, some conclusions are drawn.

2. TRANSPORT AND SEDIMENTATION MODEL

The transport-sedimentation model described herein was developed by Soares et al. (1982a). These authors made the following assumptions to derive the governing equations (Fig. 1): (1) the channel slope is small, consequently the pressure distribution in the vertical is hydrostatic (2) the sediment-water mixture is homogeneous and the concentration of suspended sediment C over a cross-section can be expressed as the average concentration of the equivalent uniform flow; (3) the concentration of deposited sediments at bed level C_b is a function of x only, and not of time. In other words, consolidation effects are considered separately from the sedimentation process; (4) there is no lateral inflow and the losses by evaporation are negligible; (5) the geometry of the cross-sections are expressed as a function of the stage, y; (6) the equation of motion is written neglecting the variation of density of laden water and assuming a gradually varied unsteady flow (Chow, 1959).

Thus, equation of motion (momentum equation) is given by:

$$\partial U/\partial t + U\partial U/\partial x + g\,\partial D/\partial x + g\partial Z/\partial x + gS_f = 0 \quad (1)$$

where, U is the flow velocity, D is the flow depth, Z is the bed coordinate; g is the gravity acceleration, $S_f = (\eta_c U/R_h^{2/3})^2$ is the friction factor, η_c is the Manning's coefficient and R_h is the hydraulic radius of the cross-section.

Equation of continuity for the mixture is expressed as:

$$\partial A/\partial t + U\partial(UA)/\partial x - [A/C_b - C](\partial C/\partial t + U\partial C/\partial x) = 0 \quad (2)$$

and conservation of solid mass equation for the mixture is given by:

$$\partial(CA)/\partial t + \partial(CUA)/\partial x + C_b\partial A_b/\partial t = 0 \quad (3)$$

These equations involve four unknowns U(x,t), D(x,t), C(x,t) and Z(x,t) (or, A_b(x,t)). The

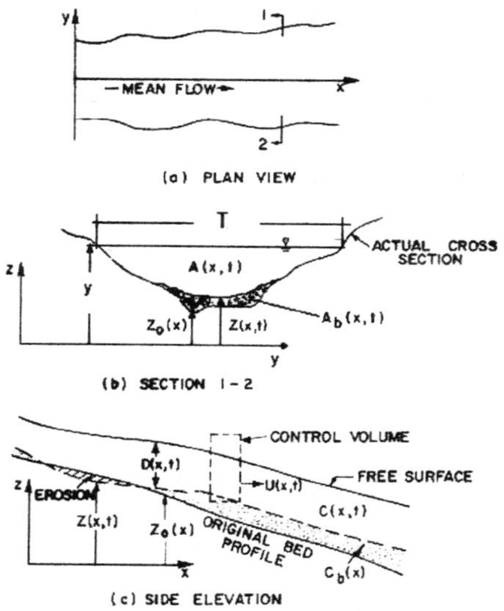

Fig. 1 - Schematization of ideal stream with control volume.

fourth equation necessary for the solution is the sediment transport capacity of the flow, relating the concentration C(x,t) with the velocity U(x,t) and depth of the flow D(x,t).

Soares et al. (1982a) deduced the following equation for the suspended-sediment load capacity of transporting a concentration of particles sizing d^j in a section i+1:

$$CT^j_{i+1} = \Omega^j \cdot C^j_b \quad (4)$$

where

$$\Omega^j = \frac{C^j_b \eta_c U}{\sqrt{S_b}\,D^{2/3}}(1+2.5\alpha)\left[\frac{1-\exp(-15\beta/\alpha)}{15\beta/\alpha}\right] + 2.5\alpha\int_0^1 \ln\eta\exp[-15(\beta/\alpha)\eta]d\eta \quad (5)$$

where $\alpha = \eta_c \frac{\sqrt{g}}{D^{1/6}}$ and $\beta = \omega^j/U$.

$\omega^j = 0{,}9(d^j)^2$ is the settling velocity of sediment-particle of size d^j.

As the sediment concentration varies with x, the balance of mass of sediment of a given size d^j, neglecting $\partial C/\partial t$, is expressed as:

$$Q(\partial C^j/\partial x)\Delta x = (-q^j_d + q^j_e)T\Delta x \quad (6)$$

where q_d^i and q^j_e are the flux of deposition and the flux of erosion of sediment of size d^j and Q is the flow discharge, respectively.

Two cases may occur (noticing that superscript relates to particle size and subscript to section along x:

(a) $C_i^j > CT^j_{i+1}$: in this case ***deposition will occur***. Since the flow is able to transport CT^j, the rate of deposition is given by $q_d^j = -\omega^j (C^j - CT^j)$ and $q_d^j = 0$. With these values and after integration of eq. (6) between sections i and (i+1) the concentration becomes:

$$C^j_{i+i} = CT^j_{i+1} = (C_i^j - CT^j_{i+1}) \cdot \exp(-\omega^j \Delta x T / Q) \qquad (7)$$

(b) $C_i^j > CT^j_{i+1}$: in this case ***erosion will occur*** depending on the availability of sediment of size d^j at the stream bed. The rate of erosion is given by $\lambda^j(C_b^j - C^j)$, i.e. the erosion is proportional to the difference between the availability on the bed and the concentration already carried by the flow. The parameter $\lambda^j = \frac{\Omega^j}{1-\Omega^j}\omega^j$ has the dimension of velocity. In spite of erosion there will still be deposition of sediment at the rate $\omega^j C^j$. With these values and after integration of eq. (6) between sections i and (i+1) the concentration becomes:

$$C^j_{i+1} = CT^j_{i+1} + (C_i^j - CT^j_{i+1}) \cdot \exp\left[-\frac{T\Delta x}{Q}(\lambda^j + \omega^j)\right] \qquad (8)$$

Letting π_i^j be the fraction of sediment of size d^j in suspension in section i, where the concentration is known to be C_i, then $C_i^j = \pi_i^j C_i$. With γ_i^j expressing the fraction of sediment of this size on the bed in the reach i, i + 1, it can be stated that $C_b^i = \gamma_i^j C_b$ and, from eq. 4, CT^j_{i+1} is calculated. If $C_i^j \rangle CT^j_{i+1}$, the concentration C^j_{i+1} is given by eq. 7, otherwise C^j_{i+1} is computed from eq. 8. After the values C^j_{i+1} are computed for all j's, the average concentration of suspended sediment in section (i + 1) is obtained as:

$$C_{i+1} = \sum_j C^j_{i+1} \qquad (9)$$

and the fraction of sediment in suspension of size d^j at this section is given by $\pi^j_{i+1} = C^j_{i+1} / C_{i+1}$. Knowing these values the procedure may be carried out for section (i + 2) and so on.

The set of equations is solved numerically by the following procedure: (a) the concentration of suspended sediment C is computed as described above (eq. 9); (b) assuming that the variation of $\partial Z/\partial x$ during a step in time is small and neglecting the term $\partial C/\partial t$ in eq.2, the pair of eqs. 1 and 2 is solved for $U(t + \Delta t)$ and , $D(t + \Delta t)$ using an implicit finite-difference scheme with appropriate boundary conditions; (c) the concentration of sediment at time $(t + \Delta t)$ is calculated as in item (a) and eq. 3 is solved numerically for $A_b(t + \Delta t)$. The area of sediment deposited during time step Δt, $[A_b(t + \Delta t) - A_b(t)]$, is distributed throughout the cross-section proportionally according to the depth of flow at each point; (d) the new stage is then computed as:

$$y(t + \Delta t) = y(t) + [A(t + \Delta t) - A(t)]/T + + [A_b(t + \Delta t) - A_b(t)]/T \qquad (10)$$

(e) the geometry of the cross-sections is altered and the procedure continues for following steps in time starting at item (b).

3. CONSOLIDATION MODEL

Gibson et al. (1967) using equations of equilibrium of the mixture and continuity of the liquid, together with Darcy-Gersevanov law (Schiffman, 1986) and the effective stress principle, derived an equation for 1D finite-strain consolidation of a layer with constant volume of solids due to the self weight of the material and/or surcharges. Lately, Schiffman (1986) modified the equations to allow analyses of layers in which the solids volume increased with time, suitable to analyze reservoir fillings. The final equation derived by Schiffman (1986) is:

$$\frac{1}{h_z} \cdot \frac{\partial}{\partial Y}\left[\frac{k}{(1+e).\gamma_w.(\partial e/\partial \sigma')}\frac{\partial e}{\partial Y}\right] + \left[Y.r_z - (1-\frac{\gamma_s}{\gamma_w}).\frac{\partial}{\partial e}(\frac{k}{1+e})\right].\frac{\partial e}{\partial Y} = h_z \frac{\partial e}{\partial t} \qquad (11)$$

where e, σ', k, γ_w and γ_s are the void-ratio, the effective stress, permeability, unit weight of water and unit weight of solids, respectively. h_z is the solids height and r_z is the rate of solids accumulation. Finally, Y is the non-dimensional reduced coordinate:

$$Y(a) = \frac{z(a,t)}{h_z(t)} \qquad (12)$$

where z(a,t) is the reduced coordinate of a point whose lagrangian coordinate is 'a', defined by (McNabb, 1960):

$$z(a) = \int_0^a \frac{1}{1+e(\zeta,0)}\, d\zeta \qquad (13)$$

Equation (11) was solved with an implicit finite difference algorithm (Azevedo and Sado, 1990) and used in conjunction with the transport-sedimentation model described before as follows: (a) for a given time step Δt, the areas of sediment deposited in each section along x are calculated with the transport-sediment model (items a, b and c); (b) with these areas, rates of accumulation of solids in each section are calculated to serve as input data to the 1D finite strain consolidation procedure that, in turn, calculates new consolidated areas of sediment deposited; (c) with these new areas, the algorithm returns to step (d) of the transport-sedimentation procedure.

4. PARAMETRIC STUDY

This section presents a parametric study based on data of the John Martin Reservoir located on the Arkansas River in Bent County, Colorado, U.S.A. (Soares et al., 1982c). The main purpose is to show the different deposition pattern according to the flow characteristics and the grain-size distribution both of the sediments and the reservoir bed.

Two flow characteristics are considered: case A corresponds to a low water level in the reservoir (higher flow velocities) whereas case B corresponds to a high water level (smaller flow velocities). In each case, two grain-size distributions are assumed for the sediments: a well-graded one (WG) and a uniform distribution of clay sizes particles (C). For the reservoir bed material also two uniform distributions are adopted: one corresponding to a fine sand (FS) (more difficult to remove) and another corresponding to a clay (C) (easier to remove).

The total time for all analyses was set equal to 3 months. The inflow was 11,65 m^3/s with concentration of 0,0036 for case A, and 60,17 m^3/s with concentration of 0,0077 for case B. The geometry and other data of the reservoir are found in Soares et al.(1982c).

Figure 2 shows the water level (a) and the flow velocity (b) along the reservoir length for case A. Figures 3 to 6 present the sediment deposit area distribution for this same case A.

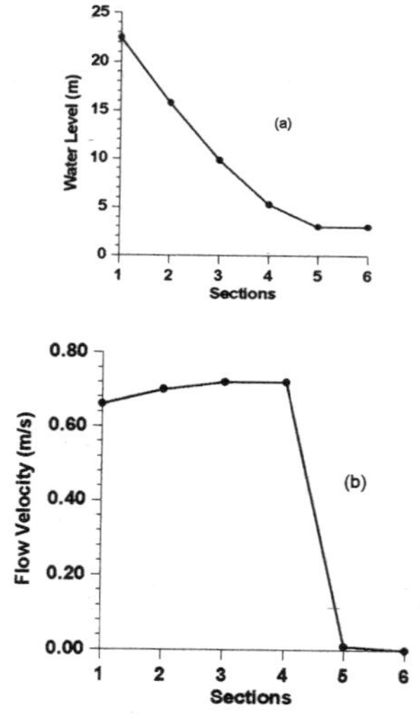

Figure 2. Case A

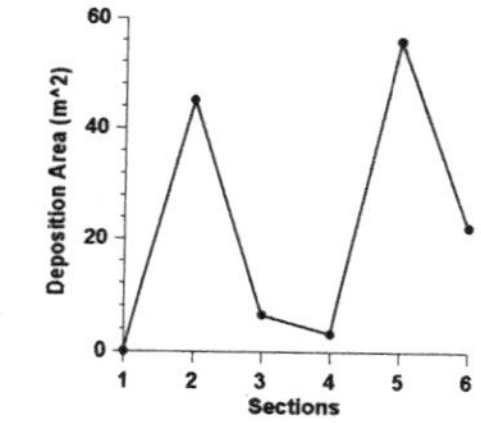

Fig. 3 - Inflow WG, bed FS

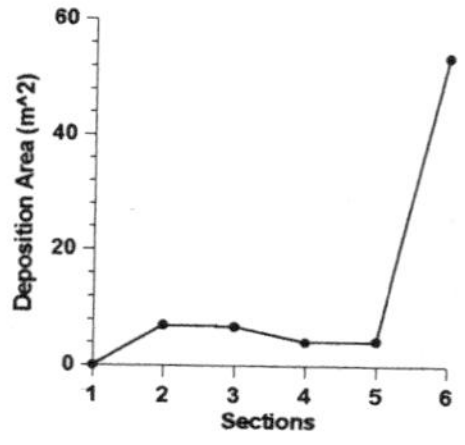

Fig. 4 - Inflow C, bed FS

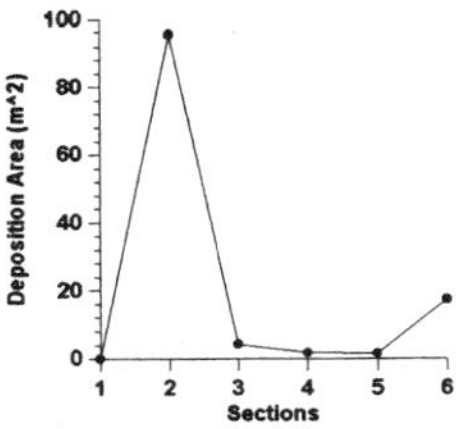

Fig. 5 - Inflow WG, bed C

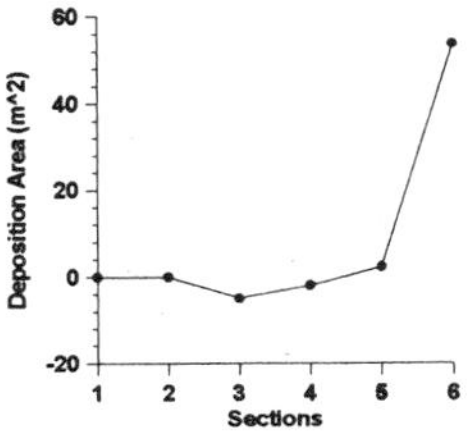

Fig. 6 - Inflow C, bed C

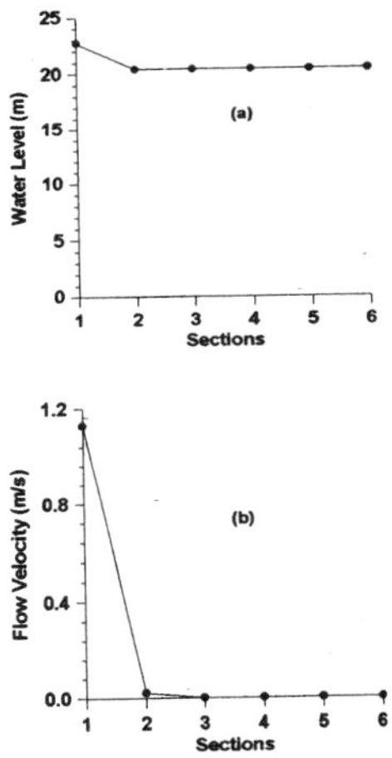

Fig. 7 - Case B

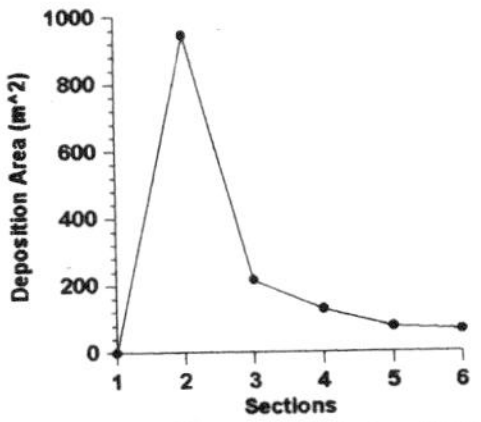

Fig. 8 - Inflow WG, bed FS

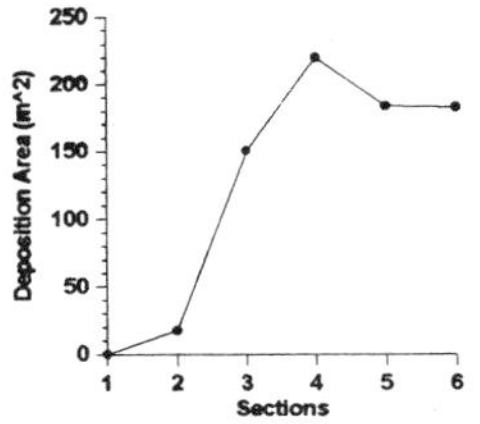

Fig. 9 - Inflow C, bed FS

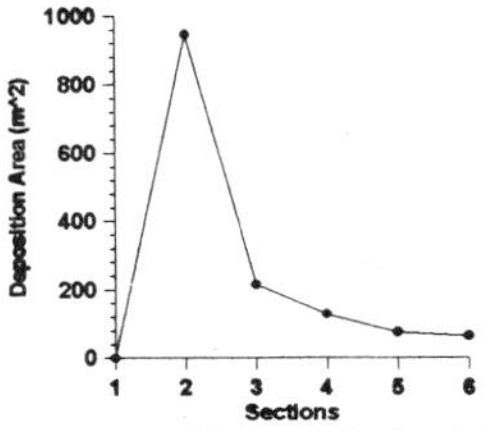

Fig. 10 - Inflow WG, bed C

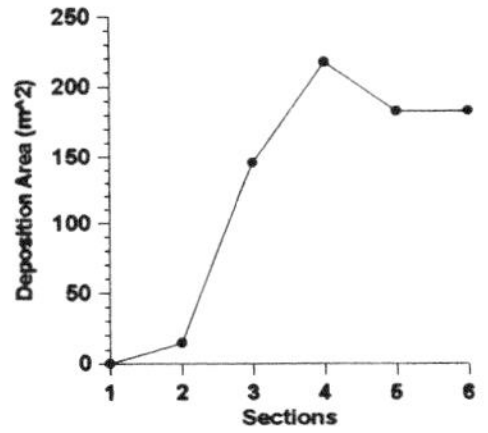

Fig.11 - Inflow C, bed C

Figures 7 shows the water level and the flow velocity along the reservoir length for case B. Figures 8 to 11 present the sediment deposit area distribution for this same case B.

Table1 and 2 present the total volume deposited in each situation.

Table 1- Flow conditions Case A

Bed \ Slurry	W. G.(m^3)	Clay(m^3)
Fine Sand	546224	220086
Clay	503162	130898

Table 2. Flow Conditions Case B

Bed \ Slurry	W. G.(m^3)	Clay(m^3)
Fine Sand	6333061	2981357
Clay	6281011	2932172

5. CONCLUSIONS

This paper presented a model for the deposition of sediments in water reservoirs considering their transport, sedimentation and consolidation. The transport-sedimentation theory was described as well as the consolidation theory used.

The parametric analysis made showed that the grain-size distribution of the sediment has a strong influence on the pattern of the deposited material as well as in the volume of deposition. On the other hand, the grain-size distribution of the bed only influences the pattern of the deposited material and the volume of deposition when the flow velocity is more intense (case A).

The new model is being used to analyze the filling of a tailing dam reservoir and comparisons between field monitored results and analytical results may be available at the Conference.

6. ACKNOWLEDGMENTS

This research was carried out at the Civil Engineering Department of Federal University of Viçosa, the authors thank the support received. The authors are also indebted to the Brazilian Government Agency for Improvement of Personnel with University Degree Level (CAPES) for the scholarship provided to Mr. E.C. Silva.

7. REFERENCES

1. Azevedo, R.F. & Sado, J.S. (1990), "Análise Uni-Dimensional do Enchimento de Reservatórios de Barragens de Rejeito através de uma Teoria de Adensamento com Grandes Deformações.", IX COBRAMSEF , vol. 1,Salvador, Brazil, pp. 71-78 (in portuguese).
2. Borland, W. M. and Miller, C. R. (1960), "Distribution of Sediment in Large Reservoirs", Trans. ASCE, 125 (3019), pp. 166-180.
3. Chow,V. T. (1959), "Open Channel Hydraulics. McGraw-Hill, New York.
4. Consoli, N. C. (1991), "Numerical Modeling of the Sedimentation and Consolidation of Tailings", Ph.D. Thesis, Concordia Univ., Quebec, Canada.
5. Gibson, R. E.;England, G. L. and Hussey, M. J. L. (1967), "The Theory of One-Dimensional Consolidation of Saturated Clays: I. Finite Non-Linear Consolidation of Thin Homogeneous Layers", Geotechnique, 17, pp. 261-273.
6. Kynch, G. J. (1952), "A Theory of Sedimentation", Trans. Faraday Society, 48, pp. 166-176.
7. Lamb, H. (1932), "Hydrodynamics", Dover Publications, New York.
8. McNabb, A.(1960), "A Mathematical Treatment of One-Dimensional Soil Consolidation", Quarterly of Applied Mathematics, 17, pp. 337-347.
9. Paulet, M. (1971), "An Interpretation of Reservoir Sedimentation as a Function of Watershed Characteristics", Ph.D. Thesis, Purdue University, USA
10. Richards, D. L. And Chang, F. M. (1975), "Deposition in Transient Flow", J. Hydraulic Div. ASCE, Vol. 97, June, pp. 837-849.
11. Schiffman, R. (1986), "Short Course o the Consolidation of Soft Clays", Internal Report, CA 01/87, PUC-Rio, Brazil.
12. Soares, E. F.; Unny, T. E. And Lennox, W. C. (1982a), "Conjunctive Use of Deterministic and Stochastic Models for Predicting Sediment Storage in Large Reservoirs. Part 1: A Stochastic Sediment Storage Model", J. Hydrology, 59, pp. 49-82.
13. Soares, E. F.; Unny, T. E. And Lennox, W. C. (1982b), "Conjunctive Use of Deterministic and Stochastic Models for Predicting Sediment Storage in Large Reservoirs. Part 2: Deterministic Model for the Sediment Deposition Process", J. Hydrology, 59, pp. 83-105.
14. Soares, E. F.; Unny, T. E. And Lennox, W. C. (1982c), "Conjunctive Use of Deterministic and Stochastic Models for Predicting Sediment Storage in Large Reservoirs. Part 3: Application of the Two Models in Conjunction", J. Hydrology, 59, pp. 107-121.
15. Terzaghi, K. (1925), "Modern Concepts Concerning Foundation Engineering", J. Boston Society of Civil Engineering, 12, No. 10, pp. 397-423.

Numerical Models in Geomechanics, Pietruszczak & Pande (eds) © 1997 Balkema, Rotterdam, ISBN 90 5410 886 X

Study on interaction between rocks and worn PDC's cutter: Numerical approach

H.Geoffroy & D.Nguyen Minh
Laboratoire de Mécanique des Solides, Ecole Polytechnique, Palaiseau, France

ABSTRACT: Two mechanisms take place during cutting of the rock by a blunt PDC cutter. One of them is the frictional contact between the cutter's wear flat and the rock which is not very well understood. In order to identify frictional forces acting on the wear flat, experiments are conducted on a single worn cutter with a constant applied vertical force on a rotating sample. They show a relationship between imposed vertical load on the cutter and penetration angle α, which is related to the helicoidal movement of the tool. In the same way, the mechanical model is a flat rigid punch sliding parallel to a surface with inclination α. The numerical model uses a steady algorithm which allows, in a reference moving with the cutter, to treat the stationary problem on a fixed geometry. The previous relationship was confirmed and it was shown that the frictional mechanism is related to a deformation of elastoplastic type.

1 INTRODUCTION

Polycrystalline Diamond Compact (PDC) bits were introduced in petroleum industry in the early 1970's. A PDC bit is composed of a multiplicity of individual PDC cutters mounted at the surface of the bit body. In general, PDC drill bits show excellent performance in soft and medium formations, but such performance will gradually decrease with increasing wear. In the same way, PDC bits remain often unreliable in hard and abrasive formations because of accelerated wear. A study of the drilling response of a PDC bit has to take into account evolution of its shape over its life, due to progressive wear. In normal conditions, abrasion is the main cause for wear, which results in a wear flat grow at the bottom of the cutter, because of his increasing bearing area (Figure1), and is responsible for degradation in ROP (rate of penetration) (Glowka, 1987), (Sinor & Warren, 1989), (Brett *et al.*, 1989).

Contrary to cutting forces for a sharp PDC cutter, which can be considered as related to yield load mechanisms, the frictional forces on wear flat are still badly understood, although they may represent 50% of total amount of boring energy. The object of this study is concerned with this specific point. This will then allow to determine the global forces acting on the bit, to improve understanding of wear processes of cutters and therefore to improve bit life.

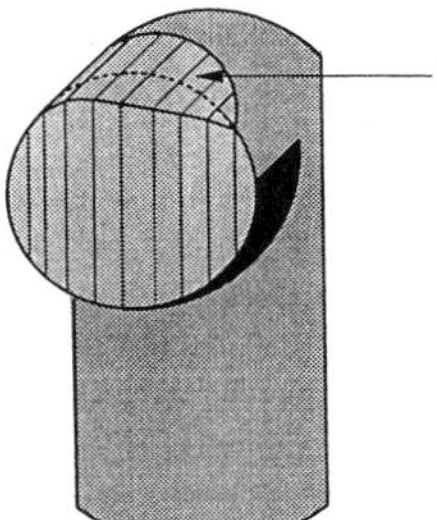

Figure 1: Schematic of single cutter with wear flat.

2 PDC CUTTER MODEL

From experimental evidence, the cutting process on the cutting face and the frictional mechanism under the wear flat can be considered as independent mechanisms (Sellami, 1987), (Kuru, 1990), (Detournay & Defourny, 1992).

Using Detournay's formulation, let us decompose the cutter force F into two vectorial components, F^c transmitted by the cutting face and F^f acting on the wear flat (Figure 2) :

$$\vec{F} = \vec{F^c} + \vec{F^f} \tag{1}$$

and let us decompose each force into its components in the direction normal and parallel to the rock surface respectively :

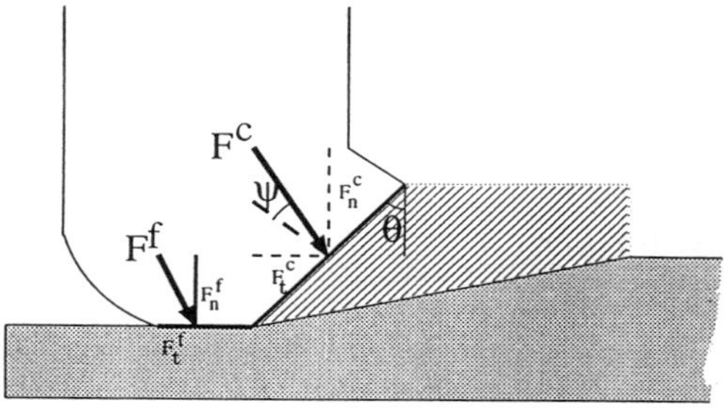

Figure 2: Forces acting on the worn cutter.

Cutting force $\vec{F^c}$: The tangential component F_t^c is proportional to the cross-sectional area A_c and $F_t^c = \epsilon\, A_c$, where ϵ is the specific energy, defined as the amount of energy necessary to cut a unit volume of rock. Its value is supposed to be constant for all cutters.

The normal cutting component F_n^c writes then as $F_n^c = \zeta\, F_t^c$, where ζ is given by $\zeta = \tan(\theta + \psi)$ and θ is the back rake angle and ψ the interfacial friction angle.

Frictional force on the wear flat F^f : The components F_t^f and F_n^f are related by the friction law :

$$\mu = \frac{F_t^f}{F_n^f} \tag{2}$$

where μ is a coefficient of friction.

In order to improve understanding of "frictional mechanism", we can recall experimental observations done by Nguyen Minh (1974). The author performed on a model material (sand mixed with paraffin standing for a plastic rock) some drag tests on a parallelepipedic cutter with a flat bottom. When submitted to a vertical applied load, the tool penetrates into the rock with an inclination angle α. A linear relationship was established between the vertical applied load and α available unless the vertical load reaches the bearing capacity of the rock surface.

These tests have been the starting point of this study ; indeed, such behavior can be assumed to represente the cutter's wear flat effect.

In our case, α can be related to the advance a per revolution of a flat bit in stabilised motion. A cutter on such a bit, will have a helicoidal movement with an inclination α with respect to the flat horizontal borehole bottom, with α defined by the relation $\tan\alpha = \frac{a}{2\pi r}$, where r represents the position of the cutter on the bit. Usually a typical penetration rate ranges between 15 m/h to 60 m/h. For an average rotary speed of 120 rev/min, the advance per revolution a vary between 2 and 8 mm. So, for a 216 mm diameter bit, $3.10^{-3} < \tan\alpha < 6.10^{-2}$.

However, tests conducted by Nguyen Minh were insufficient. Indeed, they may not be representative of actual behaviour of the wear flat because they were not performed on a natural rock. Moreover, the "bearing limit" may not be evident for certain types of rocks. Considering these tests, the frictional mechanism seems to be related to an elastoplastic deformation. This had to be verified, and was studied by an experimental and a numerical approach.

3 LABORATORY DRILLING RESULTS

Cutting tests were performed with the aim of obtaining a relationship between F_n^f and α. Tests were conducted with constant normal load and not with constant depth of cut as usual. Remark that tests performed with a constant depth of cut would not allow to show the importance of α. The difficult point was to measure reliable α values, despite their low value.

The sample is rotated by an electric motor. The cutter is mounted on a shaft fixed on a rigid frame. The shaft can be placed at different radii on the sample in order to simulate different positions of the cutter on the bit. A three dimensional piezoelectric force gage is placed near the cutter to record tangential force, normal force and side force acting on the cutter. The depth of cut is recorded by a LVDT. The rotating sample allows to measure average penetration of the cutter on several revolutions, i.e., over a long distance.

The rock is a chalk from Liege (Belgium). Its behaviour is complex and exhibits important plastic deformations due to the gradual destruction (collapse and distortion) of the porous structure, both under hydrostatic and deviatoric loadings. This rock was selected because of its good homogeneity, with a small grain size, and its low strength. The cutter is an iron parallelepiped 10 mm long, 3 mm large and 6 mm thick with 0° back rake angle θ. This special geometry allows to consider the frictional mechanism alone without action of the cutting mechanism for low values of the depth of cut (Geoffroy, 1996). A relationship between the contact stress acting on the cutter $\frac{F_n^f}{A_w}$ (A_w is the wear flat area) and α was then established (Figure 3). Measured angles vary between 2.10^{-3} rad and $1.20.10^{-2}$ rad. Although low, these values belong to the ranges of values described here above.

On this Figure 3, the points linked together by a line represent tests conducted in the same groove on the sample. For each group of points, the intercept with the vertical axis is different. This dispersion can be explained by the variations of initial operating conditions when the drill string is moved on the frame. The cutter may change slightly its position, so the initial inclination of the wear flat is modified.

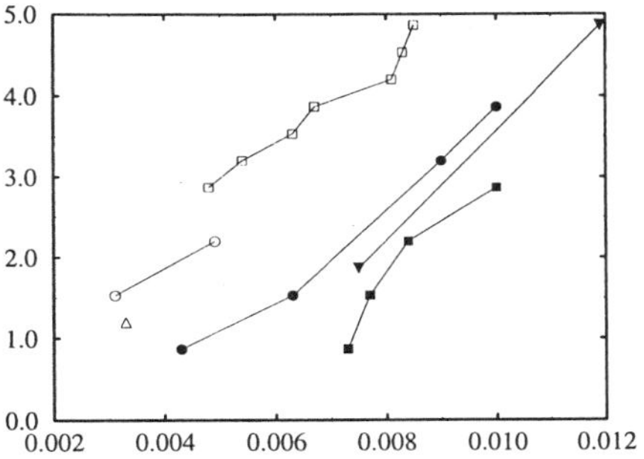

Figure 3: Relationship between $\frac{F_n^f}{A_w}$ (MPa) and α.

4 NUMERICAL ANALYSIS

4.1 Formulation

Within the hypothesis of two independent mechanisms of the worn tool - cutting and friction - the frictional contact between the cutter wear flat and the rock can be idealised by a rigid slider moving on the surface of a half plane. The slider is inclined with respect to the boundary of the half-plane with an angle α. The length of the contact between the slider and the rock beneath it is $2a$. A friction coefficient between the tool and the material is prescribed. The slider moves across the surface with a constant velocity (Figure 4).

The normal displacement $v(x)$ beneath the slider is given. On the interval $[-a, a]$, $v'(x) = -\tan\alpha$. The normal and tangential pressures $p(x)$ and $\tau(x)$ are related by the Coulomb's law, $\tau(x) = \mu p(x)$. The surface boundary outside the contact area is traction free. It is a half plane problem. The contact is applied on the length $2a$ which is unknown. This is a frictional unilateral contact problem, defined by the condition $p(x) \geq 0$ on $[-a, a]$. This condition will allow to determine the contact area by means of an algorithm of contact.

Numerical study is divided into three parts :

Elastic case : results obtained with the numerical model allow to validate it, comparing it with the analytic solution for the elastic case.

Elastoplastic case with a perfect plastic material : Von Mises law is used. These calculations allow to confirme the relationship between α and F_n^f revealed by tests and to verify whether the "frictional mechanism" may be related to a deformation of elastoplastic type.

Elastoplastic case with a strain hardening material : it was interesting to precise plastic volumic strains

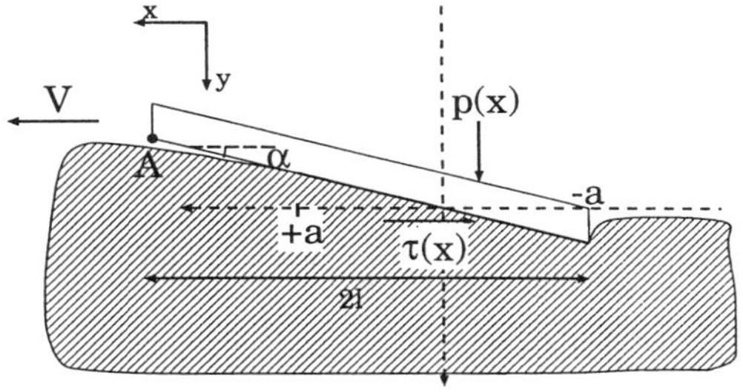

Figure 4: Wear flat model.

induced by the slider inside the rock medium. So, the Cam-Clay model, which takes into account strain hardening and softening, pressure dependency, dilatancy and pore collapse of the chalk was implemented.

4.2 Description of the numerical model

In order to simulate motion of the slider, a specific algorithm is used. Steady motion of the wear flat is supposed. This problem is well adapted to the treatment by a plane strain Finite Element Program developed in the laboratory, (Nguyen & Rahimian, 1981), (Maitournam, 1989) (Guo, 1995) and which uses a steady algorithm and allows, in a reference moving with the cutter, to treat the stationary problem on a fixed geometry. The method is based on the change of rates into gradients along the movement direction. The material derivative of a quantity A with respect to time which expresses as :

$$\dot{A} = \frac{\partial A}{\partial t} + \vec{v}.\vec{grad}A \Rightarrow \dot{A} = -v\frac{\partial A}{\partial x} \qquad (3)$$

since ($\frac{\partial A}{\partial t} = 0$) due to steady state. The time integration is converted, for any material point, into an integration along the direction x of the movement, from the initial state, far ahead of the front face, up to far behind. For this purpose, the finite element mesh is made of rectangular meshes, resulting into two set of strips, one perpendicular and the other parallel to direction of motion x.

In order to determine the contact length $2a$, an algorithm was developed as follows. An arbitrary length $2l$ was chosen. For each node of the slider the normal displacement $v(x)$ is given by :

$$v(x) = (l - x)\tan\alpha \qquad (4)$$

$v(x)$ is imposed by using the Lagrange multipliers (more details are given in (Geoffroy, 1996)).

At the point A (Figure 4), $v(x)$ is arbitrarily imposed to be null. First calculation will give negative pressure acting on some nodes of the punch. But, $p(x)$ must be positive or null on the çontact zone. So, for these nodes no more displacement will be imposed. Finally, the real contact length will be lower or equal to $2l$.

4.3 *Elastic case : validation*

Adachi (1996) has expressed the analytical solution of the normal pressure distribution for the problem of the inclined slider with frictional interface.

So, in the elastic case, a relationship between α and the average value of the distribution of pressure P_0 may be established :

$$P_0 = \pi\eta\frac{E'}{2} \tan\alpha \quad (5)$$

$$\text{avec } \eta = \frac{1}{2} - \frac{1}{\pi}\arctan\frac{1}{\lambda} \quad (6)$$

$$\lambda = -\frac{2(1-\nu)}{\mu(1-2\nu)} \quad (7)$$

$$\text{et } E' = \frac{E}{1-\nu^2} \quad (8)$$

The numerical results are represented with a contact length within the interval $[-1, 1]$ by introducing $(X = \frac{x}{a})$. In the numerical experiment, the slider has an inclination $\alpha = 10^{-2}$ rd and $\mu = 0$. Calculation give us $p(x)$ and $2a$. Figure 5 shows the plot of the normal distribution of pressure $p(x)$ obtained numerically. Analytical solution is very close to the numerical one. The algorithm of contact is validated.

4.4 *Elastoplastic case*

In the frictionless case, $2a$ and $p(x)$ are calculated for an angle α imposed ($10^{-3} \leq \alpha \leq 5.10^{-2}$ rd). An approximation is made for the frictional case ($\mu = 0.5$). $p(x)$ which was obtained from frictionless case is imposed on the contact length $2a$ together with $\tau(x) = \mu p(x)$; there results a new angle α in the frictional case.

Simulations with Von Mises law : A plot of the distribution of normal pressure $p(x)$ for different values of the penetration angle α is shown in Figure 6.

For low values of α, the distribution of pressure $p(x)$ is similar to the elastic solution. The increase of α leads the distribution of pressure $p(x)$ to "become flat". Finally, normal stresses acting on the punch reach the same value when the contact zone is completely plastic (except at the point $(+a)$ where normal pressure is null).

P_0 is the normal contact stress applied by the slider. P_0 versus α is plotted in Figure 7. In the frictionless case, elastoplastic curve is very different from elastic one.

For α lower than 2.10^{-2} rad, P_0 increases continuously with α. Once α becomes greater than 2.10^{-2} rad, P_0 reaches a stationary value. One can notice that the value of P_0 is very close to the bearing capacity of a vertical foundation.

In the frictional case, similar results are obtained. However, for a given mean applied normal stress P_0 one obtains more important α value.

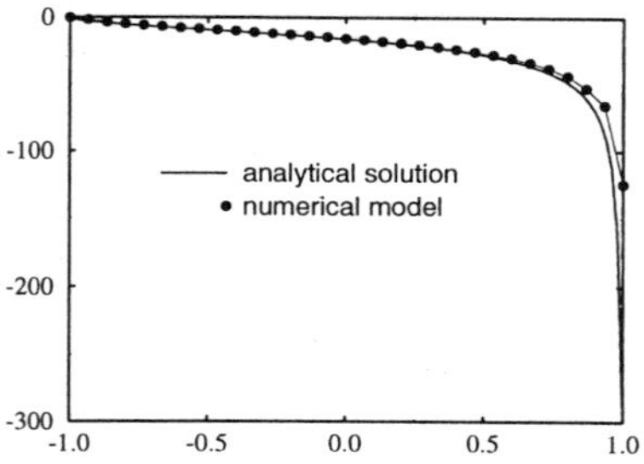

Figure 5: Distribution of normal pressure $p(X)(MPa)$ with respect to the coordinate X.

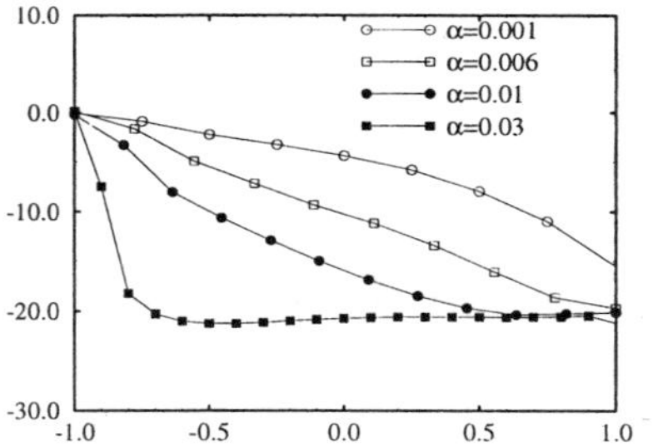

Figure 6: Evolution of the distribution of normal pressure beneath the punch for different values of penetration angle α.

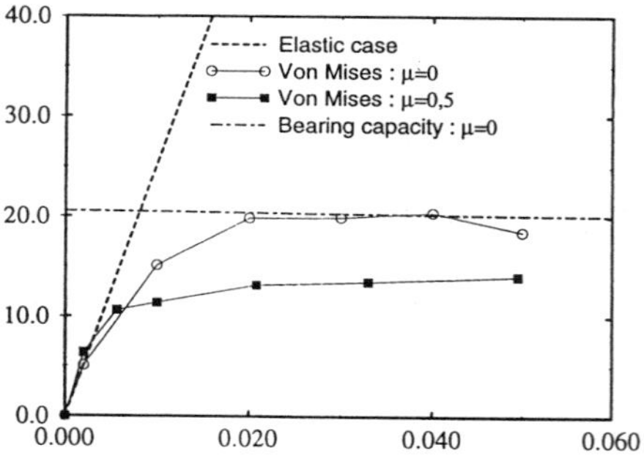

Figure 7: $P_0(MPa)$ versus α : Von Mises law.

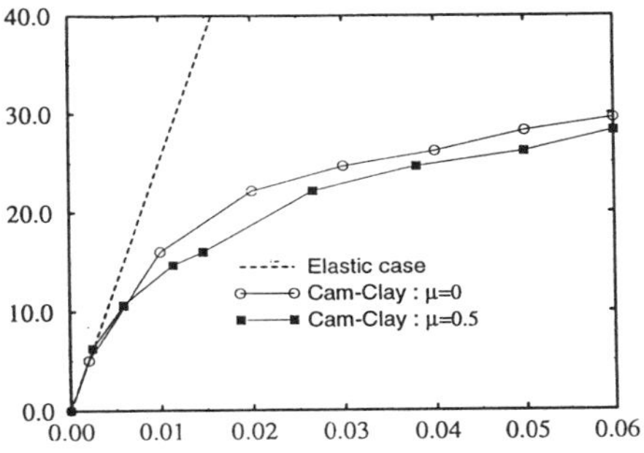

Figure 8: $P_0(MPa)$ versus α : Cam-Clay law.

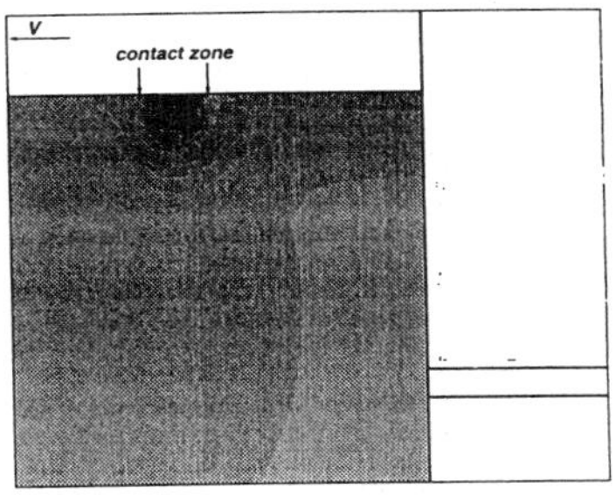

Figure 9: Von Mises law : contour plots of first invariant I_1 ; $\mu = 0.5$; $\alpha = 0.05$ rd.

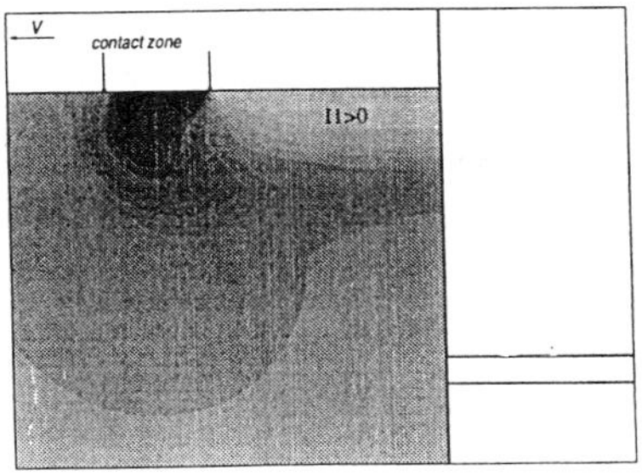

Figure 10: Cam-Clay law : contour plots of first invariant I_1 ; $\mu = 0.5$; $\alpha = 0.05$ rd.

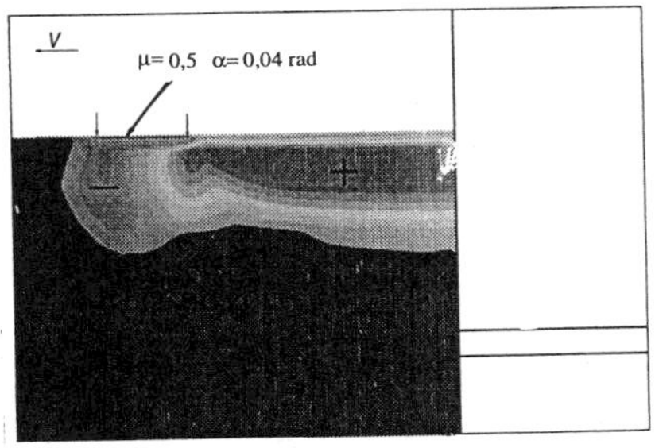

Figure 11: Cam-Clay law ; plastic volumetric zones ; $\mu = 0.5$; $\alpha = 0.05$ rd.

Simulations with Cam-Clay law : For both cases ($\mu = 0.5$ and $\mu = 0$), a relationship between α and P_0 is found too (Figure 8). But, with Cam-Clay model, P_0 value increases continuously. The ultimate bearing limit notion does not appear in this case.

Contour plots of first invariant I_1 : Figures 9 and 10 show contour plots of the first invariant I_1 in the frictional case with $\alpha = 5.10^{-2}$ rad. Both laws are considered. Contact zone on these figures is delimited by two vertical arrows. The punch is dragged on the surface from right to left. Largest stresses prevail below the contact area. The observation of the stress field within the half plane show that is always compressive for the perfect plastic material (Von Mises law) case whereas for Cam-Clay law, a zone under tensile stress appears behind the punch. It is interesting to notice that this zone also appeared in the elastic case with a frictional contact (Adachi, 1996).

Plastic zones description : Plastic volumetric strains have been represented for Cam-Clay law in the frictionless case (Figure 11). This law allows to distinguish plastic contractant zones from dilatant ones. Contractant zones appear beneath the punch (pale zone, water drop shape). Behind the punch, near the surface, there is a thin contractant zone too, overlying a a more important dilatant region (dark zone) in the rockmass. These results can be compared to the elastic case. However, with frictional contact, all the domain turns to be contractant.

5 CONCLUSION

The experimental and numerical analysis have allowed to improve understanding of the "frictional mechanism". In both cases, a relationship between the friction force acting on the wear flat and the inclination angle α of the advancing cutter has been established. It has been found that α increases with the friction force .

Numerically, the frictional contact between the wear flat and the rock surface has been idealized by a slightly inclined cutter moving over an half plane and with friction at the contact. This is an unilateral contact problem. Numerical model developed using the Lagrange multipliers has been validated in the elastic case. This validation was done by comparing the results of the model with the analytical solution.

In the elastoplastic case, sliding motion of the cutter has been conveniently simulated with the steady state algorithm. Two different rockmass hypothesis have been chosen, depending on the rockmass properties such as strain hardening and perfect plasticity or dilatancy and contractancy. It has been shown that the frictional mechanism, after an elastic phase can be related to a deformation of elastoplastic type when α increases. However for a perfect plastic material, the average normal stress acting on the slider reaches a limit value while for a strain-hardening one, this contact stress increases continuously.

Finally, evolution law for α can be derived depending of the rheology of the rockmass, so that friction forces that acting on the blunt cutter can be predicted. Global forces acting on the bit are then obtained by integration of total forces (cutting force added to friction force) on individual cutters.

Moreover, elastoplastic behavior in frictional mechanisms let expect that initial stress state in the well bottom may have an influence, i.e., may be controlled by e.g. bottom shape. Usually, such effect is neglected in the study of cutting process, as cutting is generally associated to yield load mechanisms. This point has to be verified.

Acknowledgments :
The authors are grateful to Dr H. Maitournam for his contribution to numerical developments in the code and to M. Bergues for his contribution to experiments.

References

J.I. Adachi, 1996. *Frictional contact in rock cutting with blunt tools.* Master's thesis, University of Minnesota.

J.F. Brett, T.M. Warren & S.M. Behr, 1989. Bit Whirl : A New Theory of PDC Bit Failure. *SPE* 521–536.

E. Detournay & P. Defourny, 1992. A phenomenological model for the drilling action of drag bits. *Int. J. Rock Mech. Min. Sci. and Geomech. Abstr.* 29(1):13–23.

H. Geoffroy, 1996. *Etude de l'interaction roche-outil de forage : influence de l'usure sur les paramètres de coupe.* Ph.D. thesis, Ecole Polytechnique, Paris.

D.A. Glowka, 1987. Development of method for predicting the performance and wear of PDC drill bits. Sandia Report 86–1745, Sandia national laboratories.

C. Guo, 1995. *Calcul des tunnels profonds soutenus. Méthode stationnaire et méthode approchée.* Ph.D. thesis, Ecole Nationale des Ponts et Chaussées.

E. Kuru, 1990. *Effects of rock/cutter friction on PDC bit drilling performance-an experimental and theoretical study.* Ph.D. thesis, Louisiana State Univ. Baton Rouge, La.

M.H. Maitournam, 1989. *Formulation et résolution numérique des problèmes thermoviscoplastiques en régime permanent.* Ph.D. thesis, Ecole Nationale des Ponts et Chaussées, Paris.

Quoc S. Nguyen & M. Rahimian, 1981. Mouvement permanent d'une fissure en milieu élastoplastique. *Journal de Mécanique Appliquée* 5(1).

H. Sellami, 1987. *Etude des pics usés. Application aux machines d'abattage.* Ph.D. thesis, Ecole Nationale Supérieure des Mines de Paris.

A. Sinor & T.M. Warren, 1989. Drag bit wear model. *SPE* 128–136.

7 Supplement

Numerical Models in Geomechanics, Pietruszczak & Pande (eds) © 1997 Balkema, Rotterdam, ISBN 90 5410 886 X

Numerical limit analysis of rock mechanics problems

L.G.Araujo, J.Macías, E.A.Vargas Jr & L.E.Vaz
Department of Civil Engineering, PUC-Rio, Rio de Janeiro, Brazil
P.O.Faria
Department of Civil Engineering, University College of Swansea, UK

ABSTRACT: Limit Analysis is a technique that has been very little explored as a method to obtain solutions to stability problems in geotechnical engineering such as slope stability, bearing capacity of foundations, earth pressure, underground excavations among others. The present work initially presents an outline of the theory of Limit Analysis as applied to general problems of stability. Subsequently, a specialization of the formulation is developed in order to apply it to stability problems in fractured media which are of interest to rock engineering. Continuum or 'equivalent' continuum can be modeled by discretizing the rock mass into finite elements. Furthermore, discrete blocks can also be considered. The paper presents results obtained with the implemented formulations.

1 INTRODUCTION

Some of the relevant stability problems in geotechnical engineering such as bearing capacity of foundations, earth pressures and slope stability have been solved mainly by established techniques such as limit equilibrium, slip line theories, plasticity theory and incremental finite element anlysis. Limit analysis, on the other hand, based on the limit theorems of plasticity theory provides a framework for the solution of various above mentioned problems although seldom used in practice. A number of analytical solutions based in this approach are available, such as found in Chen (1975) for example. A less used approach however is to obtain limit analysis solutions numerically. In relation to limit equilibrium, numerical limit analysis has the advantage of not having to establish a-priori a failure mechanism and one does not have to define a factor of safety. Moreover, incremental finite element analysis have an intrinsic difficulty in precisely determining failure.

The present paper explores a numerical solution, via finite elements, of limit analysis problems in rock masses. In the present paper, an equilibrium, lower bound approach based on the virtual work principle is derived. The derived equations, when added to the plastic admissibility constrains constitute an optimization problem. The constraints are generally nonlinear and in the present work they were linearized so that a linear programming problem could be obtained.

The above formulation, normally associated with continuum problems, was extended to fractured rock masses. This was done by using the concept of a multilaminate material as proposed by Zienkiewicz and Pande (1977) and as a further development, the formulation is extended to analyse assemblages of rigid, discrete blocks. Examples follow to illustrate and validate the formulation.

2 AN OUTLINE OF LIMIT ANALYSIS

The basic framework of limit analysis consists of the upper and lower bound theorems of plasticity theory. These theorems have been proven (Kachanov, 1971) for ideal elastoplastic materials, under conditions of associated flow and small deformations. The lower bound theorem, relevant to the present work can be stated as:

Lower Bound theorem - "If in a given structure a stress field can be found that is in equilibrium with the applied loads,satisfies the boundary condition of traction and does not violate the yield condition then the structure will support the loads".

2.1 *Continuum and Equivalent Continuum Problems*

A numerical implementation of the condition stated above can be done by interpolating stresses, acting directly on the equations of the problem on what is known as a strong formulation. For example, a strong formulation of the lower bound theorem can be stated as (Faria, 1992):

maximize μ
subjected to

$$\mathbf{B}\,\sigma_a = \mu\,\mathbf{f}_c \text{ (equilibrium)} \tag{1}$$

$$\mathbf{g}_a(\sigma_a) < 0 \text{ (yield condition)}$$

$$\mathbf{Q}^t\,\sigma_a < \sigma_a^* \text{ (linearized yield condition)}$$

where

μ = collapse load factor
$\mathbf{B}$ = matrix relating nodal stresses to nodal forces
$\mathbf{f}_c$ = vector of nodal external loads
$\mathbf{Q}$ = matrix formed by the normals to the planes of the linearized yield surface
σ_a = vector of stresses in the medium
σ_a^* = vector of distances to the planes of the linearized surface to the origin
$\mathbf{g}_a$ = vector containing the yield functions
a - points where stresses are defined
c - points where nodal forces are defined

Eqs. 1 define a mathematical programming problem. By representing the yield function as a polyhedron in hyperspace (Faria, 1992), the problem becomes a linear programming problem. This is convenient as standard linear programming solvers can be used for its solution.

Amongst other approaches (Sloan,1987 for example), an alternative approach, the one that has been used in the context of the present work, is the so called mixed formulation equations (Fonseca and Neves, 1986). Here, the equilibrium equations are weighed on the domain by functions which for convenience are the velocities (v). This means that the equilibrium equations are valid in an average sense as in standard displacement based finite elements. An obvious advantage of this formulation in relation to the strong formulations is that all the preprocessing such as mesh definition is done with standard finite element preprocessors. No discontinuities or jumps in the stress field are necessary as in the other formulations. The basic equations can be written as:

$$\int_v \mathbf{v}^t\left(\nabla^t\sigma - \mu\mathbf{f}\right)d\mathbf{V} = \mathbf{0} \tag{2}$$

where

∇ = differential operator which relates the stress to body forces

the finite discretization:

$$\mathbf{v} = \mathbf{N}_v\,\mathbf{v}_c \tag{3}$$
$$\sigma = \mathbf{N}_\sigma\,\sigma_a$$

where the subscripts c and a stands for points where velocities and stresses are prescribed respectively, substituted into (2) with some manipulation (Faria, 1992) leads to the following equations:

$$\int \mathbf{N}_v^t\nabla^t\mathbf{N}\sigma\, d\mathbf{V}\sigma_a - \mu\int_s \mathbf{N}_v^t\mathbf{f}\, d\mathbf{S} = \mathbf{0} \tag{4}$$

or

$$\mathbf{B}^t\,\sigma_a = \mu\,\mathbf{f}_c \tag{5}$$

where

$$\mathbf{B}^t = \int_v \mathbf{N}_v^t\nabla^t\mathbf{N}\sigma d\mathbf{V} \tag{6}$$

and

$$\mathbf{f}_c = \int_s \mathbf{N}_v^t\mathbf{f}d\mathbf{S} \tag{7}$$

A statement for the mixed formulation is then obtained:

maximize μ
subjected to

$$\mathbf{B}^t\,\sigma_a = \mu\,\mathbf{f}_c \tag{8}$$

$$\mathbf{g}_a(\sigma_a) \le 0 \tag{9}$$

where $\mathbf{B}$ and $\mathbf{f}_c$ are defined by (6) and (7) respectively.

The present paper is focused on numerical implementations of the mixed formulation and the Mohr-Coulomb yield criterion. Details of the linearization of the yield surface in the stress space can be found in Faria (1992).

An extension of the mixed formulation to the so called multilaminar medium which simulates the rock mass as an equivalent continuum (Zienkiewicz and Pande, 1977) is easily done by incorporating a further yield condition for the planes of the joint families.

The stress in a joint plane can be calculated as

$$\sigma_j^t = [\ \tau\ \sigma_n\ ] = \mathbf{T}\ \sigma \qquad (10)$$

where $\mathbf{T}$ is a rotation matrix for stresses, τ and σ_n are shear and normal stress components respectively in the ith joint plane. A Coulomb criterion for yield of the joint can be written as

$$|\tau| \leq \mathbf{c}_i + \sigma_n \tan \phi_i \qquad (11)$$

where $\mathbf{c}_i$ and ϕ_i are respectively cohesion and friction angle for the ith family of joints. Equations (10) and (11) can be written as

$$\mathbf{Q}_i^t \sigma \leq \mathbf{S}_i \qquad (12)$$

where

$$\mathbf{Q}_i^t = \begin{bmatrix} 1 & \tan \phi_i \\ 1 & -\tan \phi_i \end{bmatrix} \mathbf{T} \qquad (13)$$

and

$$\mathbf{S}_i^t = [\mathbf{c}_i \quad \mathbf{c}_i] \qquad (14)$$

Additional numerical details of the implementation can be found in Faria (1992) and Araujo (1996). One important point to emphasize is the necessity of using efficient algorithms for the solution of the linear programming problem. LINDO, a commercially available software package (Schrage, 1991), showed considerable efficiency in the context of the present work.

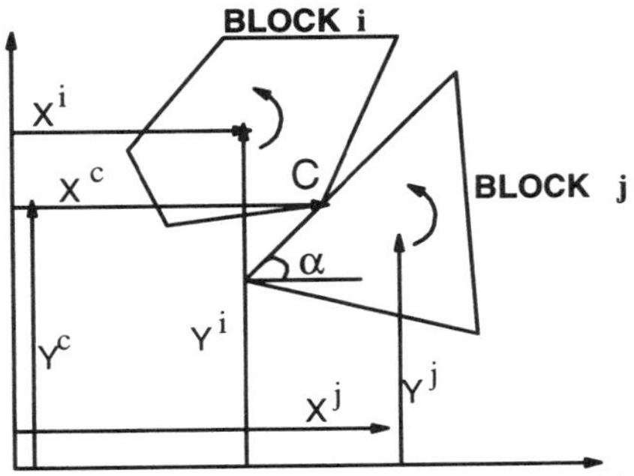

Figure 1. Global Coordinates system for Blocks

2.2 *Discrete Blocks*

The present paper constitutes an extension to the described formulation to incorporate the existence of assemblages of discrete, rigid blocks. Livesley (1978), presented a similar formulation to the one presented here but in another context; the one of the stability of stone arches.
In the case of rigid blocks in contact (fig. 1), lower bound can be simply stated as (Faria, 1992):

$$\text{Max } Y = \langle \mathbf{1} \quad \mathbf{0}^t \rangle \begin{Bmatrix} \lambda \\ \mathbf{r} \end{Bmatrix}$$

subjected to

$$\begin{bmatrix} -\mathbf{P} & \mathbf{H} \\ \mathbf{0} & \mathbf{R} \end{bmatrix} \begin{Bmatrix} \lambda \\ \mathbf{r} \end{Bmatrix} \begin{matrix} = \\ \leq \end{matrix} \begin{Bmatrix} \mathbf{0} \\ \mathbf{r}^u \end{Bmatrix} \qquad (15)$$

where

Y = objective function
λ = load factor (to be determined)
$\mathbf{r}$ = contact forces (normal and tangencial contact forces in local coordinates, to be determined)
$\mathbf{P}$ = presultant of external applied forces in global coordinates at block centroids
$\mathbf{H}$ = equilibrium matrix for each block
$\mathbf{R}$ = constraints at each contact
$\mathbf{r}^u$ = cohesion at each contact

Fig. 1 shows the global coordinates for blocks i and j in contact, as well as the global coordinates for contact C. The derivation of matrices $\mathbf{H}$ and $\mathbf{R}$ can be found in Faria (1992) and Macías (1997).

Having determined **H** and **R** for each block, the overall system for all blocks as implicit in equation 15 has to be assembled.

3 SOME EXAMPLE PROBLEMS

3.1 *Foundation on fractured rock*

The geometry and loads for this example are shown in Fig. 2. The foundation is composed of a fractured rock mass. Only one joint family is considered. The exercise consists in numerically determining the relationship between the load factor and the inclination of the joints. Properties of both the intact rock and the joints are considered. Davis (1980) obtained a solution for this problem by using slip line theory. Recently, Yu and Sloan (1994) obtained for the same problem a numerical limit analysis solution with a strong lower bound formulation. Fig. 3 shows schematically the problem and the results obtained with the three solutions the ones by Davis (1980), Yu and Sloan (1994) and the present work. The results shown in the figure correspond to

	cohesion	friction angle
intact rock	1	35°
joints	0.1	35°

One notices that for w (angle for inclination of the joints) larger than 50 degrees, some discrepancies arise between the solutions but generally one recognizes that a reasonable match is obtained.

Numerical Models in Geomechanics, Pietruszczak & Pande (eds) © 1997 Balkema, Rotterdam, ISBN 90 5410 886 X

Author index